PHOTOBIOLOGY
The Science and Its Applications

PHOTOBIOLOGY
The Science and Its Applications

Edited by
Emanuel Riklis

Israel Atomic Energy Commission
Beer-Sheva, Israel

PLENUM PRESS • NEW YORK AND LONDON

Library of Congress Cataloging-in-Publication Data

International Congress on Photobiology (10th : 1988 : Jerusalem)
    Photobiology : the science and its applications / edited by
  Emanuel Riklis.
        p.   cm.
    "Proceedings of the Tenth International Congress on Photobiology,
  held October 30-November 6, 1988, in Jerusalem, Israel"--T.p. verso.
    Includes bibliographical references and indexes.
    ISBN 0-306-43830-5
    1. Photobiology--Congresses.   I. Riklis, Emanuel.   II. Title.
    [DNLM: 1. Biology--congresses.   2. Light--congresses.
  3. Photochemistry--congresses.   4. Ultraviolet Rays--congresses.
  QH 515 I61p 1988]
  QH515.I57   1988
  574.19'153--dc20
  DNLM/DLC
  for Library of Congress   91-3013
                                                                CIP

Proceedings of the Tenth International Congress on Photobiology,
held October 30–November 6, 1988, in Jerusalem, Israel

ISBN 0-306-43830-5

Cover illustration: Figure 4a from article in
European Journal of Cell Biology, Vol. 46,
227–232 (1988).

© 1991 Plenum Press, New York
A Division of Plenum Publishing Corporation
233 Spring Street, New York, N.Y. 10013

Printed in the United States of America

Organizing Committee

Organizing Committee

E. Riklis, Chairman

M. Avron
S. Malkin
M. Ottolenghi

Honorary President
R. Latarjet, France

Board of the Association
Internationale De Photobiologie

President
K. Rohatgi-Mukherjee, India

Vice Presidents
L.O. Bjorn, Sweden
E. Riklis, Israel
W. Shropshire, USA
T. Yoshizawa, Japan

Secretary General
R.M. Tyrrell, Switzerland

Treasurer
H. Honigsmann, Austria

Immediate Past President
F. Urbach, USA

Israeli Program Committee

E. Riklis, Beer Sheva, Chairman

M. Avron, Rehovot
E. Azizi, Tel Hashomer
N. Bejerano-Roth, Beer Sheva
M. Belkin, Tel Hashomer
R. Ben-Ishai, Haifa
D. Canaani, Tel Aviv
N. Degani, Beer Sheva
B. Ehrenberg, Ramat Gan
M. Faraggi, Beer Sheva
Y. Feitelson, Jerusalem
G. Gabor, Nes Ziona
Z. Gromet-Elhanan, Rehovot
A. Haim, Haifa
B.A. Horwitz, Haifa
J. Jortner, Tel Aviv
R. Kol, Beer Sheva
A. Kushelevsky, Beer Sheva
Z. Livneh, Rehovot
S. Malkin, Rehovot
R. Margalit, Tel Aviv
I. Ohad, Jerusalem
M. Ottolenghi, Jerusalem
A. Shafrir, Haifa
G. Simchen, Jerusalem
H. Slor, Tel Aviv
J. Sperling, Rehovot
E. Waldstein, Tel Aviv

International Scientific Committee

Foreword

It is not always the case that the subject of a scientific book and its relevance to everyday life are so timely. Photobiology and its sister subject Radiobiology are now a must for understanding the environment we live in and the impact light, ultraviolet light, and radiation have on all aspects of our life. Photobiology is a true interdisciplinary field. Photobiology research plays a direct role in diverse fields, and a glance at the topics of the symposia covered in this book by over 100 articles shows the breadth and depth of knowledge acquired in fundamental research and its impact on the major issues and applied problems the world is facing.

Half a century of photobiology research brought about an understanding of the importance of light to life, both as a necessary source of energy and growth as well as its possible dangers. Research in photochemistry and photobiology led to the discoveries of cellular repair mechanisms of UV induced damages to DNA and this led to understanding of the effects of hazardous environmental chemicals and mutagenecity, and to the development of genetic engineering. This topic was given due emphasis in several symposia and chapters in this book. The similarity and differences in the effects of ultraviolet and ionizing radiations are noted in a special chapter and so are all aspects of the involvement of light and radiation in biological processes occurring in the world around us, the beneficial and the detrimental, the risks and the modes of protection, from molecules, through the plant and animal kingdom, to man. As in the field of ionizing radiation, ultraviolet light causes damage to biological systems, but is also being used to remedy deleterious effects and light is used for therapy in treating skin diseases like psoriasis as well as cancer by photodynamic therapy.

Bringing together the many experts and students researching these fields, in the framework of the International Congress of Photobiology which takes place every four years, was made possible by the support of several organizations listed on a separate page, yet special mention for generous support should be given to the U.S. Department of Energy, the Food and Drug Administration, the American Society for Photobiology, Givaudan S.A. and Laboratoire Pharmaceutique Bergaderm.

I am grateful for the efforts of the Israeli organizing and program committees, the Board of the AIP, the Association Internationale de Photobiology. International scientific committee members helped to organize and chaired the over thirty symposia, workshops and round table discussions, which were organized for this congress, in addition to sessions of contributed papers and poster presentations.

Among organizers of these were (in order of presentation): Richard B. Setlow, A half century of UV photobiology — The Alexander Hollaender Memorial Symposium; J.Z. Beer, Ultraviolet and Ionizing Radiations: From Molecules to Cells; I. Ohad, Light Regulation in Photosynthesis; G. Jori and J. Moan, Photodynamic Cancer Therapy; L. Björn, Effects of Ultraviolet Radiation on Land Plants; F. Dall'Acqua and D. Averbeck, Psorelens; R.M. Tyrrell, cellular effects of UVA and UVB radiations; H. Scheer, organization and interactions of chromophore protein complexes in the photosynthetic membrane; J. Feitelson, The Use of Optical Probes in Biological Systems; M.L. Kripke, Photoimmunology; L. Grossman, DNA Damage and Repair: A Quarter Century of Progress; C. Helene, Site-directed Photochemical Reactions in Nucleic Acids and Protein-Nucleic Acid Complexes; E. Azizi, Sunlight and Melanoma; T. Ito, Vacuum UV; E.D. Jacobson and J.C. van der Leun, UV risks and regulations; K.K. Rohatgi-Mukherjee and E. Riklis, Photobiology in Developing Countries; H. Mohr, Photomorphogenesis; H.V. Gelboin, DNA damage and Repair: Shuttle Vectors; I.Ashkenazi and A. Haim, Chronobiology: Cicadian Rhythms; M. Ottolenghi, Visual and Bacterial Rhodopsins; P.S. Song, Phytochrome; F. Urbach and E. Riklis, UV and the Environment; F. Urbach, R. Sayre and M. Pathak, Physiological Effects of UV Radiation — A Franz Greiter Memorial Symposium; G.C. Walker, UV and Light Mutagenesis; D. Forbes, Pharmaceuticals and Photobiology: Issues of Safety and Efficacy; R.E. Mascotto, Sunscreens; E. Riklis, Repair Deficiency Diseases.

Four photobiology school lectures were delivered by T.G. Truscott, J.E. Hearst, D. Shugar and S. Malkin. The prestigious Edna Roe lecture and the Finsen lecture were presented by Ethel Moustacchi and Ron Ley, respectively. Most of the invited lectures are presented in this volume, allowing the reader a view of the wide spectrum of this exciting science.

Mrs. Erna Philip is thanked for able reformatting and typing of the chapters in this book. Ms. Judie Copeland is thanked for secretarial assistance with the Congress correspondence, and "Kenes" is thanked for able general organization. Dr. Noah Dagani is thanked for advice in shaping the order of presentations in the book, and the members of the organizing committee, Professors Avron, Malkin and Ottolenghi, as well as those who were active members of the program committees are thanked for helping me shape the scientific program of the congress.

I am particularly grateful for continuous support and advice to Prof. Frederick Urbach, Past President of AIP, and to Richard J. Burk, Executive Secretary of the ASP.

The Congress participants have asked me to express their thanks to my son Eran for showing them a glimpse of this beautiful country in his film "Israel at Forty" shown at the opening ceremony, and finally, my wife Ruth and children, Liatt and Eitan are thanked for encouragement and support in many aspects of the organization.

E. RIKLIS
EDITOR & CONGRESS PRESIDENT

Acknowledgements

The Organizing Committee acknowledges with deep appreciation and thanks the generous support by the following:

Allergan Pharmaceuticals, Inc., USA
American Society for Photobiology
L.D. Caulk, USA
Department of Energy, USA
Estee Lauder Inc., USA
Food and Drug Administration, USA
Greiter AG, Switzerland
Israel Academy of Sciences and Humanities
Israel Atomic Energy Commission, Nuclear Research Center-Negev
Israel Cancer Association
Israel Discount Bank
L Givaudan & Cie. SA., Switzerland
Laboratoire Pharmaceutique Bergaderm SA, France
Meshulam Riklis, USA
Owen/Allercreme, USA
Revlon, USA
Rhone-Poulenc SA, France
Schering-Plough, USA
Technad Inc., USA
Technion–Israel Institute of Technology
The Ministry of Science and Development, Israel
The Ministry of Tourism, Israel
The Weizmann Institute of Science, Israel

Contents

xviii

Opening of the Congress

R. LATARJET

Institut Curie-Biologie
26 rue d'Ulm
75231 Paris cedex 05
France

Professor Emanuel Riklis kindly asked me to open the Congress with a few words. Why me? Because I am probably the oldest participant in international photobiology meetings. In fact, my first one was the third in 1936, at Wiesbaden (Germany) 52 years ago, almost two generations, a long time is it not? At that time, the meeting was called "Congrès International de la Lumière" (in French). The main topics were focused on the therapy of rickets, bone and skin tuberculosis, and of dermatosis, by ultraviolet rays, either natural from the sun (heliotherapy) or artificial from sources whose champion was the medium pressure mercury vapour lamp.

In the course of the following 15 years, and despite the World War, two events overturned the situation:

a) Chemistry took the place of ultraviolet in the therapy of rickets and of tuberculosis. Sanatoria were sold by apartments. Stations, especially designed for heliotherapy, were transformed into resorts for winter sports and summer holidays.

b) Conversely, radiations established themselves as a remarkable — in certain respects, unique — tool for research in biochemistry, genetics and virology. They were recognized as an environmental factory of both benefits and risks. Carried by the newly-born molecular biology, photobiology soared.

In 1951, the Comité de la Lumière held its first post-war meeting, a small-scale one in Paris. One evening, four of us were discussing the matter in a "bistro" of Montmartre: Wilhelm Morikofer (Switzerland), Alexander Hollaender (USA), Boris Rajewsky (Germany), and I. We recognized the change, and decided to turn the Comité International de la Lumière into Comité International de Photobiologie (which has since become the Association Internationale de Photobiologie — the official title is still French. For sometimes, very seldom indeed, the French language remains the international language, such as in cooking, in classical dance, and maybe in love ...).

Today, the 10th Congress of this Association begins, the 13th overall.

Photobiology, Edited by E. Riklis
Plenum Press, New York, 1991

On looking at the program, one gets a good idea of the diversity and of the wealth of the new paths along which you progress: one measures the vitality of today's photobiology, and its growing importance among the disciplines of fundamental and applied biology. The youth of that running science is well-symbolized in the youth of the State of Israel which receives us. I am glad to pay my tribute to its energetic, courageous and intelligent citizens, who, within a few decades, have achieved splendid feats — one may even say miracles - in spite of the great difficulties which they encounter.

On behalf of all of you, I thank Professor Riklis and his colleagues for their efforts in all they have done for the Congress so far. Its success will be their reward.

So — 10th Congress of Photobiology, go and be a great success.

Themes and Trends in Photobiology

K. K. ROHATGI-MUKHERJEE

Physical Chemistry Laboratory
Jadavpur University
Calcutta,700032 India

With recent excitement on high temperature superconductors, I was trying to educate myself on the intricacies of superconductivity by reading an article in the Journal of Chemical Education (Matsen, 1987). To explain the mechanism the article stated: "The 'sudden polarisation' theory provides a novel, high speed, long distance electron-pair transport mechanism. The mechanism is adapted from a theory of vision in which transmission of optical signals from the retina to the brain is by 'sudden polarisation.' Here an optically excited state of retinal is 'suddenly polarised' by an external magnetic field during the change in nuclear coordinates. In the application of 'sudden polarisation' to superconductivity, we replace retinal by a linear chain of two M-sites, ionic and valence bond states...."

Another example pointing to the underlying oneness of scientific knowledge for all basic phenomena was the subtitle to an article (Adams, 1987) which stated: "sunburnt guinea pig throws light on arthritis drug." From the knowledge that there was a relationship between the ability of aspirin to inhibit UV erythema in the guinea pig and its apparent favourable effects in rheumatoid arthritis, a correlation could be visualised. The mode of action of aspirin is due to its ability to prevent synthesis of prostaglandins which are shown to be released in many conditions of injury and disease to cause inflammation, pain and fever and also to be present in the lining of the joints in various arthritic conditions. Prostaglandins are liberated in guinea pig's skin in UV erythema which led to the discovery of brufen. The article further added: "Who could have foreseen in those early days a link between sunburn in the guinea pig and headache, toothache and the dreaded disease rheumatoid arthritis?"

These are just two examples which project the recurring structural and mechanistic themes underlying all basic phenomena in Nature. Even though the simplest cells contain more than 5000 kinds of protein, the structural and mechanistic themes seen in one protein frequently recur in others. For example, there is a close relationship between the enzymes thrombin active for blood clotting and chymotripsin necessary for digestion. The structures of many proteins have stayed the same over long evolutionary periods. There is surprisingly little difference between human and mouse haemoglobins. Enzymes

Photobiology, Edited by E. Riklis
Plenum Press, New York, 1991

work in the same way in simple organisms as in complex organisms. This knowledge has helped to perform experiments on animals and extrapolate them for human systems "to unravel disease mechanisms, devise new diagnostic tests and develop novel drugs and therapeutic strategies." (Pimental et al., 1987).

Recent X-ray and absorption spectroscopic studies of Fe in PS I preparation from spinach and thermoplastic cyanobacteria *Synchococcus sp.* at Lawrence Berkeley Laboratory have established that the structures of several ferridoxins are essentially indistinguishable. (McDermott et al., 1988). Similarly, the electron donor side of the Mn-bound protein complex, the catalyst for O_2-evolution in PS II centre is largely conserved across the evolutionary divers O_2-evolving photosynthetic species across estimated two billion years of evolutionary gap separating spinach and *Synchococcus sp.* (Biochemistry, 1988). There is substantial sequence homology of H, M and L proteins in Rhodopseudomonas, *R. capsulates*, *R. viridis* and *R. spheroides* indicating common folding of subunits and also that they are derived from common ancestors. (Deisenhofer and Michel, 1987). Some important amino acids, mainly glycines, are conserved as also proteins at the end of helices, at turns of peptide chains and as ligands to the pigments.

The organised complexity of biological phenomena has evaded complete understanding and the macromolecular level is the lowest level to which biological events can be reduced. Complete reductionism does not seem to be a valid proposition. There is a deep unity in the simple physical concepts common in a large variety of complex situations but still it is not possible to reduce biological organisation to simple sets of physical rules. It is the interaction between these rules, the perturbations and the resultant chaos, which itself is organised, that seems to be a better description. An essential feature of the complexity underlying biological organisation is the ability to regulate and correlate the various activities in both space and time according to the system's past history and to the environmental conditions, i.e., "to store past experience as well as to generate or exchange information and to evolve towards new forms and functions." (Careri and Nicolas, 1987).

The biological macromolecules are small and floppy bodies and the structures of biomolecules are such that spontaneous fluctuations are to be expected near thermodynamic equilibrium. The first evidence that rapid structural fluctuations in the nanosecond time scale are present in globular proteins and enzymes comes from experiments by Lackowics and Weber (1973) on quenching of protein fluorescence by oxygen. Such observations have been confirmed by recent studies of protein dynamics using more sophisticated techniques such as ODMR at zero field.(Ghosh et al., 1988). The ability to time-correlate the fluctuations of certain relevant variables such as relaxation of the bound water, side chain rotational correlation, proton transfer reactions of certain ionizable groups etc. which lie within the time domain of 10^{-1} and 1s, could be the characteristic kinetic property which allows macromolecules to work as an enzyme. The important feature underlying organised complexity is the possibility to use these correlations to build up new unexpected modes of behaviour associated with abrupt transitions of large amplitudes, thus achieving a role in the functional order of the living cell. This seems to be Nature's solution to the problem of faithfully generating significant biochemical events in thermal bath, i.e., noise. (Careri, 1987).

Thus, large-scale biological processes can be expressed by appropriate correlations of many small-scale events. For example, the cooperative effect involved in respiratory transport in haemoglobin is remarkable in the sense that in it Nature exploits a difference of only 13% in the size between the covalent and ionic forms of iron. The tertiary structure of the haemoglobin molecule as a whole oscillates rapidly and continuously between oxy- and deoxy-conformations.(Perutz, 1970). This brief motion is believed to trigger a whole series of subtle changes in the arrangement of the subunits and to be responsible for many remarkable aspects of oxygen-haemoglobin equilibrium. Dynamics in the time domain of femto and pico seconds have been studied (Petrich et al., 1988) to give a greater insight into the structure and functions of the respiratory chain. Similarly, the conductive regime of hydrated lysozyme powder can be described as series of correlated single proton transfers along a random thread of water molecules. Since each single proton transfer can be considered as a small-scale event caused by local fluctuations, the large-scale event results from the correlation of a great many of these small-scale events. (Careri et al., 1985). An apt example lies in laser physics, where several microscopic events, i.e., transitions among atomic or molecular states become cooperatively time-correlated to produce a giant electromagnetic pulse, a macroscopic event. Life exists only insofar as it evolves in time as never-ending streams of events.

On the other hand, the organised complexity of biological phenomena with self-interacting subunits can be appreciated only when the system is driven far away from its state of equilibrium. The system does not tend towards its equilibrium value asymptotically by entropy production. But the rate laws are complicated nonlinear functions of sets of variables describing the instantaneous state of the system in a given environment. For the non-equilibrium situation as obtained for stationary states far from equilibrium, dissipative structures are formed in which few modes or states couple strongly among each other so that the energy tends to remain localised among them rather than being distributed among many degrees of freedom, which are only weakly coupled. The hidden potentialities of nonlinear phenomena such as the ability of certain kinds of molecules to perform autocatalytic and other regulatory functions appear as a natural consequence. For all such systems connectivity and cooperativity develop at a certain threshold value specific for a given system, giving rise to specific properties such as enzyme catalysis or membrane transport or the coherence of dissipative cycles away from equilibrium as observed in glycolysis. Such behaviour cannot be directly extrapolated from the description of smaller systems. The fundamental problem of the clear understanding of all physico-chemical-biological phenomena, therefore, is that "the observed system is required to be isolated in order to be defined, yet interacting in order to be observed." (Nicolas, 1987). In such systems, any change in one region is bound to have consequences for the whole system, a phenomenon manifested very clearly on a macroscale in the biosphere regime. In the organised biological system static and dynamic aspects are well balanced when the system is poised to express its functionality.

In the transition state the system has optimal adaptibility towards the environment while maintaining its integrity. This state has been characterised as that of the highest hierarchic order. (Gutman and Resch, 1983).

An apt example is found in the theory of vision. (Bonacic-Koutecky et al., 1987; Salem, 1976). The first excited singlet state of the pigment rhodopsin has a barrierless potential surface. On photoexcitation of rhodopsin, the protonated Schiff base of 11-cis retinal, cis-trans isomerisation occurs in a few picoseconds in a concerted mechanism involving simultaneous twisting of single and double bonds. The coupling of a twisting motion with proton translocation in the environment from a strong acceptor at one end to strong donor at the other end of the chain at very near 90° twist angle creates a state of 'sudden polarisation.' Consequent changes (i.e., charge separation) in hydrogen bonding and proton configuration trigger the stimulus of vision. This system is a fascinating example of conversion of photon into a high energy electrical signal. (Schnaff and Baylor, 1987).

Thus photobiological phenomena triggered by light are systems far removed from equilibrium and functionality appears because of the non-equilibrium situation. Through ages Nature has learnt to catch the fleeting rays of the sun and make them do work as they relax towards the state of equilibrium, either the original one or a new one. The extensive cooperativity in photon-induced phenomena appears because photons are bosons. On the other hand, electrons being fermions are moved around to do specific jobs in the form of photoredox reactions and build hierarchic orders of complexity. Long distance energy transfer and electron transfer (McLendon, 1988; Khairutdinov and Briekenstein, 1986) are key phenomena, basic steps in many photobiological processes. The patterns for these processes are coded in the DNA of the species. With the help of genetic engineering, sequences of light-responsive genes of many developmental blueprints have been isolated and characterised. Light-responsive gene sequences for photomorphogenesis, a phenomenon which controls the development of plants, have recently been identified (Moses and Chau, 1988).

Many reconstitution experiments have been attempted to understand the organised complexity and for systematic exploration of basic structure-function questions through modification of various components. In some cases recovery of enzymatic functionality has been observed on such reconstitution. Recently, pigment-protein complex B 873, light-harvesting system from pure protein and pure BChl, has been reported (Parkes-Loach et al., 1988). Reconstitution of rhodopsin functionality and cGMP cascade in polymerised bilayer membranes has proved to be a convenient method for the study of protein-lipid interactions and their role in protein functionality (Tyminski et al., 1988). Mimicry of biomolecules at the molecular level has been attempted to create artificial models for natural systems such as photosynthesis (Brune et al., 1987), and develop processes for storage and utilisation of solar energy. Calvin's (1983) "synthetic leaf" for photosplitting of water has nearly reached the stage of possibility.

The stereochemical aspects of the placement of active molecules in the network of enzyme proteins and membranes are crucial to efficient biochemical processes and are being consistently established by X-ray crystallographic studies. Recent X-ray picture of the crystals of photosynthetic reaction centre (RC) of photosynthetic bacteria, *Rhodopseudomonas viridis*, by Deisenhofer and his collaborators (1984) has given insight into the whole architecture of the prosthetic groups in minute detail at 3A resolution in the framework of protein structure. The most interesting aspect of the arrangement of the pigment in the RC is the geometry and the symmetry of placement of the prosthetic groups with BChl *b* dimer, the special pair (SP), as the point of symmetry. Further study has revealed that slight asymmetry in the distribution of polar groups in the L and M protein framework on either side of the dyad, has been used by Nature to guide the electron movement only on the right arm of the dyad. The function of the left arm is still unclear. Intense activity generated around the problem has led to the observation that asymmetry appears even in the distribution of the photoexcited electron density in the primary electron donor, the special pair (Lendzian et al., 1988). The most striking results of calculations performed on the radical cation P_{960}^+ is unequal distribution of spin densities in the ratio of 0.74:0.26, in favour of BChl *b* molecule in the dimer bound to the L-branch. This seems to be the essential requirement for the high efficiency of the primary charge separation PI $\rightarrow$ P$^+$I$^-$ where I$^-$ is the intermediate acceptor BPh. It is most likely that asymmetric arrangments of proteins and pigment dimer is repeated in PS I reaction centre also. The fragment of DNA that bears most of the genes necessary for the light reaction of photosynthesis, including genes for all the proteins within the reaction centre as well as those necessary for light-harvesting antenna pigment structure, has been isolated and nucleotide sequence determined (Eckes et al., 1987).

These pioneering studies in photosynthetic bacteria have raised many more questions and have given new directions to researchers in pursuit of their answers. What is the function of spectator pigments? What is the principle that guides the geometric placement of prosthetic groups which induce fast electron transfer chain in picosecond time domain? Does the sequential near-orthogonality of the planes of several chromophores point to some basic principles of electron transfer cascades? How does the new knowledge fit in with the dynamic aspects of supramolecular architecture and functional organisation of five major protein complexes, viz., PSI, PSII, cytochrome k/f, light-harvesting complex and ATP synthetase in the thylakoid membrane? Nature must have discovered some specific advantage in introducing slight asymmetry in an otherwise symmetric arrangement of prosthetic groups as she has found in the use of chirality at the molecular level or asymmetry in DNA helix.

While the nature of the polypeptides constituting the catalytic centres of these systems has changed comparatively little during the course of phylogenetic evolution, the number of regulatory units has increased. In higher plants a 17-kDa polypeptide and a 24-kDa polypeptide, which do not appear in cyanobacteria, are well characterised (Stewart et al., 1985). The marked variations of the regulatory subunits during the course of phylogenetic development points to Nature's way for constant 'fine tuning' of the functions and the activities of these centres with increasing complexity and stress for survival in ever-changing biosphere. The centres retain the memory of the past, whereas the regulatory units invent and shape the future. The biological membranes with specific lipids can be described as appropriate correlators of events. The fast-developing areas of molecular dynamics will, perhaps, lead to a deeper mechanistic understanding of these natural phenomena. The possibility that systems consisting of interacting units may present new patterns of dynamical behaviour transcending the individual subunits and characterised by coherence, cooperativity and rich space-time correlation specturm has attracted the interest of many scientists from the disciplines of physical sciences and mathematics. No single experimental technique can suffice to unravel completely the mystery of biological structure-function symbiosis from all aspects. Techniques of divers kind are needed to glean the knowledge of, not only the molecular architecture of the funtional units and mode of cooperativity among them, but also the genes that direct such functionality. Computer simulation can provide the visual appreciation of the working model (Fruhbeis et al., 1987).

The present intense activity and breakthroughs in the appreciation of photobiological phenomena are primarily due to the development of more and more sophisticated high-intensity light sources like lasers (Kaufmann and Rentzepis, 1975; Pratesi, 1987) and synchrotrons (1986) and very sensitive detection devices. Synchrotron radiation and pulsed lasers in pico and even femto second ranges have provided probes for events ranging from femto second to seconds over wide spectrum from far UV to near IR. Every available technique has been commissioned to study delicate biological events triggered by light in femto second time scale. Various types of spectroscopy have helped to probe into the intricate patterns of biomolecules and their structure and mode of functioning. Sophisticated technologies for measurements of absorption, fluorescence (Scweckenburger et al., 1988) and even phosphorescence (Ghosh et al., 1986) in static and time-resolved modes and flash photolysis techniques have been continuously upgraded.

Genetic manipulation of plant tissue by direct injection of DNA by microperforation technique (Weber et al., 1988) has become possible by using UV laser microbeam. The UV laser microbeams can be aimed very accurately under microscopic control. They have been used successfully to induce selective cell fusion, to micro-dissect chromosomes and to introduce DNA into chloroplasts. Laser beam from Nd-YAG system has been used to manipulate subcellular structures and to introduce DNA into mammalian cells also.

With the advent of lasers, resonance Raman (RR) spectroscopy (Turner and El-Sayeed, 1985) has become a very powerful tool for the study of the ground state and the excited state structures of key molecules involved in photobiological phenomena. While the vibrational frequencies are used to determine the ground state chromophore structure, the intensities provide the information about the excited state lifetimes and geometry changes. The time-resolved resonance Raman technique has the unique ability to obtain high-quality structural data on transient intermediates which facilitate interpretation because scattering from the prosthetic group is selectively enhanced. High-quality Raman data have been obtained for the visual pigment rhodopsin and bacteriorhodopsin. The normal coordinate analysis of the vibrational spectra has made possible *in situ* determination of retinal chromophore structure (Smith et al., 1987). Another technique, surface-enhanced resonance Raman spectroscopy, (SERRS) is used to study molecules and radicals on surfaces such as plant photosynthetic membrane and TPP-coated electrodes. It may be of interest to note that exactly 60 years ago Raman effect was discovered. This year in India, the birth centenary of C. V. Raman and diamond jubilee of the discovery of Raman effect are being celebrated from November 2 to 7 at the Indian Association for the Cultivation of Science in Calcutta where he worked.

Photobiology forms a discipline by itself. Since the energy from the sun in the form of light has played a very crucial role in the evolutionary processes, from the early synthesis of biomolecules and origin of life to the development of intricacies of photobiological phenomena as observed in chemical, structural and functional relationships in evolution, development and metabolism of living organisms, the discipline of photobiology promotes multi-disciplinary and trans-disciplinary philosophy. Fine examples of such collaboration have been observed in the unravelling of the molecular mechanism of photosynthesis in which groups of investigators (Youvan and Marrs, 1987) from three fields of research, namely spectroscopy, X-ray crystallography and genetics, which are often considered to be almost unrelated, have examined the problem each from its own distinct angle and pooled their efforts eventually to unfold the whole picture.

Photobiology is concerned not only with the constructive role of light but also its destructive aspects, as well as its power to heal. The subject of photomedicine is fast developing (eds Favre et al., 1987). Photodynamic therapy of cancer may prove to be one of the most important treatments for the dreaded disease (1987). Topics such as UV light-induced mutagenesis, photodermatology and skin cancer, vision and development of cataract, action of ionising and non-ionising radiations on biological systems, atmospheric photochemistry and photochemistry of polluted atmosphere, perturbation of ozone shield and its biological consequences, are integrated within the science of photobiology. Applications to the problem of human health and development of better agricultural practices also are within its bounds. The modern methods of gene transfer to plant cells have provided the most important applications of genetic engineering, that is, the production of plant cells with new characteristics.

There exists a deep unity in the physical concepts in a large number of complex organisations and different levels of descriptions seem to be necessary in order to express their essential features. Nevertheless, a gap exists between the description of a complex biochemical event by a chemist and a biologist, which cannot be bridged by the language of biochemistry alone. The greatest advances in the past few years have been in molecular biology which is unifying the whole field of physical sciences, biological sciences and medicine into a single discipline. No wonder there is an increasing awareness for the need to promote a common platform, a task which is so successfully taken up by AIP through organising Congresses of the type we have assembled to attend.

The history of AIP goes back to the year 1928 when the first committee to coordinate international cooperation in photobiology was set up in Lausanne under the title Comité Internationale de Lumiere (CIL). The name was changed to Comité Internationale de Photobiologie (CIP) at the Paris Conference in 1951 and later in 1976 to the present form, Association Internationale de Photobiologie (AIP). Therefore, this is the Diamond Jubilee year of AIP. It would have been a great honour for me to share the dais with Professor K. Latarjet who has seen the birth and growth of the AIP since its inception. I pay my humble tribute to him and to those who have made the AIP what it is today.

References

Adams, S.S. (1987) Sunburnt guineapig throws light on arthritis drug, *Chem. in Britain*, 23, 1193.

Bonacic-Koutecky, V., Koutecky, J. and Michel, J. (1987) Neutral and charged biradicals, zwitterions, funnels in S_1 and proton translocation: Their role in photochemistry, photophysics and vision, *Angew. Chem. Int. Ed. (Eng.)*, 26, 170-189.

Brune, D.C., Nozawa, T. and Blankenship, R.E. (1987), Antenna organisation in green photosynthetic bacteria- Oligomeric bacteriochlorophyll c as a model for 740nm absorbing BChl c in *chloroflexus auranticus* chlorosomes, *Biochemistry*, 26, 8644–8652.

Calvin, M. (1983) Artificial photosynthesis: Quantum capture and energy storage, *Photochem. Photobiol.*, 37, 349–360.

Careri, G. (1987), Biology and Complexity: Some physical facets at the molecular scale, *Biology International Special Issue, IUBS News Magazine*, 15, 10–21.

Careri, G. and Nicolas, G. (1987) Biology and Complexity: Some physical facets, *Biology International Special Issue, IUBS News Magazine*, Eds. G. Careri and G. Nicolas, 15, 1–2.

Careri, G., Geraci, M., Giansanti, A. and Rupley, J.A. (1985) Protonic conductivity of hydrated lysozyme powders at megahertz frequencies, *Natl. Acad. Sci., USA*, 82, 5342–5346.

Deisenhofer, J. and Michel, H. (1987) The structure of the photosynthetic reaction centre from *Rhodopseudomonas viridis,* in From Photophysics to Photobiology, Eds. A. Favre, R. Tyrrell, J. Cadet, Elsevier, pp. 133–140. *Photobiochemistry and Photophysics Supplement (1987)*

Deisenhofer, J., Epp, O., Mikki, K., Huber, R. and Michel, H. (1984) Structure of the protein subunits in the photosynthesis reaction centre *Rhodopseudomonas viridis* at 3A resolution, *J. Mol. Biol.*, 180, 385–398.

Eckes, P., Donn, G. and Wengenmayer, F. (1987) Genetic engineering with plants, *Angew. Chem.*, 26, 382–402.

Favre, A., Tyrrell, R. and Cadet, C., (eds.) (1987) From Photophysics to Photobiology, *Photobiochemistry and Photobiophysics Supplement* Elsevier (1987), 301–475.

Fruhbeis, H., Klein, R. and Wallmeier, H. (1987) Computer assisted molecular design (CAMD) – An overview, *Angew. Chem.*, 26, 403–418.

Ghosh, S., Petrin, M. and Maki, A. (1986) Spin lattice relaxation in the triplet state of the buried tryptophan residue of ribonuclease T1, *Biophys. J.*, 49, 753–760.

Ghosh, S., Zang, Li-Hsin and Maki, H. (1988) Relative efficiency of long-range nonradiative energy transfer among tryptophan residues in bacteriophage T4 lysozyme, *J. Chem. Phys.*, 88, 2769–2775.

Gutman, V. and Resch, G. (1983) The hierarchic order in the solid state Pt III: Spin transitions in certain iron complexes, *Inorg. Chim. Acta.*, 72, 269–275.

Kaufmann, K.J. and Rentzepis, P.M. (1975) Picosecond spectroscopy in chemistry and biology, *Acc. Chem. Res.*, 8, 407–412.

Khairutdinov, R.F. and Briekenstein, E.Kh. (1986) Long range electron tunnelling in biological and model systems, *Photochem. Photobiol.*, 43, 339–356.

Lackowics, J.R. and Weber, G. (1973) Quenching of protein fluorescence by oxygen. Detection of structural fluctuations in proteins on the nanosecond time scale.,*Biochemistry*, 12, 4171–4179.

Lendzian, F., Lubitz, W., Scheer, H., Hoff, A.J., Plato, M., Tränkle, E. and Möbius, K. (1988) ESR, ENDOR and Triple resonance studies of the primary donor radical cation P_{960}^{+} in the photosynthetic bacterium *Rh viridis*, *Chem. Phys. Lett.*, 148, 377.

Matsen, F.A. (1987) Three theories of superconductivity, *J. Chem. Edu.*, 64, 842–846.

McDermott, Anne E., Yachandra, V.K., Guiles, R.D. Britt, R.D., Drexheimer, S.L., Sauer, K. and Klein, M.P. (1988) Low potential iron-sulfer centres in Photosystem 1: An X-ray absorption spectroscopy study, *Biochemistry*, 27, 4013–4020.

McLendon, G. (1988) Long distance electron transfer in proteins and model systems, *Acc. Chem. Res.*, 21, 160–167.

Moses, P.B. and Chau, Nam-Hai (1988) Light switches for plant genes, *Sci. Amer.*, 258, 64.

Nicolas, G. (1987) Self-organization in nonequilibrium systems: Biological complexity, *Biology International Special Issue, IUBS News Magazine*, 15, 22–37.

Parkes-Loach, P.S., Sprinkle, J.R. and Loach, P.A. (1988) Reconstitution of the B873 light harvesting complex of *Rhodospirillum rubrum* form separately isolated α- and ß-polypeptides and bacteriochlorophyll *a*, *Biochemistry*, 27, 2718–2727.

Perutz, M.F. (1970) Stereochemistry of cooperative effects in hemoglobin, *Nature*, 228, 726–734. Perutz, M.F., Fermi, G., Leisi, B., Sharma, B. and Lidding, R.C. (1987) Stereochemistry of cooperative mechanism in hemoglobin, *Acc. Chem. Res.*, 20, 309–321.

Petrich, J.W., Poyart, C. and Martin, J.L. (1988) Photophysics and reactivity of heme proteins: A femtosecond study of hemoglobin, myoglobins and protoheme, *Biochemistry*, 27, 4049–4060.

Pimental, George C. and Coonrod, Janice A. Opportunities in Chemistry — Today and Tomorrow, National Academy Press, Washington D.C., 1987.

Pratesi, R. (1987) Semiconductor (diode) lasers: basic principles and potential applications in the biomedical field, in From Photophysics to Photobiology, Eds. Favre, A., Tyrrell, R. and Cadet, J., Elsevier (1987), *Photochemistry and Photophysics Supplement*, pp. 59–75.

Renger, G. (1987) Biological exploration of solar energy by photosynthetic water splitting, *Angew. Chem. Int. Ed. (Engl)*, 26, 643–660.

Salem, L. (1976) Theory of photochemical reactions, *Science*, 191, 822–830.

Schnaff, J.L. and Baylor, D.A. (1987) How photoreceptor cells respond to light, *Sci. Amer.*, 265, 32–39.

Schweckenburger, H., Seidlitz, H.K. and Ebes, J. (1988) Time resolved fluorescence in photobiology, *J. Photochem. Photobiol. B.*, 2, 1–19.

Smith, S.O., Braiman, M.S., Meyers, A.B., Pardoen, J.A., Courtin, J.M.L., Winkel, C., Lagtenburg, J. and Mathies, R. A. (1987) Vibrational analysis of the all-trans retinal chromophore in light-adapted bacteriorhodopsin, *J. Amer. Chem. Soc.*, 109, 3108–3125.

Stewart, A.C., Siczkowski, M., and Ljungberg, U. (1985) *FEBS Lett.*, 193, 175.

Turner, J. and El-Sayeed, M.A. (1985) Time resolved resonance Raman spectroscopy of photobiological and photochemical transients, *Acc. Chem. Res.*, 18, 331.

Tyminski, P.N., Latimer, L.H. and O'Brien, D.F. (1988) Reconstitution of rhodopsin and the c-GMP cascade in polymerised bilayer membranes, *Biochemistry*, 27, 2696–2705.

Weber, G., Monajembaski, S., Greulich, K.O. and Wolfram, J. (1988) Microperforation of plant tissue with a UV laser microbeam and injection of DNA into cells, *Naturwissen*, 75, 35–36.

Youvan, D.C. and Marrs, B.L. (1987) Molecular Mechanism of Photosynthesis, *Sci. Amer.*, 256, 42–48.

A Half Century of Photobiology: The Alexander Hollaender Memorial Sympsium

Alexander Hollaender (1898-1986)

R. LATARJET

Institut Curie-Biologie
26 rue d'Ulm
75231 Paris Cedex 05, France

It is particularly fitting, and I believe it would be most pleasing to Alexander Hollaender that the setting of an International Congress on Photobiology has been chosen in which to honor him. As I mentioned at the opening of this Congress, he was one of the founding fathers of the Comité International de Photobiologie in 1951. Later, he was its President from 1954 to 1960, and he was awarded the Finsen Medal in 1968 "for fundamental contributions in the early development of photobiology, in particular radiation genetics". He was also a member of the U.S. Academy of Sciences (1957); he received the Enrico Fermi Award (1983). But, most of all, he headed the Oak Ridge National Laboratory's Biology Division for 21 years, from 1946 to 1967.

During that tenure, he was host to approximately 200 researchers from 35 countries, who after returning to their homeland, continued to expand and extend research begun at Oak Ridge. This example illustrates one main feature of Hollaender's personality, to promote exchange of scientific information among people of all nations (with special emphasis on Latin America).

Others will speak of his achievements in the laboratory; so I will concentrate more on his personal life. Hollaender was born in 1898 in Samter (Germany) near the Polish border. His primary language was German - from which he kept a few words throughout his whole life in America (he said al-zo instead of also, hundert instead of hundred, tusand instead of thousand ...).

He was young when his father, who had a fur and hide business, died. At the age of 21, he went to Berlin in order to widen the scope of his activity. Two years later, for the same reason, he went to Birmingham, Alabama, then to Saint Louis to attend Washington University. There he met Henrietta Wahlert, a student in art history, whom he married: finally, at the age of 29, he succeeded in being admitted to Wisconsin University, which he entered in 1927. Four years later, he was awarded a Ph.D. degree.

During the war, he worked as a civilian with U.S. Navy, Office of Scientific Research and Development; in this capacity, he was assigned to work on problems using ultraviolet radiation. That was the starting point of his scientific destiny.

His experimental work may not so much explain how Hollaender became a force in

Photobiology, Edited by E. Riklis
Plenum Press, New York, 1991

international dissemination of scientific information. In essence, he was a world citizen whose criteria for individual value were curiosity, industry, enthusiasm and intelligence, and who recognized no externals such as nationality, race, religion, etc.

In 1974, he moved to Washington, and became affiliated with Associated Universities where he set up the Council for Research Planning in Biological Sciences (1981), and continued to undertake, plan, and organize international conferences or workshops, mainly in the fields of genetic engineering and environmental sciences.

Alexander Hollaender died December 6, 1986, of a pulmonary embolism at the age of 87, at Georgetown University Hospital.

Il resta actif et jeune d'esprit jusqu'à la fin, sans connaître les vicissitudes de l'âge.

(This text was written with the help of Mrs. Virginia White).

Alexander Hollaender, 1968, Hanover, N.H.
5th International Congress of Photobiology

Alexander Hollaender: The Man and His Work[*]

R.B. SETLOW

Biology Department
Brookhaven National Laboratory
Upton, New York 11973, U.S.A.

Hollaender's first research dealt with the interaction of light and molecules. It was this fascination with light that was to be a common thread running through many of his experimental investigations. He entered the world of science at the age of 33, long enough before the explosions of molecular biology took place so that he had a historical perspective on genetics. Indeed, he was one of the more important and influential members of the genetics community from the 1930s through the 1980s. His influence was not only a result of exciting experimental work. It also stemmed from his organization of the Biology Division of the Oak Ridge National Laboratory around the themes of genetics and nucleic acid chemistry and biochemistry.

Hollaender made important experimental contributions to genetics. He had been trained and had the inclination to be quantitative in the his analyses of the interactions of light with molecules and living systems. Thus, he brought to the field of photobiology a high standard of quantitative excellence. The first application was to mitogenetic radiation. This field was brought to prominence by the reports of Gurewitsch in the Soviet Union that dividing cells give off small amounts of radiation that may be received by nondividing cells and so trigger the nondividers to enter mitosis. The radiation was supposed to be in the ultraviolet region of the spetrum, but was too weak to be detected by ordinary physical means and necessitated the use of biological receptors. The observations were intriguing because, if they were real and the interpretation correct, it meant that living cells could communicate with one another by a radiation field. I am sure that to a physical chemist, such as Hollaender, the observations made little sense. Nevertheless, he studied them intensely, published a paper on the subject (Hollaender and Schieffel, 1931), and went so far as to travel to the Soviet Union under the auspices of the Rockefeller Foundation to work in Gurewitsch's laboratory. After a three-month stay, he returned to the United States and continued working on the properties of mitogenetic radiation in collaboration with a number of others. By 1937 there had been close to 1,000 papers published on the subject. Most of the results were inconclusive and inconsistent. A review written by Hollaender and Claus (1937) concluded that mitogenetic radiation

---

[*]By permission of the author, Genetics 116, 1–3, 1987.

Photobiology, Edited by E. Riklis
Plenum Press, New York, 1991

could not be demonstrated and so ended this intriguing possibility. Sporadic publications continued to appear in Soviet publications through the 1950s.

Hollaender's claim to scientific fame rests on his detailed studies of the effects of light — especially different wavelengths of monochromatic light — on viruses, bacteria, fungal spores, and several higher systems. He quantitatively measured killing, mutation, induction and inhibition of mitosis. The work required an appreciation of both physics and biology. He found that the most effective wavelengths for producing mutations were those most absorbed by nucleic acid. He was confident of these results but cautious in reaching a conclusion (Hollaender and Emmons, 1941): "It is probably somewhat dangerous to overemphasize the importance of nucleic acid in the study of radiation effects on living cells. It is very well possible that in radiation produced mutations, the nucleic acid is only the 'absorbent' agent, then transfers the absorbed energy to the protein closely associated with it. It is possible that among others, the following changes take place: 1) the breaking down of the nucleic acid; 2) the breaking down of the protein part of the nuclear protein; 3) a disruption in the relation of the nucleic acid to the protein, and finally, a combination of these three effects. We know at present too little about radiation effects to distinguish between these possible functions of ultraviolet radiation. We believe that more detailed knowledge of the structure, the chemical compounds and physical organizations of the cell, especially the nucleus, may lead us to a more balanced interpretation of the functions of radiation and mutation production." The general conclusion was clear. Nucleic acids should be considered as very important components of genetic material (at a time when almost all investigators thought that proteins were the primary replicating molecules).

The similarity of mutation action and absorption by nucleic acids, first reported in 1939, is shown most clearly in Figure 4 of the 1941 publication of Hollaender and Emmons. There are two details of this figure that warrant further discussion. The first is that the wavelength axis was mistakenly misplaced by 100 Å, so that the most effective mutagenic wavelength of 2650 Å appears as 2750 Å and the absorption maximum of DNA is at 2700 Å instead of 2600 Å. It is not clear whether the error was the printer's or the authors'. Nevertheless, the conclusion that action and absorption are very similar holds. The second point of interest is that there is a slight difference of 50 Å between the action and the absorption spectrum. Probably this difference arose because the monochromator and the light source available to Hollaender at that time could give sufficient intensity only at 2537 Å, 2650 Å and 2805 Å. However, many years later more detailed studies of the action spectra for affecting DNA, using more closely spaced monochromatic wavelengths, indicated that this slight difference is real (Rosenstein and Setlow, 1982). The moral of this story is, always use a first approximation to go for the grand picture. Do not worry about fine details until one has extensive molecular information about the phenomenon. As a matter of fact, the explanation for the slight difference between the two spectra is not clear today.

The cautious lesson Hollaender drew from his action spectra studies was that mutations were associated with changes in nucleic acid and, hence, nucleic acids were important chemical entities to study. The expression of this point of view was one of Hollaender's three major contributions to biology in general and to genetics in particular. The other two contributions were: organizing the Biology Division of the Oak Ridge National Laboratory around the theme of genetics and nucleic acids; and fostering numerous symposia in both the developed and developing world, being a driving force in the creation of scientific societies, such as the Radiation Research Society and the Environmental Mutagen Society (as well as their international counterparts), and editing numerous volumes, such as *Radiation Biology* (1954, 1955, 1956), *Chemical Mutagens: Principles and Methods for Their Detection* (1971, 1973, 1977), and *Genetic Engineering: Principles and Methods* (Setlow and Hollaender, 1979, 1980, 1981, 1982, 1983, 1984, 1985, 1986, 1987).

Hollaender had a feel for knowing when an unexpected result was significant. For example, Hollaender and Curtis (1935) reported some experiments on the effects of ultraviolet radiation on bacteria that hinted at the possible existence of a mechanism for the recovery of cells from the effects of radiation. This hint had to wait for many years for others, and Hollaender himself, to extend this work on recovery phenomena to biological systems exposed to ionizing radiation and to the discovery at Oak Ridge, almost 30 years later, of nucleotide excision repair following ultraviolet irradiation of bacteria (Setlow and Carrier, 1964). Thus, when you add up all aspects of Hollaender's life, it is obvious that he was a leader and not a follower.

In its heyday in the 1950s and 1960s, the Biology Division of the Oak Ridge National Laboratory was the most prestigious biological research institution in the world. It was full of exciting geneticists, biochemists, and biophysicists working on molecules, prokaryotic organisms, lower eukaryotes, insects and mice. But the emphasis in the Division was on genetics. The list of young geneticists who worked at Oak Ridge before they had made important names for themselves is so impressive that it is worth enumerating (with apologies to those omitted by accident or ignorance on the author's part): Kim Atwood, Michael Bender, Grant Brewen, Ernest Chu, Roy Curtiss III, Fred de Serres, Udo Ehling, Ed Grell, Rhoda Grell, Richard Kimball, David Krieg, Dan Lindsley, Heinrich Malling, Eugene Oakberg, James Regan, Liane Russell, William Russell, Drew Schwartz, Jane Setlow, George Stapleton, Carl Swanson, R.C. von Borstel, William Welshons, Sheldon Wolff. In addition, there were many postdoctoral students, too numerous to mention. Although Oak Ridge was often thought of as the epitome of big science, it was, in actual fact, one of the better examples of the power of small science. There were several large projects — large because they used many animals — but the research depended on the ingenuity of individual investigators and collaborative efforts among them. The system worked because those individuals who wanted large groups under their control went elsewhere. Hollaender's role was to urge us on, to goad us to greater achievements, to help us become known nationally and internationally, and to be proud of those who left. To tell the truth, he was always annoyed if someone left Oak Ridge because he could not image that there was a place as good as his Biology Division.

I suspect that Hollaender will be remembered as much for his educational efforts as for his scientific achievements. He organized innumerable symposia and meetings in the developing world. The last that he helped organize was in Mexico in December 1986. One of the participants, Hugo F. Hoenigsberg of the Universidad de los Andes in Bogota, Colombia, wrote to him on December 5, "Here I am at the 1st Congress of Mutagenesis, Carcinogenesis and Teratogenesis of Mexico with many nice people who remember you as our great leader!!" and later in the letter, "Do remember all of us, in North and South America, who remember our great science mentor and leader!"

Alexander Hollaender left a mark on all of us, young and old. He taught us to be thoughtful and helpful, to love the world, its flora, its fauna, and its people, to love science and music and art. He will be missed by many.

References

Hollaender, A., ed. 1954, 1955, 1956. Radiation Biology, Vols. 1–3, McGraw Hill, New York.

Hollaender, A., ed. 1971, 1973, 1977. Chemical Mutagens: Principles and Methods for Their Detection. Vols. 1–4. Plenum Press, New York.

Hollaender, A. and W.D. Claus. 1937. An experimental study of the problem of mitogenetic radiation. Bull. 100, Natl. Res. Council, Nat. Acad. Sci. USA.

Hollaender, A. and J.J. Curtis. 1935. Effect of sublethal doses of monochromatic ultraviolet radiation on bacteria in liquid suspension. Proc. Soc. Exp. Biol. Med. 33:61–62.

Hollaender, A. and C.W. Emmons, 1941. Wavelength dependence of mutation production in the ultraviolet with special emphasis on fungi. Cold Spring Harbor Symp. Quant. Biol. 9:179-186.

Hollaender, A. and E. Schoeffel, 1931. Mitogenetic rays. Q. Rev. Biol. 6:215–222.

Rosenstein, B.S. and R.B. Setlow, 1982. The critical target in mammalian cells for solar ultraviolet radiation. Comments Mol. Cell Biophys. 1:223–235.

Setlow, J. and A. Hollaender, eds. 1979, 1980, 1981, 1982, 1983, 1984, 1985, 1986, 1987. Genetic Engineering: Principles and Methods, Vols. 1–9, Plenum Press, New York.

Setlow, R. and W.L. Carrier, 1964. The disappearance of thymine dimers from DNA: an error-correcting mechanism. Proc. Natl. Acad. Sci. USA 51:226–231.

Defective DNA Repair and Human Disease

A.R. LEHMANN

MRC Cell Mutation Unit
University of Sussex Falmer
Brighton
Sussex BN1 9RR, England

The identification of several human genetic disorders associated with cellular defects in the ability to repair or process damage in DNA has been of immense value in helping to understand the molecular mechanisms of DNA repair, their complexity and their importance for the maintenance of a healthy human condition. In 1968 the importance of DNA repair in man was revealed with the discovery by James Cleaver (1968) that cells cultured from individuals with the genetic disorder, xeroderma pigmentosum (XP) were defective in their ability to carry out excision-repair of ultraviolet (UV) damage from their DNA. Shortly afterwards this defect was shown to result in cellular hypersensitivity to the killing effects of UV light (Cleaver, 1970; Andrews et al., 1978). Genetic heterogeneity in the condition was revealed by the finding of initially two (De Weerd-Kastelein et al., 1972) and subsequently a total of seven (Keijzer et al., 1979) genetically distinct complementation groups, all partly or totally defective in excision repair. A variety of studies in the 1970's and early 1980's using several different techniques suggested that the defect in all nine complementation groups was in a very early stage of the repair process, namely at or prior to the incision step. Introduction into XP cells of microbial enzymes capable of carrying out this incision step restored excision repair to the level found in normal cells (Tanaka et al., 1975; De Jonge et al., 1985). Further understanding of the molecular basis of the defect in XP has been hampered by the failure, despite extensive efforts in many laboratories, either to purify the enzyme(s) or to clone the gene(s) which are defective in XP cells. Advances in these areas have however been made in the last year. Wood et al. (1988) have developed a cell free system from normal cells which is capable of repairing UV damaged plasmids. This activity was not found in extracts from several XP cells. This provides a system for attempting the purification from normal cells of the "XP enzymes". In the area of molecular biology, Tanaka et al. (1988), following a monumental transfection experiment, has isolated lambda DNA clones which in combination are able to correct the defect in XP-A cells. It is likely therefore that the XP-A gene will have been cloned by the time this volume is published[*].

The relationship between the repair defect in XP and the clinical symptoms was cast in doubt by the discovery in 1971 of a variant form of XP in which classical XP symptoms were not associated with a defect in excision-repair (Burk et al., 1971), nor were cultured cells from these XP variants particularly sensitive to the killing action of UV. This dilemma was resolved in 1975 when my colleagues and I showed that XP variants had a defect in a UV tolerance mechanism termed daughter-strand repair (Lehmann et al., 1975). XP variants had a deficiency in the ability to synthesize intact DNA molecules during DNA replication following UV-irradiation. Although the mechanism of daughter-strand repair remains unclear, our observations on XP variants meant that all XP cells had some kind of defect in DNA repair. A further unifying finding was that all XP cell strains studied, whether of the excision-deficient (Maher and McCormick, 1976) or of the variant type (Maher et al., 1976), were hypermutable by UV light as well as by a large group of chemical carcinogens. Work on XP thus provided strong support for the somatic mutation theory of carcinogenesis. Defective repair of DNA at the molecular level results in increased mutability at the cellular level, which in turn gives rise to an increased frequency of carcinogenesis in the individual.

More recent studies on two other genetic disorders suggest that this hypothesis is an oversimplification.

In 1977 Schmickel et al. discovered that cells from the genetic disorder Cockayne's syndrome (CS) were, like XP cells, hypersensitive to the lethal action of UV light. CS patients are photosensitive, but unlike XP patients, they are not cancer-prone and they display a variety of symptoms not seen in XP patients, including dwarfism, retinal atrophy, premature aging and skeletal deformities. Despite their sensitivity to UV light, CS cells showed no gross abnormalities in either excision or daughter-strand repair (Schmickel et al., 1977; Wade and Chu, 1979; Mayne et al., 1982). In spite of this apparently normal repair, we were able to identify an abnormality in an early response of CS cells to UV-radiation (Mayne and Lehmann, 1982). RNA synthesis was inhibited by UV light in all cells, presumably because UV damage blocked the progress of the transcriptional machinery. This inhibition was rapidly overcome in normal cells, and RNA synthesis recovered completely within two hours. In CS cells this recovery process failed to occur. On the basis of these observations we postulated that in normal cells there was a special excision-repair process which rapidly removed UV damage from actively transcribed genes, the damage from the bulk of the DNA being repaired much more slowly. In CS cells this postulated special repair process would be defective.

Direct evidence for preferential repair of active genes in both hamster and human cells was subsequently provided by the elegant experiments of Bohr et al. (1985) and

---

[*]Note added in proof: Tanaka et al. (Proc. Natl. Acad. Sci., USA 86, 5512–5516 (1989) have recently reported the cloning and partial characterization of the XP-A gene.

Mellon et al. (1986), and recently Mayne et al. (1988) have shown that this preferential repair of an active gene is indeed inoperative in CS cells. It now seems likely that following UV-radiation, the top priority for the cell is to remove damage from active regions of the genome, ie. from those genes whose function must be maintained for cellular metabolism to continue. CS cells appear to be deficient in this crucial specific repair process. As with XP, the deficiency results in both increased lethality (Wade and Chu, 1979) and mutability (Arlett, 1980) following UV irradiation. Despite this similarity at the cellular level, CS is not a cancer-prone syndrome.

Very recently a third UV-repair-deficient disorder has been identified. Trichothiodystropy (TTD) is a genetic disorder associated with sulphur-deficient brittle hair, physical and mental retardation, ichthyosis, and in some individuals photosensitivity. Neither freckles nor skin cancer are features of this disorder. In 1986 Stefanini and co-workers showed that four Italian TTD patients were deficient in excision-repair of UV damage. They showed furthermore that this defect was in the same gene as that of XP complementation group D. Thus fusion of TTD with XP-D cells did not result in restoration of repair levels in the heterokaryons to those in normal cells, whereas fusion with XP cells from other complementation groups did restore normal levels. We have extended the findings of Stefanini et al., (1986), and in collaboration with her group we have identified three categories of cellular response to UV-irradiation among twelve TTD patients (Lehmann, 1987; Lehmann et al., 1988 and unpublished observations). One category (two patients) showed a totally normal response to UV. Cells in a second category (eight patients) displayed a UV response identical to that of XP-D cells. Cells in the third category (two patients) showed a more complex response. Following UV-irradiation cell survival, mutagenesis and removal of pyrimidine dimers were normal, whereas the rate of repair synthesis and incision of damaged DNA were only 50% of normal. Recent results suggest that these cells remove the 6-4 photoproduct from their DNA at a reduced rate (Broughton et al., 1990).

The phenotype of TTD poses even more of a quandary than CS. Many of the TTD patients appear at the present time to have a cellular UV-response which is indistinguishable from that of XP-D cells and which seems to result from a mutation in the "XP-D gene". Yet the clinical symptoms of TTD and XP are quite different, as summarized in Table 1. A possible clue to help explain this paradox comes from the recent findings of Norris and co-workers (1990) who have carried out an immunological study of five XP, two CS and one TTD patients. The only abnormality that they observed was a reduced Natural Killer cell (NK) activity in all the XP patients, but not in the CS or TTD patients. Several years ago Bridges (1981) proposed that the increased frequency of skin cancers in XP resulted not only from the elevated sunlight-induced mutation frequency in the skin but also from decreased immune surveillance in XP patients. The results of Norris et al offer some support for this hypothesis. Bridges (1981) suggested that the putative immune deficiency resulted from the repair deficiency. In view of the normal NK activity in a TTD patient with the same repair deficiency as XP-D cells, its is conceivable that the full clinical phenotype of XP results from

Table 1. XP, CS and TTD: Clinical comparison

	XP	CS	TTD
Sun-sensitive	+	+	+or-
Freckles	+++-	-	-
Skin cancers	+++	-	-
Ichtyosis	-	-	+
Dwarfism	-	+	+
Mental retardation	+or-	+	+
Neurological degeneration	+or-	+	
Cataract	-	+or-	+
Brittle hair	-	-	+

mutations in two independent loci, one controlling DNA repair, the other being involved with NK activity. The idea that XP may result from two co-recessive mutations has been previously suggested in a different context by Lambert and Lambert (1985). In CS and TTD the repair deficiency is present without the immune deficiency. This may be the reason that skin tumours are not associated with these disorders. These ideas which will be developed in detail elsewhere are highly speculative, but they may provoke further experimentation to help unravel the complexities of the relationships between DNA repair, mutagenesis and carcinogenesis.

References

Andrews, A.D. Barrett, S.F. and Robbins J.H. (1978). Xeroderma pigmentosum neurological abnormalities correlate with colony-forming ability after ultraviolet radiation. *Proc. Natl. Acad. Sci. USA* 75, 1984–1988.

Arlett, C.F. (1980). Mutagenesis in repair-deficient human cell strains. *In:* "Progress in Environmental Mutagenesis". M. Alacevic. Ed., Elsevier, Amsterdam. pp. 161–174.

Bohr, V.A. Smith, C.A. Okumoto, D.S. and Hanawalt, P.C. (1985). DNA repair in an active gene: removal of pyrimidine dimers from the DHFR gene of CHO cells in much more efficient than in the genome overall. *Cell* 40, 359–369.

Broughton, C.B., Lehmann, A.R., Harcourt, S.A., Artlett, C.F., Sarasin, A., Kleijer, W.J., Beemer, F.A., Nairn, r. and Mitchell, D.L. (1990). Relationship between pyrimidine dimers, 6-4 photoproducts, repair synthesis and cell survival: studies using cells from paitents with trichothiodystropy. *Mutation Res.* 235, 33–40.

Bridges, B.A. (1981). How important are somatic mutations and immune control in skin cancer? Reflections on xeroderma pigmentosum. *Carcinogenesis* 2, 471–472.

Burk, P.G. Lutzner, M.A. Clarke, D.D. and Robbins, J.H. (1971). Ultraviolet-stimulated thymidine incorporation in xeroderma pigmentosum lymphocytes. *J. Lab. Clin. Med.* 77, 759–767.

Cleaver, J.E. (1968) Deficiency in repair replication of DNA in xeroderma pigmentosum. *Nature* 218, 652–656.

Cleaver, J.E. (1970). DNA repair and radiation sensitivity in human (xeroderma pigmentosum) cells. *Int. J. Radiat. Biol.* 18, 557–565.

De Jonge, A.J. Vermeulen, R.W. Keijzer, W. Hoeijmakers, J.H.J. and Bootsma, D. (1985). Microinjection of *Micrococcus luteus UV-endonuclease* restores UV-induced unscheduled DNA synthesis in cells of 9 xeroderma pigmentosum complementation groups. *Mutation Res.* 150, 99–105.

De Weerd-Kastelein, E.A. Keijzer, W. and Bootsma, D. (1972). Genetic heterogeneity of xeroderma pigmentosum demonstrated by somatic cell hybridisation. *Nature (New Biol.)* 238, 80-83.

Fischer, W., Jaspers, N.g.J., Abrahams, P.J., Taylor, A.M.R., Arlett, C.F., Zelle, B., Takebe, H., Kinmont, P.D.S., and Bootsma, D. (1979). A seventh complementation group in excision-deficient xeroderma pigmentosum. Mutation Res. 62, 183–190.

Lambert, W.C. and Lambert, M.W. (1985). Co-recessive inheritance: A model for DNA repair, genetic disease and carcinogenesis. *Mutation Res.* 145, 227–234.

Lehmann, A.R. (1987). Cockayne's syndrome and trichothiodystrophy: defective repair without cancer. *Cancer reviews* 7, 82–103.

Lehmann, A.R. Arlett, C.F. Broughton, B.C. Harcourt, S.A. Steingrimsdottir, H. Stefanini, M. Taylor, A.M.R. Natarajan, A.T. Green, S. King, M.D. MacKie, R.M. Stephenson, J.B.P. and Tolmie J.L. (1988). Trichothiodystrophy, a Human DNA Repair Disorder with Heterogeneity in the Cellular Response to Ultraviolet Light. *Cancer Res.* 48m, 6090–6096.

Lehmann, A.R. Kirk-Bell, S. Arlett, C.F. Paterson, M.C. Lohman, P.H.M. De Weerd-Kastelein, E.A. and Bootsma, D. (1975). Xeroderma pigmentosum cells with normal levels of excision repair have a defect in DNA synthesis after UV-irradiation. *Proc. Nat. Acad. Sci. USA* 72, 219–223.

Maher, V.M. McCormick, J.J. (1976). Effect of DNA repair on the cytotoxicity and mutagenicity of UV irradiation and of chemical carcinogens in normal and xeroderma pigmentosum cells. In: "Biology of Radiation Carcinogenesis", J.M. Yuhas, R.W. Tennant and J.B. Regan, Eds. Raven Press, New York pp. 129–145.

Maaher, V.M., Ouellette, L.M. Curren, R.D. and McCormick, J.J. (1976). Frequency of ultraviolet light-induced mutations is higher in xeroderma pigmentosum variant cells than in normal human cells. *Nature* 261, 593–595.

Mayne, L.V. Lehmann, A.R. (1982). Failure of RNA Synthesis to Recover after UV-irradiation: An Early Defect in Cells from Individuals with Cockayne's Syndrome and Xeroderma Pigmentosum,. *Cancer Res.* 42, 1473–1478.

Mayne, L.V. Lehmann, A.R. and Waters, R. (1982). Excision repair in Cockayne syndrome. *Mutation Res.* 106, 179–189.

Mayne, L.V. Mullenders, L.H.F. Van Zeeland, A.A. (1988). Cockayne's Syndrome: A UV sensitive disorder with a defect in the repair of transcribing DNA but normal overall excision repair. In: "Mechanisms and Consequences of DNA Damage Processing", E. Friedberg and P. Hanawalt, Eds. A. R. Liss, New York, pp. 349–353.

Mellon, I., Bohr, V.A. Smith, C.A. and Hanawalt, P.C. (1986). Preferential DNA repair of an active gene in human cells. Proc. Natl. Acad. Sci. USA 83, 8878–8882.

Norris, P.G., Limb, G.A., Hamblin, A.S., KLehmann, A.R., Arlett, C.f., Cole, T., Waugh, A.P.W. and Hawk, J.L.M. (1990). Immune function, mutant frequency, and cancer risk in the DNA repair defective genodermatoses xeroderma pigmentosum, Cockayne's Syndrome, and trichothiodystrophy. *J. Invest. Dermatol.* 94, 94–100.

Schmickel, R.D. Chu, E.H.Y. Trosko, J.E. and Chang, C.C. (1977). Cockayne Syndrome: a cellular sensitivity to ultraviolet light. *Pediatrics* 60, 135–139.

Stefanini, M. Lagomarsini, P. Arlett, C.F. Marinoni, S. Borrone, C. Crovato, F. Trevisan, G. Cordone, G. and Nuzzo, F. (1986). Xeroderma pigmentosum (complementation group D) mutation is present in patients affected by trichothiodystrophy with photosensitivity. *Hum. Genet.* 74, 107–112.

Tanaka, K. Satokata, I. Ogita, Z. Okada, Y. (1988). Toward the molecular cloning of the gene for xeroderma pigmentosum, complementation group A by DNA transfection method. *J. Cell. Biochem.* Suppl. 12A–246.

Tanaka, K. Sekiguchi, M. and Okada, Y. (1975). Restoration of ultraviolet-induced unscheduled DNA synthesis of xeroderma pigmentosum cells by the concomitant treatment with bacteriophage T4 endonuclease V and HVJ (Sendai virus). *Proc. Natl. Acad. Sci. USA* 72, 4071–4075.

Wade, M.H. and Chu, E.H.Y. (1979). Effects of DNA damaging agents on cultured fibroblasts derived from patients with Cockayne syndrome. *Mutation Res.* 59, 49–60.

Wood, R.D. Robins, P. and Lindahl, T. (1988). Complementation of the xeroderma pigmentosum DNA repair defect in cell-free extracts. *Cell* 53, 97–106.

DNA-Protein Crosslinking in UV-Irradiated Human and ICR 2A Cell Lines

BARRY S. ROSENSTEIN[1] AND [2]LI-WEN LAI

[1]*Department of Radiation Medicine*
Box G, Brown University
Providence, RI 02912, USA and
[2]*The Eleanor Roosevelt Institute for Cancer Research*
1899 Gaylord Street, Denver, CO 80206, USA

Abstract

The levels of DNA-protein crosslinks (DPC) were measured using the alkaline elution assay in normal human skin fibroblasts and ICR 2A frog cell lines following UV-irradiation. For experiments utilizing human cells, the cultures were exposed to 0-200 J/m^2 of 254 nm UV and incubated 0-24 hr. It was found for these cells, that the levels of DPC increased upon incubation. In addition, when the DPC were eliminated by treatment with proteinase K, large numbers of DNA strand breaks (SSB) were revealed. The yields of SSB also increased upon incubation of the UV-irradiated cells, with kinetics similar to that observed for DPC induction. These results are suggestive of a role for a topoisomerase in this crosslinking process. DPC induction was also investigated in ICR 2A cells and two cell lines, DRP 36 and DRP 153, derived from ICR 2A, based upon their sensitivity to the solar UV induction of non-dimer DNA damages. For these experiments, the cells were treated with 150 kJ/m^2 of sunlamp UV, which was filtered through 48A Mylar to eliminate wavelengths shorter than approximately 315 nm. In addition, the cells were exposed to photoreactivating light to eliminate most of the small yield of dimers produced by this treatment. Upon incubation of the ICR 2A cells, a rapid increase in the level of DPC was detected that reached a maximum within about 2-4 hr following irradiation. In contrast, the enhancement in DPC was much less pronounced, and decreased more rapidly in the two mutant cell lines. These results suggest that this enhancement in DPC may be indicative of a process that plays a role in cellular survival following UV-irradiation.

Abbreviations: DPC, DNA-protein crosslinks; PBS, phosphate-buffered saline; PRL, photoreactivating light; SDS, sodium dodecyl sulfate; SSB, DNA single-strand breaks.

Introduction

Exposure of cells to UV radiation results in the formation of DNA-protein crosslinks (DPC). Experiments have been performed in recent years to study the fate of these DPC. Surprisingly, it has been found that the level of DPC actually increases in cells following incubation. These results have been obtained with 254 nm UV-irradiated Chinese hamster cells (Chiu, 1984) and normal human cells exposed to sunlamp UV (Lai and Rosenstein, 1987). In addition, the levels of DNA strand breaks (SSB) have also been examined and the results obtained have been found to be dependent upon whether the SSB were examined under conditions in which any DPC that may be present in the cells were eliminated (Rosenstein 1988, 1989). For assay conditions in which they were not, the level of SSB rapidly decreased. In contrast, when SSB were measured under conditions in which the DPC were eliminated, then the level of SSB detected increased. Therefore, the SSB which formed during incubation following UV-irradiation appeared to be associated with the DPC.

In order to investigate the repair of DPC in greater detail, mutant cell lines that are hypersensitive to solar UV wavelengths and deficient in the repair of non-dimer DNA damages have been isolated (Rosenstein and Chao, 1985). This has been accomplished using the ICR 2A frog cell line which possesses two useful features that have made the isolation of this type of mutant feasible. The first is that it is highly proficient in enzymatic photoreactivation (Rosenstein and Setlow, 1980). Hence, it is possible to irradiate these cells with solar UV wavelengths, which induce few dimers relative to non-dimer damages, and then eliminate most of the dimers by exposure to photoreactivating light (PRL). This results in the production of a relatively pure population of non-dimer damages. The second feature these cells possess is that they contain a haploid complement of chromosomes which should increase the likelihood of producing a mutant cell line with a particular phenotype (Freed and Mezger-Freed,1970).

The purpose of the experiments described in this paper was to examine in greater detail the induction of these DPC in normal human cells and in the frog mutant cell lines to determine whether this type of DNA damage plays a role in the hypersensitivity to solar UV radiation exhibited by these cells.

Materials and Methods

Cell strains and culture conditions

The human skin fibroblast cell line GM 3468 was derived from the foreskin of a normal male and was obtained from the Human Genetic Repository (Institute for Medical Research, Camden, NJ). Cells were grown in Eagle's Minimum Essential Medium (GIBCO, Grand Island, NY) supplemented with 10% fetal calf serum (Hyclone, Logan, UT), 100 units/ml penicillin and 100 μg/ml streptomycin, 2 mM L-glutamine, 2 × amino acids solution, 2 × non-essential amino acids solution and 2 × vitamin solution (GIBCO). Cultures were incubated at 37°C in a 5% CO_2 humidified atmosphere. Under these conditions the cells had a doubling time of 1–2 days. The ICR 2A, DRP 36 and DRP 153 frog cells were grown at 25°C in modified Leibowitz medium (GIBCO, Grand Island, NY) supplemented with 10% fetal calf serum (Hyclone, Logan, UT), 100 units/ml penicillin and 100 μg/ml streptomycin (GIBCO). Under these conditions the cells had a doubling time of 48 hr.

Labeling and irradiation conditions

Cells were plated in either 25 cm^2 or 75 cm^2 flasks (Corning Glass Works, Corning, NY) at a density of 10^4 cells/cm^2 and either methyl-[^{3}H]-thymidine (20 Ci/mmol, New England Nuclear, Boston, MA) or 2-[^{14}C]-thymidine (59 mCi/mmol) were added to final concentrations of 0.1 μCi/ml or 0.02 μCi/ml, respectively. Following a 72 hr incubation the ^{14}C-labeled cells were washed three times with phosphate buffered saline (PBS) covered with 5 ml of PBS and irradiated while being held on ice. The 254 nm UV was produced by four GE G15T8 germicidal lamps while the sunlamp UV > 315 nm represents the UV produced by two FS40 Westinghouse fluorescent sunlamps in which the UV was passed through a sheet of 48A Mylar (DuPont, Wilmington, DE) to eliminate wavelengths shorter than approximately 315 nm (Rosenstein, 1984). The fluence rates were measured using a IL1700 radiometer and were 0.3 W/m^2 for the 254 nm irradiations and 10.4 W/m^2 for the sunlamp UV > 315 nm exposures. The PRL treatment consisted of exposure to 30 kJ/m^2 of the light emitted by two GE F40B blue lamps. The PBS was then replaced with medium and the cells incubated 0-24 hr.

Elution conditions and calculations of DPC and SSB

Cells were washed twice with PBS and gently scraped off the flasks into ice-cold PBS containing 0.2 mg/ml Na$_2$EDTA. 5 × 10^5 ^{14}C-labeled cells exposed to a particular treatment were then mixed with 5x10^5 ^{3}H-labeled cells and irradiated with 10 Gy of x-rays (250 kV Phillips Model RT 250 at a dose rate of approximately 1.1 Gy/min) while held on ice for the determination of the DPC level (Kohn et al, 1981). In addition, for each treatment, 5x10^5 ^{14}C-labeled cells not exposed to x-ray were mixed with 5x10^5 ^{3}H-labeled cells that had been irradiated with 10 Gy of x-rays in order to determine the level of SSB. Each cell mixture was then loaded on a 25 mm 2 μm pore size polyvinyl chloride filter (Millipore Corp., Bedford, MA), washed twice with cold PBS, lysed with a solution containing 2% sodium dodecyl sulfate (SDS, Gallard-Schlesinger, Carleplace, NY), 0.1 M glycine and 0.02 M EDTA, pH 10. The lysis solution was allowed to flow through the filter by gravity. For elutions to measure DPC, the cell lysates were then washed with 5 ml of 0.02 M Na$_2$EDTA, pH 10, whereas for measurement of SSB, 2 ml of lysis solution containing 0.2 mg/ml proteinase K (Scientific Products) was then added. In both cases, this was followed by the addition of an elution solution consisting of 0.1 M tetrapropyl-ammonium hydroxide and 0.02 M EDTA (acid form), pH 12.1. The elution solution was pumped through the filters at approximately 0.04 ml/min and 5 ml fractions were collected at 3 h intervals for detection of DPC, while the pump speed was increased to 0.3 ml/min and fractions collected every 30 min for measurement of SSB. Upon completion of the elutions, the fractions were made isovolumetric with water and mixed with 10 ml of Budget-Solve (Research Products International Corp., Mount Prospect IL) containing 0.3% acetic acid. Filters were processed as previously described (Kohn et al 1981). All fractions were counted in a Packard 2000 scintillation counter.

The number of DPC was calculated from the following equation (Peak et al., 1987):

$$DPC_{UV} = [(1-r)^{-1/2} (P_{bR} + P_{UV})] - [(1-r_0)^{-1/2} (P_{bR}) + P_{UV}]$$

where DPC_{UV} is the frequency of UV-induced DPC, P_{bR} is the frequency of x-ray induced SSB, P_{UV} is the frequency of UV-induced SSB and r and r_0 are the fractions of DNA eluting in the slow component of the elution profiles extrapolated to time zero for UV-treated and untreated cells, respectively.

The yield of SSB induced by the UV-irradiations was calculated from the following equation:

$$SSB/dalton = 27 \times 10^{-10} (B_{UV} - B_{unirr}) / (B_{10Gy} - B_{unirr})$$

where B equals the logarithm of the fraction of DNA retained by the filter after 0 min of elution minus the logarithm of the fraction of DNA retained by the filter after 120 min of elution. The x-ray dose used, 10 Gy, produces approximately 27 breaks/10^{10} daltons (Kohn et al., 1976).

Results and Discussion

Normal human skin fibroblasts were exposed to 20-200 J/m² of 254 nm UV, incubated either 0 or 24 h, and the level of SSB and DPC measured using the alkaline elution assay. As indicated in Figs. 1 and 2, the levels of both DNA alterations increased upon incubation. However, the increase in SSB was only detectable when the DNA on the filters was treated with proteinase K to eliminate any DPC that may be present in the cells. When this treatment was not performed, then the rate of elution of DNA from cells exposed to 100 J/m² of UV and incubated 24 h was essentially identical to the elution profile for cells not exposed to UV (Fig. 3). Nevertheless, the induction of SSB could

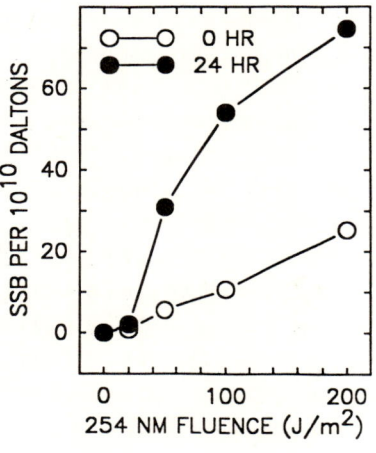

Figure 1. SSB induction in UV-irradiated human cells.

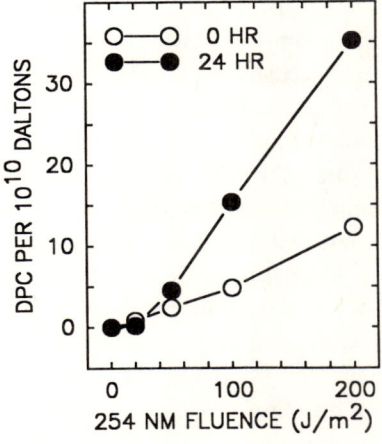

Figure 2. DPC induction in UV-irradiated human cells.

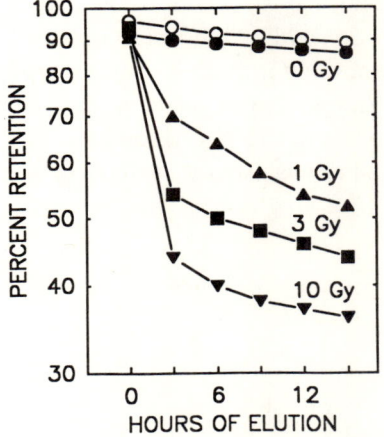

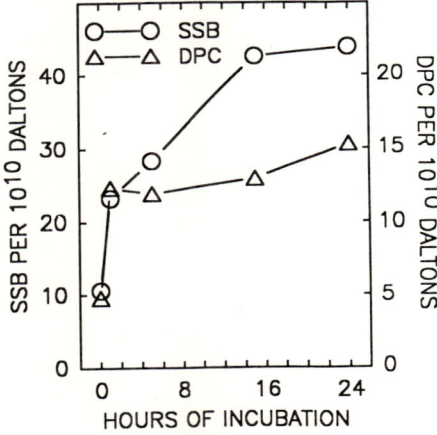

Figure 3. Elution profiles for human cells exposed to 0 (O) or 100 J/m^2 of 254 nm UV, incubated 24 h, and irradiated with either 0 (●), 1 (▲), 3 (■), or 10 (▼) Gy of x-rays.

Figure 4. Kinetics of SSB and DPC induction in human cells exposed to 100 J/m^2 of 254 nm UV.

still be manifested by exposure to x-rays. Hence, the SSB produced during incubation of UV-irradiated cells were specifically hidden by the DPC, while other SSB induced were not concealed. In addition, this result demonstrates that the increase in the rate of elution for DNA from UV-irradiated and incubated cells for experiments in which proteinase K was used, was not due to cellular lysis and general DNA breakdown.

The kinetics of the induction of SSB and DPC were also examined by exposure of cells to 100 J/m² of UV and incubation for 0-24 h following irradiation. Both the levels of SSB and DPC (Fig. 4) increased rapidly upon incubation and remained elevated for at least 24 h. Longer incubation times were not utilized as this fluence was highly lethal and cell lysis accompanied by DNA degradation occurred at longer incubation times.

ICR 2A cells and two solar UV-sensitive cell lines, DRP 36 and DRP 153, were irradiated with 150 kJ/m² of sunlamp UV > 315 nm. These cells were also exposed to PRL, resulting in the elimination of most of the small yield of dimers induced by the sunlamp irradiation. The irradiated cells were incubated for either 0–24 h and then subjected to alkaline elution. For each cell line, there was a rapid increase in the level of DPC which reached a peak within about 2 hr following irradiation and then declined. However, the increase in the level of DPC was much less pronounced in the two solar UV-sensitive cell lines. Also, the rate and extent of decrease in the level of DPC was much greater for DRP 36 and DRP 153 than that exhibited by the ICR 2A parental cell line (Fig. 5).

One explanation for these results may be that the protein involved in this crosslinking is part of a repair process or cellular response to the perturbations in DNA structure caused by the presence of damages in DNA. Possible candidates for this protein may be a type I or II topoisomerase as these proteins may recognize the alterations to DNA structure caused by the presence of damage (Liu, 1983). The results obtained are also consistent with the involvement of a topoisomerase I or II as these enzymes produce either single-strand or double-strand breaks in DNA, respectively, associated with their covalent attachment to DNA (Chen, 1984; Nelson et al., 1984; Tewey et al., 1984; Hsiang et al, 1985; Mattern, 1987). Another potential mechanism for the involvement of

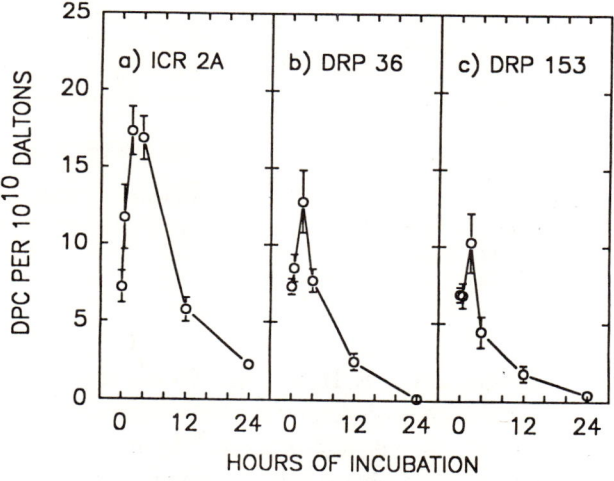

Figure 5. Kinetics of DPC induction in sunlamp UV > 315 nm-irradiated frog cells.

a topoisomerase is that the DNA damages may result in the trapping of the enzyme as it is performing its normal functions. Consistent with this hypothesis, evidence has been obtained indicating that topoisomerase I is inhibited by the induction of dimers in DNA (Pedrini, 1983, 1984). However, evidence supportive of a role for this crosslinking process in a repair or recovery system were the results obtained with the two solar UV-sensitive frog mutants which displayed a much smaller enhancement in DPC induction following solar UV-irradiation compared with the parental ICR 2A cells. This finding is consistent with the conclusion that this crosslinking process plays an important role in cellular survival.

Experiments are currently in progress to determine the nature of the proteins that are involved in formation of the DPC in these cell lines following UV-irradiation

Acknowledgements

This work was supported by DHHS grants CA45078 from the National Cancer Institute and ES04355 from the National Institute of Environmental Health Sciences.

References

Chen, G.L., L. Yang, T.C. Rowe, B.D. Halligan, K.M. Tewey and L.F. Liu (1984) Nonintercalative antitumor drugs interfere with breakage-reunion reaction of mammalian DNA topoisomerase II. *J. Biol. Chem.* 259, 13560-13566.

Chiu, S.M., N.M. Sokany, L.R. Friedman and N.L. Oleinick (1984) Differential processing of UV or ionizing radiation induced DNA-protein in Chinese hamster cells. *Int. J. Radiat. Biol.* 46, 681-690.

Freed, J.J. and L. Mezger-Freed (1970) Stable haploid cultured cell lines from frog embryos. *Proc. Natl. Acad. Sci. (U.S.A.)* 65, 337-344.

Hsiang, Y., R. Hertzberg, S. Hecht and L.F. Liu (1985) Camptothecin induces protein-linked DNA breaks via mammalian DNA topoisomerase I. *J. Biol. Chem.* 260, 14873-14878.

Kohn, K.W., L.C. Erickson, R.A.G. Ewig and C.A. Friedman (1976) Fractionation of DNA from mammalian cells by alkaline elution. *Biochemistry* 15, 4629-4637.

Kohn, K.W., R.A.G. Ewig, L.C. Erickson and L.A. Zwelling (1981) Measurement of strand breaks and cross-links by alkaline elution. In *DNA Repair* (Edited by E.C Friedberg and P.C. Hanawalt), Vol. 1, Part B, pp. 379-401. Dekker, New York.

Lai, L.-W., J.M. Ducore and B.S. Rosenstein (1987) DNA-protein crosslinking in normal human skin fibroblasts exposed to solar ultraviolet wavelengths. *Photochem. Photobiol.* 46, 143-146.

Liu, L.F., T.C. Rowe, L. Yang, K.M. Tewey and G.L. Chen (1983) Cleavage of DNA by mammalian DNA topoisomerase II. *J. Biol. Chem.* 258, 15365-15370.

Mattern, M.R., S.-M. Mong, H.F. Bartus, C.K. Mirabelli, S.T. Crooke and R.K. Johnson (1987) Relationship between the intracellular effects of camptothecin and the inhibition of DNA topoisomerase I in cultured L1210 cells. *Cancer Res.* 47, 1793-1798.

Nelson, E.M., K.M. Tewey and L.F. Liu (1984) Mechanism of antitumor drug action: Poisoning of mammalian DNA topoisomerase II on DNA by 4'-(9-acridinylamino)-methanesulfon-m-aniside. *Proc. Natl. Acad. Sci. (U.S.A.)* 81, 1361-1365.

Peak, J.G., M.J. Peak and E.R. Blazek (1987) Improved quantitation of DNA-protein crosslinking caused by 405-nm monochromatic near-UV radiation in human cells. *Photochem. Photobiol.* 46, 319-321.

Pedrini, A.M. (1984) Effect of UV induced DNA lesions on the activity of *Escherichia coli* DNA topoisomerases. In *Proteins Involved in DNA Replication* (Edited by U. Hubscher and S. Spadari), pp. 449-454, Plenum Press, New York.

Pedrini, A.M. and G. Ciarrochi (1983) Inhibition of *Micrococcus luteus* DNA topoisomerase I by UV photoproducts. *Proc. Natl. Acad. Sci. (U.S.A.)* 80, 1787-1791.

Rosenstein, B.S. (1984) Photoreactivation of ICR 2A cells exposed to solar UV wavelengths. *Photochem. Photobiol.* 40, 207-213.

Rosenstein, B.S. (1988) The induction of DNA strand breaks in normal human skin fibroblasts exposed to solar ultraviolet radiation. *Radiat. Res.* 116, 313-319.

Rosenstein, B.S. (1989) DNA strand breakage in normal and solar UV sensitive ICR 2A cell lines exposed to solar UV wavelengths. *Environ. Mol. Mutagen.* In press.

Rosenstein, B.S. and C.C.-K. Chao (1985) Isolation of a mutant cell line derived from ICR 2A frog cells hypersensitive to the induction of non-dimer DNA damages induced by solar ultraviolet radiation. *Somatic and Molecular Genetics* 11, 339-344.

Rosenstein, B.S. and R.B. Setlow (1980) DNA repair after ultraviolet irradiation of ICR 2A frog cells: Pyrimidine dimers are long acting blocks to nascent DNA synthesis. *Biophys. J.* 31, 195-205.

Tewey, K.M., G.L. Chen, E.M. Nelson and L.F. Liu (1984) Intercalative antitumor drugs interfere with the breakage-reunion reaction of mammalian DNA topoisomerase II. *J. Biol. Chem.* 259, 9182-9187.

Retroviral Based Shuttle Vectors and Endogenous Cellular Genes: Complementary Approaches for Investigations of Mutational Specificity

E. A. DROBETSKY, A.J. GROSOVSKY, A. SKANDALIS, L. R. SHEKTER
and B. W. GLICKMAN

York University
Dept. of Biology
Toronto, Canada M3J 1P3

Introduction

We have previously employed an *in vivo* recombinational technique to characterize mutational specificity at the endogenous adenine phosphoribosyl transferase (*aprt*) locus in Chinese hamster ovary (CHO) cells (Drobetsky et al, 1987; Grosovsky et al, 1988; de Jong et al, 1988; Mazur et al, 1988). The use of an endogenous cellular locus as a mutational target was an important aspect of the investigation. However, the effect of chromosomal position could not be addressed. In addition, the target gene is not readily accessible for *in vitro* manipulations that may prove useful for mechanistic studies. For example, directed sequence modifications in the target gene can be used for examining the influence of local DNA context in the appearance of hot spots.

We have thus continued to develop the *aprt* system by expanding the possibilities to include a retroviral-based shuttle vector system containing the *aprt* gene as the mutational target. A cDNA shuttle vector was created and used to transform an APRT⁻ CHO cell. An APRT⁺ derivative carrying a stably integrated, single copy vector has been obtained. To validate the system we have initially characterized a collection of UV induced mutants. In this communication we compare the results obtained with the shuttle vector to our previously published study of UV-induced mutational specificity at the endogenous *aprt* locus.

Methodology

Development of a Retroviral Shuttle Vector

The vector pRVA3, constructed by Dr P. de Jong in collaboration with our laboratory, was chosen for these studies. It is a derivative of pZip Neo SV(X) (Cepko et al, 1984) which contains an SV40 origin of replication, and the *neo* gene conferring

Photobiology, Edited by E. Riklis
Plenum Press, New York, 1991

G418 resistance in mammalian cells and kanamycin resistance in bacteria. A 2.6 kb fragment containing the entire *aprt* coding sequence along with introns and flanking regions was cloned into the vector. This insert was appropriately spliced during cellular processing of the retroviral genome and therefore, the mutational target is a cDNA version of the *aprt* gene. In addition, an M13 origin of replication and the bacterial tRNA suppressor *supF* were introduced.

Isolation of CHO Derivative Carrying Stably Integrated Retrovirus Shuttle Vector

CHO cells were chosen for these experiments since we have obtained mutational spectra using the endogenous *aprt* locus of this cell line. An APRT⁻ CHO derivative, UV 37, was transformed to APRT⁺ by infection with the retrovirus shuttle vector. One single copy infectant (UV 37-G2) was chosen for further experimentation due to its low frequency (3×10^{-5}) of reversion to the APRT⁻ phenotype. Simultaneous selection with G418 further reduced the apparent reversion frequency to approximately 5×10^{-6}. We therefore have chosen to maintain selective pressure for $G418^R$ throughout the experiment.

Recovery and Analysis of the Shuttle Vector

Recovery of plasmid constructs carrying the SV40 origin of replication is generally achieved by fusion with cell lines which constitutively express SV40 large T antigen. COS cells, a CV-1 cell derivative, are commonly used for this purpose.

UV 37-G2 are fused with COS cells and low molecular weight DNA is recovered by Hirt extraction for transformation of an appropriate bacterial host. Approximately 25% of the plasmids contain gross rearrangements visualized using agarose gel electrophoresis. The remaining 75% demonstrate the anticipated restriction enzyme cleavage pattern. Only the unrearranged plasmids were further analyzed by DNA sequencing.

UV Light Mutagenesis of the Shuttle Vector

A fluence dependent induction of APRT⁻ mutations was observed for cultures of UV37-G2 irradiated with 254 nM UV light. 5 J/m² (85% survival) resulted in a 50-fold increase over background and this dose was chosen for our mutant collection. This is the same dose used to produce the UV-light mutational spectrum in the endogenous gene.

Results and Discussion

Classes of Mutation Induced by UV-Light Within the Coding Sequence of the Chinese Hamster aprt Locus

Table 1 compares the types of UV-induced mutations recovered in the endogenous CHO *aprt* locus (Drobetsky et al, 1987), and in the spliced version of the gene carried on the retroviral shuttle vector. The endogenous data presented therefore, includes only

Table 1

	Endogenous gene	Shuttle vector
Transitions	15 (48%)	18 (72%)
G:C => A:T	14	15
A:T => G:C	1	3
Transversions	8 (26%)	3 (12%)
G:C => T:A	0	0
G:C => C:G	4	1
A:T => T:A	2	1
A:T => C:G	2	1
Frameshift	1 (3%)	0
Deletion	0	1 (4%)
Multiple Events	7 (23%)	3 (12%)
Total	31	25

coding sequence alterations; the 3 mutations occurring in intron sequences at splice junctions have been excluded from the comparison.

An examination of Table 1 reveals that both spectra are characterized by a predominance of G:C to A:T transition events. In general, there appeared to exist an overall similarity among other classes of events. However, some specific differences may be noteworthy. For example, the multiple events class may include important qualitative differences. In these mutants, more than one sequence alteration has occurred. We believe that these multiple events are unlikely to be attributable to independent mutations. In the first place, frequency considerations would argue against the idea and secondly, in every case the multiple events have only been observed as closely-clustered sequence alterations and never more than 10 base pairs apart. Among mutants of the endogenous gene 4 tandem and 3 non-tandem double mutations were recovered. In the case of the shuttle vector, multiple events included 1 tandem double, 1 tandem triple and 1 non-tandem triple sequence alteration. Triple mutations have been previously reported in UV-induced mutation of the *supF* (Hauser et al, 1986) and *lacI* (Lebkowski et al, 1985) target genes carried on extrachromosomally replicating shuttle vectors but not in an endogenous gene.

The absence of triple mutations in the endogenous gene target may reflect the fidelity of DNA replication and repair processes for genes in their native chromosomal context. Alternatively, the fidelity of these processes may vary depending on the site within the genome at which the target gene has been inserted. Investigation of additional single copy infectants might thereby be expected to yield a modified mutational spectrum. A final possibility is that vector sequences which flank the target gene directly influence the fidelity of replication and repair.

Comparative Distribution of UV-Induced Transitions

As noted above, both spectra are characterized by a predominance of G:C to A:T transitions. Additionally, these transitions frequently occur at the same sites within the coding sequence of the *aprt* gene (Table 2). In fact, approximately half of the transition sites occur in both spectra. This is remarkable in view of the relatively large size of the mutational target (540 bp coding sequence). Moreover, it is unlikely that the preferential recovery of mutation at such a small number of sites reflects a paucity of available sites. Spectra from other treatments including gamma rays (Grosovsky et al, 1988), BPDE (Mazur and Glickman, 1988), and cisplatina (de Boer and Glickman, 1988) reveal a number of other potential G:C to A:T target sites. These data are therefore interpreted as suggesting that local DNA sequence context plays an important role in the probability that a pre-mutational photoproduct will be fixed as a mutation.

Target Sequences for UV-Induced Single-Base Substitution

In order to evaluate the actual similarity of UV-induced mutation in the endogenous gene and the shuttle vector carried targets, we have compared the local DNA sequence context within which these events occur (Table 3). Most UV-induced single-base substitutions occur within pyrimidine runs. Among G:C to A:T transitions, the most common event in both spectra, all but one occurred in the middle or at the 3' end of the run. Other types of single-base substitutions were however, recovered in the middle or on the 5' side of pyrimidine runs, or at Py-T dipyrimidines.

We have previously proposed (Drobetsky et al, 1987) that the consistency of this pattern may reflect different pre-mutagenic lesions responsible for each class of base substitution. We suggested, from genetic (Glickman et al, 1986; Schaaper et al, 1987) and physical chemical (Franklin et al, 1985) evidence, that the G:C to A:T transitions arising in the middle or on the 3' side of pyrimidine runs may be attributable to 6-4 pyrimidine-pyrimidone lesions. Transversions arising in the middle or on the 5' side of pyrimidine runs may, in contrast, be attributable to cyclobutane pyrimidine dimers.

Table 2

	Endogenous	Shuttle vector
Total Transitions	15	18
Total number of sites	11	14
Total sites in common	6	6
Fraction of sites in common	.55	.43

Table 3

Type of target sequence	Endogenous gene		Shuttle vector	
	G:C => A:T	OTHERS	G:C => A:T	OTHERS
Pyrimidine Run				
3' Side	2	0	0	0
5' Side	0	4	0	1
Middle	15	2	14	2
TT or CT	0	3	0	2
Isolated				
Pyrimidine	0	0	1	1

Of particular significance are events occurring at TT or CT dipyrimidines. 6–4 pyrimidine-pyrimidone lesions are not commonly recovered at such sites (Brash and Haseltine, 1982) at the low fluence used in this investigation. The fact that G:C to A:T transitions were not recovered at CT sites, is consistent with the notion that this class of mutation is attributable to the formation of 6–4 lesions.

Two single-base substitutions observed in the shuttle vector spectrum occurred at isolated pyrimidines. One of these was a G:C to A:T transition, the other an A:T to C:G transversion. In contrast, all UV-induced single-base substitution in the endogenous gene occurred at pyrimidine runs. While the number of mutations yet analyzed is small, this may represent a significant qualitative difference; we note that single-base substitutions at isolated pyrimidines have been reported in other studies of UV mutagenesis using shuttle vectors (Hauser et al, 1986; Lebkowski et al, 1985).

Conclusions

In conclusion, the two spectra of UV-induced mutagenesis compared in this report are generally similar, although some specific, and potentially significant differences have been observed. Both spectra are predominated by G:C to A:T transitions occurring in the middle, or on the 3' end of pyrimidine runs. There is as well, a striking similarity in the sites at which these transitions occur within the 540 bp target sequence. On the other hand, the shuttle vector spectrum includes two triple-base substitutions, and two single-base substitutions at isolated pyrimidines. Neither class of event was recovered at the endogenous gene target. However, such alterations were reported in previous investigations of UV-induced mutagenesis using extrachromosomally replicating shuttle vectors. Therefore, the differences we observed between the two spectra can be accounted for by classes of events previously observed only in other shuttle vectors. This observation suggests that there may be subtle but fundamental differences between how an identical DNA sequence target behaves in its evolutionarily established location compared to it being resident elsewhere.

Acknowledgements

This work was supported by a grant from the Medical Research Council of Canada (MA - 8677)

References

Brash, D.E. and W.A. Haseltine (1982), UV-induced mutation hotspots occur at DNA damage hotspots, *Nature,* 298, 189–192

Cepko, C.L., B.E. Roberts, and R.C. Mulligan (1984), Construction and applications of a highly transmissable murine retrovirus shuttle vector, *Cell,* 37, 1053–1062

De Boer, J. and Glickman, B.W.(1988) Sequence specificity of mutation induced by the antitumor drug cis-platin in the CHO *aprt* gene, submitted

De Jong, P., Grosovsky, A.J. and B. W. Glickman (1988) Spectrum of spontaneous mutation at the APRT locus of Chinese hamster ovary cells: An analysis at the DNA sequence level, *Proc. Natl. Acad. Sci.,* 85, 3499 – 3505

Drobetsky, E.A., A.J. Grosovsky, and B.W. Glickman (1987), The specificity of UV-induced mutations at an endogenous locus in mammalian cells, *Proc. Natl. Acad. Sci.,* 84, 9103–9107

Franklin, W.A., P.W. Doetsch, and W.A. Haseltine (1985), Structural determination of the ultraviolet induced thymine-cytosine pyrimidine-pyrimidone (6–4) photoproduct, *Nucleic Acids Res.,* 13, 5317–5325

Glickman, B.W., R.M. Schaaper, W.A. Haseltine, R.L. Dunn, and D.E. Brash (1986), The C-C (6–4) photoproduct is mutagenic in *E. coli, Proc. Natl. Acad. Sci.,* 83, 6945–6950

Grosovsky, A.J., J.G. deBoer, P.J. deJong, E.A. Drobetsky, and B.W. Glickman (1988), Base substitutions, frameshifts and small deletions constitute ionizing radiation induced point mutations in mammalian cells, *Proc. Natl. Acad. Sci.,* 85, 185–188

Hauser, J., M.M. Seidman, K. Sidur, and K. Dixon (1986), Sequence specificity of point mutations induced during passage of a UV-irradiated shuttle vector plasmid in monkey cells, *Mol. Cell Biol.,* 6, 277–285

Lebkowski, J.S., S. Clancy, J.H. Miller, and M.P. Calos (1985), The lac I shuttle: Rapid analysis of of the mutagenic specificity of ultraviolet light in human cells, *Proc. Natl. Acad. Sci.,* 82, 8606–8610

Mazur, M.M. and B.W. Glickman (1988), Sequence specificity of mutations induced by benzo(a)pyrene-7,8-diol-9,10-epoxide at the endogenous aprt gene in CHO cells, *Somat.Cell Molec.Genet.,* 14: 393–400

Schaaper, R.M., R.L. Dunn, and B.W. Glickman (1987), Mechanisms of UV induced mutation: Mutational spectra in the *E. coli lacI* gene for a wild type and an excision repair deficient strain, *J. Mol. Biol.,* 198, 187–202

**Ultraviolet and Ionizing Radiations:
From Molecules to Cells**

Photobiology and Radiobiology: Divisions and Common Ground

J. Z. BEER

Center for Devices and Radiological Health
FDA, Rockville, MD 20857
USA

This symposium on "Ultraviolet and Ionizing Radiations: From Molecules to Cells" has few, if any, precedents. Comparisons of the effects of different types of radiation are made at meetings devoted to mutagenesis, carcinogenesis, DNA repair, etc. However, I am not aware of any scientific meeting *specifically* designed to compare photobiological and radiobiological effects.

This introductory paper primarily discusses those areas of photobiology and radiobiology that are close to medicine. It is mostly limited to the non-particulate ionizing radiations, i.e. x rays and gamma rays, as are the papers presented in this Symposium.

Photobiology and radiobiology both study the interactions of electromagnetic radiation with living matter. Yet, since photobiologists and radiobiologists usually go to different meetings and publish in different journals, the two diciplines are largely divided. They are like two adjacent countries sharing a common border but not a common culture. The Land of Photobiology and the Land of Radiobiology are close neighbors on the Continent of Science (Figure 1). Their common frontier is long and complex. There are many provinces in these Lands, and a complete map is not simple. Some provinces, such as the Province of DNA, stretch on both sides of the frontier. Others, such as the Provinces of Vision, Photosynthesis, Radionuclides, or RBE are located far from the frontier. Both Lands coexist fairly peacefully, unlike many other neighbors in this world. Nevertheless, they are divided.

Sources of Divisions

Reasons behind divisions, like those between photobiology and radiobiology, are often historic in nature. Indeed, the realization that electromagnetic radiation surrounds us came at different times for different types of radiation (Table 1). The presence of visible radiation has certainly been recognized from the beginnings of mankind. However, an

Photobiology, Edited by E. Riklis
Plenum Press, New York, 1991

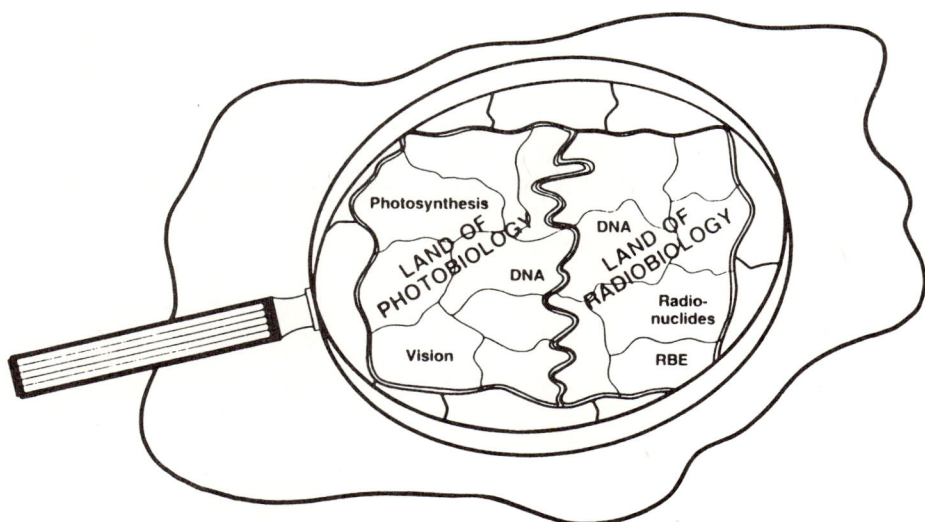

Figure 1. Photobiology and Radiobiology are neighbors on the Continent of Science.

Table 1. Milestones in radiation sciences

When	Who	What
Once Upon a Time		visible light
1666	Newton	spectrum of visible light
1800	Herschel	IR radiation
1801	Ritter	UV radiation
1895	Roentgen	x rays
1898	The Curies	radioisotopes
1900	Villard	gamma rays
1942	Fermi	nuclear reactor
1960	Maiman	laser

understanding of its nature came much later. Newton's demonstration in 1666 of the spectrum of colors in white light initiated the scientific approach to the study of optical radiation. It was, in fact, the beginning of a scientific approach to all electromagnetic radiations. The discovery of infrared radiation came in 1800. A year later ultraviolet radiation (UVR) was discovered. It was almost a century later in 1895, when Roentgen discovered x rays. In 1898 Marie and Pierre Curie discovered polonium, the first radioactive element. Gamma radiation was discovered almost exactly at the turn of the century. The first controlled nuclear chain reaction was accomplished by a team directed by Fermi in 1942. Almost two decades later, in 1960, the first ruby laser was successfully fired by Maiman. Thus, while the basic discoveries related to ionizing radiation were made within just a few decades, the discoveries related to optical radiation came at different times over centuries.

Discoveries of new physical agents have always raised hopes for their beneficial use, particularly in the healing arts. These possibilities for optical and ionizing radiations not only came at different times, but also attracted different medical specialties and different industries (Table 2). X rays were initially applied in diagnostics and cancer therapy. The original medical applications of UVR took place in dermatology. The effects of optical radiation have raised interest among ophthalmologists. Lasers were introduced to surgery soon after their development. More recently, a variety of applications of optical radiation, particularly in combination with drugs, have begun to attract oncologists. In addition, the cosmetic industry has interest in optical, but not ionizing, radiation. These distinct practical uses influenced the development of photobiology and radiobiology as separate scientific disciplines. The isolation was strengthened, on the side of photobiology, by its involvement in phenomena such as human vision and plant photosynthesis. On the other hand, the development of nuclear weapons and the nuclear industry reinforced the independent development of radiation biology. So, photobiology and radiobiology developed separately, and lost in the shuffle has been the unity of radiation science. The fact that environmentalists are interested in all radiations has been insufficient as a unifying factor.

The sun, lamps, and lasers belong to the photobiologist's arsenal. Radiobiologists use mostly x-ray machines, isotopic sources, and reactors. Differences in the ways of using these tools and in the required safety precautions deepen the divisions between the two disciplines even further.

Common Ground

The photon as a physical entity unifies photobiology and radiobiology. The interactions of photons with matter and the biological effects of radiations vary strongly in both the ionizing and the non-ionizing ranges (Figure 2). High energy photons of short wavelength radiation cause ionizations. Medium-energy photons of medium wavelength radiation cause both excitations and ionizations. Low-energy photons of long wavelength radiation cause excitations. Interactions between the effects of different electromagnetic radiations can modify the bioeffects. Interactions of low energy photons with chromophores can affect critical biomolecules and cellular structures. In practical terms, beams of *ionizing and ultraviolet radiation* carry photons of sufficiently high energy to directly modify molecules, while *visible radiation* photons can affect molecules by indirect mechanisms. Pathways and the role of cofactors are different for ionizing, ultraviolet, and visible radiation. Nevertheless, many end points are identical. Thus, at the "chemical" level, similarities and common problems should be expected for gamma rays, x rays, ultraviolet radiation, and even visible radiation.

Table 2. Some medical and non-medical communities interested in the practical applications of optical and ionizing radiation

Photobiology	Radiobiology
Dermatologists	Diagnosticians
Ophthalmologists	Oncologists
Surgeons	Nuclear industry
Oncologists	The military
Cosmeticians	Environmentalists
Environmentalists	

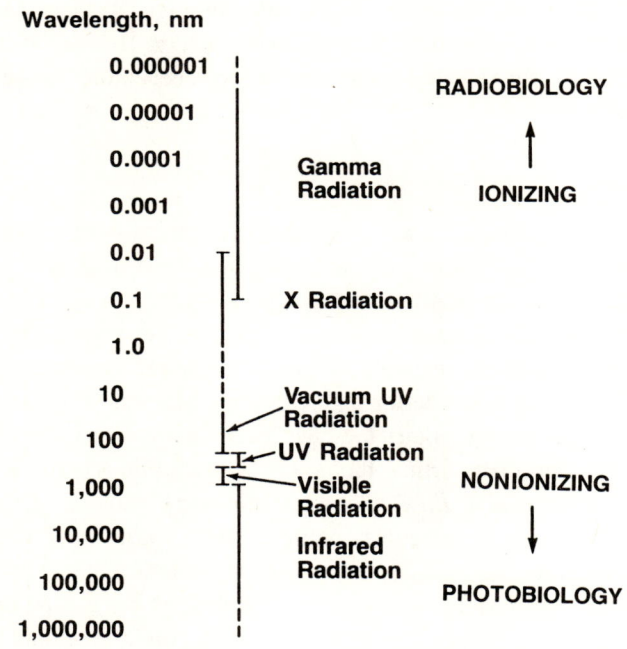

Figure 2. Photobiology, radiobiology, and the electromagnetic spectrum.

Table 3. Some bioeffects caused by optical and ionizing radiation

	Optical	Ionizing
Protein damage	+	+
DNA damage	+	+
Mutation	+	+
Cell membrane damage	+	+
Neoplastic transformation	+	+
Erythema	+	+
Cataract	+	+
"Aging"	+ (skin)	+

Notwithstanding differences in photon energy, optical and ionizing radiations both produce a variety of biological effects (Table 3). Optical and ionizing radiation can damage proteins, nucleic acids, and cell membrane, as well as induce mutations and neoplastic transformation. Both classes of radiation can cause erythema and cataracts. Both can cause phenomena resembling, or involved in, the aging process. Because of its limited penetration, optical radiation affects mostly the skin and the eye in the human organism. On the other hand, organs not easily accessible to optical radiation, such as bone marrow or intestine, are affected by the deeply penetrating ionizing radiation.

Comparisons of the phenomena caused by different radiations are difficult for a complex biological system such as the entire organism. However, such comparisons become quite feasible at the *cellular and subcellular level*. The papers presented in this symposium describe such comparisons. Both striking similarities and significant differences are noted. For example, J. Cadet, M. Berger, P.C. Joshi, A. Shaw, L. Voituriez, L.-S. Kan, and R. Wagner show that although the pathways of induction may be different, identical lesions are produced in DNA molecules by ionizing and ultraviolet radiation. H.P. Leenhouts and K.H. Chadwick demonstrate that DNA double strand breaks are critical for the cytotoxicity of ionizing radiation and UVR, and that cellular photobiological and radiobiological phenomena can be described by the same models. H.H. Evans indicates that, as mutagenic factors, both agents produce base substitutions, frameshift mutations, and deletions; however ionizing radiation induces a much higher proportion of multilocus deletions than does UVR. J.B. Little shows that the kinetics of induction of neoplastic transformation and the effects of chemical cofactors are very similar for both types of radiation. All these observations indicate how comparative photo-radiobiological studies may be useful and stimulating.

Certain studies clearly belong to both photobiology and radiobiology. Primary examples of such studies are presented in this symposium. J. Kiefer, E. Schneider, B. Bauman, F. Zölzer, and A. Schreiber describe mutagenic effects of combined exposures to

UVR and ionizing radiation and demonstrate interaction of primary lesions in yeast cells. T. Ito analyzes cytotoxicity, cell membrane damage, and DNA damage and repair in cells exposed to radiation in the so called vacuum UV range (30 to 200 nm). This range borders with the ionizing and non-ionizing regions of the spectrum. Thus, studies with vacuum UVR provide a real link between photobiology and radiobiology. The symposium material clearly demonstrates that there is a lot of common ground between photobiology and radiobiology from molecules to cells.

Perspectives

The catalog of beneficial and detrimental biomedical effects of radiations is still actively being developed. This catalog contributes to our knowledge and makes new practical applications possible. Such progress is only possible when different bioeffects are well understood. Comparisons and comparative studies can play a very special role in all areas concerned with the interaction of radiation with living matter. Such studies and comparisons should embrace a wide range of wavelengths, perhaps to include the radiofrequency/microwave radiation. They should also cover a wide range of endpoints. They should lead to the evaluation of the role of free radicals in the effects induced by different radiations and to a fuller understanding of the mechanisms involved in the induction of DNA damage in living cells. They should make it possible to establish differences in the amount and type of cell membrane damage by different radiations and to characterize the differences in the repair pathways. Further, they should make it possible to understand the effects of combined exposures to different radiations. Eventually, they should make it possible to understand relationships between human hypersensitivities to different radiations.

Joint efforts of photobiologists and radiobiologists can substantially contribute to the unification of biomedical radiation sciences. There were good pragmatic reasons to establish separate learned societies for the two disciplines and there are good reasons to organize separate meetings in photobiology and radiobiology. Nevertheless, at this time, a broad front of interactions between the two disciplines is urgently needed. Such interactions can have immediate practical implications. They may also eventually lead to a unified radiation biology that could explain the interaction of all radiation with living matter.

Comparative Effects of Ultraviolet and Ionizing Radiations on Nucleic Acids

JEAN CADET*, MAURICE BERGER*, PRAKASH C. JOSHI*, ANTHONY SHAW*, LUCIENNE VOITURIEZ*, LOU-SING KAN** and RICHARD WAGNER***

*Laboratoires de Chimie, Département de Recherche Fondamentale
Centre d'Etudes Nucléaires de Grenoble, 85X
F 38041 Grenoble Cedex, France
**Division of Biophysics
The Johns Hopkins University
Baltimore, Maryland 21205, USA
***MRC Group in the Radiation Sciences
Département de Médecine Nucléaire
Sherbrooke, Québec, Canada J1H 5N4

Abstract

This review focuses on recent aspects of the photo- and radiation chemistry of DNA model compounds. Similarities have been observed in the mode of action of ionizing radiation (direct and indirect effects) and far-UV light as well as of various types of photodynamic agents on purine and pyrimidine DNA components.

Introduction

It is well accepted that nucleic acids represent one of the major cellular targets for the deleterious effects (letality, mutagenicity, carcinogenicity) of both ionizing radiation and ultraviolet-visible light. During the two last decades significant advances have been made in the elucidation of the main decomposition pathways of biomolecules, such as nucleic acids, when exposed to continuous wave far-ultraviolet light of low intensity (Cadet et al, 1985, Cadet and Vigny, 1990) or to ionizing radiation (Téoule, 1987, von Sonntag, 1987, Teebor et al, 1988). It is clear that each of these radiations has its known specifity in terms of inducing chemical modifications of DNA constituents. Hovever, there is growing evidence that these apparently very different electromagnetic radiations, at least in terms of energy, may lead to the formation of similar radiation- or photo-induced DNA lesions which in most cases involves different initial pathways. These various aspects are

Photobiology, Edited by E. Riklis
Plenum Press, New York, 1991

reviewed in this presentation mostly on the basis of available data obtained from model studies.

Results and Discussions

Main photochemical reactions of pyrimidine and purine nucleobases induced by exposure to far-UV light of low intensity (10^{-1} W/m^2 / I /10^1 W/m^2)

Far-UV pyrimidine and purine photoproducts

Excited pyrimidine and purine nucleobases either in a singlet or a triplet state are the likely intermediates of the photoreactions induced by far-UV light. The formation of cytosine photohydrate has been rationalyzed in terms of vibrationaly excited ground S_0 state intermediate (Fisher and Johns, 1976a). An alternative mechanism has been suggested to involve the formation of a zwitterionic intermediate as the result of internal conversion of the single state S_1 (Garner and Scholes, 1985).

On the other hand cyclobutadipyrimidines, at least for nucleobases and nucleosides, are generated through the triplet state (Fisher and Johns, 1976b). It is also likely that the formation of the $5R^*$ and $5S^*$ diastereomers of 5,6-dihydro-5-(α-thymidylyl)thymidine, the so-called "spore photoproducts" involves the triplet state as inferred from recent triplet photosensitization experiments. The formation of these photoadducts is likely to be explained in terms of concerted process when thymidine is exposed to far-UV light in frozen aqueous solutions (Shaw et al, 1988a). Competitive recombination of photogenerated 5-(2'-deoxyuridylyl)methyl and 5,6-dihydrothymidyl-5-yl radicals was found to occur upon exposure of thymidine in the dry state (Shaw, 1987). This received further support from the observed formation of the (5R) and (5S) 5,6-dihydrothymidine and of two diastereomers of 5,6-dihydro-6-(α-thymidylyl)thymidine.

The pyrimidine(6-4)pyrimidone photoadducts which are much more antigenic than the cyclobutadipyrimidines constitute a fourth class of major pyrimidine photolesions (Figure 1). The mutagenic potential of these photolesions is still open to debate (for a review: Brash, 1988) despite recent findings which show that Pyr<>Pyr are the most mutagenic photolesions, particularly at TC and CC sites (Hutchinson et al, 1988). There is no available information on the excited precursors which give rise to the (6-4) adducts via unstable oxetane or azetidine intermediates. Interestingly the pyrimidine(6-4)pyrimidone adduct of thymidylyl(3'-5')thymidine monophosphate was recently found to undergo a quantitative conversion to a Dewar valence isomer upon exposure to 313 nm light (Taylor and Cohrs, 1987). It is noteworthy that the Dewar valence isomer is more unstable that the (6-4) precursor when treated under the alkali conditions used for DNA sequencing experiments (Voituriez et al, 1988).

Purine nucleobases are far less photoreactive than prymidine nucleic acid components. However adenine was shown to undergo interesting photoreaction with either a second molecule of adenine or an adjacent thymine upon exposure of appropriate dinucleoside monophosphates and DNA to far-UV light (Bose et al, 1984; Kumar et al, 1987; Kumar and Davies, 1987). The so-called "Porschke type photoadduct" was recently

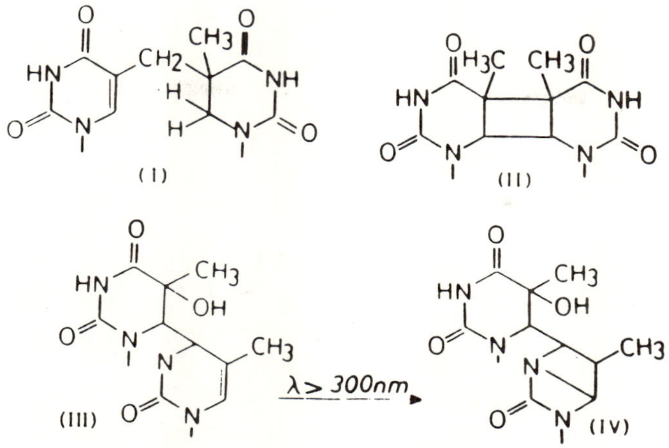

Figure 1. Structure of the main thymine photoadducts. I: 'spore photoproduct'; II cyclobutadithymine; III: pyrimidine (6-4)pyrimidone adduct; IV: Dewar valence isomer

assigned as a mixture of two diastereomers of an internal adenine adduct. The proposed mechanism of their formation involves the hydrolytic fission of an unstable azetidine intermediate which results from the covalent binding of the N(7) and C(8) atoms of the 5'-dAMP moiety to the 3'-adenine residue (Kumar et al, 1987). Far-UV irradiation of thymidylyl(3'-5')-2'-deoxyadenosine was found to generate a photoadduct whose formation involves the cycloaddition of the thymine moiety to the adenine residue through their 5,6- and 1,6-ethylenic bonds respectively (Bose et al, 1984)

Common decomposition products to far-UV radiation an ionizing radiation

It is worth noting that gamma irradiation of frozen aqueous solutions of thymidine under conditions in which direct effects are predominant was found to generate the various diastereomers of dThd<>dThd (Shaw et al, 1988b). This may be accounted for, at least partly, by a significant contribution of Cerenkov light which has been shown to induce the formation of Pyr<>Pyr in bacterial cells. However direct excitation processes cannot be ruled out. It should also be mentioned that the two diastereomers of 5,6-dihydro-(α-thymydylyl)thymidine are also generated under gamma irradiation as the result of both radical and to lesser a extent excited processes. However no detectable amount of the corresponding pyrimidine(6-4)pyrimidone photoadduct of thymidine is produced under these conditions (Figure 2).

Ionizing radiation

Direct and indirect effects are both involved in the decomposition of the base and the sugar moities of DNA upon exposure to ionizing radiation.

51

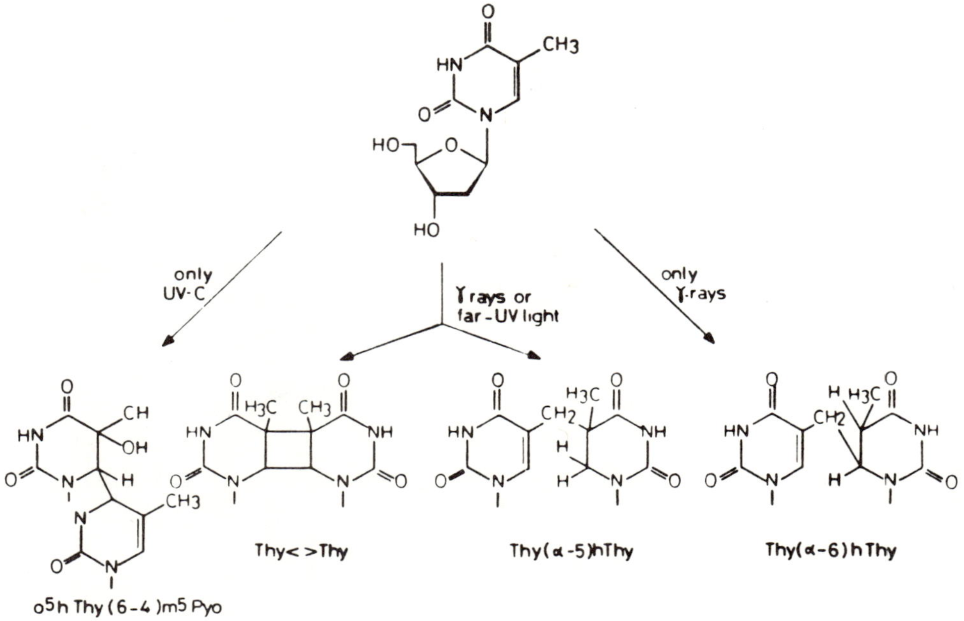

Figure 2. Main thymine type photoadducts generated by gamma irradiation and far-UV photolysis in frozen aqueous solutions

Direct effect

The nucleobase radical anions and radical cations are the main initial intermediates generated by the interaction of X- or gamma rays with nucleic acids. In frozen aqueous solutions the pyrimidine radical anions were shown to undergo protonation during the thawing process generating the corresponding 5,6-dihydro-5-yl radical. Deprotonation within the methyl group and specific hydration at position 6 are the two main competitive reactions of the thymidine radical cation as followed by ESR spectroscopy. One common feature to the nucleobase and osidic radicals in frozen aqueous solutions is their apparent lack of reactivity with dissolved oxygen as inferred from final product analysis and isotope experiments. The formation of the (5R) and (5S) diastereomers of 5,6-dihydrothymidine, the main radiation-induced decomposition products of thymidine is explained by the reduction of the transient 5,6-dihydrothymidyl-5-yl radical, followed by protonation of the resulting carbanion (Shaw et al, 1988b). Various dimeric compounds are generated by recombination of the various pyrimidine radicals including 5,6-dihydrothymidyl-5-yl, 5,6-dihydrothymidyl-6-yl, 5-(2'-deoxyuridylyl)methyl and thymidyl-3-yl radicals. Six of the eight possible diastereomers of 5',6-cyclo-5,6-dihydrothymidine are also produced by gamma irradiation of thymidine in aqueous solution (Shaw and Cadet, 1988). A likely mechanism for their formation involves

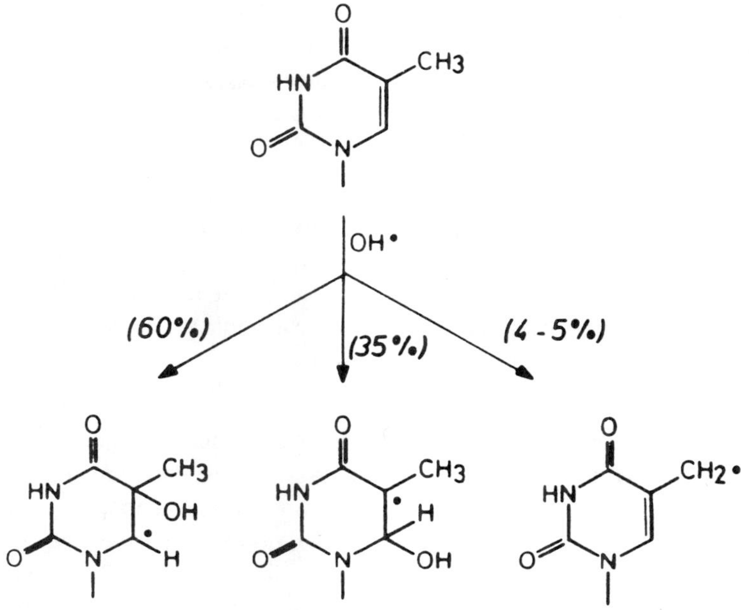

Figure 3. Main reactions of the OH° radical within the pyrimidine moiety of thymidine in aqueous solution.

intramolecular addition of the osidic 5'-yl radical to the pyrimidine carbon C(6), followed by reduction of the resulting radical and further protonation.

It is interesting to note that the pyrimidine and purine radical cations may be produced very efficiently by photosensitizers in aqueous solutions at room temperature (vide infra). As discussed below the fate of the neutral radicals which arise from competitive deprotonation and hydration is oxygen dependant under these conditions.

Indirect effect

The decomposition of both the nucleobases and the furanose ring of DNA or related model compounds following exposure to ionizing radiation in aqueous solutions is mostly mediated by the highly reactive hydroxyl radicals arising from the radiolysis of H_2O (indirect effect). Under these conditions about 80% of the hydroxyl radicals react with the pyrimidine moiety of thymidine whereas the remaining 20% abstract hydrogen atom within the osidic moiety (Téoule and Cadet, 1978, von Sonntag, 1987). The main reaction of the OH˙ radical within the thymine moiety is an addition at carbon C(5) generating a reducing radical (Figure 3). Two other relatively minor processes involve the formation of OH˙ adduct at position 6 and an hydrogen abstraction reaction within the methyl group. It is noteworthy that most of the final diamagnetic decomposition products arising from the fate of the OH˙-induced pyrimidine or purine radicals are identical to those involving the related radical cations (Cadet et al, 1987a). This is a good illustration of the occurrence of common mechanisms or at least of the formation of identical diamagnetic decomposition products upon exposure of DNA model compounds to ionizing radiation and photodynamic effects (vide infra).

Photosensitization

Various photosensitizers, mostly in a triplet excited state, are able to photooxidize purine and pyrimidine DNA components through type I mechanism as the result of electron transfer or hydrogen abstraction reaction (Cadet et al, 1986). A specific photooxidation pathway, which is only observed for the guanine derivatives, involves the participation of singlet oxygen.

Purine components

Various natural compounds and photherapeutic agents including flavin, psoralen, hematoporphyrin and phthalocyanine derivatives have been shown to oxidize purine components and in particular guanine components upon exposure to near-UV and visible light. The main photoreaction leads to the generation of a purine radical cation by electron transfer reaction. It is worthnoting that the two main final decomposition products arising from the fate of the guanine radical cation are also generated by the action of OH$^{\cdot}$ radicals in aerated aqueous solution (Cadet and Berger, 1985, Cadet et al, 1986, 1987a). This strongly suggests the occurrence of similar intermediates in both photosensitization and radiation-induced reactions as also inferred from pulse radiolysis experiments. Further insights in the initial reactions of the guanine radical cation were recently provided by a comprehensive and detailed optical and conductive pulse radiolysis study (Candeias and Steenken, 1989). The main conclusion of this interesting investigation is that the purynyl radical cation which was generated by the oxidation reaction of SO_4° and Br_2° with 2'-deoxyguanosine has a pK_a of 3.9. Therefore this charged intermediate is likely to undergo an almost complete deprotonation from N(1) in neutral aqueous solutions generating an oxyl type neutral radical. The same one-electron oxidized radical is assumed to arise by OH$^-$ (or H_2O) elimination from initially OH$^{\cdot}$ adduct to 2'-deoxyguanosine (Figure 4). Further reaction of the oxyl type radical or more likely of a tautomeric carbon centred radical with molecular oxygen is expected to generate a transient peroxy radical which may lead in subsequent steps to the formation of the final diamagnetic decomposition products.

Pyrimidine components

Several quinone derivatives including 2-methyl-1,4-naphthoquinone, a component of vitamin K_3, have been shown to photooxidize pyrimidine nucleosides in the presence of near-UV light (Decarroz et al, 1986). Again, the corresponding radical cation was found to be the predominant intermediate of these photoreactions as the result of electron transfer reaction between the pyrimidine moiety and the quinone in a triplet excited state (Fisher and Land, 1983). Hydration and deprotonation, which are the two competitive reactions involving the photogenerated radical cations, lead to the same diamagnetic

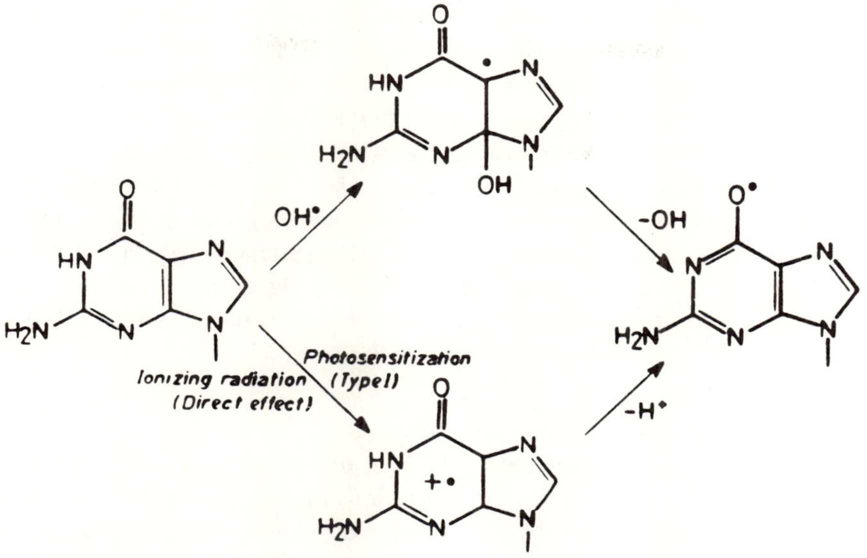

Figure 4. Proposed mechanism for the formation of the 2'-deoxyguanosine type oxyl radical following either deprotonation of the purinyl radical cation or dehydration of the OH* adduct to the guanine moiety

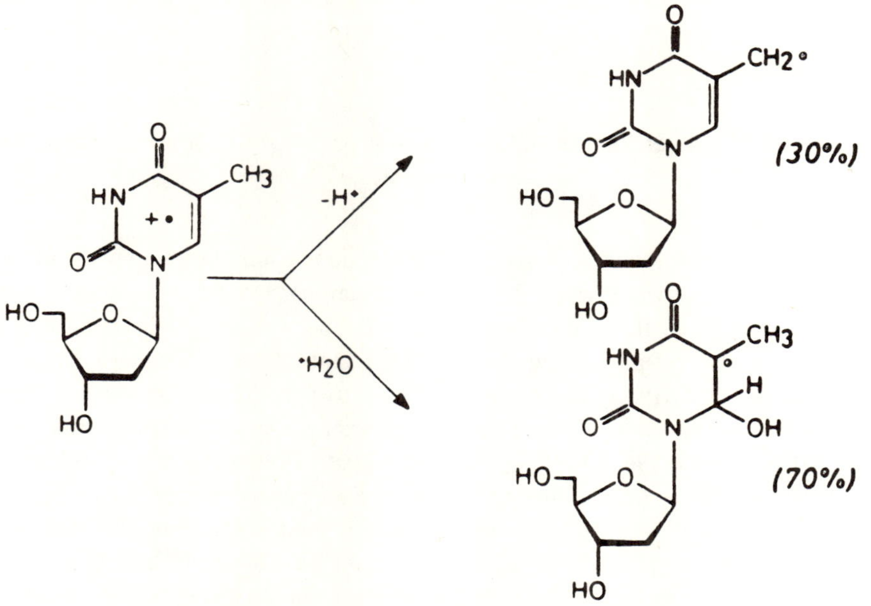

Figure 5. Hydration and deprotonation reactions of the thymidine radical cation in aqueous solution

55

degradation products than those generated by OH radicals, but in a more specific way and with a different quantitative distribution. It was inferred from various studies including final product analysis (Decarroz et al, 1986) isotopic labelling experiments with ^{18}O (Wagner et al, 1987) and spin-trapping studies (Murali Krishna et al, 1987) that hydration of the thymidine radical cation exclusively gives rise to the oxidizing hydroxy-6-dihydro-5,6-thymidyl-5-yl radical (Figure 5). It should also be noted that the competitive deprotonation reaction which occurs within the methyl group generates the 5-(2'-deoxyuridylyl)methyl radical in about 30% yield. In the same way hydration of the radical cation of 2'-deoxycytidine was found to predominantly occurs at position 6. It is worthonoting that the deprotonation of the cytosyl radical cation was shown to take place at the position 1 of the furanose ring and within the exoxyclic amino group (Decarroz et al, 1987).

Oxidation reactions mediated by hydrogen peroxide

Another point to be considered is the role of hydrogen peroxide which is an important side-product of photosensitized reactions (and also of gamma radiolysis of water) by dismutation of the initially produced unreactive superoxide radicals. Hydrogen peroxide may generate *in situ* OH˙ radicals or related reactive species when reacting with transition metals via a Fenton reaction. Evidence for the significant occurrence of this radical reaction was recently provided by the characterization of specific OH˙ decomposition products within DNA after exposure of bacterial cells to hydrogen peroxide (Cadet et al, 1989). In addition ionic oxidation leads to the specific formation of adenine N_1-oxide in cellular DNA as shown by using a sensitive ^{32}P post-labeling assay.

Other photoreactions

Vacuum-UV light and powerful laser radiation are also able, at least partly to mimic the direct and indirect effects of ionizing radiations.

Vacuum-UV light

UV light has been shown to induce radical reactions within DNA components as the result of photolysis of water when the irradiations are carried out in aqueous solutions. Release of adenine as the main photoproduct has been observed upon exposure of adenosine-5'-monophosphate (Ito et al, 1986) and 2'-deoxyadenosylyl-(3'-5')-2'-deoxyadenosine or d(ApA) (Ito and Saito, 1988) to synchroton vacuum-UV light in aqueous solutions. Vacuum-UV irradiation of guanosine-5'-monophosphate as a dry film was found to induce the release of the free base (Dodonova et al, 1982). More interestingly, formation of equimolar amounts of adenosine monophosphate, mostly as the 5'-phosphate monoester, and adenine was observed upon exposure of d(ApA) in the solid state to monochromatic 6.5-22.5 eV photons (Ito et al, 1986). This is likely accounted for by excitation processes as the result of direct absorption of vacuum-UV light by the osidic moiety.

High-powerful laser radiations

A composite system in which both conventional UV decomposition pathways and radical processes are involved is provided by far-UV nano- and picosecond laser radiations (Nikogosyan and Letokhov, 1983). The use of a high intensity laser sources was shown to induce directly, or indirectly through sensitization, the homolysis of water and the photoionization of the purine and pyrimidine nucleobases as the result of biphotonic processes. In particular, the typical decomposition products ot the thymidine radical cation were found to be generated upon exposure of this nucleoside to high intensity pulses of picosecond laser radiation (Cadet et al, 1987b). Cyclobutadithymine and radical oxidation thymine derivatives were shown to be generated following irradiation of thymidylyl-(3'-5')-thymidine with nanosecond laser far-UV pulses as the result of mono- and biphotonic excitation processes respectively (Budowsky et al, 1986). Another interesting information dealt with the specific modification of the guanine residue upon exposure of pBR322 plasmid DNA to 248 nm light provided a KrF excimer laser (Croke et al, 1988). This was rationalyzed in terms of initial formation of the guanine radical cation as the result of biphotonic photoionization. Cleavage of phosphodiester bond was found to occur within poly U exposed to 248 nm excimer laser radiation (Schulte-Frolhinde, 1986). Hydroxy-5,6-dihydrouracilyl radical which would result from the initially produced pyrimidine radical cation was suggested to abstract hydrogen atom from the vicinal osidic moiety. Similar mechanism which involves unidentified purine bases was proposed to explain the induced formation of single-strand breaks in poly A and ssDNA by excimer laser far-UV pulses (Optiz and Schulte-Frohlinde, 1987).

Conclusion

It may be concluded that some of the main decomposition reactions of DNA nucleobases induced either by ionizing radiation or by some photochemical reactions (photosensitization, powerful laser UV irradiation, vacuum UV light) give rise to identical modified compounds. Efforts are currently made to search for the formation of such base lesions in cellular DNA exposed to these different conditions.

References

Bose, S.N., Kumar, S., Davies, R.J.H., Sethi, S.K., and McCloskey, J.A., 1984, *Nucleic Acids Research*, 12, 7929-7946.

Brash, D.E., 1988, *Photochemistry and Photobiology*, 48, 59-66.

Budowsky, E.I., Kovalsky, O.I., Simukova, N.A., Tsyschewsy, V.V., Yakolev, D.Yu., and Rubin, L.B., 1986, *Lasers in the Life Sciences*, 1, 151-169.

Cadet, J., and Berger, M., 1985, *International Journal of Radiation Biology*, 47, 127-143.

Cadet, J., and Vigny, P., 1990, in Photobiochemistry Nucleic Acids, edited by H. Morrison (New York: Wiley-Interscience), in press.

Cadet, J., Shaw, A.A., Berger, M., Decarroz, C., Wagner, J.R., and van Lier, J.E., 1987a, in Radiation Research , edited by E.M. Fielden, J.F. Fowler, and J.H. Hendry (London: Taylor and Francis) Volume 2, pp.181-186.

Cadet, J., Voituriez, L., Grand, A. Hruska, F.E., Vigny, P. and Kan, L.-S., 1985, *Biochimie*, 67, 277-292.

Cadet, J., Berger, M., Decarroz, C., Wagner, J.R., van Lier, J.E., Ginot Y.M., and Vigny P., 1986, *Biochimie*, 68, 813-834.

Cadet, J., Berger, M., Decarroz, C., Shaw, A.A., Wagner, J.R., Keskinova, E., and Angelov, D.A., 1987b, in Lasers and Applications, edited by A.I. Spasov (Singapore: World Scientific Publishing Co, pp.508-526.

Cadet, J., Jouve, H., Mouret J.-F., Foray, J., Odin, F., Berger, M., and Polverelli, M., 1989, in Medical, Biochemical and Chemical Aspects of Free Radicals, edited by O. Hayaishi, E. Niki, M. Kondo and T. Yoshikawa (Amsterdam: Elsevier Publishers), pp0. 1517–1520.

Candias, L.P., and Steenken, S., 1989, *Journal of the American Chemical Society*, 111, 1094-1099.

Croke, D.T., Blau, W., OhUigin, C., Kelly, J.M., and McConnell, D.J., 1988, *Photochemistry and Photobiology*, 47, 527-536.

Decarroz, C., Wagner, J.R., and Cadet, J., 1987, *Free Radical Research and Communications*, 2, 295-301.

Decarroz, C., Wagner, J.R., van Lier, J.E., Murali Krishna, C., Riesz, P., and Cadet, J., 1986, *International Journal of Radiation Biology*, 50, 491-505.

Dodonova, N.Ya, Kiselava, M.N., Remisova, L.A., and Tsyganenko, N.M., 1982, *Photochemistry and Photobiology*, 35, 129-132.

Fisher, G.J., and Johns, H.E., 1976a, in The Photochemistry and the Photobiology of Nucleic Acids, edited by S.Y. Wang (New York: Academic Press) Volume 1, pp.169-224.

Fisher, G.J., and Johns, H.E., 1976b, in The Photochemistry and the Photobiology of Nucleic Acids, edited by S.Y. Wang (New York: Academic Press) Volume 1, pp.225-294.

Fisher, G.J., and Land, E.J., 1983, *Photochemistry and Photobiology*, 37, 27-32.

Garner, A., and Scholes, G., 1985, *Photochemistry and Photobiology*, 41, 259-265.

Hutchinson, F., Yamamoto, K., Stein, J., and Wood, R.D., 1988, *Journal of Molecular Biology*, 202, 593-601.

Ito, A., and Saito, M., 1988, *Photochemistry and Photobiology*, 48, 567-572.

Ito, A., Taniguchi, T., and Ito, T., 1986, *Photochemistry and Photobiology*, 44, 273-277.

Kumar, S., and Davies, R.J.H., 1987, *Photochemistry and Photobiology*, 45, 571-579.

Kumar, S., Sharma, N.D., Davies, R.J.H., Phillipson, D.W. and McCloskey, J.A., 1987, *Nucleic Acids Research*, 15, 1199-1216.

Murali Krishna, C., Decarroz, C., Wagner, J.R., Cadet, J. and Riesz, P., 1987, *Photochemistry and Photobiology*, 46, 175-182.

Nikogosyan, D.N., and Letokhov, V.S., 1983, *La Rivista del Nuovo Cimento*, 8, 1-72.

Optiz, J., and Schulte-Frohlinde, D., 1987, *The Journal of Photochemistry*, 39, 145-163.

Schulte-Frohlinde, D., 1986, in Mechanisms of DNA Damage and Repair, edited by M.G. Simic, L. Grossman and A.C. Upton (New York: Plenum Press) pp.19-27.

Shaw, A.A., 1987, Thesis for Doctor of Philosophy, University of Surrey, U.K.

Shaw, A.A. and Cadet, J., 1988, *International Journal of Radiation Biology*, 54, 987-997.

Shaw, A.A., Voituriez, L., and Cadet, J., 1988a, 10th International Congress of Photobiology, Jerusalem, Book of Abstracts, p.54.

Shaw, A.A., Voituriez, L., Cadet, J., Gregoli, S. and Symons, M.C.R., 1988b, *Journal of the Chemical Society, Perkin Transactions* 2, 1303-1307.

Taylor, J.-S., and Cohrs, M.P., 1987, *Journal of the American Chemical Society*, 109, 2834-2835.

Teebor, G.W., Boorstein, R.J., and Cadet, J., 1988, *International Journal of Radiation Biology*, 54, 131-150.

Téoule, R., 1987, *International Journal of Radiation Biology*, 51, 573-579.

Téoule, R., and Cadet, J., 1978, in Effects of Ionizing Radiation on DNA, edited by J. Hütterman, W. Köhnlein, R. Téoule and A.J. Bertinchamps (Berlin : Springer), pp.171-203.

Voituriez, L., Voisin, C., Kan, L.-S., and Cadet, J., 1988, 10th International Congress of Photobiology, Jerusalem, Book of Abstracts, p.83.

von Sonntag, C., 1987, The Chemical Basis of Radiation Biology (London: Taylor and Francis).

Wagner, J.R., van Lier, J.E., Decarroz, C., and Cadet, J., 1987, *Bioelectrochemistry and Bioenergetics*, 18, 155-162.

Interaction Between Ultraviolet and Ionizing Radiation in Eukaryotic Cells

J. KIEFER

Strahlenzentrum der Justus-Liebig-Universität
Giessen, W. Germany

Abstract

The combined action of ultraviolet and ionizing radiation on yeast and mammalian cells with regard to killing and mutation induction are reviewed. In most cases a positive interaction is found which takes place at the level of repair processes since the extent depends on genetically determined repair capacities. This type of study may, therefore, help unravel the complexities of repair systems in eukaryotic cells.

Introduction

Repair processes in irradiated eukaryotic cells are still not very well understood despite considerable progress in recent years. Generally speaking, more is known about the action of ultraviolet radiation, because molecular lesions like pyrimidine dimers and 6-4-photoadducts have been identified and their importance for cellular effects established. Also, the existence of excision repair in many systems — from bacteria to mammalian cells — has been clearly proven. The situation with ionizing radiations is less clear. Combination experiments employing both UV and ionizing radiation could help unravel the complicated system of damage induction and repair. This review is restricted to eukaryotic cells, references to work with bacteria may be found in the paper by Martignoni and Smith (1973).

Figure 1 gives a scheme for the classification of interactions with cell survival as an example. It is assumed that a dose of the first radiation reduces survival to a certain level beyond the shoulder region and that full survival curves are determined for the second radiation type. If they are the same as without pre-irradiation the action is "independent" (curve C) or there is "zero interaction". If the curve follows exactly that which would be expected if the first and the second irradiation were identical, in other words, if they could replace each other, the interaction is "additive" (curve A). If the first dose leads to a steeper curve than expected there is "synergism" (curve B), if survival is higher than predicted from independent action one has "antagonism" (curve D).

Interaction may take place at different levels (Fig. 2). The first radiation may alter the structure of the cellular target molecule in such a way that the probability of lesion

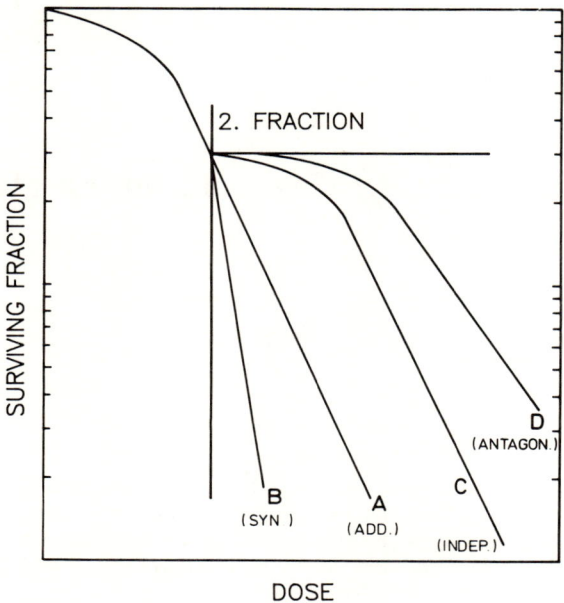

Figure 1. Interaction categories with cell survival (after Han and Elkind 1977)

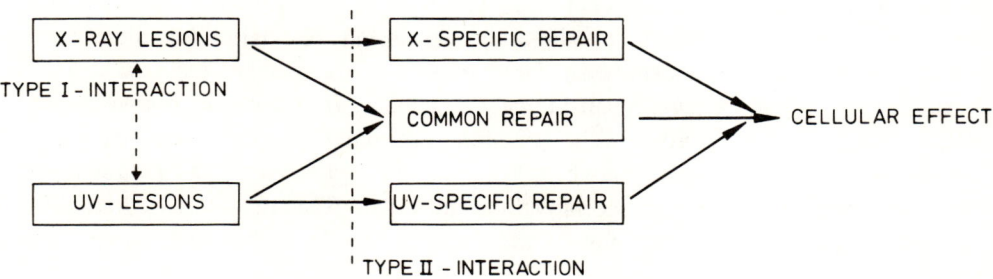

Figure 2. Classification of interaction levels

formation by the second exposure is changed ("type-I-interaction"). It is also possible that interaction occurs via repair processes ("type-II-interaction"). A discrimination between the two modes requires the use of repair deficient mutants.

Interaction with cell survival

Yeast

Yeast is a very useful test object for these kinds of experiments. There are about 100

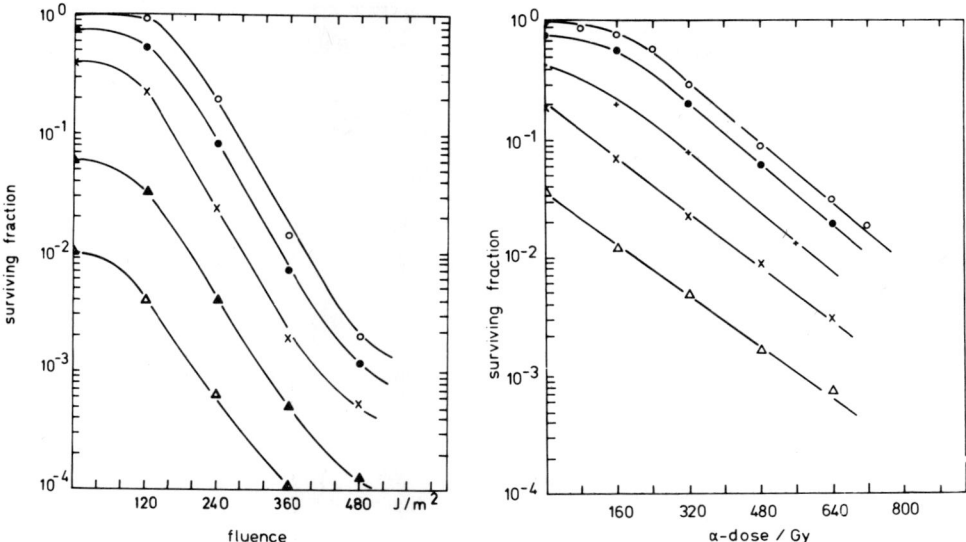

Figure 3. Interaction in diploid wild type yeast (after Schneider and Kiefer 1976). Left: UV followed by α-particles, right: α-particles followed by UV

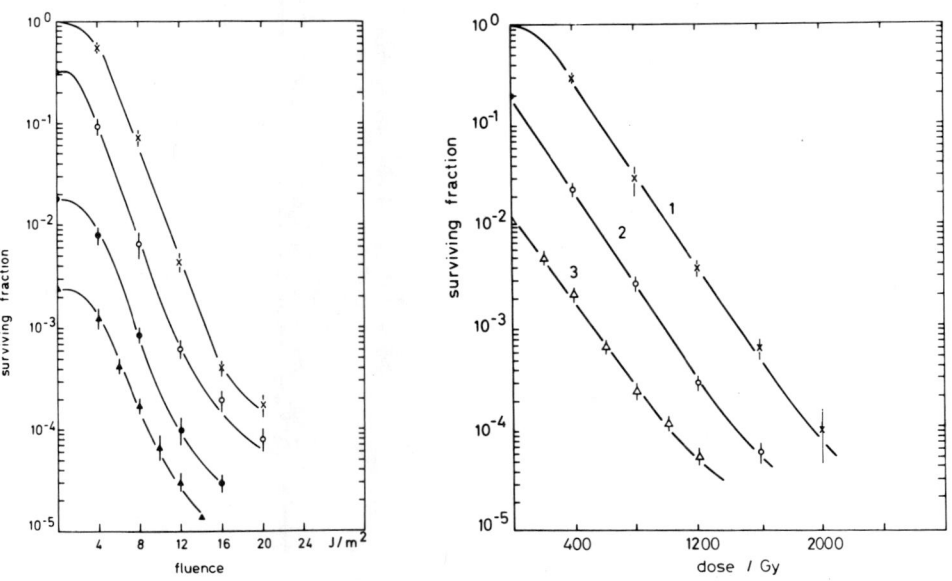

Figure 4. As in Figure 3 but for the excision-deficient strain (rad2-mutant)

63

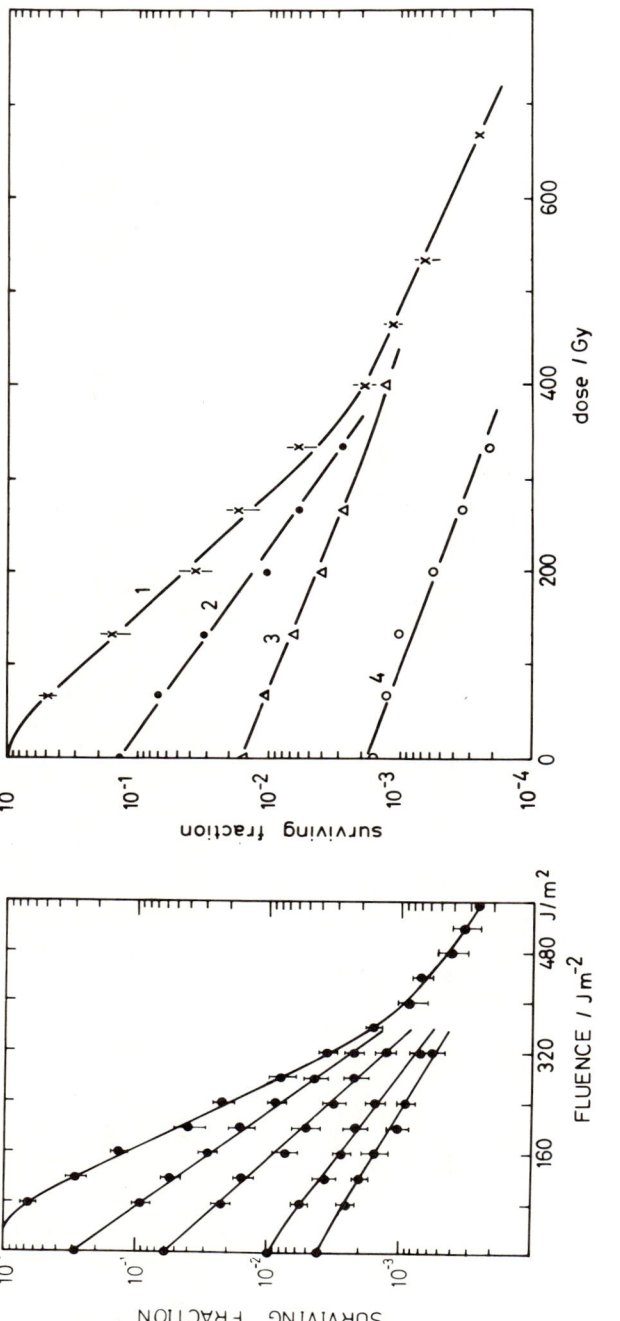

Figure 5. As in Figure 3 but for the rad9-mutant

genetic loci determining radiation sensitivity (Haynes and Kunz 1981) which can be grouped into three major repair pathways. Furthermore, radiation responses in haploid and diploid cells are different, which gives yet another parameter to be studied.

Early work by Elkind and Sutton (1959) demonstrated interaction between UV and, γ-rays both in dividing diploid and haploid yeast cells. In the latter case, a detailed analysis is hampered by the presence of resistant "tails" in the X-ray survival curve but it appears that just these cells play the most important role. In stationary haploid yeast cells, the action of UV and X-rays is independent (Uretz 1955) if given in immediate sequence. This behaviour has been interpreted to mean that radiation-induced killing proceeds via a single-hit mechanism where interaction, of course, is impossible. Mutation experiments, however, (see below) demonstrate that this interpretation is no longer tenable.

A more systematic study concerning the combined action of UV and ionizing radiation in diploid yeast was performed by Schneider and Kiefer (1976) who investigated not only the wild type but also two representative radiation sensitive strains. The results are summarized in Figs 3–5: if X- or α-irradiation is preceeded by UV, the shoulder in the survival curve is progressively lost, irrespective of the particular strain. With a reversed sequence, however, a different behaviour is found. There is a remaining shoulder in the UV survival curve both in the wild type and in the excision-deficient rad2-mutant, but not in the rad9-strain. UV-pre-exposure reduces the shoulder drastically in the wild type (the extrapolation number falls from initially 20 to finally 3) while it remains essentially unchanged in rad2. In both strains there is no effect on final slope. It is also seen from Fig. 5(b) that the rad2-mutant cannot be sensitized to subsequent UV-exposure by proceeding α–irradiation (the same is true for X-rays, data not shown). The results with the rad9-mutant are even more interesting. In this case, UV and X-rays appear to be interchangeable. The shoulder is always removed but also the slope is reduced. The biphasic survival behaviour of this strain makes the interpretation difficult. It is not due to population heterogeneity (Schneider and Kiefer 1976b) but may be caused by an inducible process. Unfortunately, this question has not been pursued.

Figure 6 summarizes the results with diploid yeast in a schematic way to facilitate later discussion.

	WILD TYPE	rad 2	rad 9
UV + IR	ADDITIVE	ADDITIVE	ADDITIVE
IR + UV	PARTLY ADDITIVE	INDEPENDENT	ADDITIVE

Figure 6. Summary of the results with diploid yeast

Mammalian Cells

Similar experiments as just reported for yeast were performed in V79 hamster cells by Han and Elkind (1977). Some of the results are shown in Fig. 7. It is immediately obvious that they resemble those obtained in wild type diploid yeast. The shoulder in the X-ray survival curve can be completely abolished by sufficiently high UV-pre-exposure but there is no change in final slope. X-rays, on the other hand, if given before UV, only reduce the shoulder, but there is always a remaining part even with high predoses. Also here no effect on final slope is seen. In a later paper (Han and Elkind 1978) showed that interaction between X-rays and UV can be found in all stages of the cell cycle although with varying extent.

Mutations

There are only few data about mutation induction after the combined action of ionizing and non-ionizing radiations in eukaryotic cells. Cleaver (1978) reported the absence of interaction between X-rays and UV in Chinese hamster cells with regard to the induction of ouabain-and thioguanine-resistance. A more extensive study was performed with wild type haploid yeast by Baumann, Zoelzer and Kiefer (1988) using resistance to canavanine as the experimental endpoint. Only the sequence UV-X-rays has been investigated so far. While there was independent action for survival (which is in agreement with earlier reports by Uretz (1955) this was not the case for mutation induction. Figure 8 displays the results: with X-rays alone the mutation frequency, i.e. mutants per survivor, is a linear function of dose, with UV alone one finds a linear-quadratic relationship. If UV is followed by a number of X-ray doses the linear

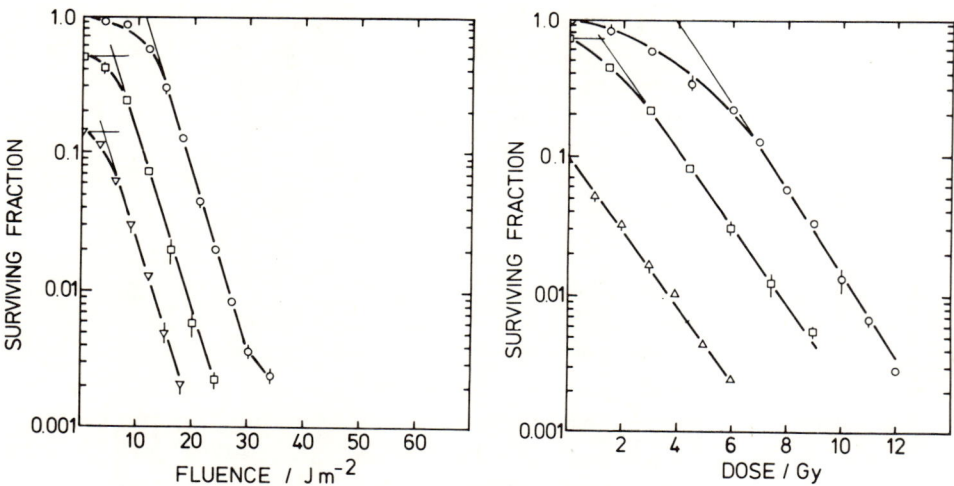

Figure 7. Interaction in V79 Chinese hamster cells (after Han and Elkind 1977)

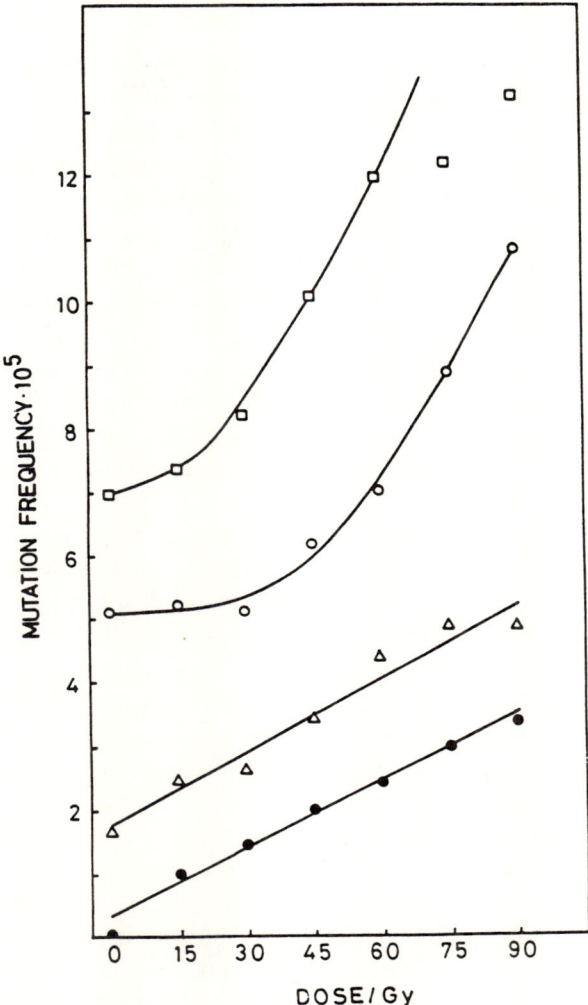

Figure 8. Interaction with mutation induction in haploid yeast (Baumann et al. 1988). Only the sequence UV + X-rays was investigated

dependency is gradually changed to an upward-concave one which can be fitted by a linear-quadratic expression. The combined treatment leads to levels of mutation frequency which cannot be achieved by X-rays alone because the induction curve levels off at higher doses. It can, therefore, not be decided whether there is really a synergistic-action or whether UV only reveals a linear-quadratic part of the X-ray induction curve which is normally not obtainable experimentally. In any case, there is a clear indication of interaction which rules out the assumption that haploid yeast is killed by a single-hit mechanism.

Discussion

The experiments reviewed show clearly that UV and ionizing radiations interact at the cellular level, both with regard to survival and mutation induction. An immediate consequence of this finding is that the damage is — at least partly — inflicted at the same site. Since action spectroscopy suggests strongly that for UV this is DNA one has to conclude that also for X-rays this molecule constitutes the critical target.

The next question to be answered is whether the interaction occurs at the level of lesion formation (type-I-interaction, see Fig. 2) or via repair-processes (type-II-interaction). At first sight, the survival data both in yeast and mammalian cells point to the first alternative. The fact that UV can always substitute for X-rays but not X-rays for UV which is reflected in the persistence of a UV-shoulder, suggests that UV might induce structural alterations in the DNA which increase its sensitivity to X-ray damage. This explanation was also proposed by Han and Elkind (1978). There are, however, a number of reasons which make this hypothesis rather improbable, both quantitatively and qualitatively.

The highest UV-fluence used in the combination experiments with diploid yeast (320 Jm^{-2}) induces about 1.5×10^5 thymine dimers in the diploid genome (Unrau et al. 1973) which consists of about 3.5×10^7 base pairs. This means that the mean distance between dimers is more than 200 base pairs. It is very unlikely that the structure is changed in such a way that the susceptibility to X-ray damage is significantly increased. With haploid cells (highest UV-fluence 48 Jm^{-2}, mean inter-dimer distance 1500 base-pairs) the probability is even less. Also, an interaction at the level of initial damage should equally show up in strains of different sensitivity unless one postulates that not only the number but also the spectrum of lesions is changed. The experiments show, however, that the interaction depends greatly on the repair capabilities of the cell. It is, therefore, suggested that "type-II-interaction" plays a predominant role.

The way of interaction is still not clear, even for yeast, let alone for mammalian cells. The scheme in Fig. 7, however, suggests that excision repair (controlled by the rad2-gene-product) is involved. This is surprising since it is generally assumed that this pathway is exclusive for UV-damage. This view is substantiated by the finding that rad2-mutants are not sensitive to X-rays (Haynes and Kunz 1981). It has to be borne in mind, however, that the rad2-mutation blocks an early step of excision-repair which is presumably not required for the repair of X-ray lesions. Later steps may be common. One could speculate, for instance, that double-strand breaks (DSB) are generated as a consequence of incision which are repaired by the same system as those directly formed by ionizing radiation. Recent results with the temperature-conditional DSB-repair-mutant rad54 are compatible with this assumption (Kiefer 1987): Incubation at the restrictive temperature increases not only sensitivity to X-rays but unexpectedly also to UV. The remaining shoulder in the wild-type UV-survival curve after pre-exposure to ionizing radiation has then to be attributed to the pathway controlled by the rad9-gene. In the wild type it has to be considered of lesser importance which becomes more significant only if excision (for UV) or DSB-repair (for ionizing radiation) are inhibited. Consequently, one would expect complete overlap between both radiation types if the rad9-pathway does not operate. This is experimentally found. One difficulty in this interpretation is that rad9-mutants are sensitive both to UV and X-rays suggesting a common mechanism. The experiments show that UV can saturate X-ray repair in all strains but that the opposite is only the case in a rad9-background. One has, therefore, to assume that UV- and X-ray damage are handled differently by the rad9-system. Obviously, more experiments are required to clarify the situation.

The scheme proposed for yeast could also be useful as a guideline for further research with mammalian cells. The similarity of the results in the "wild type" is striking. The availability of radiation-sensitive mutants opens up the possibility of unraveling the repair systems, and interaction studies could contribute to this.

Acknowledgment

I should like to thank my co-workers, B. Baumann and Dr. E. Schneider, for their permission to use their results and also for discussions.

References

Baumann, B., Zoelzer, F., Kiefer, J., Interaction between UV and X-rays for mutation induction in haploid yeast, Submitted for publication, 1988.
Cleaver, J.E., Absence of interaction between X-rays and UV light in inducing ouabain-and thioguanine-resistant mutants in Chinese hamster cells, *Mutat. Res.* 52, 247-253, 1978.
Elkind, M.M., Sutton, H., Sites of action of lethal irradiation: overlap in sites for X-ray, ultraviolet, photoreactivation, and ultraviolet protection and reactivation in dividing yeast cells, *Radiat. Res.* 10, 296-312, 1959.
Han, A., Elkind, M.M., Additive action of ionizing and non-ionizing radiations in Chinese hamster cells, *Int. J. Radiat. Biol.* 31, 275-282, 1977.
Han, A., Elkind, M.M., Ultraviolet light and X-ray damage interaction in Chinese hamster cells, *Radiat. Res.* 74, 88-100, 1978.
Haynes, R.H., Kunz, B.A., DNA repair and mutagenesis in yeast, in The molecular biology of the yeast *Saccaharomyces cerevisiae* (Eds: J.N. Strathern, E.W. Broach, J.R. Broach), Cold Spring Harbor, p. 371-414, 1981.
Kiefer, J., UV response of the temperature-conditional rad54 mutant of the yeast *Saccaromyces cerevisiae*, *Mutat. Res.* 191, 9-12, 1987.
Martignoni, K.D., Smith, K.C., The synergistic action of ultraviolet and X-radiation on mutants of *Escherichia coli* K-12, *Photochem. Photobiol.* 18, 1-8, 1973.
Schneider, E., Kiefer, J., Delayed plating recovery in diploid yeast of different sensitivities after X-ray and alpha-particle exposure, Int. *J. Radiat. Biol.* 29, 77-84, 1976b.
Schneider, E., Kiefer, J., Interaction of ionizing radiation and ultraviolet-light in diploid yeast strains of different sensitivity, *Photochem. Photobiol.* 24, 573-578, 1976.
Unrau, P., Wheatcroft, R., Cox, B., Olive, T., The formation of pyrimidine dimers in the DNA of fungi and bacteria, *Biochem. Biophys. Acta* 312, 626-632, 1973.
Uretz, R.B., Additivity of X-rays and ultraviolet light in the inactivation of haploid and diploid yeast, *Radiat. Res.* 2, 240-252, 1955.

A Molecular Model for the Cytotoxic Action of UV and Ionizing Radiation

H. P. LEENHOUTS[1] AND K H. CHADWICK[2]

[1]*L.S.O., R.I.V.M.*
P. O. Box 1
Bilthoven, The Netherlands
[2]*DG XII, C.E.C.*
200 rue de la Loi
Brussels, Belgium

Abstract

A quantitative model, which describes many aspects of cell killing by ionizing radiation, and which is based on the assumption that DNA double strand-breaks are the lethal lesions, is extended to provide a quantitative description of cell killing following UV exposure. The lethal lesion is assumed to be a pair of dimers, one on each side of a replicon terminus. The similarities and differences of the predictions from the two models are discussed. The models suggest a scheme for the interaction of ionizing radiation and UV, and an analysis of relevant data is presented. The implications of the possible interaction are discussed.

Introduction

An understanding of the hazards of low-level exposure to ionizing and ultraviolet radiations relies on a combination of epidemiological studies, experimental measurements and the extrapolation of the measurements to low exposure levels. Biophysical models used to interpret the experimental measurements can provide the quantitative methods of extrapolating to low exposure levels by making predictions about the mathematical form of the exposure-effect relationships. In this paper a mechanistic model is presented which provides a quantitative description of the cellular dose-effect relationships for ionizing radiation and has been extended to describe cell survival after ultraviolet exposure. The model also suggests that a combination of X-ray and UV exposures should lead to an interaction between the two agents. This interaction is presented and its implications are discussed.

Theoretical Background

The starting point for the development of our model has been that double stranded lesions in the nuclear DNA are the lesions which are crucial to the survival of the cell.

Photobiology, Edited by E. Riklis
Plenum Press, New York, 1991

71

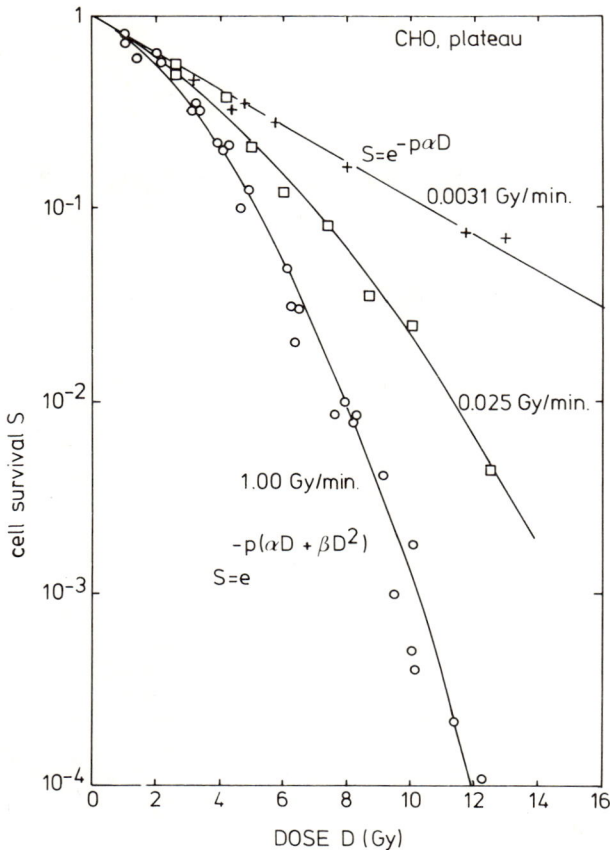

Figure 1. The effect of dose rate on cell survival analyzed according to equation (2) demonstrating that the α coefficient remains constant and the β coefficient decreases (Data from Metting et al, 1979)

Using this starting point we have proposed that DNA double strand breaks induced by ionizing radiation are the critical lesion (Chadwick and Leenhouts, 1981). We assume that the number (N) of double strand breaks induced as a function of Dose (D) is given by the equation:

$$N = \alpha D + \beta D^2 \tag{1}$$

where α is the probability per cell per unit dose that a double strand break is induced in the passage of one ionizing particle, and β is the probability per cell per unit dose squared

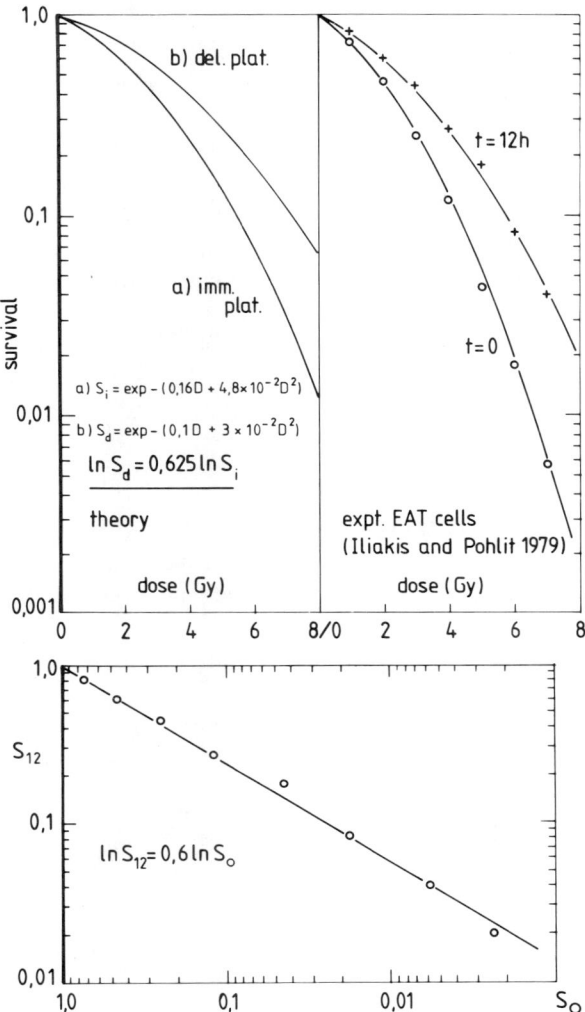

Figure 2. The theoretical prediction of the effect of post-irradiation repair on cell survival and the analysis of data from Iliakis and Pohlit (1979) according to equation (4)

that two independently induced single strand breaks combine to form a double strand break.

If p is the probability that a double strand break is lethal then cell survival (S) is given by:

$$S = \exp[-p(\alpha D + \beta D^2)] = \exp[-pN] \qquad (2)$$

73

This equation has been found to give an excellent fit to a wide variety of cell survival data (see Chadwick and Leenhouts, 1981) and Figure 1 shows data on cell survival at different dose rates fitted by equation (2). In accordance with theoretical expectations the α coefficient remains constant but as the dose rate decreases, the β coefficient decreases and this is explained on the basis of the repair of single strand breaks during irradiation.

Figure 2 presents the analysis of post-irradiation delayed plating repair. During post-irradiation delay the repair of double strand breaks can occur which means that the number (N) decreases to gN, where g is less than unity, and cell survival becomes:

$$S_2 = \exp\left[-pg(\alpha D + \beta D^2)\right] \tag{3}$$

Combining equations (2) and (3) reveals:

$$\ln S_2 = g\ln S. \tag{4}$$

Figure 3 presents two sets of extensive data on cell survival and DNA double strand breaks analysed using equation (2). The data taken from Radford (1986) and Prise et al (1987) are consistent with a correlation between the number of double strand breaks and

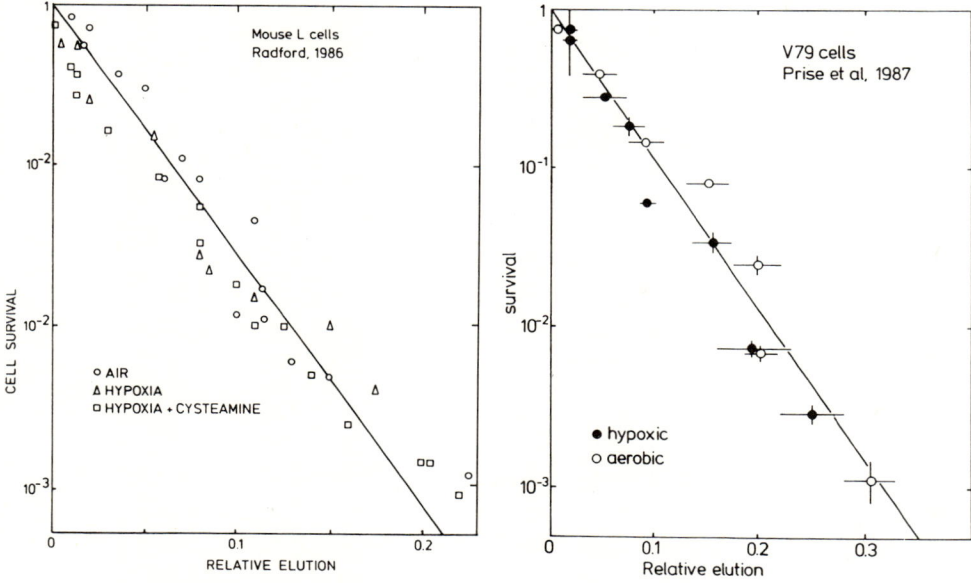

Figure 3. The relationship between DNA elution, presumed DNA double-strand breaks, and cell survival for sparsely ionizing radiation with different radiation conditions. The straight line correlations are consistent with equation (2). (Data from Radford (1986) and Prise et al (1987)

cell survival for sparsely ionizing radiation. It is worth noting that other data from Prise et al (1987) for neutron irradiation indicate a linear correlation between the logarithm of survival and number of double strand breaks, but with a different slope from that found for sparsely ionizing radiation. Even so, the results in Figure 3 may be interpreted as providing evidence of a strong connection between survival and DNA double strand breaks.

Ultra violet (UV) radiation does not cause strand breakage directly and one photon of UV cannot produce a double strand break or lesion. However, Park and Cleaver (1979) have proposed a paired dimer model for a potentially lethal lesion for UV which is schematically illustrated in Figure 4. Basically, it is proposed that a dimer can block the progression of a replication fork temporarily. Two dimers, one on each side of a replication terminus would block the two converging replication forks giving a long lived gap which could be a potentially lethal lesion. The number of pairs of dimers (N_o) induced by an exposure (X) to UV is given by:

$$N_o = \varepsilon X^2 \tag{5}$$

where ε is the probability per cell per unit exposure squared that two independently induced dimers form a pair (Chadwick and Leenhouts, 1983). Cell survival (S) is given by:

$$S = \exp\left[-p\,\varepsilon X^2\right] \tag{6}$$

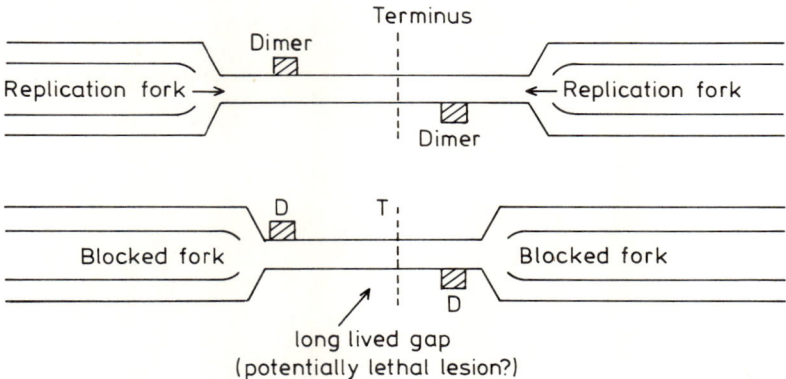

long lived gap
(potentially lethal lesion?)

Figure 4. A schematic representation of the model proposed by Park and Cleaver (1979) for the creation of a potentially lethal lesion by UV light

Figure 5 presents two survival curves analysed according to equation (6). Also in Figure 5 is some evidence from Wade and Lohman (1980) showing that pairs of dimers do form lethal lesions. The different survival curves for UV and UV and PR exposure of

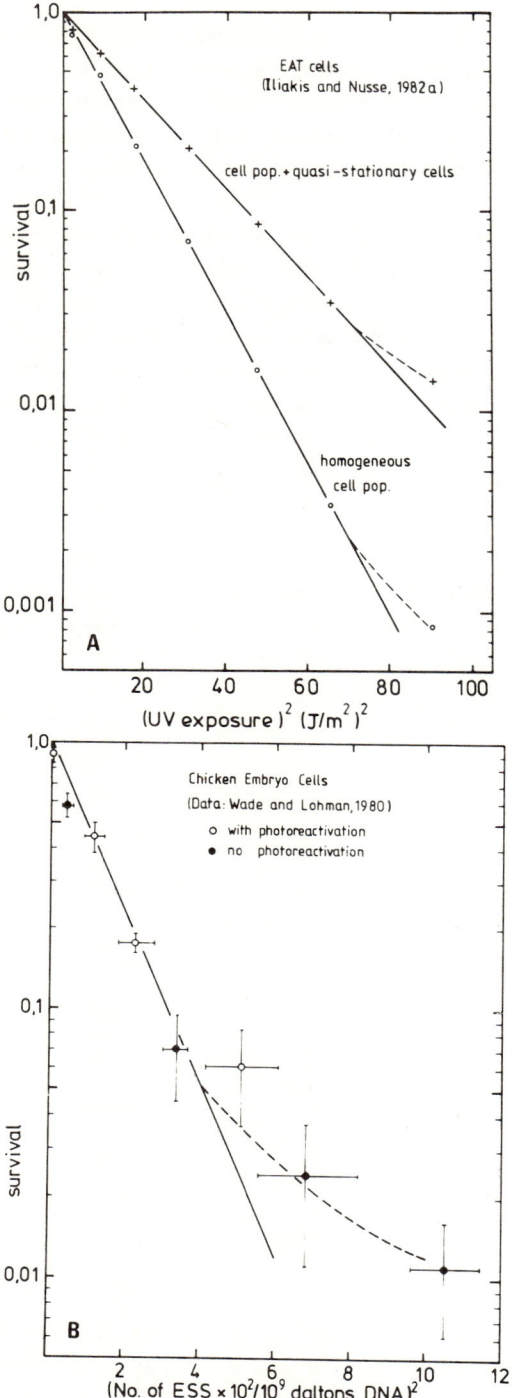

Figure 5. UV survival curves analyzed according to equation (6) (Data from Iliakis and Nusse, 1982) and an indication that pairs of dimers form potentially lethal lesions (Data from Wade and Lohman, 1980). The correlation between the cell survival, with and without photoreactivation, and the square of the numbers of endo-nuclease-sensitive sites, presumed dimers, with and without photo-reactivation. The resistant tail probably arises from G_2 phase cells.

the chick embryo cells normalize into one linear correlation when the logarithm of survival is plotted as a function of the square of the number of dimers.

The fascinating aspect of the Park and Cleaver model is that the DNA synthesis phase is absolutely crucial for the recognition of the pair of dimers as a lesion. In the period after exposure and up to the S-phase the dimers are only recognised by the cell as single-stranded damage. This means that the two strategically situated dimers form a virtual lesion after exposure up to S-phase and that the repair of either dimer of the virtual pair during and after exposure up to S-phase removes the lethal lesion. Thus, it is very difficult to show a dose-rate effect for UV because the time of exposure is only a part of the total time from the start of exposure to the next S-phase which is available for repair. The model also permits some intriguing speculation about repair in stationary cells after exposure. In this case, the repair of one of the dimers of a virtual pair in the time after exposure removes the lesion. As the dimers are 'seen' by the cell as independent single-stranded lesions, they can in principle be repaired perfectly and in stationary cells with a long enough repair time all the dimers can be repaired and no lethal lesions will be

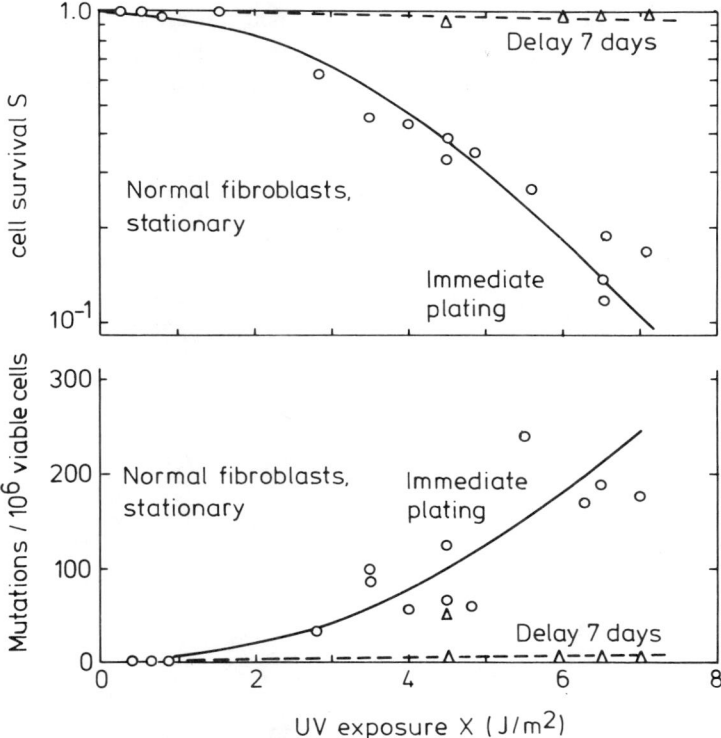

Figure 6. The influence of repair in stationary cells held for several days after exposure on survival and mutation induction. The long repair time leads to almost complete recovery (Data from Maher et al, 1979)

'seen' by the cell when it finally enters S-phase. Figure 6 presents data for human cells from Maher et al (1979) which demonstrate this effect.

When the models for ionizing radiation and for UV exposure are considered together it is relatively straightforward to think in terms of a possible interaction. In Figure 7 we present schematically an interaction process between ionizing radiation and other DNA damaging agents. In the case of UV the δ coefficient is zero. This scheme predicts that the interaction term ηXD is proportional to UV exposure and radiation dose and that the relative survival can be derived

i)   when UV exposure is followed by ionizing radiation as

$$S = \exp\left[-p\left\{(\alpha + \eta X)D + \beta D^2\right\}\right] \tag{7}$$

ii) when ionizing radiation dose is followed by UV as:

$$S = \exp\left[-p(\eta DX + X^2)\right] \tag{8}$$

Figure 8 presents an analysis of data from Han and Elkind (1977) using equations (7) and (8) which reveals the consistency of the $p\eta$ coefficient irrespective of the sequence of exposure. We are currently re-examining this interaction process using stationary cell cultures and can confirm the interaction between UV and gamma radiation.

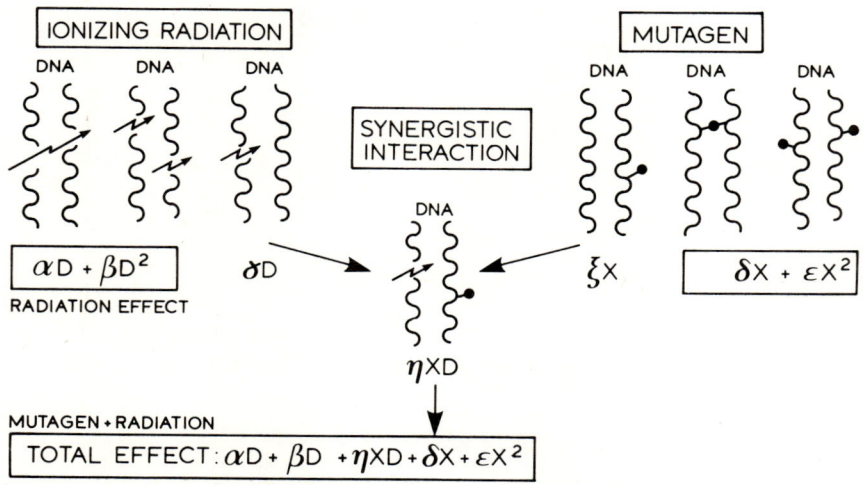

Figure 7. Schematic representation of the interaction between UV and ionizing radiation at the molecular level and the consequences for the quantitative analysis

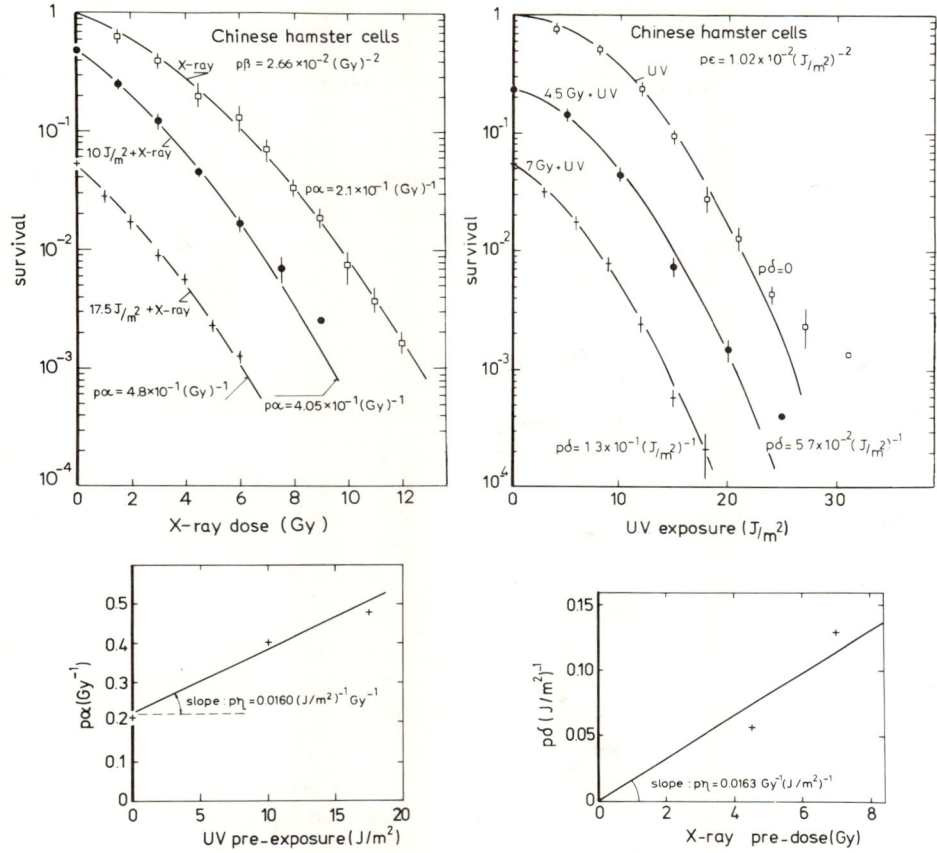

Figure 8. An analysis of cell survival data when radiation doses are preceded by a UV exposure and *vice-versa* using equations (7) and (8) and the comparability of the interaction term $p\eta$. (Data from Han and Elkind, 1977.)

Discussion

The model presented here provides a unifying concept for the cytotoxic action of ionizing radiation and UV light inasmuch as the basic assumption remains that a double stranded lesion on the DNA molecule is a potentially lethal lesion. The concept can also be extended to other DNA damaging agents (Leenhouts and Chadwick, 1984). However, the mechanisms of induction and recognition of the lethal lesions are different and have interesting implications for the effects of dose rate, fractionation and repair on cell killing. The model could be used to investigate these different implications at the cellular level and to try to correlate the repair processes with the changes in cell killing more directly.

For the case of ionizing radiation the model provides a molecular mechanism associated with the prediction that the biological effect is linearly related to radiation dose at low doses and does not have a threshold. For the case of UV light, the Park and Cleaver model, as adapted here, provides a molecular mechanism which implies that the cellular biological effect will be related to the square of the exposure, that there will be no sparing effect of dose rate, that the time from exposure to DNA synthesis is crucial for repair and that in stationary cells there is reason to expect a total repair of damage, with zero effect.

The model also suggests a mechanism for the synergistic interaction between two agents which can cause single strand damage to DNA. This interaction would eventually lead to the existence of a linear exposure relationship at low exposure levels in terms of the exposure levels of one of the two interacting agents. The interaction scheme does assume some coincidence in time of exposure to the two agents. Repair of a single-strand damage caused by one agent before the occurrence of a second damage by the second agent would eliminate the lesion if the lesion is formed when the second damage occurs, as we assume the case is for the double-strand break induced by ionizing radiation. In addition, if the lesion formed by the interactive process behaves as a UV-type lesion, then repair of either single strand damage after exposure and up to DNA synthesis would also remove the lesion. It is clearly necessary to understand more about the nature of the interactive lesion in order to be able to consider the low exposure risk.

This paper considers only a model for the cytotoxic action of ionizing radiation and UV and it is obviously interesting to investigate whether the model could also be applicable to mutation and cancer induction in connection with risk estimation of low level exposures.

Acknowledgements

This paper is publication number 2481 of the Radiation Protection Programme of the Commission of the European Communities. The work is supported by the Radiation Protection Contract B16-A-008-NL and the Dutch Ministries of Welfare, Public Health and Culture, and Housing, Physical Planning and the Environment.

References

Chadwick, K.H. and Leenhouts, H.P., The Molecular Theory of Radiation Biology, Springer Verlag, Heidelberg, 1981.

Chadwick, K.H. and Leenhouts, H.P., A model for the cytotoxic action of UV., *Phys. Med. Biol.* 28, 1369-1383, 1983.

Han, A. and Elkind, M.M., Additive action of ionizing and non-ionizing radiation throughout the Chinese hamster cell cycle, *Int. J. Radiat. Biol.*, 31, 275-282, 1977.

Iliakis, G. and Pohlit, W., Quantitative aspects of the repair of potentially lethal damage in mammalian cells, *Int. J. Radiat. Biol.*, 36, 649-658, 1979.

Iliakis, G. and Nusse, M., Conditions supporting repair of potentially lethal damage cause significant reduction of ultra-violet light induced division delay in synchronised and plateau-phase Erlich ascites tumour cells, *Radiat. Res.*, 91, 483-506, 1982.

Leenhouts, H.P. and Chadwick, K.H., A quantitative analysis of the cytotoxic action of chemical mutagens, *Mutation Res.*, 129, 345-357, 1984.

Maher, V.M., Dorney, D.J., Mendrala, A.L., Konze-Thomas, B., and McCormick, J.J., DNA excision-repair processes in human cells can eliminate the cytotoxic and mutagenic consequences of ultra-violet irradiation, *Mutation Res.*, 62, 311-323, 1979.

Metting, N.F., Braby, L.A., Roesch, W.C. and Nelson, J.M., Dose rate evidence for two kinds of radiation damage in stationary-phase mammalian cells, *Radiat. Res.*, 103, 204-218, 1985.

Park, F.D. and Cleaver, J.E., Post-replication repair: Questions of its definition and possible alteration in xeroderma pigmentosum cell strains, *Proc. Natl. Acad. Sci. USA*, 76, 3927-3931, 1979.

Prise, K.M., Davies, S. and Michael, B.D., The relationship between radiation-induced DNA double-strand breaks and cell kill in hamster V79 fibroblasts irradiated with 250 kVp X-rays, 2.3 MeV neutrons or [238]Pu alpha particles, *Int. J. Radiat. Biol.*, 52, 893-902, 1987.

Radford, I.R., Evidence for a general relationship between the induced level of DNA double strand breakage and cell killing after X-irradiation of mammalian cells, *Int. J. Radiat. Biol.*, 49, 611-620, 1986.

Wade, M.H., and Lohman, P.H.M., DNA repair and survival in UV-irradiated chicken-embryo fibroblasts, *Mutation Res.*, 70, 83-93, 1980.

Mutagenic Effects of Ultraviolet and Ionizing Radiation

HELEN H. EVANS
Case Western Reserve University
Cleveland, OH 44122 USA

Introduction

UV and ionizing radiation induce cell killing, mutation and oncogenic transformation in mammalian cells. The molecular mechanisms whereby these cellular sequelae are induced has been a subject of study for many years. Although the exact reactions involved in the induction processes remain to be elucidated, the relevant alterations responsible for these cellular events are thought to be DNA lesions.

DNA lesions induced by UV and ionizing radiation

Despite markedly different energies, UV and ionizing radiation induce a qualitatively similar spectrum of DNA lesions. This damage includes base alterations, base loss, inter- and intra-strand crosslinks, DNA-protein cross links, regions of denaturation, alkali-labile sites, single-strand breaks and double-strand breaks in the deoxyribose phosphodiester backbone (Bradley and Taylor, 1981; Hutchinson, 1985; Morgan *et al.*, 1984; Ward, 1988). Quantitatively, however, the spectrum of lesions resulting from UV vs. ionizing radiation differs markedly. For example, single-strand breaks predominate after exposure to ionizing radiation, while dipyrimidine cyclobutane dimers predominate after exposure to UV-C (210–290 nm) radiation. A majority of the detrimental effects of ionizing radiation are caused indirectly by oxidative hydroxyl radicals formed by the ionization of water molecules surrounding the DNA (Roots and Okada, 1972; Teoule and Cadet, 1978). In contrast, UV-C radiation results in the direct excitation and subsequent alteration of the purine and pyrimidine rings. UV-A (320–400 nm) and UV-B (290–315 nm) radiation cause the excitation of endogenous sensitizing molecules and in the presence of oxygen yield singlet oxygen and oxidative radicals which cause DNA damage. Cell lethality per DNA lesion is much greater for ionizing radiation than for other agents, even those yielding oxidative radicals (Ward, 1987). This efficiency in cell killing on the part of ionizing radiation may be due to the deposition in DNA of localized volumes of energy which can yield clusters of radicals and/or ionizations subsequently producing local multiply damaged sites involving *both* DNA strands (Ward, 1985). Local damage to both DNA strands may be lethal, since recovery would be hampered by the absence of an intact complementary strand to serve as a template for repair and/or the inability of some repair

Photobiology, Edited by E. Riklis
Plenum Press, New York, 1991

enzymes, such as glycosylases (Lindahl, 1982), to act on single-stranded regions (Ward, 1985).

Many different base alterations are induced by UV and ionizing radiation. These include hydrated or saturated rings such as thymine glycol and 8-hydroxyguanine, and ring cleavage products, such as the formamido pyrimidines and urea residues. Often the base alteration causes labilization of the glycosylic bond, and the entire base is then lost, yielding an apurinic or apyrimidinic site. Specific base alterations induced either *in vivo* or *in vitro* by UV-C radiation have been described by Fisher and Johns (1973), and by Cadet *et al.* (1978, 1985), while those induced by ionizing radiation have been reported by Wallace (1983), Cadet and Berger (1985), van Sonntag and Schuchmann (1986), and Teoule (1987).

Correlation of DNA lesions to cellular end points

Which of these many lesions are then responsible for the subsequent cellular lethality, mutation, or oncogenic transformation? Although it might be assumed that the most frequent lesion would be the major cause of cellular effects, this does not appear to be the case. For instance, the cyclobutane thymine-thymine dimer, the most frequently occurring lesion induced by UV radiation, does not appear to be the major cause of UV-induced mutations (Haseltine, 1983). In fact, Brash *et al.* (1987) have reported that the frequency and distribution of neither UV-induced cyclobutane dimers nor another major photoproduct, the (6–4)pyrimidine-pyrimidone dimer, are correlated to the frequency of base-substitution mutations at specific dipyrimidine sites of the *supF* tRNA gene in a UV-irradiated shuttle vector transfected into normal or excision-repair-deficient human cells. Exposure of cells to ionizing radiation induces DNA single-strand breaks in great numbers, but these lesions do not appear to be particularly detrimental to cells because of their rapid repair (Ward, 1988).

The reduction of cell killing, mutation and transformation by photoreactivating enzyme which specifically repairs cyclobutane dimers *in situ*, has traditionally been taken as evidence linking the cyclobutane dimer to these cellular events. However, the photoreversal of cyclobutane dimers may *indirectly* alleviate the damaging effects of UV light by terminating the mutagenic SOS response and/or by reducing the concentration of DNA lesions competing for repair enzymes (Witkin, 1976; Haseltine, 1983; Franklin and Haseltine, 1986). One potential explanation for the less-than-expected contribution of cyclobutane dimers to the induction of mutations may be the preferential insertion of adenine at template sites lacking coding information (Strauss *et al.*, 1982; Boiteux and Laval, 1982; Kunkel *et al.*, 1983). When adenine insertion occurs opposite sites of thymine dimers, no mutation occurs, since the normal base pair is regenerated.

The ability of a DNA lesion to produce a specific cellular effect thus depends upon a number of factors, including the frequency of the lesion, the timing of repair relative to fixation by DNA replication or the occurrence of mitosis, whether the lesion allows correct or error-prone read-through by DNA polymerase or whether it blocks replication, and whether the lesion results in a functional change in an RNA or protein product.

Cell killing and oncogenic transformation induced by UV or ionizing radiation results at least in part from chromosome aberrations, such as deletions and rearrangements (Dewey *et al.*, 1971; Croce, 1986). The aberrations in turn are thought to arise from the induction of DNA double-strand breaks (Preston, 1980). Mutations may also result from chromosomal damage, but in addition are induced by alterations in single bases leading to changes in coding properties, or by the the loss or insertion of one or more nucleotides. The balance of this paper is concerned with lesions that lead to UV and ionizing radiation-induced mutations and the influence of the target locus on the mutant frequency and the spectrum of the recovered mutations.

Miscoding by Altered Bases

In order to determine the mutational consequences of specific DNA lesions, oligonucleotide templates containing specific altered bases at unique positions have been constructed. The consequences of the alteration on the subsequent insertion of nucleotides by various DNA polymerases is then determined either in cell-free systems or by plasmid or phage replication in whole cells. In this manner it has been found that 8-hydroxy-guanine can direct the insertion of any of the four bases, and can cause miscoding by its neighboring bases as well (Kuchino *et al.*, 1987). Thymine glycol, as well as apurinic and apyrimidinic sites, commonly result in a replication block, but when bypass of these lesions occurs it has been found that an adenine nucleotide is commonly inserted (Clark and Beardsley, 1987; Loeb and Preston, 1986). Deamination of 5-methylcytosine to produce thymine, a common spontaneous reaction, results in a GC to AT transition (Coulondre *et al.*, 1978).

Specific Mutations Caused by Radiation

The *lac I* gene of *E. coli* has been used to great advantage in studying the specificity of mutagens. Base substitutions resulting in nonsense mutations at 90 different codons in the gene have been identified (Miller *et al.*, 1978). It has been shown that UV-induced mutagenic hot spots in this gene occur at sites of TC or CC sequences. These positions are the sites of (6-4)pyrimidine-pyrimidone dimers as well as cyclobutane dimers (Brash and Haseltine, 1982; Glickman *et al.*, 1986; Takimoto, 1986; Brash *et al.*, 1987). The 3'-cytosine of the photoproduct codes for an adenine nucleotide, thus resulting in a GC to AT transition. The identification of the sites of mutation in the hemizygous *aprt* gene of CHO cells has been investigated by Glickman and his group, using recombinational techniques for the cloning and sequencing of the various mutagen-induced alterations. They have found that about 75% of the recovered UV-induced mutations in this hemizygous gene are base substitutions, and that 65% of these consist of GC to AT transitions (Drobetsky *et al.*, 1987). UV-irradiation of mammalian cells harboring shuttle vectors has also been found to induce a high yield of GC to AC transitions, often at TC or CC sequences (Glazer *et al.*, 1986; Hauser *et al.*, 1986; Brash *et al.*, 1987; Drobetsky *et al.*, this volume). The GC to AT transition, one of the most common mutations which occurs following UV irradiation (Miller, 1985), thus appears to be induced to a large extent by the mispairing of adenine with the 3'-cytosine of the (6-4)photoproduct or cyclobutane dimer in both prokaryotic and eukaryotic cells. Cyclobutane pyrimidine dimers also induce transversions, frame-shifts or tandem double substitutions (Miller, 1985; Franklin and Haseltine, 1986; Protic-Sabljic *et al.*, 1986).

In contrast to UV radiation, base substitutions induced by ionizing radiation tend to occur equally at the sites of the four different bases, leading to equal proportions of transitions and transversions as determined in the *lac I* gene of *E. coli*, or in the endogenous hemizygous *aprt* gene of CHO cells (Kato *et al.*, 1985; Grosovsky *et al.*, 1988).

The Recovery of Mutants with Multilocus Deletions is Dependent upon the Target Gene

Another difference between the spectrum of mutations induced by ionizing radiation vs. UV radiation is the relatively large proportion of multilocus deletions induced by ionizing radiation. In the case of ionizing radiation, however, the recovery of mutants harboring multilocus deletions varies markedly depending upon the target locus. The recovery is low in haploid organisms and for target genes located in hemizygous regions of diploid cells. It has been suggested that this low recovery is due to the deletion of an essential gene linked to the target locus (Chu, 1971). Another copy of the essential gene would not be available in haploid cells or in diploid cells with the target gene in a hemizygous region. Alternatively, the recovery of mutants with multilocus lesions may depend on recombinational mechanisms requiring homologous DNA regions (Little *et al.*, 1988). The poor recovery of mutants harboring multilocus deletions applies to the *hprt* gene located on the X chromosome of mammalian cells (Chu, 1971; Evans *et al.*, 1986), since this chromosome is either actually monosomic (male cells) or functionally monosomic (female cells). The *aprt* gene in CHO line D422 is also located in a hemizygous chromosomal region, undoubtedly explaining why the spectrum of ionizing radiation-induced mutations obtained by Grosovsky *et al.* (1988) includes very few large deletions.

An example of the difference between the recovery of mutants harboring lesions at the site of the *hprt* locus vs. those with the target gene located on an autosome is provided by the experiments of Stankowski and Hsie (1986). These investigators compared the frequency and spectrum of mutations induced either at the *hprt* locus on the X chromosome or at the bacterial *gpt* gene, integrated into one of the CHO autosomes in cell line AS52. Although only one copy of the bacterial gene is present in this cell line, second copies of the neighboring CHO genes would be provided by the homologous CHO chromosome. The ratio of the mutant frequency at the *gpt* locus vs. the *hprt* locus was 1.4 for UV radiation, and ranged from 6 to 19 for ionizing radiation, depending on the administered dose. Analysis of the two types of mutants with regard to alteration of restriction fragments revealed that UV radiation seldom induced large deletions, whereas

86

almost all of the mutants induced by ionizing radiation involved large deletions in both the *hprt* and *gpt* genes (Table 1).

Table 1. Comparison of mutation frequency and spectra at the *gpt* and *hprt* loci[1]

Agent	Fraction showing altered RFLPs		Ratio of mutation frequencies
	gpt	*hprt*	*gpt/hprt*
UV radiation	5/26	0/26	1.4
Ionizing radiation	26/26	15/21	6–19

[1]From the data of Stankowski and Hsie (1986)

Another system which allows good recovery of mutants with multilocus deletions is that of the A_L hybrid line, consisting of CHO K1 Chinese hamster cells harboring human chromosome 11 (Waldren *et al.*, 1979). Mutants with large deletions encompassing up to four genes on the human chromosome can be recovered, since none of the genes on the human chromosome are essential for survival of the hamster cells. Ionizing radiation induced a high mutant frequency in these cells (1×10^{-3} mutants per surviving cell after exposure to the Do dose). Loss of all four chromosomal markers occurred in 16% of the mutants, while loss of 2–3 markers occurred in 64% of the mutants. Clearly mutants with multilocus mutations predominate following exposure to ionizing radiation in this system.

Mutation induction in human lymphoblast line TK-6, a cell line which is heterozygous at the *tk* locus, has been reported by Yandell *et al.* (1986). These investigators found that the entire active *tk* allele was lost in the great majority of the TK$^{-/-}$ mutants, including spontaneous mutants as well as those induced by X radiation or UV radiation. In 40% of the mutants showing loss of the entire *tk* allele, a linked gene (*erbA*) was also lost, but a more distant gene (D17S2) was still present in all but two of the TK$^{-/-}$ mutants (Little *et al.*, 1988). A fraction of the TK-6 mutants grew slowly, and a higher percentage of the slow-growing fraction showed loss of the entire *tk* allele than did the mutants with a normal growth rate. The TK-6 slow-growing mutants are thus similar to the small colony TK$^{-/-}$ mutants of L5178Y 3.7.2C lymphoblasts which exhibit cytogenetic alterations in mouse chromosome 11, the location of the mouse *tk* gene (Hozier *et al.*, 1985; Moore *et al.*, 1985).

We have observed differences in the recovery of mutants harboring multilocus lesions in strains of mouse lymphoma L5178Y cells which are either heterozygous or hemizygous for the *tk* gene (Evans *et al.*, 1986). Following treatment with ionizing radiation, the frequency of TK$^{-/-}$ mutants was approximately 50 times greater in the heterozygous strain R16 than in the hemizygous strain R83 (Table 2). The frequency obtained for ionizing radiation-induced TK $^{-/-}$ mutations in the hemizygous strain was similar to the frequency obtained for inactivation of the *hprt* gene in either of the strains, indicating that hemizygosity leads to poor recovery of such mutants at both the *tk* and *hprt* loci. After treatment with UV radiation, the frequency of mutants in strain R16 was approximately twice that observed in strain R83 at both the *hprt* and *tk* loci.

Table 2. Locus specificity in heterozygous and hemizygous strains of L5178Y cells[1]

| Target Gene | Mutant Frequency at the Do Dose[2] | | | |
| | Ionizing radiation | | UV radiation | |
	R16	R83	R16	R83
hprt	20	8	475	200
Na$^+$/K$^+$ ATPase	–	–	9	8
tk	2700	50	520	270

[1] From the data of Evans *et al.* (1986)

[2] The Do doses for both strains were 1.2 Gy for ionizing radiation and 2 J/m^2 for UV radiation.

Examination of restriction fragment patterns indicated that the entire affected *tk* gene was deleted in all ionizing radiation-induced TK$^{-/-}$ mutants (59/59). The affected *tk* allele also appeared to be deleted from all spontaneous TK$^{-/-}$ mutants. Similar results have been previously reported by Applegate and Hozier (1988) who subjected L5178Y 3.7.2C TK$^{+/-}$ cells to treatment with various chemicals. Loss of the entire *tk* allele was also observed in human TK-6 lymphoblasts (Little *et al.*, 1988. Adair (1988) has observed complete loss of the *aprt* gene in 25/52 spontaneous *aprt* $^{-/-}$ mutants of a heterozygous *aprt* $^{+/-}$ V79 cell line, while none of the *aprt* $^{-/o}$ mutants of the hemizygous CHO cell lines showed complete loss of the *aprt* gene.

The size of deletions induced by ionizing radiation is well-illustrated by the experiments of Urlaub *et al.* (1986) in which 100% of *dhfr* $^{-/o}$ mutants induced by ionizing irradiation showed major changes. Some of the deletions covered 210 kb, and all extended outside the gene in the 5' rather than the 3' direction. Inversions were observed in 3/11 mutants studied. Deletions observed in the hemizygous *aprt* gene following treatment of CHO cells with ionizing radiation also all extended to the 5' side of the gene (Breimer *et al.*, 1986; Grosovsky *et al.*, 1986; Adair, 1988). The induction of these deletions at hemizygous loci suggests that no essential genes reside within the region of the deletion.

Although poor recovery of mutants harboring multilocus deletions is obtained when *hprt* is the target gene, many more deletions and gross rearrangements have been found in the case of the *hprt* gene (45 kb) than in the case of the smaller *aprt* gene (2.5 kb) when in a hemizygous chromosomal region. Thus only 16% of recovered *aprt* $^{-/o}$ mutants showed visible changes in restriction fragment patterns (Grosovsky *et al.*, 1986), whereas over 50% of ionizing radiation-induced *hprt* $^{-/o}$ mutants showed evidence of gross changes (Fuscoe *et al.*, 1983; Turner *et al.*, 1985; Albertini *et al.*, 1985; Breimer *et al.*, 1986; Brown *et al.*, 1986; Skulimowski *et al.*, 1986). Either the large size of the *hprt* gene which allows large deletions to be contained within it or the distance of the *hprt* gene from neighboring essential genes may explain this discrepancy. The recovery of *hprt* $^{-/o}$ mutants bearing multilocus deletions also varies with the cell line, perhaps because of variations in chromosome structure and the distance between the *hprt* gene and neighboring essential genes.

Base Substitution Mutations are Induced by Ionizing Radiation at Low Frequency

The tendency for ionizing radiation to induce large deletions may explain the low frequency of the induction of point mutations in comparison to UV radiation when the assay system depends on a functional gene product. Such assays include reversions (his$^-$ $\Rightarrow$ his$^+$) or the development of drug resistance, in which the mutation involves the inactivation of a drug binding site without affecting the catalytic activity of the protein (oua^s $\Rightarrow$ oua^r). However, the ability of ionizing radiation to induce base substitution and frame-shift mutations, although at low frequency, has been well documented in both prokaryotic and eukaryotic cells (Hartman *et al.*, 1971; Phillips *et al.*, 1972; Ino and Yourno, 1974; Glickman *et al.*, 1980; Liber *et al.*, 1986; Grosovsky *et al.*, 1988; Delcourt *et al.*, 1987).

Effect of DNA Repair Capability on the Mutation Spectrum

The spectrum of mutations is also influenced by the DNA repair capacity of the cells. A greater percentage of GC to AT transitions was induced in cells obtained from individuals with xeroderma pigmentosum than found for normal human cells (Bredberg *et al.*, 1986). One hotspot for GC to AT transitions in the xeroderma cells had the same sequence (5'-TCC) as a hot spot observed in the *uvrB* strain of *E. coli* by Todd and Glickman (1982). It appears, therefore, that this mutation is caused by errors in replication rather than errors in excision repair which is deficient in these two strains (Bredberg *et al.*, 1986).

A greater proportion of multilocus lesions were found to be induced in repair-deficient *N. crassa* than in the wild-type strain (De Serres *et al.*, 1983). Similar results were oserved in L5178Y cells; thus, a greater proportion of radiation-induced slow-growing TK$^{-/-}$ colonies were observed in the case of strain LY-S (deficient in the repair of radiation-induced DNA double-strand breaks) than in strain LY-R (Evans *et al.*, 1986). Adair (1988) has reported that repair-deficient lines of CHO cells exhibit spontaneous rearrangements not seen in repair-proficient cell lines.

The spectrum of recovered mutations can also be influenced by the protocol of radiation administration. In *N. crassa* the frequency of ionizing radiation-induced mutants with multilocus deletions was found to decrease with the dose rate, whereas the frequency of mutants with base-change mutations was not influenced by a decrease in dose rate (De Serres *et al.*, 1967). In L5178Y cells the ionizing radiation-induced mutant frequency was found to decrease to a greater extent at the *tk* locus than at the *hprt* locus (Evans, unpublished results), presumably because mutants with multilocus deletions (with dose-rate dependent induction) are recovered to a much greater extent at the *tk* locus than at the *hprt* locus.

These variations in mutation spectra with the target locus and with the occurrence of DNA repair indicate that broad conclusions should not be drawn concerning the mutagenicity of a DNA damaging agent, the mutability of repair-deficient cell strains, the repairability of mutational lesions, or the effect of modifiers on mutant frequency, when the results are obtained at loci (such as *hprt*) where mutants with multilocus lesions are poorly recovered.

Effect of Radiation Energy on Mutation

The type of DNA damage and the spectrum of mutations varies with exposure to radiations of different energies. Vacuum UV radiation with a wave length of 150–180 nm causes the photolysis of water and the production of oxidative radicals (Takakura *et al.*, 1987). UV-C radiation is absorbed by the nucleic acid purines and pyrimidines, and the excitations so induced cause base alterations resulting in miscoding. UV-B and UV-A radiations are not absorbed well by nucleic acid bases, but instead are absorbed by other cellular compounds. In the presence of oxygen, excitation of these molecules can result in the production of oxidative products which cause DNA damage, cell killing, mutation, chromosome aberrations, and oncogenic transformation (Lundgren and Wulf, 1988; Lonn, 1984; Miguel and Tyrell, 1983; Matsuda and Tobari, 1988). Roza *et al.* (1985) have reported that UV-A radiation induces single-strand breaks but only a few cyclobutane dimers in the DNA of normal human fibroblasts. Addition of catalase reduced the number of single-strand breaks but did not reduce cell lethality. Xeroderma pigmentosum cells (type A) showed the same sensitivity to UV-A radiation as did the normal human fibroblasts (Roza *et al.*, 1985). The action spectrum for the induction of pyrimidine dimers, DNA single-strand breaks, lethality and mutation has been found to coincide with the absorption spectrum of nucleic acids from 254 to 313 nm. Above 313 nm the dimer and mutation frequencies decrease to a greater extent than does the occurrence of strand breaks and lethality (Peak *et al.*, 1984, 1987). The maximum number of strand breaks per lethal event occurs at 365 nm (Peak *et al.*, 1987). It is also possible that near-UV radiation causes cell killing by mechanisms other than the induction of DNA lesions (i.e., membrane damage), explaining why less mutations per lethal event are observed at the longer wave lengths (Eisenstark, 1987).

Exposure of cells to high energy ionizing radiation has generally been found to induce non-repairable cellular damage (reviewed by Blakely *et al.*, 1985). When the linear energy transfer (LET) is increased above 80 kev/μ the mutagenicity has been reported to increase to a greater extent than cytotoxicity (Cox *et al.*, 1977; Cox and Masson, 1979; Chen *et al.*, 1984; Hei *et al.*, 1988). However, Barnhart and Cox (1979) found a parallel increase in lethality and mutability at the *hprt* locus in CHO cells exposed to [238]Pu alpha particles of 4.4 Mev (100 kev/μ), as compared to 250 kVp X rays. Nakamura *et al.*

(1982) reported the induction of thioguanine-resistant mutants to be increased to a greater extent than lethality when L5178Y cells were exposed to fast neutrons (6 Mev) vs. X radiation, whereas the induction of methotrexate-resistant mutants was increased to a lesser extemt than lethality by exposure to neutron radiation. Exposure of cells to high energy ionizing radiation leads to the induction of a greater percentage of multilocus lesions than in the case of low energy radiation in *N. crassa* and in the mouse (De Serres *et al.*, 1967a; Russell, 1971). However, Thacker (1986) obtained similar percentages of *hprt* $^{-/o}$ mutants showing deletions and rearrangements after exposure of the cells to high energy vs. low energy radiation. Possibly the poor recovery of mutants containing multilocus lesions at the hemizygous *hprt* locus obscured the difference between the two types of radiation.

Conclusion

Both UV and ionizing radiation induce a large array of different DNA lesions and types of mutations. The induction of GC to AT transitions at sites of (6–4) photoproducts and cyclobutane dimers appears to play a major role in UV radiation-induced cytotoxicity and mutagenicity. Ionizing radiation induces base damage and point mutations at random throughout the genome, but the induction of large intergenic deletions and gross rearrangements probably play a major role in the induction of cellular mutation, oncogenic transformation and lethality. The recovery of mutants harboring multilocus lesions is low, however, in haploid cells or in cases in which the target gene is located in hemizygous chromosomal regions.

References

Adair, G. M. (1988) Analysis of mutation of the Chinese hamster *aprt* locus, In *Mammalian Cell Mutagenesis*, Banbury Report 28 (edited by M. M. Moore, D. M. Demarini, F. J. De Serres and K. R. Tindall) pp. 3–13, Cold Spring Harbor Laboratory, Cold Spring Harbor, N.Y.

Albertini, R. J., J. P. O'Neill, J. A. Nicklas, N. H. Heintz and P. C. Kelleher (1985) Alterations of the *hprt* gene in human *in vivo*–derived 6-thioguanine-resistant T lymphocytes. *Nature* 316, 369–371.

Applegate, M. L. and J. C. Hozier (1988) On the complexity of mutagenic events at the mouse lymphoma *tk* locus. In *Mammalian Cell Mutagenesis, Banbury Report 28* (Edited by M. Moore, D. M. Demarini, F. J. De Serres and K. R. Tindall) pp. 213–224, Cold Spring Harpor Laboratory, Cold Spring Harbor, N.Y.

Barnhart, B. J. and S. H. Cox (1979) Mutagenicity and cytotoxicity of 4.4-mev alpha particles emitted by plutonium-238. *Radiat. Res.* 80, 542–548.

Blakely, E. A., P. Y. Chang and L. Lommel (1985) Cell-cycle dependent recovery from heavy-ion damage in G1 phase cells, *Radiat. Res.* 104, S145–S147.

Boiteux, S. and J. Laval (1982) Coding propertes of poly(deoxycytidylic acid) templates containing uracil or apyrimidinic sites: *in vitro* modulation of mutagenesis by deoxyribonucleic acid repair enzymes. *Biochem.* 21, 6746–6751.

Bradley, M. O. and V. I. Taylor (1981) DNA double-strand breaks induced in normal human cells during the repair of ultraviolet light damage. *Proc. Natl. Acad. Sci. USA* 78, 3619–3623.

Brash, D. E. and W. A. Haseltine (1982) UV-induced mutation hotspots occur at DNA damage hotspots. *Nature* 298, 189–192.

Brash, D. E., S. Seetharam, K. H. Kraemer, M. M. Seidman and A. Bredberg (1987) Photoproduct frequency is not the major determinant of UV base substitution hot spots or cold spots in human cells. *Proc. Natl. Acad. Sci. USA* 84, 3782–3766.

Bredberg A., K. H. Kraemer and M. M. Seidman (1986) Restricted UV mutational spectrum in a shuttle vector propagated in Xeroderma pigmentosum cells. *Proc. Natl. Acad. Sci.USA* 83, 8273–8277.

Breimer, L. H., J. Nalbantoglu and M. Meuth (1986) Structure and sequence of mutations induced by ionizing radiation at selectable loci in Chinese hamster ovary cells. *J. Mol. Biol.* 192, 669–674.

Brown, R., A. Stretch, and J. Thacker (1986) The nature of mutants induced by ionising radiation in cultured hamster cells. II. Antigen response and reverse mutation of HPRT-deficient mutants induced by γ-rays or ethyl methanesulphonate. *Mutation Res.* 160, 111–120.

Cadet, J. and M. Berger (1985) Radiation-induced decomposition of the purine bases within DNA and related model compounds. *Int. J. Radiat. Biol.* 47, 127–143.

Cadet, J., L.-S. Kan and S. Y. Wang (1978) O^6-5'-5 cyclo-5,6 dihydro-2'deoxyuridine. Novel deoxyuridine photoproducts. *J. Am. Chem. Soc.* 100, 6715–6720.

Cadet, J., L. Voituriez, A. Grand, F. E. Hruska, P. Vigny and L.-S. Kan (1985) Recent aspects of the photochemistry of nucleic acids and related model compounds. *Biochimie* 67, 277–292.

Chen, D. J., G. F. Strniste and N. Tokita (1984) The genotoxicity of alpha particles in human embryonic skin fibroblasts. *Radiat. Res.* 100, 321–327.

Chu, E. H. Y. (1971) Mammalian cell genetics: III. Characterization of X-ray-induced forward mutations in Chinese hamster cultures. *Mutation Res.* 11, 23–34.

Clark J. M and G. P. Beardsley (1987) Functional effects of cis-thymine glycol lesions on DNA synthesis *in vitro*. *Biochemistry* 26, 5398–5403.

Cox, R. and W. K. Masson (1979) Mutation and inactivation of cultured mammalian cells exposed to beams of accelerated heavy ions. III. Human diploid fibroblast. *Int. J. Radiat. Biol.* 36, 149–160.

Cox, R., J. Thacker, D. T. Goodhead and R. J. Munson (1977) Mutation and inactivation of mammalian cells by various ionizing radiations. *Nature* 267, 425–427.

Croce, C. M. (1986) Chromosome translocations and human cancer. *Cancer Res.* 46, 6019–6023.

Coulondre, C., J. H. Miller, P. J. Farabaugh and W. Gilbert (1978) Molecular basis of base substitution hot spots in *Escherichia coli*. *Nature* 274, 775–780.

Delcourt, S. G., M. E. Kunze and P. Todd (1987) Induction of intragenic mutations in mammalian cells by ionizing radiation. In Proc. 8th Internatl. Congress of Radiat. Res. Vol. 1. (Edited by E. M. Fielden, J. F. Fowler, J. H. Hendry, and D. Scott), Edinburgh, pp. 217, Taylor and Francis, London.

De Serres, F. J., H. V. Malling, and B. B. Webber (1967) Dose rate effect on inactivation and mutation in Neurospora crassa. In *Recovery and Repair Mechanisms in Radiobiology, Brookhaven Symposium in Biology*, 20, 56–75.

De Serres, F. J., B. B. Webber and J. T. Lyman (1967a) Mutation induction and nuclear inactivation in *Neurospora crassa* using radiations with different rates and energy loss. *Radiat. Res. Supp. 7*, 160–171.

De Serres, F. J., H. Inoue and M. E. Schupbach (1983) Mutagenesis at the *ad-3A* and *ad-3B* loci in haploid UV sensitive strains of *Neurospora crassa*: VI. Genetic characterization of *ad-3* mutants provides evidence for qualitative differences in the spectrum of genetic alterations between wild-type and nucleotide excision-repair-deficient strains. *Mutation Res.* 108, 93–108.

Dewey, W. C., H. H. Miller and D. B. Leeper (1971) Chromosomal aberrations and mortality of X-irradiated mammalian cells: Emphasis on repair. *Proc. Natl. Acad. Sci. USA* 68, 667–671.

Drobetsky, E. A., A. J. Grosovsky and B. W. Glickman (1987) The specificity of UV-induced mutations at an endogenous locus in mammalian cells. *Proc. Natl. Acad. Sci. USA* 84, 9103–9107.

Eisenstark, A. (1987) Mutagenic and lethal effects of near-ultraviolet radiation (290–400 nm) on bacteria and phage. *Env. Mol. Mutagen.* 10, 317–337.

Evans, H. H., J. Mencl, M.-F. Horng, M. Ricanati, C. Sanchez, and J. Hozier (1986) Locus specificity in the mutability of L5178Y mouse lymphoma cells: The role of multilocus lesions. *Proc. Natl. Acad. Sci. USA* 83, 4379–4383.

Fisher, J. and H. E. Johns (1973) Thymine hydrate formed by UV and gamma irradiation of aqueous solutions. *Photochem. Photobiol.* 18, 23–27.

Franklin, W. A. and W. A. Haseltine (1986) The role of the (6–4) photoproduct in ultraviolet light-induced transition mutations in *E. coli. Mutation Res.* 165, 1–7.

Fuscoe, J. C., C. H. Ockey and M. Fox (1986) Molecular analysis of X-ray-induced mutants at the HPRT locus in V79 Chinese hamster cells. *Int. J. Radiat. Biol.* 49, 1011–1020.

Glazer, P. M., S. N. Sarkar and W. C. Summers (1986) Detection and analysis of UV-induced mutations in mammalian cell DNA using a lambda phage shuttle vector. *Proc. Natl. Acad. Sci. USA* 83, 1041–1044.

Glickman, B. W., K. Rietveld, and C. S. Aaron (1980) γ-ray induced mutational spectrum in the *lac I* gene of *Escherichia coli.* Comparison of induced and spontaneous spectra at the molecular level. *Mutat. Res.* 69, 1–12.

Glickman, B. W., R. M. Schaaper, W. A. Haseltine, R. I. Dunn and D. E. Brash (1986) The C-C (6-4)UV photoproduct is mutagenic in *Escherichia coli. Proc. Natl. Acad. Sci. USA* 83, 6945–6949.

Grosovsky, A. J., E. A. Drobetsky, P. J. deJong and B. W. Glickman (1986) Southern analysis of genomic alterations in gamma-ray-induced aprt⁻ hamster cell mutants. *Genetics* 113, 405–415.

Grosovsky, A. J., J. G. DeBoer, P. J. De Jong, E. A. Drobetsky and B. W. Glickman (1988) Base substitutions, frameshifts, and small deletions constitute ionizing radiation-induced point mutations in mammalian cells. *Proc. Natl. Acad. Sci. USA* 85, 185–188.

Hartman, P. E., Z. Hartman, and R. C. Stahl (1971) Classification and mapping of spontaneous and induced mutations in the histidine operon of *Salmonella. Adv. Genet.* 10, 1–34.

Haseltine, W. A. (1983) Ultraviolet light repair and mutagenesis revisited. *Cell* 33, 13–17.

Hauser, J., M. M. Seidman, K. Sidur and K. Dixon (1986) Sequence specificity of point mutations induced during passage of a UV-irradiated shuttle vector plasmid in monkey cells, *Molec. Cell. Biol.* 6, 277–285.

Hei, T. K., D. J. Chen, D. J. Brenner and E. J. Hall (1988) Mutation induction by charged particles of defined linear energy transfer. *Carcinogenesis* 9, 1233–1236.

Hozier, J., J. Sawyer, D. Clive and M. M. Moore (1985) Chromosome 11 aberrations in small colony L5178Y TK ⁻/⁻ mutants early in their clonal history. *Mutat. Res.* 147, 237–242.

Hutchinson, F. (1985) Chemical changes induced in DNA by ionizing radiation. *Progress in Nucleic Acid Res. and Mol. Biol.* 32, 116–154.

Ino, I. and J. Yourno (1974) X-ray mutagenesis: Base-pair deletion at a frameshift hotspot in *Salmonella. J. Mol. Biol.* 85, 301–307.

Kato, T., Y. Oda and B. W. Glickman (1985) Randomness of base substitution mutations induced in the *lacI* gene of *Escherichia coli* by ionizing radiation. *Radiat. Res.* 101, 402–406.

Kuchino, Y., F. Mori, H. Kasai, H. Inoue, S. Iwai, K. Miura, E. Ohtsuka and S. Nishimura (1987) Misreading of DNA templates containing 8-hydroxyguanosine at the modified base and at adjacent residues, *Nature* 327, 77–78.

Kunkel, T. A., R. M. Schaaper and L. A. Loeb (1983) Depurination-induced infidelity of deoxyribonucleic acid synthesis with purified deoxyribonucleic acid replication proteins *in vitro. Biochemistry* 22, 2378–2384.

Liber, H. L., P.-M. Leong, V. H. Terry and J. B. Little (1986) X-rays mutate human lymphoblast cells at genetic loci that should respond only to point mutagens. *Mutation Res.* 163, 91–97.

Lindahl, T. (1982) DNA Repair Enzymes. *Annu. Reviews Biochem.* 51, 61–87.

Little, J. B., D. W. Yandell and H. L. Liber (1988) Molecular analysis of mutations at the *tk* and *hgprt* loci in human cells. In *Mammalian Cell Mutagenesis*, Banbury Report 28 (edited by M. M. Moore, D. M. DeMarini, F. J. De Serres and K. R. Tindall) pp. 225–236. Cold Spring Harbor Laboratory, Cold Spring Harbor, N.Y.

Loeb, L. A. and B. D. Preston (1986) Mutagenesis by apurinic/apyrimidinic sites. *Annu. Rev. Genet.* 20, 201–230.

Lonn, U. (1984) Stability of DNA in mammalian cells irradiated with near-UV light (UV-A). *Radiat. Res.* 99, 659–664.

Lundgren, K. and H. C. Wulf (1988) Cytotoxicity and genotoxicity of UVA irradiation in Chinese hamster ovary cells measured by specific locus mutations, sister chromatid exchanges and chromosome aberrations. *Photochem. Photobiol.* 47, 559–563.

Matsuda, Y. and I. Tobari (1988) Chromosome analysis in mouse eggs fertilized *in vitro* with sperm exposed to ultraviolet light and methyl and ethyl methanesulfonate. *Mutation Res.* 198, 131–144.

Miguel, A.G. and R. M. Tyrell (1983) Induction of oxygen-dependent lethal damage by monochromatic UVB (313 nm) radiation: Strand breakage, repair and cell death. *Carcinogenesis* 4, 375–380.

Miller, J. H. (1985) Mutagenic specificity of ultraviolet light. *J. Mol. Biol.* 182, 45–65.

Miller, J. H., C. Coulondre and P. J. Farabaugh (1978) Correlation of nonsense sites in the *lacI* gene with specific codons in the nucleotide sequence. *Nature* 274, 770–775.

Moore, M. M., D. Clive, B. E. Howard, A. G. Batson and N. T. Turner (1985) *In situ* analysis of trifluorothymidine-resistant mutants of L5178Y/TK$^{+/-}$ mouse lymphoma cells. *Mutation Res.* 151, 147–159.

Morgan, T. L. J. L. Redpath and J. F. Ward (1984) Pyrimidine dimer induction in *E. coli* DNA by Cerenkov emission associated with high energy X-irradiation. *Int. J. Radiat. Biol.* 46, 443–449.

Nakamura, N., S. Suzuki, A. Ito and S. Okada (1982) Mutations induced by γ rays and fast neutrons in cultured mammalian cells. Differences in dose response and RBE with methotrexate and 6-thioguanine-resistant systems. *Muation Res.* 104, 383–387.

Peak, M. J., J. G. Peak, M. P. Moehring and R. B. Webb (1984) Ultraviolet action spectra for DNA dimer induction, lethality and mutagenesis in *Eshericia coli* with emphasis on the UV-B region. *Photochem. Photobiol.* 40, 613–620.

Peak, M. J., J. G. Peak and B. A. Carnes (1987) Induction of direct and indirect single strand breaks in human cell DNA by far and near UV radiations: Action spectrum and mechanisms. *Photochem. Photobiol.* 45, 381–387.

Phillips, S. L., S. Person, and H. P. Newton (1972) Characterization of genetic coding changes in bacteria produced by ionising radiation and by radioactive decay of incorporated ^{3}H-labelled compounds. *Int. J. Radiat. Biol.* 21, 159–166.

Preston, R. J. (1980) The effect of cytosine arabinoside on the frequency of γ ray-induced chromosome aberrations in normal human leukocytes. *Mutation Res.* 69, 71–79.

Protic-Sabljic, M., N. Tuteja, P. J. Munson, J. Hauser, K. H. Kraemer, and K. Dixon (1986) UV light-induced cyclobutane pyrimidine dimers are mutagenic in mammalian cells. *Molec. Cell. Biol.* 6, 3349–3356.

Roots, R. and S. Okada (1972) Protection of DNA molecules of cultured mammalian cells from radiation induced single strand scissions by various alcohols and SH compounds, *Int. J. Radiat. Biol.* 21, 329–342.

Roza, L., G. P. van der Schans and P. H. H. Lohman (1985) The induction and repair of DNA damage and its influence on cell death in primary human fibroblasts exposed to UV-A or UV-C irradiation. *Mutation Res.* 146, 89–98.

Russell, L. B. (1971) Definition of functional units in a small chromosomal segment of the mouse and its use in interpreting the nature of radiation-induced mutants. *Mutation Res.* 11, 107–123.

Skulimowski, A. W., D. R. Turner, A. A. Morley, B. J. S. Sanderson and M. Haliandros (1986) Molecular basis of X-ray-induced mutation at the HPRT locus in human lymphocytes. *Mutation Res.* 162, 105–112.

Stankowski, L. F., Jr., and A. W. Hsie (1986) Quantitative and molecular analyses of radiation-induced mutation in AS52 cells, *Radiat. Res.* 105, 37–48.

Strauss, B. S., S. Rabkin, D. Sagher and P. Moore (1982) The role of DNA polymerase in base substitution mutagenesis on non-instructional templates. *Biochimie* 64, 829–838.

Takakura, K., M. Ishikawa and T. Ito (1987) Action spectrum for the induction of single-strand breaks in DNA in buffered aqueous solution in the wavelength range from 150 to 272 nm: dual mechanism. *Int. J. Radiat. Biol.* 52, 667–675.

Takimoto, K. (1986) Mutational DNA base sequence changes in plasmid damaged at a specific region by ultraviolet light. *J. Radiat. Res.* 27, 310–314.

Teoule, R. (1987) Radiation-induced DNA damage and its repair. *Int. J. Radiat. Biol.* 51, 573–589.

Teoule, R. and J. Cadet (1978) Radio-induced degradation of the base component in DNA and related substances. In *Effects of Ionizing Radiation on DNA* (Edited by J. Huttermann, W. Kohnlein and R. Teoule) pp. 171–203). Springer Verlag, Berlin.

Thacker, J. (1986) The nature of mutants induced by ionising radiation in cultured hamster cells. III. Molecular characteristics of HPRT-deficient mutants induced by gamma rays or alpha particles showing that the majority have deletions of all or part of the *hprt* gene. *Mutation Res.* 160, 267–275.

Todd, P. A. and B. W. Glickman (1982) Mutational specificity of UV light in *E. coli*: Indications for a role of DNA secondary structure. *Proc. Natl. Acad. Sci.* 79, 4123–4127.

Turner, D. R., A. A. Morley, M. Haliandros, R. Kutlaca and B. J. Sanderson (1985) *In vivo* somatic mutations in human lymphoctytes frequently result from major gene alterations. *Nature* 315, 343–345.

Urlaub, G., P. J. Mitchell, E. Kas, L. A. Chasin, V. L Funanage, T. T. Myoda and J. Hamlin (1986) Effect of gamma rays on the dihydrofolate reductase locus: Deletions and inversions. *Somatic Cell and Molec. Genet.* 12, 555–566.

Van Sonntag, C. and H.-P. Schuchmann (1986) The radiolysis of pyrimidines in aqueous solutions: an updating review. *Int. J. Radiat. Biol.* 49, 1–34.

Waldren, C., C. Jones and T. T. Puck (1979) Measurement of mutagenesis in mammalian cells, *Proc Natl. Acad. Sci. USA* 76, 1358–1362.

Wallace, S. S. (1983) Detection and repair of DNA base damages produced by ionizing radiation. *Env. Mutagen.*. 5, 769–788.

Ward, J. F. (1985) Biochemistry of DNA lesions. *Radiat. Res.* 104, S103–S111.

Ward, J. F. (1987) Radiation chemical mechanisms of cell death. *Proc. 8th Int. Congress of Radiat. Res.* Vol.2, (Edited by E. M. Fielden, J. F. Fowler, J. H. Hendry and D. Scott) Edinburgh, pp.162–168, Taylor and Francis, London..

Ward, J. F. (1988) DNA damage produced by ionizing radiation in mammalian cells; identities, mechanisms of formation and reparability. *Progress in Nucleic Acid Res. and Mol. Biol.* 35, 95–125.

Witkin, E. M. (1976) Ultraviolet mutagenesis and inducible DNA repair in *Escherichia coli*. *Bacteriol. Reviews* 40, 869–907.

Yandell, D. W., T. P. Dryja and J. B. Little (1986) Somatic mutations at a heterozygous autosomal locus in human cells occur more frequently by allele loss than by intragenic structural alterations. *Somatic Cell Molec. Genet.* 12, 255–261.

Malignant Transformation by Ultraviolet and Ionizing Radiations

J. B. LITTLE

Laboratory of Radiobiology
Harvard School of Public Health
Boston, MA 02115

Introduction

There has been considerable recent interest in the use of *in vitro* techniques for studying the neoplastic transformation of mammalian cells. These are techniques whereby cells of normal growth characteristics and morphology may be converted to those with characteristics of tumor cells including the ability to form invasive cancers in syngeneic hosts. Most quantitative studies of transformation have been carried out with established mouse embryo-derived cell lines, notably the C3H10T1/2 and BALB/3T3 cell systems. Although these cell lines are already immortal and aneuploid, they can be transformed to fully malignant, tumorigenic cells *in vitro* by exposure to various physical and chemical carcinogens.

There is a considerable body of experimental data related to the induction of transformation by ionizing radiation in these cellular systems; fewer results are available for ultraviolet light. In the present report, I will review some of the existing data on the induction of transformation by these radiations. In particular, I will present evidence which suggests that the mechanisms of the initiation of transformations by x-rays and 254nm ultraviolet (UVL) may be similar, despite certain apparent differences in the cellular and molecular effects of the two types of radiation.

Results and Discussion

Dose response characteristics

The dose-response curves for the induction of neoplastic transformation by x-rays and ultraviolet light are remarkably similar. They are characterized by a rapid rise in the frequency of transformation at low radiation doses, followed by a plateau region in which the yield of transformants does not increase with further increases in radiation dose. This plateau is reached at doses greater than approximately 600 rads of x-rays (Terzaghi and Little, 1976) and 10J/m² for UVL (Chan and Little, 1976). Neither radiation is a very potent inducer of transformation; considerably higher yields of transformants are induced by polycyclic hydrocarbon carcinogens such as methylcolanthrene at doses which are much less cytotoxic (Reznikoff et al., 1973).

Both x-ray and UVL-induced transformation can be markedly potentiated by post-irradiation incubation with phorbol ester tumor promoters. This phenomenon is shown in Table 1 for BALB/3T3 cells exposed to x-rays or 254nm UVL. Phorbol ester tumor promoters are effective primarily during the proliferative phase of expression occuring 2–3 weeks after radiation exposure (Kennedy et al., 1980a). As in mouse skin, they are also effective at long times after irradiation. Their potentiating effects are particularly marked after low irradiation doses, doses which yield little if any transformation by themselves, where a 10-20 fold enhancement in the yield of transformed foci has been reported (Mondal and Heidelberger, 1976; Kennedy et al., 1978).

Table 1. Enhancement of transformation of BALB/3T3 cells by post-irradiation incubation with a phorbol ester tumor promoter (TPA)

Treatment	Surviving Fraction	Total foci Total dishes	Mean foci/ dish	Fraction dishes with foci
Control	1.0	1/31	0.03	0.03
TPA (100 ng/ml)	0.99	2/35	0.06	0.06
X-ray (50 rads)	1.0	10/29	0.34	0.28
UVL ($5J/m^2$)	0.82	3/18	0.17	0.11
X-ray + TPA	0.91	72/35	2.06	0.69
UVL + TPA	0.80	14/17	0.82	0.53

Thus, ionizing radiation and UV light show similarities both in the dose-response relationships for the induction of transformation, and for the potentiating effect of phorbol ester tumor promoters.

Role of DNA damage and mutations in transformation

It is well known that ionizing radiation and UVL induce different spectra of DNA damage. However, as Cadet et al. (1989) have pointed out in this symposium, similar DNA lesions arise in some cases following both types of irradiation, though different biochemical pathways may be involved. Indeed, Leenhouts and Chadwick (1989) and Evans (1989) in this symposium have suggested the common association of DNA double strand breaks with the cytotoxic effects of both agents.

The involvement of DNA damage in the initiation of transformation by ionizing radiation is suggested by studies with the Auger electron emitting element Iodine-125. Iodine-125 is highly effective in the induction of transformation when incorporated into cellular DNA as [125]IdUrd (LeMotte et al, 1982). Levels as low as 10–20 total disintegrations per cell led to a three to four fold enhancement in the yield of transformants. When Iodine-125 decays, it releases a shower of very low-energy electrons;

most of their energy is deposited within a sphere of 100 A° of the site of decay and is thus highly localized to cellular DNA. Little and his colleagues have compared the kinetics of the change in neoplastic transformation (Terzaghi and Little, 1975; Chan and Little, 1982a) with those for chromosomal aberrations, sister chromatid exchanges and DNA repair (Fornace et al, 1980; Nagasawa et al, 1982) during confluent holding recovery. For both ultraviolet and x-rays, the kinetics of transformation correlated with those for sister chromatid exchanges, whereas survival appeared related to chromosomal aberrations and the repair of DNA double strand breaks.

It is commonly assumed that mutagenic processes are important events in the induction of carcinogenesis *in vivo* and neoplastic transformation *in vitro*, and that indeed the initiating event may be a single gene mutation. This hypothesis has been given impetus by recent observations on the the activation of specific oncogenes in animal and human tumors. As Evans (1989) has pointed out in this symposium, the mutagenic spectrum for x-rays and ultraviolet light differ both qualitatively and quantitatively. Of particular note has been the consistent finding that ionizing radiation does not induce mutations to ouabain resistance (Na+, K+ ATPase locus).

The molecular structure of mutations induced by the two types of radiation at the autosomal tk locus in a heterozygous human cell line is shown in Table 2 (Yandell et al, 1986; Little et al, 1988). Interestingly, the spectrum of molecular changes does not differ greatly among spontaneous mutants or those induced by the two agents. Approximately 50-70% of all mutants were associated with large scale changes involving loss of the entire active tk gene, whereas a smaller fraction was caused by point mutations (molecular changes involving less than 150 base pairs, the limit of resolution of the Southern blotting technique). That x-rays do indeed induce point mutations has been demonstrated in other experiments in which the induction of mutations was examined at two other co-dominant loci which should respond only to point mutagens (Liber et al, 1986).

The fact that transformation and mutagenesis may be dissociated, however, is shown in the data in Table 3 (Chan and Little, 1978; 1982b). These experiments were carried out in parallel in mouse 10T1/2 cells. Post-irradiation incubation of the cells for 48 hrs with

Table 2. Molecular analysis of 127 mutant clones induced by X-rays or UVL in human lymphoblastoid cells

Treatment	Number	Structural changes at tk locus		
		None	Gene rearranged	Gene lost
Control (Wild type)	25	25	0	0
None (Spontaneous)	51	10	5	36 (71%)
X-ray (400 rads)	56	18	3	35 (63%)
UVL (5 J/m²)	20	10	1	9 (43%)

Table 3. Dissociation of mutagenesis (ouabain resistance) and transformation in mouse 10T1/2 cells exposed to caffeine for 48 hrs following UVL irradiation.

UVL dose (J/m^2)	caffeine conc (mM)	Transformation		Mutagenesis Mutant Fraction (X10^{-5})
		Total foci Total dishes	Fraction Dishes with Foci	
0	0	0	0	0
5	0	13/102	0.13	1.13 ± 0.30
5	1	28/90	0.23	0.51 ± 0.15
10	0	28/120	0.23	5.83 ± 0.39
10	1	75/162	0.35	2.07 ± 0.42

caffeine systematically suppressed the induction of ouabain resistant mutants by 50–70%, whereas the frequency of transformation was unchanged or elevated by ouabain treatment. Such results, however, are not conclusive; caffeine is pleomorphic in its actions and its effects might well be locus specific.

Evidence for two events in transformation

During the course of our studies on transformation of mouse 10T1/2 cells, we observed that the ultimate yield of transformed foci was largely independent of the number of irradiated cells initially seeded per dish (Kennedy et al, 1980b). The results of a typical experiment are shown in Table 4. Indeed, transformed foci arose in a relatively high fraction of dishes seeded with a single viable irradiated cell (Kennedy and Little, 1980). In these latter experiments, cell proliferation took place until a normal appearing confluent monolayer was reached; after an additional 3-4 weeks in culture, one or more transformed foci arose overlying this normal monolayer.

Table 4. Effect of number of viable cells initally seeded per dish on the yield of transformed foci in X-irradiated mouse 10T1/2 cells

No. Viable Cells/Dish	Total Foci Total Dishes	Mean No. Foci/Dish	Fraction Dishes with Foci
1-4	29/75	0.40	0.36
20-26	19/35	0.54	0.40
52-100	21/64	0.33	0.31
113-260	37/99	0.37	0.34

We interpret this phenomenon to indicate that two distinct events are involved in transformation. The first is a frequent one involving a relatively high fraction of the initial irradiated cell population, whereas the second event is a rare one and represents the actual transformation of one or more of the progeny of the irradiated cell after many rounds of cell division. Indeed, we have shown that this second event has the characteristics of a mutation, as it occurs with a constant frequency per cell per generation (Kennedy et al, 1984). A similar phenomenon has been observed for transformation by UVL (Little, unpublished data) and polycyclic hydrocarbon carcinogens (Fernandez et al, 1980). Recently, spontaneous transformants have been shown to arise in a similar manner (Grisham et al, 1988); that is, to occur with a constant frequency per cell per generation.

A summary scheme for the induction of transformation *in vitro* based on these observations is shown in the Figure below. They suggest a common pathway for the initiation of transformation by ionizing radiation and UV light. I propose that the first event in transformation represents a cellular response to non-specific DNA damage. Indeed, given the high frequency of this first event and the high efficiency of transformation by [125]IdUrd, it is highly unlikely that this first event represents a targeted gene mutation. I further propose that non-specific DNA damage induces a cellular process such as an error-prone repair mechanism that enhances the spontaneous mutation frequency among the progeny of the irradiated cells. Evidence to support this hypothesis comes from work by Stomato et al. (1987) who showed that mutations may indeed arise amongst progeny cells as a late consequence of mutagen exposure. The second event may

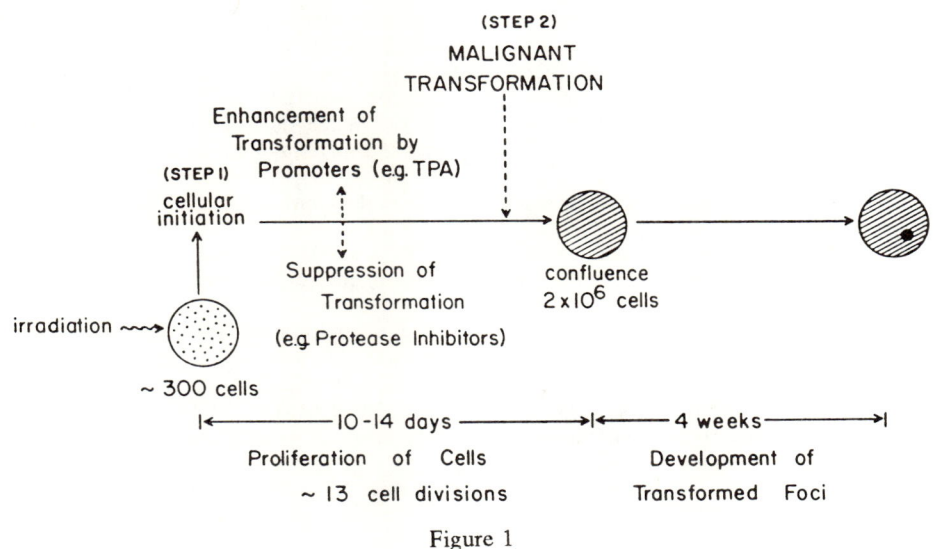

Figure 1

represent mutation occurring in a specific oncogene. We have found evidence for oncogene activation during radiation-induced transformation in transfection experiments (Little and Krolewski, 1986), but have not as yet been able to identify a specific oncogene as being rearranged or amplified.

Conclusions

Both ionizing radiation and ultraviolet light can induce neoplastic transformation in cultured mammalian cells. Despite differences in the spectrum of molecular lesions produced by the two types of radiation, the characteristics of the induction of transformation and its modification by various chemical agents are similar in many respects. Based on studies of the kinetics of transformation, we present evidence to support the hypothesis that a common pathway exists for the initiation of transformation by these two radiations. However, the nature of any induced metabolic processes as well as the involvement of specific oncogenes in transformation remain to be elucidated.

Acknowlegement

This research supported in part by Grants CA-47542 and ES-00002 from the United States Institutes of Health.

References

Cadet, J., Shaw, A., Voituriez, L. and Kan, L.-S. (1989) Comparative effects of ultraviolet and ionizing radiation on nucleic acids (This Symposium).
Chan, G. and Little, J.B. (1976) Induction of oncogenic transformation *in vitro* by ultraviolet light. *Nature* 264: 442-444.
Chan, G. L. and Little, J.B. (1978) Induction of ouabain resistant mutations in C3H 10T´ cells by ultraviolet light. *Proc. Natl. Acad. Sci. U.S.A.* 75: 3363- 3366.
Chan, G. L. and Little, J.B. (1982a) Modulation of survival and transformation during plateau-phase holding of UV-irradiated mouse cells. *Mutation Res.* 104: 183-186.
Chan, G. L. and Little, J.B. (1982b) Dissociated occurrence of single-gene mutation and oncogenic transformation in C3H 10T´ cells exposed to ultraviolet light and caffeine. *J. Cell Physiol.* 111: 309-314.
Evans, H.H. (1989) Mutagenic effecgts of ultraviolet and ionizing radiation. (This Symposium).
Fernandez, A., Mondal, S. and Heidelberger, C. (1980) Probabilistic view of the transformation of cultured C3H/10T 1/2 cells by 3-methylcholanthrene. *Proc. Natl. Acad. Sci. USA* 77, 7272-7276.
Fornace, A. J., Jr., Nagasawa, H. and Little, J.B. (1980) Relationship of DNA repair to chromosome aberrations, sister chromatid exchanges and survival during liquid holding recovery in X-irradiated mammalian cells. *Mutat. Res.* 70: 323-336.
Grisham, J.W., Smith, G.J., Lee, L.W., Bentley, K.S. and Fatteh, M.V. (1988) Spontaneous formation of foci of morphologically transformed cells in populations of C3H 10T 1/2 (clone 8) cells. *Cancer Res.* 48, 5969-5976.
Kennedy, A. R., Mondal, S., Heidelberger, C. and Little, J.B. (1978) Enhancement of X-ray transformation by 12-0-tetradecanoyl-phorbol-13-acetate in a cloned line of C3H mouse embryo cells. *Cancer Res.* 38: 439-443.
Kennedy, A. R., Murphy, G. and Little, J.B. (1980a) The effect of time and duration of

exposure to 12-0-tetradecanoyl-phorbol-13-acetate (TPA) on X-ray transformation of C3H
10T´ cells. *Cancer Res*. 40: 1915-1920.

Kennedy, A. R., Fox, M., Murphy, G. and Little, J.B. (1980b) Relationship between X-ray
exposure and malignant transformation in C3H 10T´ cells. *Proc. Natl. Acad. Sci. U.S.A.*
77: 7262-7266.

Kennedy, A. R. and Little, J.B. (1980) Investigation of the mechanism for enhancement of
radiation transformation *in vitro* by 12-0-tetradecanoylphorbol-13-acetate.
Carcinogenesis 1: 1039-1047.

Kennedy, A. R., Cairns, J. and Little, J.B. (1984) Timing of the steps in transformation of
C3H 10T´ cells by X-irradiation. *Nature* 307: 85-86.

Leenhouts, H.P. and Chadwick, K.H. (1989) A molecular model for the cytotoxic action of UV
and ionizing radiation (This Symposium).

LeMotte, P. K., Adelstein, S.J. and Little, J.B. (1982) Malignant transformation induced by
incorporated radionuclides in BALB/3T3 mouse embryo fibroblasts. *Proc. Natl. Acad. Sci.
U.S.A.* 79: 7763-7767.

Liber, H. L., Leong, P.H., Terry, V.H. and Little, J.B. (1986) X-rays mutate human
lymphoblast cells at genetic loci that should respond only to point mutagens. *Mutat. Res.*
163: 91-97.

Little, J.B. and Krolewski, B. Radiation transformation and oncogene activation. In: Primary
Changes and Control Factors in Carcinogenesis, T. Friedberg and F. Oesch, eds.,
Deutscher Fachschriften-Verlag, Wiesbaden, Germany, 1986, pp.107-112.

Little, J.B., Yandell, D.W. and Liber, H.L. (1987) Molecular analysis of mutations at the *tk*
and *hgprt* loci in human cells. Banbury Report 28, *Mammalian Cell Mutagenesis*, Cold
Spring Harbor Laboratory, pp. 235-236.

Mondal, S. and Heidelberger, C. (1976) Transformationof C3H/10T 1/2 C18 mouse embryo
fibroblasts by ultraviolet irradiation and a phorbol ester. *Nature* 260, 710-711.

Nagasawa, H., Fornace, A.J. Jr., Ritter, M.A. and Little, J.B. (1982) Relationship of enhanced
survival during confluent holding recovery in ultraviolet- irradiated human and mouse cells
to chromosome aberrations, sister chromatid exchanges, and DNA repair. *Radiat. Res.* 92:
483-496.

Reznikoff, C.A., Bertram, J.S., Brankow, D.W. and Heidelberger, C. (1973) Quantitative and
qualitative studies of chemical transformation of cloned C3H mouse embryo cells sensitive
to postconfluence inhibition of cell division. *Cancer Res.* 33, 3239-3249.

Stomato, T., Weinstein, R., Peters, B., Hu, J., Doherty, B. and Giaccia, A. Delayed mutation
in Chinese hamster cells. *Somatic Cell Molec. Gen.* 13, 57-66.

Terzaghi, M. and Little, J.B. (1975) Repair of potentially lethal radiation damage in
mammalian cells is associated with enhancement of malignant transformation. *Nature*
253: 548-549.

Terzaghi, M. and Little, J.B. (1976) X-radiation induced transformation in a C3H mouse
embryo-derived cell line. *Cancer Res.* 36: 1367-1374.

Yandell, D. W., Dryja, T.P. and Little, J.B. (1986) Somatic mutations at a heterozygous
autosomal locus in human cells occur more frequently by allele loss than by intragenic
structural alterations. *Somatic Cell and Molecular Genetics* 12: 2 55-263.

Between Ultraviolet and Ionizing Radiation: Action Spectra for Cytotoxicity in the Vacuum-UV Region

TAKASHI ITO

Institute of Physics
College of Arts & Sciences
University of Tokyo
Meguroku, Komaba 3-8-1, Tokyo 153
Japan

Introduction

Like UV-C, vacuum-UV is strongly absorbed by oxygen, so it does not reach the surface of the earth as in the case of UV-C. The vacuum-UV below 180 nm is strongly absorbed by water. This property is an important difference from UV-C. DNA as well as proteins show the first absorption peak in the vacuum-UV region around 190 nm attributable to the double bonds. For DNA such absorption is due to the π electron system of the bases. In the shorter wavelengths the absorption may be attributed largely to the sugar-phosphate moiety. The $\sigma-\sigma^*$ transition is generally thought to be responsible. Depending on the wavelength (or photon energy) the molecule may be promoted to the higher excited states (Kiseleva et al., 1975; Ito and Ito, 1986).

Two aspects must be recognized in considering vacuum-UV actions in comparison with UV-C and X-rays. According to Platzman's concept the vacuum-UV induces non-ionized, neutral excited states with a considerably high branching ratio to the direct ionization in the region Ip <E <Ip + 20 eV, where Ip is the first ionization potential (around 8 eV or 155 nm for most biological molecules) and E the energy (Platzman, 1967). The other aspect is that, since all the molecules show very strong absorption in the vacuum-UV, the penetration depth of vacuum-UV into the biological specimen of the cell size is very short. Figure 1 shows the attenuation of vacuum-UV radiation by liquid water (Ito et al, 1983). It gives an idea of the situation of living cell, since its major component is water. Although the precise absorption spectra of biological molecules over a whole range of vacuum-UV are still to be measured, it is very obvious that the experiments with living systems are very limited in the vacuum-UV region due to the low penetration. Certain systems like spores are small enough and can be dried in vacuum, therefore irradiation is possible without elaborations. Other living materials need a special device for irradiation as described later.

Besides the limitation in the irradiation conditions, the light source itself has been a long-standing problem in the vacuum-UV region. A hydrogen discharge lamp emits

Photobiology, Edited by E. Riklis
Plenum Press, New York, 1991

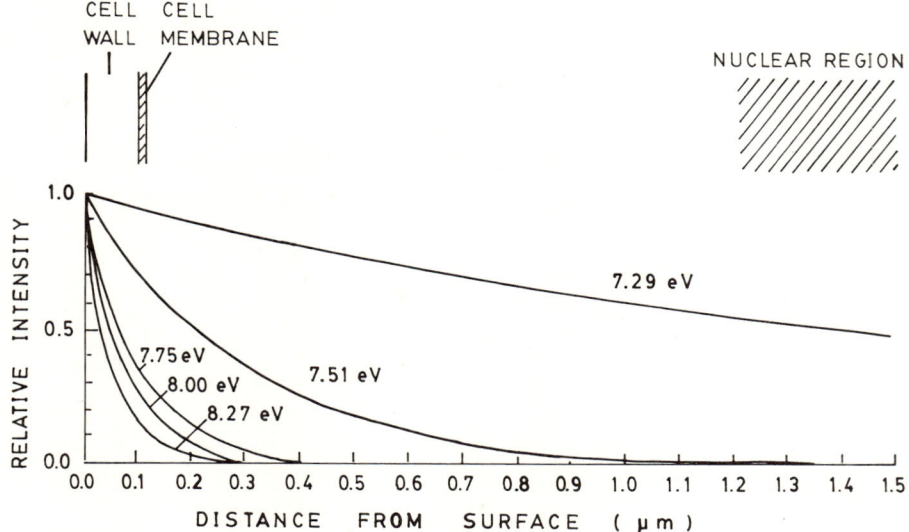

Figure 1. Attenuation of monochromatic vacuum-UV radiation in liquid water at 7.29 eV (170 nm), 7.51 eV (165 nm), 7.75 eV (160 nm), 8.00 eV (155 nm) and 8.27 eV (150 nm) as a function of the distance from the surface (reproduced from Ito, 1988)

continuous spectrum usable above 120 nm, but is weak in the photon flux if monochromatized. Rare gas discharge lamps could provide an intense light at limited spectral lines. A large-scale plasma may be a good vacuum-UV source but not well developed. Under these circumstances an excellent source for vacuum-UV radiation at the present time is the synchrotron radiation (SR) from the electron storage ring. The synchrotron radiation has continuous spectrum ranging from infra-red to X-ray region, usually very intense. It has a convenient optical geometry for adapting monochromators. Irradiation systems with SR as a light source have been constructed (Ito et al., 1984; Kobayashi et al, 1987). The only drawback is that the whole system requires large-scale instrumentation, although a small, compact storage ring (a few m in diameter) has already been in operation at several sites.

Action spectra

Cellular Vacuum-UV Actions in Wet State: Yeast and Cultured Mammalian Cells

Yeast cells were irradiated under wet conditions by vacuum-UV in the range from 254 nm down to 150 nm by using a specially designed moisture chamber (Fig. 2) (Ito and Ito, 1981). In this chamber monochromatic vacuum-UV radiation, produced in the vacuum, traverses a transparent MgF_2 window into an air-tight chamber, in which the cells are placed on a Millipore filter. H_2O vapor at saturation pressure (10–20 Torr at

106

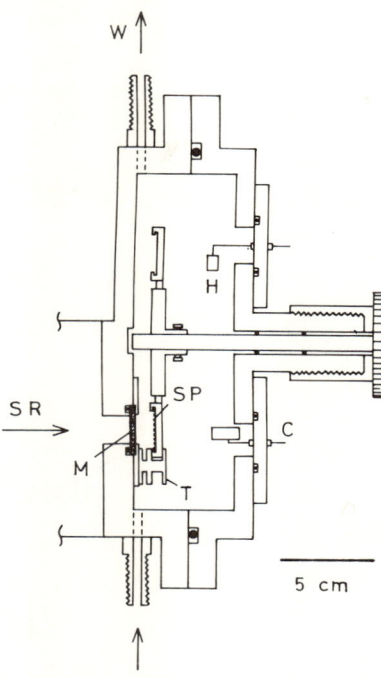

Figure 2. A sample chamber used in vacuum-UV irradiation for wet materials. SR: synchrotron radiation, M: MgF_2 window, SP: sample, T: adjustment for thickness between SP and M, C: electric connector, H: dew sensor, W: water vapor inlet and outlet (modified from Ito and Ito, 1981)

ambient temperatures) flowed into the space between the sample cells and the MgF_2 plate. The H_2O vapor (thickness, 1 mm) reduces the photon fluence reaching the cells to about 40% at most in this wavelength range. Action spectra for inactivation and for cell membrane impairment are shown in Fig. 3 (Ito et al, 1983). Before looking into the details it is to be noted that, for the radiations below 175 nm, the penetration into the cell is very limited (see Fig. 1). It can be seen that both spectra resemble the absorption of water below 175 nm, indicating that the water located a small distance from the cell surface could be responsible for the action. Then it may be reasonable to speculate that, in view of the redundance of H_2O, OH radicals produced by the photolysis of water are the most likely candidates for the action intermediate. This type of action cannot be expected in the UV-C irradiation since no absorption occurs in the water. The cross-section for the membrane damage becomes undetectably small above 190 nm while the cross-section for inactivation increases in parallel to the DNA absorption above 220 nm. Other evidence such as photoreactivation supports the transition of cellular targets from non-DNA to DNA at this wavelength. It may be summarized by saying that the UV-C type action ceases at about 190 nm and that damage localized in the cell membrane

107

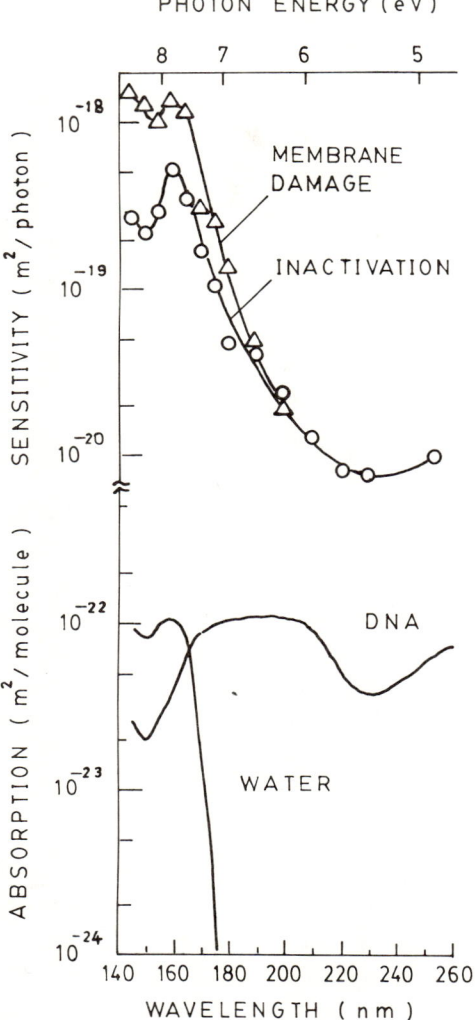

Figure 3. Action spectra for inactivation and cell membrane damage of diploid yeast cells. Absorption of water and DNA have been calculated at the depth of 1 μm water. The ordinate scale for DNA is arbitrary (modified from Ito et al, 1984)

becomes prominent in the region of vacuum-UV where H_2O shows a strong absorption. If the latter action occurs via OH radicals, it may resemble the action mode of ionizing radiations; however, an essential difference between the two is the extreme localization of OH radicals in the case of vacuum-UV, unlike ionizing radiations, where non-localized, more or less uniform production of OH radicals is generally expected. As a consequence, the cellular DNA located deep inside the cell cannot be the vacuum-UV target. Cultured mammalian cells, surprisingly withstood rather severe conditions imposed by the moisture chamber, have also been irradiated by vacuum-UV radiation. Essentially, similar results were obtained (Shinohara et al, 196). In wet systems (humidified cells and solution systems as well) all available data indicate that vacuum-UV action quite resembles the X-ray action due to the production of OH radicals from the water (see Kuwabara in this Proceedings).

Cellular Vacuum-UV Damage in Dried State

The direct damage by vacuum-UV radiation was examined on its photoreactivability, a known property for UV-C damage, by Heinmets and Taylor (1952) with bacterial cells, and later by Jagger with *Streptomyces* conidia (Jagger et al, 1967) using the bands selected by cutoff filters. It may be summarized that the photoreactivability of irradiated cells was detectable well beyond the UV-C region but tends to decrease, and finally became undetectable at somewhere around 170 nm. More recently, it was established using monochromatized SR as a light source that the induced damage in bacteriophages was photoreactivated to an appreciable extent by the host *E. coli* cells down to 170 nm (Maezawa et al, 1984). With yeast cells, irradiated under humidified conditions, the photoreactivation of genetic changes was detected only above 220 nm (Ito et al, 1983). The most informative experiments of this sort were those performed with *Bacillus subtilis* spores having different radiation sensitivities (Munakata, 1986). Figure 4 shows

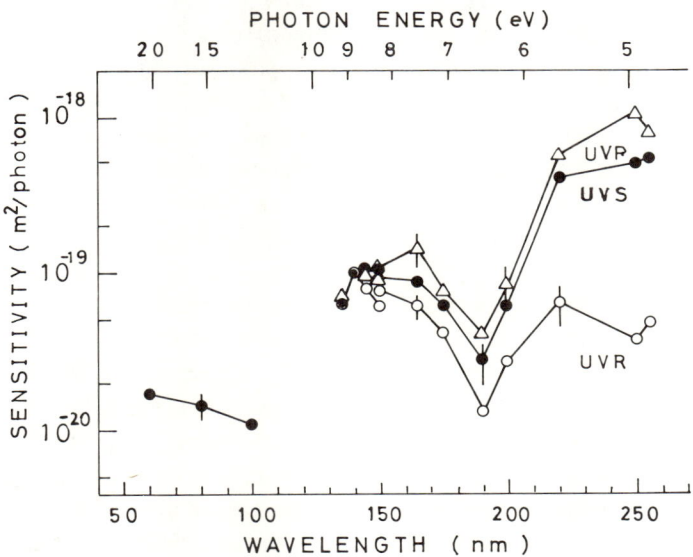

Figure 4. Action spectra for inactivation of *Bacillus subtilis* spores with different radiation sensitivities. UVR: wild type strain, UVS: UV-C sensitive strain, UVP: X-ray sensitive strain (adapted from Munakata, 1986)

action spectra for three such strains, namely, UVS being defective both in excision repair and spore repair for spore UV-photoproduct, UVP being of UVS characters plus defective in DNA polymerase I, and UVR being the wild type. One prominent feature in these spectra is that a large difference in the UV-C region between UVR and UVS (UVP as well) sharply narrowed going toward 190 nm, and they seemed to converge further toward 150 nm. The date may mean that UV-C type damage is no longer dominant at 190 nm and that vacuum-UV specific damage is replacing below that, to which neither the UV-C repair systems nor X-ray repair systems seem to respond effectively. No strain has ever been selected either for sensitive or resistant as to the possible vacuum-UV specific damage, however, it is also to be seen if the UVP strain exhibits a significantly higher sensitivity than UVS or UVR strain in the wavelength region shorter than 100 nm, where direct ionizations are expected to occur considerably. Yeast cells defective in excision repair gave essentially the same results in the region above 180 nm. However, yeast cells are too thick for the study in the shorter wavelengths (Hieda and Ito, 1986).

Comparison of vacuum-UV action with those by UV-C and ionizing radiations

Among the three wavelength regions, the UV-C is quite unique in the biological actions due to the selectively strong absorption by nucleic acids compared with other cell constituents. On the other hand, X-rays are characterized by a powerful penetration and non-specific causes of damage to the constituents of the cell, although, because of the intrinsic importance, the nucleic acid still retains the hegemony over the biological action. After a transient region (7-8 eV in photon energy) UV-C type lesions are no longer a major photoproduct and increasingly more strand breaks occur[*]. In X-ray region, where higher excited states as well as direct ionizations may occur as dominant events in the irradiated molecules, the strand breaks are clearly the major molecular damage in DNA. Interesting questions may be raised. For example, are these strand breaks induced by vacuum-UV and X-rays chemically identical? Or is there any transient region from vacuum-UV to X-rays with respect to the induced chemical changes in the DNA molecule?

Physically speaking, an X-ray action is caused predominantly by the energy dissipated by low energy electrons, and the probability (cross-section) for the energy taken up by an irradiated molecule is seen to be approximately proportional to its oscillator strength divided by the energy for such a transition (Platzman 1967). The major part of oscillator strength distribution is located in the energy region from several eV to 40 eV with the maximum around 20 eV for a large molecule. As a result, the absorption of energy in the vacuum-UV region is the predominant mode in the inelastic collision wiith a fast electron. Therefore, it can be said that the investigation with monochromatic

---

[*]For SR photobiological work at molecular level readers are referred to VUV Effect Symposium in this Proceedings.

vacuum-UV radiation should provide a dissection of otherwise mixed actions of X-rays. We may some day reconstruct the X-ray action by using the experiments performed with monochromatic vacuum-UV radiation over an appropriate energy range. This seems a basis why some vacuum-UV actions resemble more X-ray action than UV-C, and, in fact, the *raison d'être* for the vacuum-UV study in the basic understanding of radiation action in general.

References

Heinmets, F., and Taylor, Jr., W.W., Preliminary studies of electrical discharge on some chemical compounds and bacteria. *Archives Biochemistry and Biophysics*, 35, 60-62, 1952.

Hieda, K. and Ito, T., Action spectra for inactivation and membrane damage of *Saccharomyces cereviae* cells irradiated in vacuum monochromatic synchrotron ultraviolet radiation (155-250 nm). *Photochemistry and Photobiology*, 44, 409-411, 1986.

Ito, A., and Ito, T., Absorption spectra of deoxyribose, ribosephosphate. ATP and DNA by direct transmission measurements in the vacuum-UV (155-190 nm) and far-UV (190-260 nm) regions using synchrotron radiation as a light source. *Photochemistry and Photobiology*, 44, 355-358, 1986.

Ito, T. and Ito, A., Effects of 120 to 165 nm vacuum-UV light on wet yeast cells, *Radiation Research*, 85, 161-172, 1981.

Ito, T., Ito, A., Hieda, K. and Kobayashi, K., Wavelength dependence of inactivation and membrane damage to *Saccharomyces cerevisiae* cells by monochromatic synchrotron vacuum-UV radiation (145-190 nm). *Radiation Research*, 96, 532-548, 1981.

Ito, T., Kada, T., Okada, S., Hieda, K., Kobayashi, K., Maezawa, H., and Ito, A., Synchrotron system for monochromatic UV irradiation (>140 nm) of biological material. *Radiation Research*, 98, 65-73, 1984.

Jagger, J., Stafford, R.S., and Mackin, Jr., R.J., Killing and photoreactivation of *Streptomyces griseus* conidia by vacuum-ultraviolet radiation (1500 to 2700 A). *Radiation Research*, 32, 64-92, 1967.

Kiseleva, M.M., Zarochensteva, Ye. P., and Dodonova, N.Ya., Absorption spectra of nucleic acids and related compounds in the spectral region 120-280 nm. *Biophysics (Biofizika)*, 20, 571-575, 1975.

Kobayashi, K., Hieda, K., Maezawa, H., Ando, M., and Ito, T., Monochromatic X-ray irradiation system (0.08-0.4 nm) for radiation biology studies using synchrotron radiation at the Photon Factory. *Journal of Radiation Research*, 28, 243-253, 1987.

Maezawa, H., Ito, T., Hieda, K., Kobayashi, K., Ito, A., Mori, T. and Suzuki, K., Action spectra for inactivation of dry phage T1 after monochromatic (150-254 nm) synchrotron irradiation in the presence and absence of photoreactivation and dark repair. *Radiation Research*, 98, 227-233, 1984.

Munakata, N., Vacuum-UV action spectra for inactivation of *Bacillus subtilis* spores. Activity Report of Synchrotron Radiation Laboratory, Institute for Solid State Physics, University of Tokyo, pp. 76-77, 1986.

Platzma, R.L., Energy spectrum of primary activation in the action of ionizing radiation, in: Radiation Research. Ed. G. Silini, pp. 20-42, North-Holland, Amsterdam, 1967.

Shinohara, K., Ito, A., and Ito, T., Cell surface damage iin cultured mammalian cells with synchrotron radiation at 160 nm. *Photochemistry and Photobiology*, 44, 405-407, 1986.

UV Light and DNA Damage Repairs and Mutagenesis

Brief Methodological Survey of the Photochemistry of Nucleic Acid Constituents and Analogues, and Some Biological Applications, Including Photo-Affinity Labeling

DAVID SHUGAR[a,b], BORYS KIERDASZUK[b] AND RYSZARD STOLARSKI[b]

[a]*Institute of Biochemistry and Biophysics*
Polish Academy of Sciences
36 Rakowiecka Str.,02-532 Warszawa
[b]*Department of Biophysics, University of Warsaw, 93 Zwirki and Wigury Str.*
02-089 Warszawa, Poland

Abstract

This brief survey describes the methodology and, in part, the mechanisms of a variety of photochemical reactions of nucleic acid constituents and some of their analogues; as well as applications to preparative photochemical synthesis of new analogues with diverse and interesting biological properties. Also included are some aspects of photoaffinity labeling, with particular emphasis on identification of the mechanisms and products involved. Reference is made to some reported, but hitherto neglected, photochemical reactions with potential biological significance, and deserving further study.

Introduction

Studies on the photochemistry of nucleic acid constituents, and nucleic acids, have for long been directed to photoproducts which may account for the lethal and mutagenic effects of UV-irradiation. And, indeed, the first major breakthrough was the elucidation of the role of pyrimidine cyclobutane photodimers, which led to clarification of the function of the PR (photo-reactivation) enzyme and, more significantly, the phenomenon of dark excision repair. The latter for the first time signaled the presence in living cells of repair enzymes, the existence of which had previously not even been suspected. This, in turn, led to further intensification of efforts to discover other photoproducts which might be involved in the biological effects of UV.

The number of such photoproducts of nucleic acid constituents, and of many related analogues, is now very large. Many of them appear at first sight to be of little significance in relation to the biological effects of UV radiation. It is the purpose of this brief review to direct attention to other useful biological applications of a few of these reactions, as well as the potential utility of a reexamination of several of them in relation to the photobiology of nucleic acids.

Photobiology, Edited by E. Riklis
Plenum Press, New York, 1991

This account also refers to some methodological aspects, such as the role of wavelength and the environment on photochemical reactions. For more detailed aspects of the nature of the excited states, and mechanisms, involved, the reader is referred to original references.

Preparative Photochemical Synthesis

Photochemical reactions have been widely employed as an efficient tool in organic syntheses (see, e.g. Turro and Schuster, 1975). We start with several illustrative examples which, although in some instances not necessarily relevent to the biological effects of UV, have led to isolation of new products of fundamental biological significance. A good example is photochemical addition of alcohols to purine.

Photoaddition of alcohols to purine

The purine moieties of DNA and RNA, and a number of their analogues, were long ago known to be much more photoresistant in aqueous medium than the pyrimidines (McLaren and Shugar, 1964), but several purine photoproducts with potential biological significance have since been identified (see below; also Duker and Gallagher, 1988, for review).

The first such reported reactions were photoadditions of alcohols (methanol, ethanol, isopropanol) to the N(1)–C(6) bond of purine (Connolly and Linschitz, 1968), by irradiation of anaerobic alcohol solutions of purine at 254 nm, with relatively high quantum yields (~0.25). With methanol as solvent, the major products were the two diasteriomers of 1,6-dihydro-6-(hydroxymethyl)purine (see upper portion of Scheme 1). However, since purine is not a constituent of nucleic acids, this reaction appeared to be of no biological significance, and was generally ignored until Evans and Wolfenden (1970)

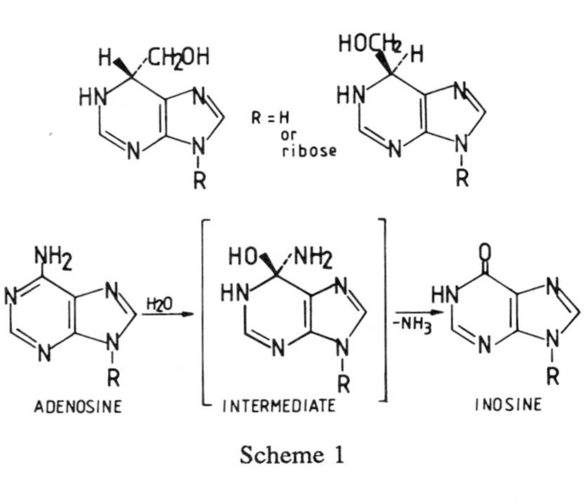

Scheme 1

demonstrated photoaddition of methanol to purine riboside (nebularine) and found that the resulting 1,6-dihydro-6-(hydroxymethyl)nebularine was a potent reversible inhibitor of adenosine deaminase, a key enzyme in the regulation of adenosine and 2'-deoxyadenosine metabolism, and of the efficacity of adenosine analogues in chemotherapy.

The foregoing led to extensive studies (see Buffel et al., 1985, for references) on a variety of additional photoproducts, but with continuing emphasis on the two diastereomers of 1,6-dihydro-6-(hydroxymethyl)nebularine, only one of which (see Scheme 1) is a potent inhibitor of adenosine deaminase ($K_i \sim 10^{-6}$ M).

Its mode of action is considered to be that of a transition-state analogue, with a tetrahedral carbon at C(6), which resembles the presumed intermediate in the pathway for hydrolytic deamination of adenosine (Scheme 1).

The photochemically prepared inhibitor has also been utilized as the starting substance for the first chemical synthesis of more potent transition-state inhibitors, e.g.coformycin (Ohno et al., 1974). It is rather odd that no report has appeared on photoaddition of methanol to nebularine-5'-phosphate, which could well prove to be a good inhibitor of another key cellular enzyme, 5'-AMP deaminase.

Pyrimidine S-nucleoside photorearrangement

With 4-thio-2-methylthiopyrimidine as a starting product, converted chemically to its S-benzyl derivative, irradiation of a solution of the latter in *t*-BuOH at 254 nm led to an intramolecular rearrangement with formation of the 5-benzyl derivative in 60% yield. Replacement of benzyl by ribose led to formation, following removal of mercapto groups, of ß-pseudouridine, a C-C nucleoside which is an important constituent of tRNA (Scheme 2; Fourrey et al., 1977).This reaction proceeded equally well when ribose was replaced by other sugars.

Scheme 2. R = benzyl or ribose

C(5)- and C(6)-allyluracil nucleosides

Another photochemical route for introduction of a C-C bond at the 5-position of a uracil nucleoside is based on irradiation at wavelengths to the red of 300 nm of an aqueous acetonitrile solution of a 5-iodouracil nucleoside in the presence of a trimethylsilylated allyl group (Scheme 3). The authors (Saito et al., 1985) were apparently unaware that such 5-allyl analogues of thymidine are potential antiviral agents. With 6-iodouracil nucleosides, a similar photoreaction leads to introduction of a C(6)-C bond.

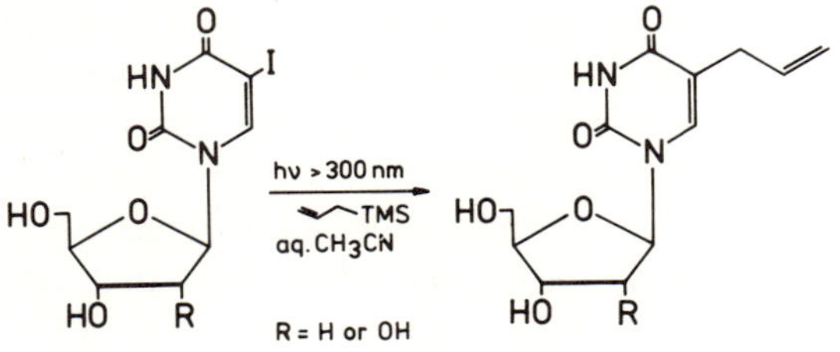

Scheme 3

Acyclonucleoside photoproducts

Irradiation of an aqueous solution of the monoanion of thymidine at 254 nm in the presence of amino acids leads to a photorearrangement with the amino acid linked to the 1-position of thymine (Saito and Matsura, 1985). This reaction has broad biological applications, discussed below. At this point we would like only to point out that the products with various amino acids, ilustrated in Scheme 4, represent a new series of acyclonucleosides of thymine. Since the discovery of the anti-viral activity of acycloguanosine (Acyclovir, an antiherpes agent licensed for clinical use), many such acyclonucleosides have found to exhibit antimetabolic activities, while acyclonucleosides of uracil and thymine are good inhibitors of uridine phosphorylase (Drabikowska et al., 1987, and references cited). The high yields (20–70%) for isolation of these photoproducts

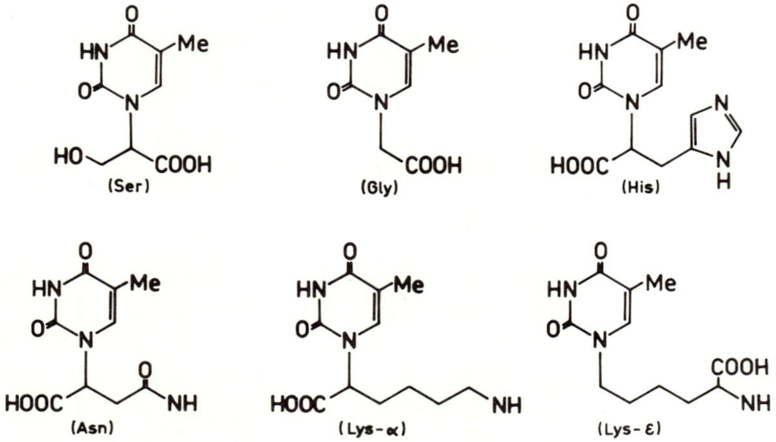

Scheme 4. Acyclonucleoside photoproducts of reaction of thymidine monoanion with amino groups of amino acids.

point to their potential applications as antimetabolites and enzyme inhibitors. They are also analogues of the naturally occurring 1-alanyluracil (willardine), a constituent of some plants. The foregoing photoreaction also proceeds with primary alkylamines (Saito and Matsura, 1985) thus making available additional series of acyclonucleoside analogues. Finally, this reaction may also lead to simultaneous preparative isolation of 2'-deoxyribosylamine (see Scheme 17, below), much simpler than chemical methods.

Photorearrangement of N(1)-oxides of adenines to isoguanines

It was initially shown by Brown et al. (1964) that irradiation of the easily prepared N(1)-oxide of adenine at 254 nm led to formation of isoguanine (2-oxoadenine), *via* an oxaziridine intermediate (Scheme 5), as for most known rearrangements of purine N-oxides. With use of a 254 nm photoreactor this reaction has been applied to the isolation on a preparative scale of isoguanosine, 2'-deoxyisoguanosine, as well as their nucleotides, and the 3':5'-cyclicphosphate of isoguanosine, a new cAMP analogue (Kazimierczuk and Shugar, 1973). These compounds subsequently proved useful for identification of isoguanosine and 1-methylisoguanosine from marine organisms. Both compounds are much more potent pharmacologically than adenosine. Furthermore, a spectroscopic study showed that, unlike the natural bases in DNA, isoguanosine exhibits very marked tautomerism (Sepiol et al., 1976), which partially accounts for its absence in nucleic acids.

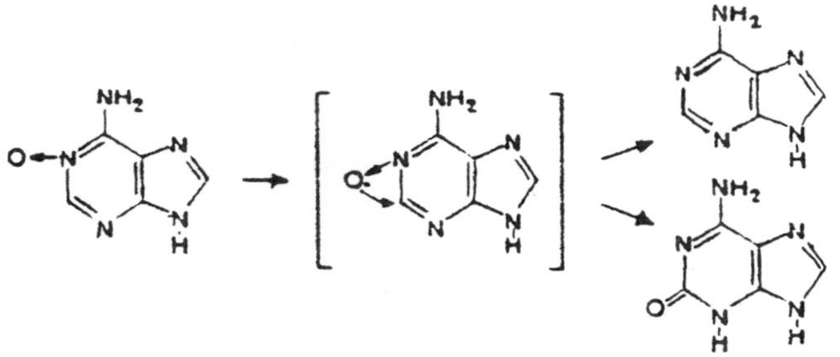

Scheme 5

Photochemical synthesis of model syn *and* anti *conformers of purine nucleosides*

The conformation of nucleosides and nucleotides about the glycosidic bond plays a key role in determining the helical structures of DNA and RNA and the mode of interaction of monomers with various enzymes.

Of various methods for examining the solution conformation of nucleotides about the glycosidic bond, the most succesful has been NMR spectroscopy (See Davies, 1978, for review). While different variants of NMR spectroscopy have occasionally led to conflicting results for purine nucleosides (Stolarski et al., 1984, and references cited) a procedure based on measurements of the chemical shifts of sugar protons and carbons, with the aid of model analogues fixed in the *syn* and *anti* conformations, has resolved these discrepancies and permited of quantitative evaluations of the relative populations of the two conformers (Scheme 6; Stolarski et al., 1984).

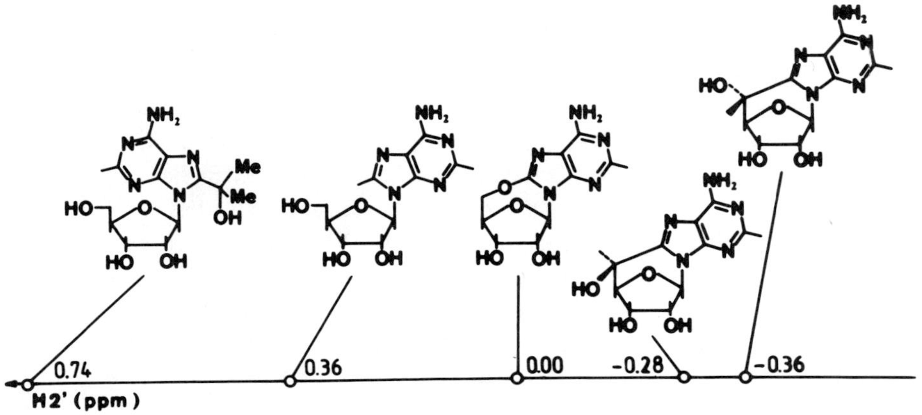

Scheme 6. Chemical shifts of ribose H(2') of adenosine (2nd from left), its 8-substituted analogue in the fixed *syn* conformation (left), and its three 8,5'-cyclo analogues in the fixed *anti* conformation. A similar relationship exists for C(2') chemical shifts. Both lead to quantitative evaluations of conformation of adenosine about the glycosidic bond.

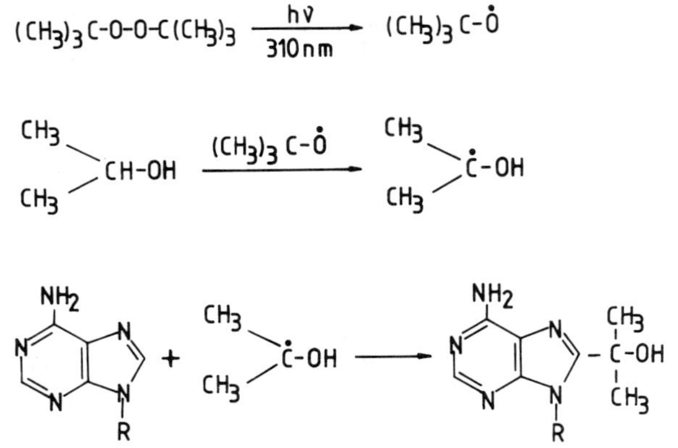

Scheme 7. Free-radical addition of isopropanol to C(8) of adenosine (R=ribose) with di-*tert.*-butyl peroxide as photoinitiator.

120

Photochemical procedures proved to be most convenient for synthesis of the model *syn* and *anti* analogues, as follows:

Syn conformers

This is based on the observation of Salomon and Elad (1973) that irradiation of an aqueous isopropanol solution of a purine nucleoside at wavelengths to the red of 300 nm in the presence of a photoinitiator such as di-*tert*-butyl peroxide (which is the only photoabsorbing species at wavelengths to the red 300 nm), leads to selective production of an 8-(α-hydroxyisopropyl)nucleoside, as shown in Scheme 7. This gives the 8-(a-hydroxyisopropyl) derivatives of adenine and guanine nucleosides and nucleotides in yields of 40–80% (Stolarski et al., 1984,). Unlike the 8-bromo nucleosides, which still exhibit a *syn-anti* equilibrium, these analogues are uniquely in the *syn* conformation (Stolarski et al., 1984), shown also by X-ray diffraction in the solid state (Birnbaum and Shugar, 1987).

It is equally of interest that photoaddition of isopropanol to C(8) of purines occurs also at the level of DNA, with interesting biological implications (Livneh et al., 1979) which have hitherto been little exploited.

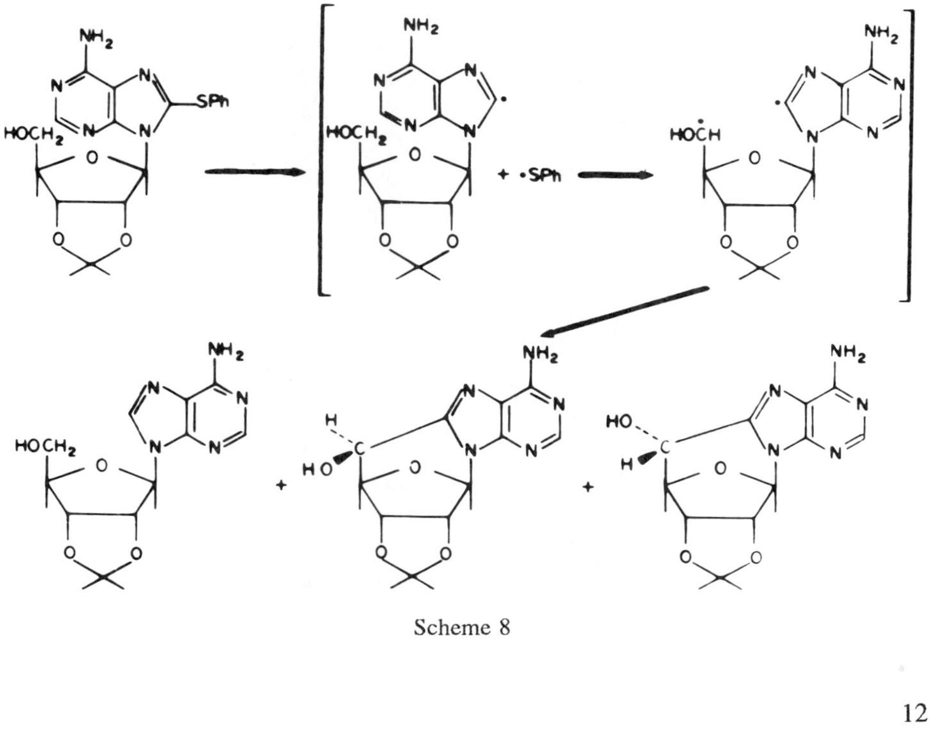

Scheme 8

Anti conformers

Apart from the appropriate model nucleoside in the *anti* conformation, such as the 8,5'-cyclo-8-oxy analogue, it proved necessary to synthesize the (*R*)- and (*S*)-epimers of an 8,5'-cyclo nucleoside in order to correct for the effect of the conformation of the exocyclic CH_2OH group on the ^1H and ^{13}C chemical shifts of the sugar protons and carbons.The two epimers of 8,5'-cycloadenosine (Scheme 8) may be prepared by two photochemical pathways, of which the simpler (Matsuda et al.,1978) is based on irradiation through a pyrex filter (λ>300 nm) of a CH_3CN solution of 8-phenylthio-2',3'-O-iso-propylideneadenosine in the presence of a photoinitiator. This gives both (*R*)-8,5'- and (*S*)-8,5'-cyclo-2',3'-O-isopropyladenosine in a single step (Scheme 8) in 10–20% yields.

Chemical synthesis of oligoribonucleotides

A key problem in the chemical synthesis of ribonucleotides is the choice of a suitable protecting group for the 2'-hydroxyl of the initial ribonucleoside monomers. A number of observers have pointed to the utility of the photosensitive *o*-nitrobenzyl group, which may be removed by irradiation at 280 nm. Originally shown to be practical by Ohtsuka et al. (1977), this procedure has now been applied to the synthesis of a tetradecamer on a preparative scale, and subsequent crystallization to furnish the most detailed analysis, by X-ray diffraction, of an A-RNA helix (Dock-Bregeon et al., 1988, and references cited).

Dewar Valence Isomer of (6–4) Product of TpT

An unusual new photoproduct of TpT, with undoubted biological significance, has been identified by Taylor and Cohrs (1987). It had long ago been shown by Johns et al. (1964) that irradiation of TpT at about 300 nm led to formation not only of the *cis-syn* (TpT1) and *trans-syn* (TpT2) cyclobutane photodimers, but to an additional product referred to as TpT4, subsequently identified as the (6–4) photoadduct of TpT (see Scheme 9). Subsequent irradiation of TpT4 at 313 nm led quantitatively, to formation of a new product, referred to as TpT3 which, on irradiation at shorter wavelengths (240–260 nm), reverted quantitatively to TpT4.

The structure of TpT3 has now been unequivocally identified by Taylor and Cohrs (1987) and its properties extensively described by the same authors (Taylor et al., 1988), as well as by Voiturez et al.(1988) (Scheme 9). Establishment of TpT3 as the Dewar valence isomer of TpT4 was based on UV, IR and NMR spectroscopy, additionaly facilitated by the earlier observations of Nishio et al. (1980) on pyrimidinones, which photoisomerize to their Dewar isomers on irradiation at wavelengths to the red of 300 nm, and revert to the parent pyrimidinones on irradiation at shorter wavelengths, or by acid catalysis or heating at neutral pH (see Scheme 10).

Such Dewar isomeric photoproducts may be more widespread and , in this context, it is worth recalling the earlier studies of Wierzchowski and Shugar (1963) on the

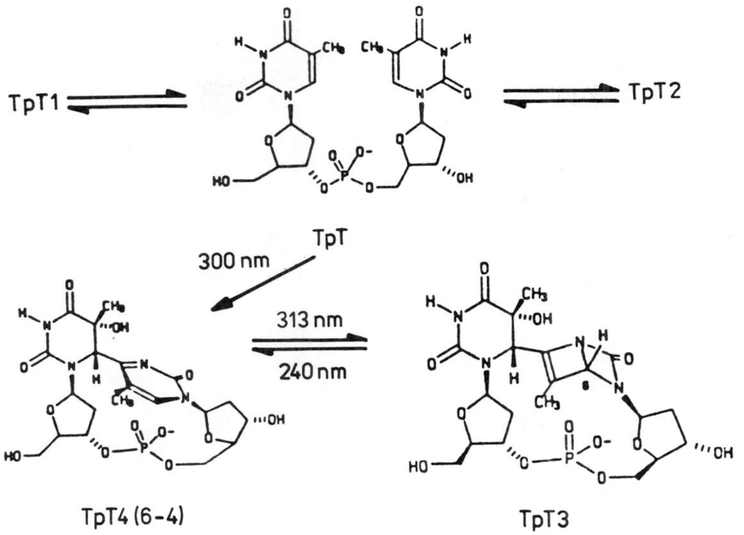

Scheme 9. See text for details

photochemistry of 4-aminopyrimidines, originally undertaken with a view to elucidating the nature of the photoproducts of 5-methylcytosines. Irradiation at 254 nm of a solution of 2,6-dimethyl-4-aminopyrimidine led to precipitation of crystals of a photoproduct (I in Scheme 11) which, following treatment with acid, gave a product (III in Scheme 11) identical with the cyanoacetylacetone chemically synthesized by Trauble (1898) many years earlier. This led to identification of I as the diaza analogue of cyanoacetylacetone and, by milder acid treatment of I, to isolation of II, the monoaza analogue. Despite numerous efforts to synthesize the diaza analogue of an acetylacetone, the above photochemical procedure is the only succesful one. The spectral properties of the foregoing products have been described in detail (Wierzchowski and Shugar,1965).

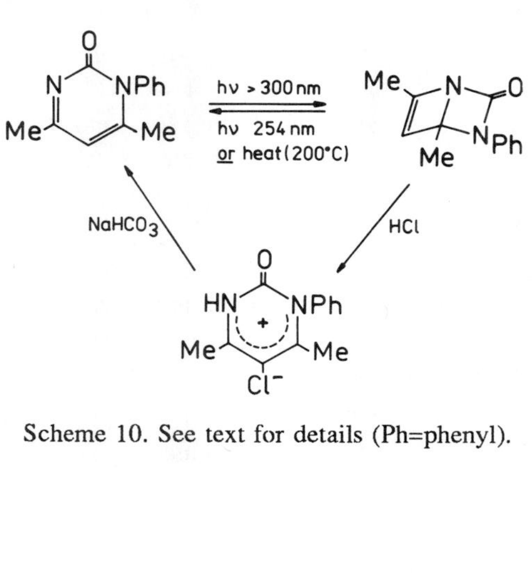

Scheme 10. See text for details (Ph=phenyl).

123

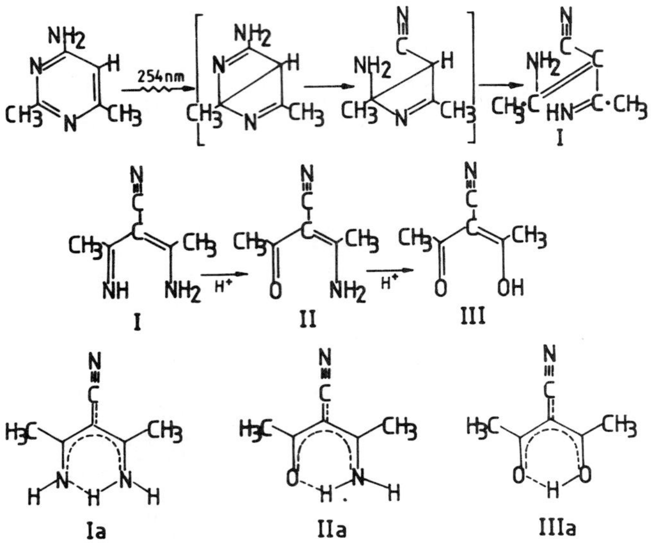

Scheme 11. Photochemical conversion of 2,6-dimethyl-4-aminopyrimidine to the diaza analogue of cyanoacetylacetone (I) *via* the Dewar intermediate (shown in brackets). I may be deaminated to the monoaza analogue II, and to cyanoacetylacetone III. All three products exist in the neutral forms as resonance stabilized intramolecularly hydrogen-bonded 6-membered rings.

Particularly relevant is the mechanism of formation of I from 2,6-dimethyl-4-aminopyrimidine, shown to be an intramolecular reaction proceeding *via* a singlet state. Since the reaction involves cleavage of two bonds (see Scheme 11), the only conceivable pathwey is *via* a C(2)–C(5) Dewar valence isomer intermediate. In retrospect, bearing in mind the findings of Taylor et al. (1988) and Nishio et al. (1980), irradiation at wavelengths to the red of 300 nm might lead to isolation of the Dewar intermediate.

Reverting once again to TpT3 (Scheme 9) it was found by Taylor and Cohrs (1987) that this is remarkably stable in solution, suggesting that such lesions in DNA might persist for long periods. Furthermore, the reversible photochemical reaction TpT4 ↔ TpT3 is reminiscent of, and may account for, the type III photoreactivation of the lethal effects of 254 nm radiation on various bacterial strains, a nonenzymatic process (Ikenaga et al., 1971), correlated with disappearence of the (6–4), i.e. TpT4, photoproduct.

During the preparation of this text, a report appeared by Takahashi et al. (1988) on the photochemistry of 5-methylpyrimidin-4-one analogues in acetic acid solution, with isolation of a variety of photoproducts formed *via* C(2)–C(5) Dewar intermediates, identified by ^{1}H NMR spectroscopy. And, in fact, one such Dewar intermediate was isolated in low yield (~14%) following irradiation of the pyrimidin-4-one in liquid NH_3-ether at –40°C under anaerobic conditions, with a high-pressure Hg lamp through quartz.

A higher yield of the Dewar intermediate might be obtained by irradiation through pyrex to cut off at wavelengths below 300 nm.

Phototransformations of 5-Alkyluracil Nucleosides

Our interest in 5-alkyluracil nucleosides stemmed initially from studies on the properties of their homopolynucleotides relative to those of poly(U) and poly(rT). It was subsequently found that 5-ethyl-2'-deoxyuridine (EtUdR), a thymidine analogue, is a good antiviral agent, with activity *vs* herpes simplex and vaccinia viruses. It is also incorporated into DNA; in the case of phages T3 and T7, it may replace up to 70% of thymidine residues (Pietrzykowska and Shugar, 1966). This prompted an investigation of its photochemical behaviour.

Irradiation of EtUdR at wavelengths to the red of 265 nm led largely to formation of cyclobutane photodimers. At 254 nm, the compound was converted to deoxyuridine in more than 80% yield, with concomitant liberation of ethylene, *via* a 5,6-cyclobutane-2'-deoxyuridine intermediate, which could be isolated (see upper portion of Scheme 12), and is the product of an intramolecular photoaddition reaction. This led to irradiation at 254 nm of deoxyuridine in the presence of ethylene (bubbled through the solution), leading to formation, as the major product, of the cyclobutane intermediate, an example of the well-known photocycloaddition reaction of olefins to aromatic molecules. The isolated cyclobutane intermediate, in turn, on irradiation at 254 nm, underwent photodissociation to deoxyuridine with a quantum yield of 0.3, hence approximately one-half the quantum yield for photodissiciation of a cyclobutane photodimer. The cyclobutane intermediate may be formally considered as a "half-photodimer".

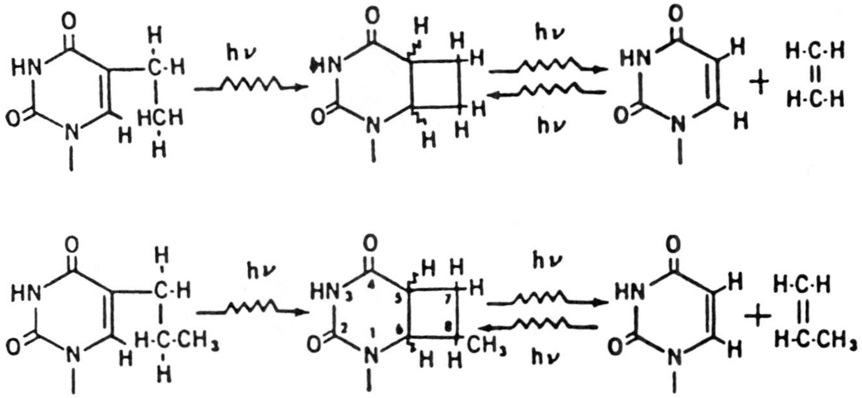

Scheme 12. See text for details

125

Similar photochemical behaviour was exhibited by 5-propyl-2'-deoxyuridine, with liberation of propylene (lower portion of Scheme 12). And, in fact, irradiation of uridine at 254 nm in the presence of propylene led to isolation of the cyclobutane intermediate as a mixture of crystals with different melting points, testifying to the presence of stereoisomers. Similar results were obtained with other 5-alkyluracil nucleosides, such as 5-isopropyl-2'-deoxyuridine and 5-hexyluridine (Sztumpf-Kulikowska and Shugar, 1974, and references cited).

Photoaddition of ethylene and propylene to thymine and thymidine was then demonstrated, with isolation of products; more significant was the finding that such photoaddition of ethylene to poly(dAT)·poly(dAT) occurs readily (Z. Zarebska and D. Shugar, unpublished), pointing to the feasability of photochemical formation of such "half-dimers" in DNA and RNA, and examining the susceptibility of these lesions to repair enzymes. Similar lesions are to be anticipated in UV-irradiated DNA of phages which contain pyrimidines with long alkyl chains, e.g. the DNA of *B. subtitlis* phage SP-15 contains, in place of thymine, 12 mole % of 5-(4',5'-dihydroxypentyl)uracil.

Pyrimidone-2 Dimers and Tetrameric Photoproduct from DNA

Some years ago Wang and Rhoades (1971) isolated a tetrameric photoproduct resulting from photodimerisation of a photoadduct of thymine and pyrimidone-2. Its structure, from X-ray diffraction of a crystal, is shown in Scheme 13, from which it will be seen that it may also be considered a dimer of a (6–4) pyrimidine photoadduct. The same product was also isolated from irradiated DNA, but its biological significance has not been examined.

Our attention was drawn to this photoproduct as an outcome of studies on the electrochemical reduction of a series of pyrimidone-2 analogues (Czochralska et al., 1986,

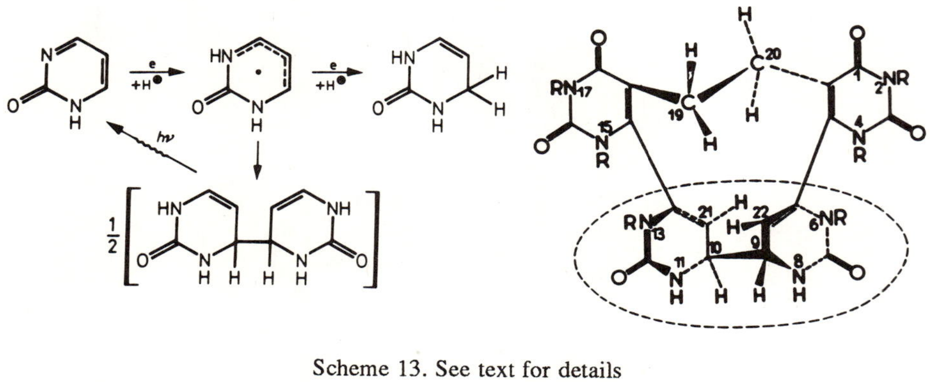

Scheme 13. See text for details

and references cited). The pathway for electrochemical reduction of pyrimidone-2 is shown in Scheme 13, the first reduction step leading to formation of a protonated radical, which dimerizes to give 6,6'-*bis*-(3,4-dihydropyrimidone-2). This dimer reduction product may be readily oxidized polarographically to regenerate the parent pyrimidone-2. On irradiation at 254 nm, the dimer is converted quantitatively to the parent pyrimidone-2 with a quantum yield of 0.1.

Similar photodissociable dimers are generated by electrochemical reduction of various methylated derivatives of pyrimidone-2, including 4,6-dimethylpyrimidone-2, obtained in crystalline form suitable for X-ray diffraction. Its structure was closely similar to that of the pyrimidone-2 dimer moiety of the tetrameric photoproduct of Wang and Rhoades (1971; see Czochralska et al., 1986)

Added interest attaches to the foregoing in the light of the finding that irradiation of 4,6-dimethylpyrimidone-2 in isopropanol at 254 nm leads to formation of a dimer (Pfoertner, 1975) similar to that obtained by electrochemical reduction. It is now pertinent to inquire whether the dimer moiety of the tetrameric photoproduct of Wang and Rhoades (1971) is susceptible to photodissociation. Furthermore, it is entirely conceivable, from the foregoing, that pyrimidine (6–4) adducts may form dimers *via* the pyrimidone-2 moieties by electrochemical reduction or by UV-irradiation in isopropanol.

Finally, it should be noted that a number of other pyrimidone, and some purine, analogues undergo electrochemical reduction to dimers which can photochemically regenerate the parent monomer (see Czochralska et al., 1986).

Methodology of Photochemical Cross-Linking

Direct photoaffinity labeling

Since cellular DNA exists in intimate contact with proteins and other molecules, it is likely that deleterious effects of UV-irradiation are due in part to cross-linking reactions. *In vitro* studies have demonstrated a variety of photoreactions between nucleic acid constituents and amino acids, e.g. thymine reacts with cysteine, on irradiation at 254 nm, with formation of five identified cross-linked products (Fisher et al, 1974; see Scheme 14). Photochemicaly induced cross-links have also been demonstrated between other bases and a variety of amino acids (Havron and Sperling, 1977; Maley et al., 1980; Varghese, 1976; Paradiso et al., 1979). For an intracellular DNA-protein complex, phototchemical formation of such a cross-link constitutes evidence for the site of binding between the two constituents. This procedure has already yielded useful results on the binding of nucleic acids with such proteins as aminoacyl tRNA synthetases, ssDNA binding proteins, DNA and RNA polymerases, and many others (see Kierdaszuk and Eriksson, 1988, for appropriate references).

The foregoing approach has also been applied to *in vitro* studies on the nature of the complexes of nucleotides and oligo- or polynucleotides with a variety of proteins and enzymes such as myosin (Maruta and Korn, 1981), ribonucleotide reductase (Kierdaszuk and Eriksson, 1988, and references cited), DNA polymerase (Biswas and Kornberg, 1984),

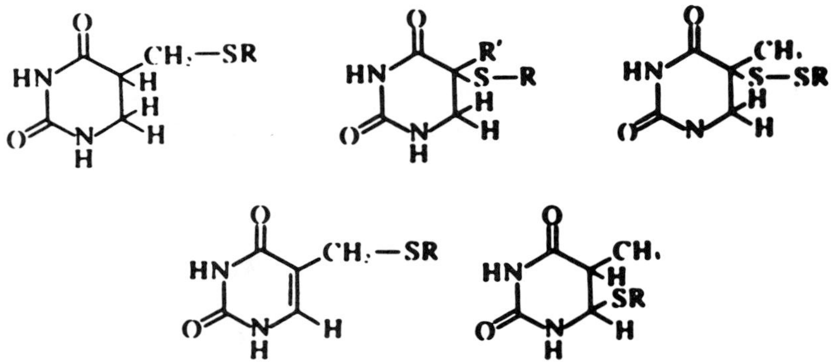

Scheme 14. Identified photoproducts of thymine with cysteine (SR).

deoxycytidine kinase (Jansson and Eriksson, 1988), etc. Occasionally the nucleotide photoligand is a photoreactive azido derivative, such as a 2- or 8-azido purine nucleotide. The ligand should, of course, exhibit a high affinity for the protein, with a life-time for the complex not shorter than the life-time of the radicals formed during photoreaction of the photoligand. Modifications of these *in vitro* techniques include elimination of free radical scavengers, the use of photosensitizers for increasing yields, and conditions for stabilizing the cross-linked photoproducts. Nonetheless, there are very few instances where direct photoaffinity labeling has led to identification of the resulting photoproducts. One interesting example is the nature of the product formed by the complex of the *fd* gene 5 protein and DNA (Paradiso et al., 1979),with identification of a tryptic peptide covalently linked *via* a cysteinylthymine bond to DNA. The absence of UV absorbance of the product at 260 nm pointed to saturation of the 5,6-bond of the thymine residue, hence to one of the 4 such products shown in Scheme 14.

   The most complete characterization of such a cross-linked product is undoubtedly that between the B1 subunit of *E. coli* ribonucleotide reductase and its allosteric activator dTTP (Kierdaszuk and Eriksson, 1988). Following irradiation at 254 nm, a dTTP-labeled tryptic peptide was isolated which, on the basis of the known sequence of the B1 subunit, was identified as the octapeptide shown in Scheme 15,with the dTTP lined to Cys[292]. The UV absorption spectrum of this product as a function of pH demonstrated that the thymine residue possessed an unsaturated 5,6-bond, so that the cross-link must involve the thymine methyl group (like the lower left-hand product in Scheme 14), further confirmed by the use of [methyl-^{3}H]dTTP. The structure of the photochemically generated cross-link is consequently as shown in Scheme 15.

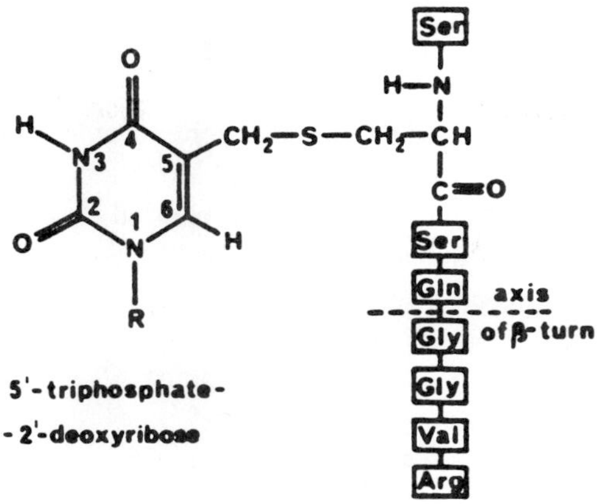

Scheme 15. See text for details

Cross-linking via transamination of cytosine photohydrates

Catalytic reduction of cytosine and its nucleosides leads initially to saturation of the 5-6 bond, and these reduced derivatives are rather labile. It was long ago reported that reduction of cytosine in the presence of amino acids led to 5,6-dihydro derivatives with considerably enhanced stability, due to transamination, as shown in Scheme 16. This transamination reaction was found to proceed efficiently with a variety of amino acids,

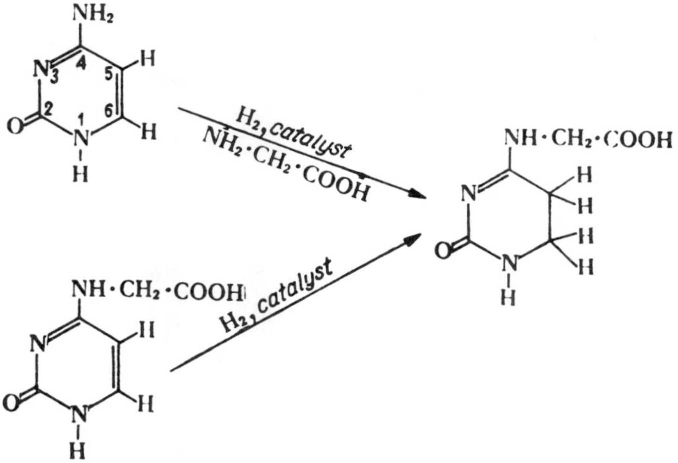

Scheme 16

129

peptides and other amines, as well as at the level of oligo(C); and attention was drawn to the enhanced stability of such transformation products of cytosine photohydrates (Janion and Shugar, 1967).

It consequently appears that, in a nucleoprotein, formation of a cytosine hydrate by UV-irradiation may lead to cross-linking by transamination *via* the α-amino group of an adjacent amino acid. There is at least one such reported example, *viz.* Budowsky et al. (1976) found that irradiation at 254 nm of phage MS2 led to cross-linking of a cytosine residue to the ε-amino group of a lysine in the coat protein, with isolation of the expected transaminated product in a hydrolysate of the phage. It is, indeed, surprising that this approach has not been more extensively exploited.

Photoreaction of thymidine monoanion with alkylamines and amino acids

This extremly interesting reaction, first described by Saito et al. (see Saito and Matsura, 1985, for review), is schematically illustrated in Scheme 17. Irradiation of thymidine in alkaline aqueous medium (pH 11, under which conditions thymidine exists as the monoanion) in the presence of an alkylamine or an amino acid, leads to nucleophilic attack by the amine at C(2), accompanied by opening of the N(1)-C(2) bond, which then undergoes ring closure with elimination of deoxyribosylamine. The ring-opened intermediate may be readily isolated and has been characterized by various physicochemical criteria. When R_1 is an amino acid, the resultant products with various amino acids are as illustrated in Scheme 4, above.

In the schematic representation of this reaction by Saito and Matsura (1985), the monoanion of thymidine is shown with the charge on O^2. However, it was long ago demonstrated that the monoanions of uracil and thymine, and their nucleosides, exist with charge distribution between O^2 and O^4 (Wierzchowski et al., 1965), as shown in our Scheme 17 (see also Scheme 18, below). This may require some revision of the mechanism of the reaction as formulated by Saito and Matsura (1985).

The foregoing reaction was demonstrated to occur at the level of DNA and, indeed, it has been proposed as a means of selective cleavage of DNA at thymine residues for sequencing purposes.

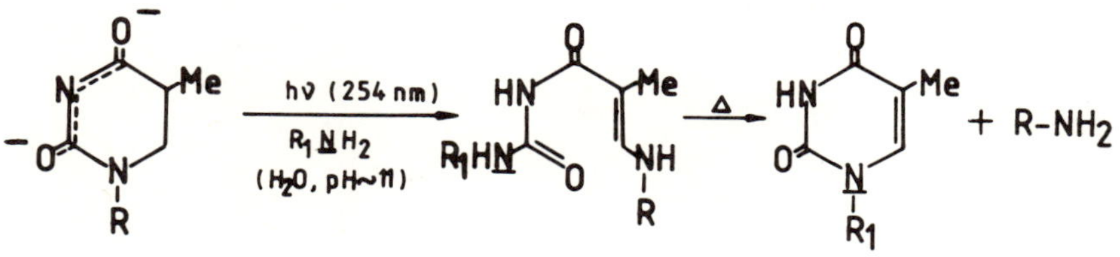

R = 2'-deoxyribose

R_1 = Me, or t-butyl, or amino acid

Scheme 17. See text for details

The same reaction has also been found to proceed at the level of nucleohistone which, when irradiated at pH 10.5 at 254 nm, led to isolation of a thymine-lysine photoadduct involving the lysine ε-amino group (see Scheme 4). This constitutes evidence for cross-linking between thymine residues in DNA and histone lysine residues in the immediate vicinity of the former. The same authors have likewise demonstrated the formation of such cross-linked products in irradiated chicken erythrocyte nuclei, pointing to the utility of these reactions in further investigations of the photobiological effects of UV radiation.

Alkaline dehydration of photohydrates of uracil nucleosides

Shetlar et al. (1984) reported that the uracil-lysine system reacts photochemically in alkaline medium like the dThd-lysine system, but at a much slower rate. However, this finding requires reexamination in the light of the results of Fikus and Shugar (1966) on the alkaline-catalyzed dehydration of photohydrates of 1-methyluracil, 1,6-dimethyluracil and uracil glycosides, illustrated in Scheme 18. It should be noted that the photohydrate of 1-methyluracil undergoes dehydration to regenerate 1-methyluracil uniquely *via* pathway A, whereas dehydration of the photohydrate of 1,6-dimethyluracil proceeds uniquely *via* pathway B. With uracil nucleosides (whatever the nature of the sugar component, and also when the sugar hydroxyls are blocked), elimination of water proceeds simultaneously *via* both pathways, and in methanolic medium (with traces of water) only by pathway B. Furthermore, it was demonstrated that pathway B proceeds *via* a ring-opened intermediate which is sufficiently stable to be isolated, and is similar to the

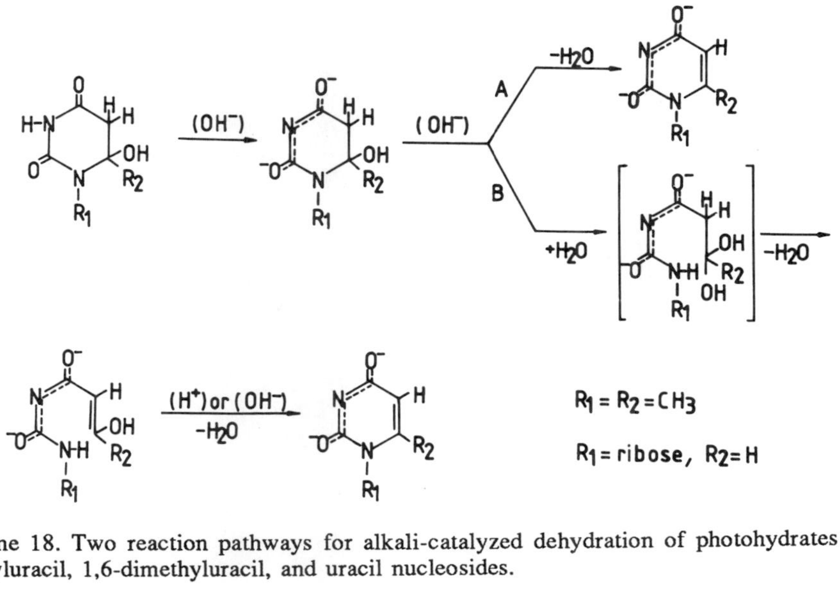

Scheme 18. Two reaction pathways for alkali-catalyzed dehydration of photohydrates of 1-methyluracil, 1,6-dimethyluracil, and uracil nucleosides.

photochemically ring-opened intermediates in the reaction of thymidine monoanion with amino acids. Finally, the alkali-catalyzed dehydration of uracil nucleosides is fully quantitative only in the presence of NH_4^+ ions, but not Na^+ or K^+. It is clear that the reaction with uracil nucleosides must take the foregoing into account.

Acknowledgements

Research supported by the Polish Academy of Sciences (CPBR 3.13) and the Polish Cancer Research Program (CPBR 5.11).

References

Birnbaum, G.I. and D. Shugar (1987) in: Topics in Nucleic Acid Structure (S. Neidle Ed., Macmillan Press, London), Part 3, 1–70.

Biswas, R. M. and A. Kornberg (1984) *J. Biol. Chem.* 259, 7990–7993.

Brown, G.B., G. Levin and S. Murphy (1964) *Biochemistry* 3, 880–883.

Budowsky E.I., N.A. Simukova, M.F. Turchinsky, I.V. Boni and Yu. M. Skoblov (1976) *Nucleic Acids Res.* 3, 261–276.

Buffel, D.K., C. McGuigan and M.J. Robins (1985) *J. Org. Chem.* 50, 2664–2667.

Connolly, J.S. and H. Linschitz (1968) *Photochem. Photobiol.* 7, 791–806.

Czochralska, B., M. Wrona and D. Shugar (1986) Topics in Crrent Chemistry (Springer-Verlag, Berlin) 130, 133–181.

Davies, D.B. (1978) *Prog. Nucl. Magn. Reson. Spectrosc.* 12, 135–225.

Dock-Bregeon, A.C., B. Chevrier, A. Podjarny, D. Moras, J.S. deBaer, G.B. Gough, P.T. Gilham and J.E. Johnson (1988) *Nature* 335, 375–378.

Drabikowska, A., M. Draminski, L. Lisowska, A. Zgit-Wroblewska and D. Shugar (1987) *Z. Naturforsch.*42c,288–296.

Duker, N.J. and P.E. Gallagher (1988) *Photochem. Photobiol.* 48, 35–39.

Evans, B. and R. Wolfenden (1970) *J. Am. Chem. Soc.* 92, 4751–4758.

Fikus, M. and D. Shugar (1966) *Acta. Biochim. Polon.* 13, 39–56.

Fisher, G.J., A.J. Varghese and H.E. Johns (1974) *Photochem. Photobiol.* 20, 109–120.

Fourrey, J.-L., G. Henry and P. Join (1977) *J. Am. Chem. Soc.* 99, 6753–6754.

Havron, A. and J. Sperling (1977) *Biochemistry* 16, 5631–5635.

Ikenaga, M., M.H. Patrick and J. Jagger (1971) *Photochem. Photobiol.* 14, 175–187.

Janion, C. and D. Shugar (1967) *Acta Biochim. Polon.* 14, 293–302.

Jansson, O. and S. Eriksson (1988) *Biochemistry* (in press).

Johns, A.E., M.L. Pearson, J.C.Le Blanc and C.W. Helleiner (1964) *J. Mol. Biol.* 9, 503–524.

Kazimierczuk, Z. and D. Shugar (1973) *Acta. Biochim. Polon.* 20, 395–402.

Kierdaszuk, B. and S. Eriksson (1988) *Biochemistry* 27, 4952–4965.

Livnoh, Z., D. Elad and J. Sperling (1979) *Proc. Natl. Acad. Sc. U.S.A.* 76, 5500–5504.

Maley, P., J. Rinke, E. Ulmer, C. Zweib and R. Brimacombe (1980) *Biochemistry* 19, 4179–4188.

Maruta, H. and E.D. Korn (1981) *J. Biol. Chem.* 256, 499–502.

Matsuda A., M. Tezuka, and T. Ueda (1978) *Tetrahedron* 34, 2449–2452.

McLaren, A.D. and D. Shugar (1964) *Photochemistry of Proteins and Nucleic Acids* (Pergamon Press, Oxford, UK).

Nair, V., D.A. Young and R. DeSilvia, Jr. (1987) *J. Org. Chem.* 52, 1344–1347.

Nishio, T., A. Kato, C. Kashima and Y. Omoto (1980) *J. Chem. Soc. Perkin. I,* 607–610.

Ohno, M., N. Yagiawa, S. Shibehara, S. Kondo, K. Maeda and H. Umezawa (1974) *J. Am. Chem. Soc.* 96, 4326–4327.

Ohtsuka, E., S. Tanaka and M. Ikehara (1977) *Chem. Pharm. Bull.* 25, 949–959.

Paradiso, P.R., Y. Nakashima and W. Konigsberg (1979) *J. Biol. Chem.* 254, 4739–4744.

Pfoertner, K.H. (1975) *Helvetica Chim. Acta.* 58, 865–869.

Pietrzykowska, I. and D. Shugar (1966) *Biochem. Biophys. Res. Commun.* 25, 567–572.

Saito, I. and M. Matsura (1985) *Acc. Chem. Res.* 18, 134–141.

Saito, I., H. Ikehira and T. Matsura (1985) *Tetrahedron Lett.* 26, 1993–1994.

Salomon, J. and Elad, D. (1973) *Biochem. Biophys. Res. Commun.* 58, 890–894.

Sepiol, J., Z. Kazimierczuk and D. Shugar (1976) *Z. Naturforsch.* 31c, 361–370.

Shetlar, M.D., J.A. Taylor and K. Hom (1984) *Photochem. Photobiol.* 40, 299–308.

Stolarski, R., C.-E. Hagberg and D. Shugar (1984) *Eur. J. Biochem.* 138, 187–192.

Sztumpf-Kulikowska, E. and D. Shugar (1974) *Acta. Biochim. Polon.* 21, 73–91.

Takahashi, T., S. Hirokami, M. Nagata and T. Yamazaki (1988) *J. Chem. Soc. Perkin. Trans. I*, No. 9, 2653–2662.

Taylor, J.-S. and M.P. Cohrs (1987) *J. Am. Chem. Soc.* 109, 2834–2835.

Taylor, J.-S. D.S. Garrett and M.P. Cohrs (1988) *Biochemistry* 27, 7206–7215.

Trauble W. (1898) *Ber. Dent. Chem. Ges.* 31, 2938–2945.

Turro, N.J. and G. Schuster (1975) *Science* 187, 303–312.

Varghese, A.J. (1976) in Aging, Carcinigenesis and Radiation Biology. The Role of Nucleic Acid Addition Reactions (Smith, K.C., Ed., Plenum Press, New York and London), pp. 207–223.

Voituriez, L., C. Voisin, L.-S. Kan and J. Cadet (1988) *10th Intern. Cong. Photobiol.* (Jerusalem, Oct. 30-Nov. 5), Abs. No. 19.

Wang, S.Y. and D.F. Rhoades (1971) *J. Am. Chem. Soc.* 93, 2554–2557.

Wierzchowski, K.L., E. Litonska and D. Shugar (1965) *J. Am. Chem. Soc.* 87, 4621–4629.

Wierzchowski, K.L. and D. Shugar (1963) *Photochem. Photobiol.* 2, 377–391.

Wierzchowski, K.L. and D. Shugar (1965) *Spectrochim. Acta.* 21, 931–941.

The Sequence Specificity of Ultraviolet Light Damage to DNA Containing Bromodeoxyuridine or Iododeoxyuridine

V. MURRAY AND R. F. MARTIN

Molecular Sciences Group
Peter MacCallum Cancer Institute
481 Little Lonsdale Street
Melbourne, Victoria 3000

Introduction

When DNA is substituted with bromodeoxyuridine (BrUdR) or iododeoxyuridine (IUdR), DNA becomes sensitive to damage by ultraviolet (UV) light (see Hutchinson, 1973; Hutchinson and Kohlein, 1980 for reviews). Damage consists of single strand breaks, alkali-labile lesions, base damage and a small proportion (1–5%) of double strand breaks. The reaction mechanism is initiated by the breaking of the carbon-halogen bond by ultraviolet light to give a uracilyl free radical and a halogen free radical. It has been proposed that the uracilyl radical abstracts a hydrogen atom from the 2' carbon of the 5' deoxyribose — this latter reaction is postulated because of the close spatial proximity of the two reaction species. The abstraction of the hydrogen from the 2' carbon is then thought to give rise to DNA damage, including single strand breaks, through unknown reaction intermediates.

In this study (Murray and Martin, 1989) the sequence specificity of UV damage to DNA containing IUdR or BrUdR was examined using a newly-developed system. This system consists of a primed single stranded M13 DNA which is replicated with iododeoxyuridine-(dIUTP) or bromodeoxyuridine triphosphate (dBrUTP) replacing dTTP. Thus the synthesised DNA is fully substituted with either IUdR or BrUdR. This DNA is then damaged with UVB light (300 nm peak) and run on thin polyacrylamide-urea DNA sequencing gels. By this technique sites of damage were located to the exact base pair.

Because versatile M13 vectors have been developed for rapid DNA sequencing, the sites of damage in any clonable DNA sequence can be examined using this system. In this study 340bp clones of alpha RI-DNA in M13 were employed. The middle repetitive human alpha RI-DNA is present at 50,000 copies per haploid genome and is found mainly as tandem repeats at the centromeres of chromosomes (Manuelidis, 1978; Darling et al., 1982).

Photobiology, Edited by E. Riklis
Plenum Press, New York, 1991

Materials and Methods

Synthesis of DNA fully substituted with IUdR or BrUdR

This method is derived from the Sanger dideoxy DNA sequencing procedure (Sanger et al., 1977; 1980) in that primed single stranded DNA is used as a template and the dideoxy sequencing reactions are used as a reference to locate the sites of damage. There were two parts to the synthesis procedure. In the first part the primer is extended with only dGTP, dCTP and [^{32}P]dATP present; and results in limited DNA synthesis (less than 10 bp) and effectively labels the DNA at the 5' end with ^{32}P. In the second part dBrUTP or dIUTP is added (along with cold dATP) and, since 4 dNTPs are present, extensive synthesis of DNA occurs (greater than 3000 bp). In this second part the synthesised DNA is fully substituted with IUdR or BrUdR instead of dTTP.

Single stranded M13 DNA (500 ng) was annealed with 15 bp primer, and in a final volume of 10ul, there was 7mM $MgCl_2$, 7mM tris-HCl (pH 7.5), 50mM NaCl, 10uM dGTP, 10uM dCTP, 2 BRESA units of Klenow polymerase I, 1uCi of [alpha ^{32}P] dATP (3000 Ci/m mole) (BRESA, Australia). After a 10 min. incubation at room temperature, 1ul of 500 µM dATP and 1ul of either 1mM dIUTP or 1mM dBrUTP was added and subsequently incubated for 20 min at 37°C. The addition of 1ul of 250mM EDTA (pH 8.0) terminated the reaction (Murray and Martin, 1989).

Three M13 clones — alpha 82,22 and 32 — were used in this study and contain 340bp inserts derived from alpha RI-DNA. Further details about these clones can be found in Murray and Martin (1985a, 1987).

Irradiation of DNA with UV light

Samples (3 ul) from the M13 DNA synthesis reaction were irradiated for 10 min in eppendorf tubes by a UVB light source (30,600 J/m^2) which had a wavelength peak of 300 nm.

Densitometer analysis of UV damage

The sequencing gel autoradiographs were scanned with a Zeineh soft laser densitometer (Biomed Instruments) and digitally analysed by Zeineh software (Biomed Instruments). After control values (non-irradiated samples) had been subtracted, the peak heights were determined for each damage site.

Results and Discussion

UV Damage to DNA containing IUdR or BrUdR

As shown in Figure 1, unirradiated DNA was mainly of high molecular weight and had a low background in the region 10 to 200 bps. When an identical sample from the same DNA synthesis reaction tube was irradiated with 300 nm UV light, extensive damage occurred. The damage sites were not of equal intensity and a wide variation in

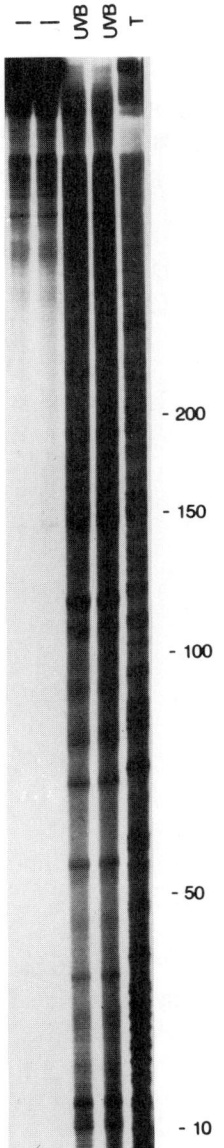

Figure 1. Autoradiograph of a polyacrylamide-urea sequencing gel showing UV damage to DNA containing IUdR
Lane T is the dideoxy T sequencing reaction. All other lanes are derived from the same reaction tube which contains IUdR-substituted DNA. Lanes UVB are irradiated by UV light while lanes (–) have not been irradiated.

degree of damage was apparent. The sites of damage were not random but were always present 2 bps faster than the dideoxy T sequencing bands. IUdR (and BrUdR) is expected to be substituted at T in the sequence. Thus all the sites of damage are caused by the presence of IUdR (or BrUdR). When the normal dTTP is used instead of dIUTP or dBrUTP, no damage is detectable above background after UV irradiation. Further confirming that the presence of IUdR or BrUdR is responsible for the UV sensitivity.

Not all sites of IUdR or BrUdR substitution give rise to detectable damage. There are several substitution sites where no significant damage is apparent.

The damage bands were generally not sharp and more "fuzzy" than the dideoxy T sequencing bands. This is presumed to reflect the presence of multiple products of the irradiation which have a slightly different mobility on the sequencing gel.

Where runs of Ts exist in the sequence (eg bp 84–88 is a run of five Ts — see Table 1), the damage site is mainly at the 5' end of the run of Ts. At present there is no completely satisfactory explanation for this phenomenon.

Table 1. The DNA sequence of M13 clone alpha 32

TGTTGTGTGT	TTTCAACTCA	CAGAGTTGAA	CGATGCTTGA	60
CACAGAGTAG	ACTTGAAACA	CTCTTTTTGT	GTAATTTGCA	100
TGTGGAGATT	TCAGCCGCTT	TGAGGTCAAT	GGTAGAAAAG	140
GAAATATCTT	CCTATAGAAA	CTAGACAGAA	TGATTGCCAG	180
AAACTCCTTT	GTGATGTGTG			

Analysis of the sites of damage by laser densitometer

A scanning laser densitometer was used to digitally analyse the degree of UV damage at each hotspot. There was a fifteen-fold range in damage intensity for IUdR (from no detectable damage to the most damaged site) and seventeen - fold for BrudR.

For clone alpha 32 from bp 30 to 188, there were 28 sites of IUdR substitution. Densitometer analysis revealed that there were 3 strong damage sites, 5 medium, 15 weak and 5 sites of no significant damage.

A comparison of the intensity of damage for IUdR and BrUdR revealed that with M13 clone alpha 32, the sites of damage were the same but variation in the degree of damage was apparent. BrUdR produced slightly more intense damage than IUdR.

DNA sequences present at the damage sites

There were eight sites where strong or medium damage occurred with clone alpha 32 with IUdR and seven sites for BrUdR. There was the same "consensus" sequence present at these sites for IUdR - RCTTG/T. (To make discussion easier, T is shown where it is expected to be substituted with IUdR or BrUdR. The damage site was at the middle base of the sequence. R is a purine).

There were five sites which had no significant damage for both IUdR and BrUdR. The "consensus" sequence present was NGTRR. (N is any base).

When a "consensus" sequence is present, it does not always result in the predicted degree of damage. On six occasions the sequence CTT appears in the region bp 30–200

(Table 1), 3 are strong damage sites, 2 medium but one is weak. Similarly with the sequence GTR, which occurs five times, four have no significant damage while one is weak. Thus the "consensus" gives a strong indication of the expected degree of damage but does not always correctly predict the degree of damage.

The effect of base substitutions on the degree of UV damage

The use of M13 clones which are closely related in DNA sequence enables the effects of random base substitutions on the degree of UV damage to be examined. Previously three 340 bp alpha RI-DNA clones were constructed in M13 and sequenced (Murray and Martin, 1987).

A substitution which removed T from the sequence, caused the damage site to disappear. A base substitution which altered the sequence so that a T was now present in the sequence, caused a damage site to appear. This data confirms that the presence of IUdR or BrUdR sensitises the DNA to UV damage.

Base substitutions in the "consensus" sequence revealed that they were important in determining the degree of cleavage. Also in one case a base substitution outside the "consensus" sequence showed that long range effects were important.

General Discussion

In this section a hypothesis is proposed to explain why different DNA sequences are associated with different degrees of UV damage to DNA containing IUdR or BrUdR.

DNA is no longer thought to be a perfect Watson-Crick double helix. The elucidation of the X-ray crystallographic structure of a dodecamer oligonucleotide revealed variations in helical twist, roll angles between base-pairs, propeller twist, relative sliding of base pairs, and conformation of the sugar-phosphate backbone (Drew et al., 1981; Dickerson and Drew, 1981). These variations in the microstructure of DNA are dependent on neighbouring sequences and principally occur to maximise base stacking energy without causing any steric clashes (see Shakked and Rabinovich, 1986 for review).

A simple hypothesis to explain the sequence dependent UV damage of DNA containing IUdR or BrUdR, is based on variation in the conformation of DNA at the site of reaction. It is proposed that the distance between the critical reactive groups determines the extent of the reaction. The UV light breaks the 5-carbon-halogen bond to give the uracilyl free radical and the halogen free radical; the uracilyl radical then abstracts a hydrogen atom from the spatially close 2' carbon of the 5'-deoxyribose; this is the critical stage in the reaction and it is proposed that the distance between the uracilyl radical (on the 5'-carbon) and the hydrogen on the 2' carbon (of the 5'-deoxyribose) determines the rate of the reaction with large distances giving a low degree of damage and short distances a high degree of damage.

The sequences CTT (high degree of damage) and GTR (low degree of damage) are theoretically expected to have very different DNA conformations because one is a run of pyrimidines and the other is alternating purine-pyrimidine-purine (Calladine, 1982). Thus

139

an obvious prediction is that the distance between the 5'-carbon and the hydrogen on the 2' carbon is larger for the sequence GTR than the sequence CTT. The importance of neighbouring sequences on the conformation of DNA can explain the effects of base substitutions in the "consensus" sequence and outside the sequence.

Conclusions

1. The presence of IUdR or BrUdR in DNA sensitises the DNA to damage by UV light.
2. The intensity of damage varies at each damage site.
3. The neighbouring sequences at the damage site are the major parameter affecting degree of damage. The "consensus" sequence RCTTG/T is present at strong and medium damage sites and the "consensus" sequence NGTRR is present at sites where there is no detectable damage.
4. The sequence dependent microstructure of DNA is the fundamental determinant of the degree of damage.

Acknowledgements

This project was supported by the National Health and Medical Research Council and research funds from the Peter MacCallum Cancer Institute. We would like to thank Larry Wakelin for helpful discussions.

References

Calladine, C.R. (1982) *J. Mol. Biol.* 161, 343–352.
Darling, S.M., Crampton, J.M., and Williamson, R. (1982) *J. Mol. Biol.* 154, 51–63.
Dickerson, R.E. and Drew, H.R. (1981) *J.Mol. Biol.* 149, 761–786.
Drew, H.R., Wing, R.M., Takano, T., Broka, C., Tanaka, S., Itakura, K. and Dickerson, R.E. (1981) *Proc. Natl. Acad. Sci. U.S.A.* 78, 2179–2183.
Hutchinson, F. (1973) *Quart. Rev. Biophys.* 6, 201–246.
Hutchinson, F. and Kohnlein, W. (1980) *Progr. Mol. Subcell. Biol.* 7, 1–42.
Manuelidis, L. (1978) *Chromosoma* 66, 23–32.
Murray, V. and Martin, R.F. (1985) *Gene Anal. Techn.* 2, 95–99.
Murray, V. and Martin, R.F. (1987) *Gene* 57, 255–259.
Murray, V. and Martin, R.F. (1989 *Nucl. Acids Res.* 17, 2675–2691.
Sanger, F. Nicklen, S. and Coulson, A.R. (1977) *Proc. Natl. Acad. Sci. U.S.A.* 74, 5463–5467. J
Sanger, F., Coulson, A.R., Barrell, B.G., Smith, A.J.H. and Roe, B.A. (1980) *J. Mol. Biol.* 143, 161–178.
Shakked, Z. and Rabinovich, D. (1986) *Prog. Biophys. Molec. Biol.* 47, 159–195.

140

UV-Light Induced Perturbation of DNA Tertiary Structure: Consequences on the Enzymes Controlling Chromosome Superhelicity

A.M. PEDRINI, F. SPIRITO, and S. TORNALETTI

Istituto di Genetica Biochimica ed Evoluzionistica del C.N.R.
Pavia
Italy

Introduction

Pyrimidine dimers, the major photodamage formed in DNA at 254 nm light, are the classic "test lesions" used for the analysis of DNA repair processes at the molecular level. Three dissimilar enzymatic systems, which initiate repair of pyrimidine dimers, have been described so far: namely photoreactivating enzyme, pyrimidine dimer DNA glycosylase, and the multiprotein Uvr ABC incision system (Grossman et al., 1988). Although the mechanism of action of these enzymes is different, all these proteins have to sense the unique structural change imposed to DNA by this damage. To obtain insight in the recognition mechanism of pyrimidine dimers by these very different enzymatic systems, it is important to know how these lesions affect B-DNA structure.

Currently, there are three basic models describing the conformational perturbation induced by pyrimidine dimers to DNA structure. The first model, based on thermal melting studies and on the increased sensitivity of UV damaged DNA to single-strand specific reagents, proposes that each dimer leads to hydrogen bond disruption of about four-five base pairs (Hayes et al., 1971; Shafranovskaya et al., 1973). The second model is the result of molecular graphic based on minimisation calculation combined with model building and of NMR studies on oligonucleotides containing dimers at a defined position (Rao et al., 1984; Kemmink et al., 1987). This model proposes that pyrimidine dimerisation does not lead to extensive denaturation nor to long range distortions of the double helix, but instead to minor conformational changes confined to the dimer region. These modifications should consist in weakening of the H-bond and in a different glycosylic orientation of the thymines forming the dimer. A third model, for the moment based only on minimisation calculation combined with model building, proposes the formation of long range distortions to the double helix as those deriving from bend or kinks, coupled with H-bond disruption and extensive unwinding (Broyde et al., 1980; Pearlman et al., 1985).

Photobiology, Edited by E. Riklis
Plenum Press, New York, 1991

DNA unwinding by pyrimidine dimers

In this communication I shall report the results of our studies on the structural changes induced by UV photoproducts to DNA in a chromatin-like structure. In fact, we have chosen as DNA substrate negatively supercoiled plasmid DNA molecules, as this DNA form better mimics DNA superstructure present within the cell. The detection and the quantitative expression of the helix perturbation associated with photodamage formation is based on the *band shift method* developed by J. Wang for the calculation of the helical repeat of DNA in solution (Wang, 1979). One of the advantages of this method is its high sensitivity, as with this experimental approach it is possible to analyze the helical deformation induced by damaging agents at doses which induce close to biologically significant level of modification (one damage every 10^3–10^4 bp). Thus, possible effects that might appear at high doses as a consequence of clustering of lesions can be avoided. Furthermore, when it is available a probe which specifically introduces a single-strand break where a lesion is formed, the use of circular closed DNA molecules offers the advantage of an easy and rapid way to quantitate low levels of DNA damage. In fact, DNA molecules containing single-strand breaks can be separated from intact molecules by a number of methods. Assuming a Poisson distribution of damage, one can calculate the number of probe sensitive sites per molecule and consequently the frequency of damage (Ciarrocchi and Pedrini, 1982).

The principle of the band shift method is based on the known topological properties of closed circular DNA molecules. In these molecules, the reciprocal relationship between the *twisting* and the *supertwisting* of the helix axis can be expressed in a quantitative form by the following equation:

$$Lk = Tw + Wr$$

$$\text{Linking number} = \text{Twist} + \text{Writhe}$$

$$Lk_{und} = Tw_{und} + Wr_{und} \qquad Lk_{dam} = Tw_{dam} + Wr_{dam}$$

$$Lk_{und} = Lk_{dam}$$

$$Tw_{und} + Wr_{und} = Tw_{dam} + Wr_{dam}$$

$$Tw_{und} - Tw_{dam} = Wr_{dam} - Wr_{und}$$

und = undamaged   *dam* = damaged

where the Linking number (Lk) is the number of times one strand goes about the other. The Lk is a topological parameter that can be modified only by subsequent breakage and rejoining of the phosphodiester bonds. Molecules which differ only in their Lk are called

DNA topoisomers. The consequences deriving from this equation are that for molecules with the same Lk, changes in the twist angle between bases cause a change of equal and opposite sign in supercoiling (intuitive expression for writhe). DNA supercoiling determines the hydrodynamic properties of closed circular DNA molecules. Agarose gel electrophoresis is the easiest and most sensitive method to study changes in the hydrodynamic properties of DNA. In fact, by this method DNA topoisomers can be separated under appropriate conditions in a series of bands, each differing from the next one by one Lk (or one superhelical turn). Changes in mobility of these bands caused by a damaging treatment indicate a change in supercoiling consequent to a change in twist (Ciarrocchi and Pedrini, 1982; Pedrini et al., 1986).

The magnitude of the helical conformational transition can be obtained by measuring the relative mobility (Rf) of each topoisomer band with respect to two undamaged markers added to the damaged topoisomer population prior to electrophoresis. By titrating the number of modified sites necessary to induce a band shift of one superhelical turn (Nd), it is possible to estimate the structural changes caused by one site. Assuming that the shift of one unit corresponds to 360° unwinding, the unwinding angle per damage will be equal to: 360°/ Nd (Pedrini et al., 1986).

Using DNA topoisomers, prepared by treatment of naturally supercoiled pAT153 DNA with DNA topoisomerase I in conditions such as upon electrophoresis on agarose gel they appeared positively and negatively supercoiled, we could observe that UV irradiation induced an upward shift of the negatively supercoiled topoisomers and a downward shift of the positively supercoiled topoisomers. The direction of the shift has indicated that UV-photodamage caused DNA unwinding. Quantitative analysis of the change in relative mobility of all visible topoisomer bands as a function of the number of dimers per molecule has shown that a dose which produces 25 dimers per plasmid pAT153 DNA molecule is required to relax negatively supercoiled topoisomers of one superhelical turn (Ciarrocchi and Pedrini, 1982; Pedrini et al., 1986), while a dose causing 42 dimers is requested to produce the same mobility change of positively supercoiled topoisomers (Pedrini et al. 1986). These findings can be explained by surmising that effects other than unwinding are influencing the electrophoretic mobility of DNA topoisomers, and consequently the extent of helical perturbation caused by dimers. In fact, although the band shift method is based on the assumption that changes in electrophoretic mobility of DNA topoisomers are exclusively due to damage induced modification in the twist angle between the damaged bases, this assumption is not always true. Alterations in flexibility, in molecular weight, or in DNA length may also modify the electrophoretic mobility of damaged topoisomers. This possibility is also suggested by the observation of a small but still detectable increase in electrophoretic mobility of the nicked circular DNA molecules damaged by UV (Ciarrocchi and Pedrini, 1982). Since in these types of molecules a change in twist is readily compensated by the rotation of the free ends, the alteration of their electrophoretic mobility has to depend from phenomena other than unwinding.

The average between the unwinding angles calculated for positively and negatively supercoiled topoisomers has allowed us to cancel off effects other than change in twist angle and to calculate an unwinding angle per dimer of 10.7°. This value is lower than the unwinding angle of damaged negatively supercoiled topoisomers and higher than that of positively supercoiled topoisomers. This finding can be explained by assuming that in addition to unwinding pyrimidine dimers distort the helix axis, and that the distortion imposes a positive writhe to negatively supercoiled topoisomers and a negative writhe to positively supercoiled topoisomers.

All figures presented so far are based on the assumption that dimers are the only photodamage responsible for DNA unwinding. However, this assumption turned out to be incorrect since complete photoreactivation with yeast photoreactivating enzyme was able to restore only up to 80% of the original electrophoretic mobility of the topoisomer bands, indicating that photodamage other than dimers affected DNA tertiary structure (Ciarrocchi et al. 1982). Thus, the correct unwinding angle per dimer is 8.6°, a value equivalent to a 25% change in twist angle between two base pairs. This small helix unwinding supports the notion that dimers do not cause significant denaturation. In fact, this value is very small when compared with 34.6° unwinding which would be produced by hydrogen bond disruption of one base pair or with 155° unwinding expected for 4-5 base pairs denaturation (Pedrini et al., 1986). Lack of dimer induced denaturation is also suggested by analysis of the UV lesions causing increased sensitivity to the single-strand specific S1 nuclease. In fact, we observed that the UV induced S1-sensitive sites represent a minor fraction among the other well characterized lesions and that these sites are not photoreactivable (Pedrini et al., 1986).

Table 1. Pyrimidine dimer unwinding angle

		–PRE	+PRE
Positively supercoiled topoisomers	360°/42 dimers	8.6°	6.5°
Negatively supercoiled topoisomers	360°/25 dimers	14.3°	10.8°

The actual unwinding angle is calculated by averaging the two values

6.5° + 10.8° / 2 = 8.6°

If the perturbation of the DNA structure described for *in vitro* irradiated DNA is an adequate basis for discrimination between damaged and undamaged sequences by DNA repair enzymes, the same type of alteration should be detected also on *in vivo* irradiated DNA. To answer this question we have carried on band shift experiments using plasmid DNA irradiated within the cell. We observed that DNA topoisomers obtained from DNA extracted from UV-irradiated cells presented the same dose dependent shift of DNA bands as that described for plasmid irradiated in solution, thus suggesting that intracellular

DNA organization does not significantly influenced the final deformation caused by UV damage to DNA tertiary structure (Pedrini et al., 1989).

DNA topoisomerase I and UV damage

Within the cell the fine tuning of DNA tertiary structure is controlled by a class of enzymes called DNA topoisomerases. These enzymes act by concomitant breakage and rejoining of the phosphodiester bonds and, by changing the DNA Lk, they modify DNA supercoiling. In bacteria there are two major DNA topoisomerases: DNA gyrase and DNA topoisomerase I (omega protein). It has been shown that these two enzymes are, either directly or indirectly, through the modulation of chromosome superhelicity, involved in a number of DNA transactions, such as replication, recombination and transcription (Wang, 1985). Since we have shown that UV-damage produces variation in DNA tertiary structure to which these enzymes are particularly sensitive, it could be suspected that they may be also involved in DNA repair processes. This possibility was tested by examining the effect of UV-damage on the activity of purified DNA topoisomerases and by studying the UV sensitivity of strains mutated in genes coding for these enzymes. While we did not observe any significant indication of a role of DNA gyrase in UV repair, we instead found indications of a possible involvement of DNA topoisomerase I in cellular recovery from UV-irradiation.

When we used an UV-irradiated DNA as substrate for DNA topoisomerase I, we observed that the apparently minor modification in DNA tertiary structure caused by DNA damage was a strong inhibitor of DNA topoisomerase I activity. When measured in processive reaction conditions, we observed an inhibition even at doses which produced one dimer per molecule. In distributive reaction conditions, we could not measure inhibition, but we observed a change in the mode of action of the enzyme (Pedrini and Ciarrocchi, 1983).

Table 2. Effect of UV light ($50 \ J/m^2$) on DNAQ topoisomerase I deletion mutant

Strains	DM 4100	DM 800
Survival	8×10^{-2}	10^{-3}
Postirradiation recovery of DNA synthesis	30 min	1-- min
Postirradiation DNA degradation		
(TCA insoluble material after 3h)	10%	50%

DM 4100	wt
DM 800	DNA topoisomerase I deletion

DNA topoisomerase I is an essential enzyme, probably involved in the advancement of the replication fork. However, it is possible to isolate viable deletion mutants of this enzyme, because some of its function can be suppressed by compensatory mutation in

DNA gyrase (Di Nardo et al., 1983). Examination of the UV sensitivity of a DNA topoisomerase I deletion mutant (DM 800) has shown that the lack of the enzyme affects UV survival. The introduction of a recA mutation in this strain abolishes this sensitivity suggesting a role of DNA topoisomerase I in the recA dependent recovery from UV damage, as already indicated by studies on UV-induced mutagenesis (Overby and Margolin, 1981). Consistent with this observation is the analysis of two other recA dependent functions, namely recovery of DNA synthesis and stability of chromosomal DNA after UV irradiation. The topoisomerase deficient cells showed a two times longer delay in the recovery of DNA synthesis, in addition to the arrest of DNA synthesis after one cycle of replication. We also found that chromosomal DNA after UV light was more unstable in this strain with respect to wild type. As observed for survival, also these two parameters were only partially impaired by the DNA topoisomerase I deletion (Pedrini, 1984). We think that this intermediate phenotype is probably due to the DNA gyrase mutation which may partially protect cells from DNA topoisomerase I deficiency.

Based on these findings, we propose that DNA topoisomerase I is requested for the advancement of the replication fork and that inhibition of its activity by UV damage is one of the factors which favour excision repair of damage over those cellular processes elicited by the proximity of DNA damage to the replication fork.

Acknowledgements

This work was supported by the C.N.R. grants: P.S. Mutagenesis and P.F. Oncology. Silvia Tornaletti was a recipient of a Fundation A. Buzzati-Traverso fellowship.

References

Broyde, S., Stellman, S., and Hingerty, B. (1980) *Biopolymers*, 9, 1965.
Ciarrocchi, G. and Pedrini A.M. (1982) *J. Mol. Biol.*, 155, 177.
Ciarrocchi, G., Sutherland, B.M. and Pedrini A.M.(1982) *Biochimie*, 64, 665.
Di Nardo, S., Voelkel, K.A., Sternglanz, R., Reynolds, A.E. and Wright, A. (1983) *Cold Spring Harbour Symp. Quant. Biol.*, 47, 779.
Grossman, L., Caron, P.L., Mazur, S.J. and OH, E.Y. (1988) *FASEB J.*, 2, 2696.
Hayes, F.N., Williams, D.L., Ratliff, R.L., Varghese, A.J. and Rupert, C.S. (1971) *J. Am. Chem. Soc.*, 93, 4940.
Kemmink, J., Boelens, R., Koning, T.M.G., Kaptein, R., van der Marel, G.A. and van Boom, J.H. (1987) *Eur. J,.,Biochem.*, 162, 37.
Overby, K.M. and Margolin, P. (1981) *J. Bacteriol.*, 146, 170.
Pearlman, D.A., Hoibrook, S.R., Pirkle, D.H. and Kim, S.H. (1985) *Science*, 227, 1304.
Pedrini, A.M. (1984) In: "Proteins Involved in DNA Replication". U. Hubscher and S. Spadari (eds.) Plenum Publishing Co., pp. 449-454.
Pedrini, A.M. and Ciarrocchi, G. (1983) *Proc. Natl. Acad. Sci. USA*, 80, 1787.
Pedrini, A.M., Tornaletti, S., Barabino, S., Menichini, P., Fronza, G. and Abbondandolo, A. (1989) In: "Cellular Responses to DNA Damage". M. Bignami, E. Dogliotti and J.M. Essigmann (eds.) *Ann. Ist. Super. Sanità*, 25, pp. 91-98.
Pedrini, A.M., Tornaletti, S., Menichini, P. and Abbondandolo, A. (1986). In: "Mechanisms of DNA Damage and Repair: Implication for Carcinogenesis and Risk Assessment". M.G. Simic, L. Grossman and A.C. Upton (eds.) *Basic Life Sci.*, 38, pp. 295–301.

Rao, S.N., Keepers, J.W. and Kollman, P. (1984) *Nucleic Acids Res.*, 12, 4789.
Shafranovskaya, N.N., Trifonov, E.N., Lazurkin, Yu.S., and Franck-Kamenetskii, M.D. (1973) *Nature New Biology*, 241, 58.
Wang, J.C. (1985) Annu. Rev. Biochem., 54, 665.
Wang, J.C.(1979) *Proc. Natl. Acad. Sci. USA*, 76, 200.

147

Excision-Repair Capacity in UV Irradiated Strains of *S. pneumoniae*

A. M. ESTEVENON[1] and N. SICARD[2]

[1] *U.F.R. Sciences Pharmaceutiques Université Paul Sabatier, Toulouse France*

[2] *Centre de Recherche de Biochimie et de Génétique Cellulaires du CNRS Toulouse France*

Abstract

We have used a rapid experiment developed in our laboratory providing an estimate of pyrimidine dimer excision repair *in vivo* in bacterial strains. It is based on the detection of the remaining dimers in a plasmid after UV irradiation and post-incubation of the host cells. We have shown that *Streptococcus pneumoniae* which is naturally deficient in photoreactivation and some SOS-like functions, has the capacity to carry out excision repair when exposed to UV light.

The analysis of the repair abilities and sensitivities to UV irradiation and chemical agents in the wild type and UV sensitive derivative strains, indicates that UV-induced pyrimidine dimers might be repaired through a system similar to the uvr-dependent system in *Escherichia coli*.

Introduction

The bacterial response to DNA damages caused by UV light, ionizing radiation and chemical agents involves the repair of DNA by specific or non-specific enzyme systems (Sancar, A. and Sancar, G.B.,1988).It has been reported that *Streptococcus pneumoniae* is deficient in photoreactivation (Goodgal, S.H. et al., 1957), and some SOS-like functions (Gasc, A.M., et al., 1980). Using a rapid experiment that we have developed in our laboratory, which provides an estimate of pyrimidine dimer excision *in vivo* based on the detection of the remaining dimers in a plasmid after UV irradiation and post incubation of the host cells, we have shown that *S. pneumoniae* has the capacity to carry out excision repair when exposed to UV light (Estevenon, A.M. and Sicard, N., in press). Excision repair can be achieved by a specific DNA glycosylase and AP endonuclease as observed in *Micrococcus luteus* (Carrier, W.L., and Setlow, R.B., 1970; Haseltine, W.A. et al., 1980) or by the uvr ABC excinuclease pathway as observed in *Escherichia coli* (Seeberg, E. et al., 1976; Yeung, A.T. et al., 1983; Sancar, A and Rupp, W.D., 1983).

We have studied the mechanism of pyrimidine dimer excision-repair in *S. pneumoniae* by the analysis of the repair abilities and sensitivities to UV irradiation and chemical agents in wild type and UV sensitive derivative strains.

Materials and Methods

Bacterial Strains

S. pneumoniae R800 was used as parental strain (Lefèvre, J.C. et al., 1979). Strain R402 is a UV sensitive mutant, isolated after transfer of the UV sensitive mutation 402 (Tiraby, G. and Sicard, A.M., 1973) into R800. The strains were transformed with pSP2 plasmid for the estimation of plasmid recovery (Prats, H. et al., 1985). Cultures of *S. pneumoniae* were grown at 37°C in complete medium (Prats H. et al., 1985).

Determination of Excision Repair Capability

The method has been previously described (Estevenon, A.M. and Sicard, N., in press).

Chemicals

cis-Diaminedichloroplatinum (Cis-platinum) and 4-Nitroquinoline-1-oxide (4NQO) were generous gifts from Dr. B. Salles and Dr. J. Alonso respectively. Mitomycin C was purchased from Sigma Chemical Company.

Survival Experiments After Irradiation and Chemical Treatments

Exponentially growing cells (5.10^7 cells/ml) were harvested, washed and resuspended in the mineral buffer of the synthetic medium (Sicard, A.M., 1964).

Samples of 5 ml were irradiated with UV from G 1518 Sylvania germicidal lamp producing an incident dose rate of $2J/m^2/s$. After irradiation the cell suspensions were diluted and plated on agar medium. Cell survival was determined after overnight incubation at 37°C.

Samples of 5 ml were treated with mutagens as follows: after addition of chemical agents at various concentrations, samples of 5 ml were kept in the dark 1 hr at 30°C for mitomycin C and 4NQO, and 1.5 hr for cis-platinum experiments. The survival estimation was performed as described above.

Results and Discussion

Sensitivity to UV Rays

The survival curves of the wild type strain R800 and its UV sensitive derivative R402 are shown in Fig 1A. R402 exhibits a high sensitivity to UV rays.

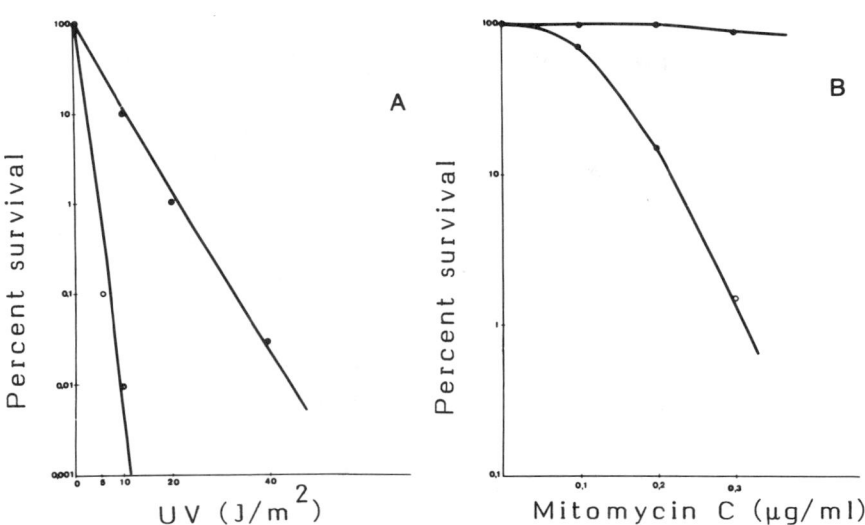

Figure 1. Effect of UV rays (A) and Mitomycin C (B) on survival of *Streptococcus pneumoniae*
∘ R800 (wild type) ∘ R402 (UV sensitive mutant)

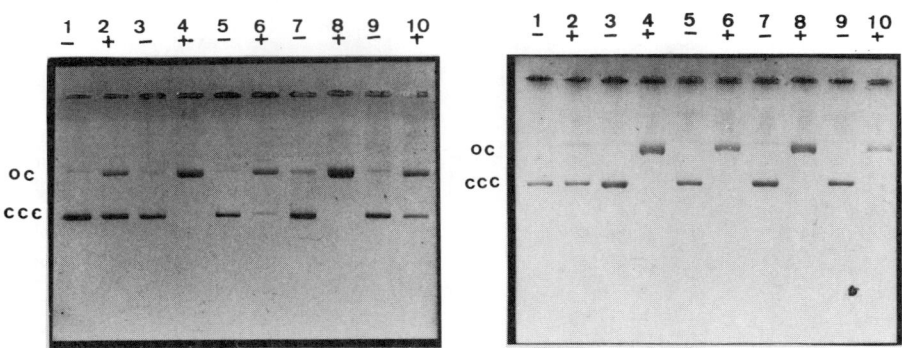

Figure 2. Incision of damaged pSP2 DNA extracted from UV irradiated *S. pneumoniae* R800(A)
and R402(B). The plasmid pSP2 has been either digested (+) or not digested (–) with UV
endonuclease from *M. luteus*. OC: open circular DNA; CCC: covalently closed circular DNA
Lane 1–2: plasmid extracted from unirradiated bacteria
Lane 3–4: irradiation A:20J/m² B:10J/m² no post incubation
Lane 5–6: irradiation A:20J/m², B:10J/m² and post-incubated 1 hr
Lane 7–8: irradiation A:40J/m², B:20J/m² no post incubation
Lane 9–10: irradiation A:40J/m², B:20J/m² and post-incubated 1 hr

Repair Capability of R800 and R402

We examined the repair capability of the wild type strain R800 bearing pSP2 plasmid. The cells were exposed to UV light and incubated 1 hr for recovery. Plasmid DNA was isolated, exposed to UV endonuclease and analysed by agarose gel electrophoresis. Pyrimidine dimers remaining in the plasmid were revealed as sites sensitive to UV endonuclease converting the covalently closed form (CCC) to open circular plasmid (OC).

For an incident dose of 20 or 40 J/m^2, the percentage of OC form over the total DNA decreases whereas some superhelical form appears (Fig. 2A). These data suggest that pyrimidine dimers are repaired. The plasmid recovery is a reflection of the repair capability of the host cell.

The same procedure was applied to the mutant R402. The results of the test following an irradiation of 10 and $20 J/m^2$ are shown in Fig 2B. The cell survival is 0.01% with $10 J/m^2$ (Fig 1A). The absence of excision-repair capacity is revealed by the persistence of the OC form in the plasmid after post-incubation and UV endonuclease treatment.

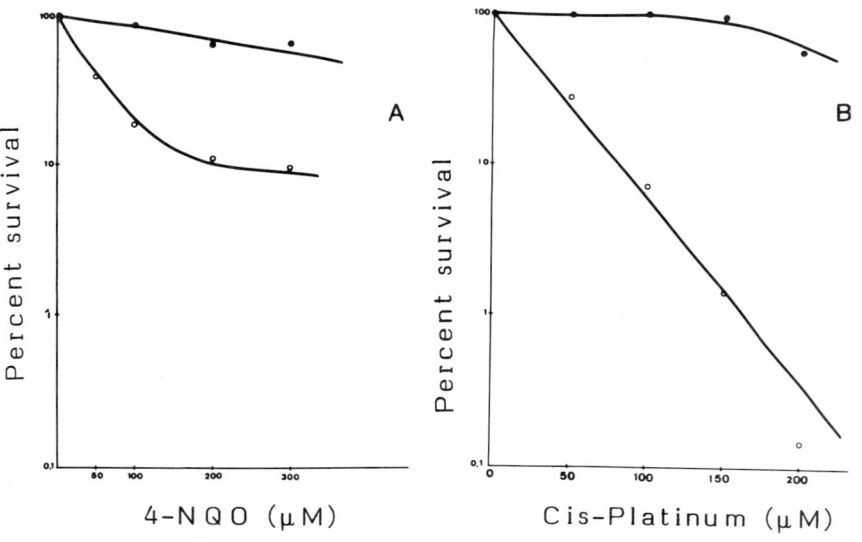

Figure 3. Effect of 4NQO(A) and cis-platinum (B) on survival of S. pneumoniae. ∘ R800 (wild type) o R402 (UV sensitive mutant). After addition of 4NQO or cis-platinum, cell samples were kept at 30°C in the dark 1 hr or 1.5 hr respectively.

Sensitivity to Chemical Agents

Pyrimidine dimers might be removed by different incision mechanisms (1). In M. Iuteus, the DNA glycosylase involves a specific recognition of the pyrimidine dimer (Haseltine, W.A. et al., 1980) whereas the uvrABC excinuclease system in *E. coli* removes through nucleotide excision repair the UV-induced pyrimidine dimers as well as the adducts produced by different chemical mutagens (Kondo, S. et al., 1970; Ikenaga, M. et al., 1975; Husain, I. et al., 1985). Interstrand crosslinks are repaired by a combined action of ABC excinuclease and recombinase (Sinden, R.R. and Cole, R., 1978). Since pneumococcus R402 is not recombinase deficient (Tiraby, G. and Sicard, A.M., 1973), an increase of sensitivity to chemical agents, should reflect nucleotide-excision deficiency. To find out whether a non-specific process promotes the restoration of UV damaged DNA in pneumococcus, we tested the sensitivity of both wild type and repair deficient strains to some chemical agents. Survival curves after mitomycin C, 4NQO and Cis-platinum treatments are shown in Fig 1B, 3A and 3B.

In all cases, the repair deficient mutant is more sensitive to these agents than the wild type strain. This suggests that a mechanism with a wide substrate specificity is lacking in this mutant.

Excision repair in *S. pneumoniae* seems to proceed through a system similar to the uvr-dependent system in *E. coli.*

References

Carrier, W.L. and Setlow, R.B., 1970, Endonuclease from *Micrococcus luteus* which has activity toward UV-irradiated DNA: purification and properties, *J. Bacteriol.* 102, 178–186.

Estevenon, A.M. and Sicard, N., Excision-repair capacity of UV-irradiated strains of *Escherichia coli* and *Streptococcus pneumoniae*, estimated by plasmid recovery, *J. Photochem. Photobiol. B*: Biology Vol. 3 No. 2 (in press).

Gasc, A.M., Sicard, N., Claverys, J.P. and Sicard, A.M., 1980, Lack of SOS repair in *Streptococcus pneumoniae, Mutation Res.*, 70, 157–165.

Goodgal, S.H., Rupert, C.S. and Herriott, R.M., 1957, In: W.D. McElroy and B. Glass, eds. The chemical basis of heredity, Johns Hopkins Press, Baltimore, p. 341–343.

Haseltine, W.A., Gordon, L.K., Lindan, C.P., Grafstrom, R.H., Shaper, N.L. and Grossman, L., 1980, Cleavage of pyrimidine dimers in specific DNA sequences by a pyrimidine dimer DNA glycosylase of *M. luteus, Nature*, 285, 634–641.

Husain, I., Chancy, S.G. and Sancar, A., 1985, Repair of cis-platinum DNA adducts by ABC excinuclease *in vivo* and *in vitro, J. Bacteriol.*, 163, 817–823.

Ikenaga, M., Ichikawa-Ryo, I. and Kondo, S., 1975, The major cause of inactivation and mutation by 4-nitro-quinoline-1-oxide in *Escherichia coli*: Excisable 4NQO purine adducts, *J. Mol. Biol.*, 92, 341–356.

Kondo, S., Ichikawa, H., Iwo, K., and Kato, T., 1970, Base-change mutagenesis and prophage induction in strains of *E. coli* with different DNA repair capacities, *Genetics*, 66, 187–217.

Lefèvre, J.C., Claverys, J.P. and Sicard, A.M., 1979, Donor DNA length and marker effect in pneumococcal transformation, *J. Bacteriol.*, 138, 80–86

Prats, H., Martin, B., Pognonec, P., Burger, C. and Claverys, J.P., 1985, A plasmid vector allowing positive selection of recombinant plasmids in *S. pneumoniae, Gene* 39, 41–48.

Sancar, A. and Rupp, W.D., 1983, A novel repair enzyme: UVR ABC excision nuclease of *E. coli* cuts a DNA strand on both sides of the damaged region, *Cell* 33, 249–260.

Sancar, A., and Sancar, G.B., 1988. DNA repair enzymes, *Ann. Rev. Biochem.*, 57, 29–67.

Seeberg, E., Nissen-Meyer, J. and Strike, P., 1976, Incision of UV-irradiated DNA by extracts of *E. coli* requires three different gene products, *Nature*, 263, 524–525.

Sicard, A.M., 1964, A new synthetic medium for *Diplococcus pneumoniae* and its use for the study of reciprocal transformations at the *amiA* locus. *Genetics,* 50, 31–44.

Sinden, R.R. and Cole, R., 1978, Topography and kinetics of genetic recombination in *E. coli* treated with psoralen and light, *Proc. Natl. Acad. Sci. USA*, 75, 2373–2377.

Tiraby, G. and Sicard, A.M., 1973, Integration efficiency in DNA-induced transformation of pneumococcus. II Genetic studies of mutant integrating all the markers with a high efficiency, *Genetics*, 75, 35–48.

Yeung, A.T., Mattes, W.B., Oh, E.Y. and Grossman, L., 1983, Enzymatic properties of purified *E. coli* uvr ABC proteins, *Proc. Natl. Acad. Sci. USA*, 80, 6157–6161

UV-Inducible Repair in Yeast

FRIEDERIKE ECKARDT-SCHUPP, ALFRED AHNE,
SABINE OBERMAIER AND SUSANNE WENDEL

GSF – Institut für Strahlenbiologie
Ingolstädter Landstr. 1
D-8042 Neuherberg, F.R.G.

Introduction

DNA is a highly dynamic material which undergoes numerous structural alterations which may or may not be accompanied by informational changes. In pro- and eukaryotic cells numerous endo- and exogenous sources are known to cause DNA damage: DNA-related processes such as replication, transcription, recombination etc. involve DNA breakage, and countless genotoxic agents induce many types of DNA damage. Opposed to these sources of genetic instability are DNA protective enzymatic processes, for example, mechanisms accounting for replication fidelity and DNA repair processes, which guarantee a steady state of genetic flexibility in cells. Genes controlling repair of spontaneous ("endogenous") and induced DNA damage also undertake functions in the control of replication, recombination, mutagenesis (for review, see Haynes and Kunz, 1981). Consequently, studies of repair processes frequently involve investigations of genetic effects.

Furthermore, repair-deficient mutants have pleiotropic phenotypes: besides being sensitive towards radiation and/or chemical agents they are either hypo- or hypermutagenic, hypo- or hyper-recombinogenic, etc. Some repair processes in yeast are probably regulated according to demand, and future research will show if they are part of a general stress responsive system. In *E. coli* such a system, the so-called SOS system has been analyzed extensively (see for review, Walker, 1985). Also for mammalian cells a complex stress response has been described, however, so far it does not include genes controlling repair processes (Karin and Herrlich, 1988).

Indirect evidences for inducibility of repair genes

For many years in the yeast *Saccharomyces cerevisiae*, a SOS-like system had been searched for. Various approaches have been undertaken, e.g. split dose experiments, inhibition of protein synthesis with cycloheximide after irradiation studying various

Photobiology, Edited by E. Riklis
Plenum Press, New York, 1991

endpoints, and dose response studies of mutation kinetics applying adequate mathematical treatment (for review, see Siede and Eckardt, 1984). The results of these different experimental designs were in favour of the idea of inducible repair, in no way could they be considered proof. However, these experiments have their value in supplying a biological basis for molecular analysis which is possible today, but was not available in former times.

Early molecular biological trials in yeast, e.g. search for inducible proteins (Schwencke and Moustacchi, 1982a,b; Rolfe, 1985a; Angulo et al., 1985) and hunting for inducible genes being identified either by their UV-responsive promoters (Ruby et al., 1983) or by their UV-enhanced level of transcripts as proven by differential colony hybridization (McClanahan and McEntee, 1984; Rolfe, 1985b) also spoke for the existence of genes responsive to DNA damaging agents in yeast. Most of the inducible proteins or genes were not identified; however, there are some interesting exceptions. One protein which had been characterized by its immunological cross-reactivity towards antibodies for the RecA protein of *E. coli* (Angulo et al., 1985), has been identified in the meantime as ribonucleotide reductase (Elledge and Davis, 1987).

These authors have demonstrated the UV inducibility of at least two of its controlling genes, *RNR1* and *RNR2*, coding for the large and the small subunit, respectively. The second protein of interest is protease (proteinase) B which is damage-inducible in repair competent yeast strains, but not in a *rad6-1* mutant (Schwencke and Moustacchi, 1982a,b). This finding is remarkable considering that the *RAD6* gene codes for a ubiquitin-conjugating enzyme E2. In yeast — as in other organisms — the coupling of ubiquitin to certain proteins supposedly results in their selective proteolysis (Jentsch et al., 1987). Whereas the cell cycle regulation and responsiveness towards DNA damage of the rad6 gene itself is still a matter of controversy (Kupiec and Simchen, 1986; Reynolds et al., 1985), the DNA damage inducibility of one of the ubiquitin genes (*UBI4*) was proven for a repair competent and a *rad6-1* strain. The *UBI4* gene was identified as one of the DDR (DNA damage responsive) genes isolated due to its UV-enhanced transcription (Treger et al., 1988). More information is required for the understanding of the role of ubiquitination for repair and induced mutagenesis which is deficient in *rad6* mutants and defective in numerous other mutants belonging to the *RAD6* epistasis group (for review, see Lawrence, 1982). Hence, one may speculate about a possible relationship between ubiquitination and inducibility of those genes (Siede, 1988).

At least five of the clones which had been isolated on the basis of their UV-enhanced expression turned out to be segments of transposable elements of yeast, so-called TY elements. It was shown that DNA damaging agents enhanced the transcription of TY elements considerably (McClanahan and McEntee, 1984; Rolfe, 1985b), and TY transposition as well (Morawetz, 1987). One may speculate whether UV-inducible genes and TY elements in yeast have controlling sequences in common as it has been proven for mammalian cells: stress-responsive genes (collagenase, methallothionein), cellular oncogenes (*c-fos*) and retrotransposons like *HIV-1* possess identical binding sites for transcription factors, "responsive elements", in their 5' control regions (Karin and Herrlich, 1988). It is not understood so far why none of the inducible genes (see Table 1) were isolated by any of these procedures but rather several TY fragments. Possibly, the level of UV-induced TY RNA is much higher than the rather moderately enhanced RNA levels of those genes.

Direct evidence for inducible genes controlling repair

Meantime, a number of yeast genes participating in the control of repair has been cloned and underly detailed molecular analysis regarding their structure, function and regulation. Those genes which are induced by DNA damaging agents are listed in Table 1.

Table 1. Cloned genes of *Saccharomyces cerevisiae* which are inducible by DNA
damaging agents

Gene	Gene product	Pathway	Cell cycle control	Agent	Damage inducibility lacZ fusions	North Analysis
RAD2	?	EXC	–	UV, 4NQO	4–6 fold	5 fold
RAD51	?	REC	–	X-rays	n.d.	3–4 fold
RAD54	?	REC	–	X-rays UV, MMS	3–12 fold	3–5 fold
CDC8	thymidilate synthetase	EXC/EPR	+	UV, 4-NQO	n.d.	2,5 fold
CDC9	ligase	EXC/REC	+	UV, γ-rays	2–3 fold	3–10 fold
POL 1	polymerase I	n.d.	+	UV	10-fold	20-fold
RNR1	ribonucleotide reductase (large subunit)	n.d.	+	UV, γ-rays MMS	10-fold	18-fold
RNR2	ribonucleotide reductase (small subunit)	n.d.	+	UV, γ-rays, MMS	n.d.	18-fold

The data has been compiled from the literature cited in the text. The inducibility of the genes by UV, X-rays, γ-rays, MMS (methyl-methane-sulfonate), or 4-NQO (4' nitroquinoline-oxide) has been determined by experimenting either for β-galactosidase activity of suitable lacZ fusions or for enhanced transcript levels by Northern analysis. Abbreviations: EXC (excision repair, *RAD3* group); REC (recombination repair, *RAD52* group); EPR (error-prone repair, *RAD6* group); n.d. (non-determined).

So far, it has not been proven whether the triggering signal in a UV, chemically or otherwise treated cell is DNA damage per se as is suggested by the frequently-used terminus "DNA damage-inducible". It may well be that disturbances in the DNA precursor pools or damages other than at the DNA level initiate the chain of signal transduction in the damaged cells.

The "damage-inducible" genes fall into two groups. One group is represented by three *RAD* genes: *RAD2* (Robinson et al., 1986), *RAD51* (Aker and Mortimer, 1986), and RAD54 (Cole et al., 1987). These repair genes had originally been identified by the characterization of radiation sensitive mutants. They belong to the *RAD3* and *RAD52* epistasis groups (Haynes and Kunz, 1981), their gene products are not known, all of them are not cell cycle controlled. Both *RAD2* and *RAD54* are induced in meiosis as well, but they do not respond to heat shock. *RAD* genes seem to have low expression levels; even after UV induction the transcription is only moderately enhanced (Friedberg, 1988). So far it is not known whether some joint regulation exists, though *RAD2* and *RAD54* have a controlling sequence element in common (W. Siede, pers. commun.). The other group contains five genes which have been characterized by their joint control in the cell cycle and their response to DNA damaging agents: *CDC8* (Sclafani and Fangman, 1984; Elledge and Davis, 1987), *CDC9* (Peterson et al., 1985; Johnson et al., 1986), *POL1* (Johnston et al., 1987), *RNR1* (Elledge and Davis, 1988) and *RNR2* (Elledge and Davis, 1987). The products of these genes are known: either they control functions in the DNA precursor synthesis like *CDC8* and the *RNR* genes, which code for thymidylate kinase and, respectively, for subunits of ribonucleotide reductase, or in DNA synthesis itself, like *CDC9* and *POL1*, which code for ligase and, respectively, for polymerase I. *CDC8* and *CDC9* have been attributed to epistasis groups of repair (Haynes and Kunz, 1981), for the others it has not been investigated so far if their mutant alleles cause sensitivity towards radiation and chemicals. The literature regarding the UV inducibility of *POL1* is controversial: Johnson et al. (1987) have given clear evidence for the UV inducibility, whereas Elledge and Davis (1987) could not find it under the experimental conditions they used. It is likely that these genes have common control elements on the level of transcription for the cell cycle control *and* for repair, since DNA synthesis is required for both processes (Elledge and Davis, 1987). The authors propose inducibility by DNA damaging agents for *CDC21* which is allelic to *TMP1* and codes for thymidylate synthetase. This gene is coordinately regulated with *CDC8*, *CDC9* and *POL1* and other genes controlling DNA precursor synthesis (Johnston et al., 1987; McEvan et al., 1987), but so far nothing is known regarding its presumed repair functions. Unpublished investigations brought no evidence for any particular sensitivity towards radiation (UV or X-rays) or a few chemicals tested in *tmp1* mutants under stress or over-supply of deoxythymidine monophosphate (M. Brendel, pers. communication; F. Eckardt-Schupp, unpublished results).

Biological and molecular analysis of the *REV2 function*

The *REV2* gene seems to meet the predictions of Elledge and Davis as far as can be guessed from the biological analysis carried out so far (Siede et al., 1983a,b; Siede and Eckardt, 1986a,b; Siede and Eckardt-Schupp, 1986a,b), and the molecular analysis being in progress. The *REV2* gene belongs to the *RAD6* epistasis group; *REV2* mutants are defective in the UV-induced reversion of certain ochre and frameshift alleles and sensitive towards various agents like UV, X-rays and nitrogen mustard (Lemontt, 1971; Siede and Brendel, 1981; Lawrence and Christensen, 1978; Lawrence et al., 1984). Remarkable for the *rev2-1* diploid were linear dose response curves for the reversion of certain *cyc1* alleles, whereas the corresponding wildtype strain had quadratic or linear-quadratic mutation kinetics (Lawrence and Christensen, 1978). Trusting in the mathematical analysis of the dose response relationship for induced mutation frequencies, M(x) [mutants per survivor], by Haynes et al (1985), the *REV2* gene was regarded to be a good candidate for an inducible repair gene controlling mutation processing as well. Biological and molecular analysis was carried out in order to test the following proposals: the inducible *REV2* function is responsible a) for the shoulder of the survival curve, and b) for the quadratic branch of the mutation kinetics obtained in selective systems for the reversion of ochre alleles like *arg4-17, his5-2, ade2-1*. Thus, it was postulated that the exponential slope of the survival curve and the purely linear dose dependence of M(x) in the *rev2* mutant is due to the defective *REV2* function.

A particular advantage for addressing these questions was a temperature sensitive *REV2* mutant. For this mutant, 23°C is the permissive and 36°C the non-permissive temperature for UV sensitivity and UV mutability, but not for growth. The mutant phenotype is exhibited without considerable delay when the cells are shifted from 23°C to 36°C; obviously, the *REV2* protein undergoes an immediate conformational change impeding its function. Temperature-shift experiments were carried out according to the following scheme: the cells, UV-treated or controls, were suitably plated and kept for various times at the permissive temperature (23°C) before being shifted to 36°C where they were incubated for up to 5 days. Two controls at both temperatures were made, and the biological endpoints, survival and mutation frequencies of various ochre alleles were tested. This experimental design allowed conclusions concerning the time dependence of the *REV2* controlled processes of repair and mutation fixation though colony formation was tested only. The experimental set-up also allowed the usage of specific inhibitors like cycloheximide which blocks protein synthesis, and hydroxyurea and aphidicolin which block DNA replication; thus, the influence of protein- and DNA synthesis on the *REV2* function was analyzed as well.

The results can be summarized as follows: Concentrating on the *REV2* controlled repair of UV damage in haploid cells by analyzing conventional survival curves and temperature shift experiments a considerable difference in the effect of the *REV2* mutation in replicating as compared to stationary cells was found: 1) Logarithmically growing *REV2* cells were twice as UV-sensitive as compared to stationary cells. This was in contrast to repair competent strains in general, and the *REV2* wildtype strain in particular, and numerous repair deficient mutants, where dividing cells were considerably more resistant than stationary cells. 2) *REV2* dependent repair was significantly faster in replicating (6–7h) than in stationary cells (8–10h) as was proven in temperature shift experiments. 3) *REV2* dependent repair could be blocked completely by 5 µg/ml cycloheximide in stationary cells, but slowed down only by concentrations up to 200 µg/ml cycloheximide in replicating cells. These results supported the proposal of replicating cells having a much higher level of Rev2 protein than stationary cells. In stationary cells, however, the low level of Rev2 protein was enhanced after UV irradiation by *de novo* synthesis. It was proposed that this effect may be due to regulation of the *REV2* gene on the transcriptional level, though the molecular evidence was missing at that time.

The analysis of the *REV2* controlled UV-induced mutagenesis by temperature shift experiments and studies of mutation kinetics, gave similar results. The supposedly higher level of Rev2 protein in replicating as compared to stationary cells was reflected by the linear mutation kinetics and high mutation frequencies of replicating cells (Eckardt-Schupp, unpublished). In stationary cells, the linear-quadratic dose-response curve was in accord with the proposed inducibility of the Rev2 protein, 5 µg/ml cycloheximide in the post-irradiation incubation medium eliminated the quadratic branch of the mutation kinetics of various ochre alleles as well as the increase in mutation frequencies in temperature shift experiments, e.g. was equal in its effect to the *REV2*^{ts} mutation at non-permissive temperature. Of special interest was the finding that excision competence is required for the inducibility of the Rev2 gene. In excision deficient strains, the *REV2* dependent mutation fixation was extremely slow (>12 h as compared to 2–4h for excision competent strains) and mainly post-replicative, though not replication dependent. The mutation kinetics were linear, and it was proposed that excision deficient strains have the constitutive level of *REV2* protein available only. In conclusion, from the biological analysis, it was postulated that the *REV2* gene controls function in repair and mutation processing and furthermore some non-essential function in DNA replication, possibly the control of replication fidelity as indicated by the mutator phenotype of *REV2* mutants. The *REV2* gene may be cell cycle controlled and inducible by DNA damaging agents in excision repair competent cells.

Molecular analysis has not yet been carried out far enough to support all these proposals, however, the *REV2* gene has been cloned (Siede and Eckardt-Schupp, 1986b). By directly ligating genomic yeast DNA restriction fragments into a YCp50 vector, a clone with a 6,7 kb insert was isolated which complemented the *REV2* phenotype (exponential survival curve, low UV mutability, linear mutation kinetics) yielding the *REV2* wildtype phenotype (survival curve with shoulder, normal UV mutability, linar-quadratic mutation kinetics). A 3,6 kb SALI fragment of the insert was hybridized to an agarose gel with well-separated individual yeast chromosomes obtained by pulsed gel electrophoresis. Both, the SALI fragment and a *ASP5* probe showed a signal at the chromosome XII, to which the *ASP5* and the *REV2* gene have been mapped in close distance by conventional genetic techniques (Mortimer and Schild, 1985).

Preliminary results of Northern analysis have shown that the transcript hybridizing with the SALI fragment is expressed in logarithmically growing cells in amounts which favor a straightforward analysis. Similarly, high levels of transcription have been described for cell cycle controlled genes, whereas "real" *RAD* genes so far analyzed seem to have only few transcripts per cell (Friedberg, 1988). Dot blot analysis has been undertaken to prove the notion of UV-enhanced expression of the *REV2* transcript in a repair competent strain and in an excision deficient mutant. Preliminary results seem to be in favor with the postulates derived from the biological results. Sequencing of the *REV2* gene is in progress. A precise sequence analysis is expected to indicate homologies

160

with sequence elements in the promoter region considered to be important for the regulation of repair and/or cell cycle controlled genes in yeast and possibly stress response genes in mammalian cells.

References

Aker, M. and Mortimer, R.K. (1986), Transcriptional patterns and nucleotide sequence of *RAD51* and its flanking regions. *Yeast* 2, (Spec. Iss.), S3.

Angulo, J.F., Schwencke, J., Moreau, P.L. and Moustacchi, E. (1985), A yeast protein analogous to *Escherichia coli* RecA protein whose cellular level is enhanced after UV irradiation. *Mol. Gen. Genet.* 201, 20–24.

Cole, G.M., Schild, D., Lovett, S.T. and Mortimer, R.K. (1987), Regulation of *RAD54*- and *RAD52*-lacZ gene fusions in *Saccharomyces cerevisiae* in response to DNA damage. Mol. Cell. Biol. 7, 1078–1084.

Elledge, S.J. and Davis, R.W. (1987), Identification and isolation of the gene encoding the small subunit of ribonucleotide reductase from *Saccharomyces cerevisiae*: DNA damage-inducible gene required for mitotic viability. *Mol. Cell. Biol.* 7, 2783–2793.

Elledge, S.J. and Davis, R.W. (1988), Identification of the genes encoding ribonucleotide reductase from yeast, a cell-cycle regulated, DNA-damage inducible enzyme required for mitotic viability. *Yeast* 4 (Spec. Iss.), S124.

Friedberg, E.C. (1988), Deoxyribonucleic acid repair in the yeast *Saccharomyces cerevisiae*. *Microbiol. Revs.* 52, 70–102.

Haynes, R.H. and Kunz, B.A. (1981), DNA repair and mutagenesis in yeast. In The Molecular Biology of the Yeast *Saccharomyces cerevisiae*: Life cycle and Inheritance (edited by J.N. Strathern, E.W. Jones and J.R. Broach), pp. 371–414, Cold Spring Harbor Laboratory, Cold Spring Harbor, N.Y.

Haynes, R.H., Eckardt, F. and Kunz, B.A. (1985), Analysis of non-linearities in mutation frequency curves. *Mutat. Res.* 15, 51–59.

Jentsch, S., McGrath, J.P. and Varshavsky, A. (1987), The yeast DNA repair gene *RAD6* encodes a ubiquitin-conjugating enzyme. *Nature*, 329, 131–134.

Johnson, A.L., Barker, D.G. and Johnston, L.H. (1986), Induction of yeast DNA ligase genes in exponential and stationary phase cultures in response to DNA damaging agents, *Curr. Genet.*, 11, 107–112.

Johnston, L.H., White, J.H.M., Johnson, A.L., Lucchini, G. and Plevani, P. (1987), The yeast DNA polymerase I transcript is regulated in both the mitotic cell cycle and in meiosis and is also induced after DNA damage. *Nucl. Acids Res.*, 15, 5017–5029.

Karin, M. and Herrlich, P. (1988), Cis- and trans-acting genetic elements responsible for induction of specific genes by tumor promoters, serum factors and stress. In Genes and Signal Transduction in Multistage Carcinogenesis (edited by N.H. Colburn), Marcel Dekker, Inc., New York.

Kupiec, M. and Simchen, G. (1986), Regulation of the *RAD6* gene of *Saccharomyces cerevisiae* in the mitotic cell cycle and in meiosis. *Mol. Gen Genet.* 203, 538–543.

Lawrence, C.W. (1982), Mutagenesis in *Saccharomyces cerevisiae*. *Adv. Genet.*, 21, 173–254.

Lawrence, C.W. and Christensen, R. (1978), Ultraviolet-induced reversion of *cyc1* alleles in radiation sensitive strains of yeast. II. *REV2* mutant strains, *Genetics*, 90, 213–226.

Lawrence, C.W., O'Brien and Bond, J. (1984), UV-induced reversion of *his4* frameshift mutations in *rad6, rev1* and *rev3* mutants of yeast. *Mol. Gen Genet.*, 195, 487–490.

Lemontt, J.F. (1971), Mutants of yeast defective in mutations induced by ultraviolet light, *Genetics*, 68,21–33.

McClanahan, T.A. and McEntee, K. (1984), Specific transcripts are elevated in *Saccharomyces cerevisiae* in response to DNA damage. *Molec. Cell. Biol.*, 4, 2356–2363.

McIntosh, E.M., Gadsden, M.G., Haynes, R.H. (1986), Transcription of genes encoding enzymes involved in DNA synthesis during the cell cycle of *Saccharomyces cerevisiae*,

Mol. Gen. Genet., 204, 363–366.

Morawetz, C. (1987), Effect of irradiation and mutagenic chemicals on the generation of *ADH2*-constitutive mutants in yeast. Significance for the inducibility of Ty transposition. *Mutat. Res.*, 177, 53–60.

Mortimer, R.K. and Schild, D. (1985), Genetic map of *Saccharomyces cerevisiae*, Ed. 9. *Microbiol. Rev.*, 49, 181–212.

Peterson, T.A., Prakash, L., Osley, M.A. and Reed, S.I. (1985), Regulation of the CDC9, the *Saccharomyces cerevisiae* gene that encodes DNA ligase. *Mol. Cell. Biol.* 5, 226–235.

Reynolds, P., Weber, S. and Prakash, L. (1985). *RAD6* gene of *Saccharomyces cerevisiae* encodes a protein containing a tract of 13 consecutive aspartates. *Proc. Natl. Acad. Sci. U.S.A.*, 82, 168–172.

Robinson, G.W., Nicolet, C.M., Kalainov, D., Friedberg, E.C. (1986), A yeast excision-repair gene is inducible by DNA damaging agents. *Proc. Natl. Acad. Sci. U.S.A.*, 83, 1842–1846.

Rolfe, M. (1985a), UV-inducible proteins in *Saccharomyces cerevisiae*, *Curr. Genet.*, 9, 529–532.

Rolfe, M. (1985b), UV-inducible transcripts in *Saccharomyces cerevisiae*, *Curr. Genet.*, 9, 533–538.

Ruby, S.W., Szostak, J.W. and Murray, A.W. (1983), Cloning regulated yeast genes from a pool of lacZ fusions. In Methods of Enzymology 101, Recombinant DNS, Part C (Edited by R. Wu, L. Grossman and K. Moldave), pp. 253–269, Academic Press, New York.

Schwencke, J. and Moustacchi, E. (1982a), Proteolytic activities in yeast after UV irradiation. I. Variation in proteinase levels in repair proficient *RAD⁺* strains. *Mol. Gen. Genet.*, 185, 290–295.

Schwencke, J. and Moustacchi, E. (1982b). Proteolytic activities in yeast after UV irradiation. II. Variation in proteinase levels in mutants blocked in DNA repair pathways. *Mol. Gen. Genet.*, 185, 296–295.

Sclafani, R.A. and Fangman, W.L. (1984), Yeast gene *CDC8* encodes thymidylate kinase and is complemented by herpes thymidine kinase gene *TK*. *Proc. Natl. Acad. Sci. U.S.A.*, 81, 5821–5825.

Siede, W. (1988), The *RAD6* gene of yeast: a link between DNA repair, chromosome structure and protein degradation, *Radiat. Environ. Biophys.*, 27, 277–286.

Siede, W. and Brendel, M. (1981), Isolation and characterization of yeast mutants with thermoconditional sensitivity to the bifunctional alkylating agent nitrogen mustard. *Curr. Genet.*, 4, 145–149.

Siede, W. and Eckardt, F. (1984), Inducibility of error-prone DNA repair in yeast, *Mutat. Res.*, 129, 3–11.

Siede, W. and Eckardt, F. (1986a), Analysis of mutagenic DNA repair in a thermoconditional mutant of *Saccharomyces cerevisiae*. III. Dose-response pattern of mutation induction in UV-irradiated rev2ts cells. *Mol. Gen. Genet.*, 202, 68–74.

Siede, W. and Eckardt, F. (1986b), Analysis of mutagenic DNA repair in a thermoconditional mutant of *Saccharomyces cerevisiae*. IV. Influence of DNA replication and excision repair on rev2 dependent UV-mutagenesis and repair. *Curr. Genet.*, 10, 871–878.

Siede, W. and Eckardt-Schupp, F. (1986a). A mismatch repair-based model can explain some features of UV mutagenesis in yeast. *Mutagenesis*, 1, 471–474.

Siede, W. and Eckardt-Schupp, F. (1986b), DNA repair genes of *Saccharomyces cerevisiae*: complementing rad4 and rev2 mutations by plasmids which cannot be propagated in *Escherichia coli*. *Curr. Genet.*, 11, 205–210.

Siede, W., Eckardt, F. and Brendel, M. (1983a), Analysis of mutagenic DNA repair in a thermoconditional mutant of *Saccharomyces cerevisiae*. II. Influence of cycloheximide on UV-irradiated exponentially growing phase rev2s cells. *Mol. Gen. Genet.*, 190. 413–416.

Treger, J.M., Heichman K.A. and McEntee, K. (1988) Expression of the yeast *UBI4* gene increases in response to DNA-damaging agents and in meiosis. *Molec. Cell. Biol.*, 8, 1132–1136.

Walker, G.C. (1985). Inducible DNA repair systems. *Ann. Rev. Biochem.*, 54, 425–457.

162

Mutant I12X86 Versus Wild-Type *Lac* Repressor Photochemistry

MÉLANIE SPOTHEIM-MAURIZOT, FRANÇOISE CULARD,
PHILIPPE GREBERT, JEAN-CLAUDE MAURIZOT AND
MICHEL CHARLIER.

Centre de Biophysique Moléculaire, C.N.R.S.
1A, Avenue de la Recherche Scientifique
45071, Orléans Cedex 02
France

Introduction

The metabolism of lactose in *E.coli* is negatively regulated through the tight interaction between a protein, the *lac* repressor, and a sequence of DNA, the *lac* operator. This binding prevents the expression of the structural genes of the lactose operon. Binding of an inducer (Isopropyl-β-D-thiogalactoside, IPTG used in most experiments), releases the operator allowing thus the expression of the genes (Muller-Hill, 1975; Bourgeois and Pfahl, 1976).

As many other DNA-interacting proteins, the repressor can bind also non-operator DNA, with a reduced affinity. This non-specific binding is not modulated by the inducer (von Hippel et al., 1974; Maurizot et al., 1974).

The repressor is organized in domains, that can be separated after proteolytic cleavage (Geisler and Weber, 1977). The tetrameric core (with each subunit composed of amino-acids 60-360) carries the binding site of the inducer. It contains the eight (two per protomer) tryptophanyl residues of the protein at positions 201 and 220 (Farabaugh, 1978). It has been shown that the four headpieces (N-terminal regions of the protein) are the major part of the DNA-interacting domain (Jovin et al. 1977; Ogata and Gilbert, 1979).

The use of altered repressor molecules with increased affinity for DNA has been a great help for studies of this regulation system (Jobe and Borgeois, 1972; Betz and Sadler, 1976; Pfahl, 1976, 1981a; 1981b; Schmitz et al., 1978; O'Gorman et al. 1981; Pfahl and Hendricks, 1984; Grebert and Maurizot, 1986; Maurizot and Grebert, 1988). The tight binding I12X86 *lac* repressor results from two mutations, the X86 mutation (Ser 61 → Leu) and the I12 mutation (Pro 3 → Tyr). When compared to wild-type repressor, this mutant shows an increased non-specific affinity for DNA, as well as large increased affinity for operator and sequences related to the operator. It has also been

Photobiology, Edited by E. Riklis
Plenum Press, New York, 1991

shown that the non-specific DNA binding of this modified repressor is IPTG dependent, contrary to the behaviour of the wild-type one (Schmitz et al., 1978) .

However, no physico-chemical method such as NMR, circular dichroism and fluorescence, could point out any differences between the wild-type and this mutant repressor except in their binding properties (unpublished results). We report here a study which uses photooxydation of tryptophan residues of *lac* repressor, as a probe to reveal such differences.

In a previous work on wild-type repressor, we have shown that U.V. irradiation (in the wavelength range 250-400 nm) induces photooxydation of tryptophan residues, leading to formation of N-formylkynurenine and consequently to the decrease of fluorescence of the protein (Spodheim-Maurizot et al., 1985). The extent of degradation of the two tryptophan residues, TRP 220 and TRP 201, can be specifically determined by a method based on the spectrofluorometric titration of the denatured protein (Spodheim-Maurizot et al., 1985). We have deduced by using this method the proximity of TRP 220 to the inducer binding site (Spodheim-Maurizot et al., 1985) and shown that only binding of operator DNA to *lac* repressor induces changes in the environment of the tryptophan residues (mainly TRP 220), but not that of non-specific DNA (Spodheim-Maurizot et. al., 1987). Thus, only specific binding of operator DNA to the headpieces, can induce conformational modifications in the vicinity of the site of the inducer situated in the core. Since for wild-type repressor IPTG binding affects only specific and not non-specific DNA complexation, this result is consistent with allosteric behaviour of the repressor.

In the present study, we applied the same type of photochemical approach to the I12X86 mutant, and compare the results to those obtained with wild-type repressor alone, in specific complexes with operator DNA, and in non-specific complex with *E.coli* DNA.

Materials and Methods

The I12X86 protein was purified as described by Rosenberg et al (1977) except that a gradient ranging from 0.12 M to 0.30 M potassium phosphate was applied to the phosphocellulose. This was done to take into account the tighter binding of the double mutant. The phosphocellulose column was followed by gel filtration on an Ultrogel AcA 34 (IBF) column. The purity of the I12X86 repressor was more than 95%, as judged by SDS-PAGE. Concentrations were determined from absorption measurements using a molecular extinction coefficient of $\varepsilon_{280} = 90800$ M^{-1}cm^{-1} for the I12X86 repressor tetramer.

Lac operator DNA fragments were isolated from plasmid pBR345, as previously described (Culard and Maurizot, 1981). The fragment is a duplex of 25 base pairs, flanked on either sides by four unpaired bases corresponding to the Eco R1 cut. The structural study of this fragment has been previously described (Culard and Maurizot, 1982). *E.coli* total DNA was prepared in our laboratory by the classical method.

All solutions were made in a buffer 10 mM Tris/HCl, 10 mM KCl, pH 7.25. In our experiments, repressor concentrations were typically 1.25×10^{-6} M. All samples were air-saturated.

Ultraviolet irradiations was carried out using a 200 Watts mercury lamp. A filter system isolated an irradiation band from 295 to 400 nm. This wavelength range were chosen in order to avoid large photolysis of DNA.

Fluorescence measurements were performed with a Kontron SFM 25 spectrofluorimeter. Spectra were recorded at 20°C, from 280 to 480 nm, with an excitation wavelength of 290 nm.

Spectrofluorimetric titrations were performed after denaturation of the protein by 6 M guanidinium chloride. The pH was adjusted using 0.05 M NaOH, and 1% acetic acid. The determination of the total number of intact TRP and of the fraction of TRP 201 and TRP 220 remaining intact is performed as previously described (Spodheim-Maurizot et al., 1985).

Results and Discussion

I12X86 repressor irradiated alone

Upon irradiation of the mutant repressor in the wavelength range 295-400 nm, the fluorescence emission of the repressor in the tryptophan band (excitation 290 nm) decreases by up to 70% after about 40 min. irradiation, without any change in the shape

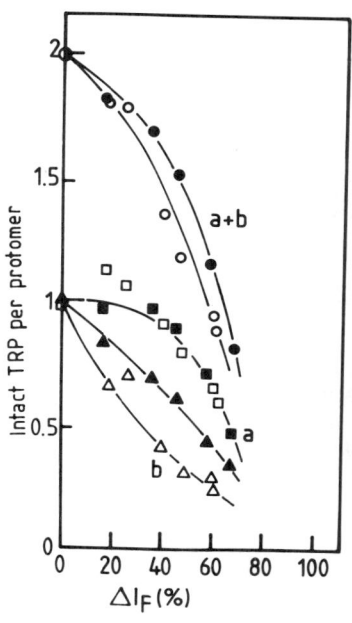

Figure 1. Variation of the total amount of intact tryptophan (a+b) – O – O –, of intact TRP 220 (a) – ▢ – ▢ –, and of intact TRP 201 (b) – Δ – Δ –, remaining in the protein, as a function of the percent decrease of fluorescence intensity of the native protein. Irradiations in the wavelength range 295-400 nm. Full symbols: wild-type repressor, open symbols: mutant I12X86 repressor.

of the spectrum. The fraction of remaining intact tryptophanyl residues 201 and 220 was determined as previously described and plotted versus loss of the protein fluorescence (Spodheim-Maurizot et al., 1985).

When comparing these results (Fig.1) to those obtained for wild-type repressor, one observes a marked difference in the photodestruction of TRP 201, no difference in that of TRP 220, and as a consequence, a difference in the total amount of intact tryptophan. We can conclude that the substitution of two amino-acids in the headpiece of I12X86 repressor may induce a structural modification of the whole protein, which changes the environment of TRP 201 in the core, and thus modifies the efficiency of the photodestruction reaction. This is the first time that a structural difference between the two repressors is observed.

Irradiation of I12X86 repressor-DNA complexes.

Two types of complexes were studied: non-specific and specific complexes. Non-specific complexes are formed after mixing of repressor with total *E.coli* DNA, in a molar ratio of 96 base pairs per protein. With DNA fragments of 25 base pairs bearing the *lac* operator sequence, specific complexes are formed. The molar ratio is two operator

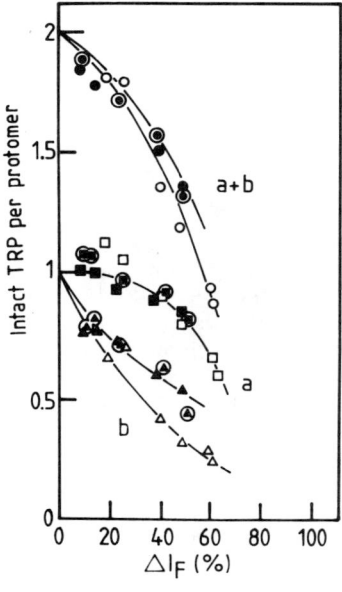

Figure 2. Variation of the amount of intact tryptophan remaining in the protein as a function of the fluorescence decrease ΔI_F. Mutant repressor irradiated in the wavelength range 295-400 nm. Open symbols: repressor irradiated alone, full symbols: repressor irradiated in a specific complex O_2R, surrounded full symbols: repressor irradiated in non-specific complex with *E.coli* DNA.

fragments per repressor molecule, to obtain the O_2R specific complex described by Culard *et al* (1987). In this complex, the two operator binding sites of the repressor are saturated, leading to identical states for the four protomers.

The determination of the fraction of intact TRP 201 and TRP 220 in the mutant repressor irradiated in the two types of complexes, leads to the results shown in Fig.2.

One observes a marked difference between the photodestruction of TRP 201 (and none for TRP 220) of I12X86 repressor irradiated alone and that of repressor irradiated in complexes with DNA. Both specific and non-specific bindings modify in the same manner the photolysis of TRP 201, and thus its environment. This result is different from that obtained with wild-type repressor, in which case only specific binding modified the environment of a tryptophan residue (TRP 220).

The similarity of effects of specific and non-specific bindings of DNA on the mutant repressor photochemistry, is in good agreement with several other results showing similarities between the two types of binding for I12X86 repressor. We have shown that the number of ion pairs liberated by complexation of mutant repressor to operator DNA and to non-specific DNA is the same (6 ion pairs) (Grebert and Maurizot, 1986). Thermodynamic parameters for both types of binding are of the same nature (Maurizot and Grebert, 1988). And the IPTG dependence of both specific and non-specific binding for I12X86 repressor is already well known (Schmitz and Galas, 1980). In the case of mutant I12X86 repressor, as in that of the wild-type one, the close relationship between the characteristics of inducer binding to the core and those of DNA binding to the headpieces of repressor, is consistent with the allosteric behaviour of this protein.

References

Betz, J.L. and Sadler, J.R. (1976). *J. Mol. Biol.*, 105, 293-319.
Bourgeious, S. and Pfahl, M. (1976). *Adv. Prot. Chem.*, 30, 1-99.
Culard, F. and Maurizot J-C. (1982). *FEBS letters*, 146, 153-156.
Culard, F. and Maurizot, J-C. (1981). *Nucleic Acids Res.*, 9, 5175-5184.
Culard, F., Charlier, M., Maurizot, J-C. and Tardieu, A. (1987). *Europ. Biophys. J.*, 14, 169-178.
Farabaugh, Ph.J. (1978). *Nature*, 274, 765-769.
Geisler, N. and Weber, K. (1977). *Biochemistry*, 16, 938-943.
Grebert, Ph. and Maurizot, J-C. (1986). *Nucleic Acids Res.*, 14, 6613-6620.
Jobe, A. and Bourgeois, S. (1972). *J. Mol. Biol.*, 72, 139-152.
Jovin, T., Geisler, N. and Weber, K. (1977). *Nature*, 269, 668-672.
Maurizot, J-C. and Grebert, Ph. (1988) *F.E.B.S. Letters*, 239, 105-108.
Maurizot, J-C., Charlier, M. and Helene, C. (1974). *Biochem. Biophys. Res. Comm.*, 60, 951-957.
Muller-Hill, B. (1975). *Prog. Biophys. Molec. Biol.*, 30, 227-252.
O'Gorman, R.B., Ferguson, L., Betz, J.L., Sadler, J.R. and Matthews, K.S. (1981). *Biochem. Biophys. Acta*, 653, 236-247.
Ogata, R.T. and Gilbert, W. (1979). *J. Mol. Biol.*, 132, 709-728.
Pfahl, M. (1976). *J. Mol. Biol.*, 106, 857-869.
Pfahl, M. (1981a). *J. Mol. Biol.*, 147, 1-10.
Pfahl, M. (1981b). *J. Mol. Biol.*, 147, 175-178.
Pfahl, M. and Hendricks, M. (1984). *J. Mol. Biol.*, 172, 405-416.

Rosenberg, J.M., Kallai, O.B., Opka M.L., Dickerson, R.E. and Riggs, A.D. (1977). *Nucleic Acids Res.*, 9, 5175-5184.

Schmitz, A. and Galas, D.J. (1980). *Nucleic Acids Res.* 8, 487-506.

Schmitz, A., Coulondre, C. and Miller, J.H.. (1978). *J. Mol. Biol.*, 123, 431-456 .

Spotheim-Maurizot, C. M. and Helene, C. (1985). *Photochem. Photobiol.*, 42, 353-359.

Spotheim-Maurizot, M., Culard, F. & Charlier, M. (1987) *Photochem. Photobiol.*, 46, 15-21.

von Hippel, P.H., Revzin, A., Gross, C.A., and Wang, A.C. (1974). in Protein-Ligands Interactions (Edited by Sund, H. and Blauer, G.), pp 270-285, De Gruyter, Berlin.

umuC-Independent, *recA*-Dependent Mutagenesis

KENDRIC C. SMITH AND NEIL J. SARGENTINI

Department of Radiation Oncology
Stanford University School of Medicine
Stanford, CA 94305-5105
USA

Abstract

Although targeted UV-radiation mutagenesis appears to require both the *recA* and *umuC* genes in *Escherichia coli*, examples of *recA*-dependent but *umuC*-independent mutagenesis exist, e.g., gamma-radiation mutagenesis (Mutat. Res. 128, 1, 1984) and streptozotocin mutagenesis (Mutat. Res. 166, 229, 1986). Most of the information on *umuC*-independent mutagenesis comes from studies on ionizing radiation mutagenesis. These results will be reviewed here. Analyses of the various suppressor and back mutations that result in *argE3* and *hisG4* ochre reversion and an analysis of *trpE9777*(+1 frameshift) reversion were performed on *umuC* and wild-type cells gamma irradiated in the presence and absence of oxygen. In wild-type cells, the presence of oxygen enhances gamma-radiation mutagenesis. Although the *umuC* strain showed the gamma-radiation induction of base substitution and frameshifts when irradiated in the absence of oxygen, the *umuC* mutation blocked all oxygen-dependent base-substitution mutagenesis, but not all oxygen-dependent frameshift mutagenesis. For anoxically-irradiated cells, the yields of GC–>AT and AT–>GC transitions were largely *umuC* independent, while the yields of (AT or GC)–>TA transversions were heavily *umuC* dependent. Therefore, the data for anoxically-irradiated cells support the hypothesis that gamma irradiation produces two kinds of DNA lesions that require *recA*-dependent misrepair to induce mutations. For base-substitution mutagenesis, one kind of lesion requires the *umuC* gene and produces transversion mutations, while a second kind of lesion produces transition mutations and does not require the *umuC* gene. For cells irradiated in the presence of oxygen, there seems to be an additional kind of lesion whose mutagenic potential for base substitutions (but not frameshifts) is completely dependent on the *umuC* gene.

Introduction

Early studies on the effect of *recA* mutations on radiation and chemical mutagenesis suggested two mechanisms of mutagenesis in *Escherichia coli:* misreplication and misrepair (e.g., Kondo, 1968; Ishii and Kondo, 1975). Misreplication mutagenesis has been defined operationally as *recA*-independent mutagenesis, and is produced by agents (e.g., ethylmethane sulfonate) that alter the coding properties of the DNA template (e.g., Miller, 1983). Conversely, misrepair [or error-prone or SOS-repair (Witkin, 1976)] mutagenesis has been synonymous with *recA*-dependent mutagenesis, and is produced by agents (e.g., ionizing and UV radiations, and 4-nitroquinoline-1-oxide) that produce noncoding sites in the template.

Photobiology, Edited by E. Riklis
Plenum Press, New York, 1991

169

   Therefore, it is not surprising that the concept has developed that the mechanism of mutagenesis (i.e., misreplication or misrepair) for any mutagen can be determined by testing whether its mutagenicity in *E. coli* depends upon the *recA* gene. However, *recA* mutations affect many phenomena in addition to mutagenesis (reviewed in Walker, 1984). With the independent discovery by Kato and Shinoura (1977) and Steinborn (1978) of the *umuC* and *umuD* mutations, which seem to only abolish misrepair, it has subsequently been considered preferable to test for *umuC* dependence rather than for *recA* dependence when one is trying to ascertain the basis of the mutagenicity of a new agent (e.g., Schendel and Defais, 1980; Shinoura et al., 1983).

   However, this preference for a *umuC*-test over a *recA*-test may be leading to incorrect conclusions. Cases in point are the following: (i) The *recA* strain was not mutated by methyl methanesulfonate (Kondo et al., 1970; Walker, 1977), while the *umuC* strain showed 30% of the mutagenesis seen in the wild-type strain when assayed by reversion of the *argE3* mutation (Schendel and Defais, 1980). However, a *umuC* strain did not show the methyl methanesulfonate-induced reversion of the *hisG4* mutation (Walker and Dobson, 1979). (ii) The *recA* strain was not mutated (a rifampicin-resistance assay) by the alkylating agent streptozotocin (a monofunctional nitrosourea), while the *umuC* strain showed the wild-type level of mutagenesis (Fram et al., 1986). (iii) The *umuC* strain was mutated by gamma radiation when assayed for Arg reversion at the same radiation doses that showed no mutagenesis of the *recA* strain (Sargentini and Smith, 1989). Therefore, it is quite clear that the *umuC* gene does not control all types of *recA*-dependent mutagenesis. Most of the information on *recA*-dependent, *umuC*- independent mutagenesis has come from studies on ionizing radiation mutagenesis. A review of this information will constitute the major emphasis of this report.

Gamma-Radiation Mutagenesis is *recA*-Dependent, but can be Either *umuC*-Dependent or *umuC*-Independent.

   Kato and Nakano (1981) reported that the *umuC* mutant was deficient in gamma-radiation mutability, when assaying for His[+], ColE[R] and Spc[R]. In contrast, however, Steinborn (1978) showed that his *uvm* mutants, which were subsequently shown by Shinagawa et al. (1983) to be *umuC* and/or *umuD* mutants, were near normal in X-ray mutability when assaying for Arg[+].

   In 1984, Sargentini and Smith attempted to resolve this apparent conflict in the literature concerning the role of the *umuC* gene in ionizing-radiation mutagenesis. Using the *umuC122*::Tn5 strain (Elledge and Walker, 1983), which is assumed to show a 'null' phenotype, and a number of different mutation assays [including some of those used by Kato and Nakano (1981) and Steinborn (1978)], they showed that the gamma-radiation mutability of this *umuC* strain varied from *no deficiency* to a 50-fold deficiency, depending upon the mutation assay and doses used (Sargentini and Smith, 1984). They concluded that both Kato and Nakano (1981) and Steinborn (1978) were correct for the mutation assays that they used, and that both *umuC*-independent and *umuC*-dependent mechanisms exist for ionizing-radiation mutagenesis, while targeted UV-radiation mutagenesis seems to depend only on the *umuC*-dependent mechanism (Sargentini and Smith, 1984).

The *recA* strain was not mutated by gamma radiation when assayed either by *arg*(Am) reversion or by assays for the production of large deletions (Kondo, 1968; Ishii and Kondo, 1975), or when assayed by *arg*(Oc) reversion (Sargentini and Smith, 1989). However, the *umuC* strain showed mutagenesis after the same radiation doses and with the same *arg*(Oc) reversion assay that failed for the *recA* strain (Sargentini and Smith, 1989). These results reaffirm that gamma-radiation mutagenesis is totally dependent on misrepair, and they suggest that the *recA* gene controls gamma-radiation mutagenesis via both *umuC*-independent and *umuC*-dependent mechanisms.

How Can the Partial Requirement for the *umuC* Gene in Certain Mutation Assays be Explained?

One complexity with many mutation assays, and especially with nonsense-reversion assays (e.g., Kato et al., 1980), is that more than one base change can lead to the same phenotypic reversion. Therefore, the partial requirement for the *umuC* gene in gamma-radiation mutagenesis may be interpreted as either a partial dependence on *umuC* at each specific mutation site, or as the net effect of an 'all-or-none' dependence on *umuC* at the several mutation sites that are scored simultaneously in one mutation assay. Since no gamma-radiation mutagenesis was observed in the *umuC* strain with the assay for resistance to spectinomycin (Sargentini and Smith, 1984), the all-or-none hypothesis is favored.

This hypothesis was supported by analyzing mutagenesis at several base-pair sites, and showing that there is a large *umuC* dependence at some base-pair sites and little or no *umuC* dependence at other base-pair sites, when assaying for Arg$^+$ and His$^+$ ochre revertants. That is, for anoxically-irradiated cells, the requirement for the *umuC* gene in gamma-radiation mutagenesis was very small at the *supB*, *supE*(Oc), *argE3*, and *hisG4* sites (where transition mutations should be produced), while the requirement for the *umuC* gene was very large at the *supC* or *M*, *supL* or *N*, and *supX* sites (where transversion mutations should be produced). In addition, all of the mutant base pairs at the suppressor loci were AT, while the mutant base pairs at the back mutation sites were GC (Table 1).

The Effect of Oxygen and the *umuC* Gene on Gamma-Radiation Mutagenesis

The *umuC* strain showed less gamma-radiation mutagenesis than wild-type cells, whether irradiated oxically or anoxically (Fig. 1a-c). The *umuC* strain did not show the oxygen enhancement of gamma-radiation-induced base-substitution mutagenesis that was seen in the wild-type strain (Figs. 1a, b); however, both strains showed the oxygen enhancement of radiation-induced frameshift reversion (Fig. 1c) and cell killing (Fig. 1d).

171

Table 1. Effects of the umuC mutation and oxic irradiation conditions on gamma radiation mutagenesis of E. coli*

Comparison of mutant frequencies	Relevant genotype	Irradiation gassing condition	Mutants per 10^8 cells induced by 30 krads							
			supB (Arg+)	supE(Oc) (Arg+)	argE3 (Arg+)	hisG4 (His+)	supC,M (Arg+)	supL,N (Arg+)	supL,N (His+)	supX (Arg+)
	WT	Air	49	34	119	88	95	80	44	125
		N$_2$	35	10	70	35	36	18	5	66
	umuC	Air	31	8	62	12	3	2	2	21
		N$_2$	29	10	56	16	2	2	1	23
WT: Air/N$_2$			1	3	2	2	3	4	9	2
umuC: Air/N$_2$			1	1	1	1	1	1	2	1
Air: WT/umuC			2	4	2	7	32	40	22	6
N$_2$: WT/umuC			1	1	1	2	18	9	5	3
Putative base charges			GC->AT Transitions		AT->GC Transitions		(AT or GC) ->TA Transversions			

*From Sargentini and Smith (1989).

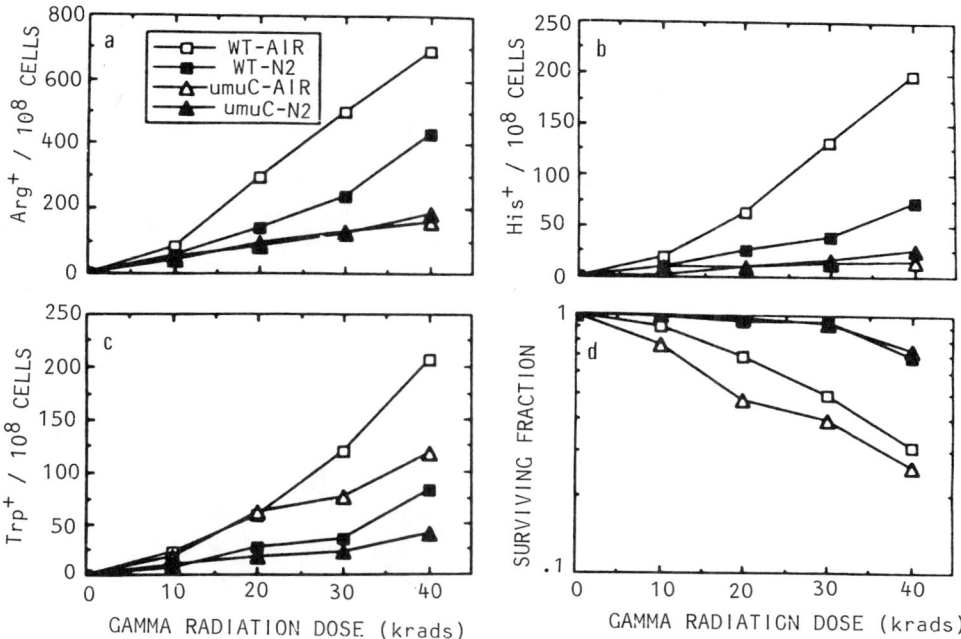

Figure 1. Gamma-radiation mutagenesis and survival for *E. coli umuC⁺* (squares) and *umuC122*::Tn*5* (triangles) cells, irradiated in the presence (open symbols) or absence (closed symbols) of O_2. Mutagenesis was assayed by reversion assays: *argE3*(Oc)->Arg⁺, (a); *hisG4*(Oc)->His⁺, (b); and *trpE9777*(+1 Fs)->Trp⁺, (c). Survival (d) was determined by plating diluted cells on mutant-selection plates. Data are the means from three experiments per strain. (From Sargentini and Smith, 1989.)

The additional mutagenic lesions that are produced in the presence of oxygen all require the *umuC* gene for the production of base substitutions. The types of alterations produced in pyrimidines by gamma irradiation are affected markedly by the presence of oxygen (Teoule, 1987). For example, seven radiolytic products of thymine are produced in DNA only in the presence of oxygen: the hydroperoxides of thymine and their degradation products (e.g., urea), and 5-hydroxymethyl uracil (reviewed in Teoule, 1987). In fact, 5-hydroxymethyl-2'-deoxyuridine produces base substitutions at AT and GC sites when it is present in bacterial-culture media, and its mutagenicity depends on the presence of the *mucAB* genes, which are analogues of the *umuDC* genes (Shirname-More et al., 1987). Thus, 5-hydroxymethyl uracil seems to be one candidate for producing *umuC*-dependent, oxygen-dependent gamma radiation-induced base substitutions.

When mutations were scored at different base-pair sites for the reversion of *argE3* and *hisG4* ochre mutations, the presence of oxygen during the gamma irradiation of wild-type cells either had no effect or it enhanced mutagenesis up to 9-fold, depending upon the specific base-pair site scored (Table 1). These results suggest that the nature of the base pair to be mutated and/or the neighboring bases have a profound effect on the role of

173

oxygen in mutagenesis. Neighboring bases have been shown to have a profound effect on chemical mutagenesis (Burns et al., 1987).

The Effect of the Absence of Oxygen on Gamma-Radiation Mutagenesis

When cells are gamma irradiated under anoxia, both umuC-independent and umuC-dependent mechanisms of mutagenesis exist. Furthermore, the yield of DNA lesions that cause umuC-independent mutagenesis is not affected by oxygen (Table 1). Among the thymine radiolysis products, only 5,6-dihydroxy-5,6-dihydrothymine (thymine glycol) is produced both in the presence and in the absence of oxygen (Teoule, 1987). However, this type of damage does not seem to be mutagenic even though it does block the replication fork (e.g., Laspia and Wallace, 1988). Another possibility is trans-5,6-dihydroxy-5,6-dihydrouracil (uracil glycol). This cytosine-derived base damage is associated with the production of C->T transitions (Ayaki et al., 1987), which were found to be umuC-independent in anoxically-irradiated cells (Table 1). Finally, the same kinds of purine radiolytic products are produced whether oxygen is present or not (R. Teoule, personal communication), which suggests that damaged purines must also be considered as a source of oxygen-independent, umuC-independent mutagenesis.

Regardless of which DNA lesions are responsible for umuC-independent mutagenesis (transitions) in anoxically-irradiated cells, another kind of lesion must be produced in anoxically-irradiated cells to explain the umuC-dependent transversions that are produced. Apurinic/apyrimidinic site mutagenesis is umuC dependent (Schaaper et al., 1982). These lesions are produced directly in DNA by gamma irradiation (Ullrich and Hagen, 1971), and they are also transiently present during the repair of gamma-radiation-induced base damage (e.g., Breimer and Lindahl, 1985). It would seem more than a coincidence that the spectral analysis shows that adenine is always part of the mutant base-pair for umuC-dependent anoxic gamma-radiation mutagenesis (Table 1).

It is known from studies on apurinic-site mutagenesis that the umuC-dependent mechanism shows a strong preference for inserting adenine when it encounters an apurinic/apyrimidinic site in the template strand (e.g., Kunkel, 1984). Furthermore, if the lesion relevant to umuC-dependent anoxic gamma-radiation mutagenesis is a damaged purine rather than an apurinic site, then the tendency for damaged purines to lead to transversions via SOS repair (Rabkin et al., 1983) provides an even better explanation for the data on umuC-dependent gamma-radiation mutagenesis in anoxically-irradiated cells.

Frameshift Mutagenesis

Even though base-substitution and frameshift mutagenesis are similar in being totally umuC dependent in UV-irradiated cells, and in being only partially umuC dependent in gamma-irradiated cells (Sargentini and Smith, 1984), the umuC gene seems to play a different role in base-substitution versus frameshift mutagenesis. The umuC gene is required for the oxygen effect on base substitutions but not for the oxygen effect on frameshifts (Fig. 1a-c). Also, the UV-radiation induction of base substitutions, but not of frameshifts, is enhanced in *umuC* cells by a delayed photoreactivation procedure (Sargentini and Smith, 1987).

Conclusions

The data for anoxically-irradiated cells support the hypothesis that gamma radiation produces two kinds of DNA lesions that require *recA*-dependent misrepair to induce mutations. For base-substitution mutagenesis, one kind of lesion requires the *umuC* gene and produces transversion mutations, while a second kind of lesion produces transition mutations and does not require the *umuC* gene. For cells irradiated in the presence of oxygen, there seems to be additional kinds of lesions whose mutagenic potential for base substitutions (but not frameshifts) is completely dependent on the *umuC* gene.

Acknowledgements

Our work reported here was supported by Public Health Service Grant CA-33738 from the National Cancer Institute, DHHS.

References

Ayaki, H., Yamamoto, O., and Sawada, S., 1987, Role of the main cytosine radiolytic product in ionizing radiation-induced mutagenesis. *Journal of Radiation Research*, 28, 254–261.

Breimer, L.H., and Lindahl, T., 1985, Enzymatic excision of DNA bases damaged by exposure to ionizing radiation or oxidizing agents. *Mutation Research*, 150, 85–89.

Burns, P.A., Gordon, A.J.E., and Glickman, B.W., 1987, Influence of neighbouring base sequence on N-methyl-N'-nitro-N-nitrosoguanidine mutagenesis in the *lacI* gene of *Escherichia coli*. *Journal of Molecular Biology*, 194, 385–390.

Elledge, S.J., and Walker, G.C., 1983, Proteins required for ultraviolet light and chemical mutagenesis. Identification of the products of the *umuC* locus of *Escherichia coli*. *Journal of Molecular Biology*, 164, 175–192.

Fram, R.J., Sullivan, J., and Marinus, M.G., 1986, Mutagenesis and repair of DNA damage caused by nitrogen mustard, N,N'–bis(2-chloroethyl)-N-nitrosourea (BCNU), streptozotocin, and mitomycin C in *E. coli*. *Mutation Research*, 166, 229–242.

Ishii, Y., and Kondo, S., 1975, Comparative analysis of deletion and base-change mutabilities of *Escherichia coli* B strains differing in DNA repair capacity (wild-type, *uvrA, polA, recA*) by various mutagens. *Mutation Research*, 27, 27–44.

Kato, T., and Nakano, E., 1981, Effects of the *umuC36* mutation on ultraviolet-radiation-induced base-change and frameshift mutations in *Escherichia coli*. *Mutation Research*, 83, 307–319.

Kato, T., and Shinoura, Y., 1977, Isolation and characterization of mutants of *Escherichia coli* deficient in induction of mutations by ultraviolet light. *Molecular and General Genetics*, 156, 121–131.

Kato, T., Shinoura, Y., Templin, A., and Clark, A.J., 1980, Analysis of ultraviolet light-induced suppressor mutations in the strain of *Escherichia coli* K-12 AB1157: An implication for molecular mechanisms of UV mutagenesis. *Molecular and General Genetics*, 180, 283–291.

Kondo, S., 1968, Mutagenicity versus radiosensitivity in *Escherichia coli*. *Proceedings of the 12th International Congress of Genetics*, 2, 126–127.

Kondo, S., Ichikawa, H., Iwo, K., and Kato, T., 1970, Base-change mutagenesis and prophage induction in strains of *Escherichia coli* with different DNA repair capacities. *Genetics*, 66, 187–217.

Kunkel, T.A., 1984, Mutational specificity of depurination, *Proceedings of the National Academy of Sciences (USA)*, 81, 1494–1498.

Laspia, M.F., and Wallace, S.S., 1988, Excision repair of thymine glycols, urea residues, and apurinic sites in *Escherichia coli*. *Journal of Bacteriology*, 170, 3359–3366.

Miller, J.H., 1983, Mutational specificity in bacteria. *Annual Review of Genetics*, 17, 215–238.

Rabkin, S.D., Moore, P.D., and Strauss, B.S., 1983, *In vitro* bypass of UV-induced lesions by *Escherichia coli* DNA polymerase I: Specificity of nucleotide incorporation. *Proceedings of the National Academy of Science (USA)*, 80, 1541–1545.

Sargentini, N.J., and Smith, K.C., 1984, *umuC*-Dependent and *umuC*-independent gamma- and UV-radiation mutagenesis in *Escherichia coli*. *Mutation Research*, 128, 1–9.

Sargentini, N.J., and Smith, K.C., 1987, Ionizing and ultraviolet radiation-induced reversion of sequenced frameshift mutations in *Escherichia coli*: a new role for *umuDC* suggested by delayed photoreactivation. *Mutation Research*, 179, 55–63.

Sargentini, N.J., and Smith, K.C., 1989, Mutational spectrum analysis of *umuC*-independent and *umuC*-dependent gamma-radiation mutagenesis in *Escherichia coli*. *Mutation Research*, 211, 193–203.

Schaaper, R.M., Glickman, B.W., and Loeb, L.A., 1982, Mutagenesis resulting from depurination is an SOS process. *Mutation Research*, 106, 1–9.

Schendel, P.F., and Defais, M., 1980, The role of *umuC* gene product in mutagenesis by simple alkylating agents. *Molecular and General Genetics*, 177, 661–665.

Shinagawa, H., Kato, T., Ise, T., Makino, K., and Nakata, A., 1983, Cloning and characterization of the *umu* operon responsible for inducible mutagenesis in *Escherichia coli*. *Gene*, 23, 167–174.

Shinoura, Y., Ise, T., Kato, T., and Glickman, B.W., 1983, *umuC*-mediated misrepair mutagenesis in *Escherichia coli*: extent and specificity of SOS mutagenesis. *Mutation Research*, 111, 51–59.

Shirname-More, L., Rossman, T.G., Troll, W., Teebor, G.W., and Frenkel, K., 1987, Genetic effects of 5-hydroxymethyl-2'-deoxyuridine, a product of ionizing radiation. *Mutation Research*, 178, 177–186.

Steinborn, G., 1978, uvm Mutants of *Escherichia coli* K12 deficient in UV mutagenesis. I. Isolation of uvm mutants and their phenotypical characterization in DNA repair and mutagenesis. *Molecular and General Genetics*, 165, 87–93.

Teoule, R., 1987, Radiation-induced DNA damage and its repair. *International Journal of Radiation Biology*, 51, 573–589.

Ullrich, M., and Hagen, U., 1971, Base liberation and concomitant reactions in irradiated DNA solutions. *International Journal of Radiaton Biology*, 19, 507–517.

Walker, G.C., 1977, Plasmid (pKM101)-mediated enhancement of repair and mutagenesis: dependence on chromosomal genes in *Escherichia coli* K-12. *Molecular and General Genetics*, 152, 93–103.

Walker, G.C., 1984, Mutagenesis and inducible responses to deoxyribonucleic acid damage in *Escherichia coli*. *Microbiological Reviews*, 48, 60–93.

Walker, G.C., and Dobson, P.P., 1979, Mutagenesis and repair deficiences of *Escherichia coli* umuC mutants are suppressed by the plasmid pKM101. *Molecular and General Genetics*, 172, 17–24.

Witkin, E.M., 1976, Ultraviolet mutagenesis and inducible DNA repair in *Escherichia coli*. *Bacteriological Reviews*, 40, 869–907.

What Is The Molecular Mechanism Of UV Mutagenesis in *Escherichia coli*?

JOHN R. BATTISTA, TAKEHIKO NOHMI, CAROLINE E. DONNELLY, and
GRAHAM C.WALKER

Biology Department
Massachusetts Institute of Technology
Cambridge, MA 02139

Abstract

Previous genetic studies have indicated that most UV mutagenesis in *Escherichia coli* requires the participation of the *umuD* and *umuC* gene products. However the mechanism of UV mutagenesis is not yet understood and the roles of the UmuD and UmuC proteins have not been elucidated. The *umuDC* operon is induced by UV irradiation and regulated as part of the SOS response. Genetic evidence now indicates that RecA-mediated cleavage activates UmuD for its role in mutagenesis. The COOH-terminal fragment of UmuD is both necessary and sufficient for this role. The RecA protein appears to have third role in UV mutagenesis besides mediating the cleavage of LexA and UmuD at the time of SOS induction. In addition, we have obtained evidence which indicates that the GroEL and GroES proteins also play a role in UV mutagenesis. Similarities of the amino acid sequence of UmuD to the sequence of gene 45 protein of bacteriophage T4 and of the sequence of UmuC to those of the gene 44 and gene 62 proteins suggest possible roles for UmuD and UmuC in mutagenesis that are supported by preliminary evidence.

Introduction

Despite many years of study, the molecular mechanism of UV mutagenesis in *Escherichia coli* has not yet been determined. Experiments of Weigle (1953) provided the first clue that, at least in *E. coli*, the process of UV mutagenesis requires the participation of host functions and that the expression of one or more of these functions was inducible. Genetic analyses have led to the identification of two host genes, *umuD* and *umuC*, whose function is required for UV mutagenesis (Kato and Shinoura 1977; Walker 1984). Mutations in either of these genes render *E. coli* largely nonmutable with UV and a variety of other chemical mutagens. The *umuD* and *umuC* genes are organized in an operon whose expression is regulated by the SOS circuitry (Walker 1984). In response to an SOS-inducing treatment, the RecA protein becomes activated and then mediates a proteolytic cleavage of LexA at its Ala^{84}–Gly^{85} bond (Walker 1984) apparently by facilitating a specific autodigestion of LexA (Little 1984). Slilaty and

Little (1987) have recently suggested that hydrolysis of the LexA Ala-Gly bond proceeds by a mechanism similar to that of serine proteases, with Ser^{119} acting as a nucleophile and Lys^{156} as an activator. LexA shares homology with the repressors of bacteriophages lambda, 434, P22, and ϕ80 and cleavage of these proteins appears to occur by an analogous mechanism (Sauer 1982, Eguchi et al., 1988). The cleavage site of all these proteins is an Ala-Gly bond except for ϕ80 repressor which has a Cys-Gly cleavage site (Eguchi et al., 1988).

The *umuD* and *umuC* genes encode proteins of approximately 15 and 47 Kd. An evolutionarily diverged but functionally analogous operon is present on the plasmid pKM101 which is present in the Ames *Salmonella* strains used for detecting mutagens (Perry and Walker 1982). The deduced amino acid sequences of the UmuD and MucA proteins are 41% homologous and those of the UmuC and MucB are 55% homologous. Our observation that each gene requires its cognate for biological function has led us to propose previously that the UmuD and UmuC proteins physically interact (Perry et al. 1985).

Despite the progress in understanding aspects of the regulation and processing of UmuD and UmuC, their biochemical role in UV and chemical mutagenesis has remained elusive as has a biochemical demonstration of the mechanism of UV and chemical mutagenesis. Our observation that overexpression of the UmuDC operon results in a cold-sensitive block to DNA replication has led us to hypothesize that UmuD and UmuC interact with components of the replication apparatus (Marsh and Walker 1985). The concept that an altered polymerase/replication apparatus might be involved in UV mutagenesis was proposed by Witkin (1969). This hypothesis was later expanded by Radman's suggestion (1974) that SOS-regulated proteins were involved in this process. Extensive analyses of DNA sequence changes resulting from UV or chemical mutagenesis have indicated that $umuD^+C^+$-dependent mutagenesis is targeted and have therefore supported the concept that a key event in such mutagenesis is misincorporation of bases opposite noncoding or potentially miscoding lesions (Walker 1984). On the basis of a set of physiological experiments involving photoreactivation of *umuC* cells, Bridges and Woodgate (1984) have proposed a two step model for mutagenesis: a $recA^+$-dependent $umuC^+$-independent step in which an incorrect base (or bases) is inserted opposite a premutagenic lesion and a subsequent $umuC^+$-dependent step in which chain elongation is continued past the misincorporated base. Based on *in vitro* studies of the behavior of DNA polymerase III holoenzyme on damaged templates, Livneh has proposed that UmuD and UmuC and possibly other SOS-induced proteins may act by helping the polymerase to reinitiate after terminating and dissociating at the site of a lesion (Livneh 1986) or by making the polymerase more processive and thus facilitating bypass (Hevroni and Livneh 1988).

UmuD Shares Homology With LexA And Phage Repressors

We recently reported that UmuD and MucA share homology with the carboxyl terminal regions of LexA and the repressors of lambda, 434, P22, and ϕ80 (Perry et al.

1985). This led us to hypothesize that UmuD and MucA might interact with activated RecA and that this interaction could result in a proteolytic cleavage of these proteins that would activate or unmask their function required for mutagenesis. The putative cleavage site of UmuD is the Cys^{24}-Gly^{25} bond. A RecA-mediated cleavage of UmuD has now been shown to occur both *in vivo* (Shinagawa et al. 1988) and *in vitro* (Burckhardt et al. 1988) and work from our lab discussed below has shown that the purpose of this cleavage is to activate UmuD for its role in mutagenesis (Nohmi et al. 1988).

Homology Of UmuD To LexA And Phage Repressors Has Functional Significance

We used site-directed mutagenesis of an $umuD^+C^+$ plasmid to create certain $umuD$ mutations that were analogous to *lexA* or lambda repressor mutations that block both RecA-mediated cleavage and autodigestion and found that all these $umuD$ mutations caused major reductions in the ability of UmuD to function in UV mutagenesis. Changing the Gly^{25} residue of the putative Cys^{24}-Gly^{25} UmuD cleavage site to Glu or Lys largely abolished the ability of UmuD to function in UV mutagenesis. A Gly→Glu change at the Ala-Gly cleavage site of lambda repressor completely blocks cleavage while a corresponding Gly→Asp change in LexA largely blocks cleavage. In addition, we found that changing either Ser^{60} or Lys^{97} to Ala also greatly reduced UmuD's ability to function in UV mutagenesis. Slilaty and Little (1987) have shown that changes of the corresponding Ser^{119} and Lys^{156} residues of LexA to Ala completely block cleavage.

Cleaved UmuD Is Functional In Mutagenesis

To test more directly the hypothesis that cleavage of UmuD is important for mutagenesis, we constructed a $umuD$ mutant of a $umuD^+C^-$ plasmid in which overlapping termination (TGA) and initiation (ATG) codons were introduced at the site in the $umuD$ sequence that corresponds to the putative cleavage site. The plasmid carrying this engineered form of UmuD encodes two polypeptides rather than one. These two polypeptides are virtually the same as those that would result from cleavage at the Cys^{24}-Gly^{25} bond of UmuD. When a plasmid carrying this engineered $umuD$ encoding two polypeptides was introduced into a nonmutable $umuD44$ strain, it restored the UV mutability of the cell to that of a $umuD^+$ strain. This result strongly indicates that at least one of the products resulting from cleavage of the UmuD at its Cys^{24}-Gly^{25} bond is capable of carrying out the role of the $umuD$ gene product in mutagenesis. Furthermore, it rules out the possibility that the purpose of UmuD cleavage is to inactivate UmuD. A plasmid that encoded only the polypeptide corresponding to the small NH_2-terminal fragment of UmuD failed to complement the UV nonmutability of a $umuD44$ strain whereas a plasmid that encoded only the large COOH-terminal polypeptide made the strain more UV mutable than a plasmid carrying $umuD^+$. These results strongly suggest that the COOH-terminal cleavage product of UmuD is both necessary and sufficient for the role of UmuD in UV mutagenesis.

The COOH-Terminal Polypeptide Of UmuD Restores Mutability To A lexA(Def) recA430 Strain

To test the physiological significance of UmuD cleavage, we introduced plasmids carrying either the engineered *umuD* encoding two polypeptides or the COOH-terminal polypeptide of UmuD into a *lexA71*::Tn5(Def) *recA430* strain. The *recA430* mutation has differential effects on RecA's ability to mediate proteolytic cleavage (Walker 1984). The RecA430 protein fails to mediate the cleavage of lambda repressor, mediates the cleavage of LexA with reduced efficiency and mediates the cleavage of $\phi80$ repressor normally. *recA430* strains are UV nonmutable. The introduction of the plasmid encoding the two engineered UmuD polypeptides partially restored the UV mutability of this strain while the plasmid encoding only the COOH-terminal UmuD polypeptide restored the UV mutability of the strain to that of a *lexA71*::Tn5(Def) *recA*[+] strain carrying a *umuD*[+] plasmid. The restoration of UV mutability to the *recA430* strains observed when we circumvented the need for UmuD cleavage strongly indicates that the primary cause for the UV nonmutability of *recA430* derivatives is an inability to mediate the cleavage of UmuD. It furthermore implies that the purpose of RecA-mediated cleavage of UmuD is to *activate* UmuD for its role in mutagenesis. Thus, it appears that RecA carries out two mechanistically related roles in UV and chemical mutagenesis — i) transcriptional derepression of the *umuDC* operon by mediating the cleavage of LexA and ii) posttranslational activation of UmuD by mediating its cleavage.

A Third Role for RecA in Mutagenesis

In contrast to the situation discussed above, the introduction of a plasmid encoding the COOH-terminal UmuD polypeptide did not suppress the nonmutability of a *lexA*(Def) *recA1730* (Dutreix et al, 1989). The *recA1730* mutation impairs the ability of RecA to mediate the cleavage of LexA but not of lambda repressor. *recA1730* is dominant to *recA*[+] with respect to the nonmutability phenotype but recessive to *recA*[+] with respect to all other functions tested. Taken together, these observations suggest that the RecA protein plays a third role in UV mutagenesis besides mediating the cleavage of LexA and UmuD.

Evidence that the GroEL and GroES Proteins Play a Role in UV Mutagenesis

We have recently observed that the *E. coli* heat-shock proteins, GroEL and GroES, appear to play roles in UV mutagenesis. This line of experimentation grew out of our previous observation that overexpression of the *umuDC* operon in a *lexA*(Def) strain results in cold-sensitive growth and that DNA replication is blocked upon a shift from 42° to 30°C (Marsh and Walker 1985). This cold sensitivity can be suppressed by various mutations affecting genes involved in the heat-shock response; recent results indicate that mutations in *groEL* and *groES* are particularly effective suppressors. These results suggest that the products of *umuDC* may interact with the replication fork and that raise the possibility that GroEL and GroES might play a role in facilitating this interaction.

This led us to the discovery that *groEL* and *groES* mutants of *E. coli* are greatly reduced in their UV mutability. It is possible that they may exert their function by playing some type of molecular chaperone role with respect to UmuD and/or UmuC since the nonmutability of these *groEL* and *groES* mutants can be suppressed by increased production of UmuD and UmuC.

Sequence Similarities Of UmuD And UmuC To DNA Accessory Proteins Of T4 Bacteriophage

We have observed that the amino acid sequences of UmuD and UmuC share limited similarity with the DNA accessory proteins 45, 44, and 62 of T4 bacteriophage. UmuD aligns with accessory protein 45 (gp45) and UmuC aligns with accessory proteins 44 and 62 (gp 44 and gp 62). The three DNA accessory proteins function as a complex and together increase the processivity of T4 DNA polymerase (Alberts 1984). The similarities between UmuD and UmuC and these DNA accessory proteins suggests the possibility that the target of the UmuD and UmuC proteins may be the replication apparatus of *E. coli* when it has been stalled by a damaged template. Such a model is consistent with suggestions by Bridges and Woodgate (1984) that *umuDC* function involves translesion bypass of DNA damage and by Hevroni and Livneh (1988) that UmuD and UmuC make the polymerase more processive.

Our approach has concentrated on developing a functional link between the DNA accessory proteins and UmuD and UmuC. This is possible because the accessory proteins are well characterized biochemically. The gp44 and gp62 proteins form a complex that exhibits a ssDNA-dependent ATPase activity which is stimulated by gp45. We have demonstrated that the UmuC protein like the gp44/62 complex binds to ssDNA. Further, binding to ssDNA appears to require ATP. UmuC binds equally well in the presence or absence of UmuD or UmuD* and there is preliminary evidence that the UmuD* protein will bind, albeit slightly, to ssDNA that has UmuC bound to it. Neither UmuD or UmuD* bind ssDNA in the absence of UmuC.

Acknowledgments

This work was supported by Public Health Service grants CA21615 and GM28988 awarded by the National Cancer Institute and National Institute of General Medical Sciences respectively. J.R.B. and C.E.D. were supported by postdoctoral fellowships from the National Institutes of Health and American Cancer Society respectively.

References

Alberts, B. M. 1984. The DNA enzymology of protein machines. *Cold Spring Harbor Symposium on Quantitative Biology*, 49, 1–12.

Bridges, B. A. and Woodgate, R. 1984. Mutagenic repair in *Escherichia coli* X. The *umuC* gene product may be required for replication past pyrimidine dimers but not for the coding error in UV mutagenesis. *Molecular and General Genetics*, 196, 364–366.

Burckhardt, S. E., Woodgate, R., Scheuermann, R. H., and Echols, H. 1988. UmuD mutagenesis protein of *Escherichia coli*: Overproduction, purification, and cleavage by RecA. *Proceedings of the National Academy of Science U.S.A.*, 85, 1811–1815.

Dutreix, M., Moreau., P.E, Bailone, A., Galibert, F., Battista., J., Walker., G.C., and Devoret, R. 1989. New *recA* mutations that dissociate the various RecA protein activities in *Escherichia coli:* Evidence for an additional role for RecA protein in UV mutagenesis. *J. of Bacteriology,* 171, 2415–2423.

Eguchi, Y., Ogawa, T. and Ogawa, H. 1988. Cleavage of phage ϕ80 cI repressor by RecA Protein. *Journal of Molecular Biology*, 202, 565–573.

Elledge, S. J., and Walker, G. C. 1983. Proteins required for ultraviolet light and chemical mutagenesis: identification of the products of the *umuC* locus of *E. coli. Journal of Molecular Biology*, 164, 175–192.

Hevroni, D., and Livneh, Z. 1988. Bypass and termination at apurinic sites during replication of single-stranded DNA *in vitro*: a model for apurinic site mutagenesis. *Journal of Biological Chemistry*, 85, 5046–5050.

Kato, T. and Shinoura, Y. 1977. Isolation and characterization of mutants of *Escherichia coli* that are deficient in induction of mutations by ultraviolet light. *Molecular and General Genetics* 156, 121–131.

Little, J. W. 1984. Autodigestion of LexA and phage λ repressors. *Proceedings of the National Academy of Science U.S.A.* 81, 1375–1379.

Livneh, Z. 1986. Mechanism of replication of ultraviolet-irradiated single stranded DNA by DNA polymerase III holoenzyme of *Escherichia coli*: implications for SOS mutagenesis. *Journal of Biological Chemistry* 261, 9526–9533.

Marsh, L., and Walker, G. C. 1985. Cold sensitivity induced by overproduction of UmuDC in *Escherichia coli. Journal of Bacteriology* 162, 155–161.

Nohmi, T., Battista, J. R., Dodson, L. A., and Walker, G. C. 1988. RecA-mediated cleavage activates UmuD for mutagenesis: Mechanistic relationship between transcriptional derepression and posttranslational activation. *Proceedings of the National Academy of Science U.S.A.* 85, 722–737.

Perry, K. L., Elledge, S. J., Mitchell, B. B., Marsh, L., and Walker, G. C. 1985. *umuDC* and *mucAB* operons whose products are required for UV light- and chemical-induced mutagenesis: UmuD, MucA, and LexA proteins share homology. *Proceedings of the National Academy of Science U.S.A.* 82, 4331–4335.

Perry, K. L., and Walker, G. C. 1982. Identification of plasmid(pKM101)-coded proteins involved in mutagenesis and UV resistance. *Nature (London)* 300, 278–281.

Radman, M. 1974. Phenomenology of an inducible mutagenic DNA repair pathway in *Escherichia coli*: SOS repair hypothesis. Molecular and Environmental Aspects of Mutagenesis, edited by L. Prakash, F. Sherman, M. Miller, C. Lawrence, and H.W. Tabor. (Springfield, Ill., Charles C. Thomas, Publisher), pp. 128–142.

Sauer, R. T., Yocum, R. R., Doolittle, R. F., Lewis, M., and Pabo, C. O. 1982. Homology among DNA-binding proteins suggests use of a conserved super-secondary structure. *Nature* 298, 447–451.

Slilaty, S. N., and Little, J. W. 1987. Lysine-156 and serine-119 are required for LexA repressor cleavage: a possible mechanism. *Proceedings of the National Academy of Science U.S.A.* 84, 3987–3991.

Walker, G. C. 1984. Mutagenesis and Inducible Responses to Deoxyribonucleic Acid Damage in *Escherichia coli. Microbiology Reviews*, 48, 60–93.

Weigle, J.J. 1953. Induction of mutation in a bacterial virus. *Proceedings of the National Academy of Science U.S.A.* 39, 628–636.

Witkin, E. 1969. Ultraviolet-induced mutation and DNA repair. Microbiological Reviews 23, 487–514.

Ultraviolet Mutagenesis of a Shuttle Vector Plasmid in Repair Proficient and Deficient Human Cells

M.M. SEIDMAN, D. BRASH, S. SETTHARAM, K.H. KRAEMER AND
A. BREDBERG

Otsuka Pharmaceutical Co. Ltd.
Rockville, MD. U.S.A. and
Laboratory of Molecular Carcinogenesis
National Cancer Institute
Bethesda, MD
U.S.A.

Studies of mutagenesis in prokaryotic systems have been performed with great sophistication for many years. There is an enormous body of literature which describes the chemistry, biochemistry, genetics, etc. of mutagenesis for a wide variety of mutagenic agents. In mammalian cells, however, it has been quite difficult to answer even the most fundamental questions about the nature of mutations and the mechanisms of mutagenesis. This is, of course, due to the complexity of the mammalian genome and the attendant difficulty of generating the specific DNA sequence data necessary for an initial consideration of these questions. The recombinant DNA revolution has spawned a variety of approaches to these problems such that answers to at least the most basic questions are available. In this paper we will discuss the development of one of these systems, the transient shuttle vector.

Shuttle vector plasmids designed for mutagenesis studies are plasmids which are able to replicate in mammalian cells and bacteria (thus the shuttle designation). In addition, they carry bacterial marker genes which permit the detection of plasmids with mutant markers when the vector population is introduced into an appropriate indicator strain. The three functional elements of the vectors represent information from three fields of study brought together by recombinant technology: the bacterial replication origin and drug resistant marker come from studies in bacterial plasmids, the mammalian replication origin and replication functions are from animal virology (SV40 in our work); the marker detection system derived from many years of study in bacterial genetics. These have been combined to develop vectors which permit mutations which arise in mammalian cells to be detected and analyzed in bacteria.

The first specific applications of shuttle vector plasmids for mutagenesis studies were reported by Calos and her colleagues (Calos et al., 1983) and Seidman and co-workers (Razzaque et al., 1983). Both papers described a most disturbing result which threatened the viability of this approach. The mutation frequency observed after passage of unmodified vectors through mammalian cells was quite high, in the neighborhood of 1%. The majority of the mutant plasmids contained deletions, some with insertions of cell DNA. Additional experiments indicated that much of the mutagenesis occurred very early in the infection, probably before plasmid replication. The two groups responded in different ways to this challenge. Calos and her group found a human cell line, Ad 293, which fortuitously had a low background mutation frequency. They have used this line in several studies (Lebkowski et al., 1985) as have others. In addition, they have developed stable extrachromosomal shuttle vectors based on Epstein Barr virus for use in lymphoid cells (DuBridge et al., 1987). The two approaches have been very useful, but may limit an investigation to specific cell types. We were interested in studying mutagenesis in a variety of cell types and so chose to modify the vector system. From the earlier work, it was clear that deletions mutations accounted for the majority of the spontaneous background (Razzaque et al., 1984). Consequently, a vector was designed such that a relatively small target gene, a suppressor tRNA, was flanked by two functional elements necessary for plasmid maintenance and replication in *E. coli* — the ampicillin resistance gene and the plasmid origin (Seidman et al., 1985). The consequence of these changes was a vector with a spontaneous background frequency, 50- to 100-fold lower than the previous constructions. This lower background permits the detection of mutations provoked by carcinogens and mutagens. Furthermore, the small size of the marker gene permits rapid sequence analysis of plasmids with mutant tRNA genes. Recently, another version of pZ189 was constructed. The plasmid, pS189, has an even smaller target size for spontaneous deletions while the mutagenesis target, the Su tRNA remains the same (Seidman, 1988). The spontaneous mutation frequency for this plasmid in Ad 293 cells is 3×10^{-5} (A. Bigger et al., submitted), as compared to 1.4×10^{-4} for pZ189 in the same cells (Yang et al., 1987).

The pZ189 plasmid was first used in studies of ultraviolet light mutagenesis in repair proficient monkey cells (by Hauser et al., 1986) and in repair proficient and deficient human cells (Bredberg et al., 1986). Sequence analysis of the mutant plasmids from both repair proficient lines revealed three types of mutations. There were single base mutations, tandem mutations, and a class of plasmids with multiple mutations scattered through the gene. The multiple mutations will not be discussed in detail here. However, we have shown that they are only indirectly related to UV damage and are not the result of UV targeted mutagenesis (Seidman et al., 1987). The majority of the single and tandem mutations (75% in both lines) were GC-AT transitions as in the case of *E. coli*. Similar results were obtained by Calos using Ad 293 cells (Lebkowski et al., 1985). There were also a variety of transversions (Table 1).

The experiment was then performed in two repair deficient lines, one from the xeroderma complementation group A, the other from the XPD complementation group (Seetharam et al., 1987). Relative to repair of proficient cell lines, these cells show reduced survival following UV exposure. Treated plasmids were passaged through these lines. They too showed a dose-dependent reduction in survival which reflects the known properties of lines from these complementation groups. Sequence analysis of the mutant plasmids again showed a preponderance of GC → AT changes. There was an interesting variability in the frequency of transversion mutations. The XPA line showed about 1/4 the number of these mutations relative to the repair proficient lines, while the XPD line had about 1/2 the repair proficient level. In other studies with repair proficient lines we have also observed this reduced transversion frequency. Although this may suggest a linkage between repair capacity and transversion mutagenesis, it is not possible at this time to come to any firm conclusions (see below).

Table 1. Transition and transversion distribution for UV mutagenesis in repair proficient CV1, human fibroplast (GM0637) and repair deficient human fibroplast XPA (XP12BE) and XPD (XPGBE)

	Mutations in Repair Proficient Lines		Mutations in Repair Deficient Lines	
	CVI	Human	XPA	XPD
Transitions (%)	73	75	94	88
GC → AT	55	59	66	57
AT → GC	5	2	1	2
Transversions (%)	27	25	4	12
GC → TA	5	8	0	5
GC → CG	3	5	1	1
AT → TA	9	6	3	0
AT → CG	5	1	0	2

The dominance of the GV → AT mutation in all of the lines is consistent with the wealth of data from studies in *E. coli*. Such observations were first made over 20 years ago by Drake (1963) and Tessman (1985). Tessman proposed that the appearance of largely single base substitutions at G:C pairs following UV treatment of DNA (which gives rise to dipyrimidine photo-products principally at T:T sites) could be explained by the famous "A rule": when in doubt put in an adenine. He suggested that this was a fundamental property of polymerases and the elegant work of Loeb (Schaaper et al., 1983) and Strauss (Rabkin et al., 1983) and their colleagues have supported this interpretation. The data from the XPA lines are one of the clearest examples of this principle at work. It is clear, however, that this is not an invariant principle as shown in the many transversion mutations which were found in the experiments with the four lines. It is tempting to suggest that the A rule is obeyed by the principle mutagenic mechanism operating in the XPA cells, and that violations of the rule reflect the function of an enzymology present in the repair proficient (or partially proficient) lines. However,

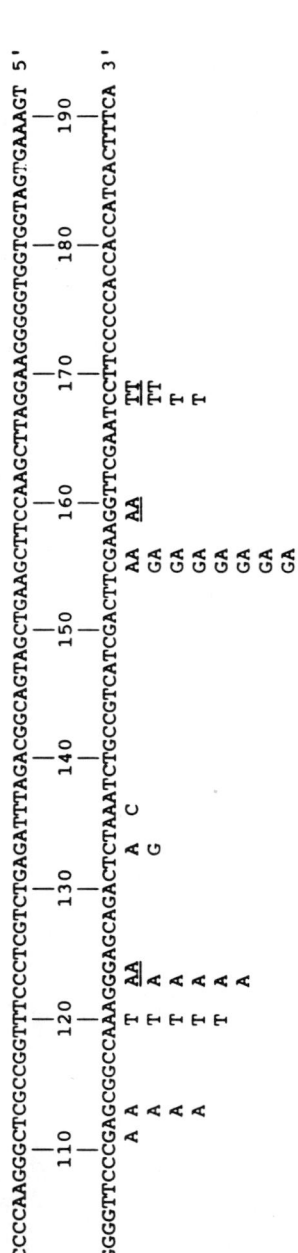

Figure 1a: Distribution of mutations in the human repair proficient line and the XPA and XPD lines

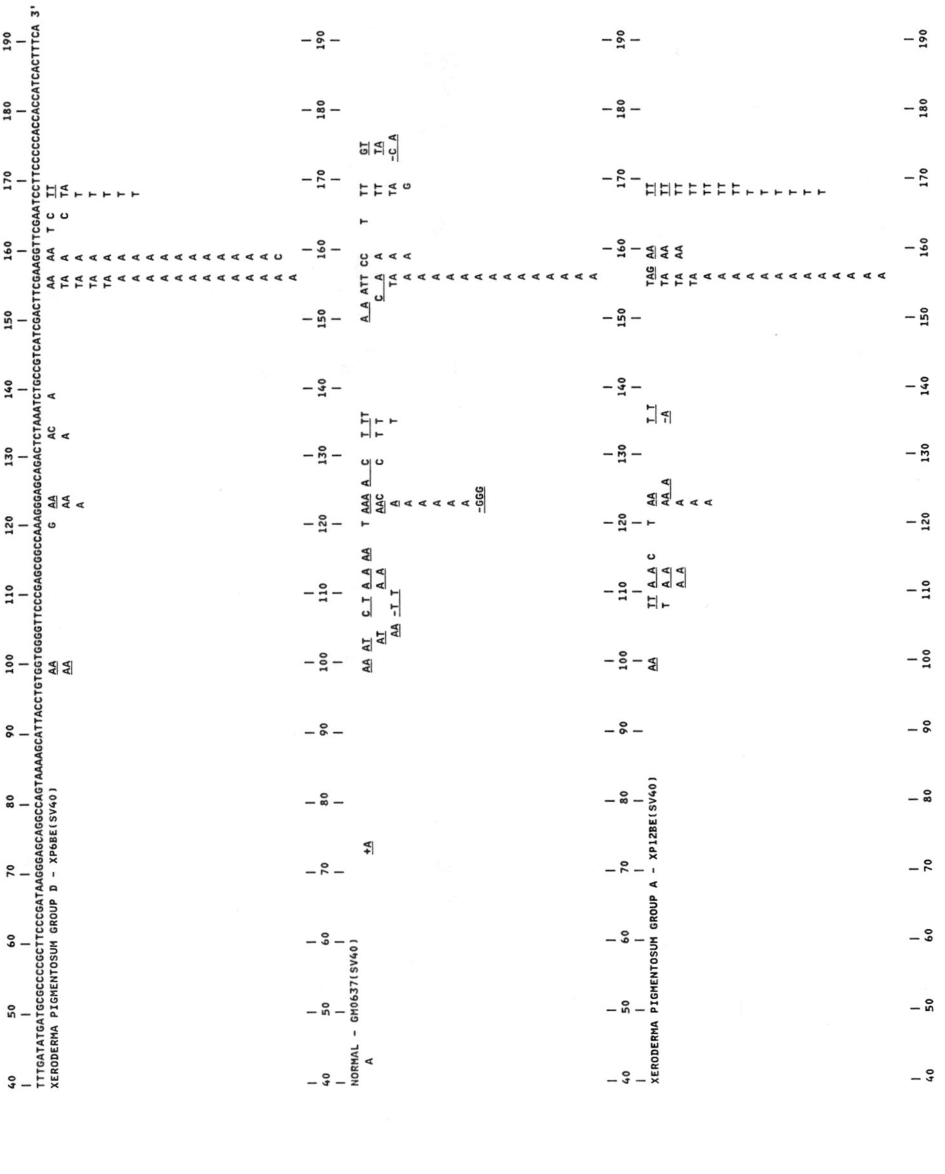

Figure 1b: Distribution of mutations after photoreactivation of the UV treated plasmid and passage through XPA cells

187

we have too little data to make this case and indeed results of our photoreactivation experiment question the absolute authority of this hypothesis (Figure 1b).

An examination of the mutation spectra from the several lines shows a number of similarities and a few striking differences. The single and tandem mutations are shown in these figures. The tandem, or closely-spaced mutations, although not the most prominent feature of the profiles, merit some comment. As predicted by the A rule, many of the tandem mutations occur at C:C sites. This is not always the case, however, and it is clear that we can not readily explain a substantial number of these mutations. The most obvious component of these displays are the hotspots. All the patterns show a major hotspot at position 156. In the repair proficient lines, there was a lesser hotspot at position 123. In the CV1 study, position 159 and 169 were also notable. At all these sites, the GC → AT transition was by far the most frequent mutation.

Comparison with the patterns from the repair deficient lines revealed some similarities and some striking differences. Both the XPA and XPD lines showed the strong hotspot at 156, hotspot at position 168, while in the XPD line position 159 emerged. Again, GC → AT transitions accounted for virtually all of the mutations at these sites.

The appearance of these hotspots leads naturally to the consideration of the basis for their occurrence. Since ultraviolet treatment of DNA does not result in uniform modification of all sites, the simplest explanation would be that a mutational hotspot is a modification hotspot. There are several different photoproducts, principally the cyclobutane dimer and the pyrimidine-pyrimidone (6-4) photoproduct. It has been suggested that the 6-4 photoproducts are the principal premutagenic lesion. If this were true, then the mutation hotspots might reflect the 6-4 distribution. In order to consider those relationships, we determined the distribution of cyclobutane dimers and 6-4 photoproducts in the Su tRNA gene (Brash et al., 1987). These measurements were made using the T4 dimer specific endonuclease (for dimers) or 1 M piperdine (for 6-4 cleavage). The results confirmed the notion that the modification was not uniform, but neither the frequency of cyclobutane dimers nor the frequency of 6-4 photoproducts correlated with the mutation frequency at individual dipyrimidine sites (Figure 2). There were sites that were lightly modified that were mutation hotspots (168) while there were other sites that were strongly modified but showed only a few mutations (43). There were also sites with strong modification levels that were mutation hotspots (156). In an effort to address the role of the 6-4 photoproducts more directly, UV treated plasmid DNA was treated with DNA photolyase prior to passage through the XPA cells. This treatment removes detectable cyclobutane dimers, but leaves 6-4 and other photoproducts. The same hotspots appeared in the mutational spectrum, suggesting that 6-4 photoproducts as well as dimers were mutagenic at this site. Interestingly, two new transversions hotspots appeared at 120 and 155. The mutations at one of these sites (155 GC → CG) clearly violate the A rule (Figure 1b).

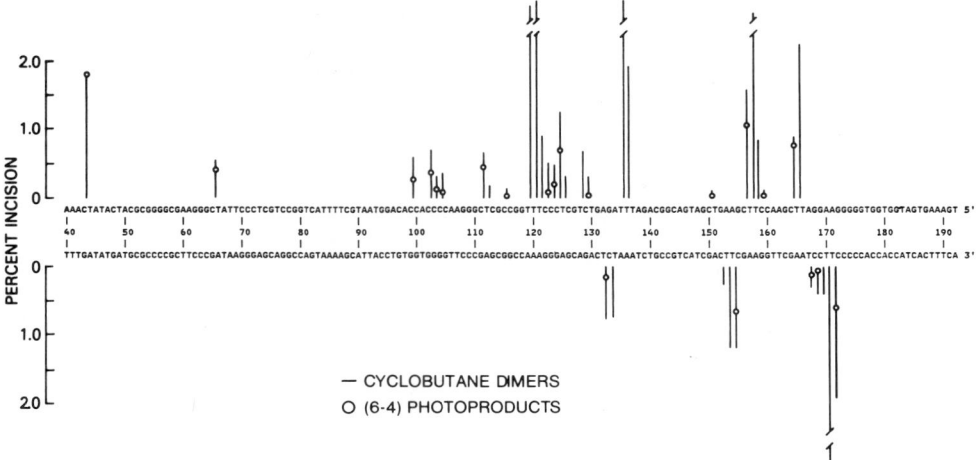

Figure 2. Site-Specific UV Photoproduct Frequencies in the SUP F Gene (at Mutation Sites)

The lack of correlation between photoproduct frequency and mutation frequency at particular sites occurred in experiments in both repair proficient and deficient cell lines (Figure 3). Although differential repair might be invoked in the repair proficient background, this is not a useful explanation in the repair deficient lines. We have suggested that the observed mutational spectra reflect the success or failure of the replication apparatus to bypass a lesion at the site in question. In this view, passing will often result in a mutation if the incorrect base is inserted while failure to bypass would result in a stalled replication, ultimately a lethal event. Each modified site would then be characterized by a particular pass/fail ratio. The pass/fail ratio would reflect the properties of the replicative enzymology, the structure of the modified DNA and the environment in which the encounter occurs. There would be, then, three classes of sites: a) sites with a bias to pass, such that sites with low levels of modification could be mutagenic hotspots (site 168, 159); b) sites with a bias to fail. Sites with high levels of modification might be mutational cold spots if they were of this type (position 43); c) finally, there would be sites whose modification frequency would be correlated with the mutation frequency (site 156).

These observations are not unique to this system. Studies with the Lac E gene in *E. coli* showed the presence of these classes of sites (Brash and Haseltine, 1982). Furthermore, Fuchs and his colleagues have described "mutation prone" sequences in their work with AAF induced frame shifts in *E. coli* (Fuchs, 1984). They have argued that the structure of the modified DNA must underlie these observations. V. Maher and her colleagues have obtained similar results with benzo- and nitro-pyrene mutagenesis (Yang et al., 1988).

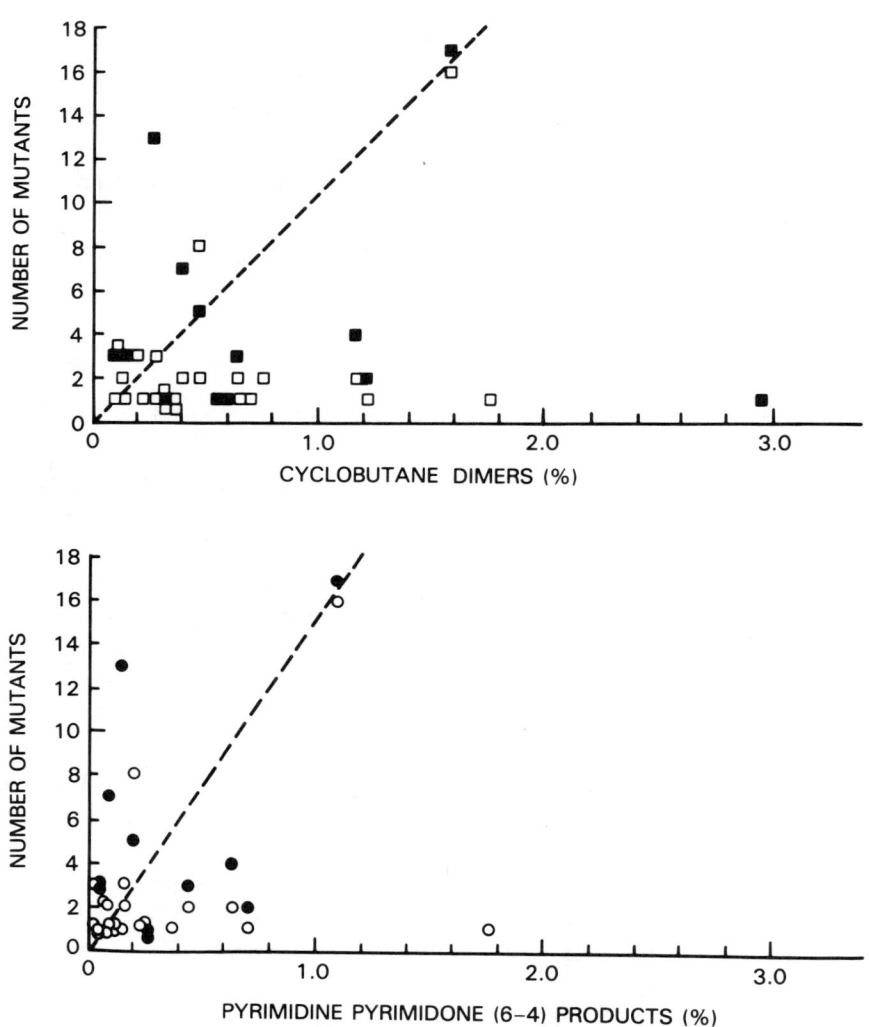

Figure 3. Relationship between cyclobutane dimer and 6-4 photoproduct frequency and the number of transition mutations at specific sites in the tRNA gene after passage through XPA cells (✱ ■) and repair proficient human cells (O ❑). Relationship of dimer and (6-4 ohotoproduct frequency to number of mutants found in pZ189 propagated in normal or xeroderma pigmentosum cells.

It is not possible at this time to go beyond the pass/fail scenario. We do not know what rules direct the molecular choice of passing or failing. Although there is a natural tendency to search the sequences of the gene in question, looking for sequences which "characterize" strong passing or failing sites (particularly sequences with secondary structural potential), it is useful to note that in these studies different hotspots appear in different cell lines. Since it is the same modified vector in all cases, there must be an important cellular component involved in this process. While it is true that the pass/fail scenario does not explain anything (as thoughtfully pointed out by V. Maher [Yang et al., 1988]), it does serve to focus thinking on the interaction of the replication apparatus and the modified DNA. This focus is not intended to slight the importance of DNA repair, but the studies in the repair deficient lines do serve to simplify slightly what must be a complex situation.

References

Brash, D.E. and Haseltine, W.A. (1982), UV induced mutation hotspots occur at DNA damage hotspots, *Nature* 298:189–192.

Brash, D.E., Seetharam, S., Kraemer, K., Seidman, M.M. and Bredberg, A. (1987), Photoproduct frequency is not the major determinant of UV based substitutions hotspots or coldspots in human cells. *Proc. Natl. Acad. Sci. USA* 84:3782–3786.

Bredberg, A., Kraemer, K.H. and Seidman, M.M. (1986), Restricted ultraviolet mutational spectrum in a shuttle vector propagated in Xeroderma pigmentosum cells. *Proc. Natl. Acad. Sci USA* 83:8273–8277.

Calos, M., Lebkowski, J.S. and Botchan, M.R. (1983), High mutation frequency in DNA transfected into mammalian cells. *Proc. Natl. Acad. Sci. USA* 80:3015–3019.

Drake, J.W. (1963), Properties of ultraviolet induced rII mutants of bacteriophage T4. *J. Mol. Biol.* 6:268–283.

DuBridge, R.B., Jang, P., Hsia, H.C., Leong, P.M., Miller, J.A. and Calos, M.P. (1987), Analysis of mutation in human cells by using an Epstein Barr virus shuttle system. *Mol. Cell. Biol.* 7:379–381.

Fuchs, R.P.P. (1984). DNA binding spectrum of the carcinogen N-acetoxy-N-2-acetylamino fluorene significantly differs from the mutation spectrum. *J. Mol. Biol.* 177:173–180.

Hauser, J., Seidman, M.M., Sidur, K. and Dixon, K. (1986), Sequence specificity of point mutations induced during passage of a UV irradiated shuttle vector plasmid in monkey cells, *Mol. Cell. Biol.* 6:277–285.

Lebkowski, J.S., Clancy, S., Miller, J.H. and Calos, M.P. (1985), The Lac I shuttle: rapid analysis of the mutagenic specificity of ultraviolet light in human cells. *Proc. Natl. Acad. Sci. USA* 82:8606–8610.

Rabkin, S.D., Moore, P.D. and Strauss, B.S. (1983), *In vitro* bypass of UV induced lesions by *E. coli* DNA polymerase I: Specificity of nucleotide incorporation. *Proc. Natl. Acad. Sci. USA* 80:1541–1545.

Razzaque, A., Chabrabarti, S., Joffe, S. and Seidman, M.M. (1984), Rearrangement and mutagenesis of a shuttle vector plasmid after passage in mammalian cells. *Proc. Natl. Acad. Sci. USA* 80:3010–3014.

Schapper, R.M., Kunkel, T.A. and Loeb, L.A. (1983), Infidelity of DNA synthesis associated with bypass of apurinic sites. *Proc. Natl. Acad. Sci. USA* 80:487–491.

Seetharam, S., Protic-Sabljic, M., Seidman, M.M. and Kraemer, K.H. (1987), Abnormal ultraviolet mutagenic spectrum in plasmid DNA replicated in cultured fibroplasts from a patient with the skin cancer prone disease Xeroderma pigmentosum.

Seidman, M.M., Dixon, K., Razzaque, A., Zagursky, R., and Berman, M.L. (1985). A shuttle vector plasmid for studying carcinogen induced point mutations in mammalian cells. *Gene* 38:233–237.

Seidman, M.M., Bredberg, A., Seetharam, S. and Kraemer, K.H. (1987), Multiple point mutations in a shuttle vector system propagated in human cells.

Seidman, M.M. (1988). The development of transient SV40 based shuttle vectors for mutagenesis studies: Problems and solutions. *Mutat. Res.* in the press.

Tessman, I. (1985). UV induced mutagenesis of phage S13 can occur in the absence of the recA and umuC proteins of *E. coli. Proc. Natl. Acad. Sci. USA* 82:6614–6618.

Yang, J.L., Maher, V.M. and McCormick, J.J. (1987), Kinds of mutations formed when a shuttle vector carrying adducts of BPDE replicates in human cells. *Proc. Natl. Acad. Sci. USA* 84:3787–3791.

Yang, J.L., Maher, V.M. and McCormick, J.J. (1988). Kinds and spectrum of mutations induced by 1-nitrosopyrene adducts during plasmid replication in human cells. *Mol. Cell. Biol.* 8:3364–3372.

Mutational Hotspots in Mammalian Cells

CHARLES R. ASHMAN

Department of Radiation and Cellular Oncology
University of Chicago
5841 S. Maryland Avenue
Box 442
Chicago, Illinois 60637, USA

Abstract

A vector, termed pZipGptNeo, has been constructed by the introduction of a DNA fragment containing the *E. coli gpt* gene into a retroviral shuttle vector. The pZipGptNeo vector was then introduced into mouse L cells to construct the A912 cell line. This cell line stably expresses the *gpt* gene and contains a single copy of the vector integrated into its chromosomal DNA. Studies utilizing the A912 cell line to determine the specificity of spontaneous and 5-bromodeoxyuridine-induced mutations will be summarized. The role of the surrounding DNA sequence in the production of these mutations will be discussed. The construction of a new retroviral shuttle vector and its introduction into the CHO-K1 cell line will be described. Preliminary experiments suggest that spontaneous *gpt* gene mutations arising in CHO cells are similar to those seen in mouse L cells. A small region of the gene which was shown to be a hotspot for spontaneous mutation in the mouse cells is apparently also a hotspot in Chinese hamster cells.

Abbreviations: gpt, genetic locus coding for xanthine (guanine) phospho-ribosyltransferase; HAT, hypoxanthine/aminopterin/thymidine; Sgua^r, 6-thioguanine-resistant; XPRTase, xanthine (guanine) phosphoribosyltransferase, BrdUrd, 5-bromode-oxyuridine; APRT, genetic locus for adenine phosphoribosyltransferase.

Introduction

The overall objective of our research effort is to gain a better understanding of the mechanisms of mutagenesis in mammalian cells. The approach which we are taking is to develop and use systems which allow one to determine the actual changes in DNA base sequence which occur when a gene mutates. Systems of this type have provided a wealth of information about the mutagenic process in *E. coli* (Miller, 1983) and, more recently, in mammalian cells (DuBridge and Calos, 1988).

For example, a knowledge of the types of mutations produced by a particular mutagen can suggest which of the lesions produced by that mutagen actually leads to mutations. Also, by carrying out these studies in the presence or absence of a particular repair system, one can get some idea of the role this repair system plays in the mutagenic process. Finally, by determining the sites in a gene where mutations are generated, one can get some idea of the roles which the DNA base sequence and the amino acid sequence of the target gene product play in determining the distribution of observed mutations within the gene.

It has been known for many years that there are certain sites within genes, referred to as hotspots, that mutate more readily than other sites. In many cases, it has been shown that a very large fraction of the mutations induced by a particular mutagen arise at a very small number of hotspots within a gene.

Clearly, we must understand why hotspots exist if we are to completely understand the mutagenic process. Today, I'd like to present some observations of hotspots which we have made and offer some speculation as to why mutations should arise so frequently at these sites.

Retroviral Shuttle Vector Systems

The systems that we have developed to study mutagenesis involve the introduction and recovery of target genes from mammalian cells through the use of shuttle vector plasmids (Ashman, et al., 1986; Ashman, 1989). Our systems differ in several respects from other shuttle vector systems so I would like to describe them in some detail before presenting our data. The systems utilize retroviral shuttle vector plasmids containing the E. coli gpt gene as a target for mutagenesis. These vectors have been introduced into mammalian cells lacking the enzyme HGPRT and cell lines have been isolated that have a single copy of the vector integrated into chromosomal DNA. The gpt gene was chosen as a target in part because of the existence of selective systems for both the GPT$^+$ (HAT medium) and GPT$^-$ (Sgua^r) phenotype in mammalian cells. Following mutagenesis of the vector-containing GPT$^+$ cells, mutant Sgua^r cell lines are isolated. Shuttle vector plasmids are recovered following fusion of the mutant cells to monkey COS cells (Cepko, et al., 1984) and the recovered plasmids are introduced into E. coli via transformation. Microgram quantities of the plasmids can then be readily isolated for direct sequencing of the mutant gpt gene.

Two retroviral shuttle vectors have been constructed for our studies. Both vectors are derivatives of the retroviral shuttle vector pZip-NeoSV(x)1 (Cepko, et al., 1984). The first vector, which we have named pZip-GptNeo, was constructed by the introduction of a 0.9 kb fragment containing the gpt gene into the pZip-NeoSV(x)1 vector. This vector was introduced via infection into the mouse fibroblast cell line A9 to create the A9I2 cell line. The second vector was constructed by the introduction of the vector pSV$_2$gpt (Mulligan and Berg, 1980) into a retroviral vector. The pZip-NeoSV(x)1 vector was first modified by the removal of an XhoI fragment containing the neo gene, pBR322 origin and SV40 origin of replication. pSV$_2$gpt was then introduced into this modified vector. This new vector, named pZip-SV$_2$gptBE had been introduced directly into the CHO cell line by CaPO4 /DNA precipitate transfection to produce a cell line which we have named CHO-10T5.

There are two major differences between our retroviral shuttle vector systems and other shuttle vector systems that have been used for mutagenesis studies. One difference is that the target gene is integrated into chromosomal DNA rather than replicating autonomously in the cell. The second difference is that mutant selection occurs in mammalian cells rather than in *E. coli.*

Spontaneous Mutations

A study of spontaneous mutation has been carried out in which a total of 77 Sguaʳ mutants were isolated in the A9I2 cell line (Ashman and Davidson, 1987). Vector sequences were recovered from 43 mutants. The remaining 34 mutants apparently arose by loss or rearrangement of vector sequences. Of the 43 mutants analyzed, 29 had deletions of all or a part of the *gpt* coding sequence.

The spontaneous deletions that we analyzed had several interesting characteristics. One was their small size. Despite the fact that deletions as large as 1250 have been recovered from A9I2 cells, most of the spontaneous deletions were less than 10 bp in length. A second interesting feature of the deletions was their nonrandom distribution throughout the *gpt* gene. Four groups of deletions had termini that either overlapped or were very close to one another. A third interesting feature was the existence of a very strong hotspot for spontaneous deletions. Of the 43 spontaneous mutations analyzed, 16 had the same 3 bp deletion.

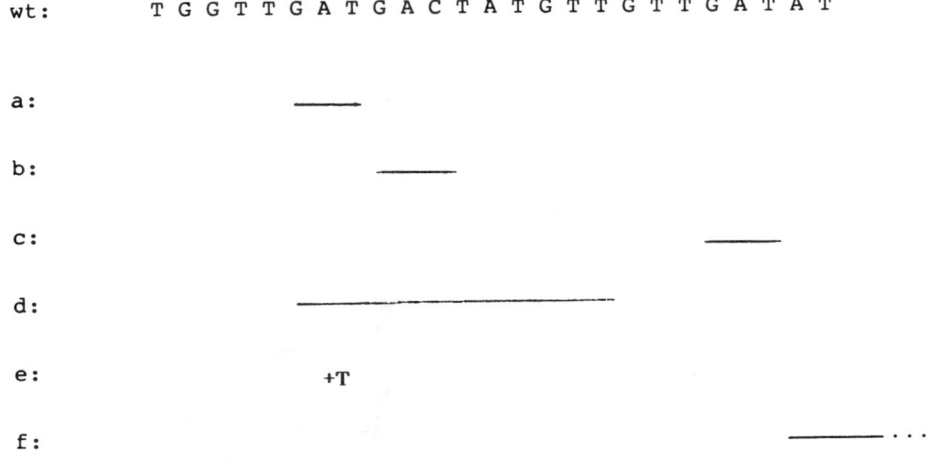

Figure 1. Location of mutations in the spontaneous hotspot region. The DNA base sequence is given for positions 362–386 of the wild-type (wt) *gpt* gene. The locations of deletion mutations in this region are indicated by solid lines beneath the wt sequence. The frameshift mutation (e) is the result of the insertion of a T between bases 369 and 370.

The DNA base sequence in the vicinity of this hotspot as well as the locations of the spontaneous mutations observed in this region are presented in Figure 1. In addition to the hotspot deletion (a), two additional 3 bp deletions (b and c) were isolated in the initial study. A 12 bp deletion (d) was isolated in the A9I2 cell line in a subsequent study (Davidson, et al., 1988). Recently, we have initiated a study of a spontaneous mutations in the CHO-10T5 cell line (C. Ashman, unpublished results). Of the first 15 mutations analyzed, 2 had the hotspot deletion observed in the A9I2 line. In addition, a +1 frameshift mutation at base 369 (e) was observed as well as a 74 bp deletion with a terminus at base 384 (f). Thus, it appears that this region is a hotspot for spontaneous mutations in cells from two different mammalian species.

Examination of the DNA base sequence of this region reveals that it contains numerous short direct repeats. Most of the deletions have short direct repeats at their termini which is characteristic of spontaneous deletions in a variety of experimental organisms (Albertini, et al., 1982; Nalbantoglu, et al., 1986). Another feature of this sequence is that it contains several copies of the consensus sequence for the strongest class of rat liver topoisomerase I breakage sites (Been, et al., 1984). While there is no evidence that it plays a role in the formation of spontaneous mutations, topoisomerase activity has been implicated in nonhomologous intrachromosomal recombination in mammalian cells (Bullock, et al., 1985) and frameshift mutagenesis in phage T4 (Ripley, et. al., 1988). Experiments are currently in progress in our laboratory to determine if this region contains strong break sites for topoisomerase I.

Another factor which may contribute to making this region a mutational hotspot is that it apparently encodes amino acids which are critical for XPRTase activity. Three of the mutations (a,b, and c) are 3 bp in-frame deletions which result in the deletion of a single aspartic acid residue from the protein. In the case of mutations a and b, the loss of one of two adjacent aspartic acid residues produces the mutant phenotype. The fact that this region is a hotspot for spontaneous mutation appears to be a consequence both of some unique aspect of the DNA base sequence (short direct repeats, topoisomerase break sites) and the extreme sensitivity of this region of the protein to changes in amino acid sequence.

A Hotspot for BrdUrd Mutagenesis

We have recently concluded a study of the specificity of BrdUrd mutagenesis in the A9I2 cell line (Davidson, et al., 1988). In this study, a total of 31 BrdUrd-induced mutants were analyzed and, of this total, 22 were found to have G:C to A:T transitions. An examination of the base sequences flanking the sites of the G:C to A:T transitions revealed a striking sequence specificity. Most of the mutations (19 of 22) occurred at the 3' G residue in the sequence 5'-G-G.

This particular sequence specificity for G:C to A:T transitions has been observed in several other studies. It has been observed for mutations induced by monofunctional alkylating agents at two genetic loci in E. coli and a human cell line (Richardson, et al., 1987; Burns, et al., 1987; DuBridge, et al., 1987). Also, it was recently reported that many spontaneous G:C to A:T transitions at the APRT locus in CHO cells had this same sequence specificity (DeJong, et al., 1988).

The similarities in sequence specificities of spontaneous and mutagen-induced G:C to A:T transitions suggests that there is a common step in the mechanisms by which these mutations are generated. Mutations induced by both BrdUrd and monofunctional alkylating agents are thought to result from base misincorporation during DNA replication. These misincorporation events may be more probable at 5'-G-G sites, possibly due to decreased polymerase fidelity or proofreading activity at these sites.

Recently, an observation was made by Gordon and Glickman (1988) which may account for the occurrence of at least some of the mutations at the 5'-G-G sites. They analyzed a large collection of G:C to A:T transitions induced by N-methyl-N'- nitro-N-nitrosoguanidine in the *lac i* gene of *E. coli*. They observed that there was a slight difference in the site specificity of G:C to A:T transitions in regions of the gene coding for different portions of the *lac i* protein. The DNA binding domain of the protein is very sensitive to amino acid substitutions. Most of the G:C to A:T transitions in the region of the gene coding for this domain occurred at 5'-purine-G sites with no preference for A or G 5' to the mutation site. In the region of the gene coding for the core domain, however, there was a very strong preference for mutations at the 3' G in the sequence 5'-G-G. They also observed that a majority of core domain G:C to A:T transitions either occurred at glycine (GGN, with N being any base) codons or produced stop codons. Both of these events would be expected to produce drastic changes in protein structure. They concluded that the distribution of observable mutations in a gene may depend to a large extent upon the nature of the protein target.

Examination of the BrdUrd-induced G:C to A:T transitions in our study showed that most (18 of 22) either occurred at glycine codons or produced stop codons. Thus, it appears that some of the BrdUrd site specificity may result from the sensitivity of the target protein to certain types of amino acid substitutions. As with the hotspot for spontaneous mutation, the BrdUrd hotspots may result from a combination of DNA sequence and amino acid substitution effects.

Acknowledgements

This work was supported by Public Health Service Grants CA31781 and CA45336 from the National Institutes of Health and a grant from the American Cancer Society.

References

Albertini, A.M., M. Hofer, M.P. Calos and J.H. Miller (1982) On the formation of spontaneous deletions: the importance of short sequence homologies in the generation of large deletions, *Cell*, 29, 319-328.

Ashman, C.R. (1989) Retroviral shuttle vectors as a tool for the study of mutational specificity. *Mutat. Res.* 220, 143- 149.

Ashman, C.R., and R.L. Davidson (1987) Sequence analysis of spontaneous mutations in a shuttle vector gene integrated into mammalian chromosomal DNA, *Proc. Natl. Acad. Sci (U.S.A.)*, 84, 3354-3358.

Ashman, C.R., P. Jagadeeswaran and R.L. Davidson (1986) Efficient recovery and sequencing of mutant genes from mammalian chromosomal DNA, *Proc. Natl. Acad. Sci. (U.S.A.)*, 83, 3356-3360.

Been, M.D., R.R. Burgess, and J.J. Champoux (1984) Nucleotide sequence preference at rat liver and wheat germ type I DNA topoisomerase breakage sites in duplex SV40 DNA. *Nucleic Acids Res.* 12, 3097-3114.

Bullock, P., J.J. Champoux, and M. Botchan (1985) Association of crossover points with topoisomerase I cleavage sites: a model for nonhomologous recombination. *Science* 230, 954-958.

Burns, P.A., A.J.E. Gordon, and B.W. Glickman (1987) Influence of neighbouring base sequences on N-methyl-N'- nitro-N-nitrosoguanidine mutagenesis in the *lac i* gene of *Escherichia coli* . *J. Mol. Biol.* 194, 385-390.

Cepko, C., B.E. Roberts and R.C. Mulligan (1984) Construction and application of a highly transmissible murine retrovirus shuttle vector, 37, 1053-1062.

Davidson, R.L., P. Broeker and C.R. Ashman (1988) DNA base sequence changes and sequence specificiy of bromodeoxyuridine-induced mutations in mammalian cells, *Proc. Natl. Acad. Sci. (U.S.A.)* 85, 4406–4410.

DeJong, P.J., A.J. Grosovsky, and B.W. Glickman (1988) Spectrum of spontaneous mutation at the APRT locus of Chinese hamster ovary cells: an analysis at the DNA sequence level. *Proc. Natl. Acad. Sci. U.S.A.* 85, 3499-3503.

DuBridge, R.B., and M.P. Calos (1988) Recombinant shuttle vectors for the study of mutation in mammalian cells. *Mutagenesis* 3, 1-9.

DuBridge, R.B., P. Jang, H.C. Hsia, P.-M., Leong, J.H. Miller and M. P. Calos (1987) Analysis of mutation in human cells by using an Epstein-Barr virus shuttle system, *Mol. Cell Biol.*, 7, 379-387.

Gordon, A.J.E., and B.W. Glickman (1988) Protein domain structure influences observed distribution of mutation. *Mutat. Res.* 208, 105-108.

Miller, J.H. (1983) Mutational specificity in bacteria, *Annu. Rev. Genet.*, 17, 215-238.

Mulligan, R.C., and P. Berg (1980) Expression of a bacterial gene in mammalian cells, *Science*, 209, 1422-1427.

Nalbantoglu, J., D. Hartley, G. Phear, G. Tear and M. Meuth (1986) Spontaneous deletion formation at the aprt locus of hamster cells: the presence of short sequence homologies and dyad symmetries at deletion termini, *EMBO J.*, 5, 1199- 1204.

Richardson, K.K., F.C. Richardson, R.M. Crosby, J.A. Swenberg and T.R. Skopek (1987) DNA base changes and aklylation following in vivo exposure of *Escherichia coli* to N-methyl-N-nitrosourea or N-ethyl-N-nitrosourea, *Proc. Natl. Acad. Aci. (U.S.A.)*, 84, 344-348.

Ripley, L.S., J.S. Dubins, J.G. DeBoer, D.M. DeMarini, A.M. Bogerd, and K.N. Kreuzer (1988) Hotspot sites for acridine- induced frameshift mutations in bacteriophage T4 corresponse to sites of action of the T4 type II topoisomerase. *J. Mol. Biol.* 200, 665-680.

Spectra of Mutations Induced by Structurally-Related Aromatic Polycyclic Carcinogens during Replication of a Shuttle Vector in Human Cells

V. M. MAHER[*], J-L. YANG, M. C-M. MAH, AND
J.J. MCCORMICK

*Carcinogenesis Laboratory – Fee Hall
Departments of Microbiology and Biochemistry
Michigan State University
East Lansing, MI 48824-1316*

Abstract

We are comparing the kinds and spectra of mutations induced when DNA containing covalently bound carcinogen residues (adducts) replicates in human cells. A shuttle vector, pZ189, carrying the *sup*F gene coding for a bacterial tyrosine suppressor tRNA as the target was treated with tritiated polycyclic aromatic carcinogens and the number of adducts per plasmid was determined. The plasmids were transfected into human cells, and after replication had occurred, the progeny plasmids were rescued and assayed for the frequency of *sup*F mutants. The carcinogens studied included the 7,8-diol-9,10-epoxide of benzo[a]pyrene (BPDE), 1-nitrosopyrene (1-NOP), N-acetoxy-2-acetylaminofluorene (N-AcO-AAF), and its trifluoroacetyl derivative (N-AcO-TFA-AF). BPDE binds principally to the N2 position of guanine. The other three carcinogens bind to the C8 position of guanine. Each agent caused a linear increase in the frequency of *sup*F mutants as a function of the number of DNA adducts formed, reaching frequencies as high as 20×10^{-4} to 40×10^{-4}, above a background frequency of 1.4×10^{-4}. When compared on the basis of adducts formed, BPDE was approximately four times more mutagenic than the other three carcinogens. This difference may reflect intrinsic differences in the nature of the adducts and their location in the DNA molecule, but it may also reflect differences in the rate of removal of particular adducts by nucleotide excision repair since we showed that the host cells excise BPDE induced adducts from genomic DNA at least three times slower than they excise 1-NOP induced adducts. Agarose gel electrophoresis and DNA sequencing analysis of mutants derived from untreated plasmids showed that the majority (70%) involved deletions, insertions, or altered gel mobility (gross rearrangements). In contrast, the majority of those from carcinogen treated plasmids were base substitutions. DNA sequencing of 86 unequivocally independent mutants derived from BPDE treated plasmids and 60 from 1-NOP treated plasmids indicated that 70% to 80% contained a single base substitution, 5%–10% had two base substitutions, and 4%–10% had small insertions or deletions (one or two base pairs). The majority (83%) of the base substitutions were transversions, predominantly $G \cdot C \rightarrow T \cdot A$. These two carcinogens produced their own spectrum

---

[*]Address for correspondence: Veronica M. Maher, Carcinogenesis Laboratory — Fee Hall, Michigan State University, East Lansing, MI 48824-1316, Telephone: 517/353-7785

Photobiology, Edited by E. Riklis
Plenum Press, New York, 1991

of mutations. Studies to date with the N-AcO-TFA-AF have shown that the AF adduct induces predominantly base subsitutions and all of these involve guanine.

Introduction

Research with oncogenes has shown that point mutations, as well as other kinds of permanent changes in the genome, are causally involved in the multistep process of changing normal cells into malignant cells (Vousden et al., 1986). As part of our investigations into the mechanisms of carcinogenesis, we are examining the specific kinds of mutations induced in human cells by a series of structurally related polycyclic aromatic carcinogens. Until recently, it has not been possible to isolate and analyze newly mutated genes from mammalian cells at the sequence level. However, several elegant systems are now available to rescue DNA containing mutations in target genes whose sequence is known. We are using two of these approaches: one involves a shuttle vector, pZ189, carrying the gene coding for a bacterial suppressor tRNA, *supF* (Yang et al., 1987, 1988a, 1988b; Maher et al., 1989); the second makes use of the polymerase catalyzed chain reaction (Saiki et al., 1985; Yang et al., 1989) to amplify the cDNA of the HPRT gene from carcinogen mutated diploid human fibroblasts that have been selected for resistance to 6 thioguanine.

Using the shuttle vector assay, we have been examining the kinds and spectra of mutations induced by four structurally related carcinogens: ($\pm$)7β, 8α-dihydroxy 9α, 10α−epoxy 7,8,9,10-tetrahydrobenzo[a]pyrene (BPDE), nitrosopyrene (1-NOP) the partially reduced intermediate metabolite of 1-nitropyrene, N-acetoxy-2-acetylaminofluorene (N-AcO-AAF), and its trifluoroacetyl-derivative (N-AcO-TFA-AF). BPDE, a direct acting metabolite of benzo[a]pyrene forms its principal DNA adduct at the N2 position of guanine (Weinstein et al., 1976). 1-NOP needs further reduction to the hydroxylamine, followed by an acid catalyzed formation of an unstable nitrenium intermediate before it can bind to DNA, principally at the C8 position of guanine (Heflich et al., 1985; Beland et al., 1986). N-AcO-AAF, a direct acting metabolite of AAF, also forms its principal DNA adduct at the C8 position of guanine (Beland et al., 1979). However, in mammalian cells, N-AcO-AAF is rapidly deacetylated so that its principal adduct is the deacetylated form of the residue (AF) on the C8 position of guanine, rather than the acetylated AAF form (Poirier et al., 1980; Heflich et al., 1988). In order to obtain deacetylated (AF) adducts on shuttle vector DNA reacted with carcinogen *in vitro*, we have made use of N-AcO-TFA-AF, in collaboration with Charles King and Thomas Reid of the Michigan Cancer Foundation (Detroit, MI). The trifluoroacetyl group on this compound rapidly dissociates in vitro, yielding the deacetylated AF form which binds to the C8 position of guanine. We are comparing the specific kinds of mutations induced by these four agents, the location of the mutations in the target gene (*supF*), as well as the biologic effectiveness of the various adducts, i.e., their ability to interfere with bacterial transformation and to induce mutations when the treated plasmid replicates in human cells. The results to date indicate that BPDE adducts are approximately four times more mutagenic than the other three carcinogens.

Materials and Methods

Treating the plasmid with radiolabeled carcinogens

The shuttle vector we use, pZ189, was constructed by Seidman et al . (1985). It is 5.5 kbp in size and contains the *supF* gene flanked by two genes essential for recovery in *E. coli*, i.e., the ampicillin gene and the bacterial origin of replication. It also carries the origin of replication and large T antigen gene from simian virus 40 and, therefore, can replicate in human cells.

For the experiments with BPDE, tritiated compound was added to a solution of DNA in Tris HC1 EDTA buffer, pH 8.0, and incubated at room temperature for 2 h protected from light. Unbound DNA was removed by three successive precipitations with ethanol. For 1-NOP, the plasmid was resuspended in helium purged Na citrate buffer, pH 5.0. Ascorbic acid (20 uM in H_2O) was added to provide needed reduction of the 1-NOP (Heflich et al., 1986), followed by tritiated 1-NOP dissolved in dimethylsulfoxide. The mixture was incubated at 37°C for 2 h and unbound 1-NOP was removed by phenol chloroform extraction and three successive ethanol precipitations. For N-AcO-TFA-AF and N-AcO-AAF, DNA dissolved in 2 mM Na citrate buffer, pH 7.0, was added to a freshly prepared ethanol solution of tritiated carcinogen and incubated at 37°C for 30 min. The unbound compound was removed by extensive ether extraction, followed by purification using phenol and finally ethanol precipitation. The number of carcinogen residues bound per mole of plasmid were calculated from the A_{260} absorption profile of the DNA and the specific activity of the carcinogens.

Transfection and rescue of replicated plasmid and identification of mutants

The human host cells used were 293 cells, an embryonic kidney cell line. They were transfected with plasmid as described by Yang et al. (1987). After 48 h, the progeny plasmids were extracted, purified extensively, and used to transform indicator bacteria. The latter, *E. coli* SY204, is ampicillin sensitive and carries an amber mutation in the β-galactosidase gene (Sarkar et al., 1984). The transformed bacteria are selected for ampicillin resistance on plates containing ampicillin, X Gal, and an inducer of the β-galactosidase gene. Transformants containing plasmids lacking a functioning *supF* gene, which is needed to suppress the amber mutation in the β-galactosidase gene of the indicator bacteria, could be identified because they form white or light blue colonies rather than dark blue colonies on X Gal plates.

Comparing the biologic effectiveness of BPDE and 1-NOP adducts

As shown in Figs. 1A and 2A, BPDE and 1-NOP induced a dose dependent increase

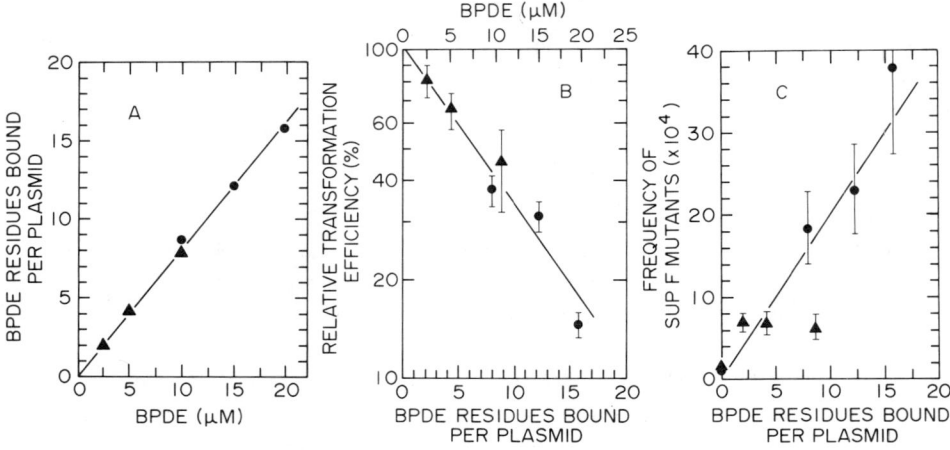

Figure 1. Number of adducts formed as a function of BPDE concentration (A); relative frequency of transformation of bacteria (B); and frequency of *sup*F mutants as a function of BPDE adducts per plasmid. (C) The symbols (■ and ▲) indicate that the plasmid cells were treated in two separate experiments (from Yang et al., 1987).

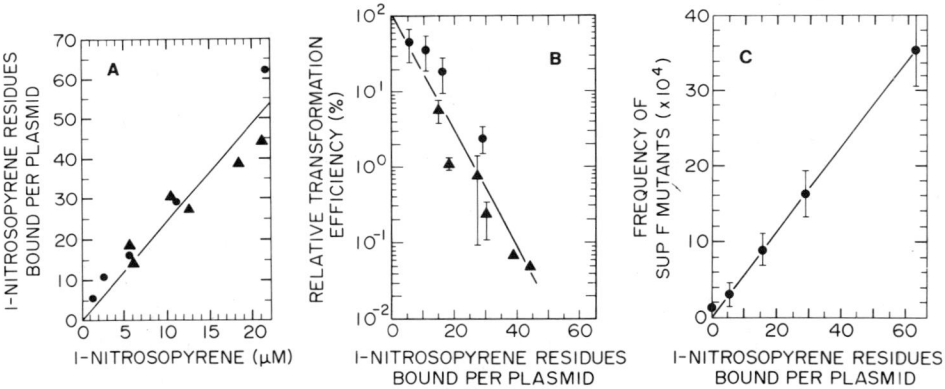

Figure 2. Number of adducts formed by 1-NOP in the presence of ascorbic acid (A); relative frequency of transformation (B) and frequency of *sup*F mutants as a function of 1-NOP adducts per plasmid. (C) The symbols (■ and ▲) indicate that the plasmid was treated in two separate experiments (from Yang et al., 1988b).

in the number of carcinogen residues bound to the DNA. 1-NOP appeared to be more reactive than BPDE, but this merely reflects the use of ascorbic acid to activate 1-NOP. The adducts of either agent were virtually equal in their ability to interfere with bacterial transformation (Figs. 1B and 2B) and induced a linear increase in the frequency of *supF* mutants during replication of the treated plasmid in human cells (Fig. 1C and 2C). However, the frequency induced per adduct was 3.7 fold higher for BPDE than for 1-NOP. For example, interpolation of the data in Figs. 1C and 2C indicates that 20 BPDE adducts per plasmid increased the mutant frequency from 1.4×10^{-4} to 40×10^{-4}, whereas 20 1-NOP adducts per plasmid increased it only to 11.6×10^{-4}.

This difference in mutagenicity may reflect the intrinsic difference in the nature of the adducts formed and their location in the DNA molecule. (The N2 position of guanine, where the major BPDE adduct is formed, is in the base pairing region of the molecule, in contrast to the C8 position where the 1-NOP is located.) However, it could reflect a difference in the rate of excision of 1-NOP and BPDE-adducts from the plasmid by the host 293 cells. Since it was impossible to determine the rate of removal of these adducts from the plasmids, we examined the ability of the host cells to remove these residues from their genomic DNA. Cellular DNA was prelabeled with ^{14}C-thymidine and the cells were treated with two concentrations of tritiated 1-NOP or BPDE. At the end of the 1 h treatment, one set of cells was assayed immediately for the number of carcinogen residues bound to DNA as described (Patton et al., 1986) and the ^{3}H to ^{14}C ratio was also determined. The rest of the cells were incubated at 37°C for 18 h or 30 h before being similarly analyzed for the number of tritiated residues remaining bound to ^{14}C-labeled parental DNA. As shown in Fig. 3, the rate of removal 1-NOP residues was significantly faster than of BPDE residues.

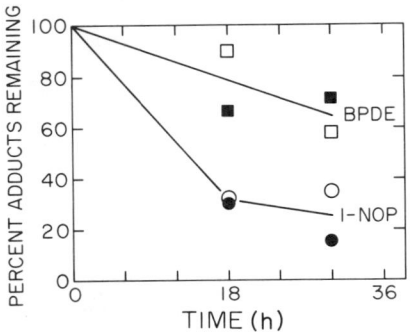

Figure 3. The rate of removal of 1-NOP or BPDE adducts by 293 cells. The cells were prelabeled with ^{14}C-TdR and treated with tritiated carcinogens. The percentage of adducts remaining was calculated by dividing the ratio of ^{3}H/^{14}C at time 18 or 30 h by the ratio of time 0. The open and closed symbols indicate that the plasmid cells were treated in two separate experiments. (From Yang et al., 1988b).

Kinds of mutations induced by BPDE and 1-NOP

A total of 40 spontaneous, 107 BPDE-induced, and 88 1-NOP induced mutants were assayed by gel electrophoresis for gross rearrangements, and those that did not show such changes, i.e., 28 spontaneous, 103 BPDE-induced, and 64 1-NOP induced mutants, were sequenced using the dideoxyribonucleotide method of Sanger et al. (1977). Polymerization was carried out from a pBR322 *Eco*RI site primer using the Klenow fragment of DNA polymerase I and ^{35}S labeled dATP as described by Yang et al. (1987). The results showed that only 30% of the spontaneous mutants contained point mutations (substitution, deletion, or insertion of 1 or 2 base pairs), compared to 73% of those from BPDE treated plasmids and 92% of those from 1-NOP treated plasmids (Yang et al., 1987, 1988b). Examination of 86 unequivocally independent mutants induced by BPDE and 60 by 1-NOP showed that the majority consisted of base substitutions (Table 1), predominantly G·C→T·A transversions (Table 2). From 87% to 90% of the base substitutions involved GC pairs, which is consistent with the DNA binding pattern of these two carcinogens (Yang et al., 1987, 1988b).

Location of the point mutations (spectra) induced by BPDE and 1-NOP

The *sup*F gene is particularly useful as a sensitive target for detecting the specificity of various mutagens, because it codes for a tRNA, rather than a protein, and a change in any one of at least 63 of the 85 bases which make up the tRNA structure results in a mutant phenotype. As shown in Fig. 4, BPDE and 1-NOP each exhibited a specific

Table 1. Comparison of sequence alterations generated in *sup*F by replication of 1-NOP- or BPDE-treated plasmid in human cells

Sequence alterations	No. of mutants sequenced		
	Control	1-NOP	BPDE
Single base substitution	3	48	51
Two base substitutions:			
Tandem	0 (24%)	2 (85%)	3 (70%)
≤ 20 bases apart	2	0	3
> 20 bases apart	0	1	3
Deletions:			
Single G·C pair	2	1	5
Single A·T pair	1	0	0
Tandem bases	0 (66%)	0 (10%)	2 (24%)
4–20 bases	4	3	3
> 20 bases	7	2	11
Insertions:			
Single A·T pair	0 (10%)	1 (5%)	4 (6%)
≤ 20 bases	2	2	1
Total	21	60	86

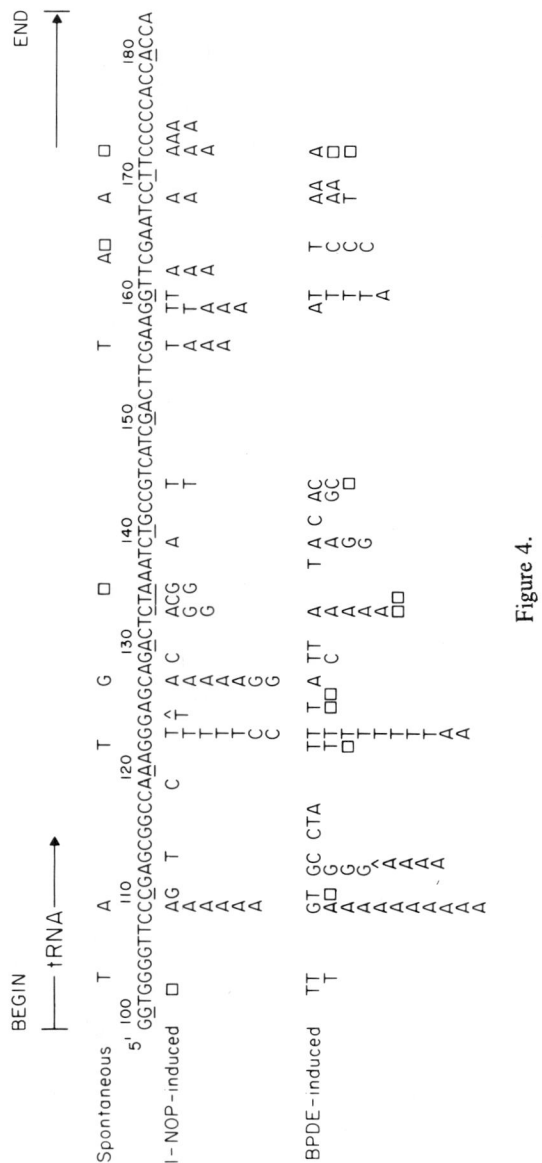

Figure 4.

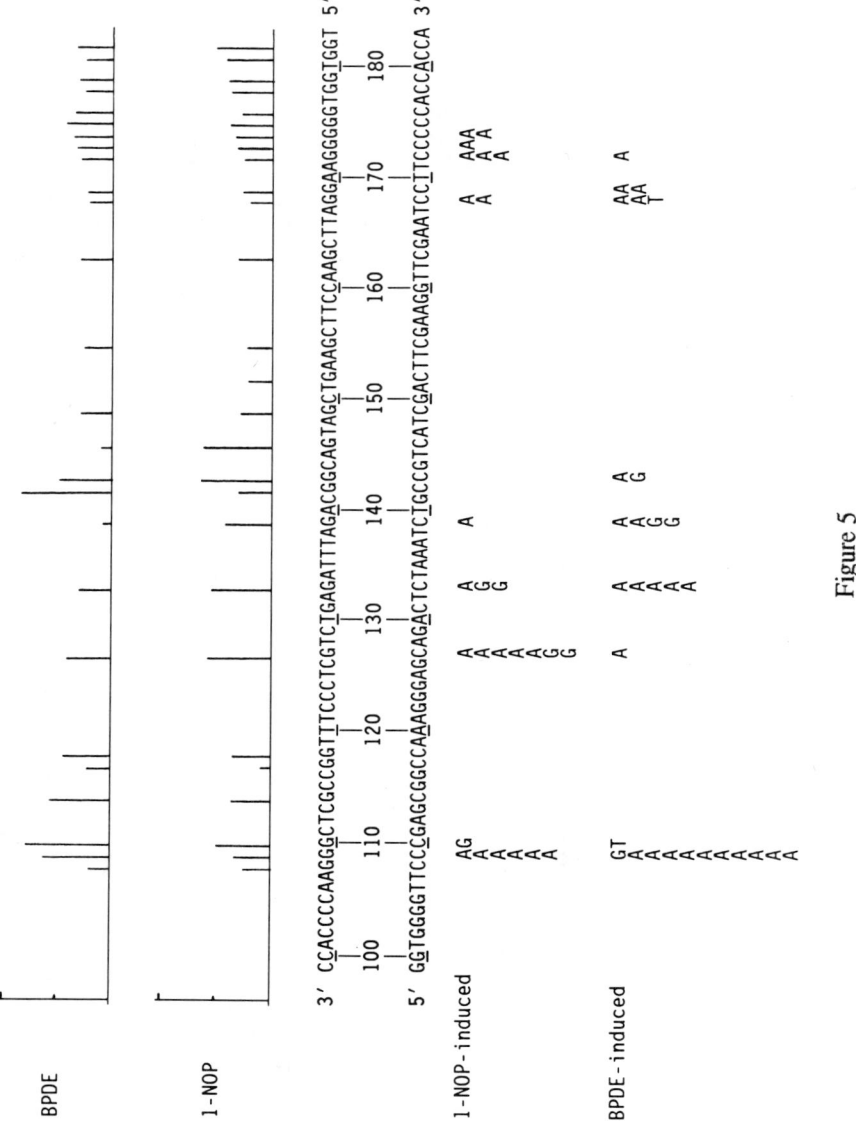

Figure 5

Table 2. Comparison of the kinds of base substitutions generated in *supF* by replication
of 1-NOP- or BPDE-treated plasmid in human cells

Base change	No. of mutations observed		
	Control	1-NOP	BPDE
Transversions:			
G·T→T·A	6	33 (61%)	45 (63%)
G·C→C·G	1	8 (15%)	13 (18%)
A·T→T·A	0	3 (5.5%)	3 (4%)
A·T→C·G	0	1 (2%)	0
Transitions:			
G·C→A·T	0	6 (11%)	6 (9%)
A·T→G·C	0	3 (5.5%)	4 (6%)
Total	7	54	71

spectrum of base changes in the *supF* gene. They exhibited two hot spots in common
(positions 109 and 123), but four of the other BPDE induced hot spots were cold spots
for 1-NOP, and the three other hot spots for 1-NOP were cold spots for BPDE. It should
be pointed out that the majority of the bases that were not changed by either compound
represent bases which have been demonstrated by other investigators to be positions that
will show a mutant phenotype, if altered.

To see if these mutation hot spots corresponded to hot spots for carcinogen binding
in the *supF* gene, we carried out the DNA synthesis stop assay of Moore and Strauss
(1979), in which bulky adducts interfere with DNA replication. Plasmid DNA containing
1-NOP or BPDE adducts (10 to 70 adducts per plasmid, 0.21 to 1.5 adducts per strand of
the *supF* gene) was denatured and annealed with the pBR322 Eco RI site primer, and
polymerization was carried out as described for the sequencing reaction, except that the
dideoxynucleotides were omitted. DNA from the four dideoxy sequencing reactions,
carried out on an untreated *supF* template, was electrophoresed on the same gel to serve
as DNA size markers. The relative intensities of the band on the autoradiograph of the gel
were determined by a laser densitometer and were corrected for position in the gene by
taking into account the number of ^{35}S-labeled adenine bases that would be present in each
length of newly synthesized DNA. The gel pattern of the bands obtained (Fig. 5)
corresponded to positions one nucleotide 5' to virtually every cytosine in the DNA
sequencing standard lane, indicating that DNA synthesis was terminated one base prior to
each guanine in the template. No bands corresponding to positions one nucleotide away
from any base other than guanine were seen, and there was no evidence of any interference
with polymerization using untreated template. If one assumes that these bands were
generated by the Klenow fragment of DNA polymerase I falling off the template at the

sites of bulky adducts, the results indicate a lack of correlation between the frequency of adduct formation in the plasmids at the time of carcinogen treatment, as judged from the intensity of the bands in the stop assay gel, and the frequency of mutations produced in the *supF* gene.

Studies with the aromatic amide derivatives N-AcO-AAF and N-AcO-TFA-AF

Similar studies are underway comparing the frequency and kinds of *supF* mutants induced when plasmids containing AAF or AF adducts replicate in human 293 cells. We have found that although the acetylated AAF residues, formed by reacting N-AcO-AAF with pZ189 in vitro, were twice as effective in interferring with bacterial transformation as the deacetylated AF residues, the mutagenic effectiveness of the two kinds of adducts was equal. Their mutation frequencies per adduct were similar to those found for 1-NOP, which also binds to the C8 position of guanine. A study of the specific spectrum of mutations induced by AAF and AF residues in the *supF* gene is in progress, but data from agarose gel electrophoresis analysis and from the sequencing analyses carried out to date indicate that the majority of the mutants induced by AF adducts contain base substitutions, predominantly G·C→T·A transversions (Mah et al., 1989).

The fact that these carcinogens cause base subsitutions mainly at G·C base pairs suggest that their mutagenesis is targeted to sites where adducts occur. The majority of the base changes were G·C→T·A transversions. This may be the result of the DNA polymerase in the human cells preferentially inserting an adenine nucleotide opposite a non instructional base containing a bulky adduct ("A rule") as suggested by Strauss et al. (1982). However, G·C→C·G transversions were also found. Another explanation for the predominance of G·C transversions, and the one which we favor, is that the presence of a bulky adduct on guanine allows purine purine base pairing to occur with some frequency and that if this misrepairing is not recognized, the result is a permanent change (mutation). Evidence of stable purine purine base pairing in a model oligonucleotide has recently been provided by Brown et al. (1986).

We did not find a correlation between the frequency of BPDE or 1-NOP induced stops in the polymerase stop assay, (putative adduct sites) and the frequency of base substitution mutations at these sites. Although the "A rule" or the purine purine mispairing mechanisms may explain G·C→T·A transversions, they cannot easily explain the location of the mutational hot spots. This lack of correlation cannot be the result of "silent" mutations, since all but six of the positions which gave frequent stops for the polymerase stop assay, but did not undergo mutations, are positions that we and others have shown will give a detectable phenotype when altered. The lack of correlation between initial adduct formation (as judged by the stop assay) and eventual base substitution during replication for the carcinogen treated plasmid in the host cells could reflect differential DNA repair. That is, nucleotide excision repair might preferentially remove adducts from particular locations in the gene.

Although both BPDE, 1-NOP, and N-AcO-TFA-AF, induced point mutations, with single base substitutions predominating and with G·C→T·A transversions being the most common mutation, BPDE adducts were 3.7 fold more mutagenic than 1-NOP or AF or AAF adducts. Since BPDE binds predominantly to the N2 position of guanine, it might well interfere more directly with base pairing than would these other guanine adducts. But the fact that 1-NOP adducts were lost from genomic DNA of the human cells at a rate three times faster than BPDE adducts suggests that following transfection, 1-NOP adducts may be removed from plasmid DNA faster than BPDE adducts and disappear before replication of the plasmids occurs. Information on this possibility is not yet available, but if this were the case, such adducts might actually be equal to BPDE adducts in their mutagenic effectiveness in human cells.

Acknowledgements

We thank Dr. F. A. Beland of the National Center for Toxicological Research for radiolabeled 1-NOP and for his helpful advice with the ascorbic acid reaction. We thank Drs. Charles M. King and Thomas M. Reid for synthesizing the N-AcO-TFA-AF and assisting with the reaction of the compound with plasmid DNA. This research was supported in part by Grant CA21253 from the National Cancer Institute and by Contract 87 2 from the Health Effects Institute (HEI), an organization jointly funded by the U.S. Environmental Protection Agency (EPA) (Assistance Agreement x812059) and automotive manufacturers. It is currently under review by the Institute. The contents of this article do not necessarily reflect the views of the HEI, nor do they necessarily reflect the policies of EPA, or automotive manufacturers.

References

Beland, F. A., M. Ribovict, P. C. Howard, R. H. Heflich, P. Kurian, and G. E. Milo (1986) Cytotoxicity, cellular transformation and DNA adducts in normal human diploid fibroblasts exposed to 1 nitrosopyrene, a reduced derivative of the environmental contaminant, 1-nitropyrene. Carcinogenesis, 7, 1279–1283.

Beland, F. A., K. L. Dooley, and D. A. Casciano (1979) Rapid isolation of carcinogen bound DNA and RNA by hydroxyapatite chromatography. *J. Chromatogr.*, 174, 177–186.

Brown, T., W. N. Hunter, G. Kneale, and O. Kennard (1986) Molecular structure of the G A base pair in DNA and its implications for the mechanism of transversion mutations. *Proc. Natl. Acad. Sci. (U.S.A.)*, 83, 2402–2406.

Heflich, R. H., P. C. Howard, and F. A. Beland (1985) 1-Nitrosopyrene: An intermediate in the metabolic activation of 1-nitropyrene to a mutagen in *Salmonella typhimurium* TA1538. *Mutat. Res.*, 149, 25–32.

Heflich, R. H., N. F. Fullerton, and F. A. Beland (1986) An examination of the weak mutagenic response of 1-nitropyrene in Chinese hamster ovary cells. *Mutat. Res.*, 161, 99–108.

Heflich, R. H., Z. Djuric, Z. Zhuo, N. F. Fullerton, D. A. Casciano, and F. A. Beland (1988) Metabolism of 2-acetylaminofluorene in the Chinese hamster ovary cell mutation assay. *Environ. Mol. Mutag.*, 11, 167–181.

Mah, M.C-M., V.M. Maher, H. Thomas, T.M. Reid, C.M. King and J.J. McCormick (1989) Mutations induced by aminofluorene DNA adducts during replication in human cells. *Carginogenesis*, 10, 2321–2328.

Maher, V. M., J. L. Yang, C. M. Mah, and J. J. McCormick (1989) Comparing the frequency and spectra of mutations induced when an SV-40 shuttle vector containing covalently bound residues of structurally related carcinogens replicates in human cells. *Mutat. Res.* 220, 83–92.

Moore, P. and B. S. Strauss (1979) Sites of inhibition of *in vitro* DNA synthesis in carcinogen- and UV-treated oX174 DNA. *Nature (London)*, 278, 664–666.

Patton, J. D., V. M. Maher, and J. J. McCormick (1986) Cytotoxic and mutagenic effects of 1-nitropyrene and 1-nitrosopyrene in diploid human fibroblasts. *Carcinogenesis*, 7, 89–93.

Poirier, M. C., G. M. Williams, and S. H. Yuspa (1980) Effect of culture conditions, cell type, and species of origin in the distribution of acetylated and deacetylated deoxyguanosine C-8 adducts of *N*-acetoxy-2-acetylaminofluorene. *Mol. Pharmacol.* 18, 581–587.

Saiki, R. K., S. Scharf, F. Faloona, K. B. Mullis, G. T. Horn, H. A. Erlich, and N. Arnheim (1985) Enzymatic amplification of globin genomic sequences and restriction site analysis for diagnosis of sickle cell anemia. *Science*, 230, 1350–1354.

Sanger, F., S. Nicklen, and A. R. Coulson (1977) DNA sequencing with chain terminating inhibitors. *Proc. Natl. Acad. Sci. (U.S.A.)*, 74, 5463–5467.

Sarkar, S., U. B. Dasgupta, and W. C. Summers (1984) Error prone mutagenesis detected in mammalian cells by a shuttle vector containing the *sup*F gene of *Escherichia coli* . *Mol. Cell. Biol.*, 4, 2227–2230.

Seidman, M. M., K. Dixon, A. Razzaque, R. J. Zagursky, and M. L. Berman (1985) A shuttle vector plasmid for studying carcinogen induced point mutations in mammalian cells. *Gene*, 389, 233–237.

Strauss, B., S. Rabkin, D. Sagher, and P. Moore (1982) The role of DNA polymerase in base substitution mutagenesis on noninstructural templates. *Biochimie*, 64, 829–838.

Vousden, K. H., J. L. Bos., C. J. Marshall, and D. H. Phillips. (1986) Mutations activating human c-Ha-*ras1* protooncogene *(HRAS1)* induced by chemical carcinogens and depurination. *Proc. Natl. Acad. Sci. (U.S.A.)*, 83, 1222–1226.

Weinstein, I. B., A. M. Jeffrey, K. W. Jennette, S. H. Blobstein, R. G. Harvey, C. Harris, H. Autrup, H. Kasai, and K. Nakanaishi (1976) Benzo[a]pyrene diol epoxides as intermediates in nucleic acid binding *in vitro* and *in vivo* . *Science*, 193, 592–595.

Yang, J.-L., V. M. Maher, and J. J. McCormick (1987) Kinds of mutations formed when a shuttle vector containing adducts of (±)-7β,8α–dihydroxy-9α,10α-epoxy-7,8,9,10-tetrahydrobenzo[a]pyrene replicates in human cells. *Proc. Natl. Acad. Sci. (U.S.A.)*, 84, 3787–3791.

Yang, J.-L., M. C. M. Mah, C. M. King, V. M. Maher, and J. J. McCormick (1988a) Ability of DNA adducts formed in a shuttle vector by reactive derivatives of 1-nitropyrene, 2-aminofluorene, and benzo[a]pyrene to induce mutations when the plasmid replicates in human cells. In: Carcinogenic and Mutagenic Responses to Aromatic Amines and Nitroarenes. C. M. King, L. J. Romano, and D. Schuetzle (eds.), Elsevier, New York, pp. 367–371.

Yang, J.-L., V. M. Maher, and J. J. McCormick (1988b) Kinds and spectrum of mutations induced by 1-nitrosopyrene adducts during plasmid replication in human cells. *Mol. Cell. Biol.*, 8, 3364–3372.

Yang, J-L., V.M. Maher and J.J. McCormick (1989) Amplification and direct nucleotide sequencing of cDNA from the lysate of low numbers of diploid human cells. *Gene*, 83, 347–354.

Twenty Years of Research on Xeroderma Pigmentosum at the National Institutes of Health

K.H. KRAEMER

Laboratory of Molecular Carcinogenesis
National Cancer Institute
National Institutes of Health
Bethesda, Maryland, 20892
U.S.A.

Abstract

Interest in studying xeroderma pigmentosum (XP) at the National Institutes of Health began in 1968 following the discovery of the DNA repair defect in XP by Dr. James Cleaver. The series of investigations initiated by Dr. Jay Robbins in the Dermatology Branch of the National Cancer Institute has continued to the present. In addition, XP has been studied by individual researchers and by collaborations among investigators in several Institutes and Laboratories. These studies demonstrated that by combining clinical and laboratory studies, XP could be a model for examining mechanisms of ultraviolet radiation-induced carcinogenesis. XP has also provided a model for human neuro-degenerative disorders. In recent years, XP has been used to evaluate DNA repair, the molecular basis of ultraviolet mutagenesis and cancer chemoprevention in humans.

The National Institutes of Health (NIH) is the major biomedical research facility of the United States Government. The intramural facility is located on 300 acres of land in Bethesda, Maryland, a suburb of Washington, D.C. There are about 60 buildings and more than 13,000 employees, including nearly 2000 with doctoral degrees, on a university campus-type setting with extensive grass and trees. There are 12 Institutes, each specializing in a organ system or disease, and the National Library of Medicine. Each Institute is divided into Branches or Laboratories which are roughly equivalent to university departments. Researchers are generally permanent staff members or post-doctoral fellows (M.D. or Ph.D.). A 540 bed research hospital, the Warren Grant Magnison Clinical Center, is the location of all inpatient and outpatient clinical studies and also contains many research laboratories.

Dr. Jay H. Robbins, a Senior Investigator in the Dermatology Branch of the National Cancer Institute (NCI) began studying xeroderma pigmentosum (XP) in 1968 following a lecture at the NIH by Dr. James E. Cleaver of the Radiobiology Laboratory of the University of California, San Francisco. Dr. Robbins studies of XP have continued to the present. Dr. Cleaver indicated that cultured skin fibroblasts from XP patients had defective excision repair of ultraviolet (UV) damaged DNA (Cleaver, 1968). Dr. Robbins reasoned that since XP was an inherited disorder, all tissues from an XP patient, including blood cells, should show the defect. Working with Dr. Peter G. Burk, a Clinical Associate, they developed a rapid assay to measure DNA repair in peripheral blood mononuclear cells. They soon were able to confirm that the DNA repair defect was present in circulating blood cells from XP patients (Burk et al, 1971a). Subsequent studies at NIH demonstrated the XP defect in every tissue examined: in cutaneous epidermal cells (Robbins et al, 1972), dermal fibroblasts (Robbins et al, 1974), cultured skin cancer cells (Robbins et al, 1975), conjunctival cells of the eye (Newsome et al, 1975), and cultured nevus cells and melanocytes (Kraemer et al, 1989). Studies performed elsewhere have shown the DNA repair defect in vivo in cutaneous epidermal cells, blood vessel walls, and smooth muscle cells (Epstein et al, 1970), and in liver cells (Dupuy et al, 1974)

Drs. Robbins and Burk chose to study XP patients with severe skin disease. The clinical research facility at NIH provided the opportunity to bring patients for study who had been referred from areas throughout the United States. The first 3 patients showed the anticipated defect in DNA repair following UV treatment of fresh blood cells (Robbins et al, 1974). The fourth patient, XP4BE, a 27 year old man with multiple skin cancers and metastatic melanoma, had clinical features equally severe as the other 3 patients. Surprisingly, he did not show the DNA excision repair defect in his cultured blood cells (Burk et al, 1971a). Additional studies revealed normal DNA excision repair in dermal fibroblasts (Burk et al, 1971b), epidermal cells and tumor cells (Robbins et al, 1975). This unusual finding was subsequently confirmed by Dr. Cleaver who called this patient an XP "variant" (Cleaver, 1972). Later studies in Japan and Europe have identified more than 40 XP variant patients - representing about 25% of the total classified XP patients (Kraemer et al, 1987a). Cleaver later studied fibroblasts from "pigmented xerodermoid" patients described by Dr. Ernst Jung of Germany (Jung, 1970) with late onset of symptoms following extensive sun exposure. Cleaver found them to have the laboratory features of XP variants (Cleaver et al, 1981).

In 1972, scientists at Erasmus University in Rotterdam, The Netherlands, under the direction of Dr. Dirk Bootsma performed cell fusion studies on fibroblasts from patients with XP (De Weerd-Kastelein et al, 1972). While unfused fibroblasts from each patient had low DNA repair (measured by autoradiography at the single cell level), fused cell heterokaryons showed increased repair in all nuclei. Thus each cell supplied components that the other was lacking. These cells were said to be in different complementation groups. Even though the precise defect was not known, these cell fusion studies demonstrated that different complementation groups had different DNA repair defects.

Dr. Robbins learned of reports of 3 XP patients with apparent neurological abnormalities and invited them to NIH for further studies (Robbins et al, 1974). Patients 5 and 6 of the NIH series, 26 and 20 year old sisters were found to have multiple skin cancers, deafness, and relatively late onset of mental retardation with normal sexual

212

development. Patient 11, a 28 year old woman had the skin changes of XP plus the features of Cockayne syndrome (dwarfism, immature sexual development, deafness, mental retardation, bony abnormalities, and pigmentary retinal degeneration). She was brought to the genetics clinic at Johns Hopkins University in Baltimore, Maryland, where the clinical diagnosis of Cockayne syndrome was confirmed, in addition to that of XP.

In 1971, as a Clinical Associate in the Dermatology Branch of the NCI working with Dr. Robbins, I was assigned the task of following Dr. Bootsma's lead and determining whether the XP patients at NIH were in different complementation groups. I learned the technique of somatic cell fusion from Dr. Hayden G. Coon, Senior Investigator in the Laboratory of Cell Biology, NCI. We fused cells from XP patients with varied clinical features and discovered 4 complementation groups (Kraemer et al, 1975a). A precedent of openness with XP research materials and cell lines was established by Dr. Robbins and Dr. Bootsma. XP cell lines were exchanged between the Rotterdam group and Bethesda. In a collaborative study 5 complementation groups were found (Kraemer et al, 1975b). These were named A-E in order of increasing levels of residual DNA synthesis.

Dr. Robbins felt strongly that continued research in XP would depend on unimpeded availability of authentic healthy XP cell lines. We had verified the clinical and laboratory diagnoses of the XP strains from patients studied at NIH. He initially arranged for contribution of the first XP fibroblast lines to the American Type Culture Collection, in Rockville, Maryland, one of the first human diseases sent there. Subsequently, he sent other lines to the Human Mutant Cell Repository at the Institute for Medical Research, Camden, N.J. The large number of papers using these cell lines and the rapid growth of investigations of XP is in large measure due to the existence of these cell repositories and the decision of investigators at the NIH and in Europe to supply XP cells freely.

Dr. Robbins realized that a large supply of XP cells for biochemical and cellular studies could more readily be obtained with Epstein-Barr virus transformed lymphoblastoid cell lines than with primary fibroblasts with limited life span. The first XP lymphoblastoid cell lines were established by Dr. Donald N. Buell, of the Immunology Branch, NCI (Andrews et al, 1974). Dr. Robbins, working with Dr. Alan Andrews, a Clinical Associate, was able to show that the lymphoblastoid cell lines have the characteristic XP UV repair deficiency (Andrews et al, 1974). These and other XP lines also were made available to the scientific community through the Human Mutant Cell Repository.

Often a series of scientific advances occur in isolation and the "big picture" only emerges after synthesis of these concepts. In 1974, Dr. Robbins organized a NIH Combined Staff Conference on XP. The transcript of this conference, published in the Annals of Internal Medicine (Robbins et al, 1974), was to become the first comprehensive review of clinical features and laboratory advances in XP and has been widely cited. There were presentations by a dermatologist, Dr. Marvin A. Lutzner, Chief of the Dermatology Branch, NCI, a neurologist, Dr. Barry W. Festoff, a Clinical

Associate in the Medical Neurology Branch, National Institute of Neurological Diseases
and Stroke, a cell biologist, Dr. Coon, Dr. Robbins and myself. Clinical features of 15
XP patients examined at the NIH were presented. This represented one of the largest
series of XP patients closely examined by one group of clinicians. The common features
of sunlight-induced damage of the skin and eyes were emphasized. A detailed analysis was
presented of the progressive neurological degeneration of some of the XP patients. A
theory was proposed linking the loss of neurons in addition to the loss of skin cells to
inefficient repair of damage to DNA. Ultraviolet photobiology and the excision repair
pathway were reviewed. The fact that different XP patients have different residual repair
levels was emphasized. The rate of repair was slower in XP than normal but after a single
UV exposure, repair continued for a longer time in the XP than in the normal. Cell
fusion studies of fibroblasts from 12 XP patients were presented. They were shown to be
in complementation groups A, B, C, and D. The clinical features of the patients in
different complementation groups were defined for the first time: group A contained
patients with severe neurological abnormalities as well as patients with minimal
neurological abnormalities, group B contained only one patient, XP11BE, the woman
with both XP and Cockayne's syndrome, group C was the most numerous and contained
patients with no neurological abnormalities, and group D had patients with neurological
abnormalities.

Several reviews of XP have subsequently been published from the NIH. These
include textbook and reference articles (Kraemer, 1980, 1987b; Kraemer and Slor, 1985;
Cleaver and Kraemer, 1989); a review in the Journal of the National Cancer Institute
(Robbins, 1978); a 1988 Grand Rounds at the NIH (Robbins, 1988a); and a literature
survey of more than 800 case reports (Kraemer et al, 1984, 1987a). The first layman's
guide to XP, "Understanding Xeroderma Pigmentosum" was recently published by the
Clinical Center Communications Office of the NIH (Kraemer et al, 1988).

Drs. Robbins and Andrews made a precise study of the post-UV colony forming
ability of fibroblasts from different XP patients (Andrews et al, 1976, 1978). They
discovered a characteristic range of UV survival for each of 5 complementation groups
and the variant form. Those cell lines that were the most sensitive to killing by UV came
from patients who had the most neurological abnormalities. Fibroblasts from patients
such as XP12BE with relatively few neurological abnormalities had better UV survival
than fibroblasts from other patients in the same complementation group with severe
neurological abnormalities. Using XP as a model, Dr. Robbins working with Dr. Fujio
Otsuka, a Visiting Fellow from Japan, have discovered cellular hypersensitivity in
several other neurodegenerative diseases including Huntington's disease, Alzheimer's
disease, Parkinson's disease, and Down's syndrome (Robbins, 1988b).

Dr. Robbins, working with Dr. Alan N. Moshell, a Clinical Associate and Dr.
Lutzner, former Chief of the Dermatology Branch, NCI who was living in Paris, found a
second patient with both XP and Cockayne Syndrome living in France. They determined
that this patient was in a separate complementation group, designated group H (Moshell
et al, 1983).

Dr. Kurt W. Kohn, Chief of the Laboratory of Molecular Pharmacology, NCI, in 1973 developed a sensitive assay for DNA strand breaks known as alkaline elution. The mechanism of excision repair, as understood at that time, required that DNA single-strand breaks would be created near the lesion sites and that breaks would reseal at a later stage of the repair process. There was, however, no unequivocal evidence for for such transient strand breaks in mammalian cells. Dr. Kohn wanted to determine the rates at which repair-induced breaks would form and reseal and whether these processes would be altered in repair deficient cells. The ultracentrifugal method available at that time had limited sensitivity. Dr. Kohn and Dr. Herbert E. Kann, Jr., a Clinical Associate, worked on improving the alkaline elution methodology. Dr. Kohn, along with Dr. Alfred J. Fornace, Jr., a Staff Associate, in 1976 applied this alkaline elution technique to XP cells (Fornace et al, 1976). The assay involves gentle alkaline lysis of radiolabeled cells on a membrane filter and the slow elution of the DNA through the filter as an alkaline solution is pumped through the filter. The rate of elution of the cellular DNA is proportional to the size of the DNA fragments. This assay successfully detected the transient DNA strand breaks introduced by the cellular excision repair system during recovery from UV damage in normal cells. In XP cells of complementation groups A, B, C and D, the formation of strand breaks following UV was greatly diminished. In 3 XP variant strains, however, strand breaks appeared in approximately normal numbers, although the rate of resealing was reduced, giving rise to the concept that these cells had an abnormality in a mechanism that allows normal cells to replicate DNA past a lesion in the template strand. These studies with XP were the first to demonstrate the extreme sensitivity of this assay.

Several investigators independently developed viral "host cell reactivation" assays with XP cells to measure DNA repair capacity of cultured cells. These assays were modeled after similar assays with prokaryotes. The damaged virus was used to infect the cells to be tested. The virus is unable to repair the damage but relies on the cells' DNA repair enzymes for recovery. With normal cells, the viral damage is repaired and the virus proceeds through replication producing progeny which eventually result in a "plaque" of infected cells on a confluent lawn of cultured cells. In cells with defective DNA repair, damage to the virus is not corrected, viral replication does not progress and plaques are not formed. Dr. Alan S. Rabson, Chief of the Laboratory of Pathology, NCI, along with Drs. Sandra A. Tyrell and Francis Y. Legallais were the first to demonstrate this effect in XP cells (Rabson et al, 1969). They infected cultured fibroblasts from an XP patient from two normal controls with UV treated herpesvirus and assayed viral production on rat kidney cells. Viral survival was more than 10-fold lower with the XP cells. Dr. Stuart A. Aaronson and Dr. C. David Lytle, in the Viral Leukemia and Lymphoma Branch, NCI used SV40 virus (Aaronson and Lytle, 1970) with similar results.

Dr. Rufus S. Day III, a Senior Investigator in the Chemistry Branch, of the NCI used UV damaged adenovirus (Day, 1974). Dr. Day was able to show that cells from 12 XP patients in 4 complementation groups were defective in this functional assay of DNA repair (Day, 1974). Dr. Robbins and I collaborated with Dr. Day to show that cell fusion

215

accomplished complementation of a survival function by demonstrating increased adenovirus host cell reactivation in complementing fused XP cells from complementation groups A and D. (Day et al, 1975). Dr. Day discovered that cells from XP4BE, the "variant" with normal excision repair, had defective DNA repair as measured by the adenovirus host cell reactivation assay (Day, 1975). This was the first laboratory demonstration of a DNA repair defect in the XP variant cells. The adenovirus host cell reactivation assay is still one of the few tests that is abnormal with every form of XP. Subsequently, studies by Dr. Alan Lehmann (Lehmann et al, 1975) in Sussex, England, described a post-replication repair defect with caffeine hypersensitivity in the XP variants.

In 1985, Dr. Miroslava Protic'-Sabljic', a Visiting Fellow from Yugoslavia and I, now a Research Scientist in the Laboratory of Molecular Carcinogenesis, NCI, described a method of adapting the viral host cell reactivation assay for use with plasmids (Protic'-Sabljic' and Kraemer, 1985). A non-replicating, plasmid expression vector had recently been constructed by Dr. Bruce Howard, Section Head in the Laboratory of Molecular Biology, NCI with Dr. Cornelia M. Gorman, a Staff Fellow, and Mrs. Raji Padmanabhan, a versatile technician (Gorman et al, 1982, 1983). This plasmid contained the bacterial gene for chloramphenicol acetyl transferase (cat) in a construction that would permit expression in human cells. We treated the plasmid with UV in vitro and introduced it into the XP and normal cells by calcium phosphate transfection. Normal cells repair the DNA damage, transcribe the RNA and produce active cat protein. Cat activity two days after transfection was measured in crude cell extracts by thin layer chromatography. XP cells are unable to repair the damage to the plasmid and exhibit low cat expression. Analysis of the target size indicated that with the XP-A cells one pyrimidine dimer per cat gene was sufficient to inactivate cat expression. Selective removal of cyclobutane dimers by use of E. coli photolyase (a gift from Dr. Aziz Sancar of the University of North Carolina) demonstrated that XP cells were also unable to repair non-dimer photoproducts (Protic'-Sabljic' and Kraemer, 1986).

Dr. Michael M. Seidman, a Senior Staff Fellow in the Laboratory of Molecular Carcinogenesis, NCI, in collaboration with Dr. Abdur Razzaque, a Visiting Fellow, Dr. Kathleen Dixon, a Senior Investigator in the Section on Viruses and Cell Biology of the National Institutes of Child Health and Human Development, and Drs. Michael L. Berman and Robert J. Zagursky of the NCI-Frederick Cancer Research Facility developed a plasmid vector designed to study mutagenesis in mammalian cells (Seidman et al, 1985). The plasmid contained SV40 sequences that permitted replication in some mammalian cells (monkey kidney, human), the prokaryotic origin of replication and ampicillin resistance genes to permit growth and selection in bacteria, and a small (150 base pair) mutagenesis marker gene (supF suppressor transfer RNA). The plasmid was treated with UV in vitro, transfected into the human host cells where repair, mutagenesis and replication occur utilizing the host cell enzymes, harvested after 2 days, and used to transform indicator bacteria to ampicillin resistance. On agar containing ampicillin, the number of bacterial colonies reflect the plasmid survival and thus the efficiency of repair in the human cells. The bacteria also contain an (amber) mutation in the gene for beta-galactosidase which can be suppressed by a functioning supF gene. The agar also contains an inducer of beta-galactosidase, IPTG, and a colorless dye, X-gal, which turns blue on cleavage by beta-galactosidase. Thus blue colonies indicate a wild type supF gene in the plasmid while light blue or white colonies are present as a result of a mutation in the supF gene in the plasmid. The mutation frequency is determined by determining the proportion of white and light blue colonies. The plasmids from these colonies can be purified and the DNA sequence of the supF gene determined thereby revealing the mutagenic consequences of the initial DNA damage.

One of the first applications of this plasmid "shuttle vector" was to examine UV mutagenesis in XP cells. Dr. Seidman, in collaboration with Dr. Anders Bredberg, a Visiting Fellow from Sweden, and I examined pZ189 mutagenesis in SV40 transformed fibroblasts from NIH patient XP12BE in complementation group A (Bredberg et al, 1986). Dr. Seidman, Protic'-Sabljic' and I subsequently performed similar studies in cells from NIH patient XP6BE along with Dr. Saraswathy Seetharam, a Visiting Fellow from India (Seetharam et al, 1987). Both of these studies indicated that, like earlier studies of cell survival and adenovirus host cell reactivation, survival of UV treated plasmids was reduced in XP cells. An increased frequency of mutations was found with the XP lines. The plasmid assay permitted, for the first time, a molecular analysis of mutations introduced by the XP cells in response to UV damage. Several striking features emerged: i. The variety and types of base sequence mutations found with the XP lines was reduced in comparison to those found with pZ189 with the normal lines. ii. Different mutational hotspots were observed with the XP and the normal lines. These observations indicated that DNA repair systems had profound effects on the types of mutations induced by UV.

The predominance of the G:C to A:T transition mutation following UV treatment of pZ189 with all cell lines indicated that photoproducts involving cytosine, rather than thymine were responsible for most of the mutations. Dr. Douglas E. Brash, a Senior Staff Fellow in the Laboratory of Human Carcinogenesis, NCI, had previously developed a technique to measure cyclobutane dimer and 6-4 photoproduct frequency at the base sequence level (Brash and Hazeltine, 1982). Dr. Brash applied this technique to analysis of the photoproduct frequency in UV treated pZ189 (Brash et al, 1987). A wide range in frequency of cyclobutane dimers and 6-4 photoproducts was found.

The question as to the identity of the mutagenic photoproduct in XP was addressed by use of the plasmid vector (Brash et al, 1987). UV treated pZ189 was reacted with E. coli photolyase (from Dr. Sancar) to selectively remove about 99% of the cyclobutane dimers but not alter other photoproducts, and then used to transfect XP-A cells. Plasmid survival increased about 90% and mutation frequency decreased about 90% indicating that cyclobutane dimers were the major contributors to plasmid killing and mutagenesis. The location of the remaining mutations was instructive. The G:C to A:T transitions again predominated and the mutagenic hotspots were found at the same sites as before photoreactivation. This indicated that in the plasmid in the XP cells, both cyclobutane dimers and 6-4 photoproducts were mutagenic.

Comparison of the frequency of transition mutations in UV treated pZ189 with the

XP-A cells to the mutation frequency revealed a surprising result: there was no correlation between photoproduct frequency and mutation frequency! Some sites with a high frequency of photoproducts had a low number of mutations and other sites with low frequency of photoproducts had high numbers of mutations. This indicated that factors other than photoproducts were of greatest importance in determining the mutagenic consequences of UV. Drs. Seidman and Brash (Brash et al, 1987) proposed a "pass-fail" rule whereby at some locations structural changes induced by photoproducts permit polymerases to pass but, since the photoproduct prevents accurate reading of the base sequence, errors are likely. At other sites, the structural changes induced by the same photoproduct would result in a block to the polymerase and cause the polymerase to fail to progress resulting in a lethal lesion.

One class of mutants, those with multiple base substitutions, was virtually absent in pZ189 passed through the XP-A cells (Bredberg at al, 1986). Dr. Seidman reasoned that since the XP cells are defective in the initial incision step of DNA repair, introduction of a single strand nick in the plasmid might alter the processing of the plasmid by the XP-A cells. There was a markedly increased frequency of multiple base substitution mutants in nicked pZ189 passed through the XP-A cells (Seidman et al, 1987). This was important evidence for action of a localized, error-prone polymerase in human cells.

Clinical studies were preceeding in parallel with the molecular biological investigations. A large literature review was initially performed by Mr. Myung Moo Lee, a medical student working in my laboratory, and analyzed in collaboration with an epidemiologist with an interest in skin cancer, Mr. Joseph Scotto, of the Biostatistics Branch, NCI. We estimated that in XP patients under 20 years of age the incidence of basal cell carcinoma, squamous cell carcinoma or melanoma of the skin was increased more than 1000 times that of the general population (Kraemer et al, 1984, 1987).

This extremely high frequency of skin cancers in XP patients enabled us to examine the possibility that an oral drug could prevent the formation of skin cancer. By studying people with XP we could obtain information on a large number of cancers in a short time with only a relatively small number of patients. In a collaborative study (Kraemer et al, 1988) with Drs. Gary L. Peck, Senior Investigator, and John J. DiGiovanna, Expert, in the Dermatology Branch, NCI, Dr. Alan N. Moshell of the National Institute of Arthritis, Musculoskeletal, and Skin Diseases, and Dr. Robert E. Tarone, in the Biostatistics Branch, NCI, we selected 5 XP patients who had a high frequency of pathologically documented skin cancers. We treated them for 2 years with high dose (2 mg/kg/da) oral isotretinoin (a retinoid) and then followed the patients for a third year without isotretinoin. The five patients had a total of 121 skin cancers in the two year interval before treatment and a total of only 25 tumors in the two years of treatment with isotretinoin. When the drug was discontinued the tumor frequency increased a mean of 8.5-fold. Although all patients experienced extensive drug related side effects, this study was one of the first demonstrations of successful prevention of cancer in humans by use of a drug.

Dr. Katherine Sanford, a Section Head in the Laboratory of Cellular and Molecular Biology, NCI, working with Dr. Ram Parshad of Howard University, Mr. Gary Jones, a talented technician, Dr. Tarone and I were for the first time able to detect carriers of XP by use of a laboratory test (Parshad et al., 1988). Dr. Sanford developed an assay to measure chromosome breakage following x-ray treatment in the G2 phase of the cell cycle. Cells from XP patients showed a delayed sealing of chromosome breaks and gaps in comparison to normal controls. Peripheral blood lymphocytes from obligate XP heterozygotes (parents of the patients) consistently showed an intermediate response.

Acknowledgements

I would like to thank Drs. Jay Robbins and Kurt Kohn for their comments on this manuscript.

References

Aaronson, S.A. and Lytle, C.D. (1970) Decreased host cell reactivation of irradiated SV40 virus in xeroderma pigmentosum. *Nature* [New Biol] 228: 359-361.

Andrews, A.D., Robbins, J.H., Kraemer, K.H., and Buell, D.N. (1974): Xeroderma pigmentosum long-term lymphoid lines with increased UV sensitivity. *J. NCI.* 53: 691-693.

Andrews, A.D., Barrett, S.F., and Robbins, J.H. (1976) Relation of DNA repair processes to pathological aging of the nervous system in xeroderma pigmentosum. *Lancet 1*: 1318-1320.

Andrews, A.D., Barrett, S.F., and Robbins, J.H. (1978) Xeroderma pigmentosum neurological abnormalities correlate with colony-forming ability after ultraviolet radiation. *Proc. Natl. Acad. Sci. USA* 75: 1984-1988.

Brash, D.E. and Hazeltine, W.A. (1982) UV-induced mutation hotspots occur at DNA damage hotspots. *Nature* 298: 189-192.

Brash, D.E., Seetharam, S., Kraemer, K.H., Seidman, M.M. and Bredberg, A. (1987): Photoproduct frequency is not the major determinant of ultraviolet mutation hotspots or coldspots in human cells. *Proc. Natl. Acad. Sci. USA* 84: 3782-3786.

Bredberg, A., Kraemer, K.H., and Seidman. M.M. (1986): Restricted mutational spectrum in an UV-treated shuttle vector propagated in xeroderma pigmentosum cells. *Proc. Natl. Acad. Sci. USA* 83: 8273-8277.

Burk, P.G., Lutzner, M.A., Clarke, D.D. et al (1971a) Ultraviolet stimulated thymidine incorporation in xeroderma pigmentosum lymphocytes. *J. Lab. Clin. Med.* 77, 759-767.

Burk, P.G., Yuspa, S.H., Lutzner, M.A., et al (1971b) Xeroderma pigmentosum and DNA repair. *Lancet 1*, 601.

Cleaver, J. E. (1968) Defective repair replication of DNA in xeroderma pigmentosum. *Nature* 218, 652-656.

Cleaver, J.E. (1972) Xeroderma Pigmentosum: variants with normal DNA repair and normal sensitivity to ultraviolet light. *J. Invest. Dermatol.* 58, 124-128.

Cleaver, J.E., Greene, A.E., Coriell, L.L. et al: (1981) Xeroderma pigmentosum variants. *Cytogenet Cell. Genet.* 31: 188-192.

Cleaver, J. and Kraemer, K.H. (1989): Xeroderma pigmentosum. In Scriver, C.R., Beaudet, A.L., Sly, W.S., and Valle, D. (Eds.) The Metabolic Basis of Inherited Disease, Sixth Edition. New York, McGraw Hill, pp 2949-2973.

Day, R.S. III (1974) Studies on repair of adenovirus 2 by human fibroblasts using normal, xeroderma pigmentosum, and xeroderma pigmentosum heterozygous strains. *Cancer Res* 34: 1965-1970.

Day, R.S.III (1975) Xeroderma pigmentosum variants have decreased repair of ultraviolet-damaged DNA. *Nature* 253: 748-749.

Day, R.S., III, Kraemer, K.H., and Robbins, J.H. (1975): Complementing xeroderma pigmentosum fibroblasts restore biological activity to UV-damaged DNA. *Mutat. Res.* 28: 251-255.

De Weerd-Kastelein, E.A., Keijzer, W.J., and Bootsma, D. (1972) Genetic heterogeneity of xeroderma pigmentosum demonstrated by somatic cell hybridization. *Nature (London)* 238, 80-83.

Dupuy J.M., Lafforet, D., Rachman, F. (1974) Xeroderma pigmentosum with liver involvement. *Helv Paediatr Acta* 29: 213-219.

Epstein, J. H., Fukuyama, K., Reed, W.B., et al (1970) Defect in DNA synthesis in skin of patients with xeroderma pigmentosum demonstrated in vivo. Science 168, 1477-1478.

Fornace A.J. Jr., Kohn, K.W., and Kann, H.E. Jr. (1976): DNA single strand breaks during repair of UV damage in human fibroblasts and abnormalities of repair in xeroderma pigmentosum. *Proc. Natl. Acad. Sci. USA* 73: 39-43.

Gorman, C.M., Moffat, L.F., and Howard, B.H. (1982) Recombinant genomes which express chloramphenicol acetyltransferase in mammalian cells. *Mol. Cell. Biol.* 2: 1044-1051.

Gorman, C.M., Padmanabhan, R. and Howard, B.H. (1983) High efficiency DNA mediated transformation of primate cells. *Science* 221: 551-553.

Jung, E.G. (1970) New form of molecular defect in xeroderma pigmentosum. *Nature* 228, 361-362.

Kraemer, K.H., Coon, H.G., Petinga, R.A., Barrett, S.F., Rahe, A., and Robbins, J.H. (1975a): Genetic heterogeneity in xeroderma pigmentosum: Complementation groups and their relationship to DNA repair rates. *Proc. Natl. Acad. Sci. USA* 72: 59-63.

Kraemer, K.H., de Weerd-Kastelein, E.A., Robbins, J.H., Keijzer, W., Barrett, S.F., Petinga, R.A., and Bootsma, D. (1975b): Five complementation groups in xeroderma pigmentosum. *Mutat. Res.* 33: 327-340.

Kraemer, K.H. (1980): Xeroderma pigmentosum. In Demis, D.J., Dobson, R.L., and McGuire, J. (Eds.): Clinical Dermatology, Vol. 4, Unit 19-7. New York, Harper and Row, pp 1-33.

Kraemer, K.H., Lee, M.M., and Scotto, J. (1984): DNA repair protects against cutaneous and internal neoplasia: Evidence from xeroderma pigmentosum. *Carcinogenesis* 5: 511-514.

Kraemer, K.H., and Slor, H. (1985): Xeroderma pigmentosum. *Clinics in Dermatol.* 3: 33-69.

Kraemer, K.H., Lee, M.M., and Scotto, J. (1987): Xeroderma pigmentosum: Cutaneous, ocular and neurologic abnormalities in 830 published cases. *Arch. Dermatol.* 123: 241-250.

Kraemer, K.H. (1987): Heritable diseases with increased sensitivity to cellular injury. In Fitzpatrick, T.B., Eisen, A.Z., Wolff, K., Freedberg, I.M., and Austen, K.F. (Eds.): Dermatology in General Medicine. New York, McGraw Hill, pp 1791-1811.

Kraemer, K.H., DiGiovanna, J.J., Moshell, A.N., Tarone, R.E., and Peck, G.L. (1988a): Prevention of skin cancer with oral 13-cis retinoic acid in xeroderma pigmentosum. *N. Engl. J. Med.* 318: 1633-1637.

Kraemer, K.H., Andrews, A.D. and Rhodes, A.R. (1988b): Understanding Xeroderma Pigmentosum U.S. Dept of Health and Human Services, Clinical Center Communications, NIH.

Kraemer, K.H., Herlyn, M., Yuspa, S.H., Clark, W.H. Jr., Townsend, G.K., Neises, G.R., and Hearing, V.J. (1989): Reduced DNA repair in cultured melanocytes and nevus cells from a patient with xeroderma pigmentosum. *Arch. Dermatol.* 125: 263-268.

Lehmann A., Kirk-Bell, S., Arlett C., Paterson, M.C., Lohman, P.H.M., de Weerd-Kastelein, E.A., and Bootsma, D. (1975) Xeroderma pigmentosum cells with normal levels of excision repair have a defect in DNA synthesis after UV-irradiation. *Proc Natl Acad Sci USA* 72: 219-223.

Moshell, A.N., Ganges, M.B., Lutzner, M.A., et al (1983) A new patient with both xeroderma pigmentosum and Cockayne syndrome establishes the new xeroderma pigmentosum complementation group H. In Friedberg E., Bridges, B. eds. Cellular responses to DNA damage. New York, Alan R. Liss pp 209-213.

Newsome, D.A., Kraemer, K.H., and Robbins, J.H. (1975): Defective repair of UV-damaged DNA in xeroderma pigmentosum conjunctiva. *Arch. Opthalmol.* 93: 660-662.

Parshad, R., Sanford, K.K., Kraemer, K.H., Jones, G.M., and Tarone, R.E. (1990): Carrier detecction in Xeroderma Pigmentosum. *J. Clin Invest.* 85:135–138.

Protic'-Sabljic', M., and Kraemer, K.H. (1985): One pyrimidine dimer inactivates expression of a transfected gene in xeroderma pigmentosum cells. *Proc. Natl. Acad. Sci. USA* 82: 6622-6626.

Protic'-Sabljic', M. and Kraemer, K. H. (1986): Reduced repair of non-dimer photoproducts in a gene transfected into xeroderma pigmentosum cells. *Photochem. Photobiol.* 43: 509-513.

Rabson, A.S., Tyrell, S.A., and Legallais, F.Y. (1969) Growth of ultraviolet-damaged herpesvirus in xeroderma pigmentosum cells. *Proc. Soc. Exp. Biol. Med.* 132, 802-806.

Robbins, J.H., Levis, W.R., and Miller, A.E. (1972) Xeroderma pigmentosum epidermal cells with normal UV-induced thymidine incorporation. *J. Invest. Dermatol.* 59, 402-408.

Robbins, J.H., Kraemer, K.H., Lutzner, M.A., Festoff, B.W., and Coon, H.G. (1974): Xeroderma pigmentosum - An inherited disease with sun sensitivity, multiple cutaneous neoplasms and abnormal DNA repair. *Ann. Intern. Med.* 80: 221-248. (A citation classic – 5/1987)

Robbins, J.H., Kraemer, K.H., and Flaxman, B.A. (1975): DNA repair in tumor cells from the variant form of xeroderma pigmentosum. *J. Invest. Dermatol.* 64: 150-155.

Robbins, J.H. (1978) Significance of repair of human DNA: Evidence from studies of xeroderma pigmentosum. *J. Natl. Cancer Inst.* 61: 645-656.

Robbins, J.H. (1988a) Xeroderma pigmentosum - Defective DNA repair causes skin cancer and neurodegeneration. *J Am Med Assn* 260: 384-388.

Robbins, J.H. (1988b) Defective DNA repair in xeroderma pigmentosum and other neurologic diseases. *Current Opinion in Neurology and Neurosurgery 1*: 1077-1083.

Seetharam, S., Protic'-Sabljic', M., Seidman, M.M. and Kraemer, K.H. (1987): Abnormal ultraviolet mutagenic spectrum in DNA replicated in cultured fibroblasts from a patient with the skin cancer-prone disease, xeroderma pigmentosum. *J. Clin. Invest.* 80: 1613-1617.

Seidman M.M., Dixon, K., Razzaque, A., Zagursky, R.J., and Berman, M.L. (1985) A shuttle vector plasmid for studying carcinogen-induced point mutations in mammalian cells. *Gene* 38: 233-237.

Seidman, M.M., Bredberg, A., Seetharam, S. and Kraemer, K.H.(1987): Multiple point mutations in a shuttle vector propagated in human cells: Evidence for an error-prone polymerase activity. *Proc. Natl. Acad. Sci. USA.* 84: 4944-4948.

The Edna Roe Memorial Lecture

Repair Control of Photoinduced Cross-Links and Monoadducts in DNA: Genetic, Molecular and Evolutionary Features

E. MOUSTACCHI

Institut Curie-Biologie
26 rue d'Ulm
75231 Paris cedex 05
France

The interest in cellular processing of DNA interstrand cross-links (CL) and monoadducts (MA) derives from a number of reasons including their induction by agents present in the human environment and the necessity to understand their genotoxic/carcinogenic potential. In the cases described so far the production of CL is accompanied by the induction of MA in different proportions according to the agent used and to the treatment conditions. Several antitumoral drugs (mitomycin C, *cis*-platinum or nitrogen mustards), environmental pollutants and products such as furocoumarins used in photochemotherapy of certain skin diseases have been shown to produce CL between either purines (alkylating agents) or pyrimidines (psoralens plus UVA). The photoaddition of furocoumarins on DNA provides attractive model systems to study DNA repair mechanisms. Indeed: a) The induction of the lesions is immediate upon absorption of the incident photons and do not follow complex kinetics as other bifunctional agents. b) Psoralens photoinduced MA and CL are stable and *in vivo* they are not "spontaneously" lost from the DNA. c) Their induction can be quanititatively controlled by the UVA fluence and a variety of biochemical methods are available in order to quantitate their amounts. d) The ratio of CL to MA, at a constant total number of adducts can be controlled by a double irradiation protocol. Moreover, in certain irradiation conditions (405 nm) only MA are formed by otherwise bifunctional psoralens. e) Metabolic activation is not required for the photoreaction of furocoumarins with DNA. Since these advantageous features are not shared by other cross-linking agents, most of our studies on MA and CL processing by normal or mutant defective cells deal with psoralen photoaddition. These tricyclic aromatic compounds contain two reactive sites,

Photobiology, Edited by E. Riklis
Plenum Press, New York, 1991

223

the 3-4 pyrone and the 4'-5' furan double bonds. They intercalate into DNA and form covalent adducts, only with pyrimidine bases, after exposure to near ultraviolet light (UVA); when activated by UVA either reactive sites of the furocoumarins can react specifically with the 5,6-double bond of pyrimidine bases forming cyclobutane-type monoadducts on the furan or pyrone side between a psoralen molecule and a pyrimidine base. Upon absorption of a second photon of 365 nm, a fraction of the 4', 5'-furan side MA (MA_f) can react with a pyrimidine base on the opposite strand by engaging its 3,4-pyrone double bond to form an interstrand CL (for review, see Cimino et al, 1985).

The isolation and characterization of mutants unusually susceptible to DNA-damaging agents in prokaryotes and lower eukaryotes considerably helped the unravelling of the genetic control and of the metabolic steps of DNA repair functions. Inherited human diseases characterized by a hypersensitivity to agents which interact with DNA provide a cellular material equivalent to such mutants. Several *Escherichia coli* mutants originally selected for their sensitivity to 254 nm ultraviolet (UV) and/or to ionizing radiations turned out to be also sensitive to the photoaddition of psoralens. The same is true for mutants isolated from the yeast *Saccharomyces cerevisiae* (*rad* mutants) or from rodent cells. New mutants have been initially selected for their sensitivity to DNA cross-linking agents from yeast (Henriques and Moustacchi, 1980), hamster (Robson and Hickson, 1986; Zdzienicka and Simons, 1987) and mouse cells (Hama-Inaba et al, 1983, 1988). In humans, such a more or less specific hypersensitivity to DNA cross-linking drugs is shared by cells derived from patients affected by Fanconi's anaemia, an inherited autosomal recessive disorder. Consequently, the induction and processing of psoralen photoadducts have been studied in a range of organisms (for review see Smith, 1988), the behaviour of mutants being compared to that of normal cells. This review will focus on these aspects.

The excision repair pathway plays a major role in the first step of processing of psoralen photoadducts

Mutants known to be defective in excision repair of UV-induced pyrimidine dimers are hypersensitive to photoaddition of both mono and bifunctional psoralens. This is the case of *uvr A* or *B* or *C* mutants of *E. coli* (Cole, 1973), of *rad3* type mutants of *S. cerevisiae* (Averbeck and Moustacchi, 1975), of at least complementation groups 1 and 4 of Chinese hamster ovary cells (Thompson et al, 1987) and of Xeroderma pigmentosum (XP) groups A and D in man (Bredberg et al, 1982; Gruenert and Cleaver, 1985; Vuksanovic and Cleaver, 1987). Biochemical studies demonstrated the persistence of CL in such excision repair deficient mutants whereas CL were efficiently removed from normal cells (Cole et al, 1976; Kaye et al, 1980; Jachymczyk et al, 1981; Magaña-Schwencke et al, 1982; Miller et al, 1982; Gruenert and Cleaver, 1985). In general, it appears that the repair of psoralen MA occurs by the same excisional pathway as that used for CL and for photoproducts produced by 254 nm UV radiation. Indeed, direct measurement using radiolabeled MA inducers such as 3-carbethoxypsoralen and 7-

methylpyrido psoralen (MPP) showed that wild type (RAD^+) yeast removed almost completely their MA within 4 hours. In contrast no removal of MPP adducts occurred in two different deletion mutants in the *RAD3* group (Magaña-Schwencke and Moustacchi, 1985). Moreover, only single-strand breaks were observed during the incubation of yeast or human cells treated with monofunctional furocoumarins (Jachymczyk et al, 1981; Kaye et al, 1980) and certain *rad3* alleles or Xeroderma pigmentosum cells were deficient in the production of these breaks.

Surprisingly, Miller et al (1984) observed that the yeast mutant *rad3-2* was proficient in incising DNA at MA but not at CL, while the mutant is sensitive to a monofunctional psoralen. This may reflect a requirement for the RAD3 protein in post-incision events. Conversely, an XP revertant, isolated from a complementation group A cell line on the basis of an acquired mutagen-induced resistance to 254 nm UV, has the unique property of being capable of removing CL but not MA. Consistent with this property, this XP revertant was found to be resistant to the cytotoxic effect of a DNA cross-linking psoralen but as sensitive as its parental cell line to a monofunctional derivative (Vuksanovic and Cleaver, 1987). This unusual property of the revertant is not confined to psoralen adducts. Indeed this cell line can excise (6-4) pyrimidine-pyrimidone photoproducts from the DNA but not cyclobutane dimers. It can be suggested that structural distortions of DNA due to the two major 254 nm UV photoproducts, to CL and to MA of the pyrone or the furan type differ in subtle ways, differently recognised in mutants. That repair efficiency varies with the nature of the damage even in normal cells is illustrated by the fact that within 24 hr after treatment of cultured human cells with 4'-hydroxymethyl-4,5',8-trimethylpsoralen, 80% of the CL but only 45% of the MA were removed from a transcribed sequence in the DHFR gene (Vos and Hanawalt, 1987).

The role of nucleotide excision repair genes in the repair of psoralen MA has been confirmed using the double irradiation protocol. When cells are treated with 8-methoxypsoralen plus UVA, a fraction of the MA (those that are potentially cross-linkable, i.e. the MA_f) can be converted into CL by further irradiation. The total yield of DNA damage remains the same but the relative proportion of MA and CL in the genome is modified by the second UVA dose. By administering this second dose of UVA at different times after the first dose, it is possible to follow the fate of cross-linkable MA (Moustacchi et al, 1983). In wild type bcteria, yeast or in normal human cells (Ben-Hur and Elkind, 1973; Chanet et al, 1983, 1985; Averbeck et al, 1988), such lesions are efficiently removed during incubation between the two doses of UVA radiation. Consistent with the role of nucleotide excision repair genes, the cross-linkable MA persist in excision-defective cells.

In vitro studies with the purified bacterial UVR ABC excinuclease complex extended our knowledge on the role of the excision repair complex (Van Houten et al, 1986). The enzymatic complex recognizes both MAs and CLs and cuts the phosphodiester bonds on both sides of these adducts. The incision of MA occurs at the eight phosphodiester bond 5' and at the fifth phosphodiester bond 3' to the furan-side thymine, generating a 13-nucleotide single strand gap. Incision of CLs occurs at the ninth phosphodiester bond 5' and the third phosphodiester bond 3' to the furan-side thymine of the CL generating a 12-nucleotides gap (Van Houten et al, 1986). Efficient incision of photoadducts in *E. coli* depends on the presence of the UVR D and POL I gene products. This indicates that helicase II and DNA Polymerase I are needed for release from the substrate DNA of the UVR ABC complex and of the fragment containing the photoadduct.

It is well established that psoralens photoinduced lesions are mutagenic in prokaryotes and eukaryotes, the efficiency of CL being higher than that of MA (Cassier et al, 1985; Bredberg and Nachmansson, 1987). In this context, it is of interest to notice that the spectrum of mutations, as determined by sequencing the *lac I⁻* mutation after treatment with psoralen plus UVA, is quite different in an excision deficient bacteria (*E. coli UVR B⁻*) from the spectrum recovered in the corresponding wild type (*Uvr B⁺*) (Yatagai et al, 1987). More precisely, the mutational spectrum of the *Uvr B-* strain reveals a failure to recover mutations frequently identified following treatment of the *Uvr+* strain. These include the spontaneous hotspot frameshift (loss or gain of 5'-CTGG-3') and the loss of an A-T base-pair at a potential CL site. Taken together with data on the frequency and nature of psoralens photoinduced lesions in relation to the DNA sequence context in the *lac I* bacterial gene (Sage and Moustacchi, 1987; Boyer et al, 1988), these results provide the basis for a) the definition of the lesions responsible for mutation; b) the definition of the strongly reactive sites as opposed to the weakly reactive sites within a DNA sequence before the intervention of the repair system(s) and c) the role of the excision repair system in the avoidance or in the fixation of mutations. Similar studies on a human gene are undertaken.

Post-incision processing of CL involves a recombinational pathway

Bacterial and yeast mutants, known to be defective in recombinational repair of lesions induced by a number of agents, are more sensitive to the cytotoxic effect of psoralen photoaddition than are the corresponding wild type cells. This is the case of the *rec A*, *rec B* and *rec F* mutants of *E. coli* (Sinden and Cole, 1978) and of the *rad52* type mutants of *S. cerevisiae* (Henriques and Moustacchi, 1980a).

DNA sedimentation analysis in alkaline sucrose gradients after treatment of wild type cells by a psoralen and UVA shows a progressive reduction in average molecular weight corresponding to about twice the average CL distance. This initial lowering in molecular weight observed in bacteria, yeast and normal mammalian cells is not seen in excision-defective cells. Upon post-treatment incubation of normal cells, the sedimentation rate of DNA increases with time, reaching an almost normal size. The *rec A* mutant of *E. coli* (Sinden and Cole, 1978), as well as the *rad51* mutant of *S. cerevisiae* (Jachymczyk et al, 1981) which are deficient in induced genetic recombination, do not demonstrate this increase in size of DNA. These results taken together with earlier studies (Cole, 1973; Cole et al, 1976) indicate that the process of CL removal involves two distinct and coordinated phases. After incision of the CL, recombinational strand exchanges involving a second intact copy of the homologous chromosome are necessary to restore the proper

DNA sequence in the gap formed by the dual first incision of the CL. After this, normal excision repair removes the residual damage on the opposite strand and the missing nucleotides are replaced using the newly exchanged DNA strand as a template. Except for the transient appearance of double-strand breaks as intermediates in CL repair in yeast as demonstrated by neutral sucrose gradient analysis (Jachymczyk et al, 1981; Magaña-Schwencke et al, 1982), there are strong similarities to the basic mechanism of CL repair in *E. coli* and yeast: the incision system acts on CLs to produce a substrate that is subject to processing by a recombinational system. In bacteria, the functional Rec A protein participates directly in the strand exchange and is needed throughout the rejoining process.

In mammalian cells, direct evidence on post-incision events in CL processing is not yet available. However, the efficient induction of sister-chromatid exchanges by CL and their increased induction by psoralen plus UVA in Fanconi's anaemia cells (Billardon and Moustacchi, 1986) suggest that recombination may be somewhat involved in the processing of CL. Moreover, studies on viruses reactivation at different multiplicities of infection (Hall, 1982) or with hamster cells at different phases of the cell cycle (Ben-Hur and Elkind, 1973) following treatment with psoralen and UVA, are consistent with a requirement for two DNA copies to fully repair CLs. The presence of psoralen MA in the replicated DNA of the human DHFR gene has been reported (Vos and Hanawalt, 1987). Two alternative mechanisms for bypass of DNA lesions in mammalian cells have been proposed: direct translesion synthesis in gaps left opposite lesions or homologous strand exchange between sister DNA molecules which would fill the gap.

Clearly, much remains to be explored in the area of post-incision processing of CL and of a fraction of MA, especially in mammalian cells.

Gene products involved in mutagenesis are also required for completion of repair of psoralen photoadducts

UV mutagenesis in *E. coli* requires the *Umu C* gene product. Strains defective in excision repair (*uvr B⁻*) and in UV-induced mutagenesis (*Umu C⁻*) are more sensitive to photoaddition of angelicin, a monofunctional furocoumarin, than the corresponding single mutant *uvr B⁻* (Miller and Eisenstadt, 1985). This is not observed after induction of pyrimidine dimers by UV treatment. It suggests that when angelicin MA is not excised from DNA, the *Umu C* gene product facilitates the bypass of lesions during replication, leading to cellular resistance. Also it appears that in cells deficient in the Uvr ABC excinuclease complex, the limited amount of CL removal which occurs, is dependent to some extent on the *Umu C* gene product (Cupido and Bridges, 1985). This minor repair pathway is *Rec A* dependent and is facilitated in cells containing the *Rep* gene product, i.e. helicase II (Bridges and Von Wright, 1981; Cupido and Bridges, 1985). Although it is clear that the processing of both MA and CL in the absence of the Uvr ABC excinuclease results in the transformation of a fraction of these lesions into less lethal forms, the precise mechanism for this minor pathway is poorly understood.

Also in *S. cerevisiae*, it appears that genes involved in a mutagenic pathway are required for the processing of MAs and CLs. Indeed, mutants belonging to the *RAD6* group characterized by sensitivity to a number of physical and chemical agents are hypersensitive to psoralens photoaddition as well (Averbeck and Moustacchi, 1975) and are immutable by these agents (Cassier et al, 1980). The *rad6* mutant has been shown to be defective in post-replication repair following UV induction of pyrimidine dimers (Prakash, 1981) and it has been suggested that the *RAD6* gene product mediates replication past noninstructional sites such as gaps formed opposite lesions in replicating DNA. Extensive DNA degradation is observed during incubation of *rad6* mutant cells after treatment with a psoralen plus UVA. The breaks observed under alkaline and neutral conditions in RAD^+ cells containing psoralen photoinduced CLs and which disappear upon incubation (see above), persist in the *rad6* mutant. In other words, the three major repair pathways genetically identified in yeast, i.e. the excision repair (*RAD3* group), the recombinational repair (*RAD52* group), and the mutagenic repair (*RAD6* group) pathways, are all involved in the processing of MA and CL.

How the gene products involved in error-prone processes interfere with repair of photo-lesions is still unknown.

Tolerance mechanisms

The presence of unrepaired bulky adducts on DNA generally blocks DNA replication. However, it appears that the bypass of a fraction of the lesions can take place *in vitro* as demonstrated in a DNA polymerase I from *E. coli* nick-translation assay (Piette and Hearst, 1983). The same is true in yeast (Chanet et al, 1983, 1985) and in human cells (Vos and Hanawalt, 1987). The reirradiation protocol which allows one to follow the fate of the cross-linkable MA, showed that in *S. cerevisiae* removal of MA is dependent only on the excision system. In excision repair deficient mutants an increased survival was observed if cells were allowed to proceed through one round of replication. Using a thermosensitive cell cycle mutant, it was possible to demonstrate that MA induced in the G1 phase were indeed bypassed after the post G2 phase. This phenomenon was independent of the RAD52 or RAD6 pathways; thus, either of them (recombinational gap-filling or special DNA translesion synthetic pathway) can overcome the block to normal replication posed by MA (Chanet et al, 1983, 1985).

Also, MA were detected in the replicated DHFR human gene. This implies that at least for actively transcribed regions of the genome, non-excised MA apparently do not constitute permanent blocks to DNA synthesis (Vos and Hanawalt, 1987). This bypass mechanism is not understood in any detail yet.

Studies with mutants more or less specifically sensitive to CL

Investigations performed in yeast (*pso* mutants) and in human cells (Fanconi's anaemia) will be briefly reviewed here.

In *S. cerevisiae* genetic and biochemical analysis has utilized mutants isolated specifically as 8-methoxypsoralen (8-MOP) plus UVA-sensitive, designated *pso1*, *pso2*, *pso3* (Henriques and Moustacchi, 1980b) and *pso4* (Henriques et al, 1988 pers. commun.). These mutants complement each other and the different *rad* mutants. The *pso1* mutant is epistatic to *rad6* (Henriques and Moustacchi, 1981; Henriques et al, 1985). More recently, the complementation test and meiotic analysis demonstrated that *pso1-1* is allelic to *rev3-1*, a mutant indeed belonging to the mutagenic *RAD6* pathway (Cassier and Moustacchi, 1988). As is true for other genes (*RAD24* and *CDC8*), *PSO1* also fits into another epistasis group; *pso1* is epistatic to *rad52* (recombinogenic pathway) in cells treated with 8-MOP or ionizing radiations. The *pso2* mutant is specifically sensitive to cross-linking agents (Cassier and Moustacchi, 1981) and as *pso1*, it is almost unmutable (Cassier et al, 1980). Mutants defective in the *PSO2* gene are able to incise DNA containing psoralen CL. However, they are defective in the repair of strand breaks (Magaña-Schwencke et al, 1982). Unlike the genes of the *rad52* group which share this last property, the *PSO2* gene is not required for generalized recombination but plays a special role in the response to CL induced recombination (Saeki et al, 1983). The *PSO2* gene is classified in the *RAD3* (excision repair) pathway and appears to interfere with step(s) following incision (Henriques and Moustacchi, 1981). The *pso3* mutant is specifically sensitive to the lethal effect of mono and bifunctional furocoumarins and is defective in mutation induction by different agents (Cassier et al, 1980; Cassier and Moustacchi, 1981). It was recently found that *pso3* is specifically defective in induced gene conversion by a variety of agents but not in mitotic crossing-over (Henriques et al, pers. commun.). This observation brings support to the notion that independent events can be at the origin of the two phenomena. Finally, the *pso4* mutant is deficient in induced mutagenesis as well as in general recombination; its phenotype recalls that of the *rec A* mutant in *E. coli* (Henriques et al, 1988).

The search for epistatic relationships taken together with the limited biochemical knowledge of the repair steps altered in mutants do not allow us to reach simple conclusions about the genetics of repair of CL in yeast. The evidence does suggest however that beside steps which are in common with the processing of other types of lesions (pyrimidine dimers, etc), the repair of CL involves some specific steps including the sequential participation of the excision-repair and the recombinational CL repair.

The mutant phenotype of more or less specific hypersensitivity to cross-linking agents is shared with human cells from the hereditary disease Fanconi's anaemia (FA). Although not completely blocked as claimed by Fujiwara (1982), several groups reported that incision of CL can take place in FA (see discussion in Papadopoulo et al, 1987). We confirmed this last observation in the two genetic complementation groups of FA (Duckworth-Rysiecki et al, 1985) but found however that the incision of CL is hampered, group A being more affected than group B. Indeed, the kinetics of incision is slowed down and the final amount of CL incised is lower in FA compared to normal cells (Papadopoulo et al, 1987). Also the recovery of a normal rate of DNA semi-conservative synthesis after treatment with a psoralen and UVA takes place in group B cells, but not in FA group A cells (Moustacchi et al, 1987). These data are correlated with the difference in clonogenic cell survival, the FA group A cells are indeed the most sensitive and FA group B cells demonstrate an intermediate response after treatment with a bifunctional psoralen plus UVA. Surprisingly, the repair of MA is also impaired in FA and group B are more sensitive to MA than group A (as seen above, the reverse is true for CL) in terms of cytotoxic effects (Averbeck et al, 1988). Furan-side MA appear to be principally involved in this effect. There are indeed indications also from studies in yeast that the mechanism of repair of different forms of psoralen MA damage may differ. For instance, the 7-methylpyrido-3,4 psoralen, a highly photoreactive monofunctional compound, is much more cytotoxic and more poorly repaired in eukaryotic cells than other psoralens (Magaña-Schwencke and Moustacchi, 1985; Nocentini, 1986).

Conclusion

Our understanding of processing of CL and of the interaction of MA on the repair of CL is still rudimentary. The cloning of genes specifically associated with processing of these lesions should in principle lead to identification of gene products and to the development of *in vitro* systems as recently done for 254 nm UV damage.

Acknowledgement

This work was supported by grants from CNRS, INSERM (Contract 852017), CEA (Saclay, France), Ligue Nationale contre le Cancer, ARC (Villejuif, France) and CEE grant BIO-151 F.

References

Averbeck, D. and Moustacchi, E. (1975), 8-methoxypsoralen plus 365 nm light effects and repair in yeast. *Biochim. Biophys. Acta*, 395, 393-404.

Averbeck, D., Papadopoulo, D. and Moustacchi, E. (1988), Repair of 4,5',8-trimethylpsoralen plus light-induced DNA damage in normal and Fanconi's anaemia cell lines. *Cancer Res.* 48, 2015-2020.

Ben-Hur, E. and Elkind, M.M. (1973), Psoralen and near ultraviolet light inactivation of cultured Chinese hamster cells and its relation to DNA cross-links. *Mutation Res.* 18, 315-324.

Billardon, C. and Moustacchi, E. (1986), Comparison of the sensitivity of Fanconi-s anaemia and normal fibroblasts to the induction of sister-chromatid exchanges by photoaddition of mono- and bifunctional psoralens. *Mutation Res.* 174, 241-246.

Boyer, V., Moustacchi, E. and Sage, E. (1988), Sequence specificity in photoreaction of various psoralen derivatives with DNA: role in biological activity. *Biochemistry*, 27, 3011-3018.

Bredberg, A., Lambert, B. and Soderhall, S. (1982), Induction and repair of psoralen crosslinks in DNA of normal human and Xeroderma pigmentosum fibroblasts, *Mutation Res.*, 93, 221-234.

Bredberg, A. and Nachmanson, N. (1987), Psoralen adducts in a shuttle vector plasmid propagated in primate cells: high mutagenicity of DNA cross-links, *Carcinogenesis*, 8, 1923-1927.

Bridges, B.A. and Von wright, A. (1981), Influence of mutations at the *rep* gene on survival of *E. coli* following UV irradiation or 8-methoxypsoralen photosensitization. evidence for a *rec A+ rep+* dependent pathway for repair of cross-lin ks. *Mutation Res.*, 82, 229-235.

Cassier, C., Chanet, R., Henriques, J.A.P. and Moustacchi, E. (1980), The effects of three *pso* genes in induced mutagenesis: a novel class of mutationally defective yeast. *Genetics*, 98, 841-857.

Cassier, C. and Moustacchi, E. (1981), Mutagenesis by mono- and bifunctional alkylating agents in yest mutants sensitive to photoaddition of furocoumarins *(pso)*. *Mutation Res.*, 84, 37-47.

Cassier, C., Chanet, R. and Moustacchi, E. (1985), Repair of 8-methoxypsoralen photoinduced cross-links and mutagenesis: role of the different repair pathway in yeast. *Photochem. Photobiol.*, 41, 289-294.

Cassier-Chauvat, C. and Moustacchi,E. (1988), Allelism between *pso-1* and *rev3-1* mutants and between *pso2-1* and *snm1* mutants in *Saccharomyces cerevisiae. Curr Genet.*, 13, 37-40.

Chanet, R., Cassier, C., Magana-Schwencke, N. and Moustacchi, E. (1983), Fate of photoinduced 8-methoxypsoralen mono-adducts in yeast. Evidence for bypass of these lesions in the absence of excision repair, *Mutation Res.*, 112, 201-214.

Chanet, R., Cassier, C., and Moustacchi, E. (1985), Genetic control of the bypass of monoadducts and of the repair of cross-links photoinduced by 8-methoxypsoralen in yeast. *Mutation res.*, 145, 145-155.

Cimino, G.D., Gamper, H.G., Isaacs, S.T. and Hearts, J.E. (1985), Psoralens as photoreactive probes of nucleic acid structure and function: organic chemistry, photochemistry and biochemistry. *Annual Rev. Biochem.*, 54, 1151-1193.

Cole, R.S. (1973), Repair of DNA containing interstrand cross-links in *Escherichia coli*: sequential excision and recombination. *Proc. Natl. Acad. Sci. USA*, 70, 1064-1068.

Cole, R.S., Levitan, D. and Sinden, R.R. (1976), Removal of psoralen interstrand cross-links from DNA of *Escherichia coli*: mechanism and genetic control, *J. Mol. Biol.*, 103, 39-59.

Cupido, M. and Bridges, B.A. (1985), Uvr-dependent repair of 8-methoxypsoralen cross-links in *Escherichia coli*: evidence of a recombinational process. *Mutation Res.*, 146, 135-141.

Duckworth-Rysiecki, G.K., Cornish, K., Clarke, C.A. and Buchwald, M. (1985), Identification of two complementation groups in Fanconi's anaemia. *Somatic Cell Mol. Genet.*, 11, 35-43.

Fujiwara, Y. (1982), Defective repair of mitomycin C cross-links in Fanconi's anaemia and loss in confluent normal and Xeroderma pigmentosum cells. *Biochim. Biophys. Acta*, 699, 217-225.

Gruenert, D.C. and Cleaver, J.E. (1985), Repair of psoralen-induced cross-links and monoadducts in normal and repair deficient human fibroblasts. *Cancer Res.*, 45, 5399-5404.

Hall, J.D. (1982), Repair of psoralen-induced cross-links in cells multiply infected with SV-40, *Mol. Gen. Genet.*, 188, 135.

Hama-Inaba, H., Hieda-Shiomi, N., Shiomi, T. and Sato, K. (1983), Isolation and characterization of mitomycin C-sensitive mouse lymphoma cell mutants. *Mutation Res.*, 108, 405-416.

Hama-Inaba, H., Sato, K. and Moustacchi, E. (1988), Survival and mutagenic responses of mitomycin C-sensitive mouse lymphoma cell mutants to other DNA cross-linking agents. *Mutation Res.*, 194, 121-129.

Henriques, J.A.P. and Moustacchi, E. (1980a), Sensitivity to photoaddition of mono- and bifuncational furocoumarins of X-ray sensitive mutants of *Saccharomyces cerevisiae. Photochem. Photobiol.*, 31, 557-563.

Henriques, J.A.P. and Moustacchi, E. (1980b), Isolation and characterization of *pso* mutants sensitive to photoaddition of psoralen derivatives in *Saccharomyces cerevisiae. Genetics*, 95, 273-288.

Henriques, J.A.P. and Moustacchi, E. (1981), Interactions between mutation for sensitivity psoralen photoaddition (*pso*) and to radiation (*rad*) in *Saccharomyces cerevisiae. J. Bacteriol.*, 148, 248-256.

Henriques, J.A.P., Da Silva, K.V.C.L. and Moustacchi, E. (1985), Interaction between genes controlling sensitivity to psoralen (*pso*) and to radiation (*rad*) after 3-carbethoxypsoralen plus 365 nm UV light treatment in yeast. *Mol. Gen. Genet.*, 201, 415-420.

Jachymczyk, W.J., Von Borstel, R.C., Mowat, M.R.C. and Hastings, P.J. (1981), Repair of interstrand cross-links in DNA of *Saccharomyces cerevisiae* requires two systems for dNA repair: the *RAD3* system and the *RAD51* system. *Mol. Gen. Genet.*, 182, 196-205.

Kaye, J., Smith, C.A. and Hanawalt, P.C. (1980), DNA repair in human cells containing photoadducts of 8-methoxypsoralen or angelicin. *Cancer Res.*, 40, 696-702.

Magaña-Schwencke, N., Henriques, J.A.P., Chanet, R. and Moustacchi, E. (1982), The fate of 8-methoxypsoralen photoinduced cross-links in nuclear and mitrochondrial yeast DNA: comparison of wild type and repair deficient strain. *Proc. Natl. Acad. Sci. USA*, 79, 1722-1726.

Magaña-Schwencke, N. and Moustacchi, E. (1985), A new monofunctional pyridopsoralen: photoreactivity and repair in yeast. *Photochem. Photobiol.*, 42, 43-49.

Miller, R.D., Prakash, L. and Prakash, S. (1982), Genetic control of excision of *Saccharomyces cerevisiae* interstrand DNA cross-links induced by psoralen plus near UV-light. *Mol. Cell. Biol.*, 2, 939-948.

Miller, S.S. and Eisenstadt, E. (1985), Enhanced sensitivity of *Escherichia coli Umu C* to photodynamic inactivation by angelicin (isopsoralen). *J. Bacteriol.*, 162, 1307-1310.

Moustacchi, E., cassier, C., Chanet, R., Magana-Schwencke, N., Saeki, T. and Henriques, J.A.P. (1983), Biological role of photoinduced cross-links and monoadducts in yeast DNA; genetic control and steps involved in their repair. In: Friedberg., E.C., Bridges, B.A. (eds) Cellular responses to DNA damage, Liss, New York, pp. 87-106.

Moustacchi, E., Papadopoulo, D., Diatloff-Zito, C. and Buchwald, M. (1987), Two complementation groups of Fanconi's anaemia differ in their phenotype response to a DNA-crosslinking treatment. *Hum. Genet.*, 75, 45-47.

Nocentini, S. (1986), DNA photobinding of 7-methylpyrido (3,4-c) psoralen and 8-methoxypsoralen. Effects on macromolecular synthesis, repair and survival in cultured human cells. *Mutation Res.*, 186, 181-192.

Papadopoulo, D., Averbeck, D. and Moustacchi, E. (1987), The fate of 8-methoxypsoralen-photoinduced DNA interstrand cross-links in Fanconi's anaemia cells of defined genetic complementation groups. *Mutation Res.*, 184, 271-280.

Piette, J. and Hearst, J.E. (1983), Termination sites of in vitro nick translation reaction of DNA that had photoreacted with psoralen. *Proc. Natl. Acad. Sci. USA*, 80, 5540-5544.

Prakash, L. (1981), Characterization of post-replication repair in *Saccharomyces cerevisiae* and effects of *rad6, rad18, rev3* and *rad52* mutations. *Mol. Gen. Genet.*, 184, 471-478.

Robson, C. and Hickson, I.D. (1986), Genetic analysis of mitomycin C-sensitive mutants of a Chinese hamster ovary cell line. *Mutation Res.*, 163, 201-208.

Saeki, T., Cassier, C. and Moustacchi, E. (1983), Induction in *Saccharomyces cerevisiae* of mitotic recombination by mono- and bifunctional agents: comparison of the *pso2* and *rad52* repair deficient mutants to the wild type. *Mol. Gen. Genet.*, 190, 255-264.

Sage, E. and Moustacchi, E. (1987), Sequence context on 8-methoxypsoralen photobinding to defined DNA fragments. *Biochemistry*, 26, 3307-3314.

Sinden, R.R. and Cole, R.S. (1978), Repair of cross-linked DNA and survival of *Escherichia coli* treated with psoralen plus lights: effects of mutation influencing genetic recombinations and DNA metabolism. *J. Bacteriol.*, 136, 538-547.

Smith, C.A. (1988), Repair of DNA containing furocoumarin adducts. In: Gasparro F.(ed), Psoralen DNA Photobiology, CRC, Boca Raton, Chap. 4.

Thompson, L.H., Salazar, E.P., Brookman, K.W., Collins, C.C., Stewart, S.A., Busch, D.B. and Weber, C.A. (1987), Recent progress with the DNA repair mutants of Chinese hamster ovary cells, in: "The molecular biology of DNa repair", *J. Cell. Sci. Suppl.*, 6, 97-110.

Van Houten, B., Gamper, H., Holbrook, S.R., Hearst, J.E. and Sancar, A. (1986), Action mechanism of position. *Proc. Natl. Acad. Sci. USA*, 83, 8077-8080.

Vos, J-M. H., and Hanawalt, P.C. (1987), Processing of psoralen adducts in an active human gene: repair and replication of DNA containing monoadducts and interstrand cross-links. *Cell*, 50, 789-799.

Vuksanovic, L. and Cleaver, J.E. (1987), Unique cross-link and monoadduct repair characteristics of a Xeroderma pigmentosum revertant cell line. *Mutation Res.*, 184, 255-263.

Yatagai, F., Horsfall, M.J. and Glickman, B.W. (1987), Defect in excision repair alters the mutational specificity of PUVA treatment in the *lac I* gene of *Escherichia coli. J. Mol. Biol.*, 194, 601-607.

Zdzienicka, M.Z. and Simons, J.W.I.M. (1987), Mutagen-sensitive cell lines are obtained with a high frequency in V-79 Chinese hamster cells. *Mutation Res.*, 178, 235-244.

Photochemistry and Photophysics: Probing the Unknown

Pulse Radiolysis and Flash Photolysis

T.G. TRUSCOTT

Department of Chemistry
University of Keele
Staffordshire ST5 5BG, UK

Photochemistry is concerned with the study of the chemical effects induced by the absorption of light (typically 200 nm–800 nm) whereas Radiation Chemistry concerns the effects induced by ionising radiations(typically X- and γ-rays, electrons, protons, α-particles, etc.). Major sources of light are discharge lamps (e.g. various pressure Hg arcs) and lasers while sources of radiation can arise with the continuous emission from isotopes (e.g. ^{60}Co) or the continuous or pulsed radiation from machines such as a Van de Graaff accelerator, the Febetron, or, more importantly, a linear-electron accelerator. The absorption of both high-energy radiation and light, can lead to similar short-lived species such as electronically excited states (singlet or triplet) and, particularly in the case of radiation chemistry, to ions, free radicals and the solvated electron. The study of these short-lived intermediates requires the use of a suitable fast-reaction technique. In photochemistry the technique of choice is flash photolysis while the complementary technique of radiation chemistry is pulse radiolysis. The time resolution of both techniques continues to shorten with new technology and pico-second (and in a few cases even faster) pulses are almost routine for studying ultra-fast processes (West, 1986). However, much of the data obtained in biological work have used nano-second pulses and this lecture will mainly concern such studies. Nevertheless, major advances in, for example, our understanding of photosynthesis, vision and haemoglobin binding have depended on pico-second and femto-second studies.

The basic arrangement for both pulse radiolysis and flash photolysis is shown in Fig. 1. The excitation source for pulse radiolysis frequently being a linear accelerator while that for flash photolysis is one of a range of pulsed lasers (e.g. ruby, YAG or dye pumped laser). Frequently, for laser flash photolysis the primary line (e.g. 1066 nm for YAG) is not useful but the harmonics are, and frequency doubling (to 532 nm), tripling to (355 nm) or quadrupling (to 266 nm) is used. For the ruby, the primary line at 694 nm is useful for some red absorbers, e.g. the bacteriochlorins and the phthalocyanines but more usually, it too is used as a frequency doubled source at 347 nm.

This paper is concerned with applications for pulse radiolysis and flash photolysis to photobiology and not with the technology of the techniques. Nevertheless, it is important to realise that interference problems and 'noise' in general are encountered in

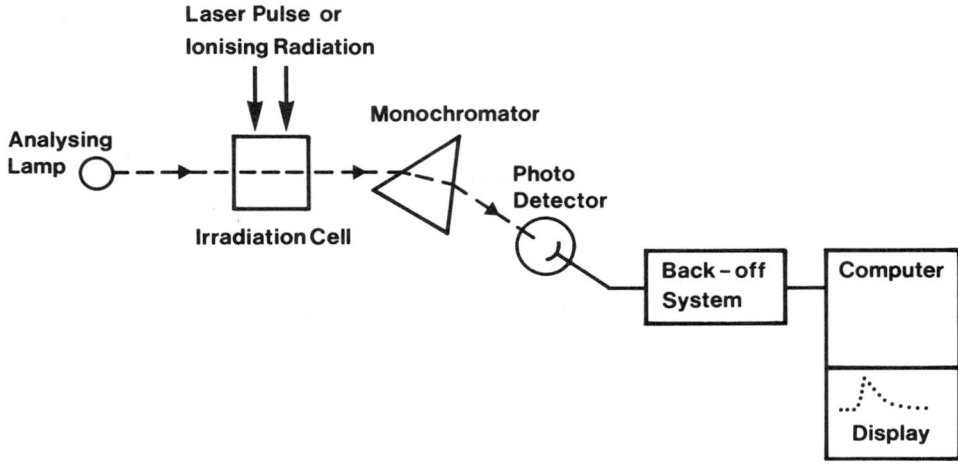

Figure 1

these pulsed techniques. For pulse radiolysis this includes Cerenkov emission from the sample (although this has been turned to advantage as a 'monitoring source' for pico-second studies) and the pick-up of the brehmsstrahlung emission from the sample, i.e. X-ray interference. Furthermore, for both techniques it is essential to maximise the signal to noise (S/N) ratio particularly because of the 'shot noise' for short-lived transient measurements. Generally, this is achieved by using a high intensity of analysing light at the photocathode (the detector). In order to avoid photomultiplier fatigue, it is necessary to only illuminate the photocathode for a short period of time. This is usually achieved by operating a Xenon arc lamp and adding a high current pulse to the steady state level while the lamp is operating in the continuous mode. This can lead to an enhancement of the monitoring intensity by up to 50 times and this enhancement 'or gain' factor can be sufficiently long-lived or 'flat' for transients with lifetimes of ≈ 50 μs to be measured. The S/N ratio for shot-noise is improved by such increased monitoring intensity because it is directly proportional to the square root of the number of photoelectrons emitted per second.

It should also be noted that other detection techniques can also prove useful. These include time-dependent esr measurements and time-dependent conductivity measurements. Generally the esr technique, which is used for studying free radicals, is very sensitive but has serious time resolution limitations while conductivity, used in pulse radiolysis to detect ions, is limited by the complexity that it measures all ions present in the solution and the monitoring of individual species is therefore not possible. For laser flash photolysis some recent measurements have used infra-red monitoring intensities which can give structural information of the excited states but to date this has not involved biological studies.

In both laser flash photolysis and pulse radiolysis much information about the absorption specra of transients and the rates of reaction of transients with other species can be obtained without knowing the dose, i.e. without measuring the absolute amount of energy deposited in the chemical system. For some purposes however, such as the measurement of quantum yields (ϕ) in photochemistry and G values for radiation chemistry, it is necessary to measure the dose. (G is defined as the number of radicals, excited states or molecular products produced or transformed in an irradiated system absorbing 100 eV of energy — the absorbed dose is often expressed in Grays [1 Gy = 100 rads]). In radiation chemistry the Fricke dosimeter, based on the Fe^{II} to Fe^{III} oxidation in water, is a well-known standard secondary dosimeter (as well as absolute dosimetry there can be beam fluctuations from the linear accelerator or pulsed laser sources and relative beam monitoring on a pulse to pulse basis is also therefore necessary) and in photochemistry there are also well-known chemical actinometers. However, frequently in practise, in both pulse radiolysis and laser flash photolysis, it is convenient to use a comparative method in which for example, the triplet quantum yield (ϕ_T) and triplet-triplet extinction coefficient (ε_T) of a standard are compared to that of the species under investigation. For example, for 355 nm excitation, anthracene is frequently used as the standard for which $\phi_T = 0.71$ and ε_T (422 nm) = 64,700 $dm^3mol^{-1}cm$ in cyclohexane and for 266 nm excitation, naphthalene is used with $\phi_T = 0.75$ with ε_T (414 nm) = 24,500 $dm^3mol^{-1}cm^{-1}$ in cyclohexane. A useful actinometer for measurements with 694 nm excitation is methylene blue with $\phi_T = 0.58$ and ε_T (370 nm) = 7,200 $dm^3mol^{-1}cm^{-1}$.

Comparison of the Consequences of Light and High Energy Radiation Absorption

The initial results of light and high energy (ionising) radiation absorption are quite different even though, as noted above, the subsequent excited states produced can be the same (e.g. excited singlet and triplet states). It is because such excited states are achieved by different routes that the two techniques of laser flash photolysis and pulse radiolysis often yield complementary data for biological molecules. With light irradiation the photons are absorbed by the solute with an efficiency based on the absorbance coefficient of the solute at the light irradiation wavelength. This usually directly generates excited singlet states which undergo vibrational deactivation and internal conversion (IC) to the first excited state (S_1). From S_1 the molecule can undergo a variety of photophysical processes particularly light emission (fluorescence) and intersystem crossing (ISC) to the triplet manifold of states. These processes, all complete in a few nano-seconds (much faster, of course, for deactivation of the upper excited states). The lowest triplet state (T_1) is inherently quite long-lived and hence the majority of photochemical reactions arise via T_1. These excited state energy levels and transitions are represented in the Jablonski diagram (Fig.2).

With high energy and relatively dilute solutions the radiation is absorbed practically exclusively by the solvent. Subsequent processes, leading mainly to solute excited states in non-polar solvents and to radical ions in polar solvents then occur. These processes in some typical solvents are as follows:

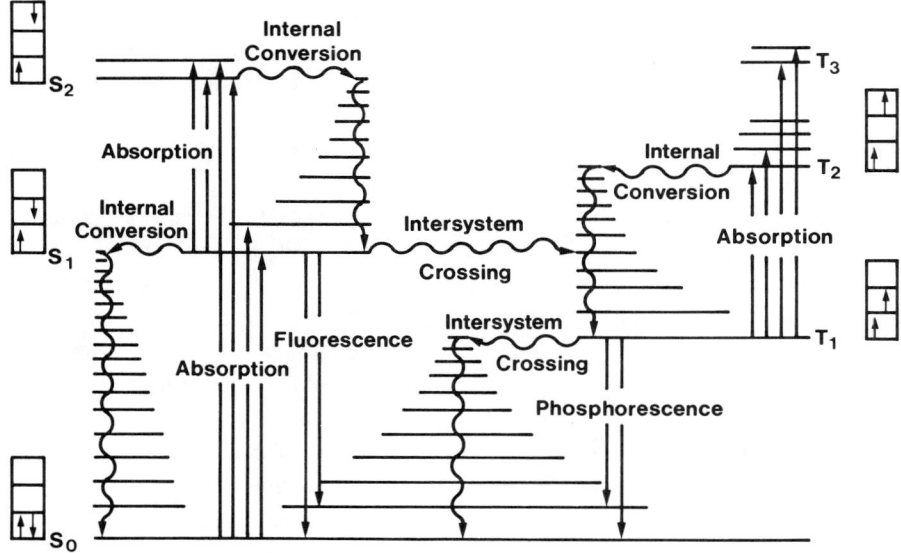

Excited state energy levels and transitions

Figure 2

Water

This is, of course, an important biological solvent. Also, one of the early contributions of pulse radiolysis was the observation of the hydrated electron (e^-_{aq}) absorption with λmax ~720 nm. The absorption of ionising radiation by water produces, in < 1 ps, excited molecules (H_2O^*), cations ($H_2O^{\cdot+}$) and electrons (e^-) ejected as quasi-free particles, these electrons can then ionise other molecules along their path. Within 1 ns of the radiation deposition the only radiolysis products are:

$$H_2O - \rightarrow e^-_{aq} + H\cdot + OH\cdot + H_2 + H_2O_2 + H_3O^+$$

The three types of radical are the most reactive and the corresponding G values are e^-_{aq} (G=2.7), H· (G=0.55) and OH· (G=2.7). The radicals e^-_{aq} and H· are reducing species while OH· is oxidising. This would lead to both oxidised and reduced solute products but fortunately these can be readily separated. Thus, often water is saturated with nitrous oxide leading to predominantly an oxidising environment by converting the e^-_{aq} into oxidising OH· radicals

$$N_2O + e^-_{aq} + H_2O \rightarrow OH\cdot + OH^- + N_2$$

Thus in the presence of a saturating concentration of N_2O (~10^{-2}mol dm^{-3}) ~90% of the water radicals formed are oxidising with ~10% being the reducing H·. The reaction of

240

N_2O with H· is too slow to be important in most situations with $k \approx 10^4 \, dm^3 mol^{-1} s^{-1}$ for:

$$N_2O + H· \rightarrow N_2 + OH·$$

However the OH· radical not only acts as an oxidising radical but often also undergoes other reactions, e.g. it (i) adds onto aromatic solutes to form hydroxy-adduct free radicals, (ii) abstracts hydrogen to give water and a free radical, (iii) reacts with another OH· radical. Also the OH· is a weak acid with pK_a for the reaction

$$OH· + OH^- = O·^- + H_2O$$

being 11.9. The reactivity of $O·^-$ (the basic form of the hydroxyl radical) can be quite different to that of OH·.

In order to avoid such complexities when studying oxidising situations, it is convenient to add high concentrations of the pseudo halide ion, SCN^-, or halide ion, Br^-, or the azide ion, N_3^-; these leading to weaker oxidising radicals which do not tend to undergo other processes, e.g.

$$OH· + 2Br^- = Br_2·^- + OH^-$$
$$OH· + 2SCN^- = (SCN)_2·^- + OH^-$$
$$OH· + N_3^- = N_3· + OH^-$$

As with N_2O the electron can be removed by scavenging with oxygen to yield the superoxide ion $O_2·^-$ and a useful way of converting the primary radicals of water radiolysis into $O_2·^-$ is to use oxygenated formate solutions, so that $CO_2·^-$ formed from OH· radicals, rapidly transfers its electron to O_2:

$$OH· + HCOO^- = CO_2·^- + H_2O$$
$$CO_2·^- + O_2 \rightarrow O_2·^- + CO_2$$

The pK_a for $O_2·^-$ is ~4.7 so that in acid solutions $HO_2·$ is produced.

Since the $CO_2·^-$ radical itself is reducing, the use of a high concentration of sodium formate leads to a reducing environment, often with e^-_{aq} reacting very quickly with the solute and $CO_2·^-$ somewhat more slowly, but both do yield the 1-electron reduction product from that solute. Alcohols also convert oxidising OH· radicals to reducing radicals, e.g. $CH_2OH·$ from methanol.

Alcohol

A rather complex mixture of radicals is formed initially on ionising irradiation of alcohols such as methanol, but many of these react rapidly with methanol to give

$CH_2OH\cdot$, e.g.

$$CH_3\cdot + CH_3OH = CH_4 + CH_2OH\cdot$$
$$H\cdot + CH_3OH\quad = H_2 + CH_2OH\cdot$$

so that, in a dilute alcohol solution, the overall 'primary effect' can be written as

$$CH_3OH \rightarrow e^-_{aq} + CH_2OH\cdot$$

with both $CH_2OH\cdot$ and e^-_{CH3OH} being reducing radicals.

Non-Polar Solvents

As examples of pulse radiolysis of dilute solutions of a solute in typical non-polar solvents we will consider anthracene (A) in hexane (H) and in benzene (B). For H, solvent ionisation occurs, and since the electrons are not readily solvated, the electrons tend to either recombine with $H\cdot^+$ or add to A:

$$H \qquad\qquad \rightarrow H\cdot^+ + e^-$$
$$e^- + A \qquad \rightarrow A\cdot^-$$
$$H\cdot^+ + A \qquad \rightarrow H + A\cdot^+$$

Fast recombination processes then lead to the A excited states

$$A\cdot^+ + A\cdot^- \rightarrow A(T_1 \text{ or } S_1) + A$$
$$A\cdot^- + H\cdot^+ \rightarrow A(T_1 \text{ or } S_1) + H$$

With benzene(B) as the solvent the situation is somewhat different because B itself has reasonably long-lived S_1 and T_1 states

$$B \qquad\qquad \rightarrow B\cdot^+ + e^-$$
$$e^- + B \qquad \rightarrow B\cdot^-$$
$$B\cdot^- + B\cdot^+ \rightarrow 2B(S_1 \text{ and/or } T_1)$$
$$B(S_1) + A \rightarrow B + A(S_1)$$
$$B(T_1) + A \rightarrow B + A(T_1)$$

Thus as a general rule for pulse radiolysis non-polar solvents lead to high yields of excited states and low yields of free radical species, whereas polar solvents tend to support low yields of excited states and high yields of radical ions. As stated above the major reason for this is that in polar solvents the ions are solvated and stabilised by the solvent and tend not to recombine to form excited states.

For organic molecules in non-polar solvents we can thus obtain information by pulse radiolysis on radical anions (non-dissociative electron attachment)

$$e^- + A = A^{\cdot-}$$

and radical cations (non-dissociative ionisation)

$$Solvent^{\cdot+} + A = Solvent + A^{\cdot+}$$

Of course laser flash photolysis can also be used to study such radicals if the excitation wavelength corresponds to sufficient energy to lead to photo-ionisation. For example, both 266 ns laser flash photolysis and pulse radiolysis have been used to ionise dihydroxyphenylalanine (dopa) and hence study the sequence of early reactions leading to melanin formation (Chedekel et al., 1984).

Perhaps less important processes as far as photobiology is concerned are dissociative electron attachment (carbanion formation)

$$e^- + (C_6H_5CH_2)_2Hg = C_5H_5CH_2^- + C_6H_5CH_2Hg$$

and dissociative ionisation (carbanion formation)

$$R-OH = R^+ + OH^-$$

One of the few such studies on biological systems concerns retinol and retinyl acetate (Lo et al., 1982).

Generally the production and spectra of radical species obtained by pulse radiolysis is quite unambiguous and such data can be used to assign spectra obtained by laser flash photolysis, thus illustrating the complementary aspects of the two techniques. There are many examples of this, one being the identification of the one-electron product of ubiquinone (a component of the mitochondrial and bacterial photosynthetic electron transport chain).

Typcial Applications in Biology and Medicine

These applications have been wide ranging and have included photosynthesis, vision, photosensitising drugs (furocoumarins and porphyrins), melanin precursors, chemically induced Parkinson's Disease, singlet oxygen, etc. As examples of the complementary application of laser flash photolysis and pulse radiolysis, we will consider next some typical results obtained on the C-40 carotenoids. These polyenes are one of the most important group of natural pigments and, as well as their wide ranging occurrence and role in photosynthesis, many are used as yellow and red colorants in the food industry and some are used as a drug to alleviate the effects of certain types of porphyric disease, e.g.

β-carotene or β-carotene/canthaxanthin mixtures are used to treat erythropoietic protoporphyria and are also used to achieve skin coloration (artificial 'tan').

The process by which photosynthetic organisms convert light energy to chemical products such as adenosine triphosphate (ATP) is, of course, one of the most studied of the photobiological processes. The role of the carotenoids in photosynthesis is considered to be mainly as protective agents against photochemical damage to chlorophylls and as accessory pigments for light harvesting in the antenna pigment-protein complexes. However, carotenoids have also been considered to have as yet undefined roles, possibly associated with the conduction of charge and as a link between electron donor and acceptor molecules. Because of this, the nature of the 1-electron oxidation and 1-electron reduction radical, and the interaction of these species with the chlorophylls are of interest. Laser flash photolysis and pulse radiolysis have been much used to characterise the triplet properties of the carotenoids and pulse radiolysis has been used to characterise the radical ions of the carotenoids (e.g. Lafferty et al., 1977) and their interaction, by electron transfer, with the chlorophylls (e.g. Chauvet et al., 1983 and Almgren & Thomas, 1980).

As an example of the use of pulse radiolysis to study such radical ions we will consider the carbonyl-containing carotenoid, canthaxanthin (Can). Flash photolysis of Can yields only an extremely weak triplet absorption and unless a triplet sensitiser (donor-D^T) is used to produce the Can triplet (Can^T) via energy transfer

$$D^T + Can \rightarrow D + Can^T$$

it is difficult to make useful measurements. On the other hand, as noted above, pulse radiolysis can be used to generate such triplets directly and, unlike laser flash photolysis, can also be used to easily generate and characterise the radical ions of such carotenoids.

Figure 3 shows the transient spectra, in the long-wavelength region, generated by pulse radiolysis of an argon-flushed solution of Can in hexane. Two of the time-resolved transient spectra shown are obtained immediately and 2.5 μs after the pulse. Also shown is the transient spectrum 2.5 μs after the pulse of a nitrous oxide flushed Can solution in hexane. As can be seen there are two absorption peaks (at ~960 nm and ~1150 nm) in the argon flushed solution and these have entirely different formation kinetics.

In hexane the radical cations are formed by positive charge capture and the radical anions by electron capture. However electron capture is extremely fast (~10^{12} $dm^3 mol^{-1} s^{-1}$) (see, e.g. Beck and Thomas, 1972). The species peaking at ~960 nm, which can also be observed in solutions bubbled with nitrous oxide, has a relatively slow growth, while that at ~1150 nm is formed immediately. The absence of the 1150 nm peak in the presence of nitrous oxide confirms that it is due to the radical anion ($Can^{\cdot-}$) produced by reaction with the electron. As noted above, nitrous oxide is an efficient scavenger of electrons and hence precludes the formation of such 1-electron reduced species. For Can the species at 960 nm is the 1-electron oxidised species, i.e. the radical cation ($Can^{\cdot+}$). A similar analysis has been made of a wide range of polyenes

244

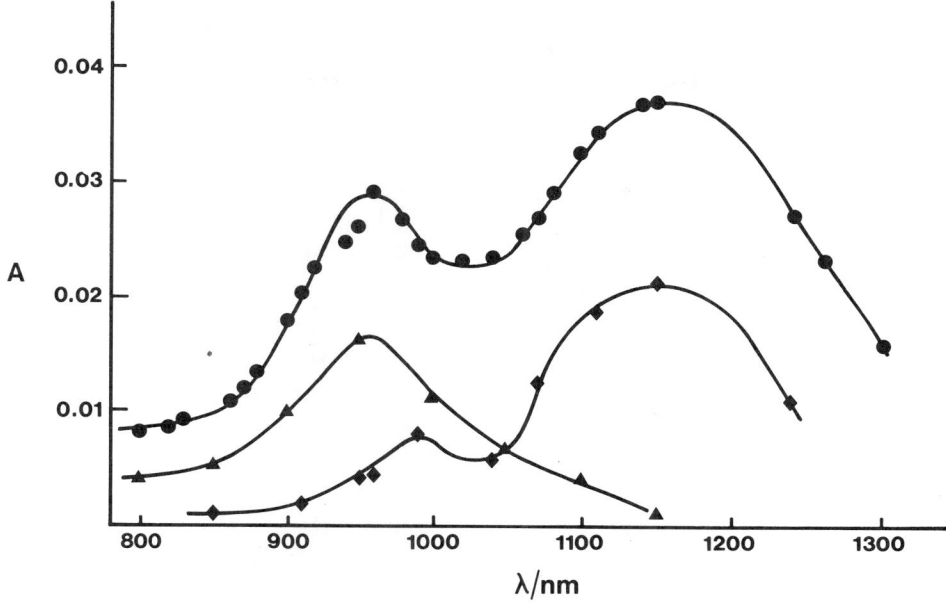

Figure 3.

(e.g. Lafferty et al., 1977). Many of the polyenes investigated in this work were also studied in methanol as solvent. As noted above in alcohol the initial products of the radiolysis can be considered as the electron and the $CH_2OH\cdot$ radical both of which are reducing. In agreement with this, studies of polyenes in methanol only give rise to a single transient which was identified as being due to polyene·⁻.

Another useful example of the complementary value of pulse radiolysis and laser flash photolysis to biological molecules came from the study of the interaction of furocoumarin triplet states (such as that of 8-Methoxypsoralen — the photochemo-therapeutic drug widely used in the treatment of psoriasis) with dopa. Thus flash photolysis allowed the triplet states of a wide range of furocoumarins to be established and the rate constant for the reaction between dopa and such triplets to be obtained (typically $\sim 2 \times 10^9$ $dm^3mol^{-1}s^{-1}$). Such reactions were found (Craw et al., 1984), by laser flash photolysis, to lead to an unidentified long-lived transient species whose spectrum could be established but not identified by the laser experiment. On the other hand, pulse radiolysis allowed the radical cation (deprotonated) spectrum of dopa to be unambiguously established and this proved to be identical to the long-lived species observed in laser flash photolysis of furocoumarin-dopa mixtures. Such complementary results allowed the establishment of reactions of the type

$$FC^T + dopa \rightarrow FC\cdot^- + dopa\cdot^+$$

Since dopa$\cdot^+$ (rapidly deprotonated) undergoes further reactions to yield melanin, such electron transfer processes could be related to the enhanced melanogenesis often observed on topical use of furocoumarins.

Singlet Oxygen

This paper is not concerned with the use of laser excitation and near infra-red emission (~1270 nm) detection of singlet oxygen although such measurements have become widespread in recent years. However, it is of interest to note that the complementary advantages of also using pulse radiolysis to generate the singlet oxygen has recently been demonstrated (Gorman et al., 1988). In this use of pulse radiolysis it was necessary to overcome the problem of the Cerenkov emission and this is still a limiting factor on the solvent system which can be used. Nevertheless, the use of pulse radiolysis to generate the singlet oxygen emission should allow biological molecules whose triplet yield is very low (e.g. the furocoumarins) to be better compared than via the usual pulsed laser excitation.

The techniques of laser flash photolysis and pulse radiolysis have now been used to study a wide range of biological systems over time domains from seconds to femtoseconds and several of these studies are reviewed in the following general references.

Acknowledgements

The author thanks Dr. E.J. Land for useful discussions.

Typical General References

Bensasson, R.V., Land, E.J. and Truscott, T.G., 1983, Flash Photolysis and Pulse Radiolysis: Contributions to the Chemistry of Biology and Medicine, (Pergamon Press).

Butler, J., Hoey, B.M. and Swallow, A.J., 1986, Radiation Chemistry. R.S.C. Ann. Report C, 129–175.

Dorfman, L.M. and Sauer Jr., M.C., 1986, Pulse Radiolysis. Chapter IX from Investigation of Rates and Mechanisms of Reactions Part II, Edited by C.F. Bernasconi, 4th Ed. (J. Wiley) pp.493–546.

Farhataziz and Rodgers, M.A.J., 1987, Radiation Chemistry (VCH Publishers).

Hughes, G., 1973, Radiation Chemistry (OUP).

Swallow, A.J., 1973, Radiation Chemistry (Longman).

West, M.A., 1986, Flash and Laser Photolysis. Chapter VIII from Investigation of Rates and Mechanisms of Reactions Part II, Edited by C.F. Bernasconi, 4th Ed. (J. Wiley), pp. 391–491.

References

Almgren, M. and Thomas, J.K., 1980, Interfacial electron transfer involving radical ions of carotene and diphenylhexatriene in micelles and vesicles. Photochem. *Photobiol.*, 31:329–335.

Chauvet, J-P., Viovy, R., Land, E.J., Santus, R. and Truscott, T.G., 1983, One-electron oxidation of carotene and electron transfers involving carotene cations and chlorophyll pigments in micelles. *J. Phys. Chem.*, 87:592–601.

246

Chedekel, M.R., Land, E.J., Thompson, A. and Truscott, T.G., 1984, Early steps in the free radical polymerisation of 3,4-dihydroxyphenylalanine (dopa) into melanin. *J. Chem. Soc., Chem. Comm.*, 1170–1172.

Craw, M., Chedekel, M.R., Truscott, T.G. and Land, E.J., 1984, The photochemical interaction between the triplet state of 8-methoxypsoralenand the melanin precursor L-3,4 dihydroxyphenylalanine. *Photochem.Photobiol.*, 39:155–159.

Lafferty, J., Roach, A.C., Sinclair, R.S., Truscott, T.G. and Land, E.J.,1977, Absorption spectra of radical ions of polyenes of biological interest. J.C.S., Faraday Trans. I, 73:416–429.

Lo, K.K.N., Land, E.J. and Truscott, T.G., 1982, Primary intermediates in the pulsed irradiation of retinoids. *Photochem. Photobiol.*, 36:139–145.

Discrimination and Coverage in Hybridization: Advantages Afforded by Crosslinkable Oligonucleotide Probes

G. D. CIMINO[*], H. B. GAMPER[*], M. FERGUSON[*], S. T. ISAACS[*] AND J.E. HEARST[**]

[*]HRI Research, Inc.
2315 Fifth Street., Berkeley , California 94710
[**]Department of Chemistry
University of California, Berkeley, CA 94720

Abstract

We describe the advantages afforded by the use of short, crosslinkable oligonucleotide probes in hybridization assays. The ability to crosslink a probe to its target facilitates solution hybridization and permits the use of probes at sufficiently high concentrations so that equilibrium is established in short times. An elementary thermodynamic theory predicts that, at equilibrium, maximal discrimination occurs at the melting temperature (T_m) of the probe/target complex. Oligonucleotides with a site-specifically incorporated psoralen monoadduct are photocrosslinkable probes which also provide an additional tier of discrimination. Photochemical crosslinkage is very rapid compared to hybridization rates. We demonstrate kinetic covalent entrapment of a short oligonucleotide (25 bases) to its complementary sequence on a longer duplex DNA (2754 bases). Conditions which eliminate the problems of renaturation of the duplex genomic DNA are defined. Additionally, increased target coverage is obtained with photocrosslinkable probes since covalent entrapment provides a means of pumping the equilibrium between an oligonucleotide probe and its target sequence.

Introduction

The ability of two polymers of nucleic acid containing complementary sequences to find each other in solution and anneal through base pairing interactions is remarkably specific and powerful. The initial observations of the "hybridization" process by Marmur and Lane (1960) and Doty et al. (1960) have been followed by the refinement of this process into an essential tool of modern genetics and molecular biology. Britten and Kohne (1968) developed a theory which explains the kinetics of reassociation of complementary DNA strands in solution. This kinetic understanding (Cot analysis) of the hybridization process led to the discovery of repetitive sequences in eukaryotic DNA. Initial studies using hybridization analysis were performed in solution (Hayashi et al., 1963; Marmur et al., 1963). Further development led to the immobilization of the target DNA or RNA on solid supports, such as nitrocellulose, nylon, and diazobenzyl-oxymethyl paper. These new immobilization techniques created a means of identifying

Photobiology, Edited by E. Riklis
Plenum Press, New York, 1991

fractionated nucleic acids of a specific length after gel electrophoresis (Southern, 1975; Alwine et al., 1977). Fractionation of genomic nucleic acid before hybridization is currently an essential step in the identification of restriction fragment length polymorphisms associated with several human disorders (Gusella, 1986). Electrophoretic fractionation spreads out the hybridization background over the entire length of a lane in the gel. This enhances the signal to noise ratio of single bands in the gel and thus permits the identification of single copy genes. Colony hybridization provides still another detection procedure where sequence separations do not arise by gel electrophoresis but by cloning and plating of living cells (Grunstein et al., 1975; Benton et al., 1977; Berent et al., 1985).

The development of chemical methods for the synthesis of oligodeoxyribonucleotides has promoted the use of short oligonucleotide probes (typically 17–40 nucleotides in length) in hybridization experiments (Studenchi et al., 1984; Torczynski et al.; Wallace, et al., 1979; Wallace, et al., 1981). Although remarkable accomplishments have been achieved, these procedures are artful and usually take long periods of time. A typical experiment involves the hybridization of a probe to an immobilized target at 10 to 20°C below the melting temperature of the probe/target complex. Probe concentrations in these experiments are approximately $1–2\times10^{-10}$ M. These concentrations are empirically derived; they minimize the use of probe and simultaneously provide enough discrimination to distinguish single copy genes utilizing probes approximately 20 nucleotides in length. Hybridization times are two to ten hours at these concentrations. After hybridization several washes of varying stringency are employed to remove excess probe, non-specifically bound probe, and probe bound to partially homologous sequences in the target genome. Careful control of these wash steps is necessary, since the signal to noise ratio of the experiment is ultimately determined by the wash procedures.

Gamper et al., (1984) have described a method of obtaining short oligonucleotide probes which contain a site-specifically bound psoralen-monoadducted thymine. The presence of the monoadduct on the probe does not effect the kinetics of hybridization, the specificity of base pairing, or the stability of the probe/target hybrid. It does, however, permit the resultant hybrid to be covalently crosslinked by a very brief exposure to actinic light (320–400 nm). Crosslinkable probes have several unique advantages over conventional probes. The most obvious advantage is the ability to use denaturing conditions to remove non-specifically bound probe. This capability facilitates solution formats and permits the use of crosslinkable probes at concentrations much higher than conventional probes. Crosslinkable probes at concentrations much higher than conventional probes. Crosslinkable probes can also be hybridized and covalently fixed to their targets at or above the melting temperature of the probe/target complex. These two advantages permit the achievement of a true equilibrium distribution between free probe and probe/target complex. Provided certain criteria are met, hybridization and crosslinkage can be carried out concurrently within a matter of minutes with good discrimination and coverage (Gamper et al., 1984).

In an effort to understand the parameters which are essential for the optimization of hybridization experiments with short, crosslinkable probes, we have developed an elementary theory that addresses the equilibrium distribution of short oligonucleotide probes to their target nucleic acids. This theory predicts that the best discrimination between true hybridization at a completely complementary sequence in the target genome relative to hybridization at sequences of partial homology occurs at the melting temperature of the probe. Furthermore, discrimination is a function of the probe concentration and it is enhanced at low concentrations. The incorporation of a crosslinkable molecule into the prove can add an additional tier of specificity which significantly reduces background. In the case of psoralen monoadducted probes, crosslinkage of the hybrid will occur only if the monoadduct is in a perfectly base paired helix.

Crosslinkable probes additionally provide a means to hybridize and crosslink in solution before fractionation by electrophoresis on a denaturing gel. An inherent problem with solution hybridization is that denatured, double-stranded target DNA anneals during the hybridization reaction. With psoralen-monoadducted probes which are present in large molar excess over denatured target nucleic acid, it is possible to irradiate and kinetically entrap probe/target complexes before renaturation of the target DNA. A further kinetic implication resulting from entrapment is the ability to drive the equilibrium thereby generating the formation of additional hybrid. This ultimately leads to high yields of covalently linked hybrid and consequently enhances the overall coverage of target molecules. Optimal conditions for photochemical entrapment and photochemical pumping will be described fully.

Theoretical Background

We refer to target by the symbol τ; probe by the symbol p; and helical duplex between target and probe by the symbol τp. The equilibrium among these three forms is represented by the chemical reaction

$$\tau + p \rightleftharpoons \tau p.$$

Elementary helix-coil transition theory for this system defines an initiation factor, β, and an equilibrium constant for the closure of a base pair once a chain has been initiated, s. The two state helix-coil theory where all base pairs in a helical run have equal stability allows us to write (Crothers et al., 1964; Crothers et al., 1965; Bloomfield et al., 1974; Tinoco et al., 1978; Cantor et al., 1980)

$$K = \beta s^n \tag{1}$$

where n+1 equals the number of base pairs in the helix and n equals the number of base-stacking interactions. The equilibrium constant, K, associated with this two state model is defined by equation 2.

251

$$K = \frac{[\tau p]}{[\tau][p]} \tag{2}$$

We now introduce the thermodynamic parameters, the enthalpy of base pair elongation, ΔS. In terms of these new parameters together with an alternate representation of the initiation parameter, the equation 2 for the equilibrium constant becomes

$$K = e^{-2.3\alpha}\, e^{-(n\Delta H/RT)}\, e^{(n\Delta S/R)} \tag{3}$$

where T is the hybridization temperature, R is the gas constant and $e^{-2.3\alpha} = \beta$.

Rearranging these equations results in an expression for the log of the ratio of free target to bound target in terms of the thermodynamic parameters

$$\log \frac{[\tau]}{[\tau p]} = -\gamma n\, (\Delta T/T) + \{\alpha - \log[p]\} \tag{4}$$

where $\gamma = -\Delta S/2.3R$ and $\Delta T = T_\infty - T$. The temperature, T_∞, is defined by the equation $T_\infty = \Delta H/\Delta S$. This is the melting temperature for a helix of infinite length.

Equation 4 indicates that the logarithm (base 10) of the ratio of the concentrations of free target to covered target is a linear function of $\Delta T/T$ where the logarithmic function is positive at high temperature (low ΔT) and negative at low temperature (high ΔT). The value of this linear function at $\Delta T = 0$ (i.e. at the melting temperature of an infinitely long helix) is $(\alpha - \log[p])$.

Theoretical and Experimental Results

Melting temperature of short probes

The melting temperature T_m is defined as in a conventional optical melting experiment. This is the temperature at which half of the target is complexed with probe and half of the target is free in solution (i.e. $\log([\tau]/[\tau p]) = 0$). For all practical purposes, a probe longer than about 200 nucleotides has a $T_m = T_\infty$. Equation 4 can be rearranged to show the dependence of melting temperature on probe length (n+1) and probe concentration, [p].

$$T_m = \frac{\gamma n}{(\alpha + \gamma n - \log[p])} \cdot T_\infty \tag{5}$$

The theoretical analysis that led to equation 4 implicitly assumes that the free energy difference between GC and AT base pair formation is zero. This situation is realizable when hybridization is performed in the presence of a chaotropic salt such as tetramethylammonium chloride (Melchior, et al., 1973; Shapiro, et al., 1969; Wood, et al., 1985.)

252

As an illustration, selecting appropriate parameters for an optically detected hybridization in a 1M NaCl solution ($\alpha=3.8$, $\gamma=4.4$, $T_\infty=368$ °K, $\log[p]=-4$) the calculated melting temperature according to Eq. 5 for a 25mer probe is about 71°C. When the probe concentration is decreased by an order of magnitude, the melting temperature correspondingly decreases by about 3°C. This 3°C decrease is approximately linear with the log of probe concentration. The change in T_m predicted here is consistent with published changes in T_m as a function of oligonucleotide concentration in the concentration ranges for which optical measurements have been made (26,27). Thus, although the T_m of the 25mer is 71°C in an optical experiment, it is only 53°C in a conventional hybridization experiment where the probe concentration is 10^{-10}M.

Discrimination under equilibrium conditions

A typical hybridization experiment involves the use of a single stranded probe of length (n + 1) to identify a specific sequence of the same length in a genome containing a total of N base pairs, which we will assume to be all composed of unique sequences of DNA. Background in such an experiment originates from two sources. Non-specific background arises when probe molecules bind to material other than nucleic acid. Specific background arises when probe molecules bind specifically to non-target regions of nucleic acid. A hybridization of p to τ which is devoid of a specific background signal requires that the accidental homologous subsequences in the target nucleic acid which might be complementary to short stretches of adjacent basis in the probe are sufficiently infrequent and have insufficient strength of interaction with the probe to represent an appreciable background. At equilibrium, specific background is inherently present. The amount of specific background, however, is governed by the hybridization conditions under which equilibrium is established. To quantitatively understand the dependence of the specific background on the various factors that contribute to hybridization, we define a mathematical parameter, ∇, which we call the decades of discrimination by equation 6.

$$\nabla = \log\{[\tau p]/[\tau]_{TOT}\} - \log\{[\tau' p]/[\tau']_{TOT}\} \tag{6}$$

where the parameter $[\tau]_{TOT}$ represents the total target concentration whether covered or free, n+1 equals the number of nucleotides in the probe, and n'+1 (which is less than n+1) equals the number of nucleotides in a sub-sequence of the probe to which complementarity may be found in nucleic acid sequences other than the true target sequence, $[\tau']_{TOT}$ is the total concentration of such complementary n'+1 mers in the solution, whether covered or free, and $[\tau'_p]$ is the concentration of probe hybridized to each (n'+1)mers.

The ratio $\{[\tau p]/[\tau]_{TOT}\}$ represents the fraction of target sites in the solution which are hybridized to probe, while the ratio of $\{[\tau' p]/[\tau']_{TOT}\}$ represents the fraction of a particular accidental homologous subsequence in the target nucleic acid which is hybridized to a segment (n'+1 long) of the probe. The ratio of these two fractions is always greater than 1 and its magnitude indicates the preference of the probe for binding to the true target over the particular accidental homologous subsequences. The logarithm of this ratio of ratios is defined as the decades of discrimination.

Figure 1 shows plots of the log of the inverse fraction of occupied targets n and n' which have been drawn using equation 4. These plots should be interpreted as logarithmic melting curves for these short probes from their targets. These curves are plotted in terms of dimensionless parameters and are therefore independent of the values of ΔH, ΔS, γ and α. They deviate from the straight line prediction of equation 4 and develop curvature near $[\tau p] = [\tau]$ because $[\tau]_{TOT}/[\tau p] = 1 + [\tau]/[\tau p]$. There are two sets of curves in Figure 1 which correspond to two different values of $[p]$. Each set contains a curve for which n = 25 and n' = 16. The $\log\{[\tau']_{TOT}/[\tau'p]\}$ is calculated by equation 4, where n' is substituted for n. The two curves of each set intersect at the melting temperature of the infinitely long helix where $\Delta T = 0$. Note that at very low $[p]$, the melting temperature of the probe from its target is low (large ΔT) and that at relatively high $[p]$, the probe melts from its target at much higher temperature. The maximum number of decades of discrimination occurs at the value of $\gamma(\Delta T/T)$ where there is a maximum difference between the n' curve and the n curve. This condition can be solved for analytically, for it

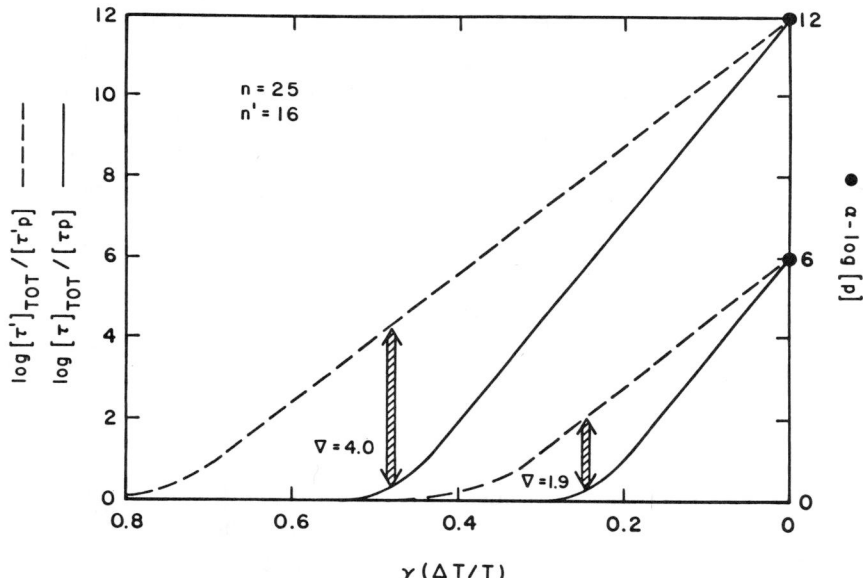

Figure 1. Plot of the logarithm to the base 10 of the ratio of total target to target hybridized with probe versus the dimensionless stability parameter $\gamma(\Delta T/T)$ for n = 25 and equal 12 and 6, ————. Plot of the logarithm to the base 10 of the ratio of total (n' + 1) mer concentration with complementarity to probe sequences to the concentration of (n' + 1) mer-probe hybrid for n' = 16 and { α - log[p] } equal 12 and 6 versus the dimensionless stability parameter, — — —. The heavy vertical arrows are equal in length to the number of decades of discrimination at the midpoint of the melting transition of probe with target for { α - log[p] } equal 12 and 6

is the point where the slopes of the two curves are equal. However, for all useful conditions, an approximate value of the abscissa is that which corresponds to the midpoint of the melting curve for the longer sequence where $[\tau]/[\tau p] = 1$. The points of maximum decades of discrimination for each of the two sets are indicated on Figure 1. Note that ∇ is much larger for the low value of $[p]$ than it is for the large $[p]$. An analytical expression for ∇ is shown in equation 7.

$$\nabla = \gamma(\Delta T/T) (n - n') + \delta \qquad (7)$$

where $\delta = \log\{[\tau]_{TOT}/[\tau p]$ at $[\tau]/[\tau p] = 1$ and is the displacement from the straight line plot in Figure 1 associated with the curvature near $[\tau]/[\tau p] = 1$. The n' curve must be linear at the midpoint value of $\gamma n(\Delta T/T)$ for equation 7 to be valid. Now if we use the approximation that the evaluation of δ should be made at $[\tau]/[\tau p] = 1$,

$$\delta = \log 0.5 = -0.3$$

At this midpoint of the transition of probe with target, equation 4 shows that

$$\gamma(\Delta T/T) = \{\alpha - \log[p]\}/n$$

The final expression for the decades of discrimination, ∇, is shown in equation 8. This is interesting, for it shows that discrimination at equilibrium depends only on the initiation parameter of equation 1 and is independent of the enthalpy of helix formation/bp and the entropy of helix formation/bp. It depends logarithmically upon the concentration of probe and it depends on the relative lengths of base sequences being compared. The best available estimate of β for polynucleotide chains comes from Pohl (1974) although several other references are consistent with this estimate. A typical range for values of β in the literature is 10^{-3} to 10^{-5} (Levin, 1974; Scheffler et al., 1968, 1970; Tinoco et al., 1973; Flavell et al., 1974; Riesner and Romer, 1973).
Using the value of $\beta = 1.6 \times 10^{-4}$ from Pohl, $\alpha = 3.8$.

$$\nabla = \{3.8 - \log[p]\}\left(\frac{n - n'}{n}\right) - 0.3 \qquad (8)$$

It should be pointed out that Eq. 8 addresses only the discrimination between true target and continuous partially homologous sequences which are $n'+1$ in length. In a hybridization experiment it is likely that the probe will also associate with $(n'+1)$ bases through may non-contiguous base pairings. For these situations, Eq. 8 will yield an underestimate of the decades of discrimination, since the free energies associated with mismatches and bulges are not considered. Theoretically, Eq. 8 does apply to the situation of $(n'+1)$ non-contiguous base pairings if the parameter α is adjusted to include the free energies of the internal mismatches and bulges.

255

Photochemical entrapment

All the equations above apply to equilibrium distributions of probe and target. It is really feasible to achieve equilibrium in these complex systems and, if so, how much time will this process take? The answers to these questions will depend upon whether hybridization is done in solution or on a membrane. We favor solution formats because hybridization reactions are approximately ten times faster when both target and probe are free in solution (Wetmur and Davidson, 1968). In addition, solution hybridization can be conducted in small volumes (~5–10 μl), facilitating the use of higher probe concentrations and further enhancing the hybridization rate. Two central points must be recognized if solution hybridization with short probes is to be practical. First, the equilibrium distribution of base pairing interactions in solution will always favor Watson (w) finding Crick (c). The genomic DNA will always be in long enough sections so that a short oligonucleotide probe, even at molar concentrations, will not be able to compete with the greater stability of wc over τp. The only way for a solution hybridization experiment to be successful is to entrap kinetically an intermediate state where the probe/target interaction is much more frequent than the perfectly renatured complementary genomic strands. Second, since time is an essential variable, the ability to freeze the distribution of interactions in a manner which does not alter the distribution is very important. We favor photochemical crosslinkage as the method for this quenching process for it can be accomplished rapidly and the resultant crosslink facilitates subsequent analysis. Nevertheless, quick cooling techniques as well as rapid mixing procedures may also be effective under approximate conditions.

The irradiation time required to crosslink a psoralen-monoadducted 13mer hybridized to a complementary oligonucleotide has been determined. The experiment was set up to allow the two oligonucleotides to reach hybridization equilibrium solution. The samples were then quick frozen and irradiated at 4°C after thawing. Irradiation at low temperature insured slow hybridization kinetics. The results show that the equilibrium distribution of the probe/target complex was photochemically fixed within two minutes of irradiation. This time scale is very rapid compared to conventional hybridization rates.

The hybridization reaction of probe with target and the annealing of complementary genomic strands are second order reactions. The rate of reaction of probe with target may be written in terms of a binary rate constant, k_p, where the subscript, p, refers to the probe.

The reaction of any ith section of Watson with its appropriate ith section of Crick may also be written in terms of a binary rate constant, k_g, where the subscript, g, refers to the genome.

$$-\frac{d[\tau]}{dt} = k_p[p][\tau] \; ; -\frac{d[w_i]}{dt} = k_g[w_i][c_i]$$

The times required for these two reactions to be half completed are

$$\Gamma_p = 0.693/k_p[p] \; ; \Gamma_g = 1/k_g[w]$$

For the first of these, the probe is considered to be in vast excess. For the second, $[w_i]$ equals $[c_i]$. In our analysis we assume the binary rate constant for hybridization to be length independent. This rate constant can be calculated from the Cot curves of Britten and Kohne (1868). Their kinetics were done on fragments of DNA approximately 400 bases long at a salt concentration of 0.18 M and a temperature approximately 20°C below the melting temperature of the DNA. The resulting rate constant is

$$k_g = k_p = 1.6 \times 10^6 \; 1/\text{mol sec}.$$

Riesner and Romer (1973) have presented data on short oligonucleotides which are in excellent agreement with this number. Their numbers range from 10^5 to 10^7, with the most frequent values between 1.5×10^6 and 5.5×10^6. These values were obtained in 1M salt at 17°C, so direct comparison might be questioned. It is striking, however, that the agreement is so good. Finally, although Wetmur and Davidson (1968) have proposed a square root dependence of the rate constant on the length of the renaturing fragments, their model also predicts a binary rate constant of 1.5×10^6 1/mol sec for 25 mer oligonucleotide probes.

    If probe hybridization is to compete kinetically with genomic renaturation, probe concentration must exceed the target concentration, since the second order rate constants are approximately equal for the two reactions. Figure 2 is a time course for the solution hybridization of a psoralen monoadducted 25mer to a denatured plasmid DNA, pUC19. Hybridization was conducted at a temperature close to the T_m of the probe/target complex utilizing a molar ratio of probe to target of 5.5 to 1. An accurate time course of hybridization was obtained by quick cooling aliquots at different hybridization times and irradiating each at 1–2°C with 360nm light. The drop in temperature stopped both probe/target hybridization and renaturation of pUC DNA. At short times, an increase in signal reflects hybridization of the 25mer to its target sequence in pUC. Later, however, that hybrid is displaced by renaturation of the complementary target strands. Maximal signal was obtained after a 30 minute hybridization and corresponded to only 3% coverage of target sequence. At this point in time, separate experiments showed that 75% of the complementary pUC strands had renatured.

    Clearly, photocrosslinkable probes can entrap a pseudo-equilibrium between probe and target. This kinetic advantage is optimized when the probe concentration exceeds the target concentration by at least a factor of 100. In this case, assuming constant binary rate constants, it will take the genome 100 times longer to renature than it will take the probe to find its target. The reaction should be stopped after four or five half-lives. If the probe concentration is 10^{-8}M, this corresponds to approximately 3–4 minutes of hybridization before the reaction is photochemically quenched. Alternatively, hybridization and photofixation can be carried out simultaneously for the same period of time. Solution hybridization of a crosslinkable probe at 10^{-8}M to a single copy sequence

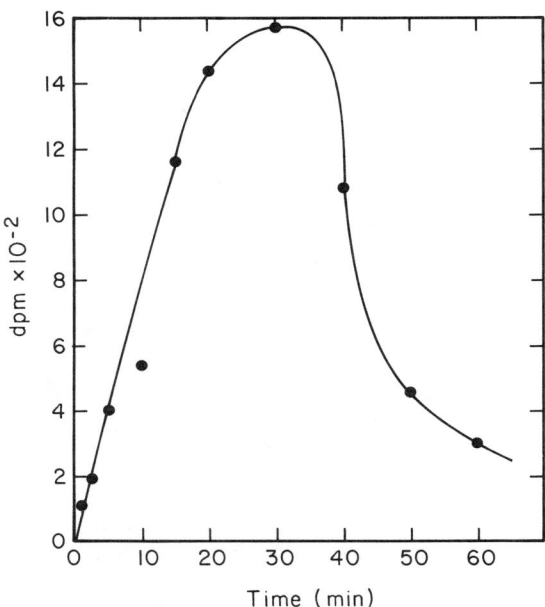

Figure 2. Kinetics of hybridization between a 4'-(hydroxymethyl)–4,5',8–trimethylpsoralen (HMT) modified 25mer probe and denatured double-stranded pUC 19 target DNA. Form III pUC 19 DNA was denatured in 100mM NaC1, 10mM Tris-HC1 pH 7.0, 1mM EDTA containing 25% formamide by heating at 100°C for 3 minutes and then quick cooling in a -70°C ethanol bath At time zero ^{32}P HMT-modified 25mer {CAGTGAATTCGAGCTCGGT(HMT)ACCCGG} was added to the thawed solution and the mixture was incubated at 45°C. Aliquots were removed at the indicated times into a -70°C ethanol bath. These aliquots were subsequently thawed in an ice bath and irradiated with 320–400 nm light for 3 minutes at 1–2°C. Formation of crosslinked probe/target hybrid was determined by electrophoresing the samples through a 1% alkaline agarose gel and counting the M13 DNA bands. [pUC DNA] = 8.47×10^{-10}M. [Probe] = 4.63×10^{-9}M. Maximum target coverage by furan side HMT-monoadducted 25mer = 2.7%. Each data point represents 15 ng pUC DNA.

in the human genome satisfies the concentration requirement. A 1.5 mg/ml solution of human DNA is 10^{-12}M in all unique human sequences. Thus, the probe is 10,000 times more concentrated than any unique sequence and genomic renaturation will not compete significantly in the pseudo-equilibrium distributions achieved after 3–4 minutes of hybridization.

Photochemical pumping

Photocrosslinkable probes provide a second advantage that is related to hybridization kinetics. This is the ability to pump an equilibrium reaction. We define "pumping" as the ability to obtain a high yield of probe/target complex when equilibrium conditions dictate a low concentration of that same complex.

$$\tau + p \; \rightleftharpoons \; p\tau \; \overset{h\nu}{\rightarrow} \; (p\tau)_{crosslinked}$$

Pumping results when a) equilibrium is reestablished after reducing the initial unfixed probe/target complex concentration by crosslinking and b) the newly formed probe/target complexes are subsequently fixed by additional crosslinking. If pumping is to occur during the time course of irradiation, three conditions must be met:

1. The half-life, τ_{XL}, required to photochemically crosslink existing probe/target complexes must be much shorter than the irradiation time.
2. The half-life for photodegradation of the probe must be much longer than τ_{XL}.
3. The half-life for hybridization, τ_p must be much shorter than the irradiation time, so that equilibrium can be reestablished many times.

The value of τ_{XL} was been measured experimentally by Gamper et al. (1987) for a monoadducted 12mer. τ_{XL} is approximately 25 seconds when preformed hybrid is irradiated with 400mW/cm^2 of 320–400nm light from a Hg source. Typically, we have used irradiation times of 2 to 20 minutes. The half-life for photodegradation was also measured. When the measured half-life is adjusted for intensity differences, a furan-side monoadducted oligonucleotide has a half-life of 15 minutes when irradiated with the 400mW/cm^2 source. Thus, conditions 1 and 2 can be achieved with psoralen monoadducted probes.

Condition 3 is determined by the probe concentration and the binary rate constant k_p. The k_p for a psoralen monoadducted 25mer binding to a single-stranded DNA near the T_m of the probe/target hybrid has been measured (Gamper et al., in press). A value of 1.5×10^6 1/mole-sec was obtained, suggesting that psoralen monoadducted probes behave kinetically as if they were conventional probes. Using a value of 1.6×10^6 1/mole-sec for k_p, the hybridization half-life, τ_p, equals 70 minutes at a probe concentration of 10^{-10}M and equals 4 seconds at 10^{-7}M. Condition 3 is satisfied when the half-life for hybridization is less than a minute, or when the probe concentration is greater than 10^{-8}M. At lower probe concentrations one can still effect pumping with a photocrosslinkable probe by cycling the sample through alternate light and dark periods. To demonstrate photochemical pumping, we have measured an apparent T_m for the crosslinking of a psoralen monoadducted oligonucleotide 13 bases to length (13mer-MA) to a complementary oligonucleotide 31 bases in length. The results are shown in figure 3. In this experiment the 13mer-MA was end labeled with ^{32}P and used as a target at 10^{-8}M. The 31mer complement, which was at 10^{-5}M, functioned as the probe. The two oligonucleotides were mixed in solution, heated to 95°C for 5 min, then cooled to a desired irradiation temperature. Duplicate samples were hybridized at each temperature for five minutes prior to irradiating for either 2 mins or 5 mins. After photofixation, the samples were run on a 20% polyacrylamide –8M urea gel. The crosslinked samples were located by autoradiography, excised, and counted. The melting profiles of Figure 3 show

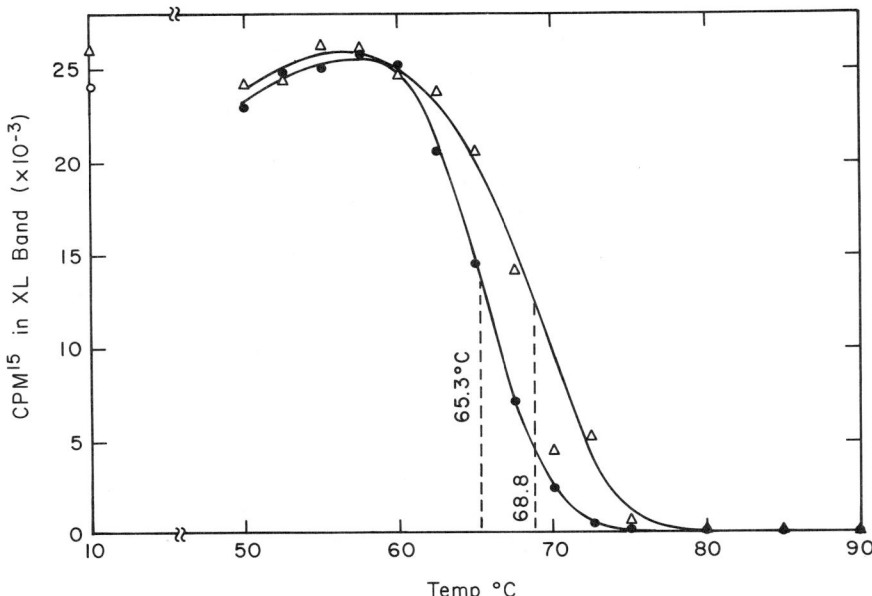

Figure 3. Effect of irradiation time on the melting profile of the 13mer HMT-MA with its complement. An oligonucleotide (31mer, 5′-GGATCCCCGGGTACCGAGCTCGAATTCACTG-3¢) containing the complement of the 13mer-MA was hybridized ^{32}P labeled 13mer-MA in 0.1M NaC1, 0.01M Tris (pH7.0). 1mM EDTA, 8% formamide. The two oligonucleotides were mixed in solution (5μ1 vol), heated to 95°C for 5 mins, then cooled to a desired irradiation temperature. Duplicate samples were hybridized at each temperature for five minutes prior to irradiating for either 2 mins of 5 mins. After photofixation, the samples were run on a 20% polyacrylamide-8M urea gel (0.04 cm x 20 cm x 40 cm) for 2 hrs at 50 watts. The crosslinked samples were located by autoradiography, excised, and counted. ● two mins of irradiation. Δ - five mins of irradiation.

that the apparent T_m is dependent upon irradiation time. The T_m equals 65.3°C when the samples are irradiated for two minutes. However, when the samples are exposed to light for 5 minutes, the apparent T_m increases to 68.8°C. From the sequence of the 13mer, a reasonable estimate for the optical T_m of this oligonucleotide under the same solvent conditions is 50–55°C. Pumping has dramatically increased the yield of probe/target complete at temperatures where equilibrium overwhelmingly favors free target over probe/target complex. At oligonucleotide concentrations where pumping is apparent, the measured T_m of a probe/target complex becomes a function of irradiation time and is substantially higher than the true T_m.

Previously, it was shown that maximal discrimination occurs at T_m, not above or below (see Fig. 3). Although pumping allows one to hybridize and crosslink above T_m, an advantage is not readily apparent since hybridization above the T_m will reduce discrimination. However, we have not yet considered all the equilibria in a real system. A more accurate view of solution hybridization with a crosslinkable probe is illustrated below:

260

$$\tau_c \rightleftharpoons \tau_o + p \rightleftharpoons \tau_o p \xrightarrow{h\nu} (\tau p)_{crosslinked}$$

Single stranded nucleic acids have secondary and tertiary structure in solution which can affect the accessibility of a target sequence for hybridization with an oligonucleotide. Open τ_o conformers which allow probe to bind to its complement exist in equilibrium with inaccessible closed conformers τ_c. Ribosomal RNAs are examples of this situation. They are known to exist in many different, stable and interconvertible conformations in solution. If a solution hybridization is performed at the T_m of the probe/target system, a low yield of probe/target complex can result if τ_c is favored over τ_o at equilibrium. This will be determined by the global free energy associated with the alternative target forms in the hybridization solution. Pumping now has the additional effect of driving the interconversion of τ_c to τ_o by LeChatelier's principle leading to improved target coverage. The above model suggests that the kinetics of hybridization at high probe concentration can be limited by the rate of interconversion of target forms. This has been experimentally confirmed and will be reported elsewhere (Gamper et al., 1987). Higher hybridization temperatures destabilize the closed conformers and increase the rates of interconversion between forms.

Comparison of the specific background of conventional probes relative to crosslinkable probes

The identification of a single unique sequence in a sample of eukaryotic DNA is technically the most challenging objective of hybridization probe technology. While examples of successful colony hybridization on cloned human libraries (Torczynski et al., 1984) and Southern hybridization in fractionated human DNA (Studenchi et al., 1984) have been reported, the goal of direct hybridization of an oligonucleotide probe to unfractionated human DNA with no significant background remains elusive. The crosslinkable oligonucleotides described here contain a site-specifically placed psoralen monoadduct. Hybridization of nucleotides adjacent to the psoralen monoadduct is a requirement for the correct positioning of the psoralen at a crosslinkable site. This correct positioning must be present during irradiation for crosslinkage to occur. On a random sequence basis, a probe 13 bases in length would be expected to be present 25–50 times in a complex eukaryotic genome. Figure 4 demonstrates this in a conventional dot blot experiment. A psoralen monoadducted oligonucleotide 13 bases in length was used to probe for the presence of its complement in a single-stranded target DNA (M13mp19). The presence of 500 vg of calf thymus DNA provided a combined signal and specific background which was equivalent to the signal given by 25 ng of target M13 DNA using conventional hybridization techniques. By irradiating this dot blot, and then washing it briefly at 100°C, 500 μg of calf thymus DNA now provided a combined signal and specific background which approximated the signal given by 10 ng of M13 DNA. The specific background was therefore reduced by photofixing the probe/target complex and subjecting it to a denaturing wash.

The kinetic advantages of crosslinkable probes are realized at concentrations in the

I3-MER MONOADDUCT

MI3mpl9 (ng)

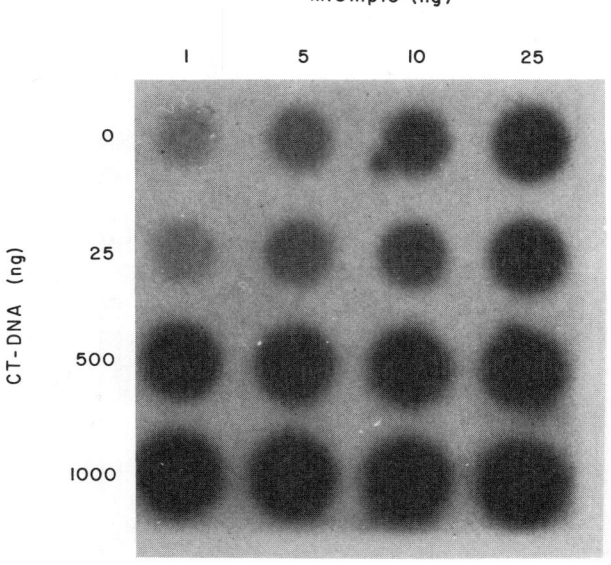

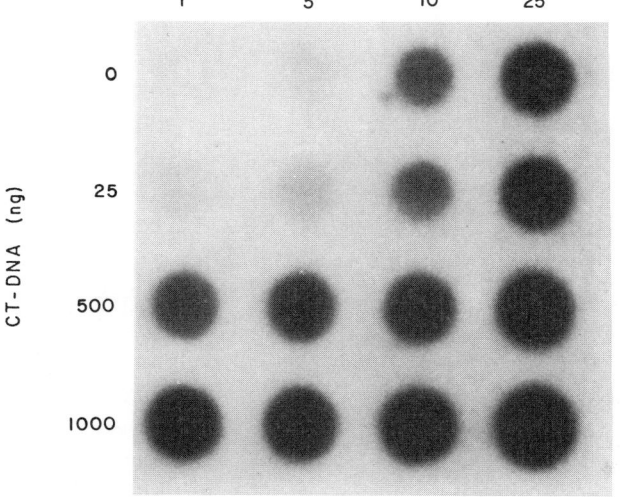

range of 10^{-7} to 10^{-8}M (see above). However, the use of probes at these concentrations results in higher non-specific and specific backgrounds. The covalent link between the photofixed probe/target complete allows one to effectively remove the non-specific background by using denaturing wash conditions (i.e. buffered solutions at 100°C containing high concentrations of formamide). A fraction of the higher *specific* background, however, will be fixed by irradiation. Under optimal hybridization conditions, is it possible to identify a single copy gene in unfractionated human DNA using a crosslinkable probe? Hearst (1988) has developed a theory for determining the minimal conditions required for effective discrimination and for estimating levels of specific background for given experimental conditions. Table 1 contains the calculated minimum length of probes and the maximum probe concentrations which may be used in probing the human genome based in this theory. Case I represents a conventional hybridization experiment with no use of a crosslinking agent. Case II represents a hybridization experiment in which the probe contains a crosslinking agent at one end. Note the advantage in reaction time of a factor of 30 provided by the crosslinkable probe over the non-crosslinkable probe.

Table 1.   Minimal conditions for oligonucleotide probing of the human genome case

Case	I	II
n_{min}	20	20
$[p]_{max}$, mol/l	7.6×10^{-10}	2.3×10^{-8}
Half life, Γ_p, sec	570	19
Specific background, B (% of signal)	17%	11%

Figure 4. Competitive hybridization of 13mer HMT-monoadducted probe to M13mp19 DNA immobilized on nitrocellulose. Increasing amounts of target M13mp19 DNA (1,5,10,25ng) in the presence of increasing amounts of calf thymus DNA (0, 25, 500, 1000ng) were probed with 13mer HMT-monoadducted DNA (5′ – ^{32}P–GCTCGGTACCCGG–3′). The target DNA and competitive DNA were immobilized to nitrocellulose. (Schleicher and Schuell, BA85) by first denaturing the DNA to 0.5M NaOH for 10 mins at 37°C. Following denaturation, the solutions were placed on ice briefly, then neutralized by addition of 1 volume of 2.0M ammonium acetate (pH7.0). The samples were placed in the appropriate minifold well, the well filled with 1.0M ammonium acetate, then suction applied at a constant rate (loading time 10–60 sec). The filter was air dried, placed in a vacuum oven, and baked for 4hrs at 80°C. The filter was prehybridized for 2 hrs at 12°C in hybridization buffer consisting of 0.9M NaCl, 0.09M Tris, pH7.5, 0.006M EDTA, 0.1% sodium dodecyl sulfate, 0.5% Nonident P–40, and 100µl homochromatography mix (Jay et al., 1974). The 13mer-MA was then hybridized a 5×10^{-10} for 16 hrs under non-stringent conditions (12°C). Following hybridization, the blot matrix was washed non-stringently for 2 mins at 12°C with a solution of 0.18M NaCl, 0.01M Na_2HPO_4 (pH=7.0), 0.001M EDT, 0.1% SDS. The blot was autoradiographed for 4 hrs, then irradiated with 320–400nm light (5 mins), stringently washed (boiling water, 5 min) and autoradiographed again.

Table 2 compares the same parameters at a constant, convenient probe length (n=25) and a constant probe concentration of 5×10^{-8}. Note the enormous advantage in calculated specific background that the crosslinkable probe provide under these conditions.

Table 2. Constant conditions for hybridization to single copy regions in the human genome

Case	I	II
n_{min}	25	25
[p], mol/1	5×10^{-8}	5×10^{-8}
Half life, Γ_p, sec	8.7	8.7
Specific background, B	102%	4.6%

For the conditions of Table 2, the specific background for a probe with a crosslinkable site at its center is 17%.

The calculations presented in Tables 2 indicate that crosslinkable probes are capable of detecting a single copy gene in unfractionated human DNA. These conclusions have been experimentally confirmed by us and reported elsewhere (Gamper et al., 1987).

Conclusions

A rapid hybridization assay for single copy unique sequences or low copy number pathogens is technically feasible using crosslinkable oligonucleotide probes. Conditions capable of eliminating the contribution of binding to non-target sequences have been defined. The necessary conditions for specific hybridization to the human genome include an oligonucleotide probe at least 20 bases long for which the maximum allowable probe concentration is 5×10^{-8}M. Longer probes provide greater discrimination and allow somewhat higher probe concentrations. Optimal discrimination is achieved by conducting the hybridization at the midpoint of the melting transition of the probe from its target. The kinetics of hybridization to accessible target is simply related to the free probe concentration. At log[p] = –6, the half-life is 0.4 second. At log[p] = –7, the half-life is 4 seconds. At log[p] = –8, the half life is 40 seconds. Use of crosslinkable oligonucleotide probes at concentrations greater than 10^{-8}M allows hybridization and photocrosslinkage to be carried out simultaneously. These conditions support sufficient photochemical entrapment of denatured double-stranded target as well as photochemical pumping. When conducted at the T_m of the probe/target hybrid, both phenomena lead to high levels of target coverage with a minimum of specific background.

We are partial to the use of photocrosslinkable probes which contain a specifically placed furan-side psoralen monoadduct at a TpA or ApT sequence. The presence of the monoadduct does not affect the kinetics of hybridization (Gamper et al., 1987) and has only a small stabilizing effect on the equilibrium interaction of probe with its target (Shi et al., 1987). As previously discussed, the monoadduct provides an additional tier of discrimination to the hybridization process and reduces specific background. Lastly, the ability to stably crosslink probe/target hybrids facilitates the use of solution formats and denaturing wash schemes (Gamper et al., 1987).

In addition to providing new procedures for hybridization diagnostics, psoralen monoadducted crosslinkable probes will permit quantitative determination of hybridization kinetics and equilibria in solution. Reaction mixtures can be quenched by rapid cooling followed by crosslinkage (Gamper et al., 1987) or simply by a short burst of saturating 320–400 nm light. The latter method should permit entrapment of hybridization distributions in times which are orders of magnitude faster than those required for the redistribution of equilibrium. These studies will provide new insights into the thermodynamic and kinetic parameters which are important for a thorough understanding of this complex process.

Acknowledgements

This study was supported by the Regents of the University of California, and by a contract from Applied Biosystems Inc. to HRI Research Inc.

References

Aboul-ela, F. Koh, D. Tinoco Jr., I. and Martin, (1985). Nucleic Acids Research 13, 4811–4824.
Alwine, J.C. Kemp, D.J. and Stark, G.R. (1977). *Proc. Nat. Acad. Sci., USA.* 74, 5350–5354.
Benton, W.D. and David, R.W. (1977). *Science,* 196, 180–182.
Berent, S.L. Mahmoudi, M. Torczynski, R.M. Bragg, P.W. and Bollen, A.P. (1985). *BioTechniques,* May/June, 208–220.
Bloomfield, V.A. Crothers, D.M. and Tinoco, T. (1974). "Physical Chemistry of Nucleic Acids". Harper & Row, NY.
Borer, P.N. Dengler, B. Tinoco, I. Uhlenbeck, O. (1974). *J. Mol. Biol.* 86, 843–853.
Breslauer, K.J. Frank, R. Blocker, H. and Uhlenbeck, O. (1974). *J. Mol. Biol.* 86, 843–853.
Britten, R.J. and Kohne, D.E. (1968). *Science* 161, 529–540.
Cantor , C.R. and Schimmel, P.R. (1980). "Biophysical Chemistry, Part III: The behavior of biological macromolecules" (W.H. Freeman & Co. - San Francisco) pp. 1183–1238.
Crothers, D.M. Kallenbach, N.R. and Zimm, B.H. (1965). *J. Mol. Biol.* 11, 802–820.
Crothers, D.M. Zimm, B.H. (1964). *J. Mol. Biol.* 9, 1–9.
Doty, P, Marmur, J. Eigner, J. and Schildkraut, C. (1960). *Proc, Nat, Acad, Sci., USA.* 461–476.
Flavell, R.A. Birfelder, E.J. Sanders, J.P..M. and Borst, P. (1974). *Eur. J. Biochem,* 47, 537–543.
Gamper, H. Cimino, G.D. and Hearst, J.E. (1987).*J. Mol. Biol.,* 197, 349–362.
Gamper, H. Piette, J. and Hearst, J.E. (1984). *Photochem. Photobiol.* 40, 29–34.
Grunstein, M. and Hogness, D.S. (1975). *Proc. Nat. Acad. Sci., USA.* 72, 3691–3695.
Gusella, J.F. (1986). *Ann. Rev. Biochem.* 55, 831–854.
Hayashi, M. Hayashi, M.N. and Spiegelman, S.(1963) *Proc, Nat. Acad. Sci., USA.* 50, 664–672.
Hearst, J. E. (1988) *Ann. Rev. Phys. Chem.* 39, 291–315.
Jay, E., Bambara, R., Padmanabhan, R. and Wu, R.(1974). *Nucleic Acids Research*, 1, 331–354.

Levine, M.D. "The Stability of Ribonucleic Acid in Solution: Model Calculations" Doctoral Dissertation, Department of Chemistry, University of California, Berkeley, 1974.

Marmur, J. and Greenspan, C.M. (1963)*Science* 142, 387–389.

Marmur, J. and Lane, D. (1960) *Proc. Nat. Acad. Sci., USA.* 46, 453–461.

Melchior, W.B. and Von Hippel, P.H.(1973) *Proc. Nat. Acad. Sci., USA.* 70, 298–330.

Pohl, F.M. (1974). *Eur. J. Biochem.* 42, 495–504.

Riesner, D. and Romer, R. (1973). In "Physico-Chemical Properties of Nucleic Acids", *vol.* 2, *ed.* J. Duchesne (London: Academic Press), p. 237–318.

Scheffler, I.E. Elson, E.L. and Baldwin, R.L. (1968). *J. Mol. Biol.* 36, 291–304.

Scheffler, I.E. Elson, E.L. and Baldwin, R.L. (1970). *J. Mol. Biol.* 48, 145–171.

Shapiro, J.T. Stannard, B.S. and Felsenfeld, G. (1969). *Biochem.* 8, 3233–3241.

Shi, Y.B. and Hearst, J.E. (1986). *Biochemistry,* in press.

Southern, E.M. (1975). *J. Mol. Biol.* 98, 503–517.

Studenchi, A.B. and Wallace, R.B. (1984).*DNA* 3, 7–15.

Tinoco, I. Borer, P.N. Dengler, B. Levine, M.D. Uhlenbeck, O.C. Crothers, D.M. Gralla, J. (1973) Nature New Biology, 246, 40–41.

Tinoco, I. Sauer, K. and Wang, J.C. (1978). "Physical Chemistry - Principles and Applications in Biological Sciences" (Prentice-Hall, Inc., Englewood Cliffs, N.J.) pp. 514–521.

Tinoco, I. Uhlenbeck, O.C. Levine, M.D. (1971). Nature, 230, 362–367.

Torczynski, R.M. Fuke, M. and Bollen, A.P. (1984) *Proc. Nat. Acad. Sci., USA.* 81, 6451–6455.

Wallace, R.B. Shaffer, M.J. Murphy, R.F. Bonner, J. Hirose, T. and Itakura, K. (1979). *Nucleic Acids Res.* 9, 879–894.

Wetmur, J.G. and Davidson, N. (1968). *J. Mol Biol.,* 31, 349–370.

Wood, W.I. Gitsbier, J. Lasky, L.A. and Lawn, R.M. (1985). *Proc. Nat. Acad. Sci., USA.* 82, 1585–1588.

The Analytical Uses of Luminescence

GABRIELLA GABOR

Israel Institute for Biological Research
Ness-Ziona 70450
Israel

Abbreviations: CL – Chemiluminescence; ACh – Acetylcholine; AChE – Acetylcholinesterase; HPTS – Hydroxypyrenetrisulfonic-acid; FIA – Fluoro Immuno Assay; PABA – Paraaminobenzoic-acid; RTP – Room Temperature Phosphorescence; HRP – Horseradish peroxidase; FITC – Fluoresceinisothiocyanate

Analytical methods based on luminescence have a long history of usefulness (Udenfriend, 1969; Guilbault, 1973). Low detection limit, variety of the measurable parameters and the excellent instrumentation developed in the last 2 decades, all contribute to the widespread use of these sparkling methods.

The different types of luminescence, depending on: a) the emitting level: fluorescence or phosphorescence and b) the means of excitation: by photons for the above, and chemical energy for obtaining chemiluminescence, are all used for analysis; the detection limits being $10^{-9} - 10^{-13}$ mole.

The measured parameters intensity, lifetime, wavelengths shifts and quenching enable determination of trace quantities as well as sites of interaction. The major disadvantage in fluorescence measurements, interference from scattered light and impurities may be overcome by measuring time-resolved spectra, (TRS) if the emitting species has a lifetime (τ) which exceeds 50ns.

Before dealing with the major topic, I would like to elaborate, on the improvements of the fluorescence (chemiluminescence) instrumentation in the last 2 decades. (Table 1).

Fluorescent substrates/leaving groups are used in enzymatic reactions to follow the kinetics and also for the quantitative determination of enzymes.

Table 2 summarizes the various types of luminescent analytical methods used in chemical and biological systems.

PH-indicators are not included in this table, however, a few examples used in fluoro-immunoassay (FIA) are given. FIA, developed to a field by itself, uses fluorescence techniques coupled with immunological selectivity, and is, as of today, the major analytical method used by immunologists (Soini and Hemmila, 1979). The most popular fluorescent label for FIA is fluorescence isocyanate, while the chemiluminescent method based on the HRP-isoluminol reaction is even more widespread due to its higher

Photobiology, Edited by E. Riklis
Plenum Press, New York, 1991

Table 1

Method/Instrument	Resulting improvement
1) Detectors based on single photon counting	Measurement of very low light intensities including chemiluminescence.
2) Laser excitation	Selectivity, both wavelength (λ) and geometry (in situ).
3) Pulsed light sources	Lifetime measurements
4) Computerized spectrofluorimeters	Measurement of all parameters easy, fast, accurate and stored (libraries of standards etc.).

When using "luminescent" analytical methods the analyte may be: 1) the fluorescent compound itself, 2) a compound that gives a fluorescent (chemiluminescent)/non-fluorescent product with a fluorescent/non-fluorescent reagent, 3) a compound labelled by a fluorescent label that enables its determination.

A fluorescent (chemiluminescent) compound may be used as an indicator in titrations; fluorescent compounds used for acidimetric titrations have suitable pKa values. Other indicators produce fluorescent complexes with positive or negatively charged colloids indicating the transformation between them. Such a reaction is used by first year students in chemistry for argentometric titrations, (indicating also the simplicity of these methods).

sensitivity (Belanger et al., 1987). Lately this method became even more popular with the introduction of chemiluminescent enhancers like iodophenol etc (Thorpe and Kricka, 1985). The nature of these enhancers confirm that at least one of the intermediates in the chemiluminescent reaction is a radical-like species and therefore, the promotion by a radical releasing compound.

Table 2

Analyte	Fluorescent/Chemilumin- escent Participant	Type of Reaction
1) Polycyclic compounds naphthalene, anthracene pyrene		"Eigen" Fluorescence
2) Salycilic Acid (Aspirin)		" "
3) Amino Acids	Fluorescamine	Schiff base[3]
4) Histidine (Histamine)	O-Phthalaldehyde	Condensation[4]
5) Thiamine	Thiochrome	Oxidation[5]
6) Ca^{++}	Quin-2	Chelate formation[6]
7) Oxidases	7-ethoxy-4-methylcoumarin and analogs, Ethoxy- fluorescein	Enzymatic[7] rxn with fluorescent leaving group.
8) Antigens, Antibodies	Fluorescein derivatives Dansyl derivatives	FIA site informative[i] " " "
9) Hydrolyses, Esterases	Esters of HPTS	Enzymatic rxn with[9] fluorescent leaving group
10) Phosphatases	Phosphate esther of Umbelliferrone	Enzymatic rxn with[1i] fluorescent leaving group
11) Acetylcholine Esterase (AChE)	N-methyl-7-carbamoyl quinolinium halides	Enzymatic[11] hydrolysis with fluorescent substrate and leaving group. their max being ∿100 nm apart.
12) Horse radishperoxidase H_2O_2	Luminol	Oxidation CL[12]
13) " " "	" Iodophenol	Enhanced CL[13]
14) AChE/Acetylcholine (ACh)	ACh/ACHE, Cholinoxidase HRP Luminol H_2O_2	Coupled CL[14]

Room Temperature Phosphorescence

Phosphorescence is a very long-lived emission, and as such easily quenched by self or other species and therefore not suitable for quantitative measurements. Until the early seventies it was detected only at low temperatures in frozen media that prevented molecular motions or encounters.

Since the seventies, room temperature phosphorescence (RTP) has been observed and quantitatively measured (Roth, 1967). The rigid surrounding "matrice" was supplied by absorbing the phosphorescent compound on a solid surface. It was shown that the phosphorescent species is hydrogen bounded to the surface of the solid (Schulman and Parker, 1977). Moisture quenches RTP by disrupting the H-bonds (Parker et al. 1979) and thus it also permits O_2 infiltration that results in additional quenching. However, by adding sodium acetate, H_3BO_3, NaF, glucose etc. the quenching became negligible.

269

These compounds have a matrix packing effect: when the phosphorescent compound is more rigidly held it is less subject to (self or impurities) quenching (Niday and Seybold, 1978).

Limit of detection by RTP is the pg-ng range. To enhance the signal a salt with a heavy atom is added. This increases the intersystem crossing and also contributes to the extent of hydrogen bounding. Fig 1 shows that the enhancement is also selective (Vo-Dinh, 1984).

Para-aminobenzoic acid (PABA), which is: a) sunscreen agent, b) a cofactor with the vitamin B complex and c) raises the salycilate level in blood and has also other therapeutic effects, has been extensively investigated and determined by RTP methods. Table 3 shows the accuracy of these measurements (Guilbault, 1973).

As the measured emission has a very long lifetime TRS can be measured. This enables the determination of a variety of compounds whose emissions overlap but their lifetimes are different. Vo-Dinh et al. (1980) used this method to identify PNA-S-benzo(a)pyrene, chrysene, fluorene, fluoranthene, phenanthrene and pyrene in synthoil. They further emphasized the differences by using selective enhancers.

This method needs very small quantities but still one has to extract the samples and the measurements are carried out by conventional instruments.

Biosensors

Recently all Analytical Chemists are looking for micro devices to carry out analyses *in situ*. The general idea is to miniaturize well-known existing techniques. The recent advances in optical fiber technology draw the attention of analytical chemists using luminescent methods. In fact ever since the beginning of the eighties there has been a burst of research on this subject (Lubbers and Opitz, 1984; Hirshfeld et al. 1984; Seitz, 1984; Peterson and Vurek, 1984; Wolfbeis, 1985, 1986, 1987).

A limited number of colored or fluorescent compounds may be determined by bare end fibers. Other non-fluorescent compounds may be determined by this method, by their reactions with fluorescent reagents or products, i.e. they involve a change of some optical property. These reactions have to be absolutely reversible, to follow quantitatively the positive and negative changes of the analyte by fluorescence. However, when using chemiluminescent methods, the reversibility of the reaction has no meaning, as at a given time the light emitted is proportional to the quantity of the analyte.

One needs only nanomolar (or less) quantities of the reagents, because of the high sensitivity of this method. To immobilize such a small quantity of the reagent on a solid surface, like beads of glass or polymers and attach it to the end of a fiber, seems easy and inviting. A device that consists of a reagent bound to a solid support (bead), held at the end of an optical fiber is named "optrode" (Fig. 2) in analogy to electrodes (Borman, 1981) used for analysis since the 1960's (Severinghaus and Bradley, 1958). The advantages of the optrode — compared to the electrode are many, as summarized in Table 4.

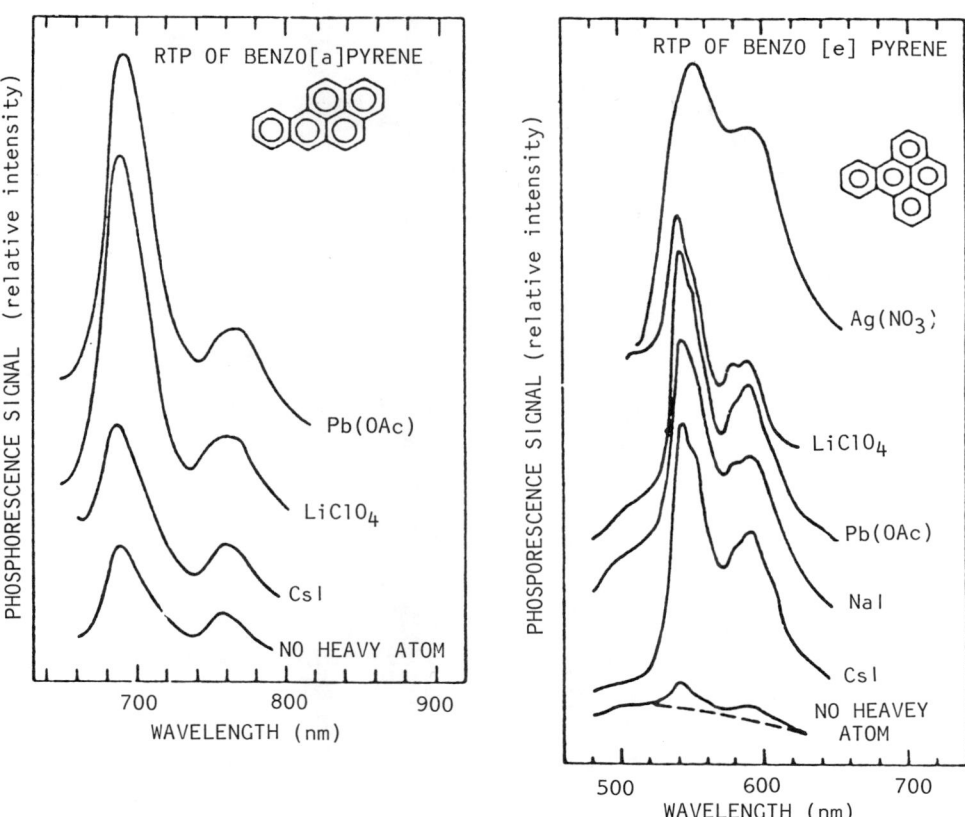

Fig.1 Enhancement of RTP of two different benzopyrenes by different heavy atom salts. Selectivity of enhancement is shown by comparing the relative enhancements by $LiClO_4$ and Pb(OAc).

Table 3. Precision of RTP measurements for p-Aminobenzoic Acid (PABA) adsorbed on paper

Amount of Analyte (ng)	Relative Intensity*	Relative Standard Deviation (%)
0.05	36	0.91
0.5	283	1.06
1	514	1.32
2	1,040	0.88
5	2,360	1.78
10	4,300	0.50
20	8,030	1.64
50	19,200	1.01

*Average relative intensity for five samples (background signal subtracted).

271

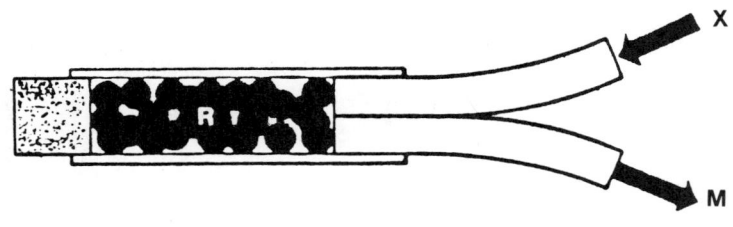

Figure 2. Optrode end with fluorescent reagent R bound to support attached to end of fibers that transfer the exciting (X) and emitted (M) light intensities.

Table 4. Advantages of optrodes

1) No electrical interference - local currents in the body

2) No reference needed

3) The flexibility of the fibers with the point like bead (d = 0.2 mm) at the edge serves as a miniature spectrofluorimeter and permits invasive sensing in clinical measurements.

4) Permits remote sensing[30]

5) Real time, in situ measurements.

6) Disposable reagent phase (reagent + support).

7) Measurements in all media: polar and non-polar solutions, gas, phase, colloids, etc.

There are some disadvantages as well, (Table 5).

Because of their very small size, optrodes are suitable for *in situ* on line measurements in body fluids. Optrodes have been designed for blood gases pO_2, pH and pCO_2. The O_2 sensors (Miller and Hirschfeld, 1986; Lubbers and Opitz, 1975; Peterson et al. 1984; Bergman, 1968; Wolfbeis et al. 1984; 1988; Lee et al. 1987;) are based on the quenching of the fluorescence by the para- magnetic oxygen molecules according to the Stern-Volmer law

$$I_0/I = I + K(O_2)_2 = I + kq_o(O_2) \tag{1}$$

Table 5. Disadvantages of optrodes

```
1)   Ambient light interference.

2)   Photo bleaching of reagents.

3)   Optrodes using indicators or chelating agents are ruled by
     the  Mass  Action  law,  and  therefore  they  have  a  limited
     dynamic range.
```

Io and I are the intensities in the presence and the absence of oxygen respectively, τ_o is the half life of the fluorescence in the absence of oxygen. Eq. 1 shows a linear relationship between the ratio of the intensities and the respective concentrations of the quencher - oxygen in this case. K in eq. 1 is the rate constant of a diffusion controlled process, therefore

$$K = k_q\tau = 10^{10} \, M^{-1}s^{-1}$$

The concentration of oxygen in aqueous solutions being 10^{-2} M

$$k_q\tau = 10^{-8}s^{-1}$$

for compounds having very short fluorescent lifetimes $\tau = 10^{-9}s$ quenching is negligible. For $\tau \, 10^{-8s}$, $I_o/I=2$. On degassing $[O_2] = 10^{-6}M$; $k_q\tau[O_2] = 10^4 \, s^{-1}$. In this case, in order to follow the changes at very low concentration of oxygen the necessary value of τ is 10^{-4} s.

pH sensors based on absorption (Peterson, 1980) or fluorescence changes (Peterson, 1980; Offenbacher et al. 1986; Saari and Seitz, 1982; Zhujun and Seitz, 1984) consist of a pH indicator with the appropriate pKa immobilized on a solid support attached to the edge of the fiber. A pH optrode based on energy transfer consists of a non-fluorescent dye with the appropriate pKa quenching the fluorescence of another dye (Jordan et al., 1987). Appropriate pKa means that the dye has a pKa value close to the pH values to be measured.

pCO_2 sensors are based on the reversible reaction between HCO_3 ions and gaseous CO_2 releasing H^+ ions (eq. 2), and their measurement by the pH optrode (Lubbers and Optiz, 1975; Wolfbeis et al. 1988;Zhujun and Seitz, 1984; Hirschfeld et al. 1987).

$$H_2O + CO_2 \quad HCO_3 + H^+$$

Remote sensing optrodes have been used for the on-line determination of uranium in nuclear reactors (Malstrom et al. 1985).

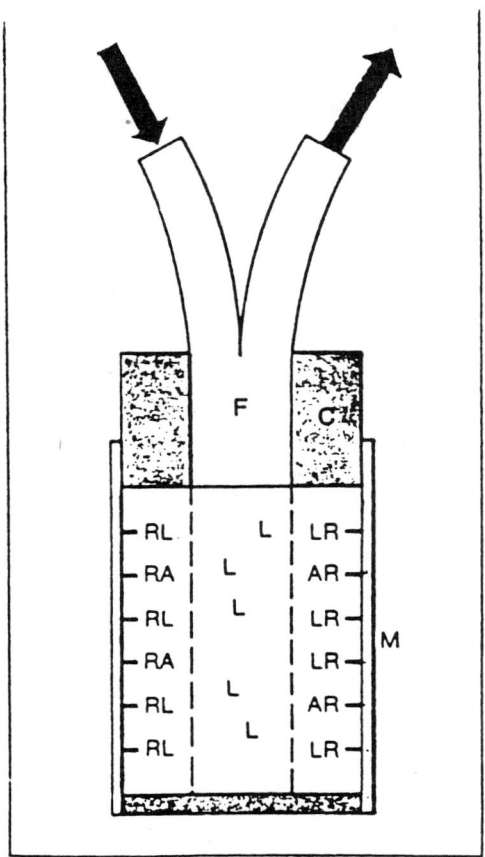

Figure 3. Bare fibers (F) follow concentration changes of fluoreseent ligand (L) — fluorescein labelled dextran — that competes for binding the reagent concanavalin — (R) with glucose (A).

Glucose sensors have been built based on: a) the enzymatic oxidation of glucose by glucose oxidase followed by the measurement of the released O_2 by an O_2-optrode (Uwira et al. 1984; Trettnak et al. 1988); b) the competitive binding of concanavalin-A (the immobilized reagent) by glucose and FITC labelled dextran (Wolfbeis, 1986). A bare fiber follows the concentration of the FITC labelled dextran in the solution (Fig.3).

Enzyme sensing optrodes consist of an immobilized lumogenic or luminescent substrate that change their optical properties during the enzymatic reaction hydrolysis, oxidation, etc. (Table 6). CAlternatively substrates may be determined in situ by immobilizing one or more enzymes whose action involves light emission; e.g. acetylcholine and choline may be determined simultaneously as shown in the following scheme.

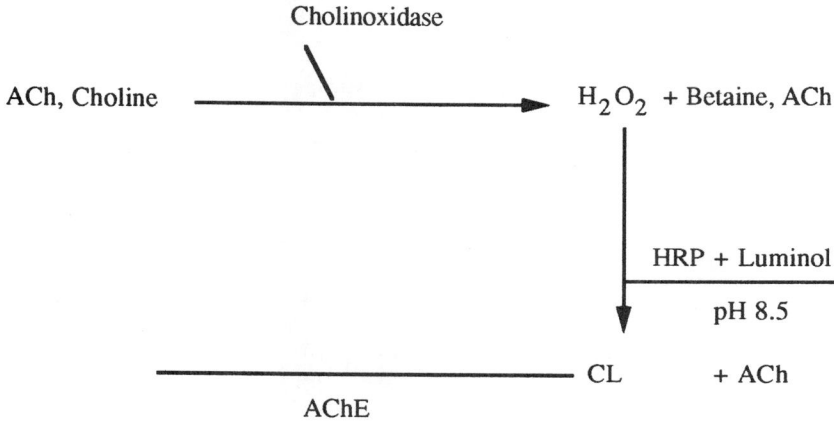

These are just a few examples of sensors that have been built and used so far.

Altogether, designing and using biosensors seems to be a very attractive field. One just has to couple the vast amount of known reactions used for the determination of simple analytes, enzymes, antibodies and antigens. (Table 2), to optical fibers to enable *in situ* on-line measurements.

It is my opinion that the TIME correspondent who estimated the contribution of biosensors in the future and defined the 21st century as the age of light (against the 20th century — the age of electronics) must have been a prophet.

Acknowledgment

The author wishes to thank Mrs. A. Raz for correcting and typing the MS.

References

Belanger, A., Brassard, P., Laquerre, S. and Merand, Y. (1987) *Com. J. Chem.* 65, 1392-1396.
Bergman, I., (1968), *Nature*, 218, 396.
Borman, S.A. (1981) *Anal. Chem.* 53, 1616A-1618A.
Eichhorn F., Rutenberg, A. and Kott, E., (1971), . *Clin. Chem.* 17, 296.
Guilbault G.G. et al (1967) *Anal. Chem.* 39, 271, (1968) ibid.
Guilbault, G.G., Practical Fluorescence (1973) Marcel Dekker.
Hirschfeld, T., Miller, F., Thomas, S., Miller, H.H., Milanovitch, F. and Gaver, R.W. (1987) *J. Lightwave Tech.* LT-5 1027.
Hirshfeld, T., Callis, J.B. and Kowalski, R.B. (1984) *Science* 226, 312.
Isacsson U. and Wettermark, G. (1974). *Anal. Clin. Acta* 68, 339.
Israel, M. and Lesbats, B. (1981) *Neurochem. Int.* 3, 81-90; B(1982) *J. Neurochem.* 39, 248-250.
Jordan, D.M., Walt, D.R. and Milanovich, F.P. (1987) *Anal. Chem.* 59, 437–439.
Koller E. and Wolfbeis, O.S. (1984) *Anal. Biochem.* 143, 146-151.
Lee E.D., Werner, T.C. and Seitz, W.R. (1987) *Anal. Chem.* 59, 279-283.
Lubbers, D.W. and Opitz, N. (1975) *Naturforsch.* 30c 532-533.
Lubbers, D.W. and Opitz, N. (1984) Sensors and Actuators, 4, 641.

Malstrom, R., Hirschfeld, T. and Deaton, T. (1985) Presented at the *184th Mtg. of the Am. Chem. Soc.*, Kansas City.

Matsoukas K.A. and Demertzis, M.A. (1988). *Analyst*, 113, 251.

Miller, H.H. and Hirshfeld, T.B. (1986) *Proc. SPIE*, 718, 39.

Niday, G.L. and Seybold, P.G. (1978) ibid 50, 1577.

Offenbacher, H., Wolfbeis, O.S. and Furlinger, E. (1986) *Sensors and Actuators* 9, 73-84.

Parker R.T., Freedlander, R.S., Schulman, E.M. and Dunlap, R.B. (1979) *Anal. Chem.* 51, 1921.

Peterson, J.I. and Vurek, G.G. (1984) *Science* 224, 123-127.

Peterson, J.I., Fitzgerald, R.V. and Buckhold,D.K. (1984) Anal. Chem. 56, 62-67.

Peterson, J.I., Goldstein, S.R. and Fitzgerald, R.V. (1980) *Anal. Chem.* 52, 864-869.

Rosenberry T.L. and Bernhard, S.A. (1971). *Biochem.* 10, 4114.

Roth, M. (1967) *J. Chromatogr.* 30, 276.

Saari, L.A. and Seitz, W.R. (1982) *Anal. Chem.* 54, 821-823

Schulman, E.M. and Parker, R.T. (1977) *J. Phys. Chem.* 81, 1932.

Seitz W.R. (1984) *Anal. Chem.* 56, 16A-34A.

Seitz W.R. and Neary, M.P. (1974) *Anal. Chem.* 46, 188A.

Severinghaus, J.W. and Bradley, A.F. (1958) *J. Appl. Physiol.* 13, 515-520.

Shore, P.A., Burkhalter, A. and Cohn, V.H. Jr. (1959), *J. Pharmacol. Exptl. Therap.*, 127, 182.

Soini F. and Hemmila, I. (1979) *Clin. Chem.* 25, 353 and other references quoted therein.

Thorpe G.H.G. and Kricka, L.J. et al (1985) *Anal. Biochem.* 145, 596-100.

Trettnak, W., Leiner, M.J.P. and Wolfbeis, O.S. (1988) *Analyst* 113, 1519-1523.

Tsien R.Y. et al (1982) , *J. Cell Biol.* 94 325; (1984) *T.I.B.S.* 263; (1985) *J.B.C* 260, 3440-3450.

Udenfriend, S., Fluorescence Assay in Biology and Medicine (1969) AP.

Udenfriend, S., Stein, S., Bohlen, P. Darman. W., Leimburger, W. and Weigele, M., (1973), *Science* 178, 871.

Uwira, N., Opitz, N. and Lubbers, D.W. (1984) *Adv. Exp. Med. Biol.* 169, 913.

Vo-Dinh, T, (1984) Room Temperature Phosphorimetry for Chemical Analysis. J. Wiley. pp 164.

Vo-Dinh, T., Gammage, R.B. and Martinez,P.R. (1980) *Anal. Chim.Acta* 118, 313.

Weigele, J.F., Blount, J.P., Tengi, R.C., Czajkowski and Leimburger, W., (1972) *J.A.C.S* 94, 4052..

Wolfbeis O.S. and Koller, E. (1983) *Anal. Biochem.* 129, 365-70

Wolfbeis O.S., et al (1984) *Microchim. Acta* 1, 153-158,

Wolfbeis, O.S (1986) Z. *Anal. Chem.* 325, 387.

Wolfbeis, O.S (1987) *Pure & Appl. Chem.* 56, 62-67.

Wolfbeis, O.S. (1985) Trends. *Anal. Chem.* 4, 184-188

Wolfbeis, O.S. (1986) *Anal. Chem.* 58, 2876-2879.

Wolfbeis, O.S. et a. (1985) *Anal. Chem.* 57, 2556.

Wolfbeis, O.S., Weis, C.J., Leiner, M.J.P. and Ziegler, W.E. (1988) *Anal. Chem.* 60, 2028-2030.

Yusem M., Delanay, W.E., Lindberg, M.A. and Fashing, E.M., (1969) *Anal. Chim. Acta.* 44, 403-409.

Zhujun, Z. and Seitz, W.R. (1984) *Anal. Chim. Acta.* 160, 47-55.

Zhujun, Z. and Seitz, W.R. (1984) *Anal. Chim. Acta.* 160, 305–309.

Electric Parameters of Purple Membrane Fragments

S.G. TANEVA AND I.B. PETKANCHIN[*]

Central Laboratory of Biophysics
Bulgarian Academy of Sciences
Sofia 1113, Bulgaria
[*]*Institute of Physical Chemistry*
Bulgarian Academy of Sciences
Sofia 1040, Bulgaria

Summary

The electric parameters (permanent and induced dipole moments) of purple membrane fragments in the reversible purple-to-blue transition and of surfactant-treated purple membrane were studied by electric light scattering (rotational electrokinetics) and transient reversal techniques.

A decrease in the values of both electric moments was observed upon deionization and treatment with relatively low concentration of the anionic surfactant. The restoration of the higher charge asymmetry upon cation addition confirms the interfacial origin of the permanent dipole moment. The relation of the membrane electric properties to the membrane function is discussed.

Introduction

Electro-optic techniques (electric light scattering (ELS), electric dichroism (ED) and electric birefringence (EB)) have a wide application in the study of biological systems. These methods could provide information on the electrical and optical properties of biological membranes — permanent dipole moment and membrane polarizability, the dynamics of the membrane charges, optical anisotropy and the direction of the chromophore transition moment.

Electro-optic methods have been successfully applied on purple membrane (PM) and have demonstrated the existence of a permanent dipole moment in direction perpendicular to the membrane plane (Keszthelyi, 1980; Kimura, et al., 1981; Kimura, et al., 1984; Todorov, et al., 1982), the effect of pH and ionic strength on the permanent dipole moment, electric polarizability and retinal tilt angle (Kimura, et al., 1984; Barabas, et al., 1983; Otomo, et al., 1986).

Recent studies concerned with the blue form of purple membrane gave emphasis to the specific functional role of the bound cations (Mowery, et al., 1979; Kimura, et al., 1984; Chang, et al., 1985; Szundi, et al., 1987). This is a particularly important question in respect to the molecular mechanism of the light-driven proton pumping in bacteriorhodopsin (bR). We have previously reported a strong change in charge

Photobiology, Edited by E. Riklis
Plenum Press, New York, 1991

asymmetry (permanent dipole moment) caused by deionization of PM (Taneva, et al., 1987).

The study reported here is an exploration of electric light scattering technique with the aim to follow changes in the orientation behaviour of PM fragments after different treatments of the membrane surface: i) cations removal from purple membrane, ii) addition of cations to deionized-blue membrane and iii) addition of the anionic surface active substance sodium dodecyl sulfate (SDS) to PM. This approach is expected to contribute to the understanding of the correlation between the membrane charge asymmetry and proton pump functioning, and the origin of the electric moments and their importance for the orientation phenomena.

Theory

Electro-optic orientational theory for non-spherical particles has been described elsewhere (Dukhin, et al., 1977).

The external electric field induces anisotropy of the dispersed system leading to changes in the optical properties (the intensity of the scattered light in the case of ELS).

The electro-optic effect (α) is given by:

$$\alpha = (I_E - I_0) / I_0 \tag{1}$$

where I_E, I_0 are the intensities of the scattered light when the external orienting field is switched on and off, respectively.

Since, the orientation of the particles is caused by the interaction between the induced dipole moment (connected with the ionic movement) and the electric field, when the frequency of the field is high enough to permit relaxation of the permanent dipole moment, one can determine the electric polarizability. Combined d.c. and a.c. fields measurements make it possible to evaluate the permanent dipole moment too.

The theoretical expression for the relative electro-optic effect of rigid thin disk in Rayleigh-Debye-Gans approximation for low degree of orientation and plane of observation perpendicular to the electric field is:

$$\alpha = \frac{A(KB)}{I_0(KB)} (\mu^2 + \delta) E^2 \tag{2}$$

where B is the diameter of the particle
$K = (2\pi / \lambda') \sin (\theta' / 2)$ λ — the wavelength of the light in solution, θ' — the angle of observation
$\mu = p / kT$, p — the permanent dipole moment
$\delta = (\gamma_1 - \gamma_2) /kT$, γ_1, γ_2 — the polarisability along and perpendicular to the particle symmetry axis. A(KB) and $I_0(KB)$ are optical functions depending on the particle dimensions and on the measuring systems characteristics (λ', θ').

The permanent dipole moment value can be evaluated using transient reversal experiments (described in Tinoko et al., (1959)) if the electric polarizability is known from ELS measurements.

The other parameter which can be determined from these measurements is the diameter of the fragments (B). The evaluation of B can be done from the disorientation relaxation time after switching off the field:

$$\alpha_t = \alpha_0 \exp(-t / \tau) \tag{3}$$

using the following expressions:

$$B = (3kT / 4\eta\, D)^{1/3} \tag{4}$$
$$\text{and} \quad D = 1/6\tau \tag{5}$$

α_t and α_0 are the values of the electro-optic effect at times $t=t$ and $t=0$, respectively; D is the rotational diffusion constant; k, T and η have their usual meaning.

B can be determined from the experimentally observed saturated value (α_∞) of the electro-optic effect (α) as well.

Experimental procedure

A combined electro-optic apparatus for ELS, ED and EB measurements was used Taneva, et al., 1987). The light source can be a He-Ne laser HNA 50 ($\lambda = 632.8$ nm) or an incandescent lamp followed by an interference filter (the wavelength of the filter depends on the absorption maximum of the sample). The scattered light was detected at 90° by a photomultiplier. The photomultiplier signal was displayed on a storage oscilloscope CB–13.

The response time of the measuring system is less than 50 μs.

Measurements were performed on suspensions of a very low bR concentration (0.5 μM) at room temperature. This concentration is much lower compared with other techniques (ED,EB), which is important for the interaction of the membrane fragments and their aggregation.

Deionized purple membranes were prepared on a cation exchange column (AG-50W, Fluka) as described in (Kimura, et al., 1984).

Recent experiments have demonstrated the induction of a blue form of bR after treatment with low concentration of SDS (Padros, et al., 1984). We carried out several sets of experiments with ionic surfactants intended to provide further information on the surfactant-induced changes in the electrical moments of PM fragments (in press). Results for concentration of SDS-0.8 mM are presented in this work.

Purple membrane was incubated with SDS for 1 h at room temperature and then centrifuged twice at 48,000 × g for 30 min.

The conductance and pH of all samples were about 0.5×10^{-5} $\Omega^{-1}\text{cm}^{-1}$ and 4.6–5, respectively.

Results and Discussion

Figure 1 represents the osciloscope traces of the ELS effect upon a.c. (A) and d.c. (B) electric fields applied to the suspension of purple, deionized and SDS-treated purple membranes. The frequency (1 kHz) of the a.c. field applied was chosen from the plateau region of the frequency dependence of the electro-optic effect, where the fragment orientation is believed to be due only to electric polarizability. The effect in d.c. field is shown for low field strength of the applied field at which the permanent dipole moment contribution for particles orientation dominates. The negative effect observed when d.c. field was applied to native PM suspension (Fig. 1B,a) indicates that the membrane fragments orient with their symmetry axis (perpendicular to the disk plant) along the field direction, hence the permanent dipole moment (or electric charge asymmetry) is in transversal direction to the membrane surface ($\gamma E \ll p$).

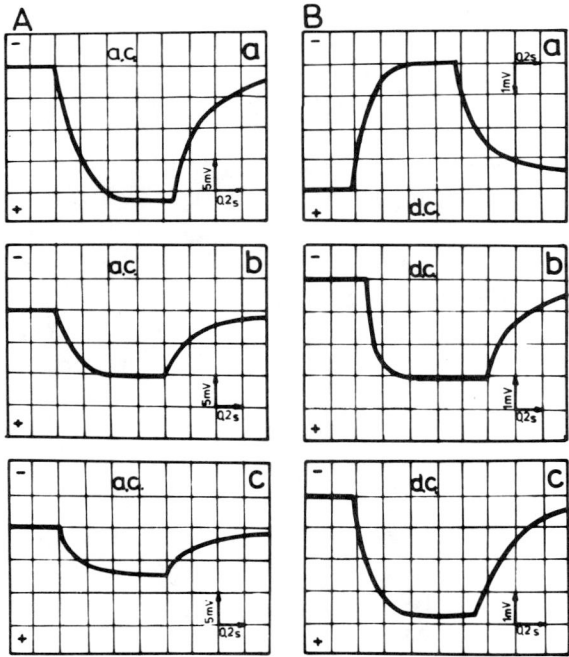

Figure 1. Time dependence of the electric light scattering effect of purple (a), blue (b) and SDS-treated (c) membranes; (A) – a.c. orienting field of 6×10^3 V/m and 1 kHz frequency, (B) – d.c. orienting field of intensity 3×10^3 V/m.

The first strong effect observed after the removal of cations (blue membrane) and SDS-treatment of PM is the sign reversal of the d.c. field electro-optic effect. In both cases (deionized and SDS-treated membranes) no negative part of the effect in d.c. orienting field was observed.

This behaviour is displayed in Fig. 2. which shows the dependence of the electro-optic effect on the square of the d.c. electric field strength. Qualitatively, one observes difference in the sign of the d.c. field effect (at low d.c. field strength) and in the initial slopes of the field dependences between PM and both — deionized and SDS-treated membranes.

The frequency dependence of the electric light scattering effect (dispersion curve — Fig. 3) confirms the existence of three distinct parts. At lowest frequencies (first part of the dispersion curve) the effect is negative for purple, whereas for deionized and SDS-treatment membranes though decreasing with frequency decreasing the effect remains positive. The slopes of the low frequency part of the dispersion curve is smaller for blue and SDS-treated membranes as compared to native PM, obviously indicating a decrease in the membrane charge asymmetry (permanent dipole moment).

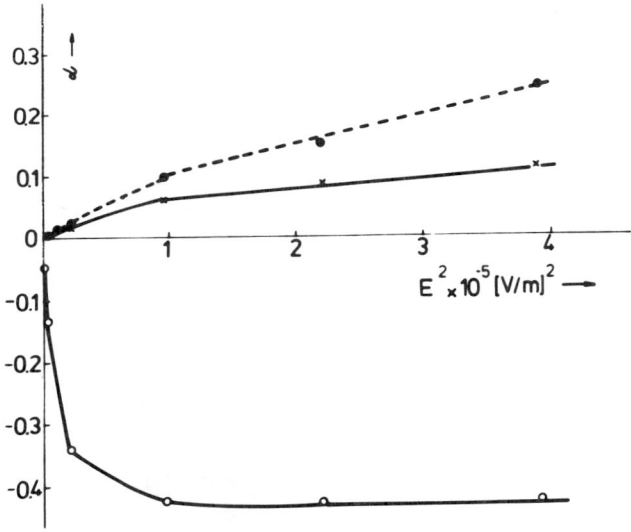

Figure 2. Dependence of the electro-optic effect on the square of the d.c. electric field in purple (O -full line), blue (● dotted line) and SDS-treated (× - full line) membranes.

The plateau region (second part of the dispersion) is shortened after both treatments of purple membrane and it is slightly shifted to higher frequencies for deionized membrane.

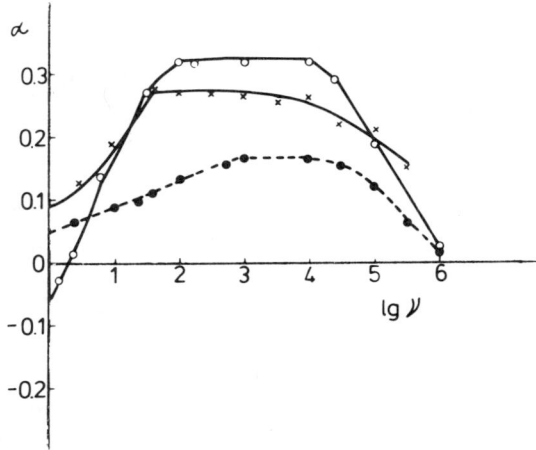

Figure 3. Dispersion dependence of the relative electro-optic effect (measured at electric field strength 3×10^3 V/m) in purple (O), blue (●) and SDS-treated (×) membranes.

There is a difference in the third (high frequency) part of the dependence between the three types of membrane under study. This difference could be due to different ions mobility if the dimension of the fragments remains unchanged after the treatments performed.

The magnitude of the electro-optic effect upon a.c. field was smaller for both deionized and SDS-treatment membranes as compared to PM within the entire range of the a.c. field strength applied (see Fig. 3 — the plateau region, the effect measured at a.c. field strength 6×10^3 V/m). This shows that the electric polarizability diminishes for both treated purple membranes.

Transient reversal measurements were carried out and transient data used to evaluate the permanent dipole moment. The reversing in the field direction leads to particles reorientation which depends on the permanent dipole moment of the particles. The change of the effect on the reversal of the field direction drastically decreases in a rather similar fashion for both deionized and SDS-treated PM (Fig. 4, b,c) as compared to native PM (Fig. 4a). The observed effect of purple membrane when the field direction is reversed (Fig. 4a) is typical for disperse particles of a large transversal permanent dipole moment. In contrast to the behaviour of purple membrane the transient data for blue and SDS-treated membranes indicate the existence of a smaller permanent dipole moment in the latter cases.

In order to verify the restoration of the electric moments we studied the black blue-to-purple transition (in press). We have found that a negative part of the effect in d.c. field, typical for PM, appeared after addition of Ca^{2+} (Fig. 5) The d.c. field dependence of regenerated membrane closely resembles that of native PM and reflects the

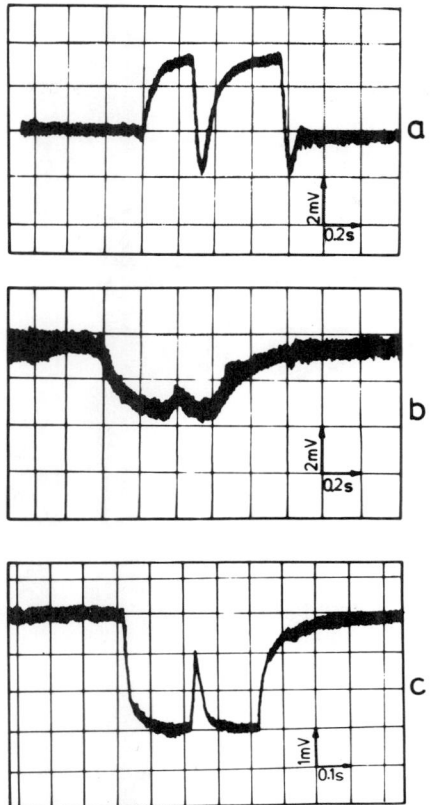

Figure 4. Time course of the electric light scattering upon orienting pulses of reversing polarity: purple membrane (a), blue (b) and SDS-treated membrane (c).

restoration of the membrane charge asymmetry (permanent dipole moment), which at low fields overcompensates the action of the electric polarizability and leads to a big negative initial slope (Fig. 2). The value of the permanent dipole moment tends to saturate at Ca^{2+}/bR ratio about 6–7 (figure not shown). This result is consistent with other observations on the cation binding to blue membrane (Kimura, et al., 1984; Ariki, et al., 1986; Chang, et al., 1986; Dunach, et al., 1988; Dunach, et al., 1988).

The values of the electric moments for native and treated purple membranes are compared in Table 1.

The interaction of PM with the anionic surfactant SDS, (at the concentration shown) exerts the same effect on the electric polarizability and even stronger effect on the permanent dipole moment as compared to deionization. It has to be noted that the pH and conductivity of the suspensions of deionized and SDS-treated membranes are the same.

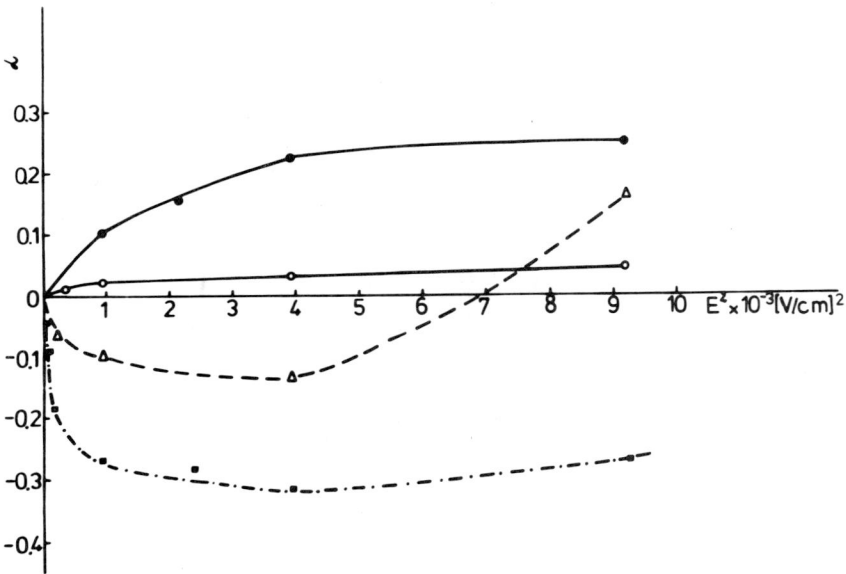

Figure 5. Dependence of the electric light scattering effect on the square of the d.c. electric field in deionized memhrane (●) and after addition of calcium to blue membrane : $Ca^{2+}/bR=1.6$ (○), 3.12 (△), 6.25 (□).

Significantly, in contrast to the permanent dipole moment value, the value of electric polarizability can not be restored during the back titration experiment. A changed counter-ions distribution in the reionized state could be the reason for the unrestored electric polarizability.

Table. 1 Permanent dipole moment and electric polarizability calculated from electric light scattering experiments for native and treated purple membrane.

Sample	$p \times 10^{24}[C. m]$	$\gamma \times 10^{28}[F.m^2]$
purple membrane	4	5.8
blue membrane	1.09	2.5
SDS-treated (0.8 mM)	0.65	2.68
regenerated-PM		
$(Ca^{2+}/bR=6.2)$	3.5	3.1

No complete restoration of the induced-type dipole moment (permanent dipoles do not contribute to the total reaction moment measured) upon addition of Mg^{2+} to cation-depleted membrane has been reported in (Tsuji, et al., 1987). Except that addition of cations to blue membrane has been shown to be insufficient for the restoration of the fluorescent spectrum — fluorescence emission and intensity (Mercier, et al., 1988). Therefore, the restoration of the purple color does not necessarily correlate with restoration of some other parameters.

The large decrease in the permanent dipole moment value could be rationalized as being caused by an increase in the negative charge of the extracellular side of the membrane as a consequence of: i. a poorer screening due to cations exchange by protons and ii. the presence of the negatively charged surfactant molecule near the membrane surface.

This explanation is based partially on ED data proving that the permanent dipole moment direction is from the intracellular to the extracellular side of the membrane (Kimura, et al., 1984). In addition, evidence for the location of more negative charges on the internal side of the membrane comes from fluorescence quenching experiments (Renthal, et al., 1984) and the prediction of the polypeptide arrangement across the membrane (Engelman, et al., 1980).

In favour of this interpretation is the 30% increase in the total negative charge of deionized membrane (measured microelectrophoretically) related to an increase of the diffuse layer electric charges (Taneva, et al., 1987). The binding of cations reduces the magnitude of the ζ-potential. This result is coincident with more recent ESR data proving that cation-depleted membrane possesses a higher negative surface potential than native PM in a wide pH range (Dunach, et al., 1988).

The difference obtained for the permanent dipole moment values between PM and either deionized or SDS-treated membrane corresponds to about 4–5 electric charges difference using the estimation of Otomo, et al., (1986) on the basis of ED measurements of papain-digested purple membrane. Moreover, a change in the net surface charge density by 5.7 negative charges/bR after binding of the lipophilic anion (tetrakis)4-fluorophenyl)boron) to purple membrane, which induces a blue form of bR has been estimated (Kamo, et al., 1987).

These findings do not contradict to the suggestions that cations effect in the purple-to-blue transition is through raising the surface pH and/or conformational change in the protein molecule (Szundi, et al., 1987; Szundi, et al., 1988).

The underlying changes of the permanent dipole moment induced by cations removal from purple membrane as well as treatment with SDS infer its interfacial origin. We also believe that the changes in the close vicinity of the membrane surface would be related to the proton translocation through alteration of the interfacial charge distribution.

References

Ariki, M., and Lanyi, J.K. (1986). Characterization of metal ion-binding sites in bacteriorhodopsin. *J. Biol. Chemistry.* 261, 8167–8174.

Barabas, K., Der, A., Dancshazy, Z., Keszthelyi, L., and Marden, M. (1983). Electro-optical measurements in aqueous suspensions of purple membrane from Halobacterium halobium, *Biophys. J.* 43, 5–11.

Chang, C.H., Chen, R., Govindjee, T., and Ebrey, T.G. (1985). Cation binding by bacteriorhodopsin. *Proc. Natl. Acad. Sci.* USA. 82, 396–400.

Chang, C.H., Jonas, R., Melchiore, S., Govindjee, R., and Ebrey, T.G. (1986). Mechanism and role of divalent cation binding of bacteriorhodopsin. *Biophys. J.* 49, 731–739.

Dukhin, S.S., Shilov, V.N., Stoylov, S., Sokerov, S., and Petkanchin, E.B. (1977). Colloid Electro-optics. Naukowa Dumka, Kiev.

Dunach, M. Padros, E.E., Seigneuret, M., and Rigaud, J.L. (1988). On the molecular mechanism of the blue to purple transition of bacteriorhodopsin. *J. Biophys. Chemistry.* 263, 7555–7559.

Dunach, M., Seigneuret, M., Rigaud, J.L., Padros, E. (1988). Influence of cations on the blue to purple transition of bacteriorhodopsin. *J. Biophys. Chemistry.* 263, 17378–17384.

Engelman, D.M., and Zaccai, G. (1980). Bacteriorhodopsin is an inside-out protein. *Proc. Natl. Acad. Sci.* USA. 77, 5894–5898.

Kamo, N., Yoshimoto, M., Kobatake, Y., and Itoh, S. (1987). Formation of blue membrane of bacteriorhodopsin by addition of tetrakis(4-fluorophenyl) boron, an hydrophobic anion. *Biochim. Biophys. Acta.* 904, 179–186.

Keszthelyi, L. (1980). Orientation of membrane fragments by electric field. *Biochim. Biophys. Acta.* 598, 429–436.

Kimura, Y., Fujiwara, M., and Ikegami, A. (1984). Anisotropic electric properties of purple membrane and their change during the photoreaction cycle. *Biophys. J.* 45, 615–625.

Kimura, Y., Ikegami, A., and Stoeckenius, W. (1984). Salt and pH-dependent changes of the purple membrane absorption spectrum. *Photochem. Photobiol.* 40, 641–646.

Kimura, Y., Ikegami, K., Ohno, S., Saigo, S., and Takeuchi, Y. (1981). Electric dichroism of purple membrane suspensions. *Photochem. Photobiol.* 33, 435–439.

Mercier, G., and Dupuis, P. (1988). The effect of deionization on the protein fluorescence of bacteriorhodopsin. *Photochem. Photobiol.* 47, 433–436.

Mowery, P.C., Lozier, R.H., Chae, Q., Tseng, Y., Taylor, W., and Stoeckenius, W. (1979). Effect of acid pH on the absorption spectra and photoreactions of bacteriorhodopsin. *Biochem.* 18, 4100–4107.

Otomo, J., Ohno, K., Takeuchi, and Ikegami, S. (1986). Surface charge movements of purple membrane during light-dark adaptation. *Biophys. J.*. 50, 205–211.

Padros, E., Dunach, M., and Sabes. (1984). Induction of the blue form of bacteriorhodopsin by low concentrations of sodium dodecyl sulfate. *Biochem. Biophys. Acta.* 769, 1–7.

Renthal, R., and Cha, C.H. (1984). Charge asymmetry of the purple membrane measured by uranyl quenching of dansyl fluorescence. *Biophys. J.* 1001–1006.

Szundi, I., and Stoeckenius, W. (1987). Effect of lipid surface charges on the purple-to-blue transition of bacteriorhodopsin. *Proc. Natl. Acad. Sci.* USA. 84, 3681–3684.

Szundi, I., and Stoeckenius, W. (1988). Purple-to-blue transition of bacteriorhodopsin in a neutral lipid environment. *Biophys. J.* 54, 227–232.

Taneva, S.G., Todorov, G., Petkanchin, I.B., and Stoylov, S.P. (1987). Electro-optic study of the deionized form of bacteriorhodopsin. *Eur. Biophys. J.* 14, 415–421.

Tinoko, I., and Yamaoka, K. (1959). The reversing pulse technique in electric birefringence. *J. Phys. Chem.* 633, 432–437.

Todorov, G., Sokerov, S., and Stoylov, S. (1982). Interfacial electric polarizability of purple membranes in solution. *Biophys. J.* 40, 1–5.

Tsuji, K., and Hess, B. (1987). Electro-optical analysis of blue and cation-regenerated bacteriorhodopsin. *Eur. Biophys. J.* 15, 231–236.

Probing the Internal Spectral Distribution of UV Radiation in Plants with Fibre Optics

J.F. BORNMAN AND *T.C. VOGELMANN

Department of Plant Physiology
University of Lund, Box 7007, S-22007 Lund, Sweden
*Department of Botany, University of Wyoming
Laramie, WY 82071, USA

Introduction

The complex internal light distribution and composition within different plant organs is relatively unknown. Therefore it is not easy to make valid assumptions on the detection by photoreceptors of light direction, quantity and quality. Light scattering and absorption complicate measurements of the internal spectral regime of plant organs such as a leaf or cotyledon. Thus, when a collimated light beam penetrates plant tissue the spectral quality and quantity are likely to be altered by wavelength-dependent absorption.

Indirect measurements of these internal light gradients have been done by determining transmittance and reflectance of light from, for example, leaves. A more direct method uses optical fibers as microprobes (Vogelmann and Björn 1984; Bornman and Vogelmann 1988; Vogelmann et al., 1988). Briefly, one end of the fibre is pulled to a fine tip (1 to 5 μm in diameter), coated with either silver or platinum, and truncated on a diamond knife in order to break the light seal at the tip of the fibre. The fibres are then calibrated in air and water to determine their acceptance angles before being used for light measurements.

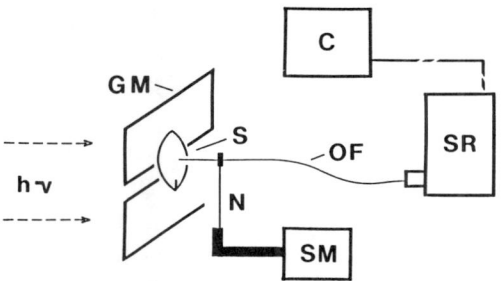

Figure 1. Schematic diagram of the experimental set-up for fibre optic measurements. C, microcomputer; GM, glass mount for sample; N, needle mount for fibre optic probe; OF, optical fibre; S, sample; SM, stepper motor; SR, spectroradiometer.

Photobiology, Edited by E. Riklis
Plenum Press, New York, 1991

Several experiments have been done on the optical effects of surface properties of leaves (Gausman et al., 1975; Sheehy 1975; Ehleringer and Björkman 1978; Ehleringer 1981). We were therefore interested in investigating the effect of pubescence on the internal distribution of UV radiation in leaves; penetration of blue light was also compared.

Effect of pubescence on internal light gradients

Leaves of *Verbascum thapsus,* which exhibit dense pubescence on both surfaces, were used as experimental material. Plants were collected from the field and kept in a greenhouse using supplemental lighting (832 μmol m^{-2} s^{-1}, metal halide lamps). Measurements were done on intact leaves and on those with the adaxial pubescence removed. Optical fibres allow only light to enter from directions that lie within their acceptance angle. One can therefore control the angle at which these directional sensors are inserted into a sample, thus making it possible to measure the spatial and angular distribution of light that is transmitted or scattered. Since plants under natural conditions will receive both collimated (transmitted) and scattered (diffuse) light, it is necessary to measure both these components. Internal light measurements were carried out with the adaxial surface of the leaves as the illuminated surface. Similar results were obtained from 2 separate sets of experiments. Each curve is the average of 4 to 6 measurements, with the curves normalized to 100.

Results and Discussion

The disribution with depth of internal UV-A (360 nm) radiation for transmitted light showed a more complicated profile (Fig. 2) than that for blue (460 nm) light (Fig. 3). The curves showing the penetration of UV radiation with depth (Fig. 2) were steeper exponential functions than those for blue light (Fig. 3), indicating that there was a greater absorption of UV-A with depth. Removal of the adaxial trichomes resulted in a uniform increased penetration of blue light with depth (Fig. 3), whereas intact leaves there was an apparent increase in the amount of UV that could penetrate the initial 20 μm (Fig. 2).

In previous experiments with spruce and fir, removal of the epidermal covering, in that case, epicuticular wax, resulted in only a ca 10% increase in blue light penetration with little change in UV penetration. Nevertheless, in intact needles, transmitted UV radiation was able to penetrate to surprising depths, especially in spruce (Fig. 4) (Bornman and Vogelmann 1988).

In the present study, light profiles of internal scattered UV were similar to those for transmitted blue light (compare Fig 3 with Fig. 6), with a greater penetration of UV radiation in the absence of adaxial trichomes. On the other hand, the presence of pubescence markedly enhanced the scattered blue light component (Fig. 5).

On a quantitative energy basis (Fig. 7, energy values, unnormalized), less collimated UV-A was able to penetrate in intact leaves than in those leaves with the adaxial

288

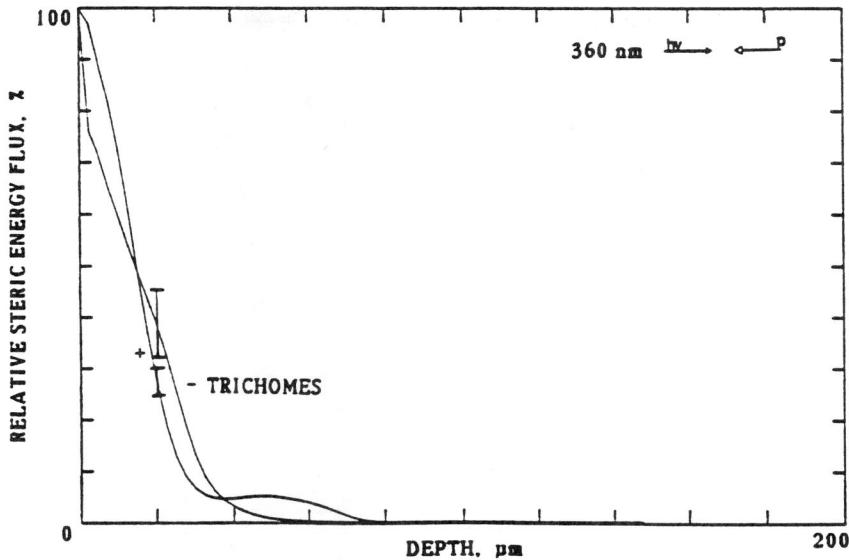

Figure 2. Distribution of *transmitted* UV-A radiation in leaves of *Verbascum*. Bars, ± SE (n = 4–6 scans). P, fibre optic probe.

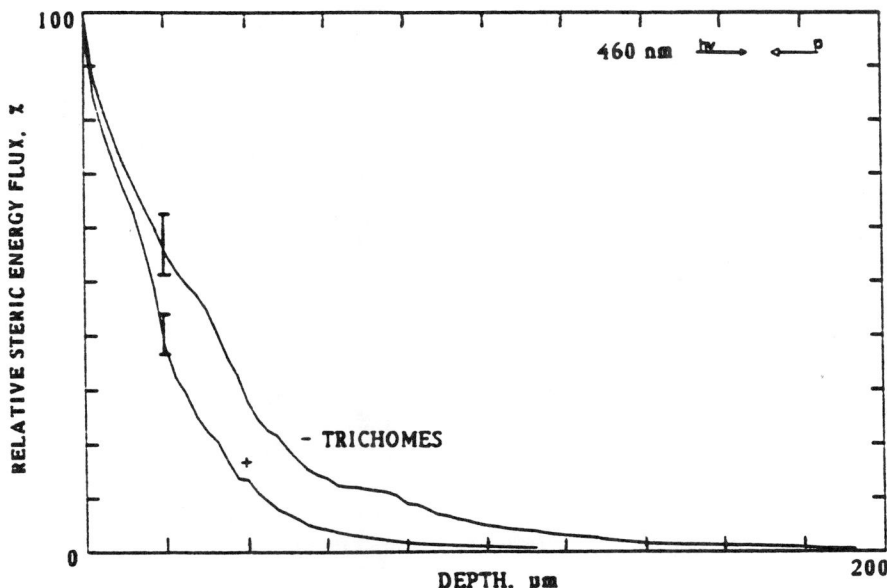

Figure 3. Distribution of *transmitted* blue light in leaves of *Verbascum*. Bars, ± SE (n = 4–6 scans). P, fibre optic probe.

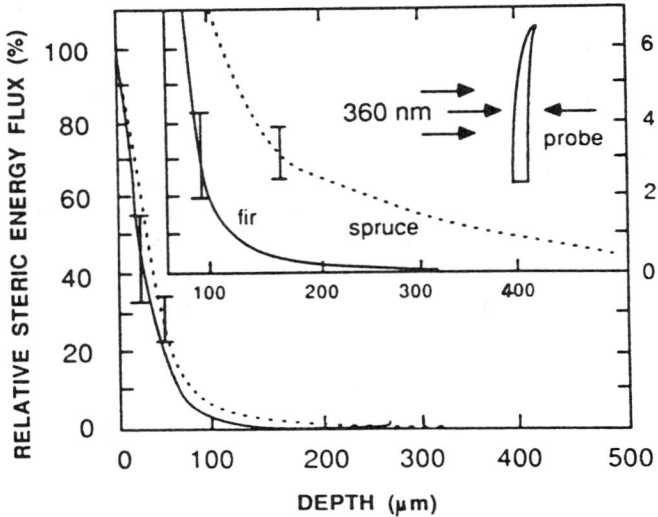

Figure 4. Distribution of *transmitted* 360 nm radiation in intact spruce and fir needles. The y-axis in the inset is expanded 10-fold. Bars, ± SE (N=4–6 scans). From Bornman and Vogelmann 1988).

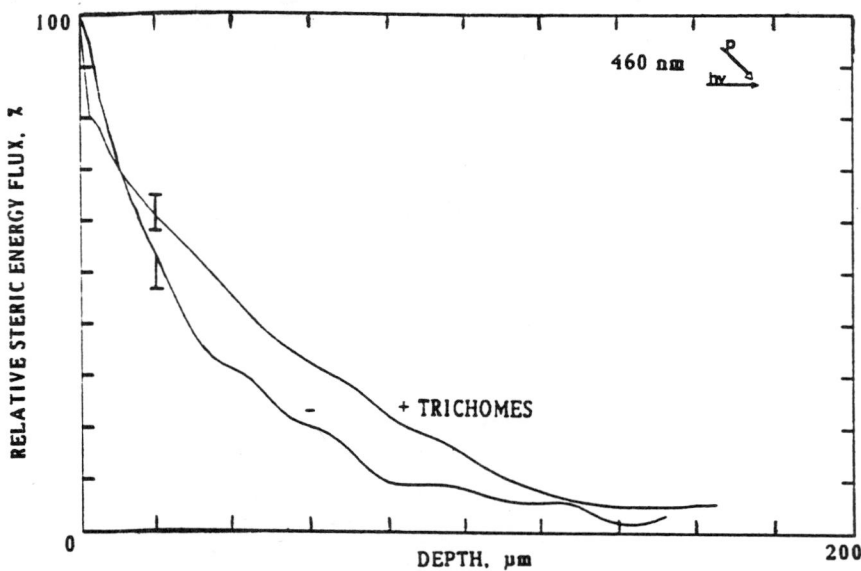

Figure 5. Distribution of *back-scattered* blue light in *Verbascum* leaves. Bars, ± SE (n = 4–6 scans). P, fibre optic probe.

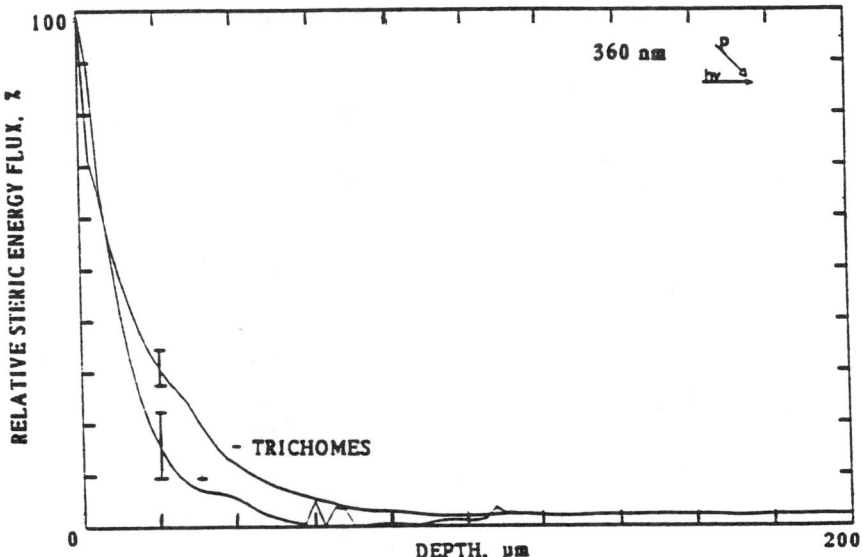

Figure 6. Distribution of *back-scattered* UV-A radiation in *Verbascum leaves*. Bars, ± SE (n = 4–6 scans). P, fibre optic probe.

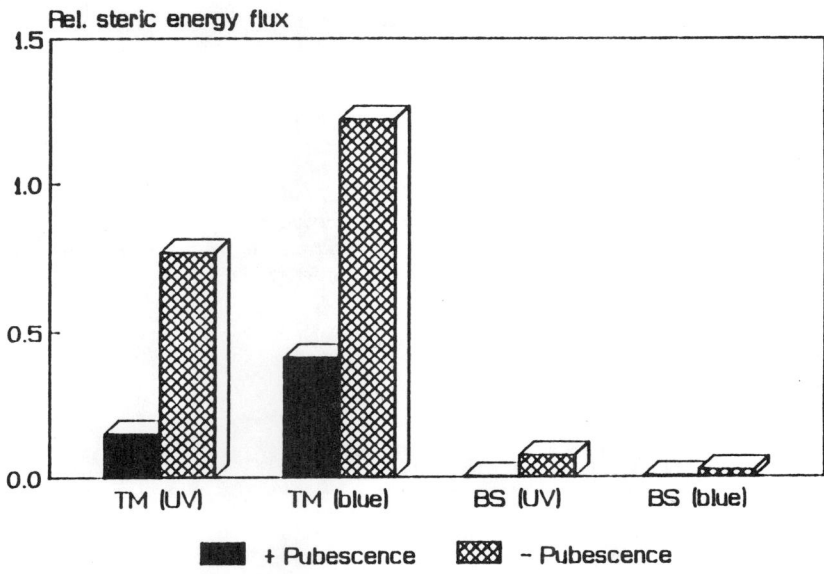

TM, transmitted; BS, back-scattered

Figure 7. Relative steric energy flux in leaves of *Verbascum*. Energy values ar derived from Figs. 2,3,5,6 except that the energy values are not normalized.

trichomes removed. The same was true for blue light. There was also more internal blue light than UV, whereas for back-scattered light, a larger amount of UV-A was measured than for blue light in leaves with adaxial trichomes removed. This could mean that apart from absorbing UV (significant quantities of flavonoids were found in extracts of the trichomes) the dense pubescence may also have scattered the UV radiation to a certain extent.

The results have shown that with the use of fibre optic microprobes it is possible to follow the penetration of radiation inside plant organs. In this respect it is also possible to follow changes in light gradients of damaged leaves from plants that have been grown under enhanced UV radiation (in preparation), since changes in pigments (e.g. protective screening pigments, chlorophyll content) and structural alterations will change the internal light profiles in UV-treated plants, and these profiles can then be compared to those from control plants. In this way certain types of internal damage can be mapped, giving an indication of changes that have taken place inside e.g. a leaf and what these changes might mean with regard to the survival of the plant.

References

Bornman, J.F., and Vogelmann T.C. (1988). Penetration of blue and UV radiation measured by fiber optics in spruce and fir needles. *Physiologia Plantarum*, 72, 699–705.

Ehleringer, J.R. (1981). Leaf absorptances of Mohave and Sonoran desert plants. *Oecologia*, 49, 366–370.

Ehleringer, J.R., and Björkman, O. (1978). Pubescence and leaf spectral characteristics in a desert shrub, *Encelia farinosa*. *Oecologia*, 36, 151–162.

Gausman, H.W., Rodriguez, R.R., and Escobar, D.E. (1975). Ultraviolet radiation reflectance, transmittance and absorptance by plant leaf epidermises. *Agronomy Journal*, 67, 720–724.

Sheehy, J.E. (1975). Some optical properties of leaves of eight temperate forage grasses. *Annals of Botany*, 39,377–386.

Vogelmann, T.C., and Björn, L.O. (1984). Measurements of light gradients and spectral regime in plant tissue with a fiber optic probe. *Physiologia Plantarum*, 60, 361–368.

Vogelmann, T.C., Knapp, A.K., McClean, T.M., and Smith, W.K. (1988). Measurements of light within thin plant tissues with fiber optic microprobes. *Physiologia Plantarum*, 72, 623–630.

Cancer Diagnosis by Fluorescence Polarization

A. WEINREB*, M. DEUTSCH AND S. CHAITCHIK+

Department of Physics
Bar Ilan University, Ramat Gan, Israel
+Department of Oncology
Tel-Aviv Medical Center, Tel-Aviv, Israel
*Permanent address: Racah Institute of Physics
Hebrew University, Jerusalem
Israel

Introduction

The work presented in the following originated in an attempt to modify the SCM test in such a way as to make it applicable as a routine method for the early detection and follow up of cancer. To do this we had first to validate the test itself which has been described by B.& L. Cercek in 1974 and later on, since until this day the test is a controversial issue. When this was done in 1980 we made it our aim to facilitate the test, to increase its accuracy and speed. This lead to the development of a novel method of cell scanning which permits the repeated measurement of a great number of cells, each one individually, after any desired number of manipulations on these cells. Although this method may have important applications in many fields of biology, we limit our discussion here to its original purpose, the diagnosis of cancer. Our presentation will thus be divided into two parts. In the first part we give a description of the SCM test as executed by us; in the second part we describe the Cytoscan, which is the method of cell scanning developed by us, and the tests performed with this equipment until present.

The SCM test

a) *Basic features of the test*

The SCM (Structuredness of Cytoplasmic Matrix) cancer test is very difficult to perform. If the test, however is carried out properly it presents an extremely powerful diagnostic tool. Since the details of the test have been published throughout the years (e.g. Cercek and Cercek 1977) we will present here only the basic steps. It will then also be appreciated why so many groups failed to execute the test and why a modification of the test was necessary. The test runs as follows:

About 20 milliliter of peripheral blood are taken from the patient. From this a particular group of lymphocytes has to be separated. These particular lymphocytes which are the only ones of the whole family of white blood cells which exhibit the SCM phenomenon could so far not be characterized by any criterion except for their mode of separation. They are separated by the exact density of the gradient of separation. For this purpose the blood is carefully placed onto a Ficoll-Triosil gradient of 1,081 g/cc at 24,5°C and an osmolality of 0.320 Osmol/Kg. The Cerceks provide a conversion table of the density for different temperatures and osmolalities. The blood gradient composite is then centrifuged for 20 minutes at an acceleration of 550g and temperature of 24.5°C. The temperature has to be maintained constant within 0.2°C. Considering the fact that the mere friction with the air in the centrifuge can easily raise the temperature by 0.3°C, this is another possible pitfall of the test. If the centrifugation has been carried our properly, a thin disc of lymphocytes will have aggregated on the interface between the gradient and the plasma. If everything has been done correctly these are the proper lymphocytes for the SCM test. They are pipetted out, washed, suspended in phosphate buffer solution (PBS) and introduced into the measuring cell. Fluorescein diacetate (FDA), a non-fluorescent coumpound, is introduced into the lymphocyte suspension. By an enzymatic reaction in the cell FDA is converted into fluorescent fluorescein. The cell suspension is now introduced into an appropriate optical system (usually a converted spectrofluorimeter) and the fluorescence of the fluorescein in the lymphocytes is excited by polarized light of 470 ± 10nm. The polarization of the resulting fluorescence is measured at 510 ± 5nm. This requirement of an exact wavelength of excitation and more so of an exact wavelength of emission for the detection of the effect to be described readily is one of the pecularities of the phenomenon, and from a physicists point of view certainly the most interesting one, to be investigated in the future. Since the polarization is measured while the fluorescence intensity is increasing (because of the ongoing conversion of FDA to fluorescein) the intensity in the two directions of polarization has to be measured continuously. During this time fluorescein is leaking out of the lymphocytes, giving rise to a (sometimes prohibitively high) background of non-polarized fluorescence. This background has to be subtracted from the measured intensities before the degree of polarization can be evaluated. This is done by filtering out the cells and measuring the fluorescence of the remaining solution.

At this stage there is no difference in the degree of polarization between the lymphocytes of healthy donors and donors with malignant diseases. By healthy we mean individuals with no malignant disease, who may suffer from other illnesses. When, however prior to the measurement the lymphocytes are incubated with a solution of PHA (phytohemagglutinin) then the degree of polarization will be lower than the control value if the individual is healthy and it will remain unchanged for individuals with a malignant disease. Conversely, if the lymphocytes are incubated with EF (encephalitogenic factor, a synthetic material) the degree of polarization will be reduced for diseased individuals and will remain unchanged for healthy donors. The implied reason for these changes is that the incubation with the reagent causes a stimulation of the lymphocytes from their resting phase which expresses itself in structural changes of the cytoplasmic matrix -

hence the name of the method. These structural changes are monitored by the fluorescence polarization which depends on the structure and fluidity of the microenvironment of the fluorescent marker molecules. By dividing the polarization after incubation with EF by the polarization after incubation with PHA we obtain a numerical parameter which is greater than unity if the donor of the lymphocytes is healthy and less than unity if he suffers from a malignant disease. We have confirmed this fact in a previous publication (Deutsch and Weinreb 1983).

b)   *The specificity test*

We later widely extended an early observation of the Cerceks, that the test also diagnoses the type of cancer. If prior to the measurement the lymphocytes are incubated (instead of PHA or EF) with the tissue or tissue extract of a tumor, the degree of polarization will decrease if the donor of the lymphocytes suffers from a malignancy of the same type or organ as that from which the tumor has been excised, and it will remain unchanged for any other kind of tumor. Some of our earlier results of the specificity

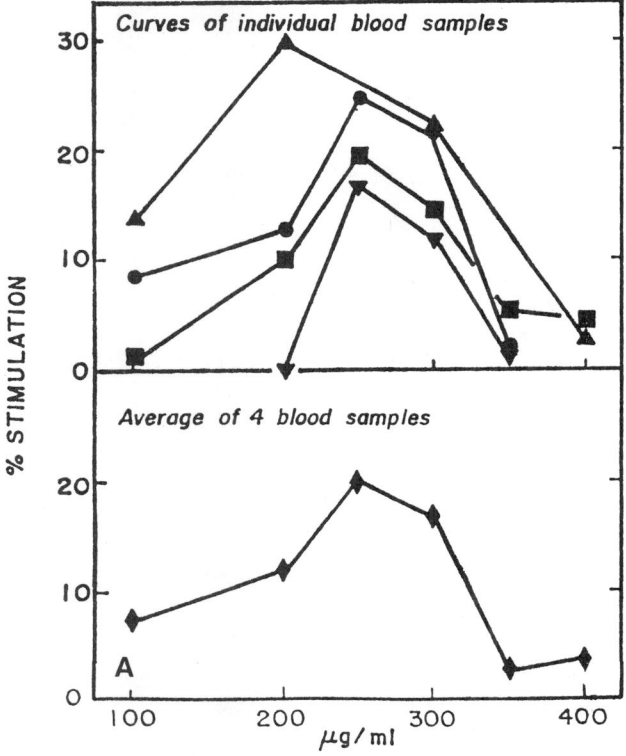

Figure 1a. Titration of lung IIO extract. Human tumor tissue

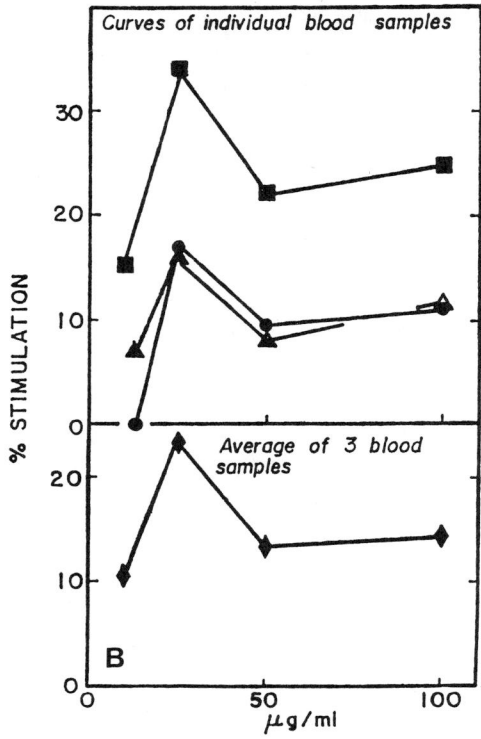

Figure 1b. Titration of breast 204-01 extract. Human tumor cell-line

test have been published elsewhere (Chaitchik et al 1985). Like many other stimulation procedures the effect is greatest for a given concentration of the stimulant. Figure 1(a) shows the calibration curve for a given tumor extract. Instead of tumor extracts we later used extracts from cell lines in order to reach a better standardization of the procedure. Fig. 1(b) shows a calibration curve for such cell line extracts. It will be noted that the optimum concentration is now lower by an order of magnitude when compared with that of the tumor extract. Later on we used supernatants of cell line media and purified them by molecular weight fractionation, with the test as the assay. This reduced the optimum concentration by five orders of magnitude. The use of specific stimulants greatly improved the reliability of the test as such.

We never performed diagnostic tests for their own sake. Our results are, therefore not the outcome of a methodical study with the aim to arrive at a statistical evaluation of the test. In course of the development of the test, however we measured many hundreds of patients before they were operated and compared the results later with the histopathologic findings.

The Cytoscan Method

The new method which we started to develop several years ago we call by the name of Cytoscan. Even from the short description of the SCM test given here it is clear that this test, with all its power, can not serve as a routine diagnostic tool. The meticulous care to be taken in the separation of the proper lymphocytes, when the slightest deviation from the protocol may yield meaningless or even false results; the measuring process with the necessity to evaluate the background intensity, and a certain arbritariness in the extrapolation procedure for the time at which the background was measured; these together with a host of other possible pitfalls, require an enormous experience and dexterity of the experimenter. An experienced worker can perform one good series of tests on a given blood sample per day. So we were searching for a different method which is based on the principles of the SCM test and is free from most of its major difficulties, and at the same time fast and manageable by trained hospital personal. We hope we are approaching this stage. The second part of my presentation deals with this equipment.

The principle

Since the lymphocytes which are active in the SCM test can so far not be separated by any conventional method (e.g. morphological parameters, monoclonal antibody markers etc.) their only means of recognition is their reaction to certain stimulants. Thus instead of separating these lymphocytes by the described procedure, running the risk of a unsuccessful separation, and later on investigating their reaction, let us investigate the whole family of lymphocytes and count those which react in the test. This requires a basic change in our approach: Instead of measuring the overall response of a million

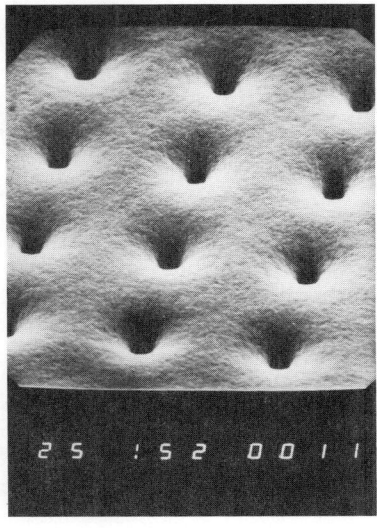

Figure 2

297

lymphocytes we have to measure each lymphocyte separately in order to find the reacting ones. This necessitates at least two measurements on the same lymphocyte, one before and one after stimulation with the relevant reagent; or generally, a method which permits consecutive measurements on individual lymphocytes after any desired manipulation.

The heart of the new method is the "grid". This is an array of conically shaped apertures, each of the size of a lymphocyte. Figure 2 shows an electron microscope picture of a section of such a grid. When a drop of blood is spread out over the grid, blood cells which are smaller than lymphocytes will pass through the holes of the grid, those greater than the holes will be rinsed away so that the lymphocytes are retained in the holes. In fact, in order to facilitate the procedure we still separate the lymphocytes out of the blood by common lymphprep separation and then spread them out over the grid. Figure 3a shows such a lymphocyte in a hole. Figure 3b shows the lymphocyte from the lower side of the grid.

The grid is fixed on a carrier. The carrier with the lymphocytes on the grid is placed on a scanning table under a microscope. This scanning system permits the fast scan of the grid in two dimensions in steps of a tenth of a micron. The grid is automatically positioned with regard to the microscope axis. Since the position of each lymphocyte is fixed on the matrix of apertures of the grid, the position of each lymphocyte is also exactly defined with regard to the axis of the microscope.

The carrier with the lymphocytes can therefore be safely removed from under the microscope, stimulation or any other manipulation can be applied to the lymphocytes, and then the carrier can be placed again under the microscope will full recognition of the

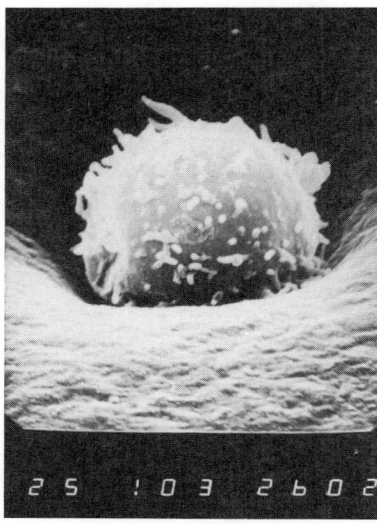

Figure 3a

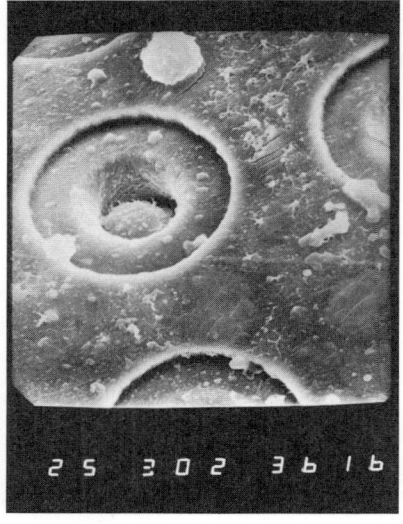

Figure 3b

identity of each individual lymphocyte. It is clear that was holds for lymphocytes can easily be made applicable for any other kind of cells.

Now the test can begin. We place the lymphocytes under the microscope and rinse them with a solution of FDA. Fluorescein which may now leak out of the cells does not accumulate around the cell. Nevertheless any trace of the external fluorescein is rinsed away by an addition of FDA solution, so that no background fluorescence exists. Light from a laser after additional filtration and polarizaton passes through the objective of the microscope onto the lymphocytes. Initially we have used an argon laser. The argon line of 472,6 nm, however, did not well respond to the required excitation wavelength of 470 nm. Recently we were informed by L. & B. Cercek of a second combination of wavelengths for excitation and detection of the fluorescence which permits the observation of the SCM phenomenon, namely excitation at 442 nm and detection at 527 nm. We thoroughly investigated this relationship and confirmed it. Since the 441,6 nm line of a He-Cd laser ideally fits the condition for excitation, we employ now this laser as our source of excitation. The average intensity of the beam at any given cell is of the order of 1 microwatt. The fluorescence passes the objective, a beam splitter, a lens, an iris, a broad-band interference filter and reaches the analyzing polarizer which separates the s from the p component of the fluorescence. Each of the two polarized beams is further split into two mutually perpendicular components. One component passes a filter of 527 nm at which wavelength the change in the degree of polarization upon stimulation is greatest and the other passes a filter of 510nm at which the effect is now nil. Each of the four beams incides on a photomultiplier which monitors its intensity. After proper calibration of the system the four intensities permit the on-line computation of the degree of polarization of the two wavelengths for each cell separately.

Many experiments were performed in order to evaluate the reproducibility of the results. For repeated measurements on a given cell the polarization remains constant within ±3%. For an ensemble of several tens of cells the mean square root deviation is about 2%. A change in the degree of polarization of such an ensemble greater than 4% is therefore considered significant for the establishment of a diagnosis. This shall be compared with the requirement of a minimum change of 10% for the original SCM test. The scan of a whole grid takes a few minutes. Thus the possibility of routine testing seems to be, in principle at least, opened.

Results

Figure 4 shows a representative result of such a test. Fig. 4a shows two consecutive control measurements. Both performed before the application of a stimulant. Each point represents the polarization of a lymphocyte at the two consecutive measurements. When such an experiment is carried out with fluorescent beads the points lie all on the 45 degree line within the accurancy of the measuring equipment. With living lymphocytes the majority of the cells still retains their degree of polarization but changes with time occur. Still the center of gravity of the points lies somewhere on the 45 degree line. Fig.

4b shows the same lymphocytes but after stimulation. The abscissa gives the polarization values before stimulation and the ordinate gives the values after stimulation. The center of gravity of the points is now markedly shifted below the 45 degree line, indicating a significant overall decrease in the microviscosity or structuredness of the cells.

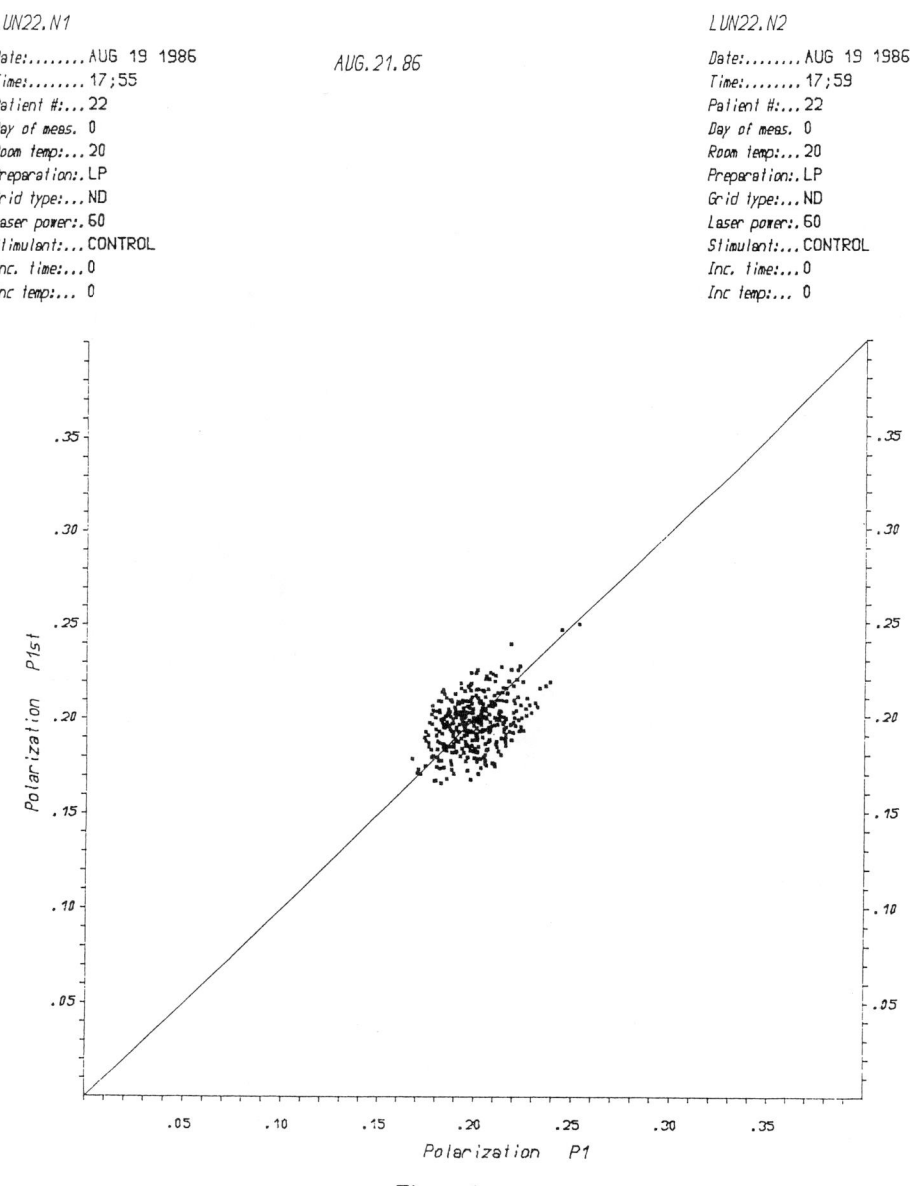

LUN22.N1

Date:........AUG 19 1986
Time:........17;55
Patient #:...22
Day of meas. 0
Room temp:...20
Preparation:.LP
Grid type:...ND
Laser power:.60
Stimulant:...CONTROL
Inc. time:...0
Inc temp:... 0

AUG.21.86

LUN22.N2

Date:........AUG 19 1986
Time:........17;59
Patient #:...22
Day of meas. 0
Room temp:...20
Preparation:.LP
Grid type:...ND
Laser power:.60
Stimulant:...CONTROL
Inc. time:...0
Inc temp:... 0

Figure 4a

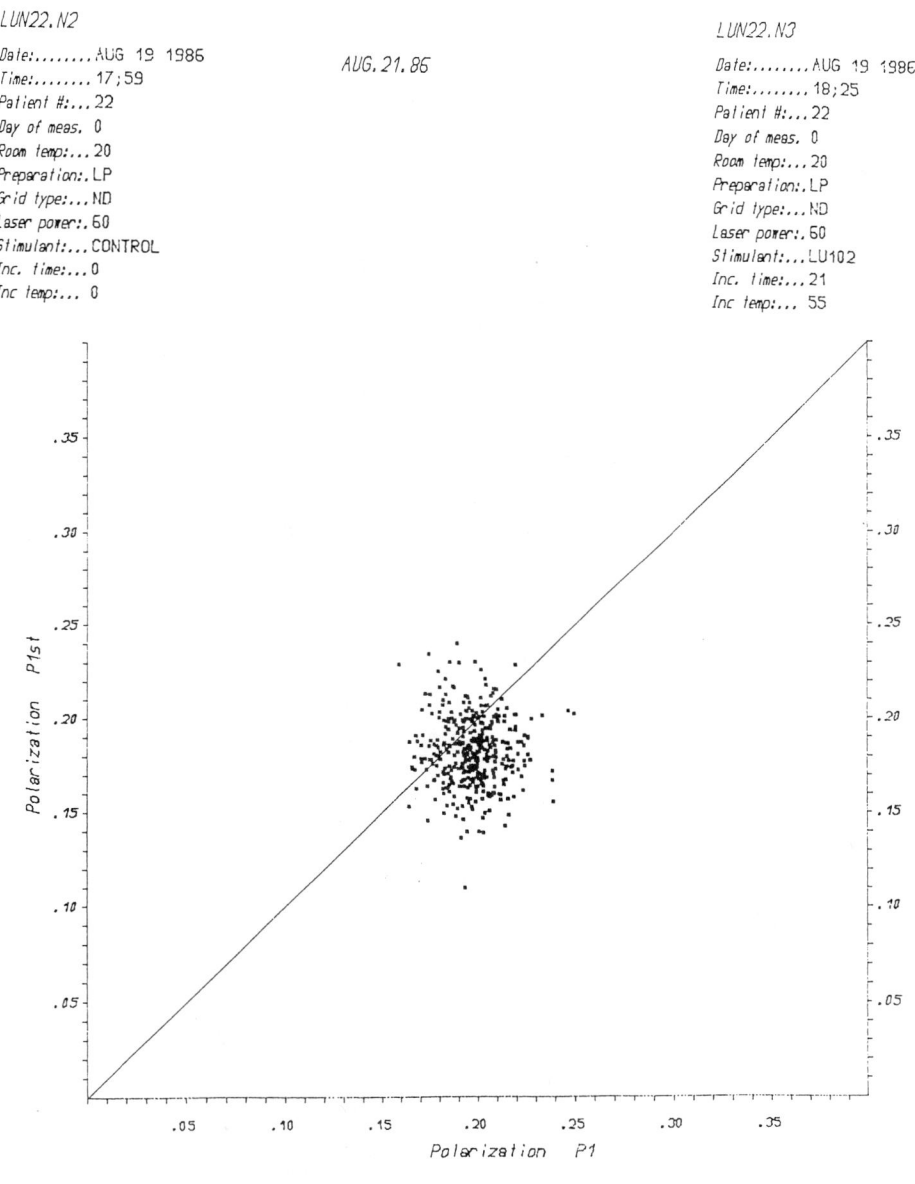

Figure 4b.

Figure 5 shows the individual behavior of a few lymphocytes picked out from a large group of lymphocytes which were measured on the same grid. The time elapse between two consecutive measurements is 5 minutes. Between time points 5 and 6 a stimulant

301

was applied. At point 7 the cells were again rinsed with the stimulant. At point 8 they were rinsed with PBS. Between points 8 and 9 they were rinsed with a solution of imidazole which turned out to erase the stimulation of some cells. Point 10 shows the effect of imidazole a few minutes later. We see that different cells have been affected differently. This shows how the Cytoscan method permits the study of the dynamical behavior of cells in response to various manipulations.

Table 1. Cytoscan: Summary of results

LAB/CLIN	BREAST −M	COLON −M	NORMAL
YES	49/51 96.1%	55/56 98.2%	
NO	2/51 3.9%	1/56 1.8%	37/37 100%
TOTAL	51	56	37

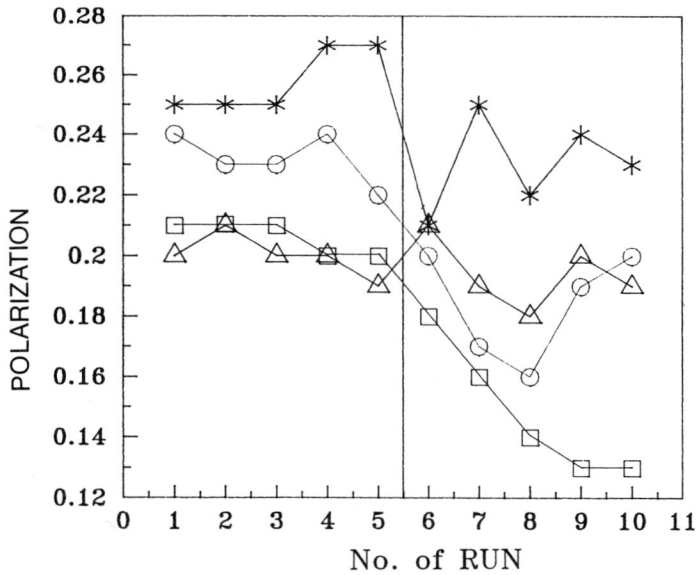

Figure 5. Sample of typical trajectories of polarization. (Data taken from patient COL211)

Let us now return to the application of the Cytoscan for cancer detection. Table 1 shows the results obtained with the test for breast and colon cancer patients and for healthy individuals (normal). M-indicates that the histopathological examination showed a malignancy; "Yes" and "No" mean that the Cytoscan test was positive or negative, respectively. The results are almost too good to be true. We have here a sensitivity and specificity not known to us for any other cancer detection method. I hope to have shown the potential of the method. Its applicability as a routine diagnostic tool has now to be established.

References

Cercek, L. and Cercek, B. 1974. *Brit. J. Cancer* 29, 345.

Cercek, L. and Cercek, B. 1977. *Eur. J. Cancer* 13, 903.

Chaitchik, S., Asher, O., Deutsch, M. and Weinreb, A. 1985 *Eur. J. Cancer Clin Oncol* 21, 1165.

Deutsch, M. and Weinreb, A. 1983. *Eur. J. Cancer Clin Oncol* 19, 187.

Microspectrofluorometry of Human Melanoma Cells and Fibroblasts Treated with Azelaic Acid

ELLI KOHEN[1], CAHIDE KOHEN[1], DIETRICH O. SCHACHTSCHABEL[2],
JOSEPH G. HIRSCHBERG[3], BURTON L. SHAPIRO[4]
AND ALINE MCHEILEH[1]

[1]*Department of Biology, University of Miami*
Coral Gables, Florida, USA
[2]*Institüt für Physiologische Chemie II der Medizinischen*
Fakültät der Philipps Universität, Marburg, Lahnberge, FRG
[3]*Department of Physics, University of Miami*
Coral Gables, Florida, USA
[4]*Department of Oral Biology*
University of Minnesota, Minneapolis, Minnesota, USA

Introduction

A variety of dicarboxylic acids (Breatnach, 1984; Hsu, 1986; Passy, 1984; Picardo, 1983; Robbins, 1985; Schachtshabel et al., 1986), such as azelaic acid and sebacic acid have been shown to be cytotoxic for melanoma cells with varying degrees of effectiveness. The mechanism of action of these drugs remains to be elucidated, but evidence from electron microscopic studies points to the mitochondria as a primary target of dicarboxylic acids. The microspectrofluorometric approach (Hirschberg, 1978; Kohen et al., 1986) which we have developed opens the possibility to investigate the intracellular actions of azelaic acids and other such compounds *in situ* at the level of organelles and metabolic compartments. An illustrative use of this method has already been made in connection with another dermatological affection: a microspectrofluorometric study was carried out recently on the effect of anthralin (Kohen et al., 1986), an antipsoriatic drug, on cellular structures and metabolism.

By the same logic used in studying the actions of an antipsoriatic drug at the level of structure and function in the intact living cell, an extension of the method was attempted to the study of antimelanoma drugs. While human melanoma cells were a primary object of these studies, a parallel investigation was made on normal and pathological human fibroblasts to determine whether the intracellular alterations due to dicarboxylic acids are specific to melanoma cells or whether they involve more general mechanisms. The studies described here are also aimed at the exploration of the usefulness of the methods applied in terms of cellular pharmacology. It is of course timely to start understanding drug effects at a level of structural, functional and temporal resolution presently attainable

Photobiology, Edited by E. Riklis
Plenum Press, New York, 1991

with new microscopic optical methods which can account for the microarchitecture of the living cell and actual rapid pace at which intracellular processes take place. The fluorescence probes used to monitor cell metabolism, organelle structure and interactions are endogenous NAD(P)H (Hirschberg et al., 1978; Kohen et al., 1986; Kohen et al., 1986) and flavin coenzymes (Kohen et al., 1982) coupled to cell bioenergetics or exogenous vital probes (e.g. rhodamine 123) (Goldstein and Korczack, 1981).

The above microspectrofluorometric studies are better interpreted if parallel biochemical studies using conventional techniques are carried out on the same cell cultures. In this way a correlation can be established between bulk biochemical changes within individual cells. Thus biochemical alterations recognized in cell extracts or homogenates may be understood in terms of drug effects evaluated at the level of cell organelles or bioenergetic and other metabolic pathways followed within intracellular metabolic compartments.

Materials and Methods

Microspectrofluorometer

The grating microspectrofluorometer (Hirschberg, 1978) which allows unidimensional linear topographic scans of total fluorescence emission in the spectral band selected throughout an array of adjacent cell regions down to the structural resolution of one to two micrometers or spectral scan of the emission from one such region in the spectral band from about 410 to 590 nm, has been extensively described elsewhere (Hirschberg, 1978). This apparatus is used in conjunction with a microelectrophoresis assembly which allows the intracellular microinjection of metabolites from one micropipette (and if required the injection of modifiers such as adenine nucleotides or cations from another micropipette). The key instrumental features allowing rapid micromanipulatory procedures are inverted microscope optics and a special custom-designed long working distance (up to three inches) high resolution phase condenser. Other characteristic instrumental features are a Ploemopak illuminator system allowing fluorescence excitation by epiillumination with dichroic arrangement to reflect excitation wavelengths and transmit emission wavelengths, a supplementary multioptional slide downstream in the light path with multiple dichroic filters to separate red light used for cell and microinstrument illumination from blue, green or yellow fluorescence emission, and fused silica optics from the mirror in front of the mercury light source to the coverglass on which the cells are grown for fluorescence excitation at wavelengths below 360 nm. For studies on NAD(P)H fluorescence, excitation was carried out at 365 nm through UG1 Schott filters. Fluorescence of rhodamine 123 was excited using the Leitz shortpass filter 450, dichroic 490 with barrier 510 for the emission.

A two-dimensional microspectrofluorometer allows total fluorescence scan of whole cells along the two coordinates (x and z) in the image plane, as shown in Fig. 1 for a fibroblast.

306

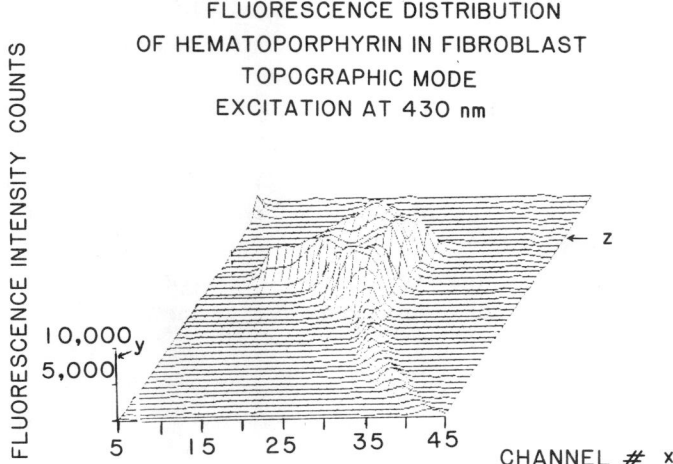

Figure 1. Two-dimensional scan of fluorescence distribution in fibroblast loaded with hematoporphyrin. The red fluorescence of hematoporphyrin is scanned along x and z dimensions. The intensity of fluorescence is shown as counts in the y coordinate. The fluorescence of porphyrin was excited at 380 nm. A similar scan can be carried out in melanoma cells for emission of porphyrins as well as other natural or exogenous fluorochromes.

Fluorescence micrographs

Fluorescence micrographs of cells vitally stained with rhodamine 123 (Goldstein and Korszack, 1981) were obtained at 1600X magnification using the above described conditions.

Cell cultures and preparations

M8255 human melanoma cells were kindly provided by Professor F. Meiskens at the University of Arizona. Human Fibroblasts (normal and cystic fibrosis) (Shapiro et al., 1982) used for parallel studies of dicarboxylic acid action were obtained from Professor B.L. Shapiro at the Department of Oral Biology, University of Minneapolis, Minneapolis, Minnesota. Melanoma cells and fibroblasts were cultured in Eagle's minimum essential medium (MEM) with 10 fetal calf serum. For cultures in the presence of azelaic acid, aliquots of azelaic acid stock solution (0.1 M) in MEM medium neutralized to pH 7.2 by addition of 1.0 M NaOH were added to a final concentration of 0.010 M.

For microspectrofluorometric studies and fluorescence micrographs using fluorescent probes, cells were grown on a 0.1 mm thick coverglass in a region circumscribed by a 20-mm-diameter stainless steel ring held onto the coverglass with aquarium cement, and provided with notches for the introduction of microinstruments. Prior to

307

microspectrofluorometric observations and microelectrophoretic injections the medium covering the cells grown in the above described open chambers was replaced by a modified Krebs Ringer buffer with triethanolamine instead of bicarbonate.

To test the reliability of metabolic responses obtained by microinjections of metabolites and modifiers, cell tolerance to microelectrophoretic currents in the range of up to 10-7 A/pl or to microinjections of volumes up to 0.1 pl was experimentally verified (Kohen et at., 1966) by injections of metabolically inert compounds (e.g. nitrates, chlorides and usually non-metabolizable intermediates such as glucose-6-sulfate or fructose-1-P). The stability of cell fluorescence provides a control of cell tolerance to the microinjection current as well as the exciting ultraviolet irradiation.

Fluorescence probes

Rhodamine 123 (Goldstein and Korczack, 1981) was used as a vital fluorescence probe to study the status of melanoma cell or fibroblast mitochondria before and at different times after the addition of dicarboxylic acids. A stock solution of rhodamine 123 was made in distilled water at 1 mg/ml and kept at 4°C in the dark. Immediately before use rhodamine was diluted in the cell incubation Ringer medium to 10 μg/ml and added to the cells. The rhodamine medium was washed after a 15 min incubation and replaced by fresh medium.

Evaluation of metabolic responses

Extensive microspectrofluorometric observations in a variety of cell lines lead to the conclusion that the microinjection of two substrates can provide essential information on the mitochondrial and extramitochondrial bioenergetic pathways. Initially glucose-6-P was injected to assess glycolytic activity. More recently it has been found that most cells can be characterized by their responses to 6-phosphogluconate (6-PG) for the hexose monophosphate shunt, and malate for the mitochondrial and extramitochondrial dehydrogenases. Metabolic responses are monitored at the sites of mitochondrial aggregates, mitochondria-free cytoplasmic and nuclear regions. In the case of the nuclear region specifically it is crucial to avoid cell sites showing a perinuclear or supra-(or infra)nuclear mitochondrial cloud. Injection of 6-PG or malate results in an NAD(P)H transient (i.e. coenzyme reduction and reoxidation). On the basis of the simplest assumption every transient with fluorescence intensity $F(t)$ at time t can be represented by a curve of the type (Kohen et al., 1982)

$$F(t) = A\,(e^{-k_2 t} - e^{-k_1 t}) \tag{1}$$

where k_2 is the rate constant of NAD(P) reoxidation in sec, k_1 is the rate constant of NAD(P)H reduction, and

$$A = \frac{k_1}{k_1 + k_s} \, [S_0] \, for \, [S_0] =$$

initial level of substrate after microinjection.

This equation is more faithfully fitted by the transients which follow the injection of glucose-6-P. In the case of 6-PG and malate injections, the reoxidation is generally slower than predicted by the equation, and tends to exhibit a prolonged tail especially after the halftime of fluorescence decay.

Results

Observations with rhodamine 123

Fluorescence observations of M8255 human melanoma and human fibroblasts treated with 10 mM azelaic acid and subsequently stained with the vital fluorochrome rhodamine 123 revealed initiation of mitochondrial damage within 30 min of dicarboxylic acid addition. More extensive mitochondrial damage was observed in M8255 cells cultured for three days in the presence of 30 mM azelaic acid (Fig. 2).

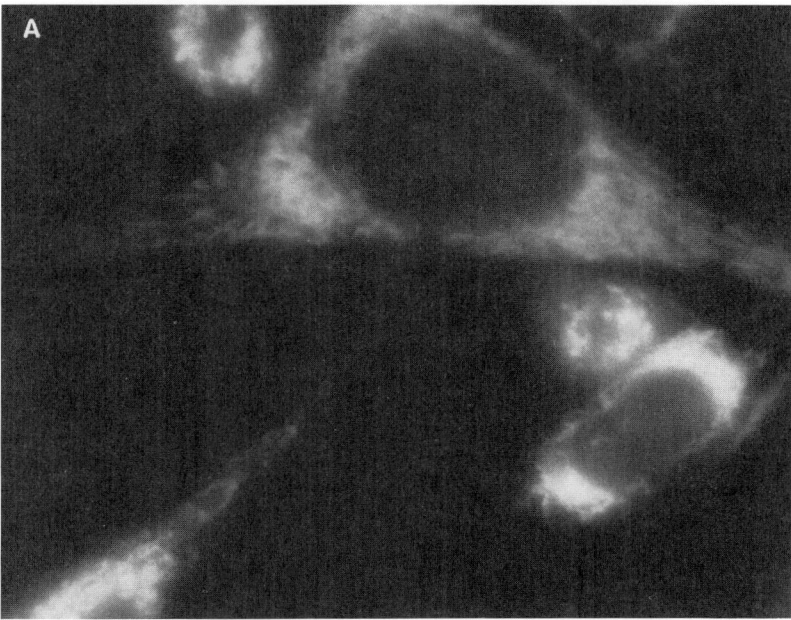

Figure 2. Fluorescence micrographs of M8255 human melanoma cells treated with the mitochondrial vital fluorescent probe rhodamine 123. The fluorescence was excited at 436 nm. A. Micrograph of control melanoma cells shows filamentous mitochondria. B. Micrograph of M8255 cells cultured for three days in presence of 30 mM azelaic acid shows structural alterations and fragmentation of mitochondria.

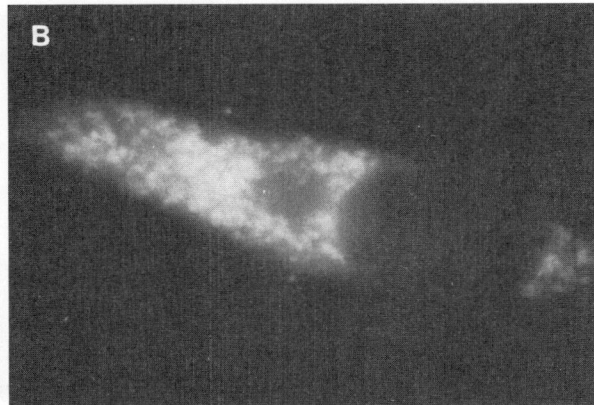

Figure 2. (cont)

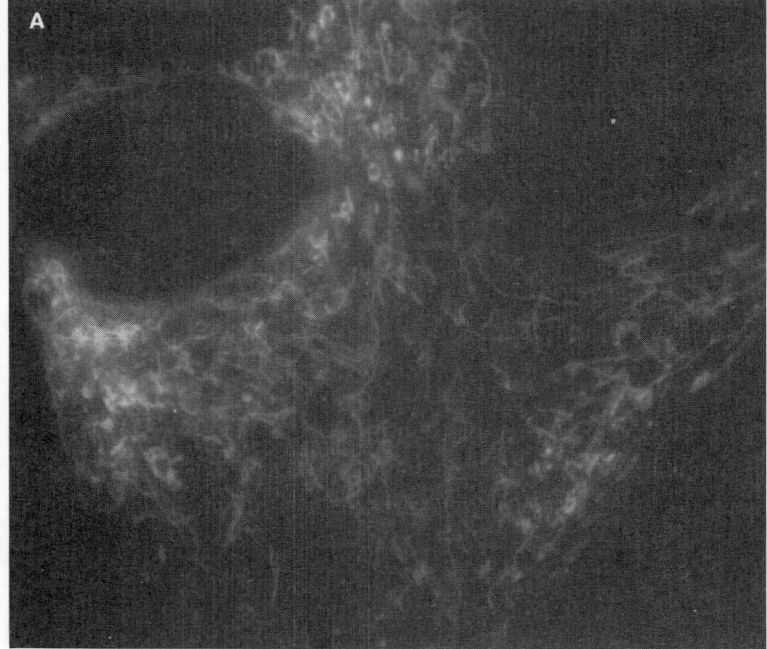

Figure 3. Fluorescence micrographs of human fibroblasts. Conditions are the same as for Fig.
2. A. Control fibroblasts maintained 4 days in the presence of 30 mM azelaic acid.

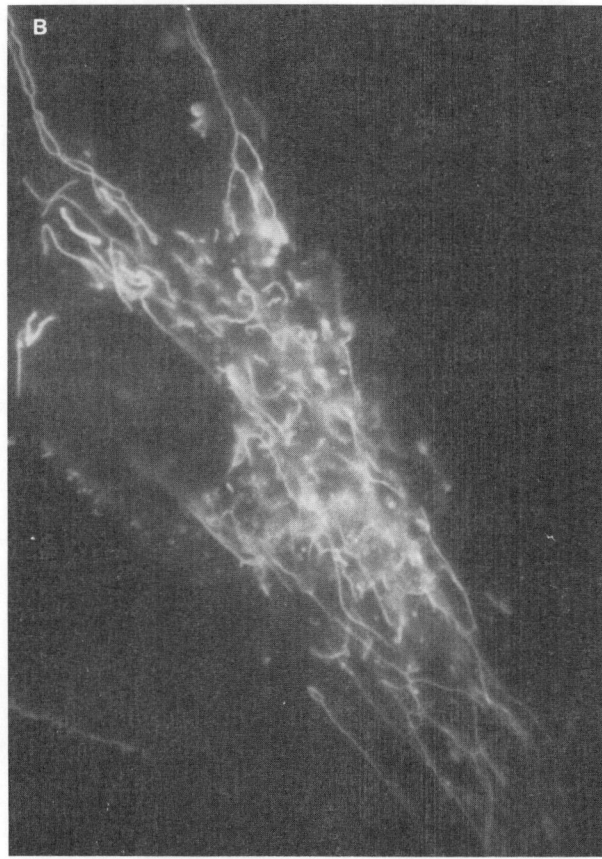

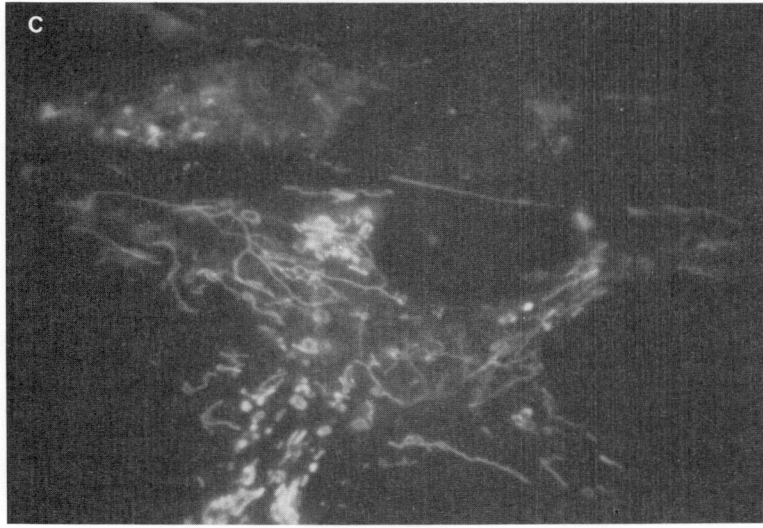

Figure 3 (cont)
B. Control fibroblast, maintained 7 days in the presence of 30 mM azelaic acid.
C. Cystic fibrosis fibroblast maintained 4 days in the presence of 30 mM azelaic acid. The micrographs show different degrees of mitochondrial damage and fragmentation. Part of the rhodamine-stained mitochondria fragments are probably incorporated within lysosome-like granules (lysosomal autophagy). The status of mitochondria in A, B and C can be compared to the rhodamine-stained filamentous mitochondria in Fig. 1a of Kohen et al., (1987) which shows the fluorescence image of normal mitochondria in a fibroblast untreated with azelaic acid.

Cultures of normal and cystic fibrosis (CF) fibroblasts maintained in presence of 10 mM azelaic acid were followed up to 7 days. Treatment with rhodamine 123 revealed considerably more severe structural damage and vesiculation of mitochondria in CF cells (Fig. 3C) as compared to control fibroblasts (Figs. 3A and 3B). Within three days mitochondria of CF cells showed considerable deterioration while in control fibroblasts it took up to a week to obtain comparable alterations (Kohen et al., 1986).

Autophagic activity of lysosomes

Extensive mitochondrial damage, more especially in azelaic acid-treated CF cells resulted in autophagic activity of lysosomes. A characteristic fluorescence image (Kohen et al., 1987) obtained with rhodamine was the coexistence of fragmented mitochondria with autophagy within lysosomes of rhodamine-stained mitochondrial fragments (see Fig. 3).

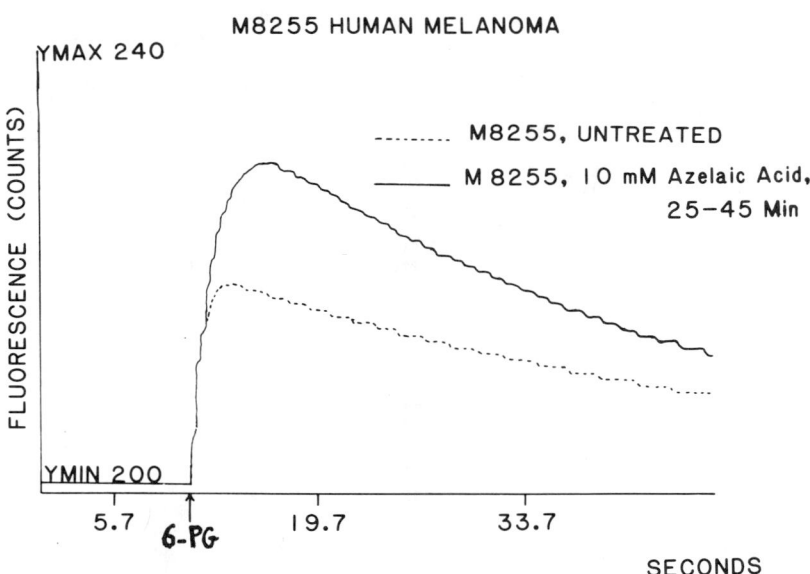

Figure 4. Averaged and computer-fitted (according to equation 1) NAD(P)H transients obtained in response to microinjected 6-phosphogluconate (6-PG) from six M8255 melanoma cells untreated with azelaic acid, and M8255 melanoma cells maintained in the presence of 10 mM azaleic acid from 25 to 45 minutes. The transient responses of azelaic acid treated melanoma cells showed variability (some were within the control range and others exhibited greater magnitude). Only cells showing the largest responses were included in the averaging. NAD(P)H fluorescence was excited at 365 nm. The ordinate shows the fluorescence intensity in counts. Increased fluorescence corresponds to NAD(P) reduction, decreased fluorescence to NAD(P)H reoxidation.

Upheavals of bioenergetic processes

Bioenergetic pathways were probed by means of 6-PG and malate injections. In melanoma cells untreated with dicarboxylic acids the metabolic responses (i.e. NADPH transients) to injected 6-PG were preferentially localized in the nuclear region or relatively larger in this region, as compared to adjacent cytoplasm. Surprisingly the responses to malate were also predominant in the same region. These findings were also repeated in CF and control fibroblasts, and seem as verified in other cultured cells (e.g. NMuLi mouse liver cells), to fit a general pattern. Due to selection of mitochondria-clear regions the responses were not attributable to a perinuclear mitochondrial cloud.

In accordance with the structural damage of mitochondria revealed by rhodamine 123, treatment with 10 mM azelaic or sebacic acid resulted within 30 min in a moderate to strong release of extramitochondrial pathways (see also Goldstein and Korczack, 1981)., evidenced by activation of NAD(P)H transient responses to both 6-PG (Fig. 4) and malate (Fig. 5) in 8255 melanoma cells. The NAD(P)H transients recorded after malate injections from the nuclear and cytoplasmic regions of untreated melanoma cells, and the

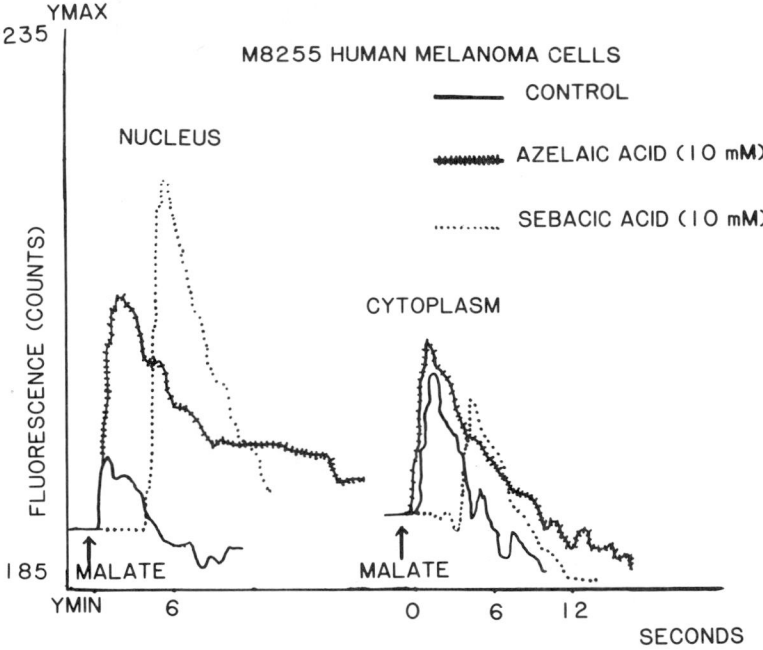

Figure 5. Representative NAD(P)H transient in response to microinjected malate, recorded from an 8255 melanoma cell control, and cells incubated for about 30 minutes in the presence of 10 mM azelaic acid or 10 mM sebacic acid. The transient curves are representative of at least five treated or untreated cells showing similar responses.

313

same cells treated with 10 mM azelaic acid are shown in Fig. 5. Activation of the NAD(P)H responses to injected malate was also observed in melanoma cells grown for 24 hours in the presence of 10 mM azelaic acid (Fig. 6).

Parallel determinations of transient responses to malate injections in azelaic acid (10 mM)-treated control and CF fibroblasts are tabulated in Table 1 (first 30 min of azelaic acid addition) and Table 2 (over 30 min) for recordings made from the nuclear region. The results are shown for paired control and CF cells in terms of amplitude A, NAD(P) reduction constant k_1 and NAD(P)H reoxidation constant k_2, the three figures required to define the time course of the NAD(P)H transient according to the equation in Methods.

As seen from the Tables, in the first 30 minutes of azelaic acid addition little difference is noticeable between control and CF fibroblasts. After 30 minutes the CF cells continue to exhibit transients, while the control fibroblasts become refractory to injections of malate. The same results are summarized in Fig. 7.

Table 1. NAD(P)H transients recorded from control and cystic fibrosis (CF) fibroblasts in response to malate injections within the first 30 min of 10 mM azelaic acid addition

A is the maximum amplitude of the transient calculated from equation:
$$F(t) = A\,(e^{-k_2 t} - e^{-k_1 t})$$
where $F(t)$ = intensity of NAD(P)H fluorescence in counts,

$A = \dfrac{k_1}{k_1 + k_s}\ [S_0]$ k_1 = NAD(P) reduction rate constant

k_2 = NAD(P)H reoxidation rate constant

$[S_0]$ = initial level of substrate after injection

Standard error of the mean indicated A_1, k_1, k_2

Cell type	A* (counts)	k_1 (sec^{-1})	k_2 (sec^{-1})
Control	24	1.4	0.06
CF	12	2.2	0.04
Control	26	2.5	0.07
CF	25	0.5	0.07
Control	43	–	0.02
CF	32	0.4	0.04
Control	47	1.3	0.06
CF	20	0.4	0.03
Control	–	–	–
CF	22	1.5	0.12
Control	–	–	–
CF	23	1.7	0.05
Control	23±8	1.7±0.4	0.05±0.01
CF	22±3	1.1±0.3	0.06±0.01

* For controls not responding A = 0

Table 2.   NAD(P)H transients recorded in response to malate injections starting from 30 min after addition of 10 mM azelaic acid

Cell type	A (counts)	k_1* (sec^{-1})	k_2* (sec^{-1})
Control	8	4.4	0.09
CF	12	2.1	0.05
Control	27	1.0	0.04
CF	8	3.7	0.07
Control	25	1.0	0.04
CF	18	1.7	0.06
Control	–	–	–
CF	100	2.2	0.06
Control	–	–	–
CF	27	1.2	0.05
Control	–	–	–
CF	25	2.2	0.04
Control	–	–	–
CF	21	1.3	0.05
Control	–	–	–
CF	14	1.3	0.08
Control	–	–	–
CF	27	1.0	0.04
Control	–	–	–
CF	26	1.0	0.03
Control	6±3	2.1±1.1	0.05±0.01
CF	28±8	1.8±0.3	0.05±0.005

*In the case of control cells, k_1 and k_2 are averaged only for cells showing a recordable NAD(P)H transient.

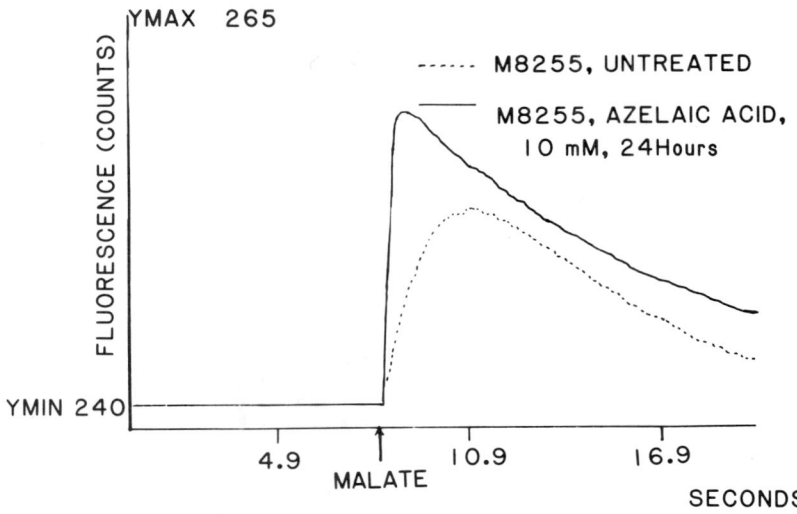

Figure 6. Averaged (from at least five cells) and computer-fitted transients obtained in response to injections of malate from control and 24 hour azelaic acid (10 mM)-grown M8255 melanoma cells.

315

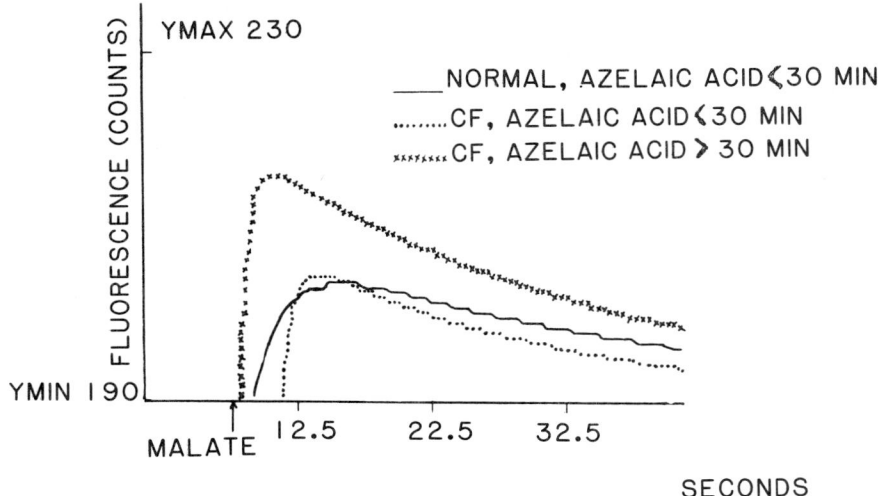

Figure 7. Simultaneously plotted averaged and computer fitted transients obtained from normal and cystic fibrosis (CF) fibroblasts maintained for up to 30 minutes in the presence of 10 mM azelaic acid, and CF fibroblasts maintained over 30 minutes in the presence of azelaic acid. Within the first 30 minutes of treatment with azelaic acid the transients recorded from normal and cystic fibrosis fibroblasts are largely superposable. However, past 30 minutes, the transients of CF fibroblasts continue to grow in amplitude while transients are no more recordable from normal fibroblasts. Each transient plot was averaged from at least five cells.

Nuclear energy metabolism.

In the absence of treatment with azelaic acid, the responses to the injected 6-PG and malate are predominant in the nuclear region of melanoma cells and fibroblasts, as compared to cytoplasmic regions (Kohen et al., 1989). Following addition of azelaic acid responses in the nuclear region are highly activated in melanoma cells and fibroblasts, and remain activated in CF fibroblasts beyond 30 min after addition of the dicarboxylic acid, while as time passes, control fibroblasts tend to become unresponsive to malate.

Discussion

The effects of dicarboxylic acids have been evaluated so far in terms of therapeutic results and mitochondrial alterations as revealed by electron microscopy. Parallel observations in 8255 melanoma cells, normal and CF fibroblasts allow some generalizations.

A first characteristic is in the topography of the transient NAD(P)H response to injected substrate, whether 6-PG or malate. The association of the highest response with the nuclear region is more easily understandable in the case of 6-PG since the nucleus and nuclear membrane have been identified as important sites of activity for the hexose monophosphate shunt (Georgiev, 1967; Kohen et al., 1986a; 1986b). A connection is suspected between the localization of this pathway and the NAD(P)H demand for nucleic acid synthesis. Our earlier studies also indicate that the center of gravity of the NAD(P)H response to substrates of the shunt can be shifted to the cytoplasm (Kohen et al., 1989) when a higher load is placed on the cell's detoxification apparatus (Kohen et al., 1988;, Kohen et al., 1988) involving the endoplasmic reticulum and other interconnected organelles, upon addition of metabolizable carcinogens or other toxic agents. Evidence from our parallel fluorescence and electron microscopic studies seems to indicate that the mobilization of organelles presumably towards detoxification activities encompasses the endoplasmic reticulum, the lysosomes, the Golgi apparatus and the nuclear membrane.

The nuclear response to injected malate seems more intriguing, although other researchers have occasionally pointed to the presence of a nuclear energy metabolism (Conover and Siebert, 1965) supported by a postulated nuclear electron transport chain or other putative mechanisms. We are probably at the very beginning of our understanding of such processes, but there is the possibility of a nuclear pump energized as postulated above, which may be involved in extrusion of toxic or other compounds (e.g. messenger RNA) from the nucleus, and in protection of the genetic apparatus, a not implausible hypothesis. Fluorescence observations which reveal the organelle interconnections described above do not suggest per se a mitochondrial connection. However, our studies on anthraline-treated cells, as well as azelaic acid treated cells, reveal in both instances extensive structural and functional mitochondrial damage. Apparently, extramitochondrial metabolic pathways, including nuclear, are under mitochondrial control. Drug-induced mitochondrial uncoupling and damage result in unleashing extramitochondrial pathways which are then out of control.

In the case of CF cells, independent studies (Shapiro, 1979) as well as our own point to abnormal activation of the NADH dehydrogenase, and a tendency of mitochondria to become uncoupled and destabilized. Thus, the mitochondria of CF cells, as also demonstrated by timed fluorescence studies with rhodamine 123, are more fragile and disintegrate more rapidly under the action of azelaic acid. As a result the upheavals of extramitochondrial metabolism, such as the striking activation of NAD(P)H transient responses to 6-PG and malate, are more noticeable and persistant in CF cells as compared to normal fibroblasts.

In the case of 8255 human melanoma cells, upheavals of nuclear energy metabolism may be relevant in understanding the antimelanoma therapeutic action of dicarboxylic acids. Our experience from microspectrofluorometric studies suggests that in following drug effects at the level of the intact living cell consideration must be given to the reciprocal interactions and interdependence of different intracellular organelles and metabolic pathways. Therefore it should not be surprising that drug-induced structural and functional alterations of mitochondria could result in severe upheavals of nuclear energy metabolism and consequently nucleic acid synthesis and mitotic activity. The reported studies on dicarboxylic acid actions in melanoma cells and fibroblasts point to a new way to assess and eventually predict drug actions at the level of cell organelles and structures, which seems more in tune with the actual pace and "modus operandi" of intracellular processes.

Acknowledgements

This work was supported by National Science Foundation grants DMB-8303691 and DMB-8705491, and Cystic Fibrosis Foundation grant G1057/01, 02.

The authors acknowledge the excellent secretarial help of Mrs. Maureen Grover and Mrs. Maria Reynardus. They are also appreciative of the help of Mr. Bill May with the art work.

References

Breatnach, A.S., 1984, *Brit. J. Dermatol.* 111: 115–120.

Conover, T.E. and Siebert, G., 1965, *Biochim. biophys. Acta* 99: 1–12.

Georgiev, G.P., 1967, *Enzyme Cytology*, edited by D.B. Roodyn (London and New York: Academic Press), pp. 27–102.

Goldstein, S. and Korczack, L.B., 1981, *J. Cell Biol.* 91: 392–398.

Hirschberg, J.G., Wouters, A.W., Kohen, E., Thorell, B., Eisenberg, B., Salmon, J.M. and Ploem, J.S., 1978, *Multichannel Image Detectors*, edited by Y. Talmi (Washington, D.C.: American Chemical Society Symposium), pp. 263–289.

Hsu, F., 1986, *Brit. J. Dermatol.* 114: 17–26.

Kohen, E., Hirschberg, J.G., Fried, M., Kohen, C., and Prince, 1988, *Chemical Carcinogens*, edited by P. Politzer and F.J. Martin Jr. (Amsterdam: Elsevier), pp. 345–361.

Kohen, E., Hirschberg, J.G., Fried, M., Kohen, C., Santus, R., Reyftmann, J.P., Morliere, P., Schachtschabel, D.O., Mangel, W.F., Shapiro B.L., and Prince, J., 1987, *Microbeam Analysis*, edited by Roy H. Geiss (San Francisco: San Francisco Press), pp. 231–240.

Kohen, E., Kohen, C., Hirschberg, J.G. and Schachtschabel, D.O. 1989, *Cell Structure and Function by Microspectrofluorometry*, edited by E. Kohen, J.G. Hirschberg and J.S. Ploem (New York Academic Press), pp. 199–228.

Kohen, E., Kohen, C., Hirschberg, J.G., Wouters, A.W., Bartick, P.R., Thorell, B., Bereiter-Hahn, J., Meda, P., Rabinovitch, A., Mintz, D. and Ploem, J.S., 1982 *Techniques in Cellular Physiology*. Part I, edited by P.F. Baker (Amsterdam: Elsevier Biomedical), pp. P103/1–28.

Kohen, E., Kohen, C. and Jenkins, W., 1966, *Exptl Cell Res.* 44: 175–194.

Kohen, E., Kohen, C., Morliere, P., Santus, R., Reyftmann, J.P., Dubertret, L., Hirschberg, J.G. and Coulomb, B., 1986b , *Cell Biochemistry and Function*, 4: 157–168.

Kohen, E., Prince, J., Kohen, C., Hirschberg, J.G. and Fried, M., (1988). *Time-Resolved Laser Spectroscopy in Biochemistry*, SPIE Vol. 909 (Bellingham, WA: The Society of Photo-Optical Instrumentation Engineering), pp. 231–240.

Kohen, E., Welch, G.R., Kohen, C., Hirschberg, J.G. and Bereiter-Hahn, J. (1986b). *The Organization of Cell Metabolism*, edited by J.S. Clegg and G.R. Welch (New York: Plenum), pp. 251–275.

Meiskens, F., personal communication.

Passy, B.S., 1984, *Biochem. Pharmacol.* 33: 103–108.

Picardo, M., 1983, *J. Investig. Dermatol.* 80: 350.

Robbins, E.J., 1985 *J. Investig. Dermotol.* 80: 216–222.

Schachtshabel, D.O., Thome, D., Salzer, V., 1986, *J. Invest. Dermatol.* 87: 423 Abstract

Shapiro, B.L., 1979, *Nature* 278: 276–277.

Shapiro, B.L., Lam, L.F. and Feigal, R.J., 1982, *Amer. J. Human Genet.* 34: 846–852.

Recent Advances in Chemical Modeling of Bacterial Bioluminescence Mechanism

SHIAO-CHUN TU* AND HUMPHREY I. X. MAGER

Department of Biochemical and Biophysical Sciences
University of Houston
Houston, Texas 77204-5500
U.S.A.

Bacterial luciferase catalyzes the oxidation of $FMNH_2$ and a long-chain aliphatic aldehyde by O_2 to yield FMN, fatty acid, water, and light. The *in vitro* emission spectrum has a peak near 490 nm and a quantum yield of 0.1-0.2, but both could vary somewhat depending on the bacterial strain from which the luciferase is obtained. The mechanism for this bioluminescent reaction has been the focus of considerable research interests. In this report, our recent studies on the bacterial bioluminescence mechanism will be summarized, with special emphases on a new radical mechanism and the use of chemical models. Some earlier findings from our and other laboratories which are of mechanistic significance will also be briefly discussed.

The Eberhard-Hastings Mechanism

An important chemical mechanism has been proposed by Eberhard and Hastings (1972) which, with modifications, is shown in Scheme 1. The enzyme-bound $FMNH_2$ (intermediate I) reacts with O_2 to form an FMN 4a-hydroperoxide intermediate II. In the absence of aldehyde, II breaks down to form FMN and H_2O_2 with very little light emission (i.e. the dark decay). In the presence of aldehyde, an FMN 4a-peroxyhemiacetal intermediate III is formed. The decay of III, following a Baeyer-Villiger type of reaction, leads to the generation of the aliphatic acid and an excited emitter proposed to be a FMN 4a-hydroxide species (Hastings and Nealson 1977, Kurfürst *et al.* 1984). After light emission, the ground state FMN 4a-hydroxide subsequently decays to produce water and the oxidized FMN.

Several features of Scheme 1 have received strong experimental supports. Foremost is the detection and identification of the 4a-hydroperoxyflavin intermediate II (Balny and Hastings 1975, Hastings and Balny 1975, Hastings *et al.* 1973, 1979, Kemal and Bruice 1976, Kurfürst *et al.* 1983, Tu 1979, 1982, 1986, Vervoort *et al.* 1986). It has also been shown that the light emission is coupled to aldehyde oxidation and is the major reaction pathway (Dunn *et al.* 1973, Hastings and Balny 1975, Tu 1982, Tu *et al.* 1987).

Photobiology, Edited by E. Riklis
Plenum Press, New York, 1991

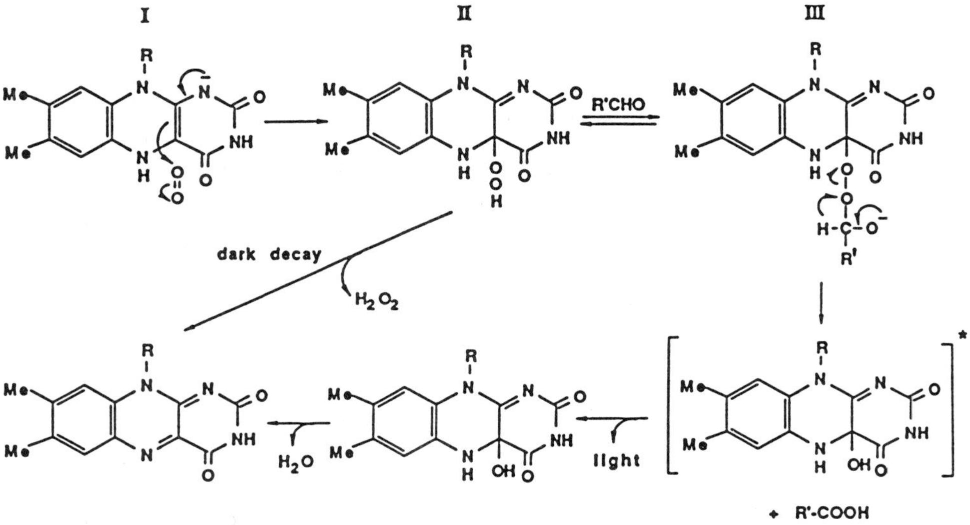

Scheme 1.

The Primary Excited Species and The Emitter

As illustrated in Scheme 1, the decay of III results in the direct formation of FMN 4a-hydroxide at the excited state as the emitter. This aspect has now been challenged by more recent findings. A blue fluorescent protein has been isolated from several strains of luminous bacteria by Lee and colleagues (Gast and Lee 1978, Koka and Lee, 1979, O'Kane *et al.* 1985), and the bound chromophore has been identified as lumazine (Koka and Lee 1979). The addition of lumazine protein (LP) to the in vitro luciferase reaction solution results in (1) an increase in the quantum yield of bioluminescence, (2) small but significant changes in emission kinetics, and (3) a blue shift of emission spectrum from 490 to 476 nm (Gast and Lee 1978). The blue-shifted bioluminescence can not be attributed to Förster energy transfer from the excited FMN 4a-hydroxide emitter to the lumazine chomophore as an acceptor. This raises the possibility that the primary excited species generated in the reaction may not be the 4a-hydroxyflavin. The chemical nature of this putative primary excited species has, however, not been specified.

A Proposed CIEEL Mechanism for Bacterial Bioluminescence

Schuster (1979) has proposed a *Chemically Initiated Electron Exchange Luminescence* (CIEEL) mechanism for chemiluminescent reactions of some organic peroxides in the presence of a catalytic "activator." The key steps of this mechanism are: (1) a one-electron transfer from the activator to the peroxide to form a radical pair P^- and

$A^{\dot{+}}$, (2) a chemical transformation of $\overset{.}{P^-}$ to a new anionic radical $\overset{.}{D^-}$ which is a powerful electron donor, (3) a second step of one-electron transfer from $\overset{.}{D^-}$ to $A^{\dot{+}}$ leading to the formation of excited state A* and (4) the radiative relaxation of A*. The initial proposal

$$P + A \rightarrow [P \text{---} A] \rightarrow \overset{.}{P^-} + A^{\dot{+}} \tag{1}$$
$$\overset{.}{P^-} \rightarrow \overset{.}{D^-} + \text{product(s)} \tag{2}$$
$$\overset{.}{D^-} + A^{\dot{+}} \rightarrow D + A^* \tag{3}$$
$$A^* \rightarrow A + \text{light} \tag{4}$$

by Schuster identifies A* as the excited state but in principle either A* or D* could be formed at step 3.

Mager and Addink (1984) have proposed the first detailed CIEEL mechanism for bacterial luciferase. We believe that this mechanism, shown in Scheme 2 with modifications (Mager and Tu 1987, Mager *et al.* 1988), may shed light on the following major unresolved questions: (1) What are the chemical steps leading to the formation of the primary excited species from III? (2) What is the chemical identity of the primary excited species? (3) Are the primary excited species and the emitter the same? (4) If different, how is the emitter formed? (5) How to explain the blue shift, enhanced quantum yield, and changed kinetics of emission by the addition of LP?

Starting with the intermediate III, this CIEEL mechanism depicts a one-electron transfer from the N5 to the peroxide bond leading to the formation of radical pair 1 and

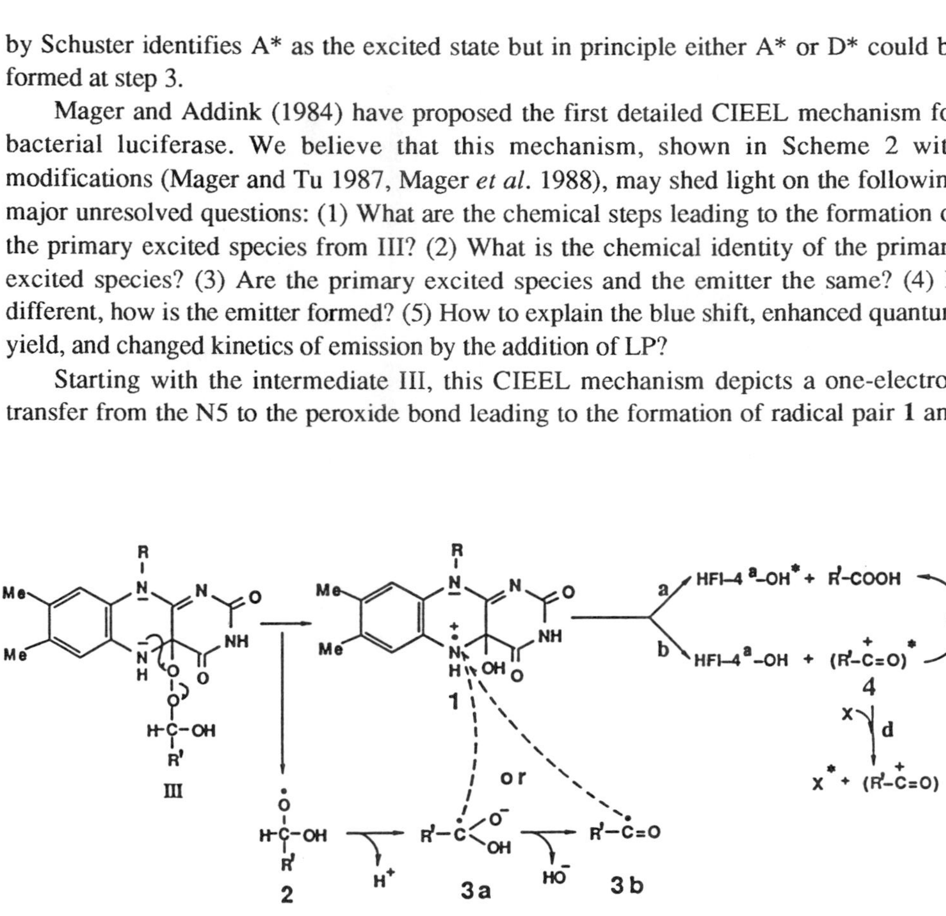

Scheme 2

2. A chemical transformation of 2 then occurs to generate a radical 3a or subsequently 3b. In the third step, a back one-electron transfer from 3a or 3b to 1 generates either an excited flavin 4a-hydroxide (Pathway a) or an excited acylium cation 4 (Pathway b). Finally, energy transfer from the excited 4 to the flavin 4a-hydroxide or to a second

chromophore X (such as LP) occurs to produce the HFl-4a-OH* (Pathway **bc**) or X* (Pathway **bd**).

At present, we favor Pathway **b** over Pathway **a** for the generation of the primary excited species. In comparison with other chemiluminescence reactions, the proposed excited acylium cation should be sufficiently energetic as an energy donor for either FMN 4a-hydroxide or LP, if present. This CIEEL mechanism can thus easily account for the blue shift caused by the addition of LP. Moreover, LP has a fluorescence quantum yield of 0.54–0.59 (Lee et al. 1985), higher than the quantum yield of 0.18 for the FMN 4a-hydroxide emitter (Tu 1982). Therefore energy transfer from the primary excited species to lumazine will result in an increase in bioluminescence quantum yield. LP also has a small effect on the bioluminescence decay rate (Gast and Lee 1978) but the underlying mechanism is still not clear. It should be noted that bioluminescence kinetics is sensitive to a number of factors. For example, a single point mutation on luciferase can markedly change the light decay kinetics (Cline and Hastings 1972). Also, for the *V. harveyi* enzyme, the decay rate of decanal-initiated emission is about 7.5-fold faster than that with dodecanal or octanal (Tu 1979). It is highly unlikely that different chemical mechanisms are involved for these aldehydes. Therefore these findings suggest that the reaction kinetics may be sensitive to the microenvironment of luciferase active site. The LP-initiated changes in decay kinetics could then be a result of luciferase conformational change triggered by the binding of LP. Alternatively, Scheme 2 can be modified to involve LP as a reactant for the formation of the primary excited species. The LP* is formed directly or in a subsequent energy (or electron) transfer step. This could also account for the changed emission kinetics. A rigorous evaluation of this possibility, however, must await further studies.

Chemical Models for The Proposed CIEEL Mechanism

Schuster (1979) has demonstrated that, for chemiluminescence of the CIEEL type, the efficiency of light generation is primarily regulated by the rate-limiting one-electron transfer from the activator to an acceptor such as a peroxide. A linear relationship can be obtained between the emission efficiency and the redox potential of the activator radical cation/activator couple, with a higher emission efficiency for an activator having a lower redox potential. In Scheme 2, the activator radical cation would correspond to the proposed key intermediate 4a-hydroxyflavin radical cation **1**. To subject Scheme 2 to experimental tests, our initial efforts were thus focused on the formation, detection, and characterization of this novel 4a-hydroxyflavin radical cation. Since many flavoprotein hydroxylases involve flavin 4a-hydroxide as an intermediate, we believe that the importance of such studies goes beyond the luciferase system. It is expected that **1** can be formed by the removal of one electron from the 4a-hydroxyflavin pseudobase **5** (Eq. 5). However, **1** has never been detected in such a reaction probably due to the instability of **5** and, most likely, **1**. The 5-ethyl-4a-hydroxy(or methoxy)-3-methyl-4a,5-dihydrolumiflavin **6** has thus been chosen as a model for its much enhanced stability.

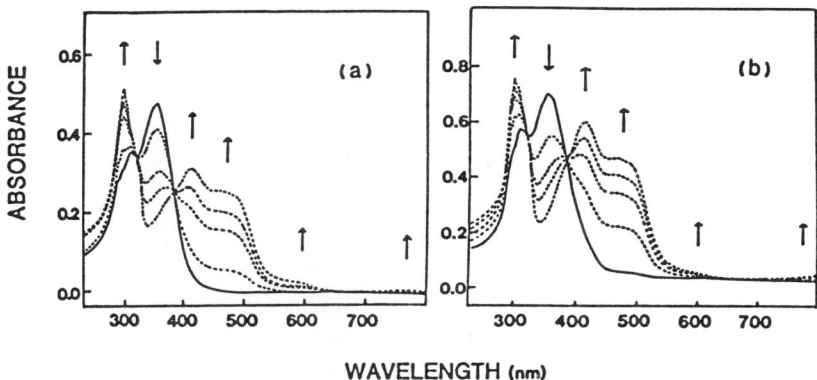

$$(5)$$

$$(6)$$

(X = OH, OMe)

Using electrochemical techniques, we have obtained unambiguous evidence (Mager *et al.* 1988) that the 5-ethyl-4a-hydroxy(or methoxy)lumiflavin radical cation **7** can indeed be generated in acetonitrile by a reversible one electron removal from **6** (Eq. 6). Time-resolved absorption spectra during controlled-potential electrolysis of **6** (X= OH, OMe) at 1.3 V are shown in Fig. 1A, B. For **6** (X = OH), the absorption band at 351 nm (ε = 9600 $M^{-1}cm^{-1}$ in acetonitrile) decreases during the transition to **7** while the shoulder at 307 nm increases and shifts to 298 nm and new absorption bands appear at 412 and 470 nm (Fig. 1A). Similar spectral results were observed for the conversion of **6** (X = OMe) to **7** (Fig. 1B). The redox potential $E_{1/2}$ for reaction shown in Eq. 6 has been determined to be 1.05 and 1.11 V for **7/6** with X as OH and OMe, respectively (Table 1). For

Figure 1. Conversion of 6 to 7 (X = OH, OMe) by controlled-potential oxidation at 1.3 V in acetonitrile. Spectra were taken (a) during the first 14 s for X = OH and (b) during the first 9 s for X = OMe. The initial spectra are shown as solid lines.

Table 1. Redox Potentials of Flavin Models in Acetonitrile

Redox Couple		$E_{1/2}$ (V)
7/6	(X = OH)	1.05
	(X = OMe)	1.11
8/9	(X = OH, OMe)	0.29
9/1 0	(X = OH; OMe)	−0.41

comparison, the ($E_{1/2}$) redox potentials for the one-electron reduction of oxidized flavin **8** to the flavosemiquinone **9** and subsequently to the dihydroflavin **10** as shown in Eq. 7 have also been determined to be 0.29 and −0.41 V, respectively (Table 1). It is obvious that the redox potential of reaction Eq. 6 is quite different from those of reactions shown in Eq. 7. Macheroux et al. (1984, 1987) have utilized several flavin derivatives as

$$(7)$$

substrate for bacterial luciferase. An apparent linear relationship was obtained between the luminescence decay rates and the redox potentials. Such a finding was considered by these authors as a strong support for a CIEEL mechanism for bacterial luciferase. However, the redox potentials used in their correlation are those corresponding to the reduction of oxidized flavins to dihydroflavins similar to that shown in Eq. 7 rather than the redox potentials of favin 4a-hydroxide radical cations.

We have explored another method for the generation of 4a-hydroxyflavin radical cations in quantities above those usually obtainable by electrochemical techniques. It is known that oxidized flavin **8** and 4a-hydroxy(or methoxy)flavin **6** (X = OH, OMe) can reach an equilibrium (Eq. 8). Therefore, a comproportionation of **8** and **6** conceivably could lead to the formation of flavin semiquinone **9** and the desired 4a-hydroxy(or methoxy)flavin radical cation **7** (X = OH, OMe) as shown in Eq. 9. This is indeed the case. Starting with either **6** (X = OH) or **8** in aqueous medium, the formation of **7** (X = OH) and **9** has been detected and quantitated by absorption measurements (Mager and Tu 1988). Since **6** and **8** reach a rapid equilibrium in water, the relative amounts of these two flavin species are pH dependent as dictated by Eq. 8 (X = OH). Consequently the rate of reaction shown in Eq. 9 is also pH dependent. When the equilibrium is shifted to either direction of the Eq. 8, one of the two reactants required in Eq. 9 would exist in trace amount thus resulting in a very slow reaction rate. Increases in the Eq. 9 reaction rate can

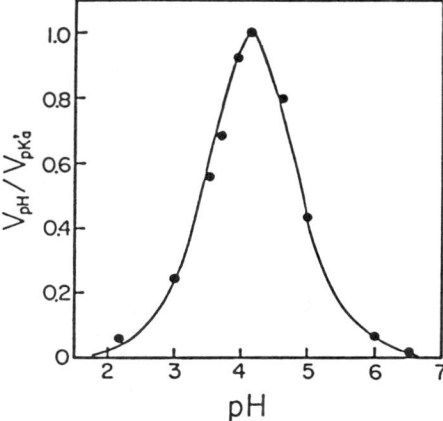

$$\mathbf{8} \quad + \text{HX} \; \rightleftharpoons \quad \mathbf{6} \quad + \text{H}^+ \tag{8}$$

$$\mathbf{6} \; + \; \mathbf{8} \; \longrightarrow \quad \mathbf{9} \quad + \quad \mathbf{7} \tag{9}$$

then be expected as the medium pH gets closer to the apparent pK'a of **8/6** (X = OH). Using a pK'a = 4.15 in 0.1 M citrate buffer, the observed relative initial rates (V_{pH} / $V_{pK'a}$) at various pH are compared with the theoretical profile and an excellent fit is obtained (Fig. 2). This provides an unambiguous evidence for the proposed mechanism shown in Eqs. 8, 9. However, the 4a-hydroxyflavin flavin radical cation formed in water is labile and can not be accumulated quantitatively (Mager and Tu 1988). When reactions shown in Eqs. 8 and 9 are carried out in benzene containing substoichiometric amount of

Figure 2. pH effect on the relative initial rate ($V_{pH}/V_{pK'a}$) of the reaction shown in Eq. 9. Solid circles are observed results and the line is the theoretical profile.

325

acid and trace amount of water, a rapid formation of flavin semiquinone **9** and the 4a-hydroxyflavin radical cation **7** (X = OH) is achieved in quantities near the theoretical 50% value (Mager and Addink, 1984). At present, this method gives the highest yield of the 5-ethyl-4a-hydroxyflavin radical cation. After the accumulation of **7** (X = OH) as described above, the subsequent addition of dibenzoylperoxide as an electron donor results in chemiluminescence emission (work in progress).

Conclusion

Several aspects of Scheme 1 are strongly supported by results from our and other laboratories. More recently, we have proposed a CIEEL mechanism for bacterial luciferase. This mechanism depicts chemical steps involved in the decay of the flavin 4a-peroxyhemiacetal intermediate III to form a primary excited species derived from the aliphatic aldehyde moiety which in turn transfers the energy to an emitter proposed to be a flavin 4a-hydroxide. This mechanism can account for the blue-shifted emission sensitized by the lumazine protein which also serves as an emitter in competition with the flavin 4a-hydroxide as an acceptor of energy from the primary excited species. A 4a-hydroxyflavin radical cation is a key intermediate of the proposed CIEEL scheme. In our recent chemical model work, this novel flavin radical cation has been generated by two methods, and characterized with respect to redox potentials, chemical reactivities, and spectral properties. Chemiluminescence was observed by reacting 5-ethyl-4a-hydroxyflavin radical cation with an electron donor in our preliminary studies.

Acknowledgment

We acknowledge the support of grants to S.-C. T. from NIH (GM25953), The Robert A. Welch Foundation (E-1030), and the Texas Advanced Technology Research Program.

References

Balny, C., and Hastings, J. W., 1975, Fluorescence and bioluminescence of bacterial luciferase intermediates. *Biochemistry*, 14, 4719–4723.

Cline, T. W., and Hastings, J. W., 1972, Mutationally altered bacterial luciferase. Implications for subunit functions. *Biochemistry*, 11, 3359–3370.

Dunn, D. K., Michaliszyn, G. A., Bogacki, I. G., and Meighen, E. A., 1973, Conversion of aldehyde to acid in the bacterial bioluminescent reaction. *Biochemistry*, 12, 4911–4918.

Eberhard, E., and Hastings, J. W., 1972, A postulated mechanism for the bioluminescent oxidation of reduced flavin mononucleotide. *Biochemical and Biophysical Research Communications*, 47, 348–353.

Gast, R., and Lee, J., 1978, Isolation of the in vivo emitter in bacterial bioluminescence. *Proceedings of the National Academy of Sciences USA*, 75, 833–837.

Hastings, J. W., and Balny, C., 1975, The oxygenated bacterial luciferase-flavin intermediate. Reaction products via the light and dark pathways. *The Journal of Biological Chemistry*, 250, 7288–7293.

Hastings, J. W., and Nealson, K. H., 1977, Bacterial Bioluminescence. *Annual Review of Microbiology*, 31, 549–595.

Hastings, J. W., Balny, C., LePeuch, C., and Douzou, P., 1973, Spectral properties of an oxygenated luciferase-flavin intermediate isolated by low–temperature chromatography. *Proceedings of the National Academy of Sciences USA*, 70, 3468–3472.

Hastings, J. W., Tu, S.-C., Becvar, J. E., and Presswood, R. P., 1979, Bioluminescence from the reaction on FMN, H_2O_2, and long chain aldehyde with bacterial luciferase. *Photochemistry and Photobiology*, 29, 383–387.

Kemal, C., and Bruice, T. C., 1976, Simple synthesis of a 4a-hydroperoxy adduct of a 1,5-dihydroflavine: Preliminary studies of a model for bacterial luciferase. *Proceedings of the National Academy of Sciences USA*, 73, 995–999.

Koka, P., and Lee, J., 1979, Separation and structure of the prosthetic group of the blue fluorescence protein from the bioluminescent bacterium *Photobacterium phosphoreum*. *Proceedings of the National Academy of Sciences USA*, 76, 3068–3072.

Kurfürst, M., Ghisla, S., and Hastings, J. W., 1983, Bioluminescence emission from the reaction of luciferase-flavin mononucleotide radical with O_2^{-}. *Biochemistry*, 22, 1521–1525.

Kurfürst, M., Ghisla, S., and Hastings, J. W., 1984, Characterization and postulated structure of the primary emitter in the bacterial luciferase reaction. *Proceedings of the National Academy of Sciences USA*, 81, 2990–2994.

Lee, J., O'Kane, D. J., and Visser, A. J. W. G., 1985, Spectral properties and function of two lumazine proteins from *Photobacterium. Biochemistry*, 24, 1476–1483.

Macheroux, P., Ghisla, S., Kurfürst, M., and Hastings, J. W., 1984, Studies on the bacterial luciferase reaction: isotope effects on the light emission. *Flavins and Flavoproteins*, edited by R. C. Bray, P. C. Engel, and S. G. Mayhew (Berlin: de Gruyter), pp.669–672.

Macheroux, P., Eckstein, J., and Ghisla, S., 1987, Studies on the mechanism of bacterial bioluminescence. Evidence compatible with a one electron transfer process and a CIEEL mechanism in the luciferase reaction. *Flavins and Flavoproteins*, edited by D. E. Edmondson and D. B. McCormick (Berlin: de Gruyter), pp. 613–619.

Mager, H. I. X., and Addink, R., 1984, On the role of some flavin adducts as one-electron donors. *Flavins and Flavoproteins*, edited by R. C. Bray, P. C. Engel, and S. G. Mayhew (Berlin: de Gruyter), pp. 37–40.

Mager, H. I. X., and Tu, S.-C., 1987, One-electron transfers in flavin systems: Relevance to the postulated CIEEL mechanism in bacterial bioluminescence. *Flavins and Flavoproteins*, edited by D. E. Edmondson and D. B. McCormick (Berlin: de Gruyter), pp. 583–592.

Mager, H. I. X., and Tu, S.-C., 1988, Spontaneous formation of flavin radicals in aqueous solution by comproportionation of a flavinium cation and a flavin pseudobase. *Tetrahedron*, 44, 5669–5674.

Mager, H. I. X., Sazou, D., Liu, Y. H., Tu, S.-C., and Kadish, K. M., 1988, Reversible one-electron generation of 4a,5-substituted flavin radical cations: Models for a postulated key intermediate in bacterial bioluminescence. *Journal of the American Chemical Society*, 110, 3759–3762.

O'Kane, D. J., Karle, V. A., and Lee, J., 1985, Purification of lumazine proteins from *Photobacterium leiognathi* and *Photobacterium phosphoreum*: Bioluminescence properties. *Biochemistry*, 24, 1461–1467.

Schuster, G. B., 1979, Chemiluminescence of organic peroxides. Conversion of ground-state reactants to excited-state products by the chemically initiated electron-exchange luminescence mechanism. *Accounts of Chemical Research*, 12, 366–373.

Tu, S.-C., 1979, Isolation and properties of bacterial luciferase-oxygenated flavin intermediate complexed with long-chain alcohols. *Biochemistry*, 18, 5940–5945.

Tu, S.-C., 1982, Isolation and properties of bacterial luciferase intermediates containing different oxygenated flavins. *The Journal of Biological Chemistry,* 257, 3719–3725.

Tu, S.-C., 1986, Bacterial luciferase 4a-hydroperoxyflavin intermediates: stabilization, isolation, and properties. *Methods in Enzymology,* 133, 128–139.

Tu, S.-C., Wang. L.-H., and Yu, Y., 1987, Applications of deuterium and tritium isotope effects to the elucidation of kinetic mechanisms of flavoprotein hydroxylases. *Flavins and Flavoproteins*, edited by D. E. Edmondson and D. B. McCormick (Berlin: de Gruyter), pp. 539–548.

Vervoort, J., Muller, F., Lee, J., van den Berg, W. A. M., and Moonen, C. T. W., 1986, Identifications of the true carbon-13 nuclear magnetic resonance spectrum of the stable intermediate II in bacterial luciferase. *Biochemistry,* 25, 8062–8067.

Use of Image Analysis in Photobiology

D-P. HÄDER

Institut für Botanik und Pharmazeutische Biologie
Universität Erlangen
D-8520 Erlangen, FRG

Introduction

Image analysis is a tool to quantitate, facilitate and automate the extraction of key parameters from data available in the form of images. The appearance of fast and inexpensive microcomputers with sufficient memory and computing capacity allows tasks to be performed which were limited to main frame computers only a few years ago (Shipton, 1979). However, it should be kept in mind that the human eye and brain far exceeds the capabilities of machine vision both in temporal and in spatial resolution (Poggio, 1984; Blake, 1987). Even though, image analysis can be used successfully to enhance, e.g., microscopic images (Serra, 1980; Kokubo and Hardy, 1982) or extract structural details of static images or a film (or video) sequence (McMillan et al., 1987; Russ and Russ, 1987; Thurston et al., 1986).

General principles of operation

Whatever the source of the image is (video tube camera, charged coupled device (CCD), video recorder, image scanner), the first step of image analysis is a digitization procedure: the image is divided into a matrix of pixels (image points) arranged in rows and columns. The number of rows and columns defines the spatial resolution of the digitized image. Each pixel is analyzed individually: in a black and white image (color images will be discussed below) each pixel is assigned a number which reflects its gray level, ranging from 0 (darkest black) to, e.g., 255 (brightest white). This process is performed by an A/D (analog to digital) converter. The numeric result for each pixel is stored in a matrix of memory cells in an electronic memory, such as a RAM (read and write memory) (Preston, 1983; Desai and Reimer, 1985). In low cost applications the sequential digitization of all pixels within an image can take up to several minutes, but in modern framegrabbers this is done by flash converters in real time, i.e. within 40 ms (50 Hz video repeat frequency) or 33 ms (60 Hz).

The hardware can be either contained in a separate unit connected with the computer

Photobiology, Edited by E. Riklis
Plenum Press, New York, 1991

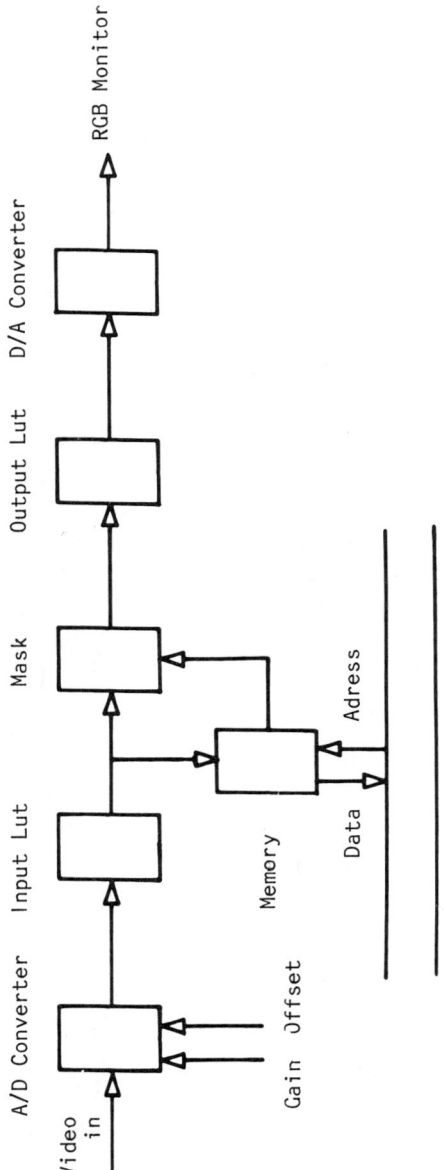

Figure 1. Sequence of hardware components to digitize, store and display a video image (after Häder, 1988).

or concentrated on a card and occupy a bus slot of the host computer. The image can be stored either in a designated video RAM or part of the general RAM of the microcomputer (Steinbach et al., 1982; Allen and Allen, 1983). The result of this first step is that the image is broken down into an array of numbers, coding gray levels, stored in an electronic memory which can be accessed by the host computer (Castillo et al., 1982). In order to visualize the digitized image, the process is reversed: the contents of the memory cells is read, converted into an analog signal (D/A conversion) and displayed on a monitor (Fig. 1).

Several hardware and software approaches can be employed to manipulate the image during the process. Most modern framegrabbers allow to alter brightness and contrast under software control. A versatile tool is a LUT (look up table) at the input side before the image is stored: each individual digitized pixel value is replaced by a value looked up in a table stored previously (Bryan et al., 1985). As an example, the image can be replaced by its negative (Table 1) by exchanging each pixel value (P) with its difference to the maximum number (e.g. 255):

$$P' = 255 - P$$

Likewise the contrast can be enhanced or certain gray levels can be stretched over the full range. An interesting application is the use of pseudocolor representation in which certain gray levels are mapped to three different output D/A converters which drive the three basic color channels (green, red and cyan) of an RGB monitor. For instance all dark grays could be mapped to blue, all bright grays to red and intermediate ones to green. Bigger systems use three individual systems to digitize and store real color images.

Software manipulation of the image

Since the image is represented by an array of digital numbers in a memory it can be easily manipulated by computer access. Each memory cell can be read or written on an individual basis. Increasing or decreasing the brightness is done by adding or substracting a constant; gradients in darkness can be imposed or removed by adding or substracting linearly or exponentially increasing values in horizontal or vertical directions. Multiplication of each pixel with a constant enhances the contrast. Thresholding techniques are used to remove a dark or bright background or to isolate areas of distinct gray levels.

Noise reduction and smoothing can be achieved by replacing each pixel with the arithmetic mean of a quadratic matrix, e.g. 3×3 pixels, surrounding and including the pixel of calculation (Julez and Harmon, 1984). Laplace and Sobel filters use more complex mathematical techniques to enhance edges. In the Laplace filter each original pixel value P is multiplied by a constant (called kernel, e.g. 8) and the values of its eight neighbors are substracted from the product, which yields the new pixel value P'. A number of other techniques can be employed to enhance the image quality or to extract

Table 1. Use of look-up tables (LUTs) to invert the image into its negative (A), increase the contrast to two levels (B) or expand the gray level of a small band (125-131) over the whole possible range (0-255) (C). Mapping certain gray level ranges to three colors results in pseudocolor representation (D).

	A	B	C	D	
0	255	0	0	0	
1	254	0	0	3	
2	253	0	0	6	blue
3	252	0	0	9	
4	251	0	0	12	
.	.	.	.	.	
.	.	.	.	.	
.	.	.	.	.	
124	131	0	31	118	
125	130	0	63	121	
126	129	0	95	124	
127	128		127	127	green
128	127	255	159	130	
129	126	255	191	133	
130	125	255	223	136	
131	124	255	255	139	
.	.	.	.	.	
.	.	.	.	.	
.	.	.	.	.	
251	4	255	255	243	
252	3	255	255	246	
253	2	255	255	249	red
254	1	255	255	252	
255	0	255	255	255	

the parameters of interest (Pavlides, 1982). Zooming and panning is achieved by mapping pixels into other memory cells (Hainfeld et al., 1982).

In the following I will discuss some real applications realized with a framegrabber used in an IBM AT compatible microcomputer. The tasks range from area detection and cell counting to densitometric analysis of absorbing and fluorescent gels. Other systems analyze motility, velocity and movement vectors of motile organisms in real time.

Area calculation and object counting

The objects analyzed can be on a microscopic or macroscopic scale. The image can be recorded using a video camera on a microscope or the object, such as a leaf, could be placed on a light box in order to enhance the contrast. Once the image is digitized and stored in memory the computer algorithm starts to scan the field line by line until it hits the object, the brightness of which differs from the background by a predefined threshold value (Fig. 2). Then a filling routine takes over which examines each of the adjacent pixels shell by shell until all connected pixels have been found (Häder, 1987a; Kim et al., 1987).

Basically the same technique can be used to count cells, colonies, organisms or other objects (Häder and Griebenow, 1987). After an object has been analyzed the scanning routine continues until the whole screen has been analyzed. However, this technique is not reliable when objects are in direct contact, since the computer regards the continuous area as belonging to one single object (Erhardt et al., 1980). This problem is solved by the erosion algorithm: one pixel wide shells are removed from the circumference of the composite object one after the other until the isthmus connecting the two adjacent objects disappears. Subsequently, the objects are inflated by filling in the shells to the original size leaving a one pixel wide free space between the objects. In other systems manual separation techniques using light stylos or cross hair indicators are employed to separate adjacent objects.

Area detection and object counting have been used successfully to calculate the cell density of, e.g., blood cells or cancer cells (Koss et al., 1983; Kaufman et al., 1987; Wittekind and Schulte, 1987). In one approach the cell density of flagellates was determined in a column of water: A suspension was filled into a plexiglass cylinder 1 m long and 90 mm in diameter which was placed vertically into a pond (Häder and Griebenow, 1988). After thermal equilibration samples were taken from 18 outlets 50 mm apart along the length of the column (Fig. 3). The cell density was determined using the technique described above.

Image anlysis has also been used to extract parameters of interest from light or electron microscope images (Spring, 1983; Caldwell, 1985; Wollmer, 1987; Gronsky, 1988). Especially in medicine it is desirable to develop a system which automatically scans and identifies abnormal cells (Preston and Dekker, 1980; Wittekind and Schulte, 1987). This task involves the recognition of subtle changes in the cellular morphological characteristics and requires a high spatial resolution of at least 10 pixel per μm (Gunzer et al., 1987). Image analysis has also been suggested and used to reconstruct cellular structures on a three-dimensional basis (Nierzwicki-Bauer et al., 1983; Jimenez et al., 1986).

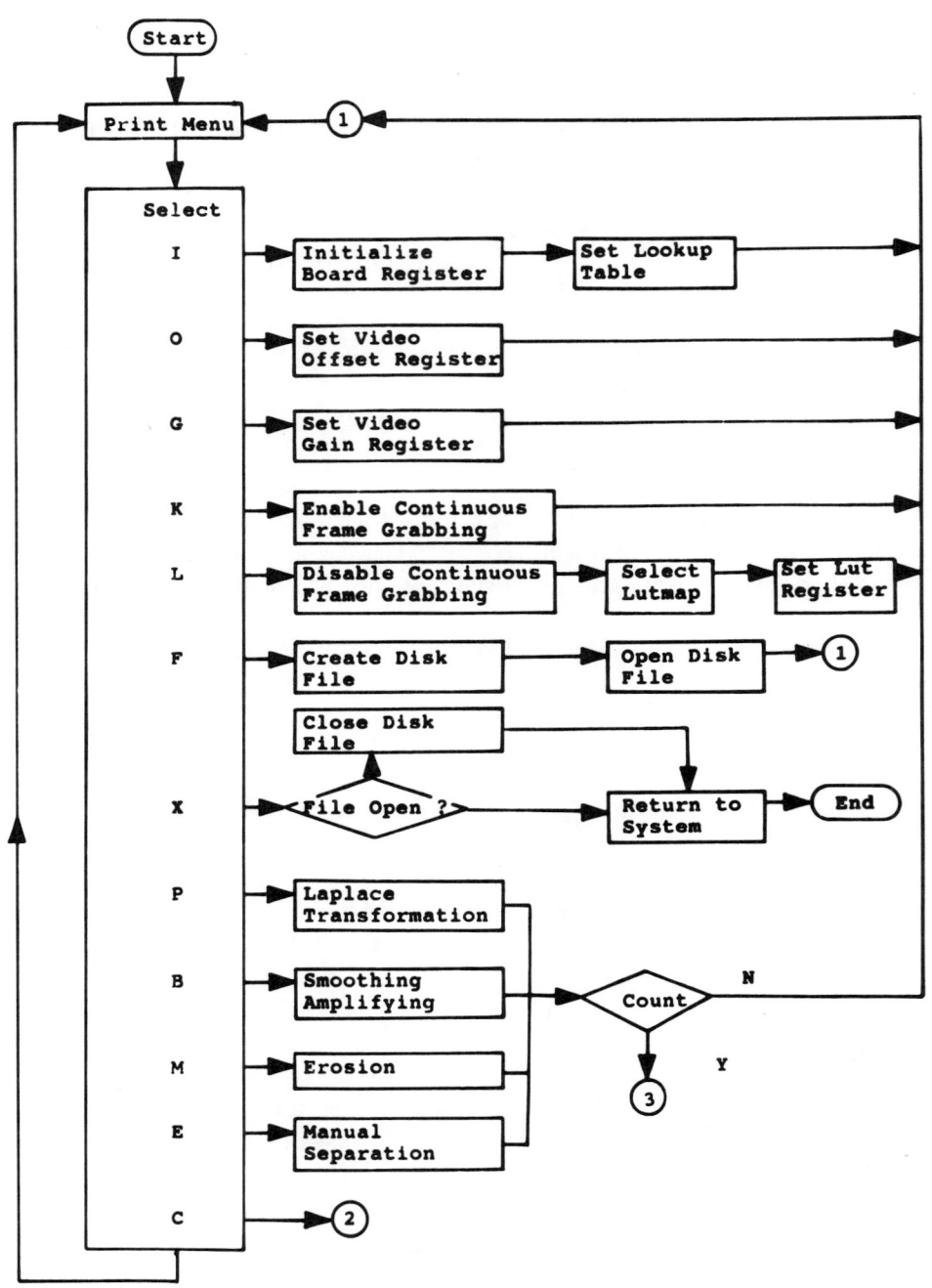

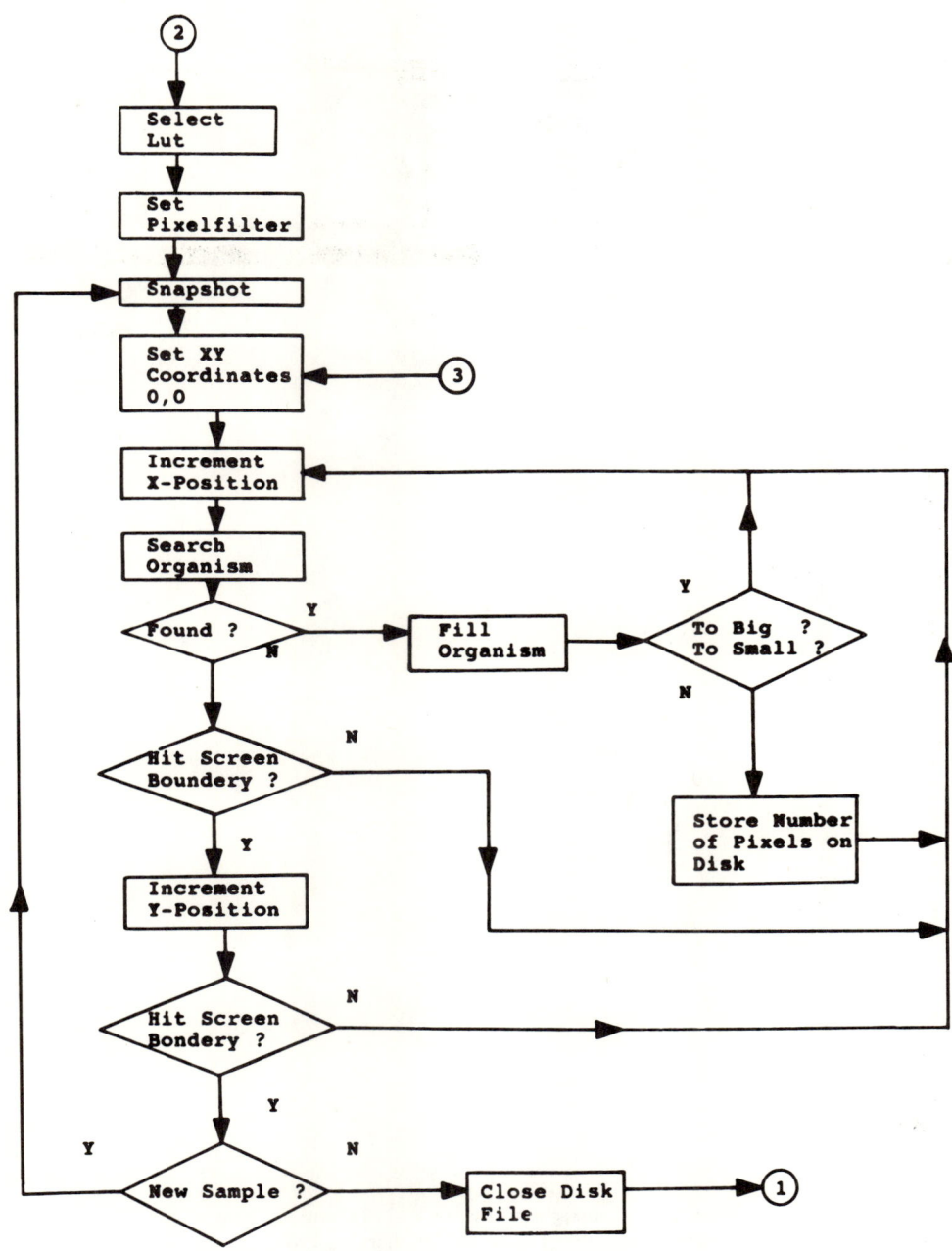

Figure 2. Flowchart of an algorithm to calculate the area of an object in an image (after Häder, 1987a).

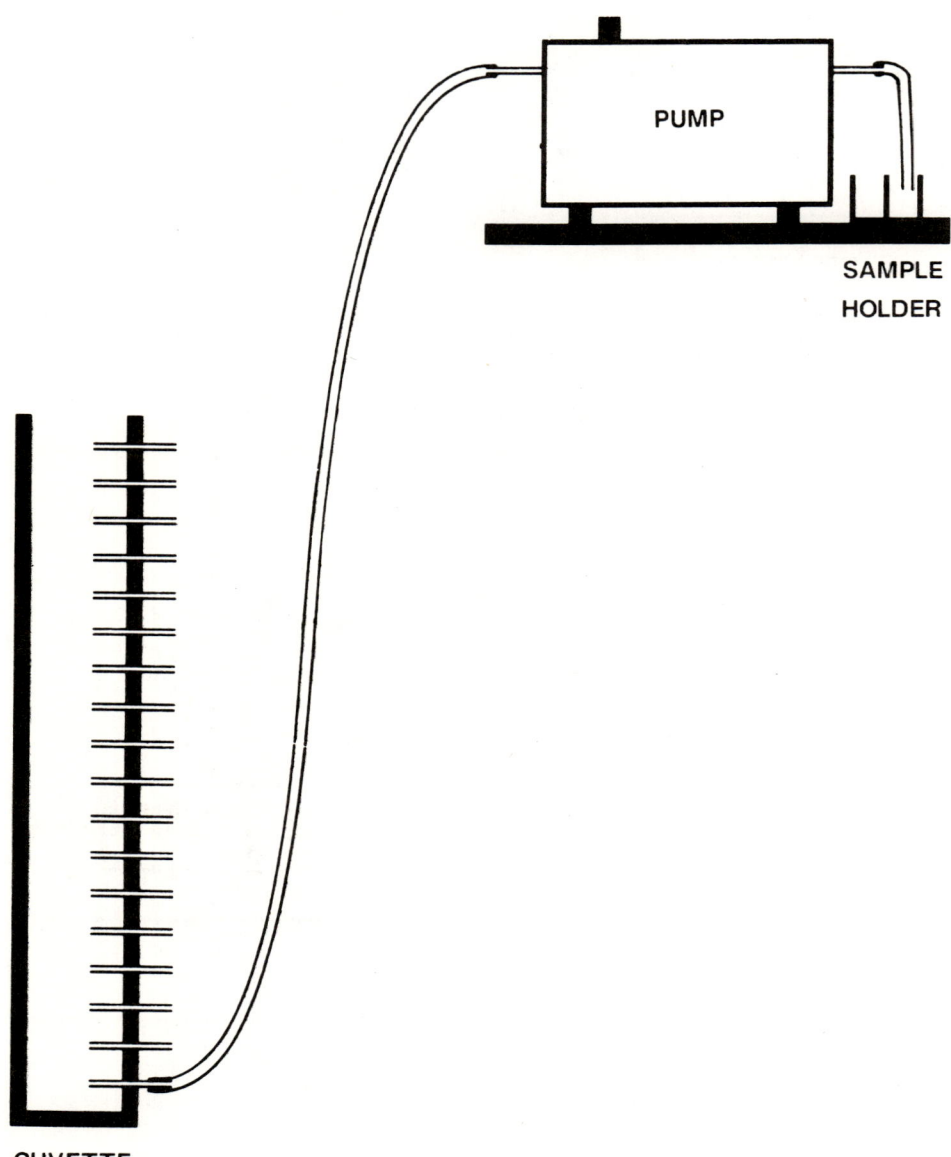

CUVETTE

Figure 3. Schematic diagram of a plexiglass cuvette with 18 outlets from which 18 samples were drawn using a peristaltic pump (after Häder and Griebenow, 1988).

Tracking of motile organisms

Analysis of movement vectors of motile microorganisms requires the comparison of a series of images taken at regular time intervals (Coates et al., 1985; Rikmenspoel and Isles, 1985; Burton et al., 1986). The objects analyzed range from unicellular organisms (Mikolajczyk et al., 1985, 1986; Spudich, 1985; Häder 1986, 1987b; Häder et al., 1986, 1987, 1988) to multicellular organisms (Dusenbery 1985a, b).

Two basic techniques have been developed to follow moving objects: In an off-line approach tracks of organisms were recorded on a tape recorder (Gualtieri et al., 1985). During playback the position of each cell was determined in each frame, so that individual tracks were reconstructed. In real time analysis of motile microorganisms a cell is selected at random from a population and its outline is determined using standard algorithms (Grant and Reid, 1981; Berns and Berns, 1982). The center of gravity (centroid) is calculated and stored in the computer memory (Fig. 4). Starting from this position the new center of gravity is determined in the next video frame and after a

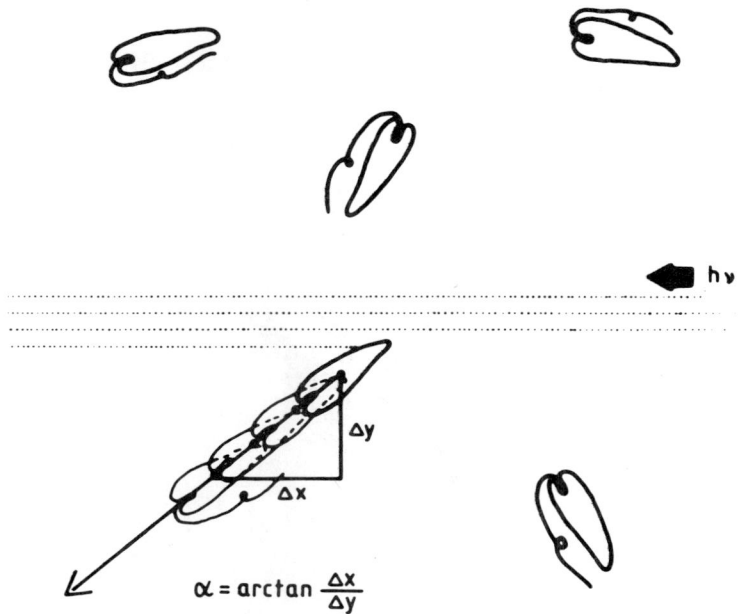

$$\alpha = \arctan \frac{\Delta x}{\Delta y}$$

Figure 4. Schematic diagram of the procedure to find and outline the position of an organism in the field of view and to follow its movement vector.

337

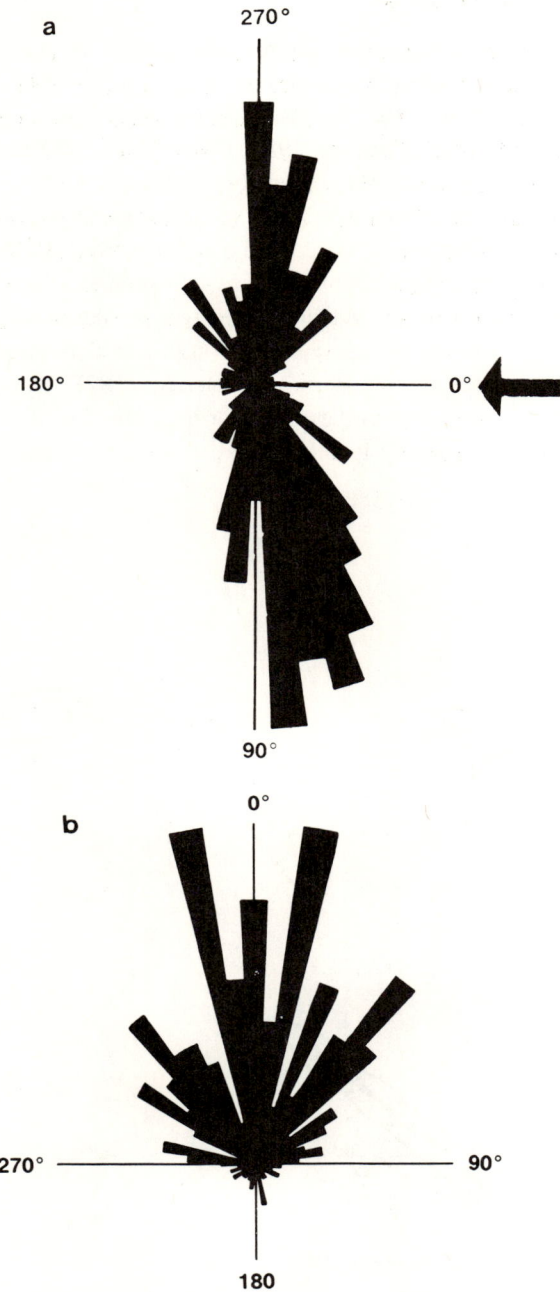

Figure 5. Circular histograms representing the orientation of *Cryptomonas* in (a) light and (b) gravity (after Rhiel et al., 1988)

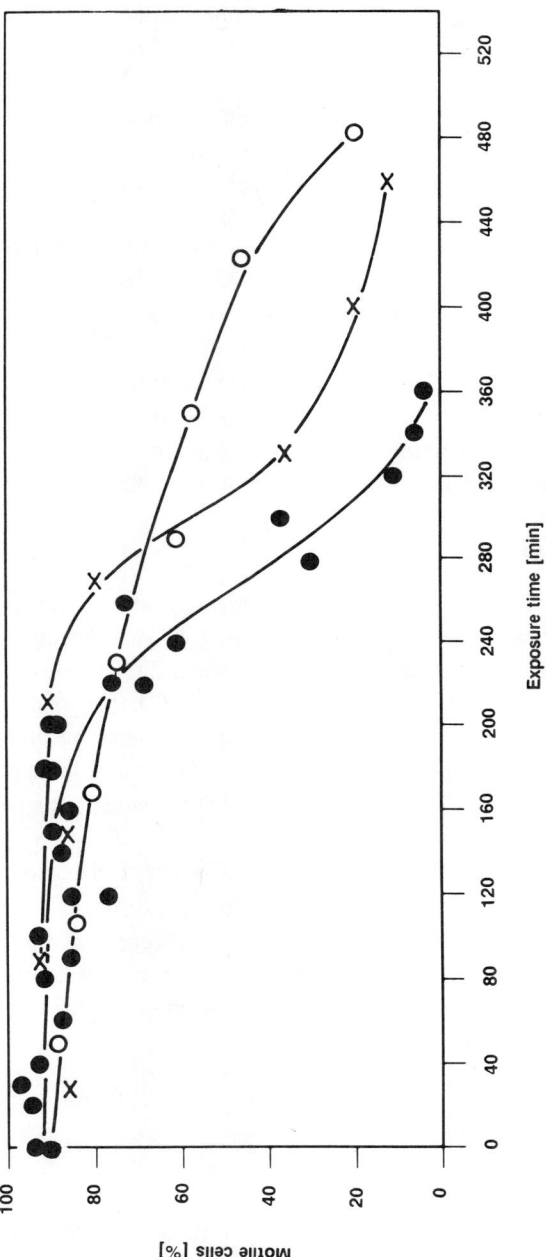

Figure 6. Percentage of motile *Euglena gracilis* after solar irradiation due to unfiltered solar radiation (filled circles), radiation filtered through a UV-B cut-off filter WG 320 (crosses) and radiation filtered by an ozone layer (open circles) (after Häder and Häder, 1988a).

predefined number of frames the movement vector is calculated from the first and final centroid position (Häder and Lebert, 1985).

The hardware consists of an infrared dark field microscope, on the stage of which the organisms move in a cuvette. The image is taken by a CCD camera and digitized in a Matrox board (Quebec, Canada), with a spatial resolution of 512×512 pixels with 256 possible gray levels each, plugged into an AT compatible microcomputer. The digitized image is displayed on an analogue monitor in pseudocolor graphics.

Angular deviations from a light direction are calculated from the displacements in x- and y-directions and stored in a disk file. These raw data are used by subsequent statistical techniques to construct circular histograms (Fig. 5) (Häder and Lipson, 1986), to calculate the degree of orientation using the Rayleigh test (Batschelet, 1981; Mardia, 1972) and to perform a Fast Fourier Analysis (Häder and Lipson, 1986).

The movement vectors can also be used to determine the percentage of motile organisms in a population (Fig. 6) defined by a minimal distance a cell must move to be regarded as motile (Häder and Häder, 1988a, b). Similarly, the individual speed of movement can be determined for the organisms in a population and broken down in the form of histograms which are finally plotted on a laser printer at a resolution of 300 dots per inch.

Analysis of electrophoresis gels

The hardware described above can easily be adapted to scan single- (Ford-Holevinski et al., 1983) or two-dimensional electrophoresis gels. Like in a densitometer the stained bands or autoradiographs (Elder et al., 1986) are scanned. The gray levels along the path are digitized and stored in memory (Häder and Truß, 1987). The spatial resolution can easily be increased by using a macro objective. In order to enhance accuracy, several scans can be averaged. The result can be shown graphically or used to calculate the areas under the peaks. Pattern recognition is used to analyze DNA sequencing gels (Mount and Conrad, 1986; Elder et al., 1986).

An attractive application is the use of image analysis on DNA gels stained with the fluorescent dye ethidium bromide. There is no need to take a photograph of the fluorescent gel for scanning but the gel is placed directly over a 302 nm transilluminator. The fluorescence intensity is digitized as the gel is scanned. Likewise, fluorescent, protein-specific stains have been used in thin-layer chromatography and analyzed using quantitative image analysis (Rees et al., 1985).

Acknowledgement

The compilation of this review was funded by a grant from the Bundesminister für Forschung und Technologie under contract number KBF 57. The author thanks H. Vieten, U. Neiß and M. Rudyk for skillful technical assistance.

References

Allen R D and Allen N S 1983 Video-enhanced microscopy with a computer frame memory; *J. Microsc.* 129 3-17

Batschelet E 1981 Circular Statistics in Biology, Academic Press, London.

Berns G S and Berns M W 1982 Computer-based tracking of living cells; *Exp. Cell Res.* 142 103-109

Blake A 1987 Real-time seeing machines? *Nature* 328 759-760

Bryan S R, Woodward W S, Griffis D P and Linton R W 1985 A microcomputer based digital imaging system for ion microanalysis; *J. Microscopy* 138 15-28

Burton J L, Law P and Bank H L 1986 Video analysis of chemotactic locomotion of stored human polymorphonuclear leukocytes; *Cell Motility Cytoskeleton* 6 485-491

Caldwell D E 1985 New developments in computer-enhanced microscopy (CEM); *J. Microbiol. Meth.* 4 117-125

Castillo X, Yorkgitis D and Preston K Jr 1982 A study of multidimensional multicolor images; *IEEE Trans. Biomed. Eng.* 29 111-120

Coates T D, Harman J T and McGuire W A 1985 A microcomputer-based program for video analysis of chemotaxis under agarose; *Computer Methods Programs Biomed.* 21 195-202

Desai V, Reimer L 1985 Digital image recording and processing using an Apple II microcomputer; *Scanning* 7 185-197

Dusenbery D B 1985a Video camera-computer tracking of nematode *Caenorhabditis elegans* to record behavioral responses; *J. Chem. Ecol.* 11 1239-1247

Dusenbery D B 1985b Using a microcomputer and videocamera to simultaneously track 25 animals; *Comput. Biol. Med.* 15 169-175

Elder J K, Green D K and Southern E M 1986 Automatic reading of DNA sequencing gel autoradiographs using a large format digital scanner; *Nucleic Acids Res.* 14 417-424

Erhardt R, Reinhardt E R, Schlipf W and Bloss W H 1980 FAZYTAN A system for fast automated cell segmentation, cell image analysis and feature extraction based on TV-image pickup and parallel processing; *Anal. Quant. Cytol. J.* 2 25-40

Ford-Holevinski T S, Agranoff B W and Radin N S 1983 An inexpensive, microcomputer-based, video densitometer for quantitating thin-layer chromatographic spots; *Analyt. Biochem.* 132 132-136

Grant G and Reid A F 1981 An efficient algorithm for boundary tracing and feature extraction; *Computer Graph. Imag. Process* 17 225-237

Gronsky R 1988 Spectroscopic information from high resolution images; *Ultramicroscopy* 24 155 -168

Gunzer U, Aus H M and Harms H 1987 Letter to the editor; *J. Histochem. Cytochem.* 35 705-706

Häder D-P 1986 Effects of solar and artificial UV irradiation on motility and phototaxis in the flagellate, *Euglena gracilis*; *Photochem. Photobiol.* 44 651 - 656

Häder D-P 1987a Automatic area calculation by microcomputer-controlled video analysis; *EDV in Med. Biol.* 18 33-36

Häder D-P 1987b Polarotaxis, gravitaxis and vertical phototaxis in the green flagellate, *Euglena gracilis*; *Arch. Microbiol.* 147 179-183

Häder D-P and Griebenow K 1987 Versatile digital image analysis by microcomputer to count microorganisms; *EDV in Med. Biol.* 18 37-42

Häder D-P and Griebenow K 1988 Orientation of the green flagellate, *Euglena gracilis*, in a vertical column of water; *FEMS Microbiol. Ecol.* 53 159-167

Häder D-P and Häder M 1988a Ultraviolet-B inhibition of motility in green and dark bleached *Euglena gracilis*; *Current Microbiol.* 17 215-220

Häder D-P and Häder M 1988b Inhibition of motility and phototaxis in the green flagellate, *Euglena gracilis*, by UV-B radiation; *Arch. Microbiol.* 150 20-25.

Häder D-P and Lebert M 1985 Real time computer-controlled tracking of motile microorganisms; *Photochem. Photobiol.* 42 509-514

Häder D-P and Lipson E 1986 Fourier analysis of angular distributions for motile microorganisms; *Photochem. Photobiol.* 44 657-663

Häder D-P and Truß M 1987 High resolution scanning of absorbing and fluorescent electrophoresis gels using video image analysis; *CABIOS* 3 339-343

Häder D-P, Lebert M and DiLena M R 1986 New Evidence for the mechanism of phototactic orientation of *Euglena gracilis*; *Curr. Microbiol.* 14 157-163

Häder D-P, Rhiel E and Wehrmeyer W 1987 Phototaxis in the marine flagellate *Cryptomonas maculata*; *J. Photochem. Photobiol. B* 1 115-122

Häder D-P, Rhiel E and Wehrmeyer W 1988 Ecological consequences of photomovement and photobleaching in the marine flagellate *Cryptomonas maculata*; *FEMS Microbiol. Ecol.* 53 9-18

Hainfeld J F, Wall J S and Desmond E J 1982 A small computer system for micrograph analysis; *Ultramicroscopy* 8 263-270

Jimenez J, Santisteban A, Carazo J M and Carrascosa J L 1986 Computer graphic display method for visualizing three-dimensional biological structures; *Science* 232 1113-1115

Julez B and Harmon L D 1984 Noise and recognizability of coarse quantized images; *Nature* 308 211-212

Kaufman A G, Nathwani B N and Preston K Jr 1987 Subclassification of follicular lymphomas by computerized microscopy; *Human Pathol.* 18 226-231

Kim N H, Wysocki A B, Bovik A C and Diller K R 1987 A microcomputer-based vision system for area measurement; *Comput. Biol. Med.* 17 173-183

Kokubo Y and Hardy W H 1982 Digital image processing: a path to better pictures; *Ultramicroscopy* 8 277-286

Koss L G, Sherman A B and Adams S E 1983 The use of hierarchic classification in the image analysis of a complex cell population. Experience with the sediment of voided urine; *Anal. Quant. Cytol.* 5 159-166

Mardia K V 1972 Statistics of Directional Data. Acad Press, London

McMillan P J, Yakush A, Frykman G, Nava P B and Ras V R 1987 Minima equalization: a useful strategy in automatic processing of microscopic images; *J. Microscopy* 148 253-262

Mikolajczyk E, Häder D-P and Nultsch W 1985 Photodynamically induced chemoresponses of the colorless flagellate, *Astasia longa*, in the presence of riboflavin; *Arch. Microbiol.* 142 397-402.

Mikolajczyk E, Nultsch W and Häder D-P 1986 Chemoaccumulation of the colorless flagellate, *Astasia longa*, in the presence of the photosensitizer methylene blue; *Acta Protozool.* 25 179-186.

Mount D W and Conrad B 1986 Improved programs for DNA and protein sequence analysis on the IBM personal computer and other standard computer systems; *Nucleic Acids Res.* 14 443-454

Nierzwicki-Bauer S A, Balkwill D L and Stevens S E Jr 1983 Three-dimensional ultrastructure of a unicellular cyanobacterium; *J. Cell Biol.* 97 713-722

Pavlides T 1982 Algorithms for graphics and image processing, Springer-Verlag, Berlin-Heidelberg

Poggio T 1984 Vision by man and machine; *Scient. Amer.* 250 68-78 Apr.

Preston K Jr 1983 Gray level image processing by cellular logic transforms; *IEEE transactions on pattern analysis and machine intelligence.* Vol. Pami-5 55-58

Preston K Jr and Dekker A 1980 Differentiation of cells in abnormal human liver tissue by computer image processing; *Anal. Quant. Cytol.J.* 2 1-14

Rees D D, Fogarty K E, Levy L K and Fay F S 1985 Computerized analysis of TV images for ultrasensitive monitoring of the reaction of fluorochrome with protein; *Analyt. Biochem.* 144 461-468

Rikmenspoel R and Isles C A 1985 Digitized precision measurements of the movement of sea urchin sperm flagella; *Biophys.J.* 47 395-410

Russ J C and Russ J C 1987 Automatic discrimination of features in grey-scale images; *J. Microscopy* 148 263-277

Serra J 1980 Digitalization; *Mikroskopie* 37 Suppl. 109-118

Shipton H W 1979 The microprocessor, a new tool for the biosciences; *Ann. Rev. Biophys. Bioeng.* 8 269-286

Spring K R 1982 Application of video to light microscopy p 15-20, In: *Membrane Biophysics II, physical methods in the study of Epithelia*, Alan R. Liss, New York

Spudich J 1985 Color-sensing by phototactic *Halobacterium halobium* p 113-118, In: *Sensory Perception and Transduction in Aneural Organisms* (eds) G Colombetti, F Lenci and P-S Song, Plenum Press, New York, London

Steinbach T, Unland F and Müller K-M 1982 Kostengünstiges Mikroprozessorsystem zur Ergänzung eines Quantimet 720-Bildanalysegerätes; *Microsc. Acta* 86 139-145

Thurston G, Jaggi B and Palcic B 1986 Cell motility measurements with an automated microscope system; *Exp. Cell Res.* 165 380-390

Wittekind C and Schulte E 1987 Computerized morphometric image analysis of cytologic nuclear parameters in breast cancer; *Anal. Quant. Cytol. Histol.* 9 480-484

Wollmer W 1987 Application of a small microcomputer to cell image analysis; *Anal. Quant. Cytol. Histol.* 9 535

Degradation of Oligonucleotides and DNA by VUV Radiation in Solids

TAKASHI ITO

Institute of Physics
College of Arts and Sciences
University of Tokyo
Meguroku, Komaba 3-8-1
Tokyo 153, Japan

Introduction

Development of synchrotron radiation (SR) technique for photobiology was a challenging attempt, since SR could cover a wide wavelength region between ultraviolet (< 190 nm) and soft X-ray region (> 0.1 nm) that have been left almost untouched. The undertaking requires a large-scale instrumentation involving an electron storage ring and a vacuum beamline with appropriate monochromators. Among the wide range of wavelength, VUV region was in great demand. There are two potentially important subjects, namely, the energy absorption processes by molecules in solid state and their chemical consequences. In this article I shall review our photobiological research with DNA and related compounds on the latter subject.

VUV Radiation Source

Since 1980 we have been using the electron storage ring operated at 0.4 GeV (Synchrotron Radiation Laboratory, University of Tokyo) as a radiation source. Figure 1 shows the irradiation system for VUV radiation in the region from 190 nm (6.5 eV) down to 50 nm (25 eV) (Hieda et al, 1986). At a reasonable wavelength width, the system provides a workable photon fluence (Figure 2) that enables us to analyze, by a standard technique, end-products of the photo-reaction induced in a relatively small molecule with the absorption cross-section of the order of 10^{-20} m^2. The irradiation of sample molecules was performed in the high vacuum in thin films because of the extremely high absorption coefficient in the VUV region.

VUV-Induced Higher Excited States and Consequences?

Unlike near- and far-UV (UV-A, B and C) radiations, VUV radiation generally

Photobiology, Edited by E. Riklis
Plenum Press, New York, 1991

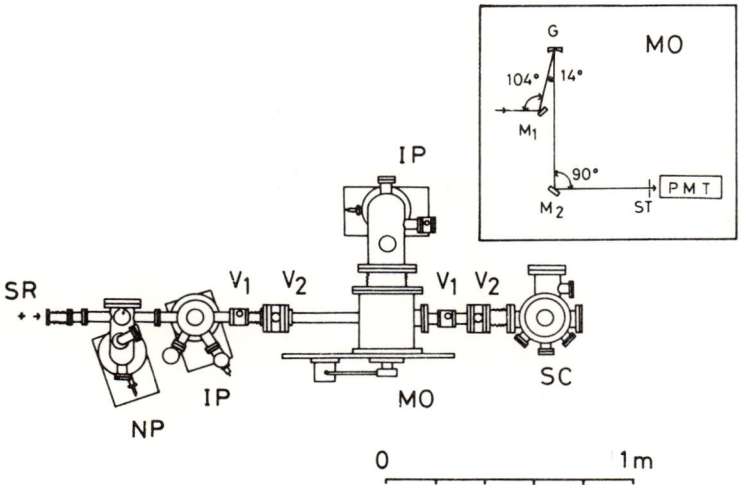

Figure 1. Schematic diagram of VUV irradiation system with a 2.2 m modified Wadsworth monochromator (inset). M_1 : collimating mirror, G: concave grating, M_2: deflecting mirror, ST: slit, PMT: photomultiplier tube, NP: noble pump, IP: ion pump, V_1 : fast closing valve, V_2 : gate valve, MO: monochromator, SC: sample chamber (reproduced from Hieda et al, 1986).

promotes the electrons to a higher excited level. With DNA or related compounds, for example, the first absorption band in the VUV region occurs around 190 nm which is attributable to the excitation of a π electron system of the bases. Then, other absorption bands occur in the shorter wavelength region. These absorptions are largely due to $\sigma-\sigma^*$ transitions in the sugar-phosphate moiety (Sontag and Weibezahn, 1975; Kiseleva et al, 1975; Ito and Ito, 1986). Without exception, the large molecule shows a large absorption around 60-80 nm (15-20 eV), part of it may be ascribed to a collective characteristic of electrons in the condensed matter (Inagaki et al, 1974; Arakawa et al, 1986).

It is noteworthy that the first ionization potential (Ip) of a large molecule is about 8 eV or 155 nm (Iwanami and Oda, 1983). The occurrence of the strong optical absorption in the region mentioned above suggests that electrically neutral excited states exist well above the ionization potential. Platzman (1962) coined "superexcited" state for such an excitation to be present as a consequence of absorption of radiation energy (E>Ip) from a charged particle. Direct evidence with optical absorption recently has begun obtaining with a simple organic molecule such as C_2H_6O by use of VUV radiation from SR (Hatano, 1987).

Since the nature of the superexcited states of biomolecules is not defined well, the relaxation processes are the matter of future research, especially in the amorphous solid state. Platzman suggested that the molecules at the superexcited state may well undergo fragmentation. It is not known whether peripheral hydrogen atoms are more likely

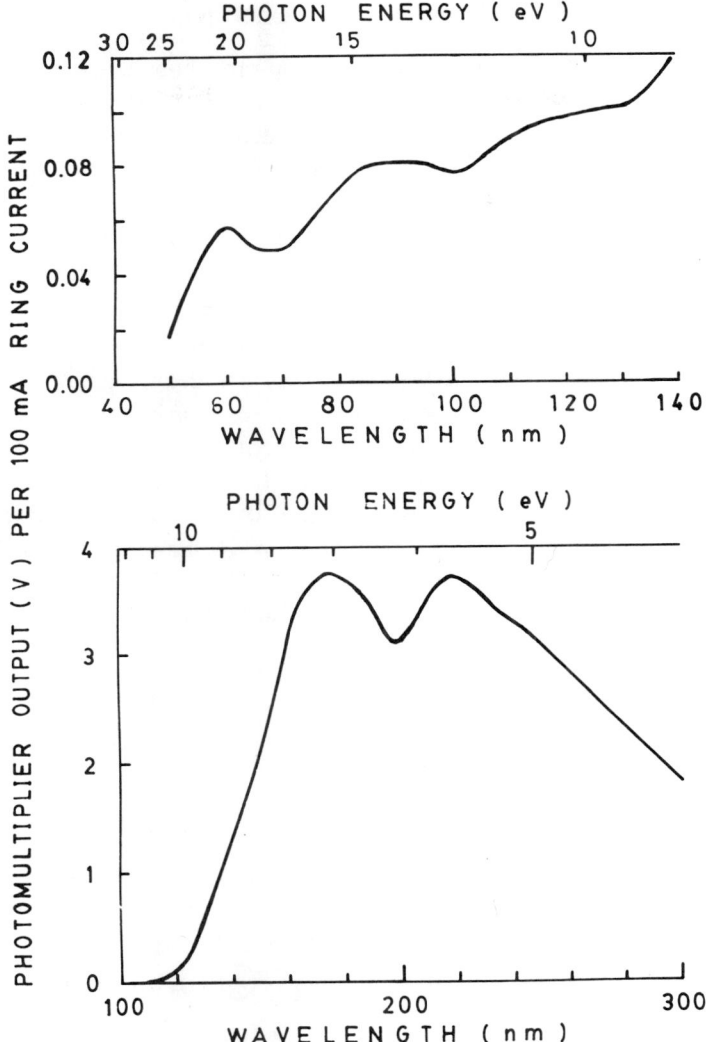

Figure 2. Output spectra of monochromatized SR in the energy region from 5-10 eV with Al-coated optical elements and 10-25 eV with Au-coated optical elements. The fluence rate at the photomultiplier output of 1 V is 2.7×10^{17} photons/m^{2}s (reproduced from Ito, 1988).

dispelled from the excited molecule or even direct rupture of C-C bond occurs. In any case, if such an event is found to occur specifically to the superexcitation, it not only reveals a new mode of molecular change induced by radiation, but also provides useful information on the relationship connecting radiation induced excited states with the associated molecular changes.

Endproducts of Irradiated Deoxyoligonucleotides

For the past few years we have attempted end-product analysis of VUV irradiated oligonucleotides in solids in order to obtain essential information on the possible molecular fragmentation from the superexcited states. The following are the results by the thin layer chromatography (TLC). A simple selectivity in the photodegradation was characteristic.

dApdA and dApdC

Photodegradation of dinucleoside monophosphate of adenine, dApdA (2'-deoxyadenylyl -(3'-5')-2'-deoxyadenosine), by VUV radiation occurs in a simple fashion, resulted in always adenine and 5'-dAMP (Scheme 1). The deoxysugar of adenylyl residue seem particularly vulnerable to VUV irradiation in solid state. The HPTLC profiles for such products were common in the energy range from 10.3 eV (124 nm) to 22.5 eV (80 nm) (Figure 3) (Ito et al, 1987). It was found that if deoxyadenosine residue is replaced with cytidine (dApdC in the same nomenclature) the degradation followed the same principle, namely, adenine and 5'-dCMP were found as the major products. However, interestingly, the cross-section of such a degradation decreased by half as compared with dApA (Ito and Saito, 1988). The cytosine residue seemed somehow to exert a stabilizing effect on the dApdC photodegradation.

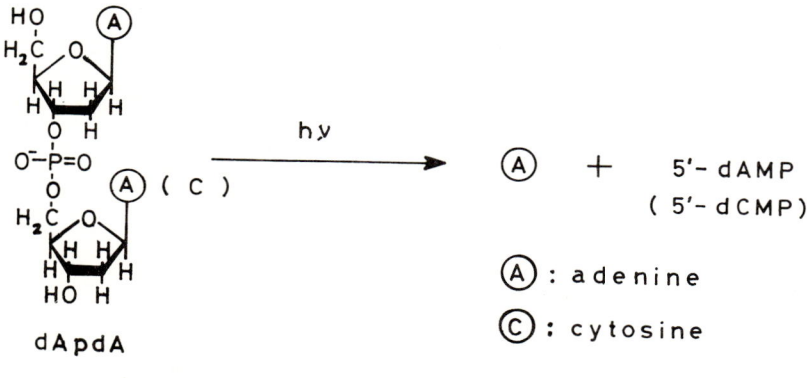

Scheme 1

dCpdA

The degradation of dCpdA by VUV irradiation was not so simple as seen in Scheme 1. Adenine and adenine- and cytosine-containing phosphate compounds (neither 5'-dAMP nor 3'-dCMP, however) were found in TLC as the product. A plausible interpretation, in considering the mode of Scheme 1 photofragmentation and also the unique effect of cytosine residue mentioned above, would be that the deoxypentose that has no phosphate

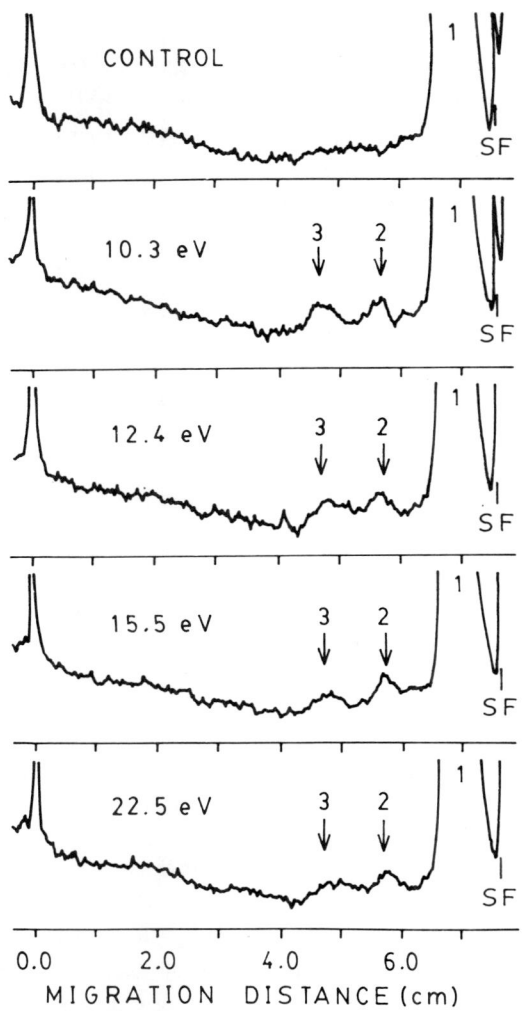

Figure 3. Chromatograms of VUV-irradiated dApdA at indicated photon energies. 1: dApdA, 2: adenine, 3: 5'-dAMP, SF: solvent front (redrawn from Ito et al, 1987).

group at the 5' position still exhibits a high sensitivity even with the suppressive cytosine attached to it, but, at the same time, the presence of adenine in the same molecule may have rendered the other deoxypentose vulnerable, resulting in the breakage of complex nature (Scheme 2). The limitation of the TLC technique did not permit us to detect more subtle changes such as base or base/sugar modifications which might have occurred as part of the fragmentation.

dTpdC and dTpdT

With these two compounds no VUV photoproducts were obtained under the same irradiation conditions. Use of more sensitive technique may well detect the release of bases or mononucleotides. However, it may be safe to say that these pyrimidine compounds have much lower sensitivity to the VUV photodegradation. It is tempted to believe that pyrimidines in general have a great stabilizing effect on the VUV photodegradation of oligonucleotides.

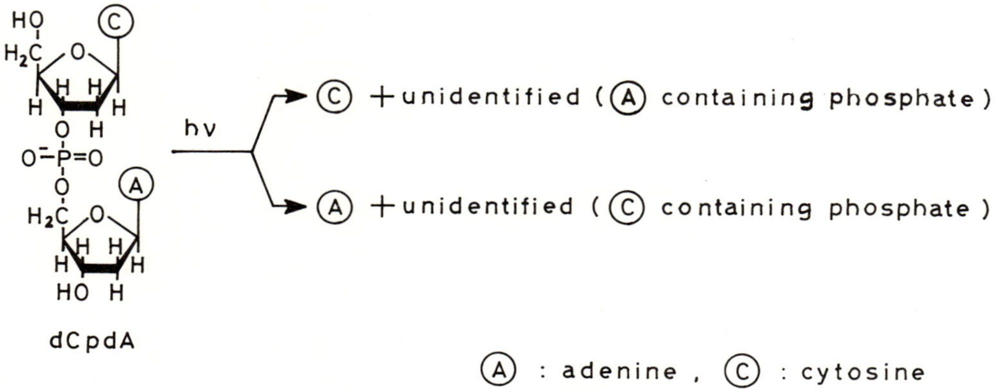

Scheme 2

Oligonucleotides with a longer chain

Recently, a series of compounds such as pdApdA, pdApdApdA and pdApdApdApdA were investigated to see how the chain length affects the mode of photofragmentation (Ito and Saito, unpublished). Figure 4 is an example of TLC profiles of VUV irradiated pdApdApdA. Adenine, 5'-dAMP and pdApdA may easily be identified. One or two other products can also be seen. A pattern of the same rule has been obtained for the longer molecules. The analysis of relative quantities of products may reveal a rule of the photodegradation in the VUV region.

Strand breaks of DNA

DNA strand breaks are induced by VUV radiation (Wirths and Jung, 1972; Hieda et al, 1986). With a plasmid DNA an action spectrum for single-strand break has been constructed in the region frrom UV-C to VUV. The cross-section for single-strand breaks spanned over 4 decades on the logarithmic scale (Hieda et al, 1986): the spectrum indicated no sign of similarity to the absorption of bases; rather it paralleled to that of a sugar-phosphate backbone. Such problems as the identification of endgroups of the broken site or the specificity for the induction as to the base or the base sequence remain to be investigated. It was noted that the nature of strand breaks induced at 160 nm

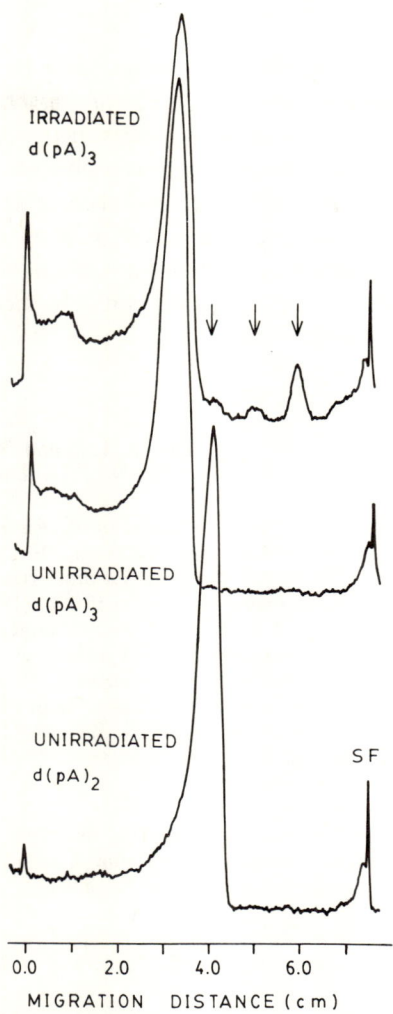

Figure 4. Chromatograms of VUV (7.59 eV or 163 nm) - irradiated pdApdApdA---d(pA)₃. Arrows indicated identified photoproducts, SF represents solvent front (reproduced from Ito, 1988).

resembled more those induced by a mild DNase treatment than X-rays with respect to the stimulative activity for poly (ADP-ribose) synthetase (Ito and Taniguchi, 1986). The release of bases from VUV-irradiated DNA was recently studied, although the quantitative relationship between the base release and the strand break was not established. The results, nevertheless, were not contradictory if the two go in parallel (Saito, unpublished).

351

Concluding remarks

We conclude that the degradation of oligonucleotides by VUV radiation may follow simple rules. The rules involve (1) the absence of phosphate group at 5' position of deoxypentose is an important factor in the sugar destruction, (2) each type of base exerts a different effect on the degradation, (3) the base sequence affects the stability of the molecule, and (4) the pyrimidine residue stabilizes the molecule. Photodegradation of tri- and tetra-oligonucleotides of adenine seems also to follow a systematic mode. We are currently working on the quantitative relationship between the products. We expect the strand break of a polynucleotide by VUV radiation is be seen in the direct extension of the behavior of these model compounds.

References

Arakawa, E.T., Emerson, L.C., Juan, S.I., Ashley, J.C. and Williams, M.W. (1986), The optical properties of adenine from 1.8 to 80 eV. *Photochemistry and Photobiology*, 44:349-353.

Hatano, Y. (1987), The chemistry of synchrotron radiation. *Radiation Research*, Proceedings of 8th International Congress of Radiation Research,, Vol. 2. ed by E.M. Fielden, J.F. Fowler, J.H. Hendry and D. Scott, pp.35-41, Taylor and Francis, London.

Hieda, K., Hayakawa, Y., Ito, A., Kobayashi, K., and Ito, T. (1986), Wavelength dependence of the formation of single-strand breaks and base changes in DNA by the ultraviolet radiation above 150 nm, *Photochemistry and Photobiology*, 44:379-383.

Hieda, K., Maezawa, H., Ito, A., Kobayashi, K., Furusawa, Y. and Ito, T. (1986), Choice of coatings for the optical elements in the irradiation system of vacuum-ultraviolet radiation above 50 nm. *Photochemistry and Photobiology*, 44:417-419.

Inagaki, T., Hamm, R.N., Arakawa, E.T., and Painter, L.R. (1974), Optical and dielectric properties of DNA in the extreme ultraviolet. *Journal of Chemical Physics*, 61:4246-4250.

Ito, A., and Ito, T. (1986), Absorption spectra of doxyribose, ribosephosphate, ATP and DNA by direct transmission measurements in the vacuum-UV (155-190 nm) and far-UV (190-260 nm) regions using synchrotron radiation as a light source. *Photochemistry and Photobiology*, 44:355-358.

Ito, T. (1989), Vacuum ultraviolet photobiology with synchrotron radiation. In: Synchrotron Radiation in Structural Biology. Ed. R.M. Sweet and A.D. Woodhead, pp. 221–241, Plenum Press, New York.

Ito, T., Saito, M. (1988), Degradation of oligonucleotides by vacuum-UV radiation in solid: Roles of the phosphate group and bases. *Photochemistry and Photobiology*, 48:567-572.

Ito, T., and Taniguchi, T. (1986), Enzymatic quantification of strand breaks of DNA induced by vacuum-UV radiation. *FEBS Letters*, 206, 151-153.

Ito, T., Taniguchi, T., Saito, M. (1987), A survey of photoproducts of an irradiated oligonucleotide by monochromatic photons with energy ranged from 6.5 to 22.5 eV. *Photochemistry and Photobiology*, 46:979-984.

Iwanami, S., and Oda, N. (1983), Photoabsorption and photoelectron yield spectra of polypeptides in the vacuum-UV region. *Radiation Research*, 95:24-31.

Kiseleva, M.M., Zarochensteva, Ye. P., and Dodonova, N.Ya.,(1975), Absorption spectra of nucleic acids and related compounds in the spectral region 120-180 nm. *Biophysics (Biofizika)*, 20:571-575.

Platzman, R.L. (1962), Superexcited states of molecules and the primary action of ionizing radiation. *The Vortex*, 23:372-385.

Sontag, W. and Weibezahn, K.F. (1975), Absorption of DNA in the region of vacuum-UV (3-25 eV). *Radiation and Environmental Biophysics*, 12:169-174.
Wirths, A. and Jung, H. (1972), Single-strand breaks induced in DNA by vacuum-ultraviolet radiation. *Photochemistry and Photobiology*, 15:325-330.

Photolysis of Water by VUV Radiation and Reactions with DNA and Related Compounds in Aqueous Systems

M. KUWABARA[a], A. MINEGISHI[b], K. TAKAKURA[c], K. HIEDA[d] AND
T. ITO[e]

[a]Faculty of Veterinary Medicine, Hokkaido University
Sapporo 060, Japan
[b]Tokyo Metropolitan Isotope Research Center
Tokyo 158, Japan
[c]College of Liberal Arts, International Christian University, Mitaka
Tokyo 171, Japan
[d]Faculty of Science, Rikkyo University
Tokyo 171, Japan
[d]College of Arts and Sciences, University of Tokyo
Tokyo 153, Japan

Abstract

The induction of OH and H radicals in vuv-exposed aqueous solutions at wavelengths from 150 nm to 200 nm was investigated by a spin trapping method using DMPO. The generation of OH and H radicals as a result of the absorption of vuv-light by H_2O was confirmed by measuring OH-DMPO and H-DMPO adducts with esr spectrometry.

Free radicals produced in aqueous uridine solution by vuv radiation at 163 nm were studied by a method combining esr-spin trapping and gel permeation chromatography as a model experiment to elucidate the chemical pathways accounting for damage formation in vuv-exposed DNA. Two base radicals were identified as precursors of the oxidative conversions of the base. Two sugar radicals were also identified as precursors to account for the strand break formation.

The induction of single-strand breaks by vuv radiation (140–200 nm) in plasmid pBR 322 DNA was studied by agarose gel electrophoresis. A good parallelism in the action spectra between the induction of the single-strand breaks and the generation of OH radicals below 170 nm was found, suggesting that photolysis of water was mainly responsible for the induction of the strand breaks.

Keywords: Vacuum UV, OH[*] generation, spin trapping, OH-induced radicals, strand breaks, pBR 322

Introduction

Since Ito and Kobayashi (1976) suggested the participation of OH radicals in the formation of DNA lesions in VUV-exposed yeast cells, our main object has been to elucidate the chemical processes from the absorption of light by H_2O to the production of

DNA damage. In the vacuum-uv region, water, which constitutes about 70% of living cells, has an absorption spectrum that increases sharply below around 180 nm as reported by Painter et al., (1969). Recently, we have constructed a new irradiation device with which aqueous biological materials can be irradiated by monochromatic synchrotron vacuum-uv radiation from 140 nm to far uv (Ito et al., 1984; Takakura et al., 1986). Using this system we have investigated, (1) the wavelength dependence of OH production was measured by esr-spin trapping with 5,5-dimethyl-1-pyrroline N-oxide (DMPO), (2) free radicals induced in a vuv-irradiated uridine by a method combining esr-spin trapping with 2-methyl-2-nitrosopropane (MNP) and gel chromatography, and (3) the wavelength dependence of the production of single-strand breaks in supercoiled DNA such as pBR 322 DNA as measured by agarose gel electrophoresis.

Materials and Methods

Chemicals

The spin traps, DMPO and MNP, were purchased from Aldrich Chemical Company. Uridine was acquired from Sigma Chemical Company. Bio-Gel P2 (200–400 mesh) was obtained from Bio-Rad Laboratories. Plasmid pBR 322 DNA was kindly donated by Dr. H. Yasuda of the University of Tokyo.

Irradiation technique (Ito et al., 1984; Takakura et al., 1986)

A cell made of polycrystal plates was specially designed for irradiation of aqueous biological materials and attached to a vacuum chamber installed at one of the beam ports of an electron storage ring at the Synchrotron Radiation Laboratory, University of Tokyo. A modified Wadsworth-type monochromator was used to get monochromatic vuv-light. The bandwidth was 6.4 nm. The aqueous cell was separated from the vacuum chamber by a CaF_2 or MgF_2 window with a thickness of 3 mm. The cell was 1.5 mm in thickness, with a 5×7 mm^2 irradiation area, and was used for all experiments under aerobic conditions. when the irradiation was carried out in the absence of O_2, another cell was designed to make continuous Ar-flow through the solution possible (Minegishi et al., 1986).

Spin trapping study of liquid water irradiated by vuv-light

Sixty μl of a solution containing a spin trap, DMPO, was placed in the cell for irradiation. The irradiation was made under aerobic or anaerobic conditions with constant stirring using a glass-coated iron rod. The anaerobic condition was achieved by Ar-flow. The concentration of DMPO was 2 to 120 mM. After irradiation, the solution was taken out of the irradiation cell and transferred to an aqueous flat cell for esr measurements. Esr measurements were performed by a JEOL JES-FE3X X-band spectrometer. In this method the DMPO converts the short-lived radicals (R·) into long-lived, esr-observable nitroxide radicals (the spin-adducts).

$$(2)$$

It has been proved that OH-DMPO adduct gives an esr spectrum consisting of a 1:2:2:1 quartet with 1.48 mT hyperfine splitting and the H-DMPO gives an esr spectrum consisting of a primary triplet of 1.62 mT and a secondary 1:2:1 triplet of 2.25 mT (Makino et al., 1983).

Spin trapping of radicals in vuv-exposed uridine and the separation of spin adducts by gel chromatography

Fifty mM of MNP and 20 mM of uridine were dissolved in aqueous solution. Vuv irradiation was carried out at 163 nm under open air in the manner described above. The solution was stirred during irradiation. In this case, MNP converts the short-lived radicals (R) into long-lived, esr-observable nitroxide radicals.

$$R^{\cdot} + tBu–N{=}0 \rightarrow tBu–N(0\ ) \rightarrow tBu–N^{+}(0^{-}) \tag{2}$$

Spin trapping usually gives several spin adducts corresponding to the number of radicals produced. Therefore, the irradiated solution was applied to a Bio-Gel P2 column (200–400 mesh, 1 cm i.d. and 90 cm long) to isolate the spin adducts. The elution was carried out with 1 mM phosphate buffer (pH 7.2) at a rate of 15 ml/h. One-ml portions were corrected and monitored for uv-absorbance at 260 nm. Each fraction was freeze-dried once, dissolved again in 200 µl of distilled water and examined be esr.

Measurements of strand-breaks of pBR 322 DNA by agarose gel electrophoresis (Takakura et al., 1986; Takakura et al., 1987)

A slab-type agarose gel of 0.7% (wt/vol) was used to measure the induction of strand-breaks in vuv-irradiated pBR 322 DNA. A buffer consisting of 40 mM Tris-acetic acid, 5 mM sodium acetate and 1 mM EDTA (pH 7.8) was used for electrophoresis. Following electrophoresis (1 V/cm, 18 h), the gel was stained with ethidium bromide (0.5 µg/ml in H_2O). The fluorescence from DNA bands was measured by a chromatoscanner (Shimazu Model CS-910). The mass of the supercoiled form I DNA was estimated by the integrated fluorescence of the specified band after correcting for reduction (factor 1.42) due to the topologically restricted uptake of ethidium bromide in form I DNA by comparison with that in form II (open circular) (Lloyd et al., 1978).

Results and Discussion

Spin trapping study of liquid water irradiated by vuv-light

When the solution containing 13.4 mM DMPO was exposed to vuv-light at 160 nm in the presence or absence of O_2, the ESR spectra in Fig. 1. were observed. The spectrum consists of a 1:2:2:1 quartet hyperfine structure and is typical for OH-DMPO adduct. Another spectrum consisting of a secondary triplet of 2.22 mT, which is typical for H-DMPO adduct, was also observed. Thus, we could get evidence that the absorption of vuv-light at 160 nm by H_2O clearly produced both OH and H radicals. Irradiation in the presence of O_2 resulted in the increase of OH-DMPO adduct and the decrease of H-DMPO adduct (the dotted line). This suggests that some H radicals react with O_2 to produce HO_2, which leads to the formation of HO_2-DMPO adduct. The transformation of HO_2-DMPO to OH-DMPO is a cause of the increase of OH-DMPO adduct.

The wavelength dependence of OH-DMPO adduct in the absence of oxygen is presented in Fig. 2A. The amount of OH-DMPO adduct decreases with increasing wavelengths. There seems to be a parallel relationship between the curve for OH-DMPO formation and the curve for the extinction coefficient of water in the vacuum-uv region as shown by the dotted line. However, the decrease of DMPO-OH adducts against the wavelength was not so rapid when compared to that of the extinction coefficient of water. This phenomenon was more pronounced when irradiation was carried out in the presence of oxygen (Fig. 2B). The amount of OH-DMPO adduct fluctuated around 175 nm. Taking the extinction coefficients of water, oxygen and DMPO, as well as wavelength-dependence of the penetration depth of light in water, into consideration, we theoretically calculated the concentration of OH-DMPO adduct (Table 1).

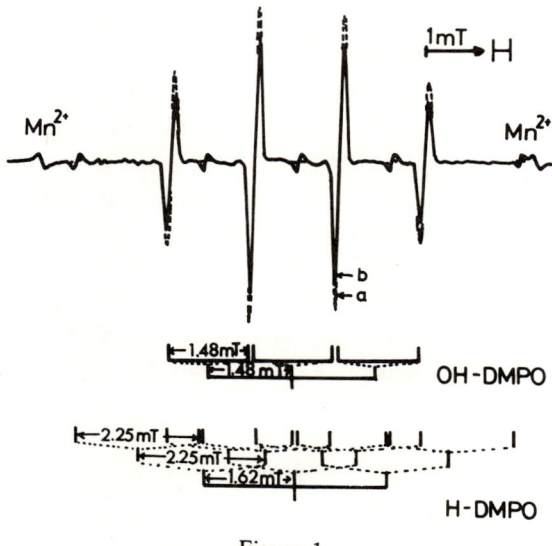

Figure 1.

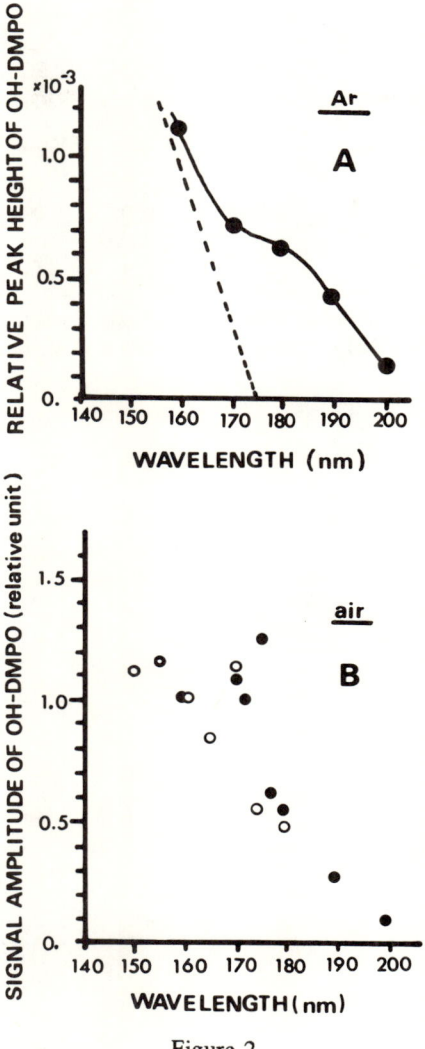

Figure 2.

Majority of OH radicals react with OH itself by OH-OH reaction and only minority of OH radicals can react with DMPO at 160 nm. As a result, about 7 mM of OH-DMPO adducts are calculated to be produced. On the other hand, the yield of OH radical generation was quite low at 180 nm, while the formation of 0.5 mM of excited DMPO molecules was estimated from the extinction coefficient of DMPO. If it can be assumed that the excited DMPO reacts with H_2O to produce OH-DMPO adduct, the concentration of OH-DMPO adduct becomes about 0.5 mM. This theoretical calculation suggests that the formation of OH-DMPO adduct at around 180 nm may be due to the direct absorption

359

Table 1.

Extinction Coefficient K (cm^{-1})

	150 nm	160 nm	170 nm	180 nm	190 nm	200 nm
H$_2$O (55M)	134,500	92,000	1,100	20	0.2	0.06
O$_2$ (0.3mM)	3.6	0.73	0.15	0.007	0.0004	–
DMPO (20mM)	27.5*			25*		

*Values roughly estimated by the extinction coefficients of TE buffer (Takakura *et al.*,1987) and PNDA (Takakura *et al.*, unpublished)

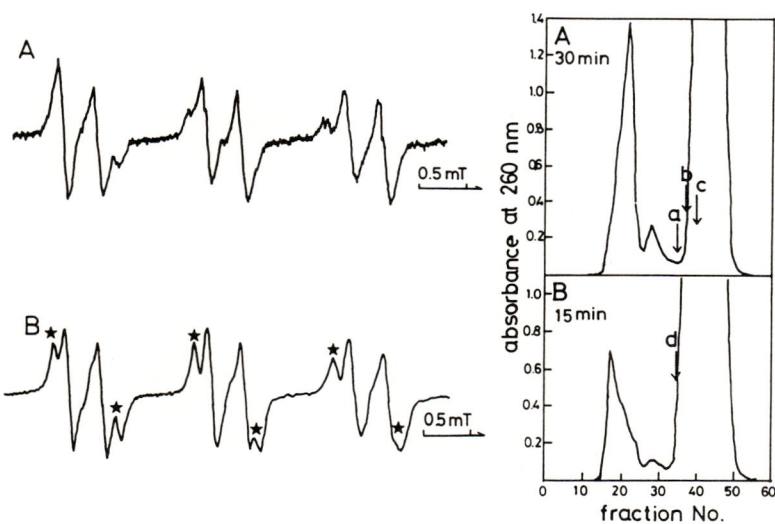

Figure 3.

of light by DMPO. Summarizing the above discussion, we can conclude that OH radicals are efficiently produced at wavelengths below 170 nm, and that the parallel relationship between the OH production and the extinction coefficient of water is maintained.

Spin trapping of the free radicals in aqueous uridine solution after exposure to vuv-light at 163 nm

Figs. 3A and B show esr spectra recorded after vuv irradiation for 30 min and for 15

360

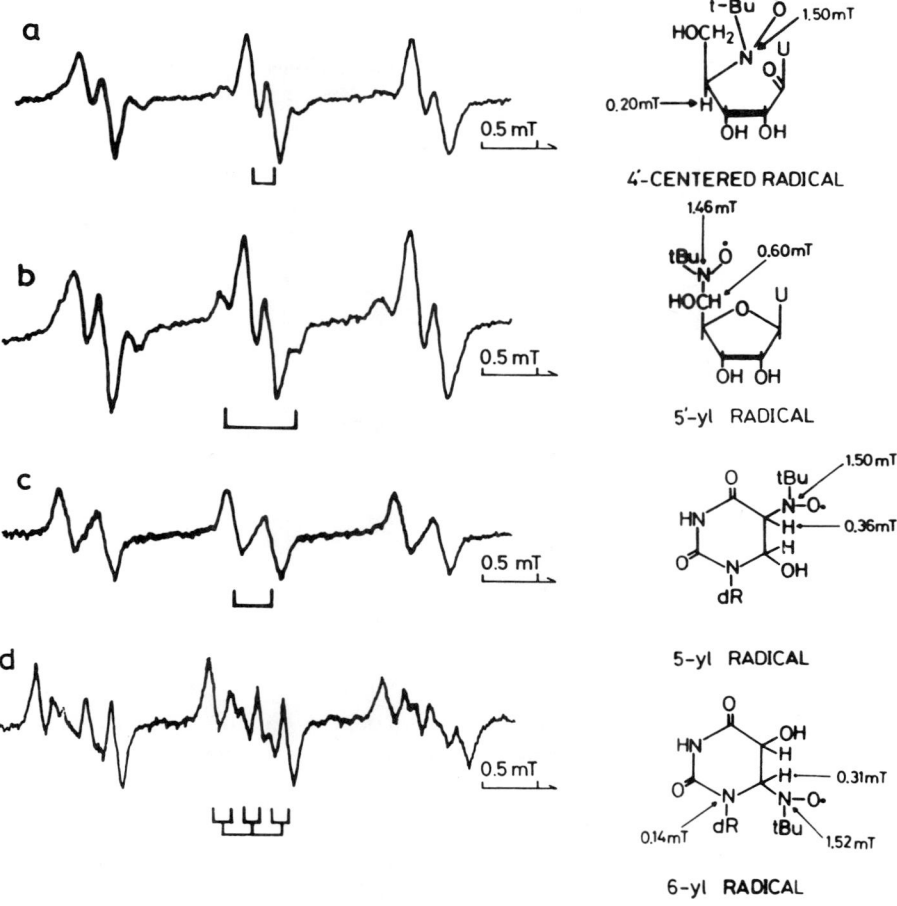

Figure 4.

min, respectively. An additional component marked with (*) was observed in the esr spectrum of Fig. 3B. This means that the esr spectrum consists of the overlapping of some signals; in other words, several spin-adducts were formed. The gel permeation chromatography was first applied to the longer-irradiated sample because this contained fewer spin-adducts than the shorter-irradiated sample did. The elution chromatogram obtained by monitoring uv-absorbance at 260 nm is shown in the right-hand side of Fig. 3A. When each fraction was examined by esr, signals were observed from three fractions, denoted a–c. The esr spectra from fractions a, b and c are presented in Figs. 4a, b and c, respectively. These esr spectra consist of 3×2 lines. Spectra a, b and c were assigned to the spin-adducts between the C4'-radical at the sugar moiety and MNP, between the

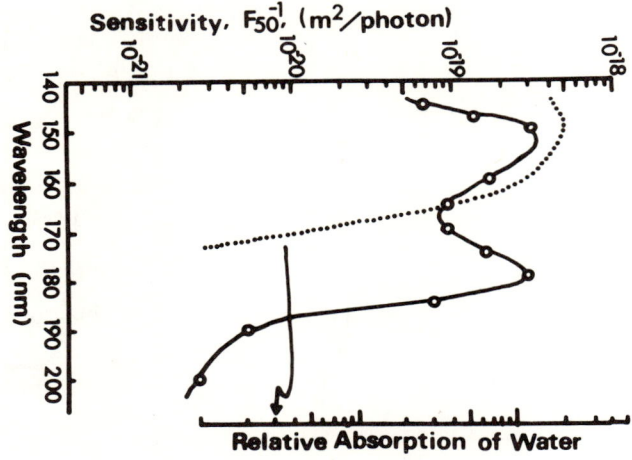

Figure 5.

C5′-radical at the sugar moiety and MNP, and between the C5-radical at the base moiety and MNP, respectively (Inanami et al., 1987). These spin-adducts are presented in the right-hand side of Fig. 4.

The gel chromatogram from the shorter-irradiated sample is presented in the right-hand side of Fig. 3B. In addition to three spin-adducts, one more adduct was newly recovered in fraction d. The spectrum due to fraction d is presented in Fig. 4d. This esr spectrum, consisting of secondary 3×2 lines, was assigned to the spin-adduct between the C6-radical of the base moiety and MNP.

The C4′- and C5′-radicals, which are induced at the sugar moiety by an H-abstraction with OH radicals, can be regarded as precursors leading to the strand break formation in poly(U) (von Sonntag, 1980), and also seem to be relevant to DNA. The C5- and C6-radicals, which are induced by OH-addition to the C5–C6 double bond of the base moiety, can be regarded as precursors of the oxidative conversions of the bases (Téoule and Cadet, 1978). Observation of these radicals gives evidence that the precursors leading to the strand breaks and to the base alteration are induced by OH-radical reactions.

The induction of single-strand breaks of plasmid pBR 322 DNA was measured by monitoring the decrease in the amount of form I DNA and the concomitant rise of form II DNA. The ratio of the mass of form I to the total mass of DNA was designated as the remaining fraction of form I, and plotted against vuv-light fluence for 145 nm to 200 nm.

The F_{50}, that is, the fluence required to give 50% of the remaining fraction, was then determined graphically for each wavelength. The reciprocal of F_{50}, the sensitivity for the induction of single-strand breaks, was plotted as a function of the wavelength as shown in Fig. 5.

362

Two maximums are present, one centered around 150 nm and the other at 180 nm. Clearly, the former maximum corresponds to that of the absorption spectrum of water. Furthermore, the addition of OH scavenger protects against the induction of strand breaks around at 150 nm and not at around 180 nm (Takakura et al., 1987). Although the maximum at 180 nm remains unexplained, these results indicate that the induction of strand breaks at around 150 nm can be explained by the action of OH radicals on DNA.

The present study proved that (1) OH radicals are generated by absorption of vuv-light by water, (2) the OH radicals produce free radicals in uridine which account for the chemical processes leading to DNA damage, and (3) the OH radicals participate in the production of single-strand breaks of pBR 322 DNA. The present study strongly suggests that the OH radical plays an essential role in the effects of vuv-radiation on biological systems.

References

Inanami, O., Kuwabara, M., and Sato, F. (1987). OH-induced free radicals in 3'-UMP and poly(U): Spin-trapping and radical chromatography. *Radiat. Res.* 112, 36–44.

Ito, T., and Kobayashi, K. (1976). Induction of lethal and genetic damage by vacuum-ultraviolet (163 nm) irradiation of aqueous suspensions of yeast cells. *Radiat. Res.* 68, 275–283.

Ito, T., Kada, T., Okada, S., Hieda, K., Kobayashi, K., Maezawa, H., and Ito, A. (1984). Synchrotron system for monochromatic uv radiation (>140 nm) of biological material. *Radiat. Res.* 98, 65–73.

Lloyd, R. S., Haidle, C.W., and Robberson, D.L. (1978). Bleomycin-specific fragmentation of double-stranded DNA. *Biochemistry* 17, 1980–1986.

Makino, K., Mossoba, M.M., and Riesz, P. (1983). Chemical effects of ultrasound on aqueous solutions. Formation of hydroxyl radicals and hydrogen atoms. *J. Phys. Chem.* 87, 1369–1377.

Minegishi, A., Kuwabara, M., Hieda, K., and Ito, T. (1986). VUV-induced OH-DMPO spin adduct in aqueous DMPO solution using 3 mm cell and continuous Ar-flow system. In: Activity Report of Synchrotron Radiation Laboratory, pp. 64–65, ISSP, The University of Tokyo.

Painter, L.R., Birkoff, R.D., and Arakawa, E.T. (1969). Optical measurements of liquid water in the vacuum ultraviolet. *J. Chem. Phys.* 51, 243.

Takakura, K., Ishikawa, M., Hieda, K., Kobayashi, K., Ito, A., and Ito, T. (1986). Single-strand breaks in supercoiled DNA induced by vacuum-uv radiation in aqueous solution. *Photochem. Photobiol.* 44, 397–400.

Takakura, K., Ishikawa, M., and Ito, T. (1987). Action spectrum for the induction of single-strand breaks in DNA in buffered aqueous solution in the wavelength range from 150 to 272 nm. *Int. J. Radiat. Biol.* 52, 667–675.

Téoule, R., and Cadet, J. (1978). Radiation-induced degradation of the base component in DNA and related substances. In: Effects of Ionizing Radiation on DNA (Ed. by J. Hüttermann et al.,), pp. 171–202. Springer-Verlag.

von Sonntag, C. (1980). Free radical reactions of carbohydrates as studied by radiation techniques. *Adv. Carbohydr. Chem. Biochem.* 37, 7–77.

Light Regulated Biological Processes

Photosynthesis

Trapping of Excitation Energy in Photosynthetic Purple Bacteria

R. VAN GRONDELLE[1], H. BERGSTRÖM[2] AND V. SUNDSTRÖM[2]

[1]Department of Biophysics,
Physics Laboratory,
Free University of Amsterdam,
1081 HV Amsterdam, The Netherlands [1]
and [2]Department of Physical Chemistry,
University of Umeå,
S 90187 Umeå, Sweden

Abstract

The transfer and trapping of excitation energy in a variety of photosynthetic bacteria was investigated using steady-state and time-resolved polarized light spectroscopy. The experiments at room temperature show that equilibration of the excitation density among different antenna pools is the dominant process before trapping or losses occur. Only at extremely high light-intensities effects due to multiphoton processes are observed, which indicate that in many of these bacterial systems the excitons can diffuse over a 1000 pigments or more before their disappearance.

The main antenna pigment-protein complex, B875, is not homogeneous, but contains a minor spectral form, B896, absorbing between 890 and 900 nm dependent on the temperature. B896 is the terminal excitation acceptor, before transfer to the reaction center occurs. At 77 K transfer from B875 to B896 and from B896 to the reaction center was studied. The former takes 20 ps, while the latter transfer step occurs within about 40 ps. This implies that B896 is situated at a distance of 2.5–3.0 nm from the special pair.

The function of B896 is probably to act as a focus for the excitations and to provide a specific entry into the reaction center. This may be a general feature in photosythetic systems.

Introduction

Photosynthesis requires the efficient collection of solar energy absorbed by pigment molecules. To this end the photosynthetic reaction centers are coupled to a large antenna of protein-bound pigment molecules, that absorb the light and transfer the excited state energy to the reaction center with efficiency of more than 90% (for reviews: see van Grondelle 1985; Holzwarth 1986; Geacintov and Breton 1987; van Grondelle and Sundström 1988; Scheer 1982; Glazer 1984). Many organisms show a variety of chemically identical, but spectroscopically different antenna molecules. All evidence suggests that these are arranged such that the pigments with a relatively high excited state energy are found at the periphery of the photosynthetic system, while those with a relatively low excited state energy are located close to the reaction centers. This architecture is found in the phycobilisome antenna, the light-harvesting antennas of Photosystem 1 and 2, and the antenna of the photosynthetic purple and green bacteria. As an example we shall discuss here our results obtained with the purple bacterium *Rhodobacter (Rb.) sphaeroides*, which contains the so-called LH2 antenna with major absorption bands at 800 and 850 nm, and the LH1 antenna with a single major absorption band at 875 nm.

Not so long ago it was generally believed that the long- wavelength antenna, LH1 in our example, provided a large network of identical antenna pigments containing at least several hundreds or maybe even close to 1000 Bchl molecules, efficiently connected via energy transfer and thus serving at least a few reaction centers (den Hollander et al. 1983; Bakker et al. 1983). The long-wavelength antenna molecules were assumed to be more or less degenerate with the excited state energy of the special pair of the reaction center, which allowed the excitation to enter and leave the reaction center at least several times before charge separation took place. This was the basis for the famous trap-limited trapping concept (Knox 1977; Pearlstein 1982) in contrast to diffusion-limited trapping.

X-ray diffraction of crystals of the photosynthetic reaction center of purple bacteria has provided us with a high resolution structure of this pigment-protein complex (Deisenhofer et al. 1984, 1985; Allen et al. 1987; Yeates et al. 1987). From the structure it is clear that the minimum distance between the special pair and an antenna complex is probably larger than 2.5 nm and even more when measured along the long axis of the protein. This implies that the rate of excitation transfer from this antenna pigment to the special pair is in the 10–100 ps range. If the excited state of all the antenna pigments and the special pair is truly degenerate, than the 10-100 ps time-constant implies an effective trapping rate of more than 400 ps or a quantum efficiency of 60% or less. This is clearly in contrast to the observations and implies that the light-harvesting system of purple bacteria is not homogeneous, but, on the contrary, must contain special pigments close to the reaction center, where the excitation energy is effectively localized before trapping takes place.

Several other groups have suggested that the antenna of purple bacteria, green plants and algae is not homogeneous (Borisov et al. 1982; Kramer et al. 1984; Vos et al. 1986; Garab and Breton 1976; Thornber 1986; Wittmershaus et al. 1985). For the purple bacteria evidence has been accumulated from a variety of spectroscopic techniques, that a special long-wavelength antenna exists (Borisov et al. 1982; Kramer et al. 1984; Vos et al. 1986), called B986, and this evidence will be reviewed below. For Photosystem 1 fluorescence measurements have indicated a strong heterogeneity of the light- harvesting antenna, which manifests itself most clearly in low temperature emission spectra, both steady-state and time-resolved (Thornber 1986, Wittmershaus et al. 1985). Models have been proposed for an arrangement of the various emitting species around the Photosystem 1 reaction center (Garnier et al. 1986). For Photosystem 2 it has been known for quite some time that a relatively red-shifted (695nm) emission occurs at low temperature with an anomalous polarization (Garab and Breton 1976). A few years ago it was suggested that this emission may originate from the reaction center pheophytin, but more recently it was shown that the 695 nm emission is associated to the CP47 antenna complex (Van Dorssen et al. 1987).

Materials and Methods

Chromatophores of *Rb. sphaeroides* were prepared from light grown cells as described elsewhere (Kingma 1983). Picosecond absorption kinetics were measured as described in (Åkesson et al. 1985). The state of open traps was simulated by fully reducing the Quinone acceptors with sodium dithionite and cooling to 77 K. Under these conditions the fluorescence yield is identical to that measured with open traps (van Grondelle et al. 1978), and charge separation results in the reaction center triplet state with high efficiency. The fast cycling of the triplet state allows the measurement of picosecond absorpton kinetics with 'open' traps at low temperatures, where a fast renewal of the illuminated material is impossible.

Results and Discussion

We shall start this section with a review of the older data concerning the existence of a minor antenna species absorbing to the red of the main long-wavelength absorption band (870–880 nm) of purple bacteria. Bolt et al. (1981) reported that in the isolated LH1 antenna complex of *Rb. sphaeroides* the polarization of the LH1 emission rose from a value of about 0.1 upon excitation in the peak to a value of close to 0.3 upon excitation in the red wing of the 870–880 nm absorption band. The low polarization of 0.1 is generally ascribed to energy transfer among similar chromophores leading to a loss of polarization in the plane of the membrane. Bolt et al. ascribed the polarization increase to exciton effects.

At about the same time Borisov and coworkers in the Soviet Union measured time-resolved absorption difference spectra and observed that upon excited state formation the maximum bleaching of the LH1 absorption band did not coincide with the absorption maximum, but was clearly red shifted (Borisov et al. 1982). Although part of the phenomenon was due to the fact that bleaching of the absorption of one BChl was always coupled to a blue shift of the absorption of at least 6 others (Nuijs et al. 1985), the tentative interpretation by Borisov et al. turned out correct, namely that a relatively high fraction of the excited state resided on a red-shifted chromophore.

Kramer et al. (1984) extended the original measurements by Bolt et al. and showed that at very low temperatures (4K) in all species of purple bacteria that were examined the fluorescence polarization increased to a value above 0.4 upon excitation in the red edge of the B875 absorption band. The simplest interpretation of their experiments was that the long-wavelength band was not homogeneous, but consisted of a major component at about 884 nm (at 4K) and of a minor component at about 896 nm. The minor component carried about 10–15% of the total absorption. The polarization of the emission upon excitation of the major component was low due to extensive energy transfer among identical molecules before the 896 pigment was reached. The high polarization upon red edge excitation suggested either a high order of the $BChl_{896}$ pigments or an isolated pigment molecule. Of course in the latter case the $BChl_{896}$ can still be connected to the reaction center if present, the emission of the latter is not observed. At room temperature in intact membranes all these polarization phenomena are not observed, mainly because thermal equilibration between major and minor pigment molecules occurs long before fluorescence takes place.

We have extensively studied the excited state dynamics in membranes and isolated pigment-protein complexes of *Rb. sphaeroides* at room temperature (Sündström et al. 1986; Bergström et al. 1986; van Grondelle et al. 1987; Bergström et al. 1988a,b: Gillbro et al. 1988). All the experiments show that the transfer of excitation energy among identical and different pigments is much faster then the time-resolution of our equipment (~1ps) and individual transfer rates may easily be of the order of 100fs. As an example we show the kinetics observed upon excitation and detection at 800nm in two different LH-2 complexes of *Rps. acidophila*: the 800/850 and the 800/820 complex (Angerhofer et al. 1986). In both complexes at room temperature the initial downward bleaching due to the formation of excited $BChl_{800}$ is almost absent, in agreement with extremely fast transfer to $BChl_{850}$ or $BChl_{820}$. The large absorption increase at later times is due to the formation of excited $BChl_{850}/BChl_{820}$. The rapid energy transfer among identical $BChl_{850}/BChl_{820}$'s is observed as an initial depolarization of the corresponding absorption changes directly after excitation (not shown). From Fig. 1 it is also clear that lowering the temperature to 77 K leads to the appearance of the $BChl_{800}$ bleaching. For the 800/850 complex the lifetime is now in the order of 1–2 ps. For the 800/820 complex the transfer time stays below 1 ps, probably due to the much better spectral overlap between $BChl_{800}$ emission and $BChl_{820}$ absorption.

At room temperature transfer between different pigment pools was dominated by equilibration of the excitation density (Sündström et al. 1986). In *Rb. sphaeroides* the absorption measurements gave no indication for the existence of a special long-wavelength component and the excited state kinetics mainly reflected equilibration between LH1, LH2 and reaction centers. In *Rs. rubrum* with traps closed, the biphasic decay of the absorption changes indicated the equilibration between $BChl_{875}$ and $BChl_{896}$ and this suggested a 2:1 equilibrium distribution. Both in *Rb. sphaeroides* and in *Rs. rubrum* the room temperature absorption changes with open traps reflected the net trapping process, with a characteristic time-constant of about 60 ps superimposed on the excitation equilibration.

Strong support for the involvement of a long-wavelength antenna component mediating energy transfer from the bulk antenna to the reaction center was obtained from low temperature picosecond absorption measurements (van Grondelle et al. 1987; Bergström et al. 1988a). In Fig. 2 we show the kinetics measured in *Rb. sphaeroides* chromatophores at 77 K under conditions that all the reaction centers were in the state with the BChl-dimer (P_{870}) oxidised. This state was simply achieved by not flushing the

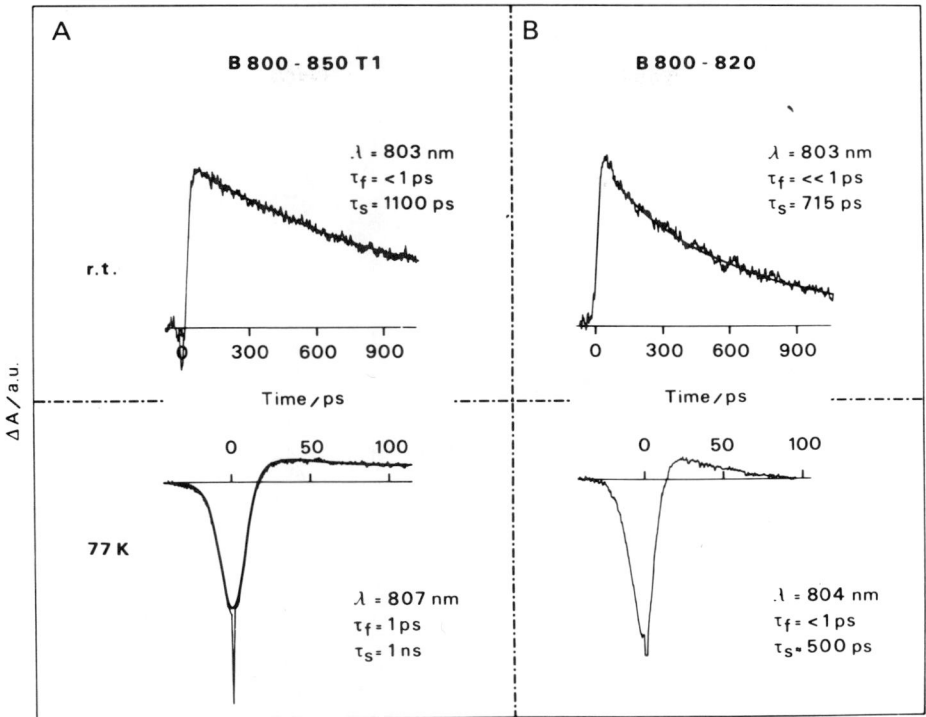

Figure 1. Isotropic absorption recovery kinetics aroung 800 nm, measured at 77 K and room temperature. (A) Isolated B800–850 type I of *Rps. acidophila*. (B) Isolated B800-820 complex of *Rps. acidophila*. For further details see ref. 32, from which this figure was taken.

sample and not tempering the intensity of the laser flashes. The integrated light intensity was more than sufficient to maintain the state P_{870}^+. The kinetics at 800 nm due to the $BChl_{800}$ excited state decay with a time-constant of about 1–2 ps, similar to what is observed in the LH2 complex of the same organism (Bergström et al. 1988b). Between 800 and 850 nm the $BChl_{850}$ excited state shows an absorption increase, above 855 nm a bleaching. The characteristic decay time is about 40–50 ps with no indication of a much faster component. Around 870 nm the signals are complicated and probably involve the absorption changes from a third LH2 pigment, $BChl_{870}$, recently identified in a mutant of *Rb. sphaeroides* (van Dorssen et al. 1988). The lifetime of this BChl870 excited state would be in the 10 ps range. Around 880 to 890 nm characteristic absorption changes can be observed due to the formation of the $BChl_{875}$ excited state. The decay is fast, between 20–30 ps, indicating that in this system the $BChl_{875}$ is not the terminal acceptor of the excitation energy. Measurements above 900 nm show the long-lived bleaching of

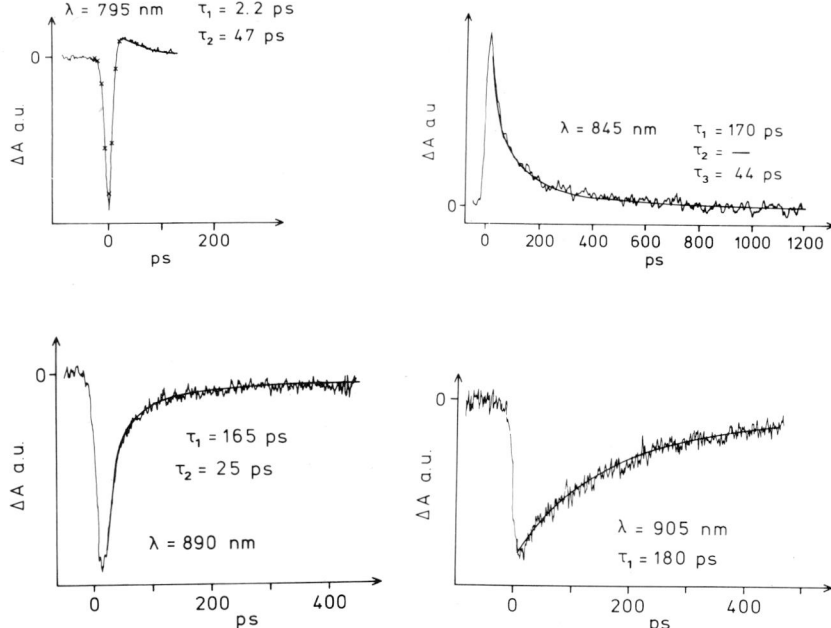

Figure 2. Isotropic absorption recovery kinetics of *Rb. sphaeroides* chromatophores at 77 K with closed reaction centers. A positive ΔA corresponds to an increased absorption relative to the initial state. Fitted kinetics are displayed with a full line or crosses, whichever stands out most clearly against the experimental curve. Maximum absolute bleaching (negative ΔA) of the kinetic curves are about $0.5 \cdot 10^{-3}$ at the peak of the absorption bands and correspondingly smaller at the other wavelengths.

$BChl_{896}$, indicating that this compound is indeed the final acceptor in the energy transfer chain. The $BChl_{896}$ excited state has a characteristic decaytime of about 150–200 ps, which is still much shorter than the lifetime observed in an isolated LH1 complex (Bergström et al. 1988a; Sebban et al. 1985), or in only LH1-containing mutants (Sebban et al. 1984) at 77 K. Also the steady-state fluorescence increase upon the formation of the state P_{870}^+ is only a factor 3–4 (van Grondelle et al. 1978), in principle corresponding rather well with the 150–200 ps lifetime. Probably, the reaction center in the state with P_{870}^+ remains a rather efficient quencher for the excitations in the antenna.

Polarisation measurements on the same system (van Grondelle et al. 1987, Bergström et al. 1988b) showed that the $BChl_{800}$ changes were highly polarised, which is not in agreement with the model originally proposed by Kramer et al. (1984). It seems that the $BChl_{800}$ pigments represent monomeric BChl molecules or dimers, but in the latter case with only a relatively small angle between the participating Q_y-transition moments. The $BChl_{850}$ and $BChl_{875}$ absorption changes were only weakly polarised at the start of the measurement and the anisotropy did not show a significant decay over the time of the measurement. This shows that already in the smallest unit rapid energy transfer has led to a depolarised excited state and further transfer does not change the polarisation any more. The smallest unit is therefore organised in such a way that the Qy- transition moments of BChl$_{850}$ in LH2 and of BChl$_{875}$ in LH1 have equal probability of being found along two orthogonal axes. In view of linear dichroism experiments these must both be parallel to the plane of the membrane. From our results it is hard to decide whether or not the excited state in such a minimal unit is a true exciton-state or a more localized Förster-state. At very early times after excitation we observe around the isosbestic point rapidly decaying, highly polarised absorption changes, which may reflect the disappearance of the initially created exciton state (van Grondelle et al. 1987, Bergström et al. 1988a,b). These phenomena are currently being investigated using spectroscopy with better time-resolution. High polarisations are observed above 900 nm, in agreement with Kramer et al. (1984). In our experiments the maximum initial anisotropy was about 0.25 and the anisotropy did not show a significant decay on the timescale of the experiment (1 ns). Compared to Kramer et al. the anisotropy is not so high but this must be due to better overlap between the bands at 77 K.

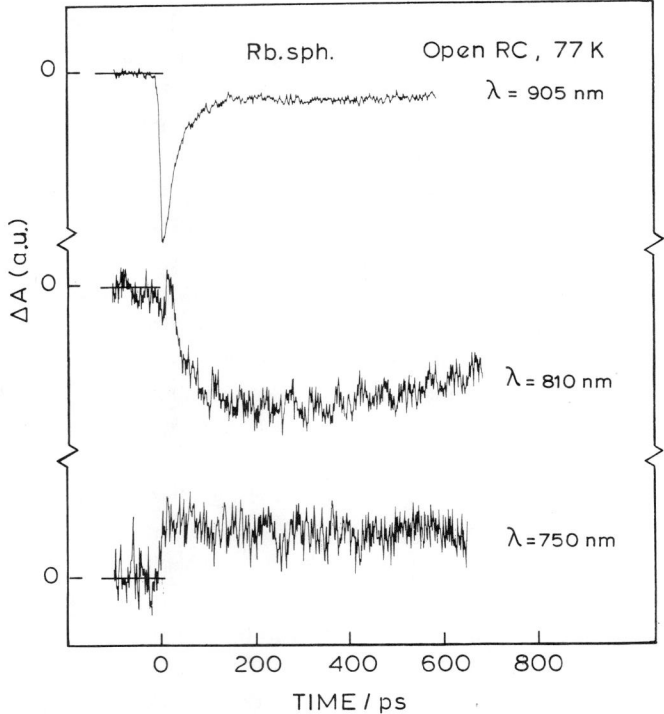

Figure 3. Isotropic absorption recovery kinetics of *Rb. sphaeroides* chromatophores at 77 K with open reaction centers, i.e. in the state with Q_A reduced with sodium dithionite. Note the long-lived bleaching at 810 nm and 905 nm and the absorption increase at 790 nm, consistent with the formation of $P_{870}^+ B_{pheo}^-$ (see for further details ref. 47).

Finally, we show the measurement at 77 K with open reaction centers in Fig. 3. It can be observed that the $BChl_{875}$ excited state decay is not affected by the state of the reaction centers and remains 20–30 ps. Only a stationary bleaching remains after the fast decay and this is due to the radical pair $P_{870}^{+}B_{pheo}^{-}$ that is formed after the excitation flash. In this particular experiment we could only maintain the reaction centers in the open state after the reduction of Q_A with sodium-dithionite. Illumination will than produce a reaction center triplet state, which even at 77 K will decay rapidly to the ground state, thus avoiding the accumulation of any photoproduct. Already more than 10 years ago it was shown that under these conditions at 77 K the fluorescence yield of *Rb. sphaeroides* membranes is identical to that of membranes with 'truly' open reaction centers (van Grondelle et al. 1978). The kinetics above 900 nm are strongly affected by the state of the reaction centers. The $BChl_{896}$ decay takes now about 30–40 ps, which most probably reflects direct transfer of the excitation to the special pair. Considering that this represents a single transfer step, 30–40 ps is a rather long time.

If the Förster R_0 is calculated using a 77 K reaction center absorption spectrum (Shuvalov et al. 1986) and a 77 K emission spectrum from a only LH1 containing mutant (Sebban et al. 1984) we calculate $R_0 = 9.7$–10.7 nm. This implies that R is at most 3.5–3.9 nm. Evidently, the calculated overlap is too large, because we have used the total emission spectrum, while we should have used only the $BChl_{896}$ contribution. Taking the same deconvolution as used by Shiwada et al. (1989) for the 77 K emission spectrum, we calculate $R_0=7.3$–8.6 nm, which implies that R=2.6–3.1 nm. In this calculation various values for the orientation parameter were used.

These measurements offer strong support for the existence of a specific entry for excitations arriving in the antenna into the reaction center. This entry consists of only a few BChl molecules per reaction center and these are identified as $BChl_{896}$. At 77 K the excited state energy is below that of the bulk LH1 antenna and close to degenerate with the excited state energy of the reaction center special pair. The relative energies at room temperature are not known, but it seems likely that the $BChl_{896}$ is still below the bulk antenna. At room temperature the main bleaching of the reaction center seems to coincide with the main LH1 absorption band, suggesting that transfer to the reaction center is an energetically uphill process, although this remains to be established. At very low temperatures in some species of purple bacteria (Rijgersberg et al. 1980), but also in Photosystem 2 of higher plants (Rijgersberg et al. 1979), the fluorescence with active reaction centers increases dramatically. At least in one specific case it has been shown that this is associated with a decrease of the quantum efficiency of trapping (Rijgersberg et al. 1980). As the actual rate of charge separation in the reaction center itself does not show a strong temperature dependence, the rate even speeds up somewhat upon cooling (Breton et al. 1988), this effect must be located in the antenna. Although it is clear that also the Förster rate of energy transfer among identical molecules is temperature dependent (Rijgersberg et al. 1980), a more plausible explanation at this moment seems to us that in some species the final transfer step from $BChl_{896}$ to the special pair becomes rate limiting. In those organisms where the entry-pigment is at lower energy than the special pair at low temperatures, the trapping will be inhibited and the fluorescence will increase.

About the significance of the arrangement in which red-shifted pigments provide the entry of excitations into the reaction center, we can only speculate. Photosynthetic antennas display an enormous variation, but they share the property that high-energy pigments are always found at the periphery, while low-energy pigments are located at the core. In the case of the phycobilisome, in which many molecular details of the energy transfer process are understood (Sauer et al. 1987; Sauer and Scheer 1988), this principle is clearly demonstrated. Even special proteins are applied in that case, so-called linker polypeptides, that serve both a structural role and at the same time fine-tune the spectra of some of the chromophores (Scheer 1982; Glazer 1984). Therefore, focussing the excitation energy at those points where an essential step has to be made to the next unit seems a universal feature and in that sense it is not surprising that we observe the phenomenon in the purple bacteria. More specifically, the role of focussing may be that in case the transfer process to the special pair is intrinsically a slow process, the system cannot allow the excitation density to be distributed equally over all antenna molecules. In that case the probability would be too low to find the excitation on a chromophore neighbouring the reaction center and the net energy trapping process would have a very low rate. In these circumstances focussing can help to speed up the trapping process (van Grondelle et al. 1988). If this is true we would expect that besides a slow forward rate into the reaction center also the back transfer rate for excitations leaving the reaction center is slow, i.e. in the 10–100 ps range. Direct excitation of the reaction center pigments will always give rise to a charge separation (3 ps vs. 10–100 ps) and the fluorescence yield upon reaction center excitation will be rather low. This is exactly what was observed already many years ago (Wang and Clayton, 1971) and remained unexplained for quite some time, especially in those days when it was believed that excitation transfer into and out of the reaction center were relatively fast processes (see for instance the discussion in Van Grondelle, 1985). In our current view the process of energy transfer is neither diffusion-limited, nor trap-limited, but some strange mixture of both concepts.

Acknowledgements

This research was supported by the Dutch Foudation of Biophysics, the Swedish Natural Research Council, Knut and Alice Wallenberg, the Magnus Bergwall Foundation and grant SC1* 0004-C(EDB) from the European Community.

References

Åkesson, E., Sundström, V., and Gillbro, T., 1985, Solvent dependent barrier heights of excited-state photoisomerization reactions. Chemical Physics Letters, 121, 513–522.

Allen, J. P., Feher, G., Yeates, T. O., Komiya, H., and Rees, D. C., 1987, Structure of the reaction center from _Rhodobacter sphaeroides_ R-26: the protein subunits. Proceedings of the National Academy of Sciences USA, 84, 6162–6166.

375

Angerhofer, A., Cogdell, R. J., and Hipkins, M. F., 1986, A spectral characterization of the light-harvesting pigment-protein complexes from *Rhodopseudomonas acidophila*. Biochimica et Biophysica Acta, 848, 333–341.

Bakker, J. G. C., Grondelle R. van, and Den Hollander, W. Th. F., 1983, Trapping, loss and annihilation of excitations in a photosynthetic system II. Experiments with the purple bacteria *Rhodospirillum rubrum* and *Rhodopseudomonas capsulata*. Biochimica et Biophysica Acta, 725, 508–518.

Bergström, H., Sundström, V., Grondelle, R. van, Åkesson, E., and Gillbro, T., 1986, Energy transfer within the isolated light-harvesting B800-850 pigment-protein complex of *Rhodobacter sphaeroides*. Biochimica et Biophysica Acta, 852, 279–287.

Bergström, H., Westerhuis, W. H. J., Sundström, V., Grondelle, R. van, Niederman, R. A., and Gillbro, T., 1988, Energy transfer within the isolated B_{875} light-harvesting pigment-protein complex of *Rhodobacter sphaeroides* at 77 K studied by picosecond absorption spectroscopy. FEBS Letters, 233, 12–16.

Bergström, H., Sundström, V., Grondelle, R. van, Gillbro, T., and Cogdell, R. J., 1988, Energy transfer dynamics of isolated B800-850 and B800-820 pigment-protein complexes of *Rhodobacter sphaeroides* and *Rhodopseudomonas acidophila*. Biochimica et Biophysica Acta, 936, 90–98.

Bergström, H., Grondelle R. van, and Sundström, V., 1989, Characterization of excitation energy trapping in photosynthetic purple bacteria at 77 K. FEBS Letters, submitted for publication.

Bolt, J. D., Hunter, C. N., Niederman, R. A., and Sauer, K., 1981, Linear and circular dichroism and fluorescence polarization of the B875 light-harvesting bacteriochlorophyll-protein complex from *Rhodopseudomonas sphaeroides*. Photochemistry and Photobiology, 34, 653–656.

Borisov, A. Yu., Gadonas, R. A., Danielius, R. V., Piskarskas, A. S., and Razjivin, A. P., 1982, Minor component B-905 of light-harvesting antenna in *Rhodospirillum rubrum* chromatophores and the mechanism of singlet-singlet annihilation as studied by difference selective picosecond spectroscopy. FEBS Letters, 138, 25–28.

Breton, J., Martin, J. -L., Fleming, G. R., and Lambry, J. -C., 1988, Low-temperature femtosecond spectroscopy of the initial step of electron transfer in reaction centers from photosynthetic purple bacteria. Biochemistry, 27, 8276–8284.

Deisenhofer, J., Epp, O., Miki, K., Huber, R., and Michel, H., 1984, X-ray structure analysis of a membrane protein complex. Electron density map at 3 Å resolution and a model of the chromophores of the photosynthetic reaction center from *Rhodopseudomonas viridis*. Journal of Molecular Biology, 180, 385–398.

Deisenhofer, J., Epp, O., Miki, K., Huber, R., and Michel, H., 1985, Structure of the protein subunits in the photosynthetic reaction centre of *Rhodopseudomonas viridis*. Nature, 318, 618–624.

Den Hollander, W. Th. F., Bakker, J. G. C., and Grondelle, R. van, 1983, Trapping, loss and annihilation of excitations in a photosynthetic system I. Theoretical aspects. Biochimica et Biophysica Acta, 492–507.

Garab, J., and Breton, J., 1976, Polarized light spectroscopy on oriented spinach chloroplasts and fluorescence emission at low temperatures. Biochemical and Biophysical Research Communications, 71, 1095–1102.

Garnier, J., Maroc, J., and Guyon, D., 1986, Low-temperature fluorescence emission spectra and chlorophyll-protein complexes in mutants of *Chlamydomonas reinhardtii*: evidence for a new chlorophyll a-protein complex related to Photosystem I. Biochimica et Biophysica Acta, 851, 395–406.

Geacintov, N. E., and Breton, J., 1987, Energy transfer and fluorescence mechanisms in photosynthetic membranes. CRC Critical Reviews in Plant Science, 5, 1–44.

Gillbro, T., Cogdell, R. J., and Sundström, V., 1988, Energy transfer from carotenoid to bacteriochlorophyll a in the B800-820 antenna complexes from *Rhodopseudomonas acidophila* strain 7050. FEBS Letters, 235, 169–172.

Holzwarth, A. R., 1986, Fluorescence lifetimes in photosynthetic systems. Photochemistry and Photobiology, 43, 707–725.

Kingma, H., 1983, Energy transfer and mechanism of carotenoid triplet formation. Effects of magnetic field. Doctoral thesis, State University of Leiden, The Netherlands.

Knox, R. S., 1977, Photosynthetic efficiency and exciton transfer and trapping. Topics in Photosynthesis, Vol. 2, edited by J. Barber, Elsevier (Amsterdam), 55–97.

Kramer, H. J. M., Pennoyer, J. D., Van Grondelle, R., Westerhuis, W. H. J., Niederman, R. A., and Amesz, J., 1984. Low temperature optical properties and pigment organization of the B875 light-harvesting bacteriochlorophyll-protein complex of purple photosynthetic bacteria. Biochimica et Biophysica Acta, 767, 335–344.

Nuijs, A. M., Van Grondelle, R., Joppe, H. L. P., Van Bochove, A. C., and Duysens, L. N. M., 1985, Singlet and triplet excited carotenoid and antenna bacteriochlorophyll of the photosynthetic purple bacterium *Rhodospirillum rubrum* as studied by picosecond absorbance difference spectroscopy. Biochimica et Biophysica Acta, 810, 94–105.

Pearlstein, R. M., 1982, Chlorophyll singlet excitons. Photosynthesis: Energy Conversion by Plants and Bacteria, Vol. 1, edited by Govindjee, Academic Press (New York), 292–331.

Rijgersberg, C. P., Amesz, J., Thielen, A. P. G. M., and Swager, J.A., 1979, Fluorescence emission spectra of chloroplasts and subchloroplast preparations at low temperature. Biochimica et Biophysica Acta, 545, 473–482.

Rijgersberg, C. P., Van Grondelle, R., and Amesz, J., 1980, Energy transfer and bacteriochlorophyll fluorescence in purple bacteria at low temperature. Biochimica et Biophysica Acta, 592, 53–64.

Sauer, K., Scheer, H., and Sauer, P., 1987, Förster transfer calculations based on crystal structure data from Agmenellum quadruplication C-phycocyanin. Photochemistry and Photobiology, 46, 427–440.

Sauer, K., and Scheer, H., 1988, Excitation transfer in C-phycocyanin. Förster transfer rate and exciton calculations based on new crystal structure data for C-phycocyanins from *Agmenellum quadruplicatum* and Mastigocladus laminosus. Biochimica et Biophysica Acta, 936, 157–170.

Scheer, H., 1982, Phycobiliproteins: molecular aspects of a photosynthetic antenna system. Light Reaction Path of Photosynthesis, edited by F.K. Fong, Springer-Verlag (Berlin), 7–45.

Sebban, P., Jolchine , G., and Moya, I., 1984, Spectra of fluorescence lifetime and intensity of *Rhodopseudomonas sphaeroides* at room and low temperature. Comparison between the wild type, th C71 reaction center-less mutant and the B800-850 pigment-protein complex. Photochemistry and Photobiology, 39, 247–253.

Sebban, P., Robert, B., and Jolchine, G., 1985, Isolation and spectroscopic characterization of the B875 antenna complex of a mutant of *Rhodopseudomonas sphaeroides*. Photochemistry and Photobiology, 42, 573–578.

Shimada, K., Mimuro, M., Tamai, N., and Yamazaki, I., 1989, Excitation energy transfer in *Rhodobacter sphaeroides* analyzed by the time-resolved fluorescence spectrum. Biochimica et Biophysica Acta, in press.

Shuralov, V. A., Shkuropatov, A. Ya., Kulakova, S. M., Ismailov, M. A., and Shkuropatova, V. A., 1986, Photoreactions of bacteriopheophytins and bacteriochlorophylls in reaction centers of *Rhodopseudomonas sphaeroides* and *Chloroflexus aurantiacus*. Biochimica et Biophysica Acta, 849, 337–346.

Sundström, V., Van Grondelle, R., Bergström, H., Åkesson, E., and Gillbro, T., 1986, Excitation-energy transport in the bacteriochlorophyll antenna systems of *Rhodospirillum rubrum* and *Rhodobacter sphaeroides*, studied by low-intensity picosecond absorption spectroscopy. Biochimica et Biophysica Acta, 851, 431–446.

Thornber, J. P., 1986, Biochemical characterization and structure of pigment-proteins of photosynthetic organisms. Encyclopedia of Plant Physiology (new series), Vol. 19, Photosynthesis III, edited of L.A. Staehelin and C.J. Arntzen: Springer-Verlag (Berlin, Heidelberg), 98–142.

Van Dorssen, R. J., Breton, J., Plijter, J. J., Satoh, K., Van Gorkom, H. J., and Amesz, J., 1987, Spectroscopic properties of the reaction center and of the 47 kDa chlorophyll protein of Photosystem II. Biochimica et Biophysica Acta, 893, 267–274.

Van Dorssen, R. J., Hunter, C. N., Van Grondelle, R., Korenhof, A. H., and Amesz, J., 1988, Spectroscopic properties of antenna complexes of *Rhodobacter sphaeroides in vivo.* Biochimica et Biophysica Acta, 932, 179–188.

Van Grondelle, R., Holmes, N. G., Rademaker, H., and Duysens, L. N. M., 1978, Bacteriochlorophyll fluorescence of purple bacteria at low redox potentials. The relationship between the reaction center triplet yield and the emission yield. Biochimica et Biophysica Acta, 503, 10–25.

Van Grondelle, R., 1985, Excitation energy transfer, trapping and annihilation in photosynthetic systems. Biochimica et Biophysica Acta, 811, 147–195.

Van Grondelle, R., Bergström, H., Sundström, V., and Gillbro, T., 1987, Energy transfer within the bacteriochlorophyll antenna of purple bacteria at 77 K, studied by picosecond absorption recovery. Biochimica et Biophysica Acta, 894, 313–326.

Van Grondelle, R., and Sundström, V., 1988, Exciton energy transfer in photosynthesis. Photosynthetic Light-Harvesting Systems. Organization and Function, edited by H. Scheer and S. Schneider, Walter de Gruyter (Berlin, New York), 403–438.

Van Grondelle, R., Bergström, H., Sundström, V., Van Dorssen, R. J., Vos, M., and Hunter, C. N., 1988, Excitation energy transfer in the light-harvesting antenna of photosynthetic purple bacteria: the role of the long-wavelength absorbing pigment B896. Photosynthetic Light-Harvesting Systems. Organization and Function, edited by H. Scheer and S. Schneider, Walter de Gruyter (Berlin, New York), 519–530.

Vos, M., Van Grondelle, R., Van der Kooy, F. W., Van de Poll, D., Amesz, J. and Duysens, L. N. M., 1986, Singlet-singlet annihilation at low temperatures in the antenna of purple bacteria. Biochimica et Biophysica Acta, 850, 501–512.

Wang, R. T., and Clayton, R. K., 1971, The absolute yield of bacteriochlorophyll fluorescence *in vivo.* Photochemistry and Photobiology, 13, 215–224.

Wittmershaus, B., Nordlund, T. M., Knox, W. M., Knox, R. S., Geacintov, N. E., and Breton, J., 1985, Picosecond studies at 77 K of energy transfer in chloroplasts at low and high excitation intensities. Biochimica et Biophysica Acta, 806, 93–106.

Yeates, T. O., Komiya, H., Rees, D. C., Allen, J. P., and Feher, G., 1987, Structure of the reaction center from *Rhodobacter sphaeroides* R-26: Membrane-protein interactions. Proceedings of the National Academy of Sciences U.S.A., 84, 6438–6442.

The Supramolecular Structure of the Light-Harvesting System of Cyanobacteria and Red Algae

E. MÖRSCHEL, G.-H. SCHATZ*, W. LANGE

Fachbereich Biologie-Botanik der Philipps-Universität
Karl-von Frisch-Str.
D - 3550 Marburg and
*Max-Planck-Institut für Strahlenchemie
Stiftstr. 34-36, D - 4330 Mülheim 1

Introduction

Cyanobacteria including prochloron like organisms were the only procaryotes, which developed oxygenic photosynthesis. Once established the photosynthetic electron transport chain has not changed essentially during evolution to algae and higher plants. However major differences were introduced by the development of specialized light harvesting pigments, which adapt the photosynthetic apparatus of the different organisms to the changing light climates, especially to changes in light quantity and quality (Anderson, 1986). These light harvesting pigments or light harvesting antennae capture light over a large range of the spectrum and transfer the excitation energy in an energetic cascade finally to the photosynthetic reaction centres where chemical events start. Thereby they increase the photosynthetic efficency of the reaction centres to fully utilize the capacity of the electron-transport chain and that of the carbon dioxide fixation system and thus allow life in ecological niches under limited and unfavourabe conditions.

The major light harvesting pigments of higher plants, green algae, euglenoids and prochlorons are the chlorophylls a and b, absorbing light in the blue and red region of the spectrum. Brown algae, diatoms, dinoflagellates, cryptomonads, red algae and cyanobacteria have developed additional light harvesting pigments as chlorophyll c, fucoxanthin, peridinin and the biliproteins, the latter absorbing light primarily in the green gap of chlorophyll (Glazer, 1983). These pigments are especially important to optimize photosynthesis, because the light window closes in water bodies, depending on depth and water quality, to a gap between 450-600 nm, a spectral range in which chlorophyll does not absorb light efficiently (Larkum and Barrett, 1983). Depending on cell structure and organisation, two different kinds of light harvesting systems have evolved: membrane integrated light harvesting systems and antennae structures, which are extramembraneous and associated with the surface of thylakoids or cytoplasmic membranes. Higher plants, green algae, prochlorons and chromophyta contain membrane

Photobiology, Edited by E. Riklis
Plenum Press, New York, 1991

integrated antennae which represent mainly light harvesting chlorophyll (chl) a/b, chl a/c and chl-fucoxanthin complexes. Examples of extramembrane light harvesting systems are the chlorosomes of green bacteria (Blankenship and Fuller, 1986) and the phycobilisomes of cyanobacteria and red algae (Mörschel and Rhiel, 1987).

Organization of biliproteins and phycobilisomes

In accordance with the extramembraneous localization of the phycobilisomes, cyanobacterial and red algal thylakoids lie singly in the cytoplasm or stroma. As the most characteristic feature, the outer surface of the thylakoids is studded with the phycobilisomes, containing mainly the biliproteins as light harvesting pigments (Gantt, 1980, Mörschel and Rhiel, 1987). They build up arrays of short and long rows with a high regular spacing. At least four types of phycobilisomes can be distinguished: i) Bundle-shaped phycobilisomes, occuring only in the atypical cyanobacterium *Gloeobacter violaceus*, which lacks thylakoids. The phycobilisomes are bound to the inner surface of the cytoplasmic membrane. ii) Hemi-ellipsoidal phycobilisomes. These phycobilisomes are preferably found in red algae. iii) Hemi-discoidal phycobilisomes, which are the light harvesting antennae of most cyanobacteria and some red algae as *Porphyridium aerugineum*. iv) Phycobilisomes, which are intermediate between hemi-discoidal and hemi-ellipsoida types and which occur in some cyanobacterial *Phormidium* and *Synechococcus* strains belonging to the LPP group.

Phycobilisomes are made up to a major part of two to three biliproteins and additional linker polypeptides that govern the assembly of phycobilisomes to active light-harvesting complexes. The biliproteins are 1) red phycoerythrins with absorption bands between 500-570 nm, 2) blue phycocyanins with absorption maxima at 610-635 nm, including phycoerythrocyanin absorbing at 575 and 590 nm and 3) allophycocyanin and the allophycocyanin B complex with an absorption at 650 nm and 670 nm respectively. The basic building blocks of the biliproteins are heterodimers composed of two genetically related polypeptides α and β with molecular weights between 17 and 22 kDa. Occasionally phycoerythrins contain a third γ-subunit (Zuber, 1987).

The extraordinary spectroscopic properties of the biliproteins are determined by covalently bound linear tetrapyrrole chromophores, the phycobilins, which are in an extended conformation (Scheer, 1981). Four different chromophores are found in biliproteins of cyanobacteria and red algae namely phycoerythrobilin, phycocyanobilin, phycourobilin and a phycobiliviolin-like chromophore which is bound to the α-subunit of phycoerythrocyanin (Glazer, 1985). The bathochromic shift in colour from red to blue is generated by increasing the number of the conjugated double bonds. However, the spectroscopic properties of the proteins are further modulated by chromophor-protein interaction and the attachment of the chromophores to the protein. The chromophores are linked through one or two thioether bonds to ring A or D or ring A and D. Thus a large spectroscopic variability of the chromophores is introduced by simple variations of the basic molecules.

The ($\alpha\beta$)-heterodimers have the fundamental property of building up cyclic trimers, 11×3 nm in size, which form hexamers and stacks of hexamers by side to side aggregation as a basic prerequisite for phycobilisome assembly (Mörschel et al., 1980a). Recently, the X-ray structures of trimeric and hexameric C-phycocyanin from *Mastigocladus laminosus* and *Agmenellum quadruplicatum* were determined and models for energy transfer within these aggregates were developed (Schirmer et al., 1986).

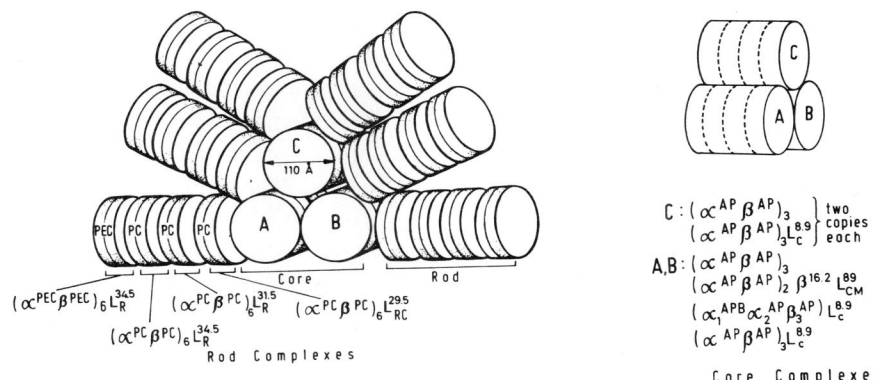

Fig. 1. Phycobilisome model of the cyanobacterium *Mastigocladus laminosus*. Allophycocyanin (AP), phycocyanin (PC), phycoerythrocyanin (PEC), linker (L) with indices rod (R), rod-core (RC), core-membrane (CM), (Zuber et al., 1987).

The hemi-discoidal phycobilisomes consist of two morphologically distinct domains: they are constructed of a core and a periphery built up of short rods (Fig. 1). The core is composed of three ring-shaped cylindrical units of 11 nm diameter arranged in a trigon. Two cylinders are localized at the base, the original attachment site to the membrane "in situ", while the third occupies the groove on top of these structures. From this core six peripheral rods radiate in a symmetrical hemi-discoidal array. They are bound to the edge of the core cylinders with their flat faces. These rods consist of stacked hexameric biliprotein double discs corresponding to the formula $(\alpha\beta)_6 L$. The peripheral biliprotein rods always contain hexameric phycocyanin aggregates proximal to the core. Depending on the organism, the distal ends are constructed of phycocyanin alone, phycocyanin-phycoerythrin or phycocyanin-phycoerythrocyanin, which are inducible or constitutive chromoproteins. Allophycocyanin is located in the cylindrical aggregates of the core (Mörschel et al., 1980b, Anderson et al., 1984, Wehrmeyer, 1983).

The assembly of such complicated aggregates as phycobilisomes into functional light harvesting aggregates at the thylakoid surface has to proceed in a regulated step by step process. Biliprotein aggregates are very similar in structure and thus additional proteins, the linker polypeptides or assembly polypeptides, are necessary to arrange the biliproteins in the right position. The linker polypeptides are genetically related to biliprotein subunits; some of them carry bilin-chromophores (Rümbeli and Zuber, 1988). Each of the hexameric biliprotein aggregates making up the phycobilisome rods is associated with a special linker polypeptide of the 25-35 kDa family. These linker polypeptides act as primers for the formation of trimeric and hexameric biliproteins and finally occupy the central channel of these aggregates. They also introduce a vectorial specificity into the hexameric biliprotein complexes and thus determine location and function within the phycobilisomes. The linker polypeptides in addition modulate the spectroscopic properties of hexameric biliproteins by interactions with the β-84 chromophores. In this way a red shift of the fluorescence maxima from the peripheral biliprotein hexamers of the rods to those proximal to the core is induced, thereby directing the energy flow towards the core and minimizing the random walk of energy within the rod subassemblies (Glazer, 1984, Holzwarth, 1986). The peripheral phycobilisome rods are an effective system to funnel light energy to the phycobilisome core. The modular design of the rods facilitates modifications due to changed environmental conditions in order to optimize light harvesting.

The core is the domain which attaches the phycobilisomes to the thylakoids and funnels the energy flow from the peripheral rods to the intramembrane chlorophyll antennae of photosystem II. With the exception of the phycobilisomes of *Synechococcus* 6301 whose cores are of only two cylinders (Glazer et al., 1983), all other investigated hemidiscoidal phycobilisomes from red algae and cyanobacteria contain three cylinder cores with a triangular symmetry. The cylindrical elements are made up mainly of allophycocyanin. Besides allophycocyanin, the core contains especially large linker biliproteins of 70-120 kDa. These large polypeptides anchor the phycobilisome on the thylakoid membrane. They carry one phycocyanobilin chromophore, and with a fluorescence maximum of 676 nm they act as a final emitter of energy within the phycobilisome. The second terminal emitter of the phycobilisomes is a special α-allophycocyanin subunit, called α-allophycocyanin B (Glazer, 1984). The isolated subunit absorbs maximally at 645 nm and fluoresces at about 680 nm. This allophycocyanin subunit is localized in each of the basal core cylinders. The tentative localization of the polypeptides within the core of *Mastigocladus laminosus* phycobilisomes is shown in Fig. 1.

No supramolecular model has been developed for hemi-ellipsoidal phycobilisomes, because of their large size that generates superimposition effects of the structures. Based on their size and electron microscopical views, these phycobilisomes may represent tentatively double versions of the hemi-discoidal type.

Phycobilisomes are dynamic structures. Their number and pigment composition is influenced by a number of developmental and environmental factors, such as light intensity and light quality. Generally an inverse correlation between biliprotein content and light intensity is observed. Some cyanobacteria are able to respond to spectral changes in light colour with the synthesis of complementary biliproteins: green light induces synthesis of red pigments, red light that of blue pigments. This phenomenon is called complementary chromatic adaptation (Tandeau de Marsac, 1983). We distinguish three groups: 1) group 1 does not adapt chromatically at all. 2) group 2 regulates phycoerythrin synthesis as a function of light quality, whereas the synthesis of phycocyanin is not affected. 3) group 3 regulates the synthesis of both biliproteins phycoerythrin and phycocyanin by light quality. The allophycocyanin level never changes and thus the core domain remains constant as an important prerequisite for energy transfer. Chromatic adaptation as an instrument to optimize the light harvesting capacity is achieved by a simple variation of the distal rod elements in a modular way, by removing, adding or replacing the biliprotein aggregates. The biliprotein variations during complementary chromatic adaptations are controlled by an unidentified photoreceptor that regulates the transcription of specific genes for biliproteins and linker proteins.

Structure of Photosystem II - Phycobilisome complexes

The well-ordered arrangement of phycobilisomes on the surface of the thylakoids affords an attachment to well aligned intramembrane particles. Thus, the phycobilisome pattern on the membrane should be mirrored by the particle pattern within the thylakoids. In many thylakoid areas we find broad lanes of densely packed protoplasmic (PF) particles of 7.0 - 9.0 nm, and on the complementary fracture face, rows of well aligned exoplasmic (EF) particles (Fig. 2a), which fit in the particle free grooves between the PF-particle lanes (Giddings et al., 1983). The minimum centre to centre distance of these EF-particles is normally 45 nm or more and is thus in register with the spacing of phycobilisome rows on the thylakoids (Neushul, 1970).

The EF-particles of cyanobacteria containing hemi-discoidal phycobilisomes measure 10×20 nm and they are aggregated linearly front to back with their longitudinal faces with a periodicity of 10 nm. Most particles reveal a central furrow perpendicular to the long axis dividing them into two side by side domains of 10 nm $\times$ 10 nm each, suggesting that each particle represents a dimer (Fig. 2a). Both domains are cleaved additionally parallel to the longitudinal face of the whole particle. The structures of the dimers are complementary to the structure of the two adjacent phycobilisome core cylinders (Mörschel and Schatz, 1987). Freeze-etched thylakoids, exposing simultaneously phycobilisomes and exoplasmic particles in the same area, show a direct alignment of both systems. The EF-particle rows are oriented towards the centre of the phycobilisome bases and they are continued by the lines of phycobilisomes on the outer surface of the thylakoids (Fig. 2b). On the basis of alignment and periodicity we proposed that hemi-discoidal phycobilisomes are associated peripherally on top of the EF-particles in a 1:1 stoichiometry (Mörschel and Mühlethaler, 1983).

The arrangement of the EF-particles of red algae containing hemi-ellipsoidal phycobilisomes is similar (Fig. 3). However, the EF-particle rows are loosened into packages of particles which are separated normally by areas of 7-10 nm. Each package is composed of two 10×20 nm EF-particles, which are aggregated with their long faces. Each EF-particle corresponds to a dimeric 10×10 nm particle as is observed in well

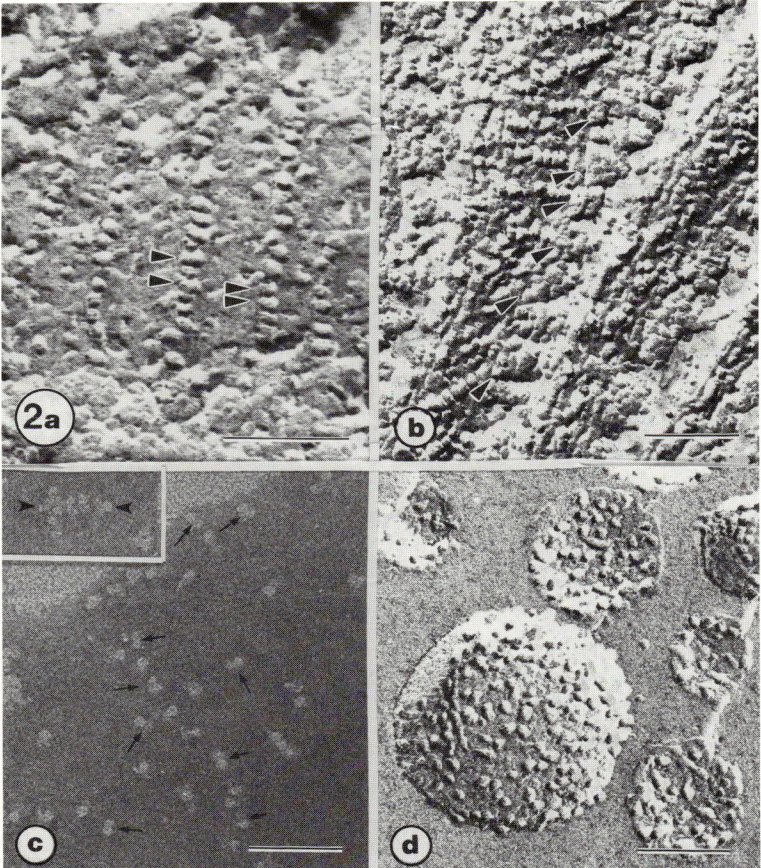

Fig. 2. a:Freeze-fracture through a thylakoid of the cyanobacterium *Synechococcus* spec. showing the dimeric 10 nm × 20 nm exoplasmic particles to which phycobilisomes are bound. b:Freeze-fractured thylakoids of *Mastigocladus laminosus*, showing the alignment of EF-particles and phycobilisomes.
c:Negatively stained PSII-particles. Monomeric and dimeric particles (arrows) are shown and a short particle row (inset). d:Freeze-fractured PSII proteoliposomes. Bars: 100 nm

resolved or partially dissociated particles and thus one particle package contains four 10 nm EF-particles. This pattern corresponds well with the pattern of the hemi-ellipsoidal phycobilisomes on the membranes, which have a basal length of 50 nm, a height of 30 nm and a width of about 22 nm (Wehrmeyer, 1983). They are separated by about the same distance as the EF-particle packages within the rows. The assembly of the EF-particles supports the hypothesis that the cores of hemi-ellipsoidal phycobilisomes are structurally double versions of the cores of hemidiscoidal phycobilisomes (Fig. 3, inset).

Because phycobilisomes transfer the captured light energy mainly to photosystem II, we proposed that the EF-particles correspond to photosystem (PS) II particles. In order to

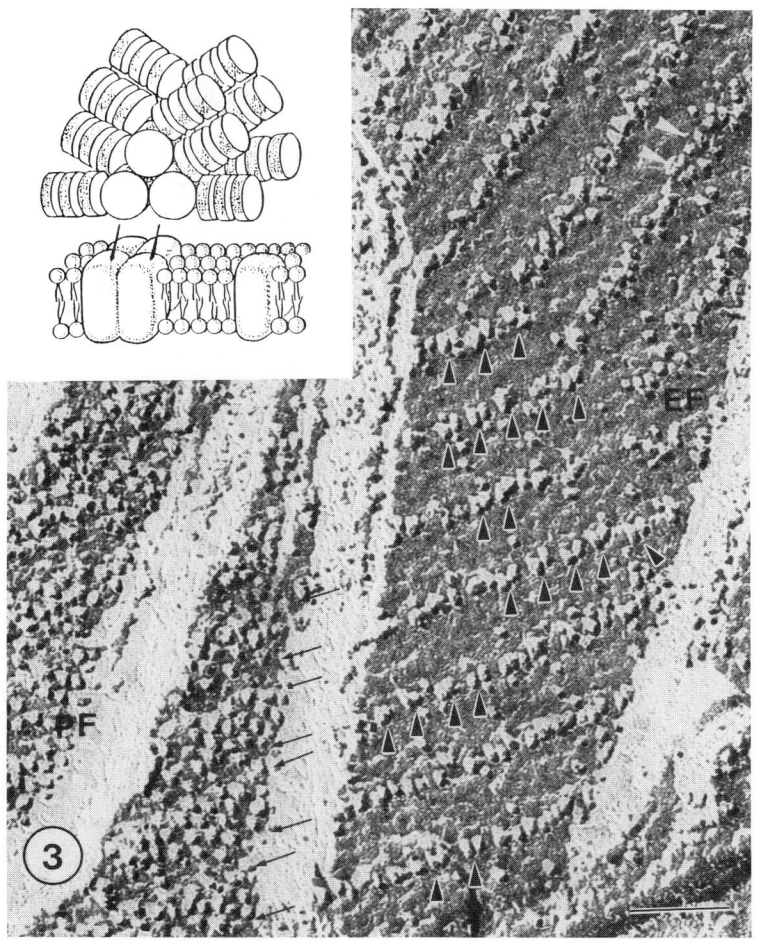

Fig. 3. Freeze-fracture showing the exoplasmic (EF) and protoplasmic fracture face (PF) of a thylakoid from *Porphyridium cruentum*. The inset shows a model of the PSII-phycobilisome: Four 10 nm PSII particles are linked to one hemi-ellipsoidal phycobilisome; the phycobilisome structure is hypothetical. Bar: 100 nm

examine this hypothesis, water splitting PSII particles were isolated from the cyanobacterium *Synechococcus* spec. by solubilization with the detergent Sulfobetain 12 without any loss in PSII activity (Mörschel and Schatz, 1987). The high degree of enrichment of PSII was indicated by chlorophyll/ PSII ratios between 70 to 90, the PSII/PSI ratio was 1500. The main pigments of the PSII complexes were phycocyanin, allophycocyanin and chlorophyll (Chl) as revealed by their absorption at 620, 652 and 680 nm, respectively. Whole cells showed a strong low temperature emission peak at 730 nm and minor peaks at 660, 685 and 695 nm when excited at 445 nm. The 730 nm emission peak was attributed to PSI. Purified PSII preparations lacked this 730 nm emission maximum and therefore PSI. When excited at 445 nm the PSII preparations showed a minor maximum at 660 nm characteristic for allophycocyanin and two major maxima at 685 and 693 nm. The fluorescence at 685 nm was attributed to an interplay of allophycocyanin B, the large membrane-phycobilisome linker (Lcm) and a chlorophyll antenna, whilst the 693 nm fluorescence belonged to a chlorophyll antenna alone. Most of the light energy captured by phycobilisomes was transferred to the final phycobilisome emitters and the PSII antennae as shown from the emission peak at 685 nm and the shoulder at 692 nm. A minor part of the fluorescence was emitted by uncoupled phycocyanin and allophycocyanin at 640 nm and 660 nm.

The chlorophyll-protein (CP) and polypeptide pattern of PSII complexes were determined by two dimensional gel electrophoresis. In the first dimension, run in the presence of LDS for optimal yield of intact chlorophyll-complexes, three chlorophyll-proteins termed CP II a,b,c as well as the biliproteins were resolved. Apparent molecular weights of 110, 79 and 45 kDa were determined for the three chlorophyll-proteins. In the second dimension, the polypeptide pattern after complete denaturation by SDS was determined. Prominent polypeptides could be attributed to the α -and β-biliprotein subunits of allophycocyanin (14, 16 kDa) and phycocyanin (15, 18 kDa) and at least four phycobilisome-linker polypeptides corresponding to 29, 31, 34 and 120 kDa. After SDS electrophoresis, the chlorophyll-complexes CP IIa and b were resolved as apoproteins of 47 kDa and CP IIc as one of 41 kDa. The chlorophyll-complexes CP II b,c contained no other proteins. Complex CP IIa revealed an additional faint band of 80 kDa. The chlorophyll-proteins were cut out of the gels and analysed by fluorescence emission spectroscopy at liquid nitrogen temperature. CP IIa and b were very similar in their spectroscopic properties and showed maxima at 685 nm and shoulders at 693 nm, whilst CP IIc exhibited only one peak at 686 nm, when excited at 445 nm. Thus the isolated chlorophyll complexes had similar emission properties as the *in situ* photosystem II antennae.

The purified photosystem II complexes were incorporated into liposomes made from soybean phosphatidylcholine. Fig. 2c shows a typical negatively stained fraction of PS II-particles solubilized by Triton X-100 from proteoliposomes which were separated from free biliproteins. Two particle classes are distinguished: spherical-ellipsoidal and binary particles. The dominant structures are the binary particles, which are preferably deposited on the support with their broad faces. They measure 20-25 × 10-14 nm and are divided perpendicular to their long axis by a lace into two spherical-ellipsoidal parts of about 10-13 nm. Aggregations of these binary particles to rows also occurred as shown in Fig. 2c (inset). Within these rows, particles are visible in "top view" and measure about 24 × 12 nm; the central division, separating the particles into two parts of about 12 nm size, is clearly visible. These binary particles are very similar in their appearance compared to the "in situ" EF-particles of Fig. 2a. The second particle class is represented by spherical-ellipsoidal particles of about 10-12 × 14-16 nm; they are supposed to be the building blocks of the binary particles.

The photosystem II proteoliposomes were analyzed by freeze-fracture analysis after incorporation into phospholipid liposomes (Fig. 2d). The freeze-fracture particles were randomly distributed and their diameter was about 10.3 nm on the concave and convex fracture faces. Some particles showed a central furrow, that was also present in the EF-particles "in situ"; side by side aggregations of two particles were observed too. Sometimes parts of phycobilisomes still bound to the incorporated photosystem II particles were observed on the outer surface of the proteoliposomes. Cross fractures running perpendicular to the plane of the liposome membrane showed that the particles were spanning the membrane. With a height of 13-16 nm they were considerably larger than the lipid bilayer. The monomeric photosystem II-particles of *P.cruentum* measured also 10 nm when incorporated into liposomes, so that a correlation with the "in situ" EF-particles can be established.

From our results we propose that the EF-particles correspond to PSII-complexes, which occur depending on organism and phycobilisome type as monomeric, dimeric or tetrameric units. The dimeric EF-particles within the rows are associated with one hemi-discoidal phycobilisome, the tetrameric EF-particles with one hemi-ellipsoidal phycobilisome as shown in the models (Figs. 3,4). Each 10 nm particle represents a PSII-complex containing at least the reaction centre complex (RC), the water splitting system, the Chl antennae with apoproteins of 41 and 47 kDa. The linker polypeptide (Lcm) is supposed to bind the phycobilisome to the Chl antennae of photosystem II. The following scheme explains the energy flow from phycobilisome to PS II:

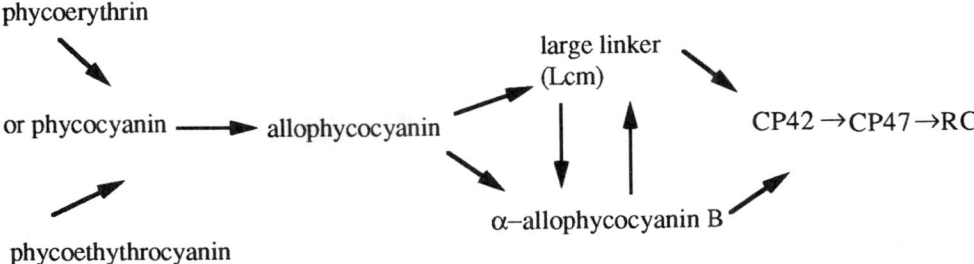

When phycobilisome-PS II complexes are tightly packed within their rows as in *Mastigocladus*, we assume that energy is transferred not only from phycobilisomes to the underlying PSII particles but also between phycobilisomes and PSII along the same row. The resulting energy conducting "fibre system" would allow for an efficient energy distribution along the plane of the thylakoid by connecting many PS II reaction centres. In the red alga *Porphyridium cruentum* only four PSII complexes are grouped. However tight packing of these particle groups to rows may occur, especially under low light conditions.

The rowed PSII-phycobilisome particles imply also a spatial separation of PS II and PS I equivalent to the lateral heterogeneity of PS I and PS II in the thylakoid system of

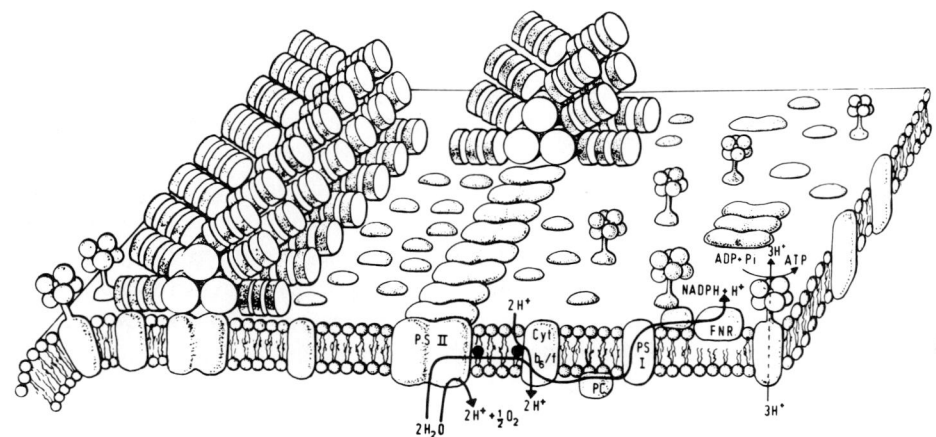

Fig. 4. Model of a thylakoid from cyanobacteria and red algae with hemi-discoidal phycobilisomes. The PSII-phycobilisomes are organized into rows. The particles located between the PSII-phycobilisome rows are supposed to correspond to PSI, cytochrome b_6/f, ATP-synthases, and complexes of the respiration chain.

grana containing chloroplasts. The mechanism of energy distribution between PSII and PSI, which is controlled by phosphorylation in higher plant chloroplasts, is not yet clear.

Acknowledgements

The study was generously supported by the Deutsche Forschungsgemeinschaft by a grant to E.M.

References

Anderson, J.M. (1986) *Ann. Rev. Plant Physiol.* 37, 93-136
Anderson, L.K., M.C. Rayner and F.A. Eiserling (1984) *Arch. Microbiol* 138, 237-243
Blankenship, R.E. and R.C. Fuller (1986) In: Encyclopedia of Plant Physiology, New Series,Vol. 19 (Edited by L.A. Staehelin and C. J. Arntzen) pp. 390-399. Springer, Berlin.
Gantt, E. (1980) *Int. Rev. Cytol.* 66, 45-79
Giddings, T.H., C. Wasman and L.A. Staehelin (1983) *Plant. Physiol.* 71, 409-419
Glazer, A.N. (1983) *Ann. Rev. Biochem.* 52, 125-157
Glazer, A.N. (1984) *Biochim. Biophys. Acta* 768, 29-51
Glazer, A.N. (1985) *Ann. Rev.Biophys. Biophys. Chem.* 14, 47-77
Holzwarth, A.R.(1986) In: Antennas and Reaction Centers of Photosynthetic Bacteria (EditedbyM.E.Michel- Beyerle) pp. 45-52 . Springer, Berlin.
Ley, A.C. and W.L. Butler (1977) *Biochim. Biophys. Acta*, 462, 290-294
Larkum, A.W.D.and J. Barrett, (1983) *Adv. Bot. Res.* 10, 1-219
Mörschel, E., K.P. Koller and W. Wehrmeyer (1980a) *Arch. Microbiol.* 125, 43-51
Mörschel, E.and K. Mühlethaler (1983) *Planta* 158, 451-457

Mörschel, E. and E. Rhiel (1987) In: Electron Microscopy of Proteins 6. Membraneous Structures (Edited by J.R. Harris and R.W.Horne) pp.209-254. Academic Press, London, New York.

Mörschel, E. and H.-G. Schatz (1987) Planta 172, 145-154 Mörschel,E., W. Wehrmeyer and K.-P. Koller (1980b) *Eur. J. Cell Biol* 21, 319-327

Neushul, M. (1970) *Am. J. Bot.* 57, 1231 -1239

Ohki, K. and Y. Fujita (1987) In: Progress in Photosynthesis Research 2 (Edited by J. Biggins) pp. 157-160 Martinus Nijhoff, Dortrecht.

Rümbeli, R. and H. Zuber (1988) In: Photosynthetic light- harvesting systems (Edited by H. Scheer and S. Schneider) pp. 61-70. De Gruyter, Berlin, New York.

Scheer, H. (1981) *Angew. Chem. Int. Ed.* 93, 241-261

Schirmer, T., R. Huber, M. Schneider, W. Bode, M. Miller and M. L. Hackert (1986) J. Mol. Biol. 188, 651-676

Tandeau de Marsac, N. (1983) *Bull. L'Inst. Pasteur* 81, 201- 254

Wehrmeyer, W. (1983) In: Photosynthetic Procaryotes: Cell differentiation and function (Edited by G.C. Papageorgiou and L. Packer) pp. 1-22. Elsevier, Amsterdam.

Zuber, H.,R. Brunisholz, and W. Sidler (1987) In: Photosynthesis (Edited by J. Amesz) pp. 231-271. Elsevier, Amsterdam.

Quinone Substituted Porphyrin Dimers: New Photosynthetic Model Systems

JONATHAN L. SESSLER,[‡] MARTIN R. JOHNSON,[‡]
STEPHEN E. CREAGER,[‡] JAMES FETTINGER,[**] JAMES A. IBERS,[**]
JUAN RODRIGUEZ,[§] CHRISTINE KIRMAIER,[§] AND DEWEY HOLTEN[§]

[‡]*Department of Chemistry, University of Texas at Austin*
Austin, TX 78712 USA
[**]*Department of Chemistry, Northwestern University*
Evanston, IL 60208 USA
[§]*Department of Chemistry, Washington University*
St. Louis, MO 63130 USA

The 1988 Nobel Prize in Chemistry was awarded to Johann Deisenhofer, Hartmut Michel, and Robert Huber for their elucidation of the X-ray crystal structure of the reaction center (RC) from the photosynthetic bacterium *Rhodopseudomonas viridis* (Deisenhofer et al. 1984). More recently, structural information for the RC of *Rhodobacter sphaeroides* has also become available (Allen et al., 1986; Chang et al., 1986). Six tetrapyrrolic subunits are found at the active sites of these two structurally similar RCs: A dimeric bacteriochlorophyll "special pair" (P), two "accessory" bacteriochlorophylls (Bchls), and two bacteriopheophytins (Bphs), all held in a well-defined but skewed geometry along a C_2 axis of symmetry. The Bchls are separated from P by center-to-center distances of ca. 11 Å and interplane angles of ca. 70°. The Bphs in turn are separated by similar distances and angles from the Bchls. Four of these six prosthetic groups are currently considered to define the relevant electron transport chain (Kirmaier and Holten, 1987). This consists in sequence of the photosensitizer (P), an "accessory" Bchl, an intermediate Bph, and a quinone acceptor (Q). In *R. sphaeroides*, Q is an ubiquinone; in *R. viridis*, it is a menaquinone. In both cases, Q lies roughly 13-14 Å away from the corresponding Bph center. Charge separation between P* and Bph entities is known to occur on a time scale of 2-4 ps with nearly 100% quantum efficiency (Woodbury et al., 1985; Martin et al., 1986; Wasielewski and Tiede, 1986; Kirmaier and Holten, 1988). Furthermore this process exhibits activationless behavior, increasing in rate by a factor of two at liquid helium temperature (Woodbury et al., 1985; Fleming et al., 1988). Subsequent electron transfer from Bph⁻ to Q to give P⁺-Bchl-Bph-Q⁻ occurs in 200 ps, also with 100% quantum yield (Kirmaier and Holten, 1987).

In spite of the availability of the above structural and kinetic data, many aspects of

Photobiology, Edited by E. Riklis
Plenum Press, New York, 1991

bacterial photosynthesis, including the rapid, activationless, and efficient nature of the initial charge separation process, remain poorly understood (Marcus and Sutin, 1985; Kirmaier and Holten, 1987; Windsor, 1986). One of the more crucial questions currently being debated is the role of the "intermediate" or "accessory" Bchl. Recent subpicosecond transient absorption experiments have failed to provide any evidence that a P^+-Bchl$^-$ state acts as a discrete intermediate in the initial P^*-Bchl-Bph $\rightarrow$ P^+-Bchl-Bph$^-$ charge separation process (Woodbury et al., 1985; Martin et al., 1986; Wasielewski and Tiede, 1986; Kirmaier and Holten 1988). Nonetheless, it seems unlikely that the electron traverses the ca. 10 Å edge-to-edge from the P to the Bph in a few ps without the Bchl playing an important role. An attractive but as yet unproven possibility is that the Bchl facilitates transfer to the Bph via a "superexchange" mechanism involving a quantum mechanical mixing of a virtual P^+-Bchl$^-$ state with the photoexcited dimer, P^* (Bixon et al., 1987; Marcus, 1987; Won and Friesner, 1987). There has been a great deal of controversy (but no definitive resolution) as to whether or not such a mechanism is consistent with all the available experimental observations. Currently extensive experimental and theoretical efforts are being made to addrress this question *in vivo*. Another potentially informative approach involves studying model systems which actually mimic important features of the RC.

Many photosynthetic model compounds have been prepared in recent years (for a listing of recent references see: Joran *et al.*, 1984; Gust et al., 1988). Some of these have proved useful in exploring how various factors, such as distance, donor-acceptor energetics, and solvent, mediate photoinduced electron transfer reactions (c.f. eg Joran et al., 1987). Others exhibit interesting charge separation properties (c.f. eg Gust et al., 1988; Wasielewski et al., 1985). Nonetheless, in essentially all cases there are substantial differences between the models and the actual RCs. For instance, with the exception of several recently reported dimers (Sakata et al., 1985; Cowan et al., 1987; Gust et al., 1988), photosynthetic model systems generally have consisted of simple monomeric porphyrins substituted with one or more acceptors. We have therefore prepared and characterized a new series of photosynthetic models: The selectively metalated, quinone-substituted "gable" and "flat" dimers **1**, **2**, **5**, and **6** (Sessler et al., 1988). These models provide the first "matched set" of photosynthetic models suitable for studying interchromophore orientation and energetic effects in biomimetic systems, and, in the case of **1** and **2**, provide the first examples wherein porphyrin-based superexchange mediated charge separation processes are observed in synthetic systems.

An essential feature of compounds **1**, **2**, **5**, and **6** is that they possess a well-defined conformational structure. In other words they are not "floppy". Initial estimates of the intersubunit orientations and distances could therefore be obtained from CPK space filling molecular models. The center-to-center distances between the unsubstituted "distal" porphyrin (MP_d) and the quinone were estimated to be 14 Å in the gable series and 20 Å in the flat compounds. Similarly, the center-to-center distances between the distal and "proximal" (MP_p) porphyrin subunits were estimated to be 10.5 Å and 12.5 Å in the gable and flat series respectively. Studies of CPK models also suggested that in both the gable and flat systems the flanking methyl substituents at positions 3 and 7 force the porphyrins to adopt a conformation that is perpendicular to the bridging phenyl subunits. Moreover, for the same type of reason, the quinone was expected to lie perpendicular to the proximal porphyrin. The available X-ray structural data are consistent with these predictions. A single crystal X-ray diffraction study of the monomeric free-base quinone **11** has now been completed; it shows that the quinone makes an angle of 84° with the porphyrin core, and reveals a center-to-center distance of 6.5 Å between the porphyrin and quinone subunits (Figure 1). Preliminary X-ray structural information is also available for the biscopper(II) chelates of the symmetrically substituted bisdimethoxyphenyl analogues (Sessler *et al.*, 1987) of **3** and **7** (Figures 2 and 3). These structures confirm the

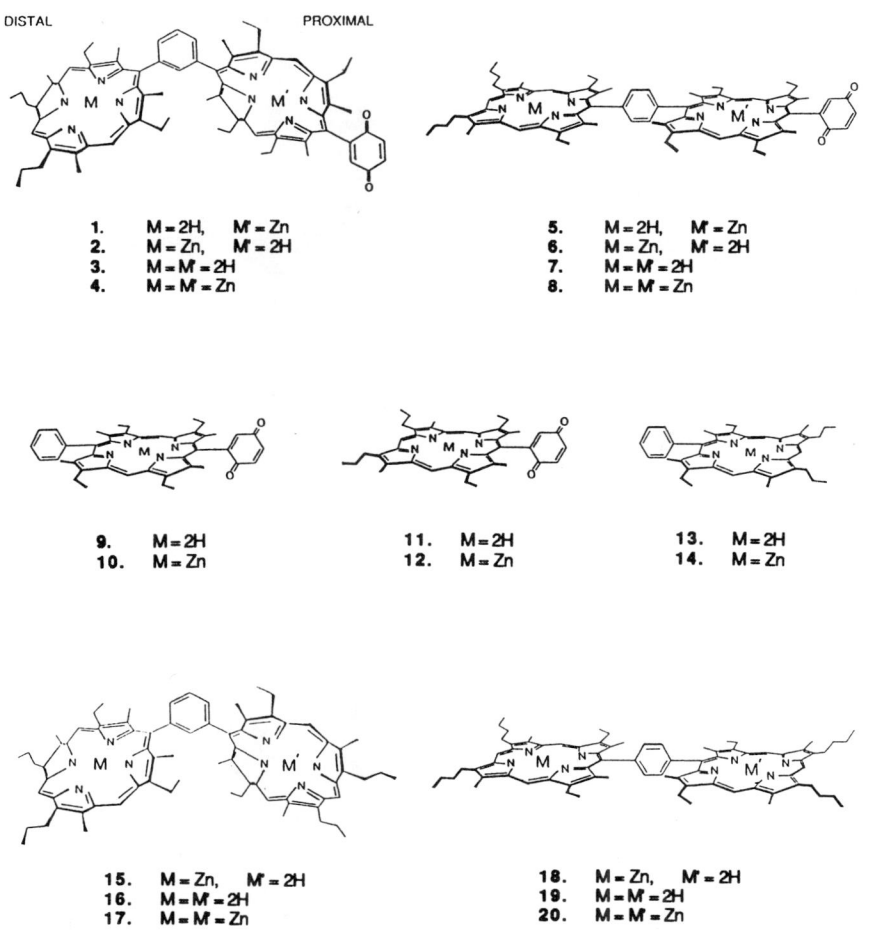

1.	M = 2H,	M' = Zn
2.	M = Zn,	M' = 2H
3.	M = M' = 2H	
4.	M = M' = Zn	

5.	M = 2H,	M' = Zn
6.	M = Zn,	M' = 2H
7.	M = M' = 2H	
8.	M = M' = Zn	

9.	M = 2H
10.	M = Zn

11.	M = 2H
12.	M = Zn

13.	M = 2H
14.	M = Zn

15.	M = Zn,	M' = 2H
16.	M = M' = 2H	
17.	M = M' = Zn	

18.	M = Zn,	M' = 2H
19.	M = M' = 2H	
20.	M = M' = Zn	

393

predicted porphyrin center-to-center distances: The intramolecular Cu-Cu separations are 10.5 Å and 12.7 Å for the gable and flat systems respectively. These structures also reveal several other interesting structural features. For instance, in the flat dimer, the two porphyrin macrocycles are found to be essentially coplanar and perpendicular to the bridging phenyl ring. In the case of the 1,3-phenyl linked system, on the other hand, the two porphyrin subunits help define what can be considered an overall "skewed" arrangement contained within the context of a gable-type configuration. The obvious structural similarity between this synthetic structure and parts of the RC (notably the Bchl and Bph pair) is a feature we consider to be of particular interest.

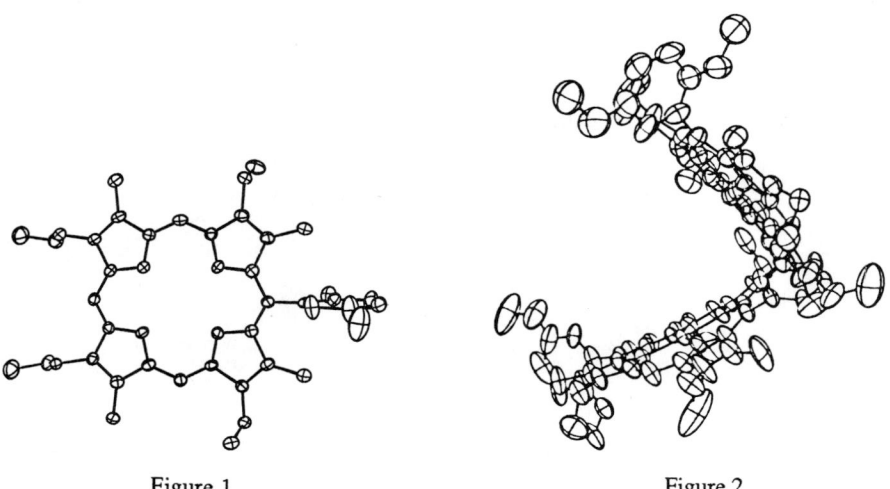

Figure 1   Figure 2

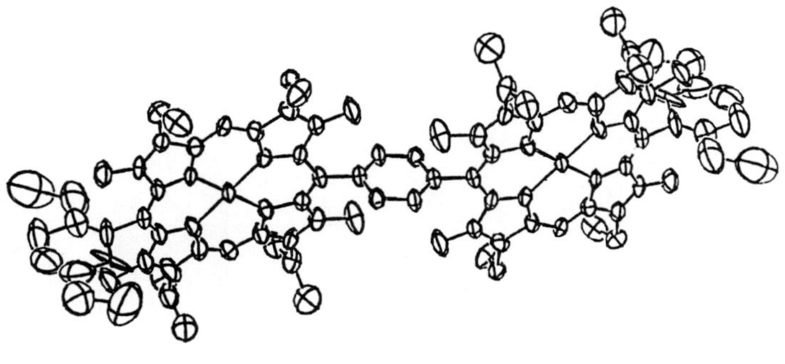

Figure 3

A second unusual feature of the selectively monometalated compounds **1, 2, 5,** and **6** is that relative subunit energetics are controlled. In these systems, the lowest excited singlet state of the zinc(II) porphyrin subunit lies ca. 0.17 eV higher in energy than that of the corresponding free-base system (Sessler *et al.*, 1988). As a result, two different energetic arrangements are defined for each regioisomeric pair **1-2,** and **5-6.** In compounds **2** and **6,** an energy gradient exists for net electron transfer from the photoexcited distal porphyrin, ZnP_d^*, through the proximal subunit $H_2P_p^*$ (or $H_2P_p^-$), to the quinone acceptor Q. In models **1** and **5,** on the other hand, the proximal ZnP_p porphyrin defines an energy barrier between the photoexcited distal subunit $H_2P_d^*$ and Q. Systems **1** and **5** thus represent photosynthetic models that mimic the apparent energetic arrangement of the P-Bchl-Bph RC chromophores.

Before considering the static and dynamic optical properties of the multi-chromophore models mentioned above, it is instructive to consider those of the quinone-free dimers **15-20.** All these compounds show evidence of optical coupling as revealed by split, broadened and/or shifted Soret bands with this interaction being most apparent for the bis-zinc complexes **17** and **20.** Irradiation of these bis-zinc complexes at the Soret maximum, however, gives rise to near normal zinc porphyrin emission bands (such as those observed for **14**) that are only slightly reduced in intensity. Similarly, irradiation of **16** and **19** gives rise to typical free-base emission spectra (e.g. that of **13**). On the other hand, irradiation of the monozinc complexes **15** and **18,** in either the H_2P or ZnP absorption bands, gives rise to strong emission bands the wavelength of which are characteristic *only* of the free-base subunit. This result implies that energy transfer takes place rapidly ($\tau \le 10$ ps) from the H_2P to the ZnP subunit.

The quinone substituted monomers **9-12** represent a second important set of controls. When these compounds are irradiated at the Soret maximum, no detectable fluorescence is observed ($\Phi_F \le 10^{-5}$). This suggests that charge separation is very rapid in these systems. This is confirmed by direct optical studies. Irradiation of these monomers in dilute toluene solution with a 350-fs 582-nm laser pulse leads to the instru-ment-limited appearance of absorption changes through the visible region which in all cases decay completely in 5-15 ps. The time evolution of the absorbance changes indicates the presence of two spectral and kinetic components, which may be ascribed to the exothermic formation of a charge separated state ($-\Delta G \cong 0.69$ and 1.03 eV for **9** and **10** respectively), followed by exothermic charge recombination ($-\Delta G \cong 1.27$ and 1.10 eV for **9** and **10** respectively) to regenerate the ground state (eq. 1).

$$MP\text{-}Q \xrightarrow{h\nu} MP^*\text{-}Q \xrightarrow{\text{electron transfer}} MP^+\text{-}Q^- \xrightarrow{\text{charge recombination}} MP\text{-}Q \qquad (1)$$

The photodynamic behavior of the distal monometalated gabled dimer **2** (and controls **3** and **4**) is basically the same as that of the monomers **9-12:** For instance, no fluorescence emission is seen upon irradiation at the Soret maximum. In addition, following a 350-fs 582-nm photoexcitation pulse, the excited states of compounds **2, 3,** and **4,** also return to the ground state very quickly (in 5-15 ps). Again, as was true for **9** and **10**, this net decay appears to consist of two kinetic components. These are assigned to the formation and subsequent decay of a photoinduced charge separated state involving the quinonone acceptor and proximal porphyin subunit. Thus both the static and dynamic optical results are consistent with a mechanism wherein rapid electron transfer from the dimer to the quinone is followed by fast recombination. In **2**, charge separation occurs either by direct excitation (to give $ZnP_d\text{-}H_2P_p^*\text{-}Q$) and subsequent charge separation, or by rapid energy migration (en. mig.) followed by fast electron transfer (E.T.) as shown in eq. 2. Here, the values in parentheses represent the state energies (in eV) of the species in question, derived from electrochemical and optical studies in CH_2Cl_2, uncorrected for any possible coulombic or solvent effects.

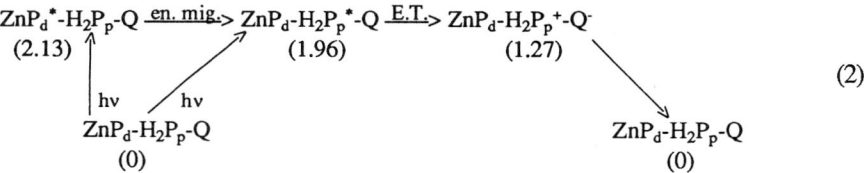

$$\begin{array}{ccccc}
ZnP_d^*\text{-}H_2P_p\text{-}Q & \xrightarrow{\text{en. mig.}} & ZnP_d\text{-}H_2P_p^*\text{-}Q & \xrightarrow{\text{E.T.}} & ZnP_d\text{-}H_2P_p^+\text{-}Q^- \\
(2.13) & & (1.96) & & (1.27) \\
\end{array}$$

$$\begin{array}{cc}
hv \quad hv & \\
ZnP_d\text{-}H_2P_p\text{-}Q & ZnP_d\text{-}H_2P_p\text{-}Q \\
(0) & (0)
\end{array}$$

$$\text{(2)}$$

Very similar behavior was also observed for the corresponding flat (1,4-phenyl linked) distal monometalated dimer **6**. In this case, however, the rates of both the fast and slow kinetic components appear to be slower by roughly 70-80% as compared to the gable compound **2**. This may reflect the fact that both charge separation and recombination are slower in the more open flat compounds, possibly because of the longer distances involved.

In contrast to the systems discussed so far, the proximal monometalated complexes **1** and **5**, containing a built-in (ZnP_p) energy barrier, display qualitatively different excited state behavior (eq. 3). This is apparent in both the static fluorescence and transient absorption measurements. Both compounds **1** and **5** show modest but detectable fluorescence emission from the distal free-base subunit in toluene at room temperature ($\Phi_F = 1.7 \times 10^{-4}$ and 7.7×10^{-4} respectively). From the ratio of emission intensities of **1** and **5** relative to **15** and **18**, respectively, net electron transfer rates (from $H_2P_d^*$ to Q) of 5.4×10^{10} s^{-1} and 1.1×10^{10} s^{-1} (k_{ET}) may be derived for **1** and **5**, respectively. These rates are ca. 100 times slower than those obtained for **2** and **6**. This indicates that the presence of a central zinc porphyin (ZnP_p), which represents an energy barrier which slows down the rate of charge separation between the free-base porphyrin and quinone subunits. Of far greater significance, however, is the realization that the net electron transfer rates between the distal H_2P and quinone subunits for **1** and **5** are still exceedingly fast. For instance, the k_{ET} value for **5** is *over 2000 times faster than that obsered for a bisbicyclooctane-derived model prepared by Joran et al.* (1984), wherein the free-base porphyrin-to-quinone separation approximates the H_2P_d to Q distance found in **5**. Clearly direct through space (or through solvent) electron transfer is not the dominant photochemical pathway in systems **1** and **5**. Rather, charge separation is most likely to be mediated by the central (or proximal) metalloporphyrin. This does not appear to be a thermally activated process involving an intermediate ZnP^* or ZnP^- state: The quantum yield for fluorescence (in 2-MeTHF) increases by roughly a factor of 3 upon cooling the sample from room temperature to 77 K; an increase of over 10^5 would be expected for a simple Arrhenius-type process with an analogous 0.17 eV barrier.

The same picture emerges from direct optical studies: Photoexcitation of solutions of **1** in toluene, benzonitrile, or 2-MeTHF, or solutions of **5** in toluene, at room temperature with a 350-fs 582-nm laser pulse, yields in addition to the fast (<15 ps) decay found for compounds **2-6**, an additional slower step having a time constant of 55-95 ps. Again, the faster of these is ascribed to the formation and subsequent decay of H_2P_d-ZnP_p^+-Q^- (obtained directly from H_2P_d-ZnP_p^*-Q). The slower process, however, appears to involve the decay of $H_2P_d^*$-ZnP_p-Q back to the ground state (as judged by the observation of stimulated emission at ca. 720 nm, and selective bleaching of the ground state absorption bands of H_2P_d). Since the observed excited singlet lifetimes of the quinone-free gable and flat analogues, **15** and **18**, in toluene are 8.6 and 10 ns respectively, the 55-95 ps deactivation (of **1** and **5**) must be occurring by a quinone-induced charge separation process mediated by the central high-energy metalloporphyrin (ZnP_p).

Low temperature time resolved studies of **1** are also in excellent agreement with the static fluorescence data and indicate that the lifetime of the long-lived state assigned to $H_2P_d^*$-ZnP_p-Q is only weakly dependent on temperature: The time constant for deactivation increases from ca. 60 ps in 2-MeTHF at 295 K to 105 ps in a 2-MeTHF glass at 77 K. This critical result provides further confirmation that decay to the ground state does not involve a thermally activated process involving an intermediate ZnP state. For example, this result precludes the possibility that deactivation of H_2P^* occurs by uphill energy transfer to the ZnP subunit, followed by quenching via electron transfer to the quinone. We suggest, therefore, that electron transfer from $H_2P_d^*$ to Q is taking place *by a direct superexchange process mediated by the proximal ZnP_p moiety* as shown in eq. 3.

$$
\begin{array}{c}
H_2P_d\text{-}ZnP_p^*\text{-}Q \xrightarrow{\text{en. mig.}} \\
(2.13) \\
\Big\uparrow hv \\
H_2P_d\text{-}ZnP_p\text{-}Q \xrightarrow{\;hv\;}
\end{array}
\Bigg\} \longrightarrow H_2P_d^*\text{-}ZnP_p\text{-}Q \xrightarrow{\text{"sup. exchg."}} H_2P_d^+\text{-}ZnP_p\text{-}Q^- \qquad (3)
$$
$$(1.96) \qquad\qquad\qquad (1.19)$$

Here, the excited distal free-base subunit ($H_2P_d^*$) could arise from both direct photo-excitation and rapid, exothermic energy transfer from ZnP_p^*. Once formed, the charge separated state $H_2P_d^+$-ZnP_p-Q^- evidently returns to the ground state by a nonradiative process at a rate that is equal to or faster than that associated with charge separation. Although not determined by the current experimental data, we consider it likely on the basis of energetics that this process occurs by fast hole migration (to give H_2P_d-ZnP_p^+ -Q^-), followed by simple charge recombination (eq. 4).

$$H_2P_d^+-ZnP_p-Q^- \longrightarrow H_2P_d-ZnP_p^+-Q^- \longrightarrow H_2P_d-ZnP_p-Q \qquad (4)$$
$$(1.19) \qquad\qquad\qquad (1.10) \qquad\qquad\qquad (0)$$

In conclusion, studies with the multichromphore model systems **1** and **5** indicate that a porphyrin molecule may serve as an effective superexchange mediator in photoinduced charge separation processes. These findings support recent suggestions that a Bchl molecule could be playing a similar role in the natural photosynthetic reaction centers.

Acknowledgment

This work was supported by the Robert A. Welch Foundation (grant F-1018 to J.L.S.), the National Science Foundation (PYI Award, 1986 to J.L.S.), and the National Institutes of Health (grant no. GM 34685 to D.H., grant no. GM 41657 to J.L.S. and grant no. HL 131572 to J.A.I.). We thank Prof. R. Friesner for helpful discussions and the staff at the Center for Fast Kinetics Research at the University of Texas at Austin (an NIH supported shared user facility) for help in recording fluorescence lifetimes.

References

Allen, J. P., G. Feher, T. O. Yeates, D. C. Rees, J. Deisenhofer, H. Michel and R. Huber (1986) *Proc. Natl. Acad. Sci. U.S.A.* 83, 8589-8593.

Bixon, M., J. Jortner, M. E. Michel-Beyerle, A. Orgodnik and W. Lersch (1987) *Chem. Phys. Lett.* 140, 626-630.

Chang, C.-H., D. Tiede, J. Tang, U. Smith, J. Norris and M. Schiffer (1986) *FEBS Lett.* 205, 82-86.

Cowan, J. A., J. K. M. Sanders, G. S. Beddard and R. J. Harrison (1987) *J. Chem. Soc., Chem. Commun.* 1987, 55-58.

Deisenhofer, J., O. Epp, K. Miki, R. Huber and H. Michel (1984) *J. Mol. Biol.* 180, 385-398.

Fleming, G. R., J. L. Martin and J. Breton (1988) *Nature (London)* 333, 190-192.

Gust, D., T. A. Moore, A. L. Moore, L. R. Makings, G. R. Seely, X. Ma, T. Trier, F. Gave (1988) *J. Am. Chem. Soc.* 110, 7567-7569.

Holten, D., C. Hoganson, M. W. Windsor, C. C. Schenck, W. W. Parson, A. Migus, R. L. Fork and C. V. Shank (1980) *Biochim. Biophys. Acta* 592, 461-477.

Joran, A. D., B. A. Leland, G. G. Geller, J. J. Hopfield, and P. B. Dervan (1984) *J. Am. Chem. Soc.* 106, 6090-6092.

Joran, A. D., B. A. Leland, P. M. Felker, A. H. Zewail, J. J. Hopfield and P. B. Dervan (1987) *Nature (London)* 327, 508-511.

Kirmaier, C. and D. Holten (1987) *Photosyn. Res.* 13, 225-260.

Kirmaier, C. and D. Holten (1988) *FEBS Lett.* 239, 211-218.

Kirmaier, C., D. Holten and W. W. Parson (1985) *Biochim. Biophys. Acta* 810, 33-48.

Marcus, R. A. and N. Sutin (1985) *Biochim. Biophys. Acta* 810, 265-322.

Marcus, R. A. (1987) *Chem. Phys. Lett.* 133, 471-476.

Martin, J.-L., J. Breton, A. J. Hoff, A. Migus and A. Antonetti (1986) *Proc. Natl. Acad. Sci. U.S.A.* 83, 957-961.

Sakata, Y., S. Nishitani, N. Nishimizu, S. Misumi, A. R. McIntosh, J. R. Bolton, Y. Kanda, A. Karen, T. Okada and N. Mataga (1985) *Tetrahedron Lett.* 1985, 5207-5210.

Sessler, J. L. and M. R. Johnson (1987) *Angew. Chem.* 99, 679-680; *Angew. Chem. Int. Ed. Engl.* 26, 678-680.

Sessler, J. L., M. R. Johnson, T.-Y. Lin and S. E. Creager (1988) *J. Am. Chem. Soc.* 110, 3659-3661.

Wasielewski, M. R., M. P. Niemczyk, W. A. Svec and E. B. Pewitt (1985) *J. Am. Chem. Soc.* 107, 5562-5563.

Wasielewski, M. R. and D. M. Tiede (1986) *FEBS Lett.* 204, 368-372.

Windsor, M. W. (1986) *J. Chem. Soc., Faraday Trans. 2* 1986, 2237-2243.

Woodbury, N. W., M. Becker, D. Middendorf and W. W. Parson (1985) *Biochemistry* 24, 7516-7521.

Won, Y. and R. A. Friesner (1987) *Proc. Natl. Acad. Sci. U.S.A.* 84, 5511-5515.

Organization of the Photosynthetic Membrane of *Rhodopseudomonas Viridis* Using Biochemical, Immunological and Electron Microscopical Methods

FRANCIS A. JAY

Sandoz AG.
CH-4002 Basel
Switzerland

Rhodopseudomonas viridis is a purple photosynthetic bacterium which has a natural econiche in the depths of stagnant water. It possesses an excellent adaptation to the energy-poor, long wavelength, light which reaches it by a repetitive, hexagonal arrangement of the light collecting antennae components and by the possession of a bacteriochlorophyll pigment (BChl b) which has a wavelength optimum at 1015–1020 nm.

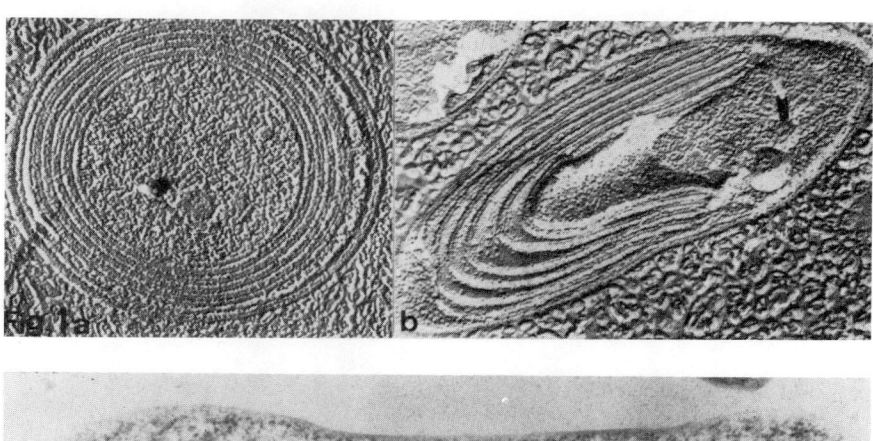

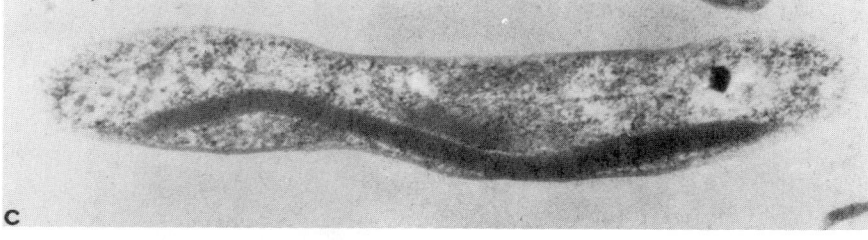

Figure 1. Freeze-etched and shadowed fracture planes of Rhodopseudomonas (Rps) viridis (transverse section in (a), longitudinal section in (b)), (c) Lowicryl embedded cells thin sectioned and stained.

Photobiology, Edited by E. Riklis
Plenum Press, New York, 1991

A large proportion of the cell interior is occupied by the photosynthetic apparatus arranged on membranes, or so-called thylakoids, the amounts of which are regulated by light intensity. Fig. 1 shows the morphology of the cell obtained by electron microscopy of thin sectioned and negatively stained, or freeze-etched and shadowed, cells. The stacked thylakoid membranes can clearly be seen and a closer examination of the longitudinally sectioned bacterium (Fig. 1 b) discerns a repetitive structure on the membrane surface. These structures represent the photosynthetic apparatus, made up of units containing the reaction centre and light harvesting components.

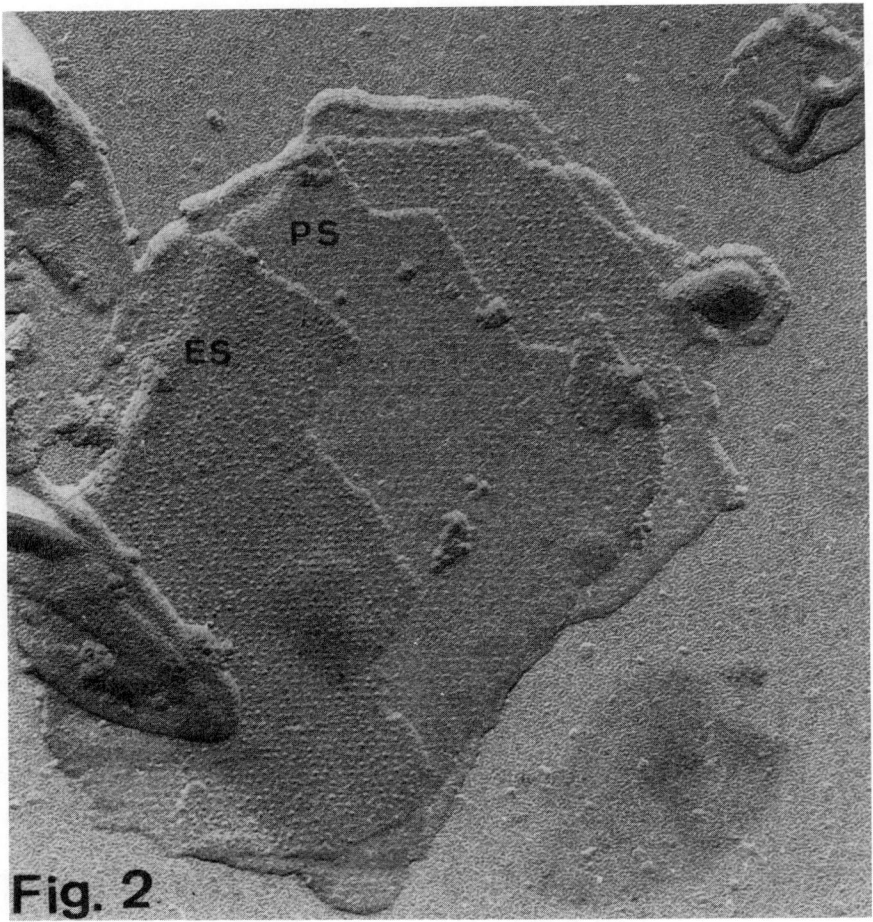

Figure 2. Electron microscopy of freeze-dried and shadowed membranes of Rps. viridis. ES = Exoplasmatic surface (exposed to the thylakoid lumen) PS = Plasmatic surface (outer thylakoid membrane)

When the cells are disruptive, the closed thylakoid stacks are broken up revealing the two surfaces of the thylakoid membrane which can be morphologically distinguished after freeze drying and shadowing (Fig. 2).

The outer plamatic surface (PS) is comparatively smooth being made up of 13 nm diameter units, whereas the inner exoplasmatic (ES) surface appears rough due to the presence of 4.5 nm protrusions.

These membranes can be induced to close, forming vesicles with the plasmatic surface uppermost (PSU vesicles), or exoplasmatic surface uppermost (ESU vesicles) which, once separated, are useful for biochemical localisation experiments (Jay and Lambillotte, 1985).

The structure of the photosynthetic units within the thylakoid membrane have been investigated using computational analyses (image processing). The highest resolution image processing information of the membranes of Rhodopseudomonas viridis were obtained by Stark et al., (1984). The photosynthetic unit is composed of a central core surrounded by a ring made up of six units (probably further subdivided on the PS surface). The outer ring of protein was shown to be closer to the centre of the ES surface than on the PS surface, which indicates an inward tilting of the outer proteins with respect to the membrane plane. Labelling the membranes with antibodies identified the central region as corresponding to the reaction centre complex (Stark et al., 1986). The size of the photosynthetic unit (13 nm diameter) compared to the dimensions of Fab fragments of antibodies (7×4 nm) used for the labelling imposed serious restrictions on the resolution of the structures within the photosynthetic unit and a clear identification of the outer ring as antennae complexes was not possible.

Due to imperfections in the structural arrays and differences in the orientation of the reaction centre with respect to the light harvesting complexes (Engelhardt et al., 1985) further structural analysis of membranes seemed limited. The most promising approach for elucidating the structure of the entire unit appeared to be to attempt crystallization. Methods had been established (Jay et al., 1984) for the isolation of photosynthetic units from Rhodopseudomonas viridis but alternative solubilizing detergents were necessary for crystallization purposes. Despite the extreme instability of the light harvesting bacteriochlorophyll (which may, by the way, account for the inability to isolate a functional light harvesting complex from this bacterium), it was possible to isolate and purify units free from accessory proteins (Fig. 3) in higher yields and more stable at 4°C than previous preparations and using detergents compatible with crystallization (Fig. 4).

Unfortunately no crystals sufficiently suitable for diffraction analysis were obtained from these preparations, but the application of image processing methods to such photosynthetic units offered a possibility of obtaining further structural information from the complexes within the photosynthetic unit.

In this method individual units were selected via corresponding analysis followed by averaging. An average from 309 negatively stained units is shown in Fig. 5(a and b). The outer ring is seen to be composed of six dimers and each reaction centre is seen to be

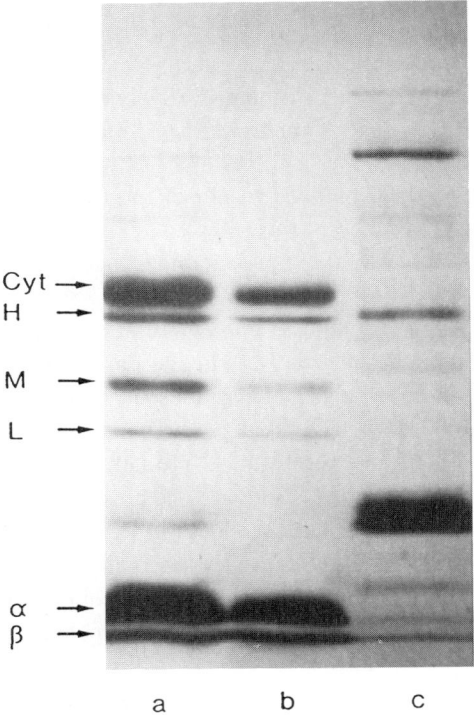

Cyt →
H →
M →
L →
α →
β →

a   b   c

Figure 3. SDS-Page of native membranes (lane a) of Rps. viridis, photosynthetic units prepared using octyl glucoside solubilization (lane b) and fraction removed during unit preparation (lane c) corresponding to denatured pigment and unidentified proteins.

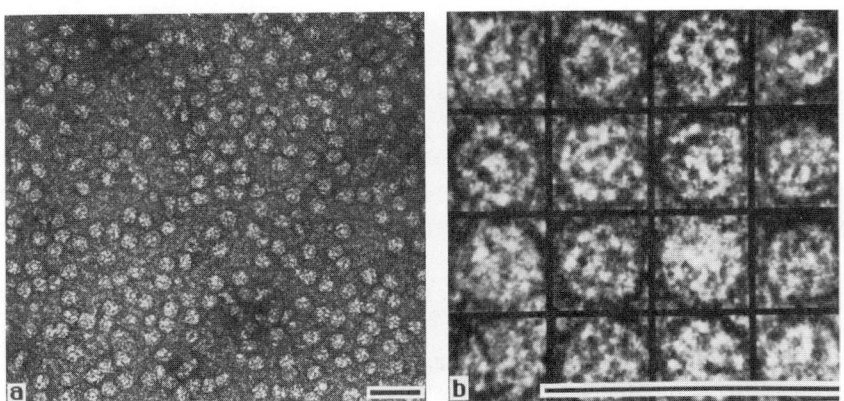

Figure 4(a). Negatively stained photosynthetic units from Rps. viridis (b) individually selected units. Bars represent 40 nm.

more closely associated with the major domain than the minor one (Jay and Wildhaber, 1988).

Biochemical information regarding the primary structure and topology of the light harvesting components were independently obtained in order to correlate with this structural mode. It was, needless to say, unnecessary to search for data on the reaction centre complex due to the success of Michel and Deisenhofer in elucidating the structure of this reaction centre complex, its location in the membrane and the sequence of the component polypeptides (Michel, 1982; Deisenhofer et al., 1984).

Methods developed in the laboratories of Loach and Zuber for the isolation of light harvesting polypeptides enabled a rapid isolation of the three light harvesting polypeptides from Rhodopseudomonas viridis (B1015-α,β, and γ) which were then sequenced (Brunisholz et al., 1985). The distribution of hydrophilic and hydrophobic amino acids in the primary structure indicate the possibility of a single hydrophobic stretch in each polypeptide, of a length sufficient to transverse the membrane once in the

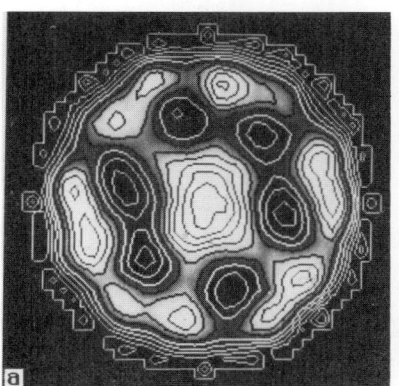 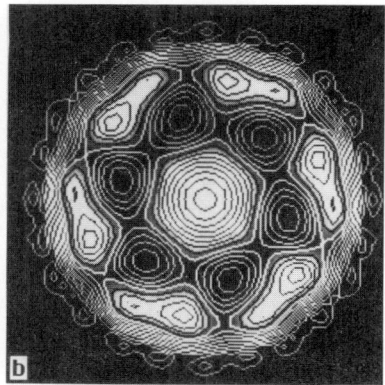

Figure 5(a) Contour map of an average of 309 negatively stained photosynthetic units selected with corresponding analysis. (b) Six-fold symmetrisation of the same map (due to the asymmetry of the reaction centre the symmetrical form is only correct for the ring structure).

form of an alpha helix. Data from Nabedryk et al., (1985) using circular dichroism of photosynthetic units support this alpha helx confrontation. Immune electron microscopy of membranes labelled with monoclonal antibodies raised against the light harvesting polypeptides confirmed a unidirectional, transmembrane, orientation of the two largest antennae polypeptides (Jay et al., 1985). An example of membranes immunolabelled with monoclonal antibody and visualised in the electron microscope using gold conjugates is shown in Fig. 6.

Epitope mapping of the antigenic determinants of the polypeptides showed that the N termini of each antennae protein is orientated toward the plasmatic surface. This was accomplished by fragmentation, separation and sequencing the polypeptides, as well as by creating synthetic peptides chosen using different epitope prediction methods. In the case

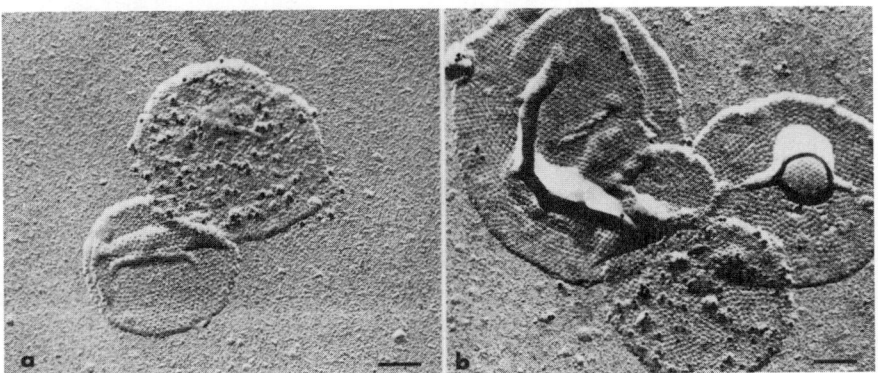

Figure 6. Immune electron microscopy with monoclonal antibody against (a) the N terminus of
B1015-β (b) the C terminus of B1015-β. Detection is with lentil lectin gold (a) or
protein A-gold (b).

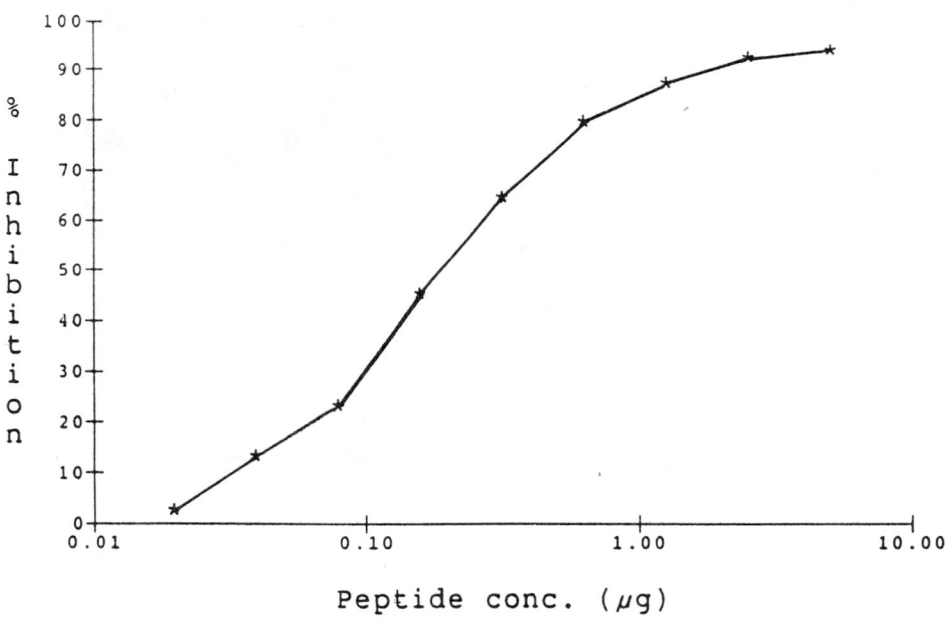

Figure 7. Inhibition ELISA using increasing amounts of peptide (Pro-Ile-Pro-Lys-Gly-Trp-Val,
corresponding to the C terminus of B1015-β) to inhibit the binding of monoclonal antibody
to membranes of Rps. viridis

of polypeptide B1015-β; two major epitopes were predicted near the N terminus (around amino acid No. 5 and in the environment of amino acid No. 15) with a number of different prediction methods (Engleman et al., 1986; Parket et al., 1986; Welling et al., 1985; Hopp and Woods, 1981; Kyte and Doolittle, 1982).

Monoclonal antibodies recognising these epitopes were indeed found. A C-terminus epitope around position 52 was only predicted using the acrophilicity method of Hopp (1984) and this proved to be a relatively dominant epitope with a number of antibodies reacting with this determinant with varying affinities. A relatively short peptide of seven amino acids (Pro-Ile-Pro-Lys-Gly-Trp-Val) completely inhibited the binding of a set of antibodies to thylakoids in the enzyme-linked-immunosorbent assay in a titratable fashion. The inhibition of binding to membranes, of one antibody (VIFF9 directed against the C terminus of B1015-β) by preincubation with increasing amounts of this peptide is shown in Fig. 7. Furthermore, the antibodies directly bound to the peptide and competed with each other i.e. the epitope corresponds to the C terminal 7 amino acids. Incidentally, this C terminal epitope, as well as the N terminal epitopes of B1015-β, are conserved in the related bacterium Rhodopsuedomonas sulfoviridis (Wyss et al. 1985).

The sequences of two of the light harvesting polypeptides (B1015-α and β) both contain histidine residues at positions suggested to act as a ligand to the bacteriochlorophyll b. In the third polypeptide no such histidine is present. The photosynthetic units contain an average of 24 bacteriochlorophyll molecules per unit (Stark et al., 1986) which, assuming these are exclusively bound on B1015-α and β, suggest that 12α and 12β polypeptides are present per unit. The polypeptides are known to be abundant in a 1:1:1 ratio (Brunisholz et al., 1985) and are all present in unit preparations; which suggests 12γ polypeptides per unit. Cross linking data has shown that the H polypeptide from the reaction centre is closely associated with B1015-β (Peters et al., 1984, Lugwig and Jay, unpublished) and that α and β are closely associated with each other (Ludwig and Jay, 1985). The association of α and β in the form of an $\alpha_2\beta_2$ complex was also suggested by partial solubilization of membranes (LDS-Page, Jay and Wildhaber, 1988).

In summary, a model is proposed on the arrangement of the light harvesting complexes in a ring surrounding the reaction centre complex. The model is made up of a concentric ring subdivided into 12 units, six of which contain $\alpha_2\beta_2$ closely associated with the reaction centre, interdiginated with six smaller units corresponding to γ_2.

References

Brunisholz, R.A., Jay, F., Suter, F., Zuber, H. *Biol. Chem. Hoppe-Seyler* 366, 87–98, 1985.

Deisenhofer, J., Epp, O., Miki, K., Huber, R., Michel, H. *J. Mol. Biol.* 180:385–398, 1984.

Engelhardt, H., Guckenberger, R., Hegerl, R., Baumeister, W. *Ultramicroscopy* 16, 395–410, 1985.

Engelman, D.M., Stetz, T.A., Goldman, A. *Ann. Rev. Biophys. Biophys. Chem.* 15, 321–352, 1986.

Hopp, T.P. p3–12 in "Synthetic peptides in biology and medicine" Eds: Alitalo, K., Partanen,
    P., Vaheri, A. Elsevier 1985.
Jay, F., Lambillotte, M. *Eur. J. Cell Biol.* 37:7–13, 1985.

Photoreceptors: Phytochrome

The Chromophore and its Role in the Phototransformation of Phytochrome: Conformational Changes Probed with bis-ANS, Monoclonal Antibodies and Phosphorylation

PILL-SOON SONG*, UMA BAI*, IN-SOO KIM*, GARRY C. WHITELAM#
AND JOHN P. MARKWELL+

*Departments of *Chemistry and +Biochemistry*
University of Nebraska
Lincoln, NE 68588, USA and
#Department of Botany
University of Leicester
Leicester LE1 7RH, UK

Introduction

Phytochrome is the photochromic receptor for a variety of photo-morphogenic and developmental responses of higher plants to red light (for a detailed review of the subject, see Song, 1983, 1988; Lagarias, 1985; Quail et al., 1986; Schaefer and Briggs, 1986; Furuya, 1987). In this paper, results on the study of the role of the chromophore on the phototransformation of 124 kDa oat phytochrome obtained in our laboratory will be reviewed.

The reversible phototransformation of 124 kDa oat phytochrome, as shown schematically below,

$$\text{Pr} \underset{730 \text{ nm}}{\overset{660 \text{ nm}}{\rightleftharpoons}} \text{Pfr} \rightarrow \text{Biological action}$$

involves observable structural changes in the chromophore and the apoprotein, including a configurational/conformational isomerization (Ruediger et al., 1983; Song, 1988, for review) and secondary/tertiary conformational changes (Chai et al., 1987; Song, 1988, for review). Our studies have suggested that there is a specific interaction between the chromophore and the amino terminus segment in the Pfr form of phytochrome, which results in a photoreversible alpha-helical folding of the amino terminus peptide chain (Chai et al., 1987). Peptide mapping (Vierstra and Quail, 1983; Lagarias and Mercurio, 1985; Grimm et al., 1988), monoclonal antibodies (Cordonnier et al., 1985) and

Photobiology, Edited by E. Riklis
Plenum Press, New York, 1991

phosphorylation (Wong et al., 1986) have also been used to probe different aspects of the conformational changes induced by the Pr → Pfr phototransformation. In this paper, we further examine the nature of the protein kinase activity of 124 kDa phytochrome preparation reported by Wong et al. (1986), as a function of perturbations on the chromophore conformation and the amino terminus peptide.

Materials and Methods

Phytochrome having a molecular mass of 124 kDa was isolated and purified from the etiolated oat seedlings, as described by Chai et al. (1987). A monoclonal antibody (LAS-41) that recognizes an epitope of the amino terminus peptide unique to the Pr form of phytochrome was produced as described by Holdsworth and Whitelam (1987). Protein kinase assay was carried out as described recently (Kim et al., 1988). Other pertinent references used in our studies have been cited elsewhere (Song, 1988).

Results and Discussion

(1) The Chromophore Topography

One of the prominent structural changes that accompany the phototransformation of phytochrome appears to be the preferential exposure of the tetrapyrrole chromophore in the Pfr form, compared to the Pr chromophore. This conclusion is derived from the fact that the rate of specific oxidation of the phytochrome chromophore is significantly faster in the former than in the latter. Results on the tetranitromethane oxidation of phytochromes are summarized in Table 1. Since the fully exposed chromophores in the Pr- and Pfr-chromopeptides are oxidized by tetranitromethane at almost identical rates, the differential rates of oxidation of the 124 kDa oat phytochrome have been attributed to the

Table 1. Relative rates of the oxidation of oat phytochrome and model chromophore (0.9 uM) by tetranitromethane (133 μM) (results taken from Hahn et al., 1984; Thuemmler et al., 1985).

Compound	Conditions	Relative rate $[(s^{-1}) \times 10^5]$	Rate ratio (Pfr/Pr)
124 kDa Pr	0.1 M KPB, pH 7.8, at 2°C	5.8	–
124 kDa Pfr	0.1 M KPB, pH 7.8, at 2°C	48.8	8.3
Pr-peptide	in methanol, 10°C	4800	–
Pfr-peptide	in methanol–5% pyridine, 10°C	5100	1.0
Bilirubin	0.1 M KPB, pH 7.8, at 2°C	1220	–
Biliverdin	0.1 M KPB, pH 7.8, at 2°C	1260	–

difference in steric accessibility of the chromophores in the two forms of phytochrome, rather than to a difference in chemical reactivity.

From a spectral analysis of the tetranitromethane oxidation of phytochrome and its proteolytic degradation fragments, position 15 of the tetrapyrrole chromophore in the Pfr form has been determined to be preferentially attacked by tetranitromethane (Farrens et al., 1989). The site-selective oxidation of the chromophore in the Pfr form further suggests that the ring D region of the chromophore becomes predominantly exposed in the Pr --> Pfr phototransformation of phytochrome.

Results similar to the tetranitromethane study have also been obtained with the borohydride reduction (Chai et al., 1987; Song and Yamazaki, 1987). Similar results have been observed using hydrophobic fluorescence probes 8-anilinonaphthalene-1-sulfonate (ANS) and bis-ANS as a probe, which force exposure of the tetrapyrrole chromophore by competitively binding to the chromophore crevice and/or the amino terminus segment (Song and Yamazaki, 1987; Song, 1988, for review). The binding of these probes to phytochrome results in spectral bleaching of the chromophore absorbance band, as a result of the disruption of the chromophore-apoprotein non-covalent bonds, presumably hydrogen bond. Fig. 1 shows an example of how ANS as a specific hydrophobic fluorescence probe preferentially bleaches the Pfr chromophore (Song, 1988, for review).

Studies of phytochromes having different degrees of truncation of the amino terminus segment, i.e., 118/114 and 60 kDa species (Hahn et al., 1984) with ANS and bis-ANS have shown that the degree of the chromophore exposure in the Pfr form of phytochrome is modulated by a specific interaction between the amino terminus segment and the chromophore (Song, 1988, for review).

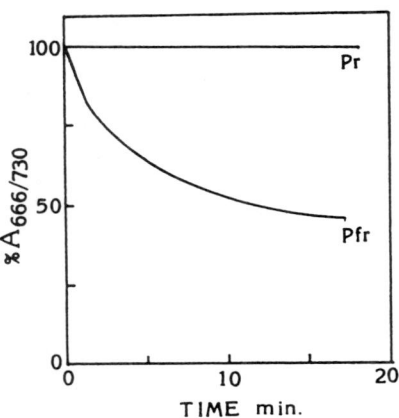

Figure 1. Time courses for the spectral bleaching of the 124 kDa oat phytochrome upon addition of 4 mM ANS at 20°C. Spectral changes were monitored by absorbance at 667 nm and 730 nm for Pr and Pfr, respectively (taken from Song, 1988).

(2) A Photoreversible Peptide Folding

Evidence for the specific interaction between the Pfr-chromophore and the amino terminus segment of the 124 kDa intact phytochrome molecule comes from a circular dichroic (CD) study (Chai et al., 1987; Song, 1988, for review). We have observed a photoreversible increase in the far-UV CD signal upon Pr → Pfr phototransformation. This CD increase has been attributed to an alpha-helical folding of the amino terminus segment in the Pfr form of phytochrome, as a result of a specific interaction between the peptide and the Pfr-chromophore. The photoreversible alpha-helical folding can be suppressed by monoclonal antibody binding near the amino terminus sequences, thus providing evidence for a Pfr-chromophore-peptide interaction (Chai et al., 1987). The photoreversible alpha-helical folding can also be abolished by sodium borohydride and tetranitromethane, which chemically modify the chromophore structure/conformation. We have recently observed that a specific binding of zinc ion to the Pfr-chromophore suppresses the photoreversible conformational change (Sommer and Song, 1990).

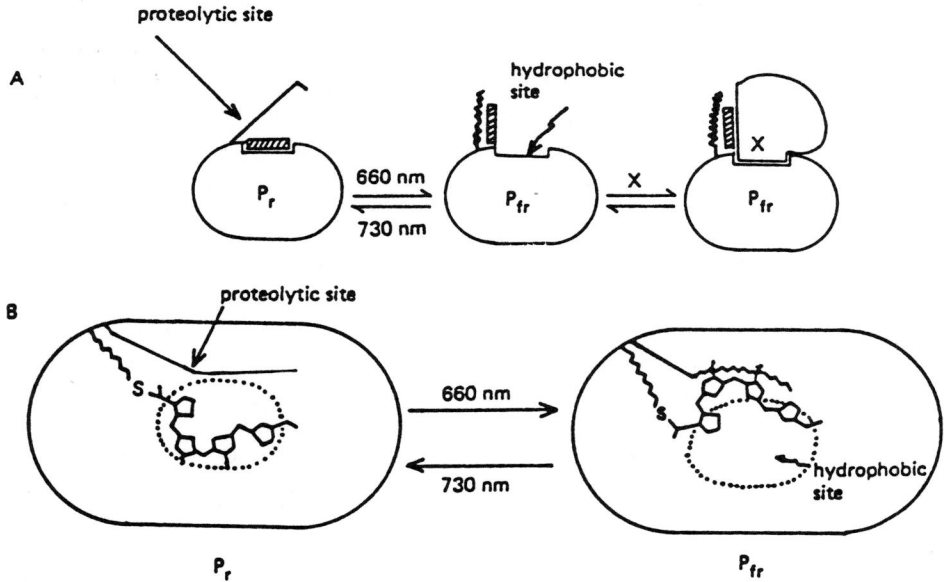

Figure 2. A schematic model for the phototransformation of 124 kDa oat phytochrome. The 6-kDa amino terminus segment is assumed to be in beta-turn and random coil conformations (Quail et al., 1987) and fully accessible in the Pr form, whereas it folds into an alpha-helical conformation in the Pfr form as a result of the interaction with the exposed Pfr-chromophore. A; side view, B; top view. X represents putative receptor/binding site (modified from Song, 1983).

Fig. 2 presents a cartoon model for the phototransformation of phytochrome, illustrating the movement/exposure of the chromophore and its topographic role in stabilizing the alpha-helical conformation of the amino terminus sequence of the Pfr-phytochrome. The hypothetical model also depicts the partially exposed chromophore binding site that contributes to the surface hydrophobicity of the Pfr form of phytochrome.

(3) Phosphorylation Studies

(a) The nature of the phytochrome autophosphorylation.

Wong et al. (1986) have used protein phosphorylation by protein kinases as a method to probe the differential conformations of the Pr vs. Pfr forms of phytochrome. It was shown that the sites of phosphorylation were different between the Pr and Pfr forms of phytochrome, indicating that a significant conformational change took place around the phosphorylation sites upon Pr → Pfr phototransformation. More intriguing was the observation that phytochrome preparation autophosphorylated both Pr and Pfr species of the oat phytochrome (Wong et al., 1986; Lagarias et al., 1987). The presence of kinase activity in the purified phytochrome preparation is interesting since the reversible phosphorylation of proteins is an important regulatory mechanism in cell metabolism. We have examined the protein kinase activity of phytochrome preparation in detail (Kim et al., 1989), and a brief account of that study is reviewed here.

In agreement with the original observation (Wong et al., 1986), highly purified phytochrome preparations showed the phosphorylating activity for phytochrome and histone. However, phytochrome preparations with different purities exhibited different levels of autophosphorylation activity. We also found that the protein kinase activity declined gradually during the storage of phytochrome at 4°C, although the characteristic absorbance spectrum of phytochrome and its phototransformability remained virtually unaltered. This observation appeared to suggest that the kinase activity is independent of the chromophore integrity. The simplest explanation is that the protein kinase activity of phytochrome preparations is due to an enzyme other than phytochrome.

SDS polyacrylamide gel electrophoresis of different phytochrome preparations and its corresponding autoradiograms of 32-P incorporation have indicated that phytochrome preparations exhibited a purity-dependent protein kinase activity (Kim et al., 1989). The protein kinase activity of the phytochrome preparation could be separated from phytochrome when the second supernatant was passed through a Bio-Gel A-0.5m column. A repeated gel filtration of the second supernatant on the column resulted in a progressive increase in the phytochrome purity as assessed by the specific absorbance ratio (SAR) value and a concomitant decline in the protein kinase activity. After three successive gel filtrations, we could obtain a phytochrome preparation of SAR=1.07 and free of detectable kinase activity (Fig. 3). The kinase-rich phytochrome fraction from the Bio-Gel A-0.5m column was mixed with an equal volume of the purest phytochrome fraction and the kinase activity of the mixture was assayed. The phosphorylating activity

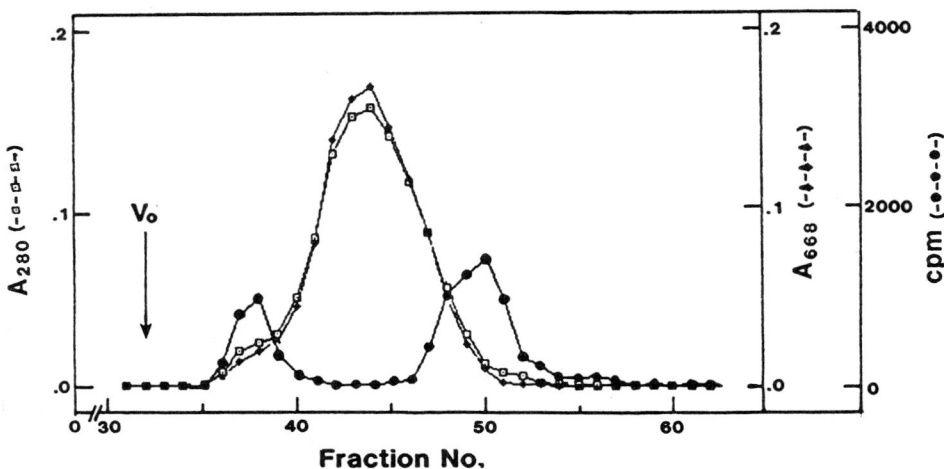

Figure 3. The elution profile of the third gel filtration of the second supernatant (SAR = 0.83) on a Bio-Gel A-0.5m column. The absorbances for phytochrome were monitored at 280 and 668 nm and the protein kinase activity was assayed by the filter paper method. Vo indicates the void volume of the column. For technical details, see Kim et al. (1989).

per unit volume of the mixture decreased below that of the kinase-rich phytochrome fraction, although the amount of phytochrome increased. When the second phytochrome preparation was analyzed by two dimensional gel electrophoresis, the major phytochrome band lacked the kinase activity, as examined by the Coomassie brilliant Blue staining and corresponding autoradiogram of the gel (Kim et al., 1989).

Our results indicated that the protein kinase activity was associated with the phytochrome preparations with SAR values up to 1.05, in spite of the fact that phytochrome preparations with SAR values greater than 0.95 were homogeneous on SDS polyacrylamide gel electrophoresis when stained with Coomassie brilliant Blue R-250. This result is consistent with the previous report (Wong et al., 1986) in which the protein kinase activity was present in highly purified phytochrome preparations with SAR values up to 1.0 and phytochrome itself was a good substrate for the kinase. Our results are explicable in terms of the protein kinase being a co-purified and/or associated contaminant, although the suggestion that phytochrome possesses an intrinsic kinase activity (Wong et al., 1986; Lagarias et al., 1987) cannot be ruled out completely at this time. Ruediger and his coworkers (private commun.) have also arrived at the conclusion that phytochrome and the protein kinase activity can be chromatographically separated from each other.

The kinase activity present in the purified phytochrome preparations exhibited a polycation-dependent enhancement of its activity and preferential phosphorylation of the Pr form over the Pfr form of phytochrome (Table 2). Polylysine and histone enhanced the

416

kinase activity by 10–20% and 43–48% for the Pr and Pfr forms of phytochrome, respectively. However, this enhancement of kinase activity by polycations is far below than that observed by Wong et al. (1986). The reproducible preference of the Pr form for phosphorylation suggests that the Pr form has more accessible phosphorylatable serine residues probably in the amino terminus segment than the Pfr form, vide infra.

Table 2. The polycation effects on the protein kinase activity of purified phytochrome preparations (data taken from Kim et al., 1989).

Phytochrome	SAR	Polycation (PC) added	^{32}P-Incorp. (cpm)		Incorp. ratio	
			Pr	Pfr	Pfr/Pr	Pr-PC/Pr[*]
2nd supernat.	0.83	none	19666	14418	0.73	
		polylysine	23992	22023		1.22
		histone	20978	20582		1.07
Bio-Gel A-0.5m fraction	1.01	none	4067	2921	0.72	
		polylysine	4851	4334	0.89	1.19
		histone	4796	3839	0.80	1.18

[*]The ratio of ^{32}P-incorporation into the Pr-phytochrome in the presence (Pr-PC) vs. in the absence (Pr) of polycations, polylysine and histone.

(b) Effects of bis-ANS, Zn ion and a monoclonal antibody (LAS-41) specific to the amino terminus of the Pr form.

Bis-ANS complexes with phytochrome, exhibiting a higher affinity for the Pfr form than for the Pr form (Choi, 1987). The effect of bis-ANS on phosphorylation of phytochrome is shown in Table 3. Bis-ANS inhibited phosphorylation of both forms of phytochrome to the same degree, although the absorbance of the Pfr form was preferentially bleached by bis-ANS than that of the Pr form. This result suggests that either the kinase activity is not associated with phytochrome chromophore or it is due to a contaminant kinase, since the effect of bis-ANS on kinase activity does not coincide with the absorbance changes of phytochrome.

Zn ions chelate with the phytochrome chromophore and change its protein conformation destroying its spectral integrity, particularly in the Pfr form (Sommer and Song, 1990). Zn ions enhanced the phosphorylation of both forms of phytochrome to the same extent at concentrations at which absorbance of the Pfr form was selectively bleached. For example, at 50–100 uM Zn^{2+} where most of the Pfr absorbance at 730 nm was bleached (Lisansky and Galston, 1974), the protein kinase activity in "autophosphorylating" both forms of the phytochrome molecule increased by 10% at most (Table 4). The lack of inhibition of the kinase activity by Zn ions suggests that the chromophore conformation/topography is not important in the kinase activity, with phytochrome either as a substrate or as a catalytic unit.

Table 3. Effect of bis-ANS on phosphorylation of phytochrome by a phytochrome preparation (SAR=0.85, 15 ug of phytochrome). For the experimental details, see Kim et al. (1989).

Conc. of bis-ANS (μM)	% Incorporation	
	Pr	Pfr (Pfr/Pfr_o)*
0	100	94 (100)
12	82	72 (76)
25	68	58 (61)
50	55	46 (49)
100	40	33 (35)
200	29	17 (18)
500	4	3 (3)

* Calculated on the basis of ^{32}P-incorporation into the Pfr form in the absence of bis-ANS (Pfr_o) as 100%.

Table 4. Effect of $ZnCl_2$ on polylysine-activated phosphorylation of oat phytochrome. The experimental conditions were the same as described in Table 3.

Conc of ZnCl$_2$ (μM)	% Incorporation	
	Pr	Pfr (Pfr/Pfr_o)*
0	100	94 100)
25	109	97 (103)
50	107	96 (102)
100	104	92 (97)
200	97	89 (94)
500	91	83 (89)
1000	7	6 (6)

*Refer to Table 3. Pfr_o represents ^{32}P-incorporation into the Pfr form without $ZnCl_2$.

A monoclonal antibody (LAS-41) raised against the amino terminus segment of the Pr form (Holdsworth and Whitelam, 1987) inhibited the phosphorylation of phytochrome by the kinase present in partially purified phytochrome preparations. Inhibition was more significant with the Pr form than with the Pfr form, as expected from the fact that this antibody is unique to the Pr form (Table 5). As the ratio of mAb to phytochrome increased, the Pr phosphorylation was proportionately inhibited, while the degree of the Pfr phosphorylation was unaffected. The initial decrease in the Pfr phosphorylation can be considered to be due to the Pr form remaining in photoequilibrium. This observation confirms the proposal of Wong et al. (1986) in that the predominant phosphorylation site is available near the amino terminus of the Pr form. The amino terminus peptide chain apparently becomes less accessible to the kinase attack in the Pfr form, as it interacts with the Pfr-chromophore and folds into an alpha-helical conformation (Chai et al., 1987)

Table 5. Inhibition of phytochrome phosphorylation by a monoclonal antibody, LAS-41

mAb added (μg)	[mAb]/[P]	% Incorporation	
		Pr	Pfr (Pfr/Pfro)*
0	0	100	68 (100)
5	0.26	95	59 (87)
10	0.52	90	58 (87)
20	1.04	40	47 (69)
40	2.08	35	53 (78)

*Refer to Table 3. Pfr_o represent ^{32}P-incorporation into the Pfr form in the absence of LAS-41.

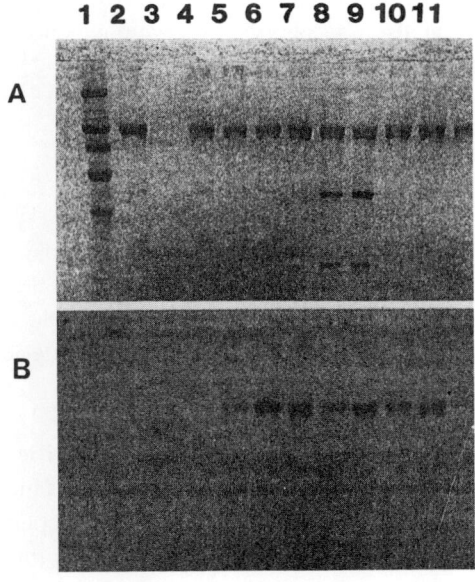

Figure 4. Phosphorylation of large oat phytochrome by a copurified kinase in the purified phytochrome preparation. Experimental conditions were the same as described in Kim et al. (1989) except that a partially purified intact phytochrome preparation was digested with endogeneous proteases to produce large phytochrome. Panel A; Coomassie brilliant Blue staining. Panel B; autoradiogram of ^{32}P-incorporation. Lane 1, MW standard; Myosin (200,000), β-Galactosidase (116,250), Phosphorylase b (97,400), BSA (66,200), Ovalbumin (42,699). Lanes 2 and 4, Phytochrome only (SAR=0.82). Lane 3, incubation mixture without phytochrome. Incubation mixtures with boiled phytochrome (lane 5), Pr-phytochrome (lane 6) and Pfr-phytochrome (lane 7). Incubation mixtures with the Pr (lane 8) and the Pfr (lane 9) forms of phytochrome (15 μg) in the presence of monoclonal antibody of LAS-41 (20 μg). Incubation mixtures with the Pr (lane 10) and the Pfr (lane 11) of phytochrome in the presence of 100 uM bis-ANS.

(c) Phosphorylation of degraded large (114 kDa) phytochrome by the purified phytochrome preparation.

The phosphorylation of large phytochrome by the copurified kinase in the phytochrome preparation is shown in Fig. 4. A comparison of Coomassie Blue staining and autoradiogram of the gel indicates that the Pfr form of large phytochrome was phosphorylated more efficiently than the Pr form (lanes 6 and 7 in Fig. 4). This may be due to the fact that the large phytochrome lost a 10 kDa amino terminus segment of the intact phytochrome, which includes one of the major phosphorylation sites in the Pr form (Wong et al., 1986). Monoclonal antibody LAS-41 and bis-ANS also inhibited the phosphorylation of large phytochrome. Bis-ANS (100 μM) inhibited the both forms of phytochrome to the same extent, but the monoclonal antibody (LAS-41) at a 1:1 (phytochrome) molar ratio inhibited the phosphorylation of Pr more effectively than the phosphorylation of Pfr. The antibody effect suggests that monoclonal antibody LAS-41 can still interact with the Pr form of large phytochrome, although its epitope is probably partially truncated. This result indicates that the phytochrome-copurified kinase in the phytochrome preparation also phosphorylates large phytochrome in the same manner as in the phosphorylation of intact phytochrome.

Conclusions

The phytochrome molecule has well been characterized with respect to its structure and spectroscopic/photochemical properties, but its molecular action mechanism largely remains to be elucidated. Our attention has been focussed on the chromophore topography and its relationship to the secondary structure of the amino treminus sequence of the intact pnytochrome. We have proposed a schematic model of these features and its implication in interactions between phytochrome and its putative receptor(s)/binding partner(s).

Wong et al. (1986) reported that the highly purified phytochrome possess polycation-dependent protein kinase activity. This is a most intriguing observation since reversible phosphorylation is involved in a variety of metabolic regulations in cells. The effects of several reagents having charasteristic binding properties with phytochrome indicate that the kinase activity present in the purified phytochrome preparation did not corelate with the chromophore integrity of phytochrome. The kinase activity was also separable from phytochrome by repeated gel filtrations and phytochrome of a purity of SAR=1.07 was free of kinase activity. These results indicate that phytochrome has no intrinsic kinase activity. The kinase in the phytochrome preparation was not specific to intact phytochrome but degraded large (114 kDa) phytochrome as well as other proteins were phosphorylated by the kinase. This suggests that this kinase may not be specific enough to play a significant physiological role in the phytochrome photoresponses.

Acknowledgements

The author's work described in this chapter has been supported by an NIH grant GM-36956.

References

Chai, Y.G., Song, P.S., Cordonnier, M.M. and Pratt, L.H. (1987) A photoreversible circular dichroism spectral change in oat phytochrome is suppressed by a monoclonal antibody that binds near its N-terminus and by chromophore modification. *Biochemistry* 26: 4947-4952.

Choi, J.K. (1987) Protein topography and binding properties of phytochrome. Ph. D. Dissertation, Texas Tech University, Lubbock, Texas.

Cordonnier, M.M., Greppin, H. and Pratt, L.H. (1985) Monoclonal antibodies with differing affinities to the red-absorbing and far-red absorbing forms of phytochrome. *Biochemistry* 24: 3246-3253.

Farrens, D., Song, P.S., Ruediger, W. and Eilfeld, P. (1989) *J. Plant Physiol.*, 134:269-275.

Furuya, M. (Ed.) (1987) Phytochrome and Photoregulation in Plants, Academic Press, Tokyo and New York.

Grimm, R., Eckerskorn, C., Lottspeich, F., Zenger, C. and Ruediger, W. (1988) Sequence analysis of proteolytic fragments of 124-kilodalton phytochrome from etiolated Avena sativa L.: Conclusions on the conformation of the native protein. *Planta* 174: 396-401.

Hahn, T.R., Song, P.S., Quail, P.H. and Vierstra, R.D. (1984) Tetranitromethane oxidation of phytochrome chromophore as a function of spectral form and molecular weight. *Plant Physiol.* 74: 755-758.

Hahn T.R. and Song, P.S. (1981) Hydrophobic properties of phytochrome as probed by 8-anilinonaphthalene-1-sulfonate fluorescence. *Biochemistry* 20: 2602-2609.

Holdsworth, M.L. and Whitelam, G.C. (1987) A monoclonal antibody specific for the red-absorbing form of phytochrome. *Planta* 172: 539-547.

Kim, I.S., Bai, U. and Song, P.S. (1988) The 124-kDa oat phytochrome does not possess protein kinase activity. *Photochem. Photobiol.*, 49:319-323.

Lagarias, J.C. (1985) Progress in the molecular analysis of phytochrome. *Photochem. Photobiol.* 42: 811-820.

Lagarias, J.C. and Mercurio, F.M. (1985) Structure-function studies of phytochrome. Identification of light-induced conformational changes in 124-kDa Avena phytochrome *in vitro*. *J. Biol. Chem.* 260: 2415-2423.

Lagarias, J.C., Wong, Y.S., Berkelman, T.R., Kidd, D.G. and McMichael, Jr. R.W. (1987) Structure-function studies on Avena phytochrome. In: Phytochrome and Photoregulation in Plants (M. Furuya, Ed) Academic Press, Tokyo and New York, pp. 51-61.

Lisansky, S.G. and Galston, A.W. (1974) Phytochrome stability *in vitro*. 1. Effect of metal ions. *Plant Physiol.* 53: 352-359.

Quail, P.H., Colbert, J.T., Peters, N.K., Christensen, A.H., Sharrock, R.A. and Lissemore, J.L. (1986) Phytochrome and the regulation of the expression of its genes. *Phil. Trans. R. Soc. London*, B314: 469-480.

Ruediger, W., Thuemmler, F., Cmiel, E. and Schneider, S. (1983) Chromophore structure of the physiologically active form (Pfr) of phytochrome. *Proc. Natl. Acad. Sci. USA*, 80: 6244-6248.

Schaefer, E. and Briggs, W.R. (1986) Photomorphogenesis from signal perception to gene expression. *Photobiochem. Photobiophys.* 12: 305-320.

Song, P.S. and Sommer, D. (1990) The chromophore topography and scondary structure of 124-kDA *Avena* Phytochrome probed by Zn^{2+}-induced chromophore modification. *Biochemistry*, 29:1934–1948.

Song, P.S. (1983) Protozoan and related photoreceptors: molecular aspects. *Annu. Rev. Biophys. Bioengin.* 12: 35-68.

Song, P.S. (1988) The molecular topography of phytochrome: Chromophore and apoprotein. *J. Photochem. Photobiol. Pt B*, 2:43-57.

Song, P.S. and Yamazaki, I. (1987) Structure-function relationship of the phytochrome chromophore. In: Phytochrome and Photoregulation in Plants (M. Furuya, Ed), Tokyo and New York, pp. 139-156.

Thuemmler, F., Eilfeld, P., Ruediger, W., Moon, D.K. and Song, P.S. (1985) On the chemical reactivity of phytochrome chromophore in the Pr and Pfr form. *Z. Naturforsch.* 40 c: 215-218.

Vierstra, R.D. and Quail, P.H. (1983) Purification and initial characterization of 124 kilodalton phytochrome from Avena. *Biochemistry*, 22: 2498-2505.

Wong, Y.S., Cheng, H.C., Walsch, D.A. and Lagarias, J.C. (1986) Phosphorylation of Avena phytochrome *in vitro* as a probe of light-induced conformation changes. *J. Biol. Chem.* 261: 12089-12097.

Molecular Properties of Phytochrome

WOLFHART RÜDIGER

Botanisches Institut der Universität
Menzinger Straße 67
D-8000 München 19, FRG

Phytochrome is the best known photoreceptor for light-dependent development and differentiation in higher plants. A general, but simplified scheme for light action through phytochrome is given in Fig. 1. The light stimulus hits at first the *chromophore* which transfers any change produced by absorption of the photon to the *protein*. The protein can only communicate with the rest of the cell if such a change implies also the surface

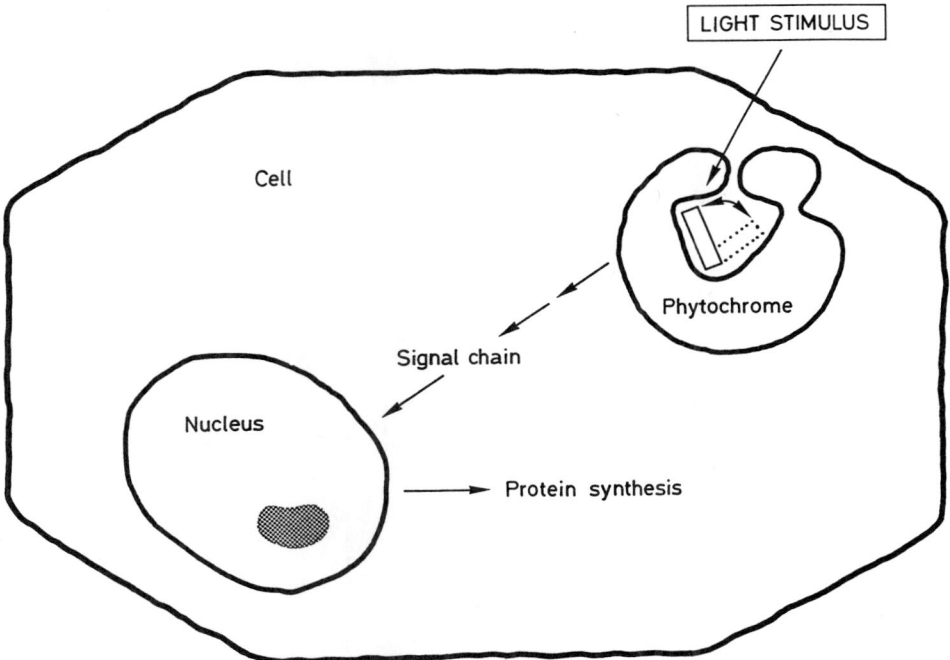

Figure 1. General scheme for action of the light stimulus through phytochrome in the plant cell.

Photobiology, Edited by E. Riklis
Plenum Press, New York, 1991

of the protein. We can define this change at the surface as formation of an "active center". This does not necessarily mean a binding site for a substrate, i.e. enzymatic activity of phytochrome, but could as well mean a binding site for a receptor molecule or other reaction partner of phytochrome. The process of binding would in this case be the start of the subsequent *signal chain*. Although such signal chains are still hypothetical items, it is clear that they must start at the "active center" of the phytochrome molecule. One important signal chain leading to the cell nucleus and differential gene expression is indicated in Fig. 1. Other signal chains starting from phytochrome are those which lead to processes like chloroplast movement in *Mougeotia* or leaflet movement in *Albizzia*; gene expression is probably not involved in this type of signal chain. It is possible (but not compelling) that different signal chains use different "active centers" of the phytochrome molecule. Neither the direct reaction partners nor "active centers" of phytochrome have so far been detected. Our research on molecular properties of phytochrome has been initiated in order to understand the dynamics of this photoreceptor; the final aim which is not yet achieved is the definition of starting points for subsequent signal chains.

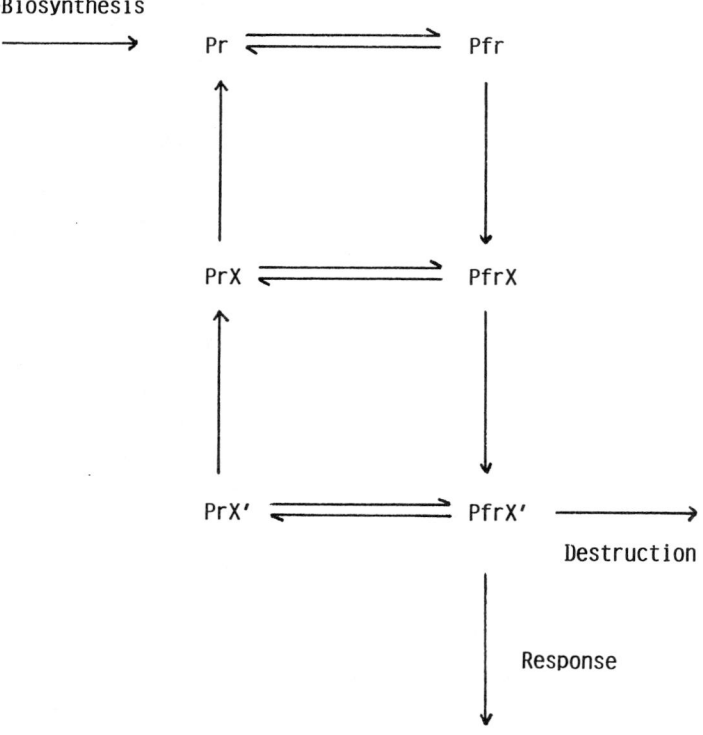

Figure 2. Common notation for phytochrome action (modified after Schäfer 1976). X = hypothetical receptor for Pfr; X' = modified receptor after Pfr binding.

A common notation for phytochrome action is given in Fig. 2. The Pr form of phytochrome is the form which is biosynthesized in the dark. Absorption of a photon leads via several intermediate steps (see below) to Pfr. The model given in Fig. 2 assumes binding of Pfr to a hypothetical receptor X which is then modified to X' (eventual conformational change?) by the binding process. The PfrX' complex would be the form from which the physiological signal chain starts. Since the Pr → Pfr transformation is photoreversible, the physiological response does not occur if the Pfr form is reconverted to Pr before the signal chain has been started. The "point of escape" from Pfr control is different for different physiological responses; it may be several seconds up to many minutes after irradiation. This model has been derived from studies of phytochrome in etiolated plants in which a redistribution of phytochrome after irradiation is observed. It is doubtful whether the model can be applied unchanged for phytochrome in green plants in which its localization might be quite different (see Whitelam and Smith, 1988). Likewise, phytochrome destruction after irradiation is typical for phytochrome in etiolated plants but not in green plants.

Studies on the phytochrome chromophore

Our first approach for investigation of molecular properties of phytochrome was the investigation of the chromophore. Since these studies have been summarized several several times (Rüdiger and Scheer, 1983; Rüdiger, 1983; Rüdiger et al. 1985; Rüdiger, 1986; 1987 a,b; Whitelam and Smith, 1988, Eilfeld and Haupt,1990), only the main line of investigation and some major results shall be mentioned here.

Spectral and chemical properties of the chromophore are dramatically modified by the native protein. The first aim was therefore to cleave the chromophore — either entirely or in fragments — from the protein. These studies revealed the tetrapyrrole nature of the chromophore and a thioether band as its covalent linkage with the protein. The mildest method for removal of the influence of the protein proved to be proteolysis under carefully controlled conditions. Small chromopeptides containing the tetrapyrrole chromophore and 8–11 amino acids (including the binding cysteine) were obtained and investigated with several methods, especially UV–Vis and NMR spectroscopy. Surprisingly, digestion of the Pfr form of phytochrome yielded a chromopeptide different from that obtained by digestion of the Pr form. Digestion and isolation of the "Pfr chromopeptide" has to be carried out in the dark, however, because irradiation transforms the "Pfr chromopeptide" immediately into the "Pr chromopeptide". Chemical and spectroscopic properties of both chromopeptides led to the chemical structures given in Fig. 3. The structures were confirmed by detailed studies on synthetic model tetrapyrroles.

The only detectable chemical difference between the Pr chromophore and the Pfr chromophore is the configuration at the C15 –C16 double bond: it is 15Z for the Pr form and 15E for the Pfr form. The 15E chromophore is relatively unstable; it is converted into the 15Z isomer either slowly in the dark (thermal reaction, proton-

425

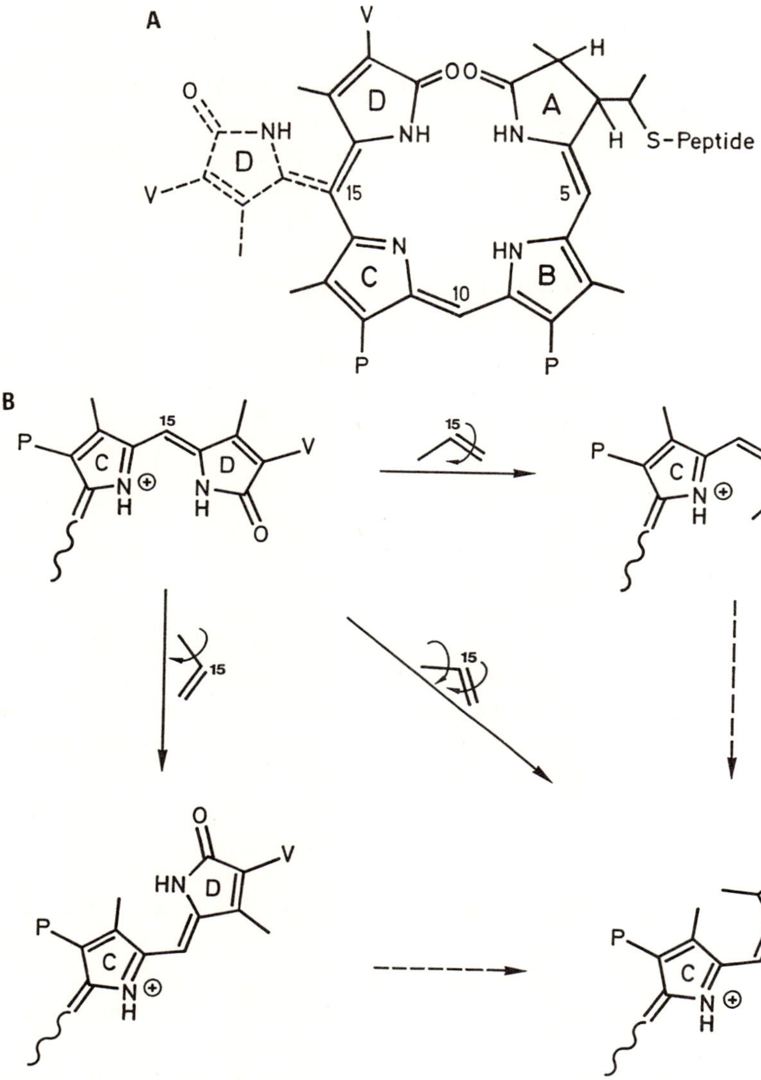

Figure 3 A. Chemical structures of the Pr chromophore (solid line, 15Z configuration, 14 syn conformation) and Pfr chromophore (dashed line, 15E configuration, 14 anti conformation;) as present in small chromopeptides. Rings A, B, C are identical in Pr and Pfr.

Figure 3 B. Hypothetical phototransformation of the Pr chromophore (upper end, left) to the Pfr chromophore (lower end, right). Z,E-isomerization (at C15–C16) and conformational change (syn, anti at C14–C15) can occur sequentially or with higher probability simultaneously.

catalyzed) or rapidly in the light (photochemical reaction). This lability shows that the 15E chromophore cannot be formed artifactually during proteolyis but that the 15E configuration must already be present or at least preformed in native Pfr.

It is reasonable to assume that the structural difference between the Pr chromophore and the Pfr chromophore is produced by a photochemical reaction, i.e. by the primary reaction of the Pr → Pfr photoconversion. This assumption is supported by studies on phytochrome intermediates (see below). We assume a simultaneous change of configuration (Z,E-isomerization at C15–C16) and conformation (syn, anti at C14–C15) as indicated by the diagonal pathway in Fig. 3B. This requires the minimum space for chromophore isomerization within the protein pocket since pyrrole ring D would only be shifted within a plane in this case. Otherwise (sequential pathways of Fig. 3B) pyrrole ring D must entirely be turned around during each step. If the protein pocket were large enough to allow free rotation of ring D, it were hard to imagine that any conformational change of the protein follows from chromophore isomerization. But conformational changes of the protein have already been demonstrated (see below).

Photochemical Z,E-(or cis,trans-) isomerization is believed to be also the primary reaction of other photoreceptors, namely of retinal in rhodopsin and of cinnamic acid in a presumed UV-B photoreceptor. It is not surprising that nature chooses cis,trans isomerization at C,C double bonds as primary reaction of photoreceptors since this is a typical photochemical reaction with a high thermal activation barrier; a thermal back reaction occurs therefore with low probability. This is important for a high signal/noise ratio. It is nevertheless surprising that this principle seems to be realized in such different classes of compounds, namely tetrapyrroles (phytochrome), isoprenoids (rhodopsin) and aromatic acids (UV-photoreceptor).

The spectral properties of the chromophore in the phytochrome chromopeptide are distinct from the those of the same chromophore in the native protein. The main difference is a relatively high oscillator strength for the red absorption band in the native protein compared with that of the chromopeptide; the oscillator strength of the blue band remains nearly unchanged while going from the native protein to the chromopeptide (see Rüdiger 1987a). Besides this difference, a significant hypsochromic shift can be seen during preparation of the Pfr chromopeptide from native Pfr (see section on intermediates for an explanation); such a shift is not observed in the Pr form.

Similar changes in the oscillator strength of red and blue bands have been observed for phycocyanin; they have been explained as conformational changes of the chromophore, namely a quasi circular shape in the chromopeptide and a more extended conformation in the native protein (Scheer 1981). The chromophore structure of phycocyanobilin is identical with that of the Pr chromophore including the thioether linkage (see Fig. 3), the only difference is an ethyl side chain at C-18 instead of the vinyl side chain in Pr. The chromophore conformation in native phycocyanin has been elucidated by X-ray analysis of the crystallized protein; it is indeed an extended conformation which can be derived from the cyclic-helical form (cf. the Pr chromophore in the chromopeptide, Fig. 3) by a twist of roughly 180° at both C5–C6 and C14–C15

	317	322
Phytochrome Avena	A L R A P H S C H L Q Y M E	
Phytochrome Cucurbita	T L R A P H S C H L Q Y M E	
Phycocyanin M.[a] α-chain	D A R G K S K C A R D I G H	
Phycocyanin M.[a] ß-chain	R N R R M A A C L R D M E I	
Phycocyanin A.[b] α-chain	D N R G K D K C A R D I G Y	
Phycocyanin A.[b] ß-chain	T N R R M A A C L R D M E I	
	79	84

[a]Mastigocladus laminosus

[b]Agmenellum quadruplicatum

Figure 4. Amino acid sequences around the chromophore in phytochromes and phycocyanins. The chromophore-binding cysteine and the arginine 5 positions ahead are underlined.

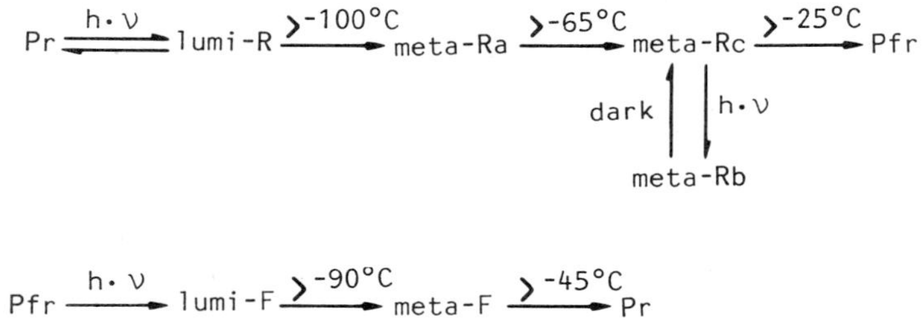

Figure 5. Phytochrome intermediates deduced from low temperature studies (after Eilfeld and Rüdiger 1985). Intermediates with the same spectral properties are also formed at physiological temperature.

single bonds (Schirmer et al. 1987, 1988). Since phytochrome has not yet been crystallized, conclusions on the chromophore conformation can only be drawn by analogy. The similarity of the spectral properties of native Pr and phycocyanin suggests a similar chromophore conformation in both cases. One of the factors which are responsible for the chromophore conformation in phycocyanin is an ionic interaction between the propionic acid side chains and lysine and arginine residues (Schirmer et al. 1988). It may be relevant in this connection that phytochrome contains an arginine residue at the same distance from the binding cysteine as phycocyanin. This arginine (R-79) interacts with the acid side chain of ring B of phycocyanobilin (Schirmer et al.1987); it is conserved in the α- and ß-chains of phycocyanins and also in the known phytochrome sequences (see Fig. 4).

Phytochrome intermediates

Phytochrome resembles rhodopsin with regard to intermediates: the first (photochemical) step is followed by a series of dark relaxation steps. Since there is a mutual interaction between protein and chromophore conformation, most (if not all) of these steps can be detected as spectral changes.

We have investigated all spectral changes in the visible and near ultraviolet region and calculatd the absolute absorption spectra for the single intermediates (Eilfeld and Rüdiger 1985). The scheme for intermediates in native phytochrome which was derived from our studies is shown in Fig. 5. I wish to mention only two points here which consider the first and last steps of the Pr→ Pfr transformation

For the formation of the first detectable intermediate, lumi R, from Pr, a small activation energy (3.6 ± 0.5 kJ/mol) was determined from the Arrhenius plot (Eilfeld et al. 1986). In accordance with these data, no photoreaction was detectable at temperatures <15 K even with intense irradiation, e.g. with a He-Ne-laser at 632 nm (Eilfeld, 1987). The same is true for the reverse reaction, lumi-R $\rightarrow$ Pr (activation energy 5.7 ± 0.7 kJ/mol). The fact which I want to discuss in this connection is the lack of any deuterium isotope effect for these activation energies (Eilfeld et al. 1986). This is a strong argument against earlier suggestions of a proton transfer or chromophore tautomerization for the primary photoprocess (Song et al. 1979; Sarkar and Song, 1981). Another argument comes from measurement of circular dichroism. Upon photoconversion of Pr to lumi-R at −110°C, the rotational strength values change sign and drastically increase from R_{666nm} band = −2.9 DBM (Pr) to R_{694nm} band = +7.5 DBM (lumi R) (Eilfeld and Eilfeld, 1988). This means inversion of the helical charge circulation around the transition moment under conditions under which the protein matrix is considered to lack flexibility. These findings again favor *cis,trans*-isomerization rather than proton transfer as the primary reaction. If any proton transfer would occur during (see discussion for rhodopsin, Becker, 1988) or after *cis,trans*-isomerization (see Rüdiger et al. 1985) it cannot be a rate-limiting step since no deuterium isotope effect has been observed during kinetic measurements (Ruzcicska et al. 1985). Thus all available data are in accordance with the assumption of the *cis,trans*-isomerization for the primary photoprocess. The small activation barrier might indicate that relaxation of the S_1 state of Pr to the S_0 state of lumi-R might involve some conformational changes of the chromophore. The situation might anyhow be somewhat more complex than this simple scheme since heterogeneity has been detected as well for lumi-R formation (Eilfeld et al. 1986) and decay (Ruzcicska et al. 1985; Inoue, 1987).

The bathochromic shift of the red absorption band of Pfr appears only at the latest stage of the dark relaxation process, i.e. on the path from meta-Rb to Pfr via meta-Rc. It is therefore clear that the typical Pfr absorption band must be created by a specific chromophore-protein interaction and cannot directly be related to the primary photoprocess. A hypothetical model has been proposed according to which the chromophore-protein interaction involves a point charge in a hydrophobic environment within the protein (Eilfeld and Rüdiger, 1984; Rüdiger et al. 1985). The smallest absorption is found for the intermediate meta-Rb which resembles very much a "free" tetrapyrrole with cyclic-helical conformation. Compared with the native forms Pr and Pfr, it can be considered as a "bleached" intermediate although it has a definite red absorption band. Some authors term all intermediates between Pr and Pfr "bleached intermediate", i.e. do not distinguish between meta-Ra, meta-Rb and meta-Rc although definitely different absorption properties have been demonstrated at low temperature (Eilfeld and Rüdiger, 1985) and at room temperature (Inoue and Rüdiger, unpublished results).

Studies on the protein

The fact that the far-red absorption band of Pfr is caused by a particular chromophore-protein interaction has been confirmed by studies on limited proteolysis of the protein. Loss of an amino-terminal 4 to 10 kDa piece from the intact 124 kDa peptide chain causes only a small shift of the absorption maximum from 730 to 722–724 nm (Vierstra and Quail, 1982). This maximum stays at 722 nm upon further proteolysis, down to a fragment size of 59 kDa. Such fragments are fully photoreversible and have been known as "small phytochrome" since many years. Further proteolysis yields a 39 kDa fragment which is still photoreversible but lacks the typical Pfr absorption band entirely (Reiff et

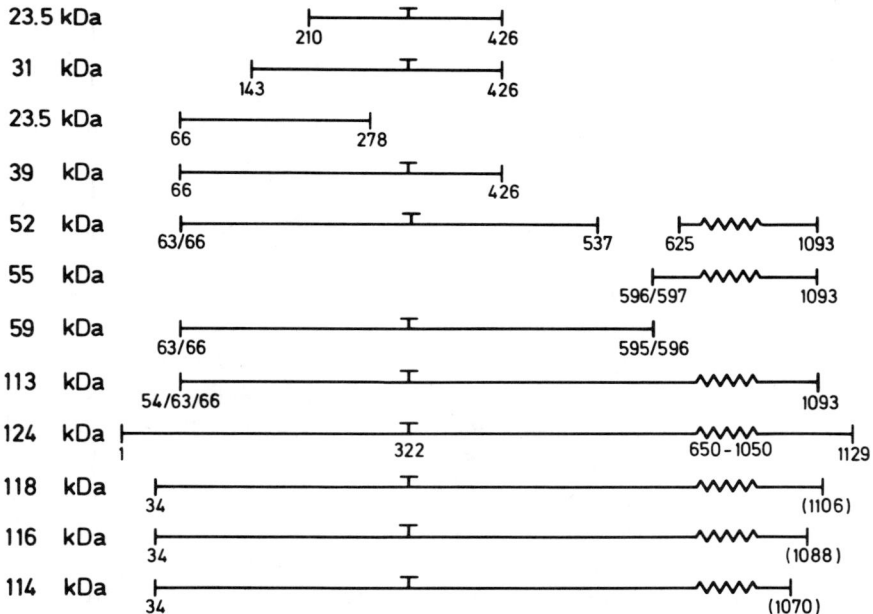

Figure 6. Localization of proteolytic fragments within the complete amino acid sequence of oat phytochrome (after Grimm et al. 1986). Upper part: fragments obtained with trypsin; lower part: fragments obtained with proteases from oat.

al. 1985). We have determined the amino-terminal sequences of this and other fragments (Grimm et al. 1986). The fragments were then localized within the complete peptide chain (see Fig. 6) for which the amino acid sequence had been determined via DNA sequencing (Hershey et al. 1985). It is clear from this investigation that the amino acid sequence which is present in the 59 kDa fragment but not in the 39 kDa fragment must (either directly or indirectly) be responsible for the chromophore-protein interaction leading to the far-red absorption band. As shown in Fig. 6, this is the region between amino acid residues 426 and 595. Likewise, the typical Pr absorption caused by extended conformation of the chromophore (see above) requires another part of the peptide chain. This part is present in the 39 and 59 kDa fragments (starting at residue 66) but not in a 23.5 kDa fragment (starting at residue 210). It should be mentioned, however, that a 16 kDa fragment obtained by digestion of phytochrome with subtilisin (Quail et al. 1987) instead of trypsin (see Fig. 6) possesses a "normal" Pr absorption although its sequence starts with amino acid 212 (Grimm and Rüdiger, unpublished results).

Another interesting aspect of limited proteolysis is the finding of different early cleavage sites for Pr and Pfr which clearly indicate that conformational changes of the protein occur during the photoconversion (Jones et al. 1985; Lagarias and Mercurio 1985). We have localized also these different early cleavage sites in Pr and Pfr by microsequencing (Grimm et al. 1988). As shown in Fig. 7, different parts of the peptide chain are exposed in Pr and Pfr. It has long been known that the amino-terminal part of phytochrome is particularly exposed in the Pr form (Cordonnier et al. 1985; Jones et al. 1985; Lagarias and Mercurio, 1985). According to our analysis, early cleavage is found in this region from amino acid 22 to amino acid 70 in the Pr form.

Two regions were identified which are more exposed in Pfr than in Pr. One of these regions is relatively near to the tetrapyrrole chromophore which is linked to cysteine-322

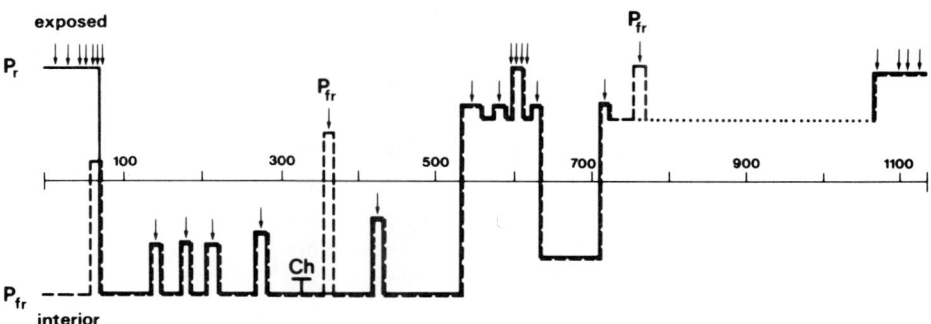

Figure 7. Schematic drawing for exposed and interior parts of the peptide chain of oat phytochrome (after Grimm et al. 1988). Vertical arrows indicate early cleavage sites for various proteases localized by microsequencing. Ch = tetrapyrrole chromophore.

(see Fig. 7). The chromophore — although shielded by the protein — is somewhat more exposed in the Pfr form than in the Pr form (Thümmler et al. 1985). It is possible, therefore, that the chromophore moves together with this part of the peptide chain from an interior to a more exterior position. The cleavage site itself at glutamate-352 is interesting because it is part of a cluster of acidic residues (6 glutamic and 2 aspartic acid residues in oat phytochrome). One can predict that exposure of these residues should increase the negative netto charge at the protein surface. We could indeed confirm, by electrophoresis and isoelectric focusing of native phytochrome, that the Pfr form carries more negative charges (or less positive charges) at the surface than the Pr form (Schendel and Rüdiger, 1988). The region from residues 323 to 360 has been described as socalled "PEST" sequence, i.e. a signal sequence for rapid protein degradation (Rogers et al. 1986). These authors considered phytochrome as a protein with rapid turnover without considering the difference between Pr and Pfr: the Pr form is indeed a stable protein; phytochrome destruction occurs only after formation of Pfr. In connection with our finding of differential exposure of this region, a general question arises: Must a "PEST" sequence be exposed at the surface of a protein in order to exhibit its signal function?

Exposure of a cluster of negative charges could still have another effect: it can be predicted that aggregation will be facilitated in the presence of divalent cations in this case. It is indeed known that Pfr (contrary to Pr) shows such aggregation phenomenon in the presence of Mg^{2+} or Ca^{2+} ions; this phenomenon is known as "*in vitro* pelletability" of Pfr (Pratt, 1982). It is not yet known whether any correlation exists with "*in vivo* pelletability" and "sequestring" of Pfr in the cell (Quail, 1983; Schäfer, 1987) processes which are probably related to phytochrome destruction. It is interesting to note that for both, pelletability and destruction, Pfr formation seems to have a signal function; after initial Pfr formation, also Pr can be implied in the reaction. It can be speculated that the peptide region from residues 323 to 360 plays an important role as signal sequence for this and eventually other responses, e.g. as part of an "active center" as defined in the Introduction.

Recently, protein kinase activity was discussed as the possible biochemical reactivity of phytochrome (Wong et al. 1986; Lagarias et al. 1987). We could confirm the presence of protein kinase acitivity in our phytochrome preparations. Whereas phytochrome itself was always phosphorylated by ATP in solutions of these preparation, incubation of a "native" gel (which gave an excellent separation for these preparations) with labeled ATP yielded radioactive phosphate only in a contaminating band but not in the phytochrome band itself (Grimm et al., 1989). Since protein kinases are autophosporylated under these conditions, we had to conclude that not phytochrome but the contaminating protein is the protein kinase of these preparations. It is not yet clear whether phytochrome regulates the activity of this kinase.

Acknowledgement

The cited work of the author was supported by the Deutsche

Forschungsgemeinschaft, Bonn-Bad Godesberg, and the Fonds der Chemischen Industrie, Frankfurt.

References

Becker, R.S. (1988) *Photochem.Photobiol.* 48, 369–399

Cordonnier, M.M., H. Greppin, L.H. Pratt (1985) *Biochemistry* 24, 3246–3253

Eilfeld, P., W. Rüdiger (1984) *Z.Naturforsch.* 39c, 742–745

Eilfeld, P., W. Rüdiger (1985) *Z.Naturforsch.* 40c, 109–115

Eilfeld, P.G., P.H. Eilfeld, W. Rüdiger (1986) *Photochem.Photobiol.* 44, 761–769

Eilfeld, P.G., W. Haupt (1990) in: Photomorphogenesis (M.G.Holmes, ed.) Academic Press, London in press

Eilfeld, P.H. (1987) Dissertation Univ. München

Eilfeld, P.H., P.G. Eilfeld (1988) *Physiol.Plant.* 74 169–175

Grimm, R., Ch. Eckerskorn, F. Lottspeich, C. Zenger and W. Rüdiger (1988) *Planta* 174, 396–401

Grimm, R., F. Lottspeich, Hj.A.W. Schneider and W. Rüdiger, (1986) *Z.Naturforsch.* 41c, 993–1000

Grimm, R., D. Gost and W. Rüdiger (1989) *Planta* 178, 199–206

Hershey, H.P., R.F. Barker, K.B. Idler, J.L. Lissemore, P.H. Quail (1985), *Nucleic Acid Res.* 13, 8543–8559

Inoue, Y. (1987) in: Phytochrome and Photoregulation in Plants (M. Furuya, ed.) pp. 117–126. Academic Press Tokyo 1987

Jones, A.M., R.D. Vierstra, S.M. Daniels, P. Quail (1985) *Planta* 164, 501–506

Lagarias, J.G., F.M. Mercurio (1985) *J. Biol. Chem.* 260, 2415–2423

Lagarias, J.G., Y.G. Wong, T.R. Berkelman, D.G. Kidd, R.W. McMichael Jr. (1987) in: Phytochrome and Photoregulation in Plants (M. Furuya, ed.) pp. 51–61, Academic Press Tokyo

Pratt, L.H. (1982) *Annu.Rev.Plant Physiol.* 33, 557–582

Quail, P. (1983) in: Photomorphogenesis, (W. Shropshire and H. Mohr) Vol. 16A, pp. 178–212, Springer Verlag Berlin, Heidelberg

Quail, P.H., C. Gatz, H.P. Hershey, A.M.J ones, J.L. Lissemore, B.M. Parks, R.A. Sharrock, R.F. Barker, K. Idler, M.G. Murray, M. Koornneef, R.E. Kendrick (1987) in: Phytochrome and Photoregulation in Plants, (M. Furuya, ed.) pp. 23–37, Academic Press Tokyo

Reiff,U., P. Eilfeld and W. Rüdiger (1985) *Z.Naturforsch.* 40c, 693–698

Rogers, S., R. Wells, M. Rechsteiner (1986) *Science* 234, 364–368

Ruzsicska, B.P., S.E. Braslavsky, K. Schaffner (1985) *Photochem. Photobiol.* 41, 681–688

Rüdiger, W. (1983) *Phil.Trans.Roy.Soc. London B* 303, 377–386

Rüdiger, W. (1986) in Photomorphogenesis in plants (R.E.Kendrick and G.H.M.Kronenberg, eds.), pp. 17–33, Dr.W.Junk Publ., Den Haag

Rüdiger, W. (1987a) *Photobiochem.Photobiophys. Suppl.*, 217–227

Rüdiger, W. (1987b) in: Phytochrome and Photoregulation in Plants (M.Furuya, ed.), pp. 127–137. Academic Press, Tokyo

Rüdiger, W., H. Scheer (1983) in Encyclopedia of Plant Physiology Vol. 16A (W. Shropshire and H. Mohr, eds.) pp. 119–151,Springer Verlag Berlin, Heidelberg

Rüdiger, W., P. Eilfeld, F. Thümmler (1985) in: Optical Properties and Structure of Tetrapyrroles (G. Blauer and H. Sund, eds.) pp. 349–366, W.de Gruyter, Berlin, New York

Sarkar H.K., P.S. Song (1981) *Biochemistry* 20, 4315–4320

Schäfer, E. (1987) in: Phytochrome and Photoregulation in Plants, (M. Furuya, ed.) pp. 279–287, Academic Press Tokyo

Schäfer, E. (1976), in: Light and Plant Development Development, (H. Smith ed.) pp. 45–59, Butterworths London

Scheer, H. (1981) *Angewandte Chemie Intern. Ed.* 20, 241–261

Schendel, R., W. Rüdiger (1989) Z. *Naturforsch.* 44c, 12–18

Schirmer, T., W. Bode, R. Huber (1987) *J.Mol.Biol.* 195, 677–695

Schirmer, T., W. Bode, R. Huber (1988) in: Photosynthetic Light-Harvesting Systems (H. Scheer and S. Schneider, eds.) pp. 195–199. De Gruyter Berlin, New York

Song, P.S., Qu. Chae, J.D. Gardner (1979) *Biochim.Biophys.Acta* 567, 479–495

Thümmler,F.,P. Eilfeld, W. Rüdiger, D.K. Moon and P.S. Song, (1985) *Z.Naturforsch.* 40c, 215–218

Vierstra, R.D., P.H. Quail (1982) Planta 156, 158–165

Whitelam, G., H. Smith (1988) in: Plant Pigments (T.W.Goodwin, ed.) pp.257–298, Academic Press London

Wong, V.-S., C.-C. Cheng, D.A. Walsh, J.C. Lagarias (1986) *J.Biol.Chem.* 261, 12089–12097

Studies of the Mechanism of the Phototransformation of Phytochrome

C. BONAZZOLA, G. VALDUGA, P. LINDEMANN, Y. KAJII,
S.E. BRASLAVSKY AND K. SCHAFFNER

Max-Planck-Institut für Strahlenchemie
D-4330 Mülheim/Ruhr
FRG

Introduction

The phototransformation $P_r \to P_{fr}$ of 124-kDa phytochrome from etiolated *Avena* was further studied using two approaches. One was the monitoring of the P_{fr} appearance after excitation of P_r with ms pulses. The absorbance increase in the region 720–750 nm after ca. 10 ms of excitation occurs in two steps (as already reported by Pratt, et al., 1982), with lifetimes (percentages) in the range of 30 ms (ca. 50%) and 1 s (ca. 50%) at 275 K. Both rate constants were only slightly affected by temperature in the range 275–293 K. In contrast, the relative amplitude of the two components of the kinetics was influenced by temperature. The small D/H isotope effect on the time constants of P_{fr} appearance and on the relative percentage of the components make it difficult to decide whether the isotope substitution affects the last or the previous to last, or both steps of P_{fr} appearance. The results indicate that a solvent-assisted activated step occurs previous to the last, while the last step is better interpreted as a conformational change without proton exchange. Parallel ways leading from P_r to P_{fr} include an initial isomerization of the chromophore and solvent-assisted chromophore and protein conformational changes. The results show that the solvent-assisted processes involve only changes in hydrogen bridges rather than full proton transfer. These data supplement those from ps fluorescence (Brock, et al., 1987) and µs transient absorbance (Aramendia, et al., 1987). A simplified picture of the transformation is shown in the scheme.

Another approach was the study by flash photolysis of the influence of ubiquitin (Ub), a protein involved in the *in vitro* proteolysis of phytochrome (Shanklin, et al., 1987), on the kinetics of the transient absorbance. Photoreversibility was not affected by Ub. After excitation of the $P_r Ub_n$ (n ca. 5 to 13) complex with 15-ns laser pulses, an absorbance in the red was observed which decayed following a biexponential law similar to that for the absorbance due to the two I_{700} µs transients of P_r in buffer (Aramendia, et al., 1987 and references therein). The absorbance maximum of one of the I_{700} transients was selectively blue shifted to 688 nm with respect to that of non-complexed I_{700} at 695

Photobiology, Edited by E. Riklis
Plenum Press, New York, 1991

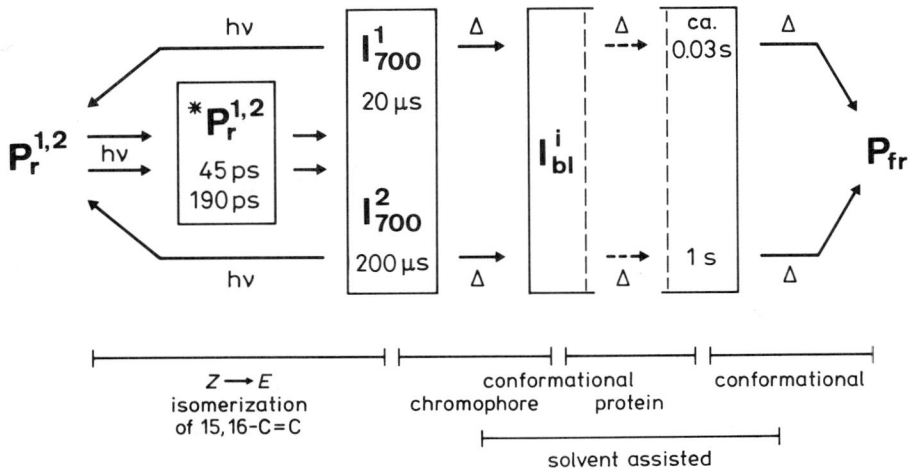

Lifetimes: 275 K

Scheme. Working hypothesis of the mechanism of the $P_r \rightarrow P_{fr}$ transformation. The number of steps between the bleached intermediate(s) I_{bl} and P_{fr} (---) is still unknown and the thermal reversion of the intermediates to P_r is not included (see Heihoff, et al., 1987).

nm. This, as well as some influence of the complexation on the activation parameters for the transients decay, points to a selective interaction of Ub with the dimeric protein playing a role during the I_{700}^i decay.

References

Aramendia, P.F., Ruzsicska, B.P., Braslavsky, S.E., and Schaffner, K. (1987). Laser flash photolysis of 124-kilodalton oat phytochrome in H_2O and D_2O solutions: Formation and decay of the I_{700} intermediates. *Biochem.* 26, 1418–1422.

Brock, H., Ruzsicska, B.P., Arai, W., Schlamann, W., Holzwarth, A.R., Braslavsky, S.E., and Schaffner, K. (1987). Fluorescence lifetimes and relative quantum yields of 124-kilodalton oat phytochrome in H_2O and D_2O solutions. *Biochem.* 26, 1412–1417.

Heihoff, K., Braslavsky, S.E., and Schaffner, K. (1987). Study of 124-kilodalton oat phytochrome photoconversion *in vitro* with laser-induced optoacoustic spectroscopy. *Biochem.* 26, 1422–1427.

Pratt, L.H., Shimazaki, Y., Inoue, Y., and Furuya, M. (1982). Spectral analysis of phototransformation n intermediates in the pathway from the red-absorbing to the far-red-absorbing form of *Avena* phytochrome. *Photochem. Photobiol.* 36, 471–477.

Shanklin, J., Jabben, M., and Vierstra, R.D. (1987). Red light-induced formation of ubiquitine-phytochrome conjugates: Identification of possible intermediates of phytochrome degradation. *Proc. Natl. Acad. Sci. USA.* 84, 359.

The Significance of Mutants in Phytochrome Research

R.E. KENDRICK[1], P. ADAMSE[1], E. LOPEZ-JUEZ[2], M. KOORNNEEF[3],
J.L. PETERS[1] and J.C. WESSELIUS[1]

[1]Plant Physiol. Res.
Gen. Foulkesweg 72, NL-6703 BW and
[3]Dept. of Genetics, Dreijenlaan 2, NL-6703 HA
Agricultural University
Wageningen, The Netherlands and
[2]Lab. of Plant Physiol., Faculty of Biology
Complutense University, 28040 Madrid, Spain

Genotypes (often as induced mutants) in which certain parts of the photomorphogenetic pathway(s) are eliminated, will exhibit a different and often simpler photomorphogenesis than their wild type. The relevance of the deleted part in the mutant is directly indicated by its difference in response compared to its isogenic wild type. Mutations may not only lead to the absence of a particular photoreceptor or response, but may result in an amplified response. Such high-response mutants would facilitate the study of a photoresponse. Phytochrome (P) mutants can be divided simplistically into three groups.

Photoreceptor mutants	Transduction chain mutants	Response mutants
hv		$\rightarrow$ R_1
$\rightarrow$	$\rightarrow$ $\rightarrow$ $\rightarrow$	$\rightarrow$ R_2
		$\rightarrow$ R_3

Photoreceptor and transduction chain mutants would be expected to have pleiotropic defects for responses R_1, R_2 and R_3, whereas the defect in a response mutant would be restricted to modification of one particular response (Adamse et al., 1988c). True photoreceptor mutants will lack the photoreceptor or contain the modified photoreceptor, which is non-functional. Mutants that contain a particular photoreceptor, but lack all photomorphogenetic responses associated with it, are most probably transduction chain mutants.

Using the spectrophotometric and immunological assays it is relatively easy to test for the existence of the bulk light-labile pool of P. However, the possible existence of

Photobiology, Edited by E. Riklis
Plenum Press, New York, 1991

multiple photoreceptor types and the presence of several structural genes for P makes the analysis of the P system more complex (Pratt and Cordonnier, 1987; Nagatani et al., 1987). Perhaps, in practice, it is only possible to obtain leaky mutants, since a plant must achieve the P-regulated de-etiolation to survive. Even so, such mutants would enable the concentration of P to be modified without irradiation. A mutant lacking P would enable the importance of the blue light (BL)/UV photoreceptor (cryptochrome) to be assessed and help understand their mode of co-action, since cryptochrome and P both absorb BL. A mutant recognized as lacking one of the P types would help in understanding the importance of light-labile and light-stable P in different responses.

Arabidopsis

The long-hypocotyl mutants at the *hy-1* and *hy-2* loci were the first recognized P photoreceptor mutants of higher plants (Koornneef et al., 1980). These mutants were isolated on the basis of their increased hypocotyl growth, relative to wild type, when grown in white light (WL). Both of these mutants lack spectrophotometrically detectable P in seeds and etiolated seedlings. More recently Chory et al. (1989) have discovered an additional locus, *hy-6*, which is even more extreme in phenotype than *hy-1* than *hy-2*. Mutants at all three loci have similar pleiotropic effects: reduced chlorophyll (Chl) level, higher Chl *a/b* ratio, smaller leaves, less cell expansion, fewer chloroplasts per cell and reduced granal stacking.

A study of the photocontrol of seed germination of mutants (Cone, 1985) revealed that the germination remained under P control despite the apparent lack of the photoreceptor as indicated by spectrophotometry. However, in most cases sensitivity towards red light RL was reduced when compared to wild type. It was speculated that the P controlling seed germination was not the bulk labile pool which is involved in the inhibition of hypocotyl growth during de-etiolation. Light-stable P is a possible contender for the photo-induction of seed germination and is probably P that has been synthesized within the seed during maturation on the mother plant. The light conditions during the latter stages of seed development are known to influence the subsequent dark (D) germination behaviour and sensitivity to RL, presumably via the stable type of P (Gettens-Hayes and Klein, 1974).

The molecular background of the loci described is not yet clear. At least one structural gene for P has been mapped to chromosome 1 (Chang et al., 1988), a different chromosome to where the hy-1 and hy-2 loci are situated. It cannot be excluded that the *Arabidopsis hy-3* mutants which contain light-labile P (Koornneef et al., 1980), show reduced levels of P in mature seeds (Cone, 1985) and physiological characteristics resembling P deficiency, are of a similar type to that of the *lh* cucumber mutant discussed below.

The *hy-4 Arabidopsis* long-hypocotyl mutants (Koornneef et al., 1980), appear to be defective with respect to BL-absorbing pigment(s). Such mutants provide a direct indication that a photoreceptor, other than P, absorbing in the BL region of the spectrum plays a role in the regulation of growth, since they are modified independently in the BL region, while response in the RL and far-red (FR) spectral region is retained.

Tomato

During selection of gibberellin (GA) mutants, a mutant was isolated that required GA for germination but in contrast to GA-deficient mutants was characterized by a long hypocotyl and a marked reduction in Chl content when grown in WL (Koornneef et al., 1985). A genetic analysis revealed that this mutant was allelic with the *aurea (au)* locus located on chromosome 1. Another mutant at the *au* locus, also with similar physiological characteristics to the previously described *au* mutants was recently isolated in the progeny of tomato plants derived from tissue culture by Lippuci di Paola et al. (1988).

Spectrophotometric and immunological (Koornneef et al., 1985; Parks et al., 1987) studies have resulted in the *au* mutant being the best characterized example of a photoreceptor mutant; having a P content of < 5% of that of the wild type. In addition, this mutant has no, or a strongly reduced, photoregulation of Chl *a/b*-binding (*cab*) protein synthesis (Sharrock et al., 1988), chloroplast development, anthocyanin synthesis, seed germination and hypocotyl growth inhibition (Koornneef et al., 1985). This true pleiotropic phenotype, coupled with lack of P is precisely that predicted for a photoreceptor mutant.

The *au* mutant, which lacks light-labile P, has been shown spectrophotometrically to possess about 50% of the P content of the wild type in tissues of light-grown plants, suggesting that it accumulates light-stable P. The cause of the lesion in the *au* mutant is not yet clear although instability of the protein has been suggested. Work in P.H. Quail's laboratory (Parks et al., 1987; Sharrock et al., 1988) has indicated that the P-mRNA is produced and is functional in an *in vitro* translation system, yet *in vivo* the protein fails to accumulate. In addition, there appear to be multiple genes coding for P in tomato. One gene has been mapped to chromosome 10, whereas the *au* locus is situated on chromosome 1. While the other genes have not been mapped and there is no knowledge as to which gene(s) is/are expressed it seems unlikely that the *au* lesion is the result of a single amino acid mutation of the protein.

The consequences of severe P deficiency can also be studied at the molecular biological level. A number of enzymes have been shown to appear or be dramatically increased as a result of *de-novo* synthesis following photoconversion of P (Schäfer, 1987; Tobin, 1987; and references therein). The *au* mutant demonstrates directly that light-labile P is functional in the control of gene expression. Modulation of the amount of *cab* mRNA is severely reduced in the *au* mutant. In at least one other species the induction of *cab* has been demonstrated to be, in part, under the control of a very low fluence response (VLFR), which by definition is a response not reversible by far-red light (FR), in contrast to the classic P responses (Kaufman et al., 1984). However, in the *au* mutant, it appears that the low level of *cab* mRNA induction by RL is reversible by

FR, while in the wild type, *cab* mRNA is strongly induced by FR. This result provides the first indication that light-labile P is the functional photoreceptor in VLFR's (Sharrock et al., 1988).

Mutants with a phenotype similar to that of *aurea*, although less extreme, are the *yellow green-2 (yg-2)* or auroid mutants that have been mapped to chromosomes 12 (for references see Koornneef et al., 1985).

While circumstantial evidence implicates that the FR-absorbing form of P (Pfr) is the physiologically active form, the possibility that it is the loss of the RL-absorbing form (Pr) which is the active process has also been postulated. Mutants lacking P provide the first direct evidence to support the hypothesis that Pfr is the active form of P. These mutants should clearly have a reduced level of Pr in D. However, such seedlings, of the *au* tomato mutant and the *Arabidopsis hy-1* and *hy-2* mutants, grow long in D, just as those of wild type, whereas if removal of Pr was the active process in P action they would be expected to be short. Mutants possessing different levels of P, a situation until now only attainable by pre-irradiations, which may selectively influence subsequent response sensitivity, would also be very useful in determining whether a response is due to the [Pfr] or Pfr/Pr+Pfr ratio (ø).

The response of plants to an end-of-day FR pulse results in a dramatic stimulation of elongation growth in plants grown in a WL/D cycle (Downs et al., 1957). It has been recently observed that the *au* mutant responds to such treatments, almost quantitatively the same as the wild type. It has been speculated that this is a response that is under the control of the light-stable type of P in adult plants (Table 1) (Adamse et al., 1988b). However, the deficiency in Chl synthesis remains the most striking feature of the *au* mutant and suggests that labile P remains functional in adult plants.

Anthocyanin synthesis in seedlings is a process which is strongly light regulated and in which P plays an important role (Beggs et al., 1986). While many of the anthocyanin deficient mutants are response mutants, P mutants would also be expected to lack photocontrol of anthocyanin synthesis. A tomato mutant (Kerr, 1965; Wettstein-Knowles, 1968) called *high pigment (hp)*, has been provisionally considered as an anthocyanin response mutant. It has about a 10-fold higher level of anthocyanin than

Table 1. The influence of 20 daily treatments with end-of-day FR (+FR) or without (–) on tomato Moneymaker wild type and *au* mutant, and cucumber wild type and lh mutant seedlings (14 h WL [35 W m^{-2}] / 10 h D) (Unpublished data E. López-Juez and J.C. Wesselius)[1].

Measurement	Tomato				Cucumber			
	Wild type		*au* mutant		Wild type		*lh* mutant	
	–	+FR	–	+FR	–	+FR	–	+FR
Plant height (cm)	22.2	30.1*	20.3	30.3*	9.5	19.7*	92.4	91.2
Leaf/stem dry wt	2.4	2.0*	2.1	1.5*	5.1	3.2*	1.2	1.2

[1]Age at start 28 d tomato and 14 d cucumber. *Significant difference

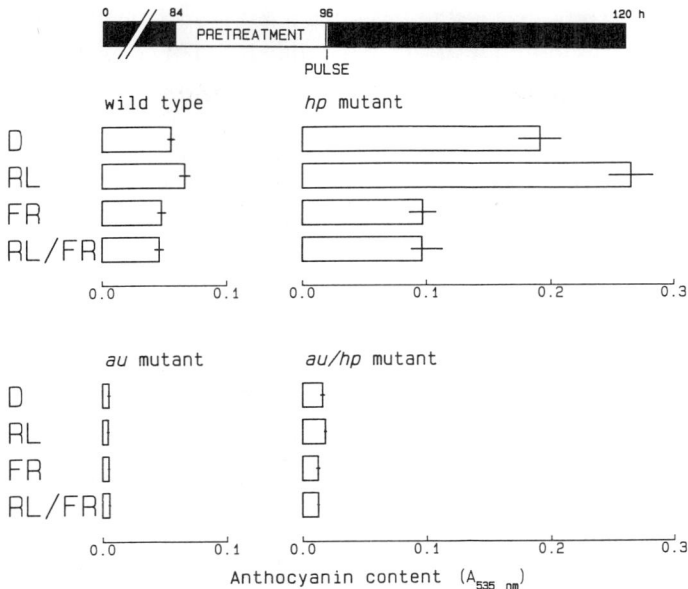

Figure 1. Anthocyanin content of tomato seedlings Moneymaker *au*, *hp*, *au/hp* and wild type after treatment with a pulse of RL; 5 min, (2.1 μmol m^{-2} s^{-1}), FR; 15 min, (2.1 μmol m^{-2} s^{-1}) or RL followed by FR terminating a 12 h BL pretreatment (2.1 μmol m^{-2} s^{-1}). Anthocyanin extracted after a further 24 h D period (Adamse et al., 1989).

wild type under P control (Adamse et al., 1989). Since the *hp* mutant shows pleiotropic effects that may also be associated with P, such as a reduced plant height under RL and yellow light (Mochizuki and Kamimura, 1985) and higher Chl levels particularly in fruit tissue, two characteristics that are the opposite of the *au* phenotype, it is likely that the *hp* mutant has enhanced sensitivity to Pfr. A study of hypocotyl elongation growth supports this conclusion (J.L. Peters pers. com.). Using the double mutant *au/hp* a reduction of the level of P by > 95% leads to a reduction of the P regulated anthocyanin synthesis to 97% of that under P control in the *hp* mutant (Fig. 1).

Cucumber

A long-hypocotyl mutant *(lh)* contains the same level of P as its isogenic wild type in D-grown seedlings and exhibits some P control of hypocotyl growth in etiolated seedlings, although de-etiolation appears to be retarded compared to wild type. However, particularly striking is the absence of P control in de-etiolated seedlings and on the basis of these observations it was proposed that the *lh* mutant was modified with respect to light-stable P (Adamse et al., 1987; 1988a). A mutant having a low level of Pfr in light-grown plants would be similar to a WL-grown wild type to which supplementary high irradiance FR has been added to lower the ø level. Such treatments have dramatic effects in many species, resulting in a massive stimulation of elongation growth (Smith, 1986). If this hypothesis is correct, the difference between mutant and wild type can be attributed to light-stable P being absent in the *lh* mutant. It is interesting to note that whereas wild type exhibits a clear end-of-day FR response the *lh* mutant fails to respond (Table 1). The *lh* mutant therefore appears to be an example of a mutant complementary to the *au* tomato mutant. This tall growing mutant, resembles the wild type that has been grown under supplementary FR to lower the ø value.

References

Adamse, P., Jaspers, P.A.P.M., Kendrick, R.E. and Koornneef, M., 1987, Photomorphogenetic responses of a long-hypocotyl mutant of *Cucumis sativus* L. *J. Plant Physiol.*, 127, 481–491.

Adamse, P., Jaspers, P.A.P.M., Bakker, J.A., Kendrick, R.E. and Koornneef, M., 1988a, Photophysiology and phytochrome content of long-hypocotyl mutant and wild-type cucumber seedlings. *Plant Physiol.*, 87, 264–268.

Adamse, P., Jaspers, P.A.P.M., Bakker, J.A., Wesselius, J.C., Heeringa, G.H., Kendrick, R.E., and Koornneef, M., 1988b, Photophysiology of a tomato mutant deficient in labile phytochrome. *J. Plant Physiol.*, 133, 436–440.

Adamse, P., Kendrick, R.E. and Koornneef, M., 1988c, Photomorphogenetic mutants of higher plants. *Photochem. and Photobiol.*, 48, 833–841.

Adamse, P., Peters, J.L., Jaspers, A.P.M., Van Tuinen, A., Koornneef, M., and Kendrick, R.E., 1989, Photocontrol of anthocyanin synthesis in tomato seedlings: A genetic approach. *Photochem. Photobiol.*, 50, 107–111.

Beggs, C.J., Wellmann, E., and Grisebach, H., 1986, Photocontrol of flavonoid biosynthesis. In: Photomorphogenesis in Plants (Edited by R.E. Kendrick and G.H.M. Kronenberg) pp. 467–499. Martinus Nijhoff Publ., Dordrecht.

Chang, C., Bowman, J.L., Dejong, A.W., Lander, E.S., Meyerowitz, E.M., 1988, Restriction fragment length polymorphism linkage map for *Arabidopsis thaliana. Proc. Natl. Acad. Sci. USA*, 85, 6856–6860.

Chory, J., Peto, C.A., Saganich, R., Pratt, L., and Ausubel, F., Ashbaugh, M., 1989, Different roles of phytochrome in etiolated and green plants deduced from characterization of *Arabidopsis thaliana* mutants. *The Plant Cell* 1, 867–880.

Cone, J.W., 1985, Photocontrol of seed germination of wild type and long-hypocotyl mutants of *Arabidopsis thaliana.* Ph.D. Thesis, Agricultural University, Wageningen.

Downs, R.J., Hendricks, S.B., and Borthwick, H.A., 1957, Photoreversible control of elongation in pinto beans and other plants under normal conditions of growth. *Bot. Gaz.*, 118, 199–208.

Gettens-Hayes, R. and Klein, W.H., 1974, Spectral quality influence of light during development of *Arabidopsis thaliana* plants in regulating seed germination. *Plant Cell Physiol.*, 15, 643–653.

Kaufman, L.S., Thompson, W.S., and Briggs, W.R., 1984, Different light requirements for phytochrome induced accumulation of *cab* RNA and *rbcS* RNA. *Science*, 226, 1447–1449.

Kerr, E.A., 1965, Identification of high-pigment, *hp*, tomatoes in the seedling stage. Can. J. Plant Sci., 45, 104–105.

Koornneef, M., Rolff, E., and Spruit, C.J.P., 1980, Genetic control of light-inhibited hypocotyl elongation in *Arabidopsis thaliana* L. Heynm. *Z. Planzenphysiol.*, 100, 147–160.

Koornneef, M., Cone, J.W., Dekens, R.G., O'Herne-Robers, E.G., Spruit, C.J.P., and Kendrick, R.E., 1985, Photomorphogenic responses of long hypocotyl mutants of tomato. *J. Plant Physiol.*, 120, 153–165.

Lipucci Di Paola, M., Collina Grenci, F., Caltavuturo, L., Tognoni, F., and Lercari, B., 1988, A phytochrome mutant from tissue culture of tomato. *Advances in Horticultural Science.*, 1, 30–32.

Mochizuki, T. and Kamimura,S., 1985, Photoselective method for selection of hp at the cotyledon stage. *Tomato Genet. Coop. Rpt.*, 35, 12–13.

Nagatani, A., Lumsden, P.J., Konomi, K., and Abe,H., 1987, Application of monoclonal antibodies to phytochrome studies. In: Phytochrome and Photoregulation in Plants (Edited by M. Furuya) pp. 95–114. Academic Press, Tokyo.

Parks, B.M., Jones, A.M., Adamse, P., Koornneef, M., Kendrick, R.E., and Quail, P.H., 1987, The aurea mutant of tomato is deficient in spectrophotometrically and immunocytochemically detectable phytochrome. *Plant Mol. Biol.*, 9, 97–107.

Pratt, L.H., and Cordonnier, M-M., 1987, Phytochrome from green *Avena*. In: Phytochrome and Photoregulation in Plants (Edited by M. Furuya) pp. 83–94. Academic Press, Tokyo.

Schäfer, E., 1987, Primary action of phytochrome. In: Phytochrome and Photoregulation in Plants (Edited by M. Furuya) pp. 279–287. Academic Press, Tokyo.

Sharrock, R.A., Parks, B.M., Koornneef, M., and Quail, P.H., 1988, Molecular analysis of the phytochrome deficiency in an *aurea* mutant of tomato., *Molec. Gen. Genetics*, 213, 9–14.

Smith, H., 1986, The perception of light quality. In: Photomorphogenesis in Plants (Edited by R.E. Kendrick and G.H.M. Kronenberg) pp. 187–217. Martinus Nijhoff Publ., Dordrecht.

Tobin, E. M., 1987, Photocontrol of gene expression. In: Phytochrome and Photoregulation in Plants (Edited by M. Furuya) pp. 39–50. Academic Press, Tokyo.

Wettstein-Knowles, P. Von, 1968, Mutations affecting anthocyanin synthesis in the tomato. II. Physiology. *Hereditas*, 61, 255–275.

Mode of Coaction between Phytochrome and Blue/UV Photoreceptors

H. MOHR AND H. DRUMM-HERREL

Biological Institute II
University of Freiburg
Germany

Abbreviations: D, darkness; B, blue light; FR, medium far-red light; P, total phytochrome (Pr + Pfr); Pfr, far-red absorbing form of phytochrome (physiologically active); Pfr/P, Pfr/P ratio in photoequilibrium (Pr $\rightleftarrows$ Pfr), often designed by φ λ; Pr, red absorbing form of phytochrome (physiologically inactive); R, red light.

Introduction

Plants are capable of adapting to the light conditions of their particular habitat. In order to respond properly, a plant has to monitor the light conditions in its environment continuously and accurately. This implies that a plant must be capable of sensing the quality and quantity of light throughout the sun's spectrum as far as sunlight leads to electronic excitations (290-800 nm). For reasons of molecular physics it is improbable that a single photoreceptor can fulfil this task. Rather, we might expect that a higher plant will use several sensor pigments to monitor the whole spectral range with the sensitivity and accuracy required.

Sensor Pigments

As far as we know today, three different sensor pigments occur in higher plants (Mohr, 1984). These are *phytochrome* (a photochromic photoreceptor, Pr $\rightleftarrows$ Pfr operating predominantly in the red and far-red spectral range), *cryptochrome* (operating in the blue/UV-A spectral range), and a *UV-B photoreceptor*. The action spectrum related to the latter photoreceptor shows a single intense peak at 290 nm and no action at wavelengths longer than 350 nm (Yatsuhashi et al. 1985).

Previous studies to establish a model of coaction between the different photosensors

One must be prepared to find different evolutionary solutions of the problem of coaction in different species and even with different photoresponses of the same plant.

Photobiology, Edited by E. Riklis
Plenum Press, New York, 1991

Moreover, the study of coaction is complicated by the fact that B/UV always operates on phytochrome, since B/UV inevitably converts Pr to Pfr.

A number of case studies have contributed to a unifying model (see Mohr et al. 1984, Mohr 1986, for references). The major points can be summarized as follows:

(i)   Phytochrome (Pfr) can, in some cases, act on growth, development and gene expression without any requirement for B/UV. An example is light-mediated synthesis of anthocyanin in the mustard (*Sinapis alba* L.) seedling cotyledons where red light can replace white light fully (Fig. 1).

(ii)   The other extreme is anthocyanin synthesis in the mesocotyl of the milo (*Sorghum vulgare* Pers.) seedling where R or FR alone have no effect (Table 1). experiments with dichromatic irradiation (simultaneous irradiation with two kinds of light to strongly modulate the level of Pfr on a constant background of B/UV) it was found that the B/UV photoreaction as such is not affected by the presence or virtual absence of Pfr during the B/UV treatment (Drumm and Mohr, 1978).

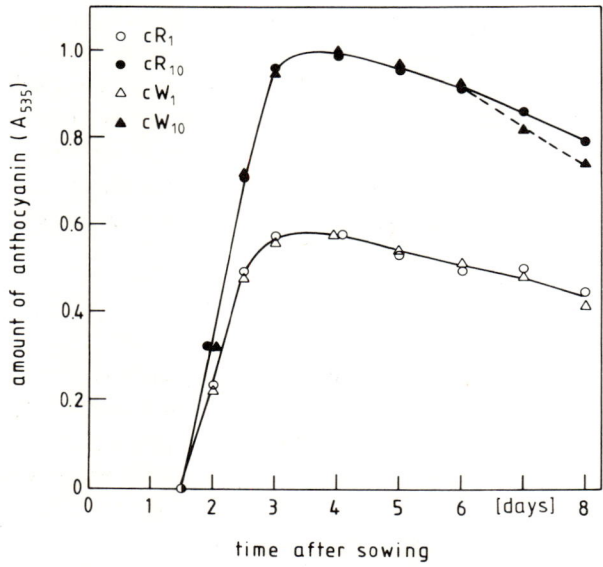

Figure 1. Time course of anthocyanin contents in mustard (*Sinapis alba* L.) cotyledons in continuous red (cR) and white light (cW) of different fluence rates (cR1, 0.68 W m^{-2}; cR10, 6.8 W m^{-2}; cW1, 0.86 W m^{-2}; cW10, 8.6 W m^{-2}). W was applied with approximately the same photon fluence rate of R as present in the corresponding R fields. This requires an approximately 30% higher fluence rate in the W fields as compared to the corresponding R fields. Photoequilibrium Pfr/P in W is determined essentially by the R part of the spectrum and is thus almost the same as in pure R (Pfr/P > 0.7).

The interpretation of the data obtained with milo was that phytochrome (Pfr) is the effector which causes anthocyanin synthesis while the B/UV effect was considered as establishing responsiveness towards Pfr.

(iii)   In many cases, it was found that responsiveness towards Pfr, established by single light pulses, is extremely weak in dark-grown material, while prolonged light treatments lead to a dramatic increase of responsiveness (degree of response per unit Pfr).

An example is induction of plastidic glyceraldehyde-3-phosphate dehydrogenase in the primary leaf of the milo seedling (Fig. 2). The data show that long-term light causes a rapid and strong increase of responsiveness (responsiveness amplification) which tends to saturate after approximately 6 h. B and UV are equally effective and far more effective than R. Since light-mediated changes of total phytochrome levels are the same in R, B and UV, it is clear that B and UV cause a several times higher responsiveness than R. However, it is equally obvious that even R alone, operating exclusively through phytochrome, exerts a considerable effect on responsiveness to Pfr.

A series of case studies of this kind has led us to suggest a general model of coaction between B/UV and light absorbed by phytochrome (Fig. 3) which generalizes the more specific model derived previously from the experiments on anthocyanin formation in the milo mesocotyl (Oelmüller and Mohr 1985).

Table 1. Induction (or lack of induction) of anthocyanin in the mesocotyl of milo seedlings (*Sorghum vulgare* Pers.) by light of different qualities (W: Xenon arc light, similar to sunlight, 250 W m^{-2}). In the case of a 3 h light treatment, the seedlings were kept in the dark for 24 h before extraction of anthocyanin. Red light (R, Pfr/P = 0.8), medium far-red light (FR, Pfr/P = 0.03), long wavelength FR (756 nm, Pfr/P < 0.01). A 5 min light pulse suffices to establish the photoequilibrium. After Drumm and Mohr (1978).

Treatment (onset 60 h after sowing)	Amount of anthocyanin (measurement 87 h after sowing) (A at 510 nm)
27 h dark	0
27 h W	1.85
27 h R	0
27 h FR	0
3 h W	0.19
3 h B/UV	0.19
3 h W + 5 min R	0.19
3 h W + 5 min 756 nm	0.06
3 h W + 5 min 756 nm + 5 min R	0.20
3 h B/UV + 5 min R	0.19
3 h B/UV + 5 min 756 nm	0.05
3 h B/UV + 5 min 756 nm + 5 min R	0.19

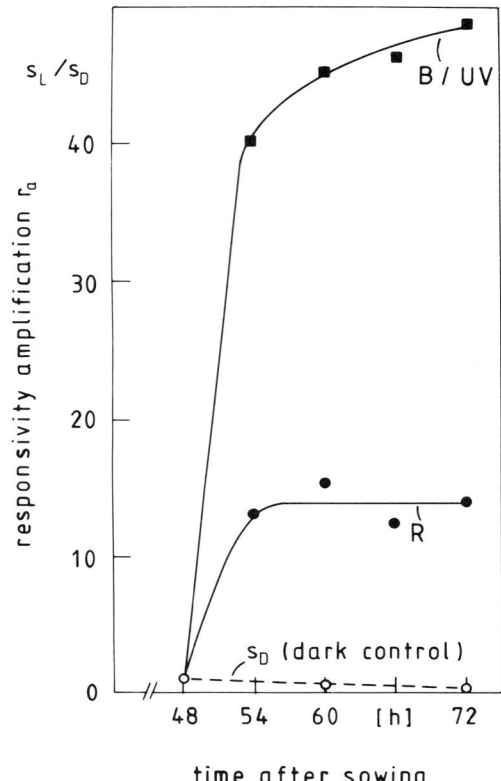

Figure 2. Time course of increase of responsiveness in long term red (R) and blue/UV light (B/UV) for phytochrome-mediated glyceraldehyde-3-phosphate dehydrogenase (GPD) induction in the milo (*Sorghum vulgare* Pers.) shoot. The light treatment commences 48 h after sowing. SL-responsiveness to Pfr in light treated material, SD-responsiveness to Pfr in dark-grown material, SD at 60 and 72 h refers to SD at 48 h = 1. After Oelmüller and Mohr (1984).

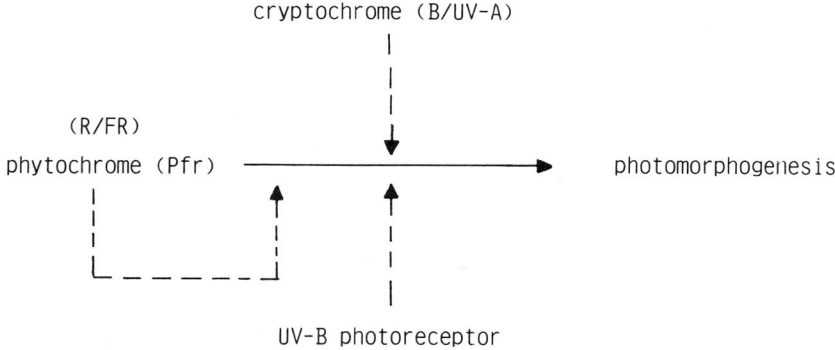

Figure 3. Suggested mode of coaction between phytochrome and the photoreceptors absorbing blue/UV light. After Mohr (1986).

Recent studies to test the model

Cucumber seedlings have often been used to study rapid inhibition of stem elongation by B (see Shinkle and Jones 1988, for reference). The response is considered *not* to be mediated by phytochrome, and similar reactions are assumed to occur in other species.

In order to understand the special position of cucumber we must briefly consider threshold control of axis elongation in other species. In the mustard seedling where only phytochrome operates (see Fig. 1), threshold controld of hypocotyl elongation is clearly expressed (Fig. 4).

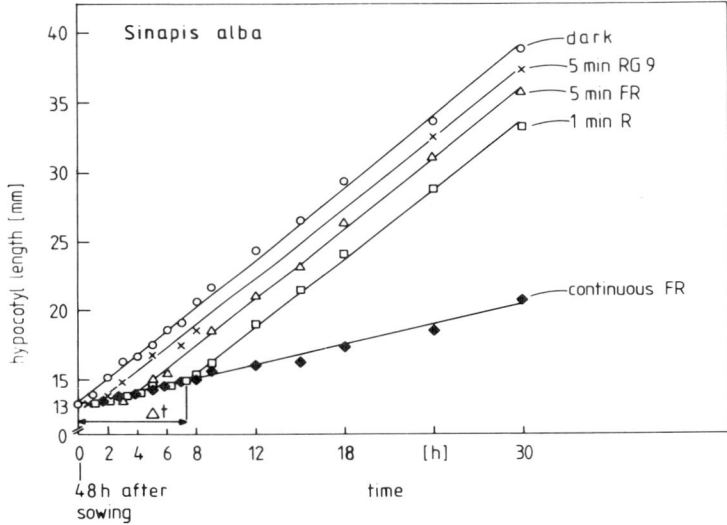

Figure 4. Time courses of hypocotyl elongation in intact mustard (*Sinapis alba L.*) seedlings in darkness (O), continuous far-red light (◆) and after saturating light pulses, which establish different Pfr levels: (✖) φ RG9 = 0.14%; (Δ) φ FR = 2.3%; (□) φ R = 80%; onset of light treatment at 48 h after sowing. Data for phytochrome photoequilibria φ λ *in vivo* as a function of wavelength are from Schäfer et al. (1975, Fig. 2). Δ t, extrapolated duration of the time between the light pulse and the point of resumption of growth. After Oelze-Karow and Mohr (1989).

Under our experimental conditions, the growth rate of the mustard hypocotyl in the dark can be considered to be constant between 48 and 78 h after sowing. Continuous FR (cFR, 3.5 W m⁻², operating through phytochrome, see Mohr 1984) results in a marked reduction in the growth rate. Growth in FR can be considered a kind of 'base line' since this growth rate cannot be decreased further by any light treatment. The same reduction in growth rate as seen in cFR is *transiently* observed when light pulses are applied at 48 h after sowing, and the seedling is then returned to darkness. The different light pulses (R, FR, RG9-light) establish different phytochrome photoequilibria (see legend to Fig. 4). Apparently, the amount of Pfr established by a light pulse determines the length of time (Δ t) before the growth rate characteristic for darkness is restored.

The data in Fig. 4 indicate that Pfr control of hypocotyl elongation in mustard acts as a threshold response. This implies that as long as the level of Pfr remains above the threshold, elongation is inhibited (i.e. elongation occurs along the 'base line').

As soon as the level of Pfr decreases below the threshold level, the dark rate of elongation is immediately restored. Phytochrome degradation kinetics, measured in the hook part of the hypocotyl, are in *quantitative* agreement with this concept (Oelze-Karow and Mohr, 1989).

The same type of control was observed in sesame (*Sesamum indicum* L.) seedlings. In this species the rate of hypocotyl elongation is controlled by W and B while R

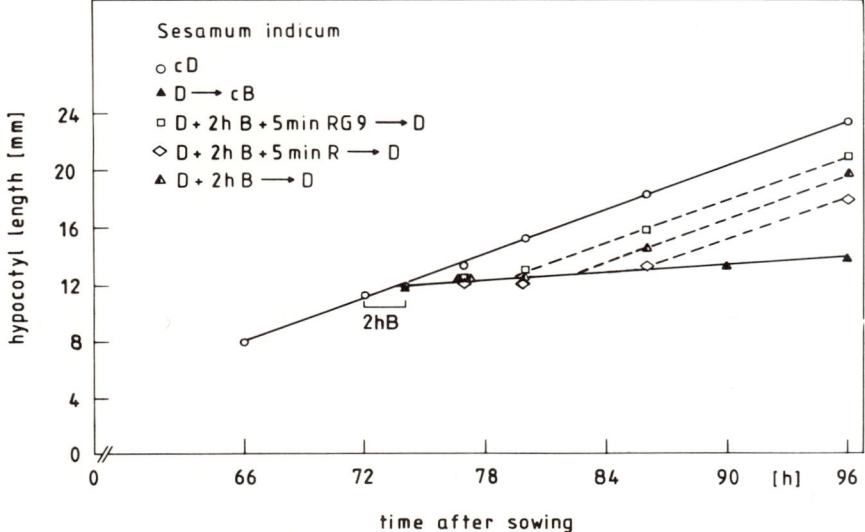

Figure 5. Time courses of hypocotyl elongation in the sesame (*Sesamum indicum* L.) seedling. Explanation of symbols: cD, continuous darkness; cB, continuous blue light (7 W m^{-2}, onset at 72 h after sowing); → D, transfer to darkness; FR (RG9), long wavelength far-red light, obtained with glass filter RG9 (10 W m^{-2}); R, red light (6.8 W m^{-2}).

(continuous R as well as repeated R pulses) is totally ineffective beyond 60 h after sowing. However, when seedlings are kept in B for 3 d and then treated with R or FR pulses growth rate responds strongly (Drumm-Herrel and Mohr 1984). If 2 h B is given to a dark-grown seedling 72 h after sowing growth rate in darkness is controlled by the level of Pfr over at least 6 h (Fig. 5). These results are explained by a phytochrome threshold control with the Pfr/P ratios established in the hook part of the sesame hypocotyl by R (0.8), long-wavelength FR (< 0.01), and B (0.38). These and further experimental data show that B is required to establish and to maintain threshold responsiveness toward Pfr. An important finding was that the fluence rate of B must itself exceed a certain threshold which is far above the fluence rates required to elicit phototropism (Drumm-Herrel and Mohr 1984, Woitzik and Mohr 1988).

In case of cucumber (Fig. 6) phytochrome exerts a strong effect even in the absence of B and a threshold control of the growth rate by Pfr is indicated by the growth kinetics

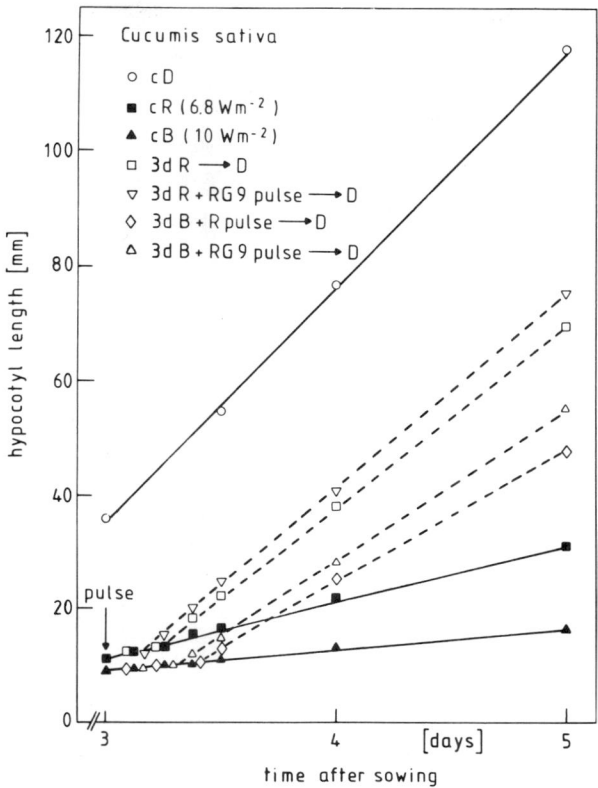

Figure 6. Time course of hypocotyl elongation in the cucumber (*Cucumis sativa* L.) seedling. Explanation of symbols: cD, continuous darkness; cB, continuous blue light (10 W m⁻²), onset at the time of sowing; cR, continuous red light (6.8 W m⁻²); - - -, growth kinetics after transfer to darkness. The R or B treatment was terminated with either a 5 min R (□ ◇)or a 5 min FR (RG9) pulse (Δ ▽).

following a light → dark transfer. It appears that there is no specific effect of B on the threshold control mechanism. The larger Δ t values in B grown seedlings are to be expected since B grown tissues contain twice as much P compared to R grown tissues (Pfr/P in B = 0.38 and Pfr/P in R = 0.8). Moreover, there is no specific B effect once growth is restored in darkness (the tendency of decreasing growth in darkness with increasing Δ t is also observed in the R grown seedlings). A specific effect of B can be described as follows: B determines the 'base line' for the phytochrome threshold control. This is the same action of B we have encountered above in case of sesame (Fig. 5).

A brief look at phototropism

Recent studies (Woitzik and Mohr 1988) have confirmed that sesame as well as mustard seedlings respond very sensitively to unilateral B (e.g. 1,4,8 or 16 mW m⁻²)

while with regard to straight growth (in omnilateral light) both seedlings are totally 'blue light-blind' at these low fluence rates. In other words, a strong phototropic response can be elicited by unilateral blue light which does not have any effect on straight growth if applied omnilaterally.

On the other hand, if the sesame seedlings were kept in red light for some period, their rate of phototropic response towards unilateral B was much higher than with dark-grown seedlings (Fig. 7). Phototropism in sesame is particularly interesting since it

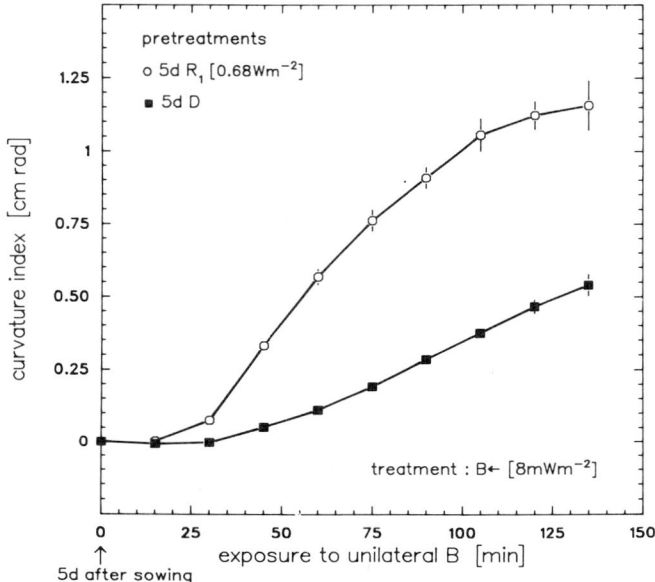

Figure 7. Phototropic curvature of the hypocotyl of the sesame (*Sesamum indicum* L.) seedling in continuous unilateral blue light (B). The 'curvature index' is a sensitive gauge of the phototropic response which considers not only the angle of curvature but also the length of the responding zone of the axis. The seedlings were either kept in darkness or in weak red light until exposure to unilateral blue light. After Woitzik and Mohr (1988).

demonstrates unambiguously that the phytochrome-mediated effect of a red light pretreatment on phototropism is unrelated to the control of straight growth by phytochrome. As mentioned above, hypocotyl straight growth in sesame is not affected by R beyond 60 h. However, a R pre-treatment exerts a strong effect on the rate of the phototropic curvature. Thus, phytochrome strongly affects differential growth of the hypocotyl flanks (elicited by unidirectional blue light) even though it does not affect straight growth of the hypocotyl. Apparently, the effect of R on the growth potential of the hypocotyl cells remains cryptic until unilateral blue light is given.

Conclusion

Comparing photomorphogenesis and phototropism, it can be stated that phytochrome (Pfr) is the effector proper in bringing about photomorphogenesis in higher plants while cryptochrome and UV-B photoreceptor (together with phytochrome) determine the plant's responsiveness towards Pfr. Phototropism, on the other hand, can only be elicited by B/UV. In this case, it is phytochrome which modulates the rate of the response.

References

Drumm, H. and Mohr, H. (1978), The mode of interaction between blue (UV) light photoreceptor and phytochrome in anthocyanin formation of the *Sorghum* seedling. *Photochem. Photobiol.*, 27, 241-248.

Drumm-Herrel, H. and Mohr, H. (1984), Mode of coaction of phytochrome and blue light photoreceptor in control of hypocotyl elongation. *Photochem. Photobiol.* 40, 261-266.

Mohr, H. (1984), Criteria for photoreceptor involvement. In *Methods in Photomorphogenesis, Biological Techniques Series*, (ed. by H. Smith and M.G. Holmes), p. 13, Academic Press, London.

Mohr, H. (1986), Coaction between pigment systems. In *Photomorphogenesis in Plants* (ed. by R.E. Kendrick and G.H.M. Kronenberg), p. 547, M. Nijhoff, Dordrecht, The Netherlands.

Mohr, H., Drumm-Herrel, H. and Oelmüller, R (1984), Coaction of phytochrome and blue/UV light photoreceptors. In *Blue Light Effects in Biological Systems* (ed. by H. Senger), p. 13, Springer, Heidelberg.

Oelmüller, R. and Mohr, H. (1984), Responsivity amplification by light in phytochrome-mediated induction of chloroplast glyceraldehyde-3-phosphate dehydrogenase (NADP-dependent, EC 1.2.1.13) in the shoot of milo (*Sorghum vulgare* Pers.) *Plant Cell Environ.*, 7, 29-37.

Oelmüller, R. and Mohr, H. (1985), Mode of coaction between blue/UV light and light absorbed by phytochrome in light-mediated anthocyanin formation in the milo (*Sorghum vulgare Pers.*) seedling. *Proc. Natl. Acad. Sci.*, 82, 6124-6128.

Oelze-Karow, H. and Mohr, H. (1989), An analysis of phytochrome-mediated threshold control of hypocotyl growth in mustard (*Sinapis alba L.*) seedlings. *Photochem. Photobiol.*, 50, 133–141.

Shinkle, J.R. and Jones, R.L. (1988), Inhibition of stem elongation in *Cucumis* seedlings by blue light requires calcium. *Plant Physiol.*, 86, 960-966.

Woitzik, F. and Mohr, H. (1988), Control of hypocotyl phototropism by phytochrome in a dicotyledonous seedling (*Sesamum indicum L.*) *Plant, Cell and Environment*, 11, 653–661.

Yatsuhashi, H., Hashimoto, T., and Shimizu, S. (1982), Ultra-violet action spectrum for anthocyanin formation in broom *Sorghum* first internode. *Plant Physiol.*, 70, 735-741.

Phytochrome Regulated Expression of Cab Genes in Higher Plants

F. NAGY[1,2], E. FEJES[1], A. PAY[1] AND E. ADAM[1]

[1]*Biological Research Center*
Hungarian Academy of Sciences
Szeged, H-6701, Hungary
[2]*The Rockefeller University*
New York, NY 10021, USA

Abstract

In higher plants the chlorophyll a/b binding protein is encoded by small multigene families (1). We have isolated and characterized the Cab-1 gene from wheat (2). We have shown that the Cab-1 gene is expressed in a tissue-specific, light-inducible manner. The regulated expression of the Cab-1 gene is controlled by phytochrome and manifested at the level of transcription (3). Recently we have demonstrated that the expression of the Cab-1 gene is also controlled by an endogenous circadian rythm (4). Both the circadian rythm and the phytochrome mediated light response of the Cab-1 gene can be recapitulated in transgenic tobacco plants. *In vitro* DNA sequence manipulation coupled with expression in transgenic plants revealed *cis*-acting elements for the regulated expression of this gene. One of these elements, a ca. 260 bp upstream enhancer-like sequence of the Cab-1 gene confers phytochrome controlled, tissue-specific expression to constitutive promoters (5). Detailed analysis of this element has revealed a complex array of multiple regulatory sequences for the for the expression of the Cab-1 gene.

Introduction

During the last decade the molecular biology of plant gene expression has shown a remarkably fast development. Different methodologies have been worked out for transferring DNA sequences into plant cells and the number of isolated and characterized genes has also increased steadily (Kuhlemeier et al., 1987). The development of methods for the production of transgenic plants has provided a new way of investigating the expression of plant genes and the impact of this new approach on our knowledge about the regulatory circuits of plant gene expression is already obvious. Many genes of higher plants are expressed in a highly regulated fashion. Certain genes are expressed only at specific stages of development only in certain cell types, or their expression is regulated by different environmental stimuli, such as light, heat, salt, wounding. Higher plants developed several signal transducing systems to react to changes in their environment. One such system, which responds to fluctuations in ambient light quality and fluences,

Photobiology, Edited by E. Riklis
Plenum Press, New York, 1991

has as its receptor a pigment known as phytochrome (Quail et al., 1986). Recent studies have indicated that a variety of developmental responses are controlled by phytochrome. It was postulated that many morphogenic changes elicited by phytochrome involve alterations in gene expression (Vierstra and Quail, 1986). The expression of a number of nuclear genes has been shown to be controlled by phytochrome.

Prominent light-induced, phytochrome regulated genes include those encoding the small subunit (rbcS) of ribulose-1,5-bisphosphate carboxylase and the chlorophyll a/b binding (Cab) protein of the light-harvesting complex (Silverthorne and Tobin, 1984; Mosinger et al., 1985). In addition to their phytochrome regulated expression the rbcS and Cab genes were shown to be expressed in a tissue-specific manner (Kuhlemeier et al. 1987). Because of their complex regulation, these two photosynthetic genes were particularly attractive as model systems to study regulation of gene expression in plants. In recent years several laboratories have begun to analyse those signal transduction pathways which lead to the regulated expression of these genes.

Our approach was aimed at delineating the sequence of events leading from light reception by phytochrome to the activation of target (Cab) gene expression. We were particularly interested in the terminal steps of these signal transduction pathways, therefore we focused on the identification of cis-acting elements that mediate light-induced, tissue-specific gene expression. To this end we have combined in vitro DNA sequence manipulation with gene expression in transgenic plants. We have shown so far that the regulation of the wheat Cab genes is excercised at the transcriptional level and a short (241 bp) promoter fragment contains the majority of *cis*-regulatory elements. (Nagy et al., unpublished results)

Materials and Methods

In vitro DNA manipulations were carried out according to Maniatis et al. (1982). Preparation of RNA and 3'or 5' S1 nuclease protection assays were done as described (Nagy et al. 1987). Chimeric gene constructs were inserted into a pMON 505 binary cloning vector and transferred into a disarmed *Agrobacterium tumefaciens* (GV3111SE) by triparental crosses. *Agrobacterium tumefaciens* cells containing chimeric gene constructs were co-cultured with leaf-discs of *Nicotiana tabacum* on a medium containing kanamycin and carbenicillin for selection and phytohormones for plant regeneration (Horsch and Klee, 1986). Transgenic plants were transferred to soil and grown to flowering in a greenhouse. Light treatments of etiolated transgenic seedlings and of mature transgenic plants were done as described (Nagy et al. 1987).

Results and Discussion

mRNA encoding the Cab protein is a major transcript in leaf mesophyll cells. In both monocots and dicots the Cab mRNA is transcribed from a multigene family in the nucleus (Kuhlemeier et al. 1987). A genomic clone for a major Cab protein has been isolated and sequenced from wheat by Lamppa et al. (Lamppa et al. 1985). This gene designated Cab-1 encodes a 70 nucleotide 5' nontranslated leader, a 34-amino acid N-terminal extension (transit peptide) and a coding region for a mature protein of 232 amino acid residues. No intervening sequences are found in this gene. The expression pattern of the Cab-1 gene was characterized at different developmental stages.

Expression of the Cab-1 gene in etiolated seedlings

The mRNA level of the Cab-1 gene is undetectable in dark-grown wheat seedlings. Upon brief illumination with red light the transcript level of the Cab-1 gene is increased considerably (20-fold) and the stimulating effect of red light can be abrogated by a subsequent far-red illumination. In fact we have shown that far-red light alone can elicit a small increase in the Cab-1 mRNA level (Nagy et al. 1986). The extreme sensitivity of the Cab-1 gene transcription to red light and its induction by far-red light alone are characteristic features of the Cab gene regulation by phytochrome (Kauffman et al. 1984). These modulating effects of red and far-red light provide evidence that the wheat Cab-1 gene is regulated by phytochrome at this developmental stage.

Expression of the Cab-1 gene in green tissue

Kloppstech (1985) reported that the Cab mRNA level of barley undergoes diurnal oscillation under light-dark conditions. We extended and confirmed these observations. We have shown that the oscillation persists even under constant environmental conditions. These results indicate that the wheat Cab gene family as a whole, and the Cab-1 gene in particular are under the control of an endogenous circadian rhythm. Although the oscillation of Cab-1 transcript level is continuous with a 24 h periodicity under extended darkness, the amplitude is dampened with time indicating that light also plays a role in regulating the expression level. Piechulla (1988) has reported similar findings by studying Cab gene expression in tomato leaves. We have shown also that the dampening in continuous darkness can be prevented by red illumination and the stimulating effect of red light can be erased by far-red treatment. These light treatments affect only the amplitude but do not elicit a phase shift in the expression of the Cab-1 gene. Our results provide evidence that phytochrome is involved in the regulation of the Cab-1 gene expression within the periodicity of the clock at this developmental stage (Nagy et al. 1988). The interaction of phytochrome with the circadian clock is reminiscent of a similar interaction reported for the photoperiodicity of higher plants (Vince-Prue, 1983).

Expression of the wheat Cab-1 in transgenic tobacco plants

To define *cis*-acting DNA sequences for light-induced, phytochrome mediated gene expression we had to monitor the transcription of *in vitro* assembled chimeric genes. No cell free system has yet been established for studying regulated plant gene expression, therefore transcription of different chimeric genes was evaluated in transgenic plants. The optimal host plant for our experiments would obviously be wheat; however, the

457

regeneration of plants from undifferentiated, transformed wheat cells is not yet possible. For practical reasons (i.e. tissue-culture methods, seed production, and relatively short vegetative period) we chose *Nicotiana tabacum* as a host plant for these studies. First we transferred a 6.6 kb wheat genomic fragment into tobacco. This fragment contained 4.4 kb of 5' upstream sequences, the entire coding region and a 600 bp 3' untranslated region of the Cab-1 gene. It was found that the expression pattern of the Cab-1 gene can be recapitulated in transgenic tobacco plants (Lamppa et al. 1985). The transferred wheat gene was expressed in a light-regulated, tissue specific manner. S1 nuclease protection experiments provided evidence that the Cab-1 gene was expressed mainly in chloroplast containing, fast growing tissues. This type of tissue specificity is reminiscent of the one found in wheat where the highest levels of mRNA were detected in the basal segments of leaves while roots did not contain measurable amounts of Cab-1 specific RNA. In addition to its tissue-specific expression we demonstrated that the transferred wheat gene maintained its phytochrome responsiveness in green as well as in etiolated tissues. The circadian rhythm was also maintained in green tissues of transgenic plants (Nagy et al. 1988).

These results indicate that the regulatory mechanism of the circadian clock and the signal transduction pathway for phytochrome regulation are conserved between monocots and dicots. Since these two groups of flowering plants are thought to have diverged in the Cretaceous (Dahlgreen et al. 1985), these experiments bridge a gap of 110×10^6 years in evolutionary time. Taken together, these observations clearly show that the transgenic tobacco system can be used as an assay system for the identification of *cis*-regulatory elements which are responsive to phytochrome.

Cis-regulatory elements for phytochrome controlled, tissue-specific Cab-1 gene expression

To delineate sequences for phytochrome regulation and for tissue specific gene expression we analyzed a series of 5' deletion mutants as well as chimeric gene constructs comprising different regions of the Cab-1 gene. Analysis of a chimeric construct revealed that a 1.8 kb 5' flanking fragment (–1816 to +31) of the Cab-1 gene can confer phytochrome and circadian rhythm controlled leaf-specific expression on a heterologous (CAT) coding sequence. The analysis of yet another chimeric gene showed that the coding region of Cab-1 (from +31 to +1100) is not involved in the regulation of this wheat gene.

Taken together, these results indicate that the circadian control and the phytochrome responsiveness of Cab-1 gene expression is largely, if not exclusively, a transcriptional phenomenon, and the stability of the Cab mRNA is not a key factor in this aspect. Run-off experiments using isolated nuclei confirmed our conclusion by demonstrating that the phytochrome regulation of Cab genes is indeed exercised at the level of transcriptional activation in dark grown seedlings (9, l0.) and in fully developed green plants (Kanevsky et al., unpublished results)

To define more precisely *cis*-regulatory sequences residing upstream of + 31, we constructed a 5' deletion mutant carrying only 357 bp 5' upstream sequences and showed that it still exhibits full regulation. Two conlusions were drawn from these results: (i) sequences upstream of –357 of the Cab-1 gene are not required and (ii) a 289 bp fragment between –90 and –357 contains elements for expression and for regulation. We confirmed these findings by several experiments in which various Cab-1 upstream fragments were evaluated for their ability to potentiate transcription from a truncated CaMV 35S promoter. We found that only one proximal subfragment (–89 to –357) of the 5' upstream region is active in these assays. More importantly, this proximal subfragment retains its activity when placed in the inverted orientation suggesting that it has a property expected of a transcription enhancer (Nagy et al. 1987). Chimeric genes containing only this 289 bp sequence from the Cab-1 gene were found to be expressed in a phytochrome regulated, tissue-specific manner either in etiolated or in green transgenic tobocco plants (Nagy et al., unpublished results). Castresana et al. (1988) have also identified a light regulatory element within the promoter region of a tobacco Cab gene extending from –396 to –186 which confers photoregulated expression when fused to a constitutive nopaline synthase promoter. Our findings are consistent with several other observations. Simpson et al. (Simpson et al. 1985) reported that a 247 bp 5' upstream fragment of a pea Cab gene can confer white light inducibility on a truncated NOS promoter. Moreover, this fragment is also active when placed in the inverted orientation. A 280 bp fragment from the corresponding region of the pea rbcS-3A gene (–48 to –327) was shown to confer phytochrome inducible transcription on a truncated 35S promoter (Fluhr et al. 1986). These results clearly show that (i) different light-induced, phytochrome controlled genes contain a ca. 280 bp enhancer element and (ii) this element is essential for their regulated expression.

Analysis of these enhancer-like elements has been started and some information about the organization of these regulatory elements is already available. Kuhlemeier et al. (Kuhlemeier et al. 1987) reported that the above described upstream region of the pea rbcS-3A gene contains multiple copies of negative and positive regulatory elements for light inducibility. Castresana et al. (1988) have described two positive and one negative *cis*-acting regulatory elements in the far upstream region of a tobacco Cab gene for maximum levels of photoregulated expression. Strittmatter and Chua (1988) analysed chimeric genes containing an artificial combination of two *cis*-regulatory elements and found that different negative regulatory sequences are responsible for the repressed transcript level of the rbcS-3A gene in roots and in dark-adapted leaves. Similar results have been reported by Simpson et al. (1986) describing a root specific silencer in the upstream region of a pea Cab gene. Taken together, these results indicate that several positive and negative *cis*-acting regulatory elements contribute to the light-induced tissue-specific, phytochrome responsive gene expression in higher plants. Further experiments are needed to define the exact number, position and relative importance of these elements for the regulated expression of different Cab genes.

Acknowledgements

The majority of this work was done in the laboratory of Professor N.-H. Chua at the Rockefeller University where F. Nagy worked as a postdoctoral fellow in 1983–1987. During this time the work was supported by grants to N.-H. Chua from the Monsanto Company. Experiments carried out since 1987 have been supported by grants to F. Nagy from a OKKFT Program (Tt) in Hungary.

References

Castresana, C., Garcia-Luque, I., Alonso, E., Malik, S. V. and Cashmore, R. A. (1988) *EMBO J.*, 7, 1929–1936.

Dahlgreen, R. M. T., Clifford, H. T. and Yeo, P. (1985) The Families of the Monocotyledons. Springer-Verlag, Berlin.

Fluhr, R., Kuhlemeier, C., Nagy, F. and Chua, N.-H. (1986) *Science* 232, 1106–1112.

Horsch, R. B. and Klee, H. J. (1986) *Proc. Natl. Acad. Sci. USA*,83, 4428–4432.

Kauffman, L. S., Thompson, W. F. and Briggs, W. R. (1984) *Science* 226, 1447–1449.

Kloppstech, K. (1985) *Planta* 165, 502–506.

Kuhlemeier, C., Fluhr, R., Green, P. J. and Chua, N.-H. (1987) *Genes and Dev.* 1, 247–255.

Kuhlemeier, C., Green, P. and Chua, N.-H. (1987) *Annu. Rev. Plant Physiol.*, 38, 221–257.

Lamppa, G., Morelli, G. and Chua, N.-H. (1985) *Mol. Cell Biol.*, 5, 1370–1378.

Lamppa, G., Nagy,. F. and Chua, N.-H. (1985) *Nature*, 316, 750–752.

Maniatis, T., Fritsch, E. F. and Sambrook, J. (1982) Cold Spring Harbor Laboratory

Mosinger, E., Batschauer, A., Schafer, E. and Apel, K. (1985) *Eur. J. Biochem.* 147, 137–142.

Nagy, F., Boutry, M. Hsu, M.-Y. Wong, M. and Chua, N.-H. (1987) *EMBO J.*, 6, 2537–2542.

Nagy, F., Kay, S. and Chua, N.-H. (1988) *Genes and Development.* 2, 376–382.

Nagy, F., Kay, S., Boutry, M. and Chua, N.-H. (1986) *EMBO J.*, 5, 1119–1124.

Piechulla, B. (1988) *Plant Molecular Biology* 11, 345–353.

Quail, P. H. et al. (1986) *Phil. Trans. R. Soc. London Ser. B.* 314, 469–480.

Rogers, S. G., Horsch, R. B. and Fraley, R. T. (1986) *Methods Enzymol.*, 118, 627–640.

Silverthorne, J. and Tobin, E. M. (1984) *Proc. Natl. Sci. USA*, 81, 1112–1116.

Simpson, J., Schell, J., Van Montagu. M. and Herrera-Estrella (1986) *Nature* 323, 551–554.

Simpson, J., Timko, M. P., Cashmore, A. R., Schell, J., Van Montagu, M. and Herrera-Estrella, L. (1985) *EMBO J.*, 4, 2723–2729

Strittmatter, G. and Chua, N.-H. (1988) *Proc. Natl. Acad. Sci. USA* 84, 8986–8990.

Tobin, E. M. and Silverthorne, J. (1985) *Annu. Rev. Plant Physiol.*, 36, 569–593.

Vierstra, R. D. and Quail, P. H. (1986) in Photomorphogenesis of Plants (Kendrick, R. E. and Kronenberg, G. H. M., eds), pp. 35–59, Martinus Nijhoff

Vince-Prue, D. (1983) Photomorphogenesis and flowering. In: Encyclopedia of Plant Physiolgy, 16B, Shropshire, W., R. and Mohr, H., eds. pp. 458–490. Springer-Verlag, Berlin.

Phytochrome Structure/Function Relationships as Probed by Monoclonal Antibodies

LEE H. PRATT, MARIE-MICHELE CORDONNIER
and LYLE CROSSLAND

Department of Botany
University of Georgia, Athens, Georgia
CIBA-GEIGY Biotechnology
Research Triangle Park
North Carolina

Abstract

Monoclonal antibodies (MAbs) are ideal for elucidation of the structure/function relationships of low abundance, relatively labile macromolecules such as phytochrome, in part because of their precise epitope specificity. MAbs that react differentially with the inactive, red-absorbing (Pr) and active far-red-absorbing (Pfr) forms of phytochrome have been obtained, as have been MAbs that recognize highly conserved domains. Similarly, MAbs that can discriminate among different pools of phytochrome from the same plant have been characterized. The locations of many of the epitopes recognized by these MAbs have now been mapped, permitting functional domains on this chromoprotein to be correlated with its primary structure. In particular, the epitope recognized by MAb Pea-25, which detects a highly conserved domain found on phytochrome from angiosperms to algae, is resolved by analysis of fusion proteins to a sequence of seven amino acids (-pro-ile-phe-gly-ala-asp-glu-, residues 765-771) on the carboxy-terminal domain of this chromoprotein.

Introduction

The exquisite epitope specificity of monoclonal antibodies (MAbs) makes them ideal for the elucidation of structure/function relationships of a complex macromolecule such as phytochrome, which is a globular chromoprotein with two similar, if not identical, monomers of about 124 kDa in size (Vierstra and Quail, 1986). In the case of phytochrome, an epitope comprises only about 0.5% of its monomer composition (5–7 amino acids out of more than 1100). Since an antibody can be sensitive to (i) relatively minor modification of a protein, such as addition or removal of a phosphate moiety (Smith et al., 1987), (ii) a single amino acid substitution (Getzoff et al., 1987), or (iii) even a relatively modest conformational change (Hansen and Beavo, 1986), it can be an equally sensitive indicator of its structure/function relationships. In addition, since immunochemical assays can require only femtomol amounts of antigen, they can be performed readily with low abundance proteins such as phytochrome. The objective of this contribution is to summarize our own efforts to utilize monoclonal antibodies for structure/function studies of phytochrome, with an emphasis on an epitope mapping

Photobiology, Edited by E. Riklis
Plenum Press, New York, 1991

461

strategy that permits quick, precise and unambiguous assignments by immunoblot analysis of fusion proteins.

Monoclonal antibodies that detect potentially important domains on phytochrome

Pr versus Pfr

Although most antibodies do not discriminate between Pr and Pfr, at least two domains that are differentially recognized by MAbs have been identified (Cordonnier et al., 1985; Shimazaki et al., 1986; Thomas and Penn, 1986; Holdsworth and Whitelam, 1988). By ELISA, one set of MAbs (Oat-23, Oat-24, Oat-25) detects Pr better than Pfr, while a second set (Oat-9, Oat-16) detects Pfr better than Pr. MAbs that prefer Pr have proven to be particularly interesting (Cordonnier et al., 1985). When bound to Pr, Oat-25 has no perceptible effect on its absorbance spectrum. When bound to Pfr, however, extinction at the far red absorbance maximum is decreased and shifted 8 nm towards the blue. Moreover, when bound first to Pr, the Pfr that is produced upon phototransformation is unstable. Unlike undegraded phytochrome from etiolated oats, Pfr with Oat-25 bound to it reverts in darkness back to Pr. Apparently, Oat-25 either binds to a domain that is essential for Pfr to assume its stable conformation, or its presence on phytochrome interferes in steric fashion with the formation of Pfr.

Evolutionarily conserved domains

Identification of evolutionarily conserved domains is of interest because their existence implies that there has been some selective pressure maintaining them, which in turn indicates that they might be important to the molecular function of phytochrome. An initial attempt to identify conserved epitopes by ELISA did select two MAbs (Oat-12 and Oat-20) that reacted equally well with phytochrome from three monocotyledons and three dicotyledons (Cordonnier et al., 1984). Unfortunately, neither detected phytochrome well by immunoblot assay, making it difficult to identify their epitopes with precision. Subsequent screening by immunoblot assay, however, did identify a MAb (Pea-25) that detects a polypeptide the size of phytochrome from angiosperms to algae, regardless of whether the plant or alga was grown in the dark or in the light (Cordonnier et al., 1986a, b). The amino acids recognized by this MAb therefore seem likely to be important to its function.

Discrimination among different populations of phytochrome

For the same reason that MAbs are useful for the identification of highly conserved domains, they are also well suited to discriminating among different populations of phytochrome. By ELISA, for example, MAb Oat-22 readily detects phytochrome from three monocotyledons, but essentially fails to detect phytochrome from three dicotyledons (Cordonnier et al., 1984). MAb Pea-1 (originally designated I-3b2) is even more specific. It detects by ELISA pea phytochrome to the exclusion of that from zucchini or lettuce, as well as that from three monocotyledons (Cordonnier et al., 1984). In addition, MAbs permit ready discrimination among different pools of phytochrome from the same plant. Not only do they distinguish between the pools of phytochrome that are abundant in etiolated and green oat shoots respectively (Tokuhisa et al., 1985; Shimazaki and Pratt, 1985; Cordonnier et al., 1986a; Pratt and Cordonnier, 1987), but initial indications are that they also discriminate among different subpopulations of the phytochrome that is abundant in green oat shoots (Cordonnier and Pratt, elsewhere in this volume).

Mapping of antigenic domains on phytochrome

The most common approach to epitope mapping is via immunoblot analysis of sodium dodecyl sulfate, polyacrylamide gels following electrophoretic separation of proteolytically derived fragments. While this approach has been used with phytochrome (e.g., Grimm et al., 1986; Pratt et al., 1988), it typically does not define an epitope with precision because it relies upon chance production of peptides for analysis. Methods that utilize predetermined amino acid sequences are thus preferred. We summarize here the use of nested sets of fusion proteins for epitope mapping. Resolution of the epitope for Pea-25 to a sequence of seven amino acids is presented as an illustration of the method (Thompson et al., 1989), which differs from earlier applications of fusion proteins for epitope mapping in two critical respects. First, it eliminates chance as an element in the assignment by utilizing fusion proteins of predetermined, rather than randomly derived, sequence. Second, interpretation of the data depends solely upon the ability of a MAb to detect a putative fusion protein, rather than its inability to do so. Since failure of a MAb to detect a putative fusion protein can occur for a variety of reasons, apart from the complete absence of structure belonging to an epitope, negative data are inherently equivocal in interpretation.

With a large protein such as phytochrome, it is most efficient to approach fine resolution mapping in two steps. Beginning with a portion of oat phytochrome cDNA isolated from λgt11, two overlapping, nested sets of subclones were generated. The original cDNA, which encodes amino acids 464 through 1129 (the C terminus; see Vierstra and Quail, 1986), was first subcloned in frame into the *lacZα* sequence of pUC18 to produce construct *pCIB315* (Fig. 1). *pCIB315* was then digested with the appropriate restriction enzyme as indicated in Fig. 1, after which fragments of interest were ligated in frame back into pUC18 (subclones *pHindIII-5'* to *pXhoII-3'* in Fig. 1). Corresponding fusion proteins were produced in *E. coli* and assayed by immunoblotting (Fig. 2a). Because Pea-25 detects the products of both *pNciI-5'* and *pAvaI*, its epitope must be encoded by the DNA between the *AvaI* and *NciI* restriction endonuclease sites in *pCIB315* (Fig. 1).

Thus, in the second step *pCIB315* was made linear by digestion on either the 3' side of the *NciI* site or the 5' side of the *AvaI* site, after which the open ends were digested with the exonuclease, Bal31. Aliquots were harvested at time intervals approximating

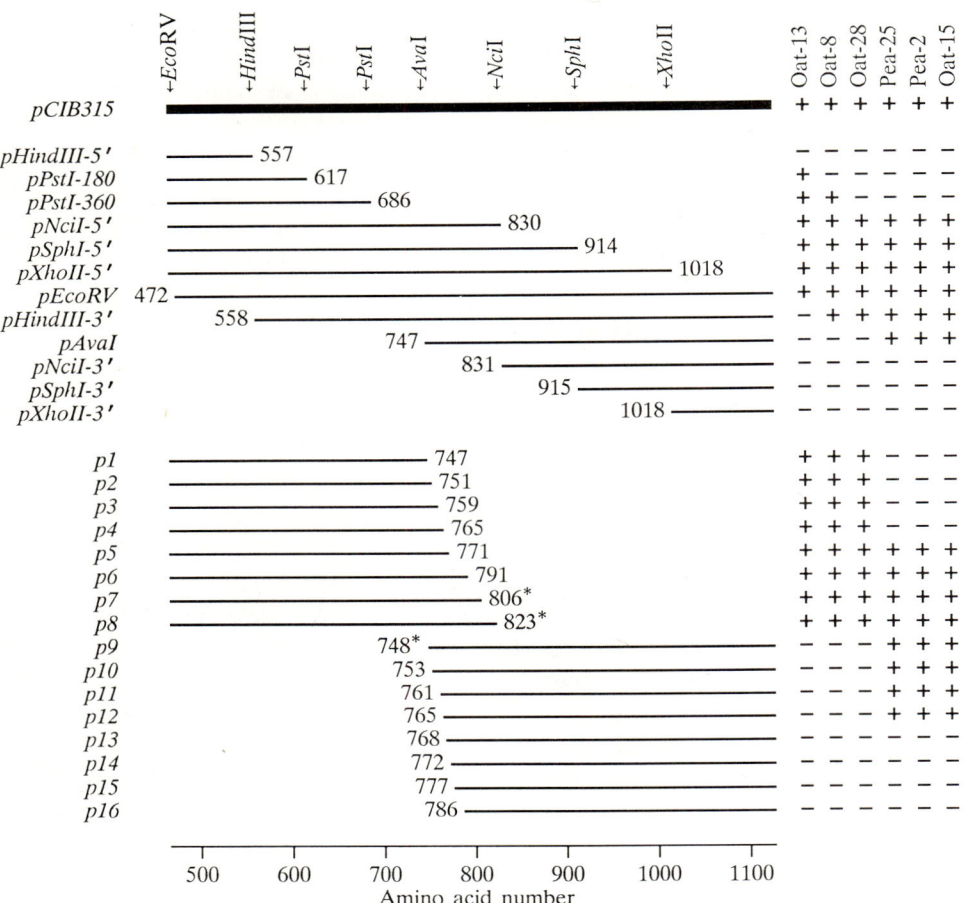

Figure 1. Diagrammatic representation of cDNA inserts in pUC18 that were used for epitope mapping. Restriction endonuclease sites are indicated at the top, corresponding amino acids at the bottom. Designations of subclones is given on the left, reactivity of expressed fusion proteins with MAbs on the right. Terminating or initiating amino acids, which were verified by sequencing except for those marked by an asterisk, are indicated for each construct.

removal of 15–20 base pairs. Thus, when the fragments of decreasing length in each direction were ligated in frame back into pUC18, two complementary, nested sets of subclones were obtained (constructs *p1-p16*, Fig. 1). Fusion proteins were again expressed in *E. coli* and recognition by MAbs assessed by immunoblotting (Fig. 2b). Because Pea-25 detects the products of subclones *p5* and *p12*, its epitope must be wholly contained on the amino acids in common to the products of these two constructs. These amino acids are residues 765–771 (-pro-ile-phe-gly-ala-asp-glu-). A competitive ELISA between oat phytochrome and a synthetic, 16-residue peptide containing this sequence (Fig. 3) confirms this epitope assignment.

464

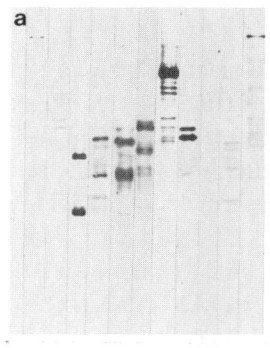

1 2 3 4 5 6 7 8 9 10 11 12

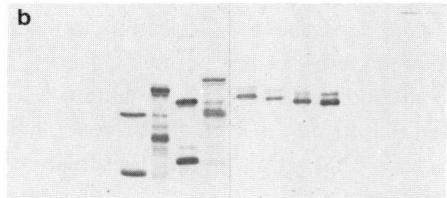

1 2 3 4 5 6 7 8 9 10 11 12 13 14 15 16

Figure 2. Immunoblot analysis of fusion proteins. *E. coli* was transformed with each pUC18 subclone of *pCIB315* as defined in Fig. 1. After expression of fusion proteins, whole extracts of *E. coli* were electrophoresed in sodium dodecyl sulfate, polyacrylamide gels. Polypeptides were electrotransferred to nitrocellulose and immunostained with MAb Pea-25. (a) Lane assignments are: 1, *pHindIII-5'*; 2, *pPstI-180*; 3, *pPstI-360*; 4, *pNciI-5'*; 5, *pSphI-5'*; 6, *pXhoII-5'*; 7, *pEcoRV*; 8, *pHindIII-3'*; 9, *pAvaI*; 10, *pNciI-3'*; 11, *pSphI-3'*; 12, *pXhoII-3'*. (b) Each lane was loaded with the subclone of the same designation (*p1* through *p16* in Fig. 1).

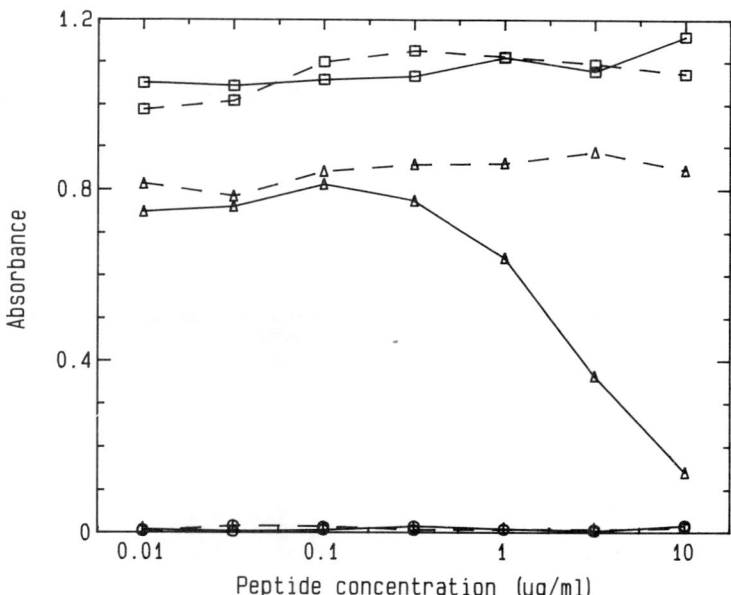

Figure 3. Competition ELISA between oat phytochrome and a synthetic, 16-residue peptide (N-pro-leu-ile-pro-pro-ile-phe-gly-ala-asp-glu-phe-gly-trp-cys-ser-C). Wells were coated with oat phytochrome at 5 μg/ml. Pea-25 (Δ), Oat-22 ($\square$), and nonimmune mouse immunoglobulins (O) were preincubated with the peptide at the indicated concentration (—) or with an equal amount of buffer (- -), after which they were added to the wells. Bound immunoglobulin was detected with alkaline phosphatase conjugated rabbit antibodies to mouse IgG.

465

Structure/function relationships

A synthesis of epitope mapping data obtained by immunoblot analysis of proteolytic digests of phytochrome (Pratt et al., 1988) with that obtained as described here by immunoblot analysis of fusion proteins (Thompson et al., 1989) identifies seven epitopes (Fig. 4). Although Oat-22 and Oat-9/Oat-16 are assigned to the same region, competition ELISA data indicate that they detect independent epitopes (Shimazaki et al., 1986; Pratt et al., 1988). Moreover, while Oat-9/Oat-16 recognize Pr and Pfr differentially, Oat-22 does not (Shimazaki et al., 1986). And, while Oat-9/Oat-16 detect some of the immunochemically distinct phytochrome isolated from green oat leaves (Shimazaki and Pratt, 1985; Cordonnier and Pratt, elsewhere in this volume), Oat-22 does not (Cordonnier et al., 1986a). These two epitopes may therefore be considered to be different. Together with the functional studies summarized above, it is possible to reach the following conclusions concerning structure/function relationships of phytochrome.

An N terminal epitope is more accessible to antibodies in the Pr conformation (Fig. 4). Although Grimm et al. (1986) have suggested that this epitope might be dependent upon structure at the C terminus, this is not the case (Pratt et al., 1988). Observations summarized above support the suggestion (Vierstra and Quail, 1986) that this domain is important for maintenance of the Pfr conformation. In addition, they lead to the further suggestion that a plant might regulate the level of Pfr during the night by making differentially available a ligand that binds to the same domain as Oat-25.

A second epitope, which is recognized by Oat-9 and Oat-16, is more accessible to antibodies in the Pfr conformation (Fig. 4). Not only is this second epitope close to the site of chromophore attachment, but because it is also found on at least a portion of the

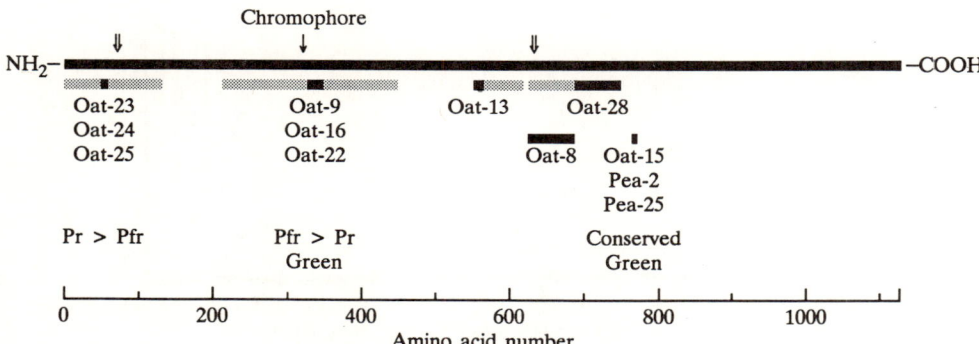

Figure 4. Summary of structure/function relationships of phytochrome as deduced by MAbs. Solid areas denote regions certain to contain amino acids that are part of an epitope. Shaded areas indicate regions that may contain amino acids that are part of an epitope, but that are tentatively excluded on the basis of negative evidence. The double arrows (⇓) mark sites of unusual susceptibility to endoproteolysis

immunochemically distinct phytochrome that predominates in green oat leaves (Shimazaki and Pratt, 1985), it is likely that it may play an important role in the molecular function of phytochrome. Moreover, it is evident that it is far removed from the N-terminal domain recognized by Oat-25 both in primary (Fig. 4) and tertiary structure. The latter conclusion derives from the observation that when bound to its antigen, an antibody occludes an area of about 3.5 nm in diameter. This domain, and that at the N terminus recognized by Oat-25, must therefore be at least this far apart in the tertiary structure of phytochrome, since competition ELISA data indicate that neither Oat-9 nor Oat-25 interfere with binding of the other to phytochrome (Shimazaki et al., 1986).

A third epitope of special interest, which has been defined with near maximal resolution, is that recognized by Pea-25 (Fig. 4). This epitope is found on some of the immunochemically distinct phytochrome from green oat shoots, and is found on phytochrome, or a polypeptide the size of phytochrome, from angiosperms, a moss, and three green algae. Since this epitope is in a region of phytochrome that is not required for its photoreversibility (Vierstra and Quail, 1986), but is presumably a region important to the function of phytochrome because of its evolutionary conservation, it may well perform an important alternative function related to the molecular mode of action of this pigment.

Acknowledgements

Research described here has been supported by a grant from the U.S. National Science Foundation (DCB 8703057).

References

Cordonnier, M.-M., Greppin, H., and Pratt, L. H., 1984, Characterization by enzyme-linked immunosorbent assay of monoclonal antibodies to *Pisum* and *Avena* phytochrome. *Plant Physiology*, 74, 123–127.

Cordonnier, M.-M., Greppin, H., and Pratt, L. H., 1985, Monoclonal antibodies with differing affinities to the red-absorbing and far-red-absorbing forms of phytochrome. *Biochemistry*, 24, 3246–3253.

Cordonnier, M.-M., Greppin, H., and Pratt, L. H., 1986a, Phytochrome from green *Avena* shoots characterized with a monoclonal antibody to phytochrome from etiolated *Pisum* shoots. *Biochemistry*, 25, 7657–7666.

Cordonnier, M.-M., Greppin, H., and Pratt, L. H., 1986b, Identification of a highly conserved domain on phytochrome from angiosperms to algae. *Plant Physiology*, 80, 982–987.

Getzoff, E. D., Geysen, H. M., Rodda, S. J., Alexander, H., Tainer, J. A., and Lerner, R.A., 1987, Mechanisms of antibody binding to a protein. *Science*, 235, 1191–1196.

Grimm, R., Lottspeich, F., Schneider, H. A. W., and Rüdiger, W., 1986, Investigation of the peptide chain of 124 kDa phytochrome: localization of proteolytic fragments and epitopes for monoclonal antibodies. *Zeitschrift für Naturforschung*, 41c, 993–1000.

Hansen, R. S., and Beavo, J. A., 1986, Differential recognition of calmodulin-enzyme complexes by a conformation–specific anticalmodulin monoclonal antibody. *The Journal of Biological Chemistry*, 261, 14636–14645.

Holdsworth, M. L., and Whitelam, G. C., 1988, A monoclonal antibody specific for the red-absorbing form of phytochrome. *Planta,* 172, 539–547.

Pratt, L. H., and Cordonnier, M.-M., 1987, Phytochrome from green *Avena.* Phytochrome and Photoregulation in Plants, edited by M. Furuya (Tokyo: Academic Press), pp. 83–94.

Pratt, L. H., Cordonnier, M.-M., and Lagarias, J. C., 1988, Mapping of antigenic domains on phytochrome from etiolated *Avena sativa* L. by immunoblot analysis of proteolytically derived peptides. *Archives of Biochemistry and Biophysics,* 267, 723–735.

Shimazaki, Y., and Pratt, L. H., 1985, Immunochemical detection with rabbit polyclonal and mouse monoclonal antibodies of different pools of phytochrome from etiolated and green *Avena* shoots. *Planta,* 164, 333–344.

Shimazaki, Y., Cordonnier, M.-M., and Pratt, L. H., 1986, Identification with monoclonal antibodies of a second antigenic domain on *Avena* phytochrome that changes upon its photoconversion. *Plant Physiology,* 82, 109–113.

Smith, S. C., McAdam, W. J., Kemp, B. E., Morgan, F. J., and Cotton, R. G. H., 1987, A monoclonal antibody to the phosphorylated form of phenylalanine hydroxylase. *Biochemical Journal,* 244, 625–631.

Thomas, B., and Penn, S. E., 1986, Monoclonal antibody ARC MAC 50.1 binds to a site on the phytochrome molecule which undergoes a photoreversible conformational change. *FEBS Letters,* 195, 174–178.

Thompson, L.K., Pratt, L.H., Cordonnier, M.-M., Kadwell, S., Darlix, J.-L., and Crossland, L., 1989, Fusion protein-based epitope mapping of phytochrome: precise identification of an evolutionary conserved domain. *The Journal of Biological Chemistry,* 264, 12426–12431.

Tokuhisa, J. G., Daniels, S. M., and Quail, P. H., 1985, Phytochrome in green tissue: spectral and immunochemical evidence for two distinct molecular species of phytochrome in light grown *Avena sativa* L. *Planta,* 164, 321–332.

Vierstra, R. D., and Quail, P. H., 1986, Phytochrome: the protein. Photomorphogenesis in Plants, edited by R. E. Kendrick and G. H. M. Kronenberg (Dordrecht: Martinus Nijhoff), pp. 35–60.

468

Phytochrome from Green *Avena* Characterized with Monoclonal Antibodies Directed to It

[1]MARIE-MICHELE CORDONNIER AND [2]LEE H. PRATT

[1]*Ciba-Geigy Biotechnology*
Research Triangle Park
North Carolina
[2]*Department of Botany*
University of Georgia, Athens
Georgia, USA

Abstract

The phytochrome that predominates in green oat (*Avena sativa* L., cv. Garry) leaves (green-oat phytochrome) differs immunochemically and spectrally from that which is most abundant in etiolated-oat shoots (etiolated-oat phytochrome). It has been difficult to characterize green-oat phytochrome because of its unusual susceptibility to modification in crude extracts and its exceedingly low abundance. Recent improvements in its purification, which not only minimize degradation but also permit its large-scale preparation, have facilitated the production of monoclonal antibodies (MAbs) directed to green-oat phytochrome. Together with the preexisting MAb Pea-25, which binds to a highly conserved region, these new MAbs recognize at least five epitopes. In addition, they detect by immunoblot assay at least two different green-oat phytochrome monomers. Reciprocal analyses of immunoprecipitates with GO–6 and GO–7 indicate that green-oat phytochrome dimers may consist in part of dissimilar monomers, neither of which is equivalent to etiolated-oat phytochrome.

Introduction

Investigations of the many physiological responses under the control of phytochrome, and of its molecular properties, have been numerous in the three decades since this chromoprotein was first isolated (see Pratt and Cordonnier, 1988, for background). Many attempts have been made to correlate phytochrome quantities with its photomorphogenic effects, both in etiolated and in green plants. Because of the presence of chlorophyll, however, spectral assay of phytochrome in light-grown, green tissues has been impossible. Moreover, the exceedingly low abundance of phytochrome in light-grown plants further complicated attempts to study it. Consequently, virtually all molecular characterization of phytochrome has been done with the pigment as isolated from etiolated seedlings (see Vierstra and Quail, 1986, for review). The many difficulties associated with characterization of the chromoprotein as it exists in green plants prompted attempts to develop highly sensitive immunochemical assays for phytochrome, with which chlorophyll would not interfere (Hunt and Pratt, 1979).

Photobiology, Edited by E. Riklis
Plenum Press, New York, 1991

Initial measurements of phytochrome in green oat leaves by radioimmunoassay, using rabbit antibodies directed to phytochrome from etiolated oats, gave values that generally agreed with expectations (Hunt and Pratt, 1979). Subsequent development of a more sensitive sandwich ELISA, however, led to the observation that phytochrome from green oat leaves was not quantitated accurately by antibodies to phytochrome from etiolated oats (Shimazaki et al., 1983). The explanation for this difficulty was provided by Tokuhisa and Quail (1983), who found that antibodies to phytochrome from etiolated oats cross-react poorly with phytochrome from green oats. Subsequent work (Tokuhisa et al., 1985; Shimazaki and Pratt, 1985) confirmed that the phytochrome that is most abundant in green oat leaves (green-oat phytochrome) is immunochemically distinct from that which predominates in etiolated oat shoots (etiolated-oat phytochrome).

These observations have reemphasized the need to investigate phytochrome from green plants, independently of the phytochrome that has been well characterized previously. The goals of this contribution with respect to green-oat phytochrome are (i) to summarize recent improvements in its purification, (ii) briefly to describe it from a physicochemical perspective, (iii) to characterize a panel of seven newly produced monoclonal antibodies (MAbs) directed to it, and (iv) to indicate what these MAbs are beginning to reveal about its properties.

Purification and properties of green-oat phytochrome

Purification of green-oat phytochrome and production of MAbs

Initial efforts to purify green-oat phytochrome and generate MAbs directed to it suffered from two difficulties, in addition to those already mentioned. (i) Green-oat phytochrome is unusually susceptible to an apparent cleavage that results in rapid loss of a *ca.* 8-kDa peptide (Cordonnier et al., 1986b). (ii) Immunodominant contaminants of about the same size as this resultant, slightly degraded phytochrome copurified with it. Thus, when phytochrome prepared by hydroxyapatite (HA) chromatography as described by Shimazaki and Pratt (1985) was purified further by DEAE chromatography (Fig. 1a, lane 3) and SDS PAGE and used to immunize mice, virtually all MAbs obtained were to these contaminants (unpublished data). Recent improvements in purification methods have not only eliminated both difficulties, but have permitted scaling up to large quantities of tissue (unpublished data). Phytochrome purified by this improved protocol is not only larger (Fig. 1b, lanes 1 *vs.* 3), but is free of similarly sized contaminants, such as that recognized by MAb G3-12H8 (Fig. 1c, lanes 2 and 3), and is significantly purer (Fig. 1a, lanes 1 *vs.* 3). This phytochrome has been further purified by SDS polyacrylamide gel electrophoresis and used to generate five new MAbs to green-oat phytochrome (GO-4, GO-5, GO-6, GO-7, and GO-8; unpublished data). These MAbs complement two that had been obtained previously (GO-1 and GO-2, referred to as G1 and G2 in Pratt and Cordonnier, 1987).

Although the initial modification of green-oat phytochrome (Fig. 1b, lanes 1 *vs.* 3) is superficially similar to the rapid N-terminal cleavage of etiolated-oat phytochrome (Vierstra and Quail, 1986), there are significant differences. (i) Phenylmethylsulfonyl fluoride, ε-aminocaproic acid, and benzamidine, inhibitors that prevented N-terminal cleavage of etiolated-oat phytochrome (Vierstra and Quail, 1986), are ineffective with green-oat phytochrome (Cordonnier et al., 1986b). Fortunately, Tokuhisa (1986) found that iodoacetamide was effective. Therefore, this inhibitor is included in the revised purification procedure. (ii) This initial cleavage of green-oat phytochrome occurs much more rapidly than does the N-terminal cleavage of etiolated-oat phytochrome (Cordonnier et al., 1986b). (iii) The difference in cleavage rates results from differences between the two types of phytochrome, not the extracts in which they are found (Cordonnier et al., 1986b). (iv) And, whereas the N-terminal degradation of etiolated-oat phytochrome occurs more rapidly for Pr than for Pfr, the initial cleavage of green-oat phytochrome occurs at the same rate for both forms (Cordonnier et al.,1986b). Direct comparison of proteolytic digestion patterns obtained with trypsin as a function of green-oat phytochrome form reinforces this earlier conclusion. No difference is seen between the two digest patterns when immunoblots are stained with MAbs detecting three different epitopes (unpublished data; *cf*. Fig. 2).

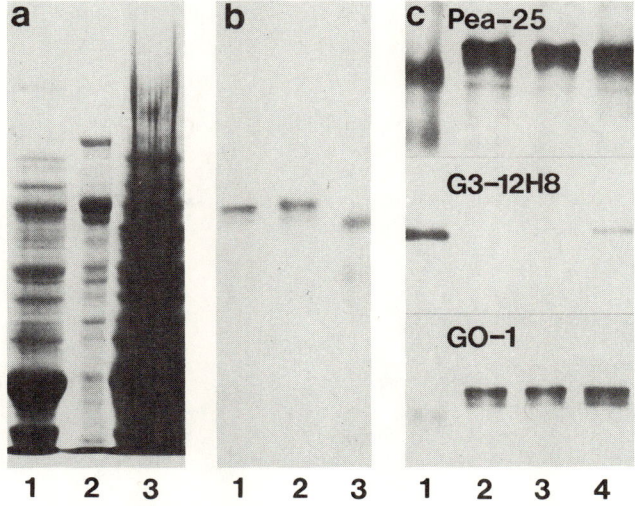

Figure 1. Comparison of phytochrome prepared by original and revised protocols. (a) Coomassie blue-stained, SDS polyacrylamide gel after electrophoresis of 0.5 µg HA-purified green-oat (lane 1, revised protocol) and etiolated-oat (lane 2) phytochrome, and 0.5 µg DEAE-purified green-oat phytochrome (lane 3, original protocol). (b) Pea-25-stained immunoblot of gel like that in (a), except each lane received 25 ng phytochrome. (c) Replica immunoblots stained with MAbs Pea-25, G3-12H8 and GO-1. Lanes received 20 ng DEAE-purified (lanes 1, original protocol) or HA-purified (lanes 2–4) green-oat phytochrome. For HA-purified phytochrome, the first ammonium sulfate precipitation was with 36% (2, 3) or 45% (4) saturation, the second 28% (2) or 32% (3, 4) saturation.

Spectral characteristics

The maximum for HA-purified green-oat phytochrome in a Pr – Pfr difference spectrum is at 656 nm, indistinguishable from that for phytochrome in light-grown, norflurazon-bleached oat leaves (Jabben and Deitzer, 1978), but blue-shifted by 11 nm as compared to etiolated-oat phytochrome (Tokuhisa et al., 1985; cf. Fig. 3). The difference minimum is also slightly blue-shifted compared to that for etiolated-oat phytochrome (727 *vs.* 731 nm), although not as much as observed for phytochrome in the norflurazon-bleached leaves (Fig. 3). The absolute absorbance maximum for green-oat phytochrome, like the difference maximum, is also blue-shifted (data not shown). The difference spectrum does not, however, differ significantly from that for etiolated-oat phytochrome in the blue/near UV region (Cordonnier et al., submitted).

Primary and quaternary structure

The monomer size of green-oat phytochrome is within about 5% of that of etiolated-oat phytochrome (Tokuhisa et al., 1985; Cordonnier et al., 1986b; *cf.* Fig. 1b, lanes 1, 2). Similarly, it is evident that both full-size and slightly degraded green-oat phytochrome behave as dimers when assayed by size exclusion chromatography (Cordonnier et al., 1986b).

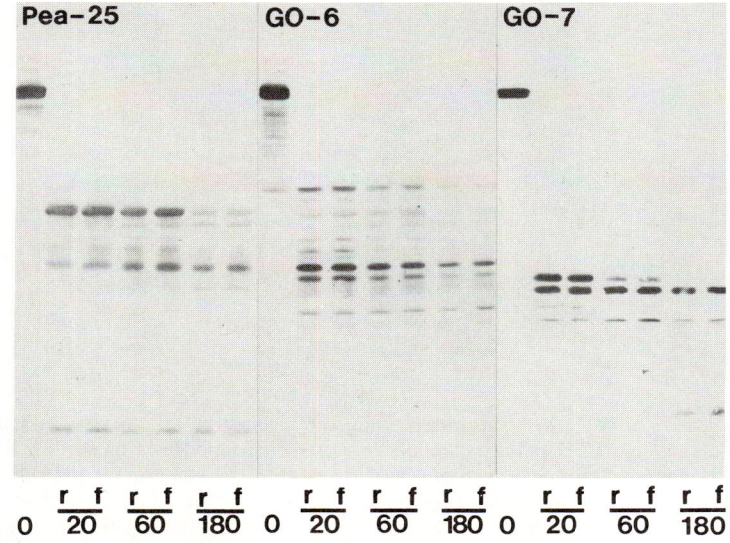

Figure 2. HA-purified, green-oat phytochrome digested as Pr (r) or Pfr (f) for 0, 20, 60 or 180 min. Aliquots of 100 ng phytochrome were electrophoresed on 7.5–15% gradient SDS polyacrylamide gels, blotted onto nitrocellulose and immunostained with Pea-25, GO-6 or GO-7 as described in Cordonnier et al. (1986b).

Green-oat phytochrome is immunochemically distinct and heterogeneous

Antibodies directed to etiolated-oat phytochrome

Polyclonal antibodies to etiolated-oat phytochrome immunoprecipitate a variable proportion of green-oat phytochrome, depending upon the antiserum used and whether the antibodies were immunopurified by a column of immobilized etiolated-oat phytochrome (Tokuhisa et al., 1985; Shimazaki and Pratt, 1985, 1986). These data are consistent with other observations indicating that etiolated- and green-oat phytochrome share only a limited number of epitopes, such as those recognized by MAbs Pea-25 and Oat-9/Oat-16 (Tokuhisa et al., 1985; Shimazaki and Pratt, 1985; Cordonnier et al., 1986a). Of perhaps greater significance are the data indicating that virtually all of the phytochrome isolated from green oat leaves differs from etiolated-oat phytochrome. In particular, of all the MAbs directed to etiolated-oat phytochrome that were tested, only two (Oat-9 and Oat-16, which detect the same epitope) immunoprecipitate a significant amount of green-oat phytochrome (Shimazaki and Pratt, 1985). Moreover, MAbs such as Oat-22 and Oat-13, which stain etiolated-oat phytochrome strongly on immunoblots, stain green-oat phytochrome only weakly, or not at all (Cordonnier et al., 1986b, and unpublished data).

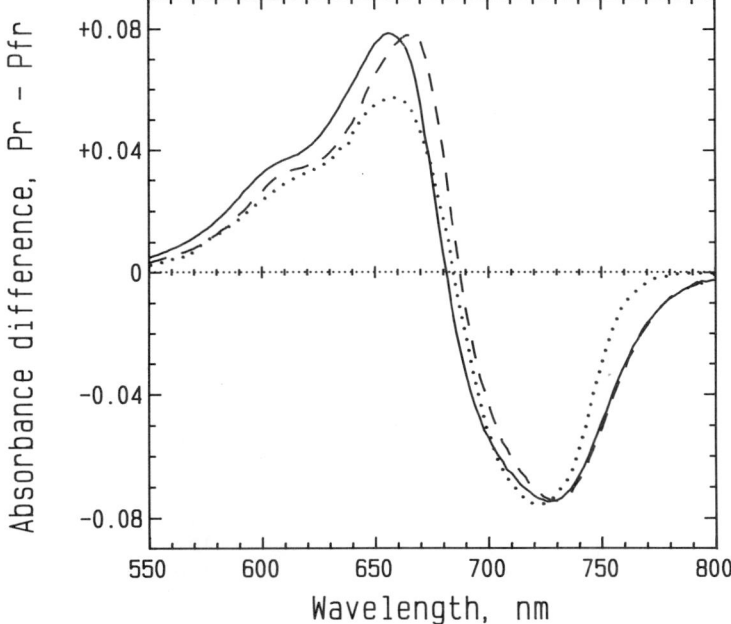

Figure 3. Pr-Pfr absorbance difference spectra for HA-purified green-oat (——) and etiolated-oat (- -) phytochrome as compared to an *in vivo* difference spectrum for norflurazon bleached, light-grown oat leaves (·····). The green-oat phytochrome spectrum and the *in vivo* spectrum have been multiplied by 1.155 and 28.4, respectively. The *in vivo* spectrum is from Jabben and Deitzer (1978).

MAbs directed to green-oat phytochrome

To determine how well the newly produced MAbs cross react with etiolated-oat phytochrome, an immunoprecipitate was prepared with Oat-3, a MAb that does not precipitate green-oat phytochrome (Shimazaki and Pratt, 1985). The precipitate was prepared at a ratio of 1 antigen-binding site per 3 phytochrome monomers, further decreasing the probability that any green-oat phytochrome would be precipitated directly. Under these circumstances, the new MAbs cross reacted weakly or not at all with etiolated-oat phytochrome by immunoblot assay (Fig. 4). Because there might be heterodimers consisting of one etiolated-oat phytochrome monomer with one green-oat phytochrome monomer, however, it is not possible to conclude with certainty that the epitopes recognized by these new MAbs exist on etiolated-oat phytochrome. These MAbs may be detecting a trace amount of green-oat phytochrome that is only fortuitously present in the immunoprecipitate.

The new MAbs to green-oat phytochrome are directed to at least four epitopes. Tryptic digests of green-oat phytochrome were electrophoresed in SDS polyacrylamide gels and immunoblotted, yielding four distinctly different peptide profiles (unpublished data). None of these four epitopes is the same as that detected by Pea-25 (*cf.* Fig. 2 for part of the data), a MAb that recognizes a highly conserved epitope found on both etiolated- and at least some green-oat phytochrome (Cordonnier et al., 1986a).

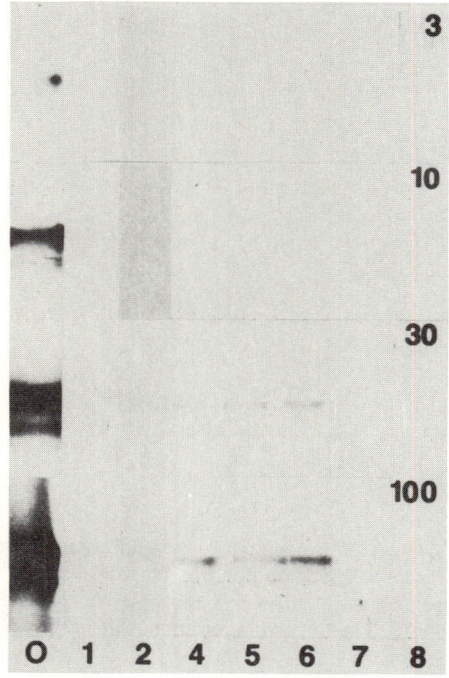

Fig. 4. Immunoblot analysis of etiolated-oat phytochrome precipitated by Oat-3, a MAb that does not immunoprecipitate green-oat phytochrome (Shimazaki and Pratt, 1985). Each lane of the SDS gel received 3, 10, 30 or 100 ng of etiolated-oat phytochrome. After electrophoresis and transfer to nitrocellulose, the blots were stained with MAbs Oat-22 (lanes O), GO-1, GO-2, GO-4, GO-5, GO-6, GO-7 and GO-8 (lanes 1, 2, 4-8). All MAbs were applied at 1 μg/ml except GO-1 and GO-2, which were used at 3 μg/ml.

Green-oat phytochrome appears heterogeneous

Immunoblot analysis of HA-purified green-oat phytochrome indicates that it consists of at least two immunochemically distinct polypeptides, neither of which is identical to etiolated-oat phytochrome (unpublished data). Data obtained with GO-6 and GO-7 serve to illustrate this point. When HA-purified green-oat phytochrome is electrophoresed in an SDS polyacrylamide gel and immunoblotted with GO-6 and GO-7, the two MAbs recognize different polypeptides. GO-6 recognizes a doublet, while GO-7 stains primarily a single polypeptide intermediate in mobility between the two stained by GO-6 (Fig. 5a). While these different polypeptides might result from a post-homogenization artifact, both GO-6 and GO-7 immunostain phytochrome in an SDS sample buffer extract of lyophilized green oat leaves that were rapidly frozen in liquid nitrogen (unpublished data). Thus, neither the epitope detected by GO-6 nor that recognized by GO-7 arises solely as a consequence of a post-homogenization artifact. Moreover, GO-6 and GO-7 do not visibly stain the polypeptide(s) stained by the other, even when a relatively heavy protein load is examined (Fig. 5a). The possibility that there are at least two immunochemically distinct green-oat phytochrome polypeptides therefore remains viable. Moreover, when either MAb is used to immunoprecipitate green-oat phytochrome, and the immunoprecipitates are reciprocally stained with the other MAb, it is evident that GO-7 immunoprecipitates not only the polypeptides stained by GO-6, but also that which it stains (Fig. 5b, lanes 7). Although not as obvious, it appears that GO-6 may also immunoprecipitate the polypeptide stained by GO-7 (Fig. 5b, lanes 6). A simple, albeit not exclusive, interpretation of these data is that heterodimers, consisting of one monomer detected by GO-6 and one by GO-7, can exist.

Green- versus etiolated-oat phytochrome

As discussed above, it has become evident in recent years that green-oat phytochrome differs in many respects from etiolated-oat phytochrome. Not only are they spectrally different (Tokuhisa et al., 1985; Fig. 3), but most MAbs to either type of phytochrome cross react poorly, if at all, with the other (Shimazaki and Pratt, 1985; unpublished data; Fig. 4). As already noted, whether some of the new MAbs to green-oat

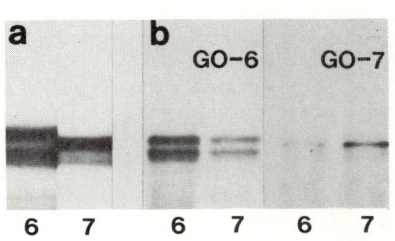

Figure 5. Immunoblot analysis of SDS polyacrylamide gels loaded with green-oat phytochrome. (a) HA-purified phytochrome electrophoresed in a single wide lane was blotted onto nitrocellulose. Immediately adjacent strips, each bearing about 180 ng of phytochrome, were cut out, stained with GO-6 and GO-7, and then repositioned precisely as they were before cutting. (b) Immuno-precipitates were prepared with GO-6 and GO-7 (lanes 6 and 7, respectively). After electrophoresis and transfer to nitrocellulose, replica blots were stained with GO-6 or GO-7.

phytochrome detect homologous epitopes on etiolated-oat phytochrome, or are merely detecting green-oat phytochrome monomers that are fortuitously coprecipitated by Oat-3, cannot be ascertained from the immunoblot data summarized here (Fig. 4). And, while both types of phytochrome have similar monomer sizes (Fig. 1b) and behave in solution as dimers (Cordonnier et al., 1986b), they are differentially sensitive to endoproteases. Not only does green-oat phytochrome undergo a more rapid N- and/or C-terminal cleavage, but partial proteolytic digestion under denaturing conditions yields different peptide patterns as assessed by immunoblotting of SDS polyacrylamide gels (Tokuhisa et al., 1985; Cordonnier et al., 1986b). Thus, while the evidence cannot be considered conclusive, it appears that these two types of phytochrome likely derive from different genes.

Of potential significance is the observation that green-oat phytochrome might consist of at least two immunochemically distinct polypeptides (e.g., Fig. 5), neither of which is equivalent to etiolated-oat phytochrome. The possibility therefore arises that green-oat phytochrome itself derives from two or more genes. Moreover, the existence of multiple species of green-oat phytchrome might explain discrepancies between the data of Tokuhisa et al. (1985) and Shimazaki and Pratt (1985) and Cordonnier et al. (1986b), as discussed by Pratt and Cordonnier (1987). Whether these different green-oat phytochrome monomers exist as heterodimers, as implied by data like those in Fig. 5, remains to be firmly established.

Acknowledgements

Research described here has been supported by a grant from the U.S. Department of Energy (DE-AC-09-81SR10925).

References

Cordonnier, M.-M., Greppin, H., and Pratt, L. H., 1986a, Identification of a highly conserved domain on phytochrome from angiosperms to algae, *Plant Physiology*, 80, 982–987.
Cordonnier, M.-M., Greppin, H., and Pratt, L. H., 1986b, Phytochrome from green *Avena* shoots characterized with a monoclonal antibody to phytochrome from etiolated *Pisum* shoots, *Biochemistry*, 25, 7657–7666.
Hunt, R. E., and Pratt, L. H., 1979, Phytochrome radioimmunoassay, *Plant Physiology*, 64,k 327–331.
Jabben, M., and Deitzer, G. F., 1978, Spectrophotometric phytochrome measurements in light-grown *Avena sativa* L., *Planta*, 143, 309–313.
Pratt, L. H., and Cordonnier, M.-M., Phytochrome from green *Avena*, *in:* Phytochrome and Photoregulation in Plants, M. Furuya, ed., Academic Press, Toyko, 1987, pp. 83
Pratt, L. H., and Cordonnier, M.-M., Photomorphogenesis, *in:* The Science of Photobiology, Second Edition, K. Smith, ed., Plenum, New York, 1988, pp. 273-304.
Shimazaki, Y., and Pratt, L. H., 1985, Immunochemical detection with rabbit polyclonal and mouse monoclonal antibodies of different pools of phytochrome from etiolated and green *Avena* shoots, Planta, 164, 333–344.
Shimazaki, Y., and Pratt, L. H., 1986, Immunoprecipitation of phytochrome from green

Avena by rabbit antisera to phytochrome from etiolated *Avena*, *Planta*, 168, 512–515.

Shimazaki, Y., Cordonnier, M.-M., and Pratt, L. H., 1983, Phytochrome quantitation in crude extracts of *Avena* by enzyme-linked immunosorbent assay with monoclonal antibodies, *Planta*, 159, 534–544.

Tokuhisa, J. G., 1986, Characterization of a distinct species of phytochrome in light-grown *Avena sativa* L. Doctoral Dissertation, University of Wisconsin, Madison, USA.

Tokuhisa, J. G., and Quail, P. H., 1983, Spectral and immunochemical characterization of phytochrome isolated from light-grown *Avena sativa*, *Plant Physiology* (Supplement), 85, 483.

Tokuhisa, J. G., Daniels, S. M., and Quail, P. H., 1985, Phytochrome in green tissue: spectral and immunochemical evidence for two distinct molecular species of phytochrome in light-grown *Avena sativa* L., *Planta*, 164, 321–332.

Vierstra, R. D., and Quail, P. H., Phytochrome: the protein, *in:* Photomorphogenesis in Plants, R. E. Kendrick and G. H. M. Kronenberg, eds., Dordrecht, Martinus Nijhoff, 1986, pp. 35–60.

Phytochrome and Cryptochrome: Coaction or Interaction in the Control of Chloroplast Orientation

W. HAUPT

Institute of Botany
University of Erlangen-Nürnberg, FRG

Introduction

Movement of cytoplasm in the cell can be controlled by light. Such a light effect can be scalar, i.e. the response has no relation to the direction of light, as found in light-controlled cytoplasmic streaming, or it can be vectorial, as in the light-oriented redistribution or rearrangement of chloroplasts. The latter response, as it is more spectacular, has been investigated more thoroughly. The present report, therefore, will be centered to the oriented chloroplast movements, but cytoplasmic streaming will be included at the beginning, because there appear to exist close causal relationships between the two kinds of responses (for references cf. Haupt, 1982).

In most of the well-investigated cases of light-controlled cytoplasmic streaming and chloroplast rearrangement the actin-myosin system is the mechanical basis of the movement (Haupt and Wagner, 1984), and the important question to be answered concerns the chain of events that is started by light and that eventually controls the actin-myosin interaction. Among these events, the first step is the most interesting one for the photobiologist, i.e. the perception of the light signal and of its direction.

Cryptochrome or phytochrome as photoreceptor pigment

In higher plants and mosses the main photosensory pigment for light-controlled cytoplasmic streaming and chloroplast orientation generally is a yellow pigment, as the responses are nearly restricted to the blue/near-UV range of the spectrum (cf. Britz, 1979). From action spectrometry, from investigations on action dichroism and from inhibitor experiments, there is good evidence that this "cryptochrome" belongs to the flavins (cf. Zurzycki, 1972). Moreover, action spectra in different systems are very similar (Lechowski, 1972; Inoue and Shibata, 1973), and if light control of both cytoplasmic streaming and chloroplast redistribution can be investigated in one system (viz., *Vallisneria spiralis*), the photosensory pigments appear to be identical (Seitz, 1967a,b; cf. also Haupt and Wagner, 1984).

Photobiology, Edited by E. Riklis
Plenum Press, New York, 1991

Yet, there are a few remarkable exceptions from this general rule: In the green algae *Mougeotia* and *Mesotaenium* and in the gametophyte of the fern *Adiantum capillus veneris*, red light via phytochrome is either the main photosensory pigment or at least it contributes substantially to the control of chloroplast orientation (for references cf. Haupt, 1987). Thus, apparently plants make use, for the control of intracellular movements, of either cryptochrome or phytochrome.

However, this "either-or" statement is too simple. Instead, there is increasing evidence that in those plants that use cryptochrome also marginal effects of red light can be found, and that in phytochrome-using plants also cryptochrome can be involved. Thus, the question arises how in these cases the two pigments or their transduction chains coact or interact.

Red-light effects in *Vallisneria*

The aquatic plant *Vallisneria spiralis* is a typical example for blue-light control of cytoplasmic streaming and chloroplast orientation, and typical flavin-like action spectra have been reported (Seitz, 1967a). However, in addition a marginal effect of red light has been found (Fig.1, solid curve). This red-light effect has been investigated in more detail for cytoplasmic streaming in the mesophyll cells (Seitz, 1967b). Here, the action spectrum is drastically changed by a treatment with potassium iodide, which is assumed —as a quencher of the triplet-excited state of flavins — to abolish the flavin effect. Accordingly, the main flavin peaks disappear, but other parts of the spectrum, especially the red region, are hardly influencd. This is shown by the dotted curve in Fig.1 as compared to the dashed curve. The red-light effect is not far-red reversible, and from additional inhibitor experiments it has been concluded that it is mediated by chlorophyll and that both blue and red light start quite different transduction chains (Seitz, 1972). Thus, in this case we are clearly dealing with a pure coaction of two independent photoreceptor systems, which even start in two different cell compartments (cytoplasm vs. chloroplasts).

Interestingly, in another species, *V. gigantea*, red light is by far most effective in inducing cytoplasmic streaming; this effect is far-red reversible and hence interpreted as mediated by phytochrome (Takagi and Nagai, 1985). Moreover, in this species also chloroplast orientation is mainly controlled by red light (Nagai, personal communication).

It appears reasonable to assume that there are no fundamental differences in two species of one genus concerning perception of a light signal, viz. chlorophyll and cryptochrome versus phytochrome. Thus, it might be predicted that both species have available a triple-pigment coaction system, viz., cryptochrome, phytochrome, and chlorophyll, and that the species specificity determines the relative contribution of each system to the final response. It is thus a challenge to find an aditional effect of phytochrome in *V. spiralis* and additional effects of cryptochrome and chlorophyll in *V. gigantea*. Mutual transfer of the experimental approaches in one species to the other is

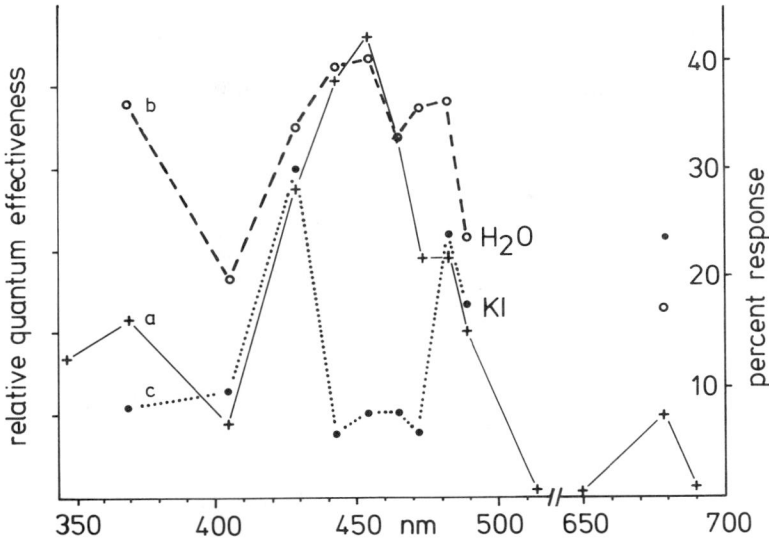

Figure 1. Action spectra for induction of cytoplasmic streaming in mesophyll cells of *Vallisneria spiralis* after 20 min irradiation. a. Solid curve: Leaf sections kept in water, relative quantum effectiveness for induction of streaming in 10% of the cells (left ordinate). b. Dashed curve: Independent experiment under the same conditions as curve a, but percent cells with streaming when induced with equal quantum flux density (quantum responsiveness, right ordinate). c. Dotted curve: As in b, but leaf sections incubated in 0.2 M potassium iodide. The two uninfluenced spectra (a and b) differ from each other in the relative height of their peaks and shoulders, as is frequently found under slightly modified conditions; but the dashed and dotted curves are strictly comparable with each other. After Seitz (1967 a,b).

recommended as well as investigation of blue-light effects on the background of strong far-red light that abolishes the phytochrome effect.

Coaction of blue and red light in *Adiantum*

We now turn to oriented responses of chloroplasts with phytochrome as an important or even as the main photosensory system and ask for coaction or interaction with a blue-light system. We have to keep in mind that in these cases always vectorial effects of light are required.

The gametophyte of *Adiantum capillus veneris* can be considered as a "balanced" type: The action spectrum (Fig.2, solid curve) has strong peaks in the blue as well as in the red range, and the latter is attributed to phytochrome because of its far-red reversibility (Yatsuhashi et al., 1985); this contribution of phytochrome is substantial, even though blue light is about double as effective as red. The blue-light effect is not at all affected by a far-red background (Fig.3), and thus it is obvious that either photosensory system can independently start a transduction chain: the two systems can fully replace each other. It

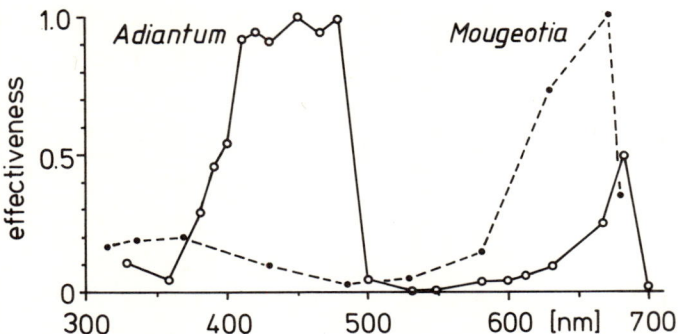

Figure 2. Action spectra of chloroplast orientation (low fluence-rate response) in *Adiantum capillus veneris* and *Mougeotia scalaris*. The ordinate denotes relative fluence effectiveness, normalized to the respective maximum. After Haupt (1987).

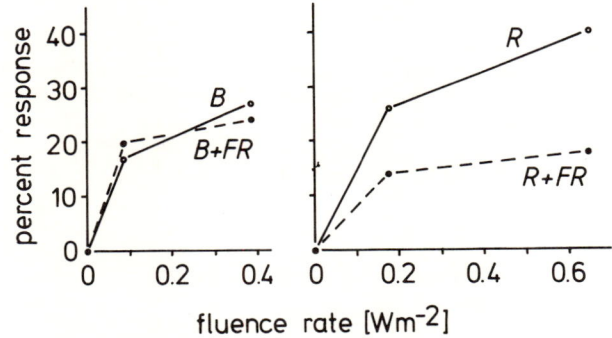

Figure 3. Chloroplast orientation in *Adiantum capillus veneris* as induced by repetitive pulses of 5 min red (R) or blue (B) light, interrupted by 5 min darkness (solid curves) or by 5 min far-red (FR) (dashed curves). The irradiation protocol was applied for 3 hours. The ordinate denotes percent protonemata with chloroplasts in diastrophe, i.e. at the wall facing the light. After Yatsuhashi et al. (1985).

might be added that for both photoreceptor pigments there is good evidence for their being localized in or close to the cell membrane (Yatsuhashi et al., 1985).

In conclusion, according to current knowledge we have, in *Adiantum*, a typical coaction of two independent systems, and it is a challenge to analyze the two transduction chains and to elucidate where they fuse to a common part.

482

Coaction or interaction of blue and red light in *Mougeotia*

A similar independence of the two photosensory systems is found in *Mougeotia scalaris*, as long as we are restricted to the so-called "low fluence-rate response", i.e. orientation of the ribbon-shaped chloroplast to the face position (cf. Haupt, 1987). In this case, red light is much more effective than blue light (Fig. 2, dashed curve). Originally, this had suggested that the blue-light effect is simply due to the minor absorption band of phytochrome in the blue region (cf. Hartmann and Cohnen Unser, 1973). Recently, it has been found, however, that this blue-light control of chloroplast orientation remains fully effective on the background of strong far-red light that completely abolishes the red-light control of the movement, even if strong red light is applied (Fig. 4; Gabrys et al., 1984). Thus, there appears to be a coaction between the two photoreceptor systems rather than an interaction, and the difference to *Adiantum* is only quantitative insofar as the effectivenesses of the two photosensory systems are shifted relative to each other. Accordingly, in *Mougeotia*, too, future research should center to elucidate differences and common parts of the transduction chains started by the two systems.

This situation changes drastically if we turn to the so-called "high fluence-rate response" in *Mougeotia*, i.e. orientation of the chloroplast's profile to the light. Originally this response was reported to be a pure blue-light effect. In fact, however, two pigment systems are involved that have to interact (cf. Schönbohm, 1966, 1980). The directional light signal, responsible for the orientation of the chloroplast, is perceived by phytochrome as in the "low fluence-rate response" referred to above. But in addition, relatively strong blue light has to be given in order to "inform" the chloroplast to reverse the sense of its orientation to the intracellular gradient of P_{fr}: If only phytochrome is

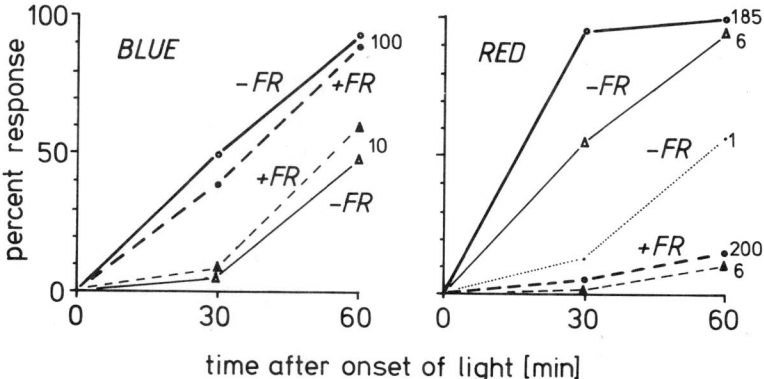

Figure 4. Chloroplast orientation in *Mougeotia scalaris* in continuous blue (449 nm) or red light (681 nm) with the fluence rates in Wm^{-2} given at the curves. Solid curves without far-red (-FR); dashed curves with far-red given simultaneously (+FR). Ordinate: percent of cells showing orientation to face position after 30 or 60 min irradiation. After Gabrys et al. (1984).

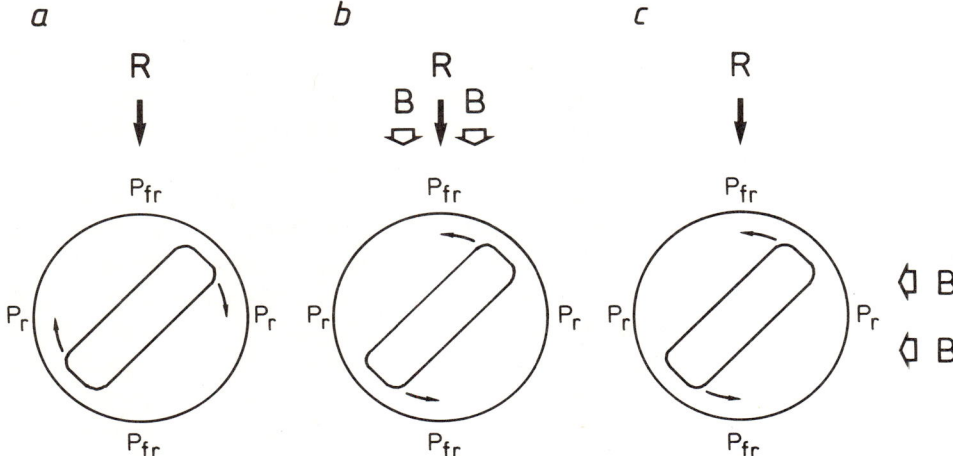

Figure 5. Schematic cross section through a *Mougeotia* cell, showing the orientation of its chloroplast in a gradient of P_{fr} that is established by red irradiation (R). a. Profile-to-face orientation to R. b,c. Face-to-profile orientation to R, if strong blue light (B) is given in addition, irrespective of its direction.

involved in light perception, the chloroplast's edges escape from the highest levels of P_{fr}, the chloroplast orients its face to the light source (Fig. 5a); but if in addition cryptochrome is sufficiently excited, the chloroplast reverses its response to the phytochrome signal, its edges now approach the highest levels of P_{fr}, and accordingly, the chloroplast orients its profile to the red light (Fig. 5b). The blue light (via cryptochrome) apparently operates a switch that determines the orientation of the response to phytochrome, and consequently both systems have to interact. Experimentally, this can be realized by either white light, containing red for phytochrome and blue for the "cryptochrome switch", or strong blue light, which is absorbed by both phytochrome and cryptochrome.

Although we have not yet full understanding about this phytochrome-cryptochrome interaction in *Mougeotia*, some details are interesting to note. Remarkably, the direction of the blue-light signal (operating the switch) has no bearing on the orientation of the chloroplast, which instead is exclusively determined by the phytochrome-absorbed light (Fig. 5b and c). Thus, phytochrome acts in a vectorial way, but cryptochrome in a scalar way (cf. Schönbohm, 1980). Moreover, the action of red and blue light can be separated in time (Gabrys et al., 1985): the reversion to high fluence-rate response can be obtained also if the two light signals are not applied simultaneously, but if the orienting red light is preceded by a blue light pulse, even with an intervening dark period of one to two minutes (Fig.6). Thus, the interacting principle of the blue-light system is not the

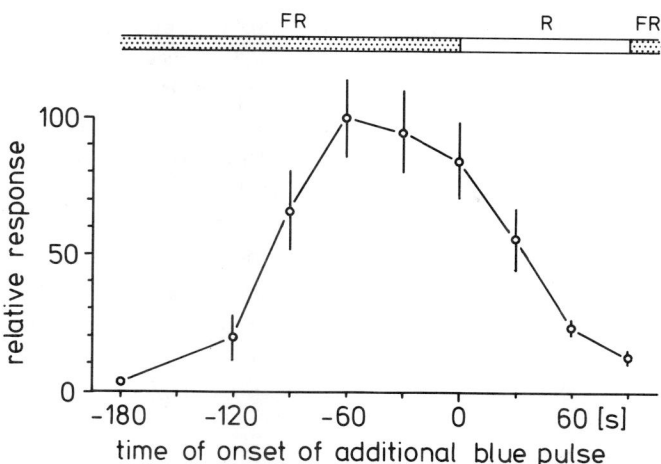

Figure 6. Chloroplast orientation in *Mougeotia scalaris* as induced by a combination of red and blue pulses. The basic irradiation protocol is indicated on top of the Figure, with a red pulse (R) given from above, preceded and followed by continuous far-red (FR). The additional blue pulse (5s), given from the side, is applied at the time indicated on the abscissa, with zero time fixed at the start of R. Ordinate: cells with chloroplasts orienting to profile position 60 min later, photometrically recorded and normalized to the maximum. After Gabrys et al. (1985).

excited photoreceptor pigment, but an early product of its action with a life time of a few minutes; this product is supposed to interact with P_{fr}.

To summarize, we have the interesting phenomenon that the coaction/interaction of cryptochrome and phytochrome appears to be fundamentally different in the two response types of one system. In the "low fluence-rate response", both systems act vectorial and independently of each other, they can substitute for each other, we may speak of a coaction. In the "high fluence-rate response", in contrast, phytochrome acts vectorial, but cryptochrome has a scalar action, and the response requires both; thus, an interaction of two transduction chains has to be postulated.

Accordingly, challenge for future research is different for the two types: for the former, the point is of interest, as in *Adiantum*, where the two alternative transduction chains fuse; for the latter, the nature of the blue-light-operated switch and its action onto the phytochrome transduction chain needs to be elucidated.

Interaction of blue and red light in *Mesotaenium*

Perhaps the most complicated system is the low fluence-rate chloroplast orientation in *Mesotaenium caldariorum* (Kraml et al, 1988). This alga is a single-cell organism in contrast to the filamentous alga *Mougeotia*, but otherwise it shares its general cell morphology. Superficially, orientation of the *Mesotaenium* chloroplast to its face position is a pure phytochrome response as in *Mougeotia*, at least as long as continuous red light is applied during the whole movement, which takes at least half an hour (Haupt and Thiele, 1961). If, however, a P_{fr} gradient in the cell is established only once by a single short pulse of red light — which is fully effective in *Mougeotia* —, almost no response can be observed. Interestingly, such a red pulse becomes highly effective in combination with blue light, which by itself cannot start the response either (Fig. 7). In this combined action of red and blue, full orientation is achieved within 10 minutes already. As in the high fluence-rate response of *Mougeotia*, this blue-light effect is independent of its direction. Thus, the directional signal is perceived by phytochrome, and cryptochrome operates a switch in a scalar way to make phytochrome effective (Kraml et al., 1988). It might be speculated that blue light via cryptochrome makes the system responsive to the phytochrome action, e.g. by making the microfilaments competent for the activating signal.

As a first approach to elucidate the interaction of the two systems, their actions have been separated in time (Kraml et al., 1988). Again, there is no interaction at the level of photoperception, as blue light still is fully effective if it follows the red pulse. Moreover, blue light can make the red-light pulse effective even if P_{fr} is reverted to P_r. This is analyzed to some detail in Fig. 8, where the 10 min blue pulse is split into two, with saturating far-red in between. It is true, a full effect of 10 min blue light, which is required to complete the response, is found only if during its first minute P_{fr} is still present —for the remaining 9 minutes blue light acts independently of P_{fr} ; but a significant response is obtained already if P_{fr} is abolished, by far-red, before the onset of blue light (zero point of the solid curve in Fig. 8). On the other hand, there is no measurable aftereffect of a blue-light pulse, as orientation movement stops as soon as blue light is switched off (cf. dashed curve). Obviously, blue light and thus excited cryptochrome mainly interacts with an early product of P_{fr} rather than with P_{fr} proper.

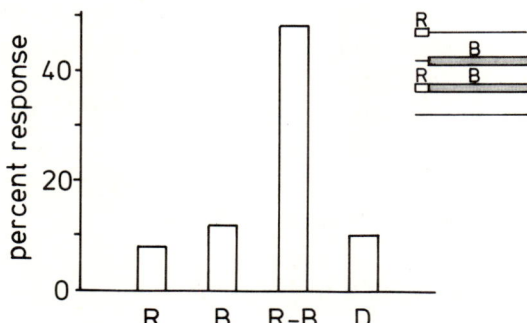

Figure 7. Chloroplast orientation in *Mesotaenium caldariorum* as induced by a sequence of 1 min red (R) and 10 min blue (B), with the corresponding controls (R only, B only, or dark control D; see inset). Ordinate: Percent of cells showing orientation to face position 10 min after termination of R or after onset of B, respectively. After Kraml et al. (1988).

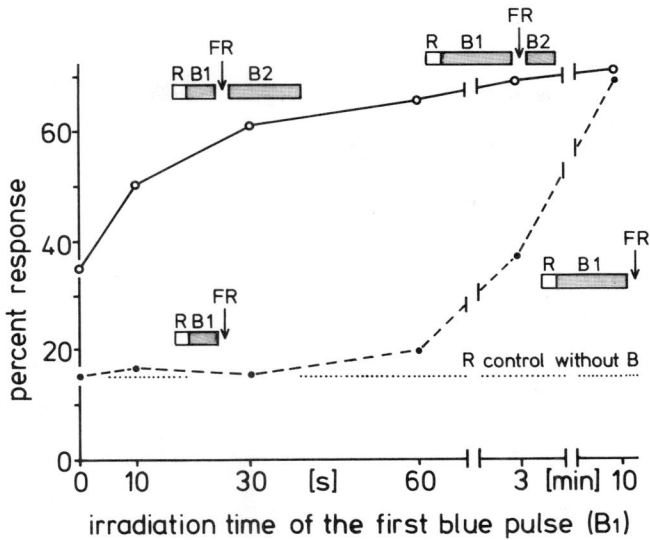

Figure 8. Chloroplast orientation in *Mesotaenium caldariorum* as induced by a sequence of 1 min red (R) and 10 min blue (B), but the blue pulse interrupted by a saturating far-red pulse (FR; 90s). Thus, the two blue pulses (B1, B2) add up to a constant total irradiation. In the lower curve, for comparison, the second blue pulse is omitted. The protocols are schematically indicated in the insets. After Kraml et al. (1988).

Although the situation appears similar to that for the high fluence-rate response in *Mougeotia*, there is an important difference: in the latter, blue light must precede the action of P_{fr}, and a product of excited cryptochrome interacts with P_{fr}; but in *Mesotaenium*, blue light must follow the P_{fr} action, and the interaction appears to concern excited cryptochrome with an early product of *Pfr* action (cf. also Fig.9, below). This interaction, then, may become a tool to elucidate the nature of primary effects of phytochrome, at least in the particular case of chloroplast orientation, and it might be worth intensifying the comparison of the interaction in *Mesotaenium* with those in other systems, particularly in *Mougeotia*.

Synopsis

A comparison of the examples for chloroplast orientation, given in this paper, is summarized in Fig. 9 (the additional complication by chlorophyll in *Vallisneria spiralis* has been omitted). The first case (a) concerns coaction of phytochrome and cryptochrome systems, with either of them acting independently of the other. This has been reported for *Adiantum* and for the low fluence-rate response in *Mougeotia*. Not included are quantitative differences insofar as either the cryptochrome or the phytochrome action prevails (*Adiantum* and *Mougeotia*, respectively). In the other two examples (b and c),

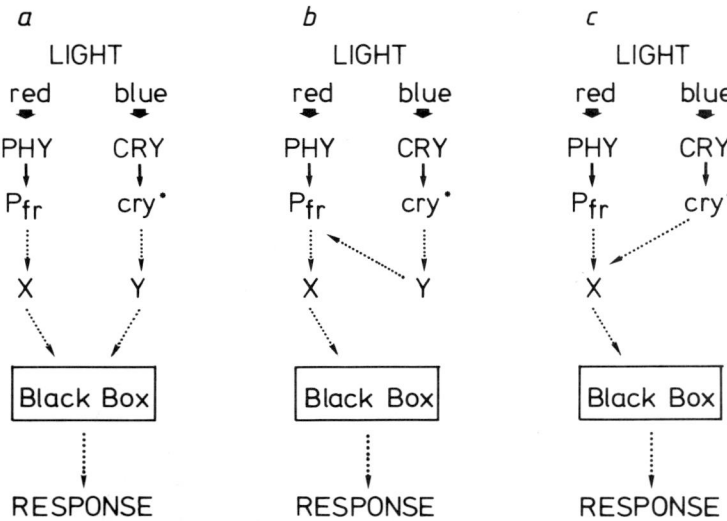

Figure 9. Summary of the possibilities of coaction and interaction of the phytochrome (PHY) and cryptochrome (CRY) system, as found in chloroplast orientation. a. Coaction in *Adiantum* and *Mougeotia* (profile-to-face orientation). b. Interaction in *Mougeotia* (face-to-profile orientation): a product (Y) of excited cryptochrome (cry*) acts on P_{fr}. c. Interaction in *Mesotaenium* (profile-to-face orientation): excited cryptochrome acts on a product (X) of P_{fr}.

instead, a true interaction has to be assumed, showing as an important difference the site of interaction: a product of cryptochrome action (Y) or excited cryptochrome itself is required to interact with P_{fr} or with a product (X) of P_{fr} action, to render the phytochrome system effective (high fluence-rate response in *Mougeotia* or low fluence-rate response in *Mesotaenium*, respectively).

In all these cases the whole transduction chain from perception of the light signal(s) to the response proceeds at the level of individual cells, and thus the complication of intercellular (or even interorgan) communication is avoided. Moreover, the responses in question are rather fast, usually requiring less than one hour. Thus, light control of intracellular movements appears to be an excellent response for further investigations of blue/red coactions or interactions.

References

Britz, S.J. (1979): Chloroplast and Nuclear Migration. In: Encycl. Plant Physiol. vol.7 (W.Haupt and M.E.Feinleib, eds.): pp 170–205. Springer, Berlin-Heidelberg-New York.

Gabrys, H., T. Walczak and W. Haupt (1984): Blue-light-induced chloroplast orientation in *Mougeotia*. Evidence for a separate sensor pigment besides phytochrome. *Planta* 160: 21–24.

Gabrys, H., T. Walczak and W. Haupt (1985): Interaction between phytochrome and the blue light photoreceptor system in *Mougeotia*. *Photochem. Photobiol.* 42: 731–734.

Hartmann, K.M., and I. Cohnen Unser (1973): Carotenoids and flavins versus phytochrome as the controlling pigment for blue-UV mediated photoresponses. *Z. Pflanzenphysiol.* 69: 109–124.

Haupt, W. (1982): Light-mediated movement of chloroplasts. *Ann. Rev. Plant Physiol.* 33: 205 –233.

Haupt, W. (1987): Phytochrome Control of Intracellular Movement. In: Phytochrome and Photoregulation in Plants (M. Furuya, ed.): pp. 225–237. Academic Press, Tokyo.

Haupt, W., and R. Thiele (1961): Chloroplastenbewegung bei *Mesotaenium*. *Planta* 56: 388–401.

Haupt, W., and G. Wagner (1984): Chloroplast Movement. In: Membranes and Sensory Transduction (G. Colombetti and F. Lenci, eds.): pp. 331–375. Plenum Press, New York-London.

Inoue, Y., and K. Shibata (1973): Light-induced chloroplast rearrangements and their action spectra as measured by absorption spectrophotometry. *Planta* 114: 341–358.

Kraml, M., G. Büttner, W. Haupt and H. Herrmann (1988): Chloroplast orientation in *Mesotaenium*: The phytochrome effect is strongly potentiated by interaction with blue light. *Protoplasma*, (Suppl. I): 172–179.

Lechowski, Z. (1972): Action spectrum of chloroplast displacements in the leaves of land plants. *Acta Protozool.* 11: 201–209.

Schönbohm, E. (1966): Der Einfluß von Rotlicht auf die negative Phototaxis des *Mougeotia*-Chloroplasten: Die Bedeutung eines Gradienten von P730 für die Orientierung. *Z. Pflanzenphysiol.* 55: 278–286.

Schönbohm, E. (1980): Phytochrome and non-phytochrome dependent blue light effects on intracellular movements in freshwater algae. In: The Blue Light Syndrome (H. Senger, ed.): pp 69–96. Springer, Berlin-Heidelberg-New York.

Seitz, K. (1967a): Wirkungsspektren für die Starklichtbewegung der Chloroplasten, die Photodinese und die lichtabhängige Viskositätsänderung bei *Vallisneria spiralis* ssp. *torta*. *Z. Pflanzenphysiol.* 56: 246–261.

Seitz, K. (1967b): Eine Analyse der für die lichtabhängigen Bewegungen der Chloroplasten verantwortlichen Photorezeptorsysteme bei *Vallisneria spiralis* ssp. torta. *Z. Pflanzenphysiol.* 57: 96–104.

Seitz, K. (1972): Primary processes controlling the light induced movement of chloroplasts. *Acta Protozool.* 11: 226–235.

Takagi, S., and R. Nagai (1985): Light-Controlled Cytoplasmic Streaming in *Vallisneria* mesophyll Cells. *Plant Cell Physiol.* 26: 941-951.

Yatsuhashi, H., A. Kadota and M. Wada (1985): Blue- and red-light action in photoorientation of chloroplasts in *Adiantum* protonemata. *Planta* 165: 43-50.

Zurzycki, J. (1972): Primary reactions in the chloroplast rearrangements. *Acta Protozool.* 11: 189-200.

Photochemistry and Photophysics of Biliprotein Chromophores: A Case of Molecular Ecology

HUGO SCHEER

Botanisches Institut der Universitaet
Menzinger Str. 67
D-8000 Muenchen, FRG

Biliproteins are widespread pigments in nature, and perform a variety of rather different functions (Kayser, 1985; Scheer, 1982; Braslavsky et al., 1983; Ruediger and Scheer, 1983): The phycobiliproteins occurring in cyanobacteria, rhodophytes and cryptophytes, function as photosynthetic light-harvesting pigments. Phytochromes are sensory photoreceptors in plants and algae, and the putative adaptochromes and photomorphochromes of cyanobacteria are probably also biliproteins. The bilirubin-serum albumin conjugates are involved in transporting this poorly water-soluble pigment, and many invertebrates contain biliproteins which play a role in their pigmentation, but may have additional functions.

According to these different functions, the photochemical and photophysical properties of these complexes are rather diverse; a fact that is surprising in view of the structural similarities of many of the chromophores involved. The relative contributions of different deexcitation pathways are regulated by specific interactions between the chromophores and the proteins. In analogy to interactions among living systems, I like to term these interactions as "molecular ecology".

In the context of studies on phycobiliproteins from cyanobacteria, we became interested in the factors regulating the contributions of the different deexcitation pathways, as well as in the influence of chromophore structure and photochemistry on protein structure and aggregation. Additional reasons for such studies are the potential buildup of background during time-resolved and other laser-spectroscopic techniques working at high intensities or repetition rates, and the possible involvement of phycobiliproteins in light perception (Bjoern and Bjoern, 1980; Scheer, 1982; Kufer, 1988).

Today, I would like to report some recent work we have done with two phycobiliproteins, phycocyanin and phycoerythrocyanin from the thermophilic cyanobacterium, *Mastigocladus laminosus,* for which primary (Zuber, 1986) and crystal structures (Schirmer et al., 1987; Duerring and Huber, private communication, 1988) are known.

Photobiology, Edited by E. Riklis
Plenum Press, New York, 1991

Energy transfer

Experimental work on this subject has been carried out in cooperation with Siegfried Schneider (Technical University, Muenchen, FRG; Schneider et al., 1988). Both pigments in different aggregate sizes, and their α- and β-subunits have been studied by two complementary picosecond techniques: The decay of fluorescence and of fluorescence polarization was measured with a repetitive streak-camera, and the ground-state de- and repopulation kinetics by absorption-recovery.

The smallest PC-unit, e.g. the monomer of the α-subunit, shows >95% of its fluorescence as a single-exponential decay with a rate constant of appx. 900 psec, an expected pattern for a single, isolated chromophore. More complex decay patterns had been observed before in subunit preparations which we now know were dimers. However, aggregation does not change the decay pattern and the depolarisation, and energy transfer among the two chromophores in the α_2-dimer appears to be negligible. The β-subunit shows a biexponential decay, with rate constants in the 30 and 1100 psec range. The first is due to energy transfer from the high-energy β-155 to the low-energy β-84, because population of the excited state of β-84 is delayed relative to the excitation of β-155 by this time. All higher aggregates require three or even more exponential components to fit the decay patterns. The most dramatic change occurs upon aggregation of the $(\alpha\beta)_1$ monomers (=heterodimers) to the $(\alpha\beta)_3$ trimers (= heterohexamers), where depolarization kinetics are increased by an order of magnitude.

In cooperation with Ken Sauer (University of California, Berkeley), we have simulated the dynamics by using several models for energy transfer, and the chromophore geometries as determined by x-ray crystallography (Sauer and Scheer, 1988). Although a model based on Foerster-type energy transfer gives results which model satisfactorily existing experimental data, there is evidence that excitonic coupling cannot be ignored in trimers and higher aggregates. α-84 on one monomeric unit, and β-84 on the adjacent one, come so close that Foerster transfer times <1 psec are calculated, and excitonic coupling energies in the range of 75cm^{-1}. This intermediate region between Foerster transfer and excitonic interaction is hitherto only little explored, and the phycobiliproteins may be useful in this respect.

Radiationless deactivation

The single-exponential lifetime of the α-subunit is in the range of 900 psec, whereas the longest fluorescence component in larger aggregates (assigned to cumulative decay of all chromophores over which excitation energy is thermally equilibrated) is 1.4–2 nsec. Since oscillator strengths of the chromophores do not change markedly upon aggregation of the phycobiliproteins, the increase in fluorescence lifetimes then corresponds to a decrease in energy losses with increasing aggregate size. This corroborates earlier steady-state results showing an increase in fluorescence yields from around 40% in subunits to close to 90% in aggregates.

As photochemistry to stable products is no major deexcitation pathway in PC (see below), and intersystem crossing is not important either, these losses are due to efficient internal conversion. Three mechanisms to this have been discussed in free bile pigments (Scheer, 1982, Braslavsky et al., 1983). The first is a high density of vibrational states, which is related to the high conformational flexibility of bile pigments. The second deexcitation mechanism is *via* photochemical channels leading to the ground-state of unstable products, which revert to the ground state of the original species. Specifically, internal H-transfer and Z/E-isomerizations at the central methine bridge (C-10/11) have been discussed. Isomerization involves rather large structural changes which are unlikely in the native or near-to-native protein environment. H-transfer is, on the other hand, even a candidate for photochemistry at temperatures close to absolute zero (Koehler et al., 1988). Little is known on the flexibility of the chromophores in their native environment, but reagents known to increase protein mobility decrease biliprotein fluorescence. Obviously, the details of radiationless deexcitation need further study.

Photochemistry

Whereas the photochemical events discussed in the context of radiationless decay are transitory, long-lived photoproducts have been observed in several phycobiliproteins. The most interesting and potentially most important type of photochemistry is observed upon partial denaturation. This photochemistry has now been found to occur under a variety of conditions. Its magnitude (defined by the ratio of the amplitude of the difference spectrum, to the maximum absorption) can be as high as 60% in the presence of 20% mercaptoethanol. It is most likely due to a Z/E interconversion of the chromophore(s) at the C-15,16 double-bond. Under such conditions, the chromophores are apparently capable to perform the same type of photochemistry as that of phytochrome.

More recently, we have studied in cooperation with W. Kufer from our laboratory a different pigment, e.g. PEC. Its α-subunit had been linked previously to photochromic activities in cyanobacterial extracts, and possibly to photomorphogenesis (Bjoern and Bjoern, 1980; Kufer, 1988). This pigment, which is structurally very similar to PC (Bryant, 1982; Duerring and Huber, private comm.), carrries a rare phycoviolobilin chromophore at cys α-84 (Bishop et al., 1987), which replaces the common phycocyanobilin chromophore present at the same location in PC. Being a component of the phycobilisome, it is commonly regarded a light-harvesting pigment. A distinct difference from other phycobiliproteins, is however its pronounced photochemistry in the native state, which is most likely again a Z/E-isomerization at the C-15 methine bridge. Th reaction would require a decreased rigidity in the environment of α-84, which has been born out in the crystal structure of PEC (Duerring and Huber, private comm.).

Effect of chromophore α-84 on aggregation

Recently, we noticed that PEC shows not only increased photochemistry upon disaggregation, but there is also a reciprocal dependence of biliprotein aggregation on photochemistry. When PEC is alternately irradiated with orange (600nm) and green light

(500nm), the two photoequilibria were enriched in the 15-E- and 15-Z-configured forms, of α-84 chromophore respectively. Ultracentrifugation showed, that at the same time there occurs a photoreversible change in aggregation: The amount of trimer increased each time the last irradiation was with green light, and decreased each time it was with orange light. This means, that the configuration of α-84 controls aggregation. This can be rationalized again from the X-ray structure: α-84 is located very close to the contact surface of monomers in trimers.

To test the sensitivity of aggregation to the structure of α-84 we have done another experiment with PC. The chromophores in isolated subunits were reduced to rubins. Modified α-subunits were then hybridized with original β-subunits and *vice versa* to yield hybrid PC. In this experiment, the hybrids containing modified α-subunit only formed monomers, whereas those hybrids containing modified β-subunits reaggregated to trimers.

This result points to an involvement of the biliprotein chromophores not only in energy transfer and photochemistry of biliproteins, but also in their structure. This effect may well be at the origin of a signal chain leading eventually to photomorphogenesis. In a more general context, it is an example for the intricate interplay of proteins with their cofactors, which leads to the stunning variety of properties of pigments with the same or very similar molecular structures.

References

Bishop, J.E., Rapoport, H., Klotz, A.V., Chan, C.F., Glazer, A.N., Fueglistaller, P., and Zuber, H., 1987, Chromopeptides from phycoerythrocyanin. Structure and linkage of the three bilin groups, .Journal of the American Chemical Society 109:875-881.

Bjoern, L.O., and Bjoern, G.S., 1980, Photochromic Pigments and photoregulation in blue-green algae, . Photochemistry and Photobiology 32:849-852.

Braslavsky, S., Holzwarth, A.R., and Schaffner, K.,1983, Solution conformations, photophysics and photochemistry of bilipigments. Bilirubin- and biliverdindimethyl-esters and related linear tetrapyrroles, Angewandte Chemie 94:670-689. International Edition (English) 22:656-674.

Bryant., D., 1982, Phycoerythrocyanin and phycoerythrin: Properties and occurence in cyanobacteria, Journal of General Microbiology 128:835-844.

Kayser,H., 1985, in Comprehensive Insect Physiology, Biochemistry and Pharmacology edited by G.A. Kerkut and L.I. Gilbert, (New York: Pergamon), pp. 367-415.

Koehler, W., Friedrich, J., Fischer, R., and Scheer, H., 1988, Low temperature spectroscopy of cyanobacterial antenna pigments, in Photosynthetic Light-Harvesting Systems: Organisation and Function edited by H.Scheer and S.Schneider, (Berlin: W. deGruyter), pp. 293-306.

Kufer, W., 1988, Concerning the relationship of light-harvesting biliproteins to phycochromes in cyanobacteria, in .us on;Photosynthetic Light-Harvesting Systems: Organisation and Function edited by H.Scheer and S.Schneider, (Berlin: W. deGruyter), pp. 89-92.

Ruediger, W., and Scheer, H., 1983, Chromophores in Photomorphogenesis, in Encyclopedia of plant physiology, Vol. 16, Photomorphogenesis edited by W. Shropshire and H. Mohr (Berlin: Springer), pp. 119-151.

Sauer, K., and Scheer, H., 1988, Excitation transfer in C-phycocyanin: Foerster transfer rate and exciton calculations based on new crystal structure data for C-phycocyanins from Agmenellum quadruplicatum and Mastigocladus laminosus, Biochimica et Biophysica Acta 936:157–170.

Scheer, H., 1982, Phycobiliproteins: Molecular aspects of photosynthetic antenna system, in Light reaction path of photosynthesis edited by F.K. Fong (Berlin: Springer), pp. 7-45.

Schirmer, T., Bode, W., Huber, R., 1987, Refined 3-dimensional structures of 2 cyanobacterial C-phycocyanins at 2.1 and 2.5Å resolution - A common principle of phycobilin-protein interaction, Journal of Molecular Biology 196:677-695.

Schneider, S., Geiselhart, P., Baumann, F., Siebzehnreubl, S., Fischer, R., and Scheer, H., 1988, Energy transfer in "native" and chemically modified C-phycocyanin trimers and the constituent subunits, in Photosynthetic Light-Harvesting Systems: Structure and Function edited by H.Scheer and S.Schneider, (Berlin: W. deGruyter), pp 469-482.

Zuber, H., 1986, Primary structure and function of the light-harvesting polypeptides from cyanobacteria, red algae, and purple photosynthetic bacteria, in Encyclopedia of Plant Physiology, Vol. 19, Photosynthesis III edited by L.A. Staehelin and C.J. Arntzen, (Berlin: Springer), pp. 238-251.

Strategy of Orientation in Flagellates

D. P. HÄDER

Institute für Botanik und Pharmazeutische Biologie
Staudtstr. 5, D-8520 Erlangen
Fed. Rep. Germany

Introduction

Motile microorganisms use external physical and chemical parameters to orient in their environment and to optimize their position in space (Häder, 1988; Nultsch and Häder, 1988). In addition to gravity (Bean, 1984; Kessler, 1985, 1986), temperature (Mizuno et al., 1984; Poff, 1985), chemical gradients (Berg, 1985; MacNab, 1985) and the earth's magnetic field (Ofer et al., 1984; Frankel, 1984; Stolz et al., 1986), light seems to play a major role to guide not only photosynthetic but also non-photosynthetic microorganisms in their environment (Foster and Smyth, 1980; Colombetti et al., 1982b; Haupt, 1983).

Most flagellates studied so far seem to use two or more antagonistic responses to optimize and adjust their position in their (aquatic) habitat with respect to the constantly changing conditions. This overview will stress light and gravity as the major external factors for orientation in flagellates, even through the other parameters may have an at least modulating effect. The following section will provide the definitions used to describe the responses to light and gravity in flagellates:

Photokinesis defines the dependence of the linear swimming velocity on the fluence rate of the incident radiation. It is compared to the speed measured in darkness; a positive photokinesis describes a movement faster than in the dark control and a negative photokinesis a slower one. Temperature, chemicals and other external factors can also influence the linear velocity, which is consequently defined as thermokinesis, chemokinesis, etc. (Diehn et al., 1977).

A *photophobic response* is elicited by a sudden change in the fluence rate; both a temporal change — e.g. when a cloud reduces the incident fluence rate — and a spatial change — e.g. when the cell swims into the shade of a floating object — can induce a phobic response. Likewise, sudden changes in other parameters may cause phobic reactions, which are then defined as, e.g., chemophobic, thermophobic etc. responses.

Phototaxis defines an oriented movement with respect to the light direction. A movement toward the light source is called a positive phototaxis and away from the light source a negative one. Recently, several organisms have been found to show a transversal phototaxis (or diaphototaxis) which describes a movement perpendicular to the incident

Photobiology, Edited by E. Riklis
Plenum Press, New York, 1991

light beam (Rhiel et al., 1988b). In analogy we can define positive and negative gravitaxis as a movement with respect to the gravity vector of the earth.

Orientation in *Euglena*

The green flagellate *Euglena gracilis* has been studied as a model organism for almost a century (Jennings, 1904; Mast, 1911; Buder, 1917). In addition to both step-up and step-down photophobic responses (Shimmen, 1980; Doughty and Diehn, 1984) and a photokinetic effect (Wolken and Shin, 1958), the flagellate shows a pronounced phototaxis (Bancroft, 1913; Diehn, 1973a,b; Colombetti et al., 1982a).

At low fluence rates up to 1.4 Wm^{-2} the organisms move toward the light source while at higher fluence rates they were found to show a negative phototaxis (Fig. 1a,b). The negative phototaxis is always much more pronounced than the positive phototaxis when quantified, e.g., by the Rayleigh test (Mardia, 1972; Batschelet, 1981). However, the sense of the direction depends on the age and culture conditions: Young cultures show a positive phototaxis even at intermediate fluence rates which reverses into a negative one when the culture grows older (Häder et al., 1987a).

The photoreceptor is assumed to be a flavoprotein localized in the paraflagellar body (PFB), a swelling at the basis of the longer flagellum which emerges from the reservoir (Benedetti and Checcucci, 1975; Doughty and Diehn, 1980; Ghetti et al., 1985). This hypothesis is supported by microspectrophotometric and fluorometric measurements and by action spectra of photoaccumulations. However, no action spectrum of phototaxis has been determined yet.

Mechanism of Photoorientation

Until recently the mechanism of orientation with respect to the light direction was explained by the so-called shading hypothesis. During forward locomotion the cell rotates around its long axis with a frequency of 1–2 Hz (Diehn, 1969), so that in lateral light the stigma periodically intercepts the light impinging on the PFB. This periodic signal causes the flagellum to swing out and move the front end of the cell toward the light source by a certain angle. This repetitive phobic response eventually causes the cell to be aligned with the impinging light beam in which case no error signal is detected by the PFB. The same reasoning holds for negative phototaxis with the only difference that the photoreceptor is additionally shaded by the chloroplasts and other cell organelles in the rear end of the cell.

This shading hypothesis has however been proven not be valid in this organism, based on behavioral and biochemical grounds: When exposed to two weak light beams perpendicular to each other the cells do not swim on the resultant as expected from the shading hypothesis but rather the population splits into two components moving towards either light source (Häder et al., 1986). A small decrease in the fluence rate of one of the light sources causes most cells to swim toward the other light source. The second proof

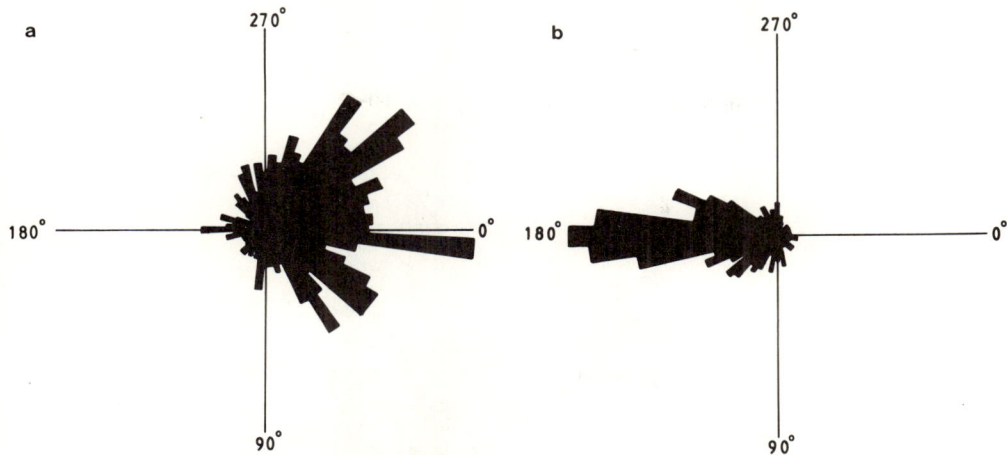

Figure 1. Circular histograms of (a) positive phototaxis at 100 lx and (b) negative phototaxis at 30 klx in *Euglena gracilis* (Häder et al., 1987a).

against the shading hypothesis is the fact that cells lacking the stigma still orient with respect to the light source, even though no shading device is present. Finally, inhibitor and ionophore studies have indicated that the biochemical sensory transduction chains of photophobic responses and phototaxis are totally different in *Euglena*. Doughty and Diehn (1979, 1982, 1983; Doughty et al., 1980) have shown that a light-modulated sodium/potassium pump and ion-dependent Na^+ and Ca^{2+} channels are involved in the transduction chain of phobic responses. Even at much higher concentrations these inhibitors and ionophores had no effect on the phototactic orientation of the cells unless they impaired motility altogether (Häder et al., 1987a).

As an alternative to the shading hypothesis, the mechanism of photoorientation has been found to be based on the dichroic orientation of the photoreceptor molecules (Creutz and Diehn, 1976; Häder, 1987a). When irradiated from above with polarized light, the cells swimming in a horizontal cuvette orient in two directions about 30° clockwise from the plane of the electrical vector (Fig. 2). Similarly, the orientation of the photoreceptor molecules has been analyzed with respect to the flagellar plane of the cell. Thus, the light direction is determined by the modulation in the absorption of the incident light beam due to the dichroic orientation of the photoreceptor molecules. The shading properties of the stigma and the rear end may modify the stimulus absorption.

Gravitaxis in *Euglena*

In addition to positive and negative phototaxis, *Euglena gracilis* shows a prominent negative gravitactic orientation in a vertical cuvette (Fig 3 Häder, 1987a). Since this response is much more pronounced than the positive phototaxis, the orientation in the three-dimensional habitat is defined by the antagonism between negative gravitaxis (plus

499

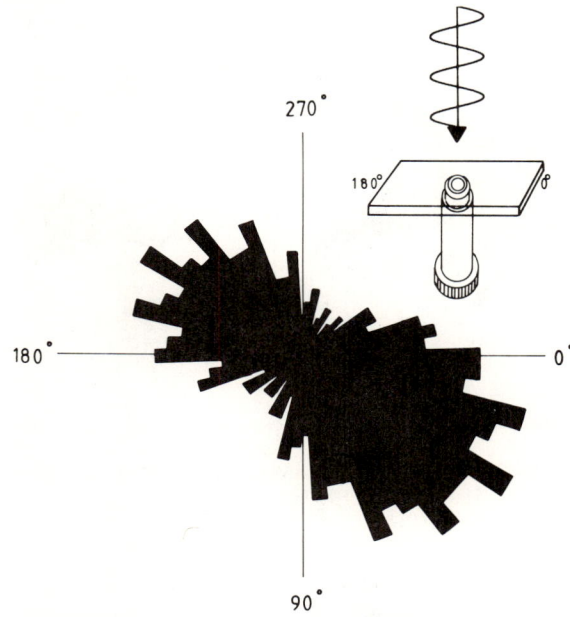

Figure 2. Orientation of *Euglena gracilis* with respect to a light beam from above, polarized between 0° and 180° (Häder, 1987a).

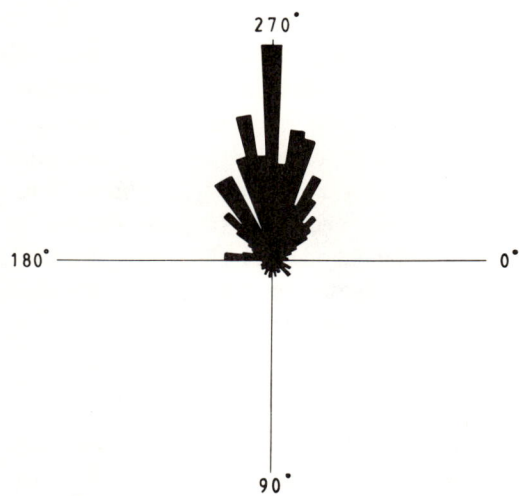

Figure 3. Circular histogram of negative gravitaxis of *Euglena gracilis* in a vertical cuvette (Häder, 1987a).

500

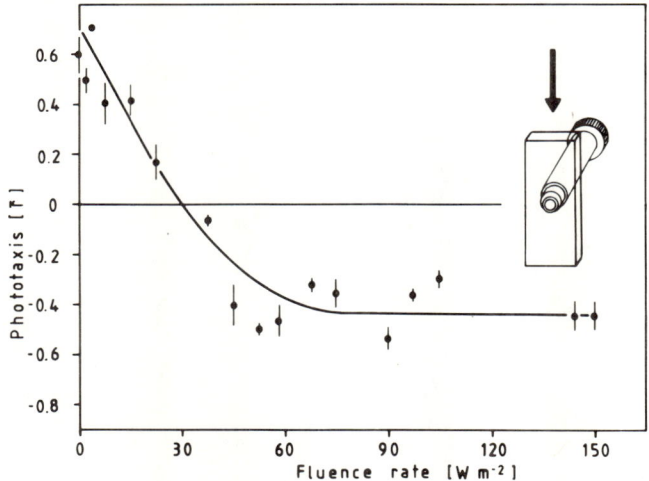

Figure 4. Direction and degree of orientation of *Euglena gracilis* in a vertical cuvette exposed to white light from above. Symbols above the horizontal zero line define upward movement, those below downward movement. Vertical bars indicate standard error of the means calculated from the Rayleigh tests of six histograms per point with 1000 organisms each. Abscissa: fluence rate of the white actinic light in [W m^{-2}], ordinate: degree of orientation as calculated from the Rayleigh test (Häder, 1987a).

a weak positive phototaxis), which causes the cells to swim upwards, and a likewise precise negative phototaxis, which guides the cells downwards, away from the too bright sunlight at the surface. The upward orientation is compensated by a downward component at higher fluence rates and the crossover point is found at a fluence rate of 30 Wm^{-2} (Fig. 4; Häder, 1987a).

This behavior was verified in an ecophysiological approach (Häder and Griebenow, 1988): A vertical plexiglass cylinder (1 m long, 90 mm inner diameter) was filled with a *Euglena* suspension and submerged in a pond. After thermal equilibration samples were taken at regular time intervals from evenly spaced (50 mm apart) openings along the length of the column using a peristaltic pump which could handle 18 samples in parallel. The cell density in the individual samples were analyzed automatically by an image analysis system which determined the cell count (Häder and Griebenow, 1987). After sunrise the cells started to move downward and formed a dense layer on the bottom of the cuvette (Fig. 5.). At night the population started to rise again; however the upward movement was far slower than the downward movement (Häder and Griebenow, 1988).

Orientation in *Cryptomonas*

Phototaxis in marine and freshwater Cryptomonas species

In an unidentified freshwater species Watanabe and Furuya (1974) have found an

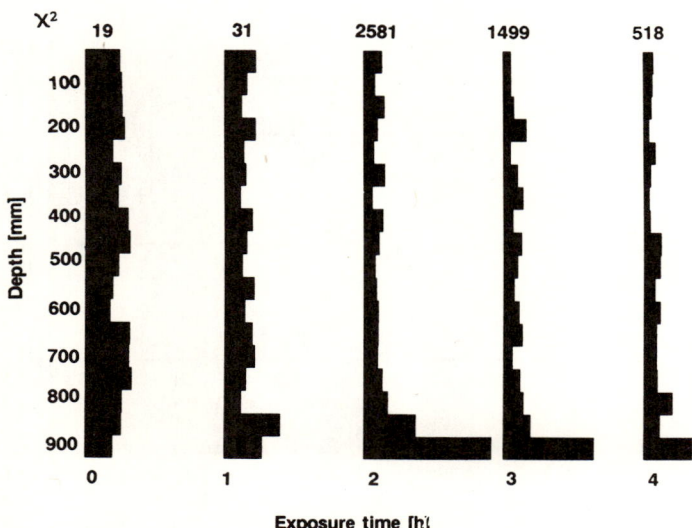

Figure 5. Cell density measured in samples taken from the 18 outlets in the submersed column at 1 h intervals. Numbers above the histograms indicate the chi square values (Häder and Griebenow, 1988).

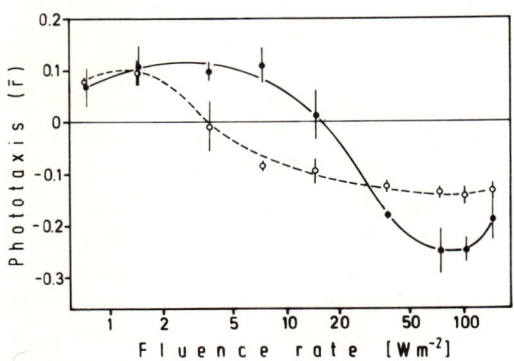

Figure 6. Fluence rate response curves of positive and negative phototaxis of normally pigmented (continuous line, closed symbols) and bleached (broken line, open symbols) *Cryptomonas maculata*. Abscissa: white actinic light fluence rate in [Wm⁻²], ordinate; phototaxis as quantified by the Rayleigh test (Häder et al., 1987b).

502

exclusive positive phototaxis. The action spectrum indicated the involvement of phycoerythrin as the major photoreceptor pigment, though the photosynthetic electron transport chain did not seem to be involved in the sensory transduction chain, since inhibitors of photosynthetic oxygen production were uneffective (Watanabe et al., 1976). The mechanism of light direction detection seems to involve a periodic modulation of the receptor pigments since repetitive light pulses were as effective as continuous irradiation; provided the dark intervals did not exceed 50% of the rotation time (450 mn) during the helical swimming (Watanabe and Furuya, 1978).

In contrast, the marine *Cryptomonas maculata* uses both positive and negative phototaxis for orientation in its habitat with a crossover point at 15 Wm^{-2} (Fig. 6; Häder et al., 1987b). Nitrogen-deficient cells which have lost most of their phycoerythrin and some of the other photosynthetic pigments showed a similar behavior, but the crossover point between positive and negative phototaxis was shifted to about 3.5 Wm^{-2}.

Another unidentified freshwater *Cryptomonas* species (S2) showed a remarkable diaphototaxis (= transversal phototaxis) both at high and low fluence rates (Fig. 7; Rhiel at al., 1988b). This behavior seems to allow the organisms to stay at a given depth in the water column. A slight change in the movement vectors upward or downward causes the population to rise or sink in accordance with the prevalent light conditions.

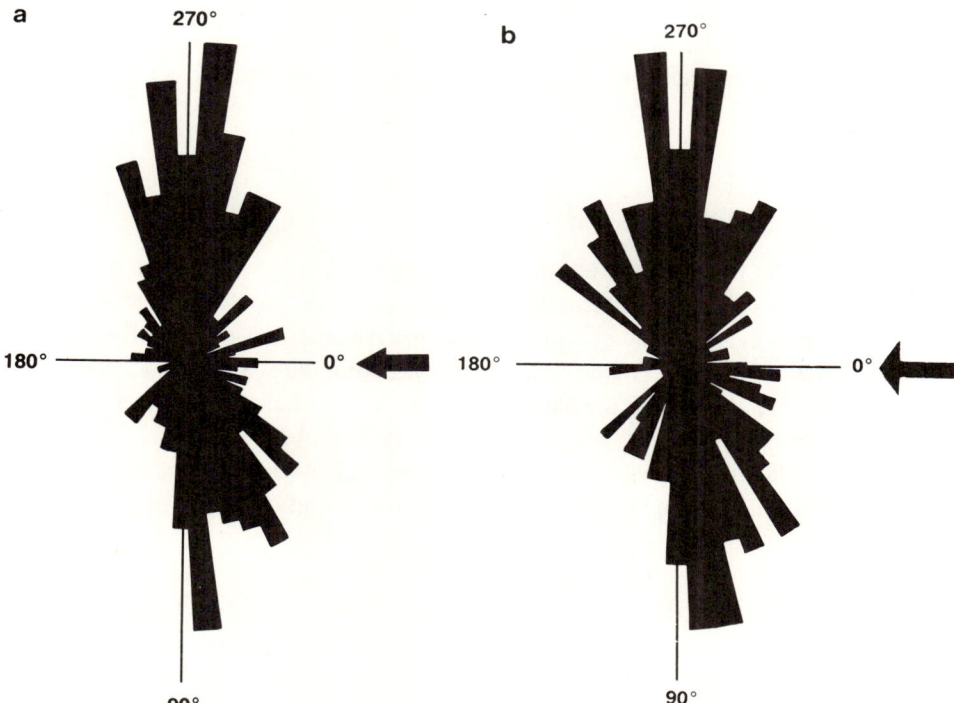

Figure 7. Circular histograms of Cryptomonas spec. (S2) when irradiated with (a) $5Wm^{-2}$ and (b) $100Wm^{-2}$ from 0° (Rhiel at al., 1988a).

Gravitaxis in Cryptomonas

No gravitaxis was detected in *Cryptomonas maculata,* but was very prominent in the freshwater species S2 (Rhiel et al., 1988a). In a vertical cuvette, the cells were found to move upward even when irradiated from above using high fluence rates. The behavior is remarkable since it seems to override negative phototaxis or diaphototaxis and cause the cells to move exclusively to the surface of the body of water. Since most *Cryptomonas* species are extremely sensitive to white solar radiation at high fluence rates, this behavior would lead to an extermination of the population on clear days. On the other hand, *Cryptomonas* species have been found to occupy lower horizons in the water column where they are well protected from the bright sunlight at the surface (Burns and Rosa, 1980). The mechanism of downward orientation still needs clarification.

Vertical Orientation in *Gyrodinium*

Phototaxis was analyzed in two species of *Gyrodinium, G. aureolum* and *G. dorsum.* Even at extremely high fluence rates only positive phototaxis was found; no negative or diaphototaxis seems to exist in either species (Ekelund and Häder, 1988). The fluence rate response curves show an increasing orientation with an optimum at about 300 Wm^{-2} (Fig. 8). The existence of gravitaxis has not been studied yet, so that the diurnal movements of these flagellates found in nature (Burns and Rosa, 1980) are not easily explicable.

Photobleaching by White Light

In spite of the fact that many flagellates depend on the availability of white solar radiation for their photosynthetic energy production, most colored flagellates have been found to be easily bleached and eventually killed when exposed to strong unfiltered solar radiation (Häder, 1987b). When irradiated by continuous white light for 24 h, suspensions of *Euglena gracilis* are bleached at high fluence rates (Häder, 1985). Up to an illuminance of 14 klx there was no increase in the absorption spectrum at all wavelengths indicating growth and pigment production. At an illuminance of 20 klx there was a drastic decrease indicative of photobleaching and inhibition of growth (Fig. 9).

Similarly, the marine *Cryptomonas maculato* is drastically bleached by continuous exposure to white light (Häder et al., 1988). Fluence rates of 422, 211 and 105 Wm^{-2} causes the cells to be totally bleached within 8, 30 and 52 h, respectively (Fig. 10). Even at 42 and 21 Wm^{-2} the cells are irreversibly bleached, though only after prolonged exposure times.

In contrast, *Gyrodinium* is less prone to photobleaching by white light (Ekelund and Häder, 1988). The bleaching experiments are in good agreement with the orientation

504

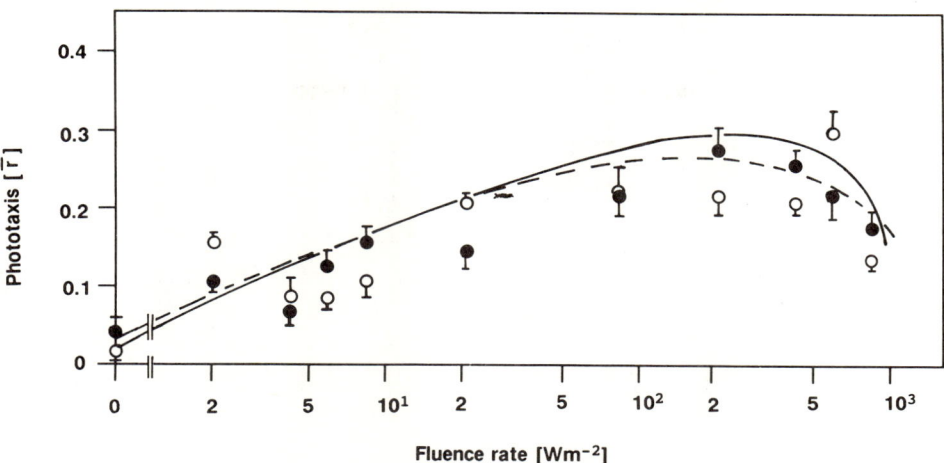

Figure 8. Fluence rate response curves of the positive phototaxis in *Gyrodinium aureolum* and *G. dorsum*.Abscissa: White actinic light in [Wm⁻²], ordinate: phototaxis as quantified by the Rayleigh test (Ekelund and Häder, 1988).

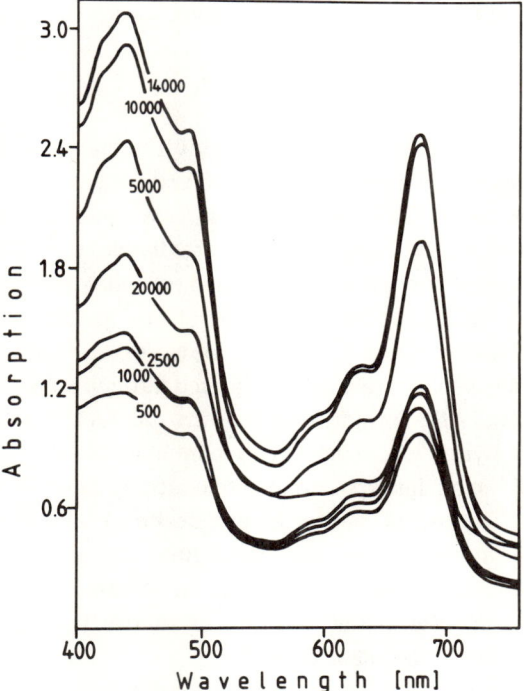

Figure 9. Absorption spectra of *Euglena gracilis* measured after 24 h white light irradiation of 5000 to 20000 lx (Häder, 1985).

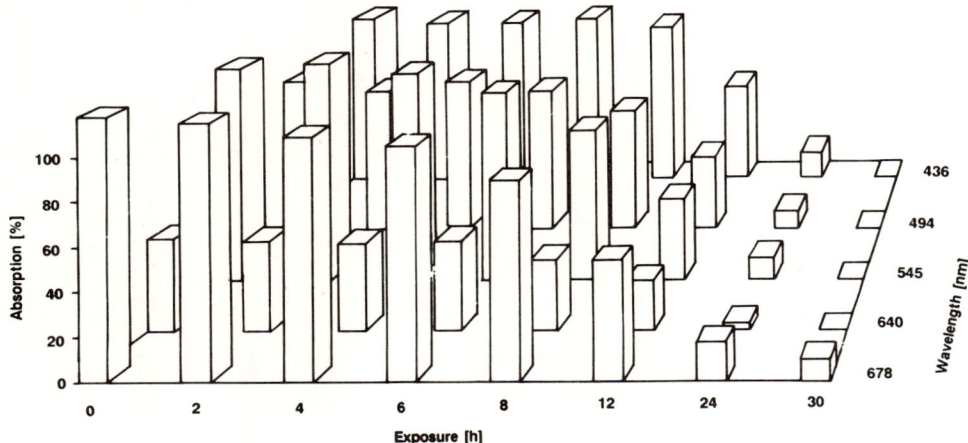

Figure 10. Effect of continuous exposure to white light at a fluence rate of 211 Wm^{-2} on normally colored cells of *Cryptomonas maculata*. The absorption was measured at 5 key wavelengths indicative of the main photosynthetic pigments (Häder et al., 1988).

within the water column. Obviously, photoorientation is an ultimate necessity for colored flagellates in order to avoid high solar fluence rates.

Effects of UV-B Radiation

Many flagellates have been found to be highly susceptible to UV-B (280–320 nm) radiation. Experiments with artificial irradiation demonstrated that both photoorientation and motility are impaired by UV-B radiation in *Euglena gracilis* (Häder, 1985). In order to evaluate the ecological significance of UV-B irradiation, the effects of solar radiation were determined in a number of photosynthetic and non-photosynthetic flagellates.

Both the average speed and the percentage of motile cells within a population of *Euglena gracilis* decreased within 2 h when exposed to direct unfiltered solar radiation (Häder, 1986). When the solar radiation was filtered through a plexiglass cuvette flooded with ozone as a natural UV-B filter, motility was unaffected for considerably longer exposure times. Likewise, when the UV-B component was gradually removed by inserting WG cut-off filters, the exposure time tolerated by the cells was increased (Häder and Häder, 1988a). The use of a fully automatic image analysis system allowed to determine velocity histograms of *Euglena* populations at short intervals. During the first few minutes of exposure the speed of movement increased by about 50% (positive photokinesis). However, after about 200 min of solar exposure the speed decreased dramatically (Fig. 11; Häder and Häder, 1988a).

This effect is not caused by the photosynthetic pigments, since both dark bleached *Euglena gracilis* and the colorless flagellate *Astasia longa* showed the same UV-B inhibition of motility (Häder and Häder, 1988b).

506

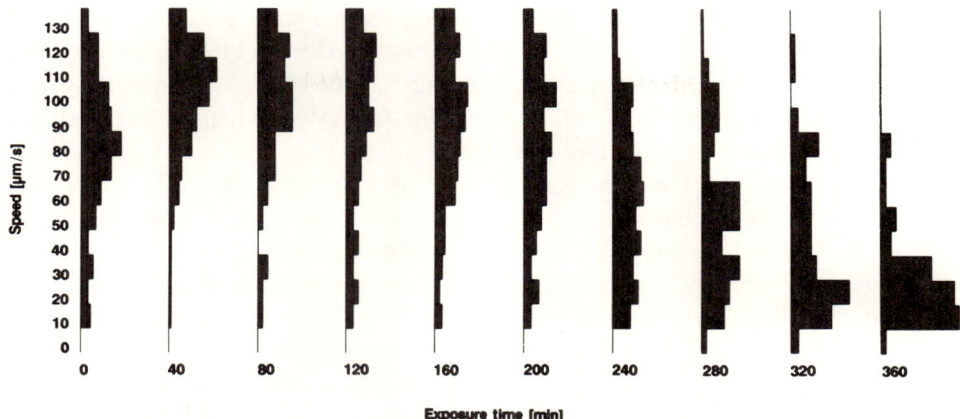

Figure 11. Histograms of the velocity distribution (ordinate) of a *Euglena gracilis* population in dependence of the solar exposure time (abscissa). At noon a total solar irradiation of 1340 Wm^{-2} was measured and the UV-B irradiation was 1.29 Wm^{-2} (Häder and Häder, 1988a).

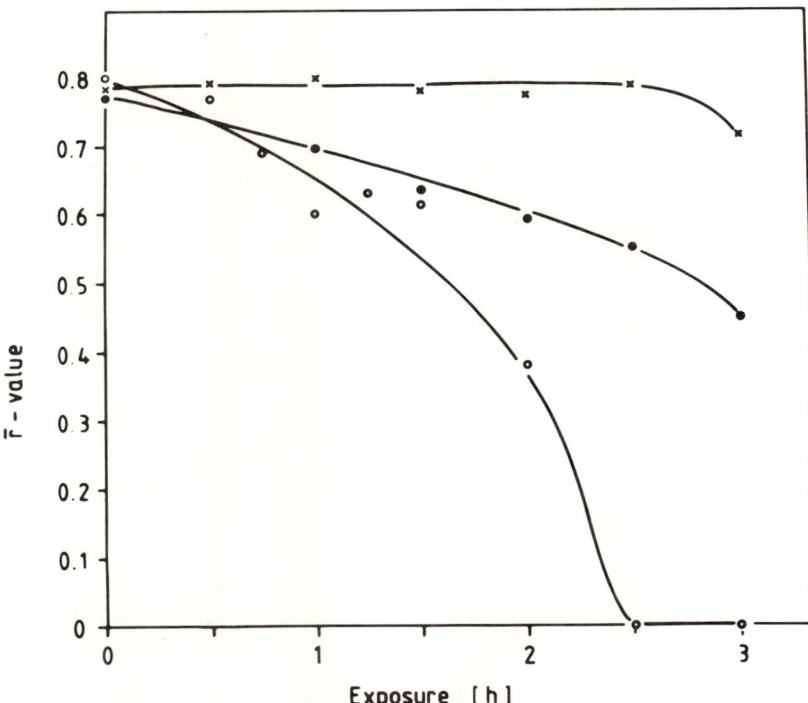

Figure 12. Degree of phototactic orientation in dependence of the unfiltered solar radiation (open circles), solar radiation filtered through the ozone cuvette (closed circles) and glass (crosses) (Häder, 1986).

507

Likewise, the precision of orientation with respect to a white 30 klx test light was drastically reduced in *Euglena gracilis* by solar radiation (Häder and Häder, 1988a). After about 2 h the cells were completely unoriented. Filtering the solar radiation by the ozone cuvette and even more by glass, which is opaque for UV-B radiation, increased the tolerated solar radiation exposure time considerably (Fig. 12).

The target for UV-B inhibition is not known. DNA does not seem to be responsible for the damage, since no photorepair mechanism could be demonstrated (Häder and Häder, 1988a). Likewise type I and type II photodynamic effects could be excluded since neither specific quenchers for singlet oxygen nor free radicals were effective in preventing UV-B inhibition of motility (Häder and Häder, 1988b).

The results indicate that flagellates are under a considerable UV-B stress at ambient solar UV-B levels. Any increase beyond current UV-B levels due to a partial decrease of the stratospheric ozone layer by manmade gaseous pollutants (Molina and Rowland, 1974; Galbally, 1976) poses a potential hazard for growth and survival of these and other phytoplankton organisms.

References

Bancroft, F.W. (1913). Heliotropism, differential sensibility and galvanotropism in *Euglena*. *J. Exp. Zool.* 15, 383–420.

Batschelet, E. (1981). Circular Statistics in Biology, Academic Press, London.

Bean, B. (1984). Microbial geotaxis. In Colombetti, G. and Lenci, F. (eds.), Membranes and Sensory Transduction, Plenum Press, New York, London, pp. 163–198.

Benedetti, P.A., Checcucci, A. (1975). Paraflagellar body (PFB) pigments studied by fluorescence microscopy in *Euglena gracilis*. *Plant Sci.* Lett. 4, 47–51.

Berg, H.C., (1985). Physics of bacterial chemotaxis, In Colombetti, G. Lenci, F. and Song, P.S. (eds.). Sensory Perception and Transduction in Aneural Organisms, Plenum Press, New York, London, pp. 19–30.

Buder, J. (1917). Zur Kenntnis der phototaktischen Richtungsbewegungen. *Jahrber, Wissenschaftl. Botanik* 58, 105–220.

Burns, N.M. and Rosa, F. (1980). *In situ* measurements of the settling velocity of organic carbon particles and ten species of phytoplankton, *Limnol. Oceanogr.* 25, 855–864.

Colombetti, G., Häder, D.P., Lenci, F., Quaglia, M. (1982a). Phototaxis in *Euglena gracilis:* effect of sodium azide and triphenylmethyl phosphonium ion on the photosensory transduction chain. *Curr. Microbiol.* 7, 281–284.

Colombetti, G., Lenci, F. (1982b). Responses to photic, chemical and mechanical stimuli. In: Buetow, D.E. (ed). The Biology of *Euglena*, Vol. III. *Acad. Press*, New York, pp. 169–195.

Creutz, C., Diehn, B. (1976). Motor response to polarized light and gravity sensing in *Euglena gracilis*. *J. Protozool.* 23, 552–556.

Diehn, B. (1969). Action spectra of the phototactic responses in *Euglena*. *Biochim. Biophys. Acta* 177, 136–143.

Diehn, B. (1973a). Phototaxis and sensory transduction in *Euglena*. *Science* 181, 1009–1015.

Diehn, B. (1973b). Phototaxis in *Euglena*. 1. Physiological basis of photoreception and tactic orientation In: Perez-Miravete, A. (ed) Behavior of Microorganisms. Plenum Press, London, New York, pp. 83–90.

Diehn, B., Feinleib, M., Haupt, W., Hildebrand, E., Lenci, F., Nultsch, E. (1977). Terminology of behavioral responses of motile microorganisms. *Photochem. Photobiol.* 26, 559–560.

Doughty, M.J., Diehn, B. (1979). Photosensory transduction in the flagellated alga, *Euglena gracilis*, I. Action of divalent cations, Ca^{2+} antagonists and Ca^{2+} ionophore on motility and photobehavior, *Biochem. Biophys. Acta* 588, 148–168.

Doughty, M.J., Diehn, B. (1980). Flavins as photoreceptor pigments for behavioral responses. *Structure and Bonding* 41, 45–70.

Doughty, M.J., Diehn, B. (1982). Photosensory transduction in the flagellated alga, *Euglena gracilis*. III. Induction of Ca^{2+}- dependent responses by monovalent cation ionophores. *Biochem. Biophys. Acta* 682, 32–43.

Doughty, M.J., Diehn, B. (1983). Photosensory transduction in the flagellated alga, *Euglena gracilis*. *IV*. Long term effects of ions and pH on the expression of step-down photobehavior. *Arch. Microbiol.* 134, 204–207.

Doughty, M.J., Diehn, B. (1984). Anion sensitivity of motility and step-down photophobic responses of *Euglena gracilis*. *Arch. Microbiol.* 138, 329–332.

Doughty, M.J., Grieser, R., and Diehn, B. (1980). Photosensory transduction in the flagellated alga, *Euglena gracilis*. II. Evidence that blue light effects alteration in Na^+/K^+ permeability of the photoreceptor membrane. *Biochim. Biophys.Acta* 602, 10–23.

Ekelund, N., and Häder, D.P. Ecological consequences of photomovement and photobleaching in two *Gyrodinium* species. *Plant Cell Physiol.*, in press.

Foster, K.W., and Smyth, R.D. (1980). Light antennas in phototactic algae. *Microbiol. Rev.* 44, 572–630.

Frankel, R.B. (1984). Magnetic guidance of organisms, *Ann. Rev. Biophys. Bioeng.* 13, 85–103.

Galbally, L.E. (1976). Man-made carbon tetrachloride in the atmosphere. *Science* 111, 619–624.

Ghetti, F., Colombetti, G., Lenci, F., Campani, E., Polacco, E., Quaglia, M. (1985). Fluorescence of *Euglena gracilis* photoreceptor pigment: an *in vivo* microspectrofluorometric study. *Photochem. Photobiol.* 42, 29–33.

Haupt, W. (1983). Photoreception and photomovement. *Phil. Trans. Roy. Soc. Lond. B.* 303, 467–478.

Häder, D.P. (1985). Effects of UV-B on motility and photobehavior in the green flagellate, *Euglena gracilis*. *Arch. Microbiol.* 141, 159–163.

Häder, D.P. (1986). Effects of solar and artificial UV irradiation on motility and phototaxis in the flagellate, *Euglena gracilis*. *Photochem. Photobiol.* 44, 651–656.

Häder, D.P. (1987a). Polarotaxis, gravitaxis and vertical phototaxis in the green flagellate, *Euglena gracilis*. *Arch. Microbiol.* 147, 179–183.

Häder, D.P. (1987b). Photomovement in eukaryotic microorganisms. *Photobiochem. Photobiophys., Suppl.* 203–214.

Häder, D.P. (1988). Ecological consequences of photomovement in microorganisms. *J. Photochem. Photobiol. B: Biol.* 1, 385–414.

Häder, D.P., and Griebenow, K. (1987). Versatile digital image analysis by microcomputer to count microorganisms. *EDV Med. Biol.* 18, 37–42.

Häder, D.P., and Griebenow, K. (1988). Orientation of the green flagellate, *Euglena gracilis*, in a vertical column of water. *FEMS Microbiol. Ecol.* 53, 159–167.

Häder D.P. and Häder, M.A. (1988a). Inhibition of motility and phototaxis in the green flagellate, *Euglena gracilis*, by UV-B radiation. *Arch. Microbiol.* 150, 20–25.

Häder D.P. and Häder, M. (1988b). Ultraviolet-B inhibition of motility in green and dark bleached *Euglena gracilis*. *Current Microbiol.* 17, 215–220.

Häder, D.P. Lebert, M. and DiLena, M.R. (1986). New evidence for the mechanism of phototactic orientation of *Euglena gracilis*. *Current Microbiol.* 14, 157–163.

Häder, D.P. Lebert, M. and DiLena, M.R. (1987a). Effects of culture age and drugs on phototaxis in the green flagellate, *Euglena gracilis*. *Plant Physiol. (Life Sci. Adv.)* 6, 169–174.

Häder, D.P., Rhiel, E., and Wehrmeyer, W. (1987b). Phototaxis in the marine flagellate *Cryptomonas maculata, J. Photochem. Photobiol.* 1, 115–122.

Häder, D.P., Rhiel, E., and Wehrmeyer, W. (1988). Ecological consequences of photomovement and photobleaching in the marine flagellate *Cryptomonas maculata. FEMS Microbiol. Ecol.* 53, 9–18.

Jennings, H.S. (1904). Reactions to light in ciliates and flagellates. In: Contributions to the study of the behavior of microorganisms. Carnegie Inst. Washington, Washington pp. 29–71.

Kessler, J.O. (1985). Hydrodynamic focusing of motile algal cells. *Nature* (London) 313, 218–220.

Kessler, J.O. (1986). The external dynamics of swimming microorganisms. In: Progress in Phycological Research, Round, F.E. and Chapman, D.J. (eds.) *Biopress Ltd.* 4, 258–307.

MacNab, R.M. (1985). Biochemistry of sensory transduction in bacteria. In Colombetti, G., Lenci, F., and Song, P.S. (eds.). Sensory Perception and Transduction in Aneural Organisms, Plenum Press, New York, London, pp. 31–46.

Mardia, K.V., (1972). Statistics of Directional Data, *Acad. Press.*, London.

Mast, S.O. (911). Light and Behavior of Organisms. John Wiley & Sons, New York, Chapman & Hall, London.

Mizuno, T., Maeda, K., and Imae, Y. (1984). Thermosensory transduction in *Escherichia coli.* In Oosawa, F., Yoshioka, T., and Hayashi, H. (eds.). Transmembrane Signaling and Sensation, *Japan Sci. Soc. Press. Tokyo and VNU Sci.* Press BV, Netherlands, pp. 147–195.

Molina, M.J., Rowland, F.S. (1974). Stratospheric sink of chlorofluoromethanes chlorine atoms catalyzed destruction of ozone. Nature (London) 249, 810–812.

Nultsch, W., and Häder, D.P. (1988). Photomovement in motile microorganisms II. *Photochem. Photobiol,* 47, 837–869.

Ofer, S., Nowik, I., Bauminger, E.R., Papaefthymiou, G.C., Frankel, R.B., and Blakemore, R.P. (1984). Magnetosome dynamics in magnetotactic bacteria. *J. Biophys.* 46, 57–64.

Poff, K.L. (1985). Temperature sensing in microorganisms. In: Sensory Perception and Transduction in Aneural Organisms (Eds.G. Colombetti, F. Lenci and P.S. Song), Plenum Press, New York, London. pp. 299–307.

Rhiel, E., Häder, D.P., and Wehrmeyer, W. (1988a). Diaphototaxis and gravitaxis in a freshwater *Cryptomonas. Plant Cell Physiol.* 29, 755–760.

Rhiel, E., Häder, D.P., and Wehrmeyer, W. (1988b). Photo-orientation in a freshwater *Cryptomonas* species. *J. Photochem. Photobiol. B: Biol.* 2, 123–132.

Shimmen, T. (1981). Quantitative studies on step-down photophobic response of *Euglena* in an individual cell. *Protoplasma* 106, 37–48.

Stolz, J.F., Chang, S.B.R., and Kirschvink, J.L. (1986). Magnetotactic bacteria and single-domain magnetite in hemipelagic sediments. *Nature* 321, 849–851.

Watanabe, M. and Furuya, M. (1974). Action spectrum of phototaxis in a cryptomonad alga. *Cryptomonas sp. Plant Cell Physiol.* 15, 413–420.

Watanabe, M. and Furuya, M. (1978). Phototactic responses of cell population to repeated pulses of yellow light in a phytoflagellate *Cryptomonas sp. Plant Physiol.* 61, 816–818.

Watanabe, M., Miyoshi, Y., and Furuya, M. (1976). Phototaxis in *Cryptomonas sp.* under condition suppressing photosynthesis. *Plant Cell Physiol.* 17, 683–690.

Wolken, J.J., Shin, E. (1958). Photomotion in *Euglena gracilis* I. Photokinesis, II. Phototaxis. *J. Protozool.* 5, 39–46.

Polarized Spectra of Immobilized Phycobilisomes Isolated from Various Cyanobacteria

D. FRACKOWIAK[1], Y. FUJITA[2], L.G. EROKHINA[3], M. MIMURO[2], Y. YAMAZAKI[4], N. TAMAI[4], M. NIEDBALSKA[1], M. ROMANOWSKI[1] and J.SZURKOWSKI[1]

[1]*Poznan Technical University*
Poland,
[2]*National Institute for Basic Biology*
Okazaki, Japan
[3]*Institute of Soil Science and Photosynthesis*
Pushchino, USSR
[4]*Institute for Molecular Sciences*
Okazaki, Japan

Phycobilisomes (PBS) big antenna complexes occuring in cyanobacteria and red algae are extremely efficient transducers of excitation energy (Glazer, 1984). It was suggested previously (Mimuro et al., 1988–1989) that are at least two independent pathways of excitation energy transfer from phycoerythrin to final excitation acceptor-allophycocyanin. It is possible that the chromophores taking part in various chains of energy transfer are differently oriented in respect to the PBS structure. Therefore we undertake the investigation of oriented PBS using polarized light spectroscopy.

PBS from cyanobacteria *Gloeotrichia* and *Tolypothrix tenuis* were isolated and introduced to polyvinyl alcohol (PVA) films as reported previously (Frackowiak et al., 1986a). The isotropic films and the films stretched to four times of its initial length were investigated.

Steady state polarized absorption, fluorescence and fluorescence excitation spectra were measured with various apparatus equipped with polarizers and film holders. The measurements of polarized photoacoustic spectra have been previously described (Frackowiak et al., 1986b).

Linear dichroism (LD) of *Tolypothrix* PBS was measured directly using Jasco J-200B apparatus. For these PBS the time-resolved fluorescence spectra in ps time range were also taken with the apparatus reported previously (Mimuro et al., 1985, Yamazaki et al., 1984, 1985). Pulse duration was 6 ps (FWHM). The pulse intensity at 546 nm used for the sample excitation was low enough to avoid nonlinear effects. Time zero was set to the time of the intensity maximum of excitation pulse. Other experimental details are descibed in Frackowiak et al. (1989).

Photobiology, Edited by E. Riklis
Plenum Press, New York, 1991

The energy absorbed by given type of biliprotein consisting PBS can be transfered to other type of chromophore, emitted as fluorescence or converted into heat. Therefore these three processes are competetive: the change in the yield of energy transfer influences the yields of thermal deactivation of excitation and the yield of fluorescence. Using the light polarized linearly parallel or perpendicular to the direction of film stretching we can observe two pools of chromophores: first — with transition moments of absorption or emission forming small angles with the deformation axis and the second having big projections of their transition moments on the plane perpendicular to PVA axis.

Supposing that it is only one chain of excitation energy acceptors and donors we have to expect the same yields of energy transfer, fluorescence and thermal deactivation in both pools of chromophores parallel and perpendicular to the orientation axis. The same result has to be expected in a case of the existence of more than one chain of acceptors and donors but having similar average orientations in respect to PBS structure. In both cases the yields of thermal deactivation (TD) obtained from the polarized photoacoustic and absorption spectra have to be the same for both polarized components.

In purpose to obtain the yields of polarized thermal deactivations of excitation occuring in various biliproteins constituend phycobilisomes polarized absorption and photoacoustic spectra were analyzed on Gaussian components. The yield of TD was calculated as a ratio of the photoacoustic and absorption Gaussian components surfaces. The analysis of fluorescence excitation spectrum gives the informations about energy transfer between biliproteins in PBS embedded in PVA. The component related with one type of biliprotein (Phycoerythrin – PE, Phycocyanin – PC or Allophycocyanin – APC) is the superposition of the contributions from different types of chromophores (sensitizing–s, fluorescent–f or medium–m) having different yields of fluorescence and TD (Wendler et al., 1986, Sauer et al., 1987, Schrimer and Vincent, 1987). Therefore Gaussians belonging to absorption, photoacoustic and fluorescence excitation spectra obtained without special supposition are mutually shifted. In used analysis it was supposed that the band positions in all three types of spectra are the same. The fact of the occurence of various (s, f and m) chromophores was taken into account by the different halfwidth of given component in various spectra.

The band belonging to PC and APC (not resoluted into components) is more pronouced in photoacoustic than in absorption spectra (Fig.1). It shows that excited states of final acceptors are depopulated efficiently not only by the fluorescence emission but also by the thermal deactivation of excitation. In purpose to fit the experimental spectra with sum of Gaussian the additional component in 497–511 nm region has to be supposed. In this region is located phycourobilin chomophores absorption. It is not excluded that such type of chromophores can be "produced" by the interacion of c-PE chromophores with PVA (Frackowiak et al., 1985). Table 1 shows the results of TD. All values are in the arbitrary but the same units, therefore they can be compared. From Table 1 it follows that for the longwave length the Gaussians the value of parallel polarized component of TD is higher than that of perpendicular one.

512

Table 1. Averaged TD of various components of phycobilisomes from *Gloeotrichia*.

Band position [nm]	Thermal deactivation	
	Parallel	Perpendicular
509	0.82	0.85
534	0.50	0.50
568	0.61	0.34
621	0.75	0.64

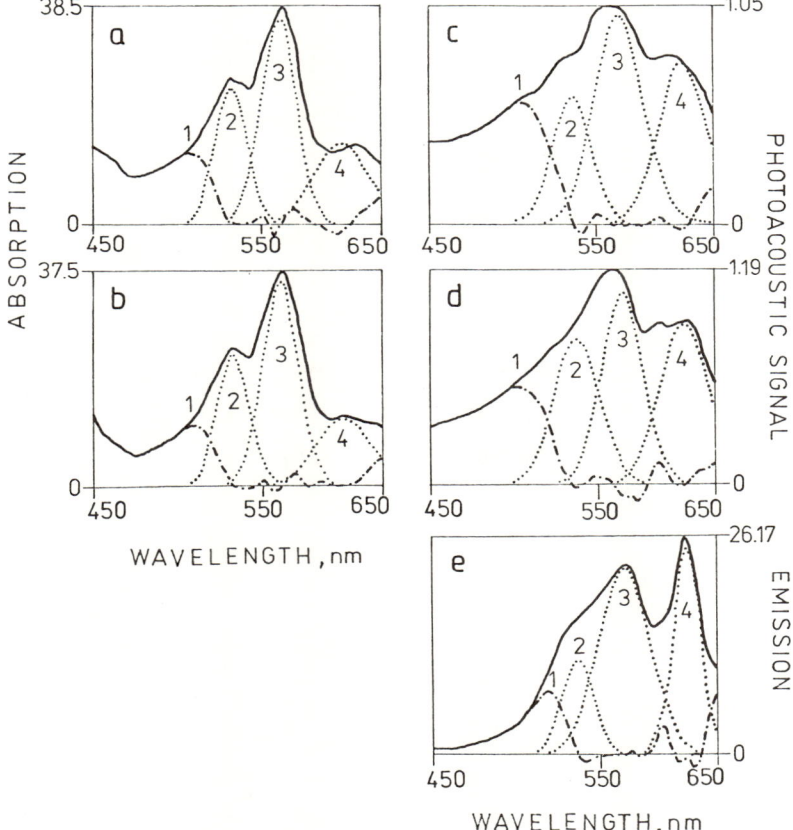

Figure 1. Gaussian analysis of absorption (a,b), photoacoustic (c,d) and fluorescence excitation spectra of phycobilisomes from Gloeotrichia in stretched (a,c parallel and b,d perpendicular component) and unstretched (e) PVA film. --- experimental results, ... Gaussian component, -.- difference between experimental and fitted curve.

It means that in this region of absorption the chromophores located parallel to the direction of film stretching are changing more efficiently their excitation into heat than a pool of chromophores located rather perpendicular to the film axis.

Anisotropy of spectral properties suggests strongly some orientation of PBS in respect to the film axis. Fig. 2 shows the linear dichroism of PBS from *Tolypothrix*. From the analysis of absorption spectra of PBS in unstretched film on the basis of Hattori and Fujita (1959) method the ratios of biliprotein concentrations were established and model shown in Fig. 3 was proposed. Similar analysis was done for the stretched film. From the analysis of the change in contribution to absorption caused by the film stretching the orientation of PBS hemidiscs diameter along PVA axis was obtained. Detail of this procedure is presented in paper by Frackowiak et al. (1989). Fig. 4 shows the example of time resolved polarized spectra of oriented PBS. The kinetics of peaks rise

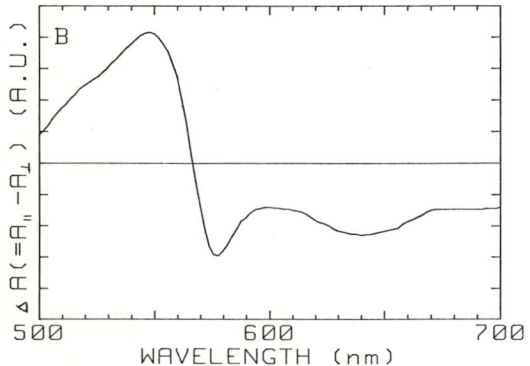

Figure 2. Linear dichroism of *Tolypothrix* PBS.

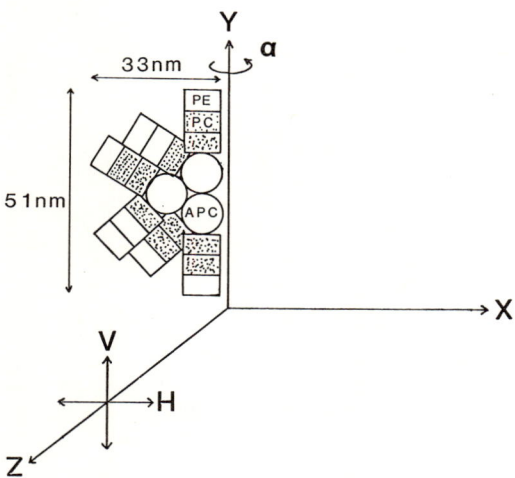

Figure 3. Model of PBS orientation in stretched PVA film. y – orientation axis, zy – PVA plane, z – direction of light beam, V – vertical, H – horizontal polarization of light.

514

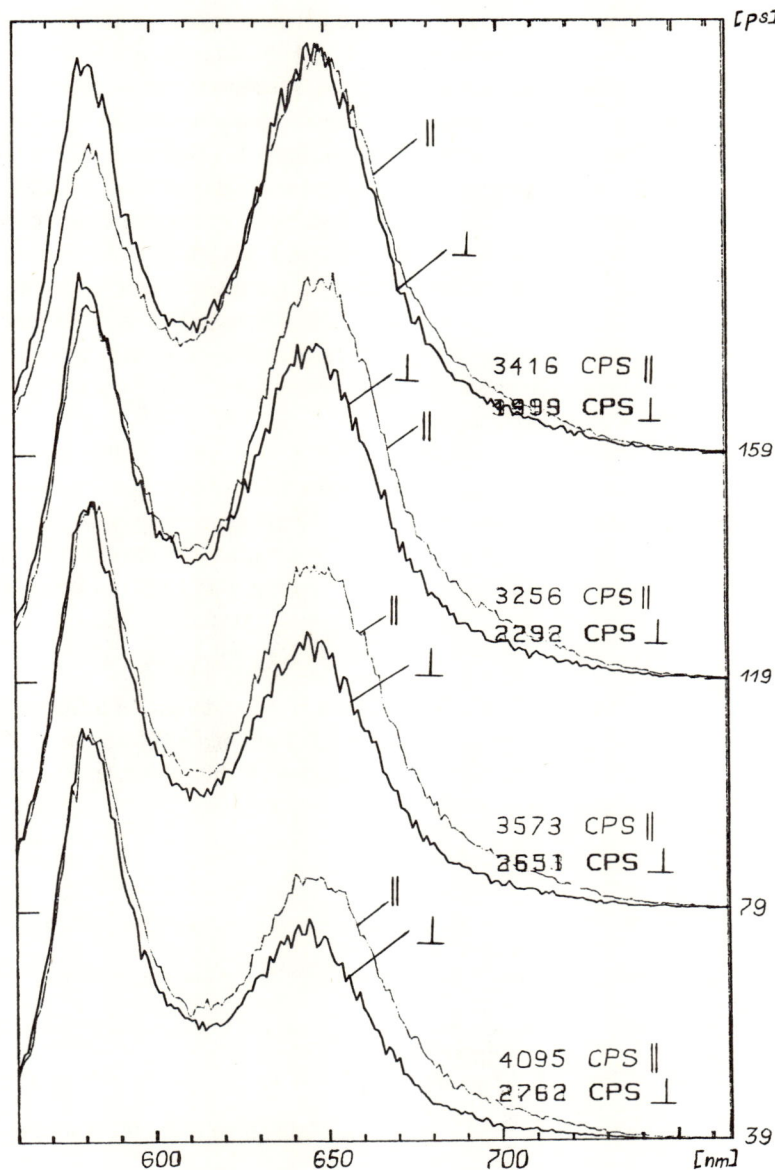

Figure 4. Example of two polarized components of time resolved fluorescence spectra of PBS from *Tolypothrix* in stretched PVA. CPS – number of counts per second. The time of taking the spectra (measured from the peak of laser pulse) is given on the graphs in ps.

and decay obtained on the basis of such spectra are different in two polarized components.For example excitation energy transfer from excited PE to APC is more efficient in parallel polarized component of stretched sample than in perpendicular one. The time of reaching the maximal value of emission of given type of biliprotein depends on the rate of energy transfer to this pigment and on the rate of depopulation of its excited state by energy transfer to other molecules, by fluorescence emission and by thermal deactivation. The times of reaching the maxima of emission of PC and APC for a sample excited in PE absorption region is much longer in immobilized PBS than in PBS in solution (Mimuro et al., 1989). The intensities ratio of parallel to perpedicular polarized components of emission measured at 660 nm APC peak is decreasing in time after laser peak. In a case of two pools of molecules differenty oriented but having similar efficiences of deactivation by energy transfer and radiative and nonradiative transitions this ratio has to be constant in time. Observed decrease in this ratio is in agreement with photoacoustic spectra results showing more prominent TD in parallel than in perpendicular component of polarized spectra.

The discussion of all set of spectra shows that light absorbed by differently oriented PE molecules is reaching different groups of final emitters. This is in agreement with previous conclusions (Mimuro et al., 1986, 1989, Gantt et al., 1988) that they are more than one independent pathway of ET in PBS. Presented results are showing that molecules taking part in various ET pathways are differently oriented in PBS structure.

Acknowledgements

D. Frackowiak wishes to thank the Yamada Science Fundation for a fund supporting her stay in National Institute for Basic Biology in Okazaki and to Professor Y. Fujita for his very kind hospitality in Bioenergetics Laboratory of this Institute. The work was partially supported by Polish Grant RPBP.2.11.

References

Frackowiak, D., L.G. Erokhina, A. Balter, L. Lorrain, J.Szurkowski and B.Szych (1986a) Biochim. Biophys. Acta 851, 173–180.
Frackowiak, D., S. Hotchandani and R.M. Leblanc (1985) Photochem. Photobiol. 42, 559–565.
Frackowiak, D., S. Hotchandani, B. Szych and R.M. Leblanc (1986b) Acta Phys. Polon. A69, 121–133.
Frackowiak, D., M. Mimuro, I. Yamazaki, N. Tamai and Y. Fujita (1989) Photochem. Photobiol. 50, 563–570.
Gantt, E., F.X. Cunningham Jr., C.A. Lipschultz and M. Mimuro (1988) Plant Physiol. 86, 996–998.
Glazer, A.N. (1984) Biochim. Biophys. Acta 768, 29–51.
Hattori, A. and Y. Fujita (1959) J. Biochem. 46, 633–644.
Mimuro, M., I. Yamazaki, T. Yamazaki and Y. Fujita (1985) Photochem. Photobiol. 41, 597–603.
Mimuro, M., I. Yamazaki, N. Tamai and T. Katoh (1989) Biochim. Biophys. Acta 973, 153–162.
Yamazaki, I., M. Mimuro, T. Murao, T. Yamazaki, K. Yoshihara and Y. Fujita (1984) Photochem. Photobiol. 39, 233–240.

Yamazaki, I., N. Tamai, H. Kume. H. Tsuchiva and K. Oba (1985) Rev. Sci. Instr. 56, 1187–1194.

Sauer, K., H. Scheer and P. Sauer, (1987) Photochem. Photobiol. 46, 427–440.

Schrimer, T. and M.G. Vincent (1987) Biochim. Biophys. Acta 893, 379–385.

Wendler, J., W. John, H. Scheer and A.R. Holtzwarth (1986) Photochem. Photobiol. 44, 79–85.

Rhodopsins

Spectra and Structures of Photorhodopsin and Bathorhodopsin Studied by Picosecond Laser Photolysis

TORU YOSHIZAWA, YOSHINORI SHICHIDA AND HIDEKI KANDORI

Department of Biophysics
Faculty of Science, Kyoto University
Kyoto 606, Japan

Introduction

The retinylidene chromophore of rhodopsin is isomerized by light from 11-*cis* to all-*trans* form. The process of the photoisomerization occurs so rapidly that rhodopsin has a high photo-sensitivity. For elucidating the mechanism of the high speed isomerization, it is essential to study the structure of the excited state of rhodopsin and its primary intermediate. For a while after discovery of bathorhodopsin, it was believed that bathorhodopsin would be the first photoproduct of photobleaching of rhodopsin (Yoshizawa and Kitô, 1958; Yoshizawa and Wald, 1963; Yoshizawa, 1972). Since bathorhodopsin has a twisted all -*trans* retinal as its chromophore (Yoshizawa and Wald, 1963; Eyring et al., 1982), it was widely accepted that the isomerization of the chromophore began with Franck-Condon state and ended at bathorhodopsin via an excited state common between rhodopsin and bathorhodopsin (Hurley et al., 1977).

In 1984, Shichida et al. detected an earlier intermediate than bathorhodopsin, which was named photorhodopsin. This intermediate has an absorption maximum at a longer wavelength side than that of bathorhodopsin. It is now essential to investigate the primary process of rhodopsin in due consideration of photorhodopsin. We have already got a result that the conversion of photorhodopsin to bathorhodopsin showed no deuterium effect (Kandori et al., 1989a) What are the structures of the chromophores in these intermediates? What happens in the process of photorhodopsin to bathorhodopsin? In this paper, two approaches were made for understanding of the primary process of rhodopsin: (1) Absolute absorption spectra of bathorhodopsin and photorhodopsin at room temperature were estimated. (2) Photochemistry of rhodopsin analogs having 11-*cis* locked retinal analogs as their chromophores was studied.

Absolute Absorption Spectra of Bathorhodopsin and Photorhodopsin.

A picosecond laser photolysis had been applied to investigate the photochemical reaction of rhodopsin at physiological temperature (Busch et al., 1972). This technique, however, might not be suitable for determining an absolute absorption spectrum of

Photobiology, Edited by E. Riklis
Plenum Press, New York, 1991

intermediate which gives us important information for elucidating the conformation of chromophore and the chromophore-protein interaction. Usually the picosecond laser pulse, regardless of excitation or monitoring pluses, is so small in diameter that only a part of the sample containing rhodopsin molecules in the sample cell can be excited or monitored. Thus it is difficult to estimate accurately a percentage of bleaching of rhodopsin in the excited spot, because the excited rhodopsin molecules can diffuse in the sample cell after the excitation. Since it is indispensable to get the accurate percentage of bleaching of rhodopsin for estimating the absolute absorption spectra of intermediates of rhodopsin, we have attempted to make rhodopsin molecules immobilized by embedding them in polyacrylamide gel and to estimate precisely the percentage of bleaching of rhodopsin, followed by computation of absolute absorption spectra of batho- and photo-rhodopsins (Kandori et al., 1989b).

Cattle rod outer segments (ROS) were isolated from retinas by a sucrose flotation method described previously (Kandori et al., 1988), followed by extraction of rhodopsin with 2% digitonin in 10mM HEPES buffer (pH 7.0). The extract was mixed with an equal volume of 30% acrylamide solution supplemented with 0.8% Bis acrylamide and then the acrylamide in the mixture was polymerized. Immediately after the start for the polymerization, the mixture was divided into several sample cells (4 mm × 60 mm × 2 mm), each of which was composed of a set of silicon rubbers (2mm thickness) with an interstitial space of 4mm which were sandwiched between two slide glasses (Fig.1). Thus a long and slender 15% polyacrylamide gel (4 mm × 60 mm × 2 mm) containing rhodopsin was prepared. The gel in the sample cell was then immersed into a neutralized 50 mM hydroxylamine solution and kept for more than 36 hrs for penetration of the hydroxylamine. This preparation is tentatively called "rhodopsin gel". All the experiments were carried out under dim red light.

Absorption spectra of the rhodopsin gel and its completely bleached gel were measured through an aperture of 2mm diameter placed in front of the sample cell. The absorption spectra and their difference spectrum were identical to those measured in 2% digitonin solution containing 50mM hydroxylamine. These results clearly show that rhodopsin molecule was not bleached by embedding in the gel and that the hydroxylamine could penetrate into the gel. The spectrum of the bleached gel never changed through the dark incubation for one hour, indicating that the rhodopsin molecule could not translate in the gel within this time range. On the other hand, no light-induced dichroism was observed in the gel, indicating that the rhodopsin molecule could rapidly rotate in the gel.

An optical setup for the picosecond laser photolysis was described previously (Shichida et al., 1984). Both excitation and monitoring pulses were 21psec in pulse widths. Wavelength of the excitation pulse was 532 nm and those of the monitoring pulse were in a range from 470 to 660 nm. Figure 1 shows a transient absorption spectrum (difference spectrum between an intermediate and rhodopsin) measured at 1 nsec after the excitation. The spectrum was almost identical in shape with that measured in 2% digitonin solution. Therefore, the photoproduct appeared in the gel should to bathorhodopsin. The percentage of bleaching of rhodopsin in a spot of the gel exposed by

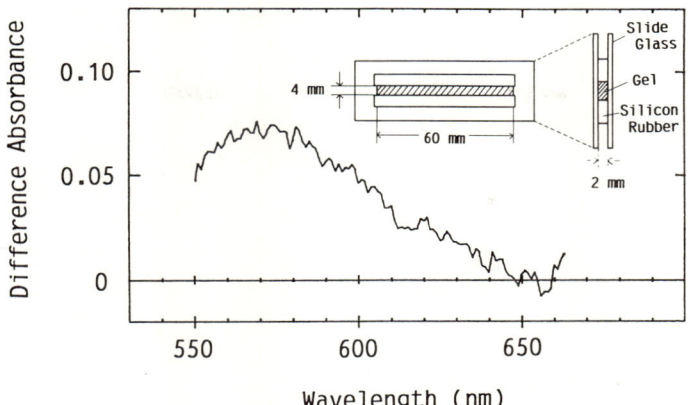

Figure 1. Transient absorption spectrum measured 1nsec after excitation of rhodopsin in the gel with a picosecond laser pulse (wavelength, 532nm; pulse width, 21psec). An optical cell for rhodopsin gel was illustrated as an insertion figure.

the pulse was estimated to be 10.9% from the absorbance change before and 5 minutes after the excitation. This value, however, is not the true percentage of bleaching of rhodopsin, because the rhodopsin molecules in the gel could rotate three-dimensionally as mentioned above, while the absorption spectra measured above gave only two-dimensional information. Thus the true percentage of bleaching was re-estimated to be $10.9 \times 3/2 = 16.4\%$.

Based upon this value, the absolute absorption spectrum of bathorhodopsin was calculated (Fig. 2). The absorption maximum was located at about 535nm which was 8nm shorter than that at low temperature. It should be noted that the absorption spectrum of bathorhodopsin was smaller in extinction coefficient and broader in spectral shape than that measured at low temperature (Yoshizawa and Wald, 1963). Furthermore, oscillator strength of bathorhodopsin at room temperature was 0.9 times smaller than that at low temperature. Thus, bathorhodopsin at room temperature was somewhat different in potential surface from that at low temperature.

The absorption spectrum of photorhodopsin (Fig. 2) was calculated from both the spectrum of bathorhodopsin at room temperature and the difference spectrum between photorhodopsin and bathorhodopsin reported previously (Shichida et al., 1984). The absorption maximum of photorhodopsin was located at about 570 nm, about 35 nm longer than that of bathorhodopsin. Moreover, it should be noted that both spectral half band-width and oscillator strength of photorhodopsin were considerably small in comparison with those of rhodopsin and bathorhodopsin. The small oscillator strength of photorhodopsin seems to be due to distortion of the retinylidene chromophore, because a value of oscillator strength is very sensitive to coplanarity of the conjugated double bond system (Sperling, 1973). Therefore, photorhodopsin may be an intermediate state on the

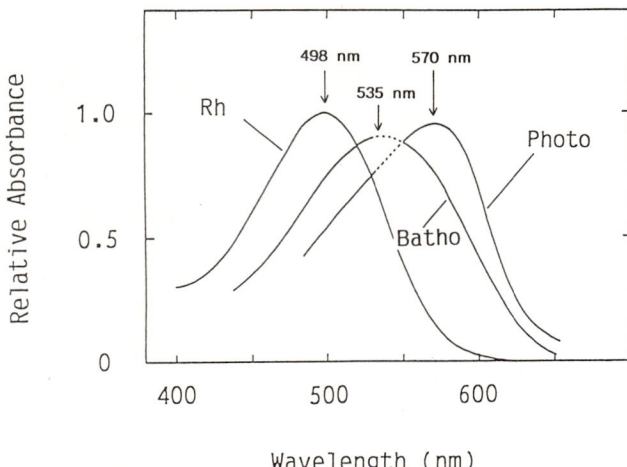

Figure 2. Absolute absorption spectra of cattle rhodopsin, bathorhodopsin and photorhodopsin at room temperature. The spectrum of bathorhodopsin was calculated from the following equation:

B = R + (Diff. Ab.)/B1,

in which B and R are spectra of bathorhodopsin and rhodopsin, respectively, Diff. Ab. is difference absorption spectrum between bathorhodopsin and rhodopsin, and B1 is the percentage of bleaching (16.4%). The spectrum at longer wavelengths than 550 nm was calculated using the difference spectrum measured in rhodopsin gel (Fig. 1). The spectrum in the range from 400 to 530 nm was calculated using the data previously reported (Shichida et al., 1984) after correcting the percentage of bleaching. The spectrum of photorhodopsin was also calculated from a difference spectrum between rhodopsin and a mixture of 70% photorhodopsin and 30% bathorhodopsin (Shichida et al., 1984). The percentages were estimated from the lifetime of photorhodopsin. Since the scattering of the excitation pulse made it impossible to measure the difference spectra in the range from 530 nm to 550 nm, the spectra in the range were estimated by extrapolating the spectra on both sides of the wavelength ranges (dotted lines).

way of isomerization. The location of the absorption maximum of photorhodopsin suggests that the proton of Schiff base may be far away from the counter charge which stabilizes the protonated Schiff base.

Studies by Use of 11–*cis*–Locked Rhodopsin Analogs

It was suggested in the previous section that the retinylidene chromophore of photorhodopsin would be twisted at C_{11}–C_{12} double bond and its neighboring single bonds. In order to confirm this suggestion, we investigated the photochemical reactions of rhodopsin analogs having cycloheptatrienylidene and cyclopentatrienylidene-11-*cis*-locked retinals (Ret7; Akita et al., 1980 and Ret5; Ito et al., 1982), respectively (Kandori et al., 1989c). These retinal analogs have 7-membered and 5-membered rings, respectively, by which the full rotation of the C_{11}–C_{12} double bonds are locked. The rhodopsin analogs (Rh7 and Rh5) displayed no spectral changes by irradiation at 77K, indicating clearly that the chromophore of bathorhodopsin is already isomerized to all-*trans* form (Mao et al., 1981; Fukada et al., 1984).

These retinal analogs are, however, different from each other in the manner of prohibition of rotation around the 11-ene. Namely, Ret5 has a perfectly planar and rigid ring, whereas Ret7 has a relatively flexible ring which enables the 11-ene or its neighboring single bonds to distort to some degree because Rh7 displayed CD in $\alpha-$ and β-bands like rhodopsin (Akita et al., 1981), while Rh5 displayed only a negative CD in β-band (Fukada et al., 1984). Therefore, we can expect to get some informations regarding the configuration and/or conformation of the chromophore of photorhodopsin by comparing the primary processes of Rh5 with that of Rh7.

Ret5 and Ret7 were chemically synthesized by Prof. Tsukida and Ito's group at Kobe Women's College of Pharmacy and Prof. Nakanishi's group at Columbia University, respectively. Rh5 was constituted from 2% digitonin extract of cattle opsin and Ret5, followed by purification with Con A-affinity column chromatography (Fukada et al., 1984). Whereas Ret7 was incorporated into cattle ROS, from which the rhodopsin analog (Rh7) was extracted with 2% digitonin by the usual method (Kandori et al., 1988). Absorbances of Rh5 and Rh7 at 532 nm were adjusted to be 0.7 for absorption measurements and 0.3 for fluorescence measurements (2 mm light path).

Picosecond absorption measurement was performed as described previously (Shichida et al., 1984). For measurement of picosecond fluorescence kinetics, a streak camera system (C–2,000, Hamamatsu) was used. An orange cut-off filter (VO 56, Toshiba) was set in front of the system for removing any scattering light owing to the excitation pulse. The time resolution of the whole system was 25 psec. The decay time of fluorescence was determined by least-squares fitting. All the experiments were performed at 18°C.

i) *Photochemistry of Rh5*

Figure 3, Left shows the transient spectral changes measured after excitation of Rh5 with weak laser pulse. Immediately after the excitation (0psec), the absorbance at about 550nm increased with concurrent decrease at about 600nm. Since Rh5 has no absorbance in the wavelength range above 600nm, the observed negative absorbance should be due to a stimulated emission signal by the monitoring pulse. The increase of absorbance at about 550nm is probably attributed to a generation of an excited state of Rh5. These absorbance changes disappeared 100psec after excitation (Fig. 3, left). Thus the life-time of the excited state of Rh5 is likely to be less than 100 psec. In fact, a measurement of fluorescence kinetics revealed that the decay time constant of the excited state was about 85psec (Fig. 4, Left). The life time of the excited state of Rh5 is about 100 times longer than that of Rh, probably owing to inhibition of rotation at the center of the linear polyene of the chromophore of Rh5. A thermally bleached Rh5 sample showed no fluorescence. There was a linear relation between the fluorescence intensity and the excitation energy in the low range of photon density (Fig. 4, Left, inset).

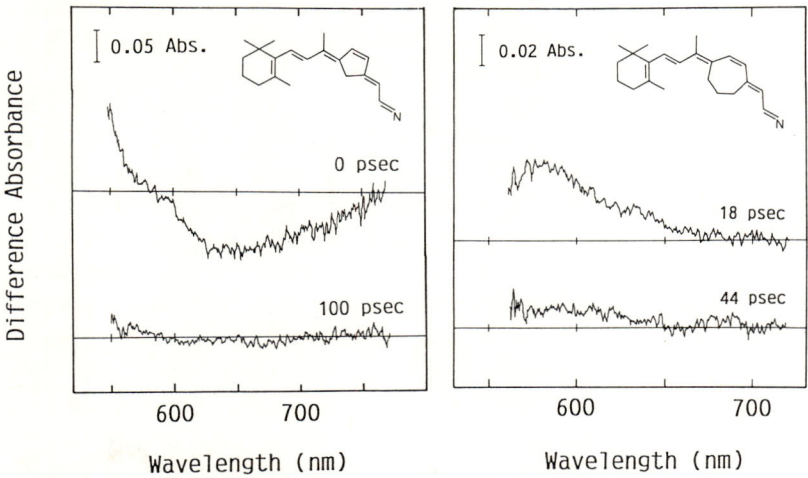

Figure 3. Left: Spectral changes of Rh5 measured at 0 and 100psec after excitation with a weak single green picosecond laser pulse (wavelength, 532nm; pulse width, 21psec;energy, 20μJ/1.8mmφ). The spectral changes correspond to difference spectra between Rh5 and its photo-transients. Right: Spectral changes of Rh7 measured at 18 and 44psec after excitation with a relatively intense single green pulse (excitation energy: 121μJ/1.8mmφ).

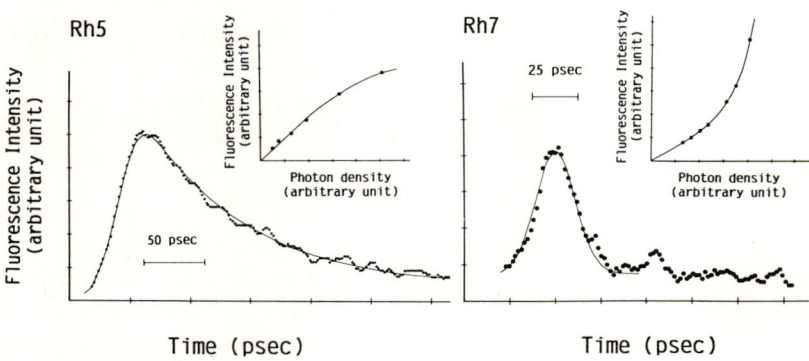

Figure 4. Left: Fluorescence kinetics of Rh5. The intensity of fluorescence emitted from Rh5 by excitation with the green pulse (532nm) was measured above 560 nm at room temperature. This is the sum of 10 data. The solid line is a fitting curve in consideration of the time resolution of the whole apparatus (25psec), giving $\tau_{1/e} = 85$psec. Inset: Relationship between fluorescence intensity and excitation energy of picosecond laser pulse. Each point is the average of 6 data. Right: Fluorescence kinetics of Rh7. No fluorescence was measured by the usual set-up because the laser pulse was too weak to detect it. For rise of photon density of the pulse, a lens was set for focusing the excitation pulse onto the sample. The fluorescence thus recorded should originate from the excited state of Rh7(580). This is the sum of 9 data. Solid line indicates the instrumental response function (half width: 25psec). *Inset*: Relationship between fluorescence intensity and excitation energy of picosecond laser pulse. Each point is the average of 4 data.

526

ii) Photochemistry of Rh7

Transient difference spectra were measured at 18 and 44ps after excitation of Rh7 with 532nm laser pulse (Fig. 3, Right). Since the maxima of the difference spectra were located at about 580nm, the product formed in this time scale was named Rh7(580). A linear relation of photon density to increase of absorbance was seen up to 70μJ/1.8mmφ, indicating that Rh7(580) was directly formed from Rh7 by absorption of one photon. Increase of absorbance at the maximum (580 nm) was estimated to be 0.008 when excited with a pulse of 20μJ/1.8mmφ, while that at 15psec after excitation of Rh was 0.05 (Matuoka et al., 1984). Since no fluorescence was detected on excitation of Rh7 under these conditions, Rh7(580) was regarded as the ground state species.

It should be noted that a photoproduct showing a different absorption maximum at about 630nm (Rh7(630) was observed when Rh7 was excited with a highly intense pulse (all the Rh7 molecules in the excited spot could absorb about 2 photons in the average). This product was similar to the photoproduct named Rh7(640) reported by Buchert et al., (1983). Thus Rh7(630) should be the product generated only by a multiphoton reaction of Rh7. Rh(630) could be detected even 40psec after excitation, indicating that its decay time constant should be larger than that of Rh7(580).

Fluorescence measurements of Rh7 also supported the interpretation that Rh7(630) was a multiphoton product. Although no fluorescence was observed from Rh7 by excitations with the weak laser pulse, fluorescence from the samples was detected by excitation with a highly intense laser pulse (Fig. 4, Right). The kinetic curve of the fluorescence displayed a Gaussian distribution function which corresponded to the resolution function of our experimental setup (25psec). Noteworthy point is that the fluorescence signal from Rh7 was not proportional to the excitation photon density (Fig. 4, Right, inset). Therefore, the origin of the fluorescence should be assigned to the excited state of Rh7(580) which was formed photochemically from Rh7. Some fraction of the excited state may convert to Rh7(630). Thus the reaction scheme of Rh7 can be drawn as Fig. 5.

As already stated, the 7-membered ring in Rh7 is more flexible than the 5-membered ring in Rh5. Thus the difference between the flexibilities of the rings can bring the difference in photophysical behavior between them. What is the structure of the chromophore of the ground state product, Rh7(580)? If Rh7(580) has a 11-*cis*-form, two kinds of 11-*cis*-forms (Rh7 and Rh7(580)) have to be present in the 7-membered chromophore.

Since the retinylidene chromophore itself would be impossible to take two kinds of *cis*-forms, a strong interaction between the chromophore and its neighboring amino-acids might be necessary.

A current finding suggests that there is little protein-induced contraint near the middle portion of the chromophore (Liu and Asato, 1985). Therefore, it is unlikely to generate a highly distorted *cis*-configuration by a strong chromophore/protein interaction. There are abundant evidence in support of an occurrence of *cis-trans* isomerization of double bonds in 6–, 7– and 8–membered rings containing 2, 3 and 4sp^2 hybridized

527

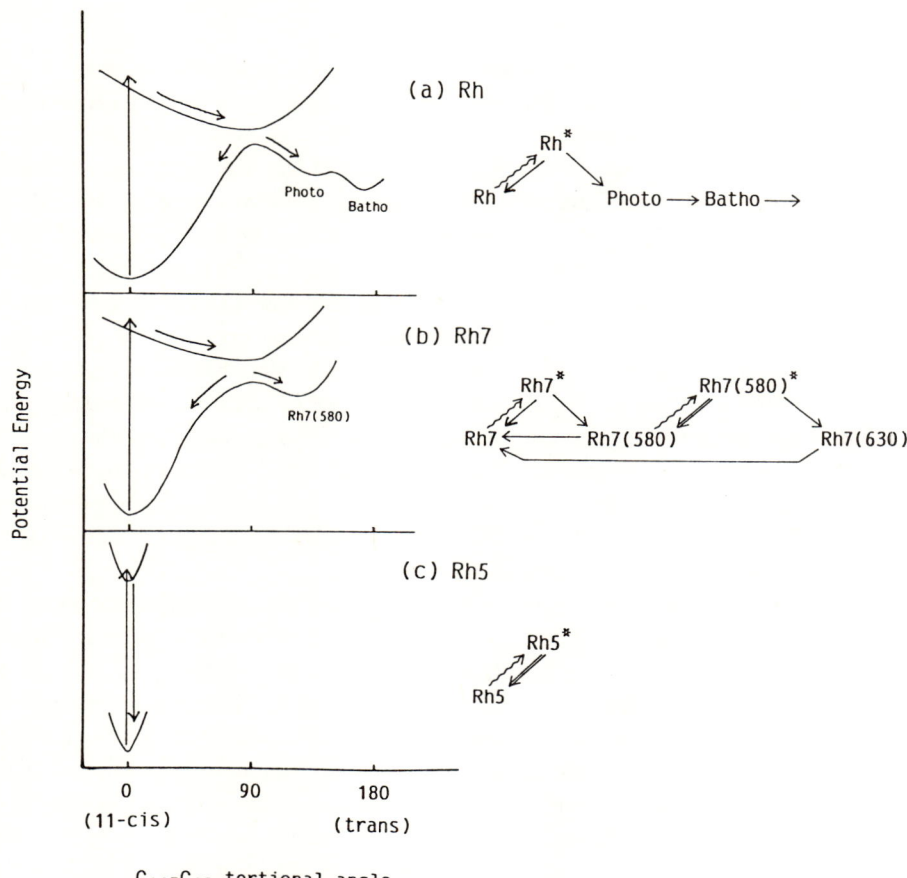

Figure 5. Left: Schematic diagrams of potential surfaces of the ground and excited state along the chromophore C_{11}-C_{12} tortional coordinate of Rh(a), Rh7(b) and Rh(5(c). Right: Primary reaction schema of Rh, Rh7 and Rh5. Solid, wavy and open arrows indicate the thermal, absorption and fluorescence processes, respectively.

carbons (Bonneau et al., 1976, 1979; Corey et al., 1965; Eaton et al., 1965; Noyori et al., 1974; Liu, 1967; Branton et al., 1965). Although the 7–membered ring of the chromophore of Rh7 is different from these compounds, above results suggest that a highly twisted *trans*–configuration at 11–ene of the 7–membered ring would be produced by the excitation.

Therefore, it is reasonable to assume that the chromophore of Rh7(580) is in a *trans*–like configuration which does not interact strongly with amino-acid residues near the 7–membered ring (Fig. 5).

528

Conclusion

Excitation of rhodopsin with picosecond laser pulse causes a formation of photorhodopsin, which decays thermally to bathorhodopsin with a time constant of about 45psec (Shichida et al., 1984; Kandori et al., 1989a). Under similar excitation conditions, no fluorescence was recorded upon excitation, indicating that photorhodopsin is a ground state species. There are remarkable resemblances between photorhodopsin and Rh7(580) in different absorption maximum (585nm and 580nm) and life time (45psec and 20–40psec). Since both the chromophores are compelled to rotate around the 11–ene in the excited states, these similarities between them may indicate that the chromophore of photorhodopsin is in a twisted *trans*–form. Probably, the distortion of the chromophore results in the small oscillator strength of photorhodopsin as shown in the first section.

Acknowledgements

This work was supported in part by Grants-in Aids for Specially Promoted Research to T.Y. (63065002), for Scientific Research on Priority Areas to Y.S. (62621004) and for Encouragement of Young Scientists to H.K. (63790474) from the Japanese Ministry of Education, Culture and Science.

References

Akita, H., Tanis, S.P., Adam, M., Balogh–Nair, V., and Nakanishi, K., 1980. Nonbleachable rhodopsins retaining the full natural chromophore. *J. Am. Chem. Soc.*, 102, 6370–6372.

Bonneau, R., Joussot-Dubien, J., Salem, L., and Yarwood, A.J., 1976. A *trans*-cyclohexene. *J.Am. Chem. Soc.*, 98, 4329–4330.

Branton, G.R., Frey, H.M., Montague, D.C. and Stevens, I.D.R. 1966. Thermal unimolecular isomerization of cyclobutenes. *Trans. Faraday Soc. (Eng.)*, 62, 659–663.

Buchert, J., Stefancic, V., Doukas, A.G., Alfano, R.R., Callender, R.H., Pande, J., Akita, H., Balogh-Nair, V., and Nakanishi, K., 1983. Picosecond kinetic absorption and fluorescence studies of bovine rhodopsin with a fixed 11–ene. *Biophys. J.*, 43, 279–283.

Busch, G.E., Applebury, M.L., Lamola, A.A., and Rentzepis, P.M., 1972. Formation and decay of prelumirhodopsin at room temperatures. *Proc. Natl. Acad. Sci. USA.* 69, 2802–2806.

Corey, E.J., Tada, M. Lamahieu, R., and Libit, L., 1965. *trans*–2–cycloheptenone. *J. Am. Chem. Soc.*, 87, 2051–2052.

Dauben, W.G., Van Riel, H.C.H.A., Hauw, C., Leroy, F., Joussot–Dubien, J., and Bonneau, R., 1979. Photochemical formation of *trans*–1–phenylcyclohexene. Chemical proof of structure. *J. Am. Chem. Soc.*, 101, 1901–1903.

Eaton, P.E., and Lin, K., 1965. *trans*–2–cycloheptenone. *J. Am. Chem. Soc.*, 87, 2052–2054.

Eyring, G., Curry, B., Broek, A., Lugtenburg, J., and Mathies, R., 1982. Assignment and interpretation of hydrogen out-of-plane vibrations in the resonance Raman spectra of rhodopsin and bathorhodopsin. *Biochemistry.* 21, 384–393.

Fukada, Y., Shichida, Y., Toshizawa, T., Ito, M., Kodama, A., and Tsukida, K., 1984. Studies on structure and function of rhodopsin by use of cyclopentatrienylidene 11–*cis*–locked–rhodopsin. *Biochemistry.*, 23, 5826–5832.

Hurley, J.B., Ebrey, T.G., Honig, B., and Ottolenghi, M., 1977. Temperature and wavelength effects on the photochemistry of rhodopsin, isorhodopsin, bacteriorhodopsin and their products. *Nature.*, 270, 540–542.

Ito, M., Kodama, A., Tsukida, K., Fukada, Y., Shichida, Y., and Yoshizawa, T., 1982. A novel rhodopsin analog possessing the cyclopentatrienylidene structure as the 11–*cis*–locked and the full planar chromophore. *Chem. Pharm. Bull.*, 30, 1913–1916.

Kandori, H., Matuoka, S., Nagai, H., Schichida, Y., and Yoshizawa, T. 1988. Dependency of apparent relative quantum yield of isorhodopsin to rhodopsin on the photon density of picosecond laser pulse. *Photochem. Photobiol.*, 48, 93–98.

Kandori, H., Matuoka, S., Shichida, Y. and Yoshizawa, T. 1989a. Dependency of photon density on primary process of cattle rhodopsin. *Photochem. Photobiol.* 49, 181–184

Kandori, H., Matuoka, S., Shichida, Y., and Yoshizawa, T. 1989b. Absolute absorption spectra of batho- and photo-rhodopsins at room temperature. Picosecond laser photolysis of rhodopsin in polyacrylamide gel., *Biophys. J.* 56, 453–457.

Kandori, H., Matuoka, S., Shichida, Y. Yoshizawa, T., Iso, M., Tsukida, K., Balogh-Nair, V., and Nakanishi, K. 1989c. Mechanism of isomerization of rhodopsin studied by use of 11–*cis*–locked rhodopsin analogs excited with a picosecond laser pulse., *Biochemistry* 28, 6460–6467.

Liu, R.S.H., 1967. Photosensitized isomerization of 1,3–cyclooctadienes and conversion to bicyclo[4.2.0] oct–7–ene. *J. Am. Chem. Soc.*, 89, 112–114.

Liu, R.S.H., and Asato, A.E., 1985. The primary process of vision and the structure of bathorhodopsin: a mechanism for photoisomerization of polyenes. *Proc. Natl. Acad. Sci. USA.* 82, 259–263.

Mao, B., Tsuda, M., Ebrey, T., Akita, H., Balogh–Nair, V., and Nakanishi, K., 1981. Flash photolysis and low temperature photochemistry of bovine rhodopsin with a fixed 11–ene. *Biophys, J.*, 35, 543–546.

Matuoka, S. Shichida, Y., and Yoshizawa, T., 1984. Formation of hypsorhodopsin at room temperature by picosecond green pulse. *Biochim. Biophys. Acta*, 765, 38–42.

Noyori, R., and Kato, M., 1974. Photo-induced polar addition of protic solvents to cycloalkenones. Evidence for the ground-state *trans* isomers as chemically-reactive intermediates. *Bul. Chem. Soc. (Japan)* , 47, 1460–1466.

Shichida, Y., Matuoka, S., and Yoshizawa, T., 1984. Formation of photorhodopsin, a precursor of bathorhodopsin, detected by a picosecond laser photolysis at room temperature. *Photobiochem. Photobiophys.*, 7, 221–228.

Sperling, W., 1973. Conformations of 11–cis retinal. in Biochemistry and Physiology of Visual Pigments (Langer, H., ed.), pp. 19–28, Springer Verlag, Heidelberg.

Yoshizawa, T., 1972. The behaviour of visual pigments at low temperatures. in Handbook of Sensory Physiology (Dartnall, H.J.A., ed.) pp. 146–179, Springer Verlag, Heidelberg.

Yoshizawa, T. and Kitô, Y., 1958. Studies on rhodopsin illuminated at low temperatures. *Ann. Rep. Sci. Works Fac. Sci. Osaka Univ.*, 56, 27–41.

Yoshizawa, T., and Wald, G., 1963. Pre-lumirhodopsin and the bleaching of visual pigments. *Nature.*, 197, 1279–1286.

The Primary Photochemical Process in Bacteriorhodopsin

W. ZINTH* AND D. OESTERHELT[+]

*Physik Department
Technische Universität München
D-8000 München
+Max-Planck-Institut für Biochemie
D-8033 Martinsried, FRG

Introduction

In the halobacterial branch of archaebacteria a special kind of retinal-based photosynthesis is found. Two light-driven ion pumps, bacteriorhodopsin (BR) as a proton pump and halorhodopsion (HR) as a chloride pump occur in the cell membrane and mediate phototrophic growth of halobacterial cells (for review see Lanyi et al., 1984).

Similar basic principles of ion transport are effective in BR and HR: In the active state the proteins contain retinal in the all-*trans* configuration linked via a protonated Schiff's base to a lysine residue of the amino acid sequence. After the absorption of a photon the chromophore retinal changes its configuration and acquired the 13-*cis* form (Nuss et al. 1985; Polland et al., 1984a, 1984b, 1986; Dobler et al., 1988; Zinth et al., 1988; Mathies et al., 1988). This isomerization induces the ion transport which must proceed via polar side groups of the protein structure (Oesterhelt and Tittor, 1989). After about 10 ms ion transport is finished and the chromoprotein has returned to its initial state. In this paper we focus on the very early steps of the photosynthetic reaction in bacteriorhodopsin prior to and during the isomerization of the retinal chromophore. We describe absorption changes seen in spectroscopic experiments with femtosecond time-resolution and discuss the related molecular processes.

Materials and Methods

Bacteriorhodopsin (purple membrane) was prepared according to the procedure published by Oesterhelt and Stoeckenius (1974). The samples were kept in the light-adapted form by appropriate background illumination. Time-resolved excite and probe experiments were performed using 80 fs pulses from a colliding pulse mode-locked (CPM) laser-amplifier system operated at a repetition rate of 7 kHz. Part of the output (10%) of the laser-amplifier system served as the exciting pulse. The excitation wavelength was λ = 620 nm. The residual of the laser output produced the probe pulses via femtosecond continuum generation. The change of transmission of the sample

Photobiology, Edited by E. Riklis
Plenum Press, New York, 1991

induced by the exciting pulses was measured with high precision as a function of time delay. The time resolution of the experiment depended on the width $\Delta t = 90$ fs were obtained from the system permitting the investigation of dynamic processes faster than 50 fs.

The choice of the probing wavelength is of major importance for the interpretation of the observed absorption transients. At short probing wavelengths, in the region of the 0-0 transition of the molecule and below, the absorption changes may be related to different processes, e.g. to cross relaxation of an inhomogeneous ground-state distribution, to excited-state processes, and to the formation of photoproducts. Working at longer wavelengths, i.e. in the fluorescent region of the molecule, the ground-state processes may be neglected, i.e. a more straight-forward interpretation of the experiment is possible (Dobler et al., 1988; Zinth et al., 1988).

Results

Time-resolved changes of absorption observed on light-adapted bacteriorhodopsin samples at room temperature are shown in Fig. 1 (circles) for three probing wavelengths in the gain region of BR. Only the very rapid processes occurring within 1.5 ps after excitation are shown. At long probing wavelengths ($\lambda = 850$ nm, Fig. 1a) a pronounced gain is found, i.e. the transmitted pulse is more intense than the incident pulse. The gain decays at later times with a time-constant of approximately 500 fs. A more careful inspection of the data points shows that a faster process ($\tau = 180$ fs, see below) also contributes to the absorption change. With decreasing probing wavelengths the 500 fs contribution diminishes and at 735 nm (Fig. 1b) the faster (180 fs) process dominates the decay of the gain. Fig. 1c shows the absorption changes at a still shorter probing wavelength of 660 nm, where, on the other hand, the S_0-S_1 absorption of BR may still be neglected. An induced absorption due to the intermediate J is built up with 500 fs. Around time zero a very short-lived ($\tau < 100$ fs) gain is found.

Discussion

The experimental data *per se* suggest a qualitative view of the primary reactions. To obtain a more quantitative picture of the ultrafast molecular processes the observed absorption changes have to be compared with the predictions from a simplifying mathematical model (Polland et al., 1984b). We assume that the reaction proceeds via several intermediate levels /i > which are characterized by their absorption spectra. The occupation of the product levels decays exponentially with decay times τ_i. While this description is well justified on the time scale of picoseconds, one should keep in mind that this model may fail when coherent motions with large amplitude along special normal coordinates take place.

A detailed analysis of the experimental data together with a precise determination of the experimental response function now reveals an interesting rapid sequence of events: Three intermediate levels appear during the first picosecond. Their decay times are $\tau_1 = 50$

532

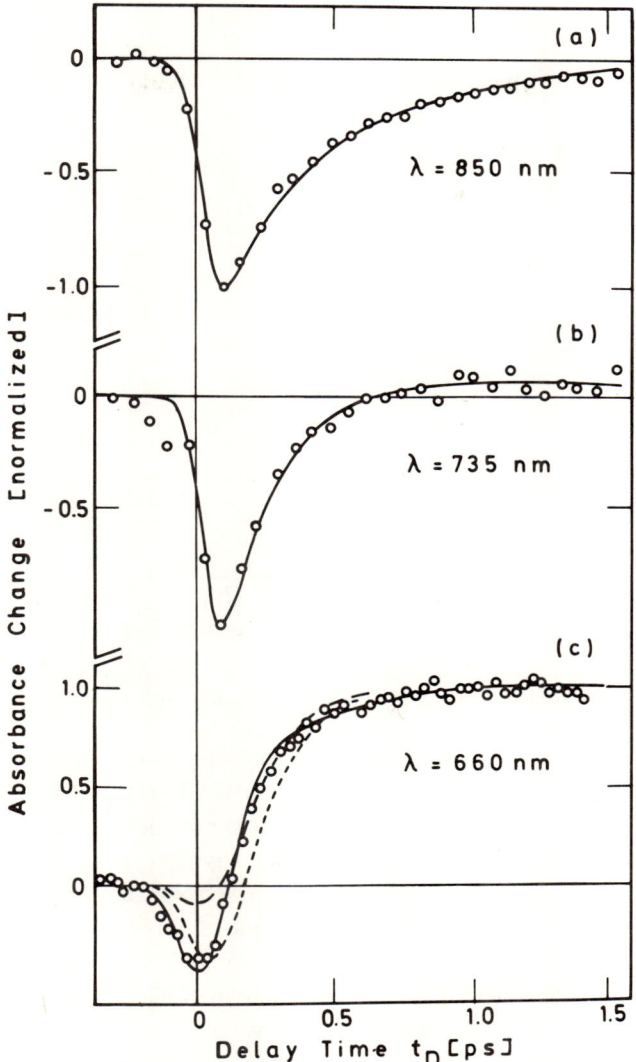

Figure 1. Time-dependent changes of absorption (negative values correspond to gain) induced by exciting femtosecond pulses at $\lambda = 850$ nm (a), $\lambda = 735$ nm (b), $\lambda = 660$ nm (c). The solid curves are calculated using the decay kinetics discussed in the text. The broken curves are calculated for two different sets of amplitude parameters excluding the 50 fs kinetics.

fs $\pm$ 30 fs, $\tau_2 = 180$ fs $\pm$ 70 fs and $\tau_3 = 500$ fs $\pm$ 100 fs. The gain related to all three levels proves that they exist in the electronically excited (S_1) state of BR.

Taking into account the spectral properties of the transient signal and the known molecular data of retinal the following microscopic picture of the very early reactions is suggested (see Fig. 2). The incident photons promote the retinal to the Franck-Condon

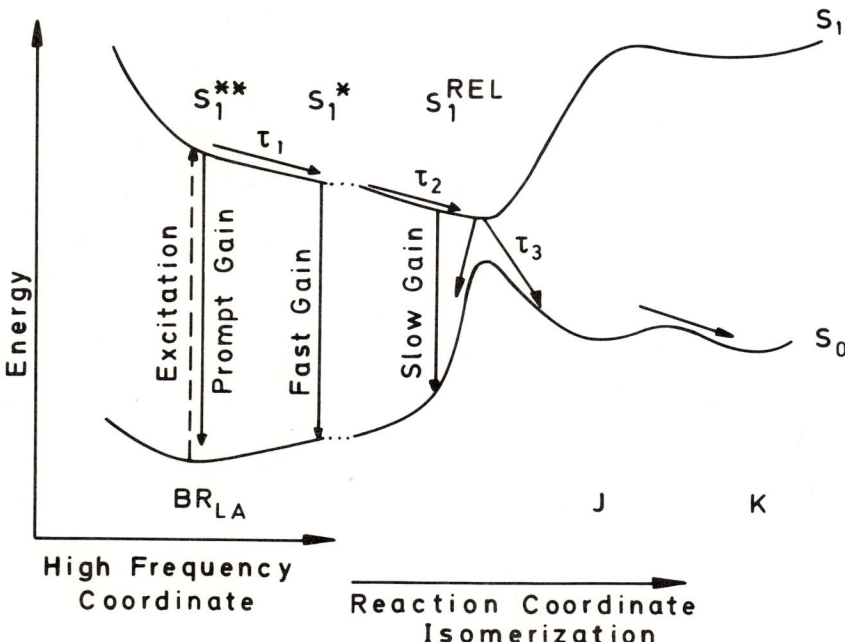

Figure 2. Scheme of the ground-state (S_0) and excited-state (S_1) potential energy surfaces as a function of the high-frequency vibrational and low-frequency reactive coordinates.

state S_1^{**} on the S_1 potential energy surface. Here a number of vibrational modes are displaced relative to the S_1 equilibrium position (Heller, 1981). Within 50 fs after light absorption an equilibration of high-frequency vibrational modes to the state S_1^* occurs. During this first reaction the molecule does not have the time to move along the coordinates of the low-frequency (reactive) modes. The following slower reactive motion of the retinal is related to the 180 fs gain kinetics. In this process, part of the isomerization (presumably a rotation by 60 to 90 degrees around the C_{13}–C_{14} double bond) takes place and the molecules arrive at the bottom of the S_1 potential surface (level S_1^{Rel}). The system leaves this area via internal conversion with a time constant $\tau_2 = 500$ fs. Two decay pathways are possible: more than 60% of the molecules form the intermediate photoproduct J, while the rest returns to the original ground state of BR. The following reaction in the active branch proceeds from J with a 3 ps time constant leading to the intermediate K which is stable on the picosecond time scale.

Conclusions

Time resolved absorption experiments on bacteriorhodopsin performed on the femtosecond time scale show strong absorption and gain dynamics which are related to the excited-state reaction of the retinal chromophore. Extremely rapid (50 fs) absorption changes reflect the relaxation of high-frequency modes. The slower kinetics are related to the isomerization of the retinal: Isomerization starts in the excited electronic state and is finished directly after the 500 fs internal conversion process during the formation of the first ground state photoproduct J.

Acknowledgements

The authors acknowledge valuable contributions by W. Kaiser, J. Dobler and K. Dressler.

References

Dobler, J. Zinth, W. Kaiser, W. Oesterhelt, D. *Chem. Phys. Lett.* 144 (1988) 215.
Heller, E.J. *Acc. Chem. Res.* 14 (1981) 368.
Lanyi, J.K. in Bioenergetics, ed. Ernster, L. (Elsevier, Amsterdam) 1984, p. 315.
Mathies, R.A. BritoCruz, C.H. Pollard, W.T. Shank, C.V.*Science* 240 (1988) 777.
Nuss, M.C. Zinth, W. Kaiser, W. Kölling, E. Oesterhelt, D. *Chem. Phys. Lett.* 117, (1985) 1.
Oesterhelt, D. Stoeckenius, W. *Meth. Enzymol.* 31A (1974) 667.
Oesterhelt, D. Tittor, J. TIBS 14 (1989) 57.
Polland, H.J. Franz, M.A. Zinth, W. Kaiser, W. Kölling, E. Oesterhelt, D. *Biochem. Biophys. Acta* 767 (1984) p. 635.
Polland, H.J. Franz, M.A. Zinth, W. Kaiser, W. Kölling, E. Oesterhelt, D. *Biophys. J.* 49 (1986). 651.
Polland, H.J. Zinth, W. Kaiser, W. in Ultrashort Phenomena, vol. 4, eds. Auston, D.H. Eisenthal, K.B. (Springer, Heidelberg, 1984) p. 456.
Zinth, W. *Naturwiss.* 751 (1988) 173.

Functional Domains in Octopus Rhodopsin

MOTOYUKI TSUDA, TOMOKO TSUDA, N.G. ABDULAEV*, AND SIGEKI
MITAKU**

Department of Life Science, Faculty of Science
Himeji Institute of Technology
Himeji 671-22, Japan
*Shemyakin Institute of Bioorganic Chemistry
USSR Academy of Science
117871 Moscow V-437, USSR
**Department of Material Systems Engineering
Faculty of Technology, Tokyo University of Agriculture and Technology
Koganei, Tokyo 184, Japan

Introduction

Photon absorption by rhodopsin triggers a transduction cascade, leading to an electrical signal in the photoreceptors of the retina (Stryer, 1986; Tsuda, 1987). Recent progress in recombinant DNA techniques provide cDNA sequences for a number of visual pigments and other related signaling receptors (Applebury and Hargrave, 1986; Dohlman et al, 1987). These studies demonstrated the gross structural and topological similarities in these receptors. Moreover, it is shown that these proteins form a superfamily of signaling receptors derived from a common ancestor. Recent evidence indicated that signaling systems of this type demonstrated a degree of structural, functional and regulatory homology that had not previously been appreciated. To date, most structural information has been available on visual photoreceptor protein, rhodopsin. Thus rhodopsin has now gained the status of a model signaling receptor of the seven helix type. We determined the nucleotide sequence of the cDNA encoding octopus rhodopsin and on the primary structure of this protein (Ovchinnikov et al., 1988). The amino acid sequence of octopus rhodopsin was compared with those of other visual pigments. It is expected that similarities in structure of the pigments which have related common functions and that differences in function to be reflected in differences in structure.

A total of three different functional domains may be defined in octopus rhodopsin; the signaling domain where the chromophore is located, an interface domain of the cytoplasmic surface where interactions with the regulatory proteins, the G-protein, kinase, and arrestin, take place. Finally an accessory domain on the C-terminal tail which is peculiar to cephalopod rhodopsin.

In this report we discuss the characterization of the various sites and functional domains in octopus rhodopsin.

Photobiology, Edited by E. Riklis
Plenum Press, New York, 1991

Topology of Octopus Rhodopsin

Figure 1 presents the optimal alignment of the deduced amino acid sequences for three invertebrate rhodopsins — octopus (Ovchinnikov et al., 1988), *Drosophila* R1 (O'Tousa et al., 1985; Zucker et al., 1985) and *Drosophila* R2 (Cowman et al., 1986), and five vertebrate visual pigments — color pigments for red, green and blue vision in human (Nathans et al., 1986) and rhodopsin in human (Nathans et al., 1984) and rhodopsin in bovine (Ovchinnikov et al., 1982). Amino acids which are identical or homologous among invertebrates and vertebrates rhodopsins were boxed and shaded, respectively. Homologies were set as G=A=S=T=P, D=N=E=Q, F=Y=W, K=R=H, and C. Using this alignment, structure and function of the visual pigment will be discussed.

The amino acid sequence is considered to have all the structural information to determine the higher order structure of the membrane protein. In order to study the secondary structure of the octopus rhodopsin, the periodicity of amino acid sequence, particulary of its hydrophobicity was studied. The amino acid sequences of octopus rhodopsin were first converted to the sequences of hydropathy index devised by Kyte and Doolittle (1982) as shown in first raw of Fig. 2. Then, the power spectral density of the sequence was calculated by the maximum entropy method (Mitaku et al., 1984). Second and third raw in Fig. 2 showed the maximum spectral density of octopus rhodopsin in the period range between 2.8 and 4.5 residues and in the range longer than 9 residues as a function of the sequence number. The average hydrophobicity and the power spectral densities of octopus rhodopsin are comparable to those of bacteriorhodopsin (Mitaku et al., 1984). This information, taken together with knowledge of the 3-dimensional folding pattern of bacteriorhodopsin (Henderson, 1977) has allowed us to predict that octopus rhodopsin probably exists as a bundle of seven helices as shown in Fig. 3.

Signaling Domain

Absorption of a photon leads to isomerization of 11-cis retinal to form all-trans retinal. The retinal is attached to the ε-amine group of Lys 306 in octopus opsin (Lys 296 in bovine opsin), by way of a protonated Schiff's base, in the transmembrane helix VII. In order to stabilized a charge in the protonated Schiff base in hydrophobic environment, a counterion is expected to locate near this linkage. Two candidates for this counterion were proposed (Honig, 1987). One is an aspartic acid residue in helix II comparable to octopus Asp 81. Every pigment is found to contain this aspartic acid residue except the human blue pigment. If the Schiff base of blue pigment is also protonated like other visual pigments, this aspartic acid residue is less likely candidate for a counterion. In helix III, a Glu-Arg or Asp-Arg pair is found in every pigment and may likewise provide a needed site of polarity in an otherwise nopolar environment. Hydropathicity analysis of octopus rhodopsin suggested that the Asp 133-Arg 134 pair is probably located in cytoplasmic loop between helix III and IV. Since the site of the retinal Schiff base is located approximately midway in the transmembrane helix VII, this ion pair seems unlikely to be a source for counterion.

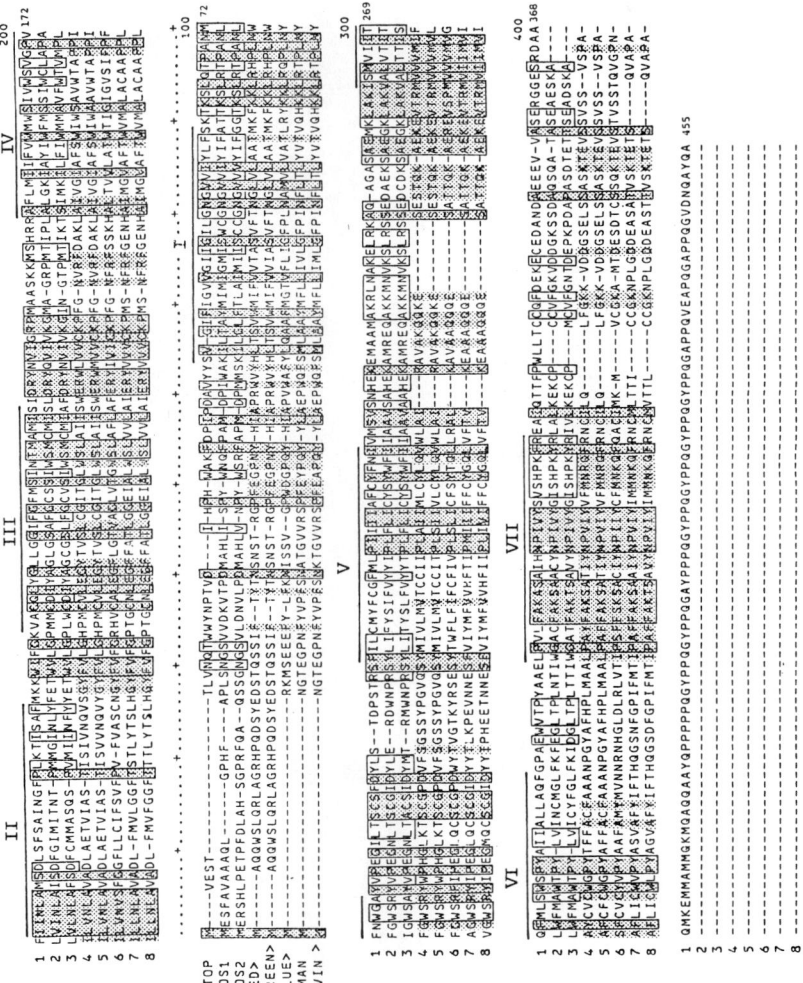

Figure 1. Alignment of amino acid sequences of eight representative visual pigments. OCTOP=octopus rhodopsin; DROS1 = *Drosophila* 1-6 pigment; DROS2 = *Drosophila* 8 pigment; RED, GREEN, BLUE = human red, green, blue pigments; HUMAN = human rhodopsin; BOVIN = bovine rhodopsin. See text for sources of amino acid sequences. The amino acids homology over all pigments were shaded and boxed whereas those among invertebrates and vertebrates were boxed and shaded, respectively.

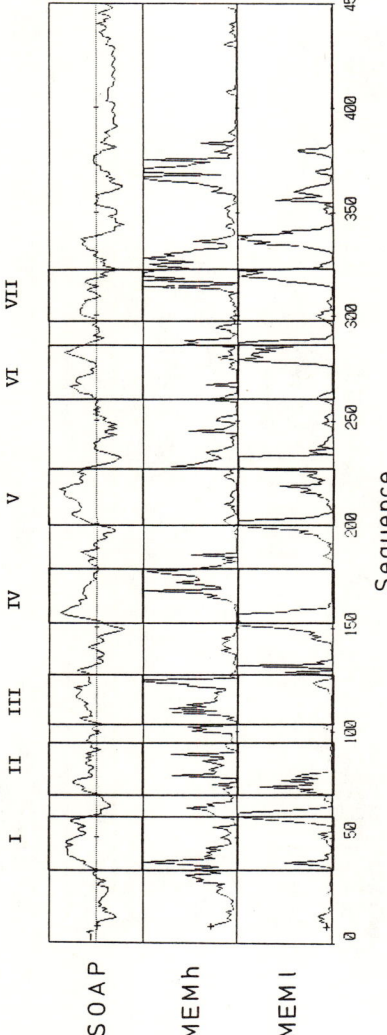

Figure 2. The average hydrophobicity (SOAP) as well as the power spectral densities of helix period [MEMh] and long period [MEN1] are plotted as a function of the sequence number of octopus rhodopsin.

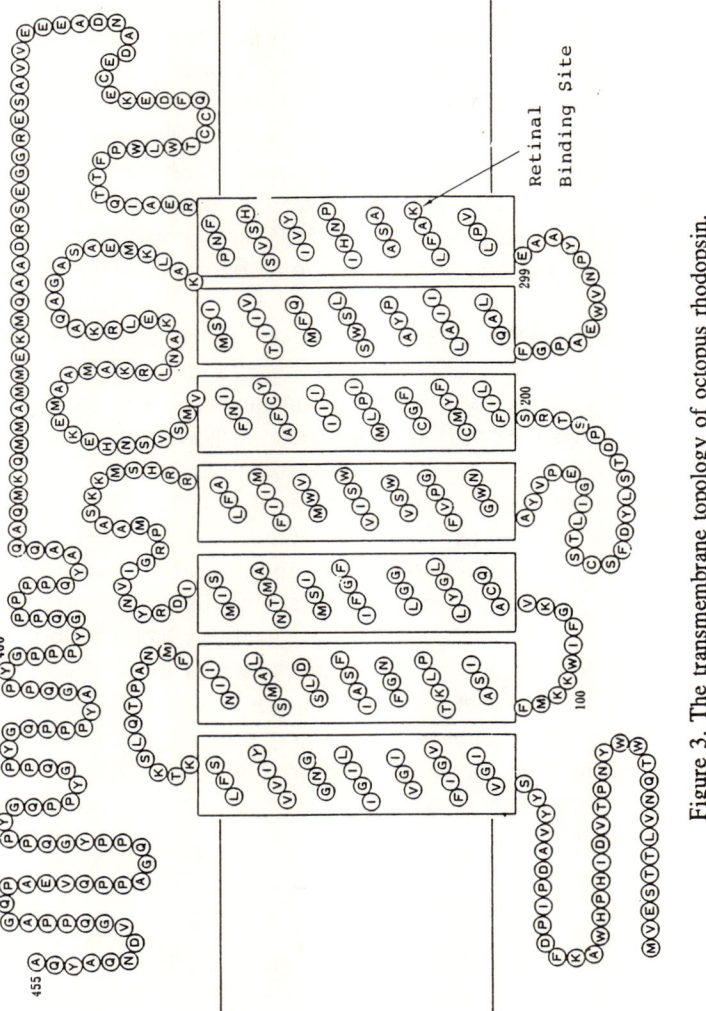

Figure 3. The transmembrane topology of octopus rhodopsin.

Recently FTIR difference spectroscopy and site-directed mutagenesis have suggested that Tyr 185 in helix VI in bacteriorhodopsin is good candidate for the counterion to the protonated Schiff base (Braimen et al., 1988). This tyrosine residue corresponds to Tyr 278 of octopus rhodopsin and strongly conserved in all visual pigments. The infrared difference spectra between bovine rhodopsin and its bathoproduct shows a line for barhorhodopsin at ca 1520 cm^{-1} (Bagley et al., 1989). This line correspond to the 1518 cm^{-1} line of K intermediate of bacteriorhodopsin, which has been assigned to a normal mode of protonated tyrosine. For octopus rhodopsin we do not resolve a bathorhodopsin line ca 1520 cm^{-1} in bathorhodopsin/rhodopsin infrared difference spectrum. However, given the lower frequency of the octopus bathorhodopsin ethylenic line (1532 cm^{-1}) compared to that for the bovine bathorhododopsin band (1536 cm^{-1}), we can not rule out the possibility that the ca 1520 cm^{-1} line is merely hidden under the more intense octopus bathorhodopsin/rhodopsin spectra suggests that a counterion to the protonated Schiff base of visual pigments could be tyrosine as has been proposed for bacteriorhodopsin.

Wavelength of chromophore absorption must be regulated by protein moeity around the retinal. The external point charge model has been used to explain the wavelength regulation on pigments (Honig et al., 1979). In this model, besides the counterion of the Schiff's base, second negative charge is used to regulate wavelength. As shown in Fig..3, the only candidate for a second negatively charged amino acid in octopus rhodopsin is Asp 81. In the bovine rhodopsin/bathorhodopsin infrared difference spectra, lines are detected for rhodopsin at 1770 cm^{-1} and for bathorhodopsin at 1773 cm^{-1} which are attributed to changes associated with -COOH group of aspartic or glutamic acid residues of the apoprotein (Bagley et al., 1989). In contrast, no discernible lines between 1700 cm^{-1} and 1800 cm^{-1} could be observed for the octopus bathorhodopsin/rhodopsin, bathorhodopsin/isorhodopsin, and isorhodopsin/rhodopsin infrared difference spectra (Bagley et al., 1989). Thus octopus Asp 81 may not be closed to the chromophore. If this is the case, there are no additional negatively charged amino acids which could contribute to the regulation of chromophore absorption. This may well be accomplished by an appropriate disposition of a large number of aromatic amino acids in the transmembrane segments.

Interface Domain

Photoisomerization of retinal induce conformational change in transmembrane α-helixes, leading to the rearrangement of polypeptide loops in the cytoplasmic surface. Activated cytoplasmic loops form interface domain which initiates transduction cascade. G-protein, a signal coupling protein, associates with this domain. The interface domain is also the site of phosphorylation by opsin kinase, which then induces the subsequent binding of a protein known as arrestin or S-antigen. This domain controls triggering and the blocking of the light activated cascade. In the invertebrate photoreceptors, a G-protein

(Gip) which is involved in the phototransduction cascade was characterized (Tsuda et. al., 1986; Tsuda, 1987). Octopus rhodopsin was shown to trigger the mamalian amplification cascade (Ebrey et al., 1980). This implies that a binding site for the G-protein should be conserved in both invertebrate and vertebrate opsins.

Based on similarities in amino acid sequence, the most conserved cytoplasmic region of the visual pigment is the loop connecting helixes I and II. Loop III-IV is also conserved in signal receptors. This homology extends into the upper part of the connecting transmembrane segment of helix IV. Especially triplet of Asp-Arg-Tyr 133-135 is completely conserved in 28 different G-protein-coupled signaling receptors examined (T. Miyata, private communication). We must consider that this conversion of structure must demonstrate the existence of some important function for this region.

The entire region of loop V-VI is highly homologous in vertebrate visual pigments. However, the loop V-VI in octopus rhodopsin, which is homologous to Drosophila, is the least conserved of the loop structure due to insertion of 13 amino acids, consistent with a potential role for this domain as a hinge or a region of conformational change. Recently, we demonstrated that trypsin digestion of octopus rhodopsin in the dark released only the C-terminal tail, but illuminated rhodopsin, metarhodopsin, was further cleaved between Arg 248-Lys 249 on loop V-VI (Tsuda, 1988). These results suggest that this region on loop V-VI is important for transferring a conformational shift induced by photoisomerization of retinal in the transmembrane helix, to the interface domain at the cytoplasmic surface.

Most interest and speculation has focused on which loop in the cytoplasmic surface is involved in the coupling of the receptor to G-protein. It was shown that cleavage between Ser-240 and Ala-241 (Kuhn and Hargrave, 1981) and a single amino acid substitution (Lys 248 - Leu) (Franke et al., 1988) in the loop V-VI of bovine rhodopsin prevent activation of G-protein. On the other hand, the binding of the antibody raised against a 14-amino-acid peptide corresponding to a sequence within loop V-VI of rhodopsin was shown to be uneffected by the presence of G-protein (Weiss et al., 1988). Moreover, the binding of antipeptide antibodies corresponding to loop III and IV and the C-terminal was significantly reduced. Thus it is still not clear which loop is important for binding G-protein.

Accessory Domain

A characteristic feature of the octopus rhodospin polypeptide chain is the unexpectedly long C-terminal tail adjacent to helix VII. This C-terminal domain of 68 amino acids contains eleven copies of a pentapeptide repeat. These repeats can be alignment for maximum homology as shown in Fig. 4. Five of the eleven repeats are completely conserved as the consensus sequence of Y-P-P-Q-G. Although these proline-rich repeats have also been found in other proteins (for example, collagen and cytokeratin), none of these contain repeats punctuated by tyrosines. The high proline and glycine content

Octopus Rhodopsin

Residue Number	Sequence
387–395	Y Q P P P P P Q G
396–400	Y P P Q G
401–406	Y P P Q G A
407–412	Y P P P Q G
413–417	Y P P Q G
418–422	Y P P Q G
423–427	Y P P Q G
428–432	Y P P Q G
433–437	A P P Q V
438–442	E A P Q G
443–447	A P P Q G
Consensus Sequence	Y P P Q G

Synexin

Residue Number	Sequence
2–7	F P P P G Q
8–13	Y P Y P S Q
14–21	F P M P G G G A
22–29	Y P P A P S S G
30–35	Y P G A G G
36–41	Y P A P G G
42–47	Y P A P G G
48–53	Y P G A P Q
54–55	Y P
Consensus Sequence	Y P P P P G

Gliadin

Residue Number	Sequence
57–62	F P P Q Q P
63–68	Y P Q P Q P
69–74	F P S Q L P
75–80	Y L Q L Q P
81–87	F P Q P Q L P
88–93	Y S Q P Q P
94–99	F R P Q G P
100–105	Y P Q P G P
Consensus Sequence	Y P P P Q P

Synaptophysin

Residue Number	Sequence
245–249	Y G P A G
250–256	Y G P G P G G
257–262	Y G P Q P S
263–268	Y G P Q G G
269–272	Y Q P A
273–277	Y G Q P A
278–282	Y S G G G G
283–288	Y G P Q G D
289–293	Y G Q Q G
294–298	Y G Q Q G
Consensus Sequence	Y G P Q G

Figure 4. Repeated sequences in the cytoplasmic tail of octopus rhodopsin, synexin, synaptophysin and gliadin. The sequence repeats are aligned for maximum homology with the residue numbers given on the left.

suggests that they form a rigid structure which is assumed to form a helix. Though these proline-rich repeats were found in cephalopod (octopus and squid) rhodopsin, no other visual pigments including *Drosophila* have such an unusual C-terminal tail.

However, it is interesting to note that proline-rich repeats was also found in the C-terminal tail of functionally distinct proteins. They are synaptophysin in synaptic vesicles, synexin in chromaffin granules and gliadin of the major wheat seed storage proteins. As shown in Fig. 4 proline-rich repeats of these proteins were aligned for maximum homology. Function of this proline-rich repeats in the protein is not known. It was suggested that the C-terminal of synaptophysin serves as a binding site for cellular function (Sudhoff et al., 1987). Moreover, the C-terminal of octopus rhodopsin contains other two interesting regions. A phosphorylated region composed of several threonine and serine residues were found in this C-terminal. A potential calcium binding site is found from Asp 341-Glu 533 where 8 negatively charged amino acids are located just like the calcium binding site of parvalbumin.

References

Applebury, M. and Hargrave, P. (1986), Molecular biology of the visual pigments. *Vision Res.* 26, 1881–1895.

Bagley, K.A., Eisenstein, L. Ebrey, T.G. and Tsuda, M. (1989). A comparative study of the infrared difference spectra for octopus and bovine rhodopsin and their bathorhodopsin iso-rhodopsin photointermediates. *Biochemistry,* in press.

Braiman, M.S, Mogi, T. Stern, L.J. Hackett, N.R. Kohrana, H.G. and Rothschild, K.J. (1988). Vibrational spectroscopy of bacterio-rhodopsin mutants: 1. Tyrosine-185 protonates and deprotonates during the photocycle. Protein: Structure, *Function and Genetics* 3, 219–229.

Cowman, A.F. Zucker S. C. and Rubin, G.M. (1986). An opsin gene expressed in only one photoreceptor cell type of the Drosophila eye. *Cell* 44, 705–710.

Dohlman, H. Caron, M. and Lefkowitz R. (1987), *Biochemistry*, 26, 2657–2662.

Ebrey, T.G. Tsuda, M. Sassenrath, G. West, J.L. and Waddell, W.H. (1980). Light activation of bovine rod phosphodiesterase by non-phyiological visual pigments. *FEBS Lett.* 116, 217–219.

Franke, R.R. Sakmar, T.P. Obrian, D.D. and Kohrana, H.G. (1988). A single amino acid substitution in rhodopsin (lysine 248 - leucine) prevents activation of transducin. *J.Biol. Chem.* 263, 2119–2122.

Honig, B. Dinur, U. Nakanishi, K. Balogh-Nair, V. Gawinowics, M. Armaboldi, M. and Motto, M. (1979). An external point-charge model for wave-length regulation in visual pigments. *J. Am. Chem. Soc.* 101, 7084–7986.

Honig, N. (1987). External point charges and amino acid sequence in retinal proteins. In Biophysical Studies of Retinal Proteins. (Eds. Ebrey et al.) p.212–218, Univ. of Illinois Press.

Henderson, R. (1977). The purple membrane from *Halobacterium halobium*. *Ann Rev. Biophys. Bioenerg.* 6, 87–109.

Kyte, J. and Doolittle, R.F. (1982). A simple method for displaying the hydropathic character of a protein. *J. Mol. Biol.* 157, 105–132.

Kuhn, H. and Hargrave, P.A. (1980). Light-induced binding of guanosinetriphosphate to bovine photoreceptor membranes; effect of limited proteolysis of the membranes. *Biochemistry* 20, 2410–2417.

Mitaku, S. Hoshi, S. Abe, T. and Kataoka, R. (1984). Spectral analysis of amino acid sequence. I. Intrinsic membrane proteins. *J. Phys. Soc. Jpn.* 53, 4083–4090.

Nathan, J. and Hogness, D.S. (1984). Isolation and nucleotide sequence of the gene encoding human rhodopsin. *Proc. Natl. Acad. Sci. U.S.A.* 81, 4851–4855.

Nathans, J. Thomas, D. and Hogness, D.S. (1986).Molecular genetics of human color vision; The genes encoding blue, green and red pigments. *Science* 232, 203–232.

O'Tousa, J.E. Baehr, W. Martin, R.L. Hirsh, I. Pak, W.L. and Applebury M.L. (1985). The Drosophila nina E gene encodes an opsin. *Cell* 40, 839–850.

Ovchinnikov, Y.A. Abdulaev, N.G. Feigine, M.Y. Artamonov, I.D. Zolotarev, A.S. Moroshinikov, A.I. Martynow, V.I. Kostina, M.B. Kudelin, A.G. and Bogachuk, A.S. (1982). The complete amino acid sequence of visual rhodopsin. *Bioorg. Khim.* 8, 1424–1427.

Ovchinnikov, Y.A. Abdulaev, N.G. Zolotarev, A.S. Artamonov, I.D. Bespalov, I.A. Dergachev, A.E. and Tsuda, M. (1988). Octopus rhodopsin amino acid sequence deduced from cDNA. *FEBS Lett.* 232, 69–72.

Stryer, L. (1986). Cyclic GMP cascade of vision. *Ann. Rev. Neurosci.* 9, 87–119.

Sudhof, T.G. Lottspeich, F. Greengard, P. Mehi, E. and Jahn, R. (1987). A syaptic vesicle protein with a novel cytoplasmic domain and four transmembrane regions. *Science* 238, 1142–1144.

Tsuda, M. Tsuda, T. Terayama, Y. Fukada, Y. Akino, T. Yamanaka, G. Stryer, L. Katada, T. Ui, M. and Ebrey, T.G. (1986). Kinship of Cephalopod photoreceptor G-protein with vertebrate transducin. *FEBS Lett.* 198, 5–10.

Tsuda, M. (1987). Photoreception and photo-transduction in invertebrate photoreceptors. *Photochem. Photobiol.* 45, 915–931.

Tsuda, M. (1987). Octopus G-protein - a signal coupling protein in invertebrate. In Retinal Proteins (Edt. Ovchinnikov) pp.393–404, *VNU Science Press*.

Tsuda, M. (1988). Signal Coupling proteins in Octopus Photoreceptors. In Molecular Physiology of Retinal Proteins. (edt. Hara) pp.167–172, *Yamada Science Foundation*

Weiss, E.R. Kellehr, D.J. and Johnson, G.L. (1988). Mapping sites of interaction between rhodopsin and transducin using rhodopsin antipeptide antibodies. *J. Biol. Chem.* 263, 6150–6154.

Zucker, C.S. Cowman, A.F. and Rubin, G.M. (1985). Isolation and structure of a rhodopsin gene from D. melanogaster. *Cell.* 40, 851–858.

Picosecond Time-resolved Spectroscopy of the Initial Events in the Bacteriorhodopsin Photocycle

GEORGE H. ATKINSON

Department of Chemistry and Optical Sciences Center
University of Arizona
Tucson, Arizona 85721

Abstract

The initial 100 ps interval of the room temperature bacteriorhodopsin photocycle is examined in terms of the molecular properties of the retinal chromophore in the K-590 intermediate. Picosecond transient absorption, picosecond time-resolved fluorescence, and picosecond time-resolved resonance Raman spectroscopy are used to characterize the electronic states and the vibrational degrees of freedom of retinal.

Introduction

The initial events in the bacteriorhodopsin (BR) photocycle have attracted considerable attention in recent years for a variety of reasons. The purple membrane of BR has been shown to sustain effective *trans*-membrane proton pumping, a process that is directly related to the synthetic functioning of the halobacterium, Oesterheldt, et al. (1973); Stoeckenius (1980). The visible chromophore contained within the purple membrane, retinal, is itself interesting since it has the same polyene structure that operates in the visual system of rhodopsin to mediate related biochemical functions, Birge (1981). In the case of BR, however, the photo-initiated changes in retinal proceed by means of a photocycle which returns on the millisecond time scale to the original all-*trans* structure. The existence of this BR photocycle makes the system particularly convenient to study experimentally. By contrast, the rhodopsin system functions only with an enzymatic step which recombines the dissociated retinal chromophore with the opsin protein, Birge (1981). The BR photocycle, therefore, not only commands interest because of its own properties, but also often stands as an experimentally accessible model to aid in the elucidation of molecular mechanisms in rhodopsin. The molecular mechanism by which retinal stores the relatively large amount of energy (~15 kcal/mole in BR, Birge and Cooper (1983)) needed to drive ATP synthesis in these two biophysical systems merits special attention since it is based in large part on changes in the retinal configuration and conformation.

The initial molecular events of the BR photocycle occur within the first 100 ps after optical excitation and consequently, studies designed to examine these processes must measure spectroscopic data with picosecond (i.e., <10 ps) time resolution while maintaining the *in vivo* functional behavior of the BR photocycle. Experimentally, these conditions suggest that low power, pump-probe instrumentation be used at repetition rates sufficiently high to obtain good signal to noise ratios. Results will be presented here that are obtained with two synchronously pumped dye lasers operating at 1 MHz repetition rates and with 5–7 ps pulsewidths. Individual laser pulses have energies of about 0.5–15 nj giving average powers of approximately 10–15 mW for pumping and 0.5–3 mW for probing. This instrumental approach has been successfully used to measure picosecond transient absorption (PTA), picosecond time-resolved fluorescence (PTRF), and picosecond time-resolved resonance Raman (PTR[3]) data, Atkinson, et al. (1989a), from flowing BR samples under *in vivo* conditions which maintain the membrane's capacity for proton pumping.

By recording all three types of data from the same sample, the spectroscopic changes occurring in the BR photocycle can be quantitatively correlated. Such information significantly aids in elucidating the overall molecular mechanism which functions during the initial 100 ps of the BR photocycle. Specifically, the results presented here describe (i) emission from the K-590 intermediate, (ii) the presence and kinetic role of vibrationally-hot K-590 (i.e., K′), and (iii) the vibrational resonance Raman (RR) spectrum of room temperature K-590.

Experimental

The instrumentation used in these measurements, shown schematically in Figure 1, has been described in detail elsewhere, Atkinson et al. (1989a) and Atkinson et al. (1989b). Briefly, the second harmonic frequency of a Nd:YAG laser is used to synchronously pump two dye lasers operating with different dye solutions. Since the dye lasers are mode-locked and cavity dumped independently, each can be optimized for either the pumping or probing process. This versatility in selecting wavelengths, pulsewidths, and pulse energies is a major experimental factor in obtaining the detailed characterization of the photocycle presented here.

The pump and probe beams are focused collinearly into a flowing jet sample stream (~400 μm diameter). In PTA measurements, the beams exiting the sample are separated by a prism before the probe laser beam intensity is detected by a photodiode operated with a lock-in amplifier. A mechanical chopper is introduced into either the pump or probe laser beam prior to its arrival at the sample in order to facilitate phase-sensitive detection of absorption changes. In PTRF measurements, the fluorescence signal is collected at 90° to the plane formed by the two laser beams and the sample jet before being focused onto the slit of a one meter monochromator. The wavelength-resolved signal is detected by a cooled photomultiplier which is operated with a lock-in amplifier. The probe laser beam is mechanically chopped before reaching the sample to permit phase-sensitive measurements of fluorescence intensities. Both spectral (i.e., scanning the

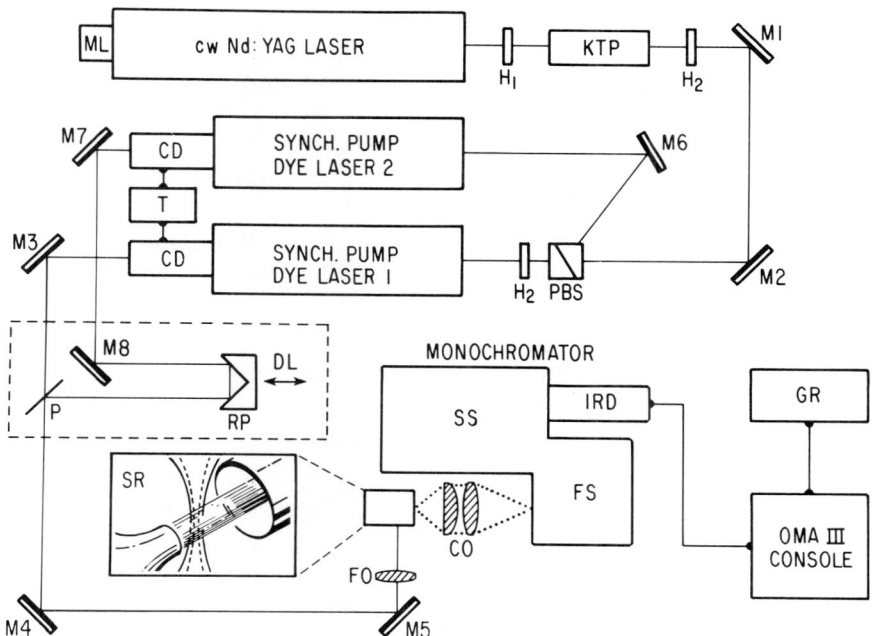

Figure 1. Picosecond time-resolved resonance Raman instrumentation: H_1 and H_2: half-wave plate; M_1 - M_{10}: dielectric mirrors; PBS: polarizing beam splitter; CD: cavity dumper; T: timing synchronization; DL: delay line; RP: retroprism; P: prism for combining beams; FO: focusing optics; CO: collection optics; SR: sample region; FS/SS: triple monochromator; IRD: intensified reticon detector; GR: graphics.

monochromator to obtain a spectrum at a fixed time delay) and time-dependent (i.e., scanning the optical delay line at a fixed spectral region of observation) data are recorded. In the PTR[3] experiments, the spontaneous resonance Raman (RR) scattering signal is collected as in the PTRF experiments and then focused onto the entrance slit of a triple monochromator. The RR signal is detected by an intensified diode array as part of an optical multichannel analyzer system.

Results

The K-590 intermediate in the BR photocycle has long been considered to be of major importance in the energy storage mechanism. First detected by PTA measurements, it has been associated with a large bathochromic shift in absorption relative to the starting species, BR-570. The results described here will focus on the spectroscopic characterization of the K-590 intermediate.

A) *Emission*

Numerous studies have attempted to detect emission from intermediates in the room temperature BR photocycle with differing results, Lewis, et al. (1982); Govindjee and Ebrey (1986); Lewis, et al. (1976); Polland, et al. (1986a); Kouyama, et al. (1985); Gillbro, et al. (1977); Sineshchekov, et al. (1977); Govindjee, et al. (1978); Kriebel, et al. (1979); Hurley and Ebrey (1978). A very small amount (quantum yield ~10^{-4}) of fluorescence has been assigned to BR-570, Lewis, et al. (1976), but there has not been a clear association of fluorescence with any of the intermediates and in fact, it was generally accepted that none existed, Lewis et al. (1982). Given the high degree to which the initial intermediates (J-625 and K-590) are photolytically-coupled with BR-570 at room temperature, the absence of detected emission may be attributable to the design of earlier experiments.

Recently, Atkinson, et al. (1989a), the fluorescence spectrum of K-590 shown in Figure 2 was reported using the PTRF techniques described here. It exhibits a 17 nm shift to the blue in its maximum relative to that of the fluorescence spectrum of BR-570. The BR-570 fluorescence spectrum also was measured here using a variety of excitation conditions designed to quantitatively minimize the presence of any intermediates, Atkinson, et al. (1989a). When viewed together with their respective absorption spectra, these fluorescence data indicate that K-590 has a substantially smaller (141 nm versus 178 nm) Stokes shift in absorption than that of BR-570, Atkinson, et al. (1989a);

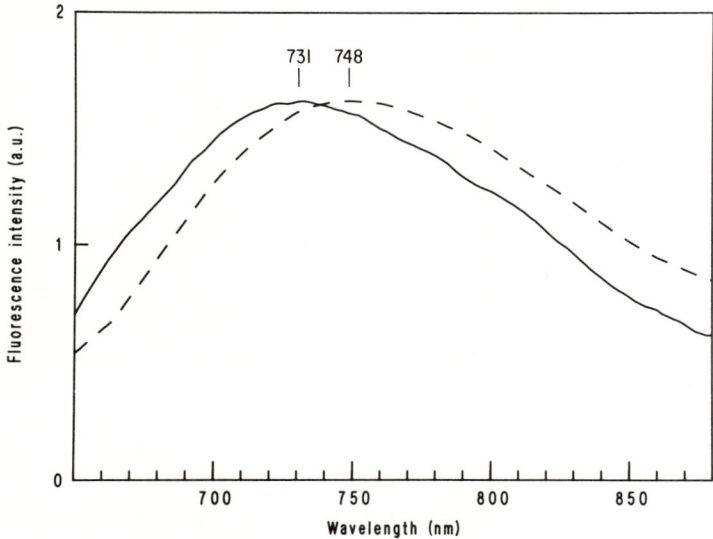

Figure 2. Fluorescence spectrum of K-590 intermediate in the BR photocycle at room temperature and 590 nm probe excitation. Fluorescence spectrum of BR-570 recorded with 590 nm excitation shown as dashed line. The maxima of the two spectra of normalized to one another for comparison purposes. Taken from Atkinson, et al. (1989a).

550

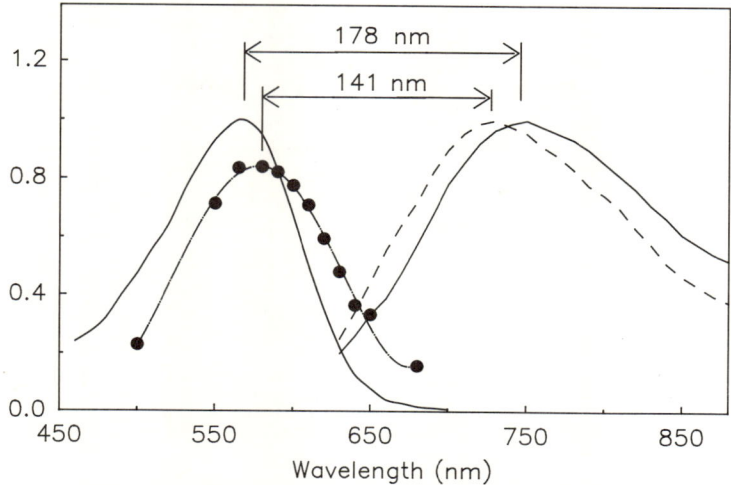

Figure 3. Absorption spectra of BR-570 (—) and K-590 (--●--) and fluorescence spectra of BR-570 (—) and K-590 (- - - -). Stokes shifts for each (141 nm for K-590 and 178 nm for BR-570) are shown.

Atkinson, et al. (1990), thus suggesting that K-590 undergoes a smaller change in geometry upon electronic-state excitation. The quantum yield of emission from K-590 also differs, being approximately twice that of BR-570, Atkinson, et al. (1989a). Both observations characterize the potential energy surfaces of K-590 as significantly different than the BR-570 species from which it is formed during the initial 10 ps of the photocycle.

The fluorescence from K-590 also can be measured as a time-dependent signal relative to the pulsed (5–7 ps) excitation of BR-570. PTA data can be measured simultaneously on the same sample providing an opportunity to quantitatively compare the absorption and fluorescence changes. Data of this type are presented in Figure 4 for the initial 100 ps of the photocycle. The probe laser wavelength of 590 nm lies very near the isobestic point of the absorption spectra of BR-570 and K-590. Data in Figure 5 show analogous results for a probe wavelengths of 630 nm, Atkinson, et al. (submitted); Blanchard (1989). The PTA data have been used to reveal the presence of the J-625 and K-590 intermediates in several studies, Atkinson, et al. (submitted); Blanchard (1989); Kaufman, et al. (1976); Applebury, et al. (1978); Ippen, et al. (1978); Dinur, et al. (1981); Shichida, et al. (1981); Gillbro, et al. (1983); Nuss, et al. (1985); Sharkov, et al. (1985); Polland, et al. (1986b), but the PTRF data have only recently been analyzed, Atkinson, et al. (1989a). One of the most important results is derived from the different

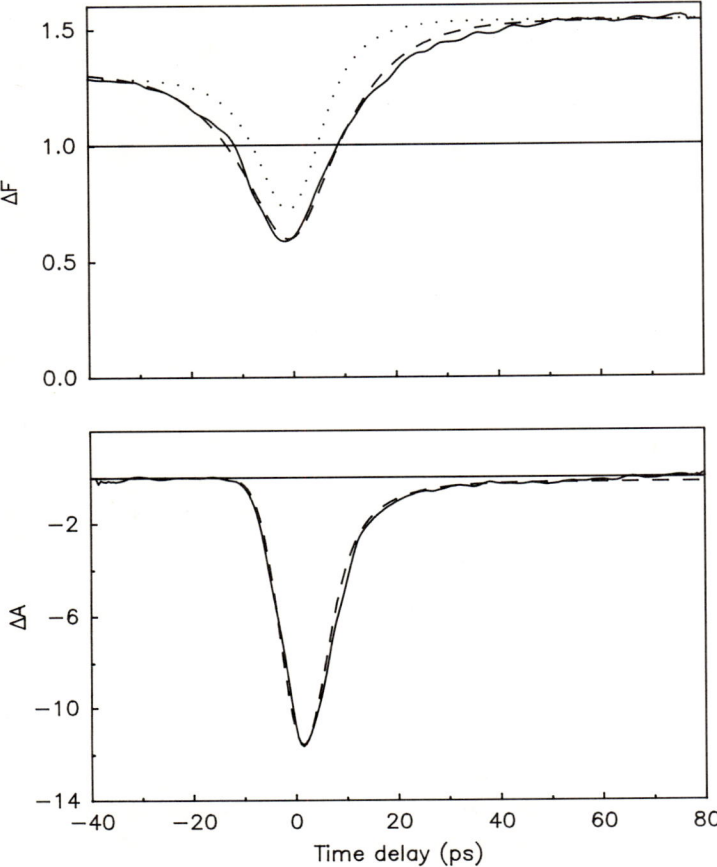

Figure 4. PTRF (top) and PTA (bottom) data recorded for 565 nm pump and 590 probe wavelengths. The dotted line represents the simulated fit to the PTRF data assuming no emission for J-625 and without the formation of K¢. The dashed line represents the simulated fit to the PTRF data assuming no emissions from J-625 and with the formation of K' with a 3.5 ps rate. From Atkinson, et al. (1990).

rates with which the PTA and PTRF signal increase over the 0–40 ps time interval (Figures 4 and 5). This difference is most evident from the 590 nm probe laser data (Figure 4) where the PTRF signal rises more slowly than that of PTA. When a mechanistic model based on the linear reaction BR-570 –> J-625 –> K-590 is used to construct a set of rate equations for simulating time-dependent populations and from these to calculate the PTA and PTRF signals, excellent fits can be obtained for all the PTA data recorded for probe laser wavelengths between 565 nm and 650 nm. Examples of these simulations are presented in Figures 4 and 5. By contrast, these same parameters yield only partially accurate fits for the PTRF data obtained over the same wavelength range and from the same BR samples. In these latter calculations, the quantum yields of

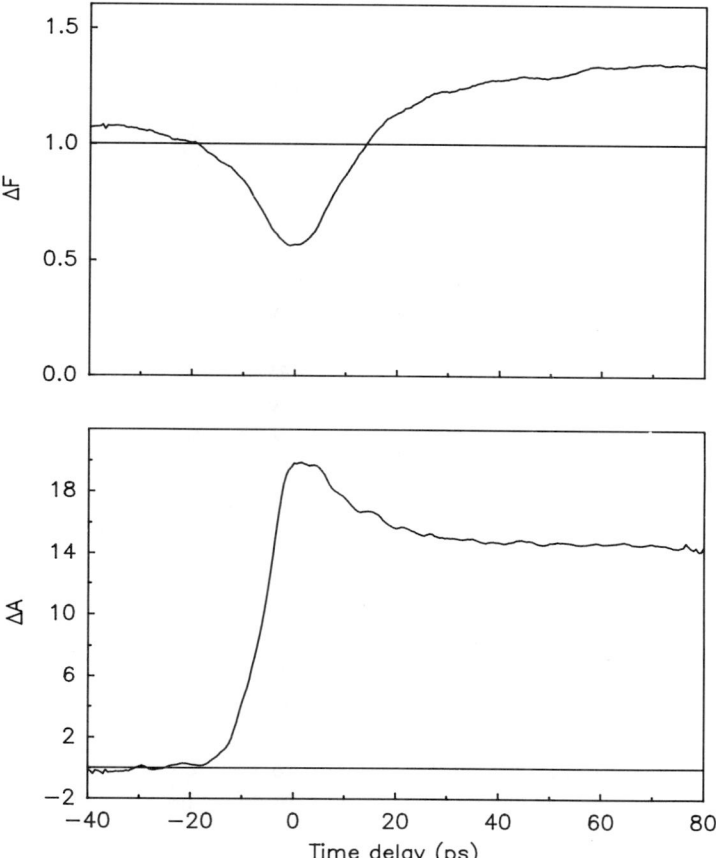

Figure 5. PTRF (top) and PTA (bottom) data recorded for 565 nm pump and 630 nm probe wavelengths. From Atkinson, et al. (1990).

fluorescence for J-625 and K-590 must be used as variables in order to test the parameterization of the model. Many features in the PTRF results are well duplicated, but the slow risetime in the 0–40 ps region cannot be fit even for widely different fluorescence yield values. An example of the simulation for the 590 nm data is shown in Figure 4. These results have led to the proposal that the linear reaction scheme be modified to: BR-570 –> J-625 –> K' –> K-590 where K' represents vibrationally-excited K-590.

The suggestion that K' is populated directly from J-625 and prior to the formation of ground-state K-590 follows reasonable expectations when the potential energy surface describing the photocycle is considered. It is evident that K' acts as an excited-state intermediate in the photocycle, but does not change the earlier kinetic scheme which involves the decay of J-625 with a rate of 3.5 ps, Atkinson, et al. (submitted); Blanchard

(1989). The vibrational relaxation of K′ to K-590 needs to occur over an interval of about 6–8 ps in order to maintain a good simulation fit to the PTRF data. The effect of including K′ can be seen in the simulation results shown in Figures 4 and 5. The inclusion of K′ also is consistent with PTA data recorded for two different pulsewidths (2.3 ps and 6–8 ps), Atkinson, et al. (1990); Blanchard (1989).

Resonance Raman Scattering

The value of absorption and fluorescence spectroscopy in characterizing electronic states, and especially the kinetics associated with their time-dependent population, is well recognized. In a molecular system such as the BR photocycle which relies strongly on rapid changes in configuration and conformation to control reactivity, however, absorption and fluorescence data are not sufficient to fully characterized photochemical mechanisms. Since the vibrational degrees of freedom are extremely sensitive to configurational and conformational properties, vibrational spectra of photocycle intermediates can be more useful in elucidating the retinal mechanism. Spontaneous RR scattering is recorded here using the same basic laser instrumentation utilized for PTA and PTRF measurements. The PTR[3] data presented here are recorded together with PTA measurements on the same BR sample in order to correlate the vibrational spectra with the changes occurring in absorption and fluorescence.

The RR spectrum of K-590 at room temperature has been sought in several studies using primarily single laser techniques, Hsieh, et al. (1981); Hsieh, et al. (1983); Stern and Mathies (1985); Terner, et al. (1979); Smith, et al. (1983). The one laser is used to both photolytically initiate the photocycle and to generate RR scattering from the resultant reaction mixture. Since neither the excitation nor the RR detection process can

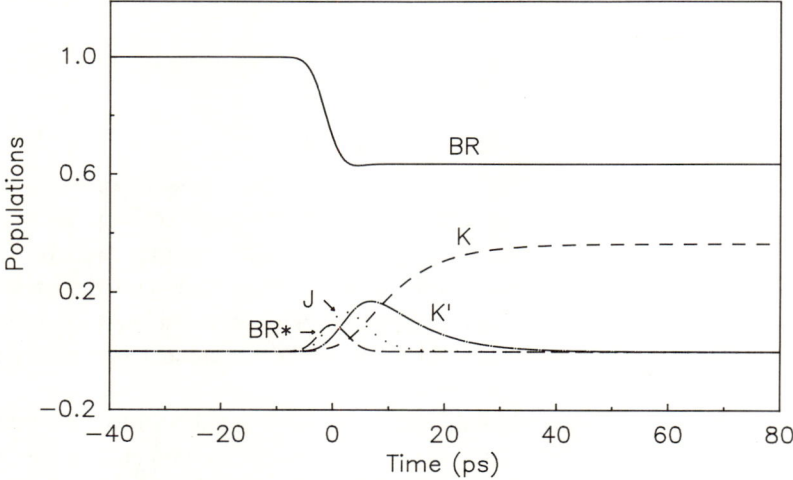

Figure 6. Time-dependent populations of BR photocycle species including electronically-excited BR-570 (BR*) and vibrationally-excited K-590 (K′). Simulations derived from a rate equation model described in Atkinson, et al. (1990).

be independently optimized using this procedure, it is difficult to separate the RR spectrum of any one species from another in the mixture. The two laser, PTR[3] technique used here addresses this limitation by permitting a high-energy, pump laser pulse to be tuned to 565 nm near the maximum of the BR-570 absorption spectrum and a low-energy (i.e., minimally perturbing) probe laser pulse to be tuned into resonance with an absorption band of the intermediate. In the case of K-590, the probe laser wavelength operates in the 570–650 nm region and at time delays of 40 ps to 1 ns. For delays longer than 40 ps, the concentration of J-625 has reached zero and the only intermediate detected by absorption to be present is K-590. Furthermore, the K-590 concentration remains unchanged over the 40 ps to 1 ns interval according to PTA data. Thus, by measuring PTR[3] data over this time interval, only one species should be detected, namely K-590.

The PTR[3] spectrum recorded at any specific time delay contains contributions from all the species in that particular, time-dependent reaction mixture. To obtain the K-590 spectrum from PTR[3] data measured at a 100 ps delay, therefore, the RR spectrum of the

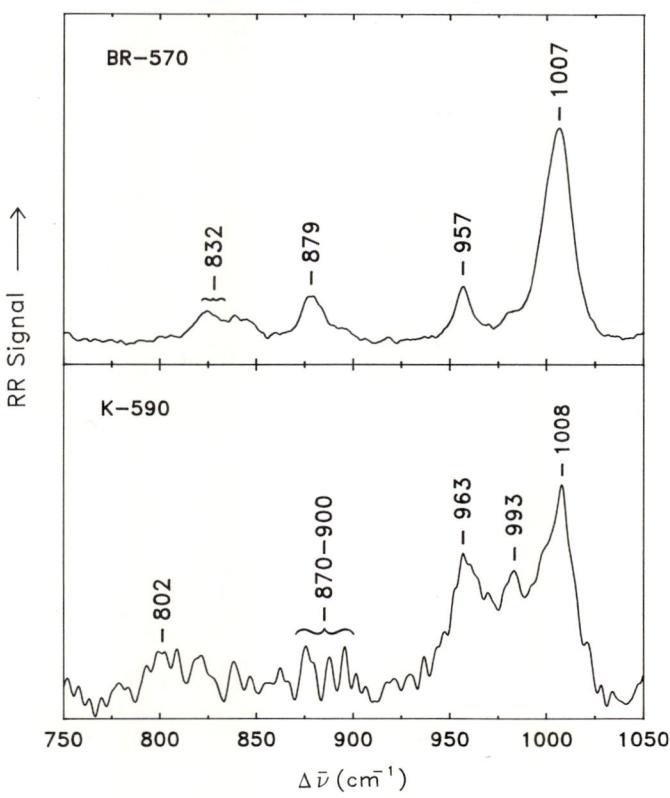

Figure 7. Resonance Raman spectra in the 750–1050 cm^{-1} region of BR-570 (top) and K-590 (bottom) recorded with 590 nm laser excitation. Taken from Brack and Atkinson (1989).

other species present (i.e., BR-570) must be subtracted. The RR spectrum of BR-570 itself has been obtained using extremely low laser power excitation, but information concerning the percentage of the reaction mixture comprised of BR-570 versus K-590 is more difficult to obtain. These relative populations of BR-570 and the various photocycle intermediates can be calculated from the rate equation model Atkinson et al. (1990). This model is validated from the time-dependent fits of PTA and PTRF data recorded over the 565 nm to 650 nm range, Atkinson, et al. (1990). From these PTA and PTRF simulations and for the experimental conditions used, BR-570 is found to comprise ≈40% of the reaction mixture at a 100 ps delay with K-590 comprising the remaining ≈60%. The time-dependent concentrations of the entire system is presented in Figure 6.

With this information, the RR spectrum of room temperature K-590 alone can be obtained by subtracting the scaled RR spectrum of BR-570 from the PTR³ data recorded over the 40 ps to 1 ns interval, Brack and Atkinson (1989). Two regions of the K-590

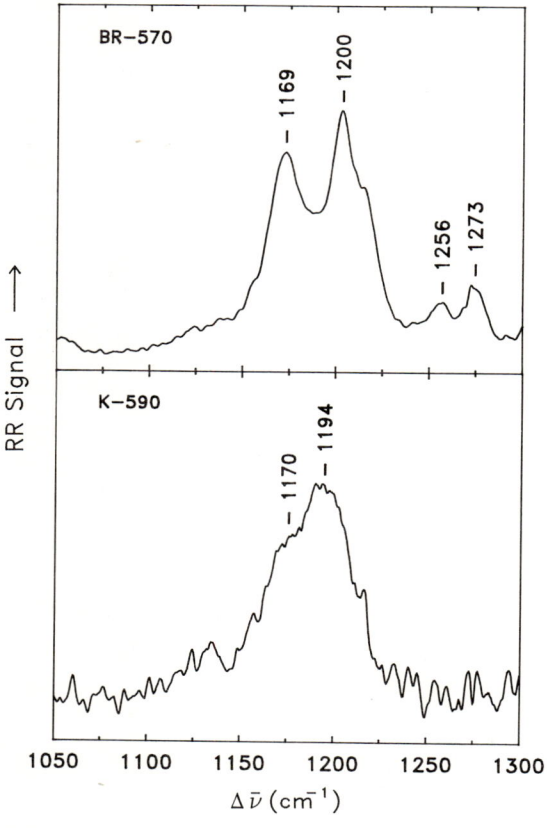

Figure 8. Resonance Raman spectra in the 1050 – 1300 cm⁻¹ region of BR-570 (top) and K-590 (bottom) recorded with 590 nm laser excitation. Taken from Brack and Atkinson (1989).

RR spectrum obtained with this analysis are shown in Figures 7 and 8. It is evident from these spectra that there are large differences between the vibrational degrees of freedom in BR-570 and K-590 at room temperature. Special note should be taken of the changing RR band intensities in the 950–995 cm^{-1} region which has been assigned to the hydrogen-out-of-plane modes and in the 1160–1210 cm^{-1} region which has been assigned to C–C stretching modes sensitive to the configuration (isomer) of the polyene backbone. The first set of changes reflect increased out-of-plane motion in K-590 such as that associated with twisting of the polyene backbone and the second indicates that the all-*trans* isomeric form of BR-570 has been altered. It is likely that K-590 contains a 13-*cis*-like isomer although the complete structural form remains unresolved, Brack and Atkinson (1989).

Concluding remarks

The results described here demonstrate the type and quality of information currently available concerning the molecular structure and electronic state changes that occur on the picosecond time scale in a complex, *in vivo* biomembrane. The vibrational degrees of freedom that can be derived from PTR^3 data are perhaps the most useful since the actual changes in molecular structure can be followed in real time. The measurement and analysis of such structure-sensitive data, however, is only beginning. It also should be noted that these measurements can be more completely analyzed if the corresponding absorption and fluorescence data are available and if the versatility of pump-probe experiments is utilized to recorded time-resolved spectroscopic signals.

Acknowledgements

The author wishes to gratefully acknowledge his coworkers on these studies, Mr. T.L. Brack, Mr. D. Blanchard, Mr. H. Lemaire, Mr. D. Gilmore, Dr. H. Hayashi, and Dr. G. Rumbles. This research was supported by a grant from the National Institutes of Health (GM 36628).

References

Applebury, M. L., Peters, K. S., and Rentzepis, P. M., 1978, Primary intermediates in the photochemical cycle of bacteriorhodopsin, *Biophysical Journal*, 23, 375–382.

Atkinson, G. H., Blanchard, D., Lemaire, H., Brack, T. L., and Hayashi, H., 1989a, Picosecond time-resolved fluorescence spectroscopy of K-590 in the bacteriorhodopsin photocycle, *Biophysical Journal,* 55, 263–274.

Atkinson, G. H., Brack, T. L., Blanchard, D., and Rumbles, G., 1989b, Picosecond time-resolved resonance Raman spectroscopy of the initial trans to cis isomerization in the bacteriorhodopsin photocycle, *Chemical Physics* 131, 1–15.

Atkinson, G. H., Blanchard, D., Gilmore, D., Brack, T., and Lemaire, H., 1990 Picosecond t-resolved absorption and fluorescence in the bacteriorhodopsin photocycle: vibrationally-excited species, *Chemical Physics* (in press).

Birge, R. R., Photophysics of light transduction in rhodopsin and bacteriorhodopsin, 1981, *Annual Review of Biophysics and Bioengineering*, 10, 315–354.

Birge, R. R., and Cooper, T. M., 1983, Energy storage in the primary step of the photocycle of bacteriorhodopsin, *Biophysical Journal*, 42, 61–69.

Blanchard, D., Ph.D. Thesis, University of Grenoble, 1989.

Brack, T. L., and Atkinson, G. H., 1989, Journal of Molecular Structure (in press).

Dinur, U., Honig, B., and Ottolenghi, M., 1981, Analysis of primary photochemical processes in bacteriorhodopsin, *Photochemistry and Photobiology*, 33, 523–527.

Gillbro, T., Kriebel, N. and Wild, V. P., 1977, On the origin of the red emission of light adapted purple membrane of Halobacterium halobium, *FEBS (Federation of European Biochemical Societies) Letters*, 78, 57–60.

Gillbro, T., and Sundstrom, V., 1983, Picosecond kinetics and a model for the primary events of bacteriorhodopsin, *Photochemistry and Photobiology*, 37, 445–455.

Govindjee, R., and Ebrey, T., 1986, Light emission from bacteriorhodopsin and rhodopsin. Light Emission by Plants and Bacteria. (Academic Press, Inc., New York), 401–419.

Govindjee, R., Becker, B., and Ebrey, T. G., 1978, The fluorescence from the chromophore of the purple membrane protein, *Biophysical Journal*, 22, 67–77.

Hsieh, C.-L., Nagumo, M., Nicol, M., and El-Sayed, M. A., 1981, Picosecond and nanosecond resonance Raman studies of bacteriorhodopsin. Do configurational changes of retinal occur in picoseconds?, 85, 2714–2717.

Hsieh, C.-L., El-Sayed, M. A., Nicol, M. Nagumo, M. and Lee, J-H., 1983, Time-resolved resonance Raman spectroscopy of the bacteriorhodopsin photocycle on the picosecond and nanosecond time scales, *Photochemistry and Photobiology*, 38, 83–94.

Hurley, J. B., and Ebrey, T. G., 1978, Energy transfer in the purple membrane of Halobacterium halobium, *Biophysical Journal*, 22, 49–66.

Ippen, E. P., Shank, C. V., Lewis, A., and Marcus, M. A., 1978, Subpicosecond spectroscopy of bacteriorhodopsin, *Science (Washington, D.C.)*, 200, 1279–1281.

Kaufmann, K. J., Rentzepis, P. M., Stoeckenius, W., and Lewis, A., 1976, Primary photochemical processes in bacteriorhodopsin, *Biochemical Biophysical Research Communications*, 68, 1109–1115.

Kouyama, T., Kinosita, K., JR., and Ikegami, A., 1985, Excited state dynamics of bacteriorhodopsin, *Biophysical Journal*, 47, 42–54.

Kriebel, A. N., Gillbro, T., and Wild, V. P., 1979, A low temperature investigation of the intermediates of the photocycle of light-adapted bacteriorhodopsin, *Biochimica et Biophysica Acta*, 546, 106–120.

Lewis, A., Spoonhower, J. P., and Perrault, G. J., 1976, Observation of light emission from a rhodopsin, *Nature (London)*, 260, 675–678.

Lewis, A., and Perreault, G. J., 1982, Emission spectroscopy of rhodopsin and bacteriorhodopsin, *Methods in Enzymology*, 88, 216–229.

Nuss, M. C., Zinth, W., Kaiser, W., Kolling, E., and Oesterhelt, D., 1985, Femtosecond spectroscopy of the first events of the photochemical cycle in bacteriorhodopsin, *Chemical Physics Letters*, 117, 1–7.

Oesterhelt, D., and Stoeckenius, W., 1973, Functions of a new photoreceptor membrane, *Proceedings of the National Academy of Science (U.S.A.)*, 70, 2853–2857.

Polland, H-J., Franz, M. A., Zinth, W., Kaiser, W., Kolling, E., and Oesterhelt, D., 1986a, Early picosecond events in the photocycle of bacteriorhodopsin, *Biophysical Journal*, 49, 651–662.

Polland, H-J., Franz, M. A., Zinth, W., Kolling, E., and Oesterhelt, D., 1986b, Early picosecond events in the photocycle of bacteriorhodopsin, *Biophysical Journal*, 49, 651–662.

Sharkov, A. V., Pakulev, A. V., Chekalin, S. V., and Matveetz, Y. A., 1985, Primary events in bacteriorhodopsin probed by subpicosecond spectroscopy, *Biochimia et Biophyscia Acta*, 808, 94–102.

Shichida, Y., Matouka, S., Hidaka, Y., and Yoshizawa, T., 1983, Absorption spectra of intermediates of bacteriorhodopsin measured by laser photolysis at room temperatures, *Biochimica et Biophysica Acta*, 723, 240–246.

Sineshchekov, V. A., and Litvin, F. F., 1977, Luminescence of bacteriorhodopsin from Halobacterium halobium and its connection with the photochemical conversions of the chromophore, *Biochimica et Biophysica Acta,* 462, 450–466.

Smith, S. O., Braiman, M., and Mathies, R., 1983, Time-resolved resonance Raman spectroscopy of the K610 and O640 photointermediates of bacteriorhodopsin Time-resolved vibrational spectroscopy, ed. G. H. Atkinson (Academic Press, NY), 219–230.

Stern, D., and Mathies, R., 1985, Picosecond and nanosecond resonance Raman evidence for structural relaxation in bacteriorhodopsin's primary photoproduct, Time-resolved Vibrational Spectroscopy, eds. M. Stockburger and A. Laubereau (Springer-Verlag, NY), 250–256.

Stoeckenius, W., 1980, Purple membrane of Halobacteria: A new light-energy converter, *Accounts of Chemical Research,* 13, 337–344.

Terner, J., Hsieh, C.-L., Burns, A. R., and El-Sayed, M. A., 1979, Time-resolved resonance Raman spectroscopy of intermediates of bacteriorhodopsin, *Proceedings of the National Academy of Science, U.S.A.,* 76, 3046–3050.

Primary Processes in Sensory Rhodopsin and Retinochrome

T. KOBAYASHI[a], H. OHTANI[a,e], M. TSUDA[b], K. OGASAWARA[a,c],
S. KOSHIHARA[a], K. ICHIMURA[a], R. HARA[d], AND M. TERAUCHI[a]

[a]Department of Physics, University of Tokyo, Bunkyo
Tokyo 113, Japan
[b]Department of Physics, Sapporo Medical College
Chuo, Sapporo 060, Japan
[c]Hamamatsu Photonics K.K. Research Div.
Hamamatsu, Shizuoka 435, Japan
[d]Department of Biology, Osaka University
Toyonaka, Osaka 560, Japan
[e]Department of Biomolecular Engineering, Tokyo Institute of Technology
Meguro, Tokyo 152, Japan

Photochemistry of two chromoproteins with retinal have been studied. One is a photoreceptor for the phototaxis of *Halobacterium halobium* and the other is a photosensitive pigment contained in cephalopod visual cells. The behaviors of the intermediates in their photocycles were clarified with picosecond and nanosecond time-resolved absorption spectroscopy apparatuses.

The kinetics of the trasient absorption in the time region from nanosecond to second was measured with the experimental apparatus shown in the previous paper (Iwai et al., 1984) with small modification. The second harmonic (532 nm, 5-ns fwhm, 0.5 mJ) of a Q-switched Nd:YAG laser (Quanta-Ray DCR-1A) was used as an excitation light pulse. Figure 1 shows the apparatus used for the measurement of the time-resolved absorption spectrum with a resolution time of 200 ns. Spectra were measured with a combined system of a Xe flash (Sugawara NP-1A, 180-ns), a polychromator, and a multichannel photodiode (Unisoku USP-450), which was interfaced to an on-site microcomputer (NEC PC9801).

For the measurement in the picosecond region, the second harmonic (532 nm, 35-ps fwhm, 0.4 mJ) of a mode-locked Nd:YAG laser (Quantel YG472) was used for the excitation of the sample. The details of the experimental apparatus are described in the previous paper (Iwai et al., 1984).

Photochemistry of Sensory Rhodopsin

Photoreceptors for the phototaxis of *H. halobium* have been found. They are retinoid pigments, sensory rhodopsin sR (third rhodopsin-like pigment, Tsuda et al., 1982; slow

Photobiology, Edited by E. Riklis
Plenum Press, New York, 1991

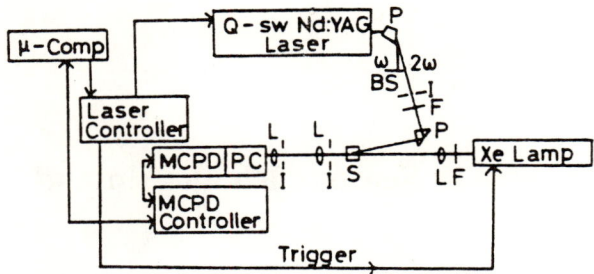

Figure 1. Block diagram of nanosecond time-resolved absorption spectroscopy apparatus. P: prism, ω: 1064-nm, 2ω: 532-nm, PC: polychromator, MCPD: multichannel photodiode, μ-Comp: microcomputer, S: sample cell, I: iris, F: filter, L: lens.

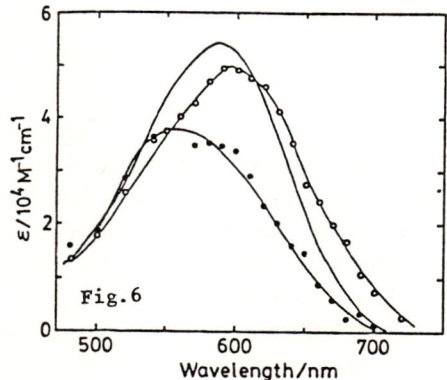

Figure 2. Absorption spectra of sR (solid line), sR_K (open circles), and sR_L (solid circles).

rhodopsin, Bogomolni and Spudich, 1982) and phoborhodopsin (Takahashi et al., 1985; Tomioka et al., 1986). It is known that sR has an absorption spectrum with an absorption maximum at 587 nm (Spudich and Bogomolni, 1984) shown in Fig. 2 and that two intermediates are sequentially formed in the photocycle of sR (Bogomolni and Spudich, 1982). Photochemical behavior of sR resembles to that of the light-adapted bacteriorhodopsin (bR). A red-shifted intermediate sR_K (sR_{680}) is formed in the early stage of the photocycle and it is converted to a blue-shifted one sR_M (sR_{370}). The former and the latter are corresponded to K (or KL) and M intermediates, respectively, in the photocycle of bR. The primary process is considered to be the all-*trans*-13-*cis* photoisomerization of retinal. sR and sR_M are the photoreceptors for the attractant and repellent phototaxis of *H. halobium* (Takahashi et al., 1985), respectively. The

562

characteristic feature of the sR photocycle is that the cycle time is slower than that of bR and that sR_K is directly converted to sR_M (Bogomolni and Spudich, 1982)

In this paper, we describe the photochemical cycle of sR clarified with nanosecond laser spectroscopy. The measured sample was prepared from membrane suspension of a carotenoid free mutant of Flx3 (Spudich and Spudich, 1983).

Intermediates in the photocycle of sR

Nanosecond spectroscopy studies show that a red-shifted intermediate sR_K is formed within 10 ns following the excitation of sR (Ohtani et al., 1986, 1988). Neither intermediate before sR_K nor sR in the excited state has not been detected. sR_K has an absorption spectrum with maximum at 595±5 nm (Ohtani et al., 1988) as shown in Fig. 1. We found that sR_K is not directly converted to sR_M (Bogomolni and Spudich, 1982) but to a newly found intermediate with a slightly blue-shifted spectrum shown in Fig. 2 (Ohtani et al., 1986). We call it sR_L on the analogy of L intermediate in the photocycle of bR. sR_L is converted to sR_M (See Fig.3).

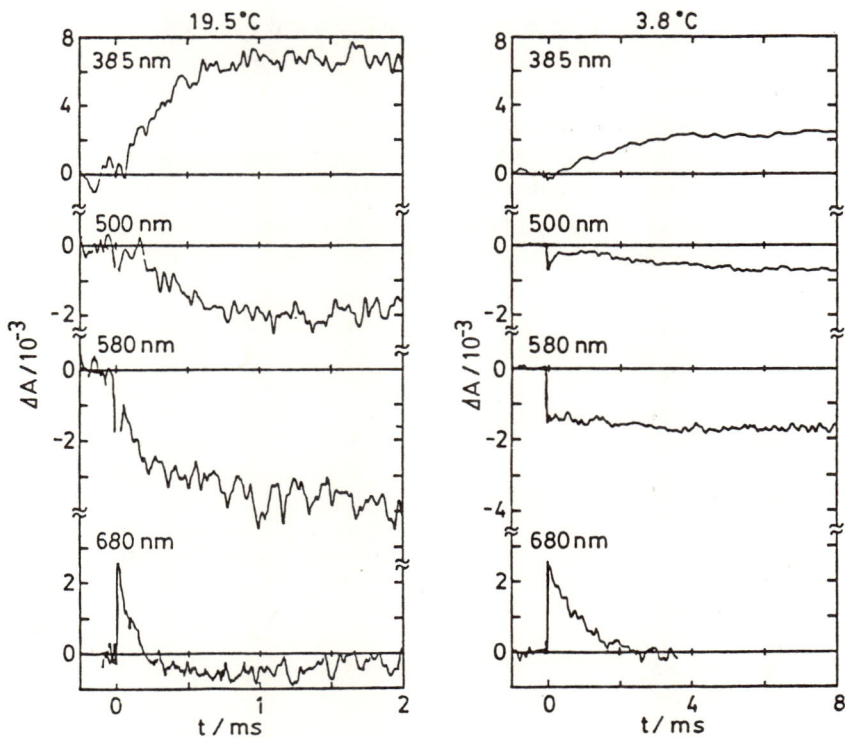

Figure 3. Kinetics of the absorbance change following the excitation of sR with 532-nm light at 19.5 and 3.8°C.

Branching Reaction

The formation yield of sR_M decreases with temperature (Hazemoto et al., 1983). We found (Ohtani et al., 1988) that the formation yields of sR_K and sR_L have no dependence on temperature (3.5–36°C). Therefore there is a branching pathway from sR_L to sR in which sR_M is not formed. The branching process was considered to obtain the absorption spectra of sR_K and sR_L shown in Fig.2.

The following photocycle of sR has been clarified.

$$sR \qquad (^1sR^*) \longrightarrow sR_K \longrightarrow sR_L \longrightarrow sR_M$$

Photochemistry of retinochrome

Retinochrome (Ret) is a photosensitive chromoprotein contained in both inner and outer segments of the cephalopod visual cells. Its properties have been studied by several groups (Hara and Hara, 1965, 1967, 1968, 1976, 1982; Sperling and Hubbard, 1975; Hamdorf, 1979; Hara et al., 1981; Seki, 1984). Ret contributes to the synthesis of squid rhodopsin in the visual cells as shown in Fig. 4. The chromophore of Ret is an all-*trans* retinal with protonated Schiff base as in rhodopsin (Hara and Hara, 1968). The absorption maximum of *Todarodes* Ret is located at 495 nm (pH 6.5) as shown in Fig. 5a and it depends on the pH value (Hara and Hara, 1965, 1968).

On irradiation with green light at room temperature (23±3°C), Ret is converted to a photoproduct, metaretinochrome (M-Ret) with an absorption maximum at 470 nm as shown in Fig.5b. On irradiation with green light (546 nm) at liquid-nitrogen temperature, Ret is converted to another intermediate, lumiretinochrome (L-Ret, λ_{max} = 475 nm). Above –20°C, L-Ret is thermally converted into M-Ret (Hara et al., 1981).

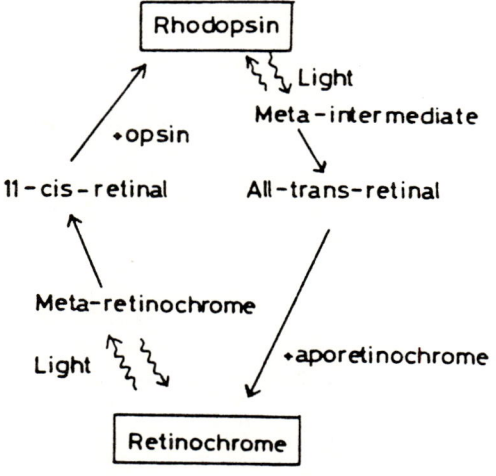

Figure 4. Photoconversions between rhodopsin and retinochrome.

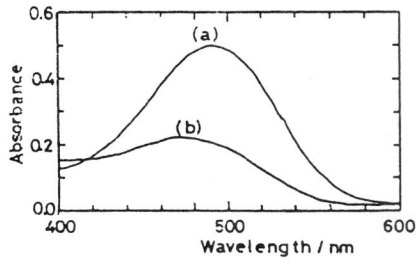

Figure 5. Absorption specbetween rhodopsin and tra of (a) retinochrome retinochrome. and (b) metaretinochrome (23°C, pH 6.5).

In this paper, we describe the photobleaching process of Ret at room temperature (23±3°C). The sample was prepared from retinas of dark-adapted squid, *Todarodes pacificus* according to the method described previously by Hara and Hara (1982).

Events in the time region from nanosecond to second

The time dependences of the absorbance changes following the excitation of Ret by the nanosecond laser at 532 nm were measured. Figure 6 shows the time-resolved difference spectra at $t_d = -1.6$ μs(a), 200 ns (b), 7 ms (c), 190 ms (d), and 1.6 s (e). The spectra at 200 ns (b) and 7 ms (c) are much the same. The difference spectra agree well with that of L-Ret minus Ret at −190°C (Hara et al., 1981).

We found that L-Ret is formed within the resolution time of the apparatus (10 ns) and decays to the next intermediate LM-Ret with a time constant of 80±15 ms at 23°C.

Figure 7 shows the difference absorption spectrum of LM-Ret which was obtained from the calculation using the data of the time-resolved absorption specta and time-dependence of the absorbance change. The differnce absorption spectrum at td = 190 ms (Fig. 6d) belongs to a mixture of L-Ret and M-Ret. The ratio of L-Ret (Fig. 7a) to M-Ret (Fig.7c) was obtained from time-dependent absorbance change and the difference absorption spectrum of LM-Ret (Fig. 7b) was calculated using this ratio.

The difference spectrum 1.6 s after excitation (Fig. 6e) agrees with that of M-Ret minus Ret at 23°C (dashed-and-dotted curve in Fig. 6e), except small discrepancy at longer wavelength than 520 nm. M-Ret is formed from LM-Ret with a time constant of 290±30 ms.

Absorbance change in picosecond region

The time-resolved difference spectra measured at td = −100 ps, 0 ps, 60 ps, and 1 ns with resolution time of 35 ps are shown in Fig. 8. The absorbance at 460 nm decreases at $t_d = 0$ ps and increases with time for the order of 100 ps (See Figs. 8b and 8c) and there is no detectable difference among the absorption spectra at $t_d = 0$ ps (Fig. 8b), 60 ps

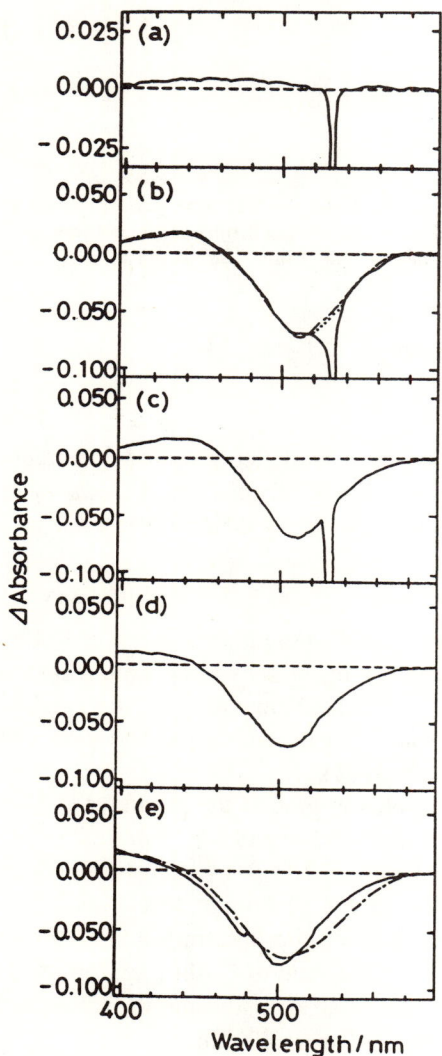

Figure 6. Time-resolved difference spectra at (a) -1.6 μs, (b) 200 ns, (c) 7 ms, (d) 190 ms, and (e) 1.6 s. Dashed-and-dotted lines in (b) and (e) are L-Ret minus L-Ret and M-Ret minus Ret difference spectra at low temperature, respectively.

(Fig. 8c), and 1 ns (Fig. 8d) between 400 nm and 640 nm. The time-resolved spectrum at $t_d = 1$ ns (Fig. 8d) agrees with the difference absorption spectrum of L-Ret minus Ret at liquid nitrogen temperature (dash-and-dotted line in Fig. 6). In the wavelength region longer than 580 nm, no absorbance change was observed in the picosecond region.

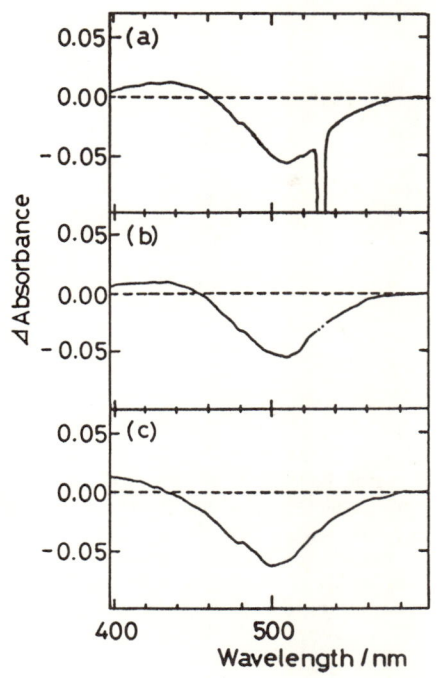

Figure 7. Difference absorption spectra at 23°C. (a) L-Ret minus Ret, (b) LM-Ret minus Ret, and (c) M-Ret minus Ret.

Photobleacing process of retinochrome

The photobleaching process of Ret at room temperature (23°C) described above are summarized as follows.

The difference absorption spectrum just after excitation with 35-ps pulse differs from L-Ret minus Ret spectrum. L-Ret is formed with a time constant of the order of 100 ps. There exists a precursor of L-Ret. The decay time of the order of 100 ps is considered to be too long to be attributed to the excited singlet state of retinochrome since the lifetimes of other retinoid pigments such as rhodopsin and bacteriorhodopsin are of the order of subpicosecond or a few picoseconds (Nuss et al., 1985; Kobayashi, 1980). Therefore we tentatively call the newly-found ground state species pre-lumiretinochrome (hereafter referred to as X-Ret).

X-Ret may not be an intermediate with a redshifted absorption spectrum such as primerhodopsin (Honig et al., 1979; Kobayashi, 1980a,1980b; Matuoka et al., 1984; Peters et al., 1977; Shichida et al., 1984) or bathorhodopsin (Yoshizawa, 1958, 1972) in rhodopsin photocycle or J (Applebury et al.,1978) or K (Polland et al.,1986) intermediate in bacteriorhodopsin photocycle. Therefore X-Ret possibly be correspondent with hypsorhodopsin (Yoshizawa,1958,1972). However further femtosecond experiment at room temperature and steady-state absorption spectrometry at helium temperature are needed for the complete clarification of the very primary processes.

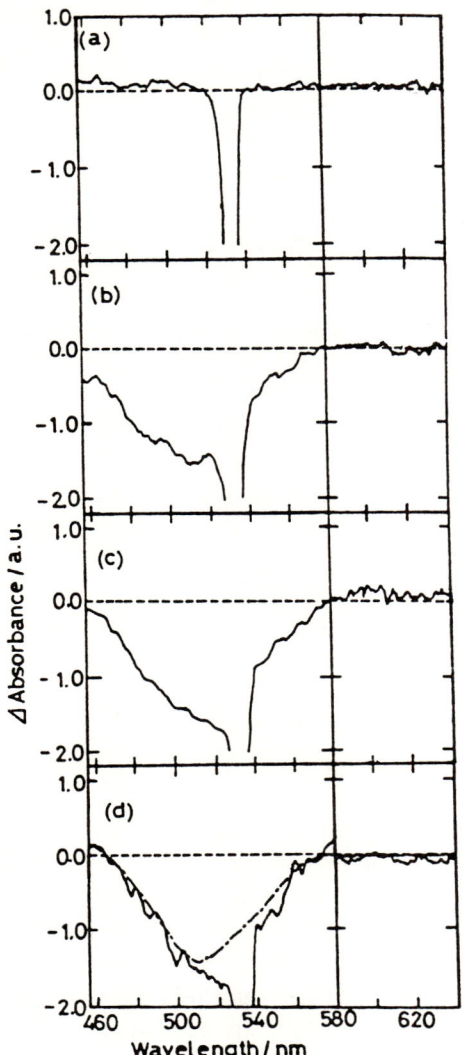

Figure 8. Time-resolved difference spectra at (a) −100 ps, (b) 0 ps, (c) 60 ps, and (d) 1 ns. Two spectra are connected at 580 nm. Dashed-and-dotted line in (d) is L-Retminus Ret difference spectrum at low temperature.

L-Ret is converted into a LM-Ret with $\tau = 80 \pm 15$ ms and M-Ret is then formed from LM-Ret with $\tau = 290 \pm 30$ ms.

The schematic model of the photobleaching process of Ret is shown in Fig. 9. There is a large difference between the photoisomerization process of retinal in rhodopsin (11-*cis* to all-*trans*) and that of Ret (all-*trans* to 11-*cis*).

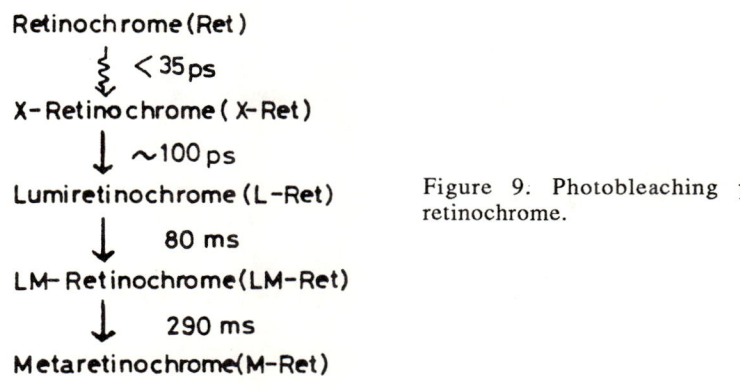

Figure 9. Photobleaching process of retinochrome.

Acknowledgment

This work was partly supported by a Grant-in-Aid for Special Distinguished Research (56222005) from the Ministry of Education, Science and Culture of Japan, and also partly by the Toray Science and Technology Foundation, and the Kurata Science Foundation.

References

Applebury, M.L., K.S. Peters and P.M. Rentzepis (1978) *Biophys. J.* 23, 375–382.

Bogomolni, R.A., and J.L. Spudich (1982) *Proc. Natl. Acad. Sci. USA.* 79, 6250–6254.

Hamdorf, K. (1979) In Handbook of Sensory Physiology (Edited by H. Autrum), Vol VII/6A, pp.145–224, Springer-Verlag, Berlin.

Hara, T., and R. Hara (1965) *Nature* 206, 1331–1334.

Hara, T., and R. Hara (1967) *Nature* 214, 572–573.

Hara, T., and R. Hara (1968) *Nature* 219, 450–454.

Hara, T., and R. Hara (1976) *J. Gen. Physiol.* 67, 791–805.

Hara, T., and R. Hara (1982) In Methods in Enzymology (Ed. by L. Packer), Vol. 81, pp.190–197.

Hara, R., and T. Hara (1984) *Vision Res.* 24, 1629–1640.

Hara, R., T. Hara, F. Tokunaga and T. Yoshizawa (1981) *Photochem. Photobiol.* 33, 883–891.

Honig, B., T. Ebrey, R.H. Callender, U. Dinur and M. Ottolenghi (1979) *Natl. Acad. Sci. USA.* 76, 2503–2507.

Hazemoto, N., N. Kamo, Y. Terayama, and M. Tsuda (1983) *Biophys. J.* 44, 59–64.

Iwai, J., M. Ikeuchi, Y. Inoue and T. Kobayashi (1984) In Protochlorophyllide Reduction and Greening (Edited by C. Sironval and M. Brouwers), pp.99–112. Martinus Nijhoff/Dr. W. Junk, The Hague.

Kobayashi, T. (1980a) *Photochem. Photobiol.* 32, 207–215.

Kobayashi, T. (1980b) *FEBS Lett.* 106, 313–316.

Matuoka, S., Y. Shichida and T. Yoshizawa (1984) *Biochim. Biophys. Acta.* 765, 38–42.

Nuss, M.C., W. Zinth, W. Kaiser, E. Kolling and D. Oesterhelt (1985) *Chem. Phys. Lett.* 117, 1–7.

Ohtani, H., T. Kobayashi, and M. Tsuda (1986) *Photobiophys. Photobiochem.* 13, 203–208.

Ohtani, H., T. Kobayashi, and M. Tsuda (1988) *Biophys. J.* 53, 493–496.

Ozaki, K., R. Hara and T. Hara (1982) *Exp. Eye Res.* 34, 499–508.

Ozaki, K., R. Hara and T. Hara (1983) *Cell Tissue Res.* 233, 335–345.

Peters, K., M.L. Applebury and P.M. Rentzepis (1977) *Proc. Natl. Acad. Sci. USA* 74, 3119–3123.

Polland, H.-J., M.A. Franz, W. Zinth, W. Kaiser, E. Kolling and D. Oesterhelt (1986) *Biophys. J.* 49, 651–662.

Seki, T. (1984) *J. Gen. Physiol.* 84, 49–62.

Shichida, Y., S. Matuoka and T. Yoshizawa (1984) *Photobiochem. Photobiophys.* 7, 221–228.

Sperling, L., and R. Hubbard (1975) *J. Gen. Physiol.* 65, 235–251.

Spudich, E.N., and J.L. Spudich (1983) *Proc. Natl. Acad. Sci. USA* 79, 4308–4312.

Spudich, J.L., and R.A. Bogomolni (1984) *Nature*, 312, 509–513.

Takahashi, T., H. Tomioka, N. Kamo, and Y. Kobatake (1985) *FEMS Microbiol. Lett.* 28, 161–164.

Tomioka, H., T. Takahashi, N. Kamo, and Y. Kobatake (1986) *Biochem. Biophys. Res. Commun.* 139, 389–395.

Tsuda, M., N. Hazemoto, M. Kondo, N. Kamo, Y. Kobatake, and Y. Terayama (1982) *Biophys. Res. Commun.* 108, 970–976.

Yoshizawa, T., and Y. Kito (1958) *Nature* 182, 1604–1605.

Yoshizawa, T. (1972) In Handbook of Sensory Physiology VII/1 (Edited by H. J. A. Dartnall), pp.146–179, Springer-Verlag, Berlin.

Circadian Rhythms

Seasonal Changes in Circadian Rhythm of Thermoregulation in Greenfinches and Siskins at Different Ambient Temperatures

SEPPO SAARELA, BERNT KLAPPER AND GERHARD HELDMAIER

Department of Zoology
University of Oulu, SF-90570 Oulu, Finland
Fachbereich Biologie, Philipps-Universität
D-3550 Marburg, Federal Republic of Germany

Abstract

The circadian variation of oxygen consumption (VO2) of greenfinches (*Carduelis chloris*, mass 27 g) and siskins (*Carduelis spinus*, mass 13.5 g) was recorded at thermoneutrality (TNZ, 26 and 29°C, respectively) and in the cold (0°C) in winter and in summer. Body temperature (Tb) was measured from greenfinches. Thermal conductance (Ct) of greenfinches was determined as well. The diurnal variation of heat production was not seasonal or temperature dependent. Nocturnal hypothermia was 2.5–3.4°C. The reduction of Ct was 41–47% in greenfinches. Daily energy savings of greenfinches and siskins was 26.8–33.0% in winter and 15.9–18.4% in summer compared with energy expenditure of birds being euthermic also at night.

Introduction

Low ambient temperature, long winter nights and restricted food supply in northern habitats require physiological adaptation of endothermic animals. As an energy saving strategy a periodic lowering of metabolism, at least partly independent of activity or food intake has been found e.g. in the chaffinch (*Fringilla coelebs*) (Aschoff and Pohl, 1970). Some endotherms are using torpor for energy and water conservation 10–88% (Hudson, 1978). Among the birds there are species from eight orders which exhibit torpor: Apodiformes, Caprimulgiformes, Coliiformes, Columbiformes, Cuculiformes, Falconiformes, Passeriformes, Strigidiformes (Reinertsen, 1983).

Seasonal changes in temperature dependent hypothermia have been demonstrated in some tit species (Reinerstsen and Haftorn, 1986). Nocturnal hypothermia has been found besides tits also with some Fringilidae birds living in Scandinavia. These results are based mainly on measurements of body temperature either in hand or after shooting the bird (Palmgren et al., 1944; Steen, 1958). Results from continuous recording of body temperature and oxygen consumption are still missing.

We used in our experiments greenfinches (*Carduelis chloris*) and siskins (*Carduelis spinus*). Both birds are typical residents in subarctic habitats. We were interested to find

Photobiology, Edited by E. Riklis
Plenum Press, New York, 1991

out whether 1) greenfinches and siskins use nocturnal hypothermia as an adaptation to the extreme condition; 2) a nocturnal hypothermia dependent on season and ambient temperature; and 3) its effect on energy conservation.

Material and methods

Adult greenfinches (*Carduelis chloris*, body mass 26.6 g) and siskins (*Carduelis spinus*, body mass 13.2 g) of both sexes were captured in Marburg area (51°N, West Germany) from their respective habitats. The birds were kept in outdoor aviaries.

Birds were measured in winter and again in summer. The first experiment was run at the birds' thermoneutral zone (TNZ, Ta = 26°C in greenfinches and Ta = 29°C in siskins) and in natural photoperiod. To find out the difference of the circadian energy expenditure of birds in winter and summer, the circadian experiment was repeated after a week at 0°C. The birds were weighted and placed into the metabolic chamber just before sunset or shortly after sunrise. During the experiment the birds were starving. The circadian rhythm of heat production was measured by indirect calorimetry. Small plastic boxes (3 l) were used as metabolic chambers. Air was sucked through these boxes into gas analyzer (Applied Electrochemistry, Model S-3A). The body temperature (Tb) was recorded continuously with temperature transmitters (Mini Mitter, Model X, weight 1.2 g) implanted into the abdominal cavity. The circadian rhythm of Tb's could be registered only in greenfinches. The body weight of birds was measured before and after each experiment. The analog output of the gas analyzer and temperature recordings were collected by a computer. Thermal conductance was calculated according to the formula C = VO2/(Tb-Ta). The values ov VO2 were converted to heat production (HP) by the equation HP = (4.44 + 1.43 * RQ) * VO2 (Heldmaier, 1975).

The three lowest values per hour of VO2 and Tb from the values recorded at 1-minute-intervals were used to calculate the hourly means for resting metabolic rate, Tb and thermal conductance.

Results

Body weight (BW)

Body weight of greenfinches and siskins are given in Table 1. BW of greenfinches was reduced during day 6.8–11.6% and at night 7.5–15.7%. The decrease of BW was smallest at 0°C in winter both during day and at night. The rate of reduction of BW was between 5.9 and 17.0% in the siskin. Similarly as in greenfinches the smallest reduction of BW occured at 0°C in winter.

Heat Production (HP)

Heat production of greenfinches and siskins was higher during day than at night (Figure 1, 2). The pattern of HP was independent on season but it was dependent on Ta.

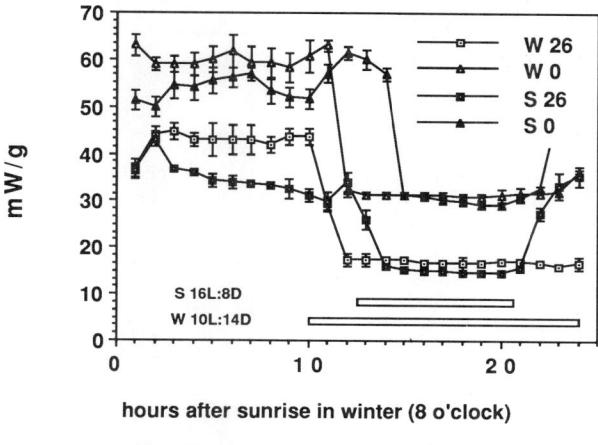

hours after sunrise in winter (8 o'clock)

Heat Production of Greenfinches

Figure 1. Circadian changes of heat production of greenfinches in winter (W) and summer (S). Ambient temperatures were 26 and 0°C. Each symbol represents average hourly means ± S.E. Number of birds were 4–6. Horizontal balks indicate the length of scotophase.

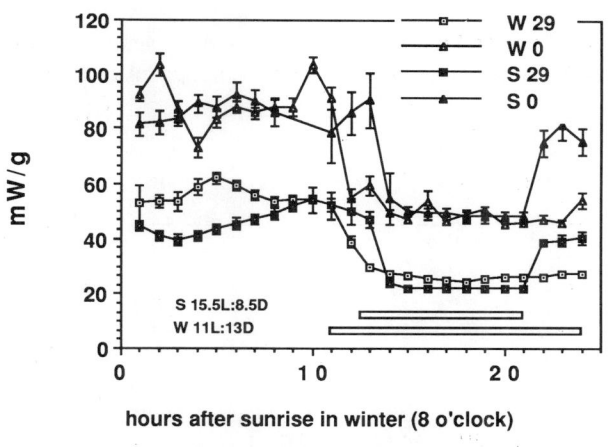

hours after sunrise in winter (8 o'clock)

Heat Production of Siskins

Figure 2. Circadian changes of heat production of siskins in winter (W) and summer (S). Ambient temperatures were 29 and 0°C. Other explanations see Figure 1.

The means of average day and night values of VO2 are given in Table 2. The reduction of VO2 was 47–57% in greenfinches and 50–55% in siskins. The reduction of VO2 of greenfinches was smaller at 0°C than at TNZ both in winter and in summer.

575

Table 1. Mean of body weight (BW, g) of greenfinches and siskins during photophase and scotophase before and after experiment. N = 4–6. Duration of day and night is indicated in hours.

Greenfinches

	Day				Night			
	Wi 26°C	Wi 0°C	Su 26°C	Su 0°C	Wi 26°C	Wi 0°C	Su 26°C	Su 0°C
BW (g) before		26.5	27.6	27.0	27.4	29.0	28.6	28.4
BW (g) after		24.5	24.4	24.9	23.2	26.8	24.6	25.0
reduction (%)		6.8	11.6	8.9	15.7	7.5	13.6	12.0
duration (h)	10	10	16	16	14	14	8	8

Siskin

	Day				Night			
	Wi 29°C	Wi 0°C	Su 29°C	Su 0°C	Wi 29°C	Wi 0°C	Su 29°C	Su 0°C
BW (g) before	13.5	13.5	13.5	13.2	13.5	13.9	13.5	13.1
BW (g) after	12.3	12.7	11.5	11.6	12.0	12.0	11.2	11.0
reduction (%)	8.9	5.9	14.8	12.9	11.1	6.7	17.0	16.8
duration (h)	11	11	15.5	15.5	13	13	8.5	8.5

Table 2. Means of day and night values of oxygen consumption (ml O_2/g'*h) of greenfinches and siskins. Nocturnal reduction is indicated as %. N = 4–6.

	Greenfinch				Siskin			
	Wi 25°C	Wi 0°C	Su 26°C	Su 0°C	Wi 29°C	Wi 0°C	Su 29°C	Su 0°C
Day	7.5	10.8	6.0	9.8	9.5	14.5	8.0	15.5
Night	3.2	5.2	2.7	5.2	4.3	7.9	4.0	8.3
Reduction (%)	57	51	55	47	55	55	50	55

Table 3. Means of day and night values and nocturnal reduction of body temperature (°C) and thermal conductance (mW/(g*°C)) of greenfinches.

	Body Temperature				Thermal Conductance			
	Wi 26°C	Wi 0°C	Su 26°C	Su 0°C	Wi 26°C	Wi 0°C	Su 26°C	Su 0°C
Day	41.5	41.7	41.5	41.8	2.68	1.40	2.12	1.28
Night	38.5	38.6	39.0	38.4	1.40	0.78	1.12	0.78
Reduction	3.0°C	3.1°C	2.5°C	3.4°C	47 %	44 %	47 %	41 %

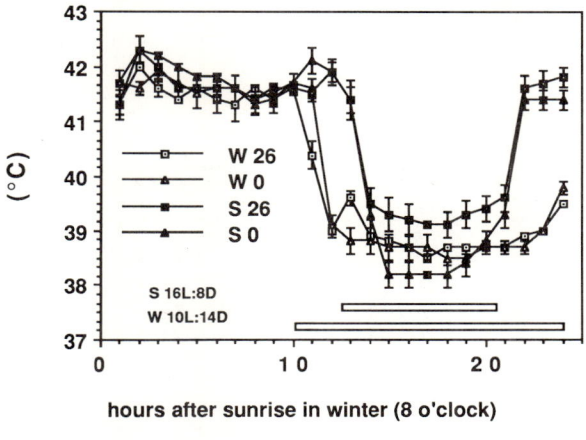

Body Temperature of Greenfinches

Figure 3. Circadian changes of body temperature of greenfinches. Other explanations see Figure 1.

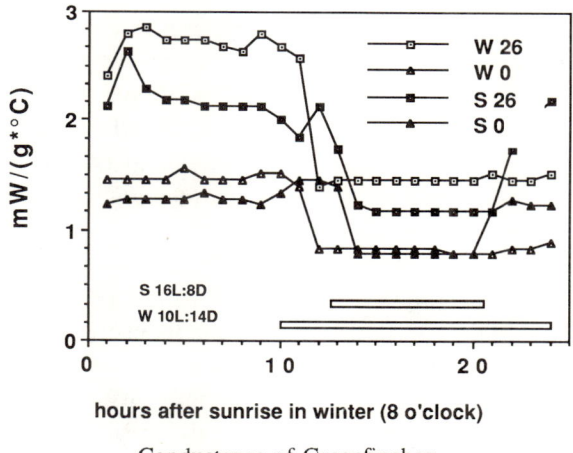

Conductance of Greenfinches

Figure 4. Circadian changes of thermal conductance of greenfinches. Other explanations see Figure 1. Body temperature (Tb)

Circadian changes of Tb are plotted in Figure 3. Interestingly, Tb of greenfinches seemed to be well regulated and cued by photoperiod. The means of average day and night values of Tb are given in Table 3. The Tb level of greenfinches was 41.5°C at Ta = 26°C, but at a Ta of 0°C even higher, 41.7 and 41.8°C in winter and in summer, respectively. The nocturnal hypothermia was 2.5–3.4°C.

577

Table 4. Energy expenditure during photophase and scotophase and daily energy savings calculated from oxygen consumption of greenfinches and siskins. Energy expenditure derived from warm-up is ignored in calculations.

Energy (kJ)	Greenfinch				Siskin			
	Wi 26°C	Wi 0°C	Su 26°C	Su 0°C	Wi 29°C	Wi 0°C	Su 29°C	Su 0°C
24 h day level	96.3	138.6	76.9	126.1	60.5	92.5	51.5	95.6
Day	40.1	57.7	51.2	84.1	26.3	40.3	32.9	61.7
Night	24.1	39.7	11.5	21.6	15.2	27.5	8.9	18.7
Saved (%)	33.0	29.7	18.4	16.2	31.4	26.8	18.0	15.9

Thermal conductance (Ct)

The pattern of Ct of greenfinches is plotted in Figure 4. Greenfinches improved their thermal insulation by 41–47% during the night (Table 3). Interestingly, the day value of Ct at 0°C was about the same than the night value at 26°C. Greenfinches were able to improve further their thermal insulation at night at 0°C.

Discussion

Greenfinches and siskins reduced their heat production by about 50% at night regardless of season or Ta. This represents daily energy savings for greenfinches and siskins by 26.8–33.0% in winter both at TNZ and 0°C (Table 4). In summer energy was saved by 15.9–18.4%. The duration of hypometabolism in winter is longer than in summer. This explains the higher amount of relative energy conservation in winter. Our results support the finding obtained previously with the willow tit that energy savings was smaller at lower Ta's (Reinertsen and Haftorn, 1986).

Body temperature of greenfinches decreased by 2.5–3.4°C at night. If the limit of nocturnal hypothermia is determined as Tb of 36°C or less (Bartholomew et al., 1983), our greenfinches did not utilize nocturnal hypothermia for energy conservation. However, as we can observe from Table 3, finches maintained their energy balance positive by reducing Tb only 2–3°C. At TNZ Ct of greenfinches remained even above minimum level. Only slight hypothermia was obtained with the great tit (*Parus major*) and the common redpoll (*Acanthis flammea*) in the cold when fed *ad libitum* (Reinertsen and Haftorn, 1986).

During nocturnal hypothermia Tb may fall by 3–10°C. The depth of hypothermia is dependent on feeding condition (Reinertsen, 1983; Reinertsen and Haftorn, 1986; Steen, 1958; Bartholomew et al., 1983; Chaplin, 1976). Small birds (body mass < 15 g) like manakins (*Manacus mentalis*, and *Pipra mentalis*) the willow tit (*Parus montanus*) enter a nocturnal hypothermia regardless of their feeding condition in the evening [Reinertsen and Haftorn, 1986; Bartholomew et al., 1983]. In laboratory experiments when the evening body weight of the common redpoll and the great tit were depleted 10–20% below the normal value, these birds also utilized nocturnal hypothermia which was significantly correlated with Ta (Reinertsen and Haftorn, 1986). In our experiments BW of greenfinches decreased at summer night 12–13.6% and at winter night 15.7% at TNZ. However, Tb was maintained above 38.4°C. We did not try to starve our birds in the evening, so the influence of BW depletion remain unsolved. Obviously both greenfinches and siskins can integrate information about physiological energy stores like manakins (Bartholomew et al., 1983), because they let a slower depletion of body weight at winter than summer night during the cold load of 0°C.

The support of Alexander von Humboldt Foundation is gratefully acknowledged.

References

Aschoff, J. and Pohl, H. (1970). Rhythmic variations in energy metabolism. *Fed. Proc.* 29(4):1541–1552.

Bartholomew, G.A., Vleck, C.M. and Bucher, T.L. (1983). Energy metabolism and nocturnal hypothermia in two tropical passerine frugivores, manacus vitellinus and pipra mentalis. *Physiol. Zool.* 56(3):370–379.

Chaplin, S.B. (1976) The physiology of hypothermia in the black-capped chickadee (*Parus atricapillus*). *J. Comp. Physiol.* 112B:335–344.

Heldmaier, G. (1975). Metabolic and thermoregulatory responses to heat and cold in the Djungarian hamster, *Phodopus sungorus. J. Comp. Physiol.* 102:115–122.

Hudson, J.W. Shallow daily torpor: a thermoregulatory adaptation. In: Strategies in Cold: Natural Torpidity and thermogenesis. (eds. L.C.H. Wang and J.W. Hudson) (1978) pp. 67–108, Academic Press, New York.

Palmgren, P. (1944). Körpertemperatur und Wärmeschutz bei einigen finnischen Vögeln. *Ornis Fennica* 21: 99–104.

Reinertsen, R. and Haftorn, S. (1986). Different metabolic strategies of northern birds for nocturnal survival. *J. Comp. Physiol. B* 156:655–663.

Reinertsen, R. Nocturnal hypothermia and its energetic significance for small birds living in the arctic and subarctic regions. A review. *Polar Res.* 1 n.s. (1983), 269–284.

Steen, J.B. (1958). Climatic adaptation in some small northern birds. *Ecology* 39:625–629.

Light and Circadian Activity in the Blind Mole Rat

[1]R. RADO, [1]H. GEV, [2]B.D. GOLDMAN, AND [1]J. TERKEL

[1]*Department of Zoology*
Tel-Aviv University, Ramat-Aviv, 69978, Israel, and
[2]*Department of Physiology and Neurobiology*
The University of Connecticut, Storrs, Connecticut, USA

Abstract

The present study describes the role of light in entraining the locomotor activity pattern of the blind mole rate, *Spalax ehrenbergi*, a rodent well adapted to a totally subterranean life. Mole rats were found to have a diurnal monophasic activity pattern under laboratory conditions with 14L/10D photoperiod. The light stimulus received by their atrophied eyes is the main zeitgeber in entraining their activity rhythm. Under constant dim light the animals free run with period lengths of close to 24 h.

Introduction

Daily variations in mammalian behavior and physiology are generated by the circadian timing system (Takahashi and Zatz, 1982). The endogenous rhythms are generally synchronized with the ambient light-dark (LD) cycle, such that physiological and behavioral events are appropriately phased with respect to environmental conditions. Under constant conditions, especially constant light (LL) or dark (DD), these rhythms free run with period lengths slightly differing from 24 h, and arE controlled by endogenous factors (Aschoff, 1981).

Photic inputs reach the mammalian circadian system through a pathway involving the retina, retinohypothalamic tract (RGT) and the suprachiasmatic nuclei (SCN) (Menaker and Binkley, 1981). Direct photic input to the brain through the skull has never been convincingly demonstrated in adult mammals (Rusak and Zucker, 1975), although in one study of neonatal rats brain photoreception was found for the first two to three weeks of their lives (Zweig et al., 1966).

Mole rats, *Spalax ehrenbergi,* are rodents that show striking behavioral and physiological adaptations to underground life (MacDonald, 1985). They excavate their own tunnel system and usually do not emerge aboveground (Nevo, 1961). Their eyes are atrophied and sealed by a thick layer of skin and dark fur. The adult retinal projections are degenerated, although connections to the SCN still exist (Bronchti et al., 1989). No visual evoked potentials can be elicited in higher visual structures such as the visual

Photobiology, Edited by E. Riklis
Plenum Press, New York, 1991

cortex (Haim et al., 1983; Gev 1984). In spite of the major atrophy of the visual system, adult mole rats show a behavioral preference to nest in dark rather than light environments (Rado, 1987). Furthermore, they are able to respond to short and long photoperiods by altering their thermoregulatory capacities (Haim et al., 1983).

The underground environment is characterized by stability of light, temperature, humidity and food availability (Nevo, 1982). The mole rat's functional blindness as well as the lack of flunctuations in its environment, presuppose that such animals would show no preference for a specific activity period during the day but, rather, would reveal a polyphasic activity pattern similar to that found for *S. leuconon* in nature (Hamer et al., 1984) and the laboratory (Savix and Mikes, 1967).

The circadian activity rhythm of *S. ehrenbergi* was studied by Nevo and his colleagues (1982). Mole rats of different karyotypes were placed in artificial tunnels for only 24 h, during which their locomotor activity was recorded. Although locomotor activity was distributed throughout this period, greater activity was recorded during the daylight hours. Unfortunately it is difficult to draw conclusions regarding circadian preference of mole rats from this study, due to the brevity of the experiment.

The aims of the present study were: (1) to determine the circadian pattern of locomotor activity of the mole rat under controlled laboratory conditions, (2) to test the role of the light-dark cycle in the entrainment of the locomotor activity, and (3) to determine the role of the atrophied eyes in detecting light stimuli.

Materials and Methods

A total of 66 adult mole rats, of both sexes, were caught near Tel-Aviv, transferred to the laboratory and housed individually in plastic cages ($33 \times 38 \times 14$ cm) in a colony room. Animals were supplied ad libitum with rodent chow and twice weekly with sufficient fresh vegetables and fruit to eliminate the need for drinking water. Room lights were programmed to a 14L/1OD cycle, with lights on at 05:00. Room temperature was maintained at $22 \pm 2°C$, and the experimental rooms were isolated from external noise. Light intensity ranged between 40–100 Lux at cage level. Feeding and maintenance were carried out twice a week, at randomly determined hours, without changing lighting conditions.

Wheel-running activity was recorded separately for each animal on an Esterline-angus even recorder in continuous operation. An actogram was made for each mole rat, by pasting each day's record immediately below that of the previous day. Onset and termination of active periods for entrained animals, and τ for free running animals, were estimated by three separate researchers on size - reduced photocopies.

Surgical or sham removal of the eyes were carried out on mole rates under equithesin anesthesia (0.3 cc/100 gr B.W). After the area overlying the atophied eye was shaved, an anterior-posterior incision, 5 cm from the end of the snout, was made. The eyes were removed and the skin was than sutured. In sham eye removal the atrophied eyes were disclosed and the skin sutured.

Experiment 1

Determination of the circadian pattern of locomotor activity.

In the first experiment we determined the circadian pattern of mole rat locomotor activity, and explored the role of the LD cycle in its entrainment. Locomotor activity of 37 mole rats exposed to a 14L/10D photoperiod was recorded.One group (n=7) was exposed to only one lighting schedule without phase shift of the LD cycle; the second group (n=22) went through one phase shift and the third group (n=6) two phase shifts of the LD cycle, all photoperiods were 14L/10D. Time of lights on was varied: 22:00, 19:00, 17:00, 13:00, 11:00 and 05:00, in order to mask potential environmental zeitgebers.

Experiment 2

Locomotor activity in constant dim light.

Entrainment to LD cycle can be a behavioral response to environment lighting conditions unrelated to an endogenous circadian timing system. On the other hand, a free running activity pattern under constant conditions is rigorous evidence for the existence of a circadian timing system with an endogenous oscillator. Two groups, of ten and six mole rats each, were exposed to constant dim (less then 1 Lux.) white (group A) and red (group B) lights. Locomotor activity was continuously recorded for two to six months, depending on period length of the oscillator. As soon as free running was detected as forming a pattern, the animal was removed from the experiment.

Experiment 3

The role of the atrophied eyes in the entrainment of locomotor activity.

In all mammals studied so far, light stimuli for the entrainment of the circadian rhythms is received through the eyes and transferred via the retinohypothalamic tract (RHT) to the suprachiasmatic nuclei (SCN). In this experiment we tested the role of the mole rat's atrophied eyes in the entrainment of locomotor activity.

Locomotor activity of 11 mole rats maintained under a 14L/10D photoperiod (lights on at 05:00) was recorded for 30–60 days until entrainment was achieved. At this stage a 6 hour phase shift of the LD cycle (lights on at 11:00) was made. After entrainment to the new schedule was achieved the atrophied eyes of four mole rats were removed while seven animals were sham operated. Surgery was carried out during the light phase of the LD cycle and immediately afterwards animals were returned to their activity cages. The ability to entrain the animals' locomotor activity was examined by phase shifting of the LD cycle every 4 weeks (14L/10D photoperiod, with light on at 17:00, 23:00, 11:00, and 17:00).

Results

Experiment 1

Thirty-five out of 37 mole rats showed circadian activity rhythm in a 14L/10D cycle. Thirty were active during the light phase of the cycle (diurnal), three during the dark phase (nocturnal), and two were initially nocturnal but become diurnal after 30–45 days of recording. Active time (α) was 15.15±0.31 h (Mean ± S.D.) and 14.57±0.95 h for the

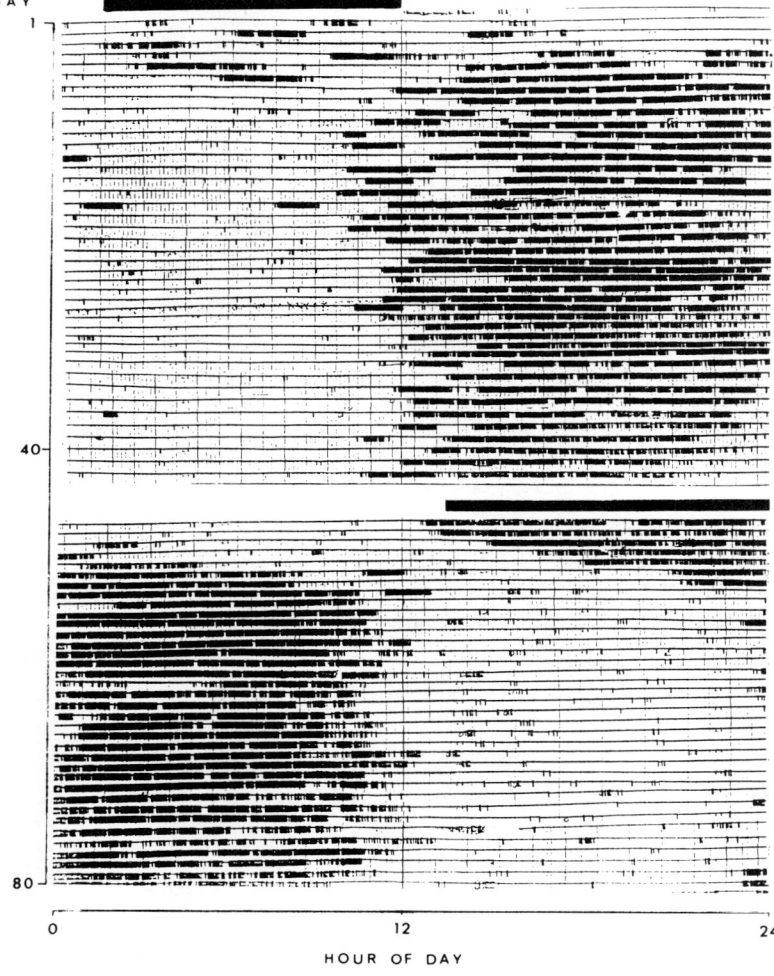

Figure 1. An eighty day actogram of mole rat No. 36 diurnal locomotor activity pattern under a 14L/10D photoperiod before and after 12 h phase shift of the LD cycle. The dark thick stripes represent the dark phase of the cycle.

diurnal and nocturnal mole rat respectively. Most of the diurnal mole rats' activity (93.8%) occurred during the light phase of the cycle and most of the nocturnal mole rats' activity time (79.4%) occurred during the dark phase. After phase shift of the LD cycle all mole rats entrained their locomotor activity rhythm to the new LD cycle within one - two weeks (Fig. 1).

Experiment 2

When mole rats were exposed to continuous illumination 56% of the animals (9 out of 16; 5 from group A – white light and 4 from group B – – red light) exhibited a free running pattern of locomotor activity (Fig. 2). The remainder (n=7) fell into three

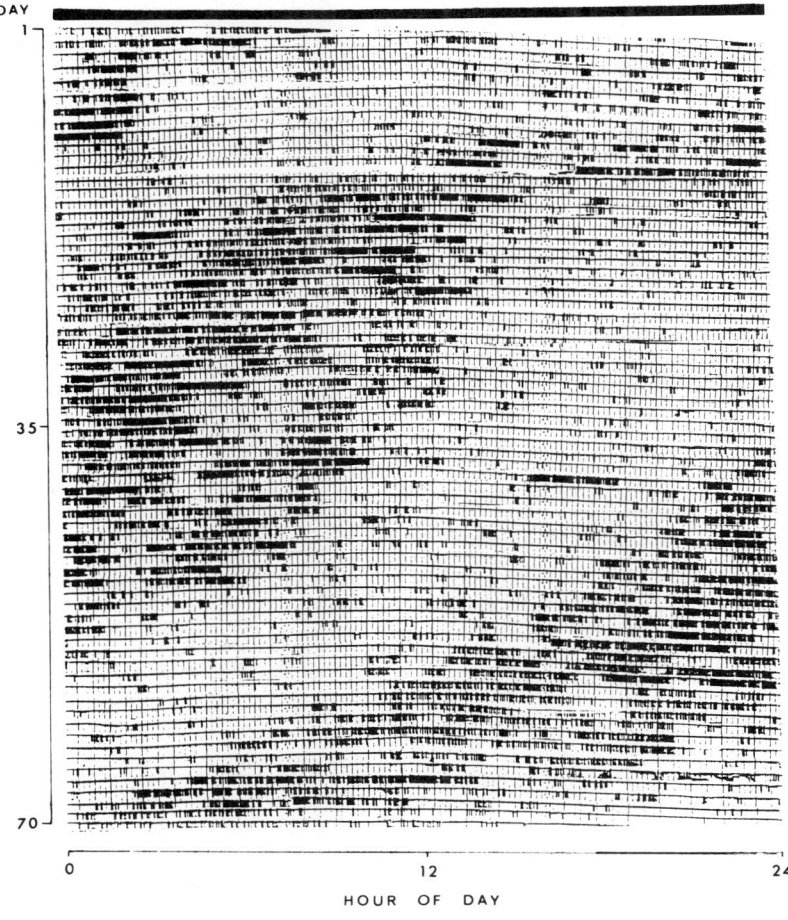

Figure 2. A seventy day actogram of free running mole rat (# 124) locomotor activity under dim (less than 1 Lux) red continuous illumination.

585

categories: three (from group A) were active during the objective night, one (from group B) was active during the objective day and three (two from group A and one from group B) did not exhibit any obvious circadian rhythm.

Period length (τ) for five mole rats (three from group A and two from group B) was longer than 24 h ($\tau = 24.24\pm0.19$ h) while that of the four other mole rats (two from each group) was shorter than 24 h ($\tau = 23.39\pm26$ h). Mean period length for all nine

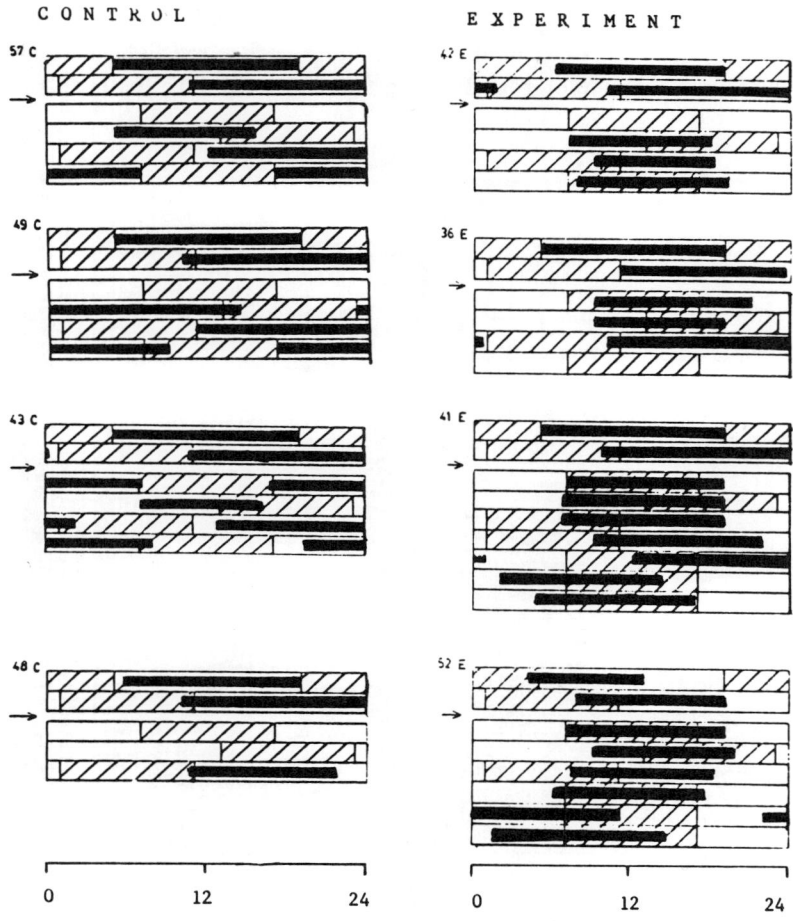

Figure 3. Locomotor activity of the eight mole rats before and after real (experimental) and sham (control) surgery. The lined area represents the dark phase of the cycle and the blank area the light phase. The dark thick stripes represent the mean locomotor activity duration during three weeks of recording under a particular lighting schedule. Five animals (57c, 49c, 48c, 42e, and 36e) recorded under a particular lighting schedule continued to exhibit locomotor activity throughout the day. In these cases no stripe representing activity time is marked. In animals 41e and 52e where a change in the time of locomotor activity occurred without change in lighting schedules, more than one stripe representing the activity period is marked on the same LD cycle.

mole rats exhibiting free running, was 23.66±0.49 h. There was no significant difference between τ of mole rats from group A or B (One way ANOVA).

Mean activity time (α) of eight mole rats exhibiting free running activity (apart from mole rate number 101 that displayed a very short activity time) was 13.67±1.86 h, similar to the activity time of diurnal mole rats exposed to a 14L/10D photoperiod (15.17±0.31 h; Exp. 1). There was no significant difference between α of mole rats with τ longer or shorter than 24 h (One way ANOVA).

Experiment 3

Before eye surgery, all 11 mole rats exhibited entrainment of locomotor activity to a 14L/10D photoperiod and to phase shift of the LD cycle (Fig. 3). All four of the mole rats that underwent eye removal lost their ability to entrain their locomotor activity to the LD cycle (Fig. 3). Six of the mole rats that underwent the sham operation continued to entrain their locomotor activity, while only one animal in this group showed free running, in response to phase shift of the LD cycle. Of these animals, three entrained their locomotor activity immediately after the sham operation, two were active around the clock at the first photoperiod and entrained after the first phase shift, and the last animal entrained only after the second phase shift.

Discussion

Most small mammals are either nocturnal or crepuscular (Ashby, 1972). Fossorial mammals, however, do not appear to conform rigidly to this generalization, as most species studied so far tend to be active polyphasically during both day and night (Nevo, 1979).

In its natural habitat, where light, temperature, humidity and food availability show little daily fluctuation, the mole rat, *Spalax ehrenbergi,* reveals a diurnal monophasic pattern of locomotor activity (Rado et al., 1988). The present study demonstrates that under controlled laboratory conditions mole rats also show a similar pattern of locomotor activity, and that the light which has been shown to be the primary stimuli to entrain the endogenous rhythm, is received through the mole rats' atrophied eyes.

The nocturnal activity pattern found in five mole rats appears to be a light avoidance response to the bright light to which they were exposed under laboratory conditions, in contrast to the complete darkness existing in their natural habitat. In a previous study, Rado (1987) found that most mole rats recorded immediately after capture revealed a nocturnal pattern of activity, but generally became diurnal after a one-two month period of adaptation to the laboratory conditions. On the other hand, when monitoring was commenced only after a six month period of adaptation to LD laboratory schedule, most of the animals revealed a diurnal pattern (Rado, 1987).

The length of the free running period of the mole rat approximates 24 h (23.86±0.49h; n=9), thus it is possible that the four animals that were active either during the objective night or objective day, entrained their locomotor activity to an external zeitgeber or, more probably, they had a free running period equal to 24 h.

The light stimulus for entraining locomotor activity of the mole rats is received by their atrophied eyes, without which such entrainment cannot be effected. Anterograde labeling with monocular injections of WGA-HRP showed that in the mole rat there is a link between the atrophied eyes and the suprachiasmatic nuclei (Bronchti et al., 1989). We suggest that the circadian system of the mole rat, like all mammals (Menaker and Binkley, 1981), involves the retina, the RHT and the SCN.

In nature mole rats show a monophasic activity pattern with most of their activity concentrated during the light hours of the day (Rado and Terkel 1989). The fact that a 20 sec bright light pulse once daily is sufficient stimulus for entrainment (Rado et al., unpublished), could explain this phenomenon. Mole rats are exposed to light pulses when excavating soil from the tunnel to the ground surface. This short light stimulus, that is transferred via the RHT to the SCN, might provide sufficient stimulation for such entrainment.

Based on these data, we propose that despite the total darkness of the mole rats' subterranean tunnels, their circadian system is similar to that of other mammals, and that the light pulses received through their atrophied eyes provide the main zeitgeber for entraining the mole rats' locomotor activity.

Acknowledgements

We wish to express our thanks to Dr. A. Terkel for critical reading of the manuscript, Ms. N. Paz for correcting and editing the paper, and to Mr. A. Shoob for the photography. This research was supported by the United-State – Israel Binational Science Foundation grant 333/87, and "The Fund for the Encouragement of Research" – Histadrut – The General Federation of Labour in Israel.

References

Aschoff, J. (1981). Free running and entrained circadian rhythms. Handbook of Behavioral Neurobiology. Vol. 4. Biological Rhythms, edited by J. Aschoff (Plenum, New-York) pp. 81–94.

Ashby, K.R. (1972). Patterns of daily activity in mammals. Mammals review, 1, 171–185.

Bronchti, G., Rado, R., Terkel, J., and Wollberg, Z. (1989). Ontogenetic degeneration of retinal projections in the blind mole rat (*Spalax ohrenbergi*). Second International Congress of Neuroethology, Berlin.

Gev., H. (1084). The role of light is entraining the circadian rhythm of the mole rat. M.Sc. Thesis. Department of Zoology, Tel-Aviv University, Israel.

Haim, A., Heth, G., Pratt, H., and Nevo, E. (1983). Photoperiodic effects on thermoregulation in a "blind" subterranean mammal. Journal of Experimental Biology, 107, 59–64.

Hamar, M.J.., Suteu, C.H., and Sutara, M. (1964). Home range and activity study of the mole rat (*Spalax leucodon* Nord.) by ^{60}C marking. Revue Roumaine de Biologie, de Zoologie., 9,6, 421–433.

MacDonald, D. (1985). The Encyclopedia of Mammals. Vol. 2 (George Allen & Unwin, London, Sydney).

Menaker, M., and Binkley, S. (1981). Neural and endocrine control of circadian rhythms in the vertebrates. Handbook of Behavioral Neurobiology. Vol. 4. Biological Rhythms, edited by J. Aschoff (Plenum, New York), pp. 243–256.

Nevo, K. (1961). Observation on Israeli population of the mole rat *Spalax ohrenbergi,* Nehring 1898. Mammalia, 25, 127–144.

Nevo, K. (1979). Adaptive convergence and divergence of subterranean mammals. Annual Reviews in Ecology, 10, 269–308.

Nevo, K. (1982). Speciation in subterranean mammals. Mechanisms of Speciation, edited by Barigozzi, C. (Alan R. Liss Publication., New York), pp. 191–218.

Nevo, K., Guttman, R., Haber, M., and Erez, K. (1982). Activity patterns in evolving mole rats. Journal of Mammalogy, 63, 453–463.

Rado, R. (1987). Circadian and circannual rhythms of the mole rat. M.Sc. Thesis, Department of Zoology, Tel-Aviv, University, Israel.

Rado, R., Gev, H., and Terkel, J. (1988). The role of light in entraining mole rats' circadian rhythm. Israel Journal of Zoology, 35, 105–106.

Rado, R., and Terkel, J. (1989). Circadian activity of the mole rat, *Spalax ehrenbergi,* monitored by radio telemetry, in seminatural and natural conditions. The Fourth International Conference on Environmental Quality and Kcosystem Stability, Jerusalem, Israel.

Rusak, B., and Zucker, I. (1975). Biological rhythms and animal behavior. Annual Review of Psychology, 26, 137–171.

Savix, V.I., and Mikes, M. (1967). Zur kenntnis des 24-stunden - rhythmus von *Spalax leucodon* Nordmann, 1840. Zurnal Saugetierkunde, 32, 233–238.

Takahashi, J.S., and Zatz, M. (1982). Regulation of circadian rhythmicity. Science, 217, 1104–1111.

Zweig, M., Snyder, S.H., and Axelrod, J. (1966). Evidence for a non-retinal pathway of light to the pineal gland of new born rats. Proceeding of the National Academy of Sciences USA, 56, 515–520.

Photoperiod Changes and Heat Production in *Meriones Crassus* — The Role of Circadian Rhythms of Body Temperature in Seasonal Acclimatization

A. HAIM, AND G. LEVI

University of Haifa
Oranim, School of Education of the Kibbutz Movement
Kiryat-Tivon 36910
Israel

Abstract

In the Arava Rift Valley of Israel and in the Negev Desert, the fat jird *Meriones crassus*, is exposed to an extreme cold climate which can sometimes be wet during winter, while in summer it is exposed to a warm and dry climate. It is reasonable to assume that due to such seasonal climatic changes, the fat jird will show seasonal acclimatization in thermoregulatory mechanisms as well as in energy demands. In this research the thermoregulatory and energy response to changes in photoperiod regimes was studied and an attempt was made to correlate the changes in these parameters with those of circadian rhythms of body temperature measured in both photoperiod regimes.

For this purpose individuals of the fat jird were acclimated to long scotophase (8L:16D) and long photophase (16L:8D) at an ambient temperature of 28°C.

The results of this study reveal that increase in scotophase increases heat production by non-shivering thermogenesis as well as the energy demands as compared with the group acclimated to long photophase. The changes of the circadian rhythms of body temperature correlates well with the thermoregulatory and energy demands under the two different photoperiod regimes.

From these results it may be concluded that extension of scotophase, without any decrease in ambient temperature, can initiate winter acclimatization of thermoregulatory mechanisms and, at the same time, adjust the circadian rhythms of body temperature.

Introduction

Rodents, like other endothermic animals, show characteristic circadian (24h) variations in body temperature. These circadian rhythms are a result of two other functions, namely heat production and heat dissipation which also show circadian rhythms (Aschoff, 1982 and Haim et al 1988).

The thermoregulatory response to changes in photoperiod regimes was recorded in several rodent species (Lynch, 1970; Haim and Fourie, 1980; Heldmaier et al. 1981; Haim, 1982; and Haim and Yahav, 1982). The results of these studies revealed that acclimation to long scotophase increased the thermoregulatory ability of the studied rodents upon exposure to cold environment.

The photoperiod regime is one of the main parameters which is manifested by seasonal acclimatization. Seasonal acclimatization of metabolism and thermoregulation has been studied in several rodent species inhabiting mesic and cold environments (Rosenmann et al. 1975; Cygan, 1985; Heldmaier et al. 1986; Feist and Feist, 1986). The fat jird *Meriones crassus*, is a nocturnal rodent distributed throughout the Palaearctic desert belt (Harrison, 1972) and is well adapted to the arid environment (Horowitz, 1971; Yahav, 1986). The fat jird in the Arava Rift Valley or the Negev desert in Israel faces pronounced seasonal climatic changes such as warm summers and cold winters.

The aim of the present research was to study the circadian rhythms of body temperature under two different photoperiod regimes (8L:16D and 16L:8D) of the fat jird at a constant ambient temperature ($T_a=28°C$), as well as to correlate the circadian rhythms of body temperature with metabolic demands and heat production by means of non-shivering thermogenesis (NST) under these photoperiod regimes.

Materials and Methods

The circadian rhythm of body temperature was measured over 30h using 12 jirds (6 males and 6 females). Jirds were acclimated for two weeks at least, prior to experiments relating to the different photoperiod regimes. During the long scotophase acclimation, lights were on between 08.30h and 16.30h while in the long photophase acclimation, lights were on between 08.00h and 24.00h. Body (rectal temperature) T_b was measured at intervals of 6h using copper-constantan thermocouples and monitored on a TH-65 Wescor digital thermometer.

Food consumption and non-shivering thermogenesis (NST) were measured in 6 individuals (2 males and 4 females) with a body mass of 120±20.1g. The jirds were kept separately in each cage on a diet of dried rat pellets and cucumber, using paper for bedding. Food consumption was measured as apparent digestible dry matter intake, the difference between the total food consumed and dry faeces, as in Haim (1987). Gross digestible energy intake was determined by bombing samples of rat pellets, cucumber and faeces using a microbomb calorimeter (Phillipson, Gentry Instruments Inc.). Food consumption was measured for two weeks after a two week period of acclimation at each photoperiod regime.

NST was measured in the jirds after completing food consumption measurements. NST was measured as the maximal oxygen consumption (Vo_2) and T_b response of nonanesthetized jirds to a noradrenaline (NA) injection (Sigma) 1.5 mg/kg body mass, using the method of Heldmaier et al (1981). Vo_2 was measured in an open flow system, Depocas and Hart (1957) and Hill (1972). Oxygen concentration was determined using an Applied Electrochemistry A3 oxygen analyzer, and a Tek-Dyn 712 recorder and Vo_2 recording was as in Haim et al (1986). At the end of Vo_2 measurements, T_b was measured and both values were taken as maximal values. Minimal Vo_2 and T_b were recorded after a period of 2h stabilizing in the metabolic chamber at an ambient temperature of 29°C. NST capacity was calculated as the ratio between maximal and minimal values of Vo_2.

All values are presented as mean ±S.D. and Student's *t*-test was used for statistical analysis.

Results and Discussion

The fat jird *Meriones crassus* under both photoperiod regimes, showed a reproducible circadian rhythm of body temperature, (Fig 1). While acclimated to long scotophase, T_b was higher for a longer period (14h) when compared to the long photophase acclimation where T_b was higher only for a short period (8h). This difference in duration of high body temperature emerges from the difference in the photoperiod regimes of the two groups. The role of circadian temporal organization in the make-up of living systems has been suggested by Aschoff (1982) and Kenagy and Vleck (1982). From the results of this study, it is obvious that circadian rhythms of body temperature vary during seasonal acclimatization and photoperiod plays a major role as a "zeitgeber" for rhythms of body temperature.

As a nocturnal species *Meriones crassus* is expected to have more hours for foraging during winter time as there is a significant increase in scotophase on the one hand, but it may be exposed to very low temperatures for a longer period each night. Will the energy demands of this species be higher in winter when compared with summer? As in other small rodents, the role of insulation in improving thermoregulation is not significant (Hart 1956). Therefore seasonal changes in mechanisms of heat production may be expected (Heldmaier et al 1986).

The results of this study indeed show a significant increase in non-shivering thermogenesis (NST) as well as a significant increase in apparent digestible dry matter intake (DDMI) and gross digestible energy intake (GDEI). Vo_2Max, T_bMax, and NST capacity were respectively: 1.536 ± 0.38 $mlO_2/g.h$, $37.8\pm0.6°C$ and 1.37 ± 0.35 for the long photophase (16L:8D) group, while for the long scotophase (8L:16D) group the values were: 2.113 ± 0.32 $mlO_2/g.h$, $38.8\pm0.6°C$, and 1.89 ± 0.22. DDMI and GDEI were respectively: 2.61 ± 0.56 $g/100gW_b.day$ and 7.49 ± 2.30 $kJ/100gW_b.day$ for the long photophase group while for the long scotophase group the values were: $5.62\pm1.06g/100gW_b.day$ and 16.75 ± 3.76 $kJ/100gW_b.day$.

An increase in NST was reported from several rodent species acclimated to long scotophase (Lynch 1970, Haim and Fourie 1980, Heldmaier et al 1981, Haim 1982 and Haim and Yahav 1982). An increase in NST is also noted during winter acclimatization, Heldmaier et al (1986). From the results of the present study, it is clear that, in the fat jird, extension of scotophase, although not accompanied by a drop in ambient temperature, initiated an increase in heat production through the NST mechanism.

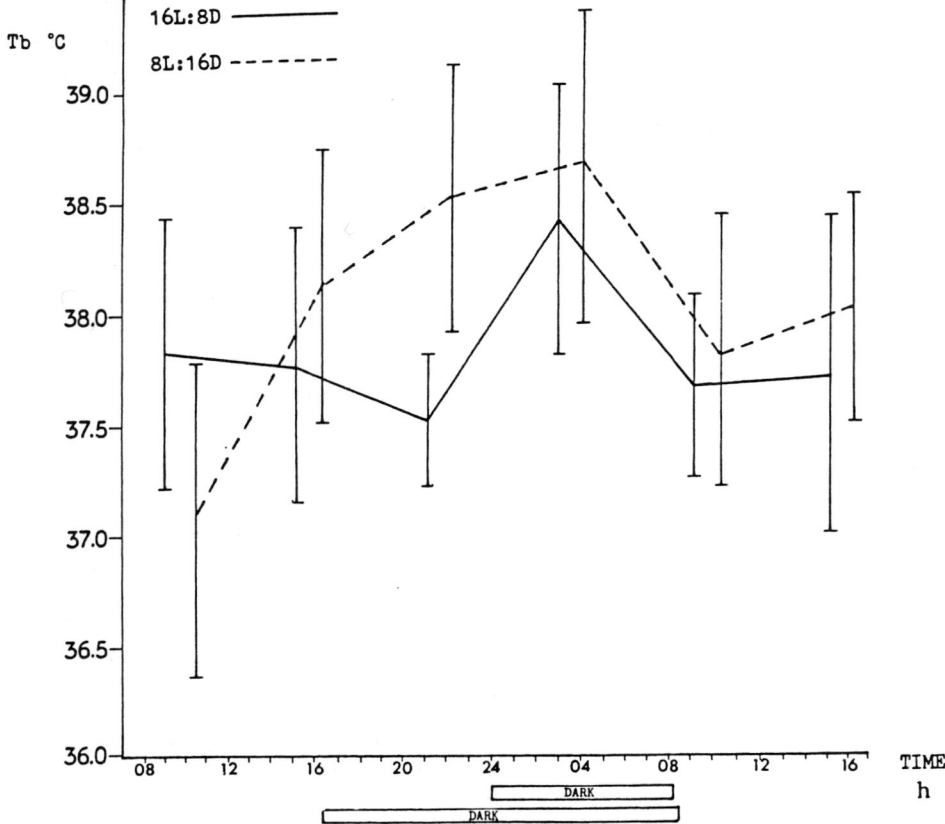

Figure 1. Body temperature (Tb°C) of 12 fat jirds *Meriones crassus*, acclimated to 16L:8D and 8L:16D at an ambient temperature of 28°C, as a function of time of the day (values are mean ± S.D.).

The increase in food and energy consumption may increase body mass or, rather, activity. As body mass was almost the same in both groups or, in other words, acclimation to either photoperiod regimes did not result in a change of body mass, it is likely that the increase in food and energy consumption increased activity. This assumption is supported by the increase in body temperature (Fig 1).

From the results of this study it may be concluded that a change in the photoperiod regime causes a change in the circadian rhythm of body temperature. This change in circadian rhythm fits well with the metabolic and thermoregulatory demands of the fat jird under experimental conditions. It is therefore reasonable to assume that the change in photoperiod regime is an essential stimulus in seasonal acclimatization.

Acknowledgements

We thank Mr. H. Kusik for mainaining the jirds, to Mrs. Yael Hollander, Mr. L. Goldzweig and Mr. S. Beris for their editing remarks, to Mrs. Tamar Luf for drawing the figure.

References

Aschoff, J., 1982, The circadian rhythm of body temperature as a function of body size. A Companion to Animal Physiology, edited by C.R. Taylor, K. Johanson and L. Bolis (Cambridge: Cambridge University Press), pp. 173–188.

Cygan, T., 1985, Seasonal changes in thermoregulation and maximum metabolism in the Yellow-necked field mouse. *Acta theriologica*, 30, 115–130.

Depocas, F., and Hart, J.S., 1957, The use of the Pauling oxygen analyzer for measurements of oxygen consumption of animals in open-circit system, and in short-lag, closed-circuit apparatus. *Journal of Applied Physiology*, 10, 388–392.

Feist, D.D., and Feist, C.F., 1986, Effects of cold, short day and melatonin on thermogenesis, body weight and reproductive organs in Alaskan red-backed voles. *Journal of Comparative Physiology B*, 156, 741–746.

Haim, A., 1982, Effects of long scotophase and cold acclimation on heat production in two diurnal rodents. *Journal of Comparative Physiology B*, 148, 77–81.

Haim, A., 1987, Thermoregulation and metabolism of Wagner's gerbil *(Gerbillus dasyurus):* A rock dwelling rodent adapted to arid and mesic environments. *Journal of Thermal Biology*, 12, 45–48.

Haim, A., and Fourie, F.le.R., 1980, Heat production in cold and long scotophase acclimated and winter acclimatized rodents. *International Journal of Biometeorology*, 24, 231–235.

Haim, A., and Yahav, S., 1982, Non shivering thermogenesis in winter acclimatized and in long scotophase and cold acclimated *Apodemus mystacinus* (Rodentia). *Journal of Thermal Biology*, 7, 193–195.

Haim, A., Ellison, G.T.H., and Skinner, J.D., 1988, Thermoregulatory circadian rhythms in the pouched mouse *(Saccostomus campastris)*, Comparative Biochemistry and Physiology. 91A, 179–181.

Haim, A., Pelaot, I., and Sela, A., 1986, Comparison of ecophysiological parameters between two *Apodemus* species coexisting in the same habitat, Environmental Quality and Ecosystem Stability, Vol. III A/B, edited by Z. Dubinsky and Y. Steinberger (Ramat-Gan: BarIlan University Press), pp. 33–40.

Harrison, D.L., 1972, The Mammals of Arabia, Vol. III (London: Benn).

Hart, J.S., 1956, Seasonal changes in insulation of the fur. *Canadian Journal of Zoology*, 34, 53–57.

Heldmaier, G., Boeckler, H., Buchberger, A., Klaus, S., Puchalski, W., Steinlechner, S., and Wiesinger, H. 1986, Seasonal variation of thermogenesis. Living in the Cold: Physiological and Biochemical Adaptations, edited by W.C. Heller et al (Amsterdam: Elsevier), pp. 361–372

Heldmaier, G., Steinlechner, S., Rafael, J., and Visiansky, P., 1981, Photoperiod control and effects of melatonin on non-shivering thermogenesis and brown adpose tissue. *Science*, 212, 917–919

Hill, R.W., 1972, Determination of oxygen consumption by use of the paramagnetic oxygen analyzer. *Journal of Applied Physiology*, 33, 261–263.

Horowitz, M., 1971, The physiological adaptations of the blood system of animals (mice) to dehydration. Ph.D thesis, submitted to the Hebrew University Of Jerusalem.

Kenagy, G.J., and Vleck, D., 1982, Daily temporal organization of metabolism in small mammals: adaptation and diversity. Vertebrate Circadian Systems, edited by J. Aschoff, S. Daan and G. Groos (Heidelberg: Springer-Verlag Berlin), pp. 322–338

Lynch, G.R., 1970, Effect of photoperiod and cold acclimation on non-shivering thermogenesis in *Peromyscus leucopus. American Zoololgist*, 10, 308.

Rosenmann, M., Morrison, P., and Fiest, D.D., 1975, Seasonal changes in the metabolic capacity of Red-Backed voles. Physiological Zoology, 48, 303–310.

Yahav, S., 1986, Bioenergtic and nitrogen metabolism in *Meriones crassus* compared to *Microtus guentheri*. Ph.D thesis, submitted to Tel Aviv University.

Physiological Responses of Melatonin-Implanted Pigeons to Changes in Ambient Temperature

T.M. JOHN and J.C. GEORGE

University of Guelph, Guelph
Ontario, Canada N1G 2W1

Introduction

Thirty years ago Lerner et al. (1958) identified and isolated from the pineal body of cattle, an unique substance, melatonin, that caused the skin of frogs and tadpoles to blanch. This substance, characterized as 5-methoxy-N-acetyltryptamine, was found to have no such activity in mammals. However, a relationship between mammalian pineal and environmental light was revealed in rats in which continuous exposure to light for several weeks caused reduction in pineal weight (Fiske et al., 1960). Such a relationship is consistent with the evolutionary history of the pineal having evolved from a photosensory structure as seen in fish to an endocrine gland in birds and mammals, with intermediate forms consisting of secretory rudimentary photoreceptor cells showing both types of activity (Bentley, 1982). The pineal, therefore, serves as a neuroendocrine transducer of photic information.

Among a number of functions ascribed to melatonin, the most extensively studied is its role in reproduction. Though melatonin has been shown to have an antigonadal influence in some seasonally breeding rodent species, information regarding its role in avian reproduction is inconsistent (Ralph, 1981). In a recent study, it was shown that subcutaneous implantation of melatonin had no significant influence on photo-induced gonadal development in the pigeon (John et al., 1986).

Several studies have shown that melatonin production in the avian pineal is regulated by the light/dark cycle and that there exists a distinct circadian rhythm in melatonin concentration (Ralph et al., 1967; Binkley et al., 1974; Pelham, 1975; Binkley, 1983; Grady et al., 1984). Since melatonin synthesis is higher and body temperature lower during night, the possibility of melatonin having a thermoregulatory role, has been investigated in the pigeon. Pinealectomy was shown to produce hyperthermia which was reversed with melatonin treatment in pigeons (John et al., 1978). It was also observed that the thermoregulatory response in pinealectomized pigeons, was dependent on photoperiod and also on the degree and duration of cold exposure. In the present study we have examined the influence of melatonin on oxygen

Photobiology, Edited by E. Riklis
Plenum Press, New York, 1991

consumption (VO$_2$), heart rate (HR), breathing frequency (BF), shivering, cloacal temperature (T$_c$) and foot temperature (T$_f$) in pigeons exposed to varying ambient temperature (T$_a$).

Materials and Methods

Adult feral pigeons (*Columba livia*) weighing between 350 and 400 g were trapped from the University of Guelph campus during November and acclimated for two months in individual cages located in a controlled environment room having a 12-h daily photoperiod (fluorescent light) and T$_a$ 23 ± 1°C. They were given commercial pigeon feed ("Purina") and water *ad libitum*. At the end of the acclimation period, they were divided into two groups, melatonin-implanted (MI) and control (MC). The MI group (4 males and 4 females) received subcutaneous implants of melatonin-beeswax (2 mg melatonin + 30 mg wax) on the dorsal side of the neck, and MC received implants of beeswax pellets (without melatonin) at the same site. Subsequently, the birds were maintained for another 12 weeks in the same controlled environment room, but with daily photoperiod increased to 16 h (16L-8D) from 12 h. Fresh implants having the same concentration of melatonin as mentioned above, were implanted every two weeks in these birds to ensure an undeclining supply of the test-substance. Although the rate of melatonin release into circulation was not measured, administration of melatonin in this manner has been found to produce thermal response in birds (John *et al.*, 1978) and endocrine responses in both pigeons (John *et al.*, 1986) as well as certain mammals tested (see Reiter et al., 1974). Melatonin used was in crystal form obtained from Sigma Chemical Company (St. Louis, Missouri, U.S.A.) and the pellets were prepared according to the procedure described by John et al. (1978).

Each pigeon, following the 12-wk treatment period, was exposed to varying T$_a$, during which different physiological parameters were monitored. The pigeon was placed in a light-proof double-walled plexiglass metabolic chamber, after properly securing the necessary electronic connections between the bird and the monitoring equipments. A thermostatically controlled "polytemp" heater-refrigerator liquid-bath with pump (Polyscience Corp., Niles, Illinois, U.S.A.) was employed to circulate an antifreeze liquid through the chamber-wall in order to regulate the T$_a$ inside the chamber. When the bird was placed inside the metabolic chamber, the T$_a$ within the chamber was set at 24°C. After allowing the bird to acclimate for 1 hr at this temperature, the chamber T$_a$ was gradually raised to 34°C (in 1 h) and was subsequently brought down to 2°C in gradual decrement (in 6 h). The readings for the physiological parameters were recorded at every 2°C drop in chamber T$_a$, beginning from 34°C.

VO$_2$ was measured with the help of a Beckman O$_2$ analyzer (model E2). Dry air, free of CO$_2$ was passed through the metabolic chamber at a rate of 480 ml/min. The air coming out from the chamber was further freed of CO$_2$ and moisture (by passing through soda lime and anhydrous calcium sulfate) and allowed to enter the O$_2$ analyzer. VO$_2$ was calculated using the equation of Hill (1972). HR, BF and shivering were monitored with a polygraph (model 7D; Grass Instruments, Quincy, Massachusetts, U.S.A.). Shivering was measured as the electromyographic (EMG) activity of the breast muscle. Integrated EMG values were obtained by using a polygraph 7P10 summating integrator. Details regarding the preparation and fixing of electrodes for the measurements of EMG, HR and BF have been described in an earlier study (John and George, 1987).

A YSI scanning telethermometer (model 47; Yellow Springs Instruments, Yellow Springs, Ohio, U.S.A.) was used for monitoring the T_a, T_c and T_f. For T_c measurements, a thermister probe was inserted approximately 4 cm deep into the cloaca, and for T_f a probe was placed in contact with the bare skin overlying the tarsometatarsus. All electrodes were firmly secured in place with pieces of adhesive tape.

Analysis of variance was used for the statistical evaluation of the data.

Results and Discussion

MI pigeons registered a significantly lower T_c than MC (Figure 1). On the other hand, T_f was significantly higher in MI pigeons (Figure 2). This drop in T_c in MI pigeons is consistent with our earlier observation with winter pigeons in which melatonin implantation nullified or reversed the hyperthermic effect of pinealectomy

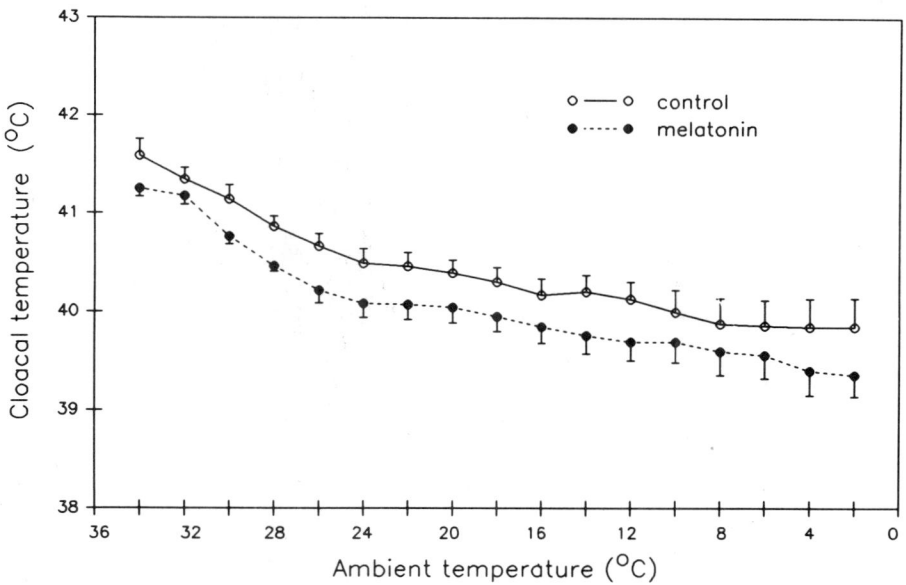

Figure 1. Cloacal temperature (T_c) in control (MC) and melatonin-treated (MI) pigeons exposed to ambient temperature (T_a) falling from 34° to 2°C. MC vs MI: P = 0.000; T_a effect: P = 0.000; inter-action: P = 1.000; n = 8 MC and 8 MI.

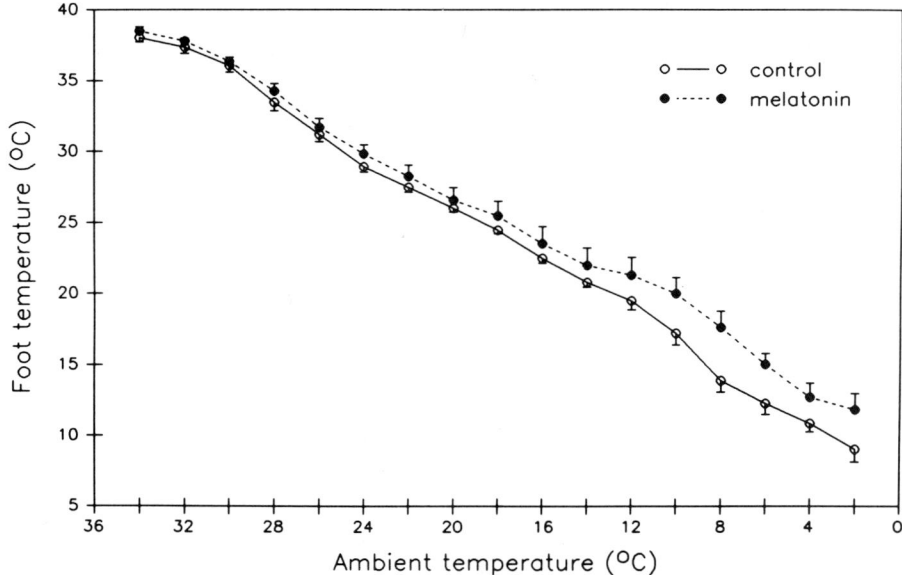

Fig. 2. Foot temperature (T_f) in MC and MI pigeons exposed to falling T_a (34° to 2°C). MC vs MI: P= 0.000; T_a effect: P= 0.000; interaction: P = 0.525; n = 8 MC and 8 MI.

(John et al., 1978). The hypothermia observed in MI pigeons may be attributed to increased heat dissipation which is indicated by higher T_f. It is well known that the leg is an important site for heat dissipation in birds (see Jones and Johansen, 1972; John and George, 1984). John and George (1984) observed a decrease in T_f in pinealectomized pigeons and concluded that the decrease in T_f was indicative of lower heat dissipation caused by the possible diminution in melatonin levels in the body. The role of melatonin, therefore, could be to enhance peripheral vasodilation and blood flow so as to facilitate greater dissipation of heat.

In the present study, both MC and MI pigeons registered gradual decreases in T_c as well as T_f in response to drop in T_a (Figures 1 and 2). Although no statistically significant differences between the MC and MI pigeons were discernible with regard to the extent of these decreases, the decrease in T_f in the latter group appeared less pronounced than that in the former as T_a continued to drop below 20°C (Figure 2). However, a similar trend was not obvious with differences in T_c.

In both MC and MI pigeons, a detectable shivering response to cold exposure was first observed when T_a fell to 26°C (Figure 3). The shivering intensity in both groups gradually increased with further drop in T_a. However, the rate at which the shivering intensity increased with continued drop in T_a was rather slow in MI pigeons. Consequently, the difference in shivering intensity between the two groups of pigeons continued to increase as T_a continued to fall (Figure 3).

The lower shivering activity observed in MI pigeons at T_a below 26°C may be interpreted to be at least partly responsible for the hypothermia in this group of pigeons.

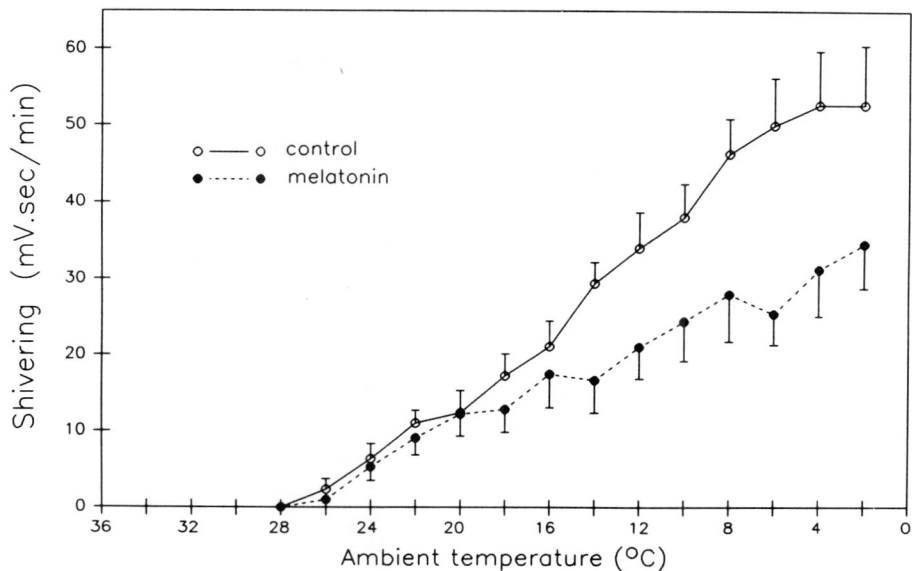

Figure 3. Shivering in MC and MI pigeons exposed to falling T_a (34° to 2°C). MC vs MI: P = 0.000; T_a effect: P = 0.000; interaction: P = 0.0289; n = 7 MC and 7 MI.

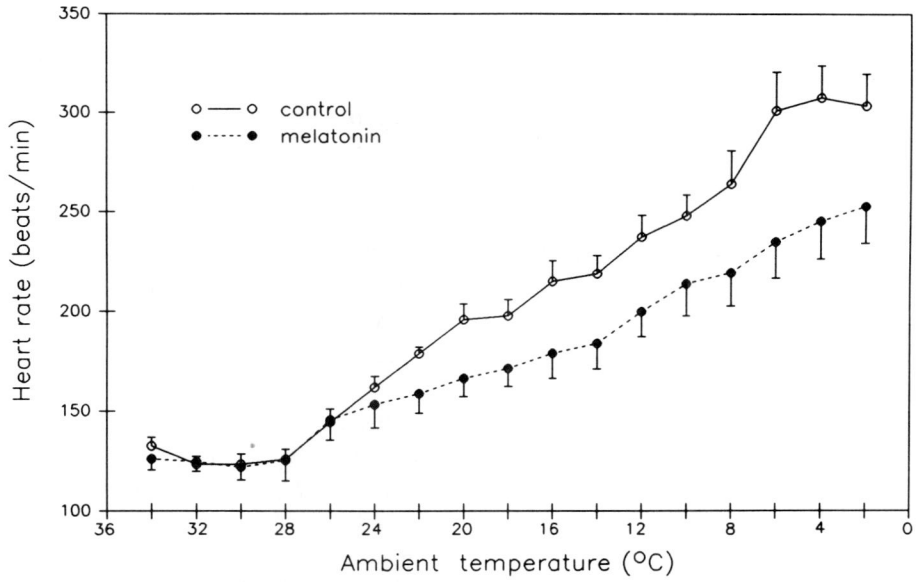

Figure 4. Heart rate (HR) in MC and MI pigeons exposed to falling T_a (34° to 2°C). MC vs MI: P = 0.000; T_a effect: P = 0.000; interaction: P = 0.033; n = 8 MC and 8 MI.

601

However, the lack of a definite correlation between the magnitude of shivering intensity increase and of T_c decrease in response to falling T_a, casts doubt on such an assumption. Moreover, hypothermia in MI pigeons was obvious even at T_a above 26°C when shivering was not registered in either MC or MI pigeons. This leads us to suggest that the action of melatonin in temperature regulation in the pigeon is by effecting a lower set-point centrally and by acting as a vasodilator peripherally. There is sufficient physiological evidence from studies on mammals to indicate the existence of melatonin receptors in the brain especially the medial preoptic and suprachiasmatic areas of the brain (for review see Zisapel, 1988). In its peripheral action, melatonin has been shown to lower vasoconstrictor effects of certain agents tested in pinealectomized rats (Cunnane et al., 1980).

HR remained fairly consistent in both MC and MI pigeons during the drop in T_a from 34 to 28°C, but began to increase gradually as T_a dropped further (Figure 4). Increase in heart rate during cold-exposure has been reported in pigeons (Barnas et al., 1984; John and George, 1987). In the present study, there were no significant differences in HR between MC and MI pigeons until T_a fell to 26°C from 34°C, after which the increase in HR became less pronounced in the MI group. As T_a continued to drop below 26°C, the difference in HR between MI and MC pigeons showed a trend to widen further (Figure 4). It appears, therefore, that shivering and HR in response to cold exposure exhibit identical profiles, and that melatonin has an inhibitory effect on HR and shivering, in a cold environment. The inhibitory effect of melatonin on HR may be attributed to its suppressive action on the sympathetic nervous system as was shown in the Syrian hamster (Viswanathan et al., 1986).

BF was significantly lower in MI pigeons than in MC when overall comparisons were made (Figure 5). The magnitude of difference between the two groups appeared less pronounced at or around T_a 26°C, and more pronounced at or below 6°C, but statistically there was no significant "interaction". BF was lowest in MC at 28°C T_a but increased as T_a hovered above or below this point, with peak response being registered at 2°C. MI pigeons also showed an identical response, but with trough at both 30 and 28°C.

Shivering is considered to be a prerequisite for increased BF in pigeons exposed to cold or subjected to spinal cooling (Barnas and Rautenberg, 1984). Similar observations were apparent in vagotomized pigeons exposed to cold (John and George, 1987). Although not conclusive, a trend indicating a similar correlation between shivering and BF is apparent in the present study as well.

That cold exposure could increase VO_2 in pigeons, has been demonstrated by Barnas et al. (1984) and John and George (1987). In the present study too, VO_2 was found to be significantly greater at lower T_a than at the higher, in both MC and MI pigeons (Figure 6). Overall, VO_2 in MI pigeons was significantly lower than that in MC. Though statistically the "interaction" was not significant, the magnitude of the difference between MC and MI pigeons showed a trend towards greater VO_2 at T_a below 12°C. Since MI pigeons had consistently shown lower BF, HR and shivering, the low VO_2 observed in these birds is not surprising. It may be mentioned here that the low VO_2 in MI pigeons

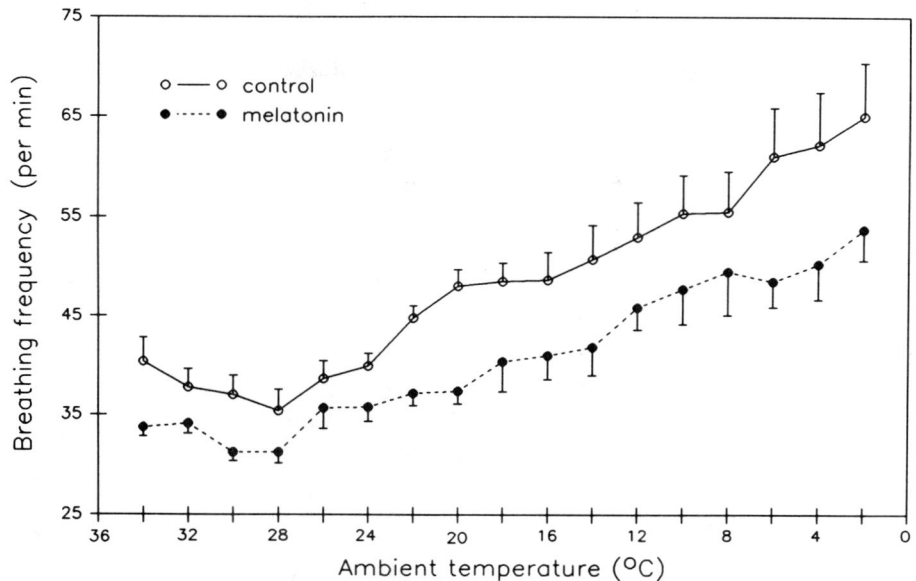

Figure 5. Breathing frequency (BR) in MC and MI pigeons exposed to falling T_a (34° to 2°C). MC vs MI: P = 0.000; T_a effect: P = 0.000; interaction: 0.910; n = 8 MC and 8 MI.

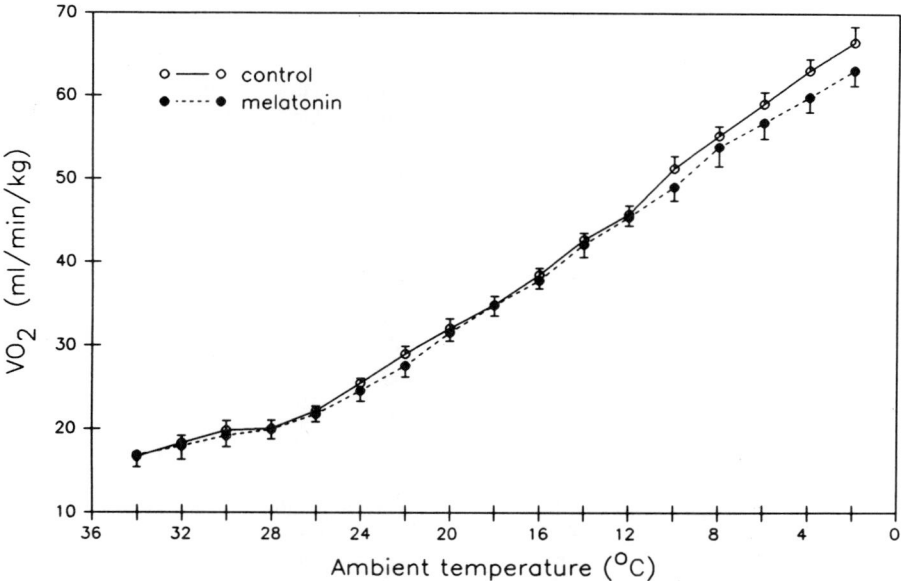

Figure 6. VO_2 in MC and MI pigeons exposed to falling T_a (34° to 2°C). MC vs MI: P = 0.016; T_a effect: P = 0.000; interaction: P = 0.994; n = 7 MC and 7 MI.

may also be due to low thyroxine levels observed in MI pigeons in another study (John et al., 1990).

Conclusions

Subcutaneous implantation of melatonin into pigeons cause lower T_c, shivering, BF and HR, but higher T_f. In both MC and MI pigeons, as T_a drops, all the monitored parameters increase, except for T_c and T_f which decrease. In general, the magnitude of differences between MC and MI pigeons tend to be greater with lower T_c, but significant only with reference to shivering and HR. The lower T_c observed in MI pigeons could be attributed to increased heat dissipation as indicated by higher T_f in these pigeons. The maintenance of higher T_f in MI pigeons, even under ambient cooling, is suggestive of greater heat dissipation in these birds. That the magnitude of shivering, HR, BF and VO_2 was consistently lower in MI than in MC pigeons with the continuing drop in T_a, suggests that melatonin has a role in temperature regulation acting centrally at the level of the hypothalamus by lowering the set-point and peripherally by facilitating heat dissipation through vasodilation.

Acknowledgements

This study was supported by funds provided by the Natural Sciences and Engineering Research Council of Canada. We are thankful to Mary Anne Finkbeiner for typing the manuscript.

References

Barnas, G., S. Nomoto and W. Rautenberg (1984) Cardiovascular and blood gas responses to shivering produced by external and central cooling in the pigeon. *Pflugers Arch.* 401, 223-227.

Barnas, G. and W. Rautenberg (1984) Respiratory responses to shivering produced by external and central cooling in the pigeon. *Pflugers Arch.* 401, 228-232.

Bently, P.J. (1982) Comparative Vertebrate Endocrinology, 2nd Edition. pp. 58-63. Cambridge University Press, Cambridge.

Binkley, S. (1983) Rhythms in ocular and pineal N-acetyltransferase: a portrait of an enzyme clock. *Comp. Biochem. Physiol.* 75A, 123-129.

Binkley, S., S.E. MacBride, D.C. Klein and C.W. Ralph (1974) Pineal enzymes: regulation of avian melatonin synthesis. *Science* 181, 273-275.

Cunnane, S.C., M. Manku, M. Oka and D.F. Horralein (1980) Enhanced vascular reactivity to various constrictor agents following pinealectomy in the rat: role of melatonin. *Can. J. Physiol. Pharmacol.* 58, 287-293.

Fiske, V.M., K. Bryant and J. Putnam (1960) Effect of light on the weight of the pineal in the rat. *Endocrinology* 66, 489-491.

Grady, R.K. Jr., A. Caliguri and I.N. Meford (1984) Day/Night differences in pineal indoles in the adult pigeon (*Columba livia*). *Comp. Biochem. Physiol.* 78C, 141-143.

Hill, R.W. (1972) Determination of oxygen consumption by use of the paramagnetic oxygen analyzer. *J. Appl. Physiol.* 33, 261-263.

John, T.M. and J.C. George (1984) Diurnal thermal response to pinealectomy and photoperiod in the pigeon. *J. Interdiscipl. Cycle Res.* 15, 57-67.

John, T.M. and J.C. George (1987) Physiological responses of vagotomized pigeons exposed to ambient temperatures gradually reduced from 34°C to 2°C. *J. Auton. Nerv. Syst.* 18, 153-161.

John, T.M., J.C. George and R.J. Etches (1986) Influence of subcutaneous melatonin implantation on gonadal development and on plasma levels of luteinizing hormone, testosterone, estradiol, and corticosterone in the pigeon. *J. Pineal Res.* 3, 169-179.

John, T.M., S. Itoh and J.C. George (1978) On the role of the pineal in thermoregulation in the pigeon. *Hormone Res.* 9, 41-56.

John, T.M., M. Viswanathan, J.C. George and C.G. Scanes (1990) Influence of chronic melatonin implantation on circulating levels of catecholamines, growth hormone, thyroid hormomes, glucose and free fatty acids in the pigeon. *Gen. Comp. Endocrinol.* (in press).

Jones, D.R. and K. Johansen (1972) The blood vascular system of birds. In Avian Biology (Edited by Farner, D.S. and King, J.R.). Vol. 2, pp. 157-285. Academic Press, New York.

Lerner, A.B., J.D. Chase, Y. Takahashi, T.H. Lee and W. Mori (1958) Isolation of melatonin, the pineal gland factor that lightens melanocytes. *J. Am. Chem. Soc.* 80, 2587.

Pelham, R.W. (1975) A serum melatonin rhythm in chickens and its abolition by pinealectomy. *Endocrinology* 96, 543-546.

Ralph, C.L. (1981) The pineal and reproduction in birds. In: The Pineal Gland. Vol. 2. Reproductive Effects (Edited by Reiter, R.J.), pp. 31-43. CRC Press, Boca Raton, Florida.

Ralph, C.L., L. Hedlund and W.A. Murphy (1967) Diurnal cycles of melatonin and bird pineal bodies. *Comp. Biochem. Physiol.* 22, 591-599.

Viswanathan, M., R. Hissa and J.C. George (1986) Suppression of sympathetic nervous system by short photoperiod and melatonin in the Syrian hamster. *Life Sci.* 38, 73-79.

Zisapel, N. (1988) Melatonin receptors revisited. *J. Neural Transm.* 73, 1-5.

Melatonin Binding Sites in Discrete Brain Areas: Coincidence with Physiological Responsiveness

NAVA ZISAPEL

Department of Biochemistry
The George S. Wise Faculty of Life Sciences
Tel-Aviv University, Tel Aviv 69978, Israel

Introduction

Much is known about the reproductive implications and regulation of the melatonin signal in mammals, but the molecular basis of melatonin action is still unknown. High affinity binding sites for ^{3}H-melatonin in membrane and cytosolic fractions of mammalian brain and peripheral organs were described in 1978-1979 by several groups (Cohen et al., 1978; Cardinali et al., 1979; Niles et al., 1979). Unfortunately, these observations could not subsequently be reproduced or extended, thus raising the possibility that the ligand used in these early studies was not authentic melatonin and raising skepticism regarding the existence of melatonin receptors. Yet, much physiological evidence points to the existence of such receptors: (a) there is a robust diurnal rhythm in sensitivity of the mammalian reproductive system to melatonin (Reiter et al. 1980; Galss and Lynch, 1982; Goldman et al. 1982; Lang et al., 1984); (b) the responsiveness of the neuroendocrine system to melatonin is not ubiquitous but rather localized at specific target sites (Tamarkin et al. 1964; Wurtman et al., 1981; Glass and Lynch, 1981; Bittman et al., 1983). The brain, especially the medial preoptic and supra-retrochiasmatic areas have been implicated as the sites of melatonin's antigonadal activity (for review see Tamarkin et al., 1985). For example, uptake of systemically administered ^{3}H-melatonin *in vivo* indicated that the hormone was concentrated in the hypothalamus, hippocampus and reproductive tract (Wurtman et al., 1981); gonadal regression in the white footed mouse was greatest when melatonin was implanted in the vicinity of the suprachiasmatic nucleus (Glass and Lynch, 1981). (c) the responsiveness to melatonin changes with age (Lang et al. 1984; Martin and Sattler, 1979); (d) the melatonin antagonist ML-23, developed in our laboratory, can prevent melatonin mediated effects in the rat *in vivo* and *in vitro* (Zisapel and Laudon 1987; Laudon et al. 1988).

We have previously introduced the use of 2-^{125}I-iodomelatonin (^{125}I-melatonin), a potent analog of melatonin in vitro in the rat brain (Laudon and Zisapel, 1985) and chicken retina (Dubocovich and Takahashi, 1987) and in the hamster fetus *in vivo* (Weaver et al., 1988) as a specific probe for melatonin binding sites (Laudon and Zisapel,

Photobiology, Edited by E. Riklis
Plenum Press, New York, 1991

1985). ^{125}I-melatonin was found to bind with high affinity to a single class of sites in rat brain synaptosomes. The binding was time-dependent and reversible. Specific ^{125}I-melatonin binding was inhibited by melatonin and by ML-23 but was almost unaffected by other structurally related compounds including serotonin (Laudon et al., 1988; Laudon and Zisapel, 1985; Zisapel et al. 1987). Specific ^{125}I-melatonin binding has recently been described in chicken retina (Dubocovich and Takahashi, 1987), suprachiasmatic nuclei of the fetal Djungarian hamster (Weaver et al., 1988), rat (Vanecek et al., 1987) and human (Reppert et al., 1988) and in the median eminence of the rat (Vanecek et al., 1987).

As previously shown, melatonin inhibits the depolarization-induced release of dopamine from discrete areas of the rat brain, predominantly the hypothalamus (Zisapel and Laudon, 1982). To elucidate the physiological relevance of ^{125}I-melatonin binding sites to melatonin receptors, the relationship between the densities of ^{125}I-melatonin binding sites and the responsiveness of the hypothalamus to the hormone *in vitro* was studied.

Methods

Animals

Rats of the CD strain were supplied by Levinstein's farm (Yokneam) and of the Sabra-Wistar strain were supplied by the animal house of Hadassa-Medical School (Jerusalem). Animals were housed under controlled temperature conditions (24+2°C) and maintained on a daily 14h light:10h darkness schedule (lights-on 05:00 h). Food and water were supplied *ad libitum*.

Preparation of tissue, labeling and stimulation of dopamine release

Rats were decapitated and their brains rapidly removed and placed on ice. The brain areas were excised manually and cut into contralateral halves. Tissue sections were incubated with 0.4 uM ^{3}H-dopamine and then subjected to two successive sets of electrical field stimulation in a superfusion chamber as described (Zisapel and Laudon, 1982). The quantity of dopamine released following stimulation was determined by scintillation counting of the effluent buffer collected in each case. The release obtained following the first train of stimuli served as an internal reference of uninhibited release. The samples were exposed to melatonin during the interval (25 min) between the two sets of stimuli. Stimulated dopamine release was expressed as the ratio between the experimental (second train, S2) and reference (first train, S1) release quantities. Percentage of inhibition by melatonin of the stimulated dopamine release was calculated from these data by the following equation:

$$\% \text{ Inhibition} = 100 \left(1 - \frac{(S2/S1) \text{ with melatonin}}{(S2/S1) \text{ without melatonin}}\right)$$

Preparation of synaptosomal fractions and ^{125}I-melatonin binding: Rats were decapitated and their brains were rapidly removed. The brain areas were excised manually and suspended in 10 ml/g ice cold 0.32 M sucrose. Crude synaptosomal pellets were prepared as described (15) and suspended in 2 vol of 50 mM Tris-HCL buffer, pH 7.4 containing 4 mM $CaCl_2$. Aliquots of the synaptosomal preparations (200 ug protein/ 20 ul) were incubated with 40 ul Tris buffer containing 5-50 nM ^{125}I-melatonin for 0-30 min at 37°C on a shaking water bath, in the absence or presence of unlabeled melatonin. Membranes were then collected by vacuum filtration using GF/C glass fiber filters and washed with 3×4 ml buffer at 4°C. The filters containing bound ^{125}I-melatonin were assayed for radioactivity using a Packard gamma counter. Specific binding was defined as that displaced by 50 uM of non-radioactive melatonin and ranged from 50 to 60% of the total binding at 50 nM ^{125}I-melatonin. The equilibrium binding data could be fitted by non-linear regression to a single site model according to the following equation: B=Bmax/(1+Kd/L), where B and Bmax are the specific binding at different ^{125}I-melatonin concentrations (L) and at saturation, respectively, and Kd is the binding dissociation constant (Laudon and Zisapel, 1985). To assess goodness of fit, the derived theoretical curves were reconstructed by computer and superimposed on the experimental data points in saturation plots. The data obtained from each experiment were analyzed separately and the mean and S.D. values of Kd and Bmax obtained from three repetitive experiments are presented (Table 1).

The results obtained from the release and binding experiments were compared by carrying out Student's t-test for differences among multiple means. Significance was determined at $p<0.05$.

Resuls and Discussion

The antigonadal effects of exogenously administered melatonin are manifested only during certain periods of the day in hamsters, white footed mice and rats (Reiter et al. 1980; Glass and Lynch, 1982; Goldman et al., 1982; Lang et al., 1984) with a peak of responsiveness late in the photophase. A counter-antigonadal effect was observed in the early photophase (Chen et al., 1980). The inhibition by melatonin of the stimulated release of preloaded ^{3}H-dopamine from the male rat hypothalamus *in vitro* exhibited a circadian rhythm with a peak response at 5 h but almost no inhibition at 15 h after the onset of light (Zisapel et al., 1985). The distribution of ^{125}I-melatonin binding sites in the brain of male rats was studied at different hours of the light-dark cycle. Our results indicated that while these sites were abundant in the rat brain, only those in the hypothalamus, medulla pons and hippocampus, exhibited diurnal variations. The density of ^{125}I-melatonin binding sites in the hypothalamus exhibited a broad peak (Zisapel et al., 1985) which overlapped the peak phase position of the maximal antigonadal and counterantigonadal responses to melatonin *in vivo* and the maximal inhibitory effect of melatonin on dopamine release *in vitro* (Table 1).

The inhibition by melatonin of dopamine release from the female rat hypothalamus varied significantly with the estrous cycles (Zisapel et. al., 1983). In long-term (2-4 weeks) ovariectomized rats, the ability of melatonin to inhibit dopamine release from the hypothalamus *in vitro* was abolished (Zisapel et al., 1987). Reinstatement of the responsiveness to melatonin was observed in the hypothalamus of the ovariectomized rats shortly (2 h) after a single subcutaneous injection of estradiol (Zisapel et al., 1987). In parallel, the effects of gonadectomy and of subsequent testosterone or estradiol treatment on the distribution of ^{125}I-melatonin binding sites in the rat brain were investigated. Ovariectomy produced a large estradiol-reversible decrease in ^{125}I-melatonin binding in the hypothalamus and medulla-pons of the female (Zisapel et al., 1987; Laudon and Zisapel, 1987; Table 1) whereas castration produced a marked testosterone-reversible decrease in ^{125}I-melatonin binding particularly in the hypothalamus and hippocampus of the male rat (Zisapel and Anis, 1988). In contrast, ^{125}I-melatonin binding in the parietal cortex, striatum and cerebellum was generally unaffected by gonadectomy (Laudon and Zisapel, 1987; Zisapel and Anis, 1988). A single injection of EDS (ethylene-1,2-

Table 1. Effects of circadian time, ovariectomy, estradiol treatment and aging on the inhibition by melatonin of dopamine release from the rat hypothalamus *in vitro* and on the equilibrium binding parameters of ^{125}I-melatonin binding sites in synaptosomes from the rat hypothalamus[a,b].

Experiment	Inhibition of release[c]		^{125}I-Melatonin binding[d]		
	%	ref.	Bmax	Kd	ref.
Male, 10:00 h	33 ± 4	23	113 ± 10	84 ± 4	24
Male, 18:00 h	1 ± 3	23	120 ± 3	78 ± 5	24
Male, 22:00 h	13 ± 4	23	59 ± 11	79 ± 4	24
Female, Estrus	52 ± 7	18	550 ± 60	100 ± 11	18
Female, OVX	-2 ± 18	18	132 ± 20	24 ± 6	18
Female, OVX+E_2	36 ± 6	18	423 ± 50	77 ± 17	18
Aged male, 10:00 h	20 ± 2	Fig 1	10 ± 4	34 ± 4	28

[a]Mean $\pm$ S.D. values from three experiments are presented.
[b]The references from which the data were quoted are depicted in "ref".
[c]Adult and aged male rats and adult intact, ovariectomized (OVX) females, or OVX females injected s.c. with estradiol (E_2; 200 ug, 2 h before sacrifice) were sacrificed between 9:00-11:00 h, if not otherwise stated. The hypothalami were incubated with ^{3}H-dopamine and subjected to stimulation in the absence and presence of 1 uM melatonin. Percent inhibition by melatonin was calculated as described in the Methods section.
[d]Rats from the various groups were sacrificed as described in (c). Synaptosomes were prepared from the hypothalami. Specific binding was determined at equilibrium. The Kd and Bmax values were obtained by best-fit analysis of the data assuming a single class of sites as described in the Methods section and are expressed in nM and fmol/mg protein, respectively.

\dimethane sulfonate) led to a marked transient decrease in ^{125}I-melatonin binding in the male rat brain concomitantly with the destruction and repopulation of the Leydig cells (Zisapel and Anis, 1988).

The inhibition by melatonin of stimulated dopamine release from the hypothalamus was markedly reduced in aged (>20 months old) male and female rats (Fig 1). The concentrations of melatonin needed to inhibit the release were about 10^4 times higher than in the adult controls (Fig 1). A dramatic age-related decline in ^{125}I-melatonin binding was observed in the hypothalamus and hippocampus of aged male rats in spite of persisting rhythm in melatonin (Table 1); ^{125}I-melatonin binding in other areas of the rat brain was relatively insensitive to aging (Laudon et al., 1988).

Taken together, the data summarized above point to the physiological relevance of ^{125}I-melatonin binding sites to melatonin receptors in the brain. Moreover, the results clearly indicate that the density of melatonin receptors in the brain does not coincide with the peak production and secretion of melatonin but is positively correlated with the timing and responsiveness of the rat neuroendocrine system to the hormone. Elevated serum melatonin concentrations therefore do not necessarily reflect potentiated biological activity.

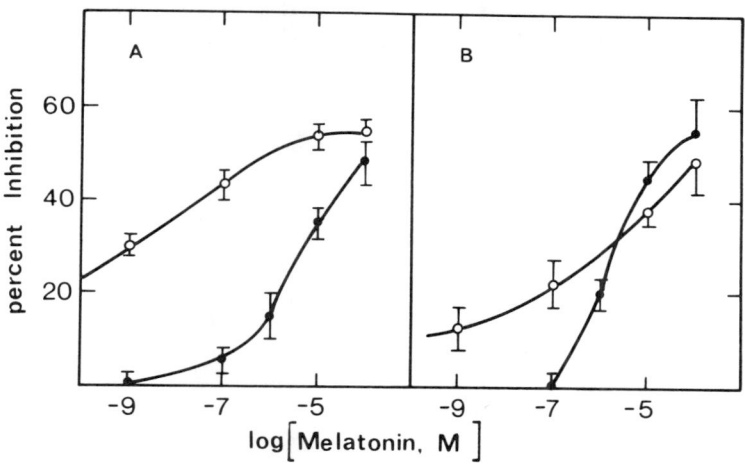

Figure 1. Inhibition by melatonin of dopamine release from the hypothalamus of the adult (○) and aged (●) female (A) and male (B) rats. The procedure described in Table 1 (d) was employed. Mean ±S.D. values of percent inhibition by various melatonin concentrations are presented (n=3).

References

Bittman, E.L., Dempsy, rSR.J. and Karsch, F.J. *Endocrinology* 113 (1983) 2276-2283.
Cardinali, D. P., Vacas, M. I. and Boyer, E.E. *Endocrinology* 105 (1979) 437-441.
Chen, H.J., Brainard, G.C. and Reiter, R.J. *Neuroendocrinology* 31 (1980) 129-132.

Cohen, M., Roselle, D. and Chabner, B. *Nature* 274 (1978) 894-895.

Dubocovich, M.L. and Takahashi, J. *Proc. Natl. Acad. Sci. USA* 84 (1987) 3916-3920.

Glass, D.T. and Lynch, G.R. *Science* 214 (1981) 821-823.

Glass, J. D. and Lynch, G.R. *Neuroendocrinology* 35 (1982) 117-122.

Goldman, B. D., Carter, D.S., Hall, V.D., Roychondhury, P. and Yellon, S.M. In: Melatonin rhythm generating systems: developmental aspects. Klein, D. ed. (Karger, Basel 1982).

Lang, U., Rivest, R.W., Schlaepfer, L.V. Bradtke, J.C. Aubert, M.L. and Sizonenko P.C. *Neuroendocrinology* 38, (1984) 261-268.

Laudon, M. and Zisapel, N. *Brain Res.* 402 (1987) 146-150.

Laudon, M. and Zisapel, N. *FEBS Lett.* 197 (1985) 9-12.

Laudon, M., Nir, I. and Zisapel, *N. Neuroendocrinology* 48 (1988) 577-584.

Laudon, M., Yaron, Z. and Zisapel, N. *J. Endocr.* 116 (1988) 43-53.

Martin, J.E. and Sattler, C. *Endocrinology* 105 (1979) 1007-1012.

Niles, L.P., Wong, Y.W., Mishra, R.K. and Brown, G.M. Eur. *J. Pharmacol.* 55 (1979) 219-233.

Reiter, R.J., Petterborg, L.J., Trakulrungsi, C. and Trakulrungsi, W.K. *J. Exp. Zool.* 212 (1980) 47-52.

Reppert, S.M., Weaver, D.R., Rivkees, S.A. and Stopa, E.G. *Science* 242 (1988) 78-81.

Tamarkin, L. Baird, C.J. and Almeida, O.F.X. *Science* 227 (1985) 714-720.

Vanecek, J., Pavlik, A. and Illnerova, H. *Brain Res.* 435 (1987) 359-362.

Weaver, D.R., Namboodiri, M.A.A. and Reppert, S.M. *FEBS Lett.* 228 (1988) 123-127.

Wurtman, R. J., ,Axelrod, J. and Potter, L.T. *J. Pharmacol. Exp. Ther.* 143 (1964) 314-318.

Zisapel, N. and Anis, Y. *Molec. Cell. Endocrinol.* 60 (1988) 119-126.

Zisapel, N. and Laudon, M. *Biochem. Biophys. Res. Commun.* 104 (1982) 1610-1616.

Zisapel, N. and Laudon, M. *Eur. J. Pharmacol.* 136 (1987) 259-260.

Zisapel, N. Egozi, Y. and Laudon, M. *Neuroendocrinology* 37 (1983) 41-47.

Zisapel, N., Egozi, Y. and Laudon, M. *Neuroendocrinology* 40 (1985) 102-108.

Zisapel, N., Nir, I. and Laudon, M. *FEBS Lett.* 232 (1988) 172-176.

Zisapel, N., Shaharbani, M. and Laudon, M. *Neuroendocrinology* 46 (1987) 207-216.

Solar UV Light Effects on Growth and Development

Signal Transduction in Blue Light-Mediated Growth Responses

T. W. SHORT[1], M. LASKOWSKI[1], S. GALLAGHER[2], and W. R. BRIGGS[1]

[1]*Department of Plant Biology*
290 Panama St.
Stanford, CA 94305 USA and
[2]*Hoefer Scientific Instruments*
654 Minnesota St.
San Francisco, CA 94107, USA

During the past decade there has appeared a large number of papers devoted to the effects of blue light on higher plants, algae, and fungi (see Senger, 1980, 1984, 1987; Senger and Schmidt, 1986). Despite intensive efforts, however, a specific blue light photoreceptor has yet to be rigorously characterized from these organisms. Indeed there is some evidence (Briggs and Iino, 1983) that there may be more than one blue light photoreceptor of physiological consequence, and there is strong evidence for an additional photoreceptor which absorbs in the UV-B region of the spectrum as well (Beggs et al., 1983; Wellmann, 1983). The absence of knowledge of the exact chemical nature of the photoreceptor(s) makes it difficult to say much either about the nature of signal perception or about early steps in the subsequent transduction chain leading to a physiological response.

One approach to identify a photoreceptor is to look for biochemical or biophysical events occurring very early following photoexcitation in intact organisms. If such changes are found, one might hope to isolate the components that change, and perhaps even develop an *in vitro* assay for the light effect obtained. Such a system could be of great value in identifying and characterizing the photoreceptor involved. Of course, finding such a change hardly constitutes proof that it lies along the transduction chain to a particular physiological response — or indeed that it has anything at all to do with it. At the very least, physiological correlations should be sought: Does the observed change precede or at least accompany the presumed physiological consequence rather than following it? If there are measurable changes leading to recovery of dark status during subsequent dark periods, do both the observed change and the physiological consequences show similar kinetics? Is the change found in responsive and not in unresponsive tissues? Are other photobiological properties — e. g. reciprocity — similar? Finally, are the fluence-response relationships for the observed change and the physiological process the same? Experiments addressing these and related questions can in principle establish strong

Photobiology, Edited by E. Riklis
Plenum Press, New York, 1991

correlative evidence for a physiological role for the change, although ultimate proof may have to depend upon the use of mutants.

Light effects on detection of phosphorylation of a 120 kD plasma membrane protein from stem tissues of etiolated pea seedlings

We have recently reported that blue light brings about a dramatic change in the detectable phosphorylation of a membrane-associated 120 kD protein from stem sections of etiolated pea seedlings (Gallagher et al., 1988). As phosphorylation and dephosphorylation of proteins is a widespread mechanism for the regulation of protein function (see Boyer and Krebs, 1986), and is a mechanism well established in higher plants (Poovaiah et al., 1987; Budde and Chollet, 1988; McFadden and Poovaiah, 1988; see Ranjeva and Boudet, 1987), we have started to characterize the system, and to apply some of the correlative tests described above to it.

Gallagher's initial experiments that lead to this study were performed on pea seedlings that were either grown in darkness or exposed to a few hours of white light (see Gallagher et al., 1988). Thereafter, growing regions of the stems were harvested and microsomal membranes purified by differential centrifugation. These membranes were then phosphorylated by addition of γ^{32}P-ATP following the technique of Shulman (1984). The membrane proteins were then separated by SDS polyacrylamide gel electrophoresis. Autoradiography was then used to detect phosphorylated proteins. One protein with a molecular weight near 120 kD appeared as a heavily phosphorylated band from the dark controls, but radioactivity at that molecular weight was essentially absent in autoradiographs of gels of membrane proteins obtained from the light-treated plants. Gallagher et al. (1988) then used sucrose density gradient separation followed by assays of marker enzymes to identify specific membrane fractions. The distribution of the 120 kD phosphorylated protein in the gradients was quite different from those of marker enzymes for mitochondria, endoplasmic reticulum, or Golgi, but very similar to that for a plasma membrane marker.

Gallagher et al. (1988) next investigated the nature of the membrane association: was it ionic, lipophilic, or the consequence of entrapment of a cytoplasmic protein into vesicles during membrane isolation? Because detectable activity was not found in cytoplasmic fractions, and neither osmotic shock nor sonication released the protein from the membranes, entrapment seemed unlikely. Because 1M KCl also failed to remove activity from the membranes, ionic binding also seemed improbable. Finally, detergent treatments sufficient to remove more than half of the total membrane protein removed only about the same fraction of the 120 kD protein. Hence these workers concluded that the association was lipophilic.

Finally they compared the relative effectiveness of red and blue light given *in vivo* in inducing the phosphorylation change noted *in vitro*. As much as 6 h of red light scarcely changed the level of phosphorylation detectable at 120 kD whereas relatively brief blue light exposures (less than 10 s) were fully effective. Hence the reaction was a response to the photoexcitation of a blue light photoreceptor, and not phytochrome.

616

Given the above information, an obvious next step is to investigate whether the photobiological properties of the phosphorylation change resemble those for any known physiological response or responses to light. By good fortune, two different physiological responses of pea seedlings to blue light that have been examined in detail recently. These are phototropic curvature in response to unilateral blue light (Baskin, 1986) and rapid suppression of epicotyl elongation (Laskowski and Briggs, 1988). In both cases, the blue light experiments were carried out on seedlings that had been grown under continuous red light even during blue light exposure. Red light irradiation was used so as to maintain most of the phytochrome as P_{fr}, the physiologically active form. Any additional effect of blue light presumably can be attributed to action through an authentic blue light photoreceptor, rather than through any P_{fr} that is formed by the action of blue light. This technique has been used recently in several other blue light studies (Baskin and Iino, 1987; Baskin et al., 1985, 1986; Iino and Briggs, 1984; Iino et al., 1985; Zeiger et al., 1985)

In the following paragraphs we discuss first pertinent known physiological and photobiological properties of these two responses, and then report our progress in examining these same properties with respect to the phosphorylation change. Finally we address briefly the possible nature of the phosphorylation change itself.

Physiology: Initial lag

Laskowski and Briggs (1988) found that rapid suppression of growth in pea seedlings following a saturating pulse of blue light or the onset of continuous blue light began after a 2-3 min lag. In continuous blue light, growth rate reached a new steady state in about 40 min. These kinetics are not unlike those reported earlier for other dicot stems (see Gaba and Black, 1983). A saturating 30 s pulse of blue light produced a decline in growth rate to a minimum at about 10 min. This minimum was followed by a transient increase and then a second minimum at about 30 min. The growth rate then gradually recovered and reached its initial level by about an hour after the light pulse. In contrast to the rapid suppression of elongation, first positive phototropic curvature in peas began only after a 10-15 minute lag period (Baskin, 1986). Maximum curvature was reached about 60 min after the onset of phototropic stimulation.

Physiology: Dark recovery

Another way that several blue light responses have been examined is to ask what effect a saturating light treatment has to subsequent sensitivity to normally effective blue light irradiations. Iino et al. (1985) found that the sensitivity of stomatal opening in response to a second blue light pulse was extremely low or absent following an initial saturating pulse, and then gradually recovered over the next 30 minutes. At present, comparable data are not available either for growth suppression or for phototropism of pea epicotyls. However, it is known that if maize or oat coleoptiles are given a unilateral pulse of light sufficient to saturate the photoreceptors on both illuminated and shaded sides of the organs, these organs are completely insensitive to normally effective phototropic stimulation. They remain insensitive for the first 5-10 minutes in darkness, and then gradually recover their phototropic sensitivity over the subsequent 30 min (Briggs, 1960).

Physiology: Tissue distribution

Laskowski and Briggs (1988) also found a distinct physiological gradient in the light response from top to bottom of the growing region of the stems of red light-grown pea seedlings. The uppermost 5 mm, just below the hook, was completely insensitive: the next 5-10 mm down responded to blue light with growth inhibition, but recovered in subsequent darkness. The lowest region of the growing zone showed strong growth inhibition, but failed to show any recovery.

A related question concerns the localization of photosensitivity for the two systems - does the responding tissue itself need to be irradiated? There are no such data either on growth suppression or phototropism for pea seedlings. However, Cosgrove (1981) concluded for cucumber hypocotyls that the sensitivity for growth suppression lies in the responding tissue itself, namely the growing region. Cosgrove (1985) failed to detect downward migration of differential growth during the development of phototropic curvature in cucumber, evidence consistent with localization of phototropic sensitivity to the growing region as well. (Such downward migration is a familiar feature of coleoptile phototropism, and reflects that the phototropic photoreceptors are in the coleoptile tip, and that there is then transmission of an auxin differential to the growing tissue below. For a more detailed discussion, see Briggs and Baskin, 1988.)

Physiology: Reciprocity

Baskin (1986) tested reciprocity for first positive phototropic curvature of red light-grown pea seedlings, and found that at least a ten-fold increase in exposure time with a concomitant ten-fold decrease in fluence rate had no effect on the magnitude of the curvature response. Likewise, Laskowski and Briggs (1988) tested reciprocity for the rapid suppression of epicotyl growth, and found that as long as the exposure time was shorter than the lag period preceding the growth decline, the reciprocity law held. It failed when illumination continued into the actual response period.

Physiology: Fluence-response relationships

The fluence-response relationships for phototropism and for light suppression of stem elongation are quite different. Phototropism is far more sensitive, showing a threshold response at about 3×10^{-2} μmol m^{-2}, and saturation at about 10^2 μmol m^{-2} (Baskin, 1986). Suppression of elongation show a threshold response well above 10^0 μmol m^{-2} and is not saturated until the fluence reaches 3×10^3 μmol m^{-2} (Laskowski and Briggs, 1988). Hence, the two kinds of light responses show photosensitivity over ranges of fluences that overlap, but are displaced by about two orders of magnitude.

Photobiological properties of the phosphorylation change

Let's examine these various properties with respect to the light- induced change in phosphorylation of the 120 kD protein. In all of the experiments summarized below, stem sections from dark-grown seedlings were harvested with red light as the safe-light (see above), and subsequently given the appropriate blue light treatments. Intact plants were not irradiated because irradiation of seedlings before the time-consuming harvest did not allow accurate measurements of kinetics on a time scale of less than 10 minutes. Where comparison of changes in irradiated sections with changes in irradiated intact seedlings have been possible (e. g. studies of dark recovery, see below), no significant differences were seen.

To begin with, the change is extremely rapid. To date, we have not been able to detect a lag between irradiation and the phosphorylation change even when the irradiation is less than 1 s in length and the sections plunged into ice-cold buffer and grinding initiated within 5 s. We plan to irradiate seedlings at temperatures near 0°C, or, alternatively, to plunge sections into liquid nitrogen immediately following irradiation in attempts to make a more precise determination of the lag time. We know, however, that the kinetics for the onset of the change are faster than those either for growth suppression (Laskowski and Briggs, 1988) or for the slower phototropic response (Baskin, 1986). Hence, as the phosphorylation change precedes either physiological change, a causal relationship with either is possible.

Second, the radioactive band is still absent when membranes are prepared from sections kept in darkness for 10 min following a saturating pulse of blue light, but gradually returns between 20 and 60 min. These kinetics for dark recovery are not incompatible with those for recovery of the dark growth rate in peas (Laskowski and Briggs, 1988), and recovery of blue light sensitivity both for phototropic sensitivity of coleoptiles (Briggs, 1960) and stomatal opening (Iino et al., 1985) following a saturating blue light treatment. Again, the available evidence does not rule out a causal relationship between the phosphorylation change and either physiological response.

Third, the strongest band of radioactivity at 120 kD is found in membranes isolated from the most apical centimeter of the epicotyl, with the signal declining sharply in more basal tissue. In addition, the radioactive band is undetectable in the combined hook and bud, excised from just above the growing region. The distribution of the capacity for change for the 120 kD protein is somewhat similar to that for the distribution of growth inhibition in pea epicotyls (Laskowski and Briggs, 1988), the distribution for growth inhibition in cucumber (Cosgrove, 1981), and that presumed for phototropic sensitivity in cucumber as well (Cosgrove, 1985). It should be noted that we had much lower resolution in the phosphorylation studies than that obtained by Laskowski and Briggs (1988) in the growth studies. The sections harvested for the phosphorylation studies could well have included both a non-responsive upper region and a strongly responsive

619

lower region. In addition, the growth studies were done with plants grown under red light, whereas the phosphorylation was on membranes from etiolated seedlings.

Finally, complete fluence-response curves have been obtained for completely etiolated seedlings and preliminary fluence-response data obtained for red light-grown seedlings — the latter for comparison with the two physiological responses for peas. In the case of the etiolated seedling, it made no difference whether time was kept constant and fluence rate was varied or vice versa, indicating that the reciprocity law probably held over a fairly wide range of fluences. Paradoxically, the fluence-response relationships in etiolated seedlings match those for phototropism of red light-grown seedlings (Baskin, 1986), although etiolated seedlings are phototropically virtually insensitive to blue light (Parker and Briggs, 1989). The fluence-response relationships for the phosphorylation change in red light-grown seedlings are the same as in etiolated seedlings. Blue light dosages sufficient for inducing phototropic curvature in red light-grown pea seedlings are effective in inducing the phosphorylation change in sections from similarly grown seedlings.

Possible explanations for the disappearance of the radiolabelled protein from membrane preparations isolated from preirradiated seedlings or sections

In principle there are at least six formal explanations for the absence of radioactivity on the gels at 120 kD following the *in vivo* light treatment: a) the protein could be released to the cytoplasm, b) a required kinase could be released to the cytoplasm (or otherwise inactivated by light), c) a specific phosphatase could be acquired by the membrane, d) a membrane-associated phosphatase could be activated by light, e) light treatment could result in phosphorylation of all available sites *in vivo*, leaving none available for subsequent phosphorylation in vitro, or f) light could alter the conformation of the protein *in vivo* so that the sites for phosphorylation or kinase association are inaccessible.

We have some preliminary results bearing on these possibilities. Irradiation of membranes obtained either from dark controls or from already irradiated stem tissue has the unexpected effect of *increasing* the capacity of a protein near 120 kD to become phosphorylated by exogenous ATP. If the 120 kD labelled proteins seen in membranes from dark-control sections and from irradiated membrane preparations are identical, then the first three possible explanations seem unlikely. Both the protein and the kinase are still present after membrane isolation, and there is no essentially cytoplasm in vitro from which to acquire a phosphatase. Light could of course activate a phosphatase *in vivo* and for some reason inactivate it *in vitro* (explanation d). However, possibilities e) or f) above seem the most viable — i. e. the protein is still present on the membrane following *in vivo* irradiation, but either does not have available sites for phosphorylation when the tissue is illuminated because they are already all phosphorylated or is driven by light into a conformation in which the sites for phosphorylation become inaccessible. If explanation e) is the case, light must be driving some cyclic phosphorylation-dephosphorylation process with the phosphorylation side predominating *in vivo* in the presence of adequate ATP, and the dephosphorylation side predominating *in vitro* in the absence of ATP. If explanation f) obtains, then light must somehow have opposite effects on the protein conformation *in vivo* and *in vitro*.

These same six formal possibilities (or some combination) could also account for the observed tissue distribution of capacity for phosphorylation. The absence of detectable phosphorylation in the hook and bud could be because a) the protein was missing, b) a specific kinase was missing or inactivated, c) a specific phosphatase is present in membranes of the hook but not in lower regions of the epicotyl, d) the phosphatase is present and active in the hook, and present but inactive except following illumination in the growing stem, e) the protein is present in the hook and bud, but is completely phosphorylated in the dark — hence light has no further effect, or f) the protein's conformation in the hook prevents the *in vitro* phosphorylation reaction whereas that in the growing stem does not. Because we have not yet attempted irradiation of membranes isolated from the hook region to determine whether capacity for phosphorylation can be photoinduced where it is otherwise absent (as was the case with membranes from irradiated sections from growing regions of the stem), we can not yet begin to distinguish between these various possibilities.

In any case, if the two proteins are indeed one and the same, then this system provides a powerful *in vitro* assay for the photoreceptor involved in the in vivo phosphorylation change. If the present correlative evidence holds up under further careful scrutiny, then one might have a good chance to identify the photoreceptor involved phototropism. Further experiments are clearly required to address these problems. Immunological techniques should help to determine whether the proteins affected by *in vitro* and *in vivo* irradiation are the same, and monoclonal antibodies may help to determine whether light-induced conformational changes are involved in the altered capacity for phosphorylation observed.

Much of the work described briefly in this article has now been published in Short, T.W. and W.R.Briggs (1990) Characterization of a rapid, blue light-mediated change in detectable phosphorylation of a plasma membrane protein from etiolated pea (*Pisum sativum* L.) seedlings. *Plant Physiol.* 92, 179–185.

Acknowledgment

The authors are grateful to Dr. William R. Eisinger for his careful review of the manuscript. This is Carnegie Institution of Washington Department of Plant Biology Publication No. 1031.

References

Baskin, T. I. (1986) Redistribution of growth during phototropism and nutation in the pea epicotyl. *Planta* 169, 406-414.

Baskin, T. I., W. R. Briggs and M. Iino (1986) Can lateral transport of auxin account for first positive curvature of maize coleoptiles? *Plant Physiol.* 81, 306-309.

Baskin, T. I., M. Iino, P. B. Green and W. R. Briggs (1985) High resolution measurements of growth during first positive phototropism in maize. *Plant Cell Environ.* 8, 595-603.

Baskin, T. I. and M. Iino (1987) An action spectrum in the blue and ultraviolet for phototropism in alfalfa. *Photochem. Photobiol.* 46, 127- 136.

Beggs, C. J., E. Wellmann and H. Grisebach (1987) Photocontrol of flavonoid biosynthesis. In Photomorphogenesis in Plants (Edited by R. E. Kendrick and G. H. M. Kronenberg), pp. 467-499. Martinus Nijhoff, Dordrecht, The Netherlands.

Boyer, P. D. and E. G. Krebs (Eds.) (1986) The Enzymes, 3rd Edit., XVII, Control by Phosphorylation, pp. 1-612. Academic Press, Orlando.

Briggs, W. R. (1960) Light dosage and the phototropic responses of corn and oat coleoptiles. *Plant Physiol.* 35, 951-962.

Briggs, W. R. and T. I. Baskin (1988) Phototropism in higher plants — controversies and caveats. *Botanica Acta* 101, 133-139.

Briggs, W. R. and M. Iino (1983) Blue light-absorbing photoreceptors in plants. *Phil. Trans. Roy. Soc. London* B 303, 347-359.

Budde, R. J. A. and R. Chollet (1988) Regulation of enzyme activity in plants by reversible phosphorylation. *Physiol. Plantarum* 72, 435-439.

Cosgrove, D. (1981) Rapid suppression of growth by blue light. Occurrence, time course, and general characteristics. *Plant Physiol.* 67, 584-590.

Cosgrove, D. (1985) Kinetic separation of phototropism from blue-light inhibition of stem elongation. *Photochem. Photobiol.* 42, 745-751.

Gaba, V. and M. Black (1983) The control of cell growth by light. In Encyclopedia of Plant Physiology NS 16A (Edited by W. Shropshire, Jr. and H. Mohr), pp. 358-400. Springer-Verlag, Berlin.

Gallagher, S., T. W. Short, P. M. Ray, L. H. Pratt, and W. R. Briggs (1988) Light-mediated changes in two proteins found associated with plasma membrane fractions from pea stem sections. *Proc. Natl. Acad. Sci. USA*, in press.

Iino, M., T. Ogawa and E. Zeiger (1985) Kinetic properties of the blue-light response of stomata. *Proc. Natl. Acad. Sci. USA* 82, 8019-8023.

Iino, M. and W. R. Briggs (1984) Growth distribution during first positive phototropic curvature of maize coleoptiles. *Plant Cell Environ.* 7, 97-104.

Laskowski, M. and W. R. Briggs (1988) Regulation of pea epicotyl elongation by blue light: fluence response relationships and growth distribution. *Plant Physiol.*, 89, 293–298.

McFadden, J. J. and B. W. Poovaiah (1988) Rapid changes in protein phosphorylation associated with light-induced gravity perception in corn roots. *Plant Physiol.* 86, 332-334.

Parker, K., T.I. Baskin and W. R. Briggs (1989) Evidence for a phytochrome-mediated phototropism in etiolated pea seedlings. *Plant Physiol.*, 89, 493–497.

Poovaiah, B. W., A. S. N. Reddy and J. J. McFadden (1987) Calcium messenger system: role of protein phosphorylation and inositol bisphospholipids. *Physiol. Plantarum* 69, 569-573.

Ranjeva, R. and A. M. Boudet (1987) Phosphorylation of proteins in plants: Regulatory effects and potential involvement in stimulus/response coupling. *Annu. Rev. Plant Physiol.* 38, 73-93.

Senger, H. and W. Schmidt (1986) Diversity of photoreceptors. In Photomorphogenesis in Plants (Edited by R. E. Kendrick and G. H. M. Kronenberg), pp. 137-183. Martinus Nijhoff, Dordrecht, The Netherlands.

Senger, H., Ed. (1980) The Blue Light Syndrome, pp. 1-655. Springer-Verlag, Berlin.

Senger, H. (Ed.) (1984) Blue Light Effects in Biological Systems, pp. 1-538. Springer-Verlag, Berlin.

Shulman, H. (1984) Phosphorylation of microtubule-associated proteins by $Ca+2$/calmodulin-dependent protein kinase. *J. Cell Biol.* 99, 11-19.

Wellmann, E. (1983) UV radiation in photomorphogenesis. In Encyclopedia of Plant Physiology NS 16B (Edited by W. Shropshire, Jr. and H. Mohr), pp. 745-756. Springer-Verlag, Berlin.

Zeiger, E., M. Iino and T. Ogawa (1985). The blue light response of stomata: pulse kinetics and some mechanistic implications. *Photochem. Photobiol.* 42, 759-763.

Potential Impacts of Increased Solar UV-B on Global Plant Productivity

ALAN H. TERAMURA AND JOE H. SULLIVAN

Department of Botany, University of Maryland
College Park, Maryland 20742 U.S.A.

Abstract

Ultraviolet-B radiation comprises only a small portion of the electromagnetic spectrum but has a disproportionately large photobiological effect. Both plants and animals are greatly affected by increases in UV-B radiation but there exists tremendous variability in the sensitivity of plant species to UV-B radiation. Approximately two out of three species tested appear sensitive and sensitivity also differs among cultivars of the same species.

Most of our knowledge of the effects of UV-B radiation on plants comes from studies of economically important crops. One species which has been extensively studied is soybean. Two cultivars grown for six seasons under enhanced UV-B radiation offered contrasting sensitivities with Essex exhibiting reductions in yield of 19 to 25% in four of the years, while Williams, was unaffected by increased UV-B radiation. The effectiveness of UV-B radiation was strongly affected by prevailing microclimatic factors such as precipitation patterns and air temperature.

In one of very few studies which examined UV-B radiation sensitivity in forest species, one important conifer species, loblolly pine, has been shown to be particularly sensitive in greenhouse tests. Ongoing field studies suggest that they also respond to UV-B radiation even under a full solar spectrum by increasing needle flavonoid content. However, the long-term effects have not yet been determined.

Plants have developed natural adaptations such as anatomical, morphological and biochemical changes which protect them from UV-B radiation. The extent of these natural adaptations may be related to the geographic origin of the species. It has been hypothesized that species originating from areas which receive high levels of UV-B radiation would be highly resistant to UV-B radiation. Plants collected along a 3000 m elevational gradient in Hawaii showed differences in sensitivity which were correlated with elevation. Most plants native to low elevations were sensitive to UV-B, but plants from the higher elevations, where UV-B is greatest, were very tolerant to UV-B radiation.

Stratospheric Ozone Depletion

Scientists have now accumulated sufficient evidence to show that human activity is rapidly changing the chemical composition of trace gases in the earth's atmosphere. Although subtle, these changes are beginning to produce global effects on the earth's ozone layer, climate and tropospheric chemistry. Of primary concern are the gases carbon dioxide (CO_2), chlorofluorocarbons (CFCs), methane (CH_4) and nitrous oxide (N_2O)

Photobiology, Edited by E. Riklis
Plenum Press, New York, 1991

(WMO, 1986). Annual rates of increase in concentrations have been measured at 0.5% for CO_2, between 5 and 7% for CFCs, 1% for CH_4, 0.2% for N_2O (WMO, 1986).

One important consequence of this rapid increase in trace gases results from the fact that they are all greenhouse gases. That is, they lead to an increase in global mean temperatures by absorbing infrared radiation. In addition to their contribution to global warming, chlorofluorocarbons (CFCs) may deplete the earth's protective stratospheric ozone layer. Their long atmospheric lifetimes (up to 150 years) allow them to be transported to the stratosphere where they are photodissociated and release chlorine, which catalytically destroys ozone.

Following an 18-month review, the International Ozone Trends Panel recently (March 1988) released its conclusive findings that significant global ozone depletion has already occurred. The panel cited a depletion of between 1.7% and 3.0% from 1969 to 1986 at latitudes between 30 and 64° North (where measurements are most extensive). The panel found that this decrease is conclusively linked to atmospheric chlorine, and is in addition to the natural variation in ozone levels (NASA, 1988).

At this time it is impossible to project future losses of ozone which will be associated with the inevitable increases in stratospheric chlorine and biogenically produced gases. Therefore, despite good prognosis for the cessation of ozone depletion based on the Montreal Protocol, which will limit CFC use, it would be premature to consider the threat of depletion to be ended.

Depletion of the ozone layer is of concern because the stratospheric ozone column is the primary attenuator of solar ultraviolet-B radiation (UV-B region, between 290 and 320 nm). A decrease in this ozone column would lead to increases in UV-B reaching the earth's surface. Though representing only a small fraction of the total solar electromagnetic spectrum, UV-B has a disproportionately large photobiological effect. One reason is that UV is readily absorbed by important macromolecules such as proteins and nucleic acids (Giese, 1964). Therefore, it is not surprising that both plant and animal life are greatly affected by increases in UV-B radiation penetrating to the earth's surface.

UV-B Effects on Crops

There exists tremendous variability in plant species sensitivity to UV-B radiation (Tevini et al., 1981; Teramura, 1983). Some species show sensitivity to present levels of UV-B radiation (Bogenrieder and Klein, 1978) while others are apparently unaffected by rather massive UV enhancements (Becwar et al., 1982). This issue is complicated further by reports of equally large response differences among cultivars of a species (Biggs et al., 1981; Teramura and Murali 1986). Approximately two-thirds of some 300 species and cultivars tested appear to be susceptible to damage from increased UV-B radiation.

Another important feature, the quality of crop yield, has only been quantitatively examined in a few species including tomato, potato (Biggs and Kossuth, 1978) and sugar beet (Ambler et al., 1978). The effectiveness of UV-B irradiation on plant growth varies seasonally and is affected by microclimate and soil fertility. For example, under water stress (Murali and Teramura, 1986) or mineral deficiency (Murali and Teramura, 1985), soybeans are less susceptible to UV-B radiation but under low levels of visible radiation sensitivity increases (Mirecki and Teramura, 1984; Warner and Caldwell, 1983). Thus field validation studies conducted over several growing seasons are crucial in any UV-B impact assessment of agricultural productivity.

In 1981, two soybean (*Glycine max* (L) Merr.) cultivars were chosen for study based upon preliminary greenhouse trials for UV sensitivity and planted into the field (Teramura and Murali, 1986). Based upon overall growth performance in the greenhouse, Essex was found to be sensitive while Williams was tolerant to UV-B radiation. Field experiments were conducted during May through October of 1981 to 1986 at the Agricultural Research Center, USDA, Beltsville, Maryland, U.S.A.

Supplemental UV-B radiation was supplied by filtered Westinghouse FS-40 sunlamps oriented perpendicular to the planted rows and suspended above the plants. Lamps were filtered either with 0.13 mm thick cellulose acetate (transmission down to 290 nm) for supplemental UV-B radiation or 0.13 mm Mylar Type S plastic films (absorbs all radiation below 320 nm) as a control. The radiation filtered through the cellulose acetate supplied a weighted daily supplemental irradiance of either 3.0 or 5.1 effective kJ m^{-2} UV-B$_{BE}$ using the generalized plant response action spectrum (Caldwell, 1971) normalized to 300 nm. Plants beneath these cellulose acetate filtered lamps received supplemental doses in addition to ambient levels of UV-B radiation. These increased levels of UV-B radiation (supplemental + ambient) were similar to those which would be received at College Park, Maryland, U.S.A. (39° N) with anticipated 16 and 25% stratospheric ozone reductions during a cloudless day on the summer solstice (Green et al., 1980). The weighted irradiance of Mylar filtered lamps was 0, so plants beneath these lamps received only ambient levels of UV-B (8.5 effective kJ m^{-2} UV-B$_{BE}$ on the summer solstice). Spectral irradiance beneath the lamps was measured with an Optronics Model 742 spectroradiometer equipped with a double monochromator with dual holographic grating and interfaced with a Hewlett Packard 85 printing calculator. The spectroradiometer was calibrated using a National Bureau of Standards traceable 1000 W tungsten halogen lamp and wavelength alignment checked with known mercury emission lines using a Hg Arc lamp.

The results of this 6-year field study demonstrate intraspecific differences in UV-B sensitivity in soybean yield and quality (Table 1). However, the expression of these sensitivity differences to UV-B radiation was altered by other prevailing microclimatic factors. For the sensitive soybean cultivar Essex, a 25% ozone reduction reduced overall yield by 19–25% during 4 of the 6 years. The 1983 and 1984 seasons were characterized as hot and dry with prolonged periods of drought. Parallel field studies have shown that the effects of UV-B radiation can be masked by drought-induced growth reduction (Murali and Teramura, 1986). For perspective, Figure 1A shows the sources of yield losses in U.S. soybean under current levels of stratospheric ozone. Figure 1B shows the relative yield loss predicted from a 25% ozone depletion, indicating that UV-B-induced

627

Table 1. Summary of UV-B radiation effects on soybean yield. A 25% ozone depletion was simulated over College Park, Maryland, U.S.A.

Year	% Change in Yield	
	Essex	Williams
1981	−25	+22
1982	−23	+14
1983*	+6	−11
1984*	−7	+10
1985	−20	+4
1986	−19	+6

*Years with prolonged drought

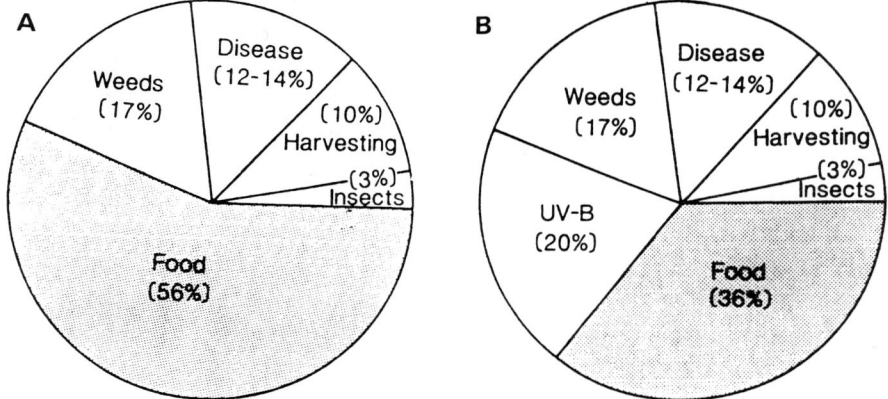

Figure 1. Sources of yield loss in U.S. soybean under current levels of stratospheric ozone (A) and predicted from a 25% ozone depletion (B).

losses may severely limit soybean productivity. In contrast, yield increased from 5 to 22% in 5 of the 6 years for the UV-B resistant cultivar Williams.

The number of precipitation events, air temperature, the number of days of low irradiance, and UV-B radiation all interact in a complex manner to affect crop yield. Although these various interactions are complex, linear models can predict crop yield reasonably well. Based on maximum improvement of R^2 (SAS 1982), the most appropriate predictive model for yield in Essex was

$$Y = 39.42 - 0.32A - 1.38B - 0.79C$$

where Y = predicted seed yield, A = total UV-B dose, B = number of days where air temperatures exceeded 35°C, and C = number of precipitation events.

The model which best approximated yield in Williams was

$$Y = 154 - 0.78B - 3.08C - 13.71D$$

where D = precipitation frequency. Expanding models beyond three variables did not significantly increase the F-statistic or the coefficient of determination. For both cultivars R^2 values exceeded 0.94 and predicted yield closely approximated actual yield. The model for Essex, the UV-susceptible cultivar, includes total UV dose while the model for the resistant Williams cultivar does not. Both models are capable of predicting crop yield within 95% confidence intervals for the five-year period (1986 was excluded from analysis because artificial irrigation was used). Through the use of such models it may be possible to more realistically assess the effects of increased levels of UV-B in concert with climatic changes than by a simple UV dose-response relationship, which does not adequately explain field observations. This interaction between UV-B dose and microclimatic variation needs further evaluation. Global climate changes altering the temperature and pattern of precipitation have been predicted due to increases in greenhouse gases in the atmosphere resulting from human activity (Titus, 1986). Such changes in climate in concert with increasing levels of UV-B radiation reaching the earth could profoundly influence the productivity of soybean and other crops, as well as that of native plant communities.

UV-B Effects on Forests

In contrast to crops, few studies have been undertaken on woody perennials (trees), which account for up to 80% of global net primary productivity and occupy as much as one-third of the land area of the United States (Whittaker, 1975; Solomon and West, 1986). Only two studies have been completed under field conditions and one of these examined the effects of solar UV-B radiation exclusion (Bogenrieder and Klein, 1982). Exclusion of naturally occurring UV-B radiation increased the growth of four broadleaf species but supplemental UV-B irradiation had no effect on growth in either Engelmann spruce or lodgepole pine (Kaufmann, 1978). Three additional studies have demonstrated deleterious effects of UV-B radiation on tree growth and physiology (Kossuth and Biggs, 1981; Sullivan and Teramura, 1988 and 1989). However, none of these studies were extended beyond a single growing season, thus the long-term effects of UV-B radiation on trees is unknown.

The examination of UV-B radiation effects on perennial species provides a unique opportunity to observe more subtle responses to protracted UV exposure which are impossible to investigate in annual species. For instance, it is presently unknown whether UV-B repair mechanisms can mitigate UV-B damage during the dormant period when ambient levels of solar UV-B are at their seasonal minimum or whether physiological changes, which accompany tissue hardening prior to entering dormancy, modify its sensitivity to subsequent UV-B exposure. The range of responses observed both inter- and intra-specifically suggests that extrapolations between annual and perennial species may not be feasible. Therefore, some direct field validation experiments on key forest species are essential before realistic estimates of the effects of increased UV-B on these species can be made.

Due to their economic importance and widespread global distribution, conifers have been selected for study in over half of the studies on trees. Some 15 species of conifers have been tested to date for susceptibility to UV-B radiation (Table 2). Of these, 7 were deleteriously affected, 5 were resistant, and 3 were favored by UV-B radiation (Sullivan and Teramura, 1988a).

Loblolly pine was one of the most susceptible species with reductions of biomass and height of 40 and 16%, respectively. Loblolly pine is the leading commercial species in the southeastern United States and accounts for a large portion of the United States pulp producing capacity (Harlow and Harbar, 1969; Walker, 1980). Therefore, there would be enormous economic consequences should increasing levels of UV-B substantially reduce loblolly pine productivity.

Kossuth and Biggs (1981) evaluated the effects of UV-B on loblolly pine growth in growth chambers and Sullivan and Teramura (1988) observed significant UV-B effects under unshaded greenhouse conditions. Both studies demonstrated that newly emergent loblolly pine seedlings were deleteriously affected by increased levels of UV-B radiation. Additionally Sullivan and Teramura (1989) have shown that established saplings are also affected in both greenhouse and field conditions by increases in UV-B radiation. In these greenhouse studies on one-year old saplings, loblolly pine responded in a dose-specific manner to UV-B radiation by increasing needle flavonoid concentrations. The rate of increase was also dose-specific, and at a low supplemental UV-B dose (11 kJ m^{-2}) the increases appeared only after photosynthetic capacity and growth had been reduced. Rapid

Table 2. Summary of effects of UV-B radiation on growth of 15 coniferous species based upon changes in biomass accumulation and height

Sensitive	Sensitivity Rating* Resistant	Favored
Abies procera	*Picea glauca*	*Abies concolor*
Pinus contorta	*Pinus edulis*	*Abies fraseri*
Pinus elliottii	*Pinus nigra*	*Picea engelmannii*
Pinus ponderosa	*Pinus strobus*	
Pinus resinosa	*Pseudotsuga menziesii*	
Pinus sylvestris		
Pinus taeda		

*Sensitive = reductions exceeding 5% of controls
 Resistant = ± 5% of controls
 Favored = increases exceeding 5% of controls

increases in flavonoid concentration initially protected plants from deleterious UV-B effects at the highest supplemental dose (19 kJ m^{-2}). However, this protection was incomplete and after six months of irradiation, needle flavonoid concentrations decreased, producing reductions in photosynthetic capacity and growth.

Due to the nature of greenhouse environments, however, caution must be exercised in the extrapolation of these results to the field setting. The effectiveness of UV-B is often exaggerated under growth chamber and greenhouse conditions due to reduced photosynthetic photon flux density (PPFD between 400 and 700 nm) levels (Mirecki and Teramura, 1984; Warner and Caldwell, 1983). This apparent increase in UV-B sensitivity in controlled environments is probably due to reductions in photoprotective and photoreactivation mechanisms. Preliminary results from a field study on loblolly pines demonstrate that needle flavonoid content was increased and growth in one of four local seed sources was reduced under supplemental UV-B radiation simulating a 25% ozone reduction. These results demonstrate that even under a complete solar spectrum, loblolly pines were sensitive to UV-B radiation and responded by increasing needle flavonoid concentrations. These increases could reduce further sensitivity but the long-term protective role of flavonoids has not yet been demonstrated. Studies extending over several growing seasons and an evaluation of the metabolic or energetic costs of producing and maintaining high flavonoid concentrations will be necessary to determine the effects of increasing solar UV-B radiation on overall tree productivity.

UV-B Effects on Natural Vegetation

Currently, almost no information exists on the degree of variability that exists in our natural vegetation and therefore we have little basis upon which to assess the potential impacts of increasing levels of UV-B on natural plant communities. To address the broad question of potential global UV effects on terrestrial communities and ecosystems, we must currently make the unlikely assumption that native perennial woody trees and shrubs respond in a fashion analogous to annual herbaceous crop species. An additional function to studying UV adaptations in natural plant species is that we might find novel or unique protective mechanisms which have not been detected in crop plants already exposed to intensive artificial selection. These might then be useful in future crop breeding programs after careful screening and genetic analysis.

Perhaps the first place to search for UV-B responsiveness in native plants is in regions where natural levels of UV-B are already quite high. Since the weighted daily UV-B dose received at low latitude, high elevation sites (i.e. tropical mountains) can be nearly six-fold greater than the maximum dose received at arctic latitudes (Caldwell et al., 1982), plants which naturally occur in such high UV environments would undoubtedly have evolved specific adaptations which protect them from the deleterious effects of UV-B (Antonovics, 1975). For example, Caldwell et al. (1982) found that arctic ecotypes (variants of the same species) of *Oxyria digyna* were consistently more sensitive to UV radiation than their counterparts collected from mountains in lower latitudes. The presence of secondary compounds such as flavonoids which may act as solar screens, absorbing UV and not allowing it to reach sensitive tissue layers in the epidermis of leaves has been shown to vary greatly among various plant species collected along a latitudinal gradient (Robberecht and Caldwell, 1978). However, a simple correlation between UV sensitivity and epidermal flavonoid concentrations does not seem to exist (Barnes et al., 1987). Caldwell et al. (1982) and Barnes et al. (1987) have also reported that the differences in sensitivity between plants collected from high UV environments (tropical mountains) were not simply due to differences in epidermal flavonoids. Instead, it appeared that the photosynthetic apparatus of these high elevation tropical plants was inherently more resistant to UV than that of plants collected from higher latitudes. However, the specific nature of these inherent differences has yet to be elucidated.

By limiting our studies to crop species in temperate environments, we may be observing only a small fraction of the range of UV protective mechanisms that plants have evolved. Therefore, it would be particularly productive to examine the types of adaptations possessed by plants naturally growing in regions of the world with the greatest solar UV-B flux. In a recent study, seeds were collected from 132 native and introduced plant species growing over a 3,000 m elevational gradient in Hawaii (23°N latitude). Seeds were brought back to the University of Maryland, U.S.A. and germinated in the greenhouse under artificial UV lamps simulating 20 and 40% ozone depletions over Honolulu, Hawaii during clear sky conditions on the summer solstice (14.6 and 22.2 kJ m^{-2} effective UV-B$_{BE}$ according to an empirical model of Green et al. 1980). In the UV-B radiation treatment simulating a 40% ozone depletion, only 8% of the species collected from sea level to 500 m were tolerant to UV, while tolerance increased markedly in species collected from higher elevations (Table 3). All species collected above 2,000 m

Table 3. Summary of UV tolerance in Hawaiian plants collected over a 3000 m elevational gradient. All plants were irradiated for 12 weeks (Teramura unpublished).

Elevation (m)	Ozone depletion*	
	20%	40%
0–500	15% favored	15% favored**
	31% tolerant	8% tolerant
	54% sensitive	77% sensitive
500–1,000	75% tolerant	75% tolerant
	25% sensitive	25% highly sensitive
>2,000	100% tolerant	100% tolerant

* ozone depletion relative to sea level at 23*N latitude (Honolulu, HI)
**favored = 5% increase in biomass
 tolerant = ±5% change
 sensitive = 5–25% decrease in biomass
 highly sensitive = >25% decrease in biomass

were found to be tolerant of UV-B radiation levels simulating a 40% ozone depletion, implying that these plants have adapted to the high levels of UV presently incident at these sites. A greater understanding of the range and types of natural UV protective mechanisms available to plants would provide a better basis upon which to make predictions of the effects of increased levels of solar UV on natural plant communities.

References

Ambler, J.E., R.A. Rowland and N.K. Maher (1978) Response of selected vegetable and agronomic crops to increased UV-B irradiation under field conditions. UV-B Biological and Climatic Effects Research (BACER), Final Report. EPA-IAG-D6-0168. U.S. EPA, Washington, D.C.

Antonovics, J. (1975) Predicting evolutionary response of natural populations to increased UV radiation. In. Climatic Impact Assessment Program (CIAP), Monograph 5 (Edited by Nachtwey, D.S., M.M. Caldwell and R.H. Biggs), pp. 8–7 — 8–27. U.S. Dept. Transport, Report No. DOT-TST-75-55, National Techn. Infor. Serv., Springfield, Virginia.

Barnes, P.W., S.D. Flint and M.M. Caldwell (1987) Photosynthesis damage and protective pigments in plants from a latitudinal arctic/alpine gradient exposed to supplemental UV-B radiation in the field. *Arctic and Alpine Research* 19,21–27.

Becwar, M.R., F.D. Moore III and M.J. Burke (1982) Effects of deletion and enhancement of ultraviolet-B (280–315 nm) radiation on plants grown at 3000 m elevation. *J. Amer. Soc. Hort. Sci.* 107,771–779.

Biggs, R.H., S.V. Kossuth and A.H. Teramura (1981) Response of 19 cultivars of soybeans to ultraviolet-B irradiance. *Physiol. Plant.* 53,19–26.

Biggs, R.H. and S.V. Kossuth (1978) Effects of ultraviolet-B radiation enhancement under field conditions on potatoes, tomatoes, corn, rice, southern peas, peanuts, squash, mustard and radish. UV-B Biological and Climatic Effects Research (BACER), Final Report, EPA, Washington, D.C.

Bogenrieder, A. and R. Klein (1978) Die abhangigkeit der UV-empfindlichkeit von der lichtqualitat bei der aufzucht (*Lactuca sativa* L.). *Angew. Botanik* 52,283–293.

Caldwell, M.M., R. Robberecht, R.S. Nowak and W.D. Billings (1982) Differential photosynthetic inhibition by ultraviolet radiation in species from the arctic-alpine life zone. *Arctic and Alpine Res.* 14,195–202.

Caldwell, M.M. (1971) Solar UV irradiation and the growth and development of higher plants. In Photophysiol. VI (Edited by A.C. Giese), pp. 131–177.

Giese, A.C. (1964) Studies on ultraviolet radiation action upon animal cells. In Photophysiology Vol. 2, (Edited by A.C. Giese), pp. 203–245. Academic Press, NY-London.

Green, A.E.S., K.R. Cross and L.A. Smith (1980) Improved analytical characterization of ultraviolet skylight. *Photochem. Photobiol.* 31,59–65.

Harlow, W.M. and E.S. Harbar (1969) Textbook of Dendrology. pp. 87–90, McGraw-Hill, ISBN 07-026569-0.

Kaufmann, M.R. (1978) The effect of ultraviolet (UV-B) radiation on Engelmann spruce and lodgepole pine seedlings. UV-B Biological and Climatic Effects Research (BACER). Final Report EPA-IAG-D6-0168. EPA, Washington, DC.

Kossuth, S.V. and R.H. Biggs (1981) Ultraviolet-B radiation effects on early seedling growth of Pinaceae species. *Can. J. For. Res.* 11,243–248.

Mirecki, R.M. and A.H. Teramura (1984) Effects of ultraviolet-B irradiance on soybean. The dependence of plant sensitivity on the photosynthetic photon flux density during and after leaf expansion. *Plant Physiol.* 74,475–480.

Murali, N.S. and A.H. Teramura (1986a) Effectiveness of UV-B radiation on the growth and physiology of field-grown soybean modified by water stress. *Photochem. Photobiol.* 44,215–220.

Murali, N.S. and A.H. Teramura (1985) Effects of ultraviolet-B irradiance on soybean. VI. Influence of phosphorus nutrition on growth and flavonoid content. *Physiol. Plant.* 63,413–416.

National Aeronautics and Space Administration (NASA) (1988) Executive Summary of the Ozone Trends Panel, NASA, Washington, D.C.

Robberecht, R. and M.M. Caldwell (1978) Leaf epidermal transmittance of ultraviolet radiation and its implication for plant sensitivity to ultraviolet-radiation induced injury. *Oecologia* 32,277–287.

Solomon, A.M. and D.C. West (1986) Simulating forest responses to expected climate change in eastern North America: Applications to decision-making in the forest industry. In Climate Change and Future Forest Management in the United States (Edited by W.E. Shands). The Conservation Foundation, Washington, D.C. In press.

Sullivan, J.H. and A.H. Teramura (1988) Effects of ultraviolet irradiation on seedling growth in the pinaceae. *Amer. J. Bot.* 75,225–230.

Sullivan, J.H. and A.H. Teramura (1989) The effects of ultraviolet-B radiation on loblolly pine: 1. Growth, photosynthesis and pigment production in greenhouse-grown saplings. *Physiol. Plant* 77, 202–207.

Teramura, A.H. and N.S. Murali (1986) Intraspecific differences in growth and yield of soybean exposed to ultraviolet-B radiation under greenhouse and field conditions. *Environmental and Experimental Botany* 26,89–95.

Teramura, A.H. (1983) Effects of ultraviolet-B radiation on the growth and yield of crop plants. *Physiol. Plant.* 58, 415–422.

Tevini, M., W. Iwanzik and U. Thoma (1981) Some effects of enhanced UV-B irradiation on the growth and composition of plants. *Planta* 153,388–394.

Titus, J.G. (1986) The causes and effects of sea level rises. In Effects of Changes in Stratospheric Ozone and Global Climate. VI. (Edited by J.G. Titus) pp. 219–248. U.S. Environmental Protection Agency. Washington, D.C.

Walker, L.C. (1980) The southern pine region. In Regional Silviculture of the United States. (Edited by J.W. Barrett). John Wiley and Sons, ISBN 0-471-05645-6.

Warner, C.W. and M.M. Caldwell (1983) Influence of photon flux density in the 400–700 nm waveband of inhibition of photosynthesis by UV-B (280–320 nm) irradiation in soybean leaves: separation of indirect and immediate effects. *Photochem. Photobiol.* 38,341–346.

Whittaker, R.H. (1975) Communities and Ecosystems. MacMillan Co., New York.

World Meteorological Organization (WMO) (1986) Atmospheric Ozone 1985. Assessment of our Understanding of the Processes Controlling its Present Distribution and Change, WMO Global Ozone Research and Monitoring Project, Report No. 16, WMO, Geneva, Switzerland.

Effects of Enhanced Solar UV-B Radiation on Growth and Function of Selected Crop Plant Seedlings

M. TEVINI, U. MARK, G. FIESER, and M. SAILE

Botanisches Institut II, Universität Karlsruhe, Kaiserstr. 12
D-7500 Karlsruhe 1, F.R.G.

Introduction

The damaging effects of enhanced UV-B radiation which is to be expected after a possible destruction of the ozone layer by chlorofluorohydrocarbons (Stolarski, 1988) have been demonstrated by several research groups. Mainly reductions of leaf area, fresh weight, lipid content and of photosynthetic activity were found in UV-B sensitive plants (Sisson and Caldwell, 1976; Bogenrieder and Klein, 1977; Brandle et al., 1977; Teramura, 1983; Teramura and Murali, 1986; Tevini and Iwanzik, 1983; Caldwell, 1981; Tevini et al., 1981, 1983 a, b, 1986; Norudeen and Kulandaivelu, 1982; Iwanzik et al., 1983). In addition structural alterations of the leaf surface (Tevini et al., 1981), the epicuticular waxes (Steinmüller and Tevini, 1985) and the diffusion of water vapour through the stomates (Teramura et al., 1983) have been observed.

These potential damages of enhanced UV-B radiation can be counteracted in many plant species by an UV protective mechanism namely the formation of UV-screening pigments, e.g. the flavonoids, especially the flavonols and the anthocyanins (Wellmann, 1982). They are located in the leaf epidermis of many plant species (Weissenböck et al., 1986), so that much of the UV-B radiation may be removed whereas most of the photosynthetically active radiation is transmitted (Caldwell et al., 1983). The red or blue anthocyanins do not absorb in the UV region and thus may not be considered as UV screens. However, they may function as screening pigments for cytochromes and phytochrome system, which is involved in the regulation of the flavonoid biosynthesis, whereas UV-B itself can induce it (Wellmann, 1974).

In all UV experiments reported so far, the plants were exposed to additional artificial UV-B radiation. Because of differences between the emission spectra of artificial lamps (emission of UV-C, low UV-A) and the solar spectrum (no UV-C, high UV-A) it is problematic to draw conclusion from experiments obtained in growth chambers and to transfer them to plants living under natural radition conditions. This problem cannot be solved by the use of cut-off-filters or plastic films which cannot exactly simulate the solar cut-off. The only solution of the basic problem is the use of the sun itself as a

Photobiology, Edited by E. Riklis
Plenum Press, New York, 1991

radiation source and ozone as a UV filter. By streaming ozone through an UV-transmissable plexiglas cuvette covering the plants, it is possible to reduce ambient UV radiation of southern latitudes to northern latitudes and to regard the latter as a "control" radiation. The ambient solar UV-B radiation of the southern location can then be regarded as enhanced compared to the "control" radiation.

This paper describes the effects of enhanced and reduced solar UV-B radiation on some agricultural plants grown from may through september in Portugal (38°N). Growth, photosynthetic gas exchange, chlorophyll fluorescence and the content of UV-screening pigments were measured in selected crop seedlings.

Materials and Methods

Plants

Sunflower seedlings (*Helianthus annuus* L.) were grown at constant temperature (20°C) and humidity (55% RH) in two identical growth chambers and used for growth analysis, gas exchange and chlorophyll fluorescence measurements. Radish (*Raphanus sativus* L.) and mustard seedlings (*Sinapis alba* L.) were taken for pigment analysis. They were grown in darkness in the lower part of the chambers.

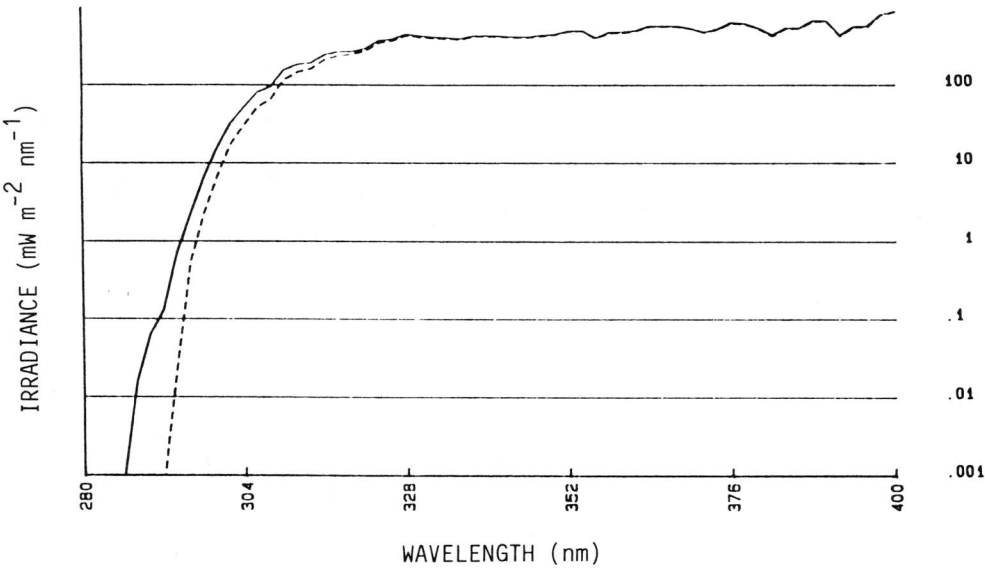

Fig.1. Spectral energy distribution of solar UV-B radiation (---) and of reduced solar UV-B radiation (- -) of Portugal (18th of August 1988, local noon, 12.40 hrs.).

Ozone cuvette

The growth chambers were covered by special plexiglass cuvettes (GS 2458, Röhm und Haas, D-6100 Darmstadt) which are permeable for visible light and UV-B radiation. One of these cuvettes contained normal air, thus plants grown under this cuvette received the full sunlight of Portugal. The second cuvette was filled with a gas mixture of O_2 and ozone which was provided by an ozone generator (Technomed, D-7504 Weingarten). Ozone concentration and atmospheric pressure inside the cuvette were controlled continuously. According to the absorbance properties of ozone the ambient UV-B radiation was shifted between 2 and 4 nm to longer wavelength and simultaneously the UV-B irradiance was reduced by 20–26% (Figure 1). Thus the UVB irradiance the plants were exposed to inside the growth chamber by this new ozone technique was, e.g. at the 18th of August at local noon (12.40 hrs) and clear sky, 2.36 $Wm^{-2}s^{-1}$ (control plants), whereas the irradiance under the 'air cuvette' was 2.97 $Wm^{-2}s^{-1}$ which is a difference of 25.6%.

Gas exchange

Sunflower seedlings up to 23 days old were fixed in a gas-tight minicuvette (Walz, D-8521 Effeltrich). This chamber contained air with a defined CO_2-concentration (ambient concentration), temperature (20°C) and humidity (55% RH). Gas exchange by the plant was detected by infrared gas analyzers (Leybold-Heraeus, D-6450 Hanau) and recorded by an IBM PC XT 286. The calculations for the gas exchange rates were made by using the equations described by von Caemmerer and Farquhar (1981). The CO_2 uptake was measured at 2811 $\mu Em^{-2}s^{-1}$.

Fluorescence induction

Fluorescence measurements were performed with a PAM fluorometer (Walz), described elsewhere (Schreiber et al., 1986). The modulated measuring light used here had an irradiance of 4.66 mWm^{-2} at a frequency of 1.6 kHz and 291.75 mWm^{-2} at 100 kHz, corresponding to 0.025 $\mu Em^{-2}s^{-1}$ at 1.6 kHz and 1.59 $\mu Em^{-2}s^{-1}$ at 100 kHz, respectively. The actinic light was provided by a Schott lamp KL 1500 at an irradiance of 54.76 Wm^{-2} (253.5 $\mu Em^{-2}s^{-1}$). Saturation pulses were also derived from a Schott KL 1500 at an intensity of 1373.7 Wm^{-2} (6800 $\mu Em^{-2}s^{-1}$). Pulses were applied at a frequency of 0.1 Hz with a duration of 0.9 seconds.

Fluorescence induction kinetics of sunflower cotyledones were measured 10 and 17 days after germination and of primary leaves 17 and 24 days, respectively. The plants were darkened for one hour prior to measurements.

The signals of fluorescence yield were recorded by the asystant+ software package (McMillan Software) via a data aquisation board (Dash 16 F, Keythley Instr., D-8000 München) within an IBM PC XT 286.

637

Pigments

Etiolated seedlings were grown for four days in darkness, then exposed to the two different radiation conditions in the growth chambers for eight hours and thereafter returned to darkness for 24 hours. The extraction of the pigments followed the method of Reznik and Krause (1968). Ten pairs of cotyledons or ten hypocotyls were used for hot extraction in distilled water. After adding either $AlCl_3$ or concentrated HCl the absorbance at 405 nm (flavonols) or 530 nm (anthocyanins) was measured in a Shimadzu UV-VIS 240 spectrophotometer. The flavonol and anthocyanin content was calculated from calibration curves of rutin and delphinidin, respectively.

The estimation of the chlorophyll content of growing seedlings followed the method described by Lichtenthaler (1987).

Growth parameters

The hypocotyl length was determined in two different ways. A transducer as described by Green and Cummins, (1974) detected the growth rate of seven days old sunflower seedlings over a period of three days. Additionally, the hypocotyl length of 25 seedlings was measured after 10, 13, 18 and 23 days.

Leaf area measurements were made with the areameter model LI-COR Li 3000.

Statistics

Each experiment was made several times so that the following results represent means and their standard deviations. For statistical calculations the Kolmogoroff-Smirnov-Test, the two-sample F-Test and the two-sample T-Test were used. The error levels, as far as the differences are significant, are indicated as follows: * = $\alpha \leq 0.05$, ** = $0.05 < \alpha \leq 0.1$, *** = $0.1 < \alpha \leq 0.2$.

Results and Discussion

Photosynthesis

Chlorophyll concentration

Green seedlings used for photosynthesis and growth analysis had always higher chlorophyll content per leaf area under enhanced UV-B radiation during development (Figure 2). A possible explanation for this effect is the decrease in the leaf area (see Figure 16) whereas the chlorophyll synthesis appeared to be uneffected by UV-B.

Net photosynthesis

Sunflower seedlings grown under enhanced UV-B radiation for 13 days showed no significant difference in the maximal CO_2 uptake based on chorophyll in comparison to control plants (reduced UV-B). A significant decline in the CO_2 fixation rate of the plants arose, however, after another 5 days under enhanced UV-B. The CO_2 uptake of the control plants amounted to 247.9 $\mu Mol\ CO_2 mMol\ chl^{-1}\ h^{-1}$ in contrast to the UV-B plants

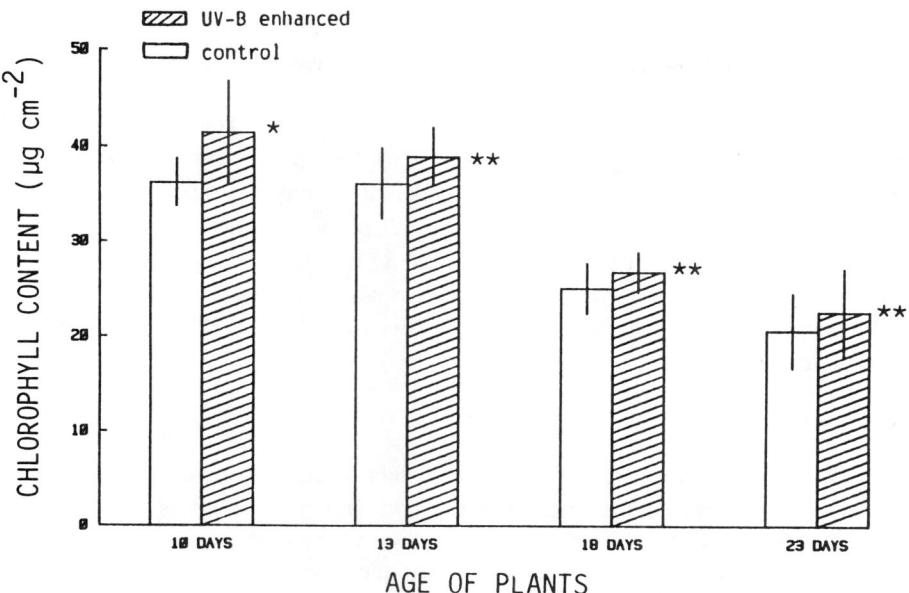

Figure 2. Chlorophyll content of leaves from sunflower seedlings at different ages.

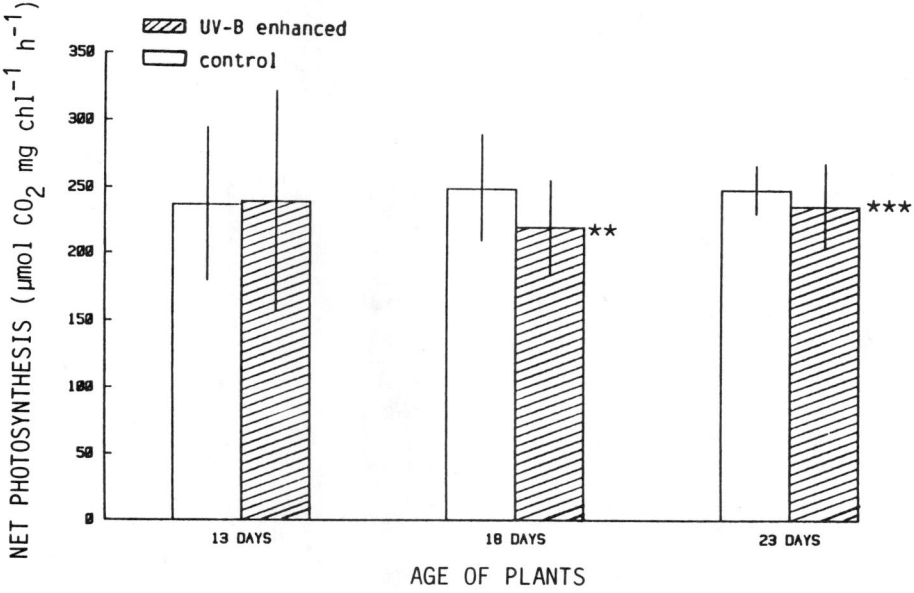

Figure 3. Net photosynthesis of sunflower seedlings at 2811 $\mu Em^{-2}s^{-1}$.

which reached only 220.7 μMol CO_2 mMol chl^{-1} h^{-1}. This represented a decline of about 11%. Twenty-three days after germination the decline was reduced to about 5%, nevertheless, the difference was still significant. The mentioned data are summarized in Figure 3.

These data are comparable to results found in plants grown under artificial radiation (Thai, 1975; Van and Garrard, 1976; Teramura, 1980; Vu, 1982). For the damaging effect of UV-B on net photosynthesis different explanations may be regarded. Firstly, the inhibition of the Ribulosebisphosphate-Carboxylase which was found in tomatoes (*Lycopersicon esculentum*; Thai, 1975) or in case of C_4-plants of Phosphoenolpyruvat-Carboxylase . Secondly, the inhibition of the photosystems, especially of PS II which was demonstrated earlier (Iwanzik et al., 1983, Tevini and Pfister, 1985). Further studies are needed to confirm one or both hypotheses.

Fluorescence

Sunflower leaves of plants grown under enhanced UV-B showed a higher ground level fluorescence, Fo (Figure 4) which correspond with earlier findings under artificial radiation (Tevini et al., 1988) and also with the higher chlorophyll content per leaf area (Figure 2). Generally, the Fo values increased with increasing age of leaves.

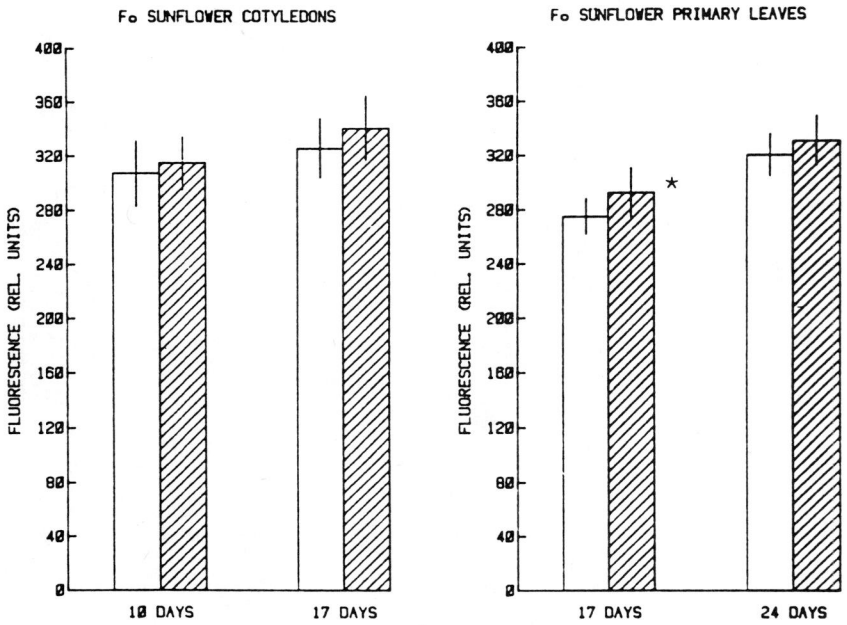

Figure 4. Ground level fluorescence Fo in sunflower leaves under different radiation conditions.

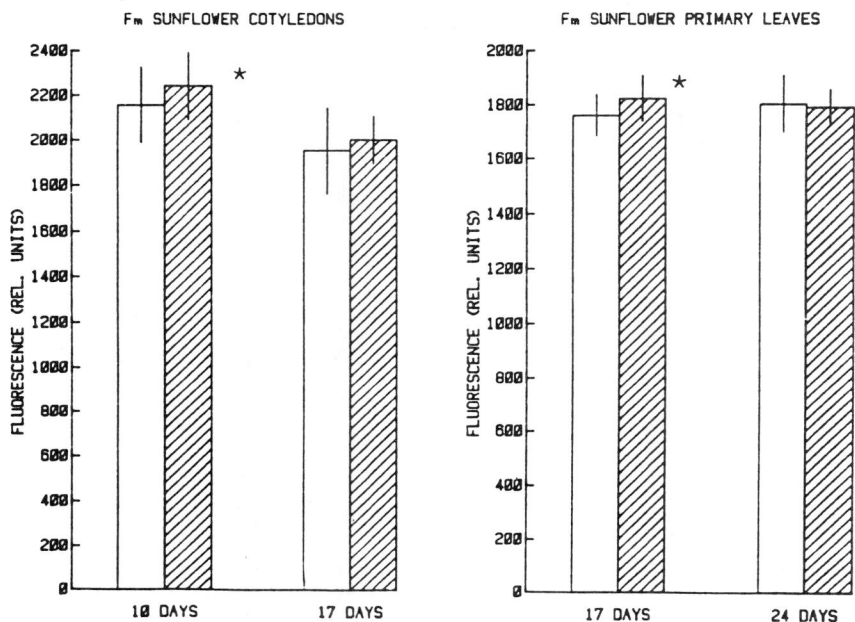

Figure 5. Maximum fluorescence Fm in sunflower leaves under different radiation conditions.

Maximum fluorescence yield, Fm decreased with increasing age of the cotyledons and increased under enhanced UV-B (Figure 5). Primary leaves (17 d) also showed an increase in Fm. However, the ratio Fm/Fo was similar for cotyledones under both radiation conditions, whereas primary leaves showed a slightly lower ratio under enhanced UV-B (Figure 6). This indicates a lower quantum yield for sunflower primary leaves grown under enhanced UV-B which is also indicated by slightly lower Fvm/Fm values (data not shown).

The area over the fluorescence induction curve under non saturating light, normalized to the variable fluorescence, Fv, was significantly lower for leaves grown under enhanced UV-B with the exception of the youngest cotyledons which showed similar values for both conditions (Figure 7). The area decreased in all cases with increasing age of the leaves.

All findings together suggest a smaller PQ-pool for plants grown under enhanced UV-B and a decreasing pool with increasing age. Under artificial light, an increase in the chlorophyll a/b ratio was induced by increasing UV-B (Tevini, unpublished). Therefore it may be possible that PS II unit size is changed under enhanced UV-B.

The q-analysis showed no differences in plants grown under the two radiation conditions in the steady-state, but higher q_Q values at the fluorescence peak in plants grown under enhanced UV-B (data not shown) indicate a more oxidized state of the

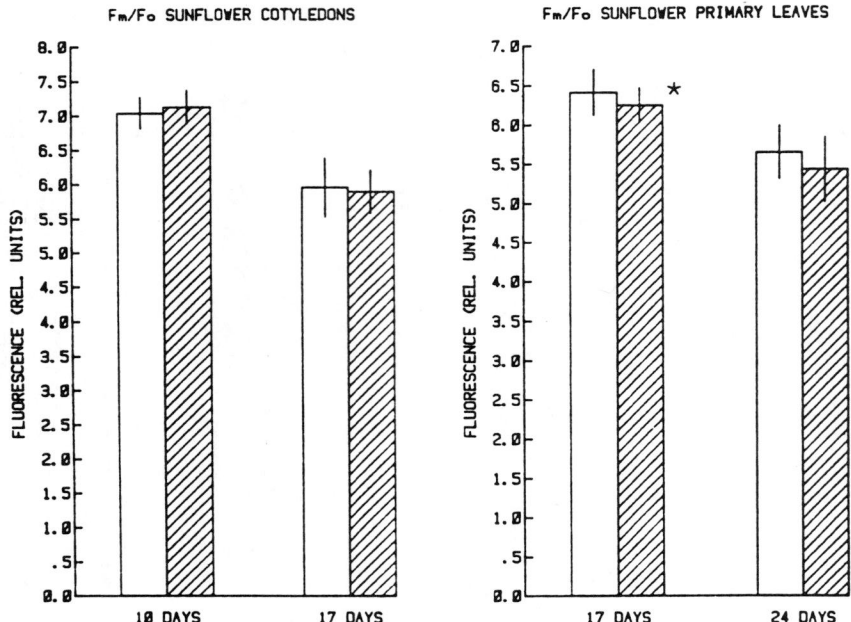

Figure 6. Ratio Fm/Fo in sunflower leaves under different radiation conditions.

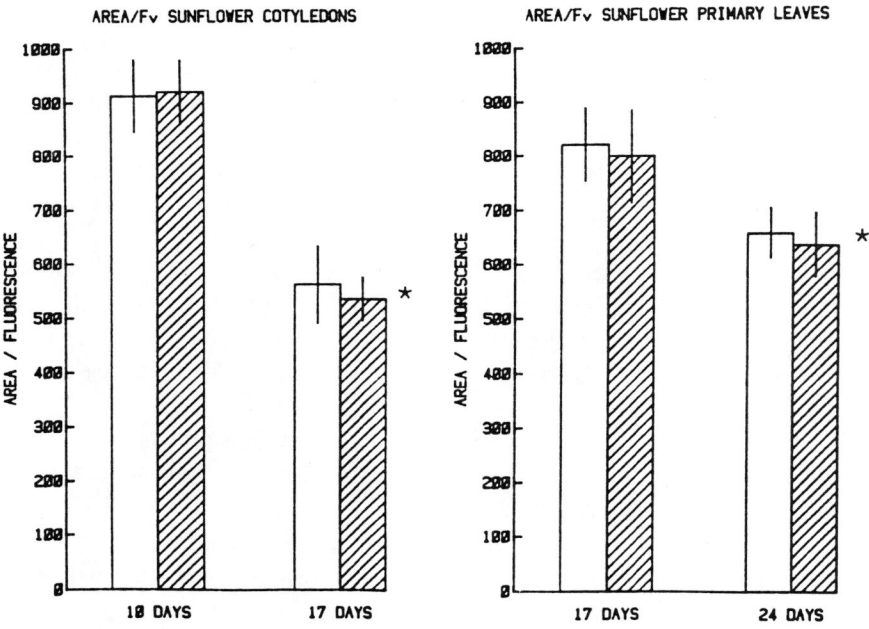

Figure 7. Area over the induction curve (normalized to Fv) under different radiation conditions.

primary electron acceptor of PS II. This could be an indication for a reduced capability to reduce PQ in these plants.

The changes in Fm, Fm/Fo, A/Fv and Fo also indicate a loss of photosynthetic activity in the examined sunflower leaves during aging. This increases the difficulty to differentiate between developmental changes and changes solely caused by enhanced UV-B radiation.

Flavonoids

In etiolated radish seedlings the irradiation with enhanced solar UV-B for 8 h caused higher flavonoid levels as shown in Figure 8 and 9.

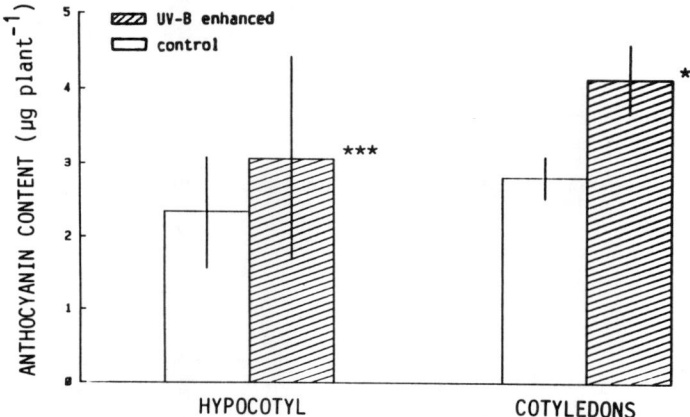

Figure 8. Anthocyanin content of etiolated radish seedlings under different radiation conditions.

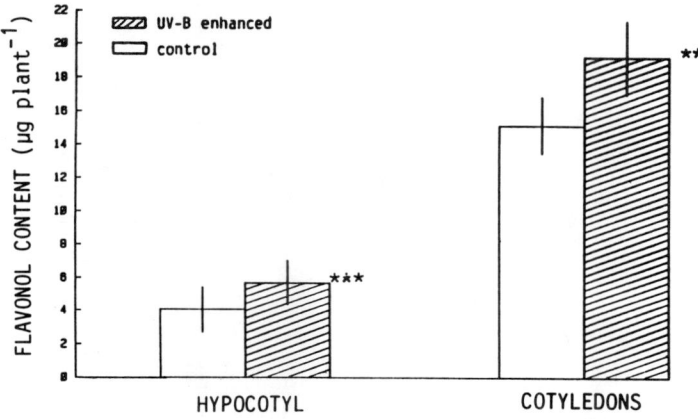

Figure 9. Flavonol content of etiolated radish seedlings under different radiation conditions.

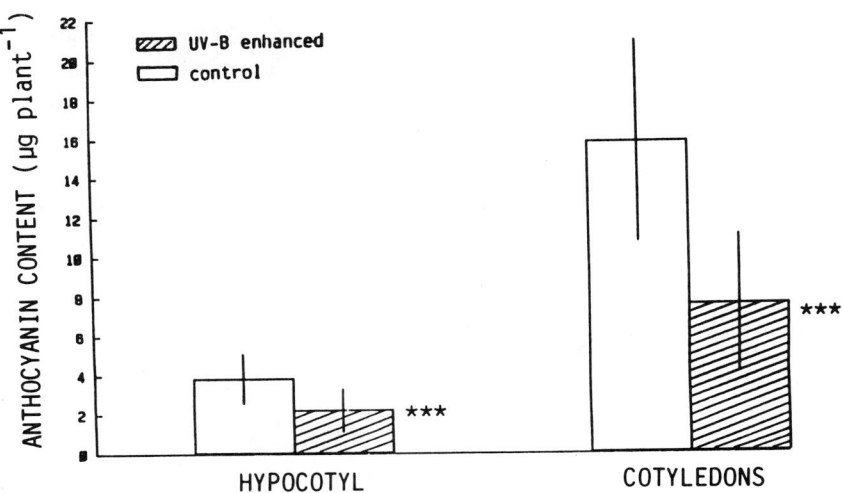

Figure 10. Anthocyanin content of etiolated mustard seedlings under different radiation conditions.

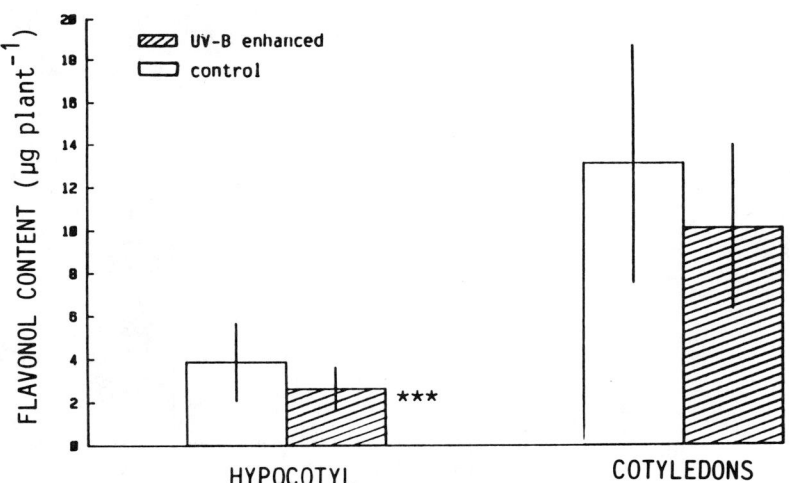

Figure 11. Flavonol content of etiolated mustard seedlings under different radiation conditions.

The anthocyanin content increased by about 31% in the hypocotyls, by 47% in the cotyledons, whereas the flavonol content increased by about 40% and 27% respectively.

These results may confirm the possible role of the flavonols as UV-screening pigments as earlier described and suggested in various studies (Wellmann, 1982, 1985; Wellmann et al., 1984). The function of the increased anthocyanin content in correlation to enhanced UV-B is so far unknown, since anthocyanins only weakly absorb in the relevant UV-B region.

An opposite effect of enhanced UV-B radiation on formation of flavonoids was observed in etiolated mustard seedlings (Figure 10 and 11). The results presented here show a definite damaging effect of enhanced UV-B radiation in that plant species, since the anthocyanin content in hypocotyls decreased by about 43%, in cotyledons by about 52%, whereas flavonol content decreased by about 32% and 22%, respectively. These results differ from the "positive" effects of UV-B radiation on flavonoid formation and may be regarded as criteria for UV sensitivity. The target for such damage is unknown, but it has been suggested that enzymes or nucleic acids, responsible for photoreactivation, may be affected (Wellmann et al., 1984).

Radish seedlings having higher flavonoid content are known to be more UV-B resistent (Tevini et al., 1983b).

Plant growth parameters

Hypocotyl length

The growth curve of sunflower seedlings showed an UV-dependent reduction of the hypocotyl length (Figure 12 and 13). During a three days period the average of the

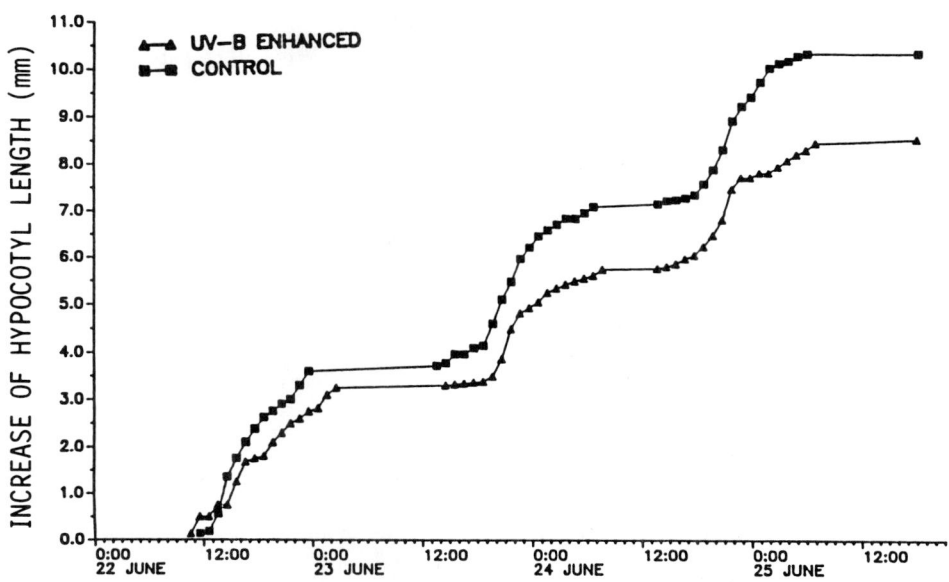

Figure 12. Growth curve of sunflower seedlings under different radiation conditions.

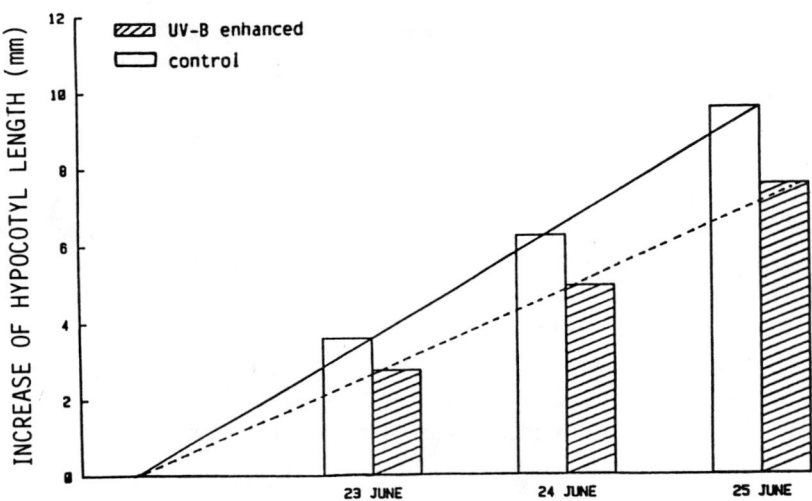

Figure 13. Growth of sunflower seedlings under different radiation conditions.

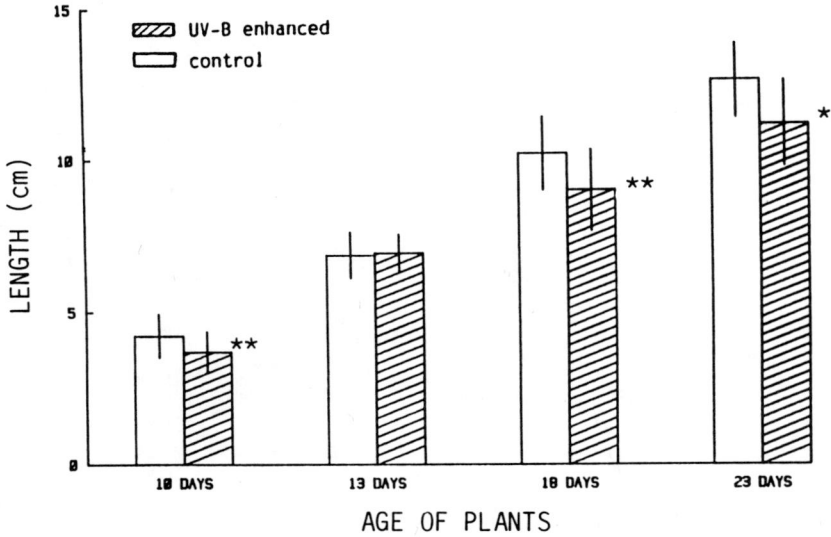

Figure 14. Hypocotyl length of sunflower seedlings at different ages under different radiation conditions.

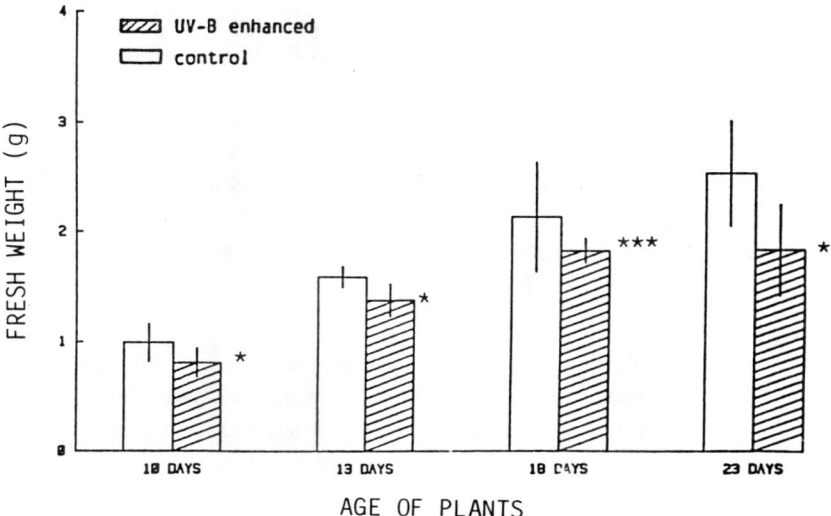

Figure 15. Fresh weight of sunflower seedlings at different ages under different radiation conditions.

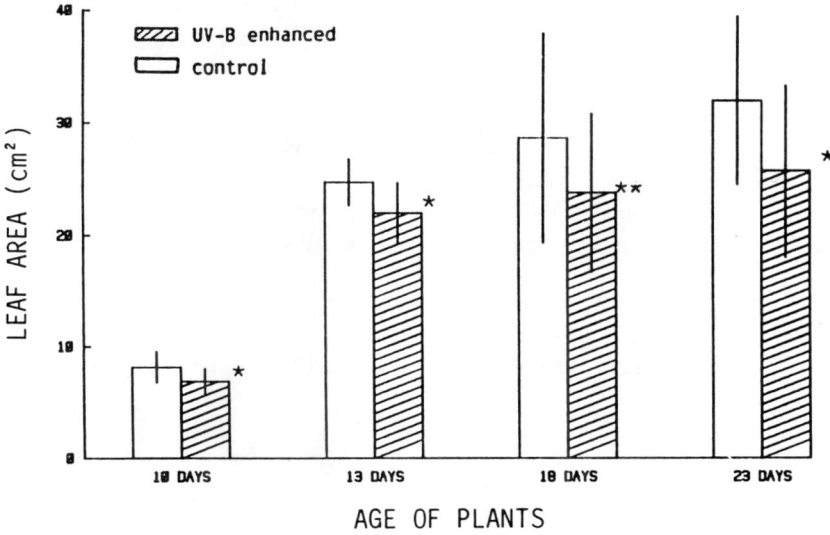

Figure 16. Leaf area of sunflower seedlings at different ages under different radiation conditions.

growth rate reduction was about 20–24% under enhanced UV-B. The hypocotyl growth during 24 hours does not follow the daily course of light intensity. The plants began to grow 6–7 hours after sunrise and continued to grow up to 3–4 hours after sunset.

Growth is regulated and controlled by phytohormones, especially by auxins e. g. indole-3-acetic acid (IAA). IAA absorbs in the UV-B range and can be destroyed *in vitro* and *in vivo* to its photooxidation products by UV-B irradiation (Ros, 1985). The photooxidation products, essentially 3-methylene-oxindole, inhibit the growth of sunflower hypocotyls, when applied exogenously (Tevini and Ros, in prep.).

The length of sunflower seedlings at different ages is presented in Figure 14 which shows that the reduction of hypocotyl length found in period of 72 hours also persists after 18 and 23 days of development.

Fresh weight

A decrease of the fresh weight of seedlings under UV-B stress was shown in laboratory studies (Tevini et al., 1986). Seedlings grown under enhanced solar UV-B had also lower fresh weight than control plants. At every age the difference was significant and increased during development of seedlings (Figure 15).

Leaf area

Similar to fresh weight leaf area was also significantly reduced under enhanced UV-B radiation (Figure 16).

Conclusions

These results show that an increase of solar UV-B radiation by about 25% has a significant impact on selected crop seedlings grown at constant temperature und humidity.

It is shown that enhanced solar UV-B radiation causes reductions in growth and photosynthesis in sunflower seedlings which may therefore be regarded as UV sensitive. Radish seedlings which were earlier shown to be less sensitive accumulate UV screening pigments under enhanced UV-B. In contrast mustard seedlings had low content in flavonoids and may therefore be classified as UV sensitive.

Acknowledgements

The financial support given by the BMFT of the FRG is gratefully acknowledged.

References

Bogenrieder, A. and R. Klein (1977) *Angew. Botanik* 51, 99–107
Brandle, J.R., W.F. Campbell, W.B. Sisson and M.M. Caldwell (1977) *Plant Physiol.* 60, 165–168
Caemmerer, S. von and G.D. Farquhar (1981) *Planta* 153, 376–387
Caldwell, M.M. (1981) In Encyclopedia of Plant Physiology 12A, Springer-Verlag, Berlin, F.R.G., 169–197
Caldwell, M.M., Robberecht, R. and S.D. Flint (1983) *Physiol. Plant.* 58, 445–450

Green, P.B. and W.R. Cummins (1974) *Plant Physiol.* 54, 863–869

Iwanzik, W., M. Tevini, G. Dohnt, M. Voss, W. Weiss, P. Grüber and G. Renger (1983) *Physiol. Plant.* 58, 401–407

Lichtenthaler, H.K. (1987) Methods in Enzymology 148, 350–382

Norudeen, A.M. and G. Kulandeivelu (1982) *Physiol. Plant.* 55, 161–166

Reznik, H. and J. Krause (1968) Staatsexamensarbeit Universität Münster

Ros, J. (1985) Staatsexamensarbeit Universität Karlsruhe

Schreiber, U., U. Schliwa and W. Bilger (1986) *Photosynth. Res.* 10, 51–62

Sisson, W.B. and M.M. Caldwell (1976) *J. Exp. Bot.* 28, 691–705

Steinmüller, D. and M. Tevini (1985) *Planta* 164, 557–564

Stolarski, R.S. (1988) *Scientific American* 258, 20–26

Teramura, A.H., M. Tevini and W. Iwanzik (1983) *Physiol. Plant.* 57, 175–180

Teramura, A.H. (1980) *Physiol. Plant.* 48, 333–339

Teramura, A.H. (1983) *Physiol. Plant.* 58, 415–427

Teramura, A.H. and N.S. Murali (1986) *Env. exp. Bot.* 26, 89–95

Tevini, M. and K. Pfister (1985) *Z. Naturforsch.* 40c, 129–133

Tevini, M., W. Iwanzik and U. Thoma (1981) *Planta* 153, 388–394

Tevini, M., U. Thoma and W. Iwanzik (1983a) *Z. Pflanzenphysiol.* 109, 435–448

Tevini, M., W. Iwanzik and A.H. Teramura (1983b) *Z. Pflanzenphysiol.* 110, 459–467

Tevini, M., D. Steinmüller and W. Iwanzik (1986) BPT-Bericht 6, GSF München

Tevini, M. and W. Iwanzik (1983) *Physiol. Plant.* 58, 395–400

Tevini, M., P. Grusemann and G. Fieser (1988) In Application of Chlorophyll Fluorescence, ed. by H. K. Lichtenthaler, Kluwer Academic Publishers, Dordrecht

Thai, V.K. (1975) Ph. D. diss., Univ. Florida, Gainesville

Van, T.K. and L.A. Garrard (1976) *Soil Crop Sci. Soc. Fla.* 35, 1–3

Vu, C.V. (1982) *Physiol. Plant.* 55, 1–16

Wellmann, E. (1974) *Ber. Deutsch. Bot. Ges.* 87, 267–273

Wellmann, E. (1982) In Biological effects of UV-B radiation, ed. by H. Bauer et al., GSF München, BPT-Bericht 5/82, 145–149

Wellmann, E. (1985) *Ber. Deutsch. Bot. Ges.* 98, 99–104

Wellmann, E., U. Schneider-Ziebert and C.J. Beggs (1984) *Plant Physiol.* 75, 997–1000

Weissenböck, G., R. Hedrich and G. Sachs (1986) *Protoplasma* 134, 141–148

649

Man and Ultraviolet Light
Photosensitization and Photocarcinogenesis

Sources of Human Exposure to Ultraviolet Radiation

B.L. DIFFEY

Regional Medical Physics Department
Dryburn Hospital
Durham DH1 5TW, England

Introduction

The sun is responsible for the development and continued existence of life on earth, with approximately 5% of terrestrial radiation in the ultraviolet region. Sunlight contributes a major fraction of human exposure to ultraviolet radiation (UVR). and prior to the beginning of the century was the only source of UVR exposure. With the advent of artificial sources of UVR the opportunity of additional exposure has increased. The exposure to UVR may be intentional, for example, cosmetic tanning from the sun or solaria, or unintentional, often as a consequence of occupational exposure.

Intentional Ultraviolet Exposure

Sunbathing

Many people believe that sunlight is beneficial to their general sense of well-being, and despite warnings by skin specialists on the dangers of overexposure, there is little doubt that nearly all of us enjoy relaxing in the sun (Wolff, 1986). Sunbathing was practised by Romans and Greek 2000 years ago, and enjoyment of the sun has persisted. For example, in the last century the English poet John Keats wrote:

> 'Give me books, fruit, french wine and fine weather and a little
> music out of doors..'

Recreational sun exposure can result in an annual UV dose (Diffey, 1986) of between 20–100 multiples of a minimal erythema dose (MED) often to a large fraction of the body surface area. In fact, the annual UV exposure of sun-loving indoor workers can result more from two or three weeks recreational exposure than from unintentional exposure during the remainder of the year (see section 3.1).

Cosmetic tanning from artificial ultraviolet sources

The social desirability of a tanned skin is apparent and many people associate a bronzed body with good health. In Northern Europe and America the lack of long periods

Photobiology, Edited by E. Riklis
Plenum Press, New York, 1991

of sunshine has led to the establishment of the 'suntanning industry' where artificial sources of ultraviolet radiation emitting almost entirely in the UVA region supplement sunlight exposure.

Most commercial tanning equipment in current use exploits the divergence in the erythema and tanning action spectra (Gange et al., 1986) and emits almost entirely UVA (315–400 nm) radiation. The sources include the so-called type 1 UVA fluorescent lamp (e.g. Philips TL09, Wotan L100.79), type 2 UVA fluorescent lamp (Philips TL10R), and optically-filtered, metal halide doped mercury arc lamps (Mutzhas, 1986). Solaria incorporating unfiltered mercury arc lamps are now less popular because of the relatively large quantities of actinic (UVB and UVC) radiation leading to the increased risk of burning and acute eye damage such as photokeratitis.

Medical therapy

Ultraviolet radiation is used in medicine mainly to treat skin diseases. The UVR may be administered on its own or in conjunction with adjunctive agents applied topically or photoactive drugs taken systemically.

Phototherapy

The skin disease which responds most readily to UVR therapy is psoriasis, whilst other conditions which are sometimes treated by this technique are acne, eczema, alopecia, pityriasis rosea, superficial ulcers and pressure sores. Standard hospital phototherapy of psoriasis normally includes the use of tar or related derivatives, or other substances such as anthralin to the skin (Parrish, 1982). The radiation may be administered as contact therapy, regional therapy or generalized therapy. The UV sources that are used include both unfiltered and optically filtered mercury arc lamps with and without metal halide additives, and fluorescent lamps with a variety of spectral emissions (Diffey and Farr, 1987).

Psoralen photochemotherapy

This form of treatment, known colloquially as PUVA, involves the combination of the photoactive drugs, psoralens (P), with longwave ultraviolet radiation (UVA) to produce a beneficial effect (Parrish et al., 1982). Psoralen photochemotherapy has been used to treat many skin diseases in the past decade, although its principal success has been in the management of psoriasis, a disorder characterised by an accelerated cycle and rate of DNA synthesis. The mechanism of the treatment is thought to be that psoralens bind to DNA in the presence of UVA resulting in a subsequent transient inhibition of DNA synthesis and cell division. The psoralens may be applied to the skin either topically or systemically; the latter route is generally preferred and the psoralens are administered as 8-methoxypsoralen (8-MOP). The patient ingests the 8-MOP tablets and, two hours later, when the photosensitivity of the skin is at a maximum, he is exposed to

UVA radiation. If the psoriasis is generalized, whole body exposure is given in a cylindrical irradiation cabinet incorporating something like 40–50 UVA fluorescent lamps mounted vertically around the walls of the cabinet.

Unintentional Ultraviolet Exposure

Most people are unintentionally exposed to ultraviolet radiation. Exposure to natural UVR is unavoidable, whereas artificial UVR exposure occurs largely in the workplace.

Natural UVR exposure

Terrestrial ultraviolet radiation has a lower wavelength cut-off in the region 290–310 nm depending on solar altitude. The path length of the sun's rays through the ozone layer is the major determinant of terrestrial UVB flux; this path length is related to the solar altitude which may be calculated from time of day, day of year and geographical latitude. Other factors which influence ultraviolet climatology are cloud cover, atmospheric pollution, albedo and altitude above sea level. The annual, unintentional natural UV exposure of indoor and outdoor workers in mid-latitudes (40°–60°N) are typically:

	UVB, MED	UVA, J/cm^2
Outdoor workers	250	3000
Indoor workers	70	1000

The ratio of UVA exposures between the two groups is less than that of UVB because window glass absorbs UVB radiation but transmits UVA.

There are groups such as the housebound elderly, some Asian immigrants and submariners who receive almost no natural UV exposure and as a consequence may suffer vitamin D deficiency.

Industrial photoprocesses

Many industrial processes involve a photochemical component. The large scale nature of these processes often necessitates the use of high power (several kilowatts) lamps such as high pressure metal halide lamps.

Photochemical syntheses

Although many organic compounds can be synthesized in the laboratory, relatively few have been translated into profitable industrial production (Clements, 1980). Probably the most successful has been the chlorination of hydrocarbons (Phillips, 1983) in which chlorine molecules absorb UVA and blue light (Gibson and Bayliss, 1933) and are photolysed to chlorine atoms, which can initiate a chain reaction. One example of photochlorination is the conversion of 1,1-dichloroethane to the solvent 1,1,1-

trichloroethane (Richtzenhain and Stephan; 1971). Other large scale syntheses involving a photochemical step are photohydrosulphonation, photosulphochlorrination and photonitrosation (Phillips, 1983).

Photopolymerization

Although bulk production of thermoplastic materials — such as PVC, polyethylene, polypropylene, nylons, polystyrene and acrylics — is possible by photopolymerization, the method is not generally used since low temperature thermal catalysts are available and normally preferred to optical radiation in the polymerizing process (Phillips, 1983). The principal industrial applications of photopolymerization include the drying of protective coatings and inks, and photoresists for printed circuit boards.

UV curing of inks and coatings

Surface coating of materials may be done for a number of reasons which include protection against chemical attack, change in the physical properties of the surface, or simply as decoration. Whatever the reason the coating is normally applied in a fluid form which must be converted to a solid form to provide the functional performance required. This 'hardening' process may be achieved by solvent evaporation or by photopolymerization — UV curing of coatings (Roffey, 1982). The curing of printing inks by exposure to UVR is now widespread, and as the cure takes only a fraction of a second it is possible to install UV drying units between the printing stations on a multicolour line, and so dry each colour before the next is applied. Another major use of UV curing has been for metal decorating in the packaging industry.

Photoresists

The action of photoresist is to protect selected areas of a substrate or support by utilising a photochemical reaction to alter the solubility of the resist material. The surface is coated with the photoresist selectively irradiated through a mask, and then developed in a suitable solvent to remove the soluble material. The technique is widely employed in the manufacture of printed circuit boards and integrated circuits in the electronic industry.

Photoageing

Artificial sources of ultraviolet radiation are often used to assess the weathering capability of many materials, especially polymers. Weathering chambers allow a controllable environment and can reduce the time required for weathering compared with natural sun exposure. Xenon arc lamps are often used as the light source due to the similarity of their emission spectrum and the spectrum of terrestrial sunlight, although some commercial weathering chambers incorporate carbon are lamps or fluorescent sunlamps (Davis and Sims, 1983).

Sterilization and disinfection

The bactericidal effects of sunlight were first noted by Downes and Blunt in 1977 and for many years this property of ultraviolet radiation has been exploited. Radiation with wavelengths in the range 260–265 nm is most effective since this corresponds with an absorption maximum in the DNA absorption spectrum. For this reason low pressure mercury discharge tubes are normally used as the radiation source as more than 90% of the radiated energy lies in the 253.7 nm line. These lamps are often referred to as 'germicidal lamps', 'bactericidal lamps', or simply 'UVC lamps'.

UVC has been used to disinfect sewage effluents (Ho and Bohm, 1981)., drinking water (Braendli et al., 1981), water for the cosmetics industry (Schenck 1979) and bathing pools (Gemne et al 1981). Germicidal lamps are sometimes used inside microbiological safety cabinets to inactivate airborne and surface microorganisms. The combination of ultraviolet radiation and ozone has a very powerful oxidizing action and is capable of reducing the organic content of water to extremely low levels (Phillips, 1983).

Welding

Welding equipment falls into two broad categories; gas welding and electric arc welding. Only the latter process produces significant levels of ultraviolet radiation, the quality and quantity of which depend primarily on the arc current, shielding gas and the metals being welded (Sliney and Wolbarsht, 1980).

Welders are almost certainly the largest single occupational group who are exposed to artificial sources of UVR. It has been estimated (Emmett and Horstman, 1976) that there may be as many as half a million welders in the USA alone. The levels of ultraviolet irradiance around electric arc welding equipment are high (Cox, 1987; Mariutti and Matzeu, 1987), and it is not surprising that most welders at some time or other experience 'welder's flash' (photokeratitis) and skin erythema.

Hospitals

Phototherapy

Many of the lamps used to treat skin diseases (see section 2.3) are unenclosed, emit high levels of actinic UV radiation, and can present a market ultraviolet exposure hazard to staff (Diffey and Langley, 1986). At 1 m from these lamps the recommended 8 h occupational exposure limits can be exceeded in less than 2 minutes. For this reason staff should always avoid the primary beam as much as possible. The personal UV doses received by staff working in phototherapy departments as a consequence of occupational exposure are around 12 MED per year (Diffey, 1988), although careless work practices can lead to much higher exposures (Diffey, 1989).

Fluorescence diagnosis

Wood's light — a source of UVA obtained by optically filtering a mercury arc lamp

657

with 'blackglass' — is used by dermatologists as a diagnostic aid in those skin conditions that produce fluorescence (Caplan, 1967). For example, the fungi responsible for *tinea capitis* fluoresce with a bright blue-green under Wood's light, whereas the bacterium responsible for erythrasma produces a porphyrin that fluoresces a bright coral-red colour.

Operating theatres

UVC lamps have been used since the 1930s to decrease the levels of airborne bacteria in operating theatres (Goldner and Allen, 1973). The technique is not widely used, filtered air units generally being preferred.

Neonatal phototherapy for hyperbilirubinaemia

Neonatal jaundice, or hyperbilirubinaemia, is found in about one half of newborn babies. The preferred method of treatment is to irradiate the baby for several hours a day for up to one week with visible light, particularly blue light (Sissons and Vogl, 1982). However, the lamps used for phototherapy, although emitting primarily visible light, also have an ultraviolet component. One commercial neonatal phototherapy unit was found to emit not only visible light and UVA, but also radiation at wavelengths down to 265 nm (Diffey and Langley, 1986).

Dentistry

Fluorescence in oral diagnosis

Irradiation of the oral cavity with a Wood's lamp will produce fluorescence which may prove useful in the diagnosis of various dental disorders,such as early dental caries, the incorporation of tetracycline into teeth, dental plaque and calculus (Hefferren et al., 1971).

Normal teeth fluoresce with a light blue colour. The application of the dye fluorescein to the teeth followed by irradiation with UVA more readily allows plaque to be visualised than with visible light. The presence of calculus on the teeth will result in a yellow-orange fluorescence under UVA illumination.

The administration of the antibiotic tetracycline can result in its incorporation into the tooth structure. Such teeth have a yellowish-brown discolouration in visible light but fluoresce with an intense yellow when exposed to Wood's light.

Polymerisation of dental resins

The restoration of pits and fissures in both deciduous and permanent teeth is often accomplished by using an adhesive resin polymerised with UVA (Buonocore, 1971).The resin is applied to the surfaces to be treated with a fine brush and is hardened by exposure to the UVA radiation at a minimum irradiance of 100 W/m^2 for 30 s or so. The restoration of teeth by resinous sealants is not only more aesthetically acceptable to patients than repair using conventional materials, but offers greater protection against tooth decay at the sites of pits and fissures.

Research laboratories

Sources of ultraviolet radiation are commonly used by most experimental scientists engaged in aspects of photobiology and photochemistry. These applications, in which the effect of ultraviolet irradiation on the biological or chemical species is of primary interest to the researcher, can be differentiated from ultraviolet fluorescence or absorption techniques where the effect is of secondary importance.

Fluorescence

The ability of ultraviolet radiation to excite electrons to higher energy states with the subsequent fluorescent emission of visible light is widely exploited. For example, detecting or differentiating materials on chromatograms is often achieved by UV fluorescence; in forensic science, examination of forged or altered documents may reveal chemically-erased writing, or fingerprints; examination of paintings under ultraviolet illumination may indicate recent restoration or addition of pigments; and so on.

Ultraviolet lasers

Lasers operating in the UV region are now commonplace. Radiation may be emitted at 337, 308, 266, 248 and 193 nm, with outputs as high as 10 W from continuous wave lasers.

Food industry

Quality assurance

Many contaminants of food products can be detected by ultraviolet fluorescence techniques. For example, the bacterium pseudomonas aeruginosa, which causes rot in eggs, meat and fish, can be detected by noting its yellow-green fluorescence under UVA irradiation. One of the longest established uses for UVA fluorescence in the public health field is to demonstrate rodent contamination. The urine of the rodent fluoresces strongly and is easily seen on bags containing flour, grain and other foods. Rodent excreta in flour may also be detected by the non-fluorescence of the faeces against the fluorescent background of the flour.

Insect traps

Many flying insects are attracted by UVA radiation, particularly in the region around 350 nm (Glick and Hollingsworth, 1955). This phenomenon is the principle of electronic insect traps in which a UVA fluorescent lamp is mounted in a unit containing a high voltage grid. The insect, attracted by the UVA lamp, flies into the unit and is electrocuted in the air gap between the high voltage grid and an earthed metal screen. Units such as these are commonly found in areas where food is prepared and sold to the public.

Leisure industry

Sunbed salons and shops

The continuing popularity of UVA sunbeds and suncanopies for cosmetic tanning has resulted in a large number of salons where the public go to use sunbeds, and shops selling sunbeds for use at home. Some shops may have 20 or more UVA tanning applicances all switched on exposing members of the public, and more importantly staff, to high levels of UVA radiation.

Discotheques

UVA 'Blacklight' lamps are sometimes used in discotheques to induce fluorescence in the skin and clothing of dancers. The UVA levels would normally be low ($< 10/Wm^2$).

Offices

Signature verification is commonly performed by exposing a signature, obtained previously with a colourless ink, to UVA radiation under which it fluoresces.

General lighting

Fluorescent lamps used for general lighting in offices and factories emit small quantities of both UVA and UVB. For typical levels of illuminance of 500 lux from bare fluorescent lamps, measurements have indicated (McKinlay and Whillock, 1987) UVA and UVB irradiances of about 30 mW/m^2 and 3 mW/m^2, respectively. These ultraviolet irradiances can be reduced appreciably by the use of plastic diffusers.

Photosensitisation

The increase in the availability of UV sources in recent times has been paralleled with a corresponding increase in the activities of the chemical and pharmaceutical industries. There now exist many substances that act as photosensitisers, and these include several component of coal tars and pitches, photoinitiators used for industrial photochemical processes, a variety of dyes and cosmetics in common use, chemicals produced by plants, and several systemic and topical therapeutic medications.

References

Braendli, G. Egger, H., and Hurlimann, R. (1981). Brit Pat 1584385 (Brown Boveri).

Buonocore, M.G. (1971). *J. Am. Dental Ass.* 82, 1090–1093.

Caplan, R.M. (1967). *J. Am. Med. Ass.* 202, 123–126.

Clements, A.D. (1980). *Chem. Brit.* 15, 464–467.

Cox, C.W.J. (1987). In: 'Human Exposure to Ultraviolet Radiation - Risks and Regulations'. (Eds. W.F. Passchier and B.F.M. Bosnjakovic), *Excerpta Medica*, Amsterdam, pp. 383–386.

Davis, A., and Sims, D. (1983). Weathering of Polymers. Applied Science Publishers Ltd., London.

Diffey, B.L. (1986). In: 'Hazards of Light' (Eds. J. Cronly-Dillon, E.S. Rosen and J. Marshall) Pergamon Press, Oxford, pp. 57–66.

Diffey, B.L., and Langley, F.C. (1986). Evaluation of Ultraviolet Radiation Hazards in Hospitals, ISPM Report 49, London.

Diffey, B.L., and Farr, P.M. (1987). *Br. J. Dermatol*, 117, 49–56.

Diffey, B.L. (1988). *Phys. Med. Biol.* 33, 1187–1193.

Diffey, B.L. (1989). *Physiotherapy*, 75, 615–616.

Emmett, E.A., and Horstman, S.W. (1976). *J. Occ. Med.* 18, 41–44.

Gange, R.W., Park, Y.K., Auletta, M. Kagetsu, N., Blackett, A.D., Parrish, J.A. (1986). In: The Biological Effects of UVA Radiation (Eds. F. Urbach and R.W. Gange) Praeger, New York, pp. 57–65.

Gemme, G., Hoffner, S., and Stenstroem, T.A. (1981). *Vatten.* 37, 265–274.

Gibson, G.E., and Bayliss, N.S. (1933).*Phys. Rev.* 44, 188–192.

Glick, P.A. and Hollingsworth, J.P. (1955). *J. Eco. Ent.* 48, 173–177.

Goldner, J.L. and Allen, B.L. (1973). *Clin. Orthop.* 96, 195–205.

Hefferren, J.J., and Cooley, R.O., Hall, J.B., Olsen, N.H., and Lyon, H.W. (1971). *J. Am. Dental. Ass.* 82, 1353–1360.

Ho, K.W.A., and Bohm, P. (1981). *Water Pollut Res, J. Can.* 16, 33–44.

Mariutti, F., and Matzeu M. (1987). In: 'Human Exposure to Ultraviolet Radiation - Risks and Regulations'. (Eds. W.F. Passchier and B.F.M. Bosnjakovic), *Excerpta Medica,* Amsterdam, pp. 387–390.

McKinlay, A.F., and Whillock, M.J. (1987). In: 'Human Exposure to Ultraviolet Radiation - Risks and Regulations'. (Eds. W.F. Passchier and B.F.M. Bosnjakovic), *Excerpta Medica,* Amsterdam, pp. 253–258.

Mutzhas, M.F. (1986). In: The Biological Effects of UVA Radiation (Eds. F. Urbach and R.W. Gange) Praeger, New York, pp. 10–23.

Parrish, J.A. (1982). In: The Science of Photomedicine'. (Eds. J.D. Regan and J.A. Parrish) Plenum Press. New York, pp. 511–531.

Parrish, J.A., Stern, R.S., Pathak, M.A., Fitzpatrick, T.B. (1982). In: The Science of Photomedicine'. (Eds. J.D. Regan and J.A. Parrish) Plenum Press. New York, pp. 595–623.

Phillips, R. (1983). Sources and Applications of Ultraviolet Radiation, Academic Press, London.

Richtzenhain, H., and Stephen, R. (1971). Ger Offen 2026671. Dynamit Nobel, AG.

Roffey, C.G. (1982). Photopolymerization of Surface Coatings, John Wiley & Sons, Chichester.

Scheneck G.O. (1979). *Parfuem Kosmet,.* 60, 433–443.

Sisson, T.R.C., and Vogl, T.P. (1982). In: The Science of Photomedicine'. (Eds. J.D. Regan and J.A. Parrish) Plenum Press. New York, pp. 477–509.

Sliney, D., and Wolbarsht, N. (1980). Safety with lasers and other optical sources. Plenum Press. New York

Wolff, F. (1986). Sunbathing Today. Naturwissenschaftliche Verlagsgesellschaft, MBH. Freiburg.

Drug Products and Photocarcinogenesis

P.D. FORBES, R.E. DAVIES AND C.P. SAMBUCO

The Center for Photobiology
Temple University *
Philadelphia, PA 19140 (USA)

Introduction

Most safety tests are designed to determine whether an agent, such as a drug or chemical, will adversely affect some normal physiological component or process. In contrast, most photobiological safety tests seek to determine whether a chemical can amplify the known noxious effects of an external agent, ultraviolet radiation (UVR). Tests for phototoxicity, for example, determine whether suberythemal doses of radiation will produce acute damage in the presence of the test agent: such tests are always preceded by a determination that the agent alone does not produce such damage. It is understood that higher doses of UVR (or, in some cases, different radiation spectra) are capable of producing erythema in the absence of the test agent, but for testing purposes the effects of low doses of UVR are regarded as the "normal" state against which toxic effects are to be evaluated. Tests for photoallergic properties may go even further in altering the reference "normal" state, by imposing various forms of trauma at either initiation or elicitation stages. The justification for introducing external trauma in a photobiological safety test is the assumption that some exposure to environmental UVR is virtually inevitable; thus an agent which can intensify trauma induced by UVR is capable of intensifying a "normal" and expected biological response.

The same type of reasoning underlies testing for modification of photocarcinogenesis. UVR is generally recognized to be the primary cause of skin cancer in man, and incidental or deliberate exposure to UVR represents a substantial hazard. Thus, a photocarcinogenesis safety test is designed to measure the ability of a test material to accelerate the process of photocarcinogenesis. In one sense this can be regarded as a specific and "practical" safety test, assessing the danger of a plausible hazard; since most humans are exposed to carcinogenic ultraviolet radiation (UVR) for much of their lives, photocarcinogenesis is a realistic risk and its acceleration is a realistic hazard. In another sense photocarcinogenesis can be regarded as representative of a class of deleterious effects of chronic UVR exposure, and acceleration of photocarcinogenesis as

---

*Address reprint requests to: The Center for Photobiology at Argus, Inc., 905 Sheehy Drive, Horsham, PA 19044, USA.

Photobiology, Edited by E. Riklis
Plenum Press, New York, 1991

663

representative of the exacerbation of such effects. The justification for selecting photocarcinogenesis as representative of chronic UVR effects lies in the fact that it is a reproducible and quantifiable phenomenon, measurable with considerable precision in established testing protocols. Most other forms of chronic damage, in contrast, can presently be elicited with only marginal reproducibility, or are describable only in qualitative terms. The question of whether photocarcinogenesis accurately represents these other changes, of course, cannot be answered for the same reasons that the changes cannot be measured.

Basis of the Test

If experimental animals are exposed repeatedly to doses of UVR, none of which produces acute trauma, the animals will develop skin tumors after a latent period. The length of the latent period and the number of exposures required are both dependent on the magnitude of the individual doses. While this response can be elicited in several species, most testing has been conducted with mice. A major impediment to photocarcinogenesis is the protection afforded by hair, an efficient sunscreen in most species. As an alternative to mechanical or chemical depilation, strains of genetically depilated mice have been employed by a number of laboratories, including ours. We use animals carrying the recessive gene "hairless" (hr) in the homozygous state, and either heterozygous or homozygous for the recessive pigment- controlling gene for albinism (c). Under defined conditions of exposure, and with specified response criteria, these animals produce tumors at a predictable time and with relatively low variability.

If irradiated animals are also exposed to certain chemicals, the production of tumors can be accelerated or delayed. Since the rate of tumor production (i.e. the time-based distribution of prevalence) is reproducibly related to the UVR dose, changes in this rate can be described as representing changes in dose, expressed as an "amplification factor. It is unlikely in most cases, that the chemicals act by altering the dose of radiation; it is useful, nevertheless, to compare their effects using a single index). In appropriate test systems it is possible to measure changes in response corresponding to changes in dose of 15% or less; this compares favorably with dose changes detectable in acute tests.

The mechanisms by which chemicals alter photocarcinogenesis are diverse, and in some cases unknown. One possibility is additive carcinogenesis, e.g., with aromatic hydrocarbon carcinogens (Epstein, 1977; Davies, 1978; Gensler, 1988). Other possibilities include direct alteration of UVR transmission (sunscreens, some vehicles; [Wulf et al. 1982; Forbes et al. 1989; Young et al. 1987]), photochemical sensitization including effects of phototoxic agents (Epstein, 1977; Davies, 1978; Young et al. 1987; Forbes and Davies, 1981; Kelly et al. 1988), and a process which may be equivalent to promotion (TPA, possibly retinoic acid; Epstein, 1977; Urbach et al. 1988; Davies and Forbes, 1988). Chemicals may also be altered by UVR, and the resulting products may influence the photocarcinogenic process (Davies, 1978). For purposes of a safety test it is not necessary to define or predict the mechanism. It is a primary objective of a general

664

testing protocol, however, to ensure that all possible mechanisms will have an opportunity to operate.

When exposure to UVR follows chemical application it is not possible to be sure how much of the radiation reaches the skin. This is a particular problem with strong absorbers (e.g. sunscreens), but can be significant with even weak absorbers because of the sensitivity of the response. The protocol described in the next section of this paper makes allowance for this problem.

A Testing Protocol for Photocarcinogenesis

The test employs albino Skh:HR-1 mice of both sexes. Groups contain a minimum of 24 or preferably 36 animals of each sex. Depending somewhat on the nature of the test material and the desired sensitivity, the "in-life" duration of the test is between 12 and 18 months.

Animals are exposed five days per week to radiation from a filtered xenon arc (solar simulator) providing both ultraviolet and visible radiation in proportions similar to those in sunlight, but at lower intensity. Typical exposure duration ranges from two to four hours per day, depending on design requirements. Two "control" or dose calibration groups receive radiation only. Dose-setting has the following basis: the *lower* weekly dose is selected to produce tumors in a period of no more than 80% of the planned duration of the study; the *higher* weekly dose might approach the maximum tolerable under chronic exposure conditions without producing subchronic irritation. The anticipated result of UVR alone is illustrated in calibration curves C-I and C-II, (Fig. 1).

For systemic administration of test articles other than phototoxic agents, the schedule for delivery of UVR to the animals can be flexible. With topical administration,

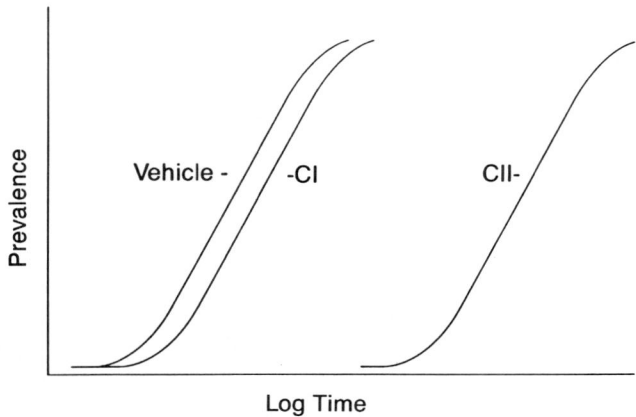

Figure 1. Idealized curves describing mortality from tumor prevalence (for 2 UVR levels) plotted against log time. The third curve illustrates a small enhancement of the UVR effect by an oily vehicle.

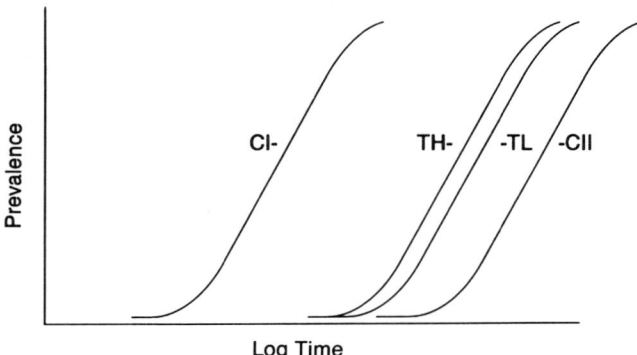

Figure 2. TH and TL represent groups receiving the lower UVR rate plus a high and low "rate" of test article respectively which enhances photocarcinogenesis.

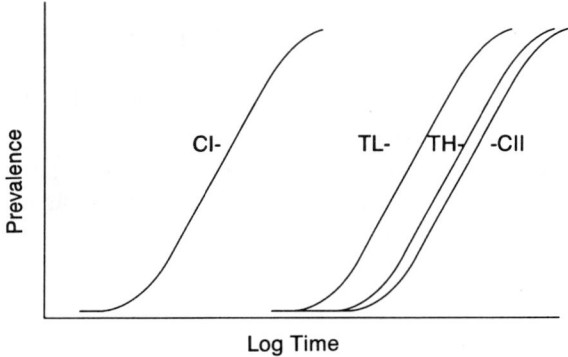

Figure 3. TH and TL represent groups receiving the higher UVR rate plus a high and low "rate" of a UVR-absorbing test article respectively which does not enhance photocarcinogenesis.

however, the timing of UVR exposure becomes critical, and the following discussion is geared to that problem.

Test groups receive daily topical applications of a test material; applications precede or follow UV exposure on alternate days. Treatments are so arranged that the most probable order in normal use occurs three days per week: prior to irradiation for sunscreens or photosensitizing medications, after irradiation for other medications.

If the test article does not absorb radiation (or absorbs it very weakly), or if it is a photosensitizer, the "effective" radiation dose to the test group equals the lower of the two calibration doses. Enhancement of photocarcinogenesis is recognized, in this situation, as acceleration relative to the lower calibration group (Fig. 2).

If the test article is an absorber (particularly an effective sunscreen) the delivered radiation dose to the test group is equal to that delivered on days when irradiation precedes chemical application, plus an unknown fraction of that delivered on the alternate days. The expected response will be some acceleration relative to the lower control group, but it is not possible in this situation to distinguish between strong absorption accompanied by enhancement, and weaker absorption. To examine this possibility we use two test groups receiving the test material at different "rates" (concentration or application volume), selected to differ measurably in acute protective efficacy. In the absence of enhancement these groups should differ in the direction that the lower "rate" produces less protection and earlier carcinogenesis. In the presence of enhancement this distinction should be reduced or absent (Fig. 4).

The significance of enhancement in the presence of protection may be questioned, especially since protection should predominate in the case of even moderately effective sunscreens. Such testing is appropriate, however, for two reasons. First, since the test detects enhancement by a variety of mechanisms, it may reveal promotion (or even carcinogenesis) attributable to test articles or their photoproducts. Secondly, enhancement would represent relative failure of long-term protection, a major reason for recommending the use of sunscreens. Thus sunscreens, like other topical agents, are reasonable candidates for testing in a general-purpose photocarcinogenesis safety test.

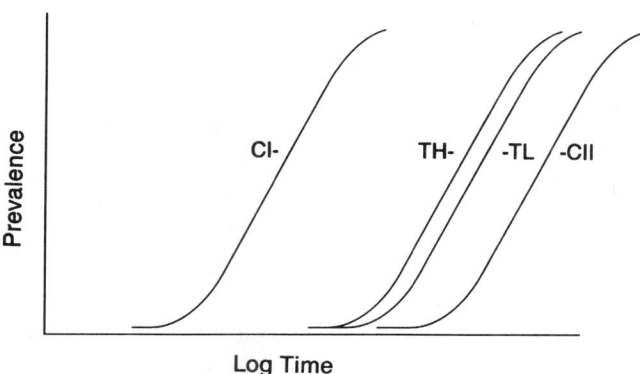

Figure 4. TH and TL, representing groups receiving the higher UVR rate plus a high and low "rate" of a UVR-absorbing test article respectively; this figure illustrates the anticipated effect from a material that provides some protection by absorbing UVR, but also enhances photocarcinogenesis in a dose-dependent manner.

Candidate Materials for the Safety Test

While there is no fixed formula for determining whether a given agent requires testing for influence on photocarcinogenesis, several circumstances may encourage such a test. These include situations in which: skin is the intended site of product application; skin is a target tissue (whatever the route of administration); the product has (or is meant to have) an effect on skin physiology; the product may be in contact with skin frequently or for prolonged periods; frequent or long-term solar exposure is likely; product photochemistry is possible or likely.

Prospects for More Rapid Safety Testing

The duration of this safety test is a direct function of the UVR calibration groups; the study need not encompass the maximal survival of the animals (about 2 years), but with current technology it cannot occupy significantly less than one year. Other endpoints as indicators of "chronic" effects are under investigation. Considering the time required for validating test methods, a replacement test of significantly reduced duration is not expected soon. *In vitro* and non-biological tests appear to be an even more remote possibility.

Acknowledgement

This paper is from the Symposium on Pharmaceuticals and Photobiology; Tenth International Congress on Photobiology; Jerusalem, Israel, 30 October – 3 November 1988. The Symposium was made possible by a grant from Givaudan SA, Geneva, Switzerland.

References

Davies, R.E. 1978. Interaction of Light and Chemicals in Carcinogenesis. National Cancer Institute Monograph 50: pp.45–50.

Davies, R.E., and Forbes, P.D. 1988. Retinoids and Photocarcinogenesis: A Review. J. Toxicology -- Cutan. and Ocular Toxicol. 7:241–253

Epstein, J.H. 1977. Chemicals and Photocarcinogenesis. Australas. J. Derm. 18: 57–61.

Forbes, P.D. and Davies, R.E. 1981. Enhancement of Photocarcinogenesis by 8-MOP. In, Psoralens in cosmetics and Dermatology. Proceedings of the International Symposium, Paris. Pergamon, Oxford, pp 365–370.

Forbes, P.D., Davies, R.E., Urbach, F., and Sambuco, C.P. 1989. Inhibition of Ultraviolet Radiation-Induced Skin Tumors in Hairless Mice by Topical Application of the Sunscreen 2-Ethyl Hexyl-p-Methoxycinnamate. J. Toxicology -- Cutan. and Ocular Toxicol. 8:209–226.

Gensler, H.L. 1988. Enhancement of Chemical Carcinogenesis in Mice by Systemic Effects of Ultraviolet Irradiation. Cancer Research 48: 620–623.

Kelly, G.E., Meikle, W.D., and Moore, D.E. 1988. Enhancement of UV-Induced Skin Carcinogenesis by Azathioprine: Role of Photochemical Sensitization. Photochem. Photobiol. 49: 59–65.

Urbach, F., Davies, R.E., and Forbes, P.D. 1988. Chemical Modifiers of Photocarcinogenesis. In, The Target Organ and the Toxic Process (Arch. Toxicol., Suppl. 12: pp. 47–51.

Wulf, H.C., Poulsen, T., Brodthagen, H., Hon-Jenson, K. 1982. Sunscreens for Delay of Ultraviolet Induction of Skin Tumors. J. Amer. Acad. Dermatol. 7: 194–202.

Young, A.R., Gibbs, N.K., Magnus, I.A. 1987. Modification of 5-methoxypsoralen phototumorigenesis by UV-B sunscreens: A statistical and histological study in the hairless albino mouse. J. Invest. Derm., 89: 611–617.

Investigating Contact Photoallergy in the Mouse

G. FRANK GERBERICK AND CINDY A. RYAN

The Procter and Gamble Co.
Miami Valley Laboratories
Cincinnati, OH 45239-8707

Abstract

Contact photoallergy is a cell-mediated immunologic reaction to chemicals which histologically and mechanistically resembles contact hypersensitivity. The critical factor differentiating these two reactions is that the chemicals that produce contact photoallergy reactions require activation by ultraviolet light in order to induce or elicit the reaction. The aim of our research is to refine a mouse model for assessment of the photoallergic potential of chemicals and to develop an understanding of the factors which influence contact photoallergy. The photoallergic potential of test chemicals is assayed in cyclophosphamide treated BALB/c mice by application of the photoallergen to clipped backs for induction and to the ears for challenge. Mice are not anesthetized for either induction or challenge. Using this mouse photoallergy model, various factors that influence the induction of photoallergy were examined. Specifically, we have studied the effects of test material dose, compound application/irradiation kinetics, and ultraviolet dose and spectrum on the induction of contact photoallergy. To date, eight known human photoallergens have been tested and successfully detected, including musk ambrette, 6-methylcoumarin, and tetrachlorosalicylanilide. We feel that this mouse ear swelling photoallergy model offers potential as a model for predictive photoallergy testing. In addition, we have recently initiated *in vitro* studies to examine the underlying mechanisms of photoallergy using murine Langerhans cell enriched epidermal cells and lymphocytes from sensitized mice. Information gained from these *in vitro* studies will enhance our ability to predict a compound's photoallergic potential.

Introduction

Contact photoallergy (CPA) is a cell-mediated immunologic reaction to chemicals which histologically and mechanistically resembles contact hypersensitivity (Epstein, 1987). The critical factor differentiating these two reactions is that the chemicals that produce CPA reactions require activation by ultraviolet radiation in order to induce and elicit the response. Clinically, CPA reactions have been detected in consumers exposed to chemicals (antimicrobials, perfumes, and sunscreens) plus ultraviolet radiation (Wennersten et al. 1986; Epstein, 1987). Although CPA reactions do not occur as frequently as allergic contact dermatitis reactions, the number of reports of chemicals eliciting possible photoallergic responses has been increasing (Wennersten et al. 1986; de Groot et al. 1987). Therefore, it is critical that valid and predictive models be employed to determine the photoallergenic potential of newly developed drugs and consumer products, especially those designed for topical application and outdoor use.

To assess whether or not a compound will elicit a photoallergic response, various animal models and human photopatch techniques have been employed (Ichikawa et al. 1981; Harber et al. 1982; Kaidbey, 1987). For ethical reasons, it is desirable to first test materials for their photoallergic potential using a valid and predictive animal model. Generally, guinea pigs are employed for this purpose (Harber et al. 1982; Jordan, 1982; Maurer, 1984). In these models, a photoallergic response is determined by subjective evaluation of the degree of erythema in the irradiated and non-irradiated sites at elicitation, in previously sensitized animals.

In addition to the guinea pig photoallergy models, investigators (Takigawa and Miyachi, 1982; Maguire and Kaidbey, 1982; Granstein et al. 1983a) have developed mouse photoallergy models. The mouse models are more quantitative than guinea pig models in that they use an objective, quantifiable ear swelling assay to measure elicitation of contact photoallergy in previously sensitized animals. In our laboratory, we have developed a mouse ear swelling model for CPA testing which is a modification of previously reported models (Takigawa and Miyachi, 1982; Maguire and Kaidbey, 1982; Granstein et al. 1983a). We have used this model to examine various parameters of CPA as well as to determine the ability of the model to detect known human photoallergens. Moreover, we have used the mouse model to investigate photoallergy *in vitro* using epidermal Langerhans cells and responder lymphocytes from sensitized mice.

Materials and Methods

Test materials

Photosensitizers used in the studies included 3,3',4',5-tetrachlorosalicylanilide (TCSA; Eastman Kodak Company, Rochester, NY), bithionol (2,2'Thio-bis-(4,6-dichlorophenol); Sigma Chemical Company, St. Louis, MO), musk ambrette (MA; Givaudan Corp., Clifton, NJ), 6-methylcoumarin (6-MC; Aldrich Chemical Company, Milwaukee, WI), chlorpromazine hydrochloride (Sigma), p-aminobenzoic acid (PABA; National Starch & Chemical Corp., Salisbury, NC) and sodium omadine (2-mercaptopyridine N-oxide, sodium salt hydrate; Aldrich). Negative control compounds tested included coumarin (Aldrich), homosalate (Kemester HMS; Humko Chemical Division, Memphis, TN), 3,4',5-tribromosalicylanilide (Pfister Chemical Inc., Ridgefield, NJ) and musk ketone (Givaudan).

Mouse photoallergy testing

Female BALB/c mice weighing 20 to 25 grams were given a single intraperitoneal injection of 200 mg/kg cyclophosphamide three days prior to the first induction. On the day of the first induction, the dorsal surface of each mouse was clipped. Animals were induced by applying 50 ul of the test material in vehicle, or vehicle alone, on days 0, 1, and 2, to the clipped area of mice in the photoallergy, contact allergy control, and vehicle/radiation control groups. Mice in the phototoxicity control group were clipped but not treated during the induction phase. In most experiments, the irradiation of the mice in the photoallergy and vehicle/radiation groups was begun 30 minutes after application of the test material or vehicle. UVA radiation (320–400nm) was supplied by a bank of eight Sylvania F40/350BL fluorescent blacklights (Figure 1), with peak emission at 350-355 nm, filtered through 3 mm plate glass to eliminate erythema producing wavelengths. UVB radiation (290–320 nm) was supplied by a bank of eight Philips F40UVB fluorescent sunlamps (Figure 1), with peak emission at 315 nm. Prior to each induction and challenge, the irradiance of both banks of lights was determined with an International Light radiometer (IL1700) fitted with either a UVA (SED 038, 300-400 nm, peak at 365 nm) or UVB (SED 240, 250-360 nm, peak at 288 nm) detector. In most experiments, the mice received a UVA dose of 10 J/cm^2 and a UVB dose of 30–45 mJ/cm^2. Seven days after the first induction, each mouse was placed in a restrainer and two measurements of ear thickness were made on each ear using an engineer's micrometer (Model D-1000; The Dyer Co., Lancaster, PA). Following baseline ear measurements, 8 ul of the test material in vehicle or vehicle alone was applied to each side of both ears of all mice. Thirty minutes after application of the test material or vehicle to the ears, the mice of the photoallergy, vehicle/radiation control, and phototoxicity control groups were irradiated as described above. At 24 hours post challenge, both ears of each mouse were measured and the increase in thickness from baseline determined.

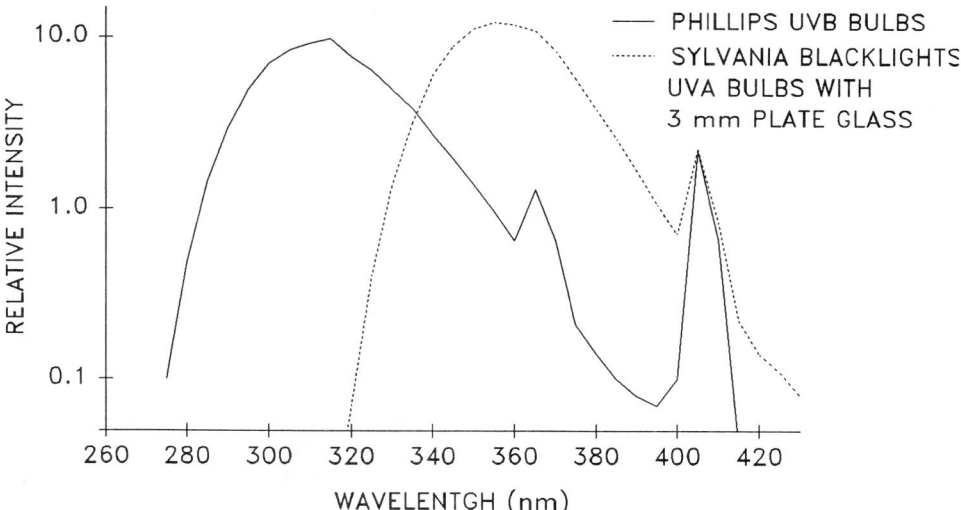

Figure 1. Emission spectra of the Sylvania blacklight bulbs with a 3 mm plate glass filter and the Philips UVB bulbs were determined with a spectroradiometer which consisted of an International Light research radiometer (model IL700A; SIW # 170 detector), monochromoeter (model GMA-201-9) and photomultiplier (model IL791).

In vitro photoallergy testing

Epidermal cells (EC) were obtained from the trunk skins and ears of normal, naive BALB/c mice. The ears were removed and split using forceps. The subcutaneous fat and cartilage was scraped from both the ears and trunk skins. The skins were floated dermal side down in petri dishes containing a solution of 0.5% trypsin (Gibco Laboratories, Grand Island, NY) for 45 minutes at 37°C in 5% CO_2. The epidermis was then gently scraped off of the dermis and was put in Hank's balanced salt solution (HBSS; Gibco) containing 0.025% DNase (Sigma) and 5% heat inactivated fetal bovine serum (FBS; Gibco). The epidermis was teased apart to release the epidermal cells from the stratum corneum so that a single cell suspension could be obtained. The suspension was then passed through a sterilized nylon filter to remove any debris. The epidermal cell suspension was washed, resuspended in culture medium (RPMI 1640 medium supplemented with 10% FBS, 2 mM L-glutamine, 50 ug/ml gentamicin; all from Gibco) containing 5×10^{-5} M 2-mercaptoethanol (Sigma), and counted. The cells were incubated four days at 37°C in 5% CO_2. Nonadherent cells were recovered from the culture plates and centrifuged on a 14.5% metrizamide (Sigma) gradient at 600 g for 30 minutes to recover the Langerhans cell (LC) enriched population. The interface cells were removed, washed, resuspended in HBSS and counted, with percent LC determined by morphology. For hapten modification, LC enriched EC were photohapten-modified by irradiating the cells in the presence of the photoallergen. TCSA was dissolved in ethanol and then diluted to the appropriate concentration in HBSS. Stimulator cells included LC enriched EC (without treatment), LC enriched EC irradiated with UVA only, LC enriched EC treated with TCSA only, and LC enriched EC treated with TCSA and UVA. For irradiation, a 1000W solar simulator (Oriel Corporation, Stratford, CT) was used to deliver a 2-5 J/cm^2 UVA dose. After treatment, the cells were centrifuged, washed and resuspended in culture medium.

Responder lymphocytes were isolated from lymph nodes of mice induced and challenged in vivo with TCSA or musk ambrette. Briefly, naive and photosensitized mice were sacrificed at 24 hours post challenge. Peripheral lymph nodes were taken and pooled for individual mice sensitized to TCSA. Nodes from mice within the naive and musk ambrette sensitized groups were pooled. Cell suspensions were prepared by gently mashing the nodes on sterilized nylon filters in culture medium. The lymph node suspensions were centrifuged, resuspended in culture medium and counted.

The responder cells and stimulator cells (variously treated LC enriched EC) were plated in quadruplicate in flat bottom, one half area, 96 well tissue culture plates (#3696, Costar, Cambridge, MA) at a responder to stimulator ratio of approximately 10:1. Control cultures of responder cells with medium only were plated to determine background proliferative activity. The cultures were incubated for 48 hours at 37°C in 5% CO_2, pulsed with 1 uCi/well ^{3}H-thymidine, incuated for another 24 hours and the harvested. ^{3}H-thymidine incorporation was then determined by liquid scintillation counting. The blastogenesis response was determined as the change in counts per minuts (ΔCPM = EC stimulated CPM — the background of CPM responder cells and medium).

Results and Discussion

The mouse ear swelling model we employed is a modification of previously described mouse models (Takigawa and Miyachi, 1982; Maguire and Kaidbey, 1982 ; Granstein et al. 1983a). In developing our methodology, we evaluated several variables which have the potential of affecting the induction of contact photoallergy (i.e. the use of cyclophosphamide, the number of inductions, radiation spectrum, application/irradiation kinetics). Initially, we examined the effect of cyclophosphamide on CPA. Cyclophosphamide has been demonstrated to augment the photoallergic response of mice to known photoallergens (Granstein et al. 1983b; Brown et al. 1986). Evidence suggests that cyclophosphamide works by elimination or inactivation of precursor T-suppressor cells (Mitsuoka et al. 1976; Granstein et al. 1983b; Takigawa et al. 1984). We determined that i.p. injection of 200 mg/kg cyclophosphamide three days prior to the first induction resulted in the strongest photoallergic response, as determined by the amount of ear swelling when using musk ambrette for induction and challenge (data not shown). Additionally, we found that by increasing the number of inductions, a corresponding increase in the photoallergic response was obtained in mice induced and challenged with MA (data not shown). Therefore, in subsequent experiments, mice were induced with the test material on three consecutive days (0, 1, and 2) and challenged on day seven. Another modification we incorporated into our mouse ear swelling testing procedure was the use of a mouse restrainer during application of the test material to the ears and measurement of the ears. By using the mouse restrainer, we avoided the loss of animals due to the repeated use of anesthetics and we found this procedure to be relatively nonstressful to the mice. Incorporating these modifications into our mouse model, we examined the photoallergic potential of a known human photoallergen, musk ambrette (MA). The results in Table 1 show that a significant photoallergic response was achieved when mice were induced and challenged with MA plus UVA radiation (Group I) as compared to the vehicle/radiation control (Group II), contact allergy control (Group III), and phototoxicity control (Group IV).

In a study designed to examine the use of UVB radiation for contact photoallergy testing, we found that successful detection of a photoallergic response with the known human photoallergen, 6-MC, was dependent upon the use of shorter ultraviolet wavelengths (Table 2). The results show that mice induced and challenged with 6-MC plus UVA and UVB radiation (Group I) demonstrated a significant photoallergic response as compared to the vehicle/radiation group (Group II), contact allergy group (Group III) and phototoxicity group (Group IV). Mice treated at induction and challenge with 6-MC plus UVA only (Group V) or 6-MC plus UVB only (Group VI), demonstrated no significant increases in ear thickness. In similar studies, we found that the photoallergic response to TCSA, MA, or the contact allergic response to oxazolone was not significantly enhanced or inhibited in mice when using both UVA and UVB radiation for induction and challenge (data not shown).

Table 1. Induction of contact photoallergy to musk abrette (MA) with UVA radiation in cyclophosphamide treated BALB/c mice

GROUP[a]	INDUCTIONS DAY 0 DAY 1 DAY 2	CHALLENGE DAY 7	EAR SWELLING ($\pm$SD) mm x 10^{-2} DAY 8
I	MA + UVA	MA + UVA	5.8 $\pm$ 2.1 [b]
II	VEHICLE + UVA	VEHICLE + UVA	0.0 $\pm$ 0.9
III	MA	MA	0.4 $\pm$ 1.0
IV	------	MA + UVA	1.3 $\pm$ 0.6

[a] MA was used at 20% for induction and 10% for challenge in acetone:corn oil (4:1) vehicle. The radiation dose was 10 J/cm^2 UVA.

[b] Significant difference from groups II, III and IV (p $\leq$ 0.002, Student's t-test).

Table 2. Induction of contact photoallergy to 6-methylcoumarin (6-MC) with UVA and UVB radiation in cyclophosphamide treated BALB/c mice

GROUP[a]	INDUCTIONS DAY 0 DAY 1 DAY 2	CHALLENGE DAY 7	EAR SWELLING ($\pm$SD) mm x 10^{-2} DAY 8
I	6-MC + UVA/UVB	6-MC + UVA/UVB	4.1 $\pm$ 1.4 [b]
II	VEHICLE + UVA/UVB	VEHICLE + UVA/UVB	0.7 $\pm$ 0.6
III	6-MC	6-MC	-0.4 $\pm$ 0.7
IV	------	6-MC + UVA/UVB	0.3 $\pm$ 0.6
V	6-MC + UVA only	6-MC + UVA only	0.9 $\pm$ 1.5
VI	6-MC + UVB only	6-MC + UVB only	0.6 $\pm$ 0.5

[a] 6-MC was used at 20% for induction and 10% for challenge in acetone:corn oil (4:1) vehicle. The irradiation dose was 10 J/cm^2 UVA and 45 mJ/cm^2 UVB.

[b] Significant difference from groups II, III, IV, V and VI (p $\leq$ 0.001, Student's t-test).

Kaidbey (1987) has shown in humans that the use of UVB radiation during induction caused a significant increase in the number of subjects becoming sensitized to 6-MC. In guinea pigs, Cripps et al. (1970) demonstrated the need for both UVA and UVB radiation for the induction of photoallergy to TCSA and bithionol. The authors proposed that skin damage induced by UVB irradiation accounted for the increased incidence of photosensitization. Supporting their explanation, TCSA photoallergy can be induced in guinea pigs with UVA only when the animals are treated with Freunds' adjuvant (Harber et al. 1982). Giudici and Maguire (1985) have demonstrated, in mice, that UVA radiation without UVB fails to induce photoallergy to systemic sulfanilamide and chlorpromazine.

In contrast to UVB enhancement of photoallergy, various investigators have demonstrated that UVB radiation can have a suppressive effect on the induction of contact photoallergy (Miyachi and Takigawa, 1982; Granstein et al. 1983a; Takigawa et al. 1984). UVB radiation has been shown to deplete Langerhans cells locally (Miyachi and Takigawa, 1982) and to be associated with the induction of antigen-specific suppressor T cells (Takigawa et al. 1984). Although there is some data which indicate that UVB radiation, at appropriate doses, can suppress the induction of contact photoallergy (Miyachi and Takigawa, 1982; Granstein et al. 1983a; Takigawa et al.1984), our results show that the doses of UVB radiation employed in our experiments were not sufficient to suppress the response and, in the case of 6-MC, UVB enhanced the photoallergic response. Since consumers are exposed to both UVB and UVA radiation when using topical products designed for outdoor use, we feel that it is important to employ both UVA and UVB radiation for photoallergy testing, especially since future test materials could have their absorption and action spectra in the UVB range.

In addition to examining the effect of UVB on the induction of contact photoallergy, we examined the effect of the amount of time allowed between application of the test material and irradiation on the photoallergic response to TCSA and MA. For induction, the mice were treated with the test material and then irradiated at either 1, 6, or 24 hours post application. The mice were challenged as previously described. The results show that allowing 24 hours between the time of test material application and irradiation for induction, caused a significant reduction in the extent of ear swelling with both TCSA and MA (Table 3). These results suggest to us the importance of understanding the penetration kinetics of the test material in the selected vehicle in relationship to the time of irradiation, since irradiating too soon or too long after the time of test material application could result in a false negative response.

Using the mouse ear swelling model as described in this paper, we have successfully detected the photoallergic potential of numerous known human photoallergens (Table 4). Included in the list are both weak (i.e., MA and PABA) and strong (i.e., TCSA) photoallergens (Wennersten et al. 1986; Kaidbey, 1987). Interestingly, PABA and musk ambrette are not detected when using a human photomaximization protocol (Kaidbey, 1987). We have also conducted experiments using analogs of positive test materials. These agents included coumarin, tribromosalicyclanilide, musk ketone and homosalate. As indicated in Table 4, no substantial increase in ear thickness was observed

Table 3. Effect of material application/irradiation time on contact photoallergy induced with musk ambrette or TCSA[a]

TIME BETWEEN APPLICATION & IRRADIATION FOR INDUCTIONS	EAR SWELLING ($\pm$SD) AT 24 HRS POST CHALLENGE mm X 10^{-2}	
	MA	TCSA
1 hour	4.6 ± 1.0	15.9 ± 2.2
6 hours	3.6 ± 1.1	13.5 ± 1.4
24 hours	0.8 ± 1.0^{b}	7.9 ± 2.1^{c}

[a] MA was used at 20% for induction and 10% for challenge in acetone. TCSA was used at 1% for both induction and challenge in acetone. UV dose for induction and challenge was 10 J/cm^2 UVA and 30 mJ/cm^2 UVB.

[b] Significantly different from MA 1 hr and 6 hr groups ($p \leq 0.01$, Student's t-test).

[c] Significantly different from TCSA 1 and 6 hr groups ($p \leq 0.001$, Student's t-test).

with these materials. We feel that the modifications made on this model have increased the sensitivity of this model, as well as its ability to detect potential photoallergens. In addition, the mouse ear swelling model offers the advantages of being quantifiable as well as requiring only two weeks for completion.

An additional advantage of the mouse model is that there is an extensive database on the mouse immune system which has allowed for investigation of the underlying mechanisms of photoallergy *in vitro*. Development of *in vitro* methods for photoallergy testing will enable us to better distinguish contact photoallergy responses from contact allergy and irritation responses. Additionally, the development of *in vitro* methods will increase our understanding of the mechanisms involved in photoallergy. This information could be of value in understanding the relationship between CPA and persistent light reactivity and/or in the development of alternative *in vitro* screening methods for photoallergens.

It has been shown that a photoallergen is produced by covalent binding of the photohapten to protein via the formation of free radicals resulting from UVA radiation (Jenkins et al. 1964; Yokozeki et al. 1985). In addition, Miyachi and Takigawa (1982) have demonstrated, in the mouse, that Langerhans cells play an important role in the induction of contact photoallergy. Recently, we have begun to examine photoallergy *in vitro* using TCSA photohapten-modified LC enriched epidermal cells and lymphocytes from TCSA photosensitized mice. When epidermal LC are cultured for 1-3 days, their antigen-presenting capacity increases markedly (Inaba et al. 1986), along with an increase

Table 4. Detection of known human photoallergens using a mouse ear swelling photoallergy model

Chemical	Mouse Ear Swelling Model	Human[a] Photomaximization Method
3,3',4',5 Tetrachlorosalicylanilide	+	+
Bithionol	+	+
Musk Ambrette	+	-
6-Methylcoumarin	+	+
Fentichlor	+	nd[b]
Chlorpromazine	+	+
p-aminobenzoic acid	+	-
Sodium omadine	+	+
Coumarin	-	-
Homosalate	-	nd
3,4',5-Tribromosalicylanilide	-	-
Musk Ketone	-	nd

[a] Kaidbey, K. 1987. In Dermatotoxicology, eds. F. N. Marzulli and H. I. Maibach, pp. 457-468.

[b] nd = not determined

in the density of class II major histocompatibility antigen (Ia) on their cell surface (Shimada et al. 1987). These Ia expressing epidermal cells, once hapten-modified, are potent stimulators of T-lymphocytes (Inaba et al. 1986; Hauser and Katz, 1988). In our studies, we observed a significant increase in the blastogenesis response of TCSA sensitized lymphocytes when cultured with epidermal cells that were photohapten-modified with TCSA and UVA radiation as compared to epidermal cells treated with TCSA or UVA alone (Figure 2). Lymphocytes from untreated mice or mice sensitized with musk ambrette demonstrated a significantly lower response to epidermal cells photohapten- modified with TCSA and UVA, indicating the specificity of the reaction.

The results of our study suggest that TCSA was successfully coupled to LC enriched EC by UVA irradiation, presumably onto cell surface proteins, and that once hapten-modified, these cells were capable of stimulating lymphocytes from TCSA sensitized animals. To our knowledge, this is the first demonstration of photohapten-modification

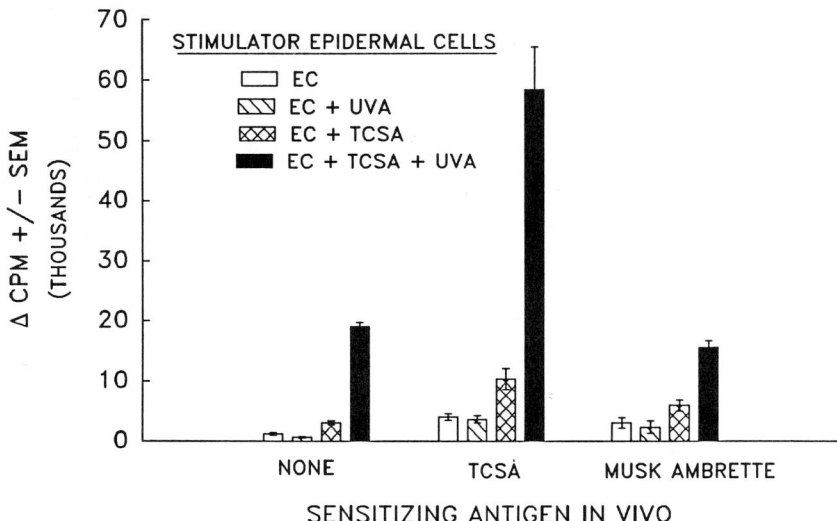

Figure 2. Lymphocyte blastogenesis response to TCSA photohapten-modified Langerhans cell enriched epidermal cells. Lymph node lymphocytes from mice induced/challenged with TCSA or MA and naive mice were cultured at 3×10^5 cells/culture in microtiter plates containing medium or medium plus stimulator epidermal cells (EC). The EC were prepared from normal, naive mice, cultured for 4 days, recovered from culture, enriched for Langerhans cells and photohapten-modified. The EC were ~48% Langerhans cells as determined by morphology. EC, EC + UVA, EC + TCSA, or photohapten-modified EC (EC+TCSA + UVA) were then added to the responder cells at a responder to stimulator ratio of approximately 10:1. Blastogenesis was assessed after 72 hr of culture. The mean background (medium only) ^{3}H incorporation for lymphocytes from antigen induced/challenged and naive control mice were as follows: Naive: 203 CPM; TCSA: 6244 CPM; MA: 5700 CPM.

of LC enriched epidermal cells. In a similar study, Tokura et al. (1988) have demonstrated photohapten-modification of TCSA with spleen cells *in vitro*. Subcutaneous injection of these photo-TCSA modified spleen cells into naive mice induced a photohapten-specific photoallergic response upon challenge with TCSA plus UVA irradiation. Using *in vitro* methodology, we plan to further investigate the photobinding of photoallergens to Langerhans cell enriched epidermal cells which may enable us to determine the relationship between photoallergy, allergy, and persistent light reactivity.

In summary, we have described the use of the mouse for investigation of photoallergy *in vivo* and *in vitro*. This mouse ear swelling photoallergy model offers potential as a method for predictive photoallergy testing as well as a model for the examination of the parameters that can influence a photoallergic response. In addition, this mouse model offers promise for elucidating the underlying mechanisms of contact photoallergy *in vitro*.

References

Brown, W.R., Ramsay, C.A., and Shivji, G.M., 1986, Dose response studies for UVA in contact photosensitivity to TCSA in the mouse. *Photodermatol.,* 3,334-339.

Cripps, D.J. and Enta, T., 1970, Absorption and action spectra studies on bithionol and halogenated salicylanilide photosensitivity. *Br. J. Dermatol.,* 82 ,230-242.

De Groot, A.C., Van Der Walle, H.B., Jagtman, B.A. and Weyland, J.W., 1987, Contact allergy to 4-isopropyl-dibenzoylmethane and 3-(4'-methylbenzylidene) camphor in the sunscreen Eusolex 8021. *Contact. Dermatitis.,*16,249-254.

Epstein, J.H., 1987, Photocontact allergy in humans. Dermatotoxicology, edited by F.N. Marzulli, and H.I. Maibach. Hemisphere, Washington, pp.441-456.

Giudici, P.A., and Maguire, H.C., JR., 1985, Experimental photoallergy to systemic drugs. *J. Invest. Dermatol.,*85,207-211.

Granstein, R.D., Morison, W.L., and Kripke, M.L., 1983a, The role of UVB radiation in the induction and elicitation of photocontact hypersensitivity to TCSA in the mouse. *J. Invest. Dermatol.,*80,158-162.

Granstein, R.D., Morison, W.L., and Kripke, M.L., 1983b, The role of suppressor cells in the induction of murine photoallergic contact dermatitis and in its suppression by prior UVB irradiation. *J. Immunol.,*130, 2099-2103.

Harber, L.C., Armstrong, R.B., and Ichikawa, H., 1982, Current status of predictive animal models for drug photoallergy and their cor4.relation with drug photoallergy in humans. *JNCI.,* 69,237-244.

Hauser, C. and Katz, S.I., 1988, Activation and expansion of hapten- and protein-specific T helper cells from nonsensitized mice. *Proc. Natl. Acad. Sci. USA,* 85, 5625-5628.

Ichikawa, H., Amstrong, R.B., and Harber, L.C., 1981, Photoallergic contact dermatitis in guinea pigs: improved induction technique using Freund's complete adjuvant. *J. Invest. Dermatol.,* 76, 498-501.

Inaba, K., Schuler, G., Witmer, M.D., Valinsky, J., Atassi, B. and Steinman, R.M., 1986, Immunologic properties of purified epidermal Langerhans cells Distinct requirements for stimulation of unprimed and sensitized T lymphocytes. *J. Exp. Med.,* 164, 605-613.

Jenkins, F.P., Welti, D., and Baines, D., 1964, Photochemical reactions of tetrachlorosalicylanilide. *Nature,* 201,827-828.

Jordan, W.P., Jr., 1982, The guinea pig as a model for predicting photoallergic contact dermatitis. *Contact. Dermatitis,* 8, 109-1167F.

Kaidbey, K., 1987, The evaluation of photoallergic contact sensitizers in humans. Dermatotoxicology, edited by F.N. Marzulli, and H.I. Maibach. Hemisphere Publishing Corporation, Washington, pp.457-468

Maguire, H.C., Jr. andKaidbey, K., 1982, Experimental photoallergic contact dermatitis: a mouse model. *J. Invest. Dermatol,* 79,147-152.

Maurer, T., 1984, Experimental contact photoallergenicity: guinea pig models. *Photodermatol.,* 1, 221-231.

Mituoka, A., Mitsuo, B., and Morikawa, S., 1976, Enhancement of delayed hypersensitivity by depletion of suppressor T cells with cyclophosphamide. *Nature,* 262, 77-78.

Miyachi, Y., and Takigawa, M., 1982, Mechanisms of contact photosensitivity in mice: II. Langerhans cells are required for successful induction of contact photosensitivity to TCSA. *J. Invest. Dermatol.,* 78, 363-365.

Shimada, S., Caughman, S.W., Sharrow, S.O., Stephany, D., and Katz, S.I., 1987, Enhanced antigen-presenting capacity of cultured Langerhans cells is associated with markedly increased expression of Ia expression. *Journal of Immunology,* 139, 2551-2555.

Takigawa, M., and Miyachi, Y., 1982, Mechanisms of contact photosensitivity in mice: I. T cell regulation of contact photosensitivity to tetrachlorosalicylanilide under the genetic restrictions of the major histocompatibility complex. *J. Invest. Dermatol.,* 79, 108-115.

Takigawa, M., Miyachi, Y., Toda, K., and Yoshioka, A., 1984, Mechanisms of contact photosensitivity in mice. IV Antigen-specific suppressor T cells induced by preirradiation of photosensitizing site to UVB. *J. Immunol.,* 132, 1124-1129.

Tokura, Y., Takigawa, M., and Yamada, M., 1988, Induction of contact photosensitivity to TCSA using photohapten-modified syngeneic spleen cells. *Archives of Dermatological Research*, 280, 207-213.

Wennerstein, G., Thune, P., Jansen, C.T., and Brodthagen, H., 1986, Photocontact dermatitis: current status with emphasis on allergic contact photosensitivity occurrence, allergens, and practical phototesting. *Seminars in Dermatology*, 5, 277-289.

Yokozeki, H., Nishioka, K., Katayama, I., Takijiri, C., and Hashoimoto, K., 1985, Induction and suppression of contact sensitivity by liposomes carrying molecules from haptenated epidermal cells. *J. Invest. Dermatol.*, 84, 33-36.

Relation Between the Molecular Structure of a Drug and its Photobiological Activity by Combination of *In vivo* and *In vitro* Research

GERARD M. J. BEIJERSBERGEN VAN HENEGOUWEN

Division of Medicinal Chemistry
Center for Bio-Pharmaceutical Sciences
Leiden University, Leiden
The Netherlands

The *in vivo* situation is very complex and information about the molecular mechanism underlying a photobiologic response is hardly gained by *in vivo* investigations only. Therefore *in vitro* experiments are performed. However, there are problems with extrapolating in vitro results to processes occurring in the whole animal. The reason is that *in vitro* research is based on models of the in vivo situation and as a result the information obtained, although abundant, often lacks the desired relevance by neglecting important parameters such as metabolism and bioavailability of a drug. Answering the question how to extrapolate *in vitro* results to the *in vivo* situation can be circumvented as follows: As the photobiologic effect to be studied is an *in vivo* problem, *in vitro* investigations should be *continuously* performed in *close* harmony with *in vivo* research; conclusions from *in vitro* research need verification in the *in vivo* system. A photobiologic problem can be solved efficiently if *in vitro* and *in vivo* research are in *continuous* interaction with each other. This means that the (experimental) approach to the problem, at any of both sides, is guided by results from the other side. Such an integration of *in vitro* and *in vivo* research was tried out with some imino-N-oxides. (Aim of the research was to find the part of the molecular structure responsible for phototoxic/photo-allergic effects. This gives the opportunity to alter the structure in such a way that the phototoxicity diminishes, whereas desired, pharmacological effects are conserved.)

Phototoxic chlordiazepoxide, an imino-N-oxide known under the trade name Librium^R is represented in Figure 1 (CDZ=A_1) with some of its analogues.

Among these are the major metabolites of CDZ, desmethyl CDZ (A_2) and demoxepam (A_3). These compounds were also investigated because phototoxic effects may be caused not only by the drug itself but also by its metabolites. A_4 is the N-oxide of diazepam (C_4). Diazepam (Valium^R) has never been reported as a phototoxic compound; its metabolites lack the N-oxide group as well, e.g. C_3. In this review only

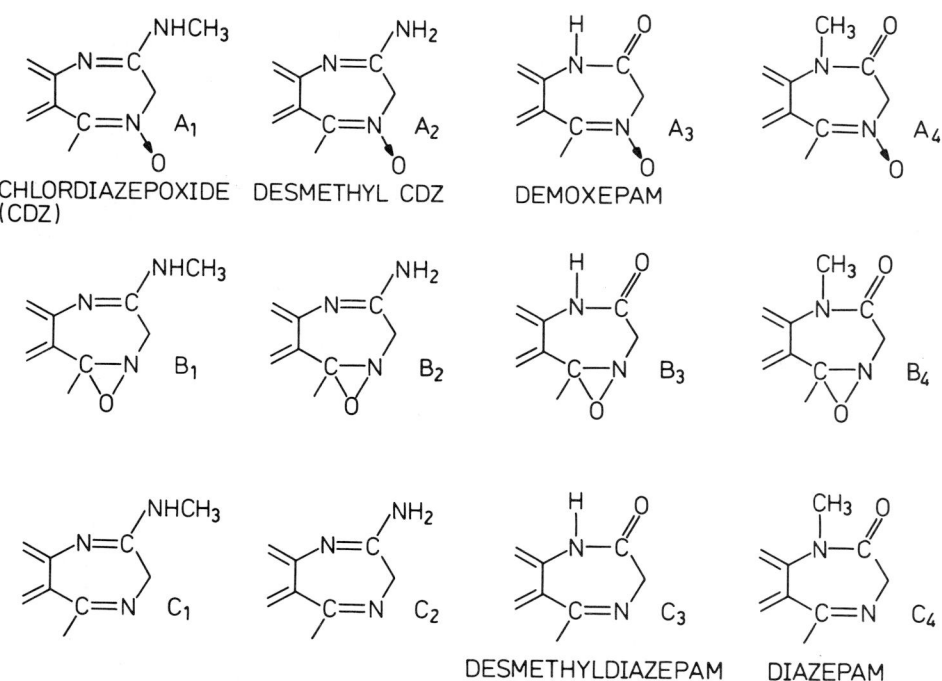

Figure 1. Structures of 7-chloro-1,4-benzodiazepines. Photoreactivity *in vitro*

some of the main results can be dealt with on which the conclusion was based that the N-oxide function is the cause of the phototoxicity of CDZ; more information and experimental details are given in the articles referred to.

Being present as the only compound in solution the N-oxides (A_1–A_4) isomerize for nearly 100% into their oxaziridine B_1–B_4 upon irradiation with UV-A (320–400 nm) (De Vries et al. 1983; Bakri et al. 1988). As oxaziridines are unstable compounds the quantity B_1–B_4 actually found depends on the conditions applied e.g. $t_{1/2}$ at pH 7.4 and 37°C: B_1 ≈140 min, B_2 ≈ 110 min, B_3 ≈ 1 min and B_4 ≈ 20 h.

Also in the presence of SH group containing compounds, such as glutathione (GSH), the photoreaction of the N-oxides proceeds via the formation of an oxaziridine. This became evident from the fact that compounds B_1 and B_2 thermochemically and A_1 and A_2 photochemically react in the same way with GSH, namely by the nonenzymatic formation of a conjugate with GSH. In the case of A_1 and B_1 this conjugate has a $t_{1/2}$ ≈ 100 min at 37°C and pH 7.4 and decomposes into the reduced form of A_1 (or B_1) namely C_1 (Cornelissen and Beijersbergen van Henegouwen 1980; De Vries et al. submitted). GSH is abundantly present in the body. The fact that both imino-N-oxides, upon

684

The toxicity curves of the oxaziridines B_1–B_4 correspond nicely with those for the phototoxicity of A_1–A_4. This indicates that formation of an oxaziridine is the cause of the phototoxicity of the N-oxides A_1–A_4. In this respect it is of importance that the reduced forms of the N-oxides (Fig.1, C_1–C_4), amongst which diazepam and its metabolite desmethyldiazepam, appeared to be nonphototoxic. (Cornelissen et al. 1980.)

Neither with strain TA 100 nor with TA 98, Cornelissen et al. (1980) found photomutagenic effects. However recently De Vries et al. (submitted) reported that the oxaziridines B_1 and B_4 can induce DNA damage with *E.coli* K-12 # 765 and # 753.

Another interesting finding from this research of De Vries is, that (photo)conjugation of A_1, A_2, B_1 and B_2 with GSH has only a small detoxificating effect (*Salmonella typhimurium* TA 100).

Phototoxicity in experimental animals

The results with bacterial test systems supported the supposition that the formation of an oxaziridine from photo-excited CDZ is the cause of the phototoxicity. Together with the data from the *in vitro* photoreactivity study they were considered as a firm bases for experiments with the rat. To verify whether the N-oxide group is the cause of the phototoxicity compound A_1, CDZ, was compared with C_1 (Bakri et al. 1983, 1985) and A_4 with C_4, diazepam (Bakri et al. 1988; Bakri and Beijersbergen van Henegouwen, submitted). A proper comparison can only be made if two factors are taken into account which quantitatively determine the eventual photobiological effect. The first is the concentration of the compound investigated in the irradiated organs, e.g. the skin, as a function of time and the second is the extent to which it absorbs the light in the spectral region of the lamp used. In the case of A_1 and C_1, the dose C_1 was 1.5 that of A_1 because of differences in these two factors. Rats of which the back was shaved got an intraperitoneal injection of the compound and were either kept in the dark or exposed to UV-A on five consecutive days for 8h/day (UV-A dose per day comparable to that on a sunny May day in Holland.)

Most of the reduced compound C_1 and its metabolites in the rat is excreted via the faeces. After deconjugation and extraction the mixture was submitted to TLC. Quantitative analysis of the spots proved that the metabolism of C_1 is not altered by UV-A. (see Table 1). The excretion of CDZ (A_1) and metabolites proceeds for $\approx 45\%$ via the urine.

Table 1. Rf and percentage of each metabolite from C_1 in the extract of faeces after deconjugation. The total quantity extracted was put at 100%.

Rf	0.27	0.33	0.53	0.46	0.59	0.71	0.90
UV-A	11.7	11.9	4.1	9.2	10.5	26.8	25.6
Dark	11.8	12.3	4.1	8.7	11.0	26.4	25.5

685

irradiation, and their first photoproducts, the oxaziridines, almost quantitatively produce the same product with this thiol indicates that other photoproducts than oxaziridines are not quite relevant for the phototoxic phenomenon.

Because of their three membered ring oxaziridines are very reactive compounds; especially with regard to photoallergy, as reported side-effect, reaction with proteins may be of interest.

The extent to which A_1 upon UV-A irradiation or B_1 in the dark irreversibly binds to human plasma proteins *in vitro* has been investigated and found to be 50% in both cases (Bakri et al.1986); for A_4 and B_4, which react far more slowly under the same conditions, the figure appeared to be 30% (Bakri et al. 1988). Under comparable conditions the compounds C_1–C_4 appeared to be photostable (Cornelissen et al. 1980, Bakri et al. 1986, 1988).

Phototoxicity in microbiogical test systems

At this stage of the research it was preferred to get an idea of the relevance of the *in vitro* data to the *in vivo* situation rather than to investigate the *in vitro* (photo)reactions of A and B in more detail. Microbiological test systems are convenient for this purpose because of their speed and simplicity. In Fig. 2 some of the results obtained with *Salmonella typhimurium* TA 100 (Cornelissen et al. 1980; De Vries et al. 1983) are presented.

The N-oxides A_1–A_4 which form an oxaziridine as primary product, are not toxic without light (not represented) but only phototoxic in this test system. Of the two metabolites of CDZ, demoxepam (A_3) appears to be less phototoxic and desmethyl CDZ (A_2) more phototoxic than CDZ.

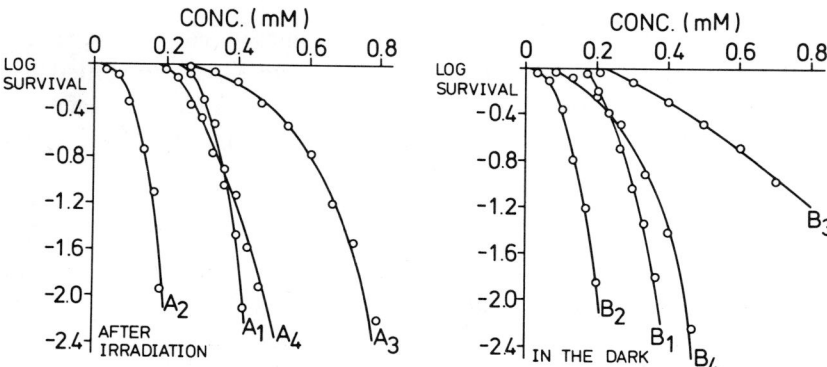

Figure 2. Concentration-dependent survival of the colony forming ability of *Salmonella thyphimurium* TA 100. Left panel: phototoxicity of the N-oxides A_1–A_4 (irradiation with UV-A, max.= 350 nm) and right panel: toxicity of the oxaziridines B_1–B_4.

In Table 2 the results are presented for urine collected 10–20h and 82–106h after adminstration of CDZ. As can be seen, the percent of N-oxymetabolites is lower for the UV-A exposed rats; this difference is much larger after 82 h: 45% *vs* 83%. The reverse was found for N-desoxymetabolites: after 82 h an increase from 6 to about 30%.

Table 2. Percentage of N-oxymetabolites (including CDZ) and N-desoxymetabolites in the extract of urine after deconjugation. The total quantity extracted was put at 100%

Metabolites	Urine (10–20 h)		Urine (82–106 h)	
	UV-A	Dark	UV-A	Dark
N-oxy	61 ± 3	87 ± 3	45 ± 3	83 ± 3
N-desoxy	17 ± 2	5 ± 2	30 ± 2	6 ± 2
Total	78 ± 3	$92 + 3$	75 ± 3	89 ± 3
Unidentified	22 ± 2	8 ± 2	25 ± 2	11 ± 2

Comparable results were obtained for diazepam (C_4) and for its N-oxide (A_4). With diazepam quantitative TLC analysis of metabolites extracted from urine and faeces after deconjugation showed that there was no difference between UV-A exposed rats and those kept in the dark. With diazepam N-oxide (A_4) the extract of urine after deconjugation appeared to contain 85% N-oxymetabolites for rats kept in the dark and 67% for UV-A exposed animals, whereas these percentages were 15% and 33% respectively for the N-desoxymetabolites.

An explanation can be proposed in the light of the reaction of oxaziridines, formed by photoisomerisation of N-oxides in the UV-A exposed skin, with SH-group containing compounds. As already mentioned CDZ upon UV-A irradiation and its oxaziridine without light react spontaneously with GSH with the N-desoxyform C_1 as the ultimate product.

Further confirmation of the responsibility of the N-oxide function in the molecule for the phototoxic effects was obtained by investigation of covalent binding of benzodiazepine fragments to biomacromolecules of, for example, the skin.

This was expected because of the *in vitro* data obtained: covalent binding to plasma proteins. With CDZ irreversible binding to biomacromolecules of the skin of the back, and to a lesser extent to those of the skin of the belly but also to those of liver and kidneys was found with rats exposed to UV-A. (Bakri et al. 1983). With the reduced form of CDZ (C_1), irreversible photobinding *in vivo* was not observed (Bakri et al. 1985), which corresponds with the *in vitro* data. It is remarkable that the N-oxide of diazepam (A_4) did not photobind *in vivo* to a measurable extent. (Bakri and Beijersbergen van Henegouwen, submitted). *In vitro* investigation had already shown that its oxaziridine (B_4) has a relatively long life time and reacts far more slowly with plasma proteins than that of CDZ (B_1). It is supposed that *in vivo* irreversible binding of the oxaziridine B_4 to proteins is suppressed by *enzymatic* reaction with GSH. This would correspond with the increased percentage of N-desoxymetabolites, at the expense of N-oxymetabolites, in the UV-A exposed rats.

Other remarkable differences between UV-A exposed rats and those kept in the dark concerned the conjugation of metabolites. These differences were observed with CDZ (A_1, Table 3, Bakri et al. 1983) but not with its reduced form (C_1, not represented, see Bakri et al. 1985) and with diazepam N-oxide (A_4) but not with diazepam (C_4) (Table 4, Bakri et al. 1988; Bakri and Beijersbergen van Henegouwen submitted).

Table 3. Percentage of glucuroconjugated metabolites of CDZ (A_1); the total quantity of metabolites, conjugated and extracted at a certain pH, was put at 100%

pH	Non-irradiated	UV-A	Non-irradiated/UV-A
2	81 ± 1	32 ± 1	2.5
4	49 ± 1	34 ± 1	1.4
7	70 ± 2	32 ± 3	2.2

Table 4: Percentage of urinary metabolites (40–50% of dose) of diazepam (C_4) and its N-oxide (A_4): I = non-conjugated; II = glucuroconjugates; III = other conjugates. I + II + III = 100%.

		I	II	III
C_4	UV–A	18 ± 2	69 ± 2	13 ± 3
C_4	Non-irradiated	19 ± 2	70 ± 3	11 ± 2
A_4	UV-A	43 ± 2	2 ± 2	55 + 3
A_4	Non-irradiated	65 ± 2	9 ± 3	26 ± 2

Perhaps the most remarkable fact is that both (a) the combination of CDZ (A_1) and UV-A, and (b) its oxaziridine (B_1) alone, caused a decrease (24% and 17% respectively) in the ratio of the weight of the liver to the total weight, whereas no change was found with C_1 and UV-A (Bakri et al. 1983, 1985).

By combination of the results from *in vitro* and *in vivo* research it could be concluded that the N-oxide function is responsible for the phototoxicity. In this respect it is interesting to mention that the reduced form of CDZ, compound C_1, shows biological activity approaching that of CDZ and the question remains whether nonphototoxic C_1 might have been an acceptable alternative to CDZ (Librium®) as tranquilizer. On the other hand as far as diazepam-N-oxide (A_4) and diazepam (C_4) are concerned this question does not arise because the nonphototoxic compound has been commercialized (Valium®).

Current research with other imino-N-oxides has features in common with the foregoing. This investigation concerns olaquindox (OX, Fig.3) which like carbadox and cyadox, is a growth-promoting substance used as an additive to pig feed. OX has two imino-N-oxide groups.

The quinoxaline-1,4-dioxides are extremely photolabile. Quindoxin caused several cases of photocontact dermatitis (Zaynoun et al. 1976) and has been removed from the market. More recently OX has also been reported to be photo-allergic in man (Fracalanci et al. 1986).

The present experiments with OX have been performed as follows. On four successive days male Wistar rats (140 g) were shaved and given 60 mg/kg OX in 0.5 ml PBS suspension by oral intubation under brief ether anesthesia. The animals were housed in small metabolism cages covered with netting (mesh 2 × 2cm). Four rats were kept in the dark and four others were exposed each day to UV-A (5 lamps, Philips TL 80W/ 10 R; UV-A = 340–400 nm, max = 370 nm) for 12 h/day; light intensity at the level of the rats is 6 mW/cm² as measured with an UV-X radiometer (UV-products Inc., San Gabriel, USA). Urine, collected in light resistant containers, was analyzed by HPLC (column: 200 mm × 3mm ID filled with Chromspher RP18 10μ; mobile phase 5% methanol/ phosphate buffer 0.01 M pH 3; detection: UV abs.260 nm; retention-time OX= 10 min).

Figure 3. Quindoxin (R_1=R_2=H); carbadox (R_1= -CH=N-NH-COOCH$_3$, R_2=H); cyadox (R_1= -CH=N-NH-COCH$_2$CN, R_2=H) and olaquindox (OX; R_1= -CO-NH-CH$_2$-CH$_2$OH; R_2=CH$_3$). OX is supposed to react according to the scheme

The complete experiment was performed twice. After four days of treatment only the OX-treated UV-A exposed rats showed severe erythema on irradiated parts of the body, e.g. skin of the back, ears, feet and tail, oedema of feet and necrosis of ears. Systemic effects were indicated by a profound change of metabolism (Fig. 4). Unchanged OX appeared to be eliminated via the urine for ~60% of total applied dose by non-UV-A exposed rats. This figure was only ~20% for the UV-A exposed rats. It appeared that one of the metabolites was excreted in a much higher quantity by the UV-A exposed rats than by those kept in dimly lighted environment (30% vs 2%, of daily dose). This metabolite was identified as OX-4-monoxide. Probably OX, photo-activated in the skin, forms a reactive oxaziridine which in the presence of, for example, GSH eventually forms the OX-4-monoxide (Fig.3, De Vries et al. 1990). *In vitro* research confirmed these results. Besides this showed that irreversible binding to and destruction of human serum albumin can take place. This occurs not only with quindoxin and olaquindox, already known as photo-allergens, but also with cyadox and carbadox (De Vries et al., 1988 and in press). For this reason the latter compounds should also be considered as potential photo-allergens.

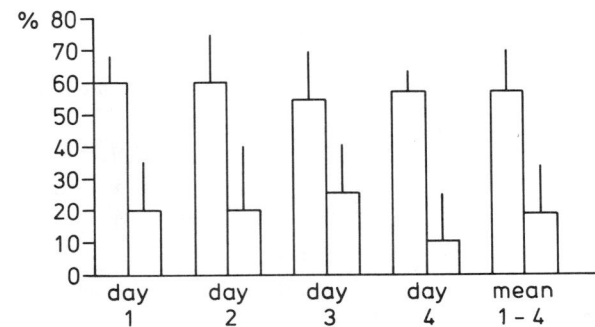

Figure 4. Urinary excretion of OX, 60 mg/kg.day, as a % of dose administered to four rats exposed to UV-A (λ_{max} = 370 nm; I=6 mW/cm^2; 12 h/day) and four rats kept in the dark. Mean is summation of % of dose found daily divided by four (Dark= 58±11%; UV-A=19±15%; p<10^{-5} Student-t-test).

References

Bakri, A. and G.M.J. Beijersbergen van Henegouwen. Photopharmacology of diazepam-N$_4$-oxide in the rat. (Submitted for publication).

Bakri, A., G.M.J. Beijersbergen van Henegouwen and J.L. Chanal (1983). Photopharmacology of the tranquilizer chlordiazepoxide in relation to its phototoxicity. *Photochem.Photobiol.*38,177-183.

Bakri, A., G.M.J. Beijersbergen van Henegouwen and J.L. Chanal (1985). Involvement of the N-4-oxide group in the phototoxicity of chlordiazepoxide in the rat. *Photodermatol.* 2,205-212.

Bakri, A., G.M.J. Beijersbergen van Henegouwen and A.G.J. Sedee (1986). Irreversible binding of chlordiazepoxide to human plasma protein induced by UV-A radiation. *Photochem.Photobiol.* 44,181-185.

Bakri, A., G.M.J. Beijersbergen van Henegouwen and H. de Vries (1988). Photobinding of some 7-chloro-1,4-benzodiazepines to human plasma protein *in vitro* and photopharmacology of diazepam in the rat. *Pharm.Weekbl. Sci.Ed.* 10,122–129.

Cornelissen, P.J.G. and G.M.J. Beijersbergen van Henegouwen (1980). Photochemical decomposition of 1,4-benzodiazepines; quantitative analysis of decomposed solutions of chlordiazepoxide and diazepam. *Pharm. Weekbl.Sci.*Ed. 2,39-48.

Cornelissen, P.J.G., G.M.J. Beijersbergen van Henegouwen and G.R. Mohn (1980). Structure and photobiological activity of 7-chloro-1,4-benzodiazepines. Studies on the phototoxic effects of chlordiazepoxide, desmethylchlordiazepoxide and demoxepam using a bacterial indicator system. *Photochem.Photobiol.* 32, 653-661.

De Vries, H., G.M.J. Beijersbergen van Henegouwen and R. Wouters (1983). Correlation between phototoxicity of some 7-chloro-1,4-benzodiazepines and their (photo) chemical properties. *Pharm.Weekbl.Sci.Ed.* 5,302-308.

De Vries, H., G.M.J. Beijersbergen van Henegouwen and A. Bakri. The effect of glutathione on the oxaziridine mediated phototoxicity of chlordiazepoxide. (Submitted for publication)

De Vries, H., G.M.J. Beijersbergen van Henegouwen, F. Kalloe and M.H.J. Berkhuysen. Phototoxicity of olaquindox in the rat. *Res. Vet. Sci.* 48, 240–244.

De Vries, H. and G.M.J. Beijersbergen van Henegouwen (1988). Photochemical reactions of quindoxin, olaquindox, carbadox and cyadox with protein, indicating photoallergic properties. *Abstracts 10th Int. Congress on Photobiol.* p.94.

Idem, *Toxicology*, in press.

Fracalanci, S., M. Gola, S. Giorgini, A. Muccinelli and A. Sertoli (1986). Occupational photocontact dermatitis from Olaquindox. *Contact Dermatitis* 15, 112–114.

Zaynoun, S., B.E. Johnson and W. Frain-Bell (1976). The investigation of quindoxin photosensitivity. *Contact Dermatitis* 2, 343–352.

Screening Materials for Photosensitization Eye Effects

PIERRETTE DAYHAW-BARKER AND FELIX M. BARKER

Pennsylvania College of Optometry
1200 West Godfrey Avenue
Philadelphia, Pa. 19141-3399

Introduction

The ocular tissues are particularly susceptible to phototoxic damage inasmuch as the eye not only allows the transmittance of possibly injurious wavelengths to deeper layers but, through its optical properties, it actually focuses the radiation onto the retina. Thus, the transmittance and absorption characteristics of the various ocular layers must be taken into account whenever photosensitization of any one eye tissue is being studied. If one also wishes to investigate the possible role of exogenous products such as pharmaceuticals, food additives or cosmetics it is necessary to consider whether or not there is a systemic distribution of the compound and the degree to which the compound will cross the blood aqueous and/or blood retinal barriers.

Lastly, perhaps the one factor that has been most frequently ignored, is that the diseased eye represents a different situation from the normal healthy eye, especially where there is a breach in the blood retinal barrier. Therefore, while it may seem perfectly appropriate to administer certain drugs (compounds) to the vast majority of the population, it may be more cautious to ascertain via experimental studies, the differences in ocular distributions of a compound that may occur whenever the retinal barriers are compromised. From these types of studies it may then be necessary to provide warning or to screen the drug's target population for the integrity of the vascular barriers before administering photosensitizing compounds.

Such are the overall facts that presently face industry, the health professions and the consumer population in regard to the marketing and distribution of potentially photosensitizing compounds.

Ocular Photosensitization

It may be said initially that any compound that exhibits photosensitizing properties in the skin must also be suspect in the eye. It has already been documented (Dayhaw-Barker et al. 1986) that the psoralens, the phenothiazines, some of the antimalarials,

Photobiology, Edited by E. Riklis
Plenum Press, New York, 1991

most notably chloroquine, and some of the tetracyclines, amongst others, adversely affect some of the ocular tissues through mechanisms of photosensitization that have not yet been experimentally determined. Nevertheless, it has been documented that there is an increase in the incidence of conjunctival/corneal alterations, lenticular opacifications and/or retinal alterations with each of these groups of compounds.

If a compound exhibits photosensitizing characteristics, the degree to which any one ocular tissue is vulnerable must next be determined. To understand this, let us first review the ocular structures and their susceptibility to phototoxicity.

Lids and Conjunctiva

The lids, conjunctiva and cornea are the first tissues upon which environmental radiation is incident. They can be directly damaged by these radiations through absorption by endogenous photosensitizers. In addition, the lids and conjunctiva are vascularized tissues susceptible to photosensitized reactions from both systemically distributed and topically applied compounds.

Cornea

An important consideration in ocular phototoxicity is the transmittance/absorption of various wavelengths in the eye. It has been shown that the cornea is highly transmissive to wavelengths greater than 295mm (Boettner and Wolter, 1962; Barker, 1979). Wavelengths less than 295mm are absorbed mainly by the corneal epithelium which is a highly regenerative tissue. Thus, an episode of photokeratitis resulting from exposure to UVB or C wavelengths causes transient pain and exfoliation of the epithelium. With even minimal clinical intervention these episodes are followed by rapid healing and seldom represent a long-term hazard to sight. Repetitive and/or long term high dosage exposure can, however, produce more serious complications such as pterygium (Lerman, 1980).

Additionally the cornea is an avascular structure. However, it is bathed anteriorly by tears and posteriorly by the aqueous humour. Photosensitizing compounds found in these fluids can theoretically affect the integrity of the corneal tissues. This possibility has been systematically investigated by Hull, Green, Csukas and colleagues (Hull et al. 1981, 1983, 1984, 1985, 1987). Using *in vitro* preparations, they ascertained that Rose Bengal, chlorpromazine, trifluoperazine and hematoporphyrin were each associated with endothelial swelling, an important finding in a tissue that regulates water transport across the cornea, thus helping to maintain transparency. Wicker (1987) has reported similar findings with changes in endothelial barrier functions.

There is, as yet, no integrated, cohesive understanding of the role and the interactions of antioxidants known to be present in the aqueous humour as these apply to the maintenance of corneal integrity. There is need for study in this area.

A variety of other alterations have been reported. For example, pigmentary deposits have been identified in the cornea of patients taking phenothiazine medications (Dayhaw-

Barker et al. 1986). Experimental studies have demonstrated dose-related opacifications and neovascularization in the corneas of unprotected mice (Barker et al. 1986) when these were administered 8-MOP and 5-MOP (with UVA exposure) by comparison to other psoralens.

These corneal effects together with the susceptibility of the lens to the cataractogenic process have guided industry and health care in providing spectacle lens protection to solar parlor clients and medicated patients.

Lens

Much more is known about the interaction of radiation and the biochemical constituents of the lens. The most obvious of these interactions is the cataractogenic process that has been shown to be related to life long exposure to ultraviolet (UV) wavelengths (295–400nm) (Pitts et al. 1986). The degree to which the absorption/transmittance of these wavelengths is correlated with age has been extensively studied in our laboratory and highlights the fact that different conditions exist at different ages. In an extension of some of the earlier studies of Boettner and Wolter (1962) and Lerman (1980), our laboratory will shortly publish the result of transmittance studies on the lenses of more than 100 donor eyes ranging in age from birth to 90 years. The data clearly show age related increases in short wave absorbance. This is, at first, restricted to the UV range but later impinges on the visible spectrum. The process is presumably due to UV-induced denaturation (Zigman, 1973) of lens proteins (Bose et al. 1986)

The studies of numerous investigators (Pitts et al. 1986) have shown that incident radiation is absorbed by endogenous lenticular photosensitizers and that through a number of reactions many of which appear to involve the amino acid tryptophan, a series of photochemical reactions occur leading to evidence of conformational changes in some lenticular proteins (Mandal et al. 1986; Bose et al. 1986; Andley and Chylack, 1986; Dillon et al. 1987) and photopolymerization (Roberts et. al. 1985) of others. This can eventually lead to opacification.

Pharmacologically, it has been documented that PUVA (Psoralen UVA) therapy is associated with cataractogenesis both in animal and human tissue (Dayhaw-Barker et al. 1986). The evidence points to protein photoproduct formation, the initiation of unscheduled DNA repair synthesis, and membrane disruption as possible mechanisms of damage (Dayhaw-Barker et al. 1986).

It is clinically well-recognized that some of the phenothiazines induce dose-related opacifications. The star-shaped chlorpromazine cataract can be seen with the biomicroscope to be made up of numerous dot-like opacities organized in a star pattern. Merville et al (1984) have demonstrated that five promazines including chlorpromazine in the presence of oxygen promote crystallin protein cross-linking. Roberts and Dillon (1984–5) have described photo-dynamically-mediated lens protein polymerization in the presence of chlorpromazine and have compared the reactivity of several known photosensitizers in promoting lens protein photopolymerization. Chlorpromazine was intermediate in this effect i.e. between tetracycline and fluorescein.

Such studies, while they do not fully describe the sequence of photochemical processes leading up to the clinically-observed cataract, do present evidence that a photosensitization process could be responsible for the opacities.

Since the lens is limited in its repair processes and since opacities on the whole are retained for the rest of the lifetime of the individual, it must be emphasized that this tissue is particularly susceptible to long-term effects of even limited exposure to photosensitization.

Retina

A similar body of information is accumulating as regards the retina. A few radioactive tracer studies have been conducted using suspect compounds. The results were analyzed at the light microscope level and demonstrated an accumulation of the radioactive-labelled compounds in the ocular pigmented structures. The exact relationship between the melanin-bound compounds, photosensitization, and clinically observable alterations in visual functions is not known. However the following associations for example have been documented as far back as the middle 1950's.

Phenothiazines

NP 207 was associated with loss of vision presumably due to a rod degeneration pattern. Pigmentary retinopathy was part of the clinical pattern as were reductions in acuity and constrictions of visual fields (Boet, 1969; Potts, 1962; Kirk et al. 1970; Burian and Fletcher, 1988; Goar and Fletcher, 1986).

Thioridazine showed similar but lesser effects (Davidorf, 1973) while chlorpromazine exerted only minor effects.

Chloroquine

The typical bull's-eye pigmentary retinopathy associated with long-term use of chloroquine has the following characteristics:

1) The retinopathies are related to increasing total dosage of the drug as well as relatively high daily dosages (Scherkel et al. 1965; Elmon et al. 1976)
2) The chloroquine is bound to melanin and remains bound for some time (even years) after cessation of intake (Lindquist and Ulberg, 1972).
3) Functional alterations include decreased EOG and ERG responses, often abnormal color vision and occasionally changes in dark adaptation (Hobbs et al. 1959; Earnshaw et al. 1966; Okun et al. 1963; Carr et al. 1966).
4) The visual changes were found to be related to the extent of retinal damage (Zinn and Marmor, 1979).

Other known photosensitizers have been reported to have retinal effects, though overall, less is known about these compounds.

One of the reasons we know comparatively less about ocular photosensitization than dermal effects is that in many instances thorough ocular studies have not been mandated by regulating agencies, though recommendations to this effect were made at a conference sponsored by the FDA in 1983 (Waxler and Hitchins, 1986). It must also be emphasized that appreciable loss of vision with certain drugs leads to fairly quick reactions by the industry and the regulating agencies, while more subtle losses in the quality of vision, especially if these occur over a span of time may not even be recognized as drug-related effects.

Therefore, we suggest that certain studies be undertaken for any compound that is found to exhibit photosensitizing characteristics.

Safety Evaluation for Suspect Compound

1) Distribution of chemical compound in ocular tissues — Perhaps the most important point to ascertain is whether or not the potentially photosensitizing compound is distributed to any of the ocular tissues. This can best be determined using radioactive tracers and autoradiography, thereby establishing the degree to which the compound crosses the blood ocular barriers as well as those tissues most susceptible to phototoxicity.

It is best to select an animal model with a vascularized retina similar to man. The rat and mouse have been frequently chosen for these purposes. Since the lenses of these animals transmit radiation greater than 300nm (Brainard et al. 1986), this animal model offers the advantage that retinal damage might be evaluated later on in the protocol.

Light microscopy/autoradiography will be adequate in most circumstances to determine distribution; electron microscopy may be needed in special circumstances.

Should the blood ocular barriers effectively obstruct the diffusion of the compound, similar autoradiographic studies using "compromised RPE systems" should be repeated to insure that such a condition does not alter the distribution pattern.

If the industry and the eye researcher work in close cooperation, availability of labelled compounds should not present a problem.

2) Animal models — Next, the studies need to be carried out in appropriate animal models utilizing ranges of dosages and radiant exposures of sufficient magnitude to identify the type of damage that could occur after long term use. Such studies must involve animal models that have both rods and cones because color vision may be affected. Since many skin studies are initiated on rodents, this group should be studied first, followed by studies on the eyes of rabbit, pig or monkey. The design should include experiments on pigmented as well as albino tissues.

3) Cell culture — Cell-culturing techniques can be utilized to study the differences in the cellular responses of pigmented versus non-pigmented cell lines.

4) Clinical evaluation — Analysis of the results in the animal models must include clinical evaluation of lids, conjunctiva, cornea, lens and retina with photographic documentation of abnormalities and electro-retinographic assessment (ERG).

697

5) Histology — The eyes should then be harvested for biochemical and/or histological evaluation. Ultrastructural level studies may be necessary to identify the initial lesion. Morphometric studies are often critical in quantitative analysis of more subtle cellular changes. The relationship of any and all changes to any visual alteration or loss should be addressed.

6) Lastly, even after the compound has been judged to be appropriate for human use, monitoring of patients in clinical trials should also include scrutiny of the ocular tissues.

Human Clinical Studies

In addition to basic science and animal studies, it is important to conduct ongoing clinical studies as a safety determination measure. This is due to species differences which may prevent detection of human eye effects. The clinical monitoring should extend for the full testing period given that compounds such as chloroquine were not identified as producing damaging effects until after 2 years of administration. Ideally patients should be pre-examined using the appropriate tests and then receive long-term periodic surveillance and assessment utilizing some of the following procedures.

Vision

The most obvious and often the simplest procedure is to monitor the visual acuity (VA) of the patient. VA can be affected if the macular (central) photoreceptors are involved. More subtle changes in acuity function can be detected through the use of contrast sensitivity testing (Arden and Jacobson, 1978) and the Amsler grid test of macular integrity.

Visual Fields

Aside from central acuity (VA), the sensitivity of the retina can be plotted over its surface using a perimeter (Hart et al. 1984). Modern computer driven perimeters enable the determination of retinal sensitivity threshold up to 30° from the macula in a time span of about 15 minutes per eye. This is a very sensitive and clinically convenient way to assess retinal function. It has proven to be a standard for the long term clinical surveillance of chloroquine users.

Color Vision

Specific effects on the optic nerve and/or the cone population(s) are often detectable by measurement of the color vision sense (Grutzner, 1969). The screening tests for alterations in color vision are very rapid to perform while the more definitive measures are quite time consuming. Nonetheless, since a number of compounds affect color vision very early on, these tests can prove to be most worthwhile.

Retinal Examination

Dilated retinal exams using indirect ophthalmoscopy, biomicroscopy and photography are also a standard that should be utilized with all patients about to undergo drug treatment. These forms of examination should then be repeated periodically with particular attention being paid to macular integrity and any changes in pigmentary patterns.

Electrophysiology

These more sophisticated testing regimens allow the objective determination of ocular potentials that are indicative of retinal function.

The Electrooculogram (EOG) (Arden et al. 1962) is a measure of the standing potential of the eye which is an indicator of Retinal Pigment Epithelium (RPE) cell health. This test has proven its use in the early diagnosis of chloroquine retinopathy.

The Electroretinogram (ERG) (Arden et al. 1983) and its variations (e.g. Pattern electroretinogram) (PERG) (Arden et al. 1982) are also very important indicators of retinal health. With these procedures it is possible to follow the generation of the visual impulses in the retina and thus to objectively determine the relative % of cells functioning as well as to more accurately pinpoint classes of cells that might be affected within the retina (Barker et al. 1987) .

Dark Adaptation

Retinal tissues normally adapt via the rhodopsin cycle to changing lighting conditions including the absence of light. Dark adaptation (Dieterle and Gordon, 1956) is clinically measurable. It is a sensitive indicator of even minor changes in retinal photochemistry, and should be included in the routine battery of tests utilized to monitor potential photosensitizers.

Summary

Overall a comprehensive series of experimental and clinical procedures are available and should be performed on animal models as well as human patients when a new compound is being assessed. The tests will provide sufficient information for the careful marketing of the new products. The procedures are neither monetarily prohibitive nor are they necessarily complicated. They do however require experienced personnel who can more easily recognize early subtle changes indicative of possible damage.

References

Andley, V.P. and L.T. Chylack Jr. Changes in sulfhydryl group microenvironment of calf lens α−crystallin by 300nm light. *Photochem. Photobiol.* 43: 175–181, 1986.

Arden, G.B. and J.J. Jacobson. A Simple Grating Test for Constant Sensitivity. *Invest. Ophthalmol. and Vis. Sci.* 17(1):23–32, 1978.

Arden, G.B., Vaegan, and C.R. Hogg. Clinical and Experimental Evidence that the Pattern Electroretinogram (PERG) is Generated in More Proximal Layers than the Focal Electroretinogram (FERG). *Ann. NY Acad. Sci.* pp. 580–601, 1982.

Arden, G.B., A. Barroda. and J.H. Kelsey. New Clinical Test of Retinal Function Based Upon the Standing Potential of the Eye. *Br. J. Ophthalmol.* 46:449, 1962.

Arden, G.B., R.M. Carter, C.R. Hogg, D.J. Powell, W.J.H. Ernst, G.M. Clover, A.L. Liyness, and M.P. Quinlan. A Modified ERG Technique and the Results Obtained in X–linked Retinitis Pigmentosa. *Br. J. Ophthalmol.* 67:419,–430, 1983.

Barker, F.M., The Transmittance of the electromagnetic spectrum Rom 200nm — 2500nm through the optical tissues of the pigmented rabbit eye. Master of Science Thesis. Univ. of Houston, 1979.

Barker, F.M., G.G. Arden, A.C Bird, J.L.M. Hawk and P.G. Norris. The ERG in Canthaxanthin therapy. *Invest. Opthalmol. Vis. Sci.* 28 (Suppl.): 304, 1987.

Barker, F.M., P. Dayhaw-Barker, P. Donald Forbes and R.E. Davies. Ocular Effects of Treatment With Various Psoralen Derivatives and Ultraviolet-A (UVA) Radiation in HRA/Skh Hairless Mice. *Acta. Ophthalmol.*, 64, pp. 471–478, 1986.

Boettner, E.A. and J.R. Wolter. Transmission of the Ocular media. *Invest. Ophthalmol.* 1(6): 776–83, 1962.

Boet, D.J., Phenothiazine retinopathy. Ophthalmologia, 158:576, 1969.

Bose, S.K., K. Mandal and B. Chakrabati. Sensitizer induced conformational changes in lens crystallin: II. Photodynamic action of riboflavin on bovine α–crystallins. *Photochem. Photobiol.* 43:525–528, 1986.

Brainard, G.C., P.L. Podolin, S.W. Leivy, M.D. Rollag, C. cole and F.M. Barker. Near ultraviolet radiation suppresses Pineal Metatonin content. *Endocrinol.* 119:2201–5, 1986.

Burian, H.M. and M.C. Fletcher. Visual functions in patients with retinal pigmentary degeneration following the use of NP 207 *Arch. Ophthalmol.*, 60: 612– 29,1988.

Carr, R.E., P. Gouras and R.D. Gunkel. Chloroquine retinopathy: early detection by retinal threshold test. *Arch. Ophthalmol.* 75: 171–179, 1966.

Davidorf, F.H. Thioridazine pigmentary retinopathy. Arch. Ophthalmol. 90: 251–255, 1973.

Dayhaw-Barker, P., Forbes, D.,Fox, D., Lerman, S., McGinness, J., Waxler, M., and R. Felton. Drug Phototoxicity and Visual Health, pp. 147–175, in Optical Radiation and Visual Health ed. by M. Waxler and V.M. Hitchins, CRC Press, Boca Raton, 1986.

Dieterle, P. and E. Gordon. Standard Curve and Physiological Limits of Dark Adaptation by Means of the Goldmann-Weekers Adaptometer. *Br. J. Ophthalmol.* 40:652–655, 1956.

Dillon, J., R. Chiesa and A. Spector. The photochemistry of specific tryptophan residues in proteins as analyzed by the fluorescent scanning of tryptic peptide maps. *Photochem. Photobiol.* 45: 147–150., 1987

Earnshaw, E.R., D.O. Miles and T.W. Stewart. Screening for chloroquine retinopathy. Br. *J. Dermatol.* 78: 669–674, 1966.

Elmon, A., A.E.R. Guelberg, E. Niesson, J. Phendahl and L. Wachtmeister. Chloroquine retinopathy in Patients with Rheumatoid Arthritis. *Scand. J. Rheumatol.* 5: 161–166, 1976.

Goar, E.L., and M.C. Fletcher. Toxic chorioretinopathy following the use of NP 207. *Trans. Am. Ophthalmol. Soc.*, 54: 603–608, 1986.

Grutzner, P. Acquired Color vision Defects Secondary to Retinal Drug Toxicity. *Ophthalmologica* (suppl.) 158:592, 1969.

Hobbs, H.E., A. Sorsby and A. Freedman. Retinopathy following chloroquine therapy. *Lancet*, 2: 478–480, 1959.

Hart, W.M., RM. Burde, G.P. Johnston and R.C. Drews. Static Perimetry in Chloroquine Retinopathy, Perifoveal Patterns of Visual Field Depression. *Arch Ophthalmol.* 102:377–380, 1984.

Hull, D.S., S. Csukas and K. Green. Rose Bengal Induced Corneal Swelling: Relation to Inciting Wavelength, *Curr. Eye Res.*, 1, pp. 487–490, 1981.

Hull, D.S., S. Csukas and K. Green. Trifluorperazine: Corneal Endothelia; Phototoxicity. *Photochem. Photobiol.* 38: 425–428, 1983.

Hull, D.S., K. Green, L. Thomas and N. Alderman. Hydrogen Peroxide Mediated Corneal Endothelial Damage. *Invest. Ophthalmol. Vis. Sci.*, 25, pp. 1246–1253, 1984.

Hull, D.S., K. Green and D. Hampstead. Effect of Hematoporphyrin Derivative on Rabbit Corneal Endothelial Cell Functionand Ultrastructure. *Invest. Ophthalmol. Vis. Sci.*, 26, pp. 1465–1474, 1985

Hull, D.S., S. Csukas and K. Green. Chlorpromazine-Induced Corneal Endothelial Phototoxicity, *Invest. Ophthalmol. Vis. Sci.*, 22, pp. 502–508, 1987.

Kirk, L, K.B. Rasmussen and A. Faurbyl. Retinopathy following thioridazine treatment. *Acta. Psychiatr. Scand.*, 46: 56, 1970.

Lerman, S. Radiant Energy and the Eye, MacMillan, NY 1980.

Lindquist, N.G. and S. Ulberg. The melanin affinity of Chloroquine ad Chlorpromazine Studies by whole body autoradiography. *Acta. Pharmacol. Toxicol* (Suppl. II), 31: 3–32, 1972.

Mandal, K., S.K. Bose and B. Chakrabarti. Sensitizer-induced conformational changes in lens crystallins: I. Photodynamic action of methylene blue and N-formylkynurenine on bovine α−crystallin. *Photochem. Photobiol.* 43: 515–523, 1986.

Merville, M.P., J. Decuyper, J. Pietle, C.M. Calbert-Bacq and A. Van deVorst. In vitro cross linking of bovine lens proteins photosensitized by promazines. *Invest. Ophthalmol. Vis. Sci.* 25: 573– , 1984.

Okun, E., P. Gouras, H. Bernstein and L. von Sallmann. Chloroquine retinopathy: a report of 8 cases with ERG and Dark adaptation findings. *Arch. Ophthalmol.*, 69:59–71, 1963.

Pitts, D.G., L.L. Cameron, J.G. Jose S. Lerman, E. Moss, S.D. Varma, S. Zigler, S. Zigman and J. Zulich. Optical Radiation and Cataracts pp. 5–41, in: Optical Radiation and Visual Health. ed. by M. Waxler and V.M. Hitchins, CRC Press, Boca Raton, 1986.

Potts, A.M., Uveal pigment and phenothiazine compounds. *Trans. Am. Ophthalmol. Soc.*, 60:517–, 1962.

Roberts, J.E. and J. Dillon. A comparison of the photodynamic effect of photosensitizing drugs on lens protein. *Lens Res.* 2: 133–144, 1984–85.

Roberts, J.E., D. Roy and J. Dillon. The photosensitized oxidation of the calf lens main intrinsic protein (MP 26) with hematoporphyrin. *Curr. Eye Res.* 4:181–185, 1985.

Scherkel, A.L., A.H. MacKenzie, J.E. Nausch, J.E. Hausel and M. Atdjian. Ocular lesions in Rheumatoid Arthritis and Related Disorders with Particular Reference to Retinopathy. *N. England J. Med.*, 273:360–366, 1965.

Waxler, M., and V.M. Hitchins. Optical Radiation and Visual Health, CRC Press, Boca Raton, 1986.

Wicker, D.E., S.D. Gottsch, C.R. Graham Jr., and W.J. Stark. Photosensitization of Rabbit Corneal Endothelium and Protoporphyrin IX. *Invest. Ophthalmol. Vis. Sci.*, 28S, p. 171, (1987).

Zigman, S., Greiss, G., Yulo, T., et al: Ocular Protein Alterations by Near UV light. *Exp. Eye Res.*, 15: 255–64, 1973.

Zinn, K.M. and M.F. Marmor. Toxicology of the human retinal pigment epithelium, pp. 395–412 in the Retinal Pigment Epithelium. ed. by K.M. Zinn and M.F. Marmor. Harvard Univ. Press, Cambridge, 1979.

UV and Cancer

Solar Ultraviolet Radiation and Skin Cancer in Man

F. URBACH

Temple University Medical Center, Philadelphia, USA

Examination of the sun's role in the production of human skin cancer does not lend itself to direct experimentation. However, extensie astute observations have strongly suggested the etiologic significance of light energy in the induction of these tumors. Skin cancers in Caucasians in general are most prevalent in geographic areas of the greatest insolation and among people who receive the most exposure, i.e., men who work outdoors. They are rare in Negroes and other deeply pigmented individuals who have the greatest protection against UV light injury. Further, the lightest complexioned individuals, such as those of Scottish and Irish descent, appear to be most susceptible to skin cancer formation when they live in geographic areas of high UV exposure. When skin cancers do occur in the darkly pigmented races, they are not distributed primarily in the sun-exposed areas as they are in light-skinned people. The tumors in these pigmented individuals are more commonly stimulated by other forms of trauma, such as chronic leg ulcers, irritation due to the lack of wearing shoes, the use of a kangri (an earthenware pot that is filled with burning charcoal and strapped to the abdomen for warmth), the wearing of a Dhoti (loin cloth), and so on. In contrast, the distribution of skin cancer in the Bantu albino and in patients with Xeroderma pigmentosum follows sun exposure patterns.

Blum (1959), Urbach et al (1967, 1972) and Emmett (1974) have reviewed the evidence supporting the role of sunlight in human skin cancer development. In brief, the main arguments are:

1) It is clearly established that superficial skin cancer occurs most frequently on the head, neck, arms and hands: parts of the body that are habitually exposed to sunlight.
2) Pigmented races, who sunburn much less readily than people with white skin, have much less skin cancer and when it does occur, it affects those areas not frequently exposed to sunlight.
3) Among Caucasians there appears to be a greater incidence of skin cancer in those who spend more time outdoors than those who work predominantly indoors.
4) Skin cancer is more common in white-skinned people living in areas where insolation is greater.
5) Genetic diseases resulting in greater sensitivity of skin to the effect of solar UV radiation are associated with marked increases and premature skin cancer development (albinism, Xeroderma pigmentosum).

Photobiology, Edited by E. Riklis
Plenum Press, New York, 1991

6) Superficial skin cancers, particularly squamous cell carcinoma of the skin, occur predominantly on the areas receiving the maximum amounts of solar UV radiation and where histologic changes of chronic UV damage are most severe.
7) Skin cancer can be produced readily on the skin of mice and rats with repeated doses of UV radiation, and the upper wavelength limit of the most effective cancer-producing radiation is about 320 nm; that is the same spectral range that produces erythema solare in human skin.

Though these arguments do not constitute absolute proof, there is excellent epidemiologic evidence supporting the role of sunlight in non-melanoma skin cancers.

Human skin cancers, especially basal cell and squamous cell carcinomas, are closely associated with chronic repeated exposure of skin to solar ultraviolet radiation. Three types of evidence indicate that the most effective wavelengths are shorter than 320 nm; (1) extensive experiments in mice show that wavelengths longer than 320 nm are almost ineffective for induction of skin cancer. 92) Wavelengths shorter than 320 nm are highly effective in inducing photochemical changes in DNA and killing of cells in tissue culture. Furthermore, damage to DNA is considered to be one of the events leading to carcinogenesis, and a number of carcinogenic chemicals mimic UV radiation damage to DNA. (3) The effective wavelengths for human skin erythema production are below 320 nm. Also, individuals who sunburn easily and have high exposure to solar UV radiation have a much higher incidence of non-melanoma skin cancer than those who rarely sunburn and have little exposure to the sun.

This last observation has been used as a basis for assuming that the human skin erythema action spectrum and the skin carcinogenesis action spectrum are closely related.

Relationship of incidence of skin cancer in man to potential changes in stratospheric ozone

Of the total solar radiant energy that reaches earth, approximately 5% consists of ultraviolet radiation (UVR). Depending on sun angle (i.e. air mass, time of day) the biologically most effective UVR comprises 10% or less of the total ultraviolet radiation (UVB). Because ozone begins to absorb UVR significantly about 325 nm, the segment of the solar UVR spectrum that would be augmented in the event of diminution of atmospheric ozone concentration includes only the small waveband between 290 and 325 nm. Wavelengths longer than 325 nm are not appreciably influenced by changes in atmospheric ozone concentration. Although the total energy that would be added to the solar (and even solar UVR) insolation would be very small, this region has notable photochemical efficiency in biologic systems.

The balance between increasing biological effectiveness of decreasing wavelengths in the UVR and increasing absorption by ozone of the shorter wavelengths becomes critical when considering the biological implications of solar UVR changes as a function of alteration of the stratospheric ozone layer.

706

For most biological action spectra, it is clear that changes in atmospheric ozone will result in increases in biologically effective UV irradiance disproportionately greater than might be indicated by simply integrating the increasing total UVR energy flux.

Extensive studies of the relative biological effectiveness of wavelengths in the UVB (i.e. action spectra) on such diverse biological materials as purified DNA, bacteria, plant and human cells in tissue culture and human skin and eye *in vivo* have shown that wavelengths of 290 nm are one thousand to ten thousand times as effective in producing damage (Thymine dimer production, single and double strand breaks in DNA, production of mutations, cell killing, production of skin erythema and skin cancer) than wavelengths of 330 nm.

Figure 1 illustrates the most recent action spectrum for skin erythema (McKinlay and Diffey, 1987). Figure 2 illustrates the doubling of skin erythema effectiveness by light sources with spectral equivalent to clear sky conditions with a sun angle of 60° (2 mm WG 320 glass filter) and 75° (1 mm WG 320 filter) even though the additional total amount of UVR is only slightly changed.

Based on these observations, it has been estimated that a 1% decrease in total ozone column would increase the biologically effective (action spectrum weighted) UVB radiation effects by between 1.7 and 2% by use of the 1935 CIE action spectrum and between 1.25 and 1.3 by use of the McKinlay-Diffey action spectrum. This has been referred to as the "optical" amplification factor. (AF_o)

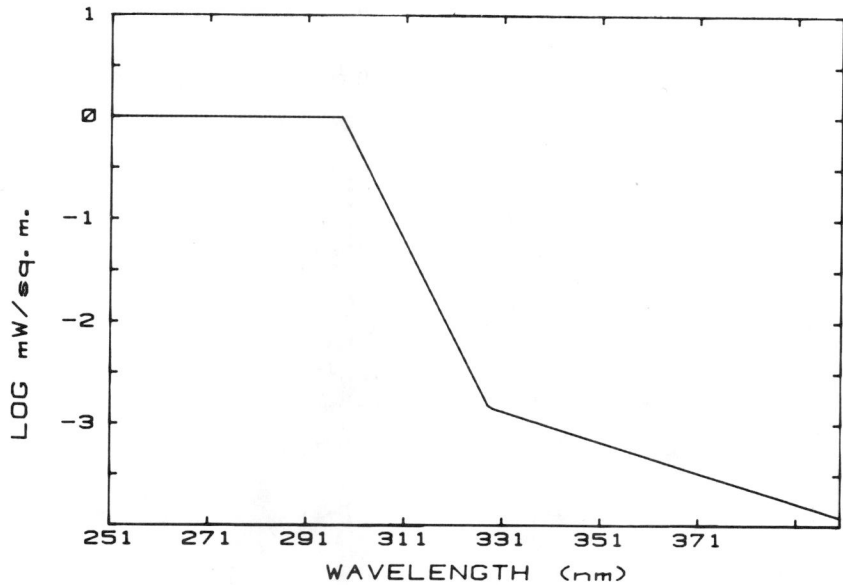

Figure 1. Spectral irradiance (McKinlay and Diffey).

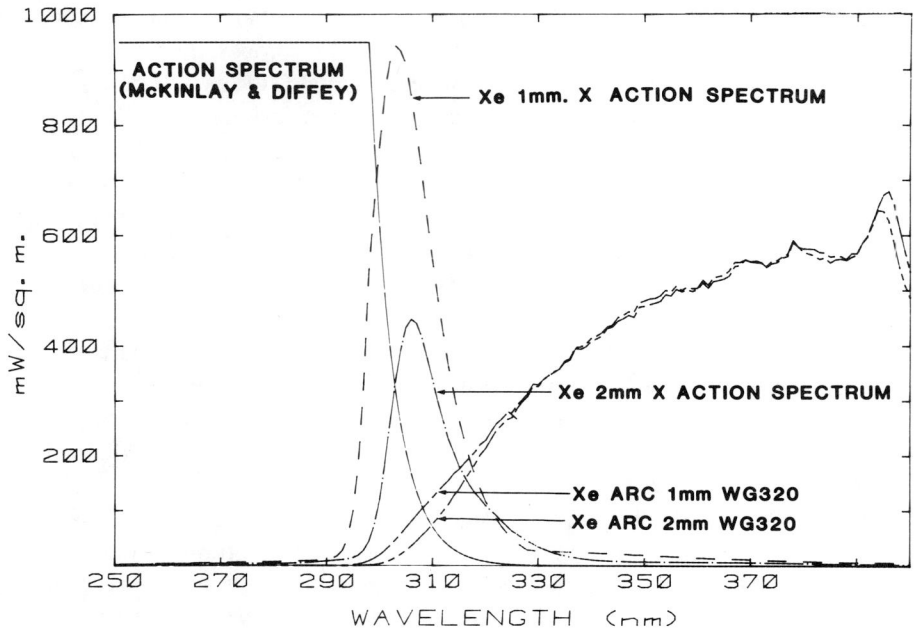

Figure 2.

Table 1.

	Latitude Band		
MONTH	53°-64°N	40°-52°N	30°-39°N
JAN	- 8.3±2.2	- 2.6±2.1	- 2.2±1.5
FEB	- 6.7±2.8	- 5.0±2.2	- 1.2±1.9
MAR	- 4.0±1.4	- 5.6±2.3	- 3.5±1.9
APR	- 2.0±1.4	- 2.5±1.7	- 1.7±1.3
MAY	- 2.1±1.2	- 1.3±1.1	- 1.7±0.9
JUN	+1.1±0.9	- 1.8±1.0	- 3.3±1.0
JUL	+0.0±1.1	- 2.2±1.0	- 1.3±1.0
AUG	+0.2±1.2	- 2.4±1.0	- 1.0±10
SEP	+0.2±1.1	- 2.9±1.0	- 1.0±0.9
OCT	- 1.1±1.2	- 1.5±1.5	- 0.9±0.8
NOV	+1.5±1.8	- 2.4±1.3	- 0.1±0.8
DEC	- 5.8±2.3	- 5.5±1.7	- 2.1±1.1
ANNUAL AVERAGE	- 2.3±0.7	- 3.0±0.8	- 1.7±0.7
WINTER AVERAGE	- 6.2±1.5	- 4.7±1.5	-2.3±1.3
SUMMER AVERAGE	+0.4±0.8	- 2.1±0.7	- 1.9±0.8

Table 2

Location	MED Highest Day	MED/yr. et ground	Average decrease O_3	Average increase in UVR		Increase in NMSC		Increase in Melanoma	
				AF∘1.8	AF∘1.25	AF∘1.8	AF∘1.25	AF∘1.8	0.6 AF∘ 1.25
								AFB	
Albuquerque (35°)	21.7	4530	− 1.81%	+3.26%	+2.26%	AF_B 3.0 +9.78%	+6.78%	+1.96%	+1.36%
Oakland (39.8°)	15.3	3079	− 2.05%	+3.69%	+2.56%	AF_B 2.0 +7.38%	+5.12%	+2.22%	+1.54%
Bismarck (46.8°)	15	2544	− 2.39%	+4.3%	+2.9%	AF_B 1.44 +6.19%	+4.18%	+2,58%	+1.74%
Davos (46.8°)	14.6	2389	−2.52%	+4.54	+3.15%	AF_B 1.44 +6.59%	+4.54%	+2.58%	+1.89%
BElsk (52°)	9.5	1330	− 2.27%	+4.1%	+2.8%	AF_B 1.44 +5.9%	+4.0%	+2.46%	+1.70%
Norrkoping (59°)	10	1206	− 0.5%	+0.92%	+0.64%	AF_B 1.44 +1.33%	+0.92%	+0.55%	+0.38%
Barrow (78°)	10	782	− 0.49%	+0.88%	+0.61%	AF_B 1.44 +1.27%	+0.88%	+0.53%	+0.37%

Based on RB meter network 1984
- 440 RB counts = 1 MED
- Ozone Trends Panel Report – Table 1. (March 1988)
Assuming AF_B 3.0 (Albuquerque), 2.0 (Oakland) 1.44 all others (based on average BCC/SCC 4:1 for Seattle 47 N)
(From Scotto et al., Table 81 and 82, 1983)

Actual increase in incidence of Malignant Melanoma in Denmark
1968–1985: males +83%, females +68%
(From Osterlind and Jensen, 1986; personal communication, 1988)
Actual increase 1971-1977-BCC +15-20%, SCC +5%
(From Scotto et al, Table 77, 1983)

A recent report of the Ozone Trends Panel (a joint effort of WHO, NASA, NOAA, EPA, CSIRO and various universities) has reported that in the 20-year period 1967–1986, total column ozone has decreased by 2 to 3 percent, depending on latitude (Table 1).

Since the decreases were greater in the winter months (when very much less UVB reaches the ground) than in the summer, the annual average increases are somewhat deceptive as far as biologically effective UVB is concerned.

From ground-based (biologically effective) UVB measurements performed with Robertson-Berger meters at Davos, Switzerland (Lat. 46.8°N) and Norrkoping, Sweden (Lat 58.75°N) for 1984, it is possible to calculate the expected UVB increases, based on the WMO/NASA report.

At Davos (Lat. 46.8°N), assuming a decrease in column ozone of 3%, and an AF. of 1.8, the (biologically effectivve) UVB would have been expected to increase by 5.4%. Utilizing the monthly changes from that report, UVB would have increased 4.54%. At Norrkoping (Lat. 58.75°N) the increase in UVB would have increased 4.54%. At Norrkoping (Lat. 58.75°N) the increase in UVB, based on an average decrease of 2.3% of 03 should have been 3.9%. Based on the monthly changes, it would have been only 0.92%. If the AF. were 1.25 (as in the McKinlay-Diffey action spectrum) the UVB increases would be +3.15% for Davos and +0.64% for Norrkoping. It is thus important that estimates of biologically effective UVB be based on monthly changes (greater in winter) than on annual averages. (Table 2).

Finally, we (Scotto et al. 1988) have reported that in the 11-year period (1974–1985) there has been *no increase* in ground level, biologically effective UVB as measured by Robertson-Berger meters, which have a response spectrum very similar to the biologic hazard spectrum of McKinlay and Diffey.

It is most likely that the lack of UVR increase measured by the R-B meters is due to metereological, climatic and environmental factors taking place in the troposphere, which attenuate UVB.

In the preceding section, the importance of weighting solar UVB radiation with a biologic effectiveness action spectrum has been discussed. Extensive studies on action spectra for two biologic effects — acute skin erythema and skin carcinogenesis — have been performed in man (experimental and epidemiologic) and in animal models (hairless mouse). Based on these studies, it has been shown that there is a precipitous, practically semilogarithmic drop in biologic effectiveness of wavelengths from 290 to 330 nm.

A modern skin erythema action spectrum has been proposed by McKinlay and Diffey (1987) and accepted by both CIE and IEC. It has been found to predict accurately the erythemal effectiveness of several polychromatic light sources differing greatly in spectral composition (Urbach, 1987).

Review of extensive photocarcinogenesis experiments in animals, performed with light sources of differing UVB content by Cole et al (1986) suggest that, as long as a light source contains 2% or more UVB, radiation of wavelengths longer than 330 nm have no measurable effect on photocarcinogenesis. That action spectrum closely

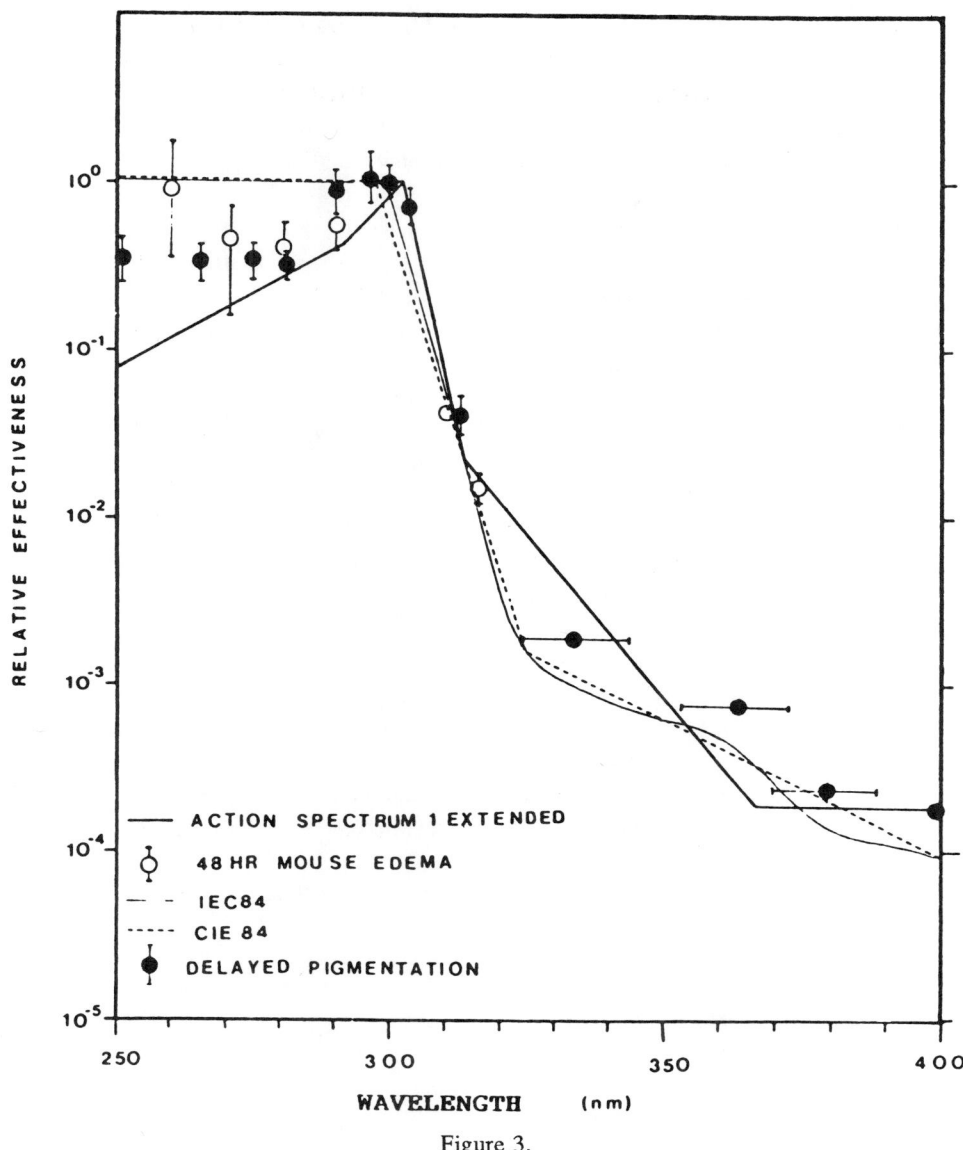

Figure 3.

simulates the older erythema action spectra. However, detailed experimental work in animals by Sternborg and van der Leun (1987) showed that the action spectrum for mouse skin cancer carcinogenesis is very similar to the new (McKinlay-Diffey) human erythema, action spectrum, i.e. wavelength longer than 330 nm are measurably effective for skin cancer induction (Figure 3).

Calculations of biologically effective UVR for the summer solstice with these two (differing in UVA effect) action spectra, performed by Prof. John Frederick (1988) show that with the action spectrum used by Cole the AF_O is 1.8, and by that used by Sterenborg and van der Leun the AF_O is 1.25, or a difference of approximately 30%. Table 2 shows the effect of use of thse two AF_O of the potential increase in skin cancer due to the ozone reduction reported in the recent Ozone Trends Panel report (1988, Table 1).

Epidemiologic studies of human non-melanoma skin cancer (NMSC) in the US (Scotto et al. 1974; Vitaliano and Urbach, 1980) have shown that there is indeed a progressive relationship of NMSC incidence with latitude and amount of outdoor sunlight exposure. Furthermore, there is a different relationship between Basal Cell and Squamous Cell Carcinoma, the latter being much more influenced by solar exposure.

Based on an extensive epidemiologic study of non-melanoma skin cancer (NMSC), Scotto et al (1983) estimated the relative increase in incidences Basal Cell (BCC) and Squamous Cell Carcinoma (SCC) of the skin in man associated with a 1% increase in UVB by geographic location in the United States.

For purposes of illustration, the biologic amplification factor (AF_B) was estimated for Albuquerque, New Mexico (Lat. 35°N), Oakland, California (Lat. 39.8°N) and for more northerly latitude based on Seattle, Washington (Lat. 47°N), assuming that the BCC/SCC ratio was 4:1.

This leads to approximate AF_B 3 for Albuquerque, 2 for Oakland and 1.44 for latitudes greater than 47°N. Utilizing the ozone reduction data from Ozone Panel report, the AF_O and AF_B reported above, and an AF_B of 0.6 for malignant melanoma (Scotto and Fears, 1987), the data shown in Table 2 are found.

In most previous reports by the US National Academy of Sciences an AF_O of 2, and AF_B of 2 were used for estimation of skin cancer increases, potentially due to ozone decreases (at equilibrium).

Comparing those estimates with those based on new data, one can estimate as follows:

Latitude 30°N — Albuquerque, NM 35°N
Assume average O_3 decrease — 1.7% and $Af_O = 2$, $AF_B = 2$
UVB + 3.4%
NMSC + 6.8%
MM + 1.0%

Actual O_3 decrease (Table 2) – 1.8%
$AF_O = 1.8$, $AF_B = 3$ UVB + 3.25%
NMSC + 9.78%
MM + 1.96%

$AF_O = 1.25$, $AF_B = 3$ UVB + 2.26%
NMSC + 6.78%
MM + 1.36%

Latitude 40° – 52°N — Davos 46.8°

Assume Average O_3 decrease -3% and $Af_O = 2$, $AF_B = 2$

UVB $+6\%$

NMSC $+12\%$

MM $+3.6\%$

Actual O_3 crease (Table 2) -2.52%

$Af_O = 1.8$, AF_B 1.44 UVB $+4.54\%$

NMSC $+6.54\%$

MM $+2.72\%$

AF_O 1.25, AF_B 1.44 UVB $+3.15\%$

NMSC $+4.54\%$

MM $+1.89\%$

Latitude 53 – 64°N — Norrkoping 58.75°N

Assume Average O_3 decrease -2.3%, $AF_O = 2$, $AF_B = 2$

UVB $+4.6\%$

NMSC $+9.2\%$

MM $+2.76\%$

Actual O_3 decrease (Table 2) -0.5%

AF_O 1.8, AF_B 1.44 UVB $+0.92\%$

NMSC $+1.33\%$

MM $+0.55\%$

AF_O 1.25, AF_B 1.44 UVB $+0.64\%$

NMSC $+0.92\%$

MM $+0.38\%$

It is interesting to compare these calculated increases in skin cancer incidence with reported epidemiologic data.

In the 20-year period 1965–1985, malignant melanoma has increased by at least 5% per year, the actual increase in some areas of the US has been 100% or more in this period. In the last decade, malignant melanoma incidence in the US has been 7% per year. In Denmark, malignant melanoma has increased by 83% in males and by 68% in females between 1968 and 1985 (Osterlind and Jensen, 1986).

The Australians believe that the increase in incidence of malignant melanoma began with people born at the end of the 19th century and has continued since. Since there is evidence that stratospheric ozone either remained stable or even increased prior to 1965, it is clear that factors other than increasing solar UVR must operate in the genesis of malignant melanoma.

In conclusion, I would like to most strongly emphasize that my colleagues and I are extremely concerned by the probability that stratospheric ozone will be decreased by man-made effluents. Although changes to date appear to be small and may not yet have caused major problems for living things on earth, the potential for an inexorable catastrophe is great.

There is every reason to believe that the effects of increased biologically effective UVR on plants and plankton and other living materials at the beginning of the food chain may be much greater than that on people, who after all can take measures to protect themselves, while stationary living things cannot do so.

It is thus of the utmost importance that the Treaties now proposed be put into effect and even strengthened and expanded to include chemicals other than CFC's. Changes in the stratosphere develop slowly, but can be reversed only in hundreds of years.

References

Blum, H.F. (1959), Carcinogenesis of ultraviolet light, Princeton University Press, Princeton, New Jersey.

Emmett, E.A. (1973), Ultraviolet radiation as a cause of skin tumors, CRC Critical Reviews in Toxicology, 2:211.

Executive Summary of the Ozone Trends Panel, NASA, Washington, DC, March, 1988.

McKinley, A.F. and Diffey, B.J. (1987), A reference action spectrum for ultraviolet induced erythema in human skin. In: Human Exposure to Ultraviolet Radiation: Risks and Regulations. Passchier, W.R. and Bosnajokovic, B.F.M. (eds.), Elsevier Science Publishers, Amsterdam 1987.

Osterlind, S. and Jensen, D. Moller (1986), Trends in incidence of malignant melanoma of the skin in Denmark 1943–1982. In: Recent Results in Cancer Research 102:8–17.

Scotto, J., Kopf, A.W. and Urbach, F. (1974), Non-melanoma skin cancer in four areas of the United States. *Cancer* 34: 1333–1338.

Scotto, J., Fears, T. and Fraumeni, J.F. (1983), Incidence of non-melanoma skin cancer in the United States. DHEW Publication NIH 83-2433, National Cancer Institute, Bethesda, MD.

Scotto, J. and Fears, T.R. (1987), The association of solar ultraviolet and skin melanoma incidence among Caucasians in United States. Cancer Investigations, in press.

Scotto, J., Cotton, G., Urbach, F., Berger, D. and Fears, T. (1988), Biologically effective ultraviolet radiation: Surface measurements in the United States 1974–1985, *Science* 239:762–764.

Slaper, H. and van der Leun, J.C. (1987), Human exposure to ultraviolet radiation: Quantitative modelling of skin cancer incidence. In: Human Exposure to Ultraviolet Radiation: Risks and Regulations. Passchier, W.R. and Bosnajokovic, B.F.M. (eds). Elsevier Science Publishers, Amsterdam. pp. 155–177.

Sterenborg, H.J.C.M. and van der Leun, J.C. (1987), Action spectra for tumorigenesis by ultraviolet radiation. In: Human Exposure to Ultraviolet Radiation: Risks and Regulations. Passchier, W.R. and Bosnajokovic, B.F.M. (eds), Elsevier Science Publishers, Amsterdam, pp. 173–190.

Urbach, F. (1969), Geographic pathology of skin cancer. In: The Biologic Effects of Ultraviolet Radiation (F. Urbach, ed.), Pergamon Press, Oxford, pp. 635–650.

Urbach, F., Rose, B.B. and Bonnem, M. (1972), Genetic and environmental interaction in skin carcinogenesis. In: Environmental Cancer, William and Wilkins, Baltimore, pp. 353–371.

Urbach, F. (1987), Man and ultraviolet radiation. In: Human Exposure to Ultraviolet Radiation: Risks and Regulations. Passchier, W.R. and Bosnajokovic, B.F.M. (eds), Elsevier Science Publishers, Amsterdam, pp. 3–17.

Vitaliano, P.P. and Urbach, F. (1980), The relative importance of risk factors in non-melanoma carcinoma. *Arch. Derm.* 116: 454–456.

Presentation of the Finsen Medal to D.I. Arnon

LARS OLOF BJÖRN

Lund, Sweden

Daniel Israel Arnon was born on November 14, 1910 in Warsaw, Poland. He emigrated to the United States at an early age, and came to study under D.R. Hoagland at the University of California. In 1936 he received his Ph.D. on "The role of trace elements in nitrogen metabolism of plants." Before becoming a photobiologist, he also made other contributions to the science of plant nutrition and the roles of metals in plants. Most importantly he discovered that vanadium is an essential element for the growth of the green alga *Scenedesmus*. We now know that vanadium is necessary also for other plants.

Arnon became professor at the Department of Soils and Plant Nutrition at the University of California at Berkeley, but during the forties and fifties he often was a guest scientist in various countries, including Sweden. One particularly important sabbatical was one spent with Otto Warburg in Berlin-Dahlem.

In 1954, soon after I had started my university studies, I saw a big headline in a Swedish newspaper: "American Professor Solves the Mistery of Life". At that time I understood even less of life than I do now (this holds for plant life, human life and a newspaper life). I did not know that in the coming years, I would encounter practically the same headline over and over again. So I thought, "Wow, solves the mystery of life, that's really something". Thus, due to this headline and a few other circumstances, I soon found myself as Professor Arnon's research assistant. I also came to realize the importance of the discovery of photosynthetic phosphorylation by isolated chloroplasts. (Inspired by the success of the Arnon group, A.W. Frenkel soon thereafter demonstrated photosynthetic phosphorylation with particles from photosynthetic bacteria.) We now take photosynthetic phosphorylation for granted, but Arnon had to go to great lengths to convince scientists that phosphorylation was really driven by light. For instance, he used sodium chloride instead of sucrose in his isolation medium to avoid the criticism that energy for phosphorylation might be derived from the oxidation of sugar. When I arrived, Arnon and coworkers had already demonstrated that the isolated chloroplast is capable of the complete light-driven carbon dioxide reduction. They went on to distinguish between cyclic and non-cyclic photophosphorylation, and later pseudocyclic photophosphorylation.

My own particular research assignment was inspired by the finding by Otto

Photobiology, Edited by E. Riklis
Plenum Press, New York, 1991

Warburg, whom Arnon very much admired, of a very high quantum efficiency for overall photosynthesis. My task was to demonstrate a correspondingly high quantum efficiency for photophosphorylation, and I think that Arnon expected *at least* 1 ATP per photon. I did not succeed with this, but had the opportunity to watch a number of other interesting projects develop in the truly multinational group. There was Bob Whatley from Britain, Achim Trebst from Germany, Manuel Losada from Spain, Joseph and Colette Bové from France, Waiwit Buddhari from Thailand, Shoitsu Ogata and Toshiro Yamada from Japan, Hans Müller from Switzerland, myself from Sweden, and in fact, even three native Americans: Mary Belle Allen, Harry Tsujimoto and Gail Sargent. One project concerned the necessity of chloride for oxygen evolution, another one was the tracking down of the mysterious "chloroplast extract factor", which was eventually characterized as ferredoxin (the first iron-sulfur protein found in chloroplasts); still other projects concerned the flavoprotein transferring electrons to NADP (TPN in those days) and photosynthetic nitrogen fixation.

I cannot go through all of Dan Arnon's achievements in photosynthesis research, as he has been very active for a long time after the period described here, and, in fact, is still very active. During the past few years he has been critically examining the so-called z-scheme, which most photosynthesis researchers regard as the basic paradigm. When I got in touch with him over the phone last night he was in the lab and, at the age of 78, he was obviously so busy following up some new idea that he could not make it here to Jerusalem. But I know that he appreciates the recognition, and he sends his thanks and thoughts to us here tonight.

Presentation of the Finsen Medal to I.A. Magnus

A.R. YOUNG

United Kingdom

Ladies and Gentlemen, I have the honor and the pleasure of saying a few words about Professor I.A. Magnus on his award of the Finsen Medal.

This award recognizes Prof. Magnus's pioneering and outstanding work on skin photobiology, especially on the photodermatoses.

Indeed, he may be regarded among the first, if not the first of the clinical photobiologists of modern times. However, his interests were not restricted to clinical aspects; his approach was scientific. He was among the first to recognize the importance of spectral delineation in the investigation and definition of the photodermatoses.

To this end, he built a monochromator — which I must say always reminded me of a coffin.

One of his special interests was in the porphyrias and he provided the first comprehesive description of the condition known as Erythropoietic Protoporphyria.

Professor Magnus was the founder of what he termed the 'Cockney School of Photobiology;. Cockney, because of its location in East London, an area, where people tend to drop their hs, known as 'ackney.

His work has been an inspiration to three generations of clinical and non-clinical photobiologists in the UK, Europe and I believe the rest of the world. Indeed from a very early time he was always keen to foster international collaboration, not least in Eastern Europe.

Of course his interests are not restricted to photobiology. He has considerable knowledge of music; opera in particular, and I hope he won't mind if I reveal his interests in the applied sciences of beer and wine making.

I am sure all those who know his outstanding work will agree that this award is both timely and most highly deserved.

Photobiology, Edited by E. Riklis
Plenum Press, New York, 1991

The Finsen Lecture

Monodelphis domestica: An Animal Model for Studies in Photodermatology Including the Induction of Melanoma

R. D. LEY

Photocarcinogenesis Program
Biomedical Research Division
Lovelace Medical Foundation
Albuquerque, NM, 87108 USA.

Studies in our laboratory have used the marsupial *Monodelphis domestica* as an animal model for investigating the effects of ultraviolet radiation (UVR) on mammalian skin. *Monodelphis domestica* (South American opossum) is an attractive model for photodermatological studies as it possesses DNA repair pathways for pyrimidine dimers similar to those found in human skin.

One pathway, photoreactivation (PR) repair, requires the presence of a photoreactivating enzyme (PRE) which recognizes and binds specifically to pyrimidine dimers in DNA. Illumination of the PRE-pyrimidine dimer complex with wavelengths in the range of 300-500 nm results, upon absorption of a photon, in the return of the dimerized pyrimidines to their monomeric form (Cook, 1970). Photoreactivation has been shown to occur in prokaryotes and certain eukaryotes (Setlow, 1966) including cells from placental mammals (Sutherland, 1974; Sutherland et al, 1974), fish (Shima et al., 1981; Mano et al., 1982; Shima and Setlow, 1984; Regan et al., 1982; Kator, 1984; Applegate and Ley, 1988), human skin *in vivo* (Sutherland et al, 1980b; D'Ambrosio et al., 1981; Eggset et al., 1983) and in marsupials (Cook and Regan, 1969; Ley, 1984; Wilkins, 1983). Photoreactivation of dimers is not measurable in mouse epidermis and has only been reported to occur in the dermis of newborn, but not adult, mice (Ley et al., 1978; Ananthaswamy and Fisher, 1981).

A second DNA repair pathway, excision repair, removes the pyrimidine dimer and some adjoining nucleotides and the integrity of the DNA molecule is restored through repair replication utilizing the undamaged complementary strand as a template. Human cells exhibit efficient kinetics of excision repair, whereas rodent cells have markedly

Photobiology, Edited by E. Riklis
Plenum Press, New York, 1991

lower rates of excision repair (Ley et al., 1977; Elliot and Johnson, 1983; Vijg et al., 1984). *Monodelphis domestica* possess both DNA repair pathways, photoreactivation and excision repair (Applegate and Ley, 1987).

The specificity of the PR repair pathway has been used to identify pyrimidine dimers as a major photoproduct involved in lethal and mutagenic (Harm, 1976), tumorigenic (Hart et al., 1977) and transformational (Sutherland et al., 1977) events. In addition, we have used the specificity of PR repair to show that dimers are involved in the induction of a number of pathological changes in UVR exposed skin of *M. domestica* following acute and chronic exposures.

When a single exposure to UVR was immediately followed by exposure to wavelengths that drive the PR repair pathway (320-500 nm) the capacity of UVR to induce erythema (Ley, 1985), edema (Ley and Applegate, 1987), and desquamation (Ley and Applegate, 1987) was substantially reduced. Similarly, the induction of histopathological changes in the skin, sunburn cells and hyperplasia, also was diminished by the post-UVR exposure to photoreactivating wavelengths (Ley and Applegate, 1985). These data indicate that pyrimidine dimers are the initiating lesion for pathological changes in mammalian skin following a single exposure to UVR. How does the induction of pyrimidine dimers in dermal or epidermal DNA result in the appearance of pathological changes hours to days later? One plausible explanation is that cell death is an intermediate event in these processes and pyrimidine dimers are lethal events in the absence of PR repair.

We also have observed that the induction of cutaneous tumors by multiple exposures to UVR could be suppressed if each UVR exposure was followed by exposure to PR light. Fifty per cent of the animals exposed to UVR alone (3 times/week for 70 weeks) had one or more tumors by 73 weeks following initiation of exposures. The median latency period for tumors in animals exposed to PR light immediately following UVR was >105 weeks. Only one tumor was observed in those animals exposed to PR light alone. In addition to the delayed time to appearance of tumors in animals exposed to PR light immediately after UVR, the yield of tumors (ave. number of tumors/surviving opossum) also was less in this group. At the time exposures were terminated (70 weeks), the yield of tumors in groups exposed to UVR alone or UVR followed by PR light was 2.0 and 0.2, respectively. These results indicate that pyrimidine dimers are the initiating lesion for the induction of cutaneous tumors in this animal.

During the course of this long-term photocarcinogenesis study, the first macroscopically observable change which occurred in the skin of irradiated opossums was the appearance of areas of focal hyperpigmentation approximately 1-2 mm in diameter. Lesions first appeared on animals at 12 weeks (36 exposures) and 14 weeks (42 exposures) in groups of animals exposed to UVR alone or PRL preceding UVR, respectively. Histologically, these areas appeared to be foci of dermal melanocytic hyperplasia (MH) without involvement of the epidermis. In those animals exposed to UVR followed by PRL or PRL alone, MH was not observed until 42 and 44 weeks of exposure, respectively. In addition to the delayed time of appearance of MH, the incidence of MH was less in those animals exposed to PRL after UVR or PRL alone. The incidence of MH observed in animals exposed to UVR alone was significantly (Chi-square analysis, $P=0.02$) higher than in animals exposed to UVR immediately followed by PRL. Because the PR enzyme acts only on pyrimidine dimers, the delayed appearance and lower incidence of focal melanocytic hyperplasia in the animals exposed to PRL after UVR suggest that UVR-induced pyrimidine dimers initiate the induction of MH.

Of 46 animals that survived 100 weeks after initiation of treatment, areas of melanocytic hyperplasia on 10 animals exposed to UVR alone or in combination with PRL continued to increase in size during the course of irradiation and following termination of exposures. These lesions have been classified as melanotic tumors with three of the 10 progressing to *malignant* melanomas based on metastasis to lymph nodes. Although the number of animals with melanotic tumors in each group is too small to determine with statistical significance whether exposure to PRL has suppressed the capacity of UVR to induce these tumors, if melanocytic hyperplasia is a precursor to melanotic tumors, then we might infer that the observed suppression of MH by PRL would likely result in decreased induction of melanoma. This represents, to our knowledge, the first study in which UVR alone has been used to induce malignant melanoma. Thus, our observations with *Monodelphis domestica* indicate that UVR alone can act as a complete carcinogen for the initiation and promotion of malignant melanoma and suggest that UVR-induced DNA damage, more specifically, pyrimidine dimers, are involved in melanoma induction.

In addition to the studies summarized above, additional investigations currently are underway to measure photoimmunological responses, life-shortening effects and ocular damage induced by exposure of *M. domestica* to UVR.

The prospect of stratospheric ozone depletion and the accompanying increase in human exposure to UVR makes it imperative that we understand the acute and chronic responses of mammalian skin to UVR. *Monodelphis domestica* represents a unique experimental animal model with which to study the role of a specific UVR-induced DNA lesion, the pyrimidine dimer, in the photobiology of mammalian skin. Furthermore, *M. domestica* is unique in that it is susceptible to UVR-induced malignant melanoma. This discovery extends the usefulness of the animal model to include studies on the exact role of UVR in the etiology of malignant melanoma.

Acknowledgements

These studies were supported from funding from the Lovelace Medical Foundation and by PHS grant AR35442 awarded by the National Institute of Arthritis and Musculoskeletal and Skin Diseases.

References

Ananthaswamy, H.N., Fisher, M.S. *Cancer Res.* 41:1829-1833 (1981).

Applegate, L.A., Ley, R.D. *Mutation Res.* 198:85-92 (1988).

Applegate, L.A., Ley, R.D. *Photochem Photobiol.* 45:241-245 (1987).

Cook, J.S., Regan, J.D.*Nature.* 233:1066-1067 (1969).

Cook, J.S.*Photophysiol.* 5:191-223 (1970).

D'Ambrosio, S.M., Whetstone, J.W., Slazinski, L., Lowney, E.*Photochem Photobiol.* 34:461-464 (1981).

Eggset, G., Volden, G., Krokan, H.*Carcinogen.* 4:745-750 (1983).

Elliot, G.C., Johnson, R.T.*Journ Cell Sci.* 60:267-288 (1983).

Harm, H.in Wang, Photochemistry and photobiology of nucleic acids, pp. 219-263 (Academic, New York 1976).

Hart, R.W., Setlow, R.B., Woodhead, A.D.*Proc Natl Acad Sci USA.* 74:5574-5578 (1977).

Ley, R.D., Sedita, B.A., Grube, D.D., Fry, R.J.M.*Cancer Res.* 37:3243-3248 (1977).

Ley, R.D., Sedita, B.A., Grube, D.D.*Photochem Photobiol.* 27:483-485 (1978).

Ley, R.D.*Photochem Photobiol.* 40:141-143 (1984).

Ley, R.D., Applegate, L.A.*Journ Invest Dermatol.* 85:365-367 (1985).

Ley, R.D., Applegate, L.A.*Arch Dermatol* 123:1032-1035 (1987).

Ley, R.D.*Proc Natl Acad Sci USA.* 82:2409-2411 (1985).

Mano, Y., Kator, K., Egami, N.*Mutation Res.* 90:501-508 (1982).

Setlow, J.K.*Radiat Res Suppl.* 6:141-155 (1966).

Shima, A., Setlow, R.B.*Photochem Photobiol.* 39:49-56 (1984).

Shima, A., Ikenaga, O., Nikaido, H., Takebe, H., Egami, N.*Photochem Photobiol.* 33:313-316 (1981).

Sutherland, B.M.*Nature.* 248:109-112 (1974).

Sutherland, B.M., Runge, P., Sutherland, *J.C.Biochem.* 13:4710-4715 (1974).

Sutherland, B.M., Cimino, J.S., Delihas, N., Shih A.G., Oliver, R.P.*Cancer Res.* 40:1934-1939 (1980a).

Sutherland, B.M., Kochevar, I.E., Harber, L.*Cancer Res.* 40:3181-3185 (1980b).

Vijg, J., Mullaart, E., van der Schans, G.P., Lohman P.H.M., Knook, D.L.*Mutation Res.* 132:129-138 (1984).

Wilkins, R.J.*Mutation Res.* 111:263-276 (1983).

Prolonged UVR-Induced Erythema and Melanoma Risk Factors

E. AZIZI M.D.[1], Y. WAX PH.D.[2], A. LUSKY M.SC.[2],
A. KUSHELEVSKY PH.D.[3], AND M. SCHEWACH-MILLET M.D.[1]

The Department of Dermatology[1] and Clinical Epidemiology[2],
The Chaim Sheba Medical Center and Sackler Faculty of Medicine,
Tel Aviv University;
The Department of Nuclear Engineering[3],
Ben Gurion University, Beer Sheva, Israel

Introduction

Cumulative epidemiological evidence suggests that exposure to solar radiation plays an important role in the pathogenesis of cutaneous malignant melamona (CM) (Kopf et al. 1984). This evidence is derived primarily from the fact that CM is most frequent in fair-skinned, red-haired persons who tend to freckle, burn easily, and tan poorly following exposure to sunlight. Direct experimental evidence that CM is correlated with increased susceptibility to skin injury induced by ultraviolet radiation (UVR) is rather limited.

The main purpose of the present study was to compare the recovery rate of UVR-induced erythema in a series of melanoma patients, to that of normal individuals with similar threshold skin sensitivity to UVR. Secondly, to find out whether a prolonged UV-erythema response is related in any way to phenotypical features that have been associated both with increased sensitivity to solar radiation and increased melanoma risk, or is it an independent marker of melanoma susceptible persons.

Material and Methods

Forty-seven patients with histologically proven CM who had agreed to participate in the study were selected from the Melanoma Clinic at the Sheba Medical Center. They all had surgical removal of the primary tumors, in disease stage I, and were free of metastases upon entry into the study. The majority of the tumors were located on the limbs, head and neck. At least half of them were superficial lesions, less than 1.5 mm thick. The control group consisted of 48 healthy hospital staff members and paid volunteers without a history of skin cancer or any skin lesions suspicious of malignancy.

Photobiology, Edited by E. Riklis
Plenum Press, New York, 1991

They were group matched to the melanoma patients by orignal hair and eye color, and presence or absence of freckles, as major matching criteria. Secondary matching criteria were: age, sex and origin.

The test procedure has been described in detail elsewhere (Azizi et al. 1990). Briefly, after determination of the minimal erythema dose (MED), three separate 2×2 cm skin patches of each subject, adjacent to the site on which the MED was previously determined, were irradiated simultaneously with either 2,4, or 6 times the subject's MED. The erythematous responses (ER) ensuing at these sites and their evolution were observed at 24 hr, one, two and three weeks post exposure. Responses were graded, by the same person, under identical light conditions using a scale of 1–6, where 1 implies no erythema, and 2 to 6 imply various degress of positive ER.

Results

Various degrees of ER occurred in all subjects in both groups 24 hrs after each of the three challenge doses. The evolution of the ER induced by UVR challenges of 2 and 4 MEDs were very similar in the two study groups: At the 2 MED sites all ER have almost completely faded by 3 weeks post exposure. At the 4 MED sites prolonged ER was evident at the end of 3 weeks in about 30% of the subjects in both groups. At the 6 MED sites, along the follow up period, ER persisted in the majority of subjects (>50%) from both groups but the proportion of CM patients with positive ER at each time point was higher than the controls. The differences between the two groups — at each UVR challenge and time point — were not statistically significant.

The proportion of subjects recovering from ER was then determined in each group by the percent of persons in whom ER graded ≥ 2 have faded at each time period, out of the number of persons at risk for prolonged ER, namely those showing positive ER at a preceeding time point. The differences between the two groups were significant at $p=0.045$ only when accumulated by Mantel Haenszel (1959) test, indicating that the proportion of subjects recovering from ER was lower in the melanoma group. Namely, these melanoma patients were more susceptible to prolonged UVR-induced skin damage than their controls.

Evaluation by logistic regression analysis of the net effect of age, sex, origin, hair and eye color, freckles, degree of sun-sensitivity and MED indicates that no other variables, except decreased MED and freckles independently contribute to the odds of persistent erythema in the CM patients group. MED lower than the median value of 75 sec. increased the odds of prolonged ER 11.3 times in comparison to MED ≥ 75 sec., and the presence of freckles had an additional 5.5 fold increase in odds, compared to absence of freckles. Data pertaining to the control group were excluded from this analysis, because this group was selected primarily to match the CM patients and does not by itself represent the normal population.

Discussion

The results of this study clearly demonstrate that prolonged ER, persisting over a period of 3 weeks or longer, has occurred in the majority of CM patients and healthy controls following a single UVR challenge by the highest dose of 6 MEDs. The proportion of persons recovering from this ER was lower in the CM patients than in the controls (p=0.045).

In contradistinction from previous works, in the present study an attempt was made to compare the recovery rates of UVR-induced ER between CM patients and healthy controls matched by several risk factors of increased sensitivity to UVR, namely — light hair and eye color and the presence of freckles, which are also major risk factors to develop CM (Azizi et al. 1988; Elwood et al. 1984; Elwood et al. 1986; Rhodes et al. 1987). The additional similarity of the study groups in the mean MED values, a non-matched objective parameter, further indicates that they were closely matched in regard to their threshold skin sensitivity to UVR.

The group of 47 cases in the present report is the largest series of CM patients followed so far for prolonged UVR erythema, and provides data indicating that decreased MED and the presence of freckles are independent predictors of increased risk of prolonged UVR erythema in these patients. It remains to be proven whether the significant association between low MED, freckles and increased risk of prolonged ER in the CM group occurs also in the general fair-skin population. In any case, the close similarity of the two groups in the present study in the distribution of freckles and proportions of subjects with decreased MED may account for the relatively small differences in the proportion of subjects recovering from the ER, between the CM patients and controls. Higher significant differences at the p<0.01 level in the recovery rate of UVR ER have been previously reported in CM and non-melanoma skin cancer patients, in comparison to normal controls, who were only randomly selected Caucasians with negative history of skin cancer (Tannenbaum et al. 1976), or age-matched normal individuals (Jung et al. 1981).

The overall findings of this study thus lead to the conclusion that although prolonged UVR-erythema is not a unique feature of melanoma patients, it probably reflects an excess of phenotypical risk factors to UVR-induced skin injury in these patients, (Rhodes et al. 1987).

The precise explanation for this type of association between decreased MED, or freckles and increased susceptiblity to prolonged UVR-induced ER remains unknown. A plausible hypothesis is that the apparent predominance of pheomelanin-forming melanocytes in the excessively freckled skin of CM patients and/or within melanocytic nevi (Breathnach et al. 1964), may contribute to the cellular damage following exposure to UVR, due to the photosensitizing properties of this pigment molecule (Chedekel et al. 1982). In a preceeding study we have also demonstrated in white healthy individuals with skin types I and II a significant correlation between freckles and low MED (Azizi et al. 1988).

725

The fact that heavy freckling has been reported by several investigators as an independent risk factor for CM suggests that common mechanisms in the skin of excessively freckled persons may account for their decreased MED, persistent erythema as well as increased melanoma risk. The production of a series of reactive oxidizing species by photodestruction of pheomelanin and its precursor 5-S-cysteinyldopa evidently plays a major role both in the induction of UV erythema and skin cancer, particularly malignant melanoma, due to their highly mutagenic and carcinogenic properties (Oberley et al. 1979, Harsanyi et al. 1980). Further studies are warranted to substantiate *in vivo* the potential role of pheomelanin photosensitizing properties in the pathogenesis of prolonged UV erythema and CM.

References

Azizi, E., Lusky, A., Kushelevsky, A.P., and Schewach-Millet, M., 1988, Skin type, hair color and freckles are predictors of decreased minimal erythema UV-radiation dose. Journal of the American Academy of Dermatology, 19:32–38.

Azizi, E., Wax, Y., Lusky, A., Kushelevsky, A.P., and Schewach-Millet, M., 1990, The recovery from UVR-induced erythema and melanoma risk factors — A case study. J. Am. Acad. Dermatology, in press.

Beral, V., Evans, S., Shaw, H., and Milton, G., 1983, Cutaneous factors related to the risk of malignant melanoma. British Journal of Dermatology, 109:165–172.

Breathnach, A., and Wyllie, L., 1964, Electron microscopy of melanocytes and melanosomes in freckled human epidermis. Journal of Investigative Dermatology, 42:389–394.

Chedekel, M.R., 1982, Photochemistry and photobiology of epidermal melanins. Yearly Review. Photochemistry Photobiology, 35:881–885.

Elwood, J.M., Gallagher, R.P., Hill, G.B., Spinelli, J.J., Pearson, J.C.G., and Thralfall, W., 1984, Pigmentation and skin reaction to sun as risk factors for cutaneous melanoma: Western Canada Melanoma Study. British Medical Journal, 288:99–102.

Elwood, J.M., Williamson, D., and Stapleton, P.J., 1986, Malignant melanoma in relation to moles, pigmentation and exposure to fluorescent and other lighting sources. British Journal of Cancer, 53:65–74.

Harsanyi, Z.P., Post, P.W., Brinkmann, J.P., Chedekel, M.R., and Deibel, R.M., 1980, Mutagenicity of melanin from human red hair. Experientia 36:291–292.

Jung, E.G., Gunthart, K., Metzger, R.F.G., and Bohnert, E., 1981, Risk factors of the cutaneous melanoma phenotype. Archives of Dermatological Research, 270:33–36.

Kopf, A.W., Kripke, M.L., and Stern, R.S., 1984, Sun and malignant melanoma. Journal of the American Academy of Dermatology, 11:674–684.

Mantel, N., and Haenszel, W., 1959, Statistical aspects of the analysis of data from retrospective studies of disease. Journal of the National Cancer Institute, 22:719–748.

Oberley, L.W., and Buettner, G.R., 1979, Role of superoxide dismutase in cancer: A review. Cancer Research, 39:1141–1149.

Rhodes, A.R., Weinstock, M.A., Fitzpatrick, T.B., Mihm, M.C., and Sober, A.J., 1987, Risk factors for cutaneous melanoma. Journal of the American Medical Association,258:3146–3154.

Tannenbaum, L., Parrish, J.A., Haynes, H.A., Fitzpatrick, T.B., and Pathak, M.A., 1976, Prolonged ultraviolet light induced erythema and the cutaneous carcinoma phenotype. Journal of Investigative Dermatology, 67:513–517.

Ultraviolet Radiation Induces Cytoskeletal Damage in Human Cells

GLEN B. ZAMANSKY AND IIH-NAN CHOU

Department of Microbiology
Boston University School of Medicine
Boston, Massachusetts, 02118, U.S.A.

Introduction

Solar ultraviolet (UV) radiation is the major cause of skin cancer and contributes to the deterioration of dermal connective tissue associated with aging. It is therefore important to identify the UV induced lesions which alter normal cellular functions. Although genetic damage may be an important initiating factor in the carcinogenic process, epigenetic phenomena are also likely to play a role in the development of neoplasia. Current evidence indicates that lesions other than pyrimidine dimers are involved in the carcinogenic, mutagenic and lethal effects of UV radiation which reaches the earth's surface (Zamansky, 1986 and references therein). Although investigators continue to seek additional DNA lesions which may contribute to these pathogenic phenomena, the importance of UV induced alterations of other cellular components must also be considered.

The cytoplasm of eukaryotic cells contains an intricate network of filamentous structures which are collectively referred to as the cytoskeleton. The three major components of the cytoskeleton are microtubules, microfilaments and intermediate filaments (Alberts et al, 1983). Microtubules are assembled from a cellular pool of tubulin dimers composed of partially homologous α and β polypeptides. Polymerization appears to originate at microtubule organizing centers close to the nuclear membrane and results in filaments extending throughout the cytoplasm. Microfilaments, composed primarily of F-actin, are often found in bundles, referred to as stress fibers. Intermediate filaments are composed of more heterogeneous proteins and are grouped into 5 sub-classes:keratin (epithelium), vimentin (mesenchyme), desmin (muscle), glial fibrillary acidic protein (glial cells) and neurofilament proteins (neurons). Since the cytoskeleton is an important participant in the control of normal cell growth and other essential cellular functions, we have begun to explore UV radiation induced cytoskeletal damage in human skin fibroblasts and keratinocytes.

Photobiology, Edited by E. Riklis
Plenum Press, New York, 1991

727

Materials and Methods

Cells AG1522, obtained from the Coriel Institute for Medical Research (Camden, New Jersey), is a normal diploid human skin fibroblast cell strain. NHEK-14, obtained from Clonetics (San Diego, California) is a normal human epidermal keratinocyte (NHEK) cell strain. Fibroblasts were grown as previously described (Zamansky, 1986). NHEK cells were grown in a modified MCDB 153 medium supplemented with 0.15 mM calcium, 10 ng/ml epidermal growth factor, 5.0 µg/ml insulin, 0.5 µg/ml hydrocortisone, 50 µg/ml bovine pituitary extract, 0.1 mg/ml streptomycin, 100 u/ml penicillin, and 0.25 µg/ml amphotericin B.

UV Irradiation General Electric G8T5 germicidal lamps, Westinghouse FS40 lamps, and Sylvania FR40T12 lamps were our sources of UVC, sun lamp, and UVA radiation respectively. The spectra of light transmitted by the Westinghouse FS40 and Sylvania FR40T12 lamps through polystyrene culture dish covers have been published (Zamansky et al, 1985). Cultures were irradiated under conditions in which cytoskeletal structures are stable in mock-irradiated controls.

Cytoskeletons Eighteen hours after plating AG1522 cells onto glass coverslips in 35 mm culture dishes, cultures were rinsed with Hank's balanced salt solution (HBSS) containing 15mM HEPES and UV irradiated in the presence of HBSS. Fixation, extraction and fluorescent staining of microtubules in AG1522 cells were performed as described (Zamansky and Chou, 1987).

Approximately five days after plating NHEK-14 cells onto glass coverslips, cultures were rinsed with Solution A (10 mM glucose, 3 mM KCl, 130 mM NaCl, 1 mM Na_2HPO_4, 30mM HEPES, pH 7.4) containing 0.15 mM $CaCl_2$ and UV irradiated in the presence of supplemented Solution A. Following irradiation cultures were either fixed immediately or reincubated in MCDB 153 containing 0.15 mM or 1.05 mM $CaCl_2$. Two hours later NHEK-14 cultures were fixed in methanol at –20°C for five minutes. Following fixation coverslips were incubated in acetone for 3 minutes at –20°C, and rinsed with cold water and cold phosphate buffered saline (PBS). In order to fluorescently label NHEK intermediate filaments, the fixed cytoskeletons were incubated with rabbit anti-keratin antibodies, washed with PBS and treated with goat antirabbit antibodies conjugated to rhodamine.

Results and Discussion

The microscopic appearance of microtubules in control and sun lamp irradiated AG1522 cells is shown in Figure 1. As expected the microtubules of unirradiated cells appear to emanate from microtubule organizing centers and extend throughout the cytoplasm. Such a distinct network of microtubules, however, is not consistently observed in sun lamp irradiated cells. Although the microtubule organizing center usually

728

Table 1. UV Induced Disruption of Microtubules[+]

Sun Lamp		UVA	
Dose (J/m^2)	Intact MT (%)	Dose (kJ/m^2)	Intact MT (%)
0	95.6 ±1.2	0	95.7 ± 0.7
750	94.3 ± 0.8	25	92.2 ± 3.1
1500	90.8 ± 1.4	50	80.4 ± 2.9
3000	61.6 ± 12.8	75	68.0 ± 5.1
5000	45.6 ± 8.5	100	44.2 ± 3.5
7500	27.5 ± 10.2	150	33.0 ± 6.7

[+]The percent of cells with intact microtubules (mean ± 1 standard error of 3–5 independent experiments at each UV dose). 200 cells were examined in randomly selected microscopic fields of coded coverslips in each experiment.

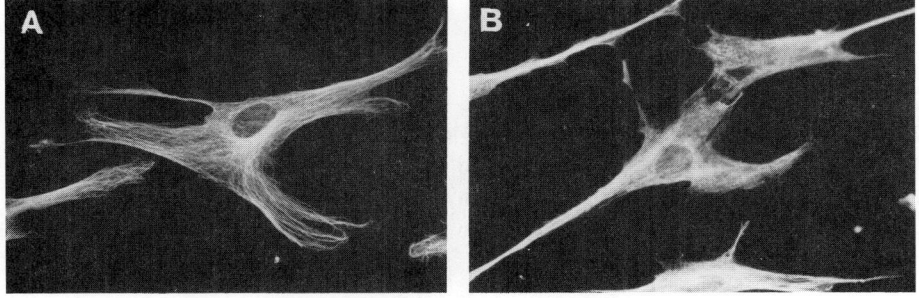

Figure 1. Microtubules in (A) control and (B) sun lamp (5000 J/m^2) irradiated AG1522 cells.

remains detectable, much of the antitubulin stained cytoplasm has a fine powdery appearance in these cells. The extent to which this apparent disassembly of microtubules occurs depends on the UV dose (Table 1) and varies from cell to cell. As can be seen in Table 1, exposure to a sun lamp dose of 3000 J/m^2 is sufficient to reduce appreciably the number of cells with intact microtubules. Experiments with UVA radiation indicated that it too causes a dose dependent loss of organized microtubules. No damage to microtubules was found in cells irradiated with UVC doses as high as 100 J/m^2, a dose at which only one out of 10^{16} AG1522 cells is expected to survive. When cells were double labelled to visualize microtubules and microfilaments, no perceptible alterations of microfilaments occurred in UVC, sun lamp or UVA irradiated cells.

We have investigated the survival of AG1522 and several other human skin fibroblast cell strains exposed to UV radiation (Zamansky et al, 1985; Zamansky, 1986). Our studies indicate that the induction of microtubule disassembly does not correlate with the cytotoxicity of UV radiation of varying composition (Zamansky and Chou, 1987). This lack of correlation between toxicity and microtubule damage does not preclude the possibility that cytoskeletal damage induced by long wavelength UV radiation contributes to its lethal effects. UVC and UVA radiation appear to kill cells by different mechanisms. Furthermore, exposure to increasing doses of UVA results in increased levels of microtubule disruption and increased cell death. Thus, microtubule disruption by UVA may be an important early event preceding cell death. In view of these findings, it will be of interest to investigate whether the disruption of cytoskeletal elements may contribute to the inactivation of cells which are hypersensitive to UV radiation.

Our cytoskeletal studies were initiated using fibroblasts because of the convenience of growing them, their immediate availability from patients with a broad range of photosensitive diseases and the depth of earlier studies into UV induced damage in this cell type. Chronic exposure to sunlight damages dermal connective tissue and hastens the aging of human skin (Gilchrest, 1984; Oikarinen, 1985). Actinically damaged connective tissue characteristically exhibits qualitative and quantitative changes in its two most prevalent components, collagen and elastin, which are synthesized by dermal fibroblasts. Since the cytoskeleton appears to participate in the control of protein synthesis (Ornelles et al, 1986), and the secretion of collagen and elastin (Uitto et al, 1976), our investigations with fibroblasts may result in a better understanding of cutaneous photoaging. However, in order to more closely examine the potential relationship between cytoskeletal damage and UV induced skin cancer, we have initiated studies utilizing normal human epidermal keratinocytes, the *in vivo* target for solar carcinogenesis.

We have chosen to grow NHEK cells in a low calcium (0.15 mM) medium which has been optimized for the proliferation of keratinocytes in the presence of very low levels of differentiation. Originally described by Ham and his colleagues, an important feature of this culture system is that terminal differentiation can be induced at higher calcium concentrations (Boyce and Ham, 1983). Recent studies have indicated that an early change induced by calcium in cultured keratinocytes is the reorganization of the keratin intermediate filaments (Watt et al, 1984; Jones and Goldman, 1985). We too have observed this calcium induced redistribution of keratin filaments. As seen in Figure 2, the keratin filaments of NHEK-14 cells grown in medium containing 0.15 mM calcium are most concentrated near the nucleus. Filaments extend towards the periphery of the cytoplasm to varying degrees. Within 2 hours after incubation in 1.05 mM calcium, a reorganization of intermediate filaments has occurred at the cell surface and filaments in adjacent cells have become aligned at cell to cell borders (Figure 2B, large arrows). No such peripheral filaments are observed in isolated cells or on the sides of cells which lack neighbors (Figure 2B, small arrows). When NHEK-14 cells grown in 0.15 mM calcium are exposed to sun lamps and then incubated in 1.05 mM calcium for 2 hours, the

730

reorganization of keratin filaments appears to be inhibited in a dose dependent manner. As seen in Figure 2C, exposure to 6000 J/m^2 results in an almost complete inhibition of the formation of aligned filaments and an apparent condensation of keratin filaments into the perinuclear region. These "condensed" filaments form a whirl-like pattern in some cells. Normal redistribution of keratin filaments occurs in mock-irradiated cultures. Thus, it appears that exposure to polychromatic UV radiation composed of environmentally relevant wavelengths has a profound effect on an early step in the calcium induced differentiation of NHEK cells.

The individual components of the cytoskeleton are structurally associated with each other as well as with the cellular membrane and nuclear matrix. It has therefore been suggested that the cytoskeleton may serve as a critical means of transmitting external signals to the nucleus. The complex network of cytoskeletal structures also participates in the regulation of cell growth, shape, and motility, the spatial arrangement of

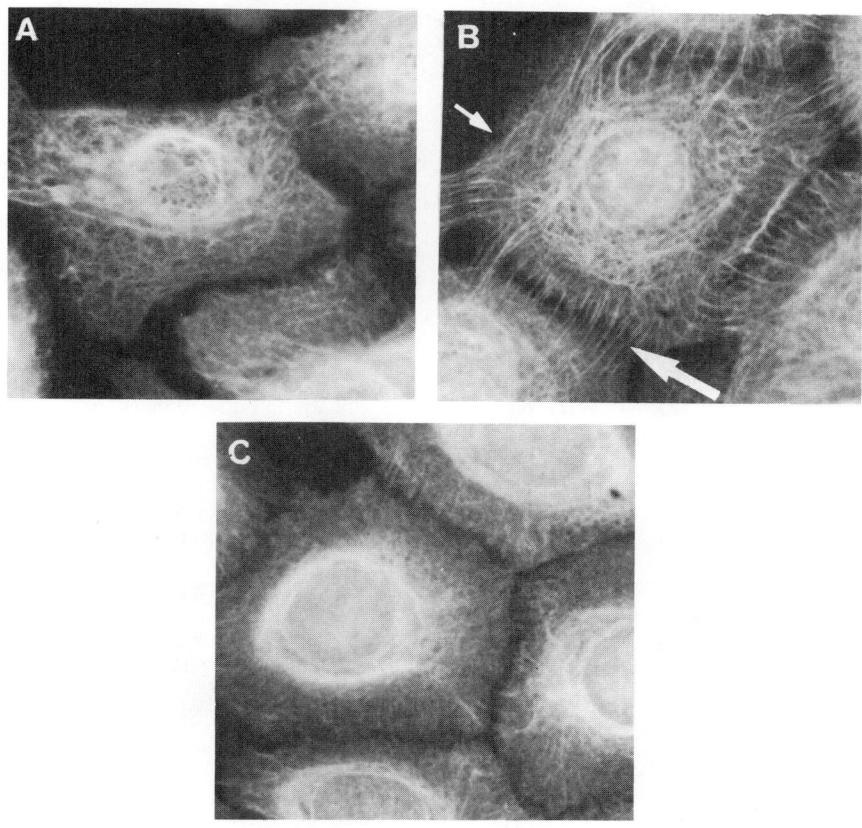

Figure 2. Keratin intermediate filaments of NHEK-14 cells (A) grown in the presence of 0.15 mM CaCl$_2$, (B) exposed to 1.05 mM CaCl$_2$ for 2 hours, or (C) sun lamp (6000 J/m^2) irradiated and then exposed to 1.05 mM CaCl$_2$ for 2 hours.

organelles, and secretory processes (Alberts et al, 1983). It is also intriguing to note that tumor promoters cause structural changes in cytoskeletal components (Rifkin et al, 1979; Schliwa, 1984). It is thus reasonable to expect that UV radiation-induced perturbations of the normal assemblage of the cytoskeleton could result in a variety of functional consequences. Our cytoskeletal studies therefore provide an important, new approach for studying cellular damage induced by UV radiation.

This work was supported in part by a grant from the National Institute of Arthritis and Musculoskeletal and Skin Diseases (U.S.A.) and by an award from the Boston University School of Medicine Biomedical Research Support Grant Award RR-05380.

References

Alberts, B., Bray, B., Lewis, J., Raff, M., Roberts, K., and Watson, J.D., 1983, Molecular Biology of the Cell, New York, Garland Press, pp. 549–609.

Boyce, S.T. and Ham, R.G., 1983, Calcium regulated differentiation of normal human epidermal keratinocytes in chemically defined clonal culture and serum-free serial culture.Journal of Investigative Dermatology, 81, 33S–40S.

Gilchrest, B.A., 1984, Skin and Aging Processes. Boca Raton, Florida, CRC Press.

Jones, J.C.R. and Goldman, R.D., 1985, Intermediate filaments and the initiation of desmosome assembly. Journal of Cell Biology, 101, 506–517.

Oikarinen, A., Karvonen, J., Uitto, J., and Hannuksela, M., 1985, Connective tissue alterations in skin exposed to natural and therapeutic UV radiation. Photodermatology, 2, 15–26.

Ornelles, D.A., Fey, E.G. and Penman, S., 1986, Cytochalasin releases mRNA from the cytoskeletal framework and inhibits protein synthesis. Molecular and Cellular Biology 6, 1650–1662.

Rifkin, D.B., Crowe, R.M. and Pollack, R., 1979, Tumor promoters induce changes in the chick embryo fibroblast cytoskeleton. Cell, 18, 361–368.

Schliwa, M., Nakamura, T., Porter, K.A., and Eutener, U., 1984, A tumor promoter induces rapid and coordinated reorganization of actin and vinculin in cultured cells. Journal of Cell Biology, 99, 1045–1059.

Uitto, J., Hoffman, H.P., and Prockop, D.J., 1976, Synthesis of elastin and procollagen by cells from embryonic aorta. Archives of Biochemistry and Biophysics, 173, 187–200

Watt, F.M., Mattey, D.L. and Garrod, D.R., 1984, Calcium-induced reorganization of desmosomal components in cultured human keratinocytes. Journal of Cell Biology, 99, 2211–2215.

Zamansky, G.B., 1986, Varying sensitivity of human skin fibroblasts to polychromatic ultraviolet light. Mutation Research, 160, 55–60.

Zamansky, G.B., and Chou, I.N., 1987, Environmental wavelengths of ultraviolet light induce cytoskeletal damage. Journal of Investigative Dermatology, 89, 603–606.

Zamansky, G.B., Minka, D.F., Deal, C.D., and Hendricks, K., 1985, The in vitro photosensitivity of systemic lupus erythematosus skin fibroblasts. Journal of Immunology, 134, 1571–1576.

Photoimmunolgy

Overview: Photoimmunology

MARGARET L. KRIPKE

Department of Immunology
University of Texas, M.D. Anderson Cancer Center
Houston, Texas, U.S.A.

Introduction

Photoimmunology is the study of the effects of photons on immunologic reactions. As a real line of investigation, it is only a little more than ten years old, although its origins can be traced back much further in the fields of dermatology and photocarcinogenesis. As a separate scientific discipline, photoimmunology arose from several independent, seemingly unrelated lines of investigation. Much of the current activity in this field was triggered by the establishment of a few key links between skin and the immune system and between the immune system and UV radiation. As a result of these inter-connections, we are developing a new understanding of the relationships among UV radiation, skin, and host defense mechanisms, and we are beginning to ask how these interrelationships might affect normal and pathological processes in the skin.

This question has turned out to be of much more significance than one might have predicted ten years ago, because of the current threat of ozone depletion. Stratospheric ozone absorbs a significant amount of UV radiation from sunlight, particularly in the UV-B (280-320 nm) and UV-C (200-280 nm) wavelength ranges. A 1% decrease in the concentration of ozone is expected to result in a 2% increase in the amount of UV-B radiation reaching the earth's surface. The ozone layer is at risk of diminishing, and may have done so already, because of the release into the atmosphere of increasing quantities of man-made chlorofluorocarbon compounds (Hoffman, ed, 1987). Thus, there has been increasing interest in determining the consequences of increased UV-B radiation on human health, including its effects on the immune system.

Immunology of Photocarcinogenesis

Historically, one aspect of photoimmunology came from studies of the antigenic properties of skin cancers induced in mice by UV radiation. Many years ago, we found that these cancers are highly antigenic. In fact, most of them are immunologically rejected upon transplantation to normal, syngeneic recipients (Kripke, 1974). This finding raised the question of how the tumors were able to survive in the original host without being destroyed by the immune system. The answer to this question turned out

Photobiology, Edited by E. Riklis
Plenum Press, New York, 1991

to be that UV radiation alters host immunity against UV-induced skin cancers. This was illustrated by transplanting these skin cancers into mice exposed for a short period of time to UV radiation. Even though these mice had not yet developed primary skin cancers, they lost their ability to reject transplants of the UV-induced tumors (Kripke and Fisher, 1976). Thus, these experiments demonstrated that UV-irradiation produced a systemic alteration that interfered with the immunologic rejection of highly antigenic skin cancers. In addition, they stimulated additional work on the nature and mechanisms of this UV-induced systemic effect.

We now know that the inability of UV-irradiated mice to reject UV-induced skin cancers results from the presence within host lymphoid tissues of suppressor lymphocytes that prevent immunologic destruction of the developing primary skin cancers (Fisher and Kripke, 1977; Daynes and Spellman, 1977; Fisher and Kripke, 1982). These suppressor lymphocytes are induced both by exposure to artificial sources of UV-B radiation, such as FS40 sunlamps (DeFabo and Kripke, 1982), and by natural sunlight (Morison and Kelley, 1985). The suppressor lymphocytes are specific for a common antigen present on UV-induced tumors (Roberts, 1986), but generally are not found on tumors induced by other agents (Kripke et al, 1979). Thus, UV radiation has at least three effects relevant to photocarcinogenesis: It induces new antigens on cells in the skin; it causes the transformation of normal cells in the skin into cancer cells; and it blocks the immune response against these cancers by inducing tumor-specific suppressor cells.

Induction of Suppressor Cells by UV-B Radiation

One of the most interesting questions raised by these studies is, how are the suppressor cells induced by exposure to UV radiation? Clues for answering this question have come from the finding that UV radiation can induce suppressor cells against other antigens, in addition to those expressed on UV-induced tumor cells. Three ways have now been described by which suppressor cells can be induced in UV-irradiated mice. First, following relatively high doses of UV radiation, antigen-specific suppressor cells are induced by subcutaneous injection of antigens that normally induce delayed type hypersensitivity (DTH; Ullrich, 1986; Jessup et al, 1978). The suppression of DTH reactions most likely results from a redistribution of antigen-presenting cells within the lymphoid organs. Evidence suggests that antigen-presenting cells leave the spleen following exposure to high doses of UV radiation and accumulate in lymph nodes draining the site of UV-induced injury, perhaps in response to a soluble mediator of inflammation (Ullrich, 1986). During this period, antigens injected subcutaneously at unirradiated sites activate the suppressor cell pathway, presumably because of an alteration in antigen presentation (Jessup et al, 1978; Greene et al, 1979; Gurish et al, 1982; Kripke and Morison, 1986).

Second, applying contact sensitizing chemicals to the unexposed skin of UV-irradiated mice also induces suppressor cells, instead of contact hypersensitivity (CHS). This suppression seems to involve a soluble mediator, but not a redistribution of antigen-presenting cells. Evidence for soluble mediators has been provided by Schwarz and colleagues (Schwarz et al, 1986), who found a suppression-inducing substance in the supernatant fluid of UV-irradiated keratinocyte cultures; by Swartz (Swartz, 1984), who demonstrated a suppressor-inducing substance in the plasma of UV-irradiated mice; by DeFabo and Noonan (DeFabo and Noonan, 1983) based on an action spectrum for inducing suppressor cells; and by Robertson et al, who mimicked the effect of *in vivo* UV-irradiation by intravenous injection of Interleukin-1, an immunologic mediator produced by keratinocytes and macrophages (Robertson et al, 1987). The identity of the other mediators has not been established, but it seems likely that several different molecules may be involved. How such mediators cause the induction of suppressor cells is completely unknown at present.

The third method for inducing suppressor cells came from an independent line of research concerned with the immunology of the skin. Bergstresser, Streilein and colleagues were interested in determining the function of a type of cell in the skin called the Langerhans cell (Toews et al, 1980). These cells are now known to be bone marrow derived, antigen-presenting cells in the epidermis, which are closely related to macrophages (Stingl et al, 1978). These investigators found that after exposure of the skin to very low doses of UV-B radiation, suppressor cells are induced to contact sensitizers applied directly onto the irradiated skin (Toews et al, 1980; Stingl et al, 1978; Elmets et al, 1983). In addition, they noted that Langerhans cells were damaged following UV-B irradiation (Toews et al, 1980). Their results suggested that an alteration in the antigen-presenting function of the Langerhans cells leads to induction of the suppressor cell pathway.

Several lines of evidence suggest that in addition to the absence of antigen presentation by Langerhans cells, a second, UV-resistant cell participates in the induction of suppressor cells. Granstein and colleagues exposed antigen-coupled epidermal cells *in vitro* to UV radiation and used these cells to induce suppressor cells. The ability of the UV-irradiated cell population to induce suppressor cells was eliminated by treatment with anti-I-J antibody and complement, suggesting the presence in the epidermis of a UV-resistant, I-J[+] cell that participates in the suppressor cell pathway (Granstein et al; Granstein et al, 1974). Sullivan et al (Sullivan et al, 1986) sorted the cells in murine epidermis using the fluorescence activated cell sorter and showed that antigen-coupled Langerhans cells induced CHS when injected into mice, whereas Thy-1[+] dendritic epidermal cells, a second type of immune cell in mouse skin, induced suppression. This study also indicated that mouse skin contains a lymphoid cell capable of activating the suppressor cell pathway.

Recent studies from our laboratory provided additional evidence of a suppressor inducing cell *in vivo* (Okamoto and Kripke, 1987). We collected cells from the lymph nodes draining the site of sensitization of mice with a contact sensitizer. These cells induced CHS when injected into the footpads of normal mice. However, lymph node cells collected from mice that were UV-irradiated before application of the antigen failed to induce CHS, but instead, induced suppressor cells. The lymph node cells that induced suppressor cells were also Thy-1[+]. This study showed that *in vivo*, as well as *in vitro*, a second lymphoid cell is involved in activation of the suppressor cell pathway.

Thus, UV-B radiation can induce suppressor cells by three apparently different mechanisms, depending on the conditions of the irradiation and the route of administration of the antigen. It is not known whether one or more of these mechanisms is involved in the induction of the suppressor cells that prevent the rejection of UV-induced skin cancers, but it is possible that these suppressor cells arise as a consequence of altered presentation of UV-associated antigens in UV-irradiated skin.

Effects of UV Radiation on Immunity to Infectious Disease

One of the most pressing questions raised by the finding that UV irradiation affects the function of the immune system is whether these effects can impair host resistanc to infectious diseases. Many disease-causing organisms enter the body through exposed skin; thus, theoretically, UVR-induced damage to Langerhans cells at the site of infection might impair the induction of host immunity and adversely alter the course of the disease. Also, immunity to some pathogenic microorganisms, notably intracellular bacteria and parasites, is thought to depend on the DTH arm of the immune response. Consequently, there is reason to believe that the ability of UVR to impair the DTH response might impair immunity to such organisms, leading to prolonged or more severe disease and a higher rate of re-infection. Because of the possibility that solar UV-B radiation is increasing, due to decreases in the stratospheric ozone layer, it is of great importance to determine whether exposure to UV-B radiation increases the incidence or severity of infectious diseases, in addition to causing cancers of the skin.

In spite of global importance of this question, there is little information available from either human or animal studies on the effects of UVR on host resistance to infectious diseases. Recent studies by Giannini using a mouse model for leishmaniasis demonstrated that the cutaneous lesions fail to develop in mice infected with the leishmania parasite through chronically UVB-irradiated skin. However, there was no decrease in the number of parasites in the UV-treated animals, their DTH response to leishmania antigens was impaired, and they failed to develop resistance to re-infection (Giannini; Giannini and DeFabo, 1987). Thus, the improvement in the cutaneous lesions failed to correlate with an improvement in health or immunity and instead, was associated with decreased resistance to systemic disease.

The infectious agent that has received the most attention in terms of its interactions with UVR is herpes simplex virus (HSV). It has long been recognized that exposure to UVR can trigger outbreaks of HSV in humans and experimental animals harboring latent virus. Studies in a mouse model of latent HSV type-1 infection have suggested that recurrence of disease may depend on suppression of local immunological mechanisms at the site of UV irradiation that permit virus replication and re-expression of disease (Harbour et al, 1983).

In support of this model, several investigators have provided evidence for an essential role of Langerhans cells in resistance to HSV infection. The severity of HSV-1 infection in mice appears to be related to the number and function of Langerhans cells at the site of infection (Sprecher and Becker, 1987). Irradiation of mouse skin with suberythemal doses of UV-B before inoculation of HSV resulted in an increase in the severity of cutaneous lesions and systemic disease (Yasumoto et al, 1987; Otani and Mori, 1987). In addition, the animals exhibited a decreased DTH response to HSV-1 and developed HSV-specific suppressor T cells (Yasumoto et al, 1987; Howie et al, 1986a; Howie et al, 1986b).

Recent studies from our laboratory have demonstrated that DTH to an opportunistic fungus, *Candida albicans*, is also abrogated by exposure of mice to UV radiation from FS40 sunlamps (Denkins et al, submitted). In these studies, mice were exposed to a single, high dose of UV radiation ($\sim$50kJ/m^2) either before or after the subcutaneous inoculation of candida at an unirradiated site. In both instances, complete suppression of the DTH response was observed in the UV-irradiated mice. Presently, we are investigating the consequences of the impaired DTH response for immunity against systemic infection.

At the present time, no information is available on the effects of UV irradiation on the development of immunity to infectious agents in humans. Clearly, this is an area where more studies are needed, both in humans and in animal models of other infectious diseases.

Conclusions

Over the past decade it has become clear that exposure to environmental and artificial sources of UV radiation has the potential to modify a variety of immunological reactions. Whether it does so in humans under the normal circumstances of everyday life is not yet known. This seems doubtful since human cells have a variety of repair mechanisms to remove UVR-induced damage in DNA, and the immune system is one of the body's homeostatic mechanisms. On the other hand, the emphasis on tanning and the advent of tanning salons, the threat of increasing UV-B radiation in sunlight, and the increasing uses of UVR for therapeutic purposes raise the possibility that UVR effects on the immune system may occur with increasing frequency.

References

Daynes, R.A. and Spellman, C.W. (1977), Evidence for the generation of suppressor cells by ultraviolet radiation, *Cell. Immunol.* 31:182-187.

DeFabo, E.C. and Kripke, M.L. (1980), Wavelength dependence and dose-rate independence of UV-radiation-induced immunologic unresponsiveness of mice to a UV-induced fibrosarcoma, *Photochem. Photobiol.* 32:183-188.

DeFabo, E.C. and Noonan, F.P. (1983), Mechanism of immune suppression by ultraviolet irradiation *in vivo*. I. Evidence for the existence of a unique photoreceptor in skin and its role in photoimmunology, *J. Exp. Med.* 157:84-98.

Denkins, Y.D., Fidler, I.J., and Kripke, M.L. (1989), Exposure of mice to UV-B radiation suppresses delayed hypersensitivity to *Candida albicans*. *Photochem. Photobiol.* 49:615–619.

Elmets, C.A., Bergstresser, P.R., Tigellar, R.E., Wood, P.J. and Streilein, J.W. (1983), Analysis of the mechanism of unresponsiveness produced by haptens painted on skin exposed to low dose ultraviolet radiation, *J. Exp. Med.* 158:781-794.

Fisher, M.S. and Kripke, M.L. (1977), Systemic alteration induced in mice by ultraviolet light irradiation and its relationship to ultraviolet carcinogenesis, *Proc. Natl. Acad. Sc. USA* 74:1688-1692.

Fisher, M.S. and Kripke, M.L. (1982), Suppressor T-lymphocytes control the development of primary skin cancers in ultraviolet-irradiated mice, *Science*, 216:1133-1134.

Giannini, S.H. and DeFabo, E.C. (1987), Abrogation of skin lesions in cutaenous leishmaniasis by ultraviolet B irradiation. In, D.T. Hart, ed. *Leishmaniasis: The First Centenary (1885-1985) New Strategies for Control.* NATA ASI Series A: Life Sciences, Plenum Publishing Co., Ltd. London.

Granstein, R.D., Askari, M., Whitaker, D. and Murphy, G.F. (1987), Epidermal cells in activation of suppressor lymphocytes: Further characterization, *J. Immunol.* 158:4055-4062.

Granstein, R.D., Lowy, A. and Greene, M.I., Epidermal antigen presenting cells in activation of suppression: Identification of a new functional type of ultraviolet radiation-resistant epidermal cell. *J. Immunol.* 132:563-565.

Greene, M.I. Sy, M.S., Kripke, M.L. and Benacerraf, B. (1979), Impairment of antigen-presenting cell function by ultraviolet radiation, *Proc. Natl. Acad. Sci. USA* 76:6592-6595.

Gurish, M.F., Lynch, D.H. and Daynes, R.A. (1982), Changes in antigen-presenting cell function in the spleen and lymph nodes of ultraviolet-irradiated mice. *Transplantation* 33:280-284.

Harbour, D.A., Hill, T.J. and Blyth, W.A. (1983), Recurrent herpes simplex in the mouse: Inflammation in the skin and activation of virus in the ganglia following peripheral stimulation. *J. Gen. Virol.* 64:1491-1498.

Hoffman John S., ed. (1987), An Assessment of the Risks of Stratospheric Modification, U.S. Environmental Protection Agency.

Howie, S.E.M., Norval, M., Maingay, J. and Ross, J.A. (1986b), Two phenotypically distinct T cells (Lyl+2- and Lyl-2+) are involved in ultraviolet-B light-induced suppression of the efferent DTH response to HSV-1 *in vivo. Immunol.* 58:653-658.

Howie, S., Norval, M. and Maingay, J. (1986a), Exposure of low-dose ultraviolet radiation suppresses delayed-type hypersensitivity to herpes simplex virus in mice. *J. Invest. Dermatol.* 86:125-128.

Jessup, J.M., Hanna, N., Palaszynski, E. and Kripke, M.L. (1978), Mechanisms of depressed reactivity to dinitrochlorobenzene and ultraviolet-induced tumors during ultraviolet carcinogenesis in BALB/c mice. *Cell. Immunol.* 38:105-115.

Kripke, M.L. and Fisher, M.S. (1976), Immunologic parameters of ultraviolet carcinogenesis, *J. Natl. Cancer Inst.* 57:211-215.

Kripke, M.L. and Morison, W.L. (1986), Studies on the mechanism of systemic suppression of contact hypersensitivity by UVB radiation II. Differences in the suppression of delayed and contact hypersensitivity in mice, *J. Invest. Dermatol.* 86:543-549.

Kripke, M.L. (1974), Antigenicity of murine skin tumors induced by ultraviolet light, *J. Natl. Cancer Inst.* 53:1333-1336.

Kripke, M.L., Thorn, R.M., Lill, P.H., Civin, C.I., Fisher, M.S. and Pazmino, N.H. (1979), Further characterization of immunologic unresponsiveness induced in mice by UV radiation: Growth and induction of non-UV-induced tumors in UV-irradiated mice. *Transplantation* 28:212-217.

Morison, W.L. and Kelley, S.P. (1985), Sunlight suppressing rejection of 280- to 320-nm UV-radiation induced skin tumors in mice, *J. Natl. Cancer Inst.* 74:525-527.

Okamoto, H. and Kripke, M.L. (1987), Effector and suppressor circuits of the immune response are activated *in vivo* by different mechanisms. *Proc. Natl. Acad. Sci. USA* 84:3841-3845.

Otani, T. and Mori, R. (1987), The effects of ultraviolet irradiation of the skin on herpes simplex virus infection: Alteration in immune function mediated by epidermal cells and in the course of infection. *Arch. Virol.* 96: 1-15.

Robertson, B., Gahring, L., Newton, R. and Daynes, R.A. (1987), In vivo administration of IL-1 to normal mice decreases their ability to elicit contact hypersensitivity responses: Prostaglandins are involved in this modification of the immune response. *J. Invest. Dermatol.* 88:380-387.

Roberts, L.K. (1986), Characterization of a cloned ultraviolet radiation (UV)-induced suppressor T cell line that is capable of inhibiting anti-UV tumor-immune responses. *J. Immunol.* 136:1908-1916.

Schwarz, T., Urbanska, A., Fritz, M.S. and Luger, T.A. (1986), Inhibition of the induction of contact hypersensitivity by a UV-mediated epidermal cytokine. *J. Invest. Dermatol.* 87:289-291.

Sprecher, E. and Becker, Y., (1987), Herpex simplex virus type 1 pathogenicity in footpad and ear skin depends on Langerhans cell density, mouse genetics, and virus strain. *J. Virol.* 61:2515-2522.

Stingl, G., Katz, S.I., Shevach, E.M., Rosenthal, A.S. and Green, I. (1978), Analogous functions of macrophages and Langerhans cells in the initiation of the immune response. *J. Invest. Dermatol.* 71:59-64.

Sullivan, S., Bergstresser, P.R., Tigelaar, R.E. and Streilein, J.W. (1986), Induction and regulation of contact hypersensitivity by resident, bone marrow-derived, dendritic epidermal cells: Langerhans cells and Thy-1+ epidermal cells, *J. Immunol.* 137:2460-2467.

Swartz, R.P. (1984), Role of UVB-induced serum factor(s) in suppression of contact hypersensitivity in mice, *J. Invest. Dermatol.* 83:305-307.

Toews, G.B., Bergstresser, P.R. and Streilein, J.W. (1980), Epidermal Langerhans cell density determines whether contact hypersensitivity or unresponsiveness follows skin painting with DNFB, *J. Immunol.* 124:445-453.

Ullrich, S.E. (1986), Suppression of the immune response to allogeneic histocompatibility antigens by a single exposure to UV radiation. *Transplantation* 42:287–291.

Yasumoto, S., Hayashi, Y. and Aurelian, L. (1987), Immunity to herpes simplex virus type 2. Suppression of virus-induced immune responses in ultraviolet B-irradiated mice. *J. Immunol.* 1139:2788-2793.

Immunology of UV-Induced Human Skin Cancer

GERDA FRENTZ

Dept of Reconstructive Surgery & Dept of Dermatology
The Finsen Institute
Copenhagen, Denmark

This paper focuses on photocarcinogenesis and non-melanoma skin cancer in man and discusses facts and theories on the role of the immune system for this process. Since photocarcinogenesis in man is closely connected with sun-exposure no details on the role of the specific wavelengths of UV radiation will be discussed.

Man has always worshipped the sun for its impact on good and evil, on fertility and wealth. That skin cancer in man is often sited on sun-exposed skin was noted about 100 years ago by the German dermatologist Unna and also by Dubreuilh in France (Unna, 1894; Dubreuilh, 1986). Since that time knowledge has expanded with increasing speed and the mechanisms involved in photocarcinogenesis have been studied thoroughly. A great deal of insight is obtained by fruitful interplay between animal experiments and clinical observations in man.

From the 1940's the concept arose that formation of cancers was a multistep process involving at least three distinct stages: initiation is an irreversible change in the genetic outfit of the cells, which is usually grossly irrecognizable; tumor promotion is a reversible, epigenetic stage often resulting in the formation of a pre-malignant visible change in the tissue; conversion designates the mutagenic conditioned change from premalignancy to overt invasive growth (Kripke, 1986). Knowledge of the nature of these processes comes mainly from experiments conducted on mouse skin. During such experiments the cancer-inducing effect of ultraviolet radiation was further studied and UV radiation was shown both to have properties as an initiating and a promoting agent. Furthermore, UV radiation has the abilities of a complete carcinogen (Epstein and Roth, 1968).

Basic parameters of interest for human skin cancer were obtained form the extended epidemiologic studies of Urbach and his associates emphasizing the importance of the cumulative sun-exposure during life, of age, ability to tan and hereditary factors (Urbach, 1969).

Before 1970 knowledge of human skin photocarcinogenesis took an important step forward on the molecular level thanks to James Cleaver. UV radiation causes DNA damages primarily by inducing thymine dimers. Normal cells have well-organized and well-functioning enzyme systems for repair of such damages to the genome. Cleaver discovered that cells from patients with Xeroderma pigmentosum were severely deficient in their ability to repair UV induced DNA damages. Hereditarily, these patients develop multiple cancers on their sun-exposed skin from childhood (Cleaver, 1968). This finding created a new wave of intense research from the idea that defective DNA repair systems might be the main cause of UV induced skin cancer. In fact, some shortcoming in the capacity for repair of UV induced DNA damages appeared in cells from patients with the pre-malignant skin condition of multiple actinic keratoses, from patients with multiple UV induced skin cancer, and less pronounced in cells from patients with multiple basal cell cancers as an expression of the dominantly inherited condition of the nevoid basal cell carcinoma syndrome (Munch-Petersen, Frentz, 1985; Frentz, et al, 1987). Later studies have shown that defective repair of certain gene loci are more important for the cell function than the DNA repair capacity as a whole (Bohr et al, 1986). Furthermore, a reduced ability of the cells to survive UV radiation seems to be more closely connected with UV induced cancers than the total capacity for repair of UV induced DNA damages measured by the size of the extra DNA synthesis (Munch-Petersen, Frentz, 1985).

For the scope of this paper, it is interesting that patients with Xeroderma pigmentosum may show signs of dysregulated T-cell mediated immune function by having a low number of helper/inducer T-lymphocytes.

Returning to the 1960's, in these years the immense role of the immune system and its importance for defence against damaging, foreign agents from the environment was recognized. Easy to understand, in cancer research this created a new wave of investigations from the idea that cancers were only allowed to grow when cancer cells escaped immune elimination, i.e., cancer formation was due to defective immuno-surveillance. This general hypothesis could not, however, be verified by sufficient, experimental evidence, and for a while many scientists working in the oncologic field felt that immunology did not influence cancer formation at all.

In the course of the 1960's renal transplantations were conducted with increasing frequency. As time went by and these renal allograft recipients were followed up, even the most doubting scientist had to admit that these immuno-suppressed renal allograft recipients developed cancers, and primarily skin cancers, at an unusually high rate (for review, see Gupta et al, 1986). Of special interest for the understanding of the interplay between immunology, exposure to UV/sun and skin cancer are the following facts: an Australian group reported a 20-fold increase in the rate of skin cancer in renal allograft recipients, sun-exposed sites were obviously most often affected, fair-skinned persons were at highest risk and tumor regression might occur after drug withdrawal. This heavily suggests that immuno-suppression implies a cancer-promoting function.

In the 1970's our knowledge of the process of photocarcinogenesis was greatly expanded through the important and comprehensive sequence of elegantly planned experiments on mice conducted by Margaret Kripke (Kripke, 1986). These experiments are of vital importance for understanding the processes which take place in the human counterpart. Briefly summarized: UV induced skin cancers have antigenic properties, UV induced skin cancers grow when transplanted to a syngenetic, conditioned mouse, but are rejected when the mouse is not conditioned, the state of susceptibility may be caused by exposure to UV radiation or by immuno-suppression, and the susceptible state is transferrable with lymphoid cells. Suppressor T-lymphocytes are responsible for this phenomenon. Furthermore, the susceptible state is associated with a decrease in cutaneous hypersensitivity reactions. Thus, the conditioning by UV radiation acts as a tumor promoting agent.

Is there any evidence that this holds true for photocarcinogenesis in man? Experiments parallel to those conducted in mice are out of the question in the human context. Short-term experiments, however, with small doses of UV-radiation comparable to those obtained by normal sunbathing or visits to tanning beds, are allowed. Such short-term experiments have ben performed by Peter Hersey and his associates in Australia (Hersey et al, 1983; Hersey et al, 1983; Hersey et al, 1987; Hersey et al, 1988). This group has studied the effects of exposure to sun/UV on the human immune system by use of monoclonal antibodies in determining the phenotypes of the lymphocytes in peripheral blood and skin infiltrates and the natural killer cell activity, too. Very briefly summarized: in the peripheral blood short-term UV-exposure caused alterations in the total number of lymphocytes, in the total number of T-lymphocytes and a decrease in the number of natural killer cells; in the lymphoid cells of the skin the number of Langerhans' cells and the number of HLA-DR+ cells decreased; a trend for decrease was seen for the total number of T-cells and the number of natural killer cells. Regarding the immune function a decrease in cutaneous hypersensitivity reactions, a decrease in the number of antigen-presenting cells, in the natural killer cell activity and in the immunoglobulin production (IgG, IgM) occurred (Hersey et al, 1988). Thus, UV-radiation on short terms influences the human immune system profoundly.

Experiments with long-lasting, intense exposure with carcinogens such as UV-radiation or X-rays are illegal in man, of course. Sometimes, however, the experiments are conducted by Mother Nature with the assistance of either human lifestyle or by chance. In the Department of Dermatology of the Finsen Institute, more than 600 persons with multiple non-melanoma skin cancer were followed up during the last 30 years. For all of them, a complete history of carcinogenic exposures was obtained. Among these patients we have been able to sort out those whose multiple cancers by clinical evidence were caused by one single, specific skin carcinogen.

We studied these patients in order to verify the following hypothesis: the mechanisms operating in photocarcinogenesis in man are analogous to those operating in mice, i.e., human UV-induced skin cancers are antigenic, and a certain state of immuno-suppression is required for the growth of these tumors. Furthermore, we hypothesized that skin cancer induction by X-rays differs basically. On this background, we studied the immune system in a group of patients with multiple skin cancers induced by sun-exposure in comparison with that of a group of patients with multiple skin cancers induced by ionizing radiation. Also a group of healthy control persons was examined.

This summarizes the results of our first study, in which none of the patients had any actual cancers present on the skin: patients who got multiple skin cancers on account of a large cumulative amount of sun-exposure during life had an increased number of T-lymphocytes in their peripheral blood, primarily of suppressor/cytotoxic cells and a low Th/Ts ratio indicating a certain state of immuno-suppression (Frentz et al, 1988). Interestingly, also the number of HLA-DR+ cells or activated cells were rather high in the UV-group. This is consistent with the findings in mice and, together with our demonstration of an inverse correlation between the number of suppressor T-cells and the ability of the lymphocytes to survive UV-radiation, gives support to the following hypothesis: heavy sun-exposure during life may cause a permanent increase in the number of suppressor T-cells, these extra T-cells may be UV-sensitive and they may protect transformed keratinocytes from immune rejection (Frentz et al, 1988).

How to approach this hypothesis further? If UV-induced skin cancers in man really have antigenic properties it could be worthwhile to study the lymphocyte subsets in the infiltrates in these cancers, always regarding basal or squamous cell cancers, no one, to my knowledge regarding the etiology of the cancers. Roughly estimated, about 90% of all skin cancers are thought to be attributed to sun-exposure. One study on infiltrates in squamous cell cancers showed a certain amount of B-cells (25%), predominance of T-cells and a rather low Th/Ts ratio of .85. Also, some cytotoxic/natural killer cells appeared. Interestingly, by far the most lymphoid cells were HLA-DR+, indicating immunological activation (Kochiyama et al, 1986). In the infiltrates of human basal cell cancers, most T-cells are activated too. Here, B-cells are sporadic, T-cells predominate, the Th/Ts ratio is above one, and cytotoxic/natural killer cells are few (Kochiyama et al, 1987; Guillen et al, 1985).

Interestingly, in a very recent study heavy infiltrates in basal cell cancers contained relatively more T-cells, primarily T-helper cells than did mild infiltrates. This might be due to variability in tumor antigenicity and/or host response (Habets et al, 1988). In any case, these studies heavily suggest that T-cells play a major role in the defence against human skin cancer proliferation and that many tumors have antigenic properties. Recent studies have also disclosed the presence of certain tumor antigens in human squamous cell cancers and high titers of squamous cell related antigens in serum from patients with advanced squamous cell cancers of the skin (Yagi et al, 1987). In some human basal cell cancers tumor cells might express the HLA-DR antigens themselves (Kochiyama et al, 1987). Also interesting in this context is the recent detection and identification of activated oncogenes in human skin cancers occurring on sun-exposed sites (Ananthaswamy et al, 1988).

Thus, the role of UV-radiation in the formation of human non-melanoma skin cancers is multi-faceted and complicated. Indeed, UV-radiation is involved in the initiation phase, perhaps activating oncogenes and certainly creating thymindimers and DNA-strand breaks, which the cell might be unable to repair. Furthermore, UV-radiation seems to transform keratinocytes and make them display antigens, to which the immune system may react, when intact, i.e. when not exerting a promoter function on account of immuno-suppression by lifelong heavy sun-exposure or by drugs or, perhaps, exhaustion from a long standing, heavy tumor burden.

This is our tentative schedule for induction of skin cancers by UV-radiation: the initiation step includes transformation of keratinocytes by damage to the DNA. The transformed keratinocytes also elicit an immune response, which may lead to immune rejection of the tumor cells. However, tumors may grow continuously if the general, systemic immunological defence is deficient, which it may be on account of repeated, heavy sun-exposure during life.

However, we need much more evidence and detailed studies before this hypothesis can be verified.

References

Ananthaswamy, H.N., Price, J.E., Goldberg, L.H., Bales, E.S. (1988), Detection and identification of activated oncogenes in human skin cancers occurring on sun-exposed body sites, *Cancer Res.* 48:3341–3346.

Bohr, V.A., Okumoto, D.S., Hanawalt, P.C.,(1986), Survival of UV-irradiated mammalian cells correlates with efficient DNA repair in an essential gene. *Proc. Natl. Acad. Sci. USA* 83:3830–3833.

Cleaver, J. (1968), Defective repair replication of DNA in Xeroderma pigmentosum, *Nature* 218:652–656.

Dubreuilh, W. (1896), Des hyperkeratoses circonscriptes, *Ann. Dermatol. Syph (Series 3)* 7:1158–1204.

Epstein, J.H., Roth, H.L. (1968), Experimental ultraviolet carcinogenesis: a study of croton oil promoting effects. *J. Invest. Dermatol.* 50:387–389.

Frentz, G, Munch-Petersen, B., Wulf, H.C., Niebuhr, E., da Cunha Bang, F. (1987), The nevoid basal cell carcinoma syndrome: sensitivity to ultraviolet radiation and X-rays, *J. Am Acad. Dermatol.* 17:637–643.

Frentz, G., dan Cunha Bang, F., Munch-Petersen, B., Lange, Wantzin, G. (1988), Increased number of circulating suppressor T-lymphocytes in sun-induced multiple skin cancers. *Cancer* 61: 294–297.

Guillen, F.J., Calvin, L. Day Jr,, Murphy, G.F., (1985), Expression of human lymphocyte antigen (HLA)-DR on tumor cells in basal cell carcinoma, *J. Am. Acad. Dermatol.* 85:203–206.

Gupta, A.K., Cardella, C.J., Habermann, H.F. (1986), Cutaneous malignant neoplasms in patients with renal transplants, *Arch. Dermatol.* 122:1288–1293.

Habets, J.M.W., Tank, B., Vuzevski, V.D., van Reede, E.C., Stolz, E., van Joost, T. (1988), Characterization of the mononuclear infiltrate in basal cell carcinoma: a predominantly T-cell-mediated immune response with minor participation of Leu-7+ (natural killer) cells and Leu-14+ (B) cells. *J. Invest. Dermatol.* 90:289–292.

Hersey, P., Haran, G., Hasic, E., Edwards, A. (1983), Alteration of T-cell subsets and induction of suppressor T-cell activity in normal subjects after exposure to sunlight, *J. Immunol.* 31:171–174.

Hersey, P., Bradley, M., Hasic, E., Haran, G., Edwards, A., McCarthy, W.H. (1983), Immunological effects of solarium exposure in human subjects, *Lancet* 1:545–548.

Hersey, P., MacDonald, M., Burns, C., Schibeci, S., Matthews, H., Wilkinson, F.J. (1987), Analysis of the effect of a sunscreen agent on the suppression of natural killer cell activity induced in human subjects by radiation from solarium lamps, *J. Invest. Dermatol.* 88:271–276.

Hersey, P., MacDonald, M., Henderson, C., Schibeci, S., d'Allesandro, G., Pryor, M., Wilkinson, F.J. (1988), Suppression of natural killer cell activity in humans b y radiation from solarium lamps depleted of UVB, *J. Invest. Dermatol.* 90:305–310.

Kochiyama, A, Oka, D., Ueki, H. (1986), Immunohistologic studies of squamous cell carcinoma: possible participation of Leu-7+ (natural killer) cells as antitumor effector cells, *J. Invest. Dermatol.* 87:515–518.

Kochiyama, A., Oka, D. Ueki, H. (1987), Expression of human lymphocyte antigen (HLA)-DR on tumor cells in basal cell carcinoma, *J. Am. Acad. Dermatol.* 16:833–838.

Kripke, M.K. (1986), Immunology and photocarcinogenesis, New light on an old problem, *J. Am. Acad. Dermatol.* 14:149–155.

Munch-Petersen, B., Frentz, G. (1985), X-ray and UV-radiation sensitivity of circulating lymphocytes in multiple epidermal cancer in relation to previous radiation exposure, *Rad. Res.* 103:432–440.

Urbach, F. (1969), Geographic pathology of skin cancer, in Urbach F., ed: The biologic effects of ultraviolet radiation, Pergamon Press, Oxford, pp. 635–650.

Unna, P.G. (1894), Die histopathologie der hautkrankheiten, Berlin, A. Hirschwald.

Yagi, H., Danno, K., Maruguchi, Y., Yamamoto, M., Imamura, S. (1987), Significance of squamous cell carcinoma (SCC)-related antigens in cutaneous SCC. *Arch. Dermatol.* 123:902–906.

Immunological Mediators Produced by UV-Irradiated Keratinocytes

T. SCHWARZ, A. URBANSKI, J. KRUTMANN, T.A. LUGER

Department of Dermatology
Hospital Vienna-Lainz
and Department of Dermatology II
University of Vienna and Ludwig Boltzmann Institute for Dermato-Venerologic
Serodiagnosis
Laboratory for Cell Biology
Vienna, Austria

Introduction

The epidermis has been identified as a place where immune responses can originate. This is primarily supported by the fact 1. that the epidermis harbors the dendritic bone marrow derived Langerhans cell which exhibits antigen presenting capacity (Stingl et. al., 1980) and 2. that recently in the murine epidermis a Thy 1-dendritic cell expressing T cell receptor γ and δ units were discovered (Stingl et al., 1987). Moreover, keratinocytes, the major constituent of the epidermis, have been shown to release a variety of immunomodulating cytokines and thus can actively participate in immune responses (Luger et al., 1988).

Cytokines are hormone like regulatory (glyco)proteins which are synthesized and released by various cells and bind to specific receptors on target cells (Cohen et al., 1983). At present there is increasing evidence for a network of interacting mediators leading to the activation, differentiation and proliferation of both immune and non-immune cells (Dinarello et al., 1987).

Interleukin 1

Similar to other mediators epidermal cell derived cytokines originally were described according to their biological and biochemical characteristics. The first cytokine definitely proved to be produced by keratinocytes was interleukin 1 (IL1). The epidermal cell product initially was described according to its property to costimulate thymocyte proliferation and thus named "epidermal cell derived thymocyte activating factor" (ETAF) (Luger et al., 1981). It soon turned out that ETAF was biologically and biochemically indistinguishable from IL1 (Ansel et al., 1988). Recently it has been shown that human and murine epidermal cells express mRNAs encoding for both IL1 α and IL1β (Kupper et

Photobiology, Edited by E. Riklis
Plenum Press, New York, 1991

749

al., 1986; Luger et al., 1983). IL1 stimulates T cells to produce other lymphokines such as IL2, IL4, interferon–γ and colony stimulating factors (CSF), it increases IL2 receptor expression on T-cells and stimulates B-cell proliferation and differentiation by enhancing the effects of B-cell stimulatory factors. IL1 induces hepatocytes to produce acute phase proteins, activates fibroblasts and osteoclasts, causes proteolysis and fever and thus is regarded as a mediator of the acute phase response. Moreover, IL1 upregulates melanocyte stimulating hormone receptor expression upon melanocytes and stimulates the release of hypothalamic and pituitary gland hormones. Thus IL1 represents a multitargeted cytokine with a broad spectrum of biological activities (Dinarello et al., 1988; Oppenheim et al., 1986).

Other Mediators Released by Keratinocytes

After the discovery that epidermal cells actively release IL1, several studies addressed the question whether other mediators are produced by keratinocytes (Luger et al., 1987; Kupper et al., 1987). It soon was demonstrated that epidermal cells synthesize a factor which stimulates the proliferation of distinct T cell lines and thus was called keratinocyte derived T cell growth factor (KTGF) (Kupper et al., 1986). KTGF subsequently was identified as granulocyte macrophage colony stimulating factor (GM–CSF) and in addition it was found that keratinocytes express the mRNA for GM–CSF (Kupper et al., 1988). GM–CSF supports the proliferation of macrophage, eosinophilic and neutrophilic colonies, it enhances biological activities of neutrophils, eosinophils and macrophages and is a growth factor for myeloid leukemic cells (Clark et al., 1987). Moreover, it was shown that GM–CSF enhances the maturation of Langerhans cells into potent immunostimulatory cells (Heufler et al., 1988).

Recently, mRNA encoding for IL6 was detected in human epidermal cells (Kirnbauer et al., 1988). IL6 has been described under a variety of other names e.g. interferon β2, B-cell stimulatory factor 2, hybridoma growth factor and hepatocyte stimulating factor (Sehgal et al., 1986). IL6 is an important regulator of B-cell growth and differentiation, stimulates the proliferation of hybridoma plasmocytoma cell lines, enhances the secretion of acute phase proteins and is pyrogenic (Wong et al., 1988). Whether IL6 exhibits antiviral activity is still a matter of debate.

Besides these multitargeted cytokines keratinocytes have been demonstrated to release also tumor necrosis factor (TNFα) (Oxholm et al., 1988; Coffey, et al., 1987) and transforming growth factor β (TGFβ) (Akhurst et al., 1988). TNFα induces hemorrhagic necrosis of tumors in animals and causes cachexia (Cerami et al., 1988). It is an important mediator of endotoxic shock and induces fever by the production of pyrogenic cytokines such as IL1 and IL6. TGFβ influences cell proliferation, promotes epithelial cell differentiation and stimulates matrix protein formation. Moreover, TGFβ seems to be a potent immunosuppressor by blocking IL1 and IL2 activity (Sporn et al., 1987; Wahl et al., 1988).

Through the capacity to release these mediators the keratinocyte may play an important role during immunologic and inflammatory reactions. Although the real causative agents in most inflammatory and immunologically mediated skin disorders is unknown, keratinocytes may participate at a very early step by releasing different cytokines. This is in accordance with the in vitro findings that the constitutive production of mediators is low, but is dramatically induced by various stimuli. This also appears to be true for the in vivo situation. Using in situ hybridization techniques it has been shown that mRNA expression e.g. for IL1 and TGFβ is weak or almost absent in normal skin, but highly induced following UVB irradiation or after application of tumor promoting agents (Oxholm et al., 1988; Akhurst et al., 1988). One may speculate that the release of a single cytokine in vivo activates the cytokine cascade inducing both the expression of specific receptors on various cells and the production of other mediators and causing the attraction of inflammatory cells which by themselves contribute different mediators not released by keratinocytes. The most potent stimuli for cytokine release by epidermal cells in vitro and in vivo include tumor promotors, endotoxin and UV–light (Luger et. al., 1988).

Ultraviolet Light and Cytokine Release

As irradiation with UVB light has various in vivo implications and as the skin is the primary target for UVB light, one can anticipate that many UV-induced effects are mediated by epidermal cell derived cytokines. This is also supported by the in vitro findings that in supernatants of UV exposed keratinocytes significantly increased IL1 activity was detected (Ansel et al., 1983) and that mRNA encoding for IL1 was significantly higher expressed than in unirradiated control epidermal cells (Kupper et al., 1986; Kupper et al., 1987). Severe sunburn results in fever, malaise and chills, effects which are mediated at least partly by IL1. Accordingly during sunburn reaction serum levels of IL1 are found significantly increased (Ansel et al., 1987). One can assume that IL1 found in the serum may originate form the epidermis. Formal proof of this challenging speculation, however, is extremely difficult.

Recently, it was also detected that epidermoid carcinoma cell lines, freshly isolated epidermal cells and long term cultured keratinocytes upon UV irradiation release significant amounts of IL6 (Kirnbauer et al., 1988). Maximum biological activity was observed within 24hr after irradiation. Accordingly IL6 mRNA expression in keratinocytes was found to be increased after UV exposure. Preliminary data also suggest that in humans after application of erythemogenic doses of UVB light IL6 serum levels are significantly increased (Urbanski et al., 1990). IL6, like IL1, causes fever and thus may be involved in the pathogenesis of systemic sunburn reaction. Whether UV light induces IL6 release directly or via IL1 which is a potent stimulus for IL6 production is still a matter of debate (Kirnbauer et al., 1988).

In addition, UV-induced release of epidermal cell derived cytokines may play a role in the recovery from bone marrow suppression. It has been demonstrated that CSF and in particular GM-CSF activity present in normal skin is enhanced by UVB-light (Birchall et al., 1988). Moreover, it could be demonstrated that 5-fluorouracil induced myelosuppression in mice can be reversed by UV exposure (Birchall et al., 1988). This observation leads to the conclusion that UVB irradiation induces the cutaneous release of CSF in amounts sufficient to reverse acute myelosuppression.

At the moment it is not known whether all epidermal cell derived cytokines can be induced by UVB light or whether this is only due for IL1, IL6, GM-CSF, TNFα and murine IL3. Further studies have to address that issue. Nevertheless, the present observations clearly demonstrate that various effects caused by UV light may be mediated by the release of cytokines from keratinocytes.

UV-Induced Suppressor Factors

When one accepts the concept that keratinocytes play a regulatory role in the epidermal immune response, one has to anticipate that the keratinocyte should also be able to release immunosuppressive factors. Until recently the only mediators with suppressor properties released by keratinocytes were prostaglandins and urocanic acid (Rola-Pleszczynski et al., 1985; DeFabo et al., 1983).

Moreover, irradiation with UVB light causes immunosuppression (Kripke et al., 1986). This immunomodulation may at least partially contribute to UV induced carcinogenesis (Kripke et al., 1986). Immunologic alterations that follow UV exposure can be divided into two forms: local and systemic immunosuppression. Local alterations result from a direct interaction between UV light and immunocompetent cells in the irradiated area. Application of a potent contact sensitizing agent e.g. dinitrofluorobenzene to a UV exposed skin site results in the inability to induce contact hypersensitivity (CHS), while administration of the sensitizer to an unirradiated area leads to a normal CHS reaction (Toews et al., 1980). The induction of this immunologic tolerance correlates with an alteration of the antigen presenting capacity of epidermal Langerhans cells and the induction of antigen specific suppressor T-cells (Elmets et al., 1983).

Higher doses of UV light alter immune responses e.g. induction of CHS at sites not exposed directly to UV light ("systemic" immunosuppression) (Jessup et al., 1978). The question, how the events occuring at the irradiated site lead to an abnormal response to an antigen applied at a distant non-UV exposed skin area, remains to be answered. The most attractive explanation for this systemic immunosuppression involves a soluble mediator (Kripke et al., 1986). If this is in fact of importance, the keratinocyte which is the main target for UVB light and endowed with immunosecretory functions, has to be considered as a primary source of such an inhibitor. Accordingly it was recently shown that the intravenous injection of supernatants derived from UV-irradiated freshly isolated murine epidermal cells into mice resulted in the inability to induce CHS to potent sensitizing agents in the recipient animals (Schwarz et al., 1986). This observation demonstrated for the first time that obviously murine epidermal cells upon UV-irradiation release an immunosuppressive factor, which blocks the induction of CHS. This mediator has to be induced by UV light, as the injection of supernatant derived from unirradiated cells did not affect CHS reaction. Keratinocytes appear to be the major source of this inhibitor since it is produced not only by epidermal cell cultures containing Langerhans cells, melanocytes and Thy 1-positive cells, but also by transformed keratinocyte cell lines, which are devoid of other cellular constituents.

Upon high performance liquid chromatography gel filtration this suppressor cytokine exhibits a molecular weight between 20kD and 50kD and thus appears to be distinct from other immunomodulators such as prostaglandins, leukotrienes and urocanic acid, a soluble compound located in the corneal layer of the epidermis, which has been suggested to be a photoreceptor for immunosuppression (DeFabro et al., 1983),

Although it is not yet clear whether this keratinocyte-derived factor is a new immunomodulator or a well known mediator with a so far undescribed function, it is possible that the inhibitor generated in UV exposed epidermal cell cultures is similar to a transferable serum suppressor factor found in UVB exposed mice, which induces suppression of CHS via generation of T-suppressor cells (Swartz et al., 1984).

Recently the concept that UV exposed keratinocytes release immunosuppressive factor(s) also has been supported by other groups. A similar epidermal cell derived inhibitor was described, which in contrast to the CHS inhibitor described above, only was able to block the induction of delayed but not of contact hypersensitivity (Kim et al., 1988). Moreover, it was demonstrated that injection of supernatant derived from UVB irradiated murine keratinocytes suppresses the induction of delayed type hypersensitivity to alloantigens and that this is mediated via the induction of antigen specific suppressor cells (Ullrich et al., 1988). These so far discrepant results might be due to different experimental conditions or due to the fact that more than one cytokine is involved in this complex system. In spite of such differences which have to be clarified in future studies these observations clearly emphasize that epidermal cells upon UV exposure release inhibitors which might be responsible for systemic immunosuppressive effects following UV irradiation.

In addition, it was recently observed that murine epidermal cells after UV exposure release a cytokine which blocks the biological activity of IL1 (Schwarz et al., 1987). This observation was both interesting and surprising, as it was well established that UV irradiation induces the release of IL1 by epidermal cells and thus supernatants derived from UV exposed epidermal cells contain high levels of IL1 (Ansel et al., 1983; Kupper et al., 1987). Consequently this IL1 inhibitor, termed EC-contra-IL1, was only detectable after chromatographic removal of the high amounts of IL1 present in the supernatants. As in all previous studies only crude supernatants were tested, this inhibitor was always overlooked. Partially purified EC-contra-IL1 blocks both natural and recombinant IL1, exhibits a molecular weight of 40kD and an isoelectric point of 8.8. EC-contra-IL1 seems to be specific for IL1, as it does not inhibit IL2 or IL3 activity and does not block spontaneous cell proliferation, thus excluding a toxic effect.

The main source of EC-contra-IL1 seems to be the keratinocyte as EC-contra-IL1 is also released by keratinocyte cell lines. EC-contra-IL1 is not constitutively produced by epidermal cells, the release has to be induced either by UVB exposure or by stimulation with the tumor promoting agent phorbol myristate acetate. The production of EC-contra-IL1 obviously is not only an in vitro phenomenon as recently an IL1 inhibitor with similar biochemical characteristics could be detected in the sera of total body UV exposed mice (Schwarz et al., 1988). This circulating factor is not found in unirradiated animals, maximum levels appear between 12 and 24hr after UV exposure. These observations support the speculation that epidermal cells release EC-contra-IL1 upon UVB irradiation which penetrates the basal lamina, enters the circulation and probably causes systemic immunosuppression.

There seems to be an obvious similarity between EC-contra-IL1 and the CHS inhibitor. Both factors are released by the same cells and cell lines under identical conditions, they are both induced by UV light and they share a similar molecular weight. However, only further biochemical characterization, sequencing and gene cloning of both mediators will reveal whether these factors are identical or two distinct cytokines released by keratinocytes. Further purification will also indicate whether these factors are new mediators or well known immunomodulators with so far underscribed activities.

Transforming growth factor beta (TGFβ) is the only keratinocyte derived cytokine with potent immunosuppressive capacities which has been fully cloned and sequenced so far (Akhurst et al., 1988; Sporn et al., 1987; Wahl et al., 1988). Moreover, TGFβ has turned out to be responsible for a variety of immunosuppressive activities previously described under various names. Accordingly a glioblastoma derived suppressor factor blocking IL1 and IL2 activity was cloned and showed complete homology to TGFβ2 (Wrann et al., 1987) and a keratinocyte derived factor blocking mixed lymphocyte reactions called keratinocyte lymphocyte inhibitory factor (KLIF) exhibited partial homology to TGFβ (Nickoloff et al., 1988). Although, TGFβ has been demonstrated to be a potent inhibitor of IL1 and IL2 activity, EC-contra-IL1 seems to be unrelated to TGFβ since its activity could not be blocked by antibodies directed against TGFβ (Urbanski et al., 1988).

Recently, EC-contra-IL1 was also described in the human system. This factor could be isolated and partially purified from either UV exposed or PMA-treated human epidermoid carcinoma cell lines (Urbanski et al., 1988). Human EC-contra-IL1 shares obvious biological and biochemical similarities with the murine IL1-inhibitor.

Ultraviolet radiation is a potent inhibitor of selected cell-mediated immune responses and this suppression is at least partially caused by an UV-induced defect in the function of accessory/antigen presenting cells required for helper T-cell activation (Krutmann et al., 1988). Since IL1, in addition to IL6, represents one of the signals provided by accessory cells to augment T-cell activation (Krutmann et al., 1988) the capacity of human EC-contra-IL1 to modulate human accessory cell function was tested. As expected, human EC-contra-IL1 down-regulated human T-lymphocyte activation in an accessory cell dependent system by specifically blocking accessory cell derived IL1 activity, whereas IL6 activity remained unaffected (Krutmann et al., 1988). These studies reveal a novel mechanism by which UV-radiation in addition to its previously defined direct effects on human accessory cell (Krutmann et al., 1988) may modulate human accessory cell activity in an indirect manner via the induction of suppressor factors such as human EC-contra-IL1.

The findings that epidermal cells and in particular keratinocytes exhibits the capacity to release immunosuppressive cytokines further supports the concept of the epidermis as an immunologic organ. It clearly demonstrates that the keratinocyte not only initiates the immune response by providing immunoenhancing mediators, but also may turn down ongoing inflammatory and immunologic reactions by secreting suppressive cytokines. The capacity to release agonistic and antagonistic factors e.g. IL1 and contra-IL1, however, might not be unique to the keratinocyte but common for all immunosecretory cells. Consequently immunosuppressive situations may originate from the keratinocyte and this seems to be true for UV-induced immunosuppression, as UV-mediated release of immunosuppressive factors by keratinocytes may at least partly explain local and in particular systemic immunosuppressive states following UV exposure.

Correspondence

Thomas Schwarz, M.D., Department of Dermatology, Hospital Vienna-Lainz, Wolkersbergenstrasse 1, A–1130 Vienna, Austria/Europe.

References

Akhurst, R.J., Fee, F., and Balmain, A. (1988). Localized production of TGF–β mRNA in tumour promoter-stimulated mouse epidermis. *Nature* 331, 363.

Ansel, J.C., Luger, T.A. Lowry, D., Perry, P. Roop, D.R., and Mountz, J.D. (1988). The expression and modulation of IL-la in murine keratinocytes. *J. Immunol.* 140, 2274.

Ansel, J.C., Luger, T.A., and Green, I. (1983). The effect of in vitro and in vivo UV irradiation on the production of ETAF activity by human and murine keratinocytes. *J. Invest. Dermatol.* 81, 519.

Ansel, J.C., Luger, T.A., and Green, I. (1987). Fever and increased IL–1 activity as a systemic manifestation of acute phototoxicity in New Zealand white rabbits. *J. Invest. Dermatol.* 89, 32.

Birchall, N., Gamba, C., and Kupper, T. (1988). Cutaneous UVB irradiation enhances recovery from bone marrow suppression. *J. Invest. Dermatol.* 90, 547.

Cerami, A., and Beutler, B. (1988). The role of cachectin/TNF in endotoxic shock and cachexia. *Immunol.* Today 9, 28.

Clark, S.C., and Kamen, R. (1987). The human hematopoietic colony stimulating factors. *Science* 236, 1129.

Coffey, R.J., Derynck, R., Wilcox, J.N., Bringman, T.S., Goustin, A.S., Moses, H.L., and Pittelkow, M.R., (1987). Production and auto-induction of transforming growth factor-α in human keratinocytes. *Nature* 328, 817.

Cohen, S., and Yoshida, T. (1983). Physiological and pathological roles of lymphokines. In Y. Amamura, H. Ayashi, T. Honjo, T. Kishimoto, M. Muramatsu, and T. Osawa, *eds.*, Humoral factors in host defense (Academic Press, Washington) pp. 245.

DeFabo, E.C., and Noonan, F.P. (1983). Mechanism of immune suppression by ultraviolet irradiation in vivo. I. Evidence for the existence of an unique photoreceptor in skin and its role in photoimmunology. *J. Exp. Med.* 157, 84.

Dinarello, C.A. (1988). Biology of interleukin 1. *FASEB J.* 2, 108.

Dinarello, C.A., and Mier, J.W. (1987). Lymphokines. *New Engl. J. Med.* 317, 940.

Elmets, C.A., Bergstresser, P.R., Tigelllar, R.E., Wood, P.J., and Streilein, J.W. (1983). Analysis of mechanism of unresponsiveness produced by haptens painted on skin exposed to low dose ultraviolet radiation. *J. Exp. Med.* 158, 781.

Heufler, C., Koch, F., and Schuler, G. (1988). Granulocyte/macrophage colony-stimulating factor and interleukin 1 mediate the maturation of murine epidermal Langerhans cells into potent immunostimulator dentritic cells, *J. Exp. Med.* 167, 700.

Jessup, M., Hanna, N., Plaszynski, E., and Kripke, M.L. (1978). Mechanisms of depressed reactivity to dinitrochlorobenzene and ultraviolet-induced tumors during ultraviolet carcinogenesis in Balb/c mice. *Cell. Immunol.* 38, 105.

Kim, T.Y., Golden, P., Ullrich, S.E., and Kripke, M.L. (1988). Effect of UV-induced epidermal cytokines on immune responses in vivo. *Clin. Res.* 36, 662A.

Kirnbauer, R., Köck, A., Schwarz, T., Urbanski, A., Krutmann, J., Borth, W., Ansel, J.C., and Luger, T.A. (1989). Interferon β2, B–Cell differentiation factor 2, hybridoma growth factor (interleukin 6) is expressed and released by human epidermal cells and epidermoid carcinoma cell lines. *J. Immunol*, 142, 1922.

Kirnbauer, R., Köck, A., Krutmann, J., Schwarz, T., Urbanski, A., and Luger, T.A. (1988). UVB irradiation stimulates expression and release of interleukin 6 (IL 6) by normal and malignant human epidermal cells. *Arch. Dermatol. Res.* 281:129.

Kripke, M.L. (1986). Immunology and photocarcinogenesis. *J. Am. Acad. Dermatol.* 14, 149.

Kripke, M.L., and Morison, W.L. (1986). Studies on the mechanism of systemic suppression of contact hypersensitivity by ultraviolet B radiation. Photodermatology 3, 4.

Krutmann, J., and Elmets, C.A. (1988). Recent studies on mechanisms in photoimmunology. Yearly Review. *Photochem. Photobiol.* 48:787.

Krutmann, J., Schwarz, T., Kirnbauer, R., Urbanski, A., and Luger, T.A. (1988). Epidermal cell-contra-interleukin 1 inhibits human accessory cell function by specifically blocking IL1 activity. *Arch. Dermatol. Res.* 281:129.

Krutmann, J., Schwarz, T., Kirnbauer, R., Urbanski, A., and Luger, T.A. (1988). Modulation of OKT3–induced activation of human T-lumphocytes by interleukin 6 and epidermal cell derived-contra-interleukin 1. *J. Invest. Dermatol.* 91, 382.

Krutmann, J., Wallis, R.S., Kahn, I.U., Zhang, F. Koehler, K.A., Rich, E.A., Jacobson, K., Ellner, J.J., and Elmets, C.A. (1990). The cell membrane is a major locus for ultraviolet-B-induced alterations in accessory cells. *J. Clin. Invest.* (in press).

Kupper, T.S., Ballard, D., Chua, A.O., McGuire, J.S., Flood, P., Horowitz, M.C., Langdon, R., Lightfood, L., and Gubler, U. (1986). Human keratinocytes contain mRNA indistinguishable from monocyte interleukin 1 mRNA. *J. Exp. Med.* 164, 2095.

Kupper, T.S., Chua, A.O., Flood, P., McGuire, J., and Gubler, U. (1987). Interleukin 1 gene expression in cultured human keratincoytes is augmented by ultraviolet irradiation. *J. Clin. Invest.* 80, 430.

Kupper, T.S., Coleman, D., McGuire, J. Goldminz, D., and Horowitz, M. (1986). Keratinocyte derived T-cell growth factor: a T-cell growth factor functionally distinct from interleukin–2. *Proc. Natl. Acad. Sci. USA.* 83, 4451.

Kupper, T.S., Horowitz, M. Lee, F., Coleman, D., and Flood, P. (1987). Molecular characterization of keratinocyte cytokines. *J. Invest. Dermatol.* 88, 501A.

Kupper, T.S., Lee, F., Coleman, D., Chodakewitz, J., Flood, P., and Horowitz, M. (1988). Keratinocyte derived T-cell growth factor (KTGF) is identical to granulocyte macrophage colony stimulating factor (GM–CSF). *J. Invest Dermatol.* 91, 185.

Luger, T.A. Stadler, B.M., Luger, B.M., Mathieson, B.J., Mage, M., Schmidt, J.A., and Oppenheim, J.J. (1983). Murine epidermal cell derived thymocyte activating factor resembles murine interleukin 1. *J. Immunol.* 128, 2147.

Luger, T.A., and Schwarz, T. (1990) Epidermal cell derived cytokines. In J. Bos, ed., Skin Immune System (CRC Press, Boca Rotan) in press.

Luger, T.A., Danner, M. Schwarz, T. Köck., and Urbanska, A. (1987). Epidermal cell derived mediators of immunity and inflammation. In R. Caputo, *ed.,* Immunodermatology (CID Edizioni Internationali, Roma) pp. 63.

Luger, T.A., Stadler, B.M. Katz, S.I. and Oppenheim, J.J. (1981). Epidermal cell (keratinocyte) derived thymocyte activating factor (ETAF). *J. Immunol.* 147, 1493.

Nickoloff, B.J., and Mitra, R.S. (1988). Transforming growth factor-beta is a keratinocyte-derived limphocyte inhibitory factor. *J. Invest. Dermatol.* 90, 592.

Oppenheim, J.J., Kovacs, E.J., Matsushima, K., and Durum, S.K. (1986). There is more than one interleukin 1. *Immunol.* Today 7, 45.

Oxholm, A., Oxholm, P., Staberg, B., and Bendtzen, K. (1988). Immunohistological detection of interleukin 1–like molecules and tumor necrosis factor in human epidermis before and after UVB-irradiation in vivo. *Brit. J. Dermtol.* 118,369.

Rola-Pleszczynski, M., (1985). Immunoregulation by leukotrienes and other lipoxygenase metabolites. *Immunol.* Today 6: 302.

Schwarz, T., Urbanski, A., Gschnait, F., and Luger, T.A. (1986). Inhibition of the induction of contact hypersensitivity by a UV-mediated epidermal cytokine. *J. Invest. Dermatol.* 87, 289.

Schwarz, T., Urbanski, A., Gschnait, F., and Luger, T.A. (1987). UV-irradiated epidermal cells produce a specific inhibitor of interleukin 1 activity. *J. Immunol.* 138, 1457.

Schwarz, T., Urbanski, A., Kirnbauer, R., Köck, A., Gschnait, F., and Luger, T.A. (1988). Detection of a specific inhibitor of interleukin 1 in sera of UVB-treated mice. *J. Invest. Dermatol.* 91:536.

Sehgal, P.B., Zilberstein, A., Ruggieri, R.M., May, L.T., Feguson-Smith, A., Slate, D.L., Revel, M., and Ruddle, F.H. (1986). Human chromosome 7 carries the β2 interferon gene. *Proc. Natl. Acad. Sci. USA.* 83, 5219.

Sporn, M.B., Roberts, A.B., Wakefield, L.M., and deCrombrugghe, B. (1987). Some recent advances in the chemistry and biology of transforming growth factor-beta. *J. Cell. Biol.* 105, 1039.

Stingl, G., Gunter, K.C., Tschachler, E., Yamada, H., Lechler, R.I., Yokoyama, E.M., Steiner, G., Germain, R.N. and Shevach, E.M. (1987). Thy-1[+] dendritic epidermal cells belong to the T-cell lineage. *Proc. Natl. Acad. Sci. USA.* 84, 2430.

Stingl, G., Tamaki, K., and Katz, S.I. (1980). Origin and function of epidermal Langerhans cells. *Immunol. Rev.* 53, 149.

Swartz, R.P. (1984). Role of UVB-induced serum factor(s) in suppression of contact hypersensitivity in mice. *J. Invest. Dermatol.* 83, 305.

Toews, G.B., Bergstresser, P.R., and Streilein, J.W. (1980). Epidermal Langerhans cell density determines whether contact hypersensitivity or unresponsiveness follows skin painting with DNFB. *J. Immunol.* 124, 445.

Ullrich, S.E. (1988). Induction of suppressor cells by a factor released by UV-irradiated epidermal cells, *Fed. Proc.* A1680.

Urbanski, A., Schwarz, T., Schneider, F.J., Adolf, G., and Luger, T.A. (1988). Release of inhibitor of IL1 activity by human epidermoid carcinoma cells. *Arch. Dermatol. Res.* 281:.

Urbanski, A., Schwarz, T., Neuner, P., Krutmann, J., Kirnbauer, R., Köck, A. and Luger, t.a. (1990). Ultraviolet light induces increased circulating interleukin 6 in humans. *J. Invest. Dermatol.* (in press).

Wahl, S.M., Hunt, D.A., Wong, H.L., Dougherty, S., McCartney-Francis, N., Wahl, L.M., Ellingsworth, L., Schmidt, J.A., Hall, G., Roberts, A.B., and Sporn, M.B. (1988). Transforming growth factor-β is a potent immunosuppressive agent that inhibits IL-1 dependent lymphocyte proliferation. *J. Immunol.* 140, 3026.

Wong, G.G., Clark, S.C. (1988). Multiple actions of interleukin 6 within a cytokine network. *Immunol.* Today 9, 28.

Wrann, M., Bodmer, S., de Martin, R., Siepl, C., Hofer-Warbinek, R., Frei, K., Hofer, E., and Fontana, A. (1987). T-cell suppressor factor from human glioblastoma cells is a 12.5kD protein closely related to transforming growth factor-beta. EMBO (*Eur. Mol. Biol. Organ*) J. 6, 1633.

The Franz Greiter Memorial Symposium:

Physiological Effects of UV Radiation on the Immune System and on the Eye

Franz Greiter — The Man and His Work

F. URBACH

Skin and Cancer Hospital,
Philadelphia, Pa. 19140,
USA

Professor Franz Greiter was born in Wittberg, Tyrol in the mountains of Austria in December 1919, and died at age 66 in 1985. He completed his early studies in Bregenz, Austria in 1945. As a devoted mountain climber and skier, Greiter frequently suffered from severe sunburn, one of the worst occurring on climbing Piz Buinm the highest mountain of the Silvretta area at the Swiss Austrian border. The experiments that led to the most effective sunscreens were begun in a small room in the paternal house. Because Greiter's sunscreens were particularly effective for mountaineers, they were sought after by climbers in the Himalayas, Andes and Mount Everest. In 1962, Greiter introduced the concept of SPF into Photobiology and industry; he was the first to develop sunscreens absorbing UVA as well as UVB, and developed water resistant products. At age 53 he returned to the University and completed work for a Ph.D. in Physiology in 1977. One of his many major contributions was the founding of the Institute for Applied Physiology in Vienna, which rapidly became an interdisciplinary team of physiologists, psychologists, chemists and biologists. In addition to his industrial pursuits, Franz Greiter was a thoughtful scientists, as can be seen in the 160 publications and 6 books. His university honored him with the title of Professor, and his country awarded him the Decoration of Merit and the Cross of Honor for Science and Art. I think he was proudest of the official gold sports badges for Austria and Germany. Franz Greiter was that most unusual combination of successful businessman and superb scientist. His contributions to photobiology and photodermatology were outstanding. We, his colleagues, collaborators and friends miss him sorely.

Photobiology, Edited by E. Riklis
Plenum Press, New York, 1991

Immune Suppression by Ultraviolet B Irradiation and Urocanic Acid

F.P. NOONAN AND E.C. DE FABO

Department of Dermatology
George Washington University Medical Center
Washington, DC, USA.

It is a pleasure to have the opportunity to speak at this Symposium in memory of Franz Greiter. I first met Professor Greiter at the 8th International Congress of Photobiology, when he came to our poster in which we proposed for the first time that UV-irradiated urocanic acid may have a role as an immune modulator. He was very curious about our work, and although our studies must have presented somewhat of an inconvenience for him as he had considered using urocanic acid in his sunscreens, he was scientific and open minded enough to support me when I worked for six months in the laboratories of Professors Stingl and Wolff at the University of Vienna.

A variety of cell mediated immune responses is altered by UV irradiation. Contact hypersensitivity responses to chemical sensitisers (Noonan et al., 1981a), delayed type hypersensitivity responses to hapten-conjugated cells (Greene et al., 1979; Noonan et al., 1981b), to viruses (Ross et al., 1986, 1987) and to alloantigens (Mottram et al., 1988) are all suppressed by UV irradiation. We have recently shown that a single dose of UV can prolong heart allografts in mice (Mottram et al., 1988).

Our own studies have been directed toward establishing the mechanism by which UV irradiation initiates immune suppression. We have used as a model the systemic suppression of contact hypersensitivity (CHS) to trinitrochlorobenzene in mice. This systemic suppression is a UVB effect, is both dose and wavelength dependent and is independent of dose-rate and dose fractionation (Noonan et al., 1981a, b, De Fabo and Noonan, 1983). Systemic suppression of CHS by UV radiation from FS40 sunlamps is proportional to $\log_{10}$ of UV dose and the dose response curves are similar for mice of different coat colors and different immunologic phenotypes (BALB/c; albino, H-2^a. C3H, brown, H-2^k; CBA/N, brown, H2^k; Sencar albino). All strains of mice so far tested with one interesting exception which will be addressed below show UV-induced systemic suppression of CHS. Most importantly, antigen-specific suppressor T cells are found when contact sensitiser is applied to a UV-irradiated mouse.

In terms of immunologic mechanism, a critical contribution was made by Greene and collaborators, including ourselves, (Greene et al., 1979; Noonan et al., 1981b) who demonstrated an antigen presenting cell defect in the spleen cells of UV irradiated

Photobiology, Edited by E. Riklis
Plenum Press, New York, 1991

animals. Using either *in vivo* or *in vitro* assays, partially purified antigen presenting cells from the spleens of UV irradiated mice were not as effective at stimulating T cells in the presence of antigen as were comparable cell preparations from normal mice. We extended these studies (Noonan et al., 1988) to show that highly purified preparations of splenic dendritic cells (potent antigen presenting cells) from UV mice were less effective at stimulating T cells than comparable cells from normal mice, indicating that the splenic antigen presenting cell defect could not be simply explained by a redistribution of cells within the body.

The finding of an antigen presenting cell alteration in UV mice allows the formulation of a central hypothesis for the mechanism of UV induced suppression (Figure 1). In this scheme, the underlying immunologic problem in UV mice is an alteration to antigen presenting cells such that, when antigen is administered to a UV-irradiated animal, suppressor T cells, which down-regulate immune responses, are formed in preference to up-regulating effector cells. The net result is a suppressed immune response. This scheme explains why antigen-specific suppressor T cells are formed when a contact sensitiser or viral antigen is applied to a UV-irradiated animal.

To address the question of whether UV-induced suppression is initiated by an interaction between UV radiation and a specific photoreceptor in the skin, an action spectrum for systemic suppression of contact hypersensitivity was determined (De Fabo and Noonan, 1983). Using a specialized source of narrow bands of UV radiation (2.5 nm half band width), dose-response curves for suppression were derived *in vivo* at 10 wavelengths from 250 to 320 nm. The action spectrum so derived has a maximum between 260 and 270 nm, a shoulder at 280–290 nm, and declines steadily to about 3% of maximum at 320 nm. The finding of such a clearly defined wavelength dependence implied the presence of a specific photoreceptor for this effect. The observation that wavelengths between 260 and 270 nm are much more effective than wavelengths

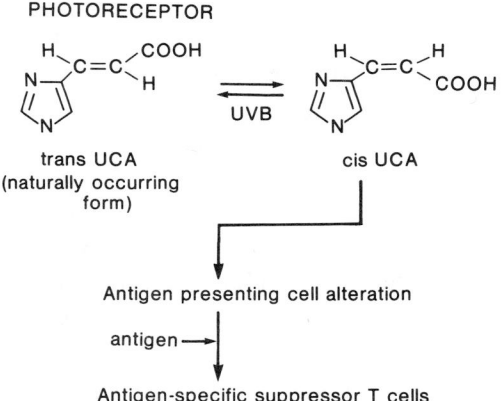

Figure 1. Hypothesis for UV-induced suppression

around 320 nm implied that such photoreceptor may be superficially located, since the shorter wavelengths are approximately 10 times more absorbed in skin than 320 nm UV.

One observation that was noted during the course of these studies, was that, in contrast to mice irradiated with UV from the broadband FS40 UV source, with one exception animals irradiated with immunosuppressive doses of narrow bands of UV did not show visible sunburn or gross skin damage. We therefore examined histologic sections of skin taken from mice at various times after administration of a UV dose sufficient to cause 50% systemic suppression. Sections were taken after irradiation with each of 10 wavelengths from 250 to 320 nm at 24 h and 72 h after UV. The slides presented were for 270, 295 and 320 nm UV. At 24 h after either 270 or 295 nm UV there was extensive killing of the epidermis, with some few inflammatory infiltrating cells. At 3 days after UV, in the skin sections from mice given 270 nm UV, considerable epidermal hyperplasia was evident, with some abnormal keratinization but very few inflammatory cells. In contrast, a well developed sunburn reaction was evident in the skin of mice which had been given 295 nm UV 3 days previously. The skin was grossly swollen, there was a huge inflammatory infiltrate, sunburn cells and vacuolation were apparent in the epidermis, and parakeratosis was evident. In the skin from mice given 320 nm UV, however, none of these features was evident - there was no evidence of epidermal killing or of gross inflammation. There were a few cells infiltrating in the dermis, but no other identifiable changes. We concluded from this investigation that neither an inflammatory response nor epidermal killing and regeneration are necessary for the generation of UV-induced immunosuppression.

As indicated above, the finding that UV of 260–270 nm was the most effective at immunosuppression, suggested that the photoreceptor initiating immunosuppression may be superficially located. This conclusion was further supported by the finding that removal of the stratum corneum prior to UV irradiation prevented the generation of immunosuppression. The implication from these studies was that the action spectrum probably showed little distortion from screening of the photoreceptor. Since the action spectrum should be congruent with the *in vivo* absorption spectrum of the photoreceptor, a comparison was made between the action spectrum and various chromophores known to be in mammalian skin. Although the action spectrum generally agreed with the action spectrum for DNA up to about 285 nm, and partially with the absorption spectra of proteins and of 7-dehydro-cholesterol, at longer wavelengths there was increasing disparity, the difference at 320 nm being more than two orders of magnitude. One substance, urocanic acid (UCA, deaminated histidine), a major-UV absorbing component of the stratum corneum (see Figure 1), has an absorption spectrum very similar to the action spectrum for immunosuppression. Accordingly, we postulated, (De Fabo and Noonan, 1983), based on the congruence between the action spectrum and the absorption spectrum of UCA, on its superficial location in the stratum corneum, and on its trans to cis isomerisation on UV irradiation (see Figure 1), that UV-induced immunosuppression is initiated by the interaction in the skin between UV and UCA.

To test this hypothesis further, with the collaboration of Drs. Henry Kacser and

Graham Bulfield of the University of Edinburgh, we tested the immunosuppressive effects of UV radiation on mice mutant in the enzyme histidase (De Fabo et al., 1983). These animals have less than 10% of normal UCA levels in their skin. Although we were able to suppress systemically the CHS response by prior UV irradiation in the wild-type animals, we could not obtain UV-induced suppression in the UCA deficient mutants. These animals have subsequently been bred for us onto an inbred background, and are currently under further study in our laboratory.

By HPLC analysis we have determined the levels of trans and cis UCA in extracts of mouse skin taken before or after UV irradiation. Before UV irradiation, trans levels are about 200 ng/mg of skin, and cis levels are less than 5 ng/mg. Immediately after 1 h of UV irradiation (27 kJ/m^2), cis levels were 60 ng/mg of skin and trans levels 130 ng/mg. 24 h after UV irradiation these levels were the same, but 4 days after UV, 50% of the cis isomer was no longer detectable, although the trans levels were similar to immediately post-irradiation levels. This finding is particularly interesting when compared with our previous observation that it takes between 1 and 3 days after a single dose of UV for the generation of UV-suppression. Animals sensitised on an unirradiated site 24 h after UV do not show systemic suppression of contact hypersensitivity.

If our hypothesis is correct that UV-induced suppression is initiated by the trans to cis isomerisation of UCA as indicated in Figure 1, it would be predicted that cis UCA should alter antigen presentation. We accordingly set up an antigen presentation assay in which the proliferative response to an antigen, DNP-ovalbumin, of purified T lymphocytes taken from mice which had previously been immunized with that antigen, was determined in the presence of antigen presenting cells from various sources (Noonan et al., 1988). We could find no effect of either isomer of UCA added directly to this assay *in vitro* at doses from 5 to 500 ug/ml, regardless of whether the antigen presenting cells were epidermal Langerhans cells, peritoneal macrophages or highly purified dendritic cells from spleen. In contrast, if the UCA was administered *in vivo*, more interesting results were obtained. Splenic dendritic cells prepared from mice which had been given i/v 100 or 200 ug/mouse (0.7 to 1.4 umoles) of cis UCA 7 days previously had severely decreased antigen presenting activity. Dendritic cells from mice given the same amount of trans UCA in contrast, had antigen presenting activity equivalent to that of dendritic cells from normal mice. This effect was not reversed by the *in vitro* addition of indomethacin, indicating lack of prostaglandin involvement, and was not due to alteration of the constitutive expression of the IA antigen on the dendritic cells (Noonan et al., 1988).

Another prediction from our hypothesis in Figure 1, is that cis UCA will systemically suppress contact sensitivity with the formation of antigen-specific suppressor T cells. Our own data, originally presented at the 9th International Congress in 1984, indicated dose-dependent suppression of contact hypersensitivity if cis UCA was administered intraperitoneally over a dose range of 50 to 200 ug per mouse. No suppression was observed with administration of equivalent amounts of trans UCA. Ross et al., (1986; 1987), also found that the delayed type hypersensitivity response to Herpes Simplex Virus Type 1 in mice was suppressed by prior administration of UV-irradiated

766

UCA (containing both isomers), but not by trans UCA alone. They were able to demonstrate further that antigen-specific suppressor T cells are formed in animals given UV-irradiated UCA and Herpes Virus Type 1.

Thus evidence has been presented for 3 steps of our hypothesis. The action spectrum study indicated that a photoreceptor for UV-induced immunosuppression existed and was most likely trans UCA; formation of cis UCA in the skin occurs following immunosuppressive doses of UV irradiation; administration of cis UCA *in vivo* in the absence of UV initiated both an alteration to antigen presenting cell function, and the generation of antigen-specific suppressor T cells. Further investigations are necessary to establish the mechanism by which cis UCA initiates the antigen presenting cell defect, and how these alterations to antigen presenting cells result in the preferential stimulation of antigen-specific suppressor T cells.

One question which naturally arises is the relevance to the human situation. Dr. De Fabo has initiated investigations of the levels of cis and trans UCA in human stratum corneum by HPLC analysis of samples from more than 50 subjects. Preliminary findings are that UCA levels vary between sites on an individual, and, as expected, cis UCA levels are higher on sun-exposed areas. Investigations are proceeding on UCA levels on both normal and diseased subjects.

Another question which often arises in these studies is how relevant the doses of UV needed to cause immunosuppression are to real life exposure to UV in sunlight. To address this question, Dr. John Frederick and his student E. Kale Haywood at the University of Chicago, convoluted our action spectrum for immunosuppression in mice described above with their model for solar UV irradiances (Frederick and Lubin, 1988). This convolution indicated that, at a latitude of 40 degrees North in either January or July at noon (clear day), most immunosuppression was initiated by wavelengths from 315 to 320 nm. Thus, although these wavelengths are the least effective at immunosuppression on a per quantum basis, this relative ineffectiveness is outweighed by their much greater preponderance in sunlight. The biologic effective irradiance for immunosuppression at noon was calculated as a function of latitude. The peak biologic irradiance was 0.32 W/m^2 at 24 degrees North in July or at 24 degrees South in January. From this data we calculate that 30 minutes exposure at noon on a clear day in July at Rockville, Maryland would be sufficient to cause approximately 50% systemic suppression of CHS in a shaved mouse. Although it is not possible at this point to extrapolate to a biologically effective human dose, it should be reemphasized that UCA is present in human stratum corneum, and has been shown to isomerize on sunlight exposure. Since the biologically effective fluences in sunlight are high, we propose that UV-induced immunosuppression may be constantly switched on at a certain level as an intrinsic control against autoimmune rejection of sunlight-altered skin cells.

References

De Fabo, E.C., and Noonan, F.P. (1983). Mechanism of immune suppression by ultraviolet irradiation *in vivo*. I. Evidence for the existence of a unique photoreceptor in skin and its role in photoimmunology. *J. Exp. Med.* 158, 84–98.

De Fabo, E.C., Noonan, F.P. Fisher, M.S., Burns, J., and Kacser, H. (1983). Further evidence that the photoreceptor mediating UV-induced systemic immune suppression is urocanic acid. *J. Invest. Dermatol.* 80, 319.

Frederick, J.E., and Lubin D., The budget of biologically active ultraviolet radiation in the earth atmosphere system. *J. Geophys. Res.* 93, 3825–3832.

Greene, M.I., Sy, M.S., Kripke, M.L., Benacerraf, B. (1979). Impairment of antigen presenting function by ultraviolet radiation. *Proc. Natl. Acad. Sci. USA.* 76, 6592–6595.

Mottram, P.M., Mirisklavos, A., Clunie, G.J.A., and Noonan, F.P. (1988). A single dose of UV radiation suppresses delayed type hypersensitivity responses to alloantigens and prolongs heart allograft survival in mice. *Immunol. Cell Biol.* 66, 377–385.

Noonan, F.P., De Fabo, E.C., and Kripke, M.L. (1981a). Suppression of contact hypersensitivity in mice by UV radiation and its relationship to UV-induced suppression of tumor immunity. *Photochem. Photobiol.* 34, 683–690.

Noonan, F.P., De Fabo, E.C., and Morrison, H. (1988). Cis-urocanic acid, a product formed by ultraviolet B irradiation of the skin initiates an antigen-presentation defect in splenic dendritic cells *in vivo. J. Invest. Dermatol.* 90, 92–99.

Noonan, F.P., Kripke, M.L., Petersen, G.M., and Greene, M.I. (1981b). Suppression of contact hypersensitivity in mice by ultraviolet radiation is associated with defective antigen presentation. *Immunology,* 43, 524–533.

Ross, J., Howie, S.E.M., Norval, M., Maingay, J., and Simpson, T. (1986). UV irradiation of urocanic acid suppresses the delayed type hypersensitivity response to H. simplex virus in mice. *J. Invest. Dermatol.* 87, 630–633.

Ross, J., Howie, S.E.M., Norval, M., Maingay, J., and Simpson, T. (1987). Two phenotypically distinct T cells are involved in ultraviolet-irradiated urocanic acid induced suppression of the efferent delayed type hypersensitivity response to Herpes Simplex Type I *in vivo. J. Invest. Dermatol.* 89, 230–233.

Effects of Radiant Energy on the Ocular Tissues

SEYMOUR ZIGMAN

Ophthalmology Research Laboratory (Box 314)
University of Rochester School of Medicine and Dentistry
601 Elmwood Avenue
Rochester, New York 14642

Introduction

Studies of the effects of radiant energy on the eye are more representative of humans when carried out using diurnal animals than nocturnal albino rodents. Our choice of an animal model is the gray squirrel (*Sciurus carolinensis*) whose eyes are large, and whose lenses contain a yellow vision enhancing filter that blocks light from 450 nm and below from reaching the vitreous humor and retina. The environment contains moderate amounts of UV energy in the UV-B (280–320nm) and UV-A (320–400 nm) ranges. As it is known that photochemical damage in the eye results from excessive ocular exposure to near UV radiation, a discussion of the types of damage that occur in two major tissues, the lens and retina, is herein provided.

Several recent publications have adequately covered the field of UV damage to the ocular tissues up to 1986 (DayHaw-Barker and Barker, 1986; Miller, 1987; Urbach and Gange, 1986; Waxler and Hutchings, 1986). This paper provides information on several very recent studies that have not as yet been covered elsewhere. It refers only to near UV (300-400 nm) radiation- induced abberations in the lens and retina of the eye of the gray squirrel, our diurnal animal model, and also considers recent epidemologic findings that are related to basic studies of ocular light damage .

Lens

Two experimental designs were used. In one procedure, freshly removed lenses of normal adult squirrels were dissected out of the eyes and were then incubated in physiological medium (TC 199 at 35°). Half of the lenses were exposed to the emission of a Woods lamp (maximum emission at 360 nm; 5mW/cm^2 intensity) up to 16 hrs, while the other half were kept as dark controls. Major changes in the lens structural proteins and the enzyme Na/K ATPase were observed.

Figure 1 shows that the proteins in the anterior lens epithelium, which is exposed to the highest irradiance of light undergo both aggregation and then degradation (PAGE gels), and Figure 2 that the distribution of crystallins in the different regions of the lens

Photobiology, Edited by E. Riklis
Plenum Press, New York, 1991

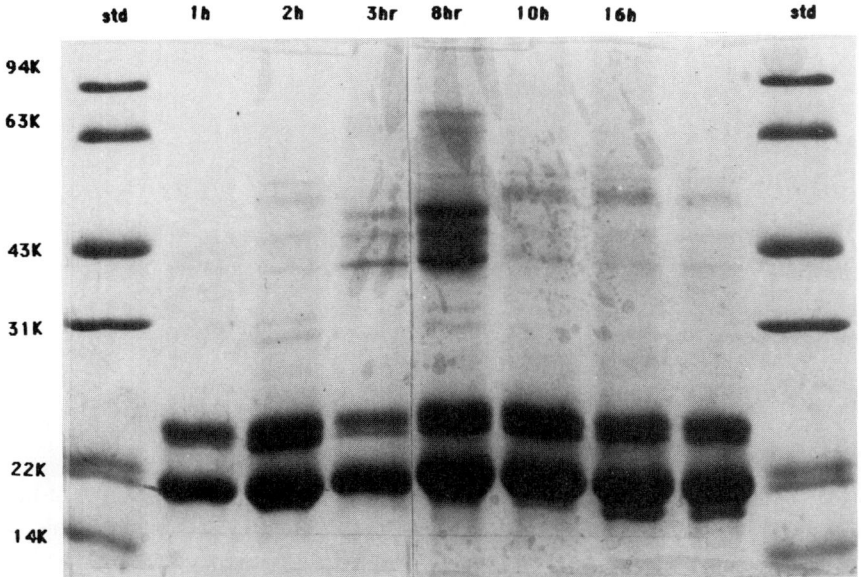

Figure 1. Polyacrylamide gel electrophoresis of the water-soluble crystallins of squirrel lens epithelium. This demonstrates the effect of *in vitro* near-UV radiation exposure. Note the higher molecular weight aggregates that increase by 6 hr of UV-exposure and their diminution by 16 hours. At 16 hrs a peptide band with a molecular weight of 16,000 daltons also appears (see Zigman et al. 1985).

Table 1. Relative rates of photopolymerization of lens protein

Sensitizer	Rates[*]
None	1
3 hydroxy kynurenine glucoside	1.5
Chorpromazine	24
Tretracycline	39
Fluorescein	7,800
Rose Bengal	11,400
Hematoporphyrin	14,100

[*]Relative to lens protein without added sensitizer. Roberts and Dillon (1984–85).

are altered by this exposure (HPLC-data). Table 1 illustrates the positive influence of various photochemical sensitizers on lens protein aggregation (Roberts and Dillon, 1984–85).

770

Another significant and important finding was that the activity of Na/K-ATPase of the lens epithelium was diminished by such exposure (Torriglia and Zigman, 1988). This loss of activity is expressed physically in the lens by a slight swelling of about 1.5% by weight (Figure 3) and by an anterior subcapsular opacity (Figure 4). This suggests that one mechanism of opacification of the lens is the inhibition of the sodium pump which leads to an osmotic type of cataract. The influence of sensitizers such as riboflavin and tetracycline, as shown in Table 2 were to enhance the degree of opacity relative to the loss of Na/K-ATPase activity of the lens.

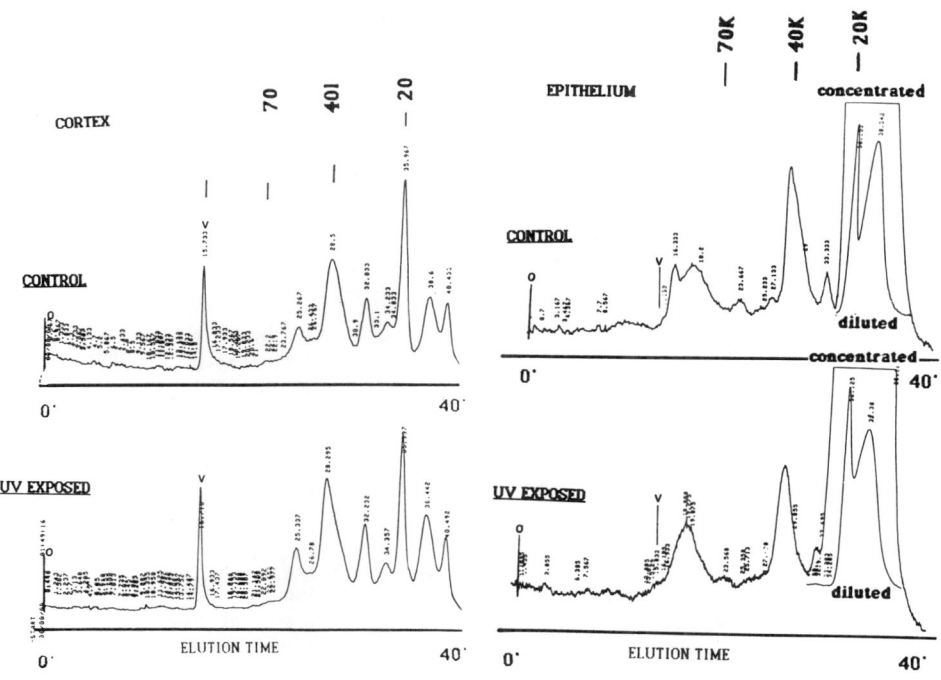

Figure 2. High pressure liquid chromatography of the soluble crystallins of the epithelium and cortex of the gray squirrel lens. Separation was on TSK 3000 gel filtration columns (conditions in 14). The effects of near-UV exposure (*in vitro* at 365 nm) on the soluble protein size distribution are shown: in epithelium, a loss in voided protein (i.e. alpha crystallin), a decrease in beta and gamma crystallins, and a marked increase in peptides with molecular weight less than 20,000 daltons; in cortex, an increase in voided protein and in beta crystallins, an increase in < 20,000 dalton peptides, and a decrease in gamma crystallin.

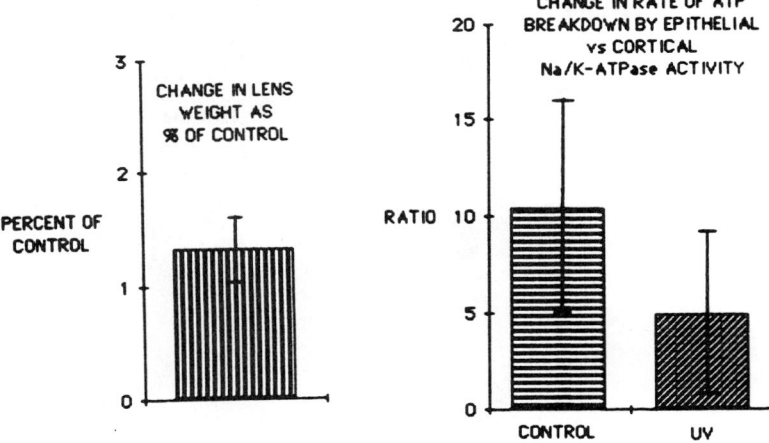

Figure 3. Effect of 16 hrs of near-UV (*in vitro* at 365 nm) on lens water content and NaKATPase activity in the epithelial cells of gray squirrel lenses. Results are shown as percent of control. As the enzyme activity in the cortex does not change, the enzyme activity is expressed as a ratio of the epithelial to the cortical activity. UV-exposure resulted in marked NaKATPase inhibition in the epithelium and a 1.5% water uptake.

Table 2. Influence of chemical agents of lens opacities due to near UV radiation of squirrel lenses

Protocol	Exposure	Result
Medium TC199	20 hrs	Dark -no effect plus UV-anterior sub-epithelial opacities
— without tryptophan	20 hrs	Same as with tryptophan
— plus 0.1 mM GSH	20 hrs	Additive diminishes UV-induced opacity slightly
— plus 1.0 mM ascorbate	20 hrs	additive diminishes UV-induced opacity slightly
— plus 0.1 mM	20 hrs	no influence
— plus 0.1 mM tetracyline	7.5 hrs	*dark*-slight posterior sub-capsular opacity
(turns, pink, brown)		*UV*-dense posterior sub-capsular opacity
— plus 0.1 mM tetracyline turns pink, brown)	20 hrs	*dark*-slight posterior opacity *UV*-dense posterior and anterior opacities
— plus 0.1 mM 3–OH KYN	20 hrs	no influence

In the other procedure, live gray squirrels were maintained for up to 2 years in mesh cages that were placed under banks of 40W BLB fluorescent lamps that provided 3 mW/cm^2 at 365 nm to the area of the animals' greatest activity. Lenses were removed from the eyes of sacrificed control and irradiated animals and aliquots of lens tissue approximately 4 mm^2 were taken from the anterior cortex, the nuclear core, and the posterior cortex. They were analyzed by HPLC using TSK 3000 size exclusion columns, after homogenization and 28,000 xg centrifugation to remove the insoluble fractions.

The findings regarding the lens proteins are shown in Figure 5 A and B. Most of the changes observed were an increase in the anterior and posterior cortex of void volume soluble proteins (> 200 kd mol. wt). and of low molecular weight peptides (< 16 kd), There was also a decrease in the nucleus of the void volume soluble proteins and an increase of nuclear low molecular weight peptides. In the nucleus, the insoluble aggregated protein level did not change, whereas it was increased in the anterior and posterior cortex (see Figure 6). Evidence is thus presented that the UV enhanced crosslinking of lens crystallins results in an increase of higher molecular weight aggregates that scatter light. Also shown is that proteolysis (enzymatic or other) increases so as to produce greater levels of low molecular weight peptides.

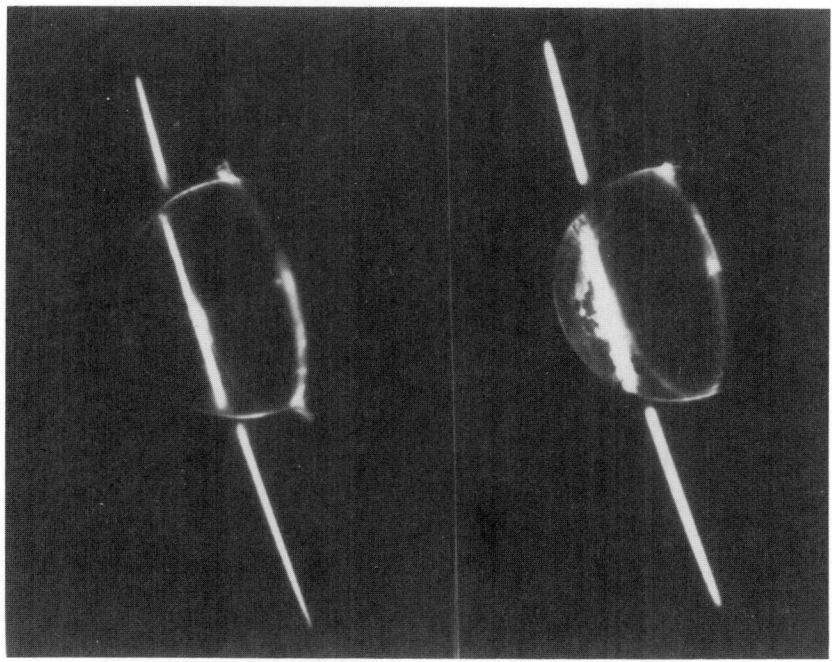

Figure 4. Development of an anterior subcapsular squirrel lens opacity due to exposure for 16 hrs to near-UV radiant energy (*in vitro* at 365 nm)

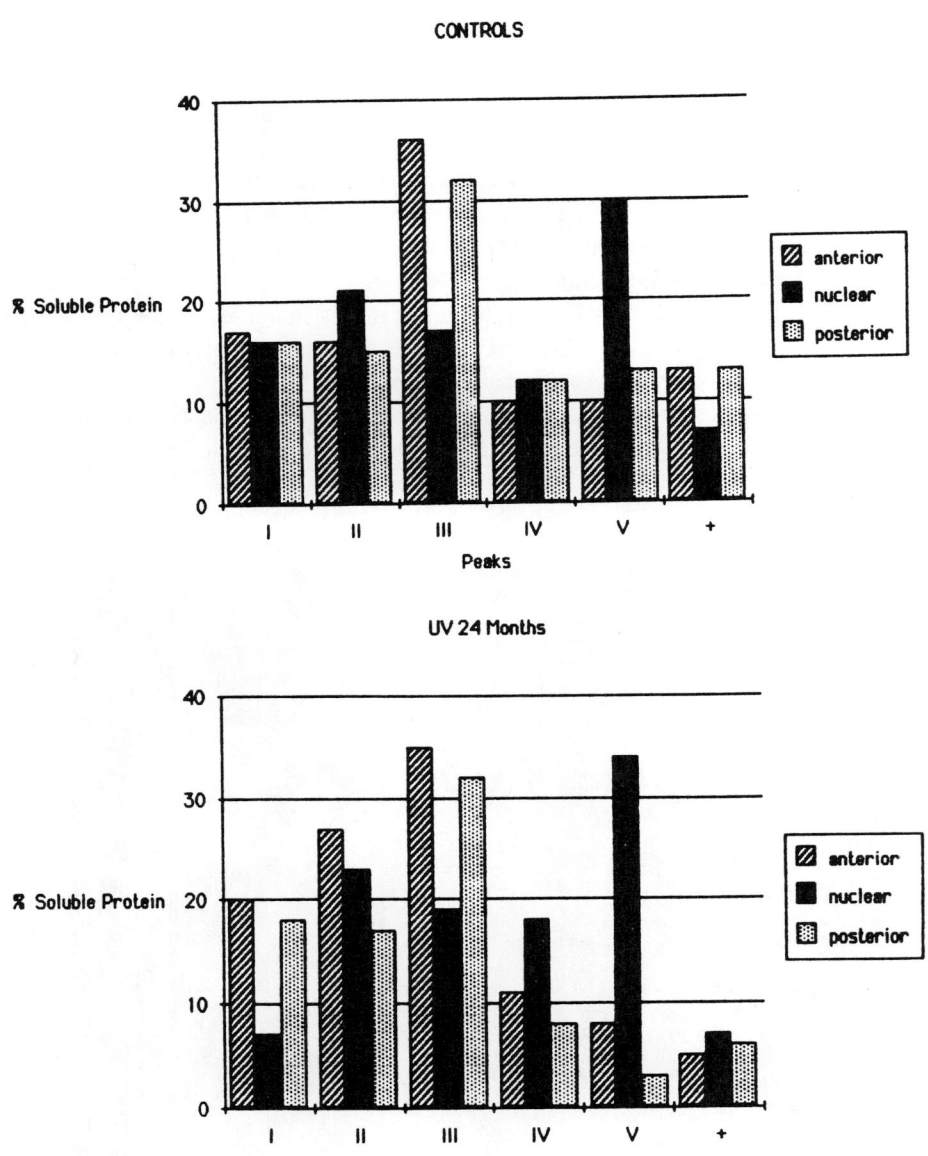

Figure 5. HPLC as in Figure 2 that illustrates changes in the crystallin levels in the anterior and posterior cortex and in the nucleus of lenses of gray squirrels due to 24 months of 12 hr/day exposure to 3 mW/cm^2 of 365 nm radiation from BLB lamps. A decrease in voided crystallins and increases in the low molecular weight crystallins and < 20,000 molecular weight peptides in the anterior and posterior cortices were observed.

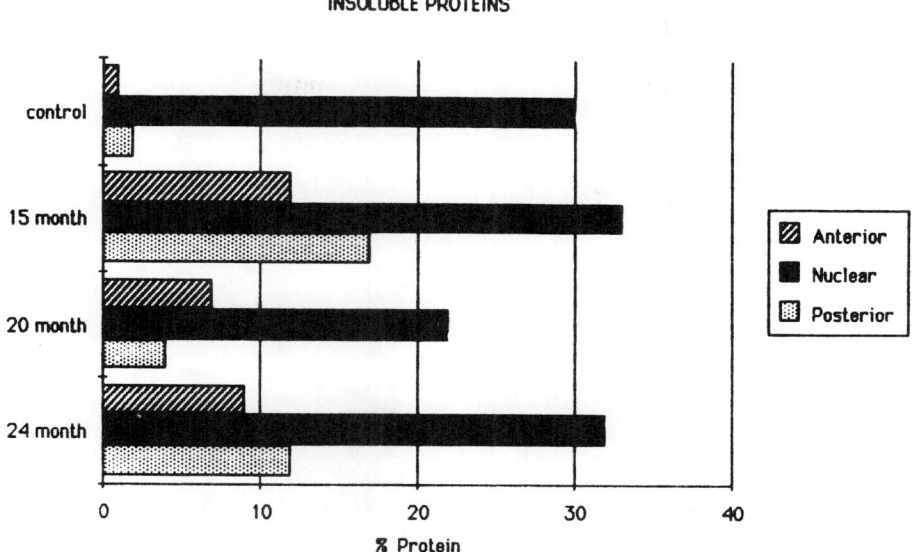

Figure 6. The enhancement with time of anterior and posterior cortical insoluble aggregated proteins in the lenses of gray squirrels exposed to ambient near-UV radiation as in the legend of Figure 5. Nuclear insoluble proteins appeared to increase much less than those in the cortex.

An epidemiology study of recent note was provided by Taylor, et al. (1988). This indicated that cortical (outer lens) cataract in a population of highly sunlight-exposed humans was related to their exposure dose of UV-B from the environment. The specificity that UV-B was the damaging wavelength range may or may not have been proven. As these recent basic studies on the squirrel lens also show more superficial damage than deep lenticular damage, it may be possible to link the results of our basic studies to the epidemiology. These suggest that the external portion of the lens, which absorbs most of the near-UV energy and in this way leads to oxidative changes which causes protein alterations that enhance cataract formation.

Retina

With regard to the damaging effects of near UV radiation on the retina, the findings of Collier and Zigman (1987, 1989) are the most recent. Both ambient unfocused and monochromatic focused radiation at 365–366 nm were used to observe the effects of short wavelength light on the squirrel retina. Gross damage could only be produced in these experiments when the UV-absorbing ocular lens was first removed surgically, as it is done for human cataract surgery. Exposures were only done after healing of the wound due to the surgery. Assessment of damage was made by histopathological studies and electroretinography.

775

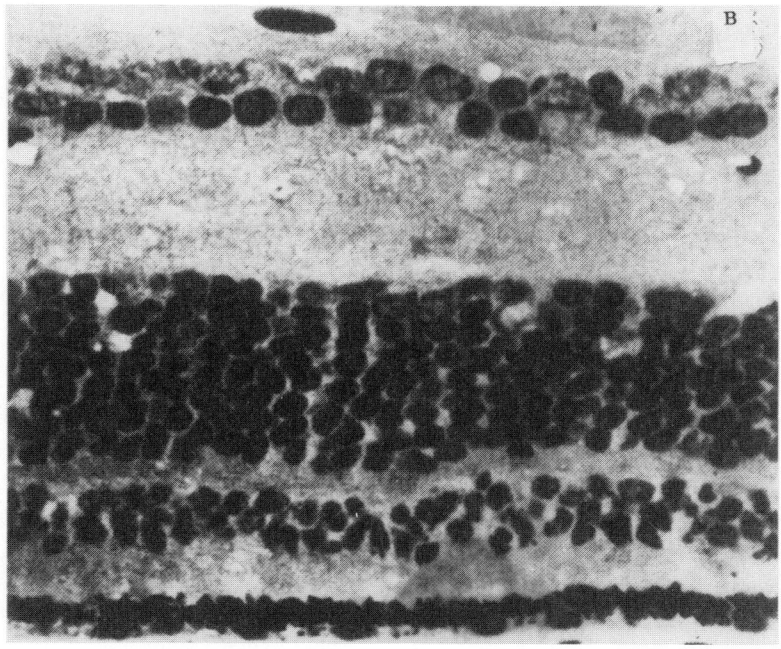

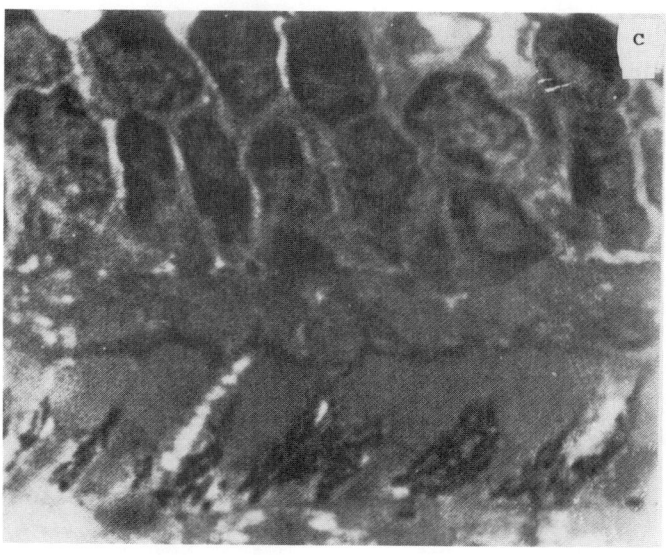

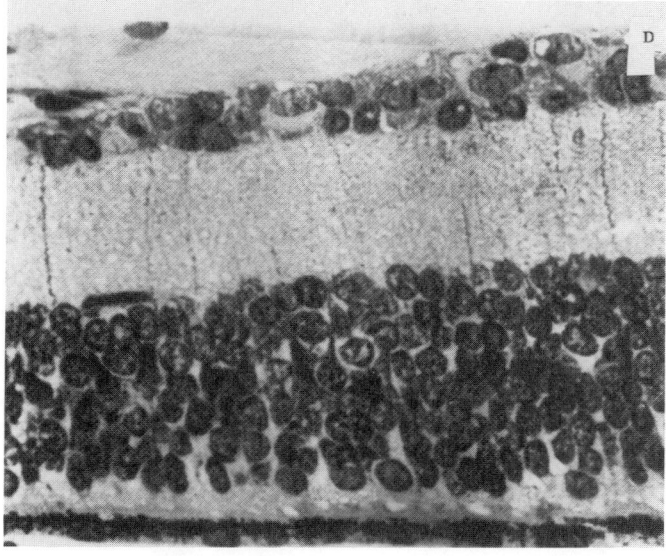

Figure 7. Effect on a gray squirrel retina target of a single 10 min. UV-exposure of 4.32 J/cm^2 (at the cornea) at 366 nm. A: control retina in eye with lens intact ; note anterior rods and posterior cones, as empahsized in the inset. B and C: aphakic retina at 24 hrs post-exposure. Photoreceptor nuclei are pyknotic, inner segments are swollen, pigment granules in apical processes of the pigment epithelium are more condensed. D: At 28 days post-exposure, the outer nuclear layer and much of the photoreceptor layer is gone. Other retina layers seem unaffected. (Data from Collier).

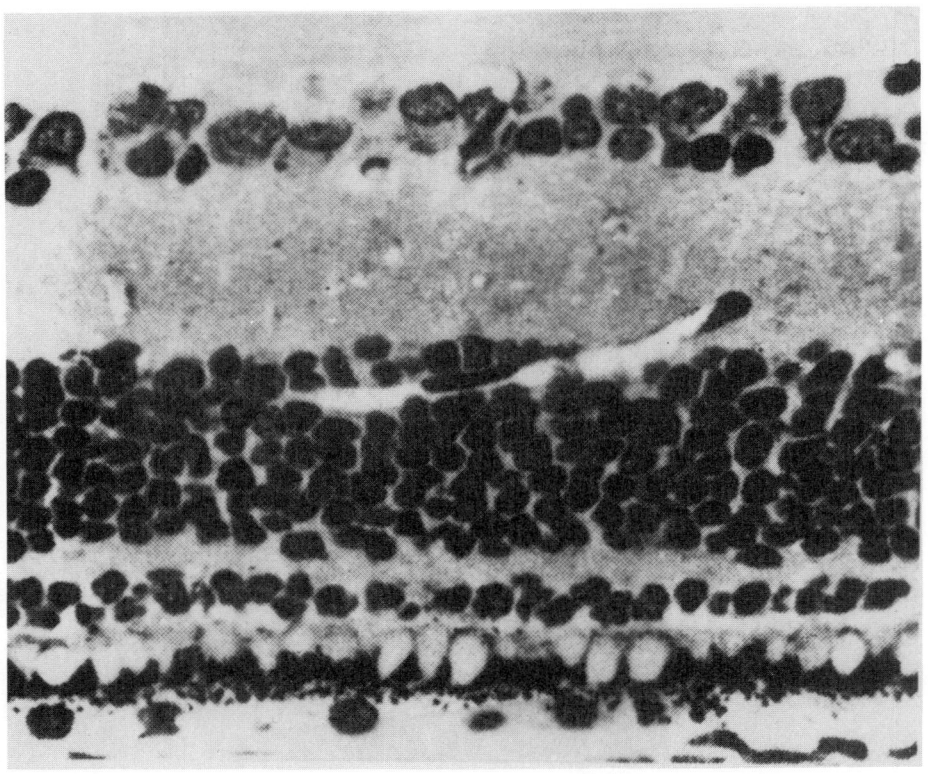

Figure 8. The effect of ambient broad-band near-UV radiation (BLB bulbs) at an irradiance at (365 nm) 2 mW/cm^2 for 3 months of 12 hrs day exposure. Use Figure 7, Frame A for control. Photoreceptor inner segments are swollen and many photoreceptors have been lost.

The damage to the retina due to near-UV exposure, as shown clearly in histological studies of short term monochromatic exposure (at 366 nm) is presented in Figure 7, and by long-term ambient exposures (BLB, 40 watt lamps, λ maximum at 365 nm) exposures, as in Figure 8. This Figure presents micro-photographs of control and of ambient UV-exposed retinas of gray squirrels above threshold irradiance levels. These photographs show early damage to rod photoreceptor inner segments (swelling, degranulation), with longer time intervals, total destruction of both rod and cone photoreceptors is induced. To summarize their effects, there was a thinning of the photoreceptors, loss of photoreceptor nuclei, decreased numbers of photoreceptors, and shorter outer segment lengths.

The use of electroretinography allowed the observation of near UV radiation damage prior to its appearance histologically. Figure 9 shows the effects of *in vivo* near UV ambient exposure in the aphakic and phakic eye. Both a and b waves have lesser amplitude and longer latency in the aphakic eye. Figure 10 summarizes the data from a

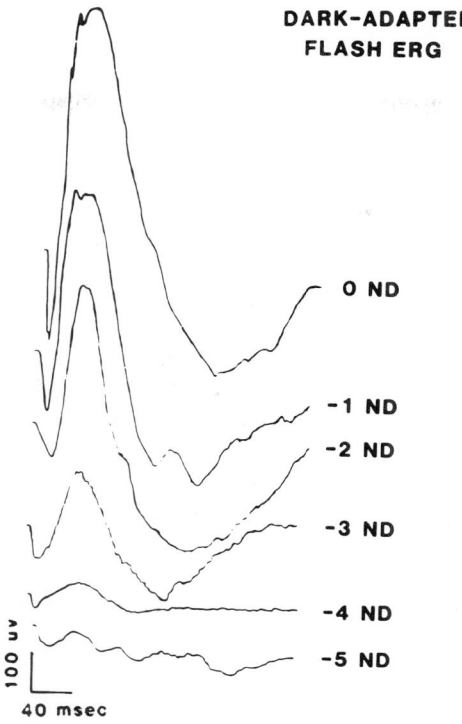

Figure 9. The electroretinogram of an aphakic eye of the gray squirrel as determined by Collier and Zigman (8) and as influenced by the addition of neutral density filters. There are clear and powerful A and B waves demonstrated.

series of such measurements of the b-wave amplitude. In the aphakic eye the b-wave amplitudes are all less than those of the phakic eyes. Figure 11 illustrates further the effects of UV *in vivo* exposure of squirrels with intact lenses. This data shows that even with strongly absorbing pigmented lenses intact, the retinas are damaged by ambient near UV exposure for 5 months.

Table 3 provides a comparision chart of the thresholds for the two types of retinal damage in aphakic gray squirrel eyes and in monkey eyes (Ham et al., 1982) using UV-A energy. As the monkey eye damage was observed was by ophthalmoscopic examination, it would be expected to require greater irradiance for damage to become apparent than when observed by histology examination.

At 366 nm, the thresholds were between 4 and 6 J/cm^2 for both animals.

In a preliminary TEM study of the mitochondria-rich myoid areas of the inner segments of the photoreceptors of gray squirrels, it appeared that there was definite damage to the mitochondria of the UV exposed retinas (see Figure 12). Inhibition of cytochrome oxidase in the retina, has been suggested to result from mitochondrial

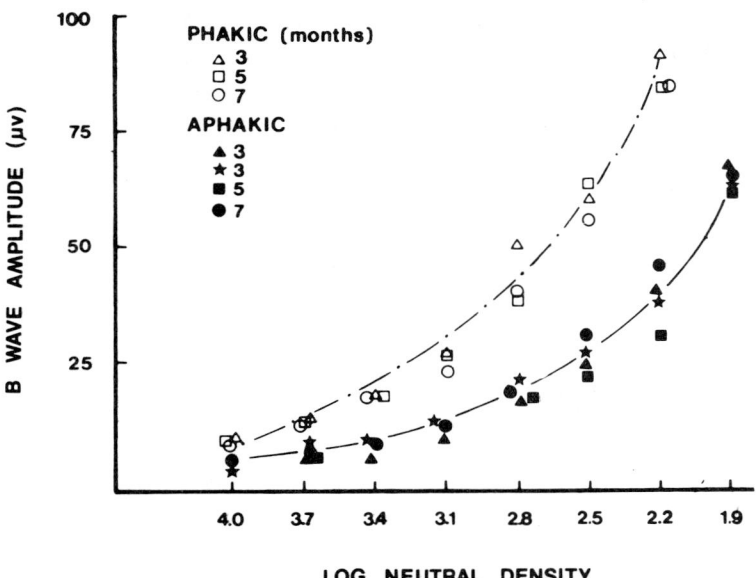

Figure 10. The B-wave amplitude as a measure of retinal function in the phakic and aphakic eyes of gray squirrels that were housed in cages illuminated with BLB lamps at 2 mW/² for 12 hours/day for increasing times. The curves represent an average of 3 determinations. It is obvious that the eye without ocular lens present is deficient in retinal response to a visible light flash compared to compared to the eye with intact lens. (Data from Collier)

Table 3. Aphakic retinal damage

Wavelength	100 sec.		Exposure 1000 sec.	
nm	J/cm^2	W/cm^2	J/mc^2	W/cm^2
325	5.0	0.05	5.1	0.0051
350	5.4	0.54	5.5	0.0055
380	8.1	0.81	9.3	0.0093

*Opthalmoscopic observation — Retinal irradiance (Xenon lamp) (Ham et al., 1982)

600 sec	
366	4.32 J/cm^2 and 0.0072 W/cm^2 (damage)
366	3.1 J/cm^2 and 0.0050 W/cm^2 (no damage)

+ERG and Histological — Corneal irradiance (Monochromator)

BLB (40 W) — Ambient 56 J/cm^2 per day (12 hrs) ERG and Histological damage at 6 wks.
(Collier and Zigman, 1987)

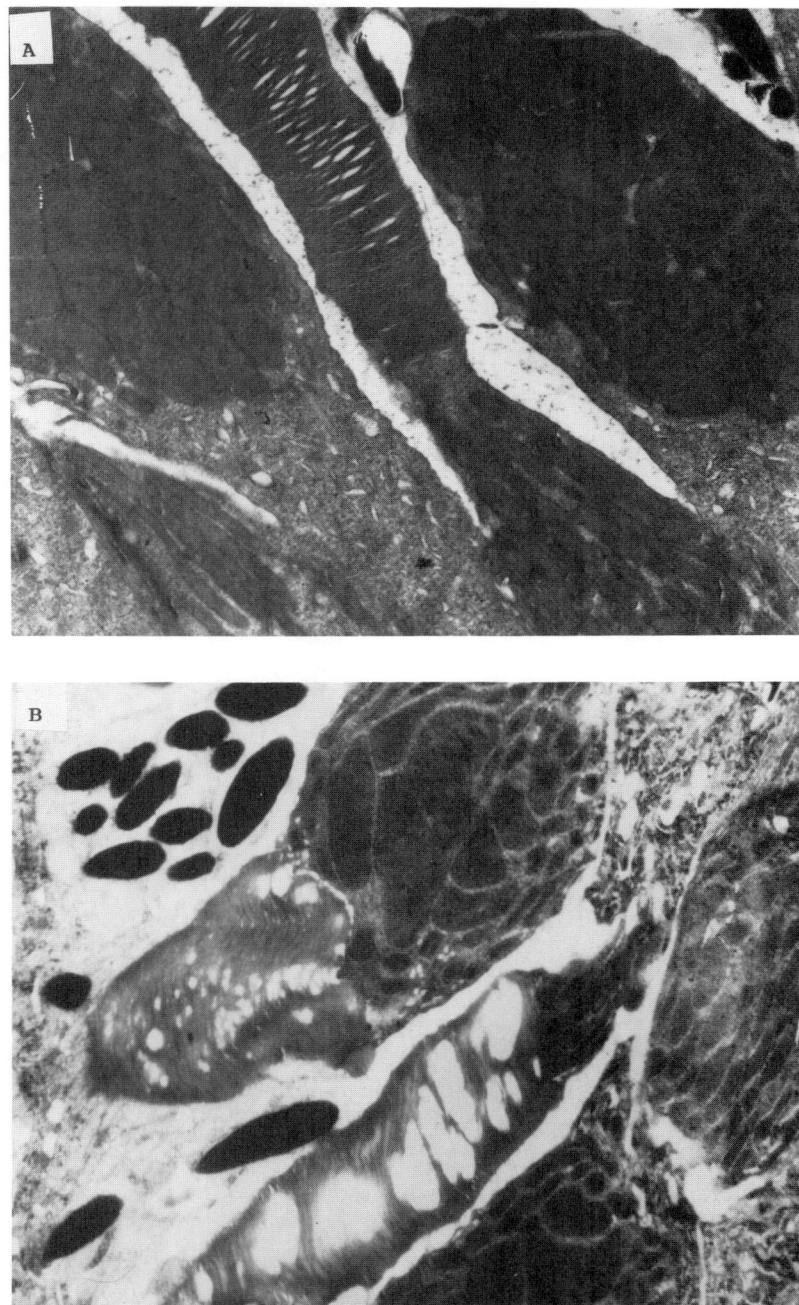

Figure 11. Electroretinograms of intact squirrel eyes as altered by ambient exposure to BLB lamps as above. Note: the depressed B-wave amplitude in the retinas of UV-exposed animals, even though the lenses were present. (Data from Collier).

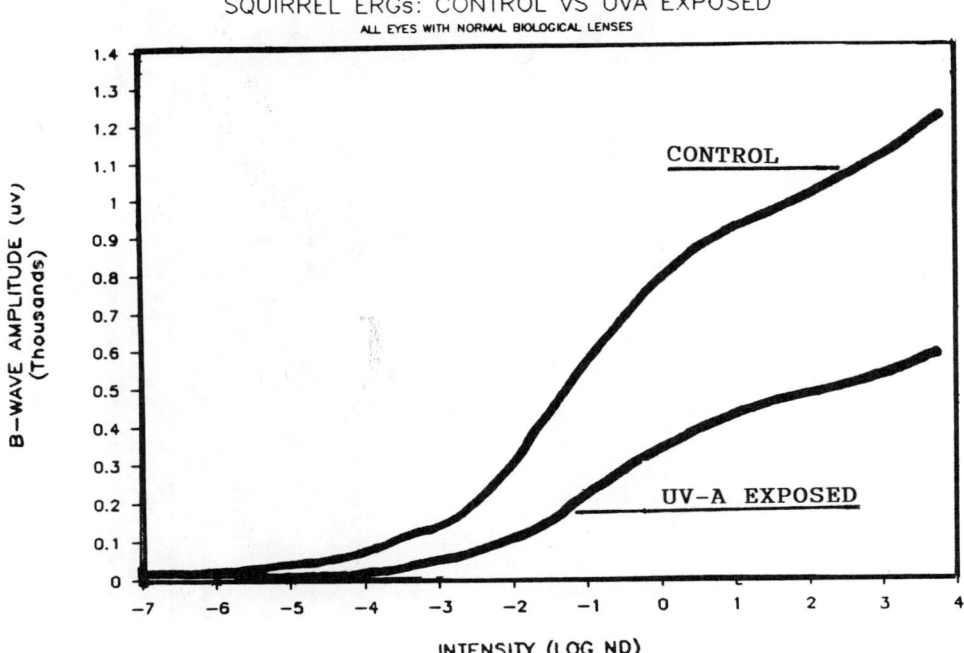

Figure 12. Transmission electron microscopy of the photoreceptors of the gray squirrel retinas in phakic (A) and aphakic (B) eyes after exposure of the animals to ambient BLB lamp radiation for weeks. While some artificats are obvious, there are distinct defects in the mitochondria of the inner segments or myoid bodies, and degenerated disk membranes in the outer segments only in the aphakic eye retina (Data from Merrill).

damage. This was confirmed experimentally. Such a mechanism would lead to photoreceptor damage in the retina. Swelling and lower metabolic energy of photoreceptors have been documented in the photoreceptors damaged by light (Sperling et al., 1980).

Just as near UV energy inhibits the activity of ATPases in the lens epithelum of the squirrel, so does it inhibit the ATPoses if the whole retina. Figure 13 shows that total ATPases are inhibited *in vitro* by exposure to near UV light (5 mW/cm^2 at 365 nm). In other experiments, the activity of retinal cytochrome oxidase (in the whole bovine retina) was inhibited by 20% to 80% by a four hour exposure *in vitro* to near UV at 3 mw/cm^2 (365 nm) without additional tryptophan in the medium. (see Table 4). The loss of photoreceptor structure and function thus may stem from the inhibition of enzymes that function by controlling salt balance and producing energy.

782

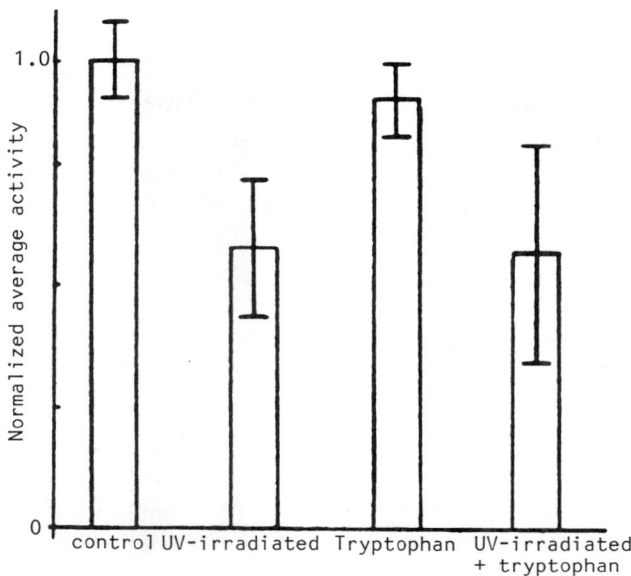

Figure 13. Inhibition of total ATPase activity in bovine retina (*in vitro*) by 4 hrs of near-UV radiation as in Figure 12. Values are expressed as percent of control. The inhibition in approximately 60%.

Table 4. Influence of phtoproducts of TRY on ocular tissue

Tissue (bovine)	Cytochrome Oxidase Conditions	Activity Units[*]
	Dark controls	4.00
Whole retina mitochondria	N-UV[+]	3.40
(After four hours of	N-UV TRY + dark	2.50
incubation of		
the whole retina)	N-UV + N-UV light	1.75

[+]Oxidation of reduced cytochrome C.
[+]N-UV light at 3 mW/cm^2 at 365 nm.

Damage of the retinas of normal phakic humans due to sunlight exposure has not been documented statistically as yet, but there is data that proves that there is a positive relationship between aphakia in humans and retinal degeneration (Cameron et al., 1983). One factor that could contribute to aphakic retinal damage is sunlight (UV).

Conclusions

This communication adds several new findings on near-UV enhanced ocular tissue damage to an already lengthy list. It shows that exposure to predominantly UV-A radiation at subsolar irradiances, both *in vivo* and *in vitro*, leads to outer and anterior lens protein changes that include aggregation and protein breakdown. These changes occur predominatly in the epithelial cells, which are exposed to radiant energy at its highest irradiance in thel ens due to their anterior position and monolayer geometry. The larger and smaller crystallins are all effected, but which are most susceptible to damage was not elucidated.

Another newly discovered type of damage to the anterior of the lens is the inhibition of the Na/K-ATPase enzyme for salt and water balance. This brings the osmotic type of cataract into the area of UV -light-induced lesions, and supports the findings that superficial cataract in humans may be related to exposure to sunlight. A common theme to be noted is that repetitive long-term exposure to sunlight or near UV-light is the condition most likely to enhance cataract formation. In the retina, the photoreceptor inner segments appear to be the first target for UV-A damage, and this defect occurs first in the more anteriorly placed rods. Only secondarily is the pigment epithelium damaged. Also it seems essential that the natural ocular lens, is the principal absorbing filter of near-UV light anterior to the retina in diurnal animals and man. Inhibition both of retinal ATPases and cytochrome oxidase, as shown *in vitro,* may explain the more basic findings that inner segment mitochondria are damaged by near-UV exposure.

Acknowledgement

Support: NIH (Ey00459) and Research to Prevent Blindness, Inc.

References

Cameron, L.L., Auer, C.L., McCormick, P.A. , Owen, S.T., Fine, S.T., and Taylor, H.R. (1983). Association of sunlight with senile macular and lens changes. *Invest. Ophthal. Vis. Sci.* 24: 202 *(abstract)* .

Collier, R.J. and Zigman, S. (1987). The Gray Squirrel Lens protects the Retina from near-UV Radiation Damage. In Degenerative Retinal Disorders. (ed. J. Holleyfield). Alan Liss, Inc. New York, NY. pp. 571-585.

Collier, R.J., Waldron, W.R., and Zigman, S. (¡989). Temporal sequence changes to the gray squirrel retina after near-UV exposure. *Invest. Ophthalmol. Vis Sci.* 30: 631-637.

DayHaw-Barker, P. and Barker F.M. (1986). Photoeffects on the eye in Photobiology of the Skin and Eye (ed. E.M. Jackson) Marcel Dekker, Inc., N pp 117-147.

Ham, W.T., Jr., Mueller, H.A., Ruffolo, J.J., Jr., Guerry, D., and Guerry, R.K. (1982). Action spectrum for retinal injury from near-ultraviolet radiation in the aphakic monkey. *Amer. J. Ophthalmol.* 93: 299-308.

Miller, D. (ed). (1987) Clinical Light Damage to the Eye. Springer-Verlage, Inc. New York, NY.

Robert, J.E., and Dillon, J. (1984-85). A comparison of the photodynamic effects of photosensitizing drugs on lens proteins. *Lens research* 2: 733-144.

Sperling, H.G., Johnson, C. and Harwerth, R.S. (1980). Differential Spectral Photo Damage to Primate Cones. *Vision Res.* 20: 1117-1125.

Taylor, H.R., West, S.K., Rosenthall, F. Munoz, Newland, H.F., Abbey, H. and Emmtt (1988). Effect of UV -radiation on cataract formation. *New England J. Med.* 319: 1429-1433.

Torriglia, A. and Zigman , S. (1988). The effects of near-UV Light on NaKATPase of the rat lens. *Current Eye Res.* 7: 539-548.

Urbach, F. and Gange, R.W. (ed) (1986). The Biological Effects of UV-A Radiation Praeger Publ. Co., New York, NY.

Waxler, M. and Hutchings, V.M. (ed) (1986). Optical Radiation and Visual Health CRC Press, Boca Raton, Fla.

Zigman, S., Paxhia, T., and Waldron, W. (1985). Biochemical features of squirrel Lens. *Invest. Ophthalmol. Vis Sci.* 26: 1075-1082.

Zigman, S., Paxhia, T., and Waldron, W. (1988). Effects of near-UV radiation on the gray squirrel lens. *Current Eye Res.* 7: 539 -548.

On the Effect of Ultraviolet Radiation on the Eye

O. HOCKWIN, J. SCHMIDT, AND C. SCHMITT

Medizinische Einrichtungen der Universität Bonn
West Germany

Electromagnetic radiation is involved in the generation of several different diseases especially those of the skin. Even the eye is subjected to the influence of electromagnetic radiation, which first reaches the cornea and the conjunctiva. These structures being superficial eye tissues, are naturally the first to be subjected to actinic influences. Additionally all other eye structures can be reached and possibly damaged by ultraviolet radiation being dependent on wavelength and intensity. In this respect, it is of crucial importance whether the damage in question is due to acute radiation or is to be considered a consequence of chronic exposure.

With respect to the acute UV-damage, there are naturally various experimental and clinical results available, while in the case of long-term exposure it is often very difficult to determine a certain correlation between the incidence and the morphological consequences.

As a first example of chronic UV-radiation damage to the eye the pterigium should be mentioned. Since the incidence of such a pterigium is typical for countries in the vicinity of the equator, a close relationship between the influence of ultraviolet light and the formation of the eye disease in question has for a considerable time been postulated. Another example of chronic UV-effect is the so-called spheroidal cornea degeneration or Labrador-keratopathia. Both are irreversible, degenerative diseases which do not react to conservative therapy, and are therefore in general subjected to surgery. Concerning the lens, there is always the so-called "brownish nuclear cataract" supposed to be associated with long-term ultraviolet radiation (Lerman et al., 1976; Zigman et al., 1979; Dilley and Pirie, 1974). Even the retina may be affected by ultraviolet light. A relationship between some kinds of retinal diseases and chronic exposure to UV-light is a controversially discussed topic in clinical ophthalmology. The so-called senile macular degenerations, in particular, are supposed to be at least partially the result of long-term UV-exposure. As opposed to the afore-mentioned effects the keratoconjunctivitis photoelectrica is due to an acute, highly-dosed UV-irradiation. All diseases mentioned so far are induced by those wavelengths of UV-light which are also absorbed by the different eye media. With regard to the cornea, this is mostly the range below 290 nm (Kinsey, 1948; Bachem, 1956; Sliney and Wolbarsht, 1980; Jose, 1982). Measurements of the cornea transmission verified that the cornea nearly completely absorbs all wavelengths below 290 nm. Therefore all intra-ocular structures of the eye are screened from the short-wave UV-portions.

The lens absorbs relatively large portions of those wavelengths which may just be transmitted by the cornea, and this absorption capacity of the lens increases with age (Bottner and Wolter, 1962; Lerman and Borkman, 1976; Laser et al., 1988). Main factors in the absorption process of UV-radiation of the lens are the aromatic amino-acids and particularly the tryptophane. In relation to the energy content of the radiation, the molecular structure of the absorbing molecules is altered. The mechanisms of energy transmission involved in these processes lead to the induction of various photochemical reactions. In particular, the catabolism of the tryptophane has grave consequences for the lens metabolism. The formation of a rather special type of lens opacity, the so-called cataracta brunescens, is particularly ascribed to the catabolic products of tryptophane (Pirie 1972, Zigman et al 1974, Lerman and Borkman 1976, Lerman 1976, Borkman et al 1977). Further, the absorbed ultraviolet radiation not only exerts a noxious effect on the lens via the tryptophane catabolism, but also directly via an alteration of the DNA (Bellows 1972, Jose and Yielding 1977, Grabner and Brenner 1982). Among the well-known phenomena are the dimerisation of pyrimidines, the hydration of certain bases, the formation of mixed aggregates of proteins and DNA, as well as disruptions of hydrogen bonds. Some characteristic changes of lens morphology following UV-B-irradiation have been mentioned.

Following UV-irradiation, the first changes in lens structure are found exclusively at the anterior pole in the form of rather discreet epithelial proliferations. In advanced stages, the outer cortex of the lens is affected as well as the epithelium. Here, occasional swelling of lens fibres is observed, as well as disruption and dehiscences in the area of the anterior lens suture. After a 6-week period of exposure to UV-irradiation with doses which do not affect the cornea, more serious changes in the epithelium were observed. Further, the disruption of lens fibres, as well as a dissolution of lens fibre substances, were seen, which as such, lead to a complete disorganization of the lens cortex. In spite of such remarkable changes in the anterior lens pole, the area of the lens equator in our experiments remained totally unimpaired during a 6-10-week observation period (Schmitt, et al., 1988).

From this, one may draw the conclusion that the influence of ultraviolet radiation on the DNA may not be that important, since in this case of direct DNA-impairment, the nuclear bow and the germinative zone, with their rather high mitotic activity, would have been the first to show pathological reactions. This indicates that UV-damage of lens proteins, finally leading to loss of transparency, might rather be the result of post-translational modifications than the result of a disturbed biosynthesis due to direct DNA-impairment. The histological changes just described all revealed alterations in rat lenses caused by exposure to UV-B-light. The main range of emission of the radiation sources used was about 300 nm. If, however, UV-A-radiation is used, only very discreet changes at the anterior lens pole were observed (Schmidt et al., 1986; Schmitt et al., 1986).

Nevertheless, it was evident that UV-A irradiation in conjunction with other noxious factors as, for instance, X-irradiation or the presence of diabetic metabolic condition, exerts an additive effect on the formation of the cataract in question. Naturally, such additive effects also occur with UV-B-irradiation in conjunction with other lens noxes. In particular, the double influence of UV-B-irradiation on the one hand and the presence of a diabetic metabolic condition on the other, leads to rather extensive changes in the lens structure (Schmidt et al., 1988). Recent investigations into this problem showed that changes in lens transparence, induced by ultraviolet radiation, may also be assessed quantitatively, by means of the Scheimpflug principle. By this procedure, it should be possible to evaluate slowly-developing UV-related changes. With respect to the reproducibility of the data obtained by Scheimpflug photography, investigations are at present in progress (Wegener, pers. comm.)

For a considerable time now in our laboratories we have also often used the UV-radiation cataract as an experimental model (Wegener and Hockwin, 1987). One of the main advantages of this cataract model is the fact that it allows a noxious mechanism to be used which is much more similar to environmental influences than any other noxious effects, such as X-rays or the presence of an untreated diabetes mellitus, and there is the further advantage in that the general state of health of laboratory animals is hardly reduced.

Besides the histological and biochemical results of animal experiments, reports of epidemiological investigations suggest a certain relationship between insolation and cataract development. A great number of investigations on this subject have been reported, but unfortunately many of them have been carried out using different and often inadequate systems for the classification and documentation of cataracts. In spite of this, the following results should be considered when discussing the effects of UV-radiation on the eye. For instance, in regions with a higher portion of ultraviolet light the rate of cataract incidence is higher than in regions with a lower intensity of ultraviolet light. As regards this relationship, a great number of results have been reported from totally different investigations, which were carried out in the United States, Australia, or in the Himalayas. The hypothesis of a cataractogenic effect of UV-light is further supported by the fact that in any given geographic district the dark brown nuclear cataract (cataracta brunescens) is much more frequently found in people who work mostly outdoors than in people who live mostly indoors (Hiller et al., 1977; Zigman et al., 1979; Taylor, 1980; Hollows and Moran, 1981; Brilliant et al., 1983).

There is no doubt, however, that in assessing these results we should not forget the fact that parameters such as the standard of living and nutritional habits, which are most closely related to social status and occupation, may also play an important role in the development of lens opacities. Taken as a whole, however, the results show that UV-radiation plays a rather important role in the multifactorial event of cataractogenesis.

Taking into consideration the results of the various experimental and epidemiological studies, it is evident that UV-radiation represents an often underestimated risk factor in the multifactorial process of cataractogenesis. This is particularly the case with technical

789

application, cosmetic irradiation, and also therapeutic irradiation with ultraviolet light, where noxious influences on the eye cannot be excluded, especially in the presence of photosensitizers like 8-Methoxypsoralen for psoriasis treatment (Lerman et al., 1977).

Summary

Biochemical results and histological findings in lenses of UV-treated animals, as well as epidemiological investigations, reveal an interdependence of ultraviolet light and the development of cataract. This paper includes some recent findings on the topic, thereby discussing the relationship between the wavelength and the corresponding histological changes and the underlying damaging mechanisms. Besides acute UV-effects on superficial eye structures even chronic changes of the eye lens should be reason for being especially careful of undesired and unhindered UV-irradiation.

References

Bachem, A., Ophthalmic Ultraviolet Action Spectrum, *A.J. Ophthalmol*. 41:969-975, 1956.

Bellows, J.G., Ultraviolet Light and the Crystalline Lens, *Ann. Ophthalmol*. 4:11-12, 1972.

Boettner, E.A., Wolter, J.R., Transmission of the Ocular Media, *Invest. Ophthalmol*. 1:776-783, 1962.

Borkman, R.F., Dalrymple, A., Lerman, S., UV Action Spectrum for Fluorogen Production in the Ocular Lens, *Photochem. Photobiol*. 26:129-132, 1977.

Brilliant, L.B., Grasset, N.C., Pokhrel, R.P., Kolstad, A., Lepkowski, J.M., Brilliant, G.E., Hawks, W.N., Parajasegaram, Associations among Cataract Prevalence, Sunlight House, and Altitude in the Himalayas, *Am. J. Epidem*. 118:250-264. 1983.

Dilley, K.J., Pirie, A. Changes to the Proteins of the Human Lens Nucleus in Cataract, *Exp. Eye Res*. 19:59-72, 1974.

Grabner, G. Brenner, W. Unscheduled DNA Repair in Human Lens Epithelium following *In-Vivo* and *In Vitro* Ultraviolet Irradiation, *Ophthalmic Res*. 14:160–166, 1982.

Hiller, R., Giacometti, L, Yuen, K. Sunlight and Cataract: An Epidemiologic Investigation, *Am. J. Epidem*. 105:450-459, 1977.

Hollows, F., Moran, D. Cataract - The UV Risk Factor, *Lancet* 2:1249-1250, 1981.

Jose, J.G. DNA Repair Synthesis Induced in Lens Epithelium Following Ultraviolet Irradiation, *Invest. Ophthalmol. Vic. Sci. Suppl*. 22:198, 1982.

Jose, J.G., Yielding, K.L. Unscheduled DNA Synthesis in Lens Epithelium Following Ultraviolet Irradiation, *Exp. Eye Res*. 24:113-119, 1977.

Kinsey, V.E. Spectral Transmission of the Eye to Ultraviolet Radiations, *Arch. Ophthalmol*. 39:508-513, 1948.

Laser, H., Hockwin, O., Schieck, A., Bialluch, A. Investigations of the Anterior Eye Segment by Scheimpflug Photography Using Visible or UV-Light with Volunteers of Different Age and with Patients with Various Type of Lens Opacifications, *Lens Research* 5:1-21, 1988.

Lerman, S. Lens Fluorescence in Aging and Cataract Formation, *Doc. Ophthalmol. Proc. Ser*. 8:241-260, 1976.

Lerman, S., Borkman, R.F. Spectroscopic Evaluation and Classification of the Normal, Aging, and Cataractous Lenses, *Ophthal. Res*. 8:335-353, 1976.

Lerman, S., Jocoy, M., Borkman, R.F. Photosensitization of the Lens by 8-Methoxypsoralen, *Invest. Ophthalmol. Vis. Sci*. 16:1065-1068, 1977.

Pirie, A. Fluorescence of N'-Formylkynurenine and of Proteins Exposed to Sunlight, *Biochem. J*. 128:1365-1367, 1972.

Schmidt, J., Wegener, A., Hockwin, O. Early Changes in Rat Lenses due to Syncataractogenic Effects of Diabetes and UV-Irradiation, *Proc. Int. Soc. Eye Research* IV:88, 1986.

Schmitt, C., Wegener, A., Hockwin, O. Implication of Biochemical, Photographic and Histological Methods for the Evaluation of Cataractogenesis Induced by X-and UV-A-Irradiation, *Proc. Int. Soc. Eye Research,* IV:89, 1986

Schmitt, C., Schmidt, J., Wegener, A., Ohrloff, S., Hockwin, O. Studie uber den Einfluss von UV-B-Strahlen auf die Entwicklung der Rontgenkatarakt, *Spektrum Augenheilk,* 1:303-309, 1987.

Sliney, D.H., Wolbarsht. Safety with Lasers and Other Optical Sources, Plenum Press, New York, 1980.

Taylor, H.R. The Environment and the Lens, *Br. J. Ophthalmol.* 64:303-310, 1980.

Wegener, A. Personal Communication (Data will be presented at the 4th Scheimpflug Club Meeting, Gamagori/Japan, April, 1989).

Wegener, A., Hockwin, O. Animal Models as a Tool to Detect the Subliminal Cocataractogenic Potential of Drugs, *Concepts. Toxicol.* 4:250-262 (Karger, Basel), 1987.

Zigman, S., Sun, M., Kalustian, A. Sensitized UV-Light Inhibition of Eye Tissue Enzymes, *Invest. Ophthalmol. Vis. Sci.* 18, Suppl. 115, 1979.

Zigman, S., Yulo, T., Schultz, J. Cataract Induction in Mice Exposed to Near UV-Light, *Ophthalmol. Res.* 6:259-270, 1974.

Photodynamic Therapy

Model System Studies on Photosensitization in Light Scattering Media

L.I. GROSSWEINER, M.J. SCHIFANO, J.L. KARAGIANNES, Z. ZHANG[*]
and Q.A. BLAN[**]

Biophysics Laboratory, Physics Department
Illinois Institute of Technology
Chicago, Illinois 60616, U.S.A., and
Wenske Laser Center
Ravenswood Hospital Medical Center
Chicago, Illinois, 60640, U.S.A.

Abstract

Biological photosensitization is being investigated in tissue models consisting of a sensitizing dye, a biological target, and light scattering particles. Results are reported for a dihematoporphyrin ether-subtilisin Carlsberg-polystyrene microsphere system in which the rate of photodynamic enzyme inactivation was measured at different wavelengths and scatterer concentrations. The optical constants were calculated with the one-dimensional diffusion approximation, from flux profiles measured with a fiber optic probe. Preliminary results arereported on lipid peroxidation in a dihematoporphyrin ether-liposome-polystyrene microsphere system. A two-channel integrating sphere spectrophotometer with on-line computer data acquisition was used to determine the optical constants of several model systems.

Introduction

Virtually all natural photosensitized processes take place in light scattering environments. The constituents of a biological photosensitizing system include the photoactive chromophore, a molecular target, and the tissue matrix. The matrix is an active component, and not merely the container of a an aqueous solution of biochemicals. Its role may include light harvesting, solubilization of reactive molecules, maintenance of spatial geometry, and product storage. This paper summarizes the results of an on-going study in which photochemical measurements on tissue models are analyzed with the methods of tissue optics. A system studied in detail consists of dihematoporphyrin ether (DHE) as the sensitizer, subtilisin Carlsberg as the target, and polystyrene

---

[*] Northwestern Telecommunication Engineering Institute, Xian, China
[**] St. Xaviar College, Chicago, Illinois, U.S.A.

Photobiology, Edited by E. Riklis
Plenum Press, New York, 1991

795

microspheres as light scattering particles (Schifano and Grossweiner 1988). Prior work with psoralens indicates that photodynamic inactivation of this enzyme is mediated by singlet oxygen (Blan and Grossweiner 1987). Measurements were made at several wavelengths to compare the energy efficiency of photosensitization by DHE with red, green, and blue light. The rate of enzyme inactivation should be proportional to the integrated energy absorption over the reaction cell. The one-dimensional diffusion approximation (1DA) was used to calculate the fractional absorption by DHE in the scattering medium. Following the approach of Grossweiner and Messina (1987), the optical constants were calculated from the flux density profiles of DHE-microsphere suspensions as measured with an inserted optical fiber technique. Preliminary results are reported on another model system consisting of DHE encapsulated in liposomes in the presence of microspheres, in which lipid peroxidation was assayed. The optical constants of this system were measured with an integrating sphere spectrophotometer. The 1DA analysis leads to the spectral dependence of the absorption and scattering coefficients over a broad wavelength range (Karagiannes, Zhang, Grossweiner and Grossweiner 1988). Additional spectral results are presented for several other model systems.

Methods

DHE-subtilisin Carlsberg-microsphere system

Aqueous suspensions of 2.0 μm polystyrene microspheres were suspended in 1% Triton X-100 containing dihematoporphyrin ether (DHE) in the form of 1:150 Photofrin II (PFII). For intensity profile measurements, the suspensions were placed in a $4 \times 4 \times 1$ cm rectangular glass cuvette and illuminated at the front face with monochromatic light from a 200 W high-pressure mercury arc. The flux profiles were measured with a 400 μm silica optical fiber with a 2 mm spherical tip inserted into the suspension from the top. The incident light was filtered by a 5 cm water filter and narrow band interference filters at 633 nm, 545 nm, and 435 nm. The photochemical system consisted of oxygen-saturated 20 μM subtilisin Carlsberg and 1:150 PFII in 1% Triton X-100. The irradiations were performed in a 2×2 cm cylindrical glass cell exposed to collimated light from the arc lamp. The 545 and 435 nm interference filters were used for green and blue light irradiations and a Corning C.Z. No. 2-63 glass filter was used for red light (> 600 nm). The incident irradiance as measured with a tunable laser power meter was 11.1 $\pm$ 1.0 mW/cm^2 at 545 nm, 7.4 $\pm$ 1.5 mW/cm^2 at 435 nm, and $\cong$ 3 mW/cm^2 from 630–635 nm. The enzymic activity was assayed with Nα-benzoyl-L-arginine ethyl ester (BAEE) as described by Grossweiner and Messina (1987). The catalytic rate constant b was calculated from the increase of OD(253 nm) with reaction time (τ) according to the following relation based on first-order kinetics:

$$OD(253) = \varepsilon_p S_o - (\varepsilon_p - \varepsilon_s)S_o Exp(-b\tau) \tag{1}$$

where S_o is the initial BAEE concentration (20 μM), ε_s is the BAEE extinction coefficient, and ε_p is the substrate product extinction coefficient. The decrease of b with irradiation time (t) led to the first-order rate constant for photodynamic inactivation of the enzyme (κ) according to:

$$b(t) = b(0)Exp(-\kappa t) \tag{2}$$

The value of κ is proportional to the rate of energy absorption in the photodynamic system.

DHE-liposome microsphere system

Liposomes were prepared by swelling a dry egg lecithin film in 1:10 PFII following the procedures of Goyal, Blum and Grossweiner (1983). The large vesicles were bath-sonicated to give a clear suspension and mixed with 2.0 μm polystyrene microspheres to provide a scattering medium. The suspensions were irradiated at 546 nm with a monochromator-illuminator system and the extent of lipid peroxidation was assayed at intervals with the iodate method following the procedures of Blan and Grossweiner (1987). The optical constants of the suspension were measured with a two-channel integrating sphere spectrophotometer providing for on-line computer data acquisition, as described by Karagiannes et al. (1989). This instrument measures total reflection (R) and total transmission (T) of layer samples from 350-1350 nm with 2 nm resolution. The linear absorption coefficient (k) and reduced scattering coefficient (s') were calculated from this data using a 1DA formulation derived from the recent work of Groenhuis, Ferwerda and Ten Bosch (1983) and Jacques and Prahl (1987).

Results

One-dimensional diffusion approximation

Radiative transfer theory relates the optical properties of a light scattering and absorbing material to optical cross sections characterizing the individual particles (Chandrasekhar 1950). Various approximate radiative transfer theories have been employed for slab geometry (Star, Marijnissen and van Gemert 1987). The diffusion approximation treats photon propagation in a turbid medium as equivalent to particle diffusion. The phase function (angular distribution of scattering) of Groenhuis et al. (1983) consists of an isotropic term plus an anisotropic term proportional to a solid angle delta function in the forward direction. The resultant diffusion equation applicable to one-dimensional slab geometry is:

$$d^2U_d(z)/dz^2 - \kappa_d^2 U_d(z) = -(3/4\pi)s'(k + s)E_oExp[-(k + s')z] \tag{3}$$

where $U_d(z)$ is the average diffuse intensity, g is the mean cosine of the scattering angle, $s' = s(1 - g)$ is the reduced scattering coefficient, $\kappa d^2 = 3k(k + s')$ is the square of the attenuation coefficient for diffuse light, and E_o is the collimated incident irradiance. In an ideal semi-infinite layer, the 1/e attenuation depth for diffuse light (δ) equals $1/\sqrt{\kappa_d}$ The boundary conditions of Groenhuis et al. (1983) are employed, in which the inward directed flux at the front surface ($z = 0$) equals the fraction of the backward-directed flux that undergoes internal reflection (r_1), and similarly for the rear surface($z = d$) with r_2. The resultant boundary conditions are:

$$U_d \pm h'(dU_d/dz) = 0 \tag{4}$$

where $h' = [(1 + r_i)/(1 - r_i)][2/3(k + s')]$, + applies for $z = d$, and – applies for $z = 0$. The complete solution to Eq. (3) is:

$$U_d(z) = C_1 Exp(\kappa_d z) + C_2 Exp(-\kappa_d z) + A Exp[-(k + s')z] \tag{5}$$

The approach of Jacques and Prahl (1987) is used to calculate the total reflection (R) and transmission (T) coefficients from Eq. (5). The total flux at depth z is expressed as:

$$I_{total}(z) = 2I(z) + 2J(z) + E_o Exp[-(k + s')z] \tag{6}$$

where I is the equivalent forward-directed diffuse flux, J is the equivalent backward-directed diffuse flux, and the factor of two accounts for the average double pathlength traveled by uniformly diffuse light compared to collimated light. These fluxes can be expressed as:

$$I(z) = 2\pi[U_d(z)/2 + B(z)] \tag{7}$$
$$J(z) = 2\pi[U_d(z)/2 - B(z)]$$

where $B(z) = - (h/2)[dU_d(z)/dz]$ and $h = 2/3(k + s')$. The diffuse transmission (T_d) and reflection (R_d) coefficients are given by:

$$Td = I(z = d)(1 - r_2)/E_o \tag{8}$$
$$R_d = J(z = 0)(1 - r_1)/E_o$$

and the experimental R and T are given by:

$$T = (1 - R_{sp1})[Td + (1 - R_{sp2})exp[-(k + s)z] \tag{9}$$
$$R = R_{sp1} + (1 - R_{sp1})Rd$$

where R_{spi} is the specular reflection coefficient. A Microsoft Fortran 4.1 program was used to calculate k, s, and g from input values of R, T, δ, r_i, and R_{spi}. A fast fitting algorithm was used based on a multi-parameter Newton's method (John C. Collins, private communication). The k and s' spectra reported in this paper were calculated from R and T with the "similarity transformation", in which the result for g = 0 leads to values of s' and k (Star et al. 1987). On a 16 Mhz Zenith "386" computer with an 80387

798

co-processor, a set of (k,s') values was calculated from input (R,T) values at about 0 results per second.

DHE-enzyme-microsphere system

Typical intensity-depth profiles obtained with the spherical fiber optic probe are shown in Figure 1. The attenuation depth (δ) was determined by exponential regression of the linear regions. The enzymic activity after each irradiation [b(t)] was calculated from Eq. (1) by non-linear regression, and the relative photoinactivation rate constant (k) was calculated from Eq. (2) by exponential regression. The values of d and the "as measured" values of k are summarized in Table 1. The data show that attenuation induced

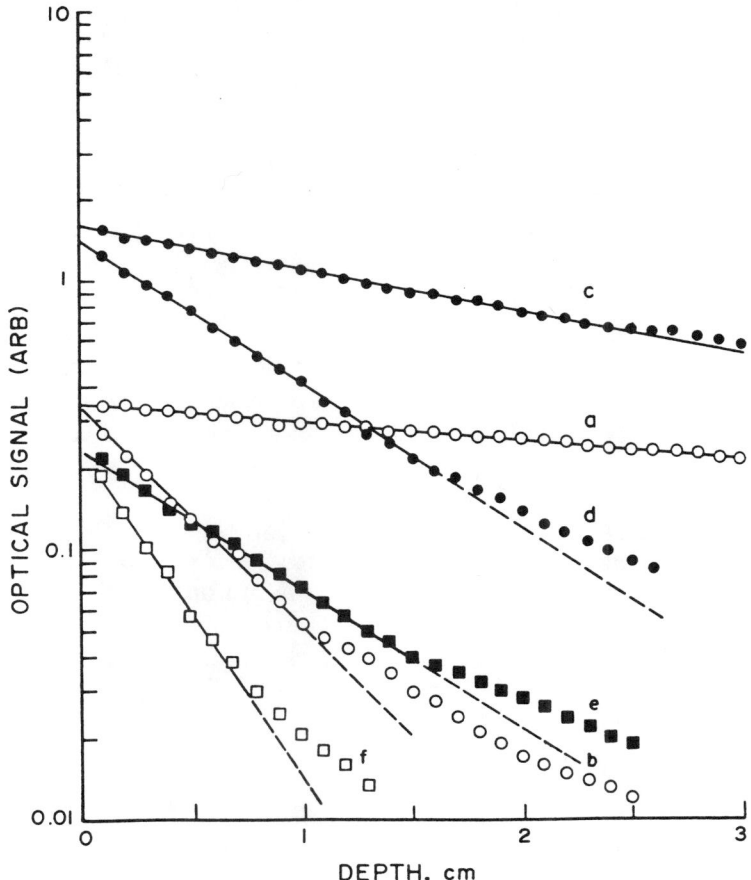

Figure 1. Intensity profiles in suspensions of 1:150 PFII (plus 1% Triton X-100) and 2.0 μm polystyrene microspheres measured with a spherical optic fiber probe: (a) 633 nm, Csc = 0; (b) 633 nm, $C_{sc} = 2.5 \times 10^8$; (c) 545 nm, Csc = 0; (d) 545 nm, $C_{sc} = 9.8 \times 10^7$; (e) 435 nm, $C_{sc} = 0$; (f) 435 nm, $C_{sc} = 2.5 \times 10^8$. The intensity scale is arbitrary for each measurement

by the microspheres had little effect on the reaction rate. This result appears to conflict with the "detour effect", where a longer light path should increase the probability of absorption. However, scattering also increases diffuse reflection and a more detailed analysis is required in order to explain the results.

The 1DA relationship between d and the optical constants is:

$$1/\delta^2 = 3k\,[k + s(1 - g)] \tag{10}$$

Since k was measured, $s' = s\,(1 - g)$ can be calculated for each sample. At a given wavelength, s' should be a linear function of C_{sc}. Fig. 2 shows a set of measurements at 633 nm at a lower range of C_{sc} than was used for the irradiations and comparative data for Mont Blanc "India Ink". The single line through the origin for two different absorbers supports the applicability of 1DA. The linear dependence also holds for the 633 nm and 545 nm data in Table 1, but not at 435 nm because of the strong DHE absorption. The computer program was used to calculate R and T for each suspension from the experimental values of s' and k, taking $r_i = 0.493$ for the water-glass-air interface (Jacques, Alter and Prahl 1987), and $R_{sp} = 0.05$. The fractional absorption in the layer is given by: $f_{abs} = 1 - (R + T)$. The results show that higher values of C_s lead to higher R and lower T such that f_{abs} is almost constant. The average values of f_{abs} are given in Table 2. The relative energy efficiency of the photodynamic process is proportional to κ_{ave}/I_o. The results in Table 2 show that red light and green light were equally efficient and blue light was two-fold more efficient than red light. The quantum efficiency (ϕ) should be independent of wavelength. The average value of ϕ is $1.3 \pm 0.3 \times 10^{-3}$ for this system, which may be compared to $V = 1.2 \times 10^{-3}$ obtained for photoinactivation of subtilisin Carlsberg at 630 nm by completely absorbing methylene blue (Grossweiner and Messina 1987).

Table 1. Photosensitized inactivation of subtilisin Carlsberg by 1:150 PFII (plus 1% Triton X-100) in the presence of 2.0 μm polystyrene microspheres. The microsphere concentration is C_{sc} (cm^{-3}), the attenuation depth is δ (cm), and the relative reaction rate is κ (min^{-1}).

	633 nm		535 nm		435 nm	
$10^{-8}C_{sc}$	δ	κ*	δ	κ	δ	κ
0.0	8.02	0.018	2.75	0.051	0.92	0.059
0.25	1.83	0.015	1.51	0.053	0.81	0.060
0.98	0.83	0.015	0.84	0.057	0.69	0.058
2.46	0.53	0.014	0.54	0.057	0.58	0.063
4.92	0.43	0.011	0.45	0.052	0.49	0.061

* red light

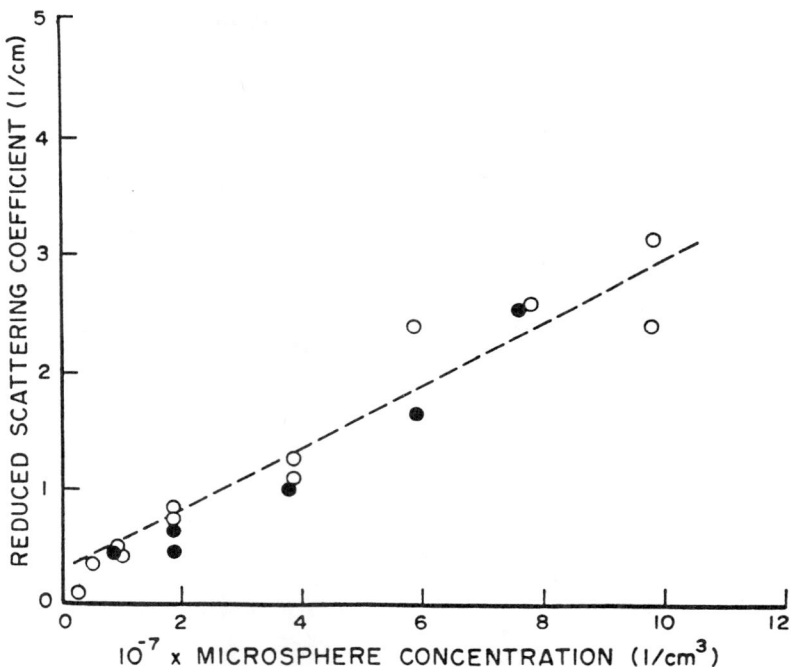

Figure 2. Dependence of s(1– g) at 633 nm as calculated from Eq. (10) on concentration of 2.0 μm polystyrene microspheres: O:150 PFII (in 1% Triton X-100); ● 1:200 "India Ink".

Table 2. Photosensitization of subtilisin Carlsberg by 1:150 PFII (with 1% Triton X-100) plus 2.0 μm polystyrene microspheres: κ_{ave} (min^{-1}) is the average "as measured" reaction rate from Table 1; I_o (mW/cm^2) is the incident irradiance; κ_{ave}/I_o is the energy efficiency; f_{abs} is the calculated integrated absorption; ϕ is the photosensitization quantum efficiency.

λ	κ_{ave}	I_o	κ_{ave}/I_o	f_{abs}	10^3 ϕ
red light	0.015±0.003	≅3	≅0.03	0.41±0.05	1–2
545 nm	0.054+0.003	11.1±1.0	0.027±0.002	0.67±0.08	1.2±0.2
435 nm	0.060±0.002	7.4±1.5	0.045±0.010	0.9±0.1	1.7+0.3

Photofrin II-liposome-microsphere system

The optical constants for this system were calculated from values of R and T measured with the integrating sphere spectrophotometer. Fig. 3 shows k and s' spectra for a liposome suspension containing 1:50 PFII in the presence of 2.0 μm microspheres ($C_{sc} = 1.1 \times 10^9$). The k spectrum has resolved absorption peaks at 400, 510, 540, 580, and 630 nm, in agreement with DHE in aqueous cetylmethylammonium bromide (Cubeddu, Keir, Ramponi and Truscott 1987). The peak value of k leads to a molar extinction coefficient of lipid bound-PFII equal to 7.3×10^4 l/mol-cm per 600 dalton porphyrin unit. The oscillations in the s' spectrum that mirror the k spectra have been observed in other model systems and tissues. Preliminary photochemical results in Fig. 4 show that the rate of lipid peroxidation was faster without microspheres, and insensitive to the microsphere concentration at high C_{sc}. Although this study is incomplete, the spectral results exemplify the usefulness of optical constant measurements on tissue model systems.

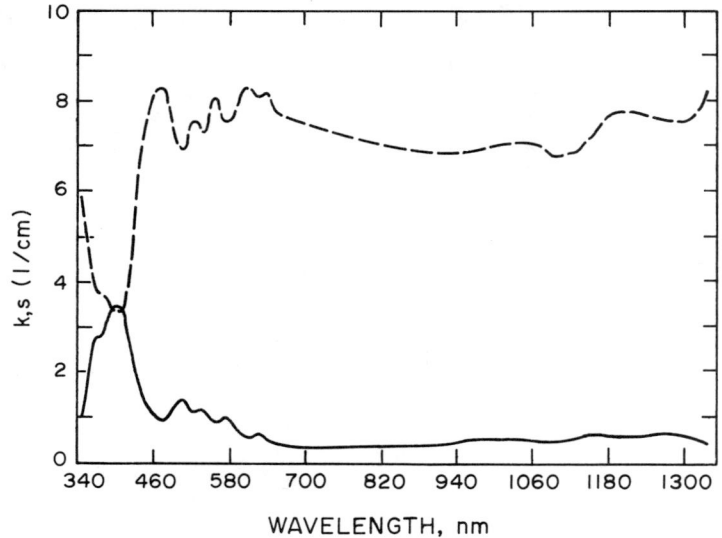

Figure 3. Optical constants of PFII (0.025 mg/ml) in sonicated egg lecithin liposomes with 2.0 μm polystyrene microspheres (1.1×10^9 cm^{-3}): *solid line*, k/cm; *dashed line*, s'/cm.

Other tissue models

Other sensitizer-target-light scattering systems are being studied in current work. The k spectrum of 38 mM methylene blue (MB) in the presence of 0.1 mg/ml artificial dairy creamer (Coffeemate) shows the MB absorption red-shifted by 4 nm (attributed to binding) and a maximum k approximately 10% lower than aqueous MB (Fig. 5). The s' spectrum is strongly perturbed by the dye absorption. However, the value at 400 nm of 16 cm^{-1} is very close to s' measured for the scatterer alone. The anaerobic photoreduction of MB is being investigated in this system. Red blood cells (rbc) are a useful scattering medium because they provide a background absorption similar to vascularized tissues. The k spectrum for 1:200 bovine rbc (Fig. 6) has maxima at 541 and 577 nm identified with oxyhemoglobin. Studies on photosensitized inactivation of enzymes by PFII are in progress.

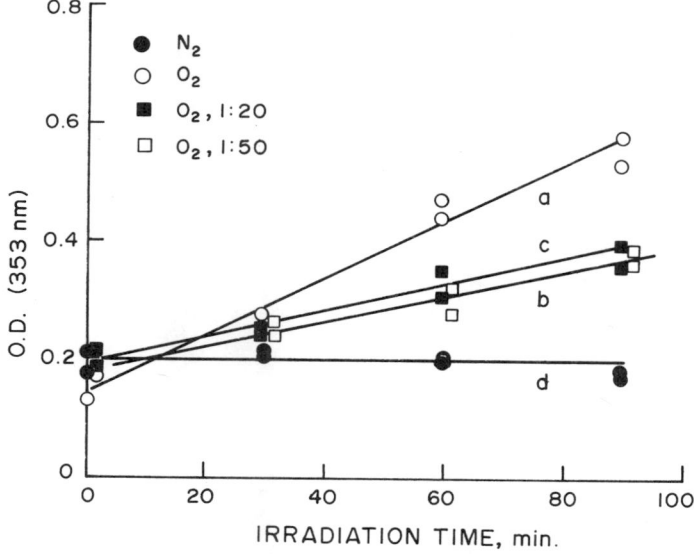

Figure 4. Photosensitized lipid peroxidation of egg lecithin liposomes by incorporated DHE at 546 nm. (a) oxygen saturated, $C_{sc} = 0$; (b) oxygen saturated, $C_{sc} = 4.4 \times 108$; (c) oxygen saturated, $C_{sc} = 1.1 \times 10^9$; (d) nitrogen saturated, $C_{sc} = 0$.

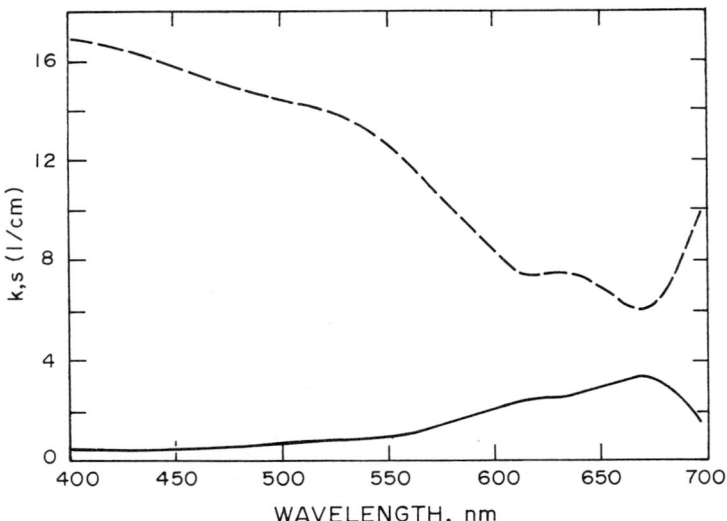

Figure 5. Optical constants of 38 μM methylene blue in the presence of 10 mg/ml artificial coffee creamer: *solid line*, k/cm; *dashed line*, s'/cm

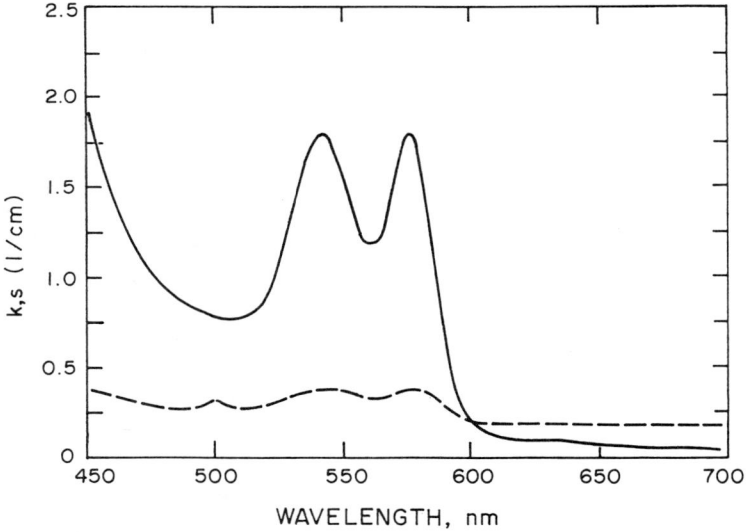

Figure 6. Optical constants of 1:200 bovine red blood cells in phosphate buffered saline. The 541 and 577 oxyhemoglobin peaks are resolved in the k spectrum (*solid line*).

Discussion

The modeling approach employed in this work has three phases:

(1)   The macroscopic optical properties of the system are measured, e.g., R and T in thin layers and δ in optically thick samples.

(2)   The optical constants k and s' are calculated with 1DA.

(3)   The integrated energy in the photochemical system is calculated with 1DA and

related to the experimental reaction rates.

This program was carried out for a DHE-enzyme-microsphere system. It was found that the reaction rate is independent of the scatterer concentration up to a 19-fold decrease in δ. The IDA calculations confirmed that scattering increases R and lowers T such that f_{abs} is almost constant. However, the apparent optical density of the layer (OD = $-$ $\text{Log}_{10}T$) was up to five-fold higher in the presence of microspheres. This is an example of the "dilution effect", where the OD of a dye is enhanced by mixing with a scattering material (Kortum 1969). Butler and Norris (1960) employed this technique to measure the spectra of weak tissue chromophores. The model system mimics photodynamic tumor therapy (PDT) to the extent that illumination of PFII in a light scattering environment damaged a non-absorbing target via the singlet oxygen pathway. Although tumor eradication and enzyme inactivation are different endpoints, the initial photophysics and photochemistry may have a similar dependence on wavelength. The results suggest that there would be no PDT dosimetry advantage with green light and perhaps a two-fold advantage with blue light.

The s' spectra do not show the true wavelength dependence of the scattering coefficients because s and g vary with wavelength. An explanation for the lower values of s' in spectral regions of high k derives from Mie scattering theory. The detailed analysis for dielectric spheres shows that absorption lowers the scattering cross section to an extent that depends on the absorption coefficient, refractive index, radius, and wavelength, e.g., Kerker (1969). Another factor may be the assumption that r_i is constant over the entire wavelength range. The higher effective refractive index in regions of strong absorption should increase r_i and therefore decrease R_d and T_d. In this case, the computer fitting with a low value of r_i would compensate by giving low values of s'. The validity of 1DA for g near unity is a subject of current interest. Recent measurements indicate that biological tissues are highly forward scattering (Bruls and van der Leun 1984; Arnfield, Tulip and McPhee 1988), which is the case also for the scatterering particles employed in this work. The phase function used in the present 1DA development may be more accurate for high g materials than Legendre polynomial expansion of Ishimaru (1978), which was used also in the recent 1DA model of Jacques and Prahl (1987). Although both 1DA models lead to the same s' and k when the similarity transformation is employed, there are significant differences in the calculated flux density profiles for g > 0. In addition to theoretical uncertainties, the measurement conditions must be consistent with the assumptions of the specific 1DA model. Finite beam size was not considered in the present analysis, which requires a beam diameter considerably larger than δ. The spectrophotometer used in the present work satisfies this condition for measurements on biological tissues, but not always for tissue models at low scatterer concentrations. In summary, tissue models are useful for the quantitative investigation of photosensitization in light scattering environments and for testing the validity of radiative transfer theories. However, experimental and analytical complications can arise, and each system must be studied in detail to ensure that the conclusions are valid.

Acknowledgements

This work was supported by NIH Grant No. 20117 and Strategic Defense Initiative Organization Medical Free Electron Laser Program on Contract No. SDIO84-88-C-0012 86-C-0188. The assistance of Dr. John C. Collins and Mrs. Bess Grossweiner with the computer programming is gratefully acknowledged.

References

Arnfield, M. R., Tulip, J., and McPhee, M. S., 1988, IEEE Trans. Biomed. Eng., 35, 372.

Blan, Q. A. and Grossweiner, L. I., 1987, Photochem. Photobiol., 45, 77.

Bohren, C. F. and Huffman, D. R., 1983, Absorption and Scattering of Light by Small Particles, (New York: John Wiley & Sons).

Bruls, W. W. G. and van der Leun, J. C., 1984, Photochem. Photobiol., 40, 231.

Butler, W. L. and Norris, K. L., 1960, Arch. Biochem. Biophys., 87, 31.

Chandrasekhar, S., 1950, Radiative Transfer (Oxford: Oxford University Press).

Cubeddu, R., Keir, W. F., Ramponi, R., and Truscott, T. G., 1987, Photochem. Photobiol., 46, 633.

Erefai, S. and Profio, S., 1985, Med. Phys., 12, 393.

Goyal, G. C., Blum, A., and Grossweiner, L. I., 1983, Cancer Res., 43, 5826.

Groenhuis, R. A. J., Ferwerda, H. A., and Ten Bosch, J. J., 1983, Applied Optics, 22, 2456.

Grossweiner, L. I. and Messina, J. W., 1987, Photochem.Photobiol., 45, 617.

Ishimaru, A., 1978 Wave Propagation and Scattering in Random Media, Vol.1 (New York: Academic Press).

Jacques, S. L. and Prahl, S. A., 1987, Lasers Surg. Med, 6, 494.

Jacques, S. L., Alter, C. A., and Prahl, S. A., 1987, Lasers Life. Scis., 1, 309.

Karagiannes, J. L., Zhang, Z., Grossweiner, B., and Grosswiener, L.I., 1989, Applied Optics, 28, 2311.

Kerker, M., 1969, The Scattering of Light, (New York: Academic Press).

Kortum, G., 1969, Reflectance Spectroscopy, (Berlin: Springer-Verlag).

Schifano, M. J. and Grossweiner, L. I., 1988, Photochem. Photobiol., in press.

Star, W. M., Marihnissen, J. P. A., and van Gemert, J. P. A., 1987, J. Photochem. Photobiol., 1B, 149.

806

In Vitro and *In Vivo* Spectral Properties of Porphyrins

R. POTTIER

Department of Chemistry and Chemical Engineering
The Royal Military College of Canada Kingston
Ontario, CANADA, K7K 5L0

Introduction

Porphyrins absorb electromagnetic radiation in both the ultraviolet and visible range, and often emit fluorescence type emission in the red. Thus conventional absorbance and emission spectroscopy becomes convenient tools to monitor the porphyrin derivative under study. For porphyrin localization phenomena, ground state absorbance and/or fluorescence spectroscopy is often used, whereas excited state lifetime and spectral measurements are most often related to the actual reactivity of the excited porphyrin. This paper will examine the ground state absorbance and first excited singlet state emission properties of porphyrins.

In vitro spectral properties

a) Porphyrin Solutions

An aqueous solution or biological preparation of a porphyrin derivative contains more than one type of molecule. In fact, several molecular species, all having different physical, chemical and spectral properties, may exist simultaneously in the solution. The various chemical species arise from ionic and monomer/dimer/aggregation equilibria that exist in solutions or biological preparations of porphyrin derivatives. Such equilibria can be represented as:

$$
\begin{array}{ccccccccc}
PH_4^{+2} & \underset{K_{a_1}}{\rightleftharpoons} & PH_3^+ & \underset{K_{a_2}}{\rightleftharpoons} & PH_2 & \underset{K_{a_3}}{\rightleftharpoons} & PH^- & \underset{K_{a_4}}{\rightleftharpoons} & P^{-2} \\
\updownarrow K_{D_1} & & \updownarrow K_{D_2} & & \updownarrow K_{D_3} & & \updownarrow K_{D_4} & & \updownarrow K_{D_5} \\
D_1 & & D_2 & & D_3 & & D_4 & & D_5 \\
\updownarrow K_{A_1} & & \updownarrow K_{A_2} & & \updownarrow K_{A_3} & & \updownarrow K_{A_4} & & \updownarrow K_{A_5} \\
A_1 & & A_2 & & A_3 & & A_4 & & A_5
\end{array}
$$

Photobiology, Edited by E. Riklis
Plenum Press, New York, 1991

PH_4^{+2} represents the dicationic species of the porphyrin ring, PH_3^+ the monocationic species, PH_2 the neutral species, PH_2 the monoanionic species and P^{-2} the dianionic species. D_1, D_2, D_3, D_4 and D_5 represent the corresponding dimeric species, whereas A_1, A_2, A_3, A_4 and A_5 represent the corresponding aggregated species. In fact, the above scheme is a simplified one, since zwitterions are possible, which would increase the number of possible chemical species. Chemical and thermal instabilities, along with solubility limits can also hamper the spectroscopic identification of some of the above species. The relative distribution of the chemical species present in a solution of porphyrin derivative is acutely dependent on the temperature, concentration, ionic strength and pH of the solution.

b) Ground state absorbance

All porphyrin spectra are characterized by two intense bands in the ultraviolet region; one in the vicinity of 280 nm (called the γ band) and another in the 400 nm region (called the Soret or B band). Less intense (approximately 1/10 Soret) bands are observed in the visible region (450–700 nm), and are called the Q bands. The number of Q bands depends on the protonation state of the imino (–N=) nitrogen, the pyrrole (–NH–) nitrogen of the tetrapyrrole nucleus, and also on metal chelation at the centre of the porphyrin ring. Both the Soret and visible bands of the free base are believed to be of a $\pi \rightarrow \pi^*$ nature, with possible contributions from $n \rightarrow \pi^*$ transitions (Gouterman 1978, Corwin et al. 1968, Chantrell et al, 1975, Pottier and Truscott 1986, Scherz and Levanon 1985). The chromophore responsible for the observed electronic transitions can be considered as an 18-membered cyclic polyene.

Conventional absorbance spectroscopy can easily discriminate between the ionic species related to the 18-membered cyclic polyene. Due to symmetry changes upon protonation, the number of observed transitions will correspondingly vary. This is easily observed in the visible, Q bands region. In acidic medium ($0 < pH < 2$), two Q bands are observed. In the pH range 3–6, three Q bands can be seen, whereas for neutral and basic solutions, four Q bands are easily distinguishable. Fig. 1 shows the absorbance spectral changes observed upon protonation of hematoporphyrin IX (Hp), for both the Soret and Q bands. In order to spectroscopically see the monocation absorbance, it is often necessary to add a surfactant to the solution. The cationic species tend to form aggregrates in aqueous solutions. While the Q bands are sensitive to protonation effects, the Soret band is very sensitive to aggregation phenomena. Upon aggregation, the Soret band near 400 nm is usually blue shifted by 20–30 nm (Pasternack et al. 1972, Pasternack 1973, Brown et al 1976, Karns et al. 1979, Moan et al. 1984 a, b Moan et al. 1985). Such a blue shifted dimer/aggregate band is shown in fig. 1E for an aqueous solution of hematoporphyrin IX. Similar blue shifts in the Soret band can be induced by an increase in the porphyrin concentration, a decrease in the solution temperature or by high ionic strength.

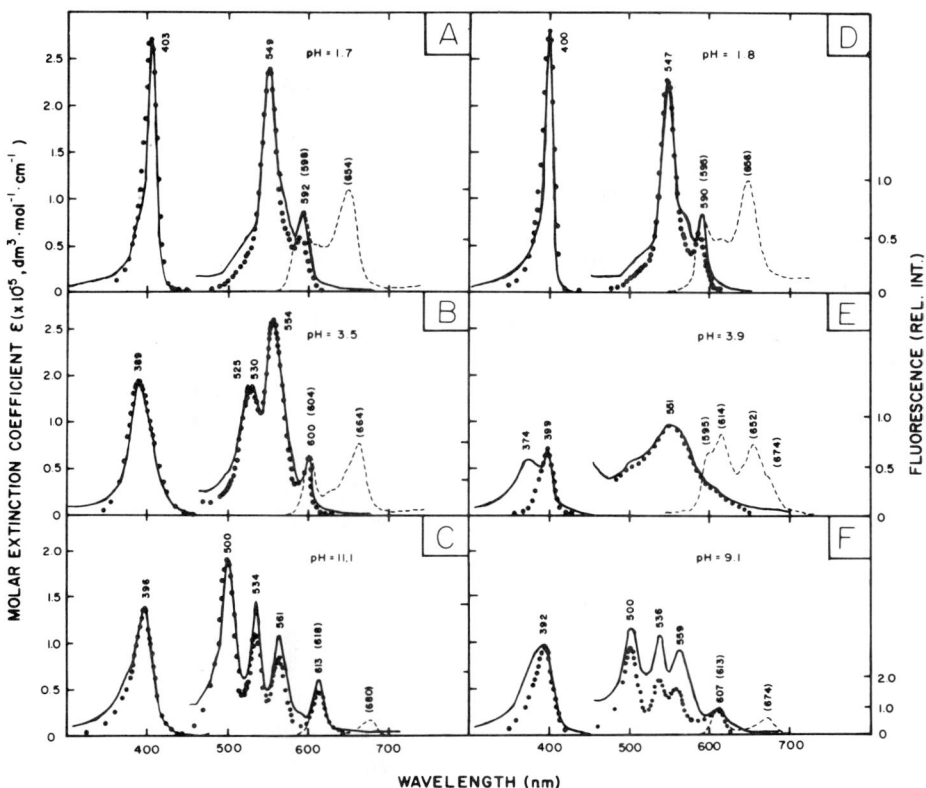

Figure 1. Absorbance (——), fluorescence emission (---) and fluorescence excitation (.....) spectra of ionic species present in water (D, E and F) and in water containing 0.4% SDS (A, B and C) at 20°C and at the indicated pH values. Solutions were unbuffered. Concentrations for absorbance spectra were 9.67×10^6 mol.dm^{-3}; for fluorescence spectra, 1.99×10^{-6}.mol.dm^{-3}. The intensity of the Q bands is amplified by a factor of 10 with respect to the Soret bands, and fluorescence intensity is normalized to absorbance intensity at the overlapping 0–0 band (except in Fig. E). Numbers shown at peaks are wavelength maxima (nm); bracketed numbers refer to the maxima of the fluorescence emission peaks.

Photoacoustic spectroscopy has recently been used to detect a new electronic band (maximum at 440 nm) in concentrated aqueous solutions of hematoporphyrin IX (Pottier et al, 1988). This new band is believed to originate from the Hp neutral dimeric species and was found in higher concentrations at or near the air-solution interface.

Aggregation phenomena observed in hydrosoluble prophyrin derivatives normally involve hydrophylic peripheral groups and the hydrophobic character of the porphyrin ring. Weaker associations between porphyrins can occur due to porphyrin ring interactions. Such interactions have recently been observed by absorbance and fluorescence emission spectroscopy for a liposoluble porphyrin, 2,7,12,17-tetrahexyl-3,8,13,18-tetramethylporphyrin (Chapados et al. 1988). Such weak associations were found to be enhanced at or near the wall of the solution cuvette.

809

Since the active component of the preparation called hematoporphyrin derivative (HPD) is believed to contain two or more porphyrins linked via covalent bonds, it is of interest to examine the spectral properties of highly purified biporphyrins. Chow et al (1988) have examined the ground and excited state properties of β,β'-biporphyrins linked via a hydrocarbon chain, $-(CH_2)_n-$, where n = 1–8 (except 7), along with porphyrin rings linked directly, i.e. n = 0 (β,β' and at meso positions). For the β,β' link, they observed a Soret band splitting for porphyrin rings separated by 0 to 3 linking carbon atoms. On the other hand, the meso linked biporphyrin did not show this splitting and behaved as two independent chromophores (due to steric hinderance).

If part of the porphyrin polyene chromophore is reduced, such as in chlorin, the resultant symmetry change induces a symmetry allowed transition in the Q band region. These new, "second generation" potential photochemotherapeutic agents, along with free base and metallophthalocyanines, hold much promise for photodynamic therapy. Their enhanced absorbance in the red or near infrared spectral region should lead to much lower dose requirements in the treatment of neoplastic disease by photosensitization.

c) Excited state fluorescence

Fluorescence spectroscopy is one of the most sensitive analytical techniques available, and has been widely used to monitor porphyrins. Fig. 1 shows typical fluorescence spectra emitted by most porphyrins. The monomeric dication fluorescence emission is characterized by two near equal intensity maxima in the vicinity of 595 and 654 nm. The monocation fluorescence is very similar to the dication emission, but both peaks are red shifted by 8–10 nm. The neutral species has two emission peaks at approximately 618 and 680 nm, the ratio of their intensities being roughly 3:1, respectively. The covalently linked biporphyrins have very similar fluorescence emission (Chow et al. 1988). The fluorescence quantum yield of these biporphyrins lie in the range of 0.08±02 which is very similar to that of 0.09 for Hematoporphyrin (Smith 1985).

When porphyrins are studied in more biologically relevant media, such as detergents, human serum and low density lipoproteins, they still show their characteristic red fluorescence, but this emission is usually red shifted by approximately 30nm (Margalit and Cohen 1983, Margalit et al. 1983, Moan and Sommer 1981, Reyftmann et al 1984, Pottier and Truscott 1987). The observed red shift has been attributed to the formation of bonds between porphyrins and proteins (Moan and Sommer 1981).

In aqueous solutions, the dimeric and/or aggregated species are either very weakly or non fluorescent (Margalit and Cohen 1983, Margalit et al 1983, Pottier et al, 1988). This can be seen by an examination of fig. 1E. The Soret absorbance band shows two maxima at 374 and 399 nm, indicating the presence of both the monomer and dimer/aggregate, respectively. However, the fluorescence excitation spectrum corresponds to the monomer absorbance only. In fact, an analysis of the fluorescence excitation spectrum in the Q band region reveals that the observed fluorescence can all be accounted for by a linear combination of the fluorescence originating from the dication and neutral species present in the solution. Further, it is found that the presence of dimers/aggregates partly quenches

810

the fluorescence emitted by both the dication and the neutral species (Pottier et al. 1988).

There have been several reports on porphyrin dimer fluorescence (Andreoni et al. 1983, Andreoni and Cubeddu 1984, Brookfield 1985, Roeder and Wabritz 1987, Kinoshita 1988) in aqueous medium, indicating a weak fluorescence similar to the neutral porphyrin, but with maxima at approximately 635 and 693 nm. Shuichi et al. (1988) have recently shown that this weak fluorescence is similar to that observed in cancerous cells. This type of fluorescence requires further investigation, since fluorescence measurements are often used to estimate the concentration of porphyrins in complex biological media and cellular systems.

In vivo spectral properties

Absorbance spectroscopy is not normally used for the monitoring and characterization of porphyrins *in vivo*. This is principally due to the low transmission and high scattering properties of tissue. Photoacoustic spectroscopy still holds some promise in this line, since PA spectra have been done on live human skin (Poulet and Chambron, 1983). For the most part, fluorescence spectroscopy has been the technique of choice for *in vivo* spectral measurements. Conventional spectrophotofluorometric instrumentation can be readily adapted to record the fluorescence spectra of tissue containing porphyrins. Several reports have shown that the porphyrin fluorescence observed *in vivo* is similar to that observed in *in vitro* preparations containing surfactants and/or proteins (Van der Putten and van Gemert 1983, Doiron et al 1984, Profio et al. 1984, Profio 1984, Gijsbers et al. 1984, Sommer et al. 1984, Fioretti et al. 1984, Pottier et al. 1976). *In vivo* fluorescence spectroscopy can be easily used to not only characterize the porphyrin, but can also yield information on the pharmacokinetics of porphyrins, such as the clearance rates of different porphyrins (Pottier, Weagle, Petryka and Kennedy, 1988). Although the results are easily obtainable, care must be exercised in the interpretation of *in vivo* spectral results, since the background fluorescence is not necessarily reproducible or totally relavent to the subject under study, and this can easily lead to artifacts. In fact, recent measurements by Weagle et al. (1988) have shown that normal mouse skin has a prominant fluorescence peak at 674 nm, and this fluorescence originates from pheophorbide a and/or pheophytin a, degradation products of chlorophyll a that are derived from mouse food. Pheophorbide a is known to selectively biodistribute towards tumor tissue and is an efficient photosensitizer (Roeder 1986).

Conclusion

The determination of the spectral properties of porphyrins *in vitro* requires the careful control of several parameters: concentration, solvent, ionic strength, pH and temperature. The simultaneous existance of several ionic, dimeric and aggregated species, regulated by two sets of mutually interactive equilibria, means that only a partial characterization of the spectral properties of porphyrins is possible. Both absorbance and fluorescence spectroscopies are applicable for *in vitro* studies, but *in vivo* studies are generally limited to emission spectroscopy. In most cases, good correlations can be obtained between *in vitro* and *in vivo* fluorescence measurements, thus establishing fluorescence spectroscopy as a powerful technique for the study of porphyrins as photochemotherapeutic agents.

References

Andreoni, A., Cubeddu, R., Desilvestri, S., Jori, G., Laporta, P., and Reddi, E., 1983, Time-resolved fluorescence studies of haematoporphyrin in different solvent systems. *Zeitschrift fur Naturforschung,* 38C,83–89.

Andreoni, A., and Cubeddu, R., 1984, Fluorescence properties of HpD and its components. Porphyrins in Tumor Phototherapy, edited by A. Andreoni and R. Cubeddu (New York: Plenum Press), pp. 11–21.

Brookfield, R.L, 1985, Time-resolved fluorescence spectroscopy of haematoporphyrin. Primary Photoprocesses in Biology and Medicine, edited by R.V. Bensasson, G. Jori, E.J. Land and T.G. Truscott (Plenum Press), pp. 329–333.

Brown, S.B., Shillock, M., and Jones, P., 1976, Equilibrium and kinetic studies of the aggregation of porphyrins in aqueous solutions. *Biochemical Journal,* 153, 279–285.

Chantrell, S.J., Mcauliffe, C.A., Munn, R.W., Pratt, A.C., and Land, E.J., 1977, Excited states of protoporphyrin IX dimethyl ester: reaction of the triplet with carotenoids. *Journal of the Chemical Society, Faraday Transactions I,* 73, 858–865.

Chapados, C., Girard, D., Pottier, R., Ringuet, M., Weagle, G., Weir, L. and Weir, M., 1989. Absorbance and fluorescence studies of liposoluble metalloporphyrins: evidence for porphyrin association in hydrophobic solvent. *Canadian Journal of Spectroscopy,* 34(4), 94–99.

Chow, Y.F.A., Dolphin, D., Paine III, J.P., MCGarvey, D., Pottier, R., and Truscott, T.G., 1988, The excited states of covalently linked dimeric porphyrins I: the excited singlet states. *Journal of Photochemistry and Photobiology, B: Biology,* 2, 253–263.

Corwin, A.H., Chivvis, A.B. Poor, R.W. Whitten, D.G., and Baker, E.W., 1968, The Interpretation of porphyrin and metalloporphyrin spectra. *Journal of The American Chemical Society,* 90, 6577–6583.

Dorion, D.R., Gomer, C.J., Fountain, S.W., and Razum, N.J., 1984, Photophysics and dosimetry of photoradiation therapy. Porphyrins in Tumor Phototherapy, edited by A. Andreoni and R. Cubeddu (New York: Plenum Press), pp. 281–291.

Fioretti, P., Facchini, V., Gadducci, A., and Cozzani, I., 1984, Monitoring of hematoporphyrin injected in humans and clinical prospects of its use in gynocologic oncology. Porphyrins in tumor Phototherapy, edited by A. Andreoni and R. Cubeddu (New York: Plenum Press), pp. 355–361.

Gijbers, G.H.M., Van Gemert, M.J.C., Breederveld, D., Longelaar, J., and Boon, T.G., 1984, *In vivo* fluorescence excitation spectra of hematoporphyrin derivative. Porphyrins in Tumor Phototherapy, edited by A. Andreoni and R. Cubeddu (New York: Plenum Press), pp. 339–345.

Gouterman, M., 1978, Optical spectra and electronic structure of porphyrins and related rings. The Porphyrins, Vol. III, edited by D. Dolphin (New York: Academic Press), pp. 1–165.

Karns, G.A., Gallagher, W.A., and Elliot, W.B., 1979, Dimerization constants of water-soluble porphyrins in aqueous alkali. *Bioorganic Chemistry,* 8, 69–81.

Kinoshita, S., Seki, T., Liv, T.F., and Kushida, T., 1988, Fluorescence of hematoporphyrin in living cells and in solutions. *Journal of Photochemistry and Photobiology, B: Biology,* 2, 195–208.

Margalit, R., and Cohen, S., 1983, Studies of hematoporphyrin and hematoporphyrin derivative equilibria in heterogeneous systems. Porphyrin-liposome binding and porphyrin aqueous dimerization. *Biochimica et Biophysica Acta,* 736, 163–170.

Margalit, R., Shaklai, N., and Cohen, S., 1983, Fluorimetric studies on the dimerization equilibrium of protoporphyrin IX and its haematoderivative. *Biochemical Journal,* 209,

547–552.

Moan, J., 1984a, Fluorescence of porphyrins in cells. Porphyrins in Tumor Phototherapy, edited by A. Andreoni and R. Cubeddu (New York: Plenum Press), pp. 109–124.

Moan, J., 1984b, The photochemical yield of singlet oxygen from porphyrins in different states of aggregation. *Photochemistry and Photobiology*, 39, 445–449.

Moan, J., and Sommer, S., 1981. Fluorescence and absorption properties of the components of hematoporphyrin derivative. *Photobiochemistry and Photobiophysics*, 3, 93–103.

Pasternack, R.K., Huber, P.R., Boyd, P., Engasser, G., Francesconi, L., Gibbs, E., Fasella, P., Venturo, G.C., and Hinds, L. DE C., 1972, On the aggregation of mesosubstituted water-soluble porphyrins. *Journal of the American Chemical Society*, 94, 4511–4517.

Pasternack, R.F., 1973, Aggregation properties of water-soluble porphyrins. *Ann. of the New York Academy of Sciences*, 206, 614–630.

Pottier, R., and Truscott, T.G., 1976, The photochemistry of haematoporphyrin and related systems. *International Journal of Radiation Biology*, 90, 421–452.

Pottier, R., Lachaine, A., Pierre, M., and Kennedy, J.C., 1988, A New electronic absorbance band in concentrated aqueous solutions of hematoporphyrin IX detected by photoacoustic spectroscopy. *Photochemistry and Photobiology*, 47, 669–674.

Pottier, R., and Nadeau, P., Unpublished results.

Pottier, R.H., Chow, Y.F.A., Laplante, J.P., Truscott, T.G., Kennedy, J.C., and Beiner, L.A., 1986, Non-invasive technique for obtaining fluorescence excitation and emission spectra *in vivo*. *Photochemistry and Photobiology*, 44, 679–687.

Pottier, R., Weagle, G., Kennedy, J., and Petryka, Z.J., unpublished results.

Profio, A.E., 1984, Laser excited fluorescence of HpD for diagnosis of cancer. *I.E.E.E. Journal of Quantum Electronics*, 20, 1502–1506.

Profio, A.E., Carvlin, M.J., Sarnaik, J., and Wuld, L.R., 1984, Fluorescence of hematoporphyrin-derivative for detection and characterization of tumors. Porphyrins in Tumor Phototherapy, edited by A. Andreoni and R. Cubeddu (New York: Plenum Press), pp. 231–337.

Poulet, P., and Chambron, J., 1983, *In vivo* photoacoustic spectroscopy of the skin. *Journal de Physique*, 44, 413–418.

Reyftmann, J.P., Morliere, P., Goldstein, S., Santus, R., Dubertret, L., and Lagrande, D., 1984, Interaction of human serum low density lipoproteins with porphyrins: a spectroscopic and photochemical study. *Photochemistry and Photobiology*, 40, 721–729.

Roeder, B., and Wabnitz, H., 1987, Time-resolved fluorescence spectroscopy of hematoporphyrin, mesoporphyrin, pheophorbide a and chlorin e_6 in methanol and aqueous solution. *Journal of Photochemistry and Photobiology, B: Biology*, 1, 103–113.

Roeder, B., 1986, Pheophorbide a — a new photosensitizer for photodynamic therapy of tumors. *Studia biophysica*, 114, 183–186.

Scherz, A., and Levanon, H., 1985, Optical transition energies of porphyrins: the application of free electron molecular orbital approach. *Molecular Physics*, 55, 923–937.

Smith, G.J., 1985, The effects of aggregation on the fluorescence and the triplet state yield of hematoporphyrin. *Photochemistry and Photobiology*, 41, 123–126.

Sommer, S., Rimington, C., and Moan, J., 1984, Formation of metal complexes of tumor-localizing porphyrins. *FEBS Letters*, 172, 267–271.

Van Der Putten, W.J.M., and Van Gemert, M.J.C., 1983, Haematoporphyrin derivative fluorescence *in vitro* and in an animal tumor. *Physics in Medicine and Biology*, 28, 633–638.

Weagle, G., Paterson, P.E., Kennedy, J., and Pottier, R., 1988, The nature of the chromophore responsible for naturally occurring fluorescence in mouse skin. *Journal of Photochemistry and Photobiology, B: Biology*, 2, 313–320

Photosensitization of Microbial Cells

Y. NITZAN, Z. MALIK, AND B. EHRENBERG*

*Department of Life Sciences and Department of Physics**
Bar-Ilan University
Ramat-Gan 52100 Israel

In this presentation we shall summarize the knowledge on the photoinactivation effects of porphyrins on bacteria. The early studies with microorganisms were done with yeasts such as *Sacharomyces cerevisiae* (Kvello-Stenstrom et al., 1980; Moan and Kvello-Stenstrom, 1981). Eukaryotic yeast cells in their exponential growth phase were inactivated when exposed to hematoporphyrin and light. Photosensitization of yeast cells was dependent on the hematoporphyrin dose, pH and intensity of illumination. Porphyrin derivatives were also shown to interact with phage g DNA or with Colicin E1 λ DNA isolated from *E. coli* K12 (Boye and Moan, 1980).

In our work with prokaryotic cells we were able to show that *Staphylococcus aureus* cultures in their logarithmic phase which were treated with small amounts (3-12 µg/ml) of hematoporphyrin derivative (HPD) and light were inhibited and their growth was stopped (Nitzan et al., 1983). The viable count of the *S. aureus* cultures show (Table 1) that more than 80% of the bacteria are killed within 3 hours of treatment by HPD and illumination. On the other hand, when these bacteria are treated at the lag phase (with HPD and light) only a minority of the bacteria are killed. Most of the cells are unaffected and may continue to multiply after the small reduction of their initial number. With deuteroporphyrin the killing rate in the logarithmic phase is even higher and faster (Table 1). After 2 hours 99% of the bacteria are dead and after 10 hours the culture is sterile and no viable counts can be detected (Nitzan et al., 1987a).

The same phenomenon of antibacterial activity of HPD and light was shown on other Gram positive bacteria such as: *Bacillus subtilis* or *Streptococcus pyogenes* (Malik et al., 1982). With other porphyrins it was shown that hematoporphyrin was acting on *Streptococcus faecalis* (Bertoloni et al., 1984) and Deuteroporphyrin was acting efficiently (Table 1) on *Streptococcus faecalis* or the spore forming *Bacillus subtilis* (Nitzan et al., 1987)

Gram negative bacteria such as E. coli or Pseudomonas aeruginosa ignore any treatment by HPD or deuteroporphyrin and light even when the concentrations of these perphyrins were 200 µg/ml (Malik et al., 1982; Nitzan et al., 1983; Nitzan et al., 1987a). The same was shown with Haemophilus influenzae type A (Malik et al., 1982)

Table 1. Survival fraction of porphyrin photosensitized bacteria. N_0 is the number of bacterial cells in the beginning of treatment and N is the number after 3 hours treatment.

Bacteria	Porphyrin	Additional conditions	N/N_0
S.aureus	HPD	Log phase	0.18
S.aureus	HPD	Lag phase	0.65
S.aureus	DT	Log phase	0.01
E.coli	DT	Log phase + nona peptide	0.03
P.aeruginosa	DT	Log phase + nona peptide	0.07
B.subtilis	HPD	Log phase	0.10
B.subtilis	DT	Log phase	0.01
S.pyogenes	HPD	Log phase	0.05
S.faecalis	DT	Log phase	0.01

or with *Klebsiella pneumoniae* (Bertoloni et al., 1984a). Various experimental conditions and reagents were tried in order to sensitize the Gram negative bacteria to porphyrins and light, all these trials failed. Recently we found that the *nonapeptide* derivative of colistin which opens channels in the bacterial membrane without having antimicrobial activity (Vaara and Vaara, 1983), enables photosensitized porphyrins to act on the Gram negative *E. coli* or *Pseudomonas aeruginosa*. The growth of both Gram negative bacteria is stopped and the viability is decreased to the same extent as the Gram positive bacteria.

There is no significant effect of the porphyrins in the dark. Even when the added amount of porphyrin was 200 µg/ml there was no significant influence in the dark.

The effect of porphyrins on *Staphylococcus aureus* cells consists of two steps: I. Absorption of the porphyrin to the bacterial cells, especially to the cell membrane as we shall see later. This step does not require light and can be carried out in the dark and is not toxic for the cells. II. Inactivation of the bacteria. In this step the porphyrin molecules which are already absorbed on the cell membrane are now activated by light and are acting and causing death to the bacterial cells. These two steps can be demonstrated by an experiment in which HPD is added in the dark to a bacterial culture without causing any inhibition. After a certain time the porphyrin was washed out in the dark. The washed bacteria that in fact absorbed a certain amount of HPD have been put in a fresh medium and exposed to light without adding any new portion of HPD. The bacterial growth was influenced and inhibited although no porphyrin was present in the medium except for the absorbed HPD (Malik et al., 1982).

The absorption of the porphyrin molecules to the bacterial membranes can be followed by fluorescene measurements (Ehrenberg et al., 1985). HPD in water gives a fluorescene band in 613-614 nm. When dissolved in a lecithin layer it gives rise to a

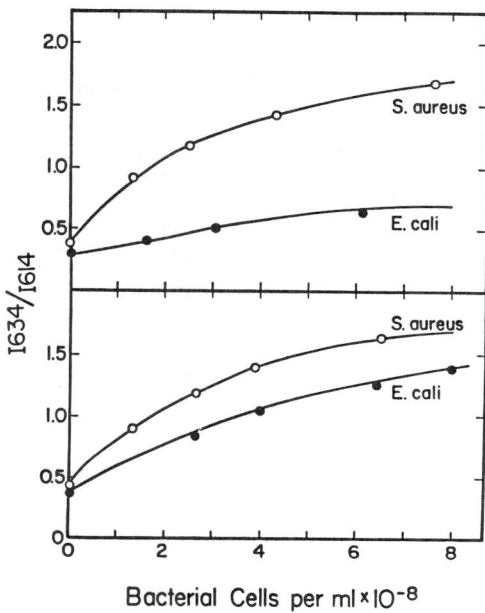

Figure 1. Ratio of the fluorescence intensities at the 634 nm band and the 614 nm band of HPD as a function of the number of *S. aureus* and *E. coli* intact cells (upper panel) or spheroplasts (lower panel).

fluorescene band at 632-634 nm while the excitation is at 458 nm. Upon examining the relative intensity of HPD fluorescence in the presence of increasing amounts of *S. aureus* or *E. coli* cells the following is seen. The fluorescence intensity at 634 nm is increased when *S. aureus* cells are added, but, on the other hand when *E. coli* cells are added only a small increase in 634 nm is obtained. The relative difference between *E. coli* and *S. aureus* is shown in Figure 1 (upper panel).

It was postulated that the binding of porphyrins to Gram positive bacteria is to the cytoplasmic membrane and as a result of the excitation the porphyrin can act and cause the death of the cell. With the Gram negative bacteria the situation is different. It seems that the porphyrin does not reach the bacterial cell membrane and the low binding indicates a low affinity of binding probably to the outer membrane of the *E. coli*. A proof for this postulation was provided when spheroplasts of these two bacteria were produced. In such an experiment the cytoplasmic membrane of both types of bacteria are exposed. As can be seen from Figure 1 (lower panel) the affinities to porphyrins of both types of bacteria with exposed cytoplasmic membranes are of similar magnitude. There is no increase in the affinity of *S. aureus* spheroplasts to the porphyrin as compared to the intact cells. On the other hand, *E. coli* spheroplasts show a high increase in affinity in comparison to the intact cells, as mentioned above. Understanding the need of porphyrin binding to the bacterial cell membrane explains the use of colistin nonapeptide mentioned before as a helper to destroy Gram negative bacteria. Colistin nonapeptide is probably acting by opening channels in the Gram negative's outer membrane and so enables the porphyrin to reach the cytoplasmic membrane. Only porphyrins reaching the cytoplasmic membrane can be photosensitized usefully and cause damage and death to the cell.

The killing effect of the porphyrins is a result of their action on the bacterial DNA and especially on it's synthesis (Nitzan et al, 1983). Synthesis of DNA is stopped immediately after the beginning of the interaction of porphyrins with bacteria. The inhibition of DNA synthesis affects obviously later the synthesis of RNA and protein. The effect on DNA can also be manifested on plasmid DNA. It can be shown with staphylococci which survived the porphyrin treatment with light that in these survivors the induction mechanism of the enzyme penicillinase, encoded on a plasmid, was damaged. As a result the resistance of these staphylococci to penicillin was reduced from 100 µg/ml to about 1 µg/ml.

The damage of the cells is mediated singlet oxygen and hydroxyl free radicals (Nitzan et al., submitted for publication). This fact was established by using singlet oxygen quenchers or hydroxyl free radical scavengers for protection from photoinactivation. It could be shown (Figure 2) that methionine tryptophan and DBCO (1.4, diaza bicyclo 2,2,2 octane) used as singlet oxygen quenchers provided 60% protection. Propyl gallate a hydroxy free radical scavanger also showed 60% protection. The same results could be seen in parallel experiments showing the growth curve and viable counts of the bacteria. Only in the presence of propyl gallate and one of the singlet oxygen quenchers complete protection from inactivation (96%) was observed. These results indicate that bacterial

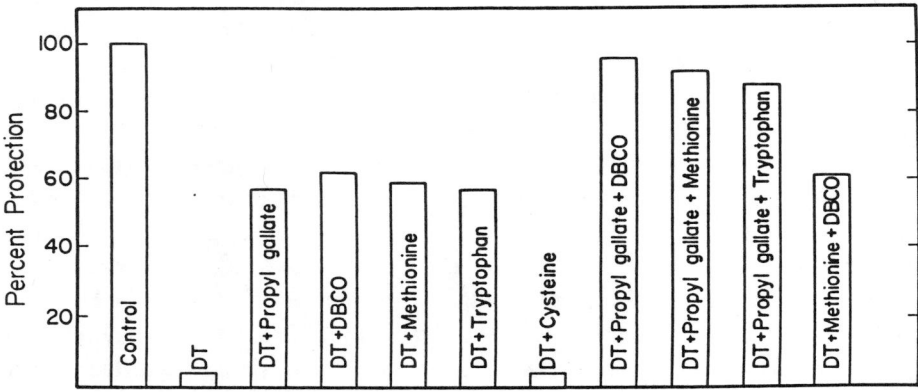

Figure 2. Protection ability of various reagents alone or in combination on *S. aureus* treated by DT and light

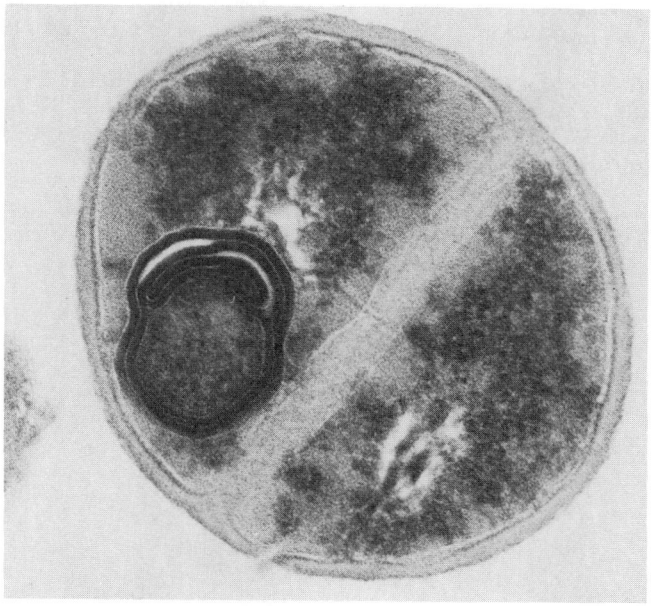

Figure 3. Ultrastructure of *S. aureus* photosensitized by DT. Appearance of mesosomes after 3 hours of treatment (×80.000)

photoactivation is mediated by both singlet oxygen and hydroxyl free radicals. In addition, from the absence of protection by cysteine it seems that the mechanism is not by oxidation as shown with iron containing porphyrins (Nitzan et al., 1987b).

One of the peculiar results of photosensitization of bacterial cells in Gram positive microorganisms is the appearance of a multilamellar structure (Figure 3) near the septum of dividing cells (Malik et al., 1982; Nitzan et al., 1983; Nitzan et al., 1987a; Malik et al., 1988). The volume of the so-called mesosomes and the frequency of their appearance in the cells was increased in a time-dependent manner. Further studies (Malik et al., 1988) have revealed the formation of membrane like structures in the mesosome. It was concluded that the photosensitization may induce disturbance in the synthesis of the membranes which now appear as a multilamellor form.

The aim of this research is to find ways to introduce porphyrins to clinical use, to overcome the problem of increased bacterial resistance to antibiotics and to prevent external and internal bacterial infections.

References

Boye, E. and J. Moan (1980) *Photochem. Photobiol.* 31, 223-228.

Bertoloni, G., M.Dall'Acqua, M. Vazzoler, B. Salvato and G. Jori (1984). In Porphyrins in Tumor Phototherapy (Edited by R. Cubeddu and A. Anderson) Plenum Press pp 177-184.

Bertoloni, G., B. Salvato, M.Dall'Acqua, M. Vazzoler and G. Jori (1984) *Photochem. Photobiol.* 39, 811-816.

Ehrenberg, B., Z. Malik and Y. Nitzan (1985) *Photochem. Photobiol.* 41, 429-435.

Kvello-Stenstrom, A.G., J. Moan, G. Brunborg and T. Eklund (1980) *Photochem. Photobiol.* 32, 349-352.

Malik, Z., S. Gozhansky and Y. Nitzan (1982) *Microbios Lett.* 21, 103-112.

Malik, Z., H. Ladan, J. Hanania and Y. Nitzan (1988) *Curr. Microbiol.* 8, 279-284.

Moan, J. and A.G. Kvello-Stenstrom (1981) *Photochem. Photobiol.* 33, 761-763.

Nitzan, Y. S. Gozhansky and Z. Malik (1983) *Curr. Microbiol.* 8, 279-284.

Nitzan, Y. B. Shainberg and Z. Malik (1987) *Curr. Microbiol.* 15, 251-258.

Nitzan, Y., H. Ladan, S. Gozhansky and Z. Malik (1987) *FEMS Microbiol. Lett.* 48, 401-406.

Nitzan, Y., B. Shainberg, S. Gozhansky and Z. Malik (1989) *curr. Microbiol.* 19, 265–269.

Vaara, M. and T. Vaara (1983) *Nature* 303, 526-528.pr

Effects of PDT on DNA and Chromosomes

J. MOAN, E. KVAM, E. HOVIG AND K. BERG

Institute for Cancer Research
The Norwegian Radium Hospital
Montebello, 0310 Oslo 3, Norway

Introduction

Many cytotoxic drugs used for cancer treatment interact with DNA. They inactivate rapidly proliferating cells with some selectivity, but may also be carcinogenic. The frequency of secondary cancer induced by such treatments may in certain cases amount to 30% of the treated patients (Schmähl et al., 1982). It is therefore highly relevant to evaluate the carcinogenic potential of new treatment modalities, such as photodynamic therapy (PDT). Most of the photosensitizing drugs used for PDT are lipophilic and localize mainly in membranes which therefore are thought to be the main targets for cell damage induced by PDT. Furthermore, the commonly used sensitizers (hematoporphyrin derivative, HpD and Photofrin II, PII) are anionic and do not readily bind to DNA in contrast to cationic dyes which have been extensively used to study photosensitized damage to DNA (Moan et al., 1989). Nevertheless, it has been reported that treatment with the anionic dye hematoporphyrin (Hp) and light may be carcinogenic (Bungeler, 1937; Santamaria, 1972).

Materials and Methods

Chemicals

All chemicals used in the present work were of the the highest purity commercially available. Two lipophilic photosensitizers were chosen: Photofrin II (PII) obtained from Photomedica, Raritan, N.J. and tetra (3-hydroxyphenyl) porphyrin (3THPP) obtained from Porphyrin Products, Logan, UT. In addition, a hydrophilic sensitizer was included in the study: Aluminium phthalocyanine tetrasulfonate ($AlPCS_4$) from Porphyrin Products.

Measurements of DNA single-strand breaks

The NHIK 3025 cell line (derived from a human cervix carcinoma) was used for the strand-break experiments. The cells were cultivated in E2a medium containing 20%

human serum and 10% horse serum. The cells were sensitized by incubation with the dyes (20 μg/ml PII, 2μg/ml 3THPP or 40 μg/ml AlPCS$_4$) for 18h in culture medium containing 3% serum. Before irradiation, the cells were trypsinized and brought into suspension in Dulbecco's phosphate buffered saline (PBS) containing 0.5% serum. The light source consisted of Philips TLD/83 fluorescent tubes with a Cinemoid 35 filter and had its main emission in the wavelength range 580–630 nm. To prevent repair of the DNA damage the cells were kept at 1°C during the irradiation. For comparison the cells were also exposed to 220kV X-rays at an exposure rate of 1.7 Gy/min. Immediately after the irradiation of $5 \cdot 10^5 - 10^6$ cells in 0.5 ml PBS, 0.5 ml 0.1N NaOH was added. Unwinding of the DNA was allowed to proceed for 30 min. at room temperature in the dark before neutralization with 0.5 ml 0.1N HCl and 0.5 ml of a solution containing 0.16% sodium lauryl sarcosinate, 150 mM K$_2$HPO$_4$, 50 mM KH$_2$PO$_4$, 40 mM disodium EDTA and 0.68 μg/ml Hoechst 33258. After neutralization, the cells were sonicated for 15 seconds. By measuring fluorescence of DNA-bound Hoechst 33258 the fraction of unbound DNA could be determined, since fluorescence decreased linearly with the proportion of single-strand DNA. Double-strand standards were obtained by mixing the alkaline solution with the neutralizing solution before addition to cells. Single-strand standards were obtained by sonicating cells during treatment with alkali. A more detailed description of the method can be found elsewhere (Kvam, 1988). We have earlier applied hydroxyapatite chromatography to separate double-strand and single-strand DNA in alkali-treated irradiated cells, a method that is comparable to the present one but more laborious (Moan et al., 1980).

Sister chromatide exchanges (SCEs)

The V79 cell line was chosen for this study since the number of chromosomes per cell is smaller for this cell line than for the NHIK 3025 cell line. The V79 cells were subcultured two times a week in MEM (Minimal Essential Medium) with Hank's salts, 100 U/ml penicillin, 100 μg/ml streptomycin and 10% foetal calf serum. The number of chromosomes per cell is 20.9 ± 0.15.

Mitotic cells were selected by shaking asynchronously growing cells with a reciprocal shaker for 10 sec. For survival experiments 100 Mitotic cells were inoculated in 25 cm^2 flasks and incubated in a sensitizer-free medium for proper binding of the cells to the substratum to occur. 1.5 hours after mitotic selection, the cells were incubated with the sensitizers (0.25 μg/ml 3THPP, 2.0 μg/ml PII or 50 μg/ml AlPCS$_4$) for 4.5 hours in the cultivation medium. At this time the cells had reached the S-phase (where the probability of inducing SCEs is supposedly at a maximum) and were exposed to light. Cells with AlPCS$_4$ were exposed to light from the Philips TLD/83 tubes described above. While cells with 3THPP and PII were exposed to light from a bank of 4 fluorescing tubes (Mod. 3026, Appl. Photophysics, London) with the highest fluence rate around 405 nm. The fluence rates of the two light sources were similar (30–40 W/m^2). After light exposure the culture medium was changed to fresh medium without sensitizer. Cell survival and SCE experiments were performed in parallel. For the

822

survival measurements, the cells (400 per 25 cm² flask) were incubated for 6 days at 37°C before fixation, staining and colony counting.

For the SCE assay 1•10⁵ mitotic cells were inoculated in 25 cm² plastic flasks and treated as described above. 5-bromodeoxyuridine (5 µg/ml), was added to the medium after the mitotic shake off and until the preparation for the SCE assay. 0.1 µg/ml colchemid was added to the medium when the cells were seen to reach the second mitosis after synchronization. After 3h in colchemid, the cells were prepared for SCE measurements as described by Alves and Jonasson (1978). The fixed cells were dripped on to cold, humidified cover slips before staining with the Giemsa solution. Coded slides were used for all experiments. 20–30 mitotic cells were scored at each dose.

Results and Discussion

DNA single-strand breaks

In earlier work PDT-induced DNA damage has been compared with X-ray-induced damage at equitoxic dose levels (Moan et al., 1980; Gomer, 1980; Evensen and Moan, 1982; Gomer et al., 1983; Ben Hur et al., 1987). This is a useful comparison, since the

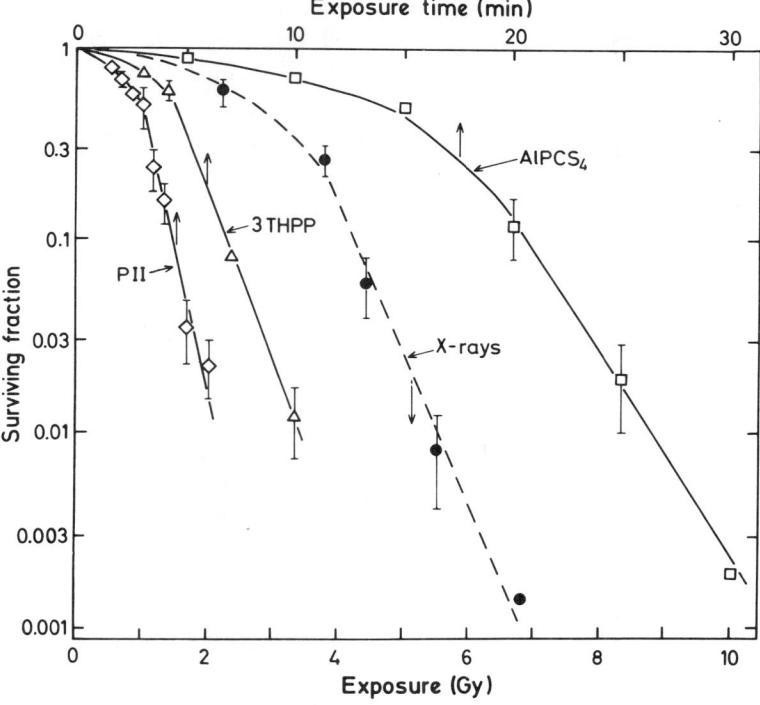

Figure 1. Survival curves for NHIK 3025 cells exposed to X-rays or to different photosensitizers and light. Incubation times with dyes: 18h at 37°C. See the text for further details.

genotoxic and carcinogenic potential of ionizing radiation has been thoroughly investigated. It is evident that data for survival should be included in such studies. Figure 1 shows survival curves for NHIK 3025 cells exposed to X-rays and to PDT. None of these sensitizers are toxic to the cells at the concentrations used here. All survival curves have a similar shape with a pronounced shoulder. Repair of sublethal damage has frequently been proposed as causing such shoulders. This is probably not always the case since curves for the lysis of liposomes exposed to PDT have similar shoulders (data not shown).

From survival curves like those shown in Fig 1, reliable D_{10}-values can be obtained. (D_{10} is the exposure needed to reduce the survival fraction to 10% of the untreated control).

All four treatments resulted in DNA single strand breaks in alkali (Fig 2). Furthermore, at an equitoxic dose level both of the lipophilic sensitizers appeared to induce less DNA damage of this type than did X-rays. This is in agreement with earlier work with lipophilic sensitizers including Hp, HpD and chloroaluminium phthalocyanine (Moan et al., 1980; Gomer, 1980; Ben-Hur et al., 1987).

The dose-response curve for the double-strand fraction (F) remaining after the treatment is logarithmic, as expected, in the case of X-rays (Fig 2). For PII and 3THPP, however, the corresponding dose-response curves bend slightly upwards. The reason for this may partly be that these sensitizers are photodegraded by large fluences (Moan et al., 1988) and partly that a certain fraction of the DNA is not reached by the photogenerated singlet oxygen ($^1O_2$). PII as well as 3THPP bind preferably to membranes, including the nuclear membrane, but do not accumulate inside the nucleus. Since $^1O_2$ does not travel more than 0.1μ during its lifetime (Moan et al., 1979), part of the DNA in the middle of the nucleus may remain unattacked during the treatment.

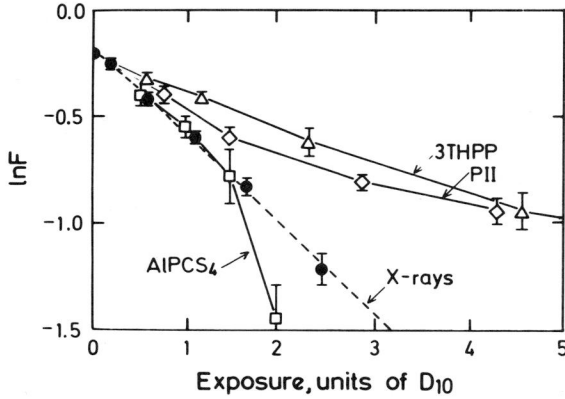

Figure 2. Induction of strand-breaks in cellular DNA by X-rays and PDT under conditons as described in the text. F is the double-stranded fraction of DNA after the alkaline treatment.

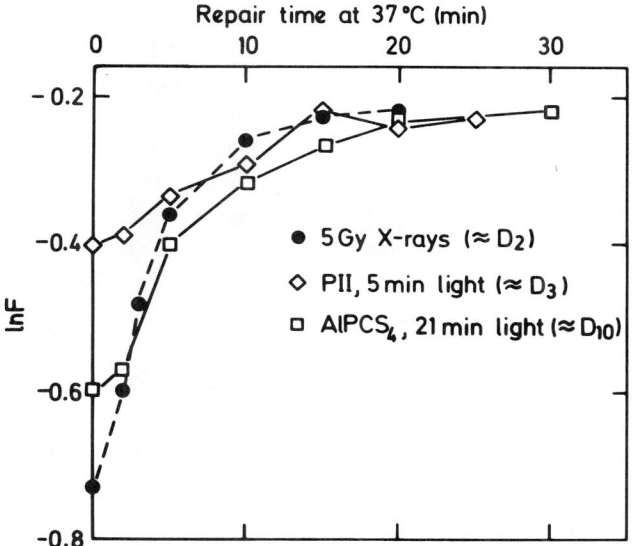

Figure 3. The repair of PDT- and X-ray-induced single-strand breaks in DNA in NHIK 3025 cells.

The dose-response curve for the hydrophilic dye AlPCS$_4$ bends downwards in the semilogarithmic plot (Fig. 2). The reason for this is probably that the radiation causes a deaggregation or a relocalization of the dye in the cells. Thus, we have observed that the fluorescence of AlPCS$_4$ in cells increases during light exposure, in contrast to what is found for PII and 3THPP. Furthermore, free AlPCS$_4$ has a higher fluorescence quantum

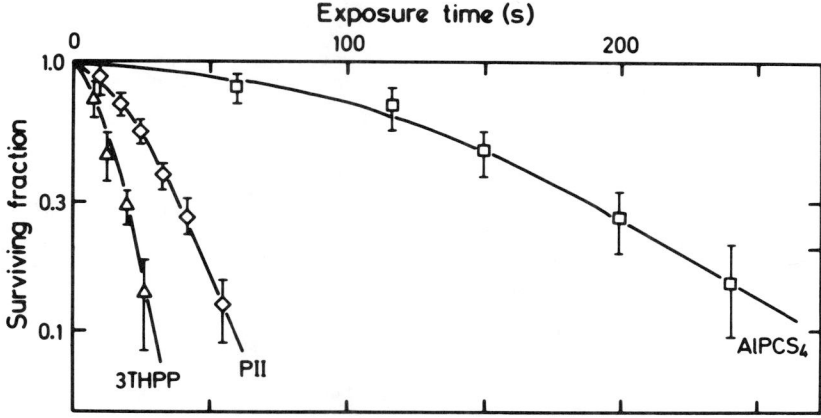

Figure 4. Survival curves for Chinese hamster V79 cells exposed to PDT under conditions as described in the text.

825

yield than protein-bound or cell bound AIPCS$_4$ (Kvam, 1988). In any case, at the same survival level, PDT with AIPCS$_4$ as the sensitizer induces a similar or even a larger amount of DNA single-strand breaks than do X-rays.

The rate of repair of DNA damage was also studied (Fig. 3). The damage induced by X-rays appear to be repaired at a similar or a slightly higher rate than the damage induced by PDT. Thus, the halflife of the damage is about 3 min for X-rays, 4 min for PDT with AIPCS$_4$ and 5 min for PDT with PII. This may indicate that the nature of the damage is different, which is not surprising since PDT is known to damage guanine specificly, while X-rays supposedly produce frank breaks randomly distributed in the DNA. The experimental method used in the present work includes both frank DNA single-strand breaks and alkaline-labile sites. We have reasons to believe that the latter type of damage plays the major role in the case of PDT (Boye and Moan, 1980).

Sister chromatid exchanges, chromosome aberrations and mutations

Survival curves for the V79 cells are shown in Fig 4. The shape of these survival curves are similar to those found for NHIK 3025 cells, although cells of the latter line are significantly less sensitive to PDT than are the V79 cells (data not shown). In agreement

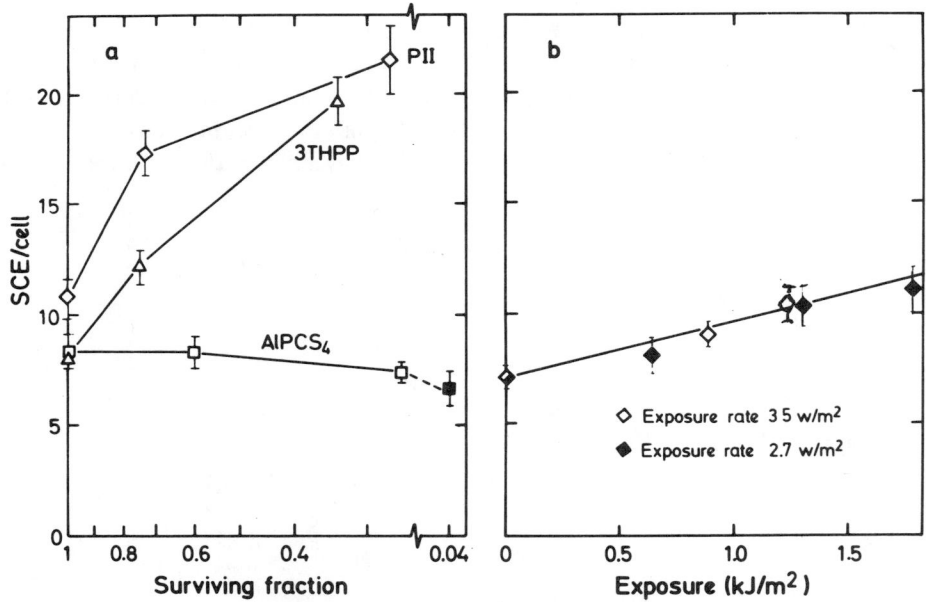

Figure 5. Induction of SCEs in V79 cells by PDT. The filled data point for AIPCS$_4$ corresponds to cells exposed to light after removal of the dye with which the cells had been incubated. This 5 min washing removed 40% of the cell-bound dye and resulted in an increase in the D$_{10}$-value from 4.5 min to 11.5 min.

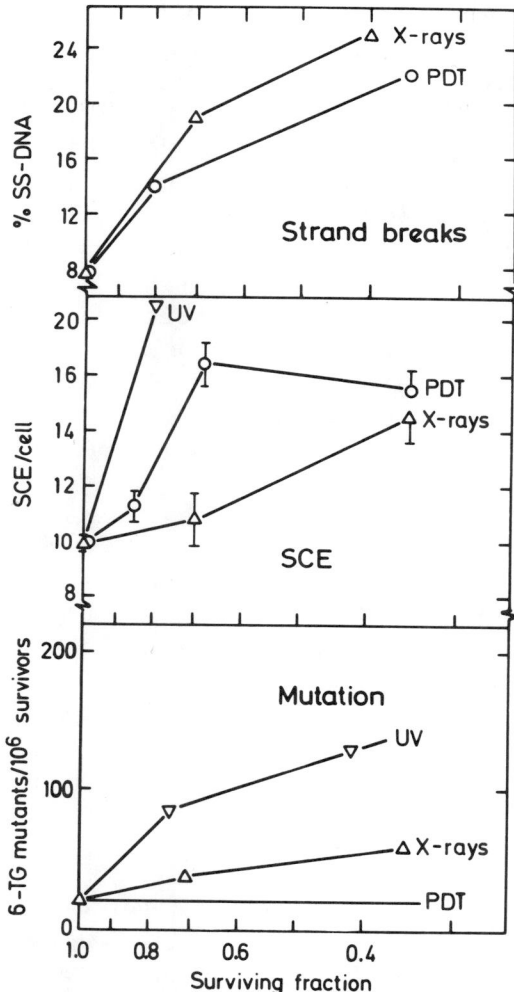

Figure 6. Upper panel: Induction of DNA single-strand breaks by X-rays (250 kV, 1.9 Gy/min) and PDT in Chinese hamster ovary (CHO) cells. Incubation for 1 h with 50μg/ml HpD in F-10 medium with 1% newborn calf serum. Data from Gomer (1980)Middle panel: Induction of SCEs in CHO cells by 254 nm UV light, HpD-sensitized PDT (1h incubation with 50 μg/ml HpD in F-10 medium with 1% FCS) or 300 kV X-rays (1.6 Gy/min, D_{10}=8Gy) (Data from Gomer et al., 1983).Lower panel: Mutations in CHO cells at the hypoxanthine-guanine phosphoribosyl transferase locus (resistance to 6-thioguanine) induced by 254 nm UV light, PDT and X-rays. Conditions as described above. Data from Gomer et al., 1983).

with earlier work (Moan et al., 1980) PDT with 3THPP and PII was found to induce SCEs (Fig. 5). Surprisingly, PDT with AlPCS$_4$, which gave the highest yield of strand-breaks (Fig. 2), did not induce SCEs. Since the exposure times for AlPCS$_4$ were much longer than those for the other drugs (Fig. 4), we thought that repair during the light exposure might explain this phenomenon. (According to figure 4, DNA repair occurs on a time-scale of minutes). It might be possible that SCEs were induced only at locations with two or more DNA single-strand breaks close to each other. However, this does not seem to be the case since the yield of SCEs was independent of the fluence rate when the latter was varied by more than a factor of 10 (Fig. 5, right panel). Therefore, SCEs and DNA single-strand breaks induced by PDT seem to be unrelated phenomena. Correspondingly, data from Gomer's group indicate that X-rays induce more single-strand breaks than does HpD sensitized PDT, in agreement with the findings described above, while the latter treatment seems to induce more SCEs, at least at low toxic levels (Fig. 6). This indicates that neither in this case is there any clear-cut correlation between single-strand DNA breaks and SCEs. The data for SCEs shown in Fig. 6 do not agree with our earlier work, where Hp sensitized PDT was found to be less potent in inducing SCEs than were X-rays (Moan et al., 1980). Correspondingly, we found that HpD sensitized PDT induced much fewer chromosome aberrations than did X-rays (Evensen and Moan, 1982). The reasons for the discrepancy between our work and that of Gomer's group may be that different incubation conditions and different cell lines were used.

Surprisingly, PDT with hydrophobic sensitizers such as HpD and chloroaluminium phthalocyanine does not induce any significant level of mutations over the control level (Fig. 6, Gomer et al., 1983; Ben-Hur et al., 1987). On the other hand, UV- light and X-rays are potent mutagens (Gomer et al., 1983; Fig. 6).

The correlation between mutant induction and SCE induction is poor (Fig. 6), an observation that complicates the choice of assay to assess the carcinogenic potential of PDT.

Conclusions

Practically all investigations carried out indicate that at a comparable cytotoxic level PDT with lipophilic and anionic sensitizers is less genotoxic than X-irradiation. These investigations include assays of DNA single-strand breaks, including alkali-labile sites, SCEs, chromosome aberrations, and mutations. Thus, PDT with such sensitizers is probably significantly less carcinogenic than X-irradiation. PDT with the hydrophilic sensitizer $AlPCS_4$ gives a high yield of DNA single-strand breaks, but no SCEs. There seems to be no correlation between the induction of DNA single-strand breaks and SCEs. Neither is there any clear-cut correlation between induction of SCEs and mutations.

Acknowledgements

The present work was supported by Norsk Hydro, by the Norwegian Cancer Society, and by the Norwegian Research Council for Science and the Humanities. The authors appreciate the critical reading of the manuscript by Professor Claude Rimington, FRS.

References

Alves, P. and J. Jonasson (1978) New staining method for the detection of sister chromatid exchanges in BrdU-labelled chromosomes. *J. Cell Sci.* 32, 185–195.

Ben-Hur, E., T. Fujihara, F. Suzuki and M.M. Elkind (1987) Gentic toxicology of the photo sensitization of Chinese hamster cells by phthalocyanines. *Photochem. Photobiol.* 45, 227–230.

Boye, E. and J. Moan (1980). The photodynamic effect of hematoporphyrin on DNA. *Photochem. Photobiol.* 31, 223–228.

Bungeler, W. (1937) Uber den Einfluss photo sensibilisierender Substanzen auf die Entstehung von Hautgeschwulsten. *Z. Krebsforsch.* 46, 130–167.

Gomer, C.J. (1980) DNA damage and repair in CHO cells following hematoporphyrin photoradiation. *Cancer Lett.* 11, 161–167.

Gomer, C.J., N. Rucker, A. Banerjee and W.F. Benedict (1983) Comparison of mutagenicity and induction of sister chromatid exchange in Chinese hamster cells exposed to hematoporphyrin derivative photoradiation, ionizing radiation, or ultraviolet radiation. *Cancer Res.* 43, 2622–2627.

Kvam, E. (1988) Photodynamic effect in NHIK 3025 cells: Survival, DNA strand-breaks and DNA-loops. Thesis, Oslo University.

Moan, J., K. Berg and E. Kvam (1989) Effects of photodynamic treatment on DNA and DNA-related cell functions. In "Photodynamic Therapy of Neoplastic Disease". (Edited by D. Kessel). CRC Press Inc. Boca Raton (in press).

Moan, J., E.O. Pettersen and T. Christensen (1979) The mechanism of photodynamic inactivation of human cells in vitro in the presence of haematoporphyrin. *Br. J. Cancer* 39, 398–407.

Moan, J., C. Rimington and Z. Malik (1988) Photoinduced degradation and modification of Photofrin II in cells *in vivo. Photochem. Photobiol.* 47, 363–367.

Moan, J., H. Waksvik and T. Christensen (1980) DNA single-strand breaks and sister chromatid exchanges induced by treatment with hematoporphyrin and light or by X-rays in human NHIK 3025 cells. *Cancer Res.* 40, 2915–2918.

Santamaria, L (1972) Further considerations on photodynamic action and carcinogenicity. In "Research Progress in Organic, Biological and Medical Chemistry", Vol 3, (Editors: U. Gallo and L. Santamaria), pp. 671–683, North Holland Publishing Company, Amsterdam.

Schmähl, D., M. Habs, M. Lorentz and I. Wagner (1982). Occurrence of second tumors in man after anticancer drug treatment. *Cancer Treatment Reviews* 9, 167–194.

829

Photophysical and Photosensitizing Properties of Phthalocyanines

JOHN D. SPIKES

Department of Biology, University of Utah,
Salt Lake City,
Salt Lake City, Utah 84112, USA

Introduction

Solid tumors selectively retain certain porphyrins; on subsequent illumination the tumors can often be destroyed with relatively little damage to the adjacent normal tissue. This modality, termed photodynamic therapy (PDT), has been used with several thousand cancer patients with generally favourable results (Dougherty, 1987). Porphyrins absorb light poorly in the red region of the spectrum where light penetration into tissues is high. However, several groups of "second generation" PDT sensitizers that absorb more effectively in the red have now been identified, among them the phthalocyanines (PCs). Phthalocyanines have a number of characteristics that recommend their application in PDT. These include strong red absorption (670–680nm), selective retention in tumors, efficient photosensitizing properties, apparent low toxicity, ease of synthesis, and resistance to chemical and photochemical degradation; several types of implanted tumors in laboratory animals have been successfully treated by PC sensitized PDT (see Spikes, 1986; Ben-Hur et al., 1987; Van Lier et al., 1988). The present paper briefly reviews the chemistry, photophysics, photochemistry and photosensitizing properties of PCs, primarily as examined under physiological conditions in aqueous solutions at pH 7.4. Phthalocyanines are blue-green compounds that are widely used as pigments for paints and inks, as light absorbers in electrocatalysis, in photovoltaic applications, as sensitizers in photocatalysis, as photoconductors, as electrochromic agents and as chemical sensors; some 4900 PCs are known (Lever, 1987).

Chemistry of Phthalocyanines

Phthalocyanine (azaporphyrin) is an analog of porphine. The PC macrocycle consists of four isoindole subunits linked by nitrogens. A large number of different metal ions can be inserted into the center of the macrocycle. Also, PCs can be prepared with a variety of different substituents on the isoindole subunits, including amino, carboxy, hydroxy, neopentoxy, nitro, sulfo, etc. groups (see the reviews listed above, also Rosenthal et al., 1987). Phthalocyanine and its metal derivatives are insoluble in water. However, zinc PC

Photobiology, Edited by E. Riklis
Plenum Press, New York, 1991

can be incorporated into phospholipid liposomes and thus be made compatible with physiological media (Valduga et al., 1988). Sulfonated PCs tend to be more water soluble, depending on the metal substituent, and have been studied extensively as sensitizers for biological systems. The mono–, di–, tri–, and tetrasulfonates of gallium PC (Brasseur et al., 1987), aluminum PC (Paquette et al., 1988; Spikes and Bommer, 1988) and zinc PC (Spikes and Bommer, 1988) have been prepared; even the four possible constitutive isomers of gallium PC disulfonate (as prepared by the condensation technique) have been purified (Brasseur et al., 1987). The water solubility of aluminum and gallium PCs tends to increase with an increasing degree of sulfonation; however, different isomers of PCs with a given degree of sulfonation can have rather different water solubilities, as in the case of the isomers of gallium PC disulfonate mentioned above. Aluminum PC tetrasulfonate (AlPCTS) dissolves in aqueous buffer at pH 7.4 and is present largely as the monomer. Zinc PC tetrasulfonate (ZnPCTS) appears to dissolve, but is actually present as dimers or aggregates; addition of 10mM cationic detergent gives complete monomerization (Darwent et al., 1982; Spikes and Bommer, 1986; 1988).

Photophysical Properties of Phthalocyanines

Monomeric aluminum and zinc PCs in aqueous media have characteristic absorption spectra with a Soret band at approximately 350nm, a small band around 600nm, and a narrow, very strong absorption peak (Q band) at 670–680nm (molar extinction coefficients in the range of $10^5 M^{-1} cm^{-1}$). Aggregated PCs have the Soret band shifted to shorter wavelengths and a broad Q band at approximately 630–640nm. Monomeric PCs fluoresce strongly, with a single peak in the 680–690nm range, while aggregated PCs do not fluoresce appreciably. The fluorescence lifetime of AlPCTS in an aqueous medium is approximately 6ns (Darwent et al., 1982; Spikes and Bommer, 1986; 1988; Svensen et al., 1988).

In aqueous buffer at pH 7.4, we find that monomeric aluminum and zinc PCs give good yields of long-lived triplet states on flashing at 355nm; the triplet-triplet difference spectra show a broad peak extending from 400–600nm. Aggregated PCs produce only very small yields of triplets. The AlPCTS triplet decays by a first order process with a lifetime of 540µs under nitrogen. With ZnPCTS in 10mM CTAB, the triplet lifetime is 240µs; little triplet is produced in the absence of detergent. Oxygen and benzoquinone quench the AlPCTS triplets efficiently with bimolecular quenching constants of 1.8 and $2.0 \times 10^9 M^{-1} s^{-1}$, respectively. The quenching constants for sodium azide, CTAB, cysteine, furfuryl alcohol, guanosine, histidine, lysozyme, methionine, tryptophan and tyrosine are less than $2 \times 10^6 M^{-1} s^{-1}$ at pH 7.4; thus these compounds would not compete significantly with oxygen for reaction with PC triplets under aerobic conditions (Spikes and Bommer, 1986; 1988). The quantum yields of triplet formation increase progressively in going from metal-free to aluminum substituted to zinc substituted PCs; PCs substituted with closed shell diamagnetic metals such as aluminum and zinc typically have long triplet lifetimes. In contrast, PCs substituted with paramagnetic transition metals have very short triplet lifetimes (in the ns range; Darwent et al., 1982). The degree of sulfonation of gallium, aluminum and zinc PCs has little effect on their photosensitizing efficiencies as long as the compounds are monomeric.

Mechanisms of Phthalocyanine-Sensitized Photoreactions

Triplet PCs react with other molecules by a variety of mechanisms. For example, in Type I (electron transfer) processes under anaerobic conditions, illuminated PCs donate an electron to methyl viologen or benzoquinone giving the PC radical cation and radical anions of the organic compounds (Ohtani et al., 1986; Ohno et al., 1983). However, they can also abstract an electron from tyrosine at high pH to give the PC radical anion and the tyrosine radical cation (Ferraudi et al., 1988). Under appropriate conditions in the presence of oxygen, illuminated PCs can produce superoxide in low efficiency reactions (Ben-Hur et al., 1985). Hydrogen peroxide can also be formed with high efficiency in the presence of certain organic substrates (Spikes and Bommer, 1986; 1988). In the Type II process, illuminated PCs react with ground state oxygen to produce singlet oxygen with quantum yields in the range of 0.3–0.5 (Rosenthal et al., 1986; Wagner et al., 1987). Phthalocyanines containing paramagnetic metal ions do not generate appreciable amounts of singlet oxygen (Rosenthal et al., 1986). In summary, PC-sensitized photoreactions can be complex, although they probably proceed largely via the singlet oxygen mechanism under aerobic conditions in aqueous solution at neutrality.

Quantum Yields, Kinetics and Mechanism of the Sensitized Photooxidation of Furfuryl Alcohol by Phthalocyanines

Furfuryl alcohol (FA) is a useful substrate for delineating the properties of photosensitizers since it is rapidly oxidized by singlet oxygen, but does not react with either hydrogen peroxide or superoxide. Further, it is miscible with water, does not change properties over the pH range 1–11, and is transparent to visible light. The quantum yields of oxygen uptake during the sensitized photooxidation of FA by monomeric PCs are 0.057 with A1PCTS and 0.15 with ZnPCTS, reflecting the higher quantum yield of triplet generation for the latter compound. The yields are independent of light intensity. They increase rapidly with increasing FA concentration, are essentially independent of sensitizer concentration, and are independent of oxygen concentration from 0.24mM (air saturated water) down to approximately 0.010mM. The yields are decreased 50% by 5×10^{-4}M sodium azide and are increased 9-fold in D_2O at low FA concentrations; since azide does not quench triplet PCs appreciably, these data suggest that the PC-sensitized photooxidation of FA is mediated by singlet oxygen. The yields are not affected by pH over the range 4–10. During the PC-sensitized photooxidation of FA, 0.8 moles of hydrogen peroxide are produced per mole of oxygen consumed (Spikes and Bommer, 1986; 1988).

The Phthalocyanine-Sensitized Photooxidation of Biomolecules

Monomeric A1PCTS and ZnPCTS sensitize the photooxidation of the susceptible amino acids in air saturated solutions. The quantum yields of oxygen uptake during these reactions, as measured at pH 7.4 with ZnPCTS, are 0.036 for cysteine, 0.074 for histidine, 0.033 for methionine, 0.059 for tryptophan and 0.0037 for tyrosine. The yield for guanosine is 0.013 and that for thymine is 0.001. With A1PCTS, the yields are approximately 50% lower, as in the case of FA photooxidation (Spikes and Bommer, 1986; 1988). The pattern of products formed during the PC-sensitized photooxidation of L-tryptophan indicates that the reaction is mediated by singlet oxygen; similarly, the photooxidation product of cholesterol (the 5–alpha–hydroperoxide) demonstrates a singlet oxygen pathway for this substrate (Langlois et al., 1986). Under anaerobic conditions, tryptophan and tyrosine interact with PC triplets by electron transfer reactions to give the one electron reduced PC and, presumably, the one electron oxidized radicals of the amino acids as initial products (Ferraudi et al., 1988). The enzyme, lysozyme, is inactivated photodynamically with monomeric ZnPCTS with a quantum yield of 0.015; this is approximately the same efficiency as observed with good porphyrin sensitizers such as tetrasulfonatophenyl porphine (Spikes and Bommer, 1986). Aluminum PC sulfonate sensitizes the photodynamic inactivation of the membrane-bound enzyme, beta-hydroxybutyrate dehydrogenase, in rat liver mitochondria and mitochondrial membrane fragments (Robinson et al., 1986). Illumination of the human serum albumin-sulfonated aluminum PC complex destroys tryptophan residues in the protein; destruction is ten times faster in D_2O as solvent than in H_2O (Svensen et al., 1988). Illumination of the aluminum PC sulfonate-bovine serum albumin complex results in the photooxidation of the sensitive amino acid residues in the protein. The binding of PC to bovine serum albumin significantly decreases its ability to sensitize the photooxidation of free tryptophan in solution (Ben-Hur, 1987).

Photobleaching of Phthalocyanines

Phthalocyanines in aqueous solution at neutrality are very resistant to photochemical degradation. For example, sulfonated aluminum PC is photobleached with the very low quantum yield of 1.7×10^{-6} in H_2O; the yield is ten times greater in D_2O, suggesting that the photobleaching process is mediated by singlet oxygen (McCubbin and Phillips, 1986). The metal substituent in a PC can have a significant effect on the light sensitivity. We find that the quantum yield of the photobleaching of ZnPCTS is over two orders of magnitude greater than that for A1PCTS (Spikes and Bommer, 1988). The presence of other molecules may also have an effect. For example, sulfonated aluminum PC bound noncovalently to bovine serum albumin photobleaches approximately seven times faster than the PC free in solution (Ben-Hur, 1987).

Naphthalocyanines

Naphthalocyanines (NPs) have strong absorption bands in the 750–780nm range where light penetration into mammalian tissues is almost twice as great as at 630nm (where porphyrins absorb). Thus there is interest in the NPs as possible sensitizers for PDT. Monomeric aluminum and zinc sulfonated NPs fluoresce with lifetimes in the 2–3ns range. On flashing, they produce triplets with absorption peaks at 580nm and lifetimes of 240 μs for the aluminum derivative and 115 μs for the zinc derivative (McCubbin and Phillips, 1986). Aluminum and zinc sulfonated NPs photobleach much more rapidly than the PCs. The rate of the photobleaching of zinc sulfonated NP is decreased by beta-carotene, suggesting the participation of singlet oxygen in the process (McCubbin and Phillips, 1986; Yates, 1988). Bis(tri–n–hexylsiloxy)(2,3 naphthalocyaninato) silicon, a lipophilic silicon derivative of NP, absorbs strongly at 776nm and fluoresces with a lifetime of 2.85ns. On flashing, a triplet is observed which peaks at 590nm and which has a lifetime of 331 μs. The triplet, which is produced with a quantum yield of 0.39, is quenched by ground state oxygen to form singlet oxygen, as observed by its near infrared luminescence. The reaction of the triplet silicon derivative with oxygen is reversible, reflecting the rather low triplet energy of this NP sensitizer (Firey and Rodgers, 1987; Firey et al., 1988).

Acknowledgements

The preparation of this paper and the original work described were supported by American Cancer Society Grant PDT-259 and by the SDIO/MFEL Program (ONR Contracts N00014-86-K-0258 and N00014-86-K-0710) and the Utah Laser Institute.

References

Ben-Hur, E. (1987). Photochemistry and photobiology of phthalocyanines: new photosensitizers for photodynamic therapy of cancer. *Photobiochemistry and Photobiophysics,* Suppl., 407–420.

Ben-Hur, E. and Rosenthal, I. (1986). Photosensitization of Chinese hamster cells by water-soluble phthalocyanines. *Photochemistry and Photobiology,* 43, 615–619.

Ben-Hur, E., Carmichael, A., Riesz, P., and Rosenthal, I. (1985). Photochemical generation of superoxide radical and the cytotoxicity of phthalocyanines. *International Journal of Radiation Biology,* 47, 837–846.

Ben-Hur, E., Rosenthal, I., Bown, S.G., and Phillips, D. (1987). The phthalocyanines: sensitizers with potential for photodynamic therapy of cancer. *Photomedicine,* edited by E. Ben-Hur and I. Rosenthal (Boca Raton, FL:CRC Press), vol 3, pp. 1–17.

Brasseur, N., Ali, H., Langlois, R., and Van Lier, J.E. (1987). Biological activities of phthalocyanines-VII. Photoinactivation of V-79 Chinese hamster cells by selectively sulfonated gallium phthalocyanines. *Photochemistry and Photobiology,* 46, 739–744.

Darwent, J.R., Douglas, P., Harriman, A., Porter, G., and Richoux, M.C. (1982). Metal phthalocyanines and porphyrins as photosensitizers for reduction of water to hydrogen. *Coordination Chemistry Reviews,* 44, 83–126.

Dougherty, T.J. (1987). Photosensitizers: therapy and detection of malignant tumors. *Photochemistry and Photobiology,* 45, 879–889.

835

Ferraudi, G., Arguello, G.A., Ali, H., and Van Lier, J.E. (1988). Types I and II sensitized photooxidation of aminoacid by phthalocyanines: a flash photochemical study. *Photochemistry and Photobiology*, 47, 657–660.

Firey, P.A., and Rodgers, M.A.J. (1987). Photo-properties of a silicon naphthalocyanine: a potential photosensitizer for photodynamic therapy. *Photochemistry and Photobiology*, 45, 535–538.

Firey, P.A., Ford, W.E, Sounik, J.R., Kenney, M.E., and Rodgers, M.A.J. (1988). Silicon naphthalocyanine triplet state and oxygen: a reversible energy transfer reaction. *Journal of the American Chemical Society*, 110, 7626–7630.

Langlois, R., Ali, H., Brasseur, N., Wagner, J.R., and Van Lier, J.E. (1986). Biological activities of phthalocyanines-IV. Type II sensitized photooxidation of L-tryptophan and cholesterol by sulfonated metallo phthalocyanines. *Photochemistry and Photobiology*, 44, 117–123.

Lever, A.B.P. (1987). The other periodic chart. *Chemtech*, August, 506–510.

McCubbin, I., and Phillips, D. (1986). The photophysics and photostability of zinc(II) and aluminium(III) sulphonated naphthalocyanines. *Journal of Photochemistry*, 34, 187–195.

Ohno, T., Kato, S., Yamada, A., and Tanno, T., (1983). Electron transfer reactions of the photoexcited triplet state of chloroaluminum phthalocyanine with aromatic amines, benzoquinones, and coordination compounds of iron(II) and iron(III). *Journal of Physical Chemistry*, 87, 775–781.

Ohtani, H., Kobayashi, T., Tanno, T., Yamada, A., Wohrle, D., and Ohno, T. (1986). Efficient photoreduction of methylviologen by metallophthalocyanine sensitizers. *Photochemistry and Photobiology*, 44, 125–129.

Paquette, B., Ali, H., Langlois, R., and Van Lier, J.E. (1988). Biological activities of phthalocyanines-VIII. Cellular distribution in V-79 Chinese hamster cells and phototoxicity of selectively sulfonated aluminum phthalocyanines. *Photochemistry and Photobiology*, 47, 215–220.

Robinson, R.S., Roberts, A.J., and Campbell, I.D. (1986). Photo-oxidation effects on beta-hydroxybutyrate dehydrogenase: studies on membrane fragments and intact mitochondria. *Photochemistry and Photobiology*, 45, 231–234.

Rosenthal, I., Murali Krishna, C., Riesz, P., and Ben-Hur, E. (1986). The role of molecular oxygen in the photodynamic effect of phthalocyanines. *Radiation Research*, 107, 136–142.

Rosenthal, I., Ben-Hur, E., Greenberg, S., Concepcion-Lam, A., Drew, D.M., and Leznoff, C.C. (1987). The effect of substituents on phthalocyanine photocytotoxicity. *Photochemistry and Photobiology*, 46, 959–963.

Spikes, J.D.(1986) Yearly Review. Phthalocyanines as photosensitizers in biological systems and for the photodynamic therapy of tumors. *Photochemistry and Photobiology*, 43, 691–699.

Spikes, J.D., and Bommer, J.C..(1986). Zinc tetrasulphophthalocyanine as a photodynamic sensitizer for biomolecules. *International Journal of Radiation Biology*, 50, 41–45.

Spikes, J.D., and Bommer, J.C. (1988). Photophysical and photosensitizing properties of sulfonated aluminum and zinc phthalocyanines, To be published.

Svensen, R., Fery-Forgues, S., MacRobert, A.J., and Phillips, D. (1988). Pulsed laser studies of aluminium phthalocyanine derivatives. *Photosensitization*, edited by G. Moreno, R.H. Pottier, and T.J. Truscott (Berlin: Springer-Verlag), pp. 445–448.

Valduga, G., Nonell, S., Reddi, E., Jori, G., and Braslavsky, S.E. (1988). The production of singlet molecular oxygen by zinc(II) phthalocyanine in ethanol and in unilamellar vesicles. Chemical quenching and phosphorescence studies. *Photochemistry and Photobiology*, 48, 1–5.

Van Lier, J.E., Brasseur, N., Paquette, B., Wagner, J.R., Ali, H., Langlois, R., and Rousseau, J. (1988). Phthalocyanines as sensitizers for photodynamic therapy of cancer. *Photosensitization*, edited by G. Moreno, R.H. Pottier, and T.G. Truscott (Berlin: Springer-Verlag), pp. 435–444.

836

Wagner, J.R., Ali, H., Langlois, R., Brasseur, N., and Van Lier, J.E. (1987). Biological activities of phthalocyanines-VI. Photooxidation of L-tryptophan by selectively sulfonated gallium phthalocyanines: singlet oxygen yields and effects of aggregation. *Photochemistry and Photobiology*, 45, 587–594.

Yates, N.C. (1988). Water-soluble metal naphthalocyanines as potential photosensitizers. *Photosensitization*, edited by G. Moreno, R.H. Pottier and T.G. Truscott (Berlin: Springer-Verlag), pp. 365–368.

Zinc(II)-Phthalocyanine as a Second-Generation Photosensitizing Agent in Tumour Phototherapy

G. JORI, R. BIOLO, C. MILANESI, E. REDDI AND G. VALDUGA

Department of Biology
University of Padova
Via Loredan 10, Padova, Italy

Keywords: Zn(II)-phthalocyanine, Photodynamic therapy, Singlet oxygen, Pharmacokinetics, Ultrastructural studies.

Introduction

The use of phthalocyanines as phototherapeutic agents for tumours was first proposed in 1985 (Ben-Hur and Rosenthal, 2985). A variety of cellular and animal studies, (for a review see Spikes, 1986; Tralau et al., 1987) have supported this proposal and it presently appears that some phthalocyanines are probable choices as second-generation tumour-localizers and -photosensitizers. At present, the phthalocyanines most frequently used in the experimental photodynamic therapy (PDT) of tumours include some sulfonated derivatives of the tetra-azaisoindole macrocycle (Brasseur et al., 1985; Ben-Hur and Rosenthal, 1986). Although these derivatives are relatively water-soluble, they are still a mixture of compounds with different degrees of sulfonation and/or showing structural isomerism.

In our laboratory, we have focused our attention on an unsubstituted phthalocyanine, Zn(II)-phthalocyanine (Zn-Pc), whose main properties of interest for phototherapeutic applications are summarized in Table 1. Besides having a high degree of purity, as assessed by elemental analysis and high-pressure liquid chromatography (Valduga et al., 1987), Zn-Pc exhibits a large molar extinction coefficient in the 670-680 nm interval, as well as a remarkable chemical stability to visible light irradiation. The latter feature guarantees against photo-induced bleaching of the photosensitizing agent during PDT, which would reduce therapeutic efficacy and possibly generate potentially toxic photoproducts. Moreover, Zn-Pc is a very hydrophobic species, and hence can only be administered to animals after incorporation into phospholipid vesicles or oil emulsions (Morgan and Garbo, 1988). Previous studies in our laboratory (Jori et al., 1986) demonstrated that the use of unilamellar liposomes of the saturated phospholipid dipalmitoyl-photophatidylcholine (DPPC) as drug carrier *in vivo* allows more efficient delivery of porphyrins to neoplastic tissues.

Photobiology, Edited by E. Riklis
Plenum Press, New York, 1991

Table 1. Selected physico-chemical properties of zinc(II)-phthalocyanine

Property	Description
Absorption maximum	673 nm (DPPC liposomes)
Extinction coefficient at λmax	153,000 M^{-1} cm^{-1}
Fluorescence maximum	680 mn (DPPC liposomes)
Fluorescence quantum yield	0.14 (DPPC liposomes)
Degree of purity	> 97%
Solubility	Insoluble in water
	Soluble in pyridine, dimethylsulfoxide
Stability to visible light irradiation	Very high

Photosensitizing Properties of Zinc(II)-Phthalocyanine

In a preliminary investigation, we ascertained that Zn-Pc can efficiently photosensitize the irreversible chemical modification of model substrates, such as the amino-acid L-tryptophan, or the heterocyclic compound 1,3-diphenylisobenzofuran (DPBF). The photoprocess was studied both in homogeneous solutions and in microheterogeneous media and was found to be oxidative in nature; in particular, the photooxidation of both substrates was exclusively carried out by electrophilic attack of $^{1}O_2$ generated through triplet-triplet energy transfer from the photoexcited Zn-Pc to ground-state dioxygen (Valduga et al., 1988). The ability of phthalocyanine to photogenerate $^{1}O_2$ has been demonstrated by several authors (Spikes, 1986; Brasseur et al., 1987), although the efficiency of the photoprocess is affected by chemical structure of the sensitizer (Rosenthal et al., 1987).

The quantum yield of $^{1}O_2$ generation by Zn-Pc was measured in both homogeneous solutions and microheterogeneous systems using DPBF as a substrate; this compound reacts with $^{1}O_2$ in a purely chemical fashion; hence, the extent of its disappearance is correlated with the amount of $^{1}O_2$ present in the system (Foote, 1988). As may be seen from the data summarized in Table 2, the efficiency of $^{1}O_2$ generation by triplet Zn-Pc and the reactivity of $^{1}O_2$ towards DPBF are closely similar in the media examined.

Table 2. Photosensitizing properties of zinc(II)-phthalocyanine toward 1,3-diphenyliso-benzofuran (DPBF).

Property	Value	
	A	B
Quantum yield for $^{1}O_2$ generation	0.53	0.70
Rate constant ($^{1}O_2$ + DPBF)	1.21	1.20

A = Data in ethanol solution; B = Data in neutral (pH 7.4) aqueous dispersions of DPPC liposomes
Zn-Pc = 5 uM; DPBF = 5-50 uM.

Under our experimental conditions, Zn-Pc was present as a monomeric species, as shown by absorbance measurements at 680 nm and time-resolved fluorescence emission studies, indicating the existence of one transient species (lifetime about 3 ns). In general, in their monomeric state, macrocyclic dyes show photosensitizing activity appreciably greater than that typical of their dimeric or oligomeric analogs, owing to the longer lifetime of the lowest excited singlet and triplet states (Jori and Spikes, 1984). The lack of any appreciable aggregation of Zn-Pc in ethanol solution or in the phospholipid bilayer of DPPC liposomes is a consequence of the low dielectric constant of these media, which ensures a thorough solvation of the hydrophobic tetraazaisoindole moiety, thus disrupting apolar intermolecular interactions. The stability of the Zn-Pc-liposome association was also verified in the presence of human serum albumin; as previously observed for hematoporphyrin (Cozzani et al., 1985), albumin is unable to promote the leakage of drugs embedded in DPPC liposomes. It therefore appears reasonable to use the liposome-bound Zn-Pc *in vivo*. The homogeneous physical state of Zn-Pc should allow a more uniform transport mechanism in the bloodstream, owing to better control of its pharamacokinetic and pharmacodynamic behaviour. Lastly, it is important to stress that the absorption and fluorescence spectra of ethanol-dissolved or liposome-bound Zn-Pc were not altered by prolonged irradiation with visible light. Zn-Pc therefore appears to be endowed with remarkable photochemical stability, whereas the corresponding sulfonated derivatives undergo rapid photobleaching (Abernathey et al., 1987). The latter process would reduce the photosensitizing efficacy of phthalocyanine *in vivo* and possible originate toxic photodegradation products.

Biodistribution of Liposome-Delivered Zinc(II)-Phthalocyanine in Tumour-Bearing Mice

The tumour-localizing activity of Zn-Pc incorporated into small unilamellar vesicles of DPPC liposomes was examined by using female Balb/c mice (weight 20-25 g) as an experimental model. The mice had a MS-2 fibrosarcoma transplanted in the right hind leg. On the seventh day after transplantation they were intravenously injected with 0.1-0.2 mg of liposome-bound Zn-Pc per kg of body weight, while the tumours had an external diameter of about 0.8 cm. At predetermined times after the administration of the photosensitizing agent, the mice were sacrificed and the Zn-Pc content was estimated in the serum and in tissue specimens: the tumour and several normal tissues were quickly removed, washed with phosphate-buffered saline, weighed and homogenized in a Potter vessel. The Zn-Pc was then extracted from the tissue homogenates by incubation with 2% aqueous SDS, whereby the phthalocyanine is incorporated into the surfactant micelle in a monomeric form. Under these conditions, the intensity of Zn-Pc fluorescence emission is linearly related with the dye concentration.

Our recovery studies were mainly focused on the tumour tissue, liver (i.e., site of metabolic elimination of poorly water-soluble phthalocyanine), muscle (representing the normal tissue where the tumour grows), and the skin, in view of the potential risk of cutaneous photosensitivity in patients subjected to photodynamic therapy (Dougherty, 1984).

The main pharmacokinetic properties of Zn-Pc are summarized in Table 3. Clearly, the disappearance of the systemically injected drug from the serum follows biphasic kinetics: the rapidly eliminated fraction, corresponding to about 80% of the totally administered Zn-Pc, is probably responsible for the accumulation of the photosensitizer by the tumour and most normal tissues. Actually, the maximum concentration of Zn-Pc in the liver and other components of the reticuloendothelial system is reached about 3 h after administration, while maximal Zn-Pc levels in the tumour are found after about 14 h. The slower uptake of Zn-Pc by the tumour compared with the normal tissues examined by us reflects the mechanism by which the drug is transported in the bloodstream and released to the neoplastic tissue. Column chromatographic analysis of serum samples taken from Zn-Pc-treated mice (Reddi et al., 1987) revealed that the liposome-associated phthalocyanine had been quantitatively transferred to serum lipoproteins already shortly after injection. The ability of DPPC liposomes to interact selectively with lipoproteins had previously been ascertained in our laboratory (Ricchelli et al., 1988). In particular, Zn-Pc is distributed among the main classes of lipoprotein family.

Several lines of evidence (Jori et al., 1984; Reyftmann et al., 1984; Zhou et al., 1988) suggest that low-density lipoproteins (LDL) play a major role in the delivery of associated photosensitizers to tumours *in vivo* via a specific receptor-mediated endocytotic process. LDL endocytosis by tumour tissues is characterized by slower kinetics than compared with drug accumulation by most normal tissues (Netland et al., 1985).

It therefore appears that the use of liposomes as drug carriers favours the transport of Zn-Pc by LDL, which would explain both the large accumulation of the phthalocyanine in the tumour and the high ratio of Zn-Pc concentration between tumour and muscle, in spite of the low injected dose. These facts should ensure an efficient response of the tumour to photodynamic treatment with minimal photoinduced damage to the surrounding normal tissues.

Experimental Photodynamic Therapy with Zinc(II)-Phthalocyanine

On the basis of the pharmacokinetic investigations described, above we developed an experimental protocol for the photodynamic treatment of the MS-2 fibrosarcoma transplanted in Balb/c mice. The main features of our phototherapeutic protocol are detailed in Table 4. Under the conditions used for irradiation, we detected no damage to the tumour tissue when no Zn-Pc was injected. Moreover, the temperature rise of the irradiated tissues never exceeded 3–4°C, indicating that no significant hyperthermal effects had taken place.

On the other hand, tumour tissues exposed to 680 nm light 24 h after injection of 0.1–0.2 mg/kg Zn-Pc displayed extensive necrosis as early as 15 hr after the completion

Table 3. Pharmacokinetic properties of Zinc(II)-phthalocyanine delivered via DPPC liposomes to Balb/c mice bearing a MS-2 fibrosarcoma (Injected dose: 0.15 mg/kg)

Tissue	Distribution of Zn-Pc
Serum	Selective transport by lipoproteins. Ca. 80% of injected Zn-Pc is cleared within $t1/2 = 6-8$ h; the slowly cleared Zn-Pc has a half-life of ca. 10 days.
Liver	Maximum accumulation (1–2 ug/g) 3 h after injection. Complete elimination after 5 days.
Skin	Maximum accumulation (0.3–0.4 ug/g) 3 h after injection. Residual Zn-Pc is below 0.1 ug/g 7 days later.
Brain	No significant accumulation.
Tumour	Maximum accumulation (0.8–1 ug/g) 24 h after injection. Very slow release: ca. 0.5 ug/g is present after 7 days. Tumour/muscle ratio of Zn-Pc concentration higher than 7 after 12 h.

Table 4. Protocol used for photodynamic therapy (PDT) of MS-2 fibrosarcoma in mice treatment with zinc(II)-phthalocyanine

Photosensitizer dose:	0.14 mg/kg body weight
Delivery system:	small unilamellar DPPC liposomes
Injection mode:	intravenous
Beginning of PDT:	24 h after injection
Irradiating light:	600–690 nm
Irradiation dose-rate:	180 mW/cm^2
Total light dose:	300 J/cm^2

of photodynamic therapy. The necrotic area gradually expanded and reached the leg bone within 3 days. The depth of tumour necrosis was evaluated by histological examination of the irradiated tissues 24 h after phototherapy and was found to increase linearly with the injected dose of Zn-Pc, at least up to 0.5 mg/kg of photosensitizer.

The extent of the tumour response also depended on the total light dose and the irradiation dose-rate. Thus, measurable tumour necrosis was found only on delivery of light doses greater than 50 J/cm^2, whereas a synergistic interaction between photochemical and thermal effects was likely to occur at dose-rates above 230 mW/cm^2. Lastly, all other experimental parameters being constant, the photoinduced necrosis of the fibrosarcoma was essentially independent of the time-interval between Zn-Pc administration and phototherapy, at least in the 12–72 h range. This is a consequence of the very slow release of LDL-delivered drugs from tumour tissues (Zhou et al., 1988) and may allow repeated phototherapeutic treatments of a given tumour after a single injection of Zn-Pc.

Ultrastructural studies of tumour samples taken at different times (3–72 h) after photodynamic treatment showed that, under our conditions, the direct photosensitized killing of neoplastic cells was by far faster and more extensive than vascular damage. Again, these observations are in agreement with the predominance of the LDL pathway for the delivery of Zn-Pc to tumours, since this mechanism involves the release of the drug from inside the cells. On the other hand, more hydrophilic photosensitizers which are chiefly carried by albumin or HDL, mainly localize in the vascular stroma and cause tumour necrosis through endothelial disruption (Zhou et al.., 1985). The photomodification of neoplastic cells was evident at only 3 h after irradiation, with the appearance of swollen mitochondria and extensive vacuolization in the cytoplasm. The photodamage gradually propagated, so that cell lysis occurred within 12–15 h; 24 h after irradiation the organized structure of the tumour tissue had almost completely disappeared. However, intact capillaries were occasionally observed even at 48 and 72 h after phototherapy. This circumstance is of the utmost importance for the efficacy of the phototherapeutic treatment, since it minimizes the possible formation of hypoxic clusters from which tumour recurrences may often originate.

References

Abernathey, C.D., Anderson, R.E., Kooistra, K.L., and Louws, E.R. (1987). Activity of phthalocyanine photosensitizers against human glioblastoma *in vitro. Neurosurgery* 21, 468–473.

Ben-Hur, E., and Rosenthal, E. (1985). The phthalocyanines: a new class of mammalian cell photosensitizers with a potential for cancer phototherapy. *Int. J. Radiat. Biol.* 47, 145–147.

Ben-Hur, E., and Rosenthal, E. (1986). Photosensitization of chinese hamster cells by water-soluble phthalocyanines. *Photochem. Photobiol..* 43, 615–619.

Brasseur, N., Ali, H., Autenrieth, D., Langlois, R., and Van Lier, J.E. (1985). Biological activities of phthalocyanines. III. Photoinactivation of V-79 chinese hamster cells by tetrasulphophthalocyanines. *Photochem. Photobiol.* 42, 515–521.

Brasseur, N., Ali, D., Langlois, R., and Van Lier, J.E. (1987). Biological activities of phthalocyanines. VII. Photoinactivation of V-79 chinese hamster cells by selectively sulfonated gallium phthalocyanines. *Photochem. Photobiol.* 46, 739–744.

Cozzani, I., Jori, G., Bertoloni, G., Milanesi, C., and Sicuro, T. (1985). Efficient photosensitization of malignant human cells *in vitro* by liposome-bound porphyrins. *Chem. Biol. Interactions* 53, 131–143.

Dougherty, T.J. (1984). Photodynamic therapy (PDT) of malignant tumours, *CRC, Crit. Rev. Oncol. Hematol.* 2, 83–116.

Foote C.S. (1988). Mechanistic characterization of photosensitizer reactions. In: Photosensitization (edited by G. Moreno, R.H. Pottier and T.G. Truscott). Springer-Verlag, Berlin. pp. 125–144.

Jori, G., Beltramini, M., Reddi, E., Salvato, B., Pagnan, A., and Tsanov, T. (1984). Evidence for a major role of plasma lipoproteins as hematoporhyrin carriers *in vivo. Cancer Lett.* 24, 291–297.

Jori, G., Reddi, E., Cozzani, I., and Tomio, L. (1986). Controlled targeting of different subcellular sites by porphyrins in tumour-bearing mice. *Br. J. Cancer* 53, 615–621.

Jori, G., and Spikes, J.D. (1983). Photobiochemistry of porphyrins. In: Topics in Photomedicine (edited by K.C. Smith), Plenum Press, New York, pp..183–319.

Morgan, A., and Garbo, G.M. (1988). Delivery systems for hydrophobic photosensitizers. *J. Photochem. Photobiol. B: Biology,* 1, 494–495.

Netland, P.A., Zetter, B.R., Via, D.P. and Vogts, J.C. (1985). In situ labelling of vascular endothelium with fluorescent acetylated low-density lipoproteins. *Histochem J.* 17, 1309–1320.

Reddi, E., Lo Castro, G., Biolo, R., and Jori, G., (1987). Pharmacokinetic studies with zinc(II)-phthalocyanine in tumour-bearing mice. *Br. J. Cancer* 56, 597–600.

Reyftmann, J., Morlière, Goldstein, S., Santus, R.C., Dubertret, L., and Lagrange, D. (1984). Interactions of human serum low-density lipoproteins with porphyrins: a spectroscopic and photochemical study. *Photochem. Photobiol.* 40, 721–729.

Ricchelli, F., Biolo, R., Reddi, E., Tognon, G., and Jori, G. (1988). Liposomes as carriers of hydrophobic photosensitizers *in vivo*: increased selectivity of tumour targeting. In: New Directions in Photodynamic Therapy (edited by D.C. Neckers), SPIE Publishers, Cambridge, Massachusetts, pp. 101–106.

Rosenthal, I., Ben-Hur, E., Greenberg, S., Conception-Lam, S., Drew, D.M., and Leznoff, C.C. (1987). The effect of substituents on phthalocyanine phototoxicity. *Photochem. Photobiol.* 46, 959–964.

Spikes, J.D. (1986). Phthalocyanines as photosensitizers in biological systems and for the photodynamic therapy of tumours. *Photochem. Photobiol.* 43, 691–699.

Tralau, C.J., MacRobert, A.J., Coleridge-Smith, P.D., Bart, H., and Bown, S.G. (1987). Photodynamic therapy with phthalocyanine sensitization: quantitative studies in a transplantable rat fibrosarcoma. *Br. J. Cancer* 55, 389–395.

Valduga, G., Nonell, S., Reddi, E., Jori, G., and Braslavsky, S.E. (1988). The production of singlet molecular oxygen by zinc(II)-phthalocyanine in ethanol and in unilamellar vesicles. Chemical quenching and phosphorescence studies. *Photochem. Photobiol.* 48, 1–5.

Valduga, G., Reddi, E., and Jori, G. (1987). Spectroscopic studies on Zn(II)-phthalocyanine in homogeneous and microheterogeneous systems. *J. Inorg. Biochem.* 29, 59–65.

Zhou, C., Milanesi, C., and Jori, G. (1988). An ultrastructural comparative evaluation of tumours photosensitized by porphyrins administered in aqueous solution, bound to liposomes or to lipoproteins. *Photochem. Photobiol.* 48, 487–492.

Zhou, C., Yang, W.Z., Ding, Z.X., Wang, Y.X., Wang, H., Shen, H., Fang, X.J., and Ha, X.W. (1985). The biological effects of photodynamic therapy on normal skin in mice: an electron microscopic study. In: Methods in Porphyrin Photosensitization (edited by D. Kessel), Plenum Press, New York, pp. 111–114.

Photosensitization by Phthalocyanines. Chemical Structure — Photodynamic Activity Relationship

IONEL ROSENTHAL[1] AND EHUD BEN-HUR[2]

[1]Department of Food Science
Agricultural Research Organization
The Volcani Center
P.O. Box 6, Bet Dagan 50250, Israel
[2]Department of Radiobiology
Nuclear Research Center-Negev
P.O. Box 9001, Beer Sheva, Israel

The very intense light absorption of phthalocyanine (PC) dyes in the spectral region of effective tissue penetration, combined with a substantial chemical stability, insignificant systemic toxicity, preferential retention in malignant tumors and, last but not least, outstanding photodynamic activity, propelled this class of dyes from photobiological obscurity to the highlights of photodynamic therapy.

The beauty of this development is the boundless research potential in revealing the yet unknown information on a new class of photosensitizers. Thus, while before 1985 the literature recorded only two publications on photobiology of PC, since then more than 60 papers have been published! Particularly attractive is the fact that with this esthetically symmetrical chromofor, virtually any element from the periodic table can be chelated, and various substituents can be attached to its periphery. Consequently, once the underlying mechanisms are deciphered, endless possibilities for tailoring a desired photobiological behavior are at hand. This gives hope for understanding the preferential affinity/retention of a chemical compound by a malignant tumor. Since PCs are stable, well defined chemical compounds, there is a viable chance to elucidate the mechanism of their complexation with various cell components and the reasons for cell specificity.

The study of the relationship between the chemical structure of phthalocyanines and their ability to act as photodynamic sensitizers indicates that two, photobiologically critical, parameters are affected by the structural diversity: the lifetime of the excited triplet state and the cellular uptake of the dye.

The nature of the central metal ligand affects intrinsically the photochemical activity and also, due to stereochemical factors, the rate of cellular uptake.

Since a long life triplet state is a prerequisite for photosensitization, a phthalocyanine dye containing a diamagnetic metal is better suited for this function than an analog compound containing a paramagnetic metal. In general, a paramagnetic ligand such as Cu, Co, Ni, Fe, VO, renders a PC photobiologically inactive, as expected from

Photobiology, Edited by E. Riklis
Plenum Press, New York, 1991

the short lifetime of the triplet state. Conversely, a diamagnetic ligand which extends the triplet life time, such as In, Al, Zn, enhances the phototoxicity (Rosenthal *et al.*, 1986). Among the dyes tested in our laboratory, ClInPC was found to be the most efficient sensitizer, as expected from its higher triplet quantum yield (0.9) as compared to ClAlPC (0.4) (Brannon and Magde, 1980). It is noted that UO_2PC sulfonate, which was the first PC shown to possess affinity to tumors (Frigerio, 1962) has only a very moderate activity (Fig.1). In apparent contradiction with this generalization, metal free-PC sulfonate although devoid of a paramagnetic ligand had only a very weak cell killing effect in experiments with human glioblastoma *in vitro* (Abernathey *et al.*, 1987) and was also inactive in sensitizing the photohemolyses of red blood cells (Sonoda *et al.*, 1987), which suggests that a central ligand is required.

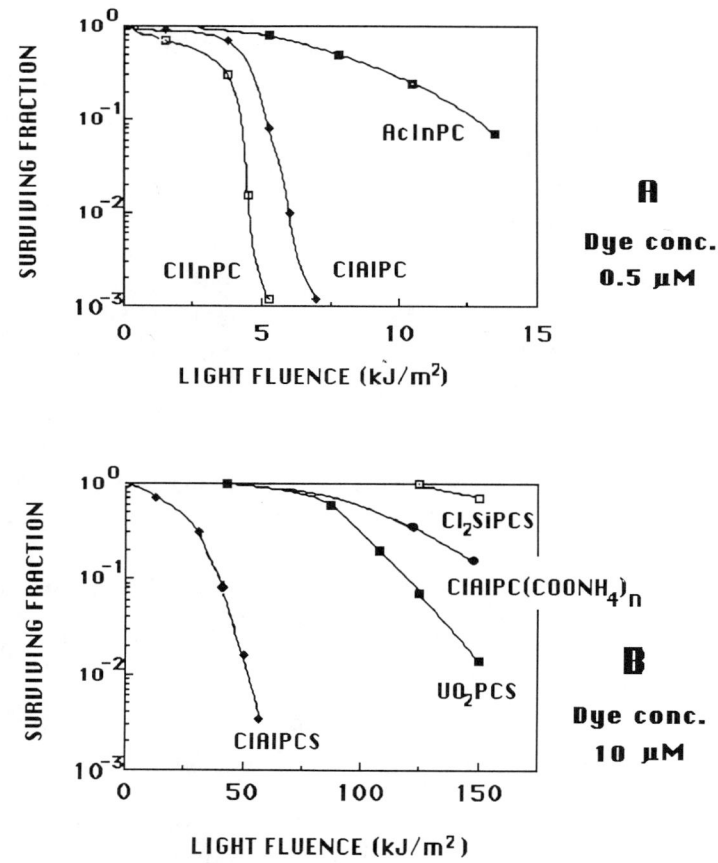

Figure 1. Survival of Chinese hamster cells photosensitized by PC derivatives.

Preliminary measurements indicate that the metal counter ion (Cl *vs*. CH_3COO) also affects the phototoxicity (Fig. 1). In this context it is noted that ClAl-PC sulfonate, which is routinely prepared by direct sulfonation with sulfuric acid to yield a mixture of isomers sulfonated to various extents, is one of the extensively used PC photodynamic sensitizers. The possibility that this preparation route might replace the chlorine atom by a hydroxyl group, apparently has not yet been considered.

In a cellular system, the central ligand also determines the rate of dye uptake and thus the overall cellular photosensitivity. UO_2-sulfonate and ClAl-sulfonate were taken up at the highest rate followed by Ni-, Zn-, Cu-, Co-, and Cl_2Si-PC sulfonate. The uptake from the growth medium containing 10% serum, in which only about 15% of the dye is not bound to serum proteins, was 5-18 fold slower than in the absence of serum, suggesting that most of the uptake is of free dye. In this context it is emphasized that the rate of uptake was unrelated to the state of aggregation of the dye. In view of the deaggregating effect of protein, which can be easily visualized *in vitro* by monitoring the change in the absorption spectrum of the dye following the addition of serum proteins, the irrelevancy of the aggregation state of dye for *in vivo* uptake should not be surprising. The temperature dependency suggests that the uptake process takes place in two steps. The first step is passive, involving binding of metallo-PC sulfonate to a receptor on the cell membrane, while the second one is active and involves internalization of the bound dye (Ben-Hur *et al.*, 1987). It results that the ability of binding to a receptor is a crucial factor in the pharmacokinetic process. Metals, higher than trivalent, which require out-of-plane counterions on both sides of the phthalocyanine ring such as Cl_2SiPC, are not taken up by cells, most probably due to the physical hindrance induced by the axial substituents which prevents the binding to the receptor. Obviously, such a compound, although photochemically active, is valueless as a photobiological sensitizer. Similarly, bulky peripheral substituents such as neopentoxy groups explain the minimal cellular uptake and the absence of photodynamic activity of heavily substituted phthalocyanines (Rosenthal *et al.*, 1987).

The stereochemistry of the metal complex affects the distribution pattern in biological tissues (Rousseau *et al.*, 1985). Thus tumor uptake and organ distribution were studied in Fischer 344/CRBL female rats bearing the 13762 mammary adenocarcinoma using sulfo PCs labelled with radioactive Ga and Tc which possess different stereochemistry. [99Tc]-tetrasulfo PC accumulated preferentially in the liver, kidney and reticuloendothelial system. The dye was also retained by the ovarian follicles and the uterus. Favorable tumor/blood ratio of 5-10 were observed for the brain, muscle and fat during the 24 hours study. The tumor activity also slightly surpassed that of the blood, colon and intestine. Substitution of ^{99}Tc by ^{67}Ga in the sulfo PC complex resulted in a shift in the tissue distribution pattern, with the bulk of material now passing through the hepatobiliary system. The kidneys retain activity at levels similar to those of the spleen, adrenals and ovaries with overall activity in kidney values at one third of those of the liver.

The peripheral substituents affect primarily the rate of uptake and consequently the photobiological activity. The binding of PCs to normal tissues has been repeatedly substantiated by histological studies. Thus basic quaternary substituted PCs have been used for many years to stain and quantify biological polyanions of all kinds. Conversely, negatively charged sulfonated PC dyes have been shown to stain tissue proteins such as collagen, proteoglycan filaments in tendons, bovine cornea, mouse lung alveoli, single muscle fibers, nerve membranes, myelin sheaths, glia fibers, erythrocytes and nucleic acids.

The ring substituents, which define the solubility features of the dye, expectedly affect the rate of uptake. The kinetics of uptake and cell retention are different for ClAl- and ClAl-PC sulfonate (Ben-Hur and Rosenthal, 1986) as expected from the different protein binding capabilities dictated by the different peripheral charges and molecular sizes. Consequently, the phototoxicities at a given time are different.

Among substituted Zn-PCs, the rate of uptake and consequently the phototoxicity decreased in the order H> OH> SO_3Na>> neopentoxy (Rosenthal *et al.*, 1987). Finally, the carboxylated ClAl-PC was much less efficient than the sulfonated counterpart (Fig. 1). Related to the different uptake rates was the observation that the photoinactivation of mammalian cells is proportional to the time of incubation of the cells with the dye prior to light exposure, and to the initial dye concentration (Ben-Hur and Rosenthal, 1986).

The activity of sulfonated ClGa-PCs was inversely related to the number of sulfonic acid groups for otherwise identical experimental conditions. Large variations in photoreactivity were observed among the four isomeric disulfonated derivatives, with the most hydrophobic isomer exhibiting the highest photoactivity (Brasseur *et al.*, 1987). The cellular uptake and distribution of ClAl-PC dyes sulfonated to different degrees were studied in an attempt to correlate hydrophobicity with cell membrane permeability and phototoxicity. The lower sulfonated derivatives (two sulfonic groups) were 25 times more efficient in photoinactivation of V-79 Chinese hamster cells, than the higher sulfonated dyes (mixed tri- and tetrasulfonated preparations) (Paquette *et al*, 1988a). However, ClAl-PC monosulfonate was reported to be inactive in sensitizing the HeLa cell photokilling, in contradistinction to the polysulfonated derivatives (Spikes and Bommer, 1987). Finally, *in vitro,* using V-79 cells, the lower sulfonated Zn-PC derivatives were the most active with the exception of the poorly water-soluble monosulfonated dye. A mixture of tetrasulfonated isomers obtained by direct sulfonation, was ten times more active than the homogeneous tetrasulfonated derivative prepared by the condensation of sulfophthalic acid. *In vivo,* testing the effect on EMT-6 mammary tumors, the latter dye was completely inactive, whereas the remainder of the sulfonated dyes exhibited a similar structure-activity pattern as observed with V-79 cells, although *in vivo* the variations are less pronounced (Brasseur *et al.*, 1988).

The peripheral ring substituents which dictate the electrical charge and solubility characteristics, also affect the tissues' distribution. Thus, the water soluble sulfonic acid derivative is cleared faster from the blood than the lipophilic nitro PC while the blood levels of the amino PC remained intermediate between those of the sulfo and nitro PC. The kinetics of tumor uptake also differ among the various derivatives. Clearance of the water-soluble PC is mainly *via* the kidneys, while the least water-soluble analogs are excreted biliarly (Van Lier et al., 1984).

In conclusion, several aspects of the relationship between chemical structure and photodynamic activity of PC have been clarified, and awaits confirmation for *in vivo* systems.

References

Abernathey, C. D., R. E. Anderson, K. L. Kooistra and E. R. Laws Jr. (1987), Activity of phthalocyanine photosensitizers against human glioblastoma *in vitro. Neurosurgery*, 21, 468-473.

Ben-Hur, E. and I. Rosenthal (1986), Photosensitization of Chinese hamster cells by water-soluble phthalocyanines. *Photochem. Photobiol.*, 43, 615-619.

Ben-Hur, E., J.A. Siwecki, H.C. Newman, S.W. Crane and I. Rosenthal (1987), Mechanism of uptake of sulfonated metallophthalocyanines by cultured mammalian cells. *Cancer Lett.*, 38, 215-222.

Brannon, J.H. and D. Magde (1980), Picosecond laser photophysics. Group 3A phthalocyanines. *J Am. Chem. Soc.*, 102, 62-65.

Brasseur, N., H. Ali, R. Langlois and J.E. van Lier (1987), Biological activities of phthalocyanines-VII. Photoinactivation of V-79 Chinese hamster cells by selectively sulfonated gallium phthalocyanines. *Photochem. Photobiol.*, 46, 739-744.

Brasseur, N., H. Ali, R. Langlois and J.E. van Lier (1988), Biological activities of phthalocyanines- IX. Photosensitization of V-79 Chinese hamster cells and EMT-6 mouse mammary tumor by selectively sulfonated zinc phthalocyanines. *Photochem. Photobiol.*, 47, 705-711.

Frigerio, N.A. (1962), Metal Phthalocyanines. U.S. Patent no. 3,027,391.

Paquette, B., H. Ali, R. Langlois and J.E. van Lier (1988a), Biological activities of phthalocyanines - VIII. Cellular distribution in V-79 Chinese hamster cells and photo-toxicity of selectively sulfonated aluminum phthalocyanines. *Photochem.Photobiol.*, 47, 215-220.

Rosenthal, I., C. Murali Krishna, P. Riesz and E. Ben-Hur (1986), The role of molecular oxygen in the photodynamic effect of phthalocyanines. *Radiat. Res.*, 107, 136-142.

Rosenthal, I., E. Ben-Hur, S. Greenberg, S. Conception-Lam, D.M. Drew and C.C. Leznoff (1987), The effect of substituents on phthalocyanine phototoxicity. *Photochem. Photobiol.*, 46, 959-964.

Rousseau, J., H. Ali, G. Lamoureux, E. Lebel and J.E. Van Lier, (1985), Synthesis, tissue distribution and tumor uptake of ^{99}mTc- and ^{67}Ga-tetrasulfophthalocyanine. *Int. J. Appl. Radiat. Isot.*, 36, 709-716.

Sonoda, M., C. Murali Krishna and P. Riesz (1987), The role of singlet oxygen in the photohemolysis of red blood cells sensitized by phthalocyanine sulfonates. *Photochem. Photobiol.*, 46, 625-632.

Spikes, J.D. and J.c. Bommer (1987), Effects of the degree of sulfonation on the photophysical and photosensitizing properties of aluminum and zinc phthalocyanines. *Photochem. Photobiol.*, 45, 79S.

Van Lier, J.E., H. Ali and J. Rousseau (1984), Phthalocyanines labeled with gamma-emitting radionuclides as possible tumor scanning agents, in "Porphyrin Localization and Treatment of Tumors", Eds. D.R. Doiron and C.J. Gomer, Alan R. Lis, Inc., pp. 315-319.

Photoprotection

A) Cellular Mechanisms

Cellular Effects of UVA: DNA Damages

MEYRICK J. PEAK AND JENNIFER G. PEAK

Molecular Photobiology Group
Biological, Environmental, and
Medical Research Division
Argonne National Laboratory
9700 South Cass Avenue
Argonne, Illinois 60439, USA

Introduction

Ultraviolet radiation between 320 nm and visible light (UVA) is a major component of both solar radiation and suntan lamps, which are being increasingly used in tanning booths. UVA has generally been considered innocuous, partially because DNA does not absorb appreciably in this region (Sutherland and Griffin, 1981). UVB radiation (290–320 nm), however, has widely been considered to be the major etiological factor in human skin carcinogenesis caused by solar UV radiation (Setlow, 1974; Parrish et al., 1978). Largely because DNA absorbs photons of UVB, which is known to produce thymine photoproducts (cyclobutane dimers and adducts). Patients with xeroderma pigmentosum are particularly prone to solar-UV-induced skin cancer, and cells derived from these people lack the ability to repair pyrimidine photoproducts by excision (Cleaver et al., 1984: Cleaver, 1987), evidence that pyrimidine photoproducts might play a role in carcinogenesis in certain specialized situations. Normal cells have the ability to repair these UVB-induced lesions (Mitchell, 1988a,b). However, UVA is considerably more penetrating and abundant than UVB, and Tyrrell and Pidoux (1987) have performed a spectral analysis claiming that 20–60% (depending upon the solar zenith angle) of the toxic biological effects of solar radiation can be attributed to UVA. The known mutagenic effects of UVA radiations (reviewed by Peak and Peak, in press) provides motivation for studying DNA changes that might be effected by this region of the electromagnetic spectrum. The following is a summary of the use of sensitive alkaline and neutral elution DNA filter assays to reveal and quantify various DNA damages resulting from exposures of human cells to isolated monochromatic UVB and UVA and visible light radiations.

Methods

The use of P3 human teratocarcinoma cells in radiation studies was discussed by Hill et al. 1988. Culture and labeling of cells and use of monochromatic UV and visible light

Photobiology, Edited by E. Riklis
Plenum Press, New York, 1991

and x-rays were described by Hill *et al.* (1988), Peak *et al.* (1985), and Peak *et al.* (in press). Elution of DNA was as described by Kohn *et al.* (1981), and Peak *et al.* (in press). The modification of the assay for measurement of DNA-to-protein crosslinking in the special case where DNA breaks are also present was described by Peak *et al.* (1987). Calibration of the neutral elution assay for measurement of double-strand DNA breaks was described by Peak *et al.* (1988).

Results and Discussion

Slowly developing alkali-labile sites: Figure 1 illustrates the different elution profiles that have been observed under various assay conditions after P3 cells were exposed to UVB and UVA, as well as visible radiations. Standard alkaline elution for detection of total strand breaks [single-strand breaks (SSB) plus double-strand breaks (DSB) plus rapidly developing alkali-labile sites (RDALS)] usually gives profiles that are exponential (Figure 1, SSB). We observed this pattern for P3 cells exposed to 365–nm radiation and green light at 512 nm (Peak *et al.* in press). In the same study, we observed that alkaline elution prifiles were down-turning (convex) after the cells were epxosed to 405 nm near-blue and 434 nm blue light (Fig. 1, ALS). After 6 h of elution at pH 12.1, the profiles beame exponential. A logical explanation is that the DNA was broken during the first 6 h elution period, i.e, that there radiations induced a class of lesion that produced SSB slowly during the first 6 h elution. This theory was tested by holding the DNA on the filters at pH 12.1 for 6 h, to allow these putative breaks to develop before commencing elution. With this treatment, profiles became entirely exponential, with exactly the same slopes as the post-6-h elution slopes, evidence that all slowly developing alkali-labile sites (SDAL) had developed by the start of elution. Table 1 shows calculated yields of SSB plus RDAL and SDAL, compared with those measured earlier (Peak *et al.* 1987) by alkaline sucrose sedimentation techniques. (Because sedimentation was for 20 h at pH 12.1, all SDAL were presumably developed in the initial part of the run, so that the DNA was fully broken during most of the sedimentation). More than 60% of lesions induced by 405–nm UVA were in the form of SDAL. It is important to discover what DNA lesion this represents and what its

Table 1. DNA Lesions Induced by 405 nm UVA

Method	Lesion	Number per 10^{10} per $J.m^{-2}$	%
Elution	FSB[*] + RDA;	1.2	35
	SDAL	1.7	65
		2.9	100
Sedimentation		3.2	

[*]Frank strand breaks

biological significance may be. Since SSB are generally assumed to be completely sealed (Painter, 1980; Ward, 1985; Ward et al. 1985), it is of interest to investigate the extent to which SDAL induced by 405 nm and 434 nm radiations are repaired.

DNA-to-protein crosslinks: The assay for DNA-to-protein crosslinks (DPC) involves first exposing DNA to a large dose of x-rays to break the DNA into small pieces that elute rapidly unless they are covalently linked to protein. This gives the biphasic profiles shown by Figure 1 (DPC). Peak *et al.* (1987) developed an improved method for quantification of DPC caused by UV radiation, which compensates for the large number of SSB induced concomitantly. We demonstrated that UVA radiations are very effective at inducing these particular lesions (Peak and Peak, in press), even more so than ionizing radiation (x-rays) when computed on the basis of events per lethal fluence (dose) per cell. The biological role of the large numbers of DPC induced in cells by UVA remains completely enigmatic. To investigate possible damage to genetic activity of

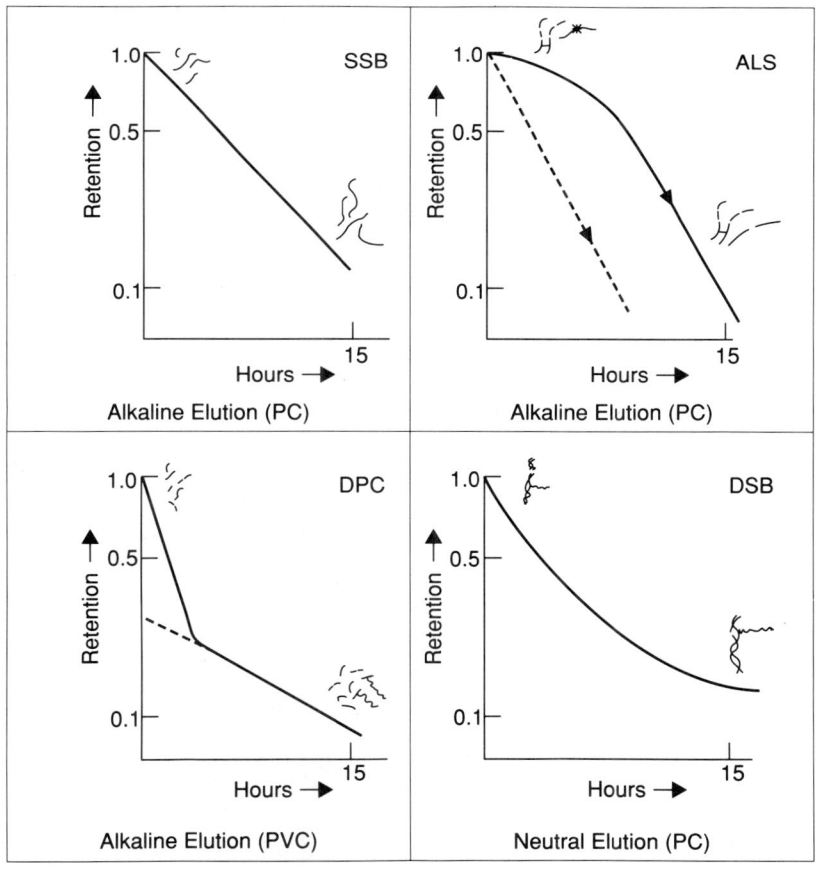

Figure 1. Elution profiles

DNA, E. R. Blazek in our laboratory has recently cross-linked a variety of proteins and amino acids to single-stranded M13 viral template DNA by using crosslinking agents (UVA plus a porphyrin sensitizer or trans-dichlorodiammine Pt 11). Possible deleterious genetic damage was studied by using a sequencing technique, described by Ito *et al.* (1988), to detect any DNA synthesis arrests. To date, few specific synthesis arrests are attributable to DPC caused by either treatment, and this finding is taken as evidence that DPC induced by these agents are linked to the DNA at sites that do not interfere with transcription (manuscript in preparation).

Neutral elution: An empirical method for the neutral elution assay (Figure 1, DSB) for measurement of DSB was described by Peak *et al.* (1988). The calibrated assay has been applied to analysis of DSB induced by UVB and UVA monochromatic radiations at 313, 365 and 405 nm (manuscript in preparation). As has been seen with ionizing radiation (Blazek *et al.* submitted), the dose response for DSB induction is not linear but quadratic, indicating that DSB induction is entirely a two-hit event. In this case, the two hits must represent two SSB close enough to cause the DNA to part. There are two explanations for the data. Either the response is correct, implying that the DSB caused by UVB and UVA are due to SSB that must be nonrandom (clustered), or else that an unexplained artifact exists in the neutral elution assay. Resolution of these alternatives is under active investigation in our laboratory.

Conclusions

Elution techniques have been successfully applied to UVB and UVA photobiology, and their use has extended our understanding of the damage that these carcinogenic radiations cause to DNA.

Acknowledgements

This work was supported in part of PHS grants R01-CR334492 and R01-CA37848 awarded by the National Cancer Institute, U.S. Department of Health, and in part of the U.S. Department of Energy, Office of Health and Environmental Research, under Contract No. W-31-109-ENG-38.

References

Cleaver, J.D., Charles, W.C., and Kong, S.H. (1984). Efficiency of repair of pyrimidine dimers and psoralen monoadducts in normal and xeroderma pigmentosum human cells. *Photochemistry and Photobiology*. 40, 621–7-629.
Cleaver, J.E. (1987). Xeroderma pigmentosum. *Photomedicine*, edited by E. Ben-Hur and I. Rosenthal (Boca Raton, Florida, CRC Press, Inc.), Vol. 11.
Hill, C.K., Holland, J., Chang-Lui, M., Buess, E.M., Peak, J.G., and Peak, M.J. (1988). Human epithelial teratocarcinoma cells (P3): Radiobiological characterization, DNA damage, and comparison with other rodent and human cell lines. *Radiation Research* 113, 278–288.

Ito, A., Robb, F.T., Peak, J.G., and Peak, M.J. (1988). Base-specific damage induced by 4-thiouridine photosensitization with 334–nm radiation in M13 phage DNA. *Photochemistry and Photobiology* 47, 231–240.

Kohn, K.W,. Ewig, R.A.G., Erickson, L.G., and Zwelling, L.A. (1981). Measurement of strand breaks and cross-links by alkaline elution. In: *DNA Repair 1. A Laboratory Manual of Research Procedures*. (Edited by E.C. Friedberg and P.C. Hanawalt), pp. 379–401, Marcel Dekker, New York.

Mitchell, D.L. (1988a). The relative cytotoxicity of (6–4) photoproduct and cyclobutane dimers in mammalian cells, Photochemistry and Photobiology 48, 51–57.

Mitchell, D.L. (1988b). The induction and repair of lesions produced by the photolysis of (6–4) photoproducts in normal and UV-hypersensitive human cells. *Mutation Research* 194, 227–237.

Painter, R.B. (1980). The role of DNA damage and repair in cell killing induced by ionizing radiation. *Radiation Biology in Cancer Research*, et. by R.E. Meyn and H.R. Withers, pp. 59–68, Raven Press, New York.

Parrish, J.A., Anderson, R.R., Urbach, F., and Pitts, D. (1978) UVA. Plenum Press, New York.

Peak, J.G., Peak, M.J., Sikorski, R.A., and Jones, C.A. (1985). Induction of DNA-protein crosslinks in human cells by ultraviolet and visible radiations: Action spectrum. *Photochemistry and Photobiology* 41, 295–302.

Peak, J.G., Peak, M.J., and Blazek, E.R. (1987). Improved quantitation of DNA-protein crosslinking caused by 405–nm monochromatic near-UV radiation of human cells. *Photochemistry and Photobiology* 46, 319–321.

Peak, J.G., Blazek, E.R., and Peak, M.J. (1988). Measurement of double-strand breaks in Chinese hamster cell DNA by neutral fiber elution: Calibration by ^{125}I decay. *Radiation Research* 115, 624–429.

Peak, M.J., Peak, J.G., Carnes, B.A., Chang-Liu, C.N., and Hill, C.K. DNA damage and repair in rodent and human cells after exposure to JANUS fission spectrum neutrons: A minor fraction of single-strand breaks as revealed by alkaline elution are refractory to repair. *International Journal of Radiation Biology*, in press.

Peak, M.J., Peak, J.G. Solar-ultraviolet-induced damage to DNA. *Photodermatology*, in press.

Setlow, R.B. (1974). The wavelengths of sunlight effective in producing skin cancer. Proceedings of the National Academy of sciences, USA. 71, 3363–3366.

Sutherland, J.C., and Griffin, K.P. (1981). Adsorption spectrum of DNA for wavelengths longer than 320 nm. Radiation Research 86, 399–409.

Tyrrell, R.M., and Pidoux, M. (1987). Action spectra for human skin cells: estimates of the relative cytotoxicity of the middle ultraviolet, near ultraviolet and violet regions of sunlight on epidermal keratinocytes. *Cancer Research* 47, 1825–1829.

Ward, J.F. (1985). Biochemistry of DNA lesions. *Radiation Research* 104, S103–S111.

Ward, J.F., Blakely, W.F., and Joner, E.I. (1985). Mammalian cells are not killed by DNA single strand breaks caused by hydroxyl radicals from hydrogen peroxide. Radiation Research 103, 383–392.

Cellular Defense Against UVA (320–380 nm) and UVB (290–320 nm) Radiations

R.M. TYRRELL, S.M. KEYSE AND E.C. MORAES

Swiss Institute for Experimental Cancer Research
CH-Epalinges/Lausanne, Switzerland

Introduction

Both acute and chronic exposure to sunlight can seriously damage human skin cells and lead to immediate effects such as erythema and sun-burn or more long-term processes including ageing and skin cancer. This overview summarises the various mechanisms by which skin cells are protected from the deleterious effects of solar radiation with emphasis on recent results from this laboratory. The sensitive basal epidermal keratinocytes and underlying fibroblasts are protected to some extent by the Stratum corneum, a physical absorption barrier whose effectiveness diminishes with increase in wavelength (as UV transmission through skin increases). An extremely important cellular protection is provided by the DNA excision repair system which constantly removes the bulk of potentially lethal and pre-mutagenic damage induced in DNA by solar UV radiation. Mechanisms of DNA repair are dealt with elsewhere in these proceedings and will only be mentioned briefly in this discussion. Suffice it to say that the importance of excision repair diminishes with increase in wavelength and appears to have little influence on the capacity of cells to survive UVA (320–380 nm) radiation damage. One reason for this is that solar radiation, and in particular UVA, generates a range of radical species and oxygen intermediates which lead to a general stress condition requiring the intervention of a series of defense mechanisms in addition to excision repair. A description as to how these active intermediates may be generated by solar radiation and their possible biological consequences will form the first part of this paper. Possible cellular anti-oxidant defense mechanisms will then be briefly considered before discussing in detail our recent findings concerning a constitutive and an inducible anti-oxidant defense mechanism in human skin fibroblasts.

The generation of active oxygen intermediates by UVA radiation

Pathways by which several oxygen species may be generated by UVA are shown in Fig 1. An important photolytic degradation product of tryptophan is hydrogen peroxide

Photobiology, Edited by E. Riklis
Plenum Press, New York, 1991

861

H_2O_2 (McCormick et al., 1976). If generated intracellularly this photoproduct can then give rise to hydroxyl radical via an iron-catalysed Fenton reaction (reaction (i))

(i) $H_2O_2 + Fe^{2+} \rightarrow Fe^{3+} + OH^- + .OH$

Superoxide ion can be generated directly by irradiation of NADH or NADPH (Czockralska et al., 1984; Cunningham et al., 1985) and can then be dismutated by

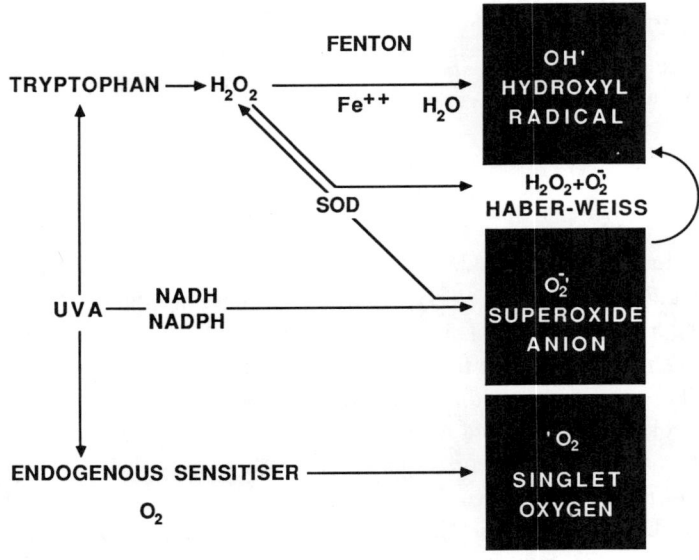

Figure 1. Possible oxygen intermediates produced by UVA (320 nm–380 nm).

superoxide dismutase to give H_2O_2. Superoxide anion may also participate in the Fenton reaction by regenerating Fe^{2+} from Fe^{3+} (Haber-Weiss reaction (ii))

(ii) $O_2^- + Fe^{3+} \rightarrow Fe^{2+} + O_2$

According to these schemes, generation of both superoxide and hydrogen peroxide would be expected to lead to the appearance of the highly reactive hydroxyl radical. In addition, cells contain significant quantities of endogenous chromophores such as flavins, quinones and porphyrins which may act as photodynamic sensitisers and lead to the generation of singlet oxygen.

Lethal consequences of active oxygen intermediates generated by UVA

Deuterium oxide enhances the lifetime of singlet oxygen and would be expected to enhance the biological action of this species. Human fibroblasts treated in the presence of deuterium are sensitised to the lethal action of both UVA (334 nm, 365 nm) and visible (405 nm) radiations (Tyrrell and Pidoux, 1989). A typical result is shown in Fig. 2 for radiation at 365 nm. In contrast to the effects of deuterium, irradiation in the presence of sodium azide, which quenches singlet oxygen, strongly protects the cell populations against the lethal action of radiation at 365 nm. Taken together, these results are a strong indication that singlet oxygen is generated intracellularly by UVA radiation and will have lethal consequences for the cells. Since oxygen intermediates apparently have lethal consequences, we would expect UVA inactivation to be dependent upon the presence of oxygen. In fact, this is known to be the case for human (HeLa) cells (Danpure and Tyrrell, 1976; Fig. 3) and from pioneering studies in bacteria by R.B. Webb (reviewed by Webb, 1977).

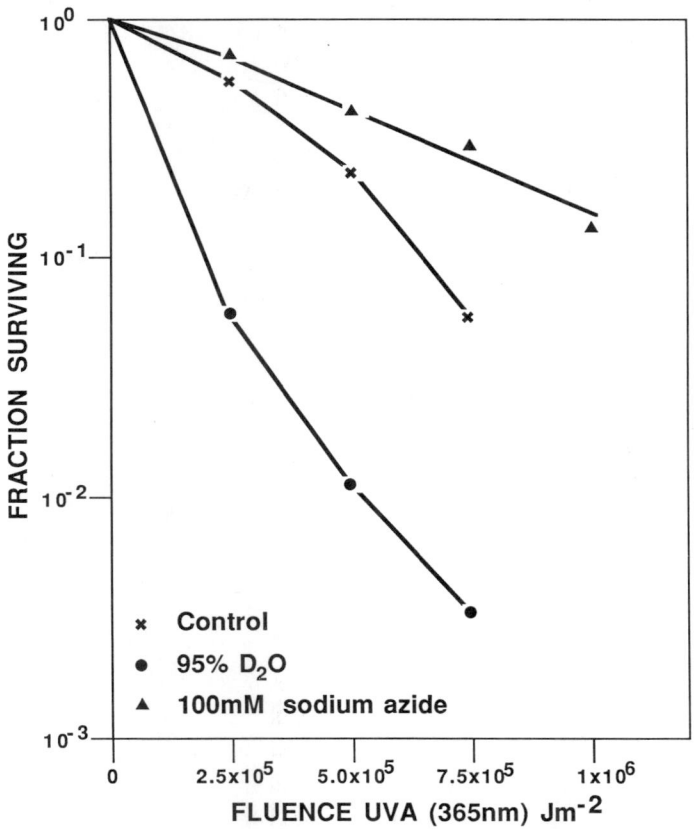

Figure 2. Human Skin Fibroblasts IBE/3. Inactivation of normal human skin cells by radiation at 365 nm in the presence of deuterium oxide and sodium azide. (Adapted from Tyrrell and Pidoux, 1989, by permission of the publishers).

Mutagenic consequences of active oxygen intermediates

Several studies have indicated that free radicals, including singlet oxygen, are mutagenic (Hsie et al., 1986; Decuyper-Debergh et al., 1987; Loeb et al., 1988). In view of the possible involvement of H_2O_2 in UVA effects, it was clearly of interest to determine the mutagenic specificity of this compound in mammalian cells. Although H_2O_2 is weakly mutagenic to bacteria (Stortz et al., 1987), it is only recently that this oxidising agent has been shown to cause phenotypic mutations in cultured populations of mammalian cells (Ziegler-Skylakakis and Andrae, 1987). We set out to confirm this finding and to obtain sequence information on induced mutants using the simian virus 40-based shuttle vector, pZ189, which contains the bacterial *supF* tRNA locus as the target for mutation and can replicate in both *Escherichia coli* and in mammalian cells . This latter property permits us to damage the plasmid either *in vitro* (extracellularly) or *in vivo* (intracellularly, following transfection) and then process the damage in a eucaryotic host. DNA is then extracted, transformed into the appropriate bacterial host and colonies containing mutant plasmid are scored and selected. The DNA from each

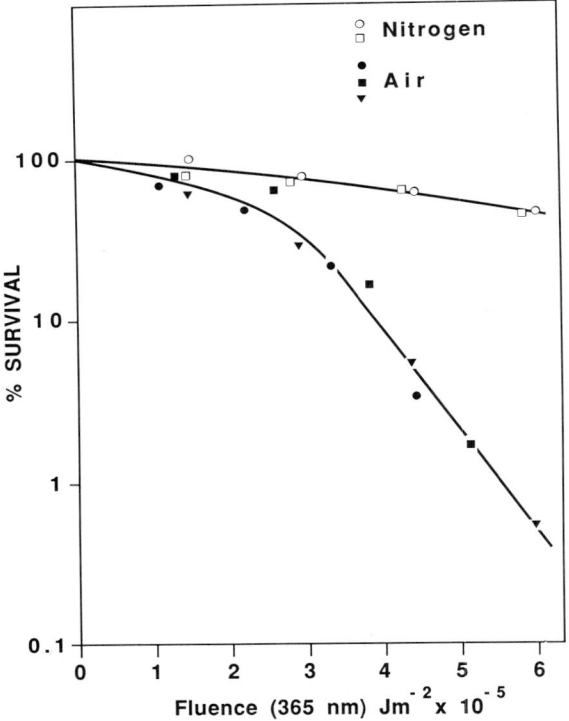

Figure 3. Hela 365 nm Killing Curves. Inactivation of human HeLa cells by radiation at 365 nm under aerobic and anerobic conditions (Adapted from Danpure and Tyrrell, 1976, by permission of the publishers).

mutant plasmid is then isolated and run on agarose gels to detect mutants with large insertions or deletions. The mutants with normal mobility are then sequenced by the dideoxy method. Using this approach we have examined H_2O_2 mutagenicity both *in vitro* (Moraes et al., 1990) and *in vivo* (Moraes et al., 1989. Using a ferric-EDTA complex together with H_2O_2 we have enhanced the mutation frequency of plasmid treated extracellularly to over 100 times over spontaneous. The mutation frequency is enhanced by a factor of up to 4 when cells previously transfected with plasmid are treated with concentrations of hydrogen peroxide in the low mMolar range. We have analysed more than 100 mutants of each type (spontaneous, H_2O_2 treatment *in vitro*, H_2O_2 treatment *in vivo*) and obtained extensive sequence information on a 150 bp region (including the supF tRNA gene and flanking sequences) which allows us to draw several conclusions concerning the mutagenicity of H_2O_2. The two main characteristics of the spectra of mutants induced *in vivo* are 1) the appearance of a high proportion of mutants with both deletions and associated base changes and 2) a majority of mutants that are small deletions (< 4 bp). Over half of these small deletions occur in runs of identical bases. Half of these appear at a single hot-spot region of 5 cytosines. These observations are consistent with an earlier model (Streisinger et al., 1966) which proposed that frameshift mutations in bacteriophage T4 are caused by slippage and mispairing of bases. Our data indicate that such a mechanism also applies to mammalian cells and that deletion formation is greatly enhanced by an agent, H_2O_2, which causes strand breakage.

The important conclusion from these mutation studies in the present context is that they provide strong evidence that both .OH and H_2O_2 are mutagenic to mammalian cells. Thus such active species have the potential to cause both cell death and mutation.

Cellular anti-oxidant defense mechanisms

Mammalian cells possess several anti-oxidant enzymes including superoxide dismutase, catalase and glutathione peroxidase. All these enzymes would be expected to be important in cellular defense. In addition, tissues contain many molecules with intrinsic antioxidant activity such as ascorbate, beta-carotene and glutathione. Glutathione is an ubiquitous tripeptide which is present at high concentrations in many cell types and the evidence from this laboratory suggests that it plays a critical role in cellular defense against the lethal action of both UVA and UVB radiations (Tyrrell and Pidoux 1986, 1988).

Constitutive cellular defense against UVA and UVB radiations: The role of endogenous glutathione

Glutathione is a powerful radical scavenger which is continually oxidised to the disulphide and regenerated by glutathione reductase. The molecule is also the unique hydrogen donor for glutathione peroxidase. Several drugs inhibit glutathione metabolism, the most specific of which is buthionine S-R Sulfoximine (BSO) which reduces cellular glutathione levels by inhibition of a critical enzyme in glutathione synthesis, gamma

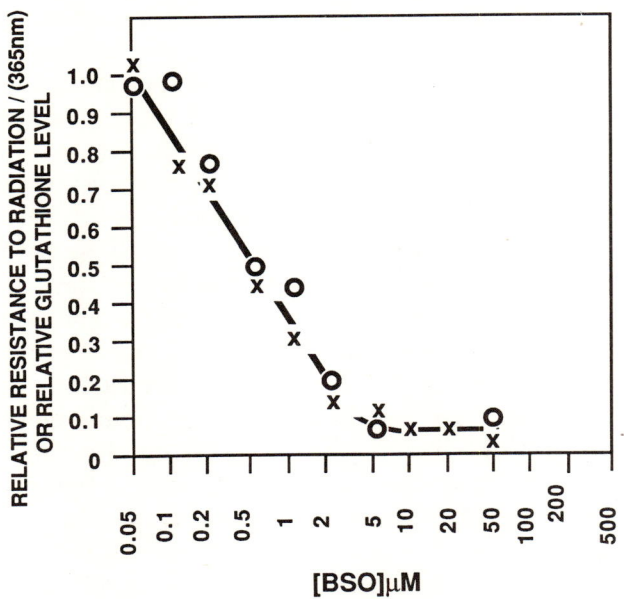

Figure 4. Correlation between resistance of colony-forming ability of human fibroblasts to radiation at 365 nm and levels of cellular glutathione (Adapted from Tyrrell and Pidoux, 1988, by permission of the publishers).

glutamyl cysteine synthetase. Concentrations of BSO as low as 5 µMolar can reduce glutathione to negligible levels within 18 h with little or no cytoxicity. We have measured cellular glutathione levels following treatment with different concentrations of BSO and correlated these values with the relative resistance of human fibroblast populations to radiation at 365 nm after comparable treatment (Fig. 4). The close correlation between the two parameters is strong evidence that glutathione plays a major role in protecting cells against the cytoxic action of radiation at 365 nm. The data in Fig 5 demonstrate that glutathione protects cells against damage induced by both UVA and near visible radiations. Most surprising is that glutathione protection extends into the UVB range (313 nm, Fig. 5; 302 nm, Tyrrell and Pidoux, 1988). These results strongly indicate that radiation in the UVB range, which is believed to contribute most to the mutagenic and carcinogenic effectiveness of sunlight, generates potentially lethal free radicals. Also included on the Figure (Fig. 5B) is a survival curve for fibroblasts derived from a patient with Xeroderma pigmentosum complementation group A. These cells are completely deficient in excision repair and the results indicate that, after treatment with radiation at 313 nm, both excision repair and glutathione are critical in cellular protection. At longer wavelengths, glutathione plays a more important role than excison repair. Additional experiments with epidermal keratinocytes derived from the same

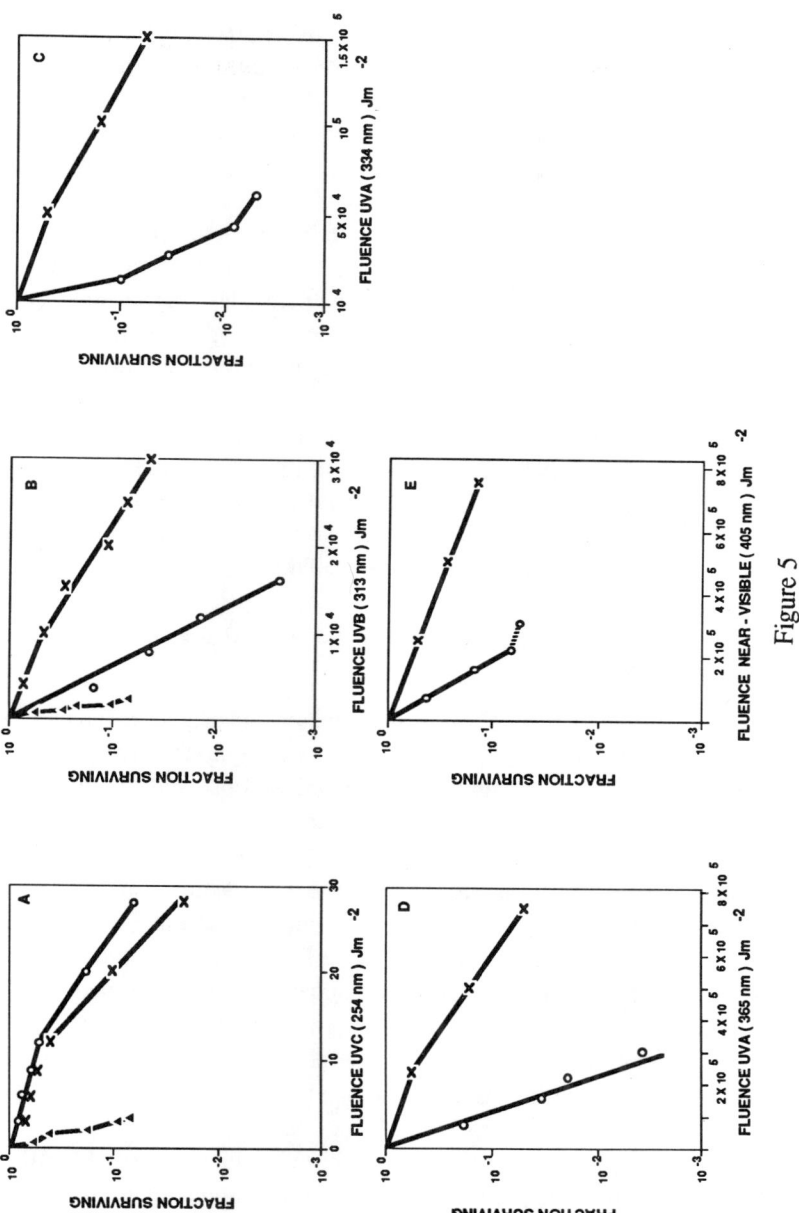

Figure 5

foreskin biopsy as the fibroblasts used in this study have shown that the cells are sensitised less by depletion of glutathione than are the fibroblasts. However, the epithelial cells contain 3 times higher levels of glutathione and even maximal concentrations of BSO only deplete the cells to 20–25 percent of the normal levels of this thiol. Higher levels of glutathione have also been seen in the epidermal tissue of mice and humans so that it appears that the tissues at greater risk of incurring solar radiation damage contain higher levels of glutathione.

The mechanism of glutathione protection is not completely clear. To determine whether the reduced free radical scavenging activity in cells depleted of glutathione is more important than a reduced activity of glutathione peroxidase (as a result of depletion of the hydrogen donor), we have added back cysteamine to glutathione depleted cells. Since cysteamine cannot act as a hydrogen donor for glutathione peroxidase, cysteamine should only restore protection if glutathione reduction leads to a critical loss of radical scavenging capacity. The experiments indicate that at a short UVB wavelength (302 nm) glutathione protects by free radical scavenging but at longer wavelengths (313 nm, 365 nm), the action of glutathione is more specific and could be related to its role as a cofactor in glutathione peroxidase activity. Whatever the mechanism, it is clear that glutathione provides a major line of defense against solar radiation damage.

Inducible cellular defense against oxidant stress: the role of heme oxygenase

Proteins are induced by both UVC (254 nm) radiation and heat shock and are believed to play a role in protecting cells against the corresponding stress. In studies using one dimensional SDS-PAGE gels to analyse total protein, we observed the induction of an unique 32 kDa protein by UVA (334 nm, 365 nm) and near-visible (405 nm) radiations but not heat shock (Keyse and Tyrrell, 1987). Detailed kinetic studies showed that the protein is induced after sub-lethal fluences and appears at a maximum level 2 h after UVA treatment. Furthermore the induction involves radical intermediates since glutathione depletion (by BSO treatment) lowers the threshold for induction. Similar studies with hydrogen peroxide together with chemical peptide mapping clearly indicated that the same protein is induced by both UVA and hydrogen peroxide. These observations led us to propose that the induction of the protein is a general response to cellular oxidant stress and encouraged us to undertake studies to identify the protein involved (Keyes and Tyrrell, 1989).

As a first step to cloning the gene involved in the induction, polyA$^+$ mRNA was prepared from human fibroblasts populations in the induced and uninduced state. *In_vitro* translation of these mRNA populations and analysis of the protein products on one dimensional gels demonstrated that the induction of the 32 kDa protein was due to enhanced mRNA levels and therefore that the inducible response was due to regulation of gene expression rather than some more indirect effect such as protein stability. In view of the high levels of mRNA induction, we used the induced mRNA population directly to prepare a cDNA library in the lambda Zap (Stratagene) vector. A sub-library of 300 individual recombinant clones was then prepared in a Bluescript phagemid by automatic excision from the lambda phage using a helper phage rescue procedure. The sub-library was screened using selective hybridisation and two clones were isolated which hybridised to an mRNA that translated into the inducible 32 kDa protein. One of the cloned inserts, which was a 550 bp fragment corresponding to the 3' untranslated region of the inducible mRNA, was used to make probes for further study.

Using the cloned DNA and peptide mapping with Staphylococcus aureus V8 protease we confirmed that the cloned insert hybridises to an mRNA that translates into the previously identified inducible 32 kDa protein. We then prepared radioactive probes from the cloned cDNA fragment and performed Northern analysis of blots prepared from RNA gels run with extracts of cells that had been treated under different inducing conditions. These studies showed that the constitutive levels of the inducible mRNA are very low and that UVA radiation, hydrogen peroxide and sodium arsenite all induce the specific mRNA within a maximum of 2–4 h after treatment. Finally, we isolated a full length (1.6 kb) clone from a second cDNA library. Information obtained from the 5' end of this clone showed that the sequence of the fragment we had isolated was 100 percent homologous with the corresponding fragment of the cDNA of human heme oxygenase (Yoshida et al., 1988).

The main function of heme oxygenase is to break down the heme in hemoglobin to produce the green pigment, biliverdin, which can then be broken down to bilirubin by biliverdin reductase (Fig 6). The enzyme may also be involved in the catabolism of hemoproteins such as cytochromes and peroxidases. Of paramount interest to our investigation is that both bilirubin and biliverdin have been shown to be powerful anti-oxidants (eg. Stocker et al., 1987). This led to the proposal that these catabolites constitute a powerful anti-oxidant defense in plasma. Since the potential cellular forms of heme catabolites are powerful anti-oxidants specifically shown to scavenge peroxyl radicals and we have observed high levels of heme oxygenase in cells from a tissue (skin)

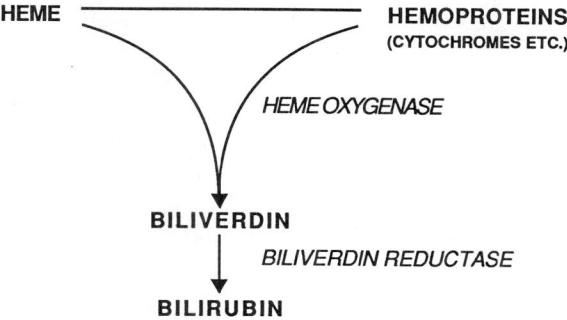

Figure 6. Catabolism of heme by heme oxygenase.

which is clearly not important in hemoglobin metabolism, we propose that the induction of heme oxygenase is critical to protecting cells against the oxidant stress resulting from exposure to UVA radiation and other oxidising agents. We are now looking for additional evidence that will confirm this role for heme oxygenase and are examining just how UVA radiation treatment is able to regulate the gene.

In summary, we believe that we have identified two pathways by which human skin cells can protect themselves against the oxidative stress imposed by solar radiation. The first is the constitutive protection provided by endogenous glutathione for which we have fairly sound functional evidence. The second is the induction of high levels of heme oxygenase which, by increasing heme catabolism, will increase the anti-oxidant capacity of cells.

Acknowledgements

This research has been supported by grants from the Swiss National Science Foundation and the Swiss League against Cancer. We gratefully acknowledge the expert technical assistance provided by Mireille Pidoux and Yvonne Tromvoukis.

References

Cunningham, M.L., Johnson, J.S., Giovanazzi, S.M. and Peak, M.J., 1985, Photosensitized production of superoxide anion by monochromatic (290–405 nm) ultraviolet irradiation of NADH and NADPH coenzymes. *Photochem. Photobiol.* 42, 125–128.

Czochralska, B., Kawiczynski, W., Bartosz, G. and Shugar, D., 1984, Oxidation of excited state NADH and NAD dimer in aqueous medium, involvement of O_2 as a mediator in the presence of oxygen. *Biochem. Biophys. Acta.* 801, 403–409.

Danpure. H.J. and Tyrrell, R.M., 1976, Oxygen dependence of near-UV (365 nm) lethality and the interaction of near-UV and X-rays in two mammalian cell lines. *Photochem. Photobiol.* 23, 171–177.

Decuyper-Deberg, D., Piette, J. and Van de Vorst, A., 1987, Singlet oxygen induced mutations in M13 lac Z phage DNA. *EMBO J.* 6, 3155–3161.

Hsie, A.W., Recio, L., Katz, D.S., Lee, M., Wagner, M. and Schenley, R.L., 1986, Evidence for reactive oxygen species inducing mutations in mammalian cells. *Proc. Natl. Acad. Sci. US.* 83, 9616–9620.

Keyse, S.M., and Tyrrell, R.M., 1987, Both near ultraviolet radiation and the oxidising agent hydrogen peroxide induce a 32 kDa stress protein in normal human skin fibroblasts. *J. Biol. Chem.* 212, 14821–14825.

Keyse. S.M. and Tyrrell, R.M., 1989, Heme oxygenase is the major 32 kDa stress protein induced in human skin fibroblasts by UVA radiation, hydrogen peroxide and sodium arsenite. *Proc. Natl. Acad. Sci. US.* 86, 99–103.

Loeb, L.A., James,.E.A., Waltersdorph, A.M. and Klebanoff, S.J., 1988, Mutagenesis by the autooxidation of iron with isolated DNA, Proc. Natl. Acad. Sci. US. 85, 3918–3922.

McCormick, J.P., Fisher, J.R., Packlatko, J.P., and Eisenstark, A., 1976, Characterization of a cell-lethal tryptophan photooxidation product: hydrogen peroxide. *Science* 191, 468–469.

Moraes, E.C., Keyse, S.M. and Tyrrell,r.M., 1990, Mutagenesis by hydrogen peroxide treatment of mammalian cells: a molecular analysis. *Carcinogenesis* 11, 283–293.

Moraes, E.C., Keyse, S.M., Pidoux, M. and Tyrrell, R.M. (1989) The spectrum of mutations generated by passage of a hydrogen peroxide damaged vector plasmid through a mammalian host. *Nucl. Acids. Res.* 17:8301–8312

Stocker, R., Yamamoto, Y., McDonagh, A.F., Glazer, A.N. and Ames,B.N., 1987, Bilirubin is an anti-oxidant of possible physiological importance, *Science* 235, 1043–1046.

Storz, G., Christman, M.F., Sies, H. and Ames, B.N., 1987, Spontaneous mutagenesis and oxidative damage to DNA in Salmonella typhimurium. *Proc. Natl. Acad. Sci. US.* 84, 8917–8921.

Streisinger, G., Okada, Y., Emrich, J., Newten, J., Tsugita, A., Terzaghi, E. and Inouye, M., 1966, Frameshift mutations and the genetic code. *Cold Spring Harbor Symp. Quant. Biol.* 31, 77–84.

Tyrrell, R.M., and Pidoux, M., 1986, Endogenous glutathione protects human skin fibroblasts against the cytotoxic action of UVB, UVA and near-visible radiations. *Photochem. Photobiol.* 44, 561–564.

Tyrrell, R.M. and Pidoux, M., 1988, Correlation between endogenous glutathione content and sensitivity of cultured human skin cells to radiation at defined wavelengths in the solar ultraviolet range. *Photochem. Photobiol.* 47, 405–412.

Tyrrell, R.M. and Pidoux, 1989, Singlet oxygen involvement in the inactivation of cultured human fibroblasts by UVA (334 nm, 365 nm) and near-visible (405 nm) radiations. *Photochem. Photobiol.* 49, 407–412.

Webb, R.B., 1977, Lethal and mutagenic effects of near-ultraviolet radiation. Photochem. Photobiol. Reviews, 2, 169–262.

Yoshida, T., Biro, P., Cohen, J., Muller, R. and Shibahara, S., 1988, Human heme oxygenase cDNA and induction of its mRNA by hemin. *Eur. J. biochem.* 171, 457–461.

Ziegler-Skylakakis, K. and Andrae, U., 1987, Mutagenicity of hydrogen peroxide in V79 Chinese hamster cells. *Mutation Res.* 192, 65–67.

Melanin is a Double-Edged Sword

JULIAN M. MENTER[1] , ISAAC WILLIS[1], MICHAEL E. TOWNSEL[3],
GEORGE D. WILLIAMSON[2], and CYRIL L. MOORE[2]

Departments of Medicine[1] and Biochemistry[2]
Morehouse School of Medicine, Atlanta, Georgia 30310
and Department of Chemistry[3], Atlanta University
Atlanta, Georgia 30314 (USA)

Abstract

Melanin is known to photoprotect by physical absorption/scattering of UV and by electron transfer/free radical scavenging. In this work we show that melanin can also facilitate potentially harmful redox reactions *in vitro*. By kinetic spectrometry and O_2 uptake, we show that both synthetic eumelanin (Sigma) and Sepia melanin extracted from cuttlefish markedly accelerate the tyrosinase-catalyzed oxygenation of the cytotoxic phenol p-hydroxyanisole MMEH to the cytotoxic 4-methoxy-1,2-benzoquinone. These studies indicate that a substrate-melanin complex transfers electrons to tyrosinase to regenerate Cu(I) necessary for reduction of molecular O_2. Prior irradiation of melanin with solar-simulating UV results in evanescent changes in the kinetics of quinone formation which are consistent with reversible photooxidation of melanin. The latter effect vis-a-vis photoprotection is ambivalent at best. Thus, melanin protects with one hand, and, perhaps unwittingly, may place surrounding cells at risk with the other.

List of Abbreviations and Symbols

Dopa –	3,4 dihydroxyphenylalanine
MMEH –	p-hydroxyanisole; monomethylether of hydroquinone
t^o –	length of induction period ("lag time") observed in the tyrosinase-catalysed oxygenation of MMEH in the absence of added melanin.
t_o –	length of induction period ("lag time") observed in the tyrosinase-catalysed oxygenation of MMEH melanin.
t_o (i) –	length of induction period ("lag time") observed using preirradiated melanin.
$t_o(u)$ –	length of induction period ("lag time") observed using "dark control" melanin.
r –	rate of quinone formation in the linear phase of tyrosinase-catalysed oxygenation of MMEH.

---

[1] Address for Correspondence: Julian M. Menter, Ph.D., Department of Medicine, Morehouse School of Medicine, 720 Westview Drive, S.W., Atlanta, Georgia 30310, (404) 752–1697.

Photobiology, Edited by E. Riklis
Plenum Press, New York, 1991

873

Introduction

Pigment melanin, particularly eumelanin, has long been considered to be a relatively "inert" biopolymer which protects the organism against solar radiation, by absorption and scattering of harmful UV (Pathak et al., (1987). Recently, there has been a new appreciation of the importance of melanin's ability to bind a wide variety of compounds, and to act as a radical scavenger in photoprotection (see, for example, Musk and Parsons, 1987; Hill et al., 1987). At the same time, however, there has been a growing awareness that these very same chemical processes could also result in "antiprotective" behavior (Hill and Hill, 1987; Cesarini, 1987; Menter and Willis, 1986). Thus, melanin may be considered a "double-edged sword" whose presence is at best somewhat ambivalent.

We have recently demonstrated that melanin can couple the anerobic oxidation of several mono-and dihydroxybenzenes to the reduction of potassium ferricyanide by binding to both reactants and acting as an electron conduit (Menter and Willis, 1986). On the one hand, the capture of electrons may be helpful in controlling cytotoxic reactions *in situ*. On the other hand, melanin may transfer the captured electron to an acceptor, which becomes chemically reactive and/or then produces other reactive species.

It turns out that melanin also transfer electrons to tyrosinase (EC 1.14.18.1), a copper-containing enzyme which, catalyzes hydroxylation of tyrosine and subsequent dehydrogenation of dopa. Although mammalian tyrosinase catalyses a variety of phenols and catechols, it is usually though of as being specific for tyrosine/dopa. However, other substrates have been reported as well, e.g. 5,6 dihydroxyindole (Pawelek and Koerner, 1982), p-hydroxyanisole (MMEH; Passi and Nazzaro-Porro, 1981) and 4-s-cysteaminyl phenol (Ito and Jimbow, 1987). All three of the latter substrates are oxidized to cytotoxic quinones by mammalian tyrosinase. These reactions may be generalized as follows:

The active form of the enzyme contains reduced copper, Cu(I), which reduces molecular O_2. The dihydroxybenzene (quinol) intermediate serves both as a substrate and as a co-factor which donates electrons to tyrosinase to maintain the active Cu(I) state. Since pigment melanin is a polymer with a large number of reversibly oxidizable quinol-quinone moieties (Hempel, 1969), and in view of its aforementioned electron-transfer capabilities, we wished to determine if eumelanin would transfer electrons to tyrosinase. As substrate, we chose MMEH (R = OCH_3 in eq I), which is well known to cause depigmentation of mammalian skin and hair (Riley, 1969, 1970). Since oxidative metabolites of MMEH are cytotoxic, this process would be antiprotective if it occurred *in vivo*. We wished to know whether or not this effect could be observed in a "biological" DOPA-melanin sample, namely sepia melanin obtained from cuttlefish. Moreover, given the non-specificity of mushroom tyrosinase, we wished to test this effect in the presence and absence of melanin using a highly purified B-16 melanoma tyrosinase preparation. Since melanins are photochemically active (Chedekel, 1982; Felix et al., 1978; Korytowski et al, 1987; Thompson et al, 1985), we wished to determine whether irradiation with solar-simulating UV would detectably alter its electron transfer properties.

In this communication, we report that melanin accelerates MMEH oxygenation by transferring electrons to tyrosinase. Prior photolysis of the pigment with solar-simulating UV results in significant alterations which are consistent with reversible photooxidation. Qualitatively similar results are obtained for both synthetic and dopa melanin samples, but their quantitative differences are enough to afford a distinction between them. Hence, these results promise a new analytical tool which can aid in the characterization of melanin samples in a non-destructive, yet meaningful way. In addition, we again mechanistic insight which is useful for explaining and predicting the phenomenological observations.

Materials and Methods

A. Reagents

The phenol p-hydroxyanisole (MMEH), obtained from Aldrich Chemical Co. and dopa (Sigma Chemical Co.) were used as received. Synthetic melanin (Sigma Chem. Co[*]), obtained from persulfate oxidation of tyrosine, was treated as previously reported (Menter and Willis, 1980). Sepia melanin extracted from cuttlefish was a kind gift from Dr. Miles R. Chedekel. L-Dopa (Sigma Chemical Co.) was used as received.

Mushroom tyrosinase (EC 1.14.18.1; Sigma Chemical Co.) was used for most experiments. Several experiments were carried out with a highly purified tyrosinase preparation obtained from B16 mouse melanoma (we thank Dr. Vincent J. Hearing, NIH,

---

[*]Although Sigma now offers a sepia melanin preparation similar to the one we used, we will use the term "Sigma melanin" to refer only to the synthetic compound prepared from persulfate oxidation of tyrosine in this paper.

for this sample). Freshly prepared stock solutions (Ca 0.1 mg/ml) were assayed spectrophotometrically at 475 nm with 300 uM dopa. One unit of activity causes an absorbance change at 475 nm of 0.001 per minute. Enzyme protein was determined by the method of Bradford (1976).

B. Tyrosinase-Catalyzed Oxygenation of MMEH in the Presence and Absence of Added Melanin

The reactions were carried out in an Aminco DW 2 Kinetic Spectrometer, in a 1.0 cm cuvette fitted with a stirring apparatus, as previously described (Menter and Willis, 1986). When mushroom or tyrosinase were used, enzyme was rapidly pipetted into a stirred solution containing 300 uM MMEH ± melanin in a total of 3.0 ml 0.1 M phosphate buffer at pH 7.4. When B-16 tyrosinase was used, 0.3 ml substrate was pipetted into tyrosinase (2.9 ml) ± melanin in buffer to make up 3.0 ml. Control experiments determined that the order of mixing did not influence the course of the reactions. Added melanin was 100 ug.

Absorption spectra were repetitively recorded on the DW 2A at 20 nm/s. Kinetic curves were then recorded using the dual beam mode, with probe wavelength 413 nm for MMEH (λ max of 4-methoxy-1,2-benzoquinone; Passi and Nazzaro-Porro, 1981; Naish et al, 1988) and reference wavelength 650 nm (no absorption). Kinetic curves were run at least three times. The precision of the reported slopes are within ± 5%. For calculation of rates of quinone formation, r, a value of 1600 $M^{-1} cm^{-1}$ was used for the molar extinction coefficient of 4-methoxy-1,2-benzoquinone (Teuber and Staiger, 1955). The "rates of quinone formation" reported below refer to those values derived from the linear part of the biphasic kinetic curve, (see Fig. 2).

Oxygen uptake experiments were carried out polarographically with YSI model 5300 oxygen analyzer. In the mushroom tyrosinase experiments, 40–80 ul of enzyme solution was rapidly injected into an air-saturated solution of MMEH ± melanin (total volume 600 ul). In the B-16 tyrosinase studies; 5 ul MMEH was injected into 40 ul enzyme ± melanin in a total of 60 ul. In both cases the O_2 content of the cell was monitored as a function of time. Oxygen uptake in the absence of any one of the other three reaction components was negligible under these conditions.

C. Static Binding Experiments.

To assess static binding of MMEH to melanin, Sigma melanin (100 mg) was incubated with 9 ml of 200 uM MMEH to 0.1 M phosphate buffer, pH 7.4. The mixture was shaken, and then centrifuged overnight at 50,000 rpm on a Beckman J-21 ultracentrifuge. Control tubes contained 100 mg melanin without MMEH or 200 uM MMEH without melanin in 9 ml buffer. Small aliquots of each were carefully pipetted from the top of the tube and placed into a 0.2 cm pathlength cuvette. Absorption of each sample which was due only to MMEH was read on the DW-2A, using the dual-mode, with probe wavelength-298 nm (λ max of MMEH) and reference wavelength 350 nm (no MMEH absorption). The readings were corrected for the small differences in melanin absorption at 298 and 350 nm (< 1%). The percent of MMEH bound to melanin was taken as $(1 - D_m/D_o) \times 100$ where D_m and D_o were the net absorbances at 298 nm in the presence and absence of melanin respectively.

To estimate binding of melanin to tyrosinase, 100 mg of sepia or Sigma melanin was mixed in a 9 ml ultracentrifuge tube with 660 units of mushroom tyrosinase in 0.1 M phosphate buffer, pH 7.4. A control tube contained the enzyme in the absence of added melanin. Both tubes were shaken thoroughly, then centrifuged as before. The supernatants were carefully decanted. The pellets, the "empty" control tubes and aliquots of the supernatants were assayed for dopa oxidase activity.

D. Photolysis of Melanin

Sepia or Sigma melanins (1.0 mg) were dispersed in 3.0 ml phosphate buffer in a 1.0 cm quartz cuvette. The stirred suspensions were then exposed to 290–400 nm broadband radiation from a 1.6 kW xenon arc solar simulator (Willis et al, 1981). The total dose (fluence) impinging on the surface of the cuvette was 150 J/cm^2. Aliquots of both the irradiated melanin samples and their unirradiated controls were analyzed for their effect on tyrosinase-catalyzed oxygenation of MMEH for post-irradiation times ranging from 5 to 300 minutes.

Results

A. Melanin-Accelerated Tyrosinase Catalyzed Oxygenation of MMEH

In aerated solution in the presence of mushroom of B-16 melanoma tyrosinase, MMEH is *hydroxylated* to 4-methoxy-1,2-diphydroxybenzene ("quinol") and then *dehydrogenated* to 4-methoxy-1,2-benzoquinone ("quinone"). Fig. 1, a repetitive rapid scan of tyrosinase catalyzed oxygenation of MMEH in the presence of Sigma melanin, clearly shows the emergence of the 413 nm band. This result is qualitatively identical to that obtained in the absence of melanin, but the reaction is faster in the presence of melanin.

The effect of added Sepia or Sigma melanin is to increase the overall rate of oxygenation. In the presence or absence of melanin, the reaction kinetics, as monitored either by O_2 uptake or increase in 413 nm (quinone) absorbance is biphasic, with an initial induction period ("lag time"), denoted by t_o, followed by a linear increase in quinone formation r. Figure 2. shows this effect for the melanin-containing system. The value of t_o is taken as the time required for the rate of quinone formation to become linear, at which point the quinol concentration reaches a steady state; it is roughly inversely proportional to the units of enzyme used, and consequently to the rate of hydroxylation (see discussion). The value of r is directly proportional to enzyme activity. Table 1 shows that the rate of reaction is roughly five times as fast with mushroom tyrosinase as it is for B-16 enzyme. Sepia melanin shortens the value of t_o (here normalized to unit activity) by 34% for B-16 melanoma tyrosinase and 70% for

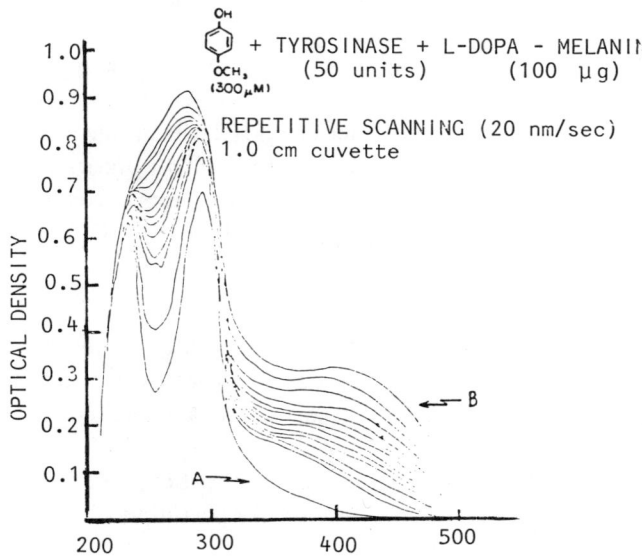

Fig. 1. Rapid repetitive scan of tyrosinase-catalyzed oxygenation of MMEH in the presence of Sigma Melanin in 0.1 M phosphate buffer, pH 7.4 [MMEH] = 100 ug/3.0 ml, mushroom tyrosinase = 50 units. Scan rate was 20 nm/sec. Several early scans were omitted from this figure for clarity. The most notable feature is the emergence of the 413 nm peak, corresponding to 4-methoxy-1,2 benzoquinone (see text). A: Initial Scan t=O B: t=10 min. Result in the absence of melanin (not shown) was qualitatively identical, but the reaction proceeded significantly slower.

mushroom tyrosinase. The normalized values r, are slightly elevated (ca 10%) for both enzyme preparations. The effect of melanin is saturable, and the values in Table 1 represent saturation values.

B. Effect of Added L-Dopa
When 30 uM L-Dopa was added to 300 uM MMEH plus tyrosinase (200 units),

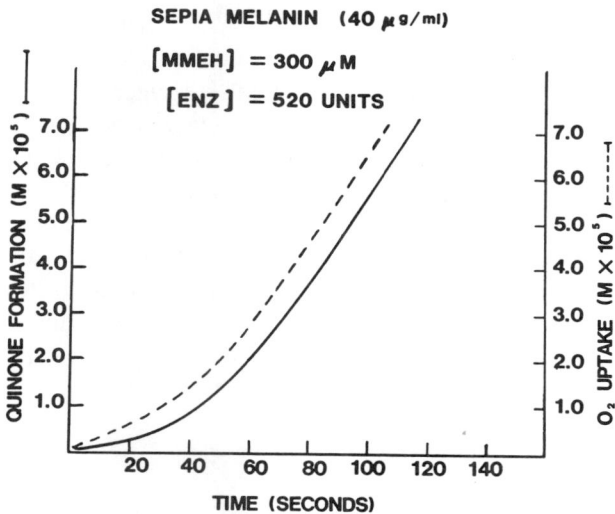

Figure 2. Time course of mushroom tyrosinase catalyzed oxygenation of p-hydroxyanisole (MMEH) in the presence of melanin. Solid line monitors 4-methoxy-1,2-benzoquinone ("quinone") formation as measured spectroscopically (left axis); dotted line is O_2 uptake, measured polarographically (right axis). [MMEH] = 300 uM; [Melanin] = 120 ug/3.0 ml tyrosinase = 520 units (see text). This figure shows results for sepia melanin; similar results were obtained for Sigma melanin. In the absence of melanin, the "lag time" was increased, but the final rate of quinone formation is similar (see text). For other experimental conditions, see text.

formation of 4-methoxy-1,2 benzoquinone was immediate, with no observable lag time. The increased rate of quinone formation could not be accounted for by oxidation of the small amount of dopa to dopachrome. Added melanin had no effect on the 4-methoxy-1,2-benzoquine formation; it also did not affect the rate of dopa oxidation to dopachrome.

C. Melanin-Binding Studies

In three separate experiments, 200˙uM MMEH was mixed with 100 mg melanin as described above. Spectroscopic analysis of the supernatant fractions indicated that under these conditions virtually all MMEH was bound to the sedimented melanin pellet, as evidenced by the lack of MMEH absorption at 298 nm in the supernatant.

On the other hand, virtually no binding was observed between mushroom tyrosinase and either Sepia or Sigma melanin. The sedimented pellet exhibited a small amount of dopa oxidase activity (Ca 0.6% of the control supernatant). However, the "empty" control tube exhibited the same amount of activity, indicating that a small amount of free enzyme sediments under these conditions, and is responsible for the activity exhibited by the sedimented pellet.

D. Irradiation of Melanin with Solar-Simulating UV

Exposure of both melanin samples to solar-simulating UV (290-400 nm) prior to assaying their effects on tyrosinase-catalyzed MMEH oxygenation revealed short-term reversible photochemical changes in both samples. Experimentally, this was manifested as an evanescent increase in the lag time of the irradiated sample $t_o^{(i)}$ relative to its unirradiated control value, $t_o^{(u)}$, as depicted in Fig. 3. The effect is more pronounced in the Sigma melanin. There, the initial $t_o^{(i)}/t_o^{(u)}$ ratio is 1.8, which decreases to unity 3 hours after irradiation. With Sepia melanin, similar effects are noted, but these are of lesser magnitude and even more evanescent, with an initial $t_o^{(i)}/t_o^{(u)}$ value of 1.4 decreasing to unity 75 minutes post-irradiation. These transient photochemical changes are super-imposed on a slow monotonic dark increase in t_o exhibited by both melanin suspensions on standing at pH 7.4.

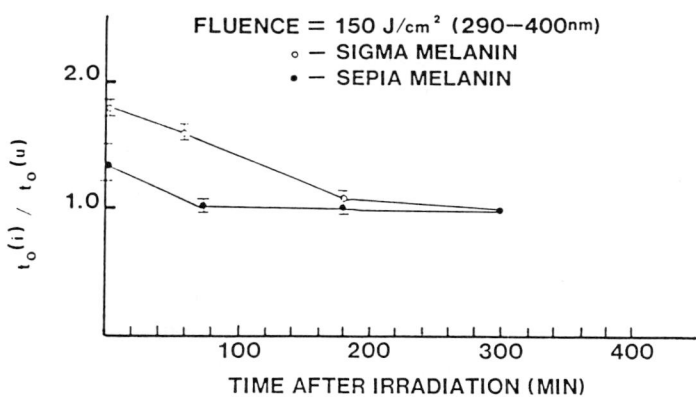

Figure 3. Effect of Solar-Simulation Radiation (290–400 nm) on subsequent melanin-mediated tyrosinase-catalyzed oxygenation of MMEH. Melanins (1.0 mg) were dispersed in 3 ml phosphate buffer, pH 7.4 in a quartz cuvette. They were irradiated with 150 J/cm^2 solar-simulating radiation, as described in the text. At various post-irradiation times, 100 ug aliquots of irradiated samples and companion dark controls were incubated with MMEH and tyrosinase as described above, and was analyzed by kinetic spectroscopy. Datapoints represent the mean ± S.D. of the ratio of lag times for irradiated: unirradiated samples, $t_o^{(i)}/t_o^{(u)}$, as functions of post irradiation time.–❼–❼– Sepia melanin; –0–0– Sigma melanin.

Discussion

In recent years, the "active" physical and chemical behavior of melanin has become well-appreciated. The net effects of binding, electron transfer, and photochemical activity may be protective or anti-protective.

In this work, we show evidence of melanin's darker side. Two different eumelanin preparation - a synthetic melanin produced from persulfate oxidation of tyrosine ("Sigma melanin") and a Sepia melanin extracted from cuttlefish are capable of markedly accelerating the tyrosinase-catalyzed oxygenation of cytotoxic phenol, MMEH to an even more cytotoxic quinone, presumably via reactions I and II, which are both catalyzed by the "substrate specific" mammalian tyrosinase from B-16 mouse melanoma. Thus, it is likely that the metabolism of xenobiotic MMEH may be accelerated *in situ* by endogenous tyrosinase, with concommitant production of cytotoxic products and intermediates.

Reactions I and II form the basis of the observed biphasic kinetics of quinol formation (Fig. 2). As in the case of tyrosine itself, hydroxylation (I) is evidently slower than dehydrogenation (II). Melanin presumably accelerates the hydroxylation by acting as a "dopa-like" co-factor which accelerates the overall oxidation of tyrosine or MMEH to the 1,2 benzoquinone. Presumably, the pigment can transfer electrons to tyrosinase in a manner analogous to, although not as efficient as, dopa, to regenerate Cu(I) necessary to reduce molecular O_2 in the slow hydroxylation reaction. Apparently 4-methoxycatechol can itself act as a dopa-like co-factor, and is evidently more efficient than melanin, since the latter did not increase the rate of 4-methoxy-1,2 benzoquinone formation in a system originally primed by a small amount of L-dopa.

From these results, it is clear that melanin significantly increases the hydroxylation reaction, but has little or no effect on the dehydrogenation. This increase is experimentally manifested by a decrease in lag time, t_o. Hydroxylation can be described by simple Michaelis-Menten Kinetics (Osaki, 1963). K_m probably not affected by the presence of melanin, since we found no evidence for a melanin-tyrosinase complex. In this case, t_o is inversely proportional to V_m, and t_o/t_o^o (where t_o^o is the lagtime in the absence of added melanin) is an approximate measure of the relative increase in rate of hydroxylation. In this instance, sepia melanin increases V_m by 3-fold and Sigma melanin increases it by 5-fold.

The experimental data are consistent with the sequence summarized in Fig. 4, where the wavy arrows represent electron transfer processes. MMEH is activated via complex formation with melanin. Melanin acts as a "middle man" by accepting electrons from bound MMEH (Menter and Willis, 1986), on the one hand, and transferring them to tyrosinase, on the other. MMEH is oxidized to 4-methoxycatechol. The catechol transfers electrons to tyrosinase, and is oxidized to quinone. Melanin has little or no effect on the dehydrogenation reaction. Tyrosinase transfers electrons to molecular O_2.

This scheme can be used estimate the rate of electron transfer from the MMEH-melanin complex (denoted by "comp") vis-a-vis that from steady-state catechol (denoted by QH_2). The active form of tyrosinase is assumed to be in a steady state, in which its rate of formation is equal to its rate of removal by reaction with molecular O_2. The resulting rate of O_2 consumption is equal to the rate of quinone formation (see Fig. 2). In the absence of melanin, we have :

$$rate_o = k_1[QH_2] \tag{1}$$

in the presence of melanin,

$$rate_{(Mel)} = k_1[QH_2] + k_D[Comp] \tag{2}$$

Since t_o is inversely proportional to the rate of hydroxylation (itself proportional to the rate of electron transfer from tyrosinase to O_2), then we can write

$$rate_o = (t_o/t_o^o) \, rate_{(mel)} \tag{3}$$

substitution of (3) into (1) and (2), and rearrangement leads to

$$k_D[Comp] = k_1[QH_2] \times [1 - (t_o/t_o^o)] / (t_o/t_o^o) \tag{4}$$

Where $k_D[Comp]$ and $k_1[QH_2]$ represent the respective rates of electron transfer to tyrosinase from the melanin complex and from the catechol. Values of t_o/t_o^o in Table 1 were used to compute $k_D[Comp1/k_1[QH_2]$ ratios, shown in Table 2. In the case of mushroom tyrosinase, the overall rate of transfer from melanin is greater than that from catechol, whereas the reverse is the case for B-16 mammalian tyrosinase. The reasons for these differences are unclear, but may lie in the greater substrate specificity of the mammalian enzyme. Sigma melanin was 1.7 times as effective as sepia melanin in transferring electrons to mushroom tyrosinase[*].

The reversible photochemical changes exhibited by UV-irradiated melanin (which are responsible for the transient increase in t_o) are consistent with reversible photooxidation, which could render the polymer less capable of electron donation. One noteworthy feature is the considerable change in t_o, observed especially with the Sigma Melanin. When inner filter effects are accounted for, the average dose "seen" by the irradiated melanin is comparable with "typical" exposure to solar UV. Clearly, melanin which has just been exposed to sunlight cannot be considered to be the same compound as the pre-irradiated polymer. In the same vein, "dark" irreversible transformations in melanin occur when it

---

[*]Note added in proof: Subsequent kinetic analysis indicated that both melanins are very efficient oat enhancing the production of oxy- from met-tyrosinase, and that Sigma melanis is approximately 3-fold more efficient than Sepia melanin (J.M. Menter et al., Pigment Cell Res., 1990, in press).

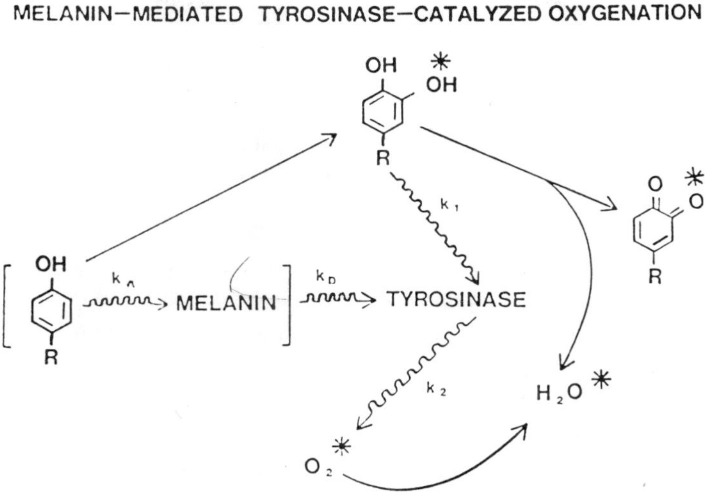

MELANIN—MEDIATED TYROSINASE—CATALYZED OXYGENATION

Figure 4. Proposed scheme for melanin-mediated tyrosinase-catalyzed oxygenation of MMEH. Wavy arrows represent electron-transfer processes. MMEH forms a complex with melanin; the activated complex transfers electrons to tyrosinase to maintain reduced Cu(I) necessary for reduction of molecular O_2. Generated 4-methoxycatechol can also transfer electrons to tyrosinase. One atom of O_2 is incorporated into the aromatic ring, the other into H_2O.

Table 1. Oxygenation of MMEH by B-16 or mushroom tyrosinases in the presence or absence of melanin

Enzyme Source	Melanin	$r \times 10^9$*	r/r_o*	$t_o \times$ units**	t_o/t_o^o**
B-16	-	4.4	=1.00	9.45	=1.00
B-16	+	5.2	1.17	6.20	0.66
Mush.	-	22.8	=1.00	1.93	=1.00
Mush.	+ (Sepia)	26.9	1.18	0.567	0.30
Mush.	+ (Sigma)	26.2	1.15	0.380	0.20

[MMEH] = 300 ug. Mel = 100 ug. B-16 tyrosinase = 70 units; Mushroom tyrosinase = 34 units in 0.1 M phosphate buffer, pH = 7.4. Other conditions as in text.
* rate of quinone formation in M^{s-1} ug protein^{-1}; r_o refers to rate in the absence of added melanin,
** lag time (seconds) normalized to unit enzyme activity (see text).
t_o^o refers to lagtime in teh absence of melanin.

Table 2. Relative efficiency of electron transfer to tyrosinase in MMEH oxygenation.

Melanin	Tyrosinase	$k_D[Comp]/k_1[OH_2]$*
Sepia	Mushroom	2.3
Sepia	B-16	0.52
Sigma	Mushroom	4.0

* Ratio of rates of electron transfer from MMEH-Melanin complex ($k_D[Comp]$) and steady-state quinol ($k_D[QH_2]$) to tyrosinase (see text).

is allowed to stand in suspension at room temperature, pH 7.4. Secondly, it is clear from the above that short and long term "light" and "dark" chemical changes in melanin may be followed in a non-destructive manner. Finally, it should be noted that, apropos photo-protection, these reactions are ambivalent at best. Immediate pigment darkening confers no significant photo-protection against erythema in human skin (Willis et al, 1972). Photolysis of eu or pheomelanin produces deleterious "active oxygen" intermediates, most notably superoxide and hydrogen peroxide (Cesarini et al, 1987; Korytowki et al, 1987, and references therein). On the other hand, the resulting oxidized melanin may be more efficient at scavenging active radicals, and it would be expected to oxidize a higher fraction of superoxide to molecular O_2, and reduce less to H_2O_2 (Korytowski et al, 1987).

Thus, melanin is a double edged sword. The very properties which make it a good photoprotection agent may also, sometimes quite unwittingly, accelerate processes which are harmful to the organism. To summarize, absorption and scattering *protects* low-lying target cells from incoming radiation. *Binding* either *protects* by localizing toxic compounds, or it may *deprotect* by retaining harmful compounds which would otherwise be excreted or metabolized. *Radical scavenging* is usually *protective*. *Electron transfer* may *protect* by defusing" active intermediates or reaction products. It may *deprotect* by inadvertently forming redox products which are more toxic than the parent compound. *Photochemical reactions* involving eu or pheomelanins are likely to lead to "active oxygen" intermediates or melanin-derived radicals which can result in cell damage. *Immediate pigment darking*, is not photoprotective against UV. Melanin is an excellent electron transfer reagent. This is usually advantageous (eg in free radical scavenging), but it may, as in the present instance, speed up the production of melanocytotoxic compounds.

Acknowledgements

The authors are very grateful to Dr. Vincent J. Hearing for providing the B-16 tyrosinase sample. We thank Betty Jean Soteres for her technical assistance. We thank Dr. Miles R. Chedekel for the Sepia melanin preparation. We thank Ms. Fran. I. Menter for the graphic illustrations and Ms. Mary Hastie and Ms. Deborah D. Smith for typing the manuscript. We gratefully acknowledge support by EPA Grants #R812605, R812542, and R814089.

References

Bradford, M.M. (1976). A Rapid and Sensitive Method for the Quantitation of Microgram Quantities of Protein Utilizing the Principle of Protein - Dye Binding. *Anal. Biochem* 72, 248–254.

Cesarini, J.P. (1988). Photo-induced Events in the Human Melanocytic System: Photoaggression and Photoprotection. *Pigment Cell Res.* 1, 223–233.

Chedekel, M.R. (1982). Photochemistry and Photobiology of Epidermal Melanins Photochem. *Photobiol.* 35, 881–885.

Felix, C.C., Hyde, J.S., Sarna, T. and Sealy, R.C. (1978). Melanin Photoreactions in Aerated Media: Electron Spin Resonance Evidence for Production of Superoxide and Hydrogen Peroxide. *Biochem Biophys Res. Comm.* 84, 335–341.

Hempel, K. (1966). Investigation on the Structure of Melanin in Malignant Melanoma with ^{3}H and ^{14}C-Dopa-Labelled at Different positions. From "Structure and Control of the Melanocyte" - G. Della and O. Muhlenbock, Eds., *Berlin-Springer Verlag*, 162–173.

Hill, H.Z., and Hill, G.J. (1987). Eumelanin Causes DNA Strand Breaks and Kills Cells. *Pigment Cell Res.* 1, 163–170.

Hill, H.Z., Huselton, C., Pilas, B., and Hill, G.J. (1987). Ability of Melanins to Protect Against the Radiolysis of Thymine and Thymidine. *Pigment Cell Res.* 1, 81–86.

Ito, Y., and Jimbow K. (1987). Selective Cytotoxicity of 4-S-Cysteamylphenol on Follicular Melanocytes of the Black Mouse: Rational Basis for its Application to Melanoma Chemotherapy. *Cancer Research*, 47, 3278–3284.

Korytowski, W., Pilas, B., Sarna, T., and Kalyanaraman, B. (1987). Photoinduced Generation of Hydrogen Peroxide and Hydroxyl Radicals in Melanins. *Photochem. Photobiol*, 45, 185–190.

Menon, I.A., and Haberman, H.F. (1978). Formation of Melanin-Tyrosinase Complex and its Possible Significance as a Model for Control of Melanin Synthesis. *Acta Dermatovener*, 58, 9–11.

Menter, J.M., and Willis, I. (1980). The interaction of L-Dopa Melanin with p-tert-Butylcatechol. *J. Invest. Dermatol*, 75, 260–265.

Menter, J.M., and Willis, I. (1986). Interaction of Several Mono- and Dihydroxybenzene Derivatives of Various Depigmenting Potencies with L – 3,4-Dihydroxyphenylalanine-Melanin. *Arch. Biochem Biophys*, 244, 846–856.

Musk, P., and Parsons, P.G. (1987). Resistance of Pigmented Human Cells to Killing by Sunlight and Oxygen Radicals. *Photochem Photobiol*, 46, 489–494.

Naish, S., Cooksey, C.J., and Riley, P.A. (1988). Initial Mushroom-Tyrosinase-Catalyzed Oxidation Product of 4-Hydroxyanisole is 4-Methoxy-Ortho-Benzoquinone. *Pigment Cell Res.* 1, 379–381.

Osaki, S. (1963). The Mechanism of Tyrosine Oxidation of Mushroom Tyrosinase. *Arch Biochem Biophys*, 100, 378–384.

Passi, S. and Nazzaro-Porro, M. (1981). Molecular Basis of Substrate and Inhibitory Specificity of Tyrosinase: Phenolic Compounds. *Br, J. Dermatol*, 104, 659–665.

Pathak, M.A., Fitzpatrick, T.B., Greiter, F., and Kraus, E.W. (1987).in "Dermatology in General Medicine" Third Edition, T.B. Fitzpatrick et al *Eds. McGraw-Hill-Ch*, 130, 1507–1522.

Pawelek, J.M., and Koerner, AM (1982). The Biosynthesis of Mammalian Melanin. *AM. Sci.* 70, 136–145.

Teuber, H.J., and Staiger, G. (1955). Reaktionen Mit Nitrosodisulfonat VIII: Ortho-Benzochinone and Phenazine. *Chem. Ber.* 88, 802–827.

Thompson, A., Land, E.J., Chedekel, M.R., Subbarao, K.V., and Truscott (1985). A Pulse

Radiolysis Investigation of The Oxidation of the Melanin Precursors 3,4-Dihydroxyphenylalanine (dopa) and the Cysteinyldopas. *Biochim Biophys. Acta,* 843, 49–57.

Willis, I., Kligman, A.M., and Epstein, J.H. (1972). The Effects of Long Ultraviolet Light on Skin-Photoprotective on Photoaugmentative? *J. Invest. Dermatol.* 59, 416–420.

Willis, I., Menter, J.M., and Whyte, H.J.. (1981). The Rapid Induction of Squamous Cell Cancer Utilizing the Principles of Photoaugmentation. *J. Invest. Derm.* (1981). 76, 404–408.

B. Protection Measures:

Sun Protection Factors

R.E. MASCOTTO

Givaudan SA
CH-1214 Vernier-Geneva

Introduction

Sunbathing and tanning, have become cosmetically desirable in the late 20th.

Almost all living species have hair, feathers or scales to protect themselves against ultraviolet (UV) radiations. Deprived of these natural protections, people must learn to avoid excess sun exposure and protect their skin to minimize photo damage (keratoses, skin cancer, premature aging).

Cosmetic scientists and regulators have recognized the importance of providing the public with effective sunscreens, which can be expected to maintain their protective capabilities under normal and abnormal conditions.

Methods to measure the protective value of sunscreen preparations have been discussed among scientists since the beginning of the century. In response to consumers' increased awareness of sunscreen benefits, biological procedures to quantify sunscreen effectiveness have formally been developed by government or private organizations in the United States of America, Germany and Australia. In general, the procedures are based on the original Schulze method. However, little has been done to provide a universal index of sunscreen protection. Significant differences still exist among the methods adopted to date. This presentation discusses and compares only the *in vivo* methods used in the U.S., Germany and Australia.

SPF Definition

In the methods employed by the three countries, the sun protection factor (SPF) is defined as the UV energy required to produce a minimal erythema dose (MED) on protected skin divided by the UV energy required to produce an MED on unprotected skin:

$$\text{SPF value} = \frac{\text{MED (protected skin)}}{\text{MED (unprotected skin)}}$$

MED, or minimal erythema dose, is the minimum quantity of radiant energy which produces the first detectable reddening of fair human skin following exposure to radiation at a specified wavelength or range of wavelengths where the radiation source has constant intensity.

Photobiology, Edited by E. Riklis
Plenum Press, New York, 1991

The MED can be stated as the time of exposure that produces the minimally perceptible erythema at 16 to 24 hours post exposure (FDA/SAA) or that produces just a perceptible reaction at 20 to 28 hours post exposure (DIN).

The MED of the unprotected skin is measured one day before the test in accordance with the FDA/SAA standards.

The DIN Norm requires the measurement on the same day. In addition the control stripes must be adjacent to test stripes.

Legislation

The U.S. Food and Drug Administration (FDA) advance notice of proposed rule making (ANPR) for over-the-counter (OTC) sunscreen drug products for human use was published in the August 25, 1978 Federal Register.

Twenty-one sunscreens have been considered as safe and effective is used at recommended concentration and provided they exhibit an SPF of at least 2.

The proposal is not a final regulation, since the FDA must consider further public comments before issuing any final rule. Nevertheless, the FDA has stated that marketing a product with a formulation or labelling that is not in accord with the ANPR, or proposed rule, may result in regulatory actions against the product, marketer, or both. Most companies currently marketing sunscreens in the U.S. use the proposal as a guide for sunscreen compositions, testing and labelling.

Unlike the German methodology, the U.S. and Australian methods define product categories in order to help consumers choose appropriate sunscreen products based on their own skin type.

Skin phototypes (M.A. Pathak - 1983)

Skin type	Unexposed skin color	MED/UV-B mj/cm^2	Sunburn and tanning history
I	White	15–30	Always burns easily, never tans
II	White	25–35	Always burns easily, tans minimally
III	White	30–50	Burns moderately, tans gradually, light brown
IV	Light Brown	45–60	Burns minimally, always tans well, moderate brown
V	Brown	60–100	Rarely burns, tans profusely, dark brown
VI	Chocolate brown, black	100–200	Never burns, deeply pigmented black

In Germany, discussion of a standardized method begins in 1980 under the auspice of the private German Standards Organization. The first draft proposal was issued for comments in March, 1984. (DIN NORM 67 501º.

As part of the comment process, industry association representatives from the U.S., U.K and Germany met during the summer of 1984 in an attempts to avoid significant

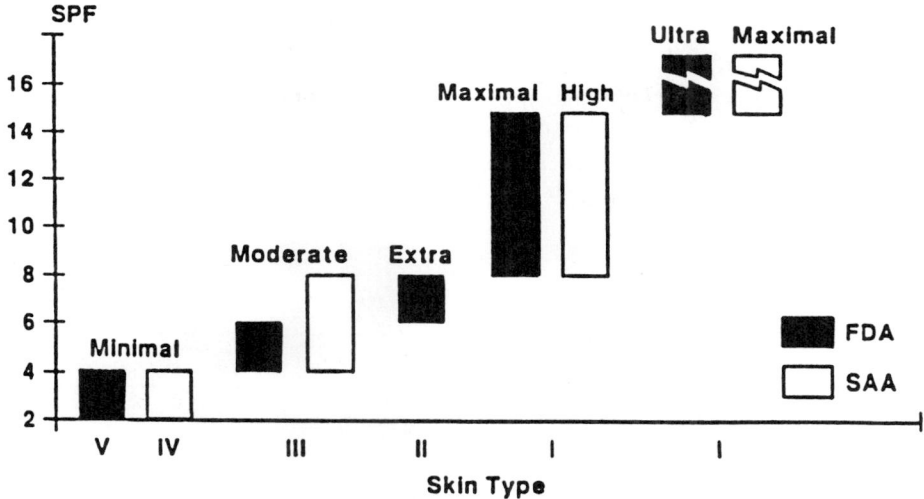

Product category designation based on SPF values (FDA vs. SAA)

differences that could result in disparate SPF values for tested sunscreen products. Several changes suggested by the industry were incorporated in the DIN Norm which was issued in September, 1984. However, key differences between Din and FDA test procedures remain.

The Standard Association of Australia (SAA) issued the Australian Standard AS 2604–1983 "Sunscreen Products Evaluation and Labelling". A revised standard AS 2604–1986 incorporating additional standards for secondary sunscreen products, broad spectrum protection products and water resistant products was issued on August 6, 1986.

Although cosmetic products may not use SPF values on their labelling and advertising, the health protection branch will permit claims that a cosmetic product will allow "X" times the normal exposure time to the sun, provided no references to sunburn or sunburn protection is made.

While Australian methodology is similar to the FDA proposal, there is no complete agreement with either the FDA standard or the DIN Norm.

The FDA may choose to publish a test methodology as a guideline that companies may use rather than a formal regulation.

The German and Australian standard, although issued by private standard organizations, may be used by the two governments to assess the validity of sunscreen product labelling. In fact, the SAA has developed a certification procedure and mark to identify products tested and labelled in conformance with the SAA standard.

Methodology

In the FDA's proposed monograph, a standard reference formulation containing 8% homomenthyl salicylate with a known SPF value is used to verify the test procedure. However, the use of homomenthyl salicylate has been criticized as a standard because its absorbing capacity is very weak as compared with the more effective and more popular ethylhexyl methoxycinnamate which is used as the standard reference formulation in the DIN Norm.

Ingredients (% w/w)

	FDA	SAA	DIN
Homosalate	8.0	8.0	0.0
Octyl methoxycinnamate	0.0	0.0	2.7
Modified O/W base	92.0	92.0	97.3
	100.0	100.0	100.0
SPF (FDA method)	4.47	4.47	--
Standard deviation	1.14	1.28	--
SPF (DIN method)	--	--	3.7 ± 0.3

More recently, it has been recommended to consider two standard reference formulations, one for SPF values ranging from two to eight and another for higher SPF values.

The test panel is well defined in the FDA and SAA standard, calling for skin phototypes I to III. In the DIN Norm, the subjects are chosen among the skin phototypes found most often in Europe.

The amount of product applied on a test site for 1 cm^2 varies by 33.3% between the DIN Norm and the FDA/SAA standards. The viscosity of the applied product will affect the thickness of the resulting film on the skin surface. For example, a very fluid preparation such as an oil will diffuse outside the test zone, resulting in a thinner film and, therefore, a less active ingredient per unit of surface.

FDA	SAA	DIN
2 mg/cm^2 or 2 ml.cm^2	2 mg/cm^2	$1.5 \pm 10\%$ mg/cm^2

Recent studies on a large scale number of consumers indicate that the average amount of product applied to the skin is 1–1.3 mg/cm^2.

In all three standards, between 15 and 20 minutes must elapse before the subjects are exposed to UV radiation.

The light source is one of the variables in the methods used for the SPF testing.

Sunlight under condition of normal or controlled use would be the obvious light source. However, several recognized factors influence the determination of SPF values under normal conditions including:

- temperature
- humidity
- direct and scattered radiation
- reflective surfaces, snow, sand, etc.
- wind velocity
- sweating

    Controlled tests that use sunlight still present serious difficulties including:

- seasons
- latitude
- altitude
- clouds, smog
- time of day

Because of the aforementioned reasons, *in vivo* human tests more commonly employ an artificial light source or a solar simulator.

Both FDA and the SAA methods recommend the use of a xenon arc solar simulator. The FDA method requires a radiation source with a continuous emission spectrum in the UV-B range of 290–320 nm. The SAA method requires a continuation of the spectrum through the UV-A region, up to 400 nm.

Standards for spectral shaping (FDA/SAA)

The required range of 290–320 nm is achieved by using appropriate filters. Extending the lower limits on this highly erythemogenic region can result in different SPF values.

In an *in vivo* study, a WG 305 filter, 2 mm thick was compared to a WG 320 filter, 1 mm thick using the same xenon arc light source. The WG 305 filter extends the range to 280 nm. The author reported a significantly higher SPF value with the WG 305 filter and PABA 4% (SPF 5.7 vs. 4.6).

However, the results were identical with the standard formulation containing 8% homosalate. This difference may be due to part to differences in the spectral absorption properties of the two sunscreens.

Specifications (FDA/SAA)

1. Less than 1% of total energy to be contributed by non-solar wavelengths (shorter than 290 nm).
2. Not more than 5% of the erythemogenically effective energy to be contributed by non-solar wavelengths.

Filters used (FDA/SAA)

Spectral shaping is achieved using filters:
WG 320/1 mm
WG 305/2 mm
Infrared filter

Specification (DIN)

Osram Ultravitalux lamp (an intermediate pressure filtered mercury vapor lamp) is the preferred light source specified in the DIN method. This source of energy exhibits a discontinuous spectrum with peaks of maximum energy at 297, 303, 313, 330 and 366 nm.

Cole et al., have investigated the performance of sunscreens using both simulators and concluded that products with a maximum absorbance in the 300–320 nm band will have a higher protection factor when tested with a xenon arc lamp. Conversely, products with maximum absorbance between 295 and 305 nm will have a higher protection factor when tested with the mercury lamp. For products exhibiting a relatively flat absorbance spectrum between 290 and 320 nm, the protection factor is approximately the same for the two light sources.

It is also important that the flux of the irradiation source be moderated. Very short irradiation time at high intensity can modify the cutaneous response and can lead to an increase in the SPF value. This can in part explain the differences in SPF values obtained under natural sunlight versus artificial light conditions.

In the FDA and SAA standards, the exposure time progresses geometrically so that the length of time for each succeeding exposure is 25% (1.25 times) longer than the previous one. The results ar read 16–24 hours after irradiation.

In the DIN Norm, the exposure time progresses geometrically so that the length of time for such succeeding exposure is 2(1.41 times), or approximately 40% longer than the previous one. The geometric time sequence is advantageous since it correlates with the approximation of most biological reactions.

However, some U.S. experts feel that the 25% stepwise increases provide for more precision and less variation of the data obtained.

Data Evaluation

In the FDA proposed standard, the SPF is the arithmetic mean of the individual SPF test values. The standard error of the mean (SEM) must not exceed 5% of the mean SPF.

In the SAA standard, the SPF is similarly calculated and the standard error of the mean (SEM) must not exceed 10% of the mean SPF.

In the DIN Norm, the SPF is the geometric mean, where the number of protected erythema-free areas are converted into binary logarithms. The geometric mean Q'(lb) is converted back into the mean protection factor Q'(log) with the aid of a table of power (See DIN NORM 67 501, page 6).

894

This method takes into considerations that the individual protective factors are not distributed in a standardized, but rather in a logarithmic fashion.

The technical advantages of either technique of calculation may be academic. The end results usually fall within the specific limits with very small differences. However, the geometric mean technique minimizes the disparity of results from a wide range of data points.

Determination of water resistance

The FDA monograph draws the line between the claims "Water Resistant" and "Waterproof". The method used determines the SPF before and after repeated periods of activity in fresh water (this condition is more severe than in salt water which has a lower dissolving power due to its salt content).

Water resistant/waterproof (FDA)

Swimming pool, 21–32° C, fresh water, indoors

20 minutes moderate activities
20 minutes rest period
20 minutes moderate activities

Repeated for waterproof claim

Water resistant (SAA)

Swimming pool, 23–28° C, indoors

20 minutes moderate activities
20 minutes rest period
20 minutes moderate activities

Drop in SPF should no exceed 50%
Indoor spa pool can also be used (see standard)

Conclusion

Discussion of appropriate methods to determine sunscreen efficacy continues. To date, testing on humans remains the only accepted method for determining the SPF values of sunscreen products. The Food and Drug Administration has recently indicated its interest to reviewing the methodology proposed in 1978 and has solicited industry comments on current industry practices. No doubt, the views of industry and academia, as well as consumer needs, will be reflected in any final FDA treatment of SPF determination testing. Similarly, discussion in Australia continues on the evaluation of "broad-spectrum" sunscreens designed to protect against UV-A and UV-B radiation.

Determination of the SPF is an evolving technique and recent or continuing studies may help point the way to a better understanding of the meaning and usefulness of SPF values, and the SPF product labelling.

References

Aarber, L. (1964). Regeneration Skin Changes Associated with Excessive UV Exposure. TGA, Proceedings of Scientific Section, No. 41, 26–31.

Cesarini, J.P. (1987). Effects of UV Radiation on the Human Skin with Emphasis on Skin Cancer. Human Exposure to UV Radiation: Risks and Regulation. Elsevier Science Publisher, Biomedical Division.

Cole, C., Davies, R.E., and Forbes, P.D. (1985). Effect of Irradiation Sources on Predicted SPF: Mercury-doped Incandescent Lamp VS Xenon Arc Simulator; Personal Communication.

Dept. of Health, Education and Welfare, U.S.A., FDA: Sunscreen Drug Products and Over-the-Counter Human Drugs. Proposed Safety, Effective and Labelling Conditions. Federal Register 43/166, 38206–38269. 1978.

Deutsches Institut Fuer Normung, Normenausschuss Lichttechnik. Experimentelle Dermatologische Bewertung des Erythemschutzes von externen Sonnenschutzmitteln für die Menschliche Haut. Deutsche Norm 67. 501, March 1984.

Groves, G. et al., (1984). Evaluation of Sunscreens. A Review of Present Methods. The University of Queensland, Australia.

Hoppe, U. et al., (1975). Statistical Evaluation of the Light Protection Factor. Arznei-Forschung (Drug Res.) 25, No. 5, 817–825.

Morton, J., and Murphy, E. (1986). Sun Protection Factors: Regulatory Issues and Methods of Determination Photobiology of the Skin and Eye. M. Dekker, Inc., N.Y.

Pathak, M.A., and Faneswlow, (1983). Photobiology of Melanin Pigmentation; Dose/response of Skin to Sunlight and its Contents. *J. Am. Acad. Dermatol.* 9, 724–733.

Serafino, G., and Frederick, J. (1986). Global Modeling of the UV Solar Flux Incident on the Biosphere; prepared for the US Environmental Protection Agency, Washington, D.C.

Standards Association of Australia: Sunscreen Products, Evaluation and Classification. Australian Standard 2604, 1986.

The Skin Cancer Foundation, Sun and Skin News, Vol. 3, No. 2, 1986.

The Reliability of Sun Protection Factor

E. AZIZI, M. MODAN, A. KUSHELEVSKY
and M. SCHEWACH MILLET

The Department of Dermatology[1] and Clinical Epidemiology[2]
The Chaim Sheba Medical Center and Sackler Faculty of Medicine
Tel Aviv University
The Department of Nuclear Engineering[3]
Ben Gurion University, Beer Sheva, Israel

Introduction

The current test procedure for determination of the sun protection factor (SPF) of sunscreen preparations (Federal Register 1978), which is based on the mean SPF of a test group, leads to an inevitable bias, since the result largely depends on the dose increments and the range of irradiation doses selected for determination of the minimal erythema dose (MED) of the sunscreen-protected skin. The use of relatively large number of high UVR dose challenges will tend to increase the mean, and vice versa. This method also ignores the considerable variability in the degree of protection afforded to different persons by the same sunscreen preparation. Thus, even when a product with a high mean SPF is applied, a considerable proportion of individuals is inadequately protected, a situation which may potentially lead to hazardous consequences. These sources of variations in several components of the test procedure have cast serious doubts on the reliability of the SPF (Farr and Diffey 1985).

Material and Methods

In order to devise a more reliable efficacy index, which can reflect and quantify these sources of variations, we have modified the current 3 days test procedure according to the following principles: on the first day, determination of the MED of the unprotected control skin (MEDc) of each individual is done employing 25% dose increments. On the second day, the MED of the skin protected by 2 ul/cm^2 of the tested sunscreen preparation is determined by exposing a number of subgroups, each to a limited, gradually increasing ranges of UVR doses, with 1 × MEDc increments. The test is terminated when threshold erythematous responses are obtained on the sunscreen-protected skin in more than 90% of the subjects. The study group as a whole is thus exposed to a wide UVR dose range. This method thus enables estimation of the inter-individual

Photobiology, Edited by E. Riklis
Plenum Press, New York, 1991

variability in the degree of protection, without exposing each individual to numerous doses.

Simultaneously, unprotected adjacent skin patches are exposed again to $1 \times$ and $2 \times$ MEDc in order to account for intra-individual variability in MEDc and to reduce observer bias by providing a concurrent reference for determination of the MED of the protected skin (MEDp).

On the third day, MEDc is re-evaluated according to the relative responses at the unprotected 1x and 2x MEDc sites, and the resulting MEDp is recorded. SPF values are calculated for each individual (SPFi) by the MEDp/MEDc ratio. For example, if no response occurs at the unprotected $1 \times$ MEDc site, and the erythema at the $2 \times$ MEDc site is more intense than the criterion for an MED, an intermediate exposure dose of 1.5 $\times$ MEDc is interpolated as the estiamted MEDc. The appropriate multiples of the MEDc administered to the protected skin sites during that test session and the resulting MEDp are recalculated accordingly.

Results and Discussion

The efficacy index of the product is determined by calculating, according to the life table method, the UVR dose, in equivalent multiples of MEDc, under which a certain percentage of the population is safely protected by the sunscreen preparation. Detailed explanation of the method and its computation has been previously described (Azizi et al. 1987). Briefly, the basic principle of this method is that all subjects in the test group are considered protected at dose levels below the MED of their sunscreen-protected skin. An important advantage of the life table method is that it takes into account individuals with MEDp levels higher or lower than their respective maximum or minimum challenge doses. The former, by considering them protected at radiation doses lower than their maximum challenge dose; the latter, by assigning them MEDp values throughout the range below their minimum challenge dose, in a weighted fashion. This achieves two purposes. Firstly, the effective sample size is increased, yielding a more precise estimate of the persons effectively protected by the sunscreen preparation (PEP), due to a smaller standard error at any dose level. Secondly, exclusion of these two categories is likely to bias the calculated PEP towards values lower or higher than the actual values, depending on the relative proportion of each category in the test series. It is notable that by this method no undue weight is given to infrequent extreme values, and unequal numbers of individuals at various doses are taken into account.

Finally, the life table method, being applicable to all sunscreens, enables statistical comparison of different products at a specific UVR dose, or over the entire dose range. It enables an assessment of the precision of the estimate of any PEP, which is inversely related to the inter-individual variability and directly related to the sample size, by calculation of the 95% confidence limits. Furthermore, comparison of the effect of other variables on sunscreen efficacy, for example swimming, sweating, ambient temperature and air humidity, may also be achieved by statistical comparison of the resulting PEP with that obtained under standard test conditions. All calculations can be done on a simple pocket calculator, or with computer programs available in standard statistical packages. Multivariate analysis included in more sophisticated packages allows simultaneous comparison of the effect of a number of factors on sunscreen efficacy.

References

Azizi, E., Modan, M., Kushelevsky, A.P., and Schewach-Millet, M., 1987, A more reliable index of sunscreen protection, based on life table analysis of individual sun protection factors. British Journal of Dermatology, 116:693–702.

Farr, P.M., and Diffey, B.L., 1985, How reliable are sunscreen protection factors? British Journal of Dermatology, 112:113–118.

Federal Register, 1978, Sunsceen drug products for over-the-counter human drugs, proposed safety, effective and labeling conditions. Washington, DC: Department of Health, Education and Welfare, Food and Drug Administration, Vol. 43, no. 166, p. 38206–38269.

Sunscreens - A Photochemical Perspective

F.P. GASPARRO

Yale University
Department of Dermatology
New Haven CT 06510, USA

I would like to thank Dr. Gesarini for asking me to address the issue of sunscreen photoreactivity. By design, sunscreens are strong absorbers of ultraviolet radiation. The absorption of UV leads to the population of excited states. The question of sunscreen safely revolves around how the excited state energy is dissipated. If collisional deactivation occurs, the molecules return to the ground state of releasing energy in the form of heat. Alternatively, emissive processes may occur. Each of these of courses is harmless. However, if a molecule is an excited state were to react with a biological moiety, then biochemical consequences ar possible. In 1985 we showed that PABA, a phototypical sunscreen, underwent a photodimerization after exposure to UVB (Gasparro, 1985). Other workers have shown that photoactivated PABA reacts with proteins (Folwks, unpublished results) and more recently in our laboratory we found that PABA reacted with DNA bases after exposure to UVB radiation (Carney, et al., unpublished results). The only other sunscreen to be studied for photo-mediated effects are cinnamate derivatives (Shimoi, et al., 1985). In this case it was found that the excision of thymine dimers had been inhibited. These studies may represent only the tip of an iceberg.

Twenty-one compounds have been approved by the United States Food and Drug Administration. Photochemical studies with two of them (see above) indicate that photobiological effects do occur. Can it be that these are the only two of the 21 that are reactive?.

Another question about sunscreens involves their *in vivo* fate. Where do sunscreens go?. Are there ways of insuring that sunscreens remain in the uppers layers of the epidermis?.

There is one other unsettling aspect of sunscreens. Although sunscreens do their primary job very well, i.e., they prevent painful sunburns, there are other light-mediated effects they do not prevent. For example, skin exposed to erythemic doses of UVB demonstrated a depression of natural killer cell activity, an important immunosurveillance role in skin. This depression was not prevented by the application of a sunscreen (8% 2-ethyl-hexyl-dimethyl p-aminobenzoate, 2% 2-hydroxy-4-methyoxy-benzophenone, 2% butyl methoxydibenzoyl methane; SPF=15) prior to UVB exposure. Thus, the possible implications for immuno-mediated effects of sunscreen protected skin merit further study.

Photobiology, Edited by E. Riklis
Plenum Press, New York, 1991

The important immune aspects of the skin have been appreciated in the past few years (Edelson, 1985). Whether sunscreens affect these processes is important to know.

In summary, the area of sunscreens photobiology has received little attention. If a photochemist were to die and then end up in a laboratory stocked with sunscreen ingredients and UV sources, I am not sure whether he would know whether he was in heaven or in hell. There would be so many places to start.

References

Carney, K., and Gasparro, F.P., unpublished results.

Edelson, R. (1985). The immunologic function of the skin. Scientific American (June) 46–53.

Folwks, W.J., unpublished results.

Gasparro, F.P. (1985). Uv-induced photoproducts of para-aminobenzoic acid. *Photodermatology*. 2, 151–157.

Shimoi, K. et al., (1985). Methyl cinnamate derivatives enhance UV-induced mutagenesis due to the inhibition of DNA excision repair in Escherichia coli B/r. *Mut. Res.* 146, 15–22.

The Protective Ability of Sunscreens

GORDON A. GROVES

Director, Unique Laboratories Pty. Ltd.
Brisbane, Australia

Since sunscreens first became available to the consuming public there has been a continual improvement, not only in their effectiveness, but also in their cosmetic appearance. No longer is it necessary for the sun-loving individual to anoint themselves with greasy concoctions such as Red Veterinary Petrolatum or products which were primarily a mixture of phenyl salicylate or PABA in soft paraffin. Today's sunscreens are, in the main, cosmetically elegant preparations which do not attract sand and dust and which are not exceptionally uncomfortable to wear even on the hottest days.

Concurrent with the appearance of more attractive vehicles has been the development of more efficient ultraviolet absorbers to be used in these products. Today we have available chemicals which can produce the same level of protection obtained with earlier absorbers such as phenyl salicylate in concentrations of less than fifty percent of what was previously employed. SPF 4 products, for example, can be obtained with a concentration of approximately three percent of either a dimethylaminobenzoate ester or a cinnamate as compared to the eight percent or higher concentration required when either phenyl or homomenthyl salicylate are employed.

A solar protective preparation or sunscreen may be defined as a product intended for application to the skin whose function is solely to reduce the intensity of ultraviolet radiation striking the skin. Recently, however, advertizing claims have appeared attributing other protective characteristics to some of these products. Other products have been made available claiming what might be considered to be exceptionally high levels of protection. In another instance sunscreens have been separated into categories on the basis of their intended use.

In this presentation an attempt will be made to examine these aspects of sunscreens and to possibly present a viewpoint on future developments in this area.

SPF Values

Both the Australian Standard for the Evaluation and Classification of Sunscreens and the FDA proposed protocol have adopted the concept of the Sun Protection Factor or SPF as the unit for depicting the effectiveness of sunscreens. In Australia, a limit is placed on the level of protection which a sunscreen may claim on its label. No sunscreen, no matter how effective it may be, is permitted to claim a protection greater than 15 plus. A product may be at the SPF 30 level of protection but the label must state only that it is SPF 15 plus. In the United States, however, no such limit exists and as a consequence there has recently appeared on the market in that country products claiming to be SPF 40 or higher. The question might be asked whether that level of protection is really necessary or whether the high SPF values are simply a marketing approach which cannot be justified on a scientific basis.

The question might be answered by considering the total amount of ultraviolet radiation to which an individual might be subjected in any period of exposure. In Table 1 are shown the maximum number of sunburn units which have been recorded in the locations indicated. Berger and Urbach (1982) have defined the sunburn unit as being equal to a minimal erythemal dose when untanned human skin is exposed to a vertical tropical sun. Robertson (1972) has stated that for most untanned skins an exposure just above 12 minutes, to a clear sky with the sun within 5° of the vertical will cause threshold erythema. This is in reference to mid-summer exposure conditions in Australia. A dose rate of 5 MED per hour is, therefore, the maximum possible.

In Table 1 it can be seen that in no instance does the maximum daily number of sunburn units exceed 25. Since these recordings were made, presumably, with a meter which operated on a 24-hour basis and which was in a stationary position throughout the recording period, it would appear to be safe to presume that the actual number of sunburn units which any individual would be receiving on any one particular site would be considerably less than the maximum. In Australia during the summer months of November to March the sun rises at approximately 5.30 a.m. and sets at approximately 6.30 p.m. These figures, therefore, for the Australia locations, represent approximately thirteen hours of recorded sunlight. It is highly improbable, therefore, that even an outdoor worker in an area such as Cloncurry would receive the recorded dose of 25 sunburn units in any one working day. Larko and Diffey (1983) have suggested that 30 to 50 percent of erythemogenic ultraviolet radiation is received between 11 a.m. and 1 p.m. Since this is the hottest time of the day it would appear to be reasonable to suggest that at least 1 hour of this period would be spent indoors. Using this criterion the total dose possible is now down to 20 sunburn units.

Table 1. Maximum daily sunburn units in summer months

Locations	1	2	3	4
Brisbane	20.9	21.5	22.7	20.7
Cloncurry	24.0	24.2	24.6	24.2
Melbourne	14.9	16.7	18.4	14.9
Townsville	21.8	24.6	23.0	20.7
El Paso	22.8	20.4	19.1	14.8
Mauna Loa	25.3	24.4	23.9	22.8

1. June or November; 2. July or December; 3. August or January; 4. September or February

The values recorded in Table 1 were obtained using a Robertson-Berger meter in a stationary position. It is highly improbable that any individual, including those addicted to sunbathing, are going to remain in a stationary or prone position for up to 13 hours daily. Thus, the actual amount of ultraviolet radiation received by different individuals will depend to a large extent upon the behaviour of the person facing exposure. Although the Robertson-Berger meter accurately measures the total amount of ambient ultraviolet radiation at any one particular time it is not capable of showing what dose of radiation is being received by individuals under different exposure conditions. Larko and Diffey (1983) have suggested that even an outdoor worker may only receive up to 25 percent of the ambient ultraviolet radiation while going about their normal chores. This would suggest that of the total daily dose of 25 sunburn units per day even those individuals who are outdoors for long periods are only being exposed to between 6 and 7 of these units on any one particular site because of the continual movement of the individual.

Using ultraviolet sensitive films to measure the anatomical distribution of sunlight Urbach (1969) has shown the dose on the horizontal surfaces such as the top of the shoulders is only 80 percent of the dose on the vertex because of shading from the rest of the body. The approximately vertical planes of the body receive about 60 percent. Thus, even a sunbather out in the full sunlight between 10 a.m. and 2 p.m. would only receive approximately 12 to 16 sunburn units at the maximum. Using the assumption of Larko and Diffey (1983) the range could be as low as 4 to 6 on any one particular site.

Considering the above it is difficult to understand why sunscreens of the level of protectiveness of SPF 40 or higher are necessary. The SPF 15 plus products available in Australia which are usually no greater than SPF 20 would appear to be capable of offering adequate protection. Even they might be providing more protection that is required.

Two approaches have been used in the formulation of the high SPF products. These are:

1. High concentrations of absorbers
2. Improved vehicles

Using the maximum concentrations of absorbers permitted by the FDA sunscreens with SPF values of between 25 and 30 are entirely feasible where a satisfactory vehicle is employed. While many absorbers, and in particular the cinnamates, have been shown to be completely innocuous, in the short term the possible long-term effects of continuous usage of products containing high concentrations of these agents must remain a consideration. A more satisfactory approach would appear to be the development of vehicles as carriers for the absorbers which would improve the level of effectiveness. During the course of doing many hundreds of SPF determinations it has been observed that the higher than predicted values obtained are, in most instances, associated with those products which leave the most uniform and compact residual film on the skin. The usual SPF values which have become the norm for the cosmetic and pharmaceutical manufacturers are often the result of too much attention being paid to getting a light cosmetic acceptable vehicle rather than a more effective sunscreen. As a consequence it has become the accepted practice to use high concentrations of ultraviolet absorbers in order to obtain values such as SPF 15. With suitable vehicles it should be possible to use considerable lower concentrations of soluble absorbers in order to obtain the necessary level of protection required.

For those individuals who appear to be hyper-sensitive to ultraviolet radiation personal observation suggests that the answer is not in creating higher and higher SPF values but in formulating products which provide a high level of protection against UV-A radiation in addition to the protection against UV-B.

Primary and Secondary Sunscreens

As a result of our increasing knowledge of the damaging effects on human skin a large number of cosmetic manufacturers are now including an ultraviolet absorber in their moisturizing creams, lipsticks and other products such as liquid makeups intended for daily use, in order to make them photoprotective. These products thus meet the accepted concept of a sunscreen. This definition was given previously.

Cosmetic preparations can be basically be divided into three categories. These are:
1. Protective
2. Decorative
3. Utilitarian

The protective cosmetics are those products designed to protect the skin from the effects of environmental exposure. This protective action is achieved in a number of ways, including the use of oils and waxes to provide a moisturizing action as well as ultraviolet absorbers. They do not serve any other than a protective purpose and, therefore, the term "multi-purpose" which has been used to describe certain absorbers, is, in the author's opinion, incorrect.

In the revised Australian Standard for sunscreen evaluation and classification which appeared in 1986 there was a division of sunscreens into classifications which would appear to be based solely on their use and not on their intended level of protection. To further add to the confusion of the consuming public in respect to sunscreens they are now asked to make a decision between what are described as "Primary" and "Secondary" sunscreens. The "Primary" sunscreen is described as being a product which is represented on the front panel as being primarily to protect the skin from certain harmful effects of the sun's rays. "Secondary" sunscreens are considered to be those which are represented on the label as protecting the skin from certain harmful effects of the sun's rays while fulfilling another primary function. We thus have the situation where the decision must be made between a product described as a sunscreen with added moisturizer and a second preparation labelled as a moisturizer with sunscreen. The differences between the two products are impossible to determine. In many instances, it might not be unfair to suggest that the only difference between these two products is the label. This arbitrary division would appear to be without any scientific foundation. A survey of the market in Australia has shown that differences in the protective ability of these products is not a criterion. Both "Primary" and "Secondary" sunscreens are commercially available which offer protection at all levels from SPF 4 to SPF 15. Both classifications also provide products with varying levels of protection with and without the incorporation of an UV-A absorber.

The classification of sunscreens should be made as simple as possible. The more the division of the products into unnecessary categories the greater the possibility of confusing the consuming public. The greater the confusion the greater the possibility that the public will begin to doubt the claims which are being made. The greater the scepticism the greater the danger that the products will not be used to the extent that is necessary. In Queensland, Australia, where there is a reported 40,000 cases of skin cancer annually the consumer needs to be encouraged to use sunscreens regardless of whether they have a moisturizing action or not.

Triplet Action

The protective action of the ordinary sunscreen has always been considered solely to prevent ultraviolet radiation striking the skin in its full intensity. A recent report by Kligman and Kligman (1983), however, suggests that perhaps attention should also be focused on preventing the effects of both infra-red radiation and visible light on the skin. The implication in this report is that all sunscreens should possess a triple blocking action in order to provide maximum protection.

One of the earliest significant studies on the effect of heat on the erythemal response of skin was carried out by Freeman and Knox (1964). Their studies showed that heat potentiates the damage to the skin caused by exposure to ultraviolet radiation. Using hairless mice these investigators were able to show that infra-red radiation enhanced both acute and chronic ultraviolet injury. Mice with elevated body and skin temperatures at the time of exposure to ultraviolet radiation were found to develop more intense erythema. In contrast, mice with lowered body temperatures during this period showed barely perceptible erythema. Significant changes were also observed in both the rate of onset and the number of tumors developing. A constant high environmental temperature (90°), which is frequently encountered by people living in tropical areas such as the northern regions of Queensland Australia, increased both the number of animals that developed tumors and the rate at which these tumors developed. These observations would appear to suggest that outdoor workers in areas such as this are at greater risk than are those in more temperate regions where, although the intensity of the ultraviolet radiation is high, the temperatures are more moderate for greater periods of the year.

Robertson (1968) has suggested that the higher incidence of skin cancers in tropical regions which is commonly attributed to the presence of much greater amounts of ultraviolet radiation is insufficient to explain the difference in the incidence of skin tumors which is evident in going from the north of Australia to the more southerly regions. This investigator's studies showed that while the total long-term exposure on horizontal surfaces varies by little more than a factor of 2, this variation is insufficient to explain the differences in the appearance of skin tumors. Robertson has observed that while the total environmental exposure to ultraviolet radiation at Cloncurry in the hot, dry, northerly interior is only 1.3 times the average near Brisbane which is located on the coast abut 1700 kilometers south, the incidence of skin tumors is approximately three times greater around Cloncurry. While no definite explanation has been provided for this discrepancy, the differences indicate support for the influence of temperature on tumor occurrence observed by Freeman and Knox (1964).

O'Brien (1978) has suggested that although ultraviolet radiation is often regarded as the sole cause of actinic elastosis (Finlayson and co-workers [1966]), infra-red radiation which is capable of penetrating more deeply into the skin may possibly be an even more dominant factor.

These studies on the possible role of infra-red in various forms of actinic damage has led to the suggestion that sunscreens should be designed to protect against the temperature effects of the sun as well as against the inroads of ultraviolet radiation. Since there are no soluble absorbers yet available capable of absorbing infra-red radiation the product would have to contain an opacifying agent of some nature. Any opacifying agent in the form of insoluble pigments such as zinc oxide or titanium dioxide is going to form a compact film on the skin which will contribute to the heating effect. The possibility exists that not only will these products be unacceptable to the average consumer because of this unpleasant temperature effect, but the heat generated in this fashion could be just as damaging to the skin over a long period as the infra-red radiation coming directly from the sun.

A recent advertisement for a sunscreen has described the product as possessing a triple block action. This product purports to protect the skin from ultraviolet, visible and infra-red radiation. The success of this product remains to be seen. Are we, however, looking at the sunscreen of the future?

References

Berger, Daniel, S., and Urbach, Frederick (1982) A climatology of sunburning ultraviolet radiation, *Photochem. Photobiol.* 35, 187-193.

Finlayson, G.R., Sams, W.M. Jr. and Smith, J.G. (1966) Erythema abigne: a pathologic study, *J. Invest Dermatol.*, 46:104-107.

Freeman, R.G. and Knox, J.M. (1964) Influence of temperature on ultraviolet injury, *Arch. Dermatol.*, 89, 858-864.

Kligman, L.H. and Kligman, A.M. (1984) Reflections on heat. *Br. J. of Dermatol.*, 110, 369-375.

Larko, O. and Diffey, B.L. (1983) Natural UVB radiation received by people with outdoor, indoor, and mixed occupations and UVB treatment of psoriasis, *Clin. Exper. Dermatol.*, 8, 279-285.

O'Brien, John P. (1978) The role of actinic damage to elastin in 'age change' and arteritis of the temporal artery in polymyalgia rheumatica, *Br. J. of Dermatol.*, 98, 1-11.

Robertson, D.F. (1968), Solar ultraviolet radiation in relation to sunburn and skin cancer, *Med. J. of Aust.*, 1, 1123-1132.

Urbach, F. (1969) Geographic pathology of skin cancer. In: The Biologic Effects of Ultraviolet Radiation with Emphasis on the Skin, Urbach, F., ed., Pergamon Press, Oxford, 635-650.

The Broad Spectrum Concept in Sunscreens

GORDON A. GROVES

Director, Unique Laboratories Pty. Ltd.
Brisbane, Australia

Abstract

Sunscreens containing both UV-A and UV-B absorbers have found extensive use in Australia in both those products which are primarily intended to act as a sunscreen and in preparations such as moisturizers, make-up and lipstick cosmetics. With the high level of intensity of both UV-A and UV-B radiation in Australia for most months of the year this action by manufacturers is to be applauded.

Introduction

The use of sunscreens to protect the skin against UV-A radiation was, for many years, only considered to be necessary for those individuals undergoing therapy or otherwise coming into contact with agents known to product phototoxic reactions. In the past fifteen years, however, considerable evidence has been gathered which shows clearly that the damage to skin produced as a result of excessive exposure to sunlight is the consequence of the action of UV-A as well as of UV-B. Radiation above 320 nm had, until these reports appeared, been discounted as a significant factor in the production of skin reactions such as erythema (Hausser, 1928). The reports by Willis et al (1972), Ying and co-workers (1974), and Spiegel et al (1978) have demonstrated quite clearly, that in the case of the erythemal response, the appearance of the erythema is the result of the potentiation of the UV-B action by the presence of UV-A. Whether the combined action is of a synergistic nature or is simply an additive effect would appear to be of academic interest only as far as the final effect on the skin is concerned. The principle significance of this combined activity is the realization that sunscreens must provide protection against UV-A and UV-B. Further support for including this added protection in sunscreens has been provided by the evidence that UV-A also is capable of carcinogenic activity (Staberg, B. et al [1983]) and has a role to play in the accelerated aging of the skin which is characteristic of white skinned individuals who spend long periods in the sun (Kligman, K.H. et al, [1985]).

Photobiology, Edited by E. Riklis
Plenum Press, New York, 1991

Sunscreens in Australia

In 1974 as a result of adverse comments in the press on the effectiveness of sunscreens on the Australian market, the government organized the publication of an annual list intended to enable the consuming public to select the sunscreen which was most suitable for their particular purpose. The initial procedure used to evaluate the sunscreens to be included in this list was the thin-film procedure developed by Groves and Robertson (1972). Subsequently, because of anomalies in this procedure a solution dilution technique was employed. Both of these procedures will be described later.

Many of the sunscreens submitted for inclusion in the annual list claimed to be capable of providing protection against both UV-A and UV-B. In order to establish a standard for products claiming this level of protection it was arbitrarily decided that the only products which would be designated "Broad Spectrum" would be those which showed less than 10 percent transmission at 360 nm and less than 3 percent transmission at 305 nm in terms of an 8 micron film on the skin. There was no scientific rationale for selecting the level of transmission at 360 nm but the Australian authorities are still employing this transmission level to designated "Broad Spectrum sunscreens.

Formulation of Broad Spectrum Sunscreens

Initially, the only "Broad Spectrum" sunscreens on the Australian market were the heavily pigmented products. Subsequently, products based on the use of soluble absorbers have become available. These products, in order to provide protection against UV-A and UV-B radiation must contain a mixture of two or more absorbers since the agents which are capable of extending the absorption into the UV-A region are only weakly absorptive in the erythemal region around 305 nm. The most frequently used UV-A absorbers are the dibenzoylmethane derivatives, Parol 1789 or Eusolex 8020. There is no restriction on the use of these derivatives in Australia and they are extensively employed where the "Broad Spectrum" activity is desired. Oxybenzone which also absorbs in the near UV-A region is not used in products categorized as such as in the commonly employed 3 percent concentration the transmission as an 8 micron layer is far greater than 10 percent at 360 nm.

When the term "Broad Spectrum" was originally introduced in Australia the designation was restricted to those products showing less than 3 percent transmission at 305 nm as stated above. When the spectrophotometric method of evaluation was replaced by the SPF method only those products having an SPF 15 or above level of protection were permitted to include an UV-A absorber in their formulation. In the latest revision of the Australia Standard, however, the use of the term "Broad Spectrum" has been extended so that any sunscreen at any SPF level may be so designated if it meets the necessary criterion.

Evaluation Procedures

With the advent of sunscreens on the commercial market which claim to provide protection against UV-A as well as against UV-B the need has arisen for a reliable procedure which can be used to assess the actual amount of protection which is being provided at wavelengths beyond 320 nm. The means of assessing these so-called "Broad Spectrum" properties had, until recently, not been considered by the regulatory authorities in either the United States or Europe. The two procedures which have been recommended for introduction as official test methods have concerned themselves only with determining the Sun Protection Factor (SPF) of the sunscreen product as a single entity. The proposed evaluation procedures do not differentiate as to whether the SPF values obtained are due to the presence of UV-A or UV-B absorbers or to a combination of the two. In this respect it might be safely concluded that a product with a high SPF as determined by any of the proposed methods is not necessarily going to provide this degree of protection against those skin reactions initiated or intensified by UV-A radiation.

The failure to provide a specific method in these protocols defining the level of activity provided by the UV-A absorber present would not appear to be due to the non-availability of procedures capable of being used for this purpose. Several procedures intended to permit the determination of the ability of a sunscreen to specifically provide protection against UV-A radiation have been described, Kooyers, W.M. (1968), Dahlen et al (1970), Barth, J. (1978), Akin et al (1979), Groves and Forbes (1982). In addition to these methods some manufacturers of UV-A absorbers have described procedures in their brochures which can be used to quantify the UV-A protective action of a sunscreen. Since UV-A does not display any immediate quantifiable effects which can be used as the end-point in determining a UV-A SPF such as the erythema used for determining the protective ability of UV-B absorbers indirect approaches must be used in these evaluations. The methods, therefore, are all based upon increasing the sensitivity of the skin to radiation between 320 nm and 400 nm by the administration of a phototoxic chemical either topically, by injection or orally. The administration of these chemicals followed by exposure to UV-A radiation using normal dosage conditions results in an erythemal response which provides a satisfactory end-point for the evaluation procedure.

Certain of the methods which have been proposed would appear to be unusually cumbersome or potentially dangerous to panel subjects. The use of human subjects where phototoxic chemicals are being employed would appear to possess unsatisfactory implications. Demethylchlortetracycline which was used to sensitize human subjects to UV-A radiation by Dahlen and co-workers (1970), for example, has been known to initiate permanent sensitivity to ultraviolet radiation, Epstein, J.H. (1971). 8-methoxypsoralen which is widely used as a photosensitizer and provides an excellent erythemal response following exposure to UV-A radiation would appear to be hazardous for use in human subjects because of its reported carcinogenic activity, Forbes, P.D. and Davies, R.N. (1981).

The use of animals such as the hairless mouse or the hairless rat to evaluate the effectiveness of sunscreens in terms of the SPF has been reported to give results consistent with those obtained using human subjects, Wolska et al (1974), Gloxhuber, W.H. (1976). The use of animals such as these, therefore, would overcome the objections associated with using potentially dangerous chemicals in evaluations of sunscreens.

Because of the potential carcinogenic activity of 8-methoxypsoralen Groves and Forbes (1982) investigated the use of anthracene as a suitable agent for evaluating the photoprotective efficiency of sunscreens against UV-A radiation. Walter and de Quoy (1978) had proposed the use of anthracene in conjunction with UV-A radiation for the phototherapy of psoriasis. Anthracene, unlike 8-methoxypsoralen, does not appear to enhance photocarcinogenesis. This lack of photocarcinogenic activity made the use of anthracene an excellent prospect for use in a photoprotective evaluation procedure. In addition it had been reported that pigskin treated with a solution of anthracene and exposed to UV-A radiation produced an erythemal response as early as two minutes after the start of irradiation, Argenbright, L.W. et al (1980). In a study by Forbes et al (1976), it had been reported that the inflammatory changes induced by anthracene in the skin of the hairless mouse were visible by 6 hours as contrasted with a time-lapse of 48 to 72 hours before any sign of edema and inflammation appeared in animals which had been treated with 8-methoxypsoralen. The utilization of the observations made in these studies in a method to evaluate UV-A radiation would, therefore, possess a number of advantages over the use of 8-methoxypsoralen.

In their study Groves and Forbes (1982) employed the concept of the SPF to compare the protective ability of sunscreens to prevent the anthracene-UV-A (ANUVA) reaction. The Protective Factor was stated to be the ratio of the time required to produce this response in the control mice to whom no sunscreen had been applied to the time required to produce the response in the sunscreen coated animal. Because of the uniformity of the response five mice were considered to be adequate for each evaluation. The studies showed that the procedure could be used to adequately define the ability of the sunscreen to protect against the specific action of UV-A radiation. Those sunscreens which contained only the UV-B absorber, PABA, did not delay the ANUVA response. In contrast, those sunscreens containing oxybenzone or dioxybenzone which absorb radiation beyond 320 nm did produce a significant delay in the appearance of the ANUVA response. The retardation of the response was also found to be directly related to the concentration of the UV-A absorber present. In contrast to the delayed response which occurs when the psoralen is used as the phototoxic agent the use of anthracene produced a definitive response with an exposure period of as little as 5 minutes.

In an unpublished study, Groves has further investigated the ANUVA response as a method for evaluating the effectiveness of the UV-A absorber portion of "Broad Spectrum" sunscreens. During the initial study which resulted in the development of the ANUVA procedure it had been observed that mice which had been treated with the anthracene solution in methanol became very irritated after a short period of exposure to the UV-A source. As a consequence of the development of this irritant action the mice began to arch their backs and to show other signs of movement. These movements had been one of the more serious disadvantages of the original ANUVA reaction where a set period of exposure was being used. With the considerable movement which occurred as the anthracene exerted its activity it was often difficult to focus the radiation uniformly on the test site. The modification of the ANUVA reaction studied by Groves used the development of this evidence of irritation as the end point of the determination. The animals were anaesthetised prior to the application of the anthracene solution by the intra-peritoneal injection of sodium pentobarbitone. This injection rendered the animal completely comatose but since it was a barbiturate the response to pain was not impaired. Fifteen minutes after the application of the anthracene the mice were exposed to the UV-A source. The time required to initiate definite movement in the animal was recorded. The animal was removed from the UV-A radiation and a layer of sunscreen was applied over the test site. This application was sufficient to form a 20 micron film over the area where the anthracene solution had been applied. After a twenty minute drying period the animal was again exposed to the UV-A source and the time required to produce definite movement was again noted. The ratio of the time required to produce movement with the sunscreen in place to the time required to produce this movement without the sunscreen was considered to be the Protective Factor of the sunscreen with respect to the action of UV-A radiation. As was the case with the ANUVA reaction the delay in the reaction was found to be entirely dependent upon the presence of an UV-A absorber in the sunscreen. This modified version of the ANUVA reaction was found to be more rapid in operation and to have a more clearly indicative end point than the original procedure.

UV-A Evaluation in Australia

In 1983, after several years of deliberation the Standards Association of Australia published an Australian Standard for the evaluation and classification of sunscreen products. This Standard was primarily intended to set forth a procedure which could be used for determining the SPF of a sunscreen product. The original publication did not consider the matter of the "Broad Spectrum" sunscreen. In a revised publication in 1986, however, the Association did include information on these sunscreens in the form of an appendix to the main publication. In the preface to this publication it was stated that because of the difficulties associated with obtaining absolute data concerning the effects of UV-A on the skin, it was decided to provide guidance only on the testing of "Broad Spectrum" products rather than specifying exact requirements for the determination of the UV-A protective activity of a sunscreen. The methods proposed for use are spectrophotometric procedures in which the transmission properties of the sunscreen are determined. The spectrophotometric procedures were selected for use for two reasons. In the first instance, the Association was reluctant to approve any procedure which involved the use of phototoxic chemicals on human subjects because of the inherent risk associated with this approach. The second reason for selecting the non-biological approach was the possible criticism which might be levelled at the Association if they were to sanction any procedure involving the use of animals.

Australian Standard 2604-1986 states that a sunscreen shall be considered as being "Broad Spectrum" if, when tested by the appropriate method, an 8 micron layer of the product does not transmit more than 10 percent of UV radiation at any wavelength between 320 nm and 360 nm . This definition changed the description of what was considered to be a "Broad Spectrum" sunscreen from the original concept in which the transmission at 360 nm only was to be less than 10 percent as an 8 micron layer as described previously. This change was made in compiling the Standard because of the desire to ensure that adequate protection against the lower region of the UV-A spectrum would be provided. Two methods are proposed for determining these transmission levels. The first method is intended to be used with sunscreens that will go completely into solution in the designated solvent. In this procedure the transmission of an appropriately diluted solution of the solution is determined using a 1 cm cell and the necessary calculations are made to show these transmission values in terms of an 8 micron film. For those products containing insoluble materials such as pigments that will not dissolve in the designated solvent the Standard specifies that the transmission of a thin-film of the product contained in an 8 micron quartz cell is to be measured.

The spectrophotometric methods designated in the Standard are comparable to methods which were originally used to evaluate sunscreens in Australia. Prior to the appearance of the first Standard in 1983, a list of all sunscreens available in Australia was published annually as a public service. This list arranged the sunscreens into categories based on the transmission shown by the sunscreen at 305 nm. The method chosen to evaluate the sunscreens was the thin-film technique developed by Groves and Robertson (1972). In this procedure the product was evaluated as a uniform layer with a thickness of 8 microns in the exact form in which it would be applied to the skin. The suggestion was made at this time that because the product was not in solution and was not diluted the transmission spectrum obtained would be more comparable to the spectrum which the preparation would present on the skin than would spectra obtained by other methods of evaluation, such as the solution-dilution procedure. In order to use this evaluation procedure a modified spectrophotometer was used in conjunction with a glass prism with an 8 micron depression precisely ground on its flat surface. The prism was fastened securely to the flat end-face of a photomultiplier tube. For the transmission study the space in the prism was filled with a sample of the sunscreen in such a manner as to avoid the inclusion of air bubbles.

Subsequent studies showed that the results obtained using the thin-film method of evaluation were not truly indicative of the protective power which the sunscreen would have on the skin (Groves et al, 1979). The conclusion was reached that the standard spectrophotometric procedures which had been so widely employed in the evaluation of sunscreens were not suitable for the purpose of predicting their biological efficacy. The authorization of a thin-film procedure by the Standards Association would, therefore, appear to be inappropriate. Not only are the results obtained by this procedure not indicative of the true effectiveness of the sunscreen, but in addition the transmission values obtained are greatly influenced by the particle size of the dispersed phase where the

914

absorber has been incorporated into an emulsion vehicle. This is particularly so where the film of the product is not in close contact with the photomultiplier. No suggestion is made in the Australian Standard that special equipment is necessary and most operators using this procedure simply place their 8 micron cell in the normal cell compartment. The results obtained would appear to be meaningless.

Photo-Protective Effectiveness

At the Ninth International Congress of Photobiology held in Philadelphia in 1984 the suggestion was made that a thorough study should be made into the various factors involved in ascertaining the effective protection of sunscreen preparations against UV-A radiation. Such a study was conducted by Kaidey and Gange (1987). The investigation considered the relative protective ability of three commonly employed UV-A absorbers, Eusolex 8021, Parsol 1789 and Oxybenzone, in the concentrations commonly employed in commercial sunscreen products. These concentrations were 3 percent in the case of Oxybenzone and 2 percent for the other two absorbers. The evaluations were made on both normal and sensitized skin. For situations where sensitized skin might be employed, a comparison was made between the results obtained with anthracene as compared to the values obtained using 8-methoxypsoralen. These two results were in turn compared with the determination on skin which had not been sensitized using either the development of pigmentation or the appearance of erythema as the end point.

The result obtained showed some interesting aspects. In the first instance, it was determined that the protective factors obtained were considerably higher when they were obtained using skin which had been sensitized with either of the two phototoxic agents. Using either the appearance of erythema or the development of pigmentation as the end points in the non-sensitized skin considerably lower protective factors were obtained. The nature of the phototoxic agent employed did not influence the protection factors, The dibenzoylmethane derivatives gave slightly higher values than did oxybenzone. The respective factors were approximately 3.0 for the former and approximately 2.0 for the oxybenzone using sensitized skin and approximately 1.8 and 1.4 respectively, using the appearance of pigmentation as the end point. Other reports have also observed that dibenzoylmethane derivatives are more effective against the skin reactions associated with exposure to UV-A than is oxybenzone. Where the development of UV-A induced erythema was used as the barometer, however, the protective factors were identical with all absorbers.

These results would appear to indicate quite clearly that the use of photosensitizing agents in the evaluation of the effectiveness of UV-A absorbers give unnaturally high results. The correct end point would appear to be the development of pigmentation.

References

Argenbright, L.W., Forbes, P.D. and Stewart, G.J. (1980), Quantitation of phototoxic hyperemia and permeability to protein, *Exper. and Molec. Path.*, 32, 154-161.

Akin, F.J., Rose, A.P. III, Chamness, T.W. and Marlowe, R., (1979), Sunscreen protection against drug-induced phototoxicity in animal models. *Toxicol. and Appl. Pharmacol.*, 49, 219-224.

Barth, J. (1978), Mouse screening test for evaluating protection to longwave ultraviolet radiation. *Br. J. Dermatol.*, 99, 357-360.

Dahlen, R.F., Shapiro, S.I., Barry, C.Z. and Schreiber, M.M. (1970), A method for evaluating sunscreen protection from longwave ultraviolet, *J. Invest. Dermatol.*, 55, 164-169.

Epstein, J.H. (1971), Adverse cutaneous reactions in the sun. In: *Yearbook of Dermatology* (ed. by F.D. Malkinson & R.W. Pearson), pp. 5-43. Yearbook Medical Publ., Chicago.

Forbes, P.D., Davies, R.E. and Urbach, F. (1976), Phototoxicity and photocarcinogenesis: comparative effects of anthracene and 8-methoxypsoralen in the skin of mice. *Fd. Cosmet. Toxicol.*, 14, 303-306.

Gloxhuber, C. (1976), Profun von Sonnenbadepraparaten an haarlosen mausen, *J. Soc. Cosmet. Chem.*, 27,399-409.

Groves, G.A. and Forbes, P.D. (1982), A method for evaluating the photoprotective action of aunscreens against UV-A radiation, *Intl. J. of Cosm. Sci.*, 4, 15-24.

Groves, Gordon A., Agin, P.P. and Sayre, Robert M. (1979), In vitro and in vivo methods to define sunscreen protection, *Austr. J. Dermatol.*, 20, 112-119.

Groves, G.A. and Robertson, D.F. (1972), The selection and use of topical sunscreens, *Med. J. Aust.*, 2, 1445-1449.

Hausser, K.W. (1928), Einfluss der Wellenlange in der Strahlenbiologie, *Strahlentherapie*, 28, 25-30.

Kaidbey, K. and Gange, R.W. (1987), Comparison of methods for assessing photoprotection against ultraviolet A in vivo, *J. Amer. Acad. Dermatol.*, 16, 346-353.

Kooyers, W.M. (1968), Method for testing the efficacy of topical sunscreen preparations, *J. Pharm. Sci.*, 57, 1236-1237.

Spiegel, F., Plewig, G., Hofman, C. and Braun-Falco, O. (1978), Photoaugmentation. *Arch. Dermatol.*, 262, 189-191.

Walter, F. and de Quoy, P.R. (1978), Anthracene with near ultraviolet light inhibiting epidermal proliferation, *Arch. Dermatol.*, 1463-1465.

Willis, I., Kligman, A.M. and Epstein, J.H. (1972), Effects of long ultraviolet rays on human skin : photoprotective or photoaugmentative? *J. Invest. Dermatol.*, 59, 416-420.

Wolska, H., Langer, A. and Marzulli, F.N. (1974), The hairless mouse as an experimental model for evaluatin g the effectiveness of sunscreen preparations, *J. Soc. Cosmet. Chem.*, 25, 639-644.

Ying, C.Y., Parrish, J.A. and Pathak, A.M. (1974), Additive erythemogenic effects of middle (280-320 nm) and Long (320-340 nm) wave ultraviolet light, *J. Invest Dermatol.*, 63, 273-278.

Trends in Sun Protection

R. S. SUMMERS AND BEVERLEY SUMMERS
School of Pharmacy
Medunsa
Republic of South Africa

Introduction

We now know that the damage caused to human skin by the sun is not limited to sunburn as was once thought. The chronic problems which result from accumulative sun exposure range from the less serious premature skin ageing, through solar keratosis and on to squamous and basal cell carcinoma and malignant melanoma (Food and Drug Administration, 1978; MacKie and Aitchison, 1982), which is life-threatening.

One of the ways in which these harmful effects can be reduced and/or prevented is by *regular* use of an effective sunscreen preparation (Pathak, 1982) as much exposure is unintentional and unconscious.

In South Africa a number of preparations of varying efficacy have been available for some years. As product quality and labelling standards for toiletries and cosmetics in South Africa did not (and still do not) require "full disclosure"[*], we reviewed the market in 1984 (Summers and Summers, 1984), in an attempt to obtain an overall picture of the situation. We have tried to document new product introductions and changes to existing products as they have occurred since.

In November 1987, we published a second review (Summers and Summers, 1987). We have observed since then that there are considerable differences between the information in the 2 reviews, separated as they are by only 3 years. Thedifferences seemed to warrant analysis, as they are indicative of trends in the market as a whole, as well as of changes in the use of active ingredients (sunscreen agents) in individual products. This paper contains the results of our analysis, and considers the trends which they indicate.

Method

Product information contained in the tables which accompanied the reviews was enumerated, analysed and compared.

---

[*]Government Notice 323 of 1986 contained a proposed 'list of UV filters which cosmetic products may contain.' These proposals are still under consideration.

Photobiology, Edited by E. Riklis
Plenum Press, New York, 1991

Results

Table 1 presents the number of companies which offered sun protection preparations, Table 2 the total number of preparations and the average per company, and Table 3 the number of preparations for which the actve ingredient(s) and/or their concentrations were **not** provided by the companies on request. The most common sunscreen agents used in the preparations are tabulated in Table 4, in rank order of frequency of use in 1987. This data also appears in Table 5 in which the occasions on which the ingredient is used are calculated as a percentage of the products available.

Table 1. Number of companies offering sunscreen preparations: South Africa: 1984 and 1987

Market Sector	1984	1987
Toiletry companies	12 (1*)	12 (3*)
Cosmetic companies	11 (1*)	14 (4*)
Total	23 (2*)	26 (7*)

Note: *South African companies

Table 2. Number of sunscreen preparations: South Africa: 1984 and 1987

Market Sector	1984			1987		
	Total number of products	Average number per company	range	Total number of products	Average number per company	range
Toiletry companies	48	4.0	1–10	58	4.8	1–19
Cosmetic companies	62	5.6	1–9	73	5.2	2–9
Total	110	4.8		131	5.0	

Table 3. Number of preparations for which 'active' agent particulars were not provided by companies[*1]: South Africa: 1984 and 1987

	1984	1987
Instances where information was **not** provided	91	36[*2]
As a percentage of possible instances (preparations × 2)	41.4%	13.7%

Note: [*1]Names and concentrations of 'active' ingredients were requested
[*2] These 36 instances all orginated from just 3 companies, all French-based.

Table 4. 'Active' ingredients used in sun protection preparations: South Africa: 1984 and 1987

Compound (Brand or approved name)	Wavelength range screened screened (nm)	Instances			
		1984		1987	
		Number	Percentage (n=125) (%)	Number	Percentage (n=221) (%)
Ethylhexyl paramethoxy (Parsol MCX)	280–320 (UV-B)	43	34.4	85	38.5
Butyl methoxy dibenzoylmethane (Parsol 1789)	320–390 (UV-A)	1	0.8	37	16.7
Octyldimethyl para-aminobenzoic acid (Padimate 0)	290–315*	15	12.0	21	9.5
Oxybenzone (Benzophenone 3) (Eusolex 4360)	270–350 (UV-B & part UV-A)	9	7.2	21	9.5
Methyl benzylidene camphor ("camphor derivative") (Eusolex 6300)	290–320 (UV-B)	6	4.8	16	7.2
Pigments	200–800 (UV & visible)	4	3.2	11	5.0
Others (specified)		22	17.6	25	11.3
Unspecified agents		25	20.0	5	2.3
Total		125	100.0	221	100.0

*A recent publication suggests some absorption in the UV-A range (Lowe et al. 1987).

Table 6 contains the number of preparations which demonstrate, in addition to the more usual UVB filtration, some degree of screening against UVA radiation. Table 7 shows the number of preparations which claim water-resistant or waterproof properties (FDA definitions).

Discussion

The main feature of the "companies" data (Table 1) is that the number of South African firms has increased markedly (from 2 in 1984 to 7 in 1987, or by 350%). The withdrawal of a (small) number of American firms is the reason for the low increase in the overall number of companies (13%). The number of preparations rose by 21 (19%) (Table 2). The average number of products per company has not changed markedly during our review period. The highest number of products offered by a single company was 19.

Table 5. Sunscreen preparations and 'active' ingredients: South Africa: 1984 and 1987

Compound (Brand or approved name)	Wavelength range screened screened (nm)	Number of preparations			
		1984		1987	
		Number	Percentage (n = 110) (%)	Number	Percentage (n = 131) (%)
Ethylhexyl paramethoxy cinnamate (Parsol MCX)	280–320 (UV-B)	43	39.1	85	64.9
Butyl methoxy dibenzoylmethane (Parsol 1789)	320–390 (UV-A)	1	0.9	37	28.2
Octyldimethyl para-aminobenzoic acid (Padimate 0)	290–315*	15	13.6	21	16.0
Oxybenzone (Benzophenone 3) (Eusolex 4360)	270–350 (UV-B & part UV-A)	9	8.2	21	16.0
Methyl benzylidene camphor ("camphor derivative") (Eusolex 6300)	290–320 (UV-B)	6	5.5	16	12.2
Pigments	200–800 (UV & visible)	4	3.6	11	8.4
Others (specified)		22	20.0	25	19.1
Unspecified agents		25	22.7	5	3.8
Total		125	113.6	221	168.6

*A recent publication suggests some absorption in the UV-A range (Lowe et al. 1987).

Table 6. Number of sunscreen preparations with screening against UV-A* in addition to UV-B protection: South Africa: 1984 and 1987

Screening Type	1984 (n = 110)	1987 (n = 132)
UV-A filter(s)	10	52
UV-A reflector(s)	4	4
UV-A filter(s) and reflector(s)	0	6
Total	14	62
As percentage of total number of preparations	12.7%	46.2%

*UV-A is particularly implicated in the long-term cumulative effects of sun exposure

Table 7.   Number of preparations with resistance against water: South Africa: 1984 and
1987

Category	1984		1987	
	No.	Percentage of total (n = 110) (%)	No.	Percentage of total (n = 132) (%)
Some degree of water-resistance from product name	11	10.0	11	8.3
Water-resistant[1]	–	–	[2]25	18.9
Water-proof[1]	3	2.7	[3]2	1.5
Total	14	12.7	38	28.8

[1]FDA definitions (Food and Drug Administration, 1978).
[2]22 of these products offer protection against UV-A and UV-B radiation
[3]Both these products offer protectin against UV-A and UV-B radiation

The original unwillingness of some companies to provide information about their products' active ingredients , which was apparent in 1984 (in 41.4% of possible instances information was **not** provided) (Table 3), has changed to the 1987 picture in which only 3 companies refused to release information (involving 13.7% of all possible instances). The 3 companies are all French-based.

The increased willingness of most companies to provide information is reflected in the data shown in Table 4, which includes the reduction of "unspecified agents" from 25 (20% of agents) in 1984 to 5 (2.3% of agents) in 1987. This tendency, combined with effective screening ability, reflects a far more scientific approach now. In 1984 ingredients like "vegetable extracts", "buckthorn" and guanin were still being used. With a few exceptions this is no longer the case.

Another trend shown by the numbers of ingredients for 1984 and 1987 respectively (Table 4) is to more "active" screening compounds per product. In 1984 there were 125 listings for 110 preparations (an average of 1.13 "actives" per product). In 1987 the figures were 221 ingredient listings for 131 products (an average of 1.69 "actives" per product).

The presentation of the active ingredient data in terms of the number of preparations (Table 5) puts their increased use into perspective. For example, although ethylhexylparamethoxy cinnamate (Parsol MCX) (290–320nm) accounted for 38,5% of active ingredient use in 1987 (Table 4), the compound was included in no less than 64.9% of the *products* available in that year. Similarly the respective 1987 figures for butylmethoxy dibenzoyl methane (Parsol 1789) (320–390nm) were 16.7% (of ingredient use) and 28.2% (of preparations). The greatly-increased use of this latter compound, a UVA filter, (one instance in 1984 and 37 in 1987) is noteworthy.

The trend to an increasing use of UVA (combined with UVB) filters shown by

Tables 4 and 5 is highlighted in Table 6 (12.7% of preparations in 1984 and 46.2% in 1987). Preparations which show some degree of water-resistance are also far more common now (28.8% of products) than they were in 1984 (12.7%) (Table 7), which indicates a greater awareness of the demands of the South African "user environment".

In essence our analysis has shown that in just 3 years there has been a trend to "more science' in sunscreen preparations. The trend is characterised by:

1. Greater use of effective UVB filters

2. Much greater use of effective UVA filters

3. More "water-resistant" products

4. Higher disclosure of information

It appears to us that no comparable sector of the market has shown such an improvement. The result will be better short-term protection against the harmful effects of the sun, and, provided that the public is informed about these effects and their long-term implications, an eventual reduction in the incidence of skin cancers.

References

Food and Drug Administration, Department of Health, Education and Welfare, 1978. Sunscreen drug products and over-the-counter human drugs: proposed safety effective and labelling conditions, Federal Register, 43 (166) Pt II: 38206-38269

Lowe NM, Domgool SH, Sefton J, Bourget T and Weingarten D, 1987, Indoor and outdoor efficacy testing of a broad-spectrum sunscreen against ultraviolet A radiation in psoralen-sensitised subjects, *J Am Acad Dermatol*, 17, 224-230

MacKie RM and Aitchison T, 1982, Severe sunburn and subsequent risk of primary cutaneous malignant melanoma in Scotland, *Br J Cancer*, 46, 955-960

Pathak MA, 1982, Sunscreens: Topical and systemic approaches for protection of human skin against harmful effects of solar radiation, *J Am Acad Dermatol*, 7, (3), 285-312

Summers Beverley and Summers RS, 1984, The general practitioner and sunscreen preparations, *SA Family Practice*,5, 344-351

Summers Beverley and Summers RS, 1987, the prescribing of sunscreen preparations : A guide for the general practitioner, *SA Family Practice*, 8, 440-446

Psoralens: Effects and Mechanisms

Photochemoprotection: Protective Effects of Psoralen-Induced Tan

P. FORLOT

Laboratoires Pharmaceutiques Bergaderm
Rungis (France)

Introduction

UVB screening chemicals incorporated in sunscreen products are considered able to provide adequate protection against acute (erythema, sunburn cell production) and probably long-term (dermatoheliosis, skin cancer) effects of solar ultraviolet radiations.

Recent evidence, however, seems to indicate that the use of such sunscreens does not eliminate completely the risk of chronic skin damage from intermittent long-term exposure to solar UV especially for people with sensitive skin unable to tan easily. In addition, sunscreen products are used during periods of heavy intentional exposure (holiday) when the remainder of the year certain parts of the body (head, neck, hands) are regularly exposed unintentionally without sunscreen protection.

Fitzpatrick (1988) was the first to describe the concept of photochemoprotection as the :"development of the UV-protection mechanism in the epidermis by the action of oral PUVA photochemotherapy". More recently, the availability of pigmentogenic derivatives with less or non-photosensitizing properties like 5-methoxypsoralen, methylangelicins or pyridopsoralens provides interesting perspective for the prophylaxis of chronic sun-induced cutaneous changes (dermatoheliosis) and skin cancer by eliciting long-lasting oral or tropical photochemoprotection.

For more than 15 years, natural citrus essential oils containing bergapten (5-methoxypsoralen) have been incorporated in sunscreen preparations to promote tanning.

Recent studies have shown that the photomutagenic/photocarcinogenic potential of 5-MOP is remarkedly suppressed by the presence of UVB filters in the citrus oils containing products. Moreover, evidence has been obtained in animals and humans that a bergapted-induced tan is far more protective than a natural tan against erythema production, sunburn cell formation and DNA damage elicited by UV radiation (Solar Simulated Radiation).

This paper will review the different aspect of photochemoprotection obtained by using a bergapten contain suntan product + filters.

Photobiology, Edited by E. Riklis
Plenum Press, New York, 1991

Photoprotection in Animals and Humans

According to Pathak (1988), human skin reacts to UV injuries by developing six defensive mechanisms:

- *Stratum corneum* thickening (keratinization process)
- Eumelanin pigmentation (absorbing and scattering of UVA)
- Accumulation of carotenoid pigments (singlet oxygen quenching)
- Urocanic acid production (absorption of UVR)
- Super Oxide Dismutase and peroxidase-reductase activation (scavenging of harmful reactive oxygen species)
- Error free DNA repair of UVA damaged DNA.

It seems evident that pigmentation is more important that *Stratum corneum* thickening in providing photoprotection (Fitzpatrick, 1986) and melanin is probably one of the most photoprotective agent in human skin by absorbing and scattering light, dissipating the absorbed energy as heat and also scavenging free radicals.

Effectively, it is well demonstrated that tanning induced by UV radiation can afford protection in animals or man against erythema protection and DNA damage (Table 1), Using endonuclease as a marker Gange et al., (1985) have shown that the UV induced tan is protective against DNA damage from a challenge dose of UVR in human volunteers capable of tanning. The same level of protection may be less likely in fair skin individuals who have difficulty in tanning (skin type I or II).

Table 1. Photoprotection studies with UVR

Type of tan	Species	Parameter	Protection	Ref.
UVA-B	Guinea pig	DNA damage (UDS)	-	(3)
UVB	Humans	DNA damage (endonuclease)	+	(10)
UVB	Humans	Erythema	+	(10)
UVA	Humans	DNA damage (endonuclease)	+	(10)
UVA	Humans	Erythema	-	(10)
SSR	Humans	DNA damage	-	(23)

− = None/minimal; + = significant

Pigmentogenic Effects of Psoralens

Mechanisms

The ability of psoralens to induce tanning in human skin has been known since 1500 B.C. and extremely well documented (Rodighiero, 1985). Psoralens, in the presence of UVA, are able to stimulate the normal pigmentogenic process in human skin by increasing the number of functional melanocytes, their dendritic arborisation, the

926

development and the melanization of the melanosomes within the melanocytes and the associated keratinocytes, enhancing tyrosinase activity as well as the migration of activated melanocytes from skin appendages (Forlot, 1988).

Despite the general agreement on the prevalent role of photosensitization in the melanogenic response of human skin to UVA radiation and psoralens, recent studies seem to indicate that induction of erythema and the subsequent enhanced pigmentation are not closely connected. In fact, some methylangelicins are able to induce dark pigmentation in human skin without causing preliminary erythema (Dall'acqua, 1988).

It appears that the mechanisms by which psoralens stimulate pigmentation may include:

a. activation of tyrosinase by MSH (Melanin Stimulating Hormone) indirectly due to DNA adduct formation by psoralens and successive repair.

b. suppression of inhibitors of tyrosinase activity by oxidation of SH groups (glutathion and cystine) (Prota, 1988).

c. photopolymerisation of melanogenic precursors due to the production of oxygen reactive species by psoralens in the presence of UV light (Joshi et al., 1987).

Clinical evidence

The numerous experiments, conducted both in animal species and in human volunteers on the capacity of various sources of psoralens to induce hyperpigmentation, have been summarized in Table 2.

Table 2. Evidence of pigmentogenic effect of psoralens

Psoralen source	Mode of administr.	Species	Source of irradiation	E	HP	Ref.
8-MOP	Oral	Humans (ST III)	Sunlight	+	+	(7)
5-MOP	Oral	Humans	UVA	-	+	(11)
Natural citrus oil + sunscreens	Topical	Humans (ST II)	Sunlight	-	+	(23)
Natural citrus oil + sunscreens	Topical	Humans (ST I,II,III)	Sunlight	-	+	(9)
Natural citrus oil + sunscreens	Topical	Humans (ST I,II,III)	Sunlight	-	+	(17)
Natural citrus oil + sunscreens	Topical	Humans (ST I,II,III)	SSR	-	+	(24) (18)
Natural citrus oil + sunscreens	Topical	Guinea pigs	SSR	-	+	(2)
Natural citrus oil + sunscreens	Topical	Mini pigs	SSR	-	+	(22)

ST: skin type; SSR: Solar simulating radiation

927

Oral and tropical psoralens are highly pigmentogenic with various sources of irradiations.

The association of natural citrus oils containing small amounts of bergapten with UVB sunscreens (ethyl-hexyl) paramethoxycinnamate and benzilidene heptanone) enhances significatively human and animal skin response to various UV sources of radiation without significant erythema protection (Young et al., 1988; Tronnier et al., 1981; Freeman, 1988; Levine et al., 1988; Kligman.et al., 1988; Cahn, 1981; Sambuco et. al., 1987). The pigmentogenic effect of psoralen is dose-dependent (Young et al., 1988; Freeman, 1988).

The available data show that, in skin type I and II volunteers, 90% of them (42 out of 47) developed tans under the influence of bergapten and sunlight, which they cannot do when they use a sunscreen or are exposed to the sun without bergapten present (9 out of 47) (Freeman, 1988; Levine et al., 1988).

Photochemoprotection of Psoralen-Induced Tan

Fitzpatrick et al., (1955) and Imbrie et. al. (1959) are the first to have shown that one single oral administration of 8-methoxypsoralen followed by sunlight exposure is able to enhance human skin pigmentation and to increase the tolerance of skin to subsequent UV exposure, doubling the dose of UVB required to produce erythema. The

Table 3. Photochemoprotection of psoralen-induced tan

Type of tan	Species	UV source	End-point of protection	Prot.	Ref.
8-MOP (oral)	Humans	Sunlight	Erythema	+	(5,12)
8-MOP (oral)	Humans	UVA (PUVA)	Erythema	+	(11)
Natural citrus oil + sunscreens	Mini pigs	SSR	Erythema	+	(22)
Natural citrus oil + sunscreens	Mini pigs	SSR	Sunburn cell damage	+	(16)
Natural citrus oil + sunscreens	Humans	SSR	Erythema	+	(16)
Natural citrus oil + sunscreens	Humans	SSR	Chemical aggression	+	(13)
5-MOP (oral)	Humans	UVA (PUVA)	Erythema	+	(13)
5-MOP (oral)	Humans	UVA (PUVA)	Sunburn cell damage	+	(13)
5-MOP (oral)	Humans	UVA (PUVA)	DNA damage (UDS)	+	(13)
Natural citrus oil + sunscreens	Humans	SSR	DNA damage (UDS)	+	(24)

Abbreviations: - none/minimal; + significant; 5-MOP 5-methoxypsoralen; 8-MOP 8-methoxypsoralen; E erythema; HP hyperpigmentation; SSR solar simulated radiation; ST skin type; UDS unscheduled DNA synthesis photoprotective effect of 5-methoxypsoralen-UVA induced tan after oral administration was later confirmed by Gschnait et al., 1978).

The different aspects of photochemoprotection induced by psoralens in various experimental conditions in human and animals are summarized in Table 3.

Tans induced by 5-methoxypsoralen (bergapten) in various forms (oral drug or topical sunscreen containing natural citrus oil) and various UV sources were able to offer remarkable protection against erythema and sunburn cell damage produced by irradiation with challenge dose of UVB. Only partial protection was observed with UV induced tan (without bergapten) alone (Imbrie et al., 1959; Gschnait et al., 1978; Kligman et al., 1988; Sambuco et al., 1987; Cripps, 1981).

In one experiment, we found that the SPF (Sun Protection Factor) of bergapten-tanned skin was in excess of 10 MED's compared to about 2 MED's of untanned caucasoïd skin and to about 4 MED's of SSR alone-tanned areas (Kligman et al., 1988). In the same experiment, the bergapten-tanned skin offered almost complete protection against application of various aggressive chemicals as measured by DMSO whealing kerosene blistering, croton oil inflammation and sodium lauryl sulfate dermatitis (Kligman et al., 1988)

In addition, using DNA synthesis as a market of DNA damage (unscheduled DNA synthesis). Young et al., (1983), have recently found that a bergapten-induced tan.offers considerably more protection against DNA damaging challenge dose of SRR (Solar Simulated Radiation) than a natural tan (induced by SSR alone) in human volunteers of skin type II (Young, 1988). The same volunteers failed to develop a tan with SSR in presence of UVB sunscreens (without bergapten) and remained unprotected. It must be underlined that, in the last described experiment, the end-point was the assessment of UV-induced and not psoralen-induced DNA damage because no psoralens were present at the stage of challenge testing.

Further aspects of this photochemoprotection are currently being investigated as onset and duration of DNA photoprotection and the effect of skin type.

Discussion

The different experiments summarized here clearly demonstrate the pigmentogenic activity of psoralens used by oral intake or topical application followed by UV exposure. The most significant recent findings are that:

1. Individuals with skin type I and skin type II who choose to expose themselves to the sun with a reasonable protective tan as an end-point have a realistic method of getting a tan with a minimum of risk of erythema and acute skin damage.
2. Substantial photochemoprotection can be obtained against challenge doses of UVR capable of producing erythema sunburn cell damage, and DNA damage in unprotected skin. This photoprotection was only achieved with a tan induced by psoralens and more specifically suntan preparations containing sunscreens and natural bergapten

(citrus oils). No photoprotection was obtained on skin areas pretreated with UVR only or with the sunscreen preparation lacking bergapten and UVR.
3. This photochemoprotection was also observed against various chemical aggressions in human volunteers.

These results indicate that melanogenesis and to a less extent *Stratum corneum* thickening are important for photoprotection and that the nature of bergapten-induced pigmentation differs from that of UVB.

As far as the potential risk of bergapten containing sunscreens is concerned as based upon the photomutagenic (Ashwood-Smith, 1981) and mouse skin photocarcinogenic (Zadjela et al., 1981; Young et al., 1983) properties of 5-MOP, recent evidence has shown that both photomutagenic (Marzin et al., 1988) and photocarcinogenic (Young et al., 1987) activity of bergapten is inhibited if UVB filters are included in topical preparations. These recent findings are important for the risk-benefit evaluation of suntan products containing bergapten.

Nowadays, despite the warnings of dermatologists, a tan is fashionable and avidly desired. Fair skinned individuals (skin type I and II) who want to achieve a tan face two alternatives which, in most cases, are unacceptable:
-   use nothing when exposed to sunlight with the inherent risk of sunburn and short/long term skin damages.
-   use conventional sun-protective products in which case they will develop no tan unless they proceed to longer sun exposures according to the SPF (Sun Protection Factor) of the product. In this case, the same UVB tanning dose will be achieved over a longer period of time of sun exposure. SPF evaluation has erythema production in exposed and unexposed skin as an end-point but does not take in account possible non visible harmful effects of sunlight with a possibly more sensible threshold than 24 h erythema (sunburn cell and DNA damage).

The judicious use of bergapten-containing sunscreens can allow fair skinned individuals to develop a more intensive tan more rapidly which provides them greater protection against erythema and DNA damage from further exposure to sunlight and hence provides the potential protection against the associated carcinogenic risk.

References

Ashwood-Smith, M.J. (1982). Comparative photobiology of psoralens. J.N.C.I., 69, (1), 189–197.
Cahn, J. (1981). Possible mechanism of psoralen induced melanogenis. *Proc. Int.* Psoralens. SIR, Pergamon Press, 31–50.
Cripps, D.S. (1981). Natural and artificial photoprotection. *J. Invest. Derm.* 76, 154–157.
Dall'acqua, F. (1988). New psoralens and analogs. In Psoralens: past, present and future of photochemoprotection. T.B. Fritzpatrick, P. Forlot, M.A. Pathak and F. Urbach edit. John Libbey Eurotext, (Paris-London).
Fitzpatrick, T.B. (1986). Ultraviolet-induced pigmentary changes: Benefits and hazards. *Curr. Probl. Derm.* 15, 25–38.

Fitzpatrick, T.B. (1988). The psoralen story: Photochemotherapy and photochemoprotection. In Psoralens: past, present and future of photochemoprotection. T.B. Fritzpatrick, P. Forlot, M.A. Pathak and F. Urbach edit. John Libbey Eurotext, (Paris-London).

Fitzpatrick, T.B., Hopkins, C.E., Blickenstaff, D.D. (1955). Augmented pigmentation and other responses of normal human skin to solar radiation following oral administration of 8-methoxypsoralen. *J. Invest. Dermatol*, 25, 187–190.

Forlot, P. (1988). Psoralen induced pigmentation in human skin: overview of clinical evidence, mechanisms of induction and future research. In Psoralens: past, present and future of photochemoprotection. T.B. Fritzpatrick, P. Forlot, M.A. Pathak and F. Urbach edit. John Libbey Eurotext, (Paris-London).

Freeman, L. (1988). A dose photo response of a sunscreen product with bergapten. In Psoralens: past, present and future of photochemoprotection. T.B. Fritzpatrick, P. Forlot, M.A. Pathak and F. Urbach edit. John Libbey Eurotext, (Paris-London).

Gange, R.W., Blackett, A.D., Matezinger, E.A., Sutherland, B.M., and Kochevar, I.E. (1985). Comparative protection efficiency of UVA and UVB tans against erythema and formation of endonuclease-sensitive sites in DNA by UVB in human skin. *J. Invest. Derm.* 85, 352–364.

Gschnait, F., Brenner, W., and Wolff, K. (1978). Photoprotective effect of a psoralen UVA induced tan. *Arch. Dermatol. Res.* 263, 181–188.

Hönigsman, H., Jaenicke, K.F., Brenner, W., Rauschmeier, W. and Parrish, A. (1981). Unsheduled DNA synthesis in normal human skin after single and combined doses of UVA, UVB and UVA with Methoxalen (PUVA). *Br. J. Derm.* 105, 491–501.

Imbrie, J.D., Daniels, F. Jr., Bergeron, L., Hopkins, C.E., and Fitzpatrick, T.B. (1959). Increased erythema threshold six weeks after a single exposure to sunlight plus oral methoxalen. *J. Invest. Dermatol.* 32, (2), 331–337.

Ishikawa, T., Kodama, K., Matsuwoto, J., and Takayama, S. (1984). Photoprotective role of epidermal melanine granules against ultraviolet damage and DNA repair in guinea pig skin. *Cancer Res.* 44, 5195–5199.

Joshi, P.C., Carraro, C., and Pathak, M.A. (1987). Involvement of reactive oxygen species in the oxidation of tyrosine and dopa to melanin and skin tanning. *Biochem. Biophys. Res. Comm.* 142, (1), 265–274.

Kligman, A.M., and Forlot, P. (1988. Comparative photoprotection in humans by tans induced either by solar simulating radiation or after a psoralen-containing sunscreen. In Psoralens: past, present and future of photochemoprotection. T.B. Fritzpatrick, P. Forlot, M.A. Pathak and F. Urbach edit. John Libbey Eurotext, (Paris-London).

Levine, N., Owens, C., Kligman, A.M., and Forlot, P. (1988). Effectiveness and safety of bergapten in sunlight induced pigmentation. In Psoralens: past, present and future of photochemoprotection. T.B. Fritzpatrick, P. Forlot, M.A. Pathak and F. Urbach edit. John Libbey Eurotext, (Paris-London).

Marzin, D., and Olivier, P. (1988). Study of the protective activity against photomutagenicity of 5-MOP on *Salmonella typhimurium* TA 102 strain by UV filters in a suntan preparation. In Psoralens: past, present and future of photochemoprotection. T.B. Fritzpatrick, P. Forlot, M.A. Pathak and F. Urbach edit. John Libbey Eurotext, (Paris-London).

Pathak, M.A. (1988). Photoprotective role of melanin (eumelanin) in human skin. In Psoralens: past, present and future of photochemoprotection. T.B. Fritzpatrick, P. Forlot, M.A. Pathak and F. Urbach edit. John Libbey Eurotext, (Paris-London).

Prota, G. (1988). Mechanisms in melanin pigmentation: an up-to-date view. In Psoralens: past, present and future of photochemoprotection. T.B. Fritzpatrick, P. Forlot, M.A. Pathak and F. Urbach edit. John Libbey Eurotext, (Paris-London).

Rodighiero, G. (1985). Hyperpigmentation induced by furocoumarins. II Farmaco. 46, (6), 173–186.

Sambuco, C.P., Forbes, P.D., Davies, R.E., and Urbach, F. (1987). Protective value of skin tanning induced by ultraviolet radiation plus a sunscreen containing bergamot oil. *J. Soc. Cosmet. Chem.* 38, 11–19.

Tronnier, H., and Agache, P. (1981). Field trial on suntan products containing bergamot oil in Tunisia. *Proc. Int.* Psoralens, SIR, Pergamon Press, 415–426.

Young, A.R., Potten, C.S., Chadwick, C.A., Murphy, G.K., and Cohen, A.J. (1988). Inhibition of UV radiation-induced DNA damage by a 5-methoxypsoralen tan in human skin. *Pigment Cell Res.* 1, 350–354.

Young, A.R., Gibbs, N.K,. and Magnus, I.A. (1987). Modification of 5-methoxypsoralen phototumerogenesis by UVB-sunscreens: A statistical and histological study in the hairless albino mouse. *J. Invest. Derm.* 89, 611–617.

Young, A.R., Magnus, I.A., Davies, A.C., and Smith, N.P. (1983). A comparison of the phototumerogenic potential of 8-MOP and 5-MOP in hairless albino mice exposed to solar simulated radiated. *Br. J. Derm.* 108, 507–518.

Zadjela, F., and Bisagni, E. (1981). 5-methoxypsoralen, the melanogenic additive in suntan preparations, is tumerogenic in mice exposed to 365 nm UV radiation carcinogenesis, 2, 121–127.

Mutagenic Effects of Psoralen-Induced Photoadducts and their Repair in Eukaryotic Cells

D. AVERBECK, M. DARDALHON AND N. MAGAÑA-SCHWENCKE

Institut Curie – Biologie, CNRS UA 1292
26 rue d'Ulm
75231 Paris Cedex 05, France

Introduction

Work on the mutagenesis of psoralens (furocoumarins) on eukaryotic cells can be taken as an example for the intimate connection between fundamental and applied aspects in psoralen research. In fact, due to the use of photoreactive psoralens in the photochemotherapy (PUVA therapy) of psoriasis and certain other skin disorders (Parrish et al., 1982), as well as in cosmetology (Suzuki et al., 1979), there has been increasing interest in evaluating the mutagenic potential of psoralens (Scott et al., 1976; IARC, 1986) with respect to long term side effects arising in humans (Stern et al., 1988) and, at the same time, to get an understanding of the underlying mechanisms. In recent years, mutagenicity studies in pro- and eukaryotic cells have not only been extremely helpful to define the genotoxicity of bifunctional psoralens in actual use, such as 8-methoxypsoralen (8-MOP), 5-methoxypsoralen (5-MOP) and 4,5',8-trimethylpsoralen (TMP) (Averbeck et al., 1981; IARC, 1986), but have also directed research work towards the development of more photoreactive and, in many cases, less genotoxic monofunctional furocoumarins for photochemotherapeutic use (Rodighiero et al., 1988).

Furthermore, the development of extremely photoreactive psoralens (Hearst et al., 1984) and progress in elucidating psoralen photoreactions with biologically important macromolecules and, especially, with DNA (Musajo and Rodighiero, 1972; Dall'Acqua, 1977; Cimino et al., 1985) have opened the possibility to use psoralens as tools in nucleic acid research (Cimino et al., 1985), as well as for asking questions on the structure (Kanne et al., 1982; Tessman et al., 1985; Cadet et al., 1986), repairability (Zolan et al., 1984; Dardalhon et al., 1988) and genotoxic consequences (Averbeck, 1985; Papadopoulo and Averbeck, 1985) of chemically well defined lesions in DNA. In this connection, it is important to note that, although psoralens may photoreact with cell constituents via photodynamic oxygen-mediated reactions (Pathak and Joshi, 1984; De Mol et al., 1986; Blan and Grossweiner, 1987; Dall'Acqua, 1988), the best known reaction is the photoaddition to DNA.

After complexation in the dark to DNA, psoralens photoreact in the presence of near-ultraviolet light (UVA) (315–400 nm), rather specifically with pyrimidine bases in DNA forming C_4-cyclobutane addition products, i.e. 3,4 (pyrone side) monoadducts (MA_p), 4',5' (furan side) monoadducts (MA_f), as well as DNA interstrand cross-links (CL) (Dall'Acqua, 1977; Ben-Hur and Song, 1984; Hearst et al., 1984; Cimino et al., 1985). If left unrepaired, psoralen induced DNA photoadducts may inhibit DNA replication and thus lead to antiproliferative effects which appear to be relevant for the photochemotherapeutic treatment of psoriasis (Anderson and Voorhees, 1980). Normal DNA synthesis can be restored by enzymatic repair or translesional synthesis (bypass) (Smith, 1988).

In repair competent cells, the repair of psoralen induced damage involves the action of gene products which are also operating on the repair of UV-induced pyrimidine dimers and other bulky lesions (Smith, 1988). The repair of DNA cross-links in eukaryotic cells necessitates a sequential action of several repair pathways (Moustacchi, 1987; Friedberg, 1988; Smith, 1988) as in prokaryotic cells (Cole et al., 1976; Van Houten et al., 1986).

However, in eukaryotes, some additional gene products are needed, for example, those involved in double strand break repair (Moustacchi, 1987; Friedberg, 1988). The processing of lesions may be, to different extents, error-free and error-prone resulting in mutagenicity.

In all pro- and eukaryotic cell systems, psoralens have been found photomutagenic (IARC, 1986, 1987). Information gained from studies in eukaryotes are important because of the chromosomal structure and involvement of specific repair systems.

The present overview attempts to highlight recent developments in the research on the photomutagenicity of psoralens and the repair of psoralen photoadducts in eukaryotic cells. New information has been provided by 1) analysing the relationship between photolesions induced in DNA and mutagenic effects by mono- and bifunctional furocoumarins; 2) the use of different activating wavelengths for the induction of either mixtures of MA and CL (at 365 nm) or monoadducts alone (at 405 nm) and reirradiation protocols (365 nm + 365 nm or 405 nm + 365 nm) to change the ratio of mono- over diadducts; 3) the use of low and high dose (fluence) rates of UVA; 3) the use of psoralen damaged (exogenous) plasmid DNA.

Materials and Methods

We used chromatographycally pure 5-methoxypsoralen and 8-methoxypsoralen as described before (Averbeck, 1985). As typical eukaryotic cells, we used the diploid strain D7 of the yeast *Saccharomyces cerevisiae* (Zimmermann et al., 1975) and cultured V-79 Chinese hamster cells (Dardalhon and Averbeck, 1988).

Treatment and culturing conditions, the determination of DNA damages and lethal and mutagenic effects induced by psoralens in yeast were the same as previously described (Averbeck, 1985; Averbeck et al., 1987).

Photofixation and the removal of monoadducts and cross-links were determined by extracting DNA containing radioactively labelled psoralen photoadducts (Averbeck et al.,

1987) and by alkaline step elution analysis (Cundari and Averbeck, 1988), respectively.

The transformation competent wild type (RAD) strain of *S. cerevisiae* was a kind gift by Dr. Giora Simchen, Jerusalem. The transformation competent strains rad1, rad6, rad18 and rad1 rad52 were constructed in the laboratory of Dr. Francis Fabre, Orsay, France. Transformation of yeast was performed according to Hinnen at al., (1978).

Treatment and culturing conditions of V-79 Chinese hamster cells were those of Dardalhon et al., (1988).

The repair of psoralen induced DNA photoadducts was followed either by alkaline denaturation and hydroxylapatite chromatography (detection of CL) (Dardalhon and Averbeck, 1988) or by extraction of DNA containing radioactively labelled psoralen adducts (^{3}H-8-MOP and ^{14}C-5-MOP) (Dardalhon et al., 1988). Irradiations were performed using either a HPW 125 Philips lamp emitting mainly at 365 nm (Averbeck, 1985) or a 2.5 kW Xenon lamp in a Schoeffel housing with a Kratos 252 high intensity grating monochromator (Kratos Analytical Instruments, Ramsay, NY 07446, USA) for monochromatic radiation at the wavelengths 365 nm and 405 nm (Averbeck et al., 1987). UVA radiation was used at high (HDR=46.4 kJm^{-2}h^{-1}) and low (LDR = 0.51 kJm^{-2}h^{-1}) dose rate (Averbeck, 1988).

Dosimetry measurements were performed as previously described (Averbeck, 1985; Averbeck et al., 1987).

Results

1. Relationship between psoralen photoadducts and mutagenic effects induced

In the past, the genotoxic (mutagenic) effects of psoralens have been determined at equal psoralen concentration and UVA doses and/or at equal survival levels. This was useful for the assessment of photobiological activities and genotoxic effects with regard to the photochemotherapeutic use of psoralens, For example, the relatively high photomutagenic activity of bifunctional psoralens such as 8-MOP was in this way recognized in several prokaryotic (Mathews, 1963; Igali et al., 1970) and eukaryotic cell systems (Alderson and Scott, 1970; Arlett, 1973; Averbeck et al., 1975; Schimmer and Hauber, 1977; Burger and Simons, 1979a, b; Schenley and Hsie, 1981). Furthermore, in comparison to 8-MOP, a relatively low mutagenic activity of the monofunctional psoralen 3-carbethoxypsoralen (3-CPs) could be shown in yeast (Averbeck and Moustacchi, 1980; Averbeck et al., (1981), in Chinese hamster V-79 (Papadopoulo et al., 1983) and in CHO cells (Loveday and Donahue, 1984). However, since psoralens differ in their dark complexing, light absorption characteristics and photoreactivity with DNA comparisons at equal numbers of photoadducts are more suitable for mechanistic studies.

In similar assays, the newly developed monofunctional pyridopsoralens, pyrido (3,4-c)psoralen (PP) and 7-methyl pyrido(3,4-c)psoralen (MPP) were found to be less photomutagenic than 8-MOP in yeast (Averbeck, 1985) but equally effective as 8-MOP in Chinese hamster V-79 cells (Papadopoulo et al., 1986) at equitoxic doses. Also 4,5'-

dimethylangelicin (4,5'-DMA) and 5-methylangelicin (5-MA), were more photomutagenic than 8-MOP in Chinese hamster V-79 cells (Pani et al., 1981; Swart et al., 1983; Loveday and Donahue, 1984). In contrast to the pyridopsoralens, 4,5'-DMA and 5-MA were more photocarcinogenic than 8-MOP when applied topically on hairless mice (Mullen et al., 1984). The reduced carcinogenicity of the two pyridopsoralens MPP and PP, in comparison to 8-MOP, and the at least equal efficiency in the topical treatment of psoriatic lesions, make them very interesting compounds for photochemotherapeutic use (Dubertret et al., 1985).

When the mutagenicity of the monofunctional compounds, 3-CPs and MPP, was studied as a function of equal number of photoadducts induced, MPP adducts were apparently more mutagenic than 3-CPs adducts (Averbeck, 1985). Since in both cases 4',5'(furan side) monoadducts are induced in DNA, the result could be explained by assuming differences in isomeric structures and/or sequential distribution of the DNA adducts. Indeed, differences in sequential binding sites have been observed (Boyer et al., 1988). However, knowing that the two compounds differ also in the capacity to undergo side reactions involving molecular oxygen, for example, 3-CPs produces singlet oxygen quite efficiently whereas MPP is less able to do so (Ronfard-Haret et al., 1987), the different mutagenic activity may also be due to differences in energy transfer reactions. This may apply also for the differences observed in terms of the induction of gene conversion. Lesions induced by MPP were apparently more convertogenic than those induced by 3-CPs (Averbeck, 1985).

The development of biochemical techniques for the detection of CL made it possible to compare the mutagenicity of bifunctional psoralens at equal levels of CL induced. Interestingly, the compounds 8-MOP and 5-MOP were found to produce per unit dose of UVA equal levels of CL in Chinese hamster V-79 cells using alkaline elution (Papadopoulo and Averbeck, 1985) and hydroxylapatite chromatography analysis (Dardalhon and Averbeck, 1988). However, at equal solubility levels, 5-MOP was always the photobiologically more active compound when compared with 8-MOP. This holds not only for Chinese hamster V-79 cells (Papadopoulo and Averbeck, 1985) but also for algae *Chlamydomonas reinhardii* (Schimmer, 1981) and yeast *S. cerevisiae* (Averbeck, 1985). As suggested earlier (Averbeck, 1985; Papadopoulo and Averbeck, 1985), the higher photomutagenicity of 5-MOP in comparison to that of 8-MOP may be, at least in part, due to a higher number and/or different isomeric type of photoadducts or to a higher ratio of mono- over diadducts induced.

2. Analysis of the photomutagenicity of bifunctional psoralens using different reirradiation regimens and activating wavelengths

For some years, it has been a major difficulty to assess the role of monoadducts and DNA cross-links in the mutagenic effects induced by bifunctional furocoumarins. Comparisons between the effects of mono- and bifunctional furocoumarins gave some hints on the possible predominant role of cross-links. Interpretations of the results, however, were unsatisfactory because all comparisons were based on two types of psoralens which may give rise to structurally different photoadducts and, consequently, to different mutagenic responses. The better understanding of psoralen photoreactions with DNA lead to two interesting approaches.

a. Comparison of the effects of cross-links at high and low proportions

After the absorption of UVA, bifunctional psoralens react with DNA under the formation of pyrone side (MA_p) and furan side (MA_f) monoadducts. By further absorption of 365 nm radiation, part of MA_f can be converted into DNA interstrand cross-links (CL) while MA_p cannot do so because they do not absorb at that wavelength (see Dall'Acqua et al., 1979; Hearst et al., 1984). Increased conversion is obtained by a reirradiation regimen consisting of a first exposure to 365 nm radiation inducing $MA_{f,p}$ and a few cross-links followed, after washing out of unbound psoralen molecules, by a second exposure to 365 nm radiation. Using such an approach with TMP in Chinese hamster cells (Ben-Hur and Elkind, 1973) and with 8-MOP in excision-deficient bacteria (Seki et al., 1978; Bridges et al., 1979) increased proportions of CL were found associated with increased lethality. 8-MOP induced mutagenicity was strongly increased by such double exposure regimens in *Aspergillus nidulans* (Belogurov and Zavilgelsky, 1981; Scott and Maley, 1981), in *C. reinhardii* (Schimmer, 1983), in yeast (Cassier et al., 1984) and in Chinese hamster cells (Babudri et al., 1981), suggesting an important role of CL in psoralen mutagenesis.

b. Comparison of the effects of monoadducts alone versus a mixture of mono- and diadducts

Recently, it has been shown that bifunctional furocoumarins, such as 8-MOP (Tessman et al., 1985; Averbeck et al., 1987) and 5-MOP (Sa E. Melo et al., 1984), predominantly form MA when using radiation at wavelengths above 380 nm. After washing out of unbound psoralen molecules, part of the furan-side adducts induced can be converted into CL by further exposure to 365 nm radiation. Using 8-MOP in combination with 405 nm radiation (inducing MA alone) and 8-MOP plus 365 nm radiation (inducing a mixture of MA and CL as demonstrated by alkaline elution analysis), it could be shown that in yeast cell killing and the induction of forward and reverse mutations are much more effective in the presence of a mixture of MA and CL than in the presence of MA alone (Averbeck et al., 1987). This appears to be valid as a function of survival and, even more strikingly, as a function of equal number of total photoadducts induced (Averbeck et al., 1987). At non toxic levels, the induction of MA alone by 8-MOP and monochromatic 405 nm radiation lead to an increase in mutation frequency above the spontaneous level (Averbeck, 1988). Treatments with 8-MOP and monochromatic 365 nm radiation resulted in a significantly higher mutation frequency due to the presence of CL induced.

Furthermore, mutagenicity could be drastically increased by a reexposure of 8-MOP-plus-405 nm radiation treated cells, after washing out of unbound 8-MOP molecules, to 365 nm radiation (Cundari and Averbeck, 1988). Since, for an equal number of adducts, the mutagenicity was higher following a 405–365 nm than a 365–365 nm reirradiation protocol, it can be assumed that more CL are induced by the former than by the latter treatment. Correspondingly, more cross-linkable MA were produced by 8-MOP using 405 nm than 365 nm radiation at first exposures (Averbeck et al., 1987).

Figure 1 shows results obtained in diploid yeast (D7) after treatment with the bifunctional furocoumarin 5-MOP and 365 nm or 405 nm radiation. It can be seen that, as a function of equal number of photoadducts induced, 405 nm radiation induced adducts (MA alone) are clearly less lethal and mutagenic than 365 nm radiation induced adducts (mixture of MA and CL). The fact that up to a certain level of adducts the effects of 405 nm and 365 nm radiation induced lesions are the same appears to indicate that, from the start, 5-MOP in combination with 365 nm radiation, induces a relatively high proportion of MA as suggested already from work in mammalian cells (Papadopoulo and Averbeck, 1985).

The results demonstrate that the induction of mutations in eukaryotic cells by bifunctional furocoumarins such as 8-MOP and 5-MOP is more dependent on the proportional amount of CL in mixtures of MA and CL than on the amount of MA alone.

3. Repair of psoralen induced DNA photoadducts

3.1 Studies with repair deficient mutants in yeast

As in bacteria (Cole et al., 1976), the existence of repair deficient mutant phenotypes has been of great help for elucidating repair pathways involved in the processing of psoralen photoadducts in eukaryotic cells (see for review Moustacchi, 1987; Friedberg, 1988; Smith, 1988). In yeast, rad mutants defective in the repair of UV or γ-ray induced DNA lesions, i.e. defective in the excision-resynthesis (epistasis group RAD3), the error-prone (mutagenic) pathway (epistasis group RAD6) and the strand break repair pathway (epistasis group RAD50) were found to be more sensitive to the photoaddition of psoralen than the repair proficient RAD strain (Averbeck and Moustacchi, 1975; Henriques and Moustacchi, 1980a). Double mutants were more sensitive than single mutants (Henriques and Moustacchi, 1981; Henriques et al., 1985) and triple mutants were extremely sensitive to psoralen and UVA treatment (Chanet et al., 1985). Thus, all three repair systems operate on MA and CL photoinduced by psoralens.

In addition, new genetic loci (pso) identified by the isolation of pso mutants were specifically required for the repair of psoralen damage in DNA but not the repair of UV induced pyrimidine dimers or ionizing radiation damage (Henriques and Moustacchi, 1980b). When examining the fate of cross-linkable MA of 8-MOP by their convertibility by a second radiation dose in reirradiation experiments in triple mutants even one or a few MA were lethal (Moustacchi et al., 1983; Chanet et al., 1985). The same authors were able to show the persistance of MA in excision deficient cells (rad2 and rad3 mutants) and their bypass through at least one round of DNA replication (Chanet et al., 1985).

In contrast to results obtained in bacteria on the repair of CL (Cole et al., 1976), it was shown by neutral and alkaline gradients that the repair of psoralen CL in yeast involves double strand breakage of DNA (Jachymczyk et al., 1981; Magaña-Schwencke et

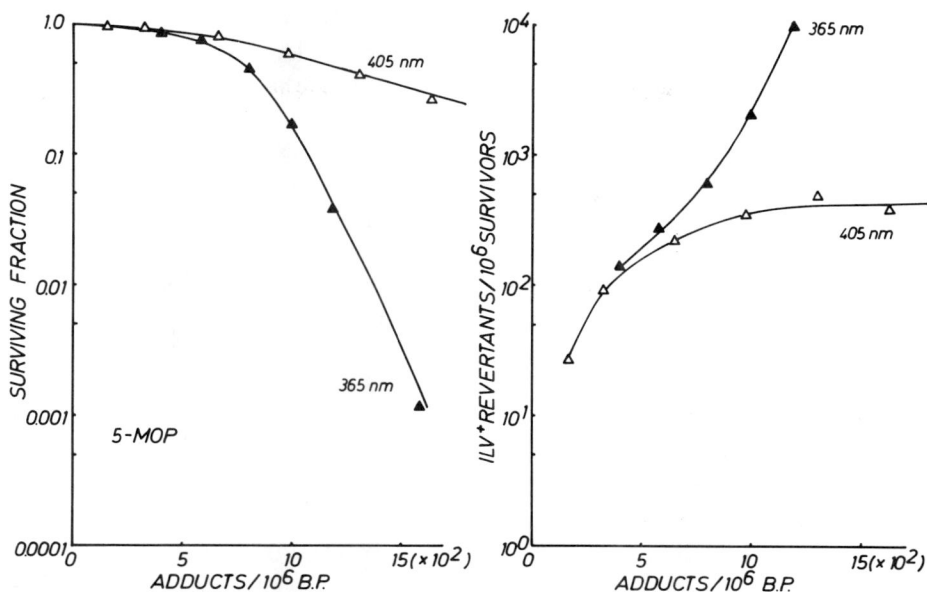

Figure 1. Induction of lethal effects (left panel) and nuclear reversions (ILV+) in the diploid strain D7 of *Saccharomyces cerevisiae* as a function of DNA photoadducts induced by 5-MOP (10 μM) and 365 nm (▲) or 405 nm (△) radiations. Points are average values from two independent experiments.

al., 1982; Miller et al., 1982a, b, 1984). Pso2 mutants, although able to incise DNA containing psoralen CL, were defective in the repair of DNA strand breaks (Magaña-Schwencke et al., 1982).

3.2 Removal of psoralen photoadducts in yeast and mammalian cells

Radioactively labelled psoralen derivatives have been used for the determination of the global rate of photofixation and the removal of photoadducts induced in DNA (Averbeck et al., 1975; Magaña-Schwencke and Moustacchi, 1985). Photobound 8-MOP, MePyPs and 3-CPs was removed quite effectively from treated repair competent wild type yeast cells during 4 hours of post-treatment incubation in complete growth medium (Magaña-Schwencke and Moustacchi, 1985). The removal was almost completely blocked in excision-defective mutants rad1-Δ again pointing to the important role of the excision-repair system in the repair of psoralen induced photoadducts.

Using a new chemical method combining high performance liquide chromatography (HPLC) and fluorescence analysis of purified and enzymatically hydrolysed DNA (Moysan et al., 1988), it could be shown in yeast cells that the two cis-syn d Th $< \frac{5\ 4'}{6\ 5'} >$ 3-CPs diastereoisomers, predominant photoadducts induced by treatments with 3-CPs and UVA in cellular DNA are removed with about the same kinetics from yeast

939

DNA during post-treatment incubation. Thus, the repair systems operating on 3-CPs plus UVA induced furan-side MA do not appear to distinguish between the two diastereoisomers (Dardalhon et al., 1988). The same appears to be true for Chinese hamster V-79 cells (Dardalhon et al., unpublished data).

As far as the removal of psoralen photoadducts from cultured mammalian cells is concerned, apparently rodent cells, green monkey cell and normal human fibroblasts were capable of removing at least part of the mono- and diadducts photoinduced by psoralens (Smith, 1988). Rodent cells appeared to be somewhat more effective in the removal of furocoumarin adducts than normal human fibroblasts. For example, guinea pig fibroblasts removed about 90% of the 8-MOP adducts (Pohl and Christophers, 1980) whereas human fibroblasts removed only 35% of these adducts in 24 hours (Nocentini, 1986). Chinese hamster cells eliminated 80% of TMP photoadducts in 8 hours (Ben-Hur and Elkind, 1973), whereas human breast carcinoma cells removed only 50% in 24 hours (Prager et al., 1983).

After treatment with 50 μM ^{3}H-8-MOP and 3.2 kJm^{-2} of UVA, 2.6 photoadducts/10^5 base pairs were induced in Chinese hamster cells. After 3 hours of post-treatment incubation, we found that 53% of total adducts were still present (Fig. 2, left panel). Using alkaline denaturation-renaturation and hydroxylapatite chromatography for detecting DNA CL induced, we observed a decrease in yield of CL by 57% after 3 hours post-treatment incubation (Fig. 2, right panel) and by 87% after 24 hours post-treatment incubation (data not shown). These data indicate that under the treatment conditions used Chinese hamster V-79 cells are quite effective in removing 8-MOP mono- and diadducts from their DNA.

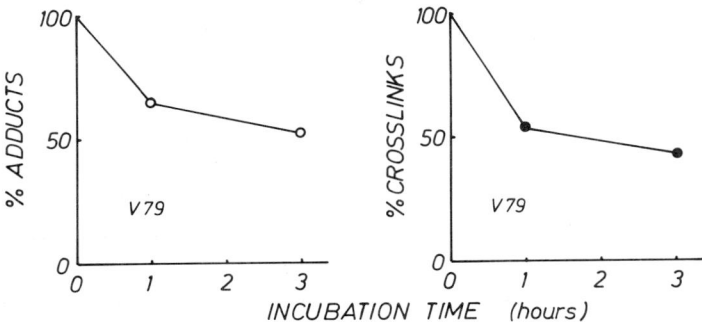

Figure 2. Removal of total photoadducts (left panel) and cross-links (right panel) after treatment of Chinese hamster V-79 cells with ^{3}H-8-MOP and UVA as a function of post-treatment incubation time at 37°C. 8-MOP concentration: 1 μM, dose of UVA: 3.2 kJm^{-2} (Removal of 8-MOP-DNA photoadducts).

Normal human fibroblasts containing angelicin plus UVA induced MA (0.3 adducts/10^5 bp) removed 80% of these adducts in 24 hours whereas excision-repair defective XP A cells (xeroderma pigmentosum complementation group A) removed only 20% in that period.(Cleaver and Gruenert, 1984). MA induced by MePyPs and UVA in normal human fibroblasts at a high level (.16 adducts/10^5 bp) were removed quite inefficiently (30% in 24 hours) (Nocentini, 1986).

An interesting situation was found when analysing the repair of TMP plus light induced adducts in normal human fibroblasts (Papadopoulo et al., 1988) and Fanconi's anemia cells (complementation group A and B) (Averbeck et al., 1988). Using the alkaline elution technique and a reirradiation protocol for the determination of the fate of CL and cross-linkable (furan-side) MA, it was demonstrated that high levels of TMP photoinduced MA_f can block the removal of CL in normal human fibroblasts (Papadopoulo et al., 1988).

After conversion of part of the furan-side TMP MA by reirradiation into CL, a substantial incision of CL (44%) in 24 hours was noted in normal human fibroblasts whereas the incision of CL was only 25% in Fanconi's anemia group A cells (Averbeck et al., 1988). Cross-linkable MA induced by TMP plus 405 nm radiation were removed by 40% in normal human fibroblasts whereas the removal was only 10% in Fanconi's anemia group A and group B cells (Averbeck et al., 1988).

Following the repair of psoralen induced lesions by S1 nuclease digestion of ^{3}H-labelled DNA adducts, Vuksanovic and Cleaver (1987) observed that normal human fibroblasts were able to remove 4'-hydroxymethyl 4,5',8-trimethylpsoralen (HMT) plus UVA induced CL and MA whereas XP group A cells failed to remove either. However, an XP revertant from a group A cell line showed the unusual feature of being able to remove psoralen CL and (6–4) pyrimidine-pyrimidine photoproducts from the DNA but not psoralen MA pyrimidine dimers (Vuksanovic and Cleaver, 1987). These results are consistent with the notion that specific repair pathways exist in human cells for psoralen induced mono- and diadducts. The processing of psoralen adducts may, however, differ in actively transcribed and non transcribed genes (Hanawalt, 1986).

3.3 Photomutagenesis and repair of psoralen damage induced at low and high dose rates of UVA

The lethal and photomutagenic effects observed with mono- and bifunctional furocoumarins in yeast were found to be dependent on the dose rate of UVA (Averbeck and Averbeck, 1978, 1979), treatments at high dose rate (HDR, 72 kJm^{-2}h^{-1}) being much more effective than treatments at low dose rate (LDR, below 1 kJm^{-2}h^{-1}). At low temperature (5°C), the effects of LDR were suppressed and similar to those of HDR. Using repair deficient mutants, it was shown that excision repair (rad3), mutagenic pathways (rad9) and other repair functions were operating during LDR (Averbeck and Averbeck, 1979).

In order to get an idea on the fate of psoralen photoadducts, we determined the number of total psoralen adducts (using a radioactively labelled psoralen) and the amount of DNA interstrand CL (measured by alkaline step elution analysis) (Averbeck et al.,

1987; Cundari and Averbeck, 1988). Fig. 3 shows the results obtained for the induction of lethal effects and the induction of ILV$^+$ revertants in diploid yeast (D7) after treatment with the bifunctional furocoumarin 5-MOP (10 μM) and 365 nm (UVA) radiation at high (HDR, 46,4 kJm^{-2}h^{-1}) and low (LDR, 0.51 kJm^{-2}h^{-1}) dose rate. It can be seen that treatments at LDR lead to very little cell killing and low mutation induction in comparison to treatment at HDR. For the same total UVA dose (12 kJm^{-2}) and using ^{14}C-labelled 5-MOP, the radioactivity photobound to DNA was 1.5 to 2 times higher in HDR than in LDR treated samples (5 independent experiments) indicating that during treatments at LDR approximately half the total photoadducts are removed. Analysis of CL induced (according to Cundari and Averbeck, 1988) revealed in two independent determinations that at LDR about 8.5 times less CL are induced than at HDR (81.8 CL per genome at HDR versus 9.6 CL per genome at LDR).

The results are consistent with the idea that during LDR treatment very efficient repair of DNA photoadducts is taking place. This repair leads to a strong decrease in phototoxic (lethal) and photomutagenic effects. The results are in line with the previous finding that at LDR the induction of intragenic mitotic recombination (gene conversion) is much less than that induced at HDR (data not shown). The processing of DNA photoadducts at LDR may also explain the fact that low dose rate effects were also observed with the monofunctional furocoumarin 3-CPs (Averbeck and Averbeck, 1979). It remains to be established whether the repair during LDR is dependent on DNA synthesis and whether this type of repair can also occur in rodent and human cells.

3.4 Repair of exogenous DNA containing psoralen photoadducts

Recently, it has been demonstrated that the excision repair system of yeast can operate on incoming plasmid DNA damaged by ultraviolet (254 cm) radiation and that to a certain extent transformation efficiency can be equated with plasmid survival and repair proficiency within the transformed cell. (White and Sedgwick, 1985, 1987).

In order to develop an assay system for the analysis of specific steps in the repair of mono- and diadducts photoinduced by psoralens, the transforming efficiency of a plasmid containing 8-MOP induced photoadducts has been determined. Table 1 shows preliminary results obtained using the plasmid YCp50 exposed to 8-MOP at 50 μM and UVA at 10 Jm^{-2}sec^{-1} and a transformation competent wild type strain of yeast. The transformation efficiency was approximately 10 transformants per ng plasmid DNA. YCp50 derives from pBR 322 and contains the yeast URA3 gene, the origin of replication ARS1 and the centromeric region CEN4 (Johnston and Davis, 1984).

It can be seen (Table 1) that in the rad1 mutant plasmid survival is considerably lowered and even more so in the rad1 rad52 double mutant whereas it is nearly unaffected in rad6 and rad18 mutants suggesting that the excision repair system of the host cell (rad1 belongs to epistasis group RAD3) is needed for the repair of plasmid damaged by 8-MOP and UVA.

Furthermore, the increased inactivation of plasmid survival in rad1 rad52 double mutant implies that also the repair system for double strand break (recombination) repair (epistasis group RAD52) is involved in plasmid repair. Since plasmid survival is nearly

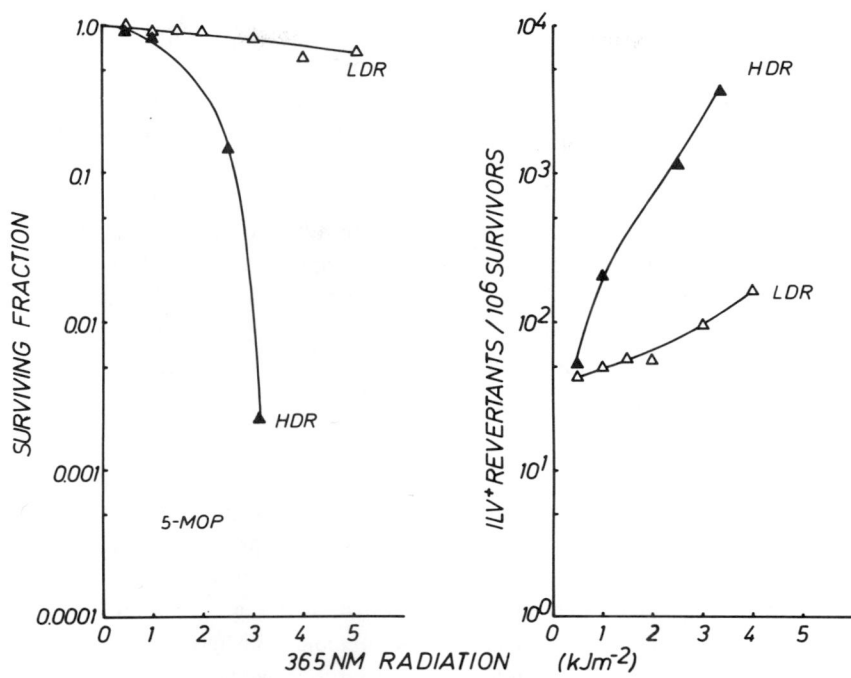

Figure 3. Induction of lethal effects (left panel) and nuclear reversions (ILV$^+$) by 5-MOP in the diploid strain D7 of *Saccharomyces cerevisiae* as a function of 365 nm (UVA) radiation dose delivered at high dose rate (HDR = 46.4 kJm^{-2}h^{-1}) and low dose rate (LDR = 0.51 kJm^{-2}h^{-1}), 5-MOP concentration: 10 μM

Table 1. Transformation of *Saccharomyces cerevisiae* by 8-MOP plus UVA treated plasmid YC$_p$50

Genotype	Inactivation parameters			DMF
	D_q(kJm^{-2})	D_{37}(kJm^{-2})	D_O(kJm^{-2})	
wild type (RAD)	0.18	0.69	0.51	1.0
rad1	0.07	0.42	0.35	1.6
rad6	0.18	0.63	0.45	1.1
rad18	0.18	0.63	0.45	1.1
rad1 rad52	0	0.27	0.27	2.6

conc. 8-MOP : 5 × 10^{-5} M, dose rate (UVA) : 10 Jm^{-2}sec^{-1}

unaffected in rad6 and rad18 mutants the mutagenic repair system of yeast does not seem to operate.

Rad1 (epistasis group RAD3) rad6 and rad18 (epistasis group RAD6) and rad52 (epistasis group RAD52) mutants have been previously shown to be sensitive to psoralen plus UVA induced damage (Friedberg, 1988). Thus, they are involved in the repair of genomic DNA of yeast. With regard to these results, the present observations with plasmid DNA on rad6 and rad18 mutants are quite unexpected. However, recent findings that the RAD6 gene encodes for a ubiquitin-conjugating enzyme and thus may be responsible for alterations of chromatine structure required for DNA repair (Jentsch et al., 1987) and that the RAD18 gene product may encode for a zinc finger like protein recognizing specific domains in DNA and regulating transcription (Chanet et al., 1988; Fabre et al., 1988; Jones et al., 1988) may provide a possible explanation. RAD6 and RAD18 gene products may fail to recognize damaged plasmid DNA if it is not an integrated part of a chromatine structure.

4. Concluding remarks

The development of new techniques for the detection of psoralen induced photoadducts in DNA offers the possibility to analyse more precisely the role of different types of adducts in the photomutagenicity of mono- and bifunctional derivatives of photochemotherapeutic and/or fundamental interest. Although psoralens are known to undergo a variety of reactions involving not only nucleic acids (see for review Cimino et al., 1985) but also proteins, membranes and their subunits (Laskin et al., 1985, 1986; Kittler and L öber, 1988), there is some hope that their contributions to important biological effects can be assessed. We have seen here that mono- and bifunctional furocoumarins exert photomutagenic effects. Bifunctional compounds appear to be more mutagenic than monofunctional ones, mostly due to the cross-linking capacity of the former. As shown by reirradiation experiments and experiments using different wavelengths not only the type (structure) and distribution of lesions is important for psoralen mutagenesis but also, in the case of bifunctional furocoumarins, the ratio of the mono- over diadducts induced. As shown in normal human fibroblasts specific types of lesions (cross-linkable MA_f) may interfere with the repair of another type of lesion (CL) (Papadopoulo et al., 1988).

The analysis of the fate of different lesions induced permits to study known biological phenomena, like dose rate effects on lethality and mutagenicity, in more detail and to approach underlying mechanisms. Finally, it appears that due to modern transformation technologies more specific questions can be asked such as whether psoralen photo-induced lesions are as accurate and effectively repaired by cellular repair mechanisms when present in exogenous (plasmid) or genomic DNA. Research work on these lines can be expected to open fascinating new insights in the interplay of genetically controlled repair systems on defined DNA adducts in eukaryotic cells.

Acknowledgements

This work was supported by the CNRS, INSERM, LNFCC (Paris) and the CEA (Saclay, France). Thanks are due to Ms S. Averbeck and D. Chardonnieras for excellent assistance.

References

Alderson, T., and Scott, B.R. (1970). The photosensitizing effect of 8-methoxypsoralen on the inactivation and mutation of *Aspergillus* conidia by near ultraviolet light. *Mutation Res.* 9, 569–578.

Anderson, T.F., and Voorhees, J.J. (1980). Psoralen photochemotherapy of cutaneous disorders. *Ann. Rev. Pharmacol. Toxicol.* 20, 235–257.

Arlett, C.F. (1973). Mutagenesis in cultured mammalian cells, *Studia Biophys.* 36/37, 139–147.

Averbeck, D., Chandra, P., and Biswas, R.K. (1975). Structural specificity in the lethal and mutagenic activity of furocoumarins in yeast cells. *Rad. Environm. Biophys.* 12, 241–252.

Averbeck, D., and Moustacchi, E. (1975). 8-methoxypsoralen plus 365 nm light effects and repair in yeast. *Biochim. Biophys. Acta.* 395, 393–404.

Averbeck, D., and Averbeck, S. (1978). Dose-rate effects of 8-methoxypsoralem plus 365 nm irradiation on cell killing in *Saccharomyces cerevisiae*. *Mutation Res.* 50, 195–206.

Averbeck, D., and Averbeck, S. (1979). Dose-rate effects of furocoumarins plus 365 nm irradiation on the cytoplasmic and nuclear genetic level in *Saccharomyces cerevisiae*.In: "Radiation Biology and Chemistry, Research Developments", Edwards, H.E., Navaratman, S., Parsons, B.J., Phillips, G.O. Eds., Elsevier Scientific Publishing Company, Amsterdam, pp. 453–466.

Averbeck, D., and Moustacchi, E. (1980). Decreased photoinduced mutagenicity of monofunctional as opposed to bifunctional furocoumarins in yeast. *Photochem. Photobiol.* 31, 475–478.

Averbeck, D., Magaña-Schwencke, N., and Moustacchi, E. (1981). Genetic effects and repair in yeast of DNA lesions induced by 3-carbethoxypsoralen and other photoreactive furocoumarins of therapeutic interest. In: "Psoralens in Cosmetics and Dermatology", Cahn, J., Forlot, B.P., Grupper, C., Meybeck, A.E., and Urbach, F. Eds., Pergamon Press, New York, pp. 143–153.

Averbeck, D. (1985). Relationship between lesions photoinduced by mono- and bifunctional furocoumarins in DNA and genotoxic effects in diploid yeast. *Mutation Press.* 151, 217–233.

Averbeck, D., Averbeck, S., and Cundari, E. (1987). Mutagenic and recombinogenic action of DNA monoadducts photoinduced by the bifunctional furocoumarin 8-methoxypsoralen in yeast (*Saccharomyces cerevisiae*). *Photochem. Photobiol.* 45, 371–379.

Averbeck, D. (1988). Photomutagenicity induced by psoralens : modulation of the photomutagenic response in eukaryotes. *Arch. Toxicol.* (in press).

Averbeck, D., Papadopoulo, D., and Moustacchi, E. (1988). Repair of 4,5'8-trimethylpsoralen plus light induced DNA damage in normal and Fanconi's anemia cell lines. *Cancer Res.* 48, 2015–2020.

Babudri, N., Pani, B., Venturini, S., Tamaro, M., Monti-Bragadin, C., and Bordin, F. (1981). Mutation induction and killing of V-79 Chinese hamster cells by 8-methoxypsoralen plus near ultraviolet light: relative effects of monoadducts and cross-links. *Mutation Res.* 91, 391–394.

Belogurov, A.A, and Zavilgelski, G.B. (1981). Mutagenic effect of furocoumarin adducts and cross-links in bacteriophage lambda. *Mutation Res.* 84, 11–15.

Ben-Hur, E., and Elkind, M.M. (1973). Psoralen plus near-ultraviolet inactivation of cultured Chinese hamster cells and its relation to DNA cross-links. *Mutation Res.* 18, 315–324.

Ben-Hur, E., Song, P.S. (1984). The photochemistry and photobiology of furocoumarins (psoralens). *Adv. Radiat. Biol.* 11, 131–171.

Blan, Q.A., and Grossweiner, L.I. (1987). Singlet oxygen generation by furocoumarins: effect of DNA and liposomes. *Photochem. Photobiol.* 45, 177–183.

Boyer, V., Moustacchi, E., and Sage, E. (1988). Sequence specificity in photoreaction of various psoralen derivatives with DNA: Role in biological activity. *Biochemistry* 27, 3011–3018.

Bridges, B.A., Mottershead, R.P., and Knowles, A. (1979). Mutation induction and killing of Escherichia coli by DNA adducts and cross-links: a photobiological study with 8-methoxypsoralen. *Chem. Biol. Interact.* 27, 221–233.

Burger, P.M., and Simons, J.W.I.M. (1979a). Mutagenicity of 8-methoxypsoralen and long-wave and ultraviolet irradiation in V-79 Chinese hamster cells. A first approach to a risk estimate in photochemotherapy. *Mutation Res.* 60, 381–389.

Burger, P.M., and Simons, J.W.I.M. (1979b). Mutagenicity of 8-methoxypsoralen and long-wave and ultraviolet light in diploid human skin fibroblasts. An improved risk estimate in photochemotherapy. *Mutation Res.* 63, 371–380.

Cadet, J.L., Voituriez, F., Gaboriau, F., and Vigny, R. (1986). Isolation and characterization of psoralen photoadducts to DNA and related model compounds. In: "The role of cyclic nucleic acid adducts in carcinogenesis and mutagenesis, Singer, B., and Bartsch, H. Eds., IARC Scientific Publ. n°70, IARC, Lyon, pp. 247–251.

Cassier, C., Chanet, R., and Moustacchi, E. (1984). Mutagenic effects of DNA cross-links induced in yeast by 8-methoxypsoralen photoaddition. *Photochem. Photobiol.* 39, 799–804.

Chanet, R., Cassier, C., and Moustacchi, E. (1983). Genetic control by the bypass of monoadducts and of the repair of cross-links photoinduced by 8-methoxypsoralen in yeast. *Mutation Res,* 145, 145–155.

Chanet, R., Magaña-Schwencke, N., and Fabre, F. (1988). Potential DNA-binding domains in the RAD18 gene product of *Saccharomyces cerevisiae. Gene* 74, 543–547.

Cimino, G.D., Gamper, H.B., Isaacs, S.T., and Hearst, J.E. (1985). Psoralens as photoactive probes of nucleic acid structure and function : organic chemistry, photochemistry and biochemistry. *Ann. Rev. Biochem.* 54, 1151–1193.

Cleaver, J.E., and Gruenert, D.C. (1984). Repair of psoralen adducts in human DNA : differences among XP complementation groups. *J. Invest. Dermatol.* 82, 311–315.

Cole, R,S., Levitan, D., and Sinden, R.R. (1976). Removal of psoralen interstrand cross-links from DNA of Escherichia coli : mechanism and genetic control. *J. Mol. Biol.* 103, 39–59.

Cundari, E., and Averbeck, D. (1988). 8-methoxypsoralen photoinduced DNA cross-links as determined in yeast by alkaline step elution under different reirradiation conditions. Relation with genetic effects. *Photochem. Photobiol.* 48, 315–320.

Dall'Acqua, F. (1977). New chemical aspects of the photoreaction between psoralen and DNA. In: "Research Photobiology", Castellani, A. Ed., Plenum Press, New York, pp. 245–255.

Dall'Acqua, F., Marciano-Magni, S., Zambon, F., and Rodighiero, G. (1979). Kinetic analysis of the photoreaction (365 nm) between psoralen and DNA. *Photochem. Photobiol.* 29, 489–495.

Dall'Acqua, F. (1988). Psoralens: a review. In: "Photosensitization. Molecular, cellular and medical aspects", Moreno, G., Pottier, R.H., Truscott, T.G. Eds., NATO ASI Series, vol. H15, Springer-Verlag, Berlin, Heidelberg, pp. 269–278.

Dardalhon, M., and Averbeck, D. (1988). Induction and removal of DNA interstrand cross-links in V-79 Chinese hamster cells by hydroxylapatite chromatography after treatments with bifunctional furocoumarins. *Int. J. Radiat. Biol.* 54. 1007–1020.

Dardalhon, M., Moysan, A., Averbeck, D., Vigny, P. Cadet, J., and Voituriez, L. (1988). Repair of two cis-syn diastereoisomers formed between 3-carbethoxypsoralen and thymidine in yeast cells, followed by a chemical method. *J. Photochem. Photobiol. Part B* 2, 389–394.

De Mol, N.J., Beijersbergen van Henegouwen, G.M.J., Weeda, B., Knox, C.N., and Truscott, T.G. (1986). Photobinding of psoralens to bacterial macromolecules *in situ* and induction of genetic effects in a bacterial test systems. Effects of singlet oxygen diagnostic aids D_2O and DABCO. *Photochem. Photobiol.* 44, 747–751.

Dubertret, L., Averbeck, D., Bisagni, E., Moron, J., Moustacchi, E., Billardon, C., Papadopoulo, D., Nocentini, S., Vigny, P., Blais, J., Bensasson, R.V., Ronfard-Haret, J.C., Land, E.J., Zajdela, F., and Latarjet, R. (1985). Photochemotherapy using pyridopsoralens. *Biochimie.* 67, 417–422.

Fabre, F., Magaña-Schwencke, N., and Chanet, R. (1988). Isolation of the RAD18 gene of *Saccharomyces cerevisiae* and construction of rad18 deletion mutants. *Mol. Gen. Genet.* 215, 425–430.

Friedberg, E.C. (1988). Deoxyribonucleic acid repair in the yeast *Saccharomyces cerevisiae.* *Microbiol. Rev.* 52. 70–102.

Hanawalt, P.C. (1986). Intragenomic heterogeneity in DNA damage processing potential implication for risk assessment. In: "Basis Life Sciences (Mechanisms of DNA damage and repair)", vol. 38, Simic M.G., Grossman, L. and Upton A. C. Eds., Plenum Press, New York, pp. 489–498.

Hearst, J.E., Isaacs, S.T., Kanne, D., Rapoport, M., and Straub, K. (1984). The reaction of the psoralens with deoxyribonucleic acid. *Quart. Rev. Biophys.* 17, 1–44.

Henriques, J.A.P. and Moustacchi, E. (1980a). Sensitivity to photoaddition of mono- and bifunctional furocoumarins of X-ray sensitive mutants of *Saccharomyces cerevisiae.* *Photochem. Photobiol.* 31, 557–563.

Henriques, J.A.P. and Moustacchi, E. (1980b). Isolation and characterization of pso mutants sensitive to photoaddition of psoralen derivatives in *Saccharomyces* cerevisiae. Genetics 75, 273–288.

Henriques, J.A.P., and Moustacchi, E. (1981). Interactions between mutations for sensitivity to psoralen photoaddition (pso) and to radiation (rad) in *Saccharomyces cerevisiae.* J. Bacteriol. 148, 248–256.

Henriques, J.A.P., Da Silva, K.V.C.L., and Moustacchi, E. (1985). Interaction between gene controlling sensitivity to psoralen (pso) and to radiation (rad) after 3-carbethoxypsoralen plus 365 nm UV-light treatment in yeast. *Mol. Gen. Genet.* 201, 415–420.

Hinnen, A., Hicks, J.B., and Fink, G.R. (1978). Transformation of yeast. *Proc. Natl. Acad. Sci. USA* 75, 1929–1933.

IARC Monograph (1986). Furocoumarins. In "IARC Monographs on the evaluation of the carcinogenic risk of chemicals to humans. Some naturally occurring and synthetic food components, furocoumarins and ultraviolet radiation". Vol. 40, IARC Lyon, France, pp. 289–376.

IARC (1987). Monographs on the evaluation of carcinogenic risks to humans. Genetic and related effects: an updating of selected IARC monographs from volumes 1 to 42". Supplement 6, IARC Lyon, France, pp. 377–385, pp. 541–544.

Igali, S., Bridges, B.A., Ashwood-Smith, M.J., and Scott, B.R. (1970). Mutagenesis in E. coli. IV: Photosensitization to near ultraviolet light by 8-methoxypsoralen. *Mutation Res.* 9, 21–30.

Jachymczyk, W.J., Von Borstel., R.C., Mowat, M.R.A., and Hastings, ,P.J. (1981). Repair of interstrand cross-links in DNA of *Saccharomyces cerevisiae* requires two systems for DNA repair: the RAD3 system and the RAD51 system. *Mol. Gen. Genet.* 182, 196–205.

Jentsch, S., McGarth, J.P., and Varschavsky, A. (1987). The yeast DNA repair gene RAD6 encodes a ubiquitin-conjugating enzyme. *Nature* 329, 131–134.

Johnston, M., and David, R.W. (1984). Sequences that regulate the divergent gall-gall0 promotor in *Saccharomyces cerevisiae. Mol. Cell. Biol.* 4, 1440–1448.

Jones, J.S., Weber, S., and Prakash, L. (1988). The *Saccharomyces cerevisiae* RAD18 gene encodes a protein that contains potential zinc finger domains for nucleic acid binding and putative nucleotide binding sequences. *Nucleic Acid Res.* 16, 7119–7131.

Kanne, D., Straub, K., Rapoport, M., and Hearst, J.E. (1982). Psoralen-deoxyribonucleic acid photoreaction. Characterization of the monoaddition products from 8-methoxypsoralen and 4,5',8-trimethylpsoralen. *Biochemistry* 21,861–871.

Kittler, L., and Löber G. (1988). Photoreactions of furocoumarins with nucleic acids, proteins, membranes and their subunits. *Studia Biophys.* 124, 97–114.

Laskin, J.D., Lee, E., Yurkow, E.J., Laskin, D.L., and Gallo, M.A. (1985). A possible mechanism of psoralen phototoxicity not involving direct interaction with DNA. *Proc. Natl. Acad. Sci. USA* 82, 6158–6162.

Laskin, J.D., Lee, E.,Laskin, D.L., and Gallo, M.A. (1986). Psoralens potentiate ultraviolet light-induced inhibition of epidermal growth factor binding. *Proc. Natl. Acad. Sci. USA* 83, 8211–8215.

Loveday, K.S., and Donahue, D.A. (1984). Induction of sister chromatid exchanges and gene mutations in Chinese hamster ovary cells by psoralens. *Natl. Cancer Inst. Monogr.* 66, 149–155.

Magaña-Schwencke, N., Henriques, J.A.,P., Chanet, R., and Moustacchi, E. (1982). The fate of 8-methoxypsoralen photoinduced cross-links in nuclear and mitochondrial yeast DNA: comparison of wild-type and repair-deficient strains. *Proc. Natl. Acad. Sci. USA* 79, 1722–1726.

Mathews, M.M. (1963). Comparative study on the lethal photosensitisation of *Sarcina lutea* by 8-methoxypsoralen and toluidene blue. *J. Bacteriol.* 85, 322–328.

Miller, R.D., Prakash, L. and Prakash, S. (1982a). Genetic control of excision of *Saccharomyces cerevisiae* interstrand DNA cross-links induced by psoralen plus near-UV light. *Mol. Cell. Biol.* 2, 939–948.

Miller, R.D., Prakash, L., and Prakash, S. (1982b). Defective excision of pyrimidine dimers and interstrand DNA cross-links in rad7 and rad23 mutants of *Saccharomyces cerevisiae.* *Mol. Gen. Genet.* 188, 235–239.

Miller, R.D., Prakash, S., and Prakash, L. (1984). Different effects of RAD genes of *Saccharomyces cerecisiae* on incision of interstrand cross-links and monoadducts in DNA induced by psoralen plus UV light treatment. *Photochem. Photobiol.* 39, 349–352.

Moustacchi, E., Cassier, C., Chanet, R., Magaña-Schwencke, N., Saeki, T., and Henriques, J.A.P. (1983). Biological role of photoinduced cross-links and monoadducts in yeast DNA: Genetic control and steps involved in their repair. In: "Cellular responses to DNA damage", Friedberg, E.C., and Bridges, B.A. Eds., Alan, R. Liss Inc. New York, pp. 87–106.

Moustacchi, E. (1987). DNA repair in yeast : genetic control and biological consequences. In: "Advances in Radiation Biology", vol. 13, Lett. J.T. Ed., Academic Press, New York, pp. 1–30.

Moysan, A., Vigny, P., Dardalhon, M,. Averbeck, D., Voituriez, L., and Cadet, J. (1988). 3-carbethoxypsoralen-DNA photolesions : Identification and quantitative detection in yeast and mammalian cells of the two cis-syn diastereoisomers formed with thymidine. *Photochem. Photobiol.* 47, 327.

Mullen, M.P., Pathak, M.A., West, J.D., Harris, T.J., and Dall'Acqua, F. (1984). Carcinogenic effects of monofunctional and bifunctional furocoumarins. *Natl. Cancer Inst. Monogr.* 66, 205–210.

Musajo, L. and Rodighiero, G. (1972), Mode of photosensitizing action of furocoumarins. In: "Photobiology", vol. VII, Giese A.C. Ed., Academic Press, New York and London, pp. 115–147.

Nocentini, S. (1986). DNA photobinding of 7-methylpyrido(3,4-c) psoralen and 8-methoxypsoralen. Effects on macromolecular synthesis, repair and survival in cultured human cells. *Mutation Res.* 161, 181–192.

Pani, B., Babudri, N., Venturini, S., Tamaro, M., Bordin, F., and Monti-Bragadin, C. (1981). Mutation induction and killing of prokaryotic and eukaryotic cells by 8-methoxypsoralen, 4,5'-dimethylangelicin, 5-methoxypsoralen, 4'-hydroxymethyl-4,5'-dimethylangelicin. *Teratog. Carcinog. Mutagen.* 1, 407–415.

Papadopoulo, D., Sagliocco, F., and Averbeck, D. (1983). Mutagenic effects of 3-carbethoxypsoralen and 8-methoxypsoralen plus 365 nm irradiation in mammalian cells. *Mutation Res,* 124, 287–297.

948

Papadopoulo, D., and Averbeck, D. (1985). Genotoxic effects and DNA photoadducts induced in Chinese hamster V-79 cells by 5-methoxypsoralen and 8-methoxypsoralen. *Mutation Res.* 151, 281–291.

Papadopoulo, D., Averbeck, D.and Moustacchi, E. (1986). Mutagenic effects photoinduced in mammalian cells *in vitro* by two monofunctional pyridopsoralens. *Photochem. Photobiol.* 44, 31–39.

Papadopoulo, D., Averbeck, D.and Moustacchi, E. (1988). High levels of 4,5',8-trimethylpsoralen photoinduced furan-side monoadducts can block cross-link removal in normal human cells. *Photochem. Photobiol.* 47, 321–326.

Parrish, J.A,. Stern, R.S., Pathak, M.A., and Fitzpatrick, T.B. (1982). Photochemotherapy of skin diseases. In: "The Science of Photomedicine", Regan, J.D., and Parrish, J.A. Eds., Plenum Press, New York, and London, pp. 595.

Pathak, M.A., and Joshi, P.C. (1984). Production of active oxygen species ($^1O_2$ and $O_2^{-\cdot}$) by psoralens and ultraviolet radiation (320–400 nm), *Biochim. Biohys. Acta* 798, 115–126.

Pohl, J., and Christophers, E. (1980). Photoinactivation and recovery in skin fibroblasts after formation of mono- and bifunctional adducts by furocoumarins-plus-UVA, *J. Invest. Dermator.* 75, 306–310.

Prager, A, Green, M. and Ben-Hur, E. (1983). Inhibition of ornithine decarboxylase induction by psoralen plus near ultraviolet light in human cells : the role of monoadducts vs DNA cross-links. *Photochem. Photobiol.* 37, 525–528.

Rodighiero, G., Dall'Acqua, F., and Averbeck, D. (1988). New psoralem and angelicin derivatives. In: "Psoralen-DNA Photobiology", vol. 1, Gasparro, F.P. Ed., CRC Press Inc. Boca Raton, pp. 37–114.

Ronfard-Haret, J.C., Averbeck, D., Bensasson, R.V., Bisagni, E., Land, E.J., and Moron, J. (1987). Correlation between the triplet photophysical properties and the photobiological action in yeast of two monofunctional pryidopsoralens. *Photochem. Photobiol.* 45, 235–239.

Sa E Melo, T., Morlière, Santus, R., and Dubertret, L. (1984). Photoactivity of 5-methoxypsoralen with calf thymus DNA upon excitation in the UV-A. *Photobiochem. Photobiophys.* 7, 121–131.

Schenley, R.L., and Hsie, A.W. (1981). Interaction of 8-methoxypsoralen and near UV light causes mutation and cytotoxicity in mammallan cells, *Photochem. Photobiol.* 33, 179–185.

Schimmer, O., and Hauber, F. (1977). Untersuchungen zur mutagenen Wirksamkeit von Cumarinderivaten in Chlamydomonas. I: Der Einfluss von Licht verschiedener Wellenlängenbereiche auf die Mutationsinduktion durch Xanthotoxin in einer arginin-bedürftigen Mutante. *Mutation Res.* 44, 21–31.

Schimmer, O. (1981). Vergleich der photomutagenen Wirkungen von 5-MOP (Bergapten) und 8-MOP (Xanthotoxin) in Chlamydomonas reinhardii. *Mutation Res.* 89, 283–296.

Schimmer, O. (1983). Effect of re-irradiation with UV-A on inactivation and mutation in arg cells of Chlamydomonas reinhardii pretreated with furocoumarins plus UV-A. *Mutation Res.* 109, 195–205.

Scott, B.R., Pathak, M.A., and Mohn, G.R. (1976). Molecular and genetic basis of furocoumarin reactions. *Mutation Res.* 39, 29–74.

Scott, B.R., and Maley, M.A. (1981). Mutagenicity of monoadducts and cross-links induced in Aspergillus nidulans by 8-methoxypsoralen plus 365 nm radiation. *Photochem. Photobiol.* 34, 63–67.

Seki, T., Nozu, K.. and Kondo, S. (1978). Differential causes of mutation and killing in Escherichia coli after psoralen plus light treatment : monoadducts and cross-links. *Photochem. Photobiol.* 27, 19–24.

Smith, C.A. (1988). Repair of DNA containing furocoumarin adducts. In: "Psoralen-DNA Photochemistry, Photobiology and Phototherapies", vol. II, Gasparro, F. Ed., CRC Press Inc. pp. 87–116.

Stern, R.S., Lange, R., and members of the Photochemotherapy Follow-up Study (1988). Non-melanoma skin cancer occurring in patients treated with PUVA five to ten years after first treatment. *J. Invest. Dermatol.* 91, 120–124.

Suzuki, H., Nakamura, K. and Iwaida, M. (1979). Detection and determination of bergapten in bergamot oil and in cosmetics. *J. Soc. Cosmet. Chem.*30, 393–400.

Swart, R.N.J., Beckers, M.A.N., and Schothorst, A.A. (1983). Phototoxicity and mutagenicity of 4',5'-dimethylangelicin and long-wave ultraviolet irradiation in Chinese hamster cells and human skin fibroblasts. *Mutation Res.* 124, 271–279.

Tessman, J.W., Isaacs, S.T. and Hearst, J.E. (1985). Photochemistry of the furan-side 8-methoxypsoralen-thymidine monoadduct inside the DNA helix. Conversion to diadduct and to pyrone-side monoadduct. *Biochemistry* 24, 1669–1676.

Van Houten, B., Gamper, H., Holbrook, S.R., Hearst, J.E., and Sancar, A. (1986). Action mechanism of ABC excision nuclease on a DNA substrate containing a psoralen cross-link at a defined position. *Proc. Natl. Acad. Sci. USA* 83, 8077–8081.

Vuksanovic, L., and Cleaver, J.E. (1987). Unique cross-links and monoadduct repair characteristics for a xeroderma pigmentosum revertant cell line. *Mutation Res.* 184, 255–263.

White, C.I. and Sedgwick, S.G. (1985). The use of plasmid DNA repair functions in the yeast *Saccharomyces cerevisiae. Mol. Gen. Genet.* 201, 99–106.

White, C.I. and Sedgwick, S.G. (1987). Repair of UV-irradiated plasmid DNA in *Saccharomyces cerevisiae.* Inability to complement mutational defects in excision repair by *in vitro* treatment with Micrococcus luteus UV endonuclease. *Mutation Res.* 183, 161–167.

Zimmermann, F.K., Kern, R., and Rasenberger, H. (1975). A yeast strain for simultaneous detection of induced mitotic crossing-over, mitotic gene conversion and reverse mutation. *Mutation Res.* 28, 381–388.

Zolan, M.E., Smith, C.A., and Hanawalt, ,P.C. (1984). Formation and repair of furocoumarin adducts in α–deoxyribonucleic acid and bulk deoxyribonucleic acid of monkey cells. *Biochemistry* 23, 63–69.

Quantification of 8-MOP Photoadducts in Lymphocytes

F.P. GASPARRO, D. WEINGOLD, E. SIMMONS, D. GOLDMINZ,
R. EDELSON

Yale University
Department of Dermatology
New Haven CT 06510, USA

Abstract

The ability of lymphocytes to respond to stimulation with a mitogen has been determined as a function of 8-methoxypsoralen concentrations (1-1000 ng/ml) and ultraviolet A doses (1–5 J/cm^2). 8-MOP concentrations as low as 5 ng/ml in combination with 3 J/cm^2 UVA completely abrogate the response. The extent of 8-MOP photoadduct formation was determined by liquid scintillometric analysis of DNA isolated from lymphocytes after treatment with 8-MOP and UVA. Greater doses of drug and light led to proportionately greater number of adducts. However, the effect on cell function as gauged by tritiated thymidine incorporation is saturated after the formation of a small number of 8-MOP photoadducts (~1 per million bases). Once this number of adducts have formed the cells lose their ability to remove the adducts.

Keywords: Lymphocyte, photopheresis, PUVA, psoralen, DNA repair, photoadduct

Abbreviations: 8-MOP, 8-methoxypsoralen; UVA, ultraviolet A (320–400 nm); PBS, phosphate-buffered saline

Introduction

8-Methoxypsoralen (8-MOP) in conjunction with long wavelength ultraviolet radiation (UVA, 320–400 nm) is used in the photochemotherapy of psoriasis (Parrish, et al., 1974) and cutaneous T cell lymphoma (Gilchrest, 1979). Administration of psoralen plus UVA (PUVA) results in the clearing of skin lesions in patients with psoriasis. The therapeutic efficacy has been attributed to the formation of psoralen photoadducts (monoadducts and crosslinks), which inhibit DNA synthesis and hence control the hyperproliferation of keratinocytes in psoriatic skin (Kraemer, et al., 1979). In our laboratory we have developed an extracorporeal modality that employs 8-MOP and UVA to treat the epidermotropic neoplasm, cutaneous T cell lymphoma (CTCL) (Edelson, et al., 1987). In extracorporeal photopheresis, a patient's blood is fractionated into three components by centrifugation; erythrocytes, plasma and leukocytes. After ~15% of the patient's leukocytes have been collected, they are combined with the 8-MOP containing plasma and passed through a UVA field in order to activate 8-MOP.

In this respect we describe methods for the quantification of 8-MOP photoadduct

Photobiology, Edited by E. Riklis
Plenum Press, New York, 1991

formation in lymphocytes. In these *in vitro* studies, liquid scintillometry has been used to detect incorporation of [^{3}H] 8-MOP in the DNA of lymphocytes. We have also studied the ability of lymphocytes to remove 8-MOP photoadducts. In these studies we have found that relatively low doses of 8-MOP and UVA are capable of inhibiting the ability of the cells to process the DNA damage.

For the analysis of *in vivo* samples obtained from photopheresis patients, we are developing ELISA methods using highly specific monoclonal antibodies that recognize 8-MOP photoadducts (Santella, et al., 1988).

Materials and Methods

Chemicals

Stock solutions were prepared by dissolving crystalline 8-MOP (Sigma, St. Louis, MO) in absolute ethanol (1 mg/ml). Intermediate dilutions with PBS were made prior to addition to the cells. All solutions contained trace amounts of 8-[methoxyl-^{3}H]methoxypsoralen ([^{3}H]8-MOP, Amersham Intl, Amersham, UK) with a specific activity of 3.07 TBq/mmol (83 Ci/mmol; 1 mCi/ml).

Methyl[^{3}H]thymidine ([^{3}H]TdR) (Amersham Intl.). Specific activity is 1.59 TBq/mmol (43 Ci/mmol; 1 mCi/ml).

4',6,4-[^{3}H]Trimethylangelicin ([^{3}H]TMA) — a gift from Professor F. Dall'Acqua, University of Padova, Italy — was dissolved in absolute ethanol (9.34 ug/ml).

Lymphocyte isolation

Whole blood was collected from normal volunteers using vacutainer tubes containing EDTA. 10 ml aliquots of blood were pipeted into prepared Leucoprep cell tubes (Becton Dickinson, Lincoln Park, NJ) for lymphocyte isolation. Prior to filling with blood, the tubes were spun for 10 min at 3000 RPM. After filling, the tubes were centrifuged at 3000 RPM for 15 mins. The buffy coat was obtained from each tube and combined and then washed gently with Dulbecco's Phosphate Buffered Saline (PBS, Gibco, Grand Island, NY) and centrifuged for 10 mins at 1500 RPM. The washing procedure was repeated three times.

Determination of cell counts and viability by trypan blue exclusion

Lymphocytes were resuspended in 2 ml of PBS. 100 uL of the cells were mixed thoroughly with 100–900 uL of 0.4% trypan blue (Sigma, St. Louis, MO) for a 1:2 or 1:10 dilution. Viable cells exclude the dye and nonviable cells take up the dye, thereby producing a visual distinction between "living" and "nonliving" cells. 40 uL of this mixture was then allowed to flow under a coverslip into the loading groove filling the grid area of the hemocytometer chamber (Warner-Lambert, Buffalo, NY).

Treatment of Lymphocytes with 8-methoxypsoralen

Lymphocytes (3×10^6 cells/ml) were treated with 8-MOP over a concentration range of 10–100 ng/ml. Trace amounts of [^{3}H]8-MOP were included in order to measure 8-MOP incorporation by liquid scintillometry. Cells suspended in PBS were incubated in the presence of 8-MOP for 30 mins in darkness at 25°C prior to UV-A irradiation. For experiments at 4°C, the cells suspended in PBS were kept at 4°C for 30 min prior to UV-A irradiation.

Ultraviolet A irradiation

After incubation with 8-MOP, the cell suspension was transferred to plastic petri dishes (15–25 nm in diameter — Corning Glass works, Corning, NY). Petri dish size was based on the volume of solution to be irradiated and was chosen so that the depth of the solution was always 0.1 cm. 80% of the cell suspension received UVA exposure and the remaining 20% were wrapped in foil and served as dark controls. Both petri dishes were then placed in a light box (Derma-Control. Dolton, IL) and irradiated with broad band UVA light (320–400 nm with peak emission at 355 nm) from black light bulbs (Derma-Control, Dolton I1). The radiation was filtered through 4 mm thick window glass to eliminate UVB radiation. UVA doses (1–5 J/cm^2) were determined by radiometry (International Light Model 700 radiometer, Newberry Port, MA) equipped with a SEE015 Probe (International Light). For experiments at 4°C, the cells were incubated at 4°C with 8-MOP and then the petri dishes were placed on ice during the irradiation. After irradiation, the cells were transferred into 15 ml conical tubes and washed with PBS and centrifuged (1500 rpm for 10 mins) between each wash. After the third wash the cells were resuspended in a culture medium consisting of RPMI medium with 10–15% FCS and 2% PCN (Gibco, Grand Island, NY) at a final concentration of 3×10^6 cells/ml and incubated at 37°C and 5% CO_2.

Treatment of cells with phytohemaglutinin

Lymhocytes suspended in culture media were stimulated with 2% PHA-M (Gibco) and incubated at 37°C in a 5% CO_2 atmosphere prior to or post treatment with 8-MOP and UVA.

Treatment of cells with [^{3}H] thymidine

To measure the extent of PHA stimulation, lymphocytes suspended in culture media were aliquoted into wells of a microtiter plate (96-Well Cluster Dish, Gibco) at a concentration of 2×10^5 cells per well. To 200 ul volume per well, 20 ul of [^{3}H]TdR was added and allowed to incubate for 4–6 hrs. in 5% CO_2 at 37°C. The assays were performed in quintuplet and harvested on to glass fibre filters using a cell harvester (Cambridge Technologies, Watertown, MA). The filters were dried and transferred to scintillation vials containing 4.5 ml of scintillation fluid. The radioactivity was measured as described above. The mean of the quintuplet cultures was used to calculate stimulation indices which were expressed as a ratio of incorporated counts for PHA stimulated cells as compared to non stimulated cells.

Isolation of DNA from lymphocytes

Lymphocytes treated with 8-MOP and UVA were washed with PBS, resuspended in culture media, and then sonicated (Heat Systems-Ultrasonics, Plainview, NY). at a low output frequency for 10 mins. 200 uL of Protease K (2 mg/ml) was added to the lysed cells which were then placed in a waterbath at 37°C for 2 hrs. After incubation, the lysed cell suspension was transferred to a 15 ml glass tube and an equal volume of phenol saturated with TNE (0.01 M Tris. 0.1 M NaC1, 0.001 M EDTA, pH 7.6) was added. After vortexing, the suspension was centrifuged at 2000 rpm for 15 mins. To the aqueous portion 5 M NaC1 was added to a final concentration of 0.5 M. Three volumes of cold 100% ethanol were added and the tube was inverted several times, and then placed in a freezer (-70°C) for 1 h. The frozen suspension was allowed to become viscous prior to centrifugation at 2000 rpm for 20 mins at a cool temperature. After centrifugation, the ethanol was carefully decanted and the tube was allowed to dry. The DNA was dissolved in 2 ml of TNE and 200 ul of ribonuclease A (Sigma, St. Louis MO) was added at a concentration of 2 mg/ml The suspension was placed in a waterbath at 37°C for 45 mins. The DNA was then treated with 200 ul of Protease K (Sigma, St. Louis MO) at 2 mg/ml before extraction with equal volumes of chloroform-isoamyl alcohol (24:1). Next, the sample was vortexed thoroughly and centrifuged at 2000 rpm for 5 mins. The aqueous layer was transferred to a glass test tube. Again, 5 M NaC1 was added to a final concentration at 0.5 M along with 3 volumes of cold 100% ethanol. To precipitate the DNA completely, the sample was incubated at −70°C for 1 hr. The cold solution was centrifuged at 2000 rpm for 30 min; the ethanol was decanted. After drying, the DNA was dissolved in 1 ml for deionized distilled H_2O.

Determination of DNA concentration and 8-MOP incorporation

The UV spectrum of the DNA solution was recorded (Ultraspec II, LKB, Piscataway, NJ). Using the absorbances at 260 nm and 280 nm, the concentration of the DNA (mg/ml) and any residual protein (mg/ml) was calculated using the following formulas:

$$[DNA] = 0.0647 \, A_{260} - 0.0389 \, A_{280} \quad [Protein] = 1.55 \, A_{280} - 0.77 \, A_{260}$$

To assess the extent of 8-MOP incorporation, duplicate samples of the DNA were assayed by liquid scintillometry (LKB 1219 Rackbeta, Gaithersburg, MD).

Results and Discussion

Figure 1 illustrates the dose range of 8-MOP required to alter the ability of normal lymphocytes to incorporate $[^3H]$-TdR. Our studies performed using a UVA dose of

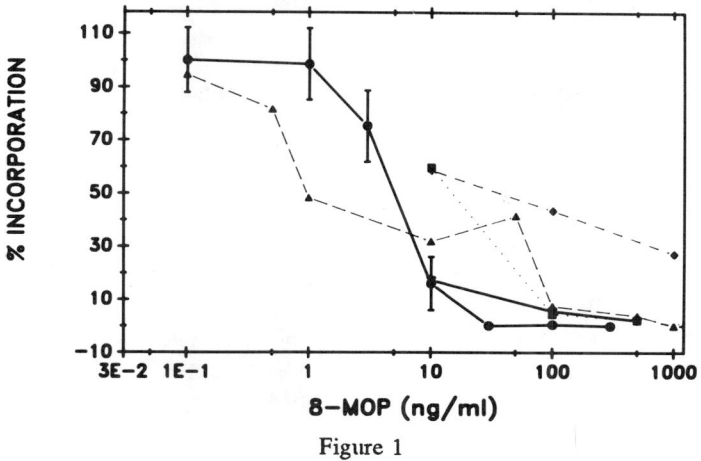

Figure 1

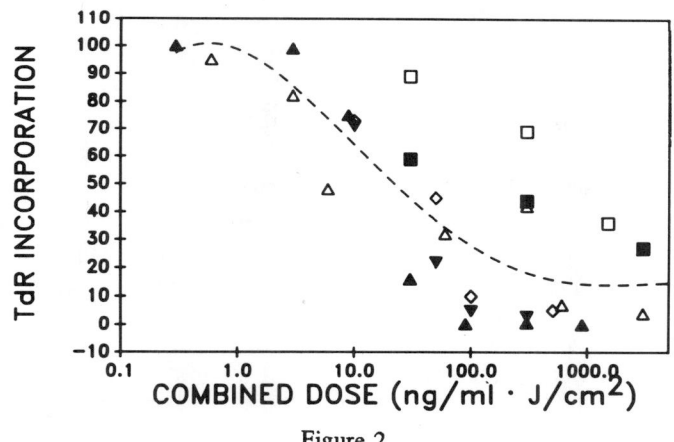

Figure 2

3 J/cm^2 are represented by the solid line. Data from other laboratories using different UVA doses are also shown for comparison. 50% suppression of [^{3}H]-TdR incorporation occurred at an 8-MOP concentration of 5 ng/ml. In Figure 2 the data is re-plotted versus the product of the UVA dose and the concentration of 8-MOP. By using this combined dose it is possible to compare results from different laboratories. The dotted line is the average of all of the data points. From this plot, the 50% inhibition occurred at a combined dose of 22. With 3 J/cm^2 of UVA this would correspond to an 8-MOP dose of 7.3 ng/ml which is in excellent agreement with the data presented in Figure 1. In fact, if the clearly deviant data is eliminated from Figure 2 the corresponding 8-MOP

concentration would be 5.2 ng/ml (50% inhibition point in Figure 3). Figure 3 also demonstrates the dose reciprocity relationship that exists between the amounts of 8-MOP and UVA used in these studies. The upright triangles represent experiments performed with a fixed dose of UVA and various amounts of 8-MOP. The inverted triangles represent data from experiments in which the 8-MOP concentration was constant and the amount of UVA was varied.

The experiments described in the preceding Figures were designed to assess the affects of 8-MOP and UVA on the functional activity of lymphocytes. In addition we wished to correlate this activity with the actual number of photoadducts formed in the DNA of the treated lymphocytes. Figure 4 illustrates the results of experiments designed to measure the extent of 8-MOP modification of (and removal from) cellular DNA isolated from lymphocytes after treatment with 8-MOP and UVA. 100 ng/ml of 8-MOP was used to treat the cells in each instance, the UVA dose and temperature were varied. These conditions were chosen initially as approximations of physiologic doses of 8-MOP and UVA (Gasparro et al., 1988).

To minimize any tendency for the lymphocytes to repair newly formed 8-MOP photoadducts, the cells were kept at 4°C when treated with 8-MOP and UVA. This temperature was maintained in all subsequent manipulations until the cells were lysed or incubated at 37°C 5% CO_2. For cells treated with 5 J/cm² at 4°C (triangles), DNA isolated on day 0, as soon after treatment with 8-MOP and UVA as was practicable, contained 11.7 photoadducts per million bases. On days 1,2, and 4 after treatment, the DNA was isolated. The extent of 8-MOP formation was approximately the same as that calculated for day 0; therefore, the photoadducts persisted over a 96 h time frame.

To study the effect of lower doses of UVA on photoadduct formation and removal, the 8-MOP treated cells were irradiated at 1 J/cm² (also at 4°C, filled circles). On day 0, the DNA contained 2.86 photoadducts per million bases. At the lower UVA dose, the

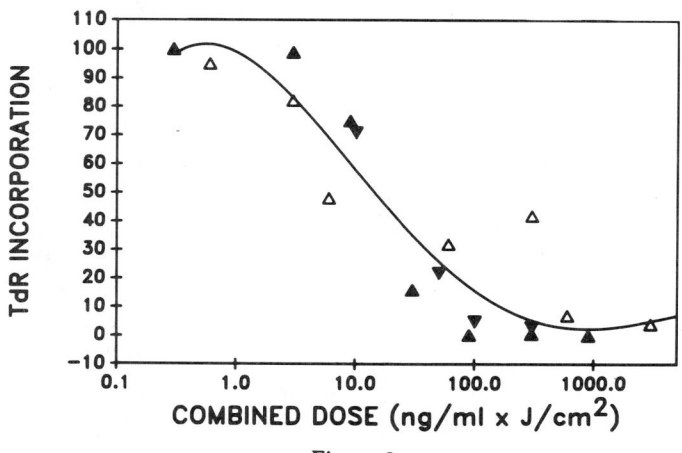

Figure 3

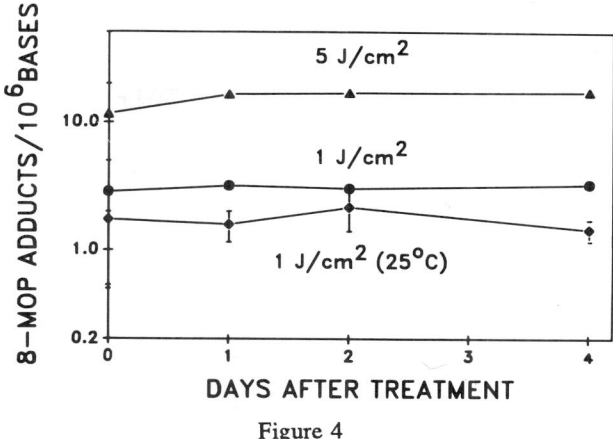

Figure 4

initial number of adducts was 4 times less than that found at 5 J/cm^2. DNA was isolated from cells that had been cultured in RPMI with fetal calf serum for 1,2 and 4 days after treatment. The number of photoadducts contained in these samples was essentially the same as that of day 0. Thus, at 1 J/cm^2 and 100 ng/ml 8-MOP the photoadducts also persisted over a 96 h period.

For comparison the same experiment was performed at 25°C (diamonds in Figure 4). DNA isolated on day 0 contained 1.63 photoadducts per million bases; the number of adducts was 1.83 times less than the adducts formed at 1 J/cm^2 and 4°C. On days 1,2, and 4, the DNA from treated cells was isolated and overall photoadduct levels on subsequent days was similar to day 0. The data shown in Figure 3 indicate that lymphocytes treated with 100 ng/ml 8-MOP and 1 J/cm^2 UVA were only capable of 20% [^{3}H]TdR incorporation. Thus, at these doses of 8-MOP and UVA the cells may be damaged to such a degree that they are not capable of repairing DNA damage.

The viability of the lymphocytes was assessed during trypan blue exclusion in all experiments on the days which DNA was isolated. Correcting for the total number of cells initially present on day 0, 51% to 72% of the lymphocytes were viable on day 4 and 74% to 92% of the original cells were present. However, it should be noted that viability as gauged by trypan blue exclusion is not an accurate measure of the functional status of a cell.

Figure 5 shows the effect of lower 8-MOP concentrations on adduct formation in the DNA of peripheral lymphocytes. 10 ng/ml of 8-MOP and 1 J/cm^2 of UVA at 4°C was used to treat the cells. Under these conditions the ability of the cells to incorporate [^{3}H]TdR after PHA stimulation was 50% of control, therefore we expected to observe repair of adducts. The data for 100 ng/ml is the same as that shown in Figure 4 and is included for comparison (upper curve). DNA isolated on days 0,1,2, and 4 after treatment

957

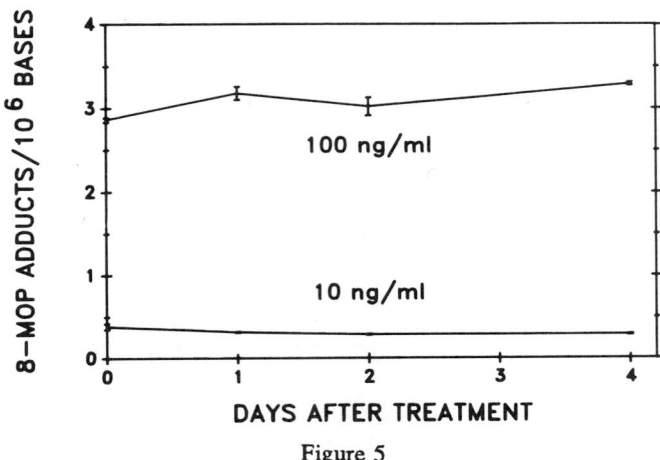

Figure 5

with 10 ng/ml is shown in the lower curve. On day 0, at the lower concentration, 0.39 photoadducts per million bases formed which was 7.4 times less than that obtained with 100 ng/ml of 8-MOP.

For cells treated with 10 ng/ml and 1 J/cm^2, photoadduct removal was measured over a 96 h period. DNA isolated from treated cells on day 1 had 18% fewer photoadducts than on day 0. On day 2, the number of adducts had decreased by 32%; by day 4, there was no evidence of further photoadduct removal. These results are shown in Figure 6 using an expanded scale. The data for two trials correlated well. On day 0, the number of adducts for trials 1 and 2 are 0.41 per million bases and 0.36 per million bases respectively. On day 1, the number of adducts had decreased by 22% in trial 1, and 14% in trials 2; and by day 2, the number of adducts had decreased by 32% in trial 1, and 20% in trial 2 without evidence of further removal by day 4. Thus, the two sets of data reveal that at these lower doses of 8-MOP and UVA lymphocytes are capable of photoadduct removal. In addition, viability assessed via trypan blue exclusion was 96% and 97% on day 4 for trials 1 and 2 respectively.

To examine the effect of a higher dose of radiation on photoadduct formation, cells were treated with 10 ng/ml of 8-MOP and 5 J/cm^2 at 4°C. On day 0, DNA was isolated and the number of photoadducts was 1.63 per million bases — 4.17 times higher than that found at 1 J/cm^2. Similarly, DNA was isolated from cells on day 1,2, and 4, and the number of adducts calculated for those days were also 4 to 5 times that found at 1 J/cm^2.

In Figure 7 the initial number of adducts formed under the preceding conditions is plotted against the combined dose of 8-MOP and UVA. Increasing doses led to an increased number of 8-MOP photoadducts with a correlation coefficient of 1.00 over a fivehundredfold dose range. It is interesting to note that at a combined dose of 100 (ng/ml × J/cm^2) 3.5 adducts/million bases are detected. It can be seen in Figure 3 that at

958

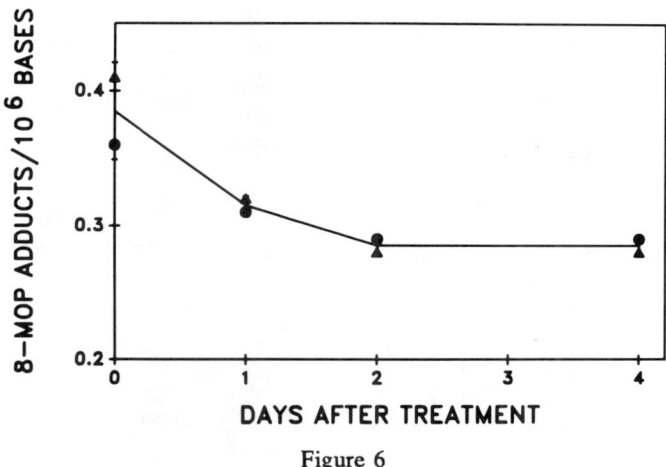

Figure 6

this dose, the cells show only a 10% response to PHA. Cells treated with greater amounts of 8-MOP are less able to incorporate [³H]TdR. These results indicate that 1 J/cm^2 and 10 ng/ml represent sub-lethal doses while 1 J/cm^2 and 100 ng/ml could be lethal. Greater doses of 8-MOP and UVA can result in a greater number of adducts as seen in Figure 7. However, once the combined dose of 8-MOP and UVA exceed the sublethal level of 10, no adduct removal is observed.

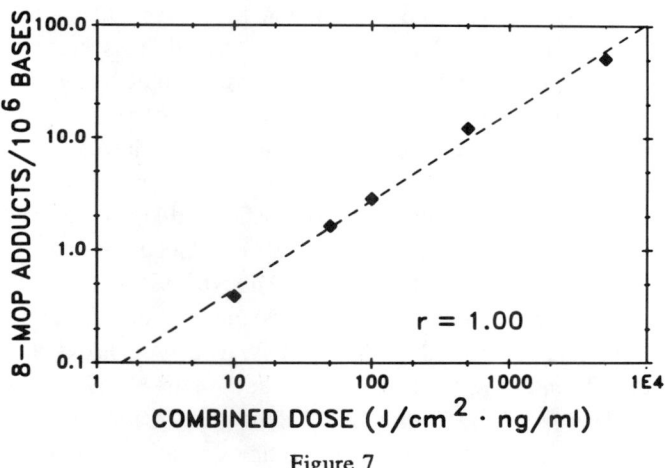

Figure 7

959

To test the hypothesis that stimulated lymphocytes could form more 8-MOP photoadducts and, perhaps be better able to repair these adducts, cells were exposed to PHA prior to 8-MOP and UVA treatment (10 ng/ml, 1 J/cm^2). Figure 8 illustrates the results for these experiments. The data from the previous experiments are provided for comparison (dotted line). The effect of pre-stimulation for 24 hrs with PHA prior to

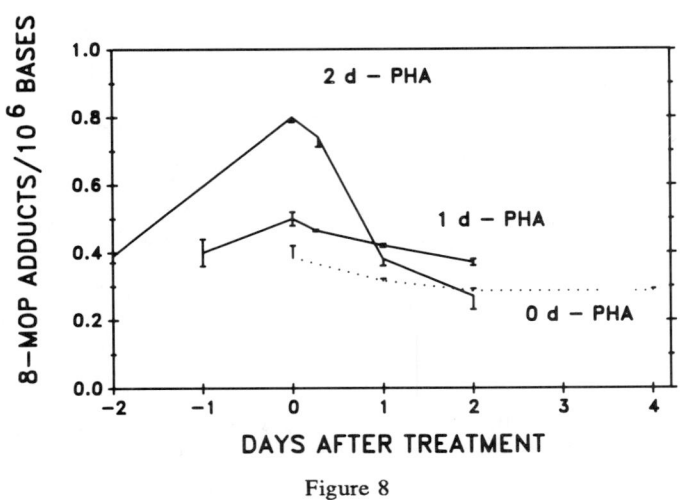

Figure 8

treatment with 8-MOP and UVA is represented by the curve labelled '1 d-PHA'. DNA isolated from cells treated with 8-MOP and UVA prior to stimulation with PHA served as a reference for each set of experiments. The number of adducts contained in this sample was 0.40 per million bases. The remaining cells were incubated with PHA for 24 hrs, then washed extensively and resuspended in PBS immediately prior to treatment with 8-MOP and UVA. DNA isolated on day 0 contained 0.5 photoadducts per million bases, a 25% increase in the number of adducts compared to those seen prior to PHA pre-stimulation. At 6.5 hrs, 1 day, and 2 days after treatment, the DNA was isolated and the number of 8-MOP photoadducts present was determined (0.47, 0.42, 0.37 per million bases respectively). Adduct removal can be detected as early as 6.5 hrs (7%). On day 1, adduct removal was 16%; and by day 2, adduct removal was 26%. The effect of pre-stimulation for 48 hrs prior to treatment with 8-MOP and UVA was more pronounced (see curve labelled '2 d-PHA'). The initial number of adducts formed was in agreement with the two previous experiments as well as the above (0.39 photoadducts per million bases). After incubation with PHA for 48 hrs the remaining cells were washed, resuspended in PBS, and treated with 8-MOP and UVA. DNA isolated on day 0 contained 0.80 adducts per million bases, a 105% increase in the number of adducts seen prior to stimulation with PHA. At 7 hrs, 1 day, and 2 days after treatment, DNA was isolated from cells. The adduct levels calculated at these times were 0.74, 0.38, 0.27 per million bases respectively. At 7 hrs, the rate of adduct removal (7.5%) approximates the removal at 6.5 hrs (7%) observed when cells were pre-stimulated for one day. However, by day 1, 54% of the adducts have been removed and by day 2, almost three times the number of adducts (67%) have been removed as compared to those seen on day 2 after 24 hrs of pre-stimulation with PHA. Thus, 48 hrs of pre-stimulation with PHA results in a) a higher level of adduct formation and b) a greater extent of adduct removal. It appears that cells which are actively processing genomic information have DNA that is more accessible to photomodification by 8-MOP. In addition, as a result of PHA stimulation, enzymes involved in the repair of DNA damage (e.g., endonucleases and polymerases) might also be present at higher levels (Ashman, 1984) and thus adduct removal would occur more efficiently.

Future Studies

One of the questions that remains to be answered involves the proportion of monoadducts and crosslinks that form during photopheresis. In Table 1 the properties of MAbs that recognize psoralen-DNA photoadducts are listed. In earlier studies, MAb 8G1 was used to measure 8-MOP photoadducts formed in the DNA of lymphocytes of patients treated by photopheresis. Adduct levels fell in the range 0.1–15 per million bases (Yang, et al., 1988). Studies using [^{3}H]8-MOP indicated that within the detection limits of the experiment MAb 8G1 recognized virtually all of the adducts present (Gasparro, et al., 1985). MAb 10B12 will be used to determine the extent of 8-MOP crosslink formation in DNA samples isolated from photopheresed lymphocytes. The combined analysis of DNA samples with these two antibodies will be used to characterize the repair of monoadducts and crosslinks.

Table 1. Monoclonal antibodies

MAb	Specificity	Application	Reference
8G1	8-MOP monoadducts (4′, 5′)	Photopheresis immunofluorescence of skin after PUVA	Edelson et al., 1987 Santella et al., 1988
10B12	8-MOP crosslinks	Characterization	Santella & Gasparro, preliminary results
7E3	TMA monoadducts (4′, 5′)	Characterization	Miolo et al., 1988

References

Ashman, E.F. (1986). Lymphocyte Activation in "Fundamental Immunology" (ed., W.E. Paul) Raven Press, NY.

Edelson, R.L., Berger, C., Gasparro, F.P. et al., (1987). Extracorporeal photochemotherapy for cutaneous T-cell lymphoma, *New Engl. J. Med.* 316, 297–303.

Gasparro, F.P., Chan, G., and Edelson, R.L. (1985). Phototherapy and photopharmacology. *Yale. J. Biol. Med.* 58, 519–534.

Gasparro, F.P., Battista, J., Song, J., Edelson, R.L. (1988). Rapid and sensitive analysis of 8-methoxypsoralen in plasma. *J. Invest. Dermatol.* 90, 234–236.

Gilchrest, B.A. (1979). Methoxsalen photochemotherapy for mycosis fungoides. Cancer Treat Rep. 63, 663–667.

Kraemer, K.H., Waters, H.L., Ellingson, O.L., Tarone, R.E. (1979). Psoralen plus ultraviolet radiation induces inhibition of DNA synthesis and viability in human lymphoid cells *in vitro. Photochem,. Photobiol.* 30, 263–270.

Miolo, G., Stefanidis, M., Santella, R.M., Dall'Acqua, F., and Gasparro, F.P. (1988). 4,5,4'-Trimethylangelicin photoadduct formation in DNA: production and characterization of a specific monoclonal antibody (in press), *J. Photochem. Photobiol).*

Parrish, J.A., Fitzpatrick, T.B., Tannenbaum, L. et al., (1974). Photochemotherapy of psoriasis with oral methoxsalen and long wavelength ultraviolet light. *New Engl. J. Med.* 29, 1207–1211.

Santella, R.M., Dharmaraja, N., Gasparro, F.P., and Edelson, R.L. (1985). Monoclonal antibodies that recognize 8-MOP-modified DNA, *Nucleic Acids Res.* 14, 2533–2544.

Yang, X.Y., Gasparro, F.P., DeLeo, V.A., and Santella, R.M. (1988). 8-Methoxypsoralen-DNA photoadducts in patients treated with 8-methoxypsoralen and ultraviolet A light. (in press, *J. Invest. Dermatol).*

New Aspects of the Mechanism of Action of Furocoumarins (Psoralens and Angelicins)

GIOVANNI RODIGHIERO

Dept of Pharmaceutical Sciences
Padua University
Via F. Marrzolo, Italy

The mechanism through which furocoumarins produce their photobiological effects has been widely investigated. Various different modes of action have been evidenced. What can occur when a furocoumarin is inserted in a biological medium and then is irradiated with long wavelength ultraviolet light is shown in Scheme 1.

Scheme 1

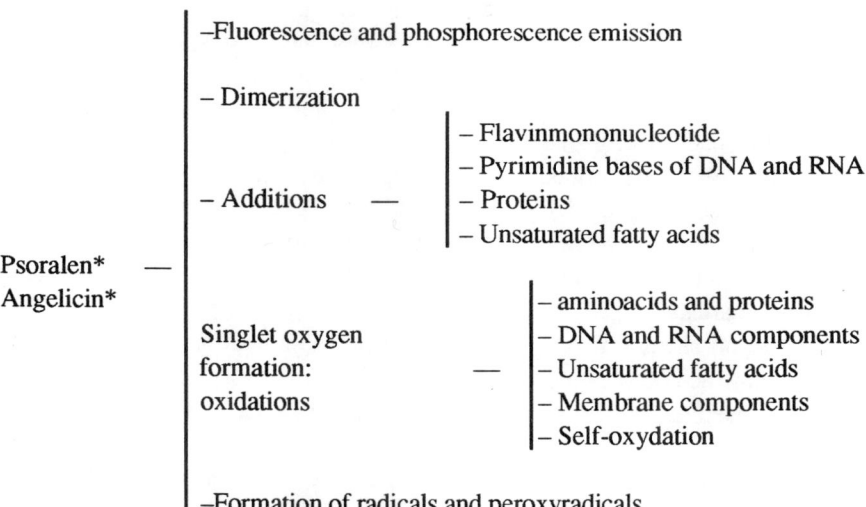

The picture appears to be rather complicated and probably it will become still more complicated in the future. However, many of these various aspects of the photochemical behaviour of furocoumarins have already been widely studied and have been summarized in several good reviews. (Musajo and Rodighiero, 1972; Pathak et al, 1974; Rodighiero et al, 1984; Cimino et al, 1985; Rodighiero et al, 1988).

In this paper only the most recently studied aspects will be considered, that is the

Photobiology, Edited by E. Riklis
Plenum Press, New York, 1991

possibility of causing damage to the cell membrane by irradiation with UV-A in the presence of furocoumarins.

Various observations have been made in the past on this possibility by various authors (Muler-Runkel and Grossweiner, 1981; Salet et al, 1982). Recently, however, two lines of research have been developed, which have obtained new, very interesting results. They are, namely, the studies on (a) the photohemolysis induced by furocoumarins, and (b) the photoreaction between furocoumarins and unsaturated fatty acids. In this paper what, at present, is known in these two fields will be summarized.

Photohemolysis induced by furocoumarins

Photohemolysis is a photosensitized effect which can be induced by many well-known photodynamic compounds, such as hematoporphyrin, chlorophyl, erythrosin, methylene blue, and so on. Therefore, since the time of the first studies on the photodynamic properties of furocoumarins, various attempt have been made to ascertain if they were able to induce this photosensitized effect, naturally making experiments in comparison with some well-known photodynamic compounds.

In a very old study, done in 1958 (Rodighiero and Caporale, 1958), furocoumarins, as well as other photodynamic compounds, were tested at a 10^{-5} Molar concentration and hemolysis was checked in the usual way soon after irradiation. The results showed that furocoumarins are practically lacking photohemolytic properties, with a dramatic difference in respect to the other photodynamic compounds. The difference in activity was made even more evident by decreasing the irradiation time to a tenth of the previous value. Some photodynamic compounds remained still very active.

Experiments were also performed by increasing the concentration of furocoumarins, using a saline saturated solution of them, that is the highest possible concentration. The results also showed no photohemolytic effect in this case; by contrast, there was a protection of red blood cells, probably due to a filter effect against UV-A radiation, exerted by furocoumarins.

Therefore, as photohemolysis was not directly obtainable with furocoumarins in the same conditions in which other photodynamic compounds were very active, the common opinion was diffused that furocoumarins are practically lacking photohemolytic properties.

In 1979, after finding that furocoumarins under UV-A radiation are able to produce singlet oxygen, Wennersten (1979) tried to observe photohemolysis using 8-MOP and perform irradiation also in D_2O solution, however, with negative results.

More recently, however Potapenko and his co-workers (Potapenko and Sukhorukov, 1984; Potapenko et al, 1986; Potapenko et al, 1986; Lisenko et al, 1988) found that a photohemolytic effect of 8-MOP can be evidenced by incubating at 37°C for 1 hour the red blood cells previously irradiated with UV-A in the presence of 8-MOP.

Studies of the same authors on the properties of the irradiated red blood cells not yet hemolyzed showed that they had a decreased osmotic resistance (Potapenko et al, 1986).

964

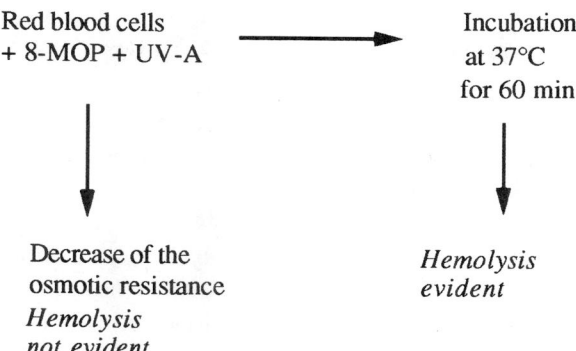

Therefore, they concluded that irradiation with UV-A in the presence of 8-MOP produce a damage of red blood cells membrane, which leads to a pre-hemolytic stage in which a decreased osmotic resistance is observable. Incubation of the so damaged red blood cells at 37°C leads to a real hemolytic condition, in which hemolysis is clearly observable.

About the mechanism through which the initial damage of the red blood cell membrane is produced Potapenko et (1986; 1988), after many studies, reached the conclusion that singlet oxygen is surely involved, however, mostly in an indirect way.

In fact, they found that if 8-MOP is previously irradiated in alcoholic solution in the presence of oxygen and then the solution is added to a red blood cell suspension, the same effects are obtained as by irradiation of red blood cells in the presence of 8-MOP (Potapenko et al, 1988).

They obtained evidence that 8-MOP, irradiated in alcoholic solution in the presence of oxygen, forms oxidation products having a not yet defined structure and indicated simply as O_2-8-MOP and O_2-(8-MOP)$_2$. These compounds are practically stable in alcoholic solution, as well as in other less polar solvents like acetone, carbon tetrachloride, benzene, but are not stable in aqueous solution. In fact, by addition of the alcoholic solution to water, they undergo decomposition with chemiluminescence emission, which intensity decays with a first order kynetic. This photo-oxidized intermediate compound could be the real agent which induces the damage to the cell membrane.

As a consequence, the mechanism could be as follows:

8-MOP $\longrightarrow$ 8-MOP* singlet

8-MOP* singlet $\longrightarrow$ 8-MOP* triplet

8-MOP* triplet + $^3O_2$ $\longrightarrow$ 8-MOP + $^1O_2$

8-MOP + $^1O_2$ $\longrightarrow$ O_2-8-MOP

O_2-8-MOP + red blood cells $\longrightarrow$ damage to membrane

Very recently, Potapenko's method for proving hemolysis has been applied by Dall'Acqua, Vedaldi and co-workers for testing other furocoumarins, both psoralens and angelicins (Dall'Acqua et al, 1987; Vedaldi et al, 1988). Moreover, as in the past Muller-Runkel and Grossweiner (1981) had found that 8-MOP and 3-carbethoxy-psoralen at high concentrations are able to induce lysis of red blood cells in the dark, experiments of dark hemolysis have also been performed (see Fig. 1).

Working out the experiments with 3-carbethoxy-psoralen and psoralen, it was found that at a concentration 10^{-5} Molar dark hemolysis was practically absend and photohemolysis was present only in 3-carbethoxy-psoralen, however at a very low level. By increasing concentration, dark hemolysis increases, but it becomes evident only at relatively high concentrations. Photohemolysis, by contrast, increases very rapidly, reaching very high levels, thus indicating a very strong photosensitizing activity.

By contrast, working with other psoralens commonly used in the photochemotherapy, that is 8-MOP and TMP, as well as with trimethyl-angelicin, both dark and photohemolysis were less pronounced, although clearly present (Fig. 1).

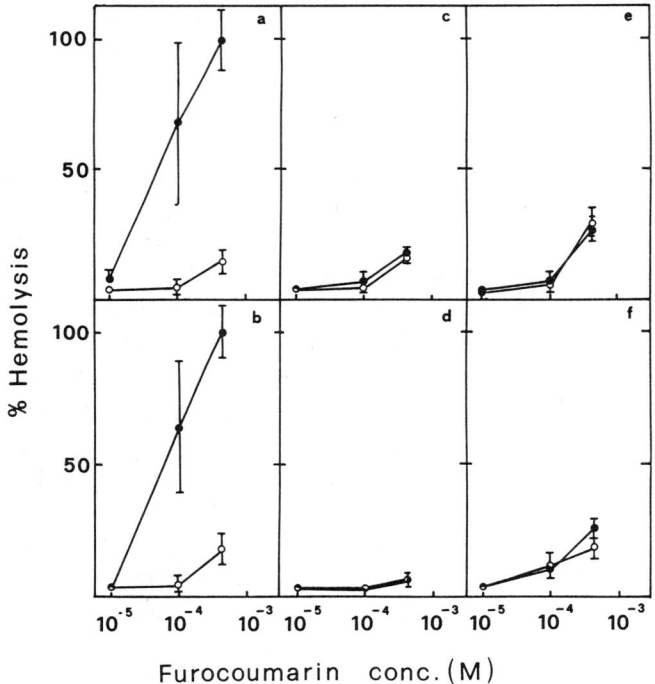

Figure 1. Photo- and dark hemolysis induced by various furocoumarins as a function of their concentrations. Irradiation at 365 nm; 59 $J \cdot sec^{-1} \cdot m^{-2}$. ✱————✱ photohemolysis: irradiation followed by incubation at 37°C for 1 hour. O————O dark hemolysis: incubation, without irradiation, at 37°C for 1 hour. a: 3-carbethoxy-psoralen; b: psoralen; c: 5-methoxypsoralen; d: 8-methoxy-psoralen; e: 4,5', 8-trimethyl-psoralen; f: 4,6,4'–trimethyl-angelicin

Extending further these experiments to some other furocoumarin derivatives, it was found that some of them have a significant dark hemolytic effect; by contrast, the photohemolytic effect was practically absent (Vedaldi et al, in press).

Experiments have also been performed on the post-irradiation dark hemolysis, that is by performing previous irradiation of the various furocoumarins, followed by their addition to a red blood cell suspension and incubation at 37°C.

The pre-irradiation of furocoumarins has been performed both in alcoholic solution, in which, according to Potapenko, the formed photo-oxidized compounds would be stable, and in aqueous saline solution, in which the same photo-oxidized compounds would be not present, because of its instability.

By examining the results reported in Figures 2 and 3 and making a comparison between irradiation in alcohol and irradiation in water, we can see that the photo-oxidized compounds seem to be formed by the main part of furocoumarins, even if to variable degrees. In fact, when the effect of post-irradiation dark hemolysis is higher when irradiation was performed in alcoholic solution than when it was performed in water, very probably the difference is due to the formation of photo-oxidized active products.

Moreover, examining the results obtained after irradiation in water, we can see that, with the exception of psoralen and 3-carbethoxy-psoralen, generally the previously irradiated furocoumarins have a lower hemolytic effect than those not irradiated.

Finally, it is possible to see that 8-MOP, which has been particularly studied by Potapenko et al, in these experiments gave results which, although in agreement with those previously obtained, are quantitatively less pronounced than those obtained with other furocoumarins, in particular with psoralen and 3-carbethoxy-psoralen. A much higher activity of psoralen in respect of 8-MOP had also been proved by Potapenko et al (1988).

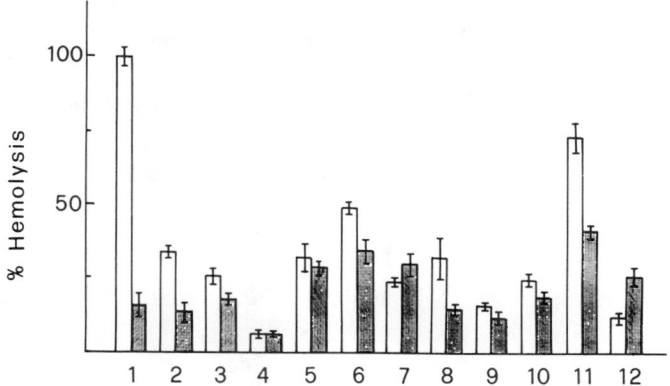

Figure 2. Post irradiation dark hemolysis induced by various furocoumarins after *irradiation in alcoholic solution* at 365 nm; 59 J•sec^{-1}•m^{-2}. ☐ PIDH: addition of the irradiated alcoholic solution of furocoumarins followed by incubation at 37°C for 1 h. ☐ Dark hemolysis: incubation, without irradiation, at 37°C for 1 h. 1: psoralen; 2: 3-carbethoxy psoralen; 3: 5-methoxy-psoralen; 4: 8-methoxy-psoralen; 5:4,5', 8-trimethyl-psoralen; 6: 4,8-dimethyl-psoralen; 7: diacetyl-psoralen; 8: 8-methyl-psoralen; 9: 8-hydroxy-psoralen; 10: 4,6,4'−trimethyl-angelicin; 11: 6,4'-diemthyl-angelicin; 12: 4,5'-dimethyl-angelicin.

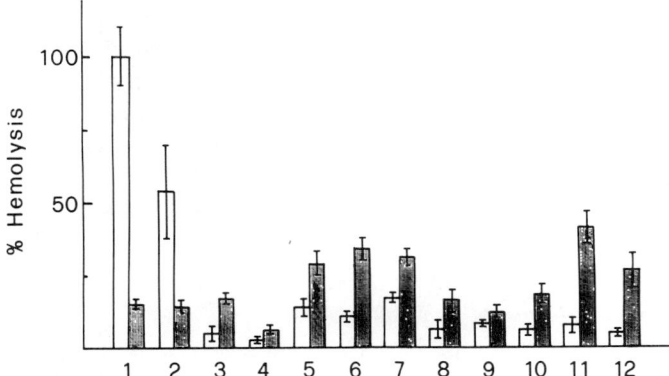

Figure 3. Post irradiation dark hemolysis induced by furocoumarins after *irradiation in aqueous saline solution*. Conditions, symbols and numbers as in Fig. 2.

Some partial conclusions can be drawn from the results now shown:

a) The photohemolytic effect is very evident in some furocoumarins, for instance psoralen and 3-carbethoxy-psoralen; however, it cannot be considered common property of all furocoumarins.

b) Both dark and photohemolysis occur only at relatively high concentrations of furocoumarins. They are much higher than those occurring in blood during PUVA therapy.

c) No correlation can be seen between the ability of the various furocoumarins to induce photohemolysis and their ability to induce skin erythema. For instance, 3-carbethoxy-psoralen which is highly photohemolytic is unable to induce skin erythema; by contrast, TMP which is strongly erythemogenic in human skin has only poor photohemolytic properties.

Of course, much more research work must be done for clarifying many aspects of this field. The recent findings of Potapenko are important because they again opened this field of research which for a very long time was considered closed.

Photoreaction between furocoumarins and unsaturated fatty acids

Two types of photodamage has been observed in unsaturated fatty acids after their irradiation with UV-A in the presence of furocoumarins: 1) formation of oxidation products and 2) formation of covalent photoadducts between fatty acids and furocoumarins.

1) Concerning the formation of oxidation products, it has been studied in particular by Potapenko et al (1982; 1983; 1988) by irradiation of an alcoholic solution of phospholipids isolated from egg yolk lecithine in the presence of 8-MOP; after irradiation an increase of the amount of products deriving from a peroxidation of unsaturated lipids was found. This fact did not take place by performing irradiation in

the absence of oxygen. By contrast, the increase was observed after addition of 8-MOP previously irradiated in alcoholic solution, therefore containing the photooxidized 8-MOP.

In this case, also, the above-mentioned authors reached the conclusion that the oxidation of the unsaturated lipids do not occur by a direct reaction of singlet oxygen, but through the formation, by the same singlet oxygen, of the intermediate oxidized product of 8-MOP.

2) Concerning the formation of covalent photoadducts between unsaturated fatty acids and furocoumarins, it has been studied until now *in vitro* by irradiation of an alcoholic solution of esters of unsaturated fatty acids, also containing furocoumarins, with long wavelength ultraviolet radiation.

The photoreaction is oxygen independent. It was found that a C_4-cyclo-addition takes place between a photoreactive double bond of furocoumarin and one of the double bonds of the unsaturated fatty acids. Until now only photoadducts involving the 3,4-double bond of furocoumarin have been obtained.

The first evidence of this photoreaction has been obtained by Kittler and Lober (1983; 1984) by irradiating a mixture of oleic, linoleic, linolenic and arachidonic acid methylesters in the presence of TMP.

Analyzing the irradiated mixture by means of reverse phase HPLC, Kittler et al (1986; 1988) obtained an elution pattern showing that after each peak of the various fatty acids a new peak was present, which, by its position, appeared to be more hydrophobic than the original fatty acid. This fact was not observed using, in the same conditions, saturated fatty acids, like palmitic and stearic acids.

Successively, Specht, Kittler et al (1988), studying in more detail the irradiated mixture of oleic acid methylester and TMP, succeeded in isolating four new products by reverse phase HPLC. The two products obtained in greater amount, examined by Mass Spectrometry showed an identical molecular weight, which exactly corresponds to the addition of the molecular weight of oleic acid methylester and that of TMP. Moreover, the UV absorption spectrum and the fluorimetric properties of the isolated compounds showed that in the formation of the photoadducts the 3,4-double bond of TMP should be involved.

The formation in this photoreaction of more than one compound is in agreement with the theoretical possibility of existence of several regio- and stereo-isomers.Further studies have been performed by Dall'Acqua, Caffieri and coworkers (1987; in press and part B) by examining the photoreactions between linolenic acid methylester and four furocoumarins, namely psoralen, 8-methoxy-psoralen, 3-carbethoxy-psoralen and 4,6,4'-trimethyl-angelicin. In all cases, by means of HPLC separation, it was possible to prove the formation of new compounds.

In the case of the photoreaction between psoralen and linolenic acid methylester (Caffieri et al, in press) (using tritium labeled psoralen) two new photoproducts have been isolated and indicated as A and B, both radioactive and having identical UV absorption spectra. (Fig. 4). The strong decrease of the absorption bands of psoralen at

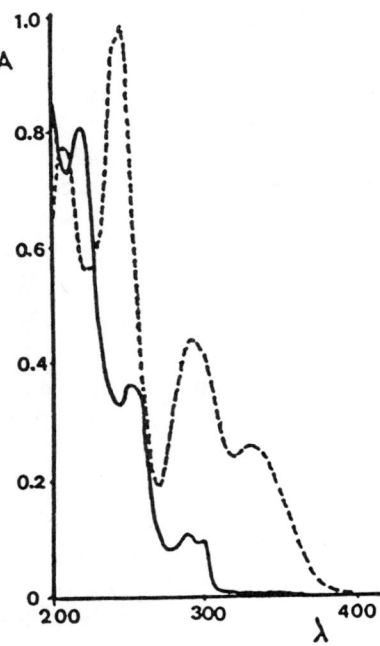

Figure 4. UV absorption spectra of compound B (————) and of psoralen (-------). Ethanolic solutions.

wavelengths longer than 300 nm is indicative of the saturation of the 3,4-double bond of the α-pyronic ring of psoralen. The spectrum of these new compounds is very similar to that of the 3,4-photoaddition product between psoralen and thymine and also to that of the 3,4-photodymer of psoralen (Musajo et al, 1967). In fact, in this spectral region the absorption is due only to the psoralen moiety, while the linolenic acid moiety absorbs only at shorter wavelengths.

Compounds A and B irradiated in alcoholic solution at 254 nm showed an evident photoreversion, that is splitting of the molecule with formation of the two parent compounds (Fig. 5).

The increase of the absorption band at 300 nm is indicative that intact psoralen, having again the 3,4-double bond in the α-pyronic ring, is gradually formed following irradiation. The effective presence of psoralen in the irradiated solution has been checked by HPLC separation.

This photoreversion is typical of the C_4-cyclo-adducts; therefore, it was concluded that really the two isolated compounds were C_4-cyclo-adducts involving the 3,4-positions of psoralen and one of the three double bonds of linolenic acid methylester. By means of NMR studies it was found that the involved double bond is that existing between the 12

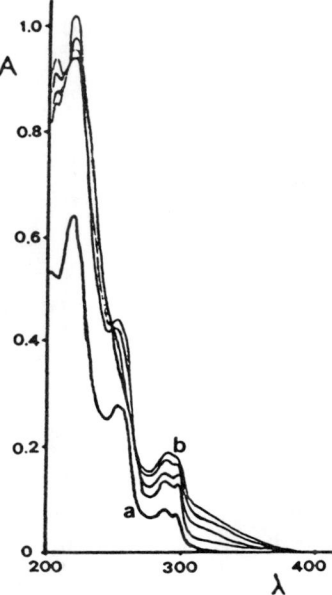

Figure 5. UV absorption spectra of an ethanolic solution of compound B submitted to irradiation at 254 nm. Times were 0 (a), 10, 20, 40 and 60 (b) minutes.

and 13 positions. The chemical structure attributed to both the new isolated photocompounds is shown in Fig. 6.

Analogous photoadducts have been isolated from the photoreaction products between psoralen and oleic or linoleic acid methylesters, as well as from the photoreaction between linolenic acid methylester and angelicin (Caffieri et al, part B). In Fig. 7 the structure of the photoadduct with angelicin is reported; also in this case the 12-13 double bond of linolenic acid is involved.

This field of research is just initiated and of course much research work is still to do. It appears extremely interesting, not only from a chemical point of view, but also for the possible biological consequences, when it would occur at the level of the cell membranes, and for the fascinating possible photochemotherapeutic applications that

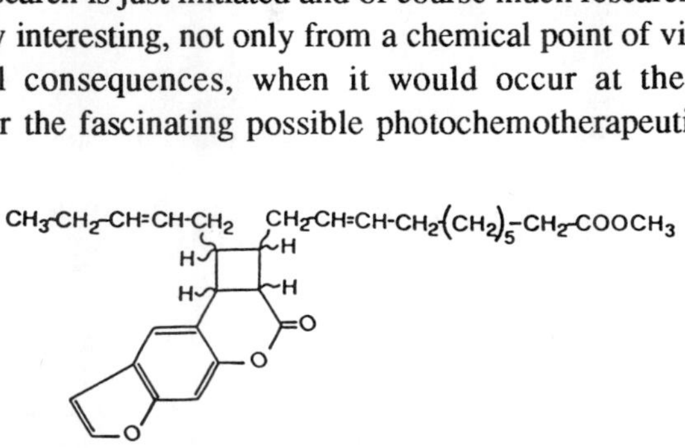

Figure 6. General structure of the C_4-cyclo-adduct between psoralen and linolenic acid methylester.

971

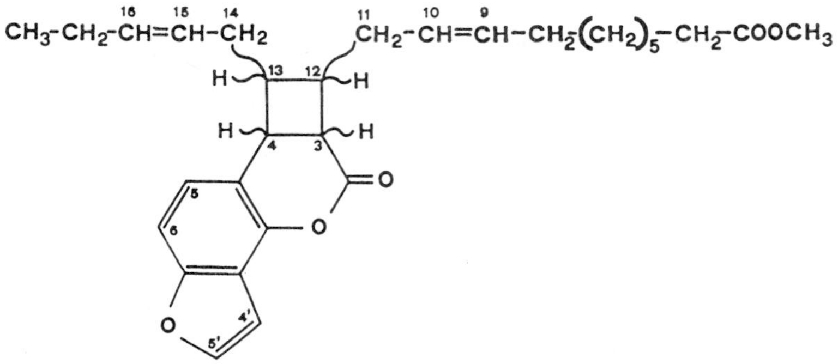

$$CH_3-CH_2-CH=CH-CH_2 \qquad CH_2-CH=CH-CH_2(CH_2)_5-CH_2-COOCH_3$$

Figure 7. General structure of the C_4-cyclo-adducts between angelicin and linolenic acid methylester.

have been suggested (Specht et al, 1988). It was suggested, in fact, that the possibility, by means of suitable modifications of the structures of psoralens and angelicins, to move the principal target of the photoreactivity of furocoumarins to the unsaturated fatty acids of the cell membranes, so producing damage to the same cells, but living intact DNA and its genetic information.

References

Akimov et al (1988), Erythrocyte damage and induction of lipid-peroxidation as a result of UV-irradiation or PUVA treatment of guinea-pig skin. *Studia Biophysica*, 124, 239-249.

Caffieri, S. et al (in press), Photosensitizing furocoumarins: photocycloaddition to unsaturated fatty acids. In: Psoralens: past, present and future, T.B. Fitzpatrick et al, eds., John Libbey Eurotext, Montrouge (France).

Caffieri, S. et al (part B), Photoaddition of angelicin to 184.inolenicacid methylester. Submitted to *J. Photochem. Photobiol.*

Caffieri, S. Tamborrino, G. and Dall'Acqua, F. (1987), Formation ofphotoadducts between unsaturated fatty acids and furocoumarins. *Med. Biol. Environ.*, 15, 11-14.

Cimino, G.P. et al. (1985), Psoralens as photoactive probes of nucleic acid structure and function: organic chemistry, photochemistry and biochemistry, *Ann. Rev. Biochem.*, 54, 1151-1193.

Dall'Acqua, F. et al (1987), Photohemolysis of erythrocytes by furocoumarins. *Med. Biol. Envir.*, 15, 43-46.

Kittler, L. and Lober, G. (1983), Furocoumarins. Biophysical investigations on their modes of action. *Studia Biophysica*, 97, 61-67.

Kittler, L. and Lober, G. (1984), Photoreactions of furocoumarins with membrane constituents. Results with fatty acids and artificial bilayers. *Studia Biophysica*,101, 69-72.

Kittler, L. and Lober, G. (1988), Photoreactions of furocoumarins with nucleic acids, proteins, membranes and their subunits. *Studia Biophysica*, 14, 97-114.

Kittler, L., Midden, W.R., and Wang, S.Y. (1986), Interactions of furocoumarins with subunit cell constituents. Photoreactions of fatty acids and aromatic aminoacids with trimethylpsoralen (TMP) and 8-methoxypsoralen (8-MOP). *Studia Biophysica*, 114, 139-148.

Lisenko, E.P. et al (1988), Photohemolysis sensitized by psoralen,angelicin and 8-methoxy-psoralen. *Studia Biophysica*, 124, 205-23.

Muller-Runkel, R. and Grossweiner, L.I. (1981), Dark membrane lysis and photosensitization by 3-carbethoxy-psoralen. *Photochem. Photobiol.*,33,399-402.

Musajo, L. and Rodighiero, G. (1972), Mode of photosensitizing action of furocoumarins. In: *Photophysiology*, Vol. VII, A.C. Giese ed. Academic Press, New York and London, pp. 115-147.

Musajo, L., Bordin, F. and Bevilacqua, R. (1967), Photoreactions at 3655 A linking the 3,4-double bond of furocoumarins with pyrimidine bases. *Photochem. Photobiol.* 6, 927-931.

Pathak, M.A., Kramer, D.M. and Fitzpatrick, T.B. (1974), Photobiology and photochemistry of furocoumarins (psoralens). In: Sunlight and Man, M.A. Pathak et al, eds., University of Tokyo Press, Tokyo, pp. 335-368.

Potapenko, A.Ya, and Sukhorukov, V.L. (1984), Photooxidative reactions of psoralens, *Studia Biophysica*, 101, 89-98.

Potapenko, A.Ya. et al (1986), Mechanism of furocoumarin-sensitized damage to biological membranes, *Studia Biophysica*, 114, 159-170.

Potapenko, A.Ya. et al (1986), Photosensitized modification of erythrocyte membranes induced by furocoumarins. *Photobiochem. Photobiophys.* 10, 175-180.

Potapenko, A.Ya. et al (1988), Hypothesis of the induction of psoralen phototoxic effects through the stage of photooxidized psoralen formation. Model studies of erythocytes. *Studia Biophysica,* 124, 205-223.

Potapenko, A.Ya. et al (1982), Dark oxidation of unsaturated lipids by the photooxidized 8-methoxy-psoralen. *Z. Naturforsch,* 37c,70-74.

Potapenko, A.Ya., Sukhorukov, V.L. and Davidov, B.V. (1983), 8-methoxy-psoralen - sensitized photooxidation of tocopherol. *Photobiochem. Photobiophys.*, 5, 113-117.

Rodighiero, G. and Caporale, G. (1958), Furocumarine ed emolisi fotodinamica. Il Farmaco, Ed. Sci., 13, 373-378.

Rodighiero, G., Dall'Acqua, F. and Pathak, M.A. (1984), Photobiological properties of monofunctional furocoumarin derivatives. In: Topics in Photomedicine, K.C. Smith, ed., Plenum Pub. Corp., pp. 319-398.

Rodighiero, G., Dall'Acqua, F., and Averbeck, D. (1988), New psoralen and angelicin derivatives. In: Psoralen DNA Photobiology, Vol. 1, F.P. Gasparro, F.P., ed. CRC Press, Boca Raton, pp. 37-114.

Salet, C., Moreno, G. and Vinzens, F. (1982), Photodynamic effects by furocoumarins on a membrane system. Comparison with hematoporphyrin, *Photochem. Photobiol.*, 36, 291-296.

Specht, K.G. et al (1988), Furocoumarin photosensitized reactions with fatty acids. In: Photosensitization. Molecular, Cellular and Medical Aspects, G. Moreno et al, eds., NATO-ASI Series, Springer Verlag, Berlin, pp. 301-303.

Specht, K.G., Kittler, L. and Midden, W.R. (1988), A new biological target of furocoumarins: photochemical formation of covalent adducts with unsaturated fatty acids. *Photochem. Photobiol.*, 47, 537-341.

Vedaldi, D. et al (1988, in press), Dark and photohemolysis of erythrocytes by furocoumarins. Z. Naturforsch. 43c.

Wennersten, G. (1979), Membrane damage caused by 8-MOP and UV-A treatment of cultivated cells. *Acta Dermato-Venereol.* (Stockholm),59, 21-26.

Photodynamic Carcinogenicity and its Prevention by Carotenoids as Oxygen Radical Quenchers: Relevance in Animals and Human Intervention

LEONIDA SANTAMARIA

"C. Golgi" Institute of General Pathology
Centro Tumori, University of Pavia
27100–Pavia, Italy

Abstract

The problem of photodynamic action and carcinogenicity as reviewed and developed with various *in vitro* experimental models in 1960, led to the demonstration that photodynamic activity *in vitro* in various polycyclic hydrocarbons is strictly associated with *in vivo* carcinogenicity. Skin carcinogenesis induced by benzo(a)pyrene (BP) in mice can be influenced by long UV radiation (300–400 nm), as well as daylight, with either dramatic enhancement or inhibition depending on total light dosage. Both phenomena were respectively interpreted as being caused by oxygen radical production compatible or otherwise with cellular life. Breast cancer induced by PUVA (8-methoxypsoralen + UVA light) in mice was interpreted as due to a two-step reaction, namely an oxygen-independent DNA-8-MOP photoadduct followed by generation of oxygen radicals and/or activated molecular oxygen (primarily $^1O_2$).

The above hypothesis relating to the mechanism of action of these photocarcinogeneses was experimentally first demonstrated in 1980 by the fact that diet supplementation with carotenoids (beta-carotene or canthaxanthin dyes, with and without provitamin A activity, respectively) as antioxidants produces remarkable prevention (up to 65–100%) against skin cancer induced in mice by BP ± long UV. Subsequently, this prevention was also active against breast cancer induced in mice by PUVA, and against direct gastric carcinogenesis induced by nitrosoguanidine in rats independently of light, but facilitated by oxygen radicals likely generated by phlogistic processes. This opened up new horizons for chemoprevention in humans for any form of cancer where oxygen radicals are involved, even independently of light excitation.

The main experiments both *in vitro* and *in vivo* carried out over a period of more than thirty years which led to the above achievements are here reviewed including a first clinical case report of carotenoid supplementation against cancer recurrencies after radical treatment.

In Vitro Experiments on Photodynamic Action and Carcinogenicity in Polycyclic Hydrocarbons

Studies on the action of long ultraviolet light (300–400 nm) on polycyclic hydrocarbon carcinogenesis *in vivo* have been reported several times in the literature with conflicting results, that is, enhancement (Findlay, 1928; Maisin and De Jonghe, 1934; Vles et al., 1935; Clark, 1964), inhibition (Doniach and Mottram, 1940; Morton et al.,

1940; Morton et al., 1942) as well as no effect on cancer induction (Kohn Speyer, 1929; Seelig and Cooper, 1933; Rusch et al., 1942). These studies were stimulated by early *in vitro* investigations on possible correlations between photodynamic action and carcinogenicity in different chemicals in order to understand the mechanism of the above-mentioned data. (Motttram and Doniach, 1938; Maltotsy and Fabian, 1943; Alexander and Fox, 1954; Gazia and Dansi, 1955). The main data emerging from these studies are reviewed here with the aim of demonstrating that the enhancement of polycyclic hydrocarbon carcinogenesis by light is dependent on a definite optimum concentration of photodynamic substance and intensity of radiation. The inhibition of cancer induction by this photodynamic treatment is due to a higher than optimum level of these factors, thus reducing the total number of potential malignant cells.

Methodologies at cellular, subcellular and molecular levels

The photodamage on paramecia was the first test used to ascertain the photodynamic activity in chemicals (Motttram and Doniach, 1938). This test at cellular level provided quick and reproducible qualitative and quantitative evaluations and proved a correlation between photodynamic action and carcinogenicity in more than 150 polycyclic hydrocarbons (Epstein et al., 1964). In this connection, however, it may be objected that there is no relation between the very different biological actions compared, namely, production of tumors in vertebrates by the above chemicals and death of protozoa caused by photodynamic effect of the same substances (Santamariam, 1960). Therefore, attention was paid in the fifties to develop photodynamic tests using systems where crystallized serum proteins or normal human blood–serum were used as substrate.

Lethal photodynamic effect on Saccharomyces cerevisiae was studied irradiating ($\lambda > 320$ nm) a suspension of washed yeast cells in the presence of a hydrocarbon (Santamaria and Prino, 1964). The survival curve, after a lag period, behaves in an exponential fashion (Fig. 2). Thus, a routine irradiation time was selected (to produce 50% survival with benzpyrene) and the activities of the compounds tested were expressed as percent values relative to benzpyrene activity, as reported in Table 1.

Electrophoretic test on normal human blood serum exposed to light ($\lambda > 320$ nm) in the presence of polycyclic hydrocarbons. In this test blood-serum was chosen as a pool of different native proteins to enhance the probability of an association of a chemical with such important proteins in their native solvent. It was assumed (and demonstrated in a few instances) that an affinity between sensitizer and substrate is fundamental to give rise to a photodynamic reaction (Santamaria and Prino, 1964). Indeed this test showed that photo-oxidation produced aggregation, diminution of the tinctorial affinity, and increased migration rate in the electric field of definite protein fractions. As an example, Fig. 1 reports the electrophoretic diagrams of blood-serum photo-oxidized in the presence

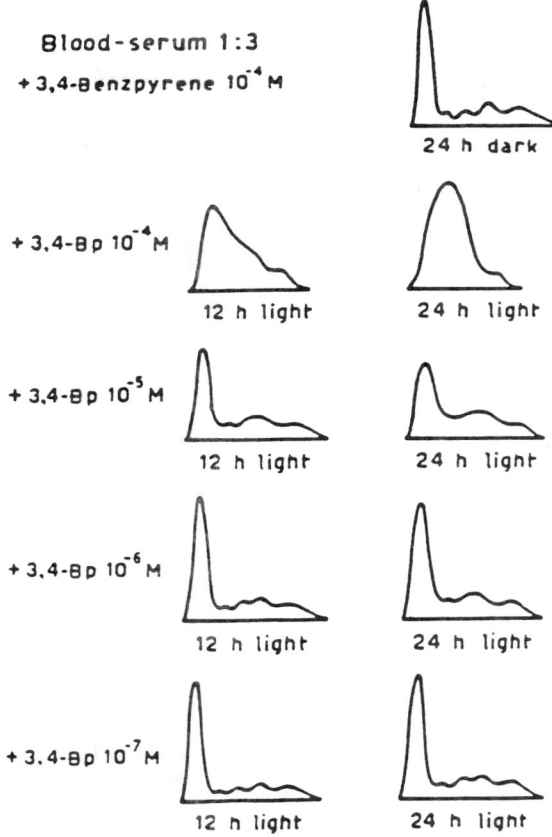

Figure 1. Alterations of the elctrophorectic pattern in normal human blood-serum + benzo(a)pyrene at different concentrations. Light source: mercury vapour lamp; radiation filtered by Wood's glass; intensity at the level of the sample *ca.* 2×10^3 erg/cm^2/sec. (After Santamaria and Prino (16)).

of BP at different concentrations. These diagrams indicated that the bands of α- and especially β-globulin are involved in the photo-oxidative process, whereas the band of γ-globulin is almost unaffected, at least at the beginning of the reaction. The chemical affinity of BP is strongest with β- and γ-globulins and weakest with α-globulin (Wunderly and Petzold, 1952). Therefore, the photo-oxidation of serum proteins, and the resulting variation of electrophoretic properties, are dependent on the rate of association of the hydrocarbon with the substrates. The different rates of band grouping or the absence of any effect observed with other polycyclic hydrocarbons stimulated an extensive investigation employing thirty-six polycyclic hydrocarbons, either carcinogenic or non-carcinogenic, many of them chemically very closely related, as reported in Table 1. The

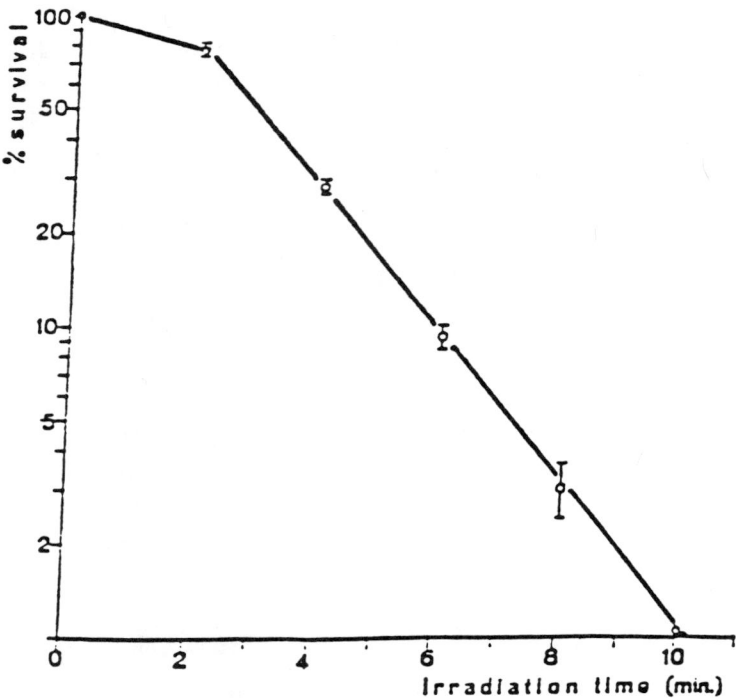

Figure 2. Survival curve of *S. cerevisiae* (5 exp.) irradiated in the presence of BP 10^{-5} M. Light source: 500-W high pressure quartz mercury lamp; radiations filtered though 4-mm common glass ($\lambda > 320$ nm). Intensity at the level of the sample *ca.* 2×10^4 ergs/cm^2/sec. (After Santamaria and Prino (16)).

assessment of this test followed early studies of photodynamic action of hematoporphyrin with different methods (Santamaria et al., 1955, 1957, 1957).

The polarographic study of blood serum photo-oxidation (Santamaria et al., 1955) used in our laboratory as a tool in the study of the photodynamic reaction showed the lowering of the oxygen wave, as well as the modifications of the "protein waves". Furthermore, it was observed that the rate of oxygen disappearance produced by 26 out of 29 polycyclic hydrocarbons follows a zero-order reaction: only in 3 of these substances such a rate behaved in an exponential fashion (Santamaria, 1960). The observation that photo-oxidation behaves according to a zero-order reaction allowed one to consider polycyclic hydrocarbons as substances acting like enzymes. Surprisingly this is consistent with experimental findings showing that BP and other polycyclic hydrocarbons play a role similar to that of peroxidase when hydrogen peroxide is present (Dansi et al., 1957, Garzia and Dansi, 1953).

Table 1. Photodynamic activity on blood serum, on mitochondria and on *S. cerevisiae* in relation to carcinogenicity. (φ_r = percent activity). After Santamaria and Prino (1964)

Substances	on blood serum	on mitochondria Δφr	on mitochondria Approx. indication	on S. cerevisiae Δφr	on S. cerevisiae Approx. indication	Carcinogenicity
3,4-Benzpyrene	++++	100	++++	100	++++	++++
Benzfluoranthrene	+++			37	+(+)	?
20-Methylcholanthrene	+++	83	+++	50	++	++++
1.2.5.6-Dibenzfluorene	+			13	−	+
1.2.5.6-Dibenzanthracene	++	6	•−	55	++	++
Anthracene	−	33	•+	2	−	−
Phenanthrene	−	0	−	8	−	−
Pyrene	−	46	++	2	−	−
1.2.5.6-Dibenzacridine	−	23	+	38	+(+)	±
2-Acetyl-amino-fluorene	•−			0	−	+++
1,2-Benzanthracene	+	38	+(+)	95	•++++	−•
1'-Methyl-1,2-benzanthracene	−	59	•++	100	•++++	−•
2'-Methyl-1,2-benzanthracene	±	82	•+++	41	•++	−•
3'-Methyl-1,2-benzanthracene	−	95	•++++	30	•+	−•
4'-Methyl-1,2-benzanthracene	±	26	•+	0	−	
3-Methyl-1,2-benzanthracene	±	84	•+++	0	−	+a +b
4-Methyl-1,2-benzanthracene	++	84	+++	100	++++	+a ++b
5-Methyl-1,2-benzanthracene	++	100	++++	66	++(+)	++ab
6-Methyl-1,2-benzanthracene	+	82	+++	22	+	+a ++b
7-Methyl-1,2-benzanthracene	±	56	++	0	−	+a −b
8-Methyl-1,2-benzanthracene	•++	85	•+++	100	•++++	+a −b
9-Methyl-1,2-benzanthracene	+(+)	38	+(+)	100	++++	+a +b
10-Methyl-1,2-benzanthracene	•±	56	++	100	++++	+++a ++b
9,10-Dimethyl-1,2-benzanthracene	•+	23	•+	100	++++	++++a ++b
1-Methyl-3,4-benzphenanthrene	++	48	++	25	+•	++ab
6-Methyl-3,4-benzphenanthrene	±	42	++	20	↓	+a −b
7-Methyl-3,4-benzphenanthrene	+	48	++	25	↓	+ab
8-Methyl-3,4-benzphenanthrene	+	17	±	3	•−	+a ++b
3-Fluoro-10-methyl-1,2-benzanthracene	•+++(+)	98	•++++	3	−	−c
4-Fluoro-10-methyl-1,2-benzanthracene	+++(+)	98	++++	60	+++	+++(+)c
6-Fluoro-10-methyl-1,2-benzanthracene	++	63	++(+)	80	+++	++c
4'-Fluoro-10-methyl-1,2-benzanthracene				0	−	
7-Fluoro-10-methyl-1,2-benzanthracene	++++	88	+++(+)	93	++++	++++c

[a]Badger (1948); [b]Von Haam (1958); [c]Miller and Miller (1963).
*Exception to the correlation between photodynamic action and carcinogenicity.

From the percent values of the oxygen disappearance after 30 min of irradiation no correlation appeared between the degree of blood serum photo-oxydation and the alteration of the serum electrophoretic pattern. In this connection, it could be assumed that the aggregation process is dependent on the oxidation of definite groups (SH) rather than on the oxidation rate of the total protein molecules.

The polarographic analysis carried out on blood-serum proteins as well as on crystallized serum-albumin photo-oxidized in the presence of BP demonstrated that polarographic "protein waves" undergo variation consistent with change in protein configuration. In fact, the total surface thiol groups (-SH and -S-S-) were increased. Amperometric – SH titration indicated that increase of the second protein wave is actually due to an oxidation leading to a decrease of -SH groups.

The photodynamic injury (λ > 320 nm) on mitochondria became a new method of study when it was possible to isolated from rat liver mitochondria which do not age at 20°C, but respond quickly to a photodynamic injury by swelling, easily detectable by fall in optical density in aerated medium (Santamaria and Fanelli, 1961). This swelling could be completely prevented when the sample is exposed to light under vacuum; addition of ATP or of reducing agents such as ascorbic acid, cysteine, glutathione or thiourea failed to prevent the swelling and even showed that in the dark they produce a drop in turbidity which is enhanced under irradiation. With mitochondria which naturally swell at 45 min of irradiation, in our experimental conditions, it was possible to distinguish clearly two classes of photoactive agents, namely, those which produce swelling (a) before 45 min of irradiation, or (b) afterwards. We called the first effect "primary" and the second effect "secondary". The "primary effect" was presented as the consequence of a prompt interaction of the photosensitizer with the mitochondria components; accordingly, the "secondary" effect was likely due to the interaction of photosensitizer and mitochondria components facilitated by the natural swelling by light. Although the mitochondria

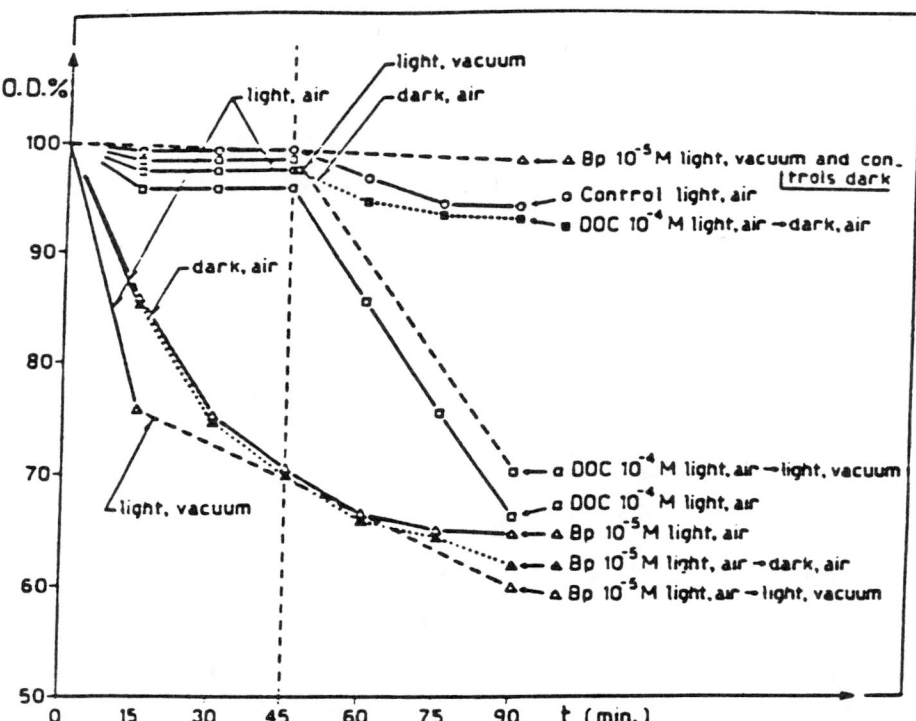

Figure 3. Swelling of rat liver mitochondria following photodynamic action of BP ("primary effect") and of desoxycorticosterone (DOC) ("secondary effect"). Evidence of the oxygen dependence of BP photoactivity. Behaviours of "primary" and "secondary" effects, with respect to degassation and replacement in the dark, where the swelling is triggered. (λ 320 nm, ca 3.3×10^4 ergs/cm^2/sec at the level of the sample). (After Santamaria and Giordano (1969).

swelling that we observed is definitely oxygen dependent, the rate of swelling — when the latter is triggered — cannot be stopped or modified by degassation or by transferring the sample in the dark. The above phenomena are expressed by an experiment reported in Fig. 3, where the photodynamic effect of BP and desoxycorticosterone on isolated mitochondria are reported. To evaluate comparatively the activities, the percent values of drop in turbidity at 45 min of exposure were chosen, with reference to that of benzpyrene, rather than the swellings at the end of the irradiation time (90 min), as reported in Table 1 (Santamaria et al., 1952, 1959).

In Vitro Experiments on Photodynamic Action and Carginogenicity

The hypothesis that photodynamic action is important in polycyclic hydrocarbons as an aspect of carcinogenic action is supported by the observation that other carcinogenic substances with different chemical structures frequently exert photodynamic action (Calcutt, 1954). Furthermore, photodynamic substances like hematoporphyrin, eosin, furocoumarins, conventionally not listed as carcinogenic agents can induce carcinogenic processes when the animals given the above substances are exposed to light under definite experimental procedures (Bungeler, 1937; Griffin et al., 1958).

Therefore, one should expect that light enhances cancer induction by hydrocarbons which are carcinogenic and photodynamic at the same time, as demonstrated in Table 1. However, early experimental attempts to show this hypothesis have provided conflicting findings (Santamaria, 1960), as reported above. A critical analysis of the papers published on this topic led to the conclusion that little attention had been paid to the experimental conditions in relation to dosage of either the sensitizer or the light intensity. As a matter of fact, the problem of an enhancement of a chemical carcinogenic process by light could be theoretically approached by assuming the phenomenon as dependent on a definite optimum concentration of photodynamic substance and intensity of radiation. This assumption was supported by a work upon erythemal and carcinogenic response in 8-methoxypsoralen (8-MOP) treated mice under the effect of light (300–400 nm) (Griffin et al., 1958). Here, a certain dosage of light produced 100% tumor incidence, whereas a higher dosage produced 35% tumor incidence or death. Similar results regarding papilloma formation in animals given 20-methylcholanthrene were also reported (Clark, 1964).

Photodynamic enhancement and inhibition of skin cancer in mice induced by benzo(a)pyrene (BP).

To underscore the above conflicting data on the influence of light on skin carcinogenesis by polycyclic hydrocarbon experiments were set up with mice painted with BP, 1,2,5,6-dibenzanthracene, and 1,2-benzanthracene. The general procedure of experiments consisted in keeping control and experimental groups in dark rooms and thick black curtains, and in exposing the experimental groups twice a week soon after painting in the same room to irradiation for different periods of time ranging from 30 min to 3 hr. The irradiation was carried out using a high-pressure Wood's 125-W Philips bulb "57236 E/70" at 30 cm. This source has its maximum output at 365 nm and is capable of exciting BP. The energy delivered by one bulb was about 1.5×10^5 ergs/cm^2/sec at 30 cm according to measurements by an "Eppley" calibrated thermopile.

With BP three experiments were carried out over a period of 3 years employing 955 female Swiss albino mice. Two of these experiments were carried out in the Institute of General Pathology of the University of Milan, and one in the Institute of Cancer Research in Naples. The experimental conditions in painting and way of keeping the

Table 2. Erythemal and carcinogenic response in Swiss male mice painted with BP (one drop each time of a 0.5 g% acetone solution = 100 μg) and esposed to Wood light. After Santamaria and Prino (16).

Group	Animal No.	Treatment 3.4-Bp	Treatment Light	Duration of experiment	Mortality death	Mortality survival	Mortality Q_m	Q_{weight}	Gross observations	Animals with epithelioma	Tumor incidence
A	50	none	2 hours twice a week for 23 weeks (total: 8 × 10¹⁰ erg/cm²)	32 weeks	25	25	0·6	+0·3	evident itching during the light exposure till the 6ᵗʰ-7ᵗʰ week	none	none
B	50	100 μg twice a week for 23 weeks (total = 5600 μg)	none*	idem	31	19	0·9	+0·3	this group was the one in the best condition. 3.4-Bp did not inhibit the growth of the hairs cut at the beginning of the experiment	2	4%
C	50	idem	3 hours twice a week for 23 weeks (total: 12 × 10¹⁰ erg/cm²)	idem	43	7	1·5	−0·2	violent itching during the light exposure till the 14ᵗʰ week. Severe erythemal response. Falling of hairs. Many a wound by bites. Cannibalism.	23	46%
D	50	idem	2 hours twice a week for 23 weeks (total: 8 × 10¹⁰ erg/cm²)	idem	35	15	1·1	−0·4	idem	29	58%
E	50	idem	1 hour twice a week for 23 weeks (total: 4 × 10¹⁰ erg/cm²)	idem	29	21	0·8	+0·2	erythemal response. Falling of hairs. Itching till the 14ᵗʰ week.	1	2%
F	50	idem	½ hour twice a week for 23 weeks (total: 2 × 10¹⁰ erg/cm²)	idem	21	29	0·5	+0·3	see group B.	none	none

*The cages were kept in a darkened room on shelves covered with a thick black curtain. These experimental conditions were adopted for all the cages of the other groups.

$$Q_m = \frac{m}{\frac{1}{2}(N+s)}$$ where m = number of death; N = initial animal number; s = survival.

$$Q_{weight} = \frac{\Delta_\omega}{\frac{1}{2}(w_1 + w_f)}$$ where Δ_ω = difference between the initial weight (w_1) and the final weight (w_f).

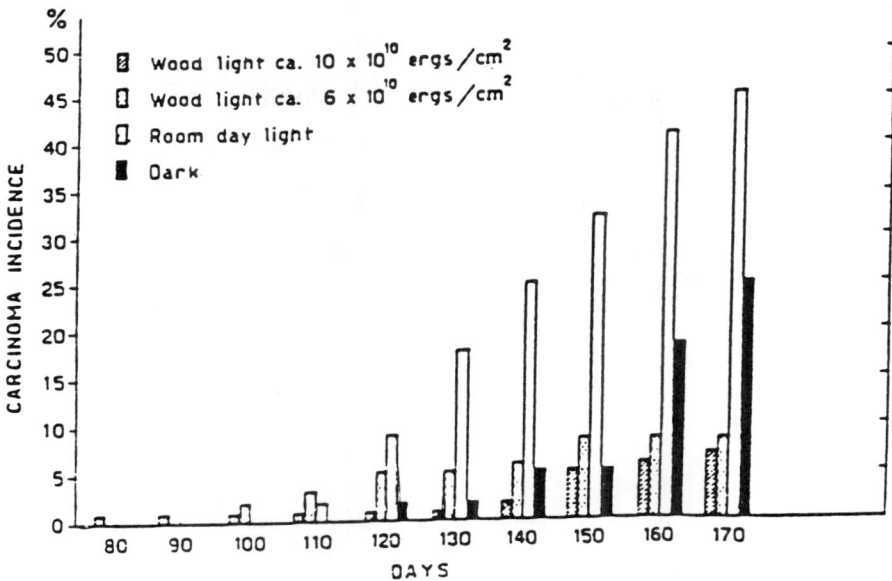

Figure 4. Enhancement of carcinoma incidence by room daylight and inhibition by long-wave ultraviolet radiation. After Santamaria et al. (26,33).

animals during irradiation (in wide or narrow containers) were different (Santamaria et al., 1966). Therefore, the results were comparable within the groups of each experiments. Table 2 and Figs. 4 and 5 depict the results of these experiments. In the experiment reported in Fig. 4, a group of animals was kept in a room with conventional windows illuminated by daylight.

The principal data of the above experiments can be summarized as follows. First experiment (Table 2): (a) very low incidence in the group of mice treated with BP and kept in the dark; (b) highest tumor incidence in the group treated with BP plus Wood's light at a dosage of about 8×10^{11} ergs/cm^2, which was not the highest dose used. Second experiment (Fig. 4): (a) Wood's light produced carcinoma earlier than in the dark. On continued irradiation the effect of Wood's light on carcinoma formation was inhibitory, (b) daylight induces an enhancement of carcinoma onset, which leads later to an increase of carcinoma induction with respect to the dark. Third experiment (Fig. 5): in this experiment attention was paid to the onset of both papilloma and carcinoma; Wood's light leads to a definite increase of both benign and malignant neoplastic processes. Histological observations have shown that when an inhibitory effect on carcinogenesis occurs an early hyperplasia of the epidermis is followed by regressive processes both in cells and intracellular materials. Such alterations are not evident when carcinogenesis is

983

stimulated by light. Therefore it seems that the inhibition of cancer induction by light is a consequence of a tissue damage by photodynamic action of BP rather than a photo-oxidation of the carcinogen as it occurs when the latter is exposed to light before application (Miller, 1951).

The observation that small doses of UV radiations enhanced the induction of tumors in rats given 9,10-dimethyl-1,2-benzanthracene (Prodi and Maltoni, 1959) can be explained assuming a mechanism as in the case of BP. The enhancement of UV (280–320 nm) carcinogenesis following a single application of 9,10-dimethyl-1,2-benzanthracene in hairless mice (Epstein, 1965) was presented as the consequence of a summation of effects.

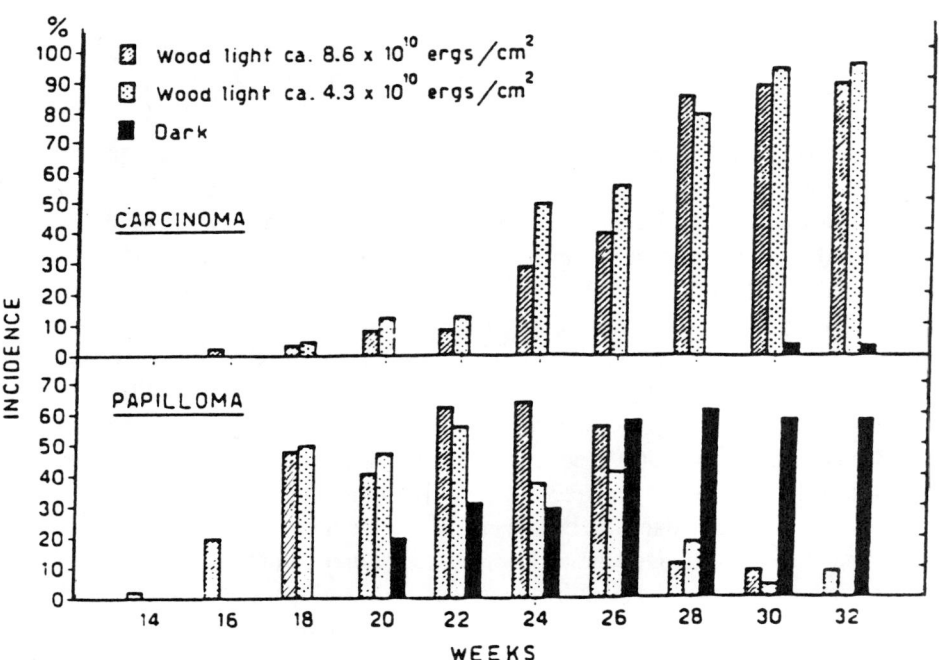

Figure 5. Enhancement of both papilloma and carcinoma formation by long-wave ultraviolet radiation. After Santamaria et al. (Santamaria et al., 1966).

Photocarcinogenicity by 8-methoxypsoralen (8-MOP), neutral red (NR), and proflavine (PF)

A combination of 8-methoxypsoralen (8-MOP), and long UV radiation 320–400 nm (PUVA) (37,38), for the treatment of psoriasis, mycosis fungoides and other skin disorders and the use of neutral red (NR) (Felber, 1971; Felber et al., 1973; Friedrich 1973) or proflavine (PF) (Kaufman et al., 1973), and visible light for the treatment of both oral and genital forms of herpes simplex virus (HSV) has been advocated. The combined "photochemotherapy" (Parrish et al., 1974) for psoriasis (Goeckerman, 1925), mycosis fungoides (Honigsmann et al., 1976; Roenigk. 1977), pustularis palmaris, pustularis plantaris (Uematza and Mizuna, 1976), and HSV infections has been reported by numerous investigators. However, data has accumulated 1) on the photocarcinogenic effect of MOP (Pasrrish, 1976; Griffin et al., 1958), and 2) on the photodynamic and carcinogenic properties of many chemicals (Santamaria, 1960; Rodighiero, 1976).

Furthermore, treatment of 1373 patients with oral MOP photochemotherapy for psoriasis produced a total of 48 basal cell and squamous cell carcinomas in 30 patients. The analysis of this data, however, failed to reveal the carcinogenic potential of long-term PUVA exposure and the role of PUVA in the development of these tumors (Wollf et al., 1976).

Therefore, attempts were made to investigate the tumorigenic effects of MOP, and the possible carcinogenic action of NR and PF, in mice by topical application under photodynamic conditions. The general criteria adopted in the experiments were similar to those previously described for investigation of the effects of light on BP carcinogenesis (Santamaria et al., 1966). The doses of both drug and irradiation were adjusted to provoke erythemal phototoxic reactions (Pasrrish et al., 1976).

The results confirmed the photocarcinogenic action of MOP with long UV irradiation and demonstrated for the first time the photocarcinogenic action of NR and PF with visible light. In no animals did the tumors develop within the clipped areas but occurred in subcutaneous sites near the neck, interscapular, and inguinoabdominal regions. The epidermis covering the neoplasms appeared to be normal but in some cases was fixed to the tumor.

Microscopically, the histologic patterns were as follows: 1) acinar or acinotubular structures with polygonal small cells, uniform in size, with scanty mitoses; these tumors were encapsulated but produced metastases; 2) semisolid cords or masses with rare acinar structures, often pierced by capillaries, lacunas, and blood extravasations; 3) dense cords and infiltrating masses of atypical cells, with atypical mitoses; 4) growths as in 3 with epidermoid-like cysts; 5) intracanalicular papillary growths with numerous mitoses; 6) solid growths with anaplastic small and large cells; 7) growths of acinotubular structures mixed with masses of polygonal sebaceous-like cells; and 8) carcino-mixo-sarcomatous growths.

Because of the morphologic similarity of the above patterns with mammary tumors in mice it appeared that almost all the tumors observed in our investigation developed from mammary glands. This breast cancer was believed to be the expression of the oncogenic properties of a virus triggered by photodynamic action, as suggested by the fact that the photodynamic inactivation of HSV-1 and HSV-2 may lead to unmasking of the oncogenic potential of these viruses (Santamaria et al., 1985). A minority of tumours apparently arose from skin appendages other than the mammary glands.

These results were unexpected, when first observed in 1966 in our laboratory. Most probably, they depended on the strain 955 of female Swiss albino mice (from Nossan, Corezzana, Milano)

The kinetics of tumor growth are illustrated in Figure 6; the percentage of mice with tumors is plotted in relation to the weeks after the initial light exposure and 20 weeks after the termination of the treatment. It appeared that MOP, PF, and NR plus light produced tumors after a latency of 15, 31, and 38 weeks, respectively. MOP was the

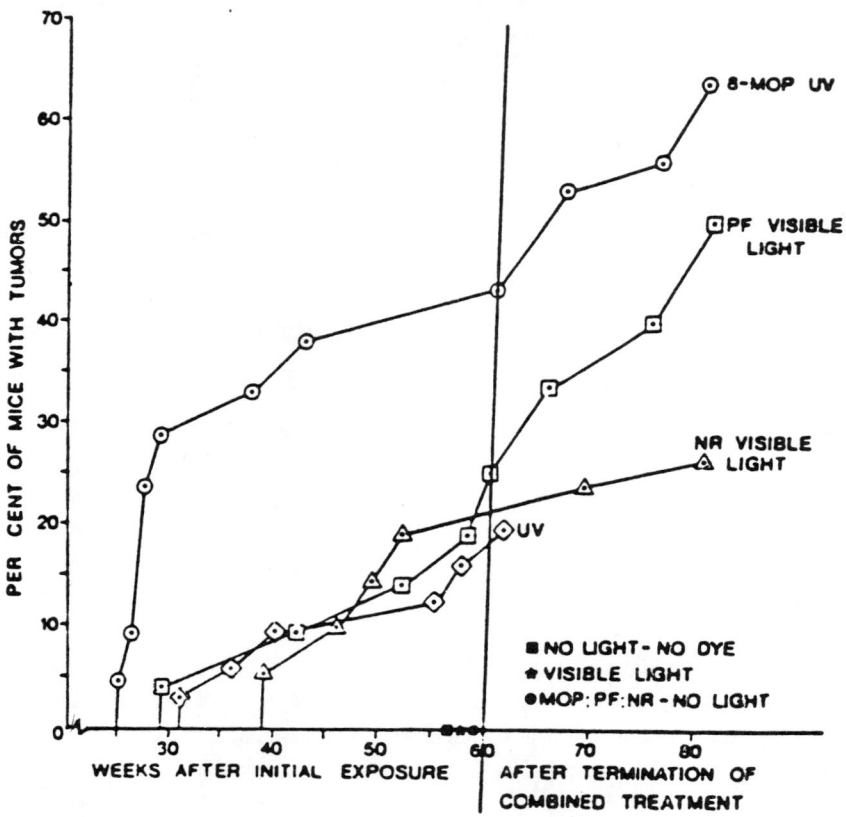

Figure 6. Breast photocarcinogenesis by 8-methoxypsoralen, Neutral red, proflavine, and long UV radiation.(After Santamaria et al. 1985).

986

most active tumor-inducer. Long UV light, 300–400 nm, apart from the negligible 2.6% of fluence at 313 nm, appeared to be carcinogenic *per se.*

After the termination of the experimental treatments, the percentage values of mice with tumors increased, with the exception of those treated with UV light alone. The control groups did not develop any tumor in the dark or in visible light, on aging up to 80 weeks after initial treatment.

Background to Oxyradical Formation and Carcinogenicity Caused by Photodynamic Action

All the above photoactivations were explained by the fact that photodynamic substances act with photooxidation mechanisms. In this connection *it was clear that the photocarcinogenic enhancement of BP, for instance, had to work by oxygen activation as eosin, hematoporphyrin, neutral red, proflavin, furocoumarins do in exerting their photocarcinogenic activity.*

The first demonstration of this oxygen activation was made possible when an oxygen dependent radical species was detected in an aqueous solution of hematoporphyrin irradiated and examined in an electron spin resonance (ESR) apparatus. In this connection it is worth reporting selected parts of the original paper by Smith, Santamaria and Smaller (1961), which was carried out applying, for the first time, an electron paramagnetic resonance instrument to detect the intermediates formed between the excited sensitizer and oxygen This paper was first published in a book of proceedings (Smith et al., 1961) and its results were later reported in a journal (Santamaria, 1962).

The model used to detect the formation of free radicals in photodynamic reactions consisted of hematoporphyrin (sensitizer) and human blood serum (substrate) in phosphate buffer at pH 7.3, as used in a previous experiment (Santamaria et al. 1957). Solutions were made up under ordinary atmospheric conditions such that hematoporphyrin was present in a concentration of 1.2×10^{-3} M and serum in a dilution of 1:3. Hematoporphyrin stock solution was prepared by dissolving hematoporphyrin in 1.0 N NaOH and adjusting the pH to 7.3 with 1 N HCl. The solutions to be irradiated and measured in the microwave spectrometer were contained in Pyrex tubes of 3 mm inside diameter.

Irradiations with a type A high-pressure mercury arc were carried out with the solutions held at –20°C or at –196°C. In all instances the samples were quickly transferred at the end of the irradiation to a Dewar flask containing liquid nitrogen. Electron spin resonance studies were carried out using a 9000 Mc electron paramagnetic resonance instrument equipped with a lock-in phase-sensitive detection system such that the second derivative of the microwave absorption was recorded. Measurements were made with the samples at the temperature of liquid nitrogen.

In some experiments the reducing agents cysteine or ascorbic acid were added to the system in concentrations of 1.2×10^{-2} M prior to irradiation. In other instances the gas content of the samples was altered prior to irradiation by degassing to high vacuum and

Table 3. Radical formation in irradiate hematoporphyrin solutions in phosphate buffer. After Santamaria et al. (1962)

System	In the presence of O_2		In the presence of He or nitric oxide	
	$-20°$ C	$-196°$ C	$-20°$ C	$-196°$ C
Buffer	0	0	0	0
Buffer + serum	0	0	0	0
Buffer + hematoporphyrin	.86	.24	0	˜.24
Buffer + hematoporphyrin + ascorbic acid	0	.24	0	˜.24
Buffer + hematoporphyrin + cysteine	0	.30	0	˜.30
Buffer + hematoporphyrin + serum	.18	.40	0	˜.40

regassing with helium or nitric oxide. The results of the experiments in which the irradiations were carried out at –20°C are summarized in Table 3.

In no case were electron spin resonance signals observed in the absence of irradiation. After irradiation, signals were obtained only from solutions containing hematoporphyrin. Signals were absent in solutions of hematoporphyrin to which either cysteine or ascorbic acid were added and they were reduced by two-thirds in samples to which serum was added prior to irradiation. They were also absent in solutions of hematoporphyrin and buffer that had been degassed to high vacuum or that had been degassed and regassed with helium or nitric oxide.

No signals were obtained from solutions of hematoporphyrin irradiated at 23°C; signals present in solutions irradiated at –20°C disappeared when the samples were warmed to 23°C.

The results of the experiments in which the irradiations were carried out with the samples at liquid nitrogen temperatures are also summarized in Table 3. It is apparent that a signal appears in hematoporphyrin solutions irradiated at a low temperature, and that this signal is not influenced by the addition of cysteine, ascorbic acid, or serum prior to irradiation. Moreover, degassing of the samples to high vacuum or replacement of the ordinary atmosphere by helium or nitric oxide prior to irradiation was followed by electron spin signals of the same character as those under ordinary atmosphere.

The signals obtained from the samples irradiated at –196°C were about one-half the size of those from the –20°C irradiations. In addition, the signals from the irradiations at the two temperature extremes were of quite different character with respect to the levels of

988

microwave power at which they became saturated, indicating that they were derived from different molecular species (Fig. 7).

It seems clear that the electron spin resonance signals recorded from the solutions of the present experiments arise only as a result of irradiation and come only from hematoporphyrin. It would appear that the signals arise from free radicals of hematoporphyrin.

The similarity between the conditions influencing signal production in the –20°C experiments and those influencing the course of photodynamic action are striking and suggest that the –20°C state allows the operation of a useful model of photodynamic action. Thus, signals are produced and photodynamic action proceeds only in the presence of oxygen, and both processes are inhibited when ascorbic acid or cysteine is introduced into the system. *We interpreted the oxygen-dependence of the signal to indicate the presence of a free radical consisting of some kind of an association between hematoporphyrin and oxygen (oxyradical).* The failure to detect signals in the presence of ascorbic acid or cysteine was thought to be due to interaction between the reducing agent and the radical, the result of which is the loss of paramagnetic characteristics of the system.

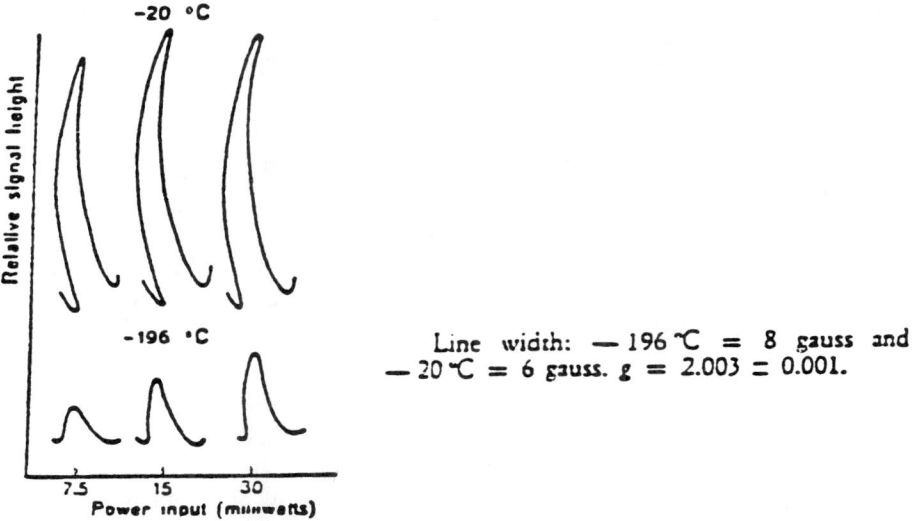

Line width: — 196 °C = 8 gauss and — 20 °C = 6 gauss. g = 2.003 ± 0.001.

Figure 7. Comparison of ESR signals in irradiated –196°C and –20°C systems with respect to dependence of microwave power. After Smith et al. (Smith et al., 1961).

The accumulation of radicals in measurable amounts may be readily explained in terms of decreased possibilities for reaction of radicals with other components of the system in the physical state at –20°C. The explanation of the results with cysteine, ascorbic acid, and serum is not so obvious, but it would appear that the –20°C state allows interaction between these substances and hematoporphyrin. The results of the 196°C experiments, however, indicate that interactions between the critical components of the system are not possible in the physical state existing at this temperature. That the physical state of the system is an important factor was indicated in preliminary experiments in which the solvent was altered. Thus, signal production was markedly decreased when glycerol was present in 10% concentration in the usual system irradiated at –20°C. When glycerol alone was used as the solvent, a signal similar to that of the usual –196°C experiments with respect to microwave power dependence appeared after irradiation at –20°C.

It was noted above that several lines of investigation have shown that substances capable of photosensitizing oxidation are excited to the triplet state upon irradiation. It is proposed that in the present –20°C experiments, hematoporphyrin in the triplet state can combine with oxygen to form an oxyradical which may then react with the substrate according to the following scheme:

$$S + h\nu \rightarrow {}^3S \qquad S = \text{sensitizer}$$

$$^3S + O_2 \rightarrow SOO'$$

$$SOO' + X \rightarrow XO_2 + S \qquad X = \text{substrate}$$

The above original report of an experiment carried out at Argonne National Laboratory during summer 1959 is to testify when the "photoperoxide" as postulated by Schenk in 1948, and by Oster et al. in 1959 was first experimentally demonstrated. However, this paramagnetic species — here first called "oxyradical" — seemed to rule out the hypothesis advanced by Schenk as to the loss of the paramagnetic character of the excited sensitizer by oxygen (Santamaria, 1962).

Later on, another important activated molecular molecular oxygen was found to occur in photodynamic action, namely the singlet oxygen ($^1O_2$) (Foote, 1968). In contrast to the oxyradical, this excited species does not involve a change in electron spin (no net unpaired spins).

At any rate, all these and other oxygen radical species or activated molecular oxygen ($O_2^{-\cdot}$, $HO_2^\cdot$, H_2O_2, HO, $RO^\cdot$, $ROOH$, ^3RO) can be quenched or scavenged by antioxidant compounds, such as carotenoids, as described by Krinsky (1979), thus inhibiting the above oxygen dependent reaction. This turned out to be the rationale of all the subsequent experimental and clinical efforts to prevent oxygen radical pathology in differenrt carcinogenic processes.

In this report clear-cut evidence is presented below to demonstrate that carotenoids are antitumorigenic, acting as anti-oxidant substances in skin and mammary photocarcinogeneses, as well as in gastric carcinogenesis independently of light excitation.

Chemoprevention by Supplemental Carotenoids of Indirect Skin Carcinogenesis induced by Mice in BP ± UV-A

In the last decade, experimental trials have suggested that a diet rich in red carrots inhibits the appearance of tumors induced by dymethyl-benzo-(a)-anthracene in mice (58) and that injection or peroral administration of carotenoids delays skin tumor induction in hairless mice exposed to UV-B (290–320 nm) irradiation (Epstein, 1977; Mathews-Roth, 1980).

Finally, an experimental trial demonstrated that β-carotene (BC) or canthaxanthin (CX), two carotenoids with and without pro-vitamin A activity respectively, when supplemented at high dosage by diet to albino mice, prevent up to about 60% skin cancer induced by benzo(a)pyrene (BP) (Santamaria et al., 1980), as shown in Fig. 8. *In this experiment, attention was paid to start carotenoid supplementation one month before carcinogenic induction and to continue this supplementation to the diet throughout the entire experiment.* Groups of animals were irradiated with long UV light (300–400 nm) twice a week half an hour after BP skin paintings. This treatment produced the marked photocarcinogenic enhancement (PCE) as first observed by Santamaria et al. (Santamaria et al., 1966), apparently through a photodynamic process involving oxygen radical species.

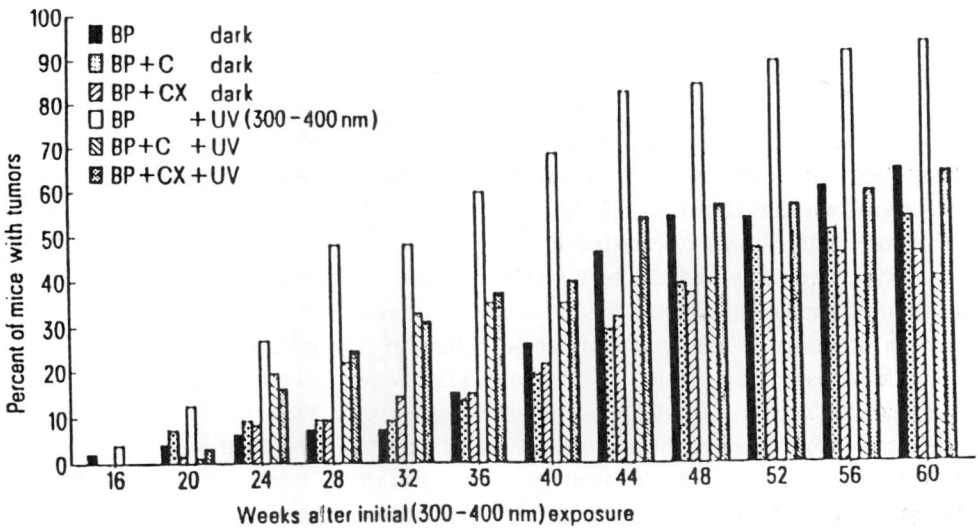

Figure 8. Percent inhibition coeffecent of BP-PCE by BC and CX against weeks after initial UV exposure. The value of 100 indicates the shifting point from the protective to the blocking effect of carotenoids on BP-PCE. After Santamaria et al. (1983).

991

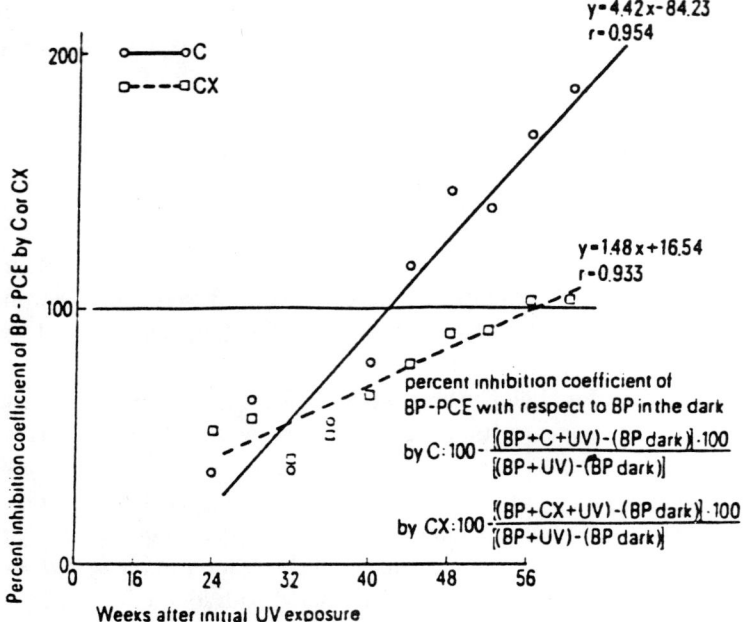

Figure 9. Percent inhibition coefficient of BP-PCE by BC (–) and CX (– –) against weeks after initial UV exposure. After Santamaria et al. (1983)

A suitable analysis of the results as in Fig. 9 pointed out that both carotenoids controlled completely the PCE. Apparently, this effect was independent of a provitamin A activity, which is absent in CX, but was highly dependent on the anti-oxidant property of carotenoids. It was interesting to point out that in this experiment the low but significant skin cancer induced by long UV light alone was also thoroughly controlled by carotenoid supplementation. The skin cancer inhibition by BC and CX observed in the dark must be due to the scavenging or quenching of oxyradicals. Indeed, these excited states are considered possible also in the dark via interaction with endogenously produced free radicals such as $O_2^-\cdot$ $OH\cdot$, or singlet oxygen ($^1O_2$). Thus, when explaining their antitumorigenic activity the main properties of carotenoids are those as free radical scavengers and singlet oxygen quenchers rather than those connected with provitamin A activity and a possible screening effect, the latter being excluded, through pigment measurement of the skin, in mice and humans fed BC (Sayre et al., 1981; Lamola and Blumberg, 1976).

These results were, to our knowledge, the first clear-cut experimental evidence that carotenoid supplementation can take care of the oxygen radical pathology involved in skin cancer inductions by an indirect carcinogen (Santamaria et al., 1983), thus preventing the tumor onset.

Chemoprevention of Indirect Breast Cancer Induced by Puva in Mice and Its Modeling Through the Two-Step 8-MOP Photomutagenesis Test

The same methodology was adapted in another experimental trial to prevent the breast and other skin appendage photocarcinogenesis in mice induced by 8-MOP (topical application) plus long UV light, practically UV-A light (P-UVA), as above reported (Samtamaria et al., 1985). Indeed, in 8-MOP photobiology it was found that singlet oxygen is generated by triplet energy transfer following a cross-linking reaction with DNA (De Mol and Beijersbergen van Henegouwen, 1981), along with oxygen-radical formation.

Also in this trial BC or CX as diet supplementation prevented both up to about 60% mammary and other skin appendage adenocarcinomas produced by PUVA (Table 4). Moreover, in this experiment the lower but evident malignancies induced by long term exposure to UV-A alone were thoroughly controlled by this diet supplementation (Santamaria et al., 1984).

Table 4. Tumour incidence in mice painted 8-MOP, fed BC or CX, kept in the dark or exposed to UV (300–400 nm) light, at 80th week after initial UV exposure

Experimental animal group (75 mice in each group)	UV light exposure	Percent of mice with tumors
8-MOP	−	0
8-MOP + BC	−	0
8-MOP + CX	−	0
8-MOP	+	38
8-MOP + BC	+	16
8-MOP + CX	+	18
Control groups	±	0
No drug	+	9
No drug	−	0
Carotenoids	±	0
Arachidic oil	±	0

As far as the mechanism of action of this chemoprevention is concerned, one should consider that BC and CX yielded results similar to those observed in the inhibition of BP-PCE. Therefore, the properties of carotenoids as radical scavengers and/or singlet oxygen quenchers must also have played a basic role here. Nevertheless, an additional experiment was carried out on the photomutagenicity of 8-MOP using *Salmonella typhimurium* TA 102, a suitable strain to study mutagenesis produced by oxidation processes.

The results indicated that 8-MOP is mutagenic after UV-A irradiation confirming previous findings on *E. coli*. But, this photomutation was prevented by BC and CX up to about 50% and 70% respectively. The same experiment carried out in a medium saturated with nitrogen (anoxia) showed that photomutagenesis by 8-MOP does occur, but to an extent up to about 65% lower than that in the presence of air and with no protection by BC or CX (Santamaria, et al., 1984). These results are reported in Fig. 10.

*For the first time, in 1984, these data demonstrated that **the photobiological reaction by 8-MOP occurs through two steps** as follows*: a first oxygen-independent step, which is apparently responsible for the well-known double strand photo-adduct between 8-MOP and DNA; a second oxygen-dependent step generates oxygen radicals, thus being responsible for the enhancement of DNA damage, up to the levels of mutagenesis as observed in the above experiments (Santamaria et al., 1984).

In terms of multiphasic carcinogenicity, it should be remarked that both BP and 8-MOP are indirect carcinogens. Actually, BP and 8-MOP can initiate a carcinogenic process when BP is oxidized in the cell to the diol-epoxy-derivative, and when 8-MOP is

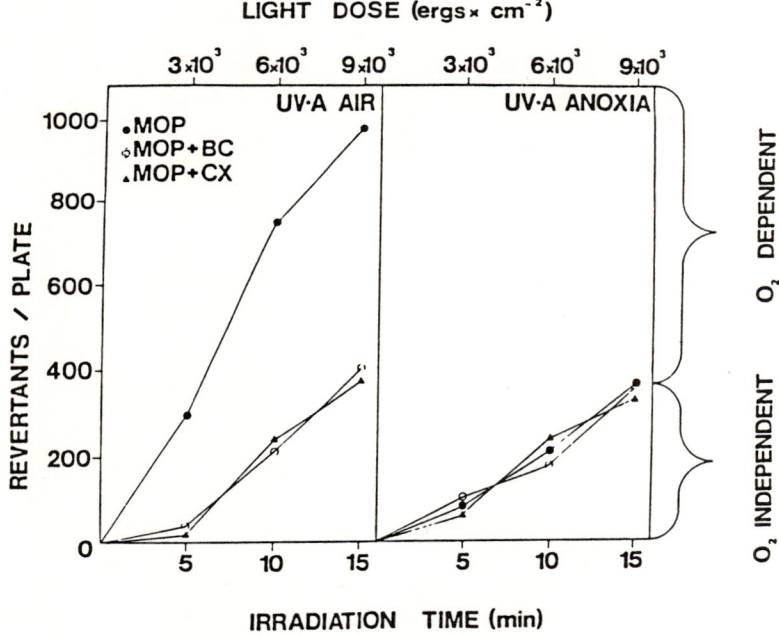

Figure 10. Photomutagenicity induced by 8-MOP in *S. typhimurium* TA 102 in aerated and anoxic conditions and its prevention by BC and CX. The latter are active only in aerated condition. (After Santamaria et al. (1984).

adducted to DNA by light excitation, thus generating oxygen radicals. Therefore, the cancer prevention exerted by carotenoids must take place at the phase of initiation in the case of BP, and at the phase of promotion in the case of 8-MOP.

Carotenoid supplementation in PUVA therapy

As regards possible correlation of the above results (the 8-MOP two-step-photoreaction)and PUVA therapy of skin disorders in humans, one could speculate that anoxic 8-MOP-DNA photobinding could take care of the therapeutic effect in clearing chronic plaque psoriasis, whereas supplemental carotenoids could scavenge or quench the oxyradicals and or $^1O_2$. This theraputic association might control completely the oxygen dependent second reaction step, thus reducing the oncogenic risk (Santamaria et al., 1984).

This prediction appeared to be somehow disproved by clinical work on PUVA therapy carried out by Mcdonald et al. in 1984 on psoriatic patients. Here supplemental carotenoids, even if administrated to patients at high dosage, did not help in preventing burning for which such treatment was attempted, but, interestingly, did not affect the therapeutic action in clearing chronic plaque psoriasis. Apparently this therapeutic action should strictly depend on the oxygen-independent step (8-MOP-DNA photoadduct).

Similar work has been recently carried out by Wolf et al. in 1988 on healthy volunteers using BC plus CX to prevent skin erythema induced by UVB, UVA, and PUVA. The results of this work demonstrated that carotenoids do not protect against both skin erythema and unscheduled DNA synthesis (UDS). On this basis, the authors concluded that one should remain very sceptical about the recommendation that oral carotenoid supplementation, instead of sunscreen, should be adopted for outdoor workers as a cancer preventive measure, as was previously advanced in our work on skin cancer and carotenoids (Santamaria et al., 1984, 1988).

Apparently, the above studies on psoriatic patients as well as on healthy volunteers lead to the conclusion that there is no place for beta-carotene/canthaxanthin in photochemotherapy for psoriasis as well as in protection for photocarcinogenesis induced by chemicals, furocoumarins included.

In this connection, let's attempt a discussion raising up the following points: a) in mice, skin BP photocarcinogenic enhancement is completely prevented by BC or CX supplementation (Santamaria et al., 1980) and breast 8-MOP photocarcinogenesis is prevented up to about 65% by the same treatment (Santamaria et al., 1984), likely through an oxygen radical quenching effect; b) the reports that carotenoids do not protect against skin erythema in healthy subjects (Wolf et al., 1988), whereas they prevent erythema by PUVA in animals (Kornhauserr et al., 1985) are questionably intriguing; c) immediate erythema is not necessarily related to oxygen effect, since it does not occur in 3-carbetoxypsoralen (3-CPS) photosensitization, although 3-CPS is very active in producing $^1O_2$ and is devoid of carcinogenicity since it bounds to DNA at a very low rate (Dubertret et al., 1979); d) erythema is not always related to carcinogenicity by furocoumarins (Dubertret et al., 1979; Pathak and Joshi, 1984); e) last but not least, the therapeutic effect of PUVA on psoriatic patients and its burning effect are apparently both independent of oxygen involvement, since both are not affected by BC/CX treatment, even at high dosage, as remarked above.

After underscoring all tese discussion points, let's consider the fact that photomutagenesis test carried out on *S. typhimurium* TA 102 disclosed a two-step-reaction of 8-MOP, as reported above. Accordingly, it appears reasonable to draw the conclusion that in Mcdonald et al. experiment (1984) the burning and the therapeutic effect depended on the anoxic step of 8-MOP photoreaction, namely on the 8-MOP-DNA photoadduct. Therefore, both burning and therapeutic effect could not be modulated by the concurrent carotenoid supplementation, that likely had to quench all the oxygen radicals produced *in vivo* by the oxygen-dependent step. On the same basis, in Wolf et al. experiments (1988), UVB, UVA, and PUVA erythema, as well as thymine dimers produced by UVB should occur throughout oxygen-independent reactions on which carotenoids cannot be active.

Then, the antitumorigenic action of carotenoid supplementation maintains its theoretical validity in both McDonald et al. and Wolf et al. experiments, but this predictable cancer prevention in humans, up to a certain percentage, can only be ascertained after several years of clinical trials. At any rate, the recommendation that oral carotenoids supplementation, instead of sunscreens, should be adopted for outdoor workers as a cancer preventive measure, is validated by the fact that current treatment with BC plus CX ameliorate photodermatosis and seems to inhibit photo-induction of labial HSV recurrencies, the latter being extremely interesting as far as the implication of viral tumorigenesis is concerned.

Finally, Gatti et al. (1989) recently demonstrated that, when psoriatic patients with skin phototype 2 are loaded with BC plus CX (Phenoro) two months before photochemotherapy, erythema and burning were significantly reduced. This permitted to increase gradually and constantly the irradiation dose, thus obtaining a better therapeutic response. The latter confirmed our assumption that the therapeutic effect is due to the oxygen independent step, but the therapeutic advantage by carotenoids, as predicted in our previous work (Santamaria, 1984) was observed only with skin type 2 and not 3. In this connection, the same authors concluded that the patients with skin type 3 and 4 are likely more exposed to the cancer risk in that they, being devoid of cutaneous side-effects, undergo more prolonged PUVA treatment.

Chemoprevention of MNNG Direct Gastric Carcinogenesis and DMH Indirect Colon Carcinogenesis in Rats and Mice

The antitumorigenic activity of carotenoids, described above, relates to substances listed as indirect carcinogenic agents, such as benzo(a)pyrene (BP) and 8-methoxypsoralen (8-MOP), their activity being dependent on light excitation and/or metabolic oxidation. Hence, tests were made on the effects of supplemental dietary carotenoids on a carcinogenic process induced by a direct agent acting independently of light excitation and/or metabolic oxidation. In this connection, the model of N-methyl-N'-nitro-N-nitrosoguanidine (MNNG) induced gastric carcinogenesis in rats according to Kunze et al. (1979) was first adopted as follows. Low doses of this carcinogen were applied over a limited period of time in drinking water to rats to allow gastric cancer to develop stepwise via several successive stages of transformation with varying biological potential, expressed by dysplasia grade I, II, III and progression to early and infiltrating cancers.

This process took place independently of erosions, ulcers, and benign-appearing proliferative or neoplastic epithelial lesions. They clearly developed stepwise via several successive stages of transformation from polypoid gastritis, gastric atrophy, intestinal metaplasia, mild dysplasia, moderate dysplasia, severe dysplasia including signet ring cell dripping and early cancer. The multifocal development of gastric cancer was expressed by the appearance of preneoplastic and neoplastic changes simultaneously in various areas of the same stomach.

From these data it appeared that supplemental carotenoids did not interfere with any pre-neoplastic lesion arising from glandular mucosa initiated by MNNG oral application. Somehow, carotenoids must have increased the severe dysplasias in the form of signet ring cell dripping. However, carotenoids clearly inhibited the progression of dysplasias to early and infiltrating carcinomas by more than 60–70%, respectively (Santamaria et al., 1987, 1988).

To explain these results, much attention was paid to the fact that reactive oxygen scavengers and detoxifier agents (superoxide-dismutase (SOD), SOD-mimetics, catalase, and phenolic antioxidants) do inhibit biochemical and biological actions of tumor promoters *in vitro* and *in vivo* (Kensler and Trush, 1984). This provided strong presumptive evidence for involvement of oxygen radicals during all the phases of neoplastic development, including progression to infiltrating malignacy.

These findings in the gastroenteric tract were recently confirmed by Temple and Basu (1987) in an experiment on mice demonstrating that dietary BC at a nutritionally relevant level (22 mg/kg diet) is active in preventing colon tumors induced by dimethylhydrazine (DMH). The incidence and multiplicity of tumors was halved. While adenomas, the predominant type of tumors, were only 40% less frequent in supplemented mice, adenocarcinomas were largely absent. Consistent with the above data, the mortality rate in BC supplemented mice was only about half that in unsupplemented mice. Significantly, BC supplementation did not alter initial mild colon mucosal hyperplasia, thus indicating, according to Temple and Basu, that the protective effect against colon cancer may have occurred at the late stage of carcinogenesis. However, considering that the promotion phase, as expressed by adenomas, already appeared to be prevented by about 50%, and that this phase is strictly dependent on the initiation of the carcinogenic process, it may be concluded that BC prevention of DMH colon carcinogenesis occurs at the initiation stage just as with indirect carcinogens. This suggests that the mild colon mucosal hyperplasia observed 12 weeks after the first dose of DMH in both BC

997

supplemented and unsupplemented groups should be interpreted as a drug effect related to the initial DMH induced injury. In this condition, the mucosa may require BC to provide more vitamin A than normal, thus stimulating proliferation (Peto et al., 1981) in both DMH treated groups.

Synoptic Table of the Basic Experimental Evidence of Carotenoid Antitumorigenic Activity

All the above data appear to answer the full demand advanced in 1981 by Peto et al. (Peto et al., 1981) for experimental work to confirm carotenoid antitumorigenic activity. In this connection, it is worth reporting in Table 4 the basic references of carotenoid effects on experimental indirect and direct carcinogeneses and tumor transplantation in mice and rats, computing the dosage of supplemented carotenoids and referring the type of the effect along with the assumed mechanism of action.

From Table 5 it is evident that the antitumorigenic effect of carotenoid is independent of the pro-vitamin activity of beta-carotene since CX (with no pro-vitamin A activity) is antitumorigenic as well. The dosage of carotenoids employed in the different experiments is quite high (50-150 mg/Kg b.w./day), but it may be also relatively low (1.8–4 mg/Kg b.w./day), thus indicating the efficacy of the drug. In this connection, it is worth recalling that BC up to the very high dosage of 1000 mg/Kg b.w./day is well tolerated by animals (Heiwood et al., 1985).

The supplementations according to Santamaria et al. started one month before carcinogenic treatment and continued throughout the experiments. The major fact in almost all this experimental work is that the mechanism of action of this antitumorigenic activity was attributed to the antioxidant property of carotenoids able to inhibit the formation of the ultimate carcinogen (as in the indirect agents) and/or the endogenous synthesis of some carcinogens (as in direct agents, like nitrosoamine). In the first case the antitumorigenesis is exerted at the level of initiation (see BP skin carcinogenesis) or promotion (see 8-MOP breast cancer); in the second case, this action is shown much later at the progression phase. Nevertheless, one should not understimate the relevant immunostimulating activity of this class of natural compounds, which certainly plays an important well known role in protection against cancer (Seifter et al., 1981). In this connection, it should be remarked that antioxidants have been proven to counteract the deterioration of the immune system induced by free radical reactions, thus stimulating the normal function of T-lymphocytes. This stimulation associated with the anticlastogenic activity of antioxidants indicated the importance of antioxidants in autoimmune diseases as, e.g., Bloom's syndrome, where chromosomal fragility, abnormal immune response, and increased frequency of tumours constitute a well known triad (see Feher et al., 1987). Therefore, when a chemopreventive effect by carotenoids has been ascertained as due to their antioxidant activity, this should include also their immunostimulating properties.

Table 5

YEAR	AUTHOR	CAROTENOIDS Type	CAROTENOIDS Dosage (*)	CARCINOGENIC MODEL	EFFECT	ASSUMED MECHANISM
1973	DOROGOKUPLA A.G. ET AL. (58)	RED CARROTS IN DIET	UNLIMITED	SKIN CANCER BY DMA IN MICE AND RATS (indirect)	PROTECTION DELAY	VITAMIN A
1977	EPSTEIN J. (59)	BETA-CAROTENE I.P.	250 mg/Kg b.w. 3 TIMES/WEEK (100 mg/Kg b.w./DAY)	UVB SKIN CANCER IN HAIRLESS MICE (indirect)	DELAY	UNDETERMINED
1980	MATHEWS - ROTH M.M. (60)	BETA-CAROTENE CANTHAXANTHIN P.O.	6680 mg/Kg b.w./DAY idem	idem OR DMBA ± CROTON OIL OR ± UVB (indirect)	DELAY PREVENTION	ANTIOXIDANT
1980	SANTAMARIA L. ET AL. (61)	BETA-CAROTENE (**) CANTHAXANTHIN supplemental P.O.	100 mg/Kg b.w./DAY (DIETARY) + 100 mg/Kg b.w./TWICE A WEEK (GAVAGE)	SKIN CARCINOGENESIS BY BP ± UVA IN MICE (indirect)	100% PREVENTION 60% TOTAL	ANTIOXIDANT ACTING AT THE INITIATION STAGE
1980	SANTAMARIA L. ET AL. (61)	idem	idem	UVA LONG TERM SKIN PHOTOCARCINOGENESIS IN MICE (indirect)	100% PREVENTION	ANTIOXIDANT
1981	SEIFTER R. ET AL. (82)	BETA-CAROTENE P.O.	1.8 mg/Kg b.w./DAY	ADENO-CARCINOMA C3HBA IN MICE (transplantation)	DELAY/PREVENTION/ REGRESSION	VITAMIN A + IMMUNOSTIMUL.
1984	SANTAMARIA L. ET AL. (64)	BETA-CAROTENE CANTHAXANTIN SUPPLEMENTAL P.O.	idem as above (1980)	BREAST PHOTO-CARCI-NOGENSIS BY 8-MOP (PUVA) IN MICE (indirect)	60% PREVENTION	ANTIOXIDANT ACTING AT THE PROMOTION STAGE
1985	SANTAMARIA L. ET AL. (66)	idem	50 mg/Kg b.w./DAY DIETARY + 100 mg/Kg b.w./3 TIMES A WEEK (GAVAGE)	MULTIPHASIC GASTRIC CARCINOGENESIS BY MNNG IN RATS (direct)	BLOCKAGE OF PROGRESSION	ANTIOXIDANT ACTING AT THE PROGRESSION STAGE
1987	TEMPLE. N.J. AND BASU. T. (70)	BETA-CAROTENE	4 mg/Kg b.w./DAY	COLON CANCER BY DMH IN MICE (indirect)	50% ADENOMA 100% ADENOCARCINOMA PREVENTION	ANTIOXIDANT + IMMUNOSTIMUL.

(*) The dosage of carotenoids was computed assuming that mice weighing 25 g eat about 5 g of pellets/day, whereas rats weighing about 250 g eat about 25 g of pellets/day

(**) All supplementation, according to Santamaria *et al.* started one month before carcinogenic treatments and continued throughout the experiments.

999

Prospect for Human Interventions with Carotenoid Supplementation Against Cancer Recurrencies After Radical Treatment

All these chemical interventions on the pathogenesis of malignancies in epithelial tissues are in keeping with the concept of cancer chemoprevention (Sporn et al., 1976), consisting in the reversion of precancerous lesions. This chemoprevention can be called primary when it scavenges the oxygen radicals capable of producing the initiator ultimate carcinogen or those involving promotion and progression phases in both direct and indirect carcinogeneses; secondary, when it can be applied in any post-surgical conditions in humans if a primary cancer in stage I (no lymphonode involvement) is completely excised by radical surgery. In such cases occurring in compartments like the lung, urinary bladder, mammary gland, stomach, colon, etc., diet supplementation with carotenoids (or other antioxidants) should prevent the expression of a "second primary neoplasia" from mucosas already initiated by chemical carcinogens. This rationale has been already approved in Italy by the Ministero della Sanità and by the Lombardy Regional Government for the Centro Tumori of the University of Pavia, where human intervention programs are in progress to prevent second primary malignancy in the lung as well as the urinary bladder after radical surgery.

First clinical case report (1980–88)

The clearcut results published in Pavia in 1980 on the antitumorigenesis of BC and CX, as demonstrated with regard to skin cancer photodynamically induced by BP in mice (Santamaria, 1980) stimulated immediate interest among GPs in a medical environment traditionally sensitive to scientific breakthroughs. This interest was particularly acute in one case, a classic story of a highly successful, highly professional middle-aged M.D. who decided to try out these findings on his family, himself and a few of his patients, naturally following ethical rules most scrupulously.

At that time, only the above experimental data provided a convincing theory as regards promising developments in the field of chemoprevention, keeping in mind that oxygen radicals played a fundamental role in carcinogenicity whenever the balance in endogenous antioxidant systems is somehow impaired, especially by phlogistic processes which generate large amounts of active oxygen-excited species. This led the colleague mentioned above to co-operate actively in clinical attempts to prevent recurrences after radical treatment of epithelial tumours. He came to agree with our theory that complete removal of a malignancy with no lymph node involvement or even with the most accurate lymphadenectomy cannot reverse the already initiated state of all the remaining epithelial tissue of a particular anatomical compartment. Any cells in this tissue were liable to undergo gene derepression by oxidation damage, thus giving rise to the expression of what is called a "second primary malignancy". Accordingly, the "saturation" dosage of carotenoids at the beginning of the treatment (as proposed elsewhere, i.e. 80 mg of carotenoids for 3 weeks) followed by a daily supplementation of

Table 6. Clinical case report (1980–88). Cancer chemoprevention with BC (40%) + CX (60%) association against cancer recurrence after radical surgery ± chemo-radiotherapy (breast, lung, urinary bladder, head and neck

Patient	Sex	Age	Neoplasm surgery date	Chemotherapy and/or radio-therapy	Chemoprevention initial date	Recurrences I II (up to 3/88)	Expected disease-free interval
R.L.	F	47 (1980)	Breast ductal infilt. Ca $T_2N_0M_0$ Mastectomy 7/80	–	1980	– –	6/10 years
A.S.	F	57 (1980)	Breast mucoid Ca $T_2N_{1(5/5)}M_0$ Mastectomy 3/80	1980 CMF 12 cycles	3/81; interrupted for 6 months in 1985 for car crash	– –	3/5 years
A.C.	M	60 (1985)	Lung epidermoid Ca right medium lobe T_3N_0 Lobectomy 7/85 II ep. Ca left upper lobe T_3N_0 Lobectomy 7/86	–	7/86	– –	1 year
L.B.	M	60 (1983)	Urinary bladder Trans cell Ca G_2 TUR 7/83	–	1985; interrupted 7-12/85 for car crash*	1985 TUR	6 Mo./2 years
G.L.	M	63 (1986)	Urinary bladder Trans cell Ca G_2 TUR 8/86	–	8/86	– –	6 Mo./2 years
F.A.	M	49 (1986)	Urinary bladder Trans cell Ca G_2 TUR 8/86	–	8/86	– –	6 Mo./2 years
B.A.	M	60 (1984)	Urinary bladder Trans cell Ca G_2 TUR 8/86	–	8/86	– –	6 Mo./2 years
P.G.	F	60 (1984)	Urinary bladder Trans cell Ca G_1 TUR 7/84	Adriblastin 20 mg/100 ml bladder washing 24 times/1 year	2/84 for one year then only vegetables and fruit enriched diet	*	6 Mo./2 years
B.C.	M	59 (1984)	Larynx epid. Ca G_1 $T_{1a}N_0M_0$ Partial laryngect. 1/84	–	2/84	– –	2 years
L.P.	M	63 (1985)	Larynx epid. Ca G_1 $T_{1b}N_0M_0$ Partial laryngect. 6/85	–	7/85	– –	2 years
A.C.	F	54 (1983)	Rhinopharynx undiff.Ca $T_2N_2M_0$ 1/83 lymphnode biopsy	radiotherapy 63+10 Gy 3-6/83	4/83 plus vegetables and fruit enriched diet	– –	6 Mo./1 years

* During this period urinary bladder catheterism caused cystitis and subsequently cancer recurrence

40 mg was adopted. The pharmaceutical preparation available in any pharmacy in nearby Switzerland or France, was a capsule containing 20 mg each of BC (40%) and CX (60%). This preparation originally made for the treatment of solar dermatosis was helpful in "sugaring the pill" i.e. in persuading patients to adhere to the treatment in that they found it produced a healthy look through the skin-tanning effect of CX.

A recent survey of these clinical trials begun 8 years ago showed an apparent limitation as far as the number of cases (eleven) was concerned, but provided highly significant results owing to the strong evidence of the total preventive effects after radical treatment (surgery ±chemo- and/or radiotherapy) well beyond the expected disease-free intervals in pathologies affecting four different compartments. This study was, therefore, considered suitable for publication (Santamaria, 1988). Here the main features are reported in Table 6 even though no controls with placebos are envisaged. The only assessable parameter is the expected disease-free interval.

These findings demonstrated that supplemental BC and CX in association produced prevention of recurrences in all of the cases selected on the basis of radical tumour excision and adherence to treatment. One case (A.S.) had serious lymphnode involvement (5/5) and underwent high dosage chemotherapy (CMF) with an expectation of a 3 year disease-free interval. Another case (P.G.) also had chemotherapy but at a very low local dosage. The A.C., M case is at present free from disease even though carotenoid supplementation was started after two subsequent lung lobectomies for two epidermoid cancers T_3N_0 (the second occurring two years ago). The L.B. case is particularly significant: the patient suffering from superficial urinary bladder carcinoma G2, started chemoprevention soon after TUR in 1983, which was interrupted after two years for six months, because of a spine traumatic fracture (due to a car crash). During this period he underwent urinary bladder catheterism with concurrent cystitis. Two months later he suffered a tumour recurrence (trans. cell ca. G1), which was removed by TUR. This clearly confirmed the theoretical expectation that activated molecular oxygen species generated by a phlogistic process can trigger the expression of a second primary cancer.

The patients suffering from lung, urinary bladder and head and neck cancers used to be heavy smokers. Apart from one (P.G.) they have succeded in abstaining from smoking.

The convincing results of this first clinical case report appear to be confirmed by the history of a case personally reported to the authors by Correa. A few years ago a middle-aged patient underwent a kidney plus ureter removal due to three carcinomas in the calyx. Soon after surgery he adhered to BC supplementation, which would seem to explain his present good health. But the most interesting fact was that soon after surgery, dysplasias of bladder mucosa were detected, which completely disappeared after only two months.

Last but not least, it should be noted that *all* the patients selected for our tentative no-recurrence cancer chemoprevention protocol are reported above, even those, such as A.C., M and A.C., F, initially considered hopeless. There is no doubt that the overall picture emerging from these results, though certainly preliminary, is extremely encouraging as regards the current human interventions with randomized methods under the supervision of the Centro Tumori of the University of Pavia.

Acknowledgements

This work was supported by the Ministero della Sanita', Roma, Direzione Generale Servizi Medicina Sociale, Div.IV, contract N. 5OO.4/ RSC/57.3/T/1719, 1986 and 500/4/RSC/57.3/T/2128, 1987. The Ministero della Pubblica Istruzione, Roma, Direzione Generale Istruzione Universitaria, is acknowledged for providing research facilities with a special contribution (1987). It was also partially supported by the Lombardy Regional Government (1986). Hoffmann-La Roche Inc., Basel is acknowledged for providing the carotenoids. Dr. A.P. Baldry is acknowledged for revising the English text. Dr. R. Pizzala and Miss Lorella Meo are acknowledged for the editorial help. Leonida Santamaria takes this opportunity to thank his many co-workers, who demonstrated their enthusiasm and skillfullness starting and developing an original research field over a period of more than 30 years.

References

Alexander, P. and Fox, M. Photodynamic degradation of macromolecules sensitizers by dyes and carcinogenic hydrocarbon. Proc. I Int. Photobiol. Congress. Amsterdam, 336–339, 1954.

Bungeler, W. Uber den Einflus Photosensibilisierender Substanzen auf die Enstehlung von Hautgeschwulsten. Z. Krebsforschung, 46, 130–167, 1937.

Calcutt, G. The photosensitizing action of chemical carcinogens. Br. J. Cancer, 8, 177–180, 1954.

Clark, J.H. The effect of long-wave ultraviolet radiation on the development of tumours induced by 20-methyl-cholanthrene. Cancer Res. 24, 207–211, 1964.

Dansi, A., Garzia, A., and Malinverni, O. Gli idrocarburi policiclici come catalizzatori di ossidazione. Ann. di Chim., 47, 110–117, 1957.

De Mol, N. J., and Beijersbergen van Henegouwen, G. M. J. Relation between some photobiological properties of furocoumarins an their extent of singlet oxygen production. Photochem. Photobiol. 33, 815–819, 1981.

Doniach, I., and Mottram, J.C. On the effect of light upon the incidence of tumours in painted mice. Am. J. Cancer, 39, 234–240, 1940.

Dorogokupla A.C., Troitzkaia, E.G., Adilgereieva, L.K., Postolnikov, S.F., Chekrigina, Z.P. Effect of carotene on the development of induced tumors. Zdravoor. Kazak. 10, 32.34, 1973.

Dubertret, I., Averbeck, D., Zaydela, F., Bisagni, F., Moustacchi, F., Touraine, N., and Latarjet, R. Photochemotherapy (PUVA) of psoriasis using 3-carbetoxy-psoralen, a compound non-carcinogenic in mice. Br. J. Dermatol., 101, 379–389, 1979.

Epstein, J.H. Comparison of the carcinogenic and cocarcinogenic of ultraviolet light on hairless mice. J. Natl. Cancer Inst., 34, 741–745, 1965.

Epstein, J.H. Effects of β-carotene on ultraviolet induced cancer formation in the hairless mouse skin. Photochem. Photobiol., 25, 211–213, 1977.

Epstein, S.S., Small, M., Falk, H.F., and Mantel, N. On the association between photodynamic and carcinogenic activities in polycyclic compounds. Cancer Res. 24, 855–862, 1964.

Feher, J., Csomos, G., and Vereckey, A. Free Radical Reactions in Medicine. Springer Verlag, Berlin, pp.52–57,1987.

Felber T.D, Smith EB, Knox JK, Wallis C, Melnick JL. Photodynamic inactivation of herpes symplex. JAMA, 223, 289, 1973.

Felber T.D. Photoinactivation may find use against herpes virus. JAMA, 217, 270, 1971

Findlay, G.M. Ultraviolet light and skin cancer. Lancet, 2, 1070, 1928.

Foote, C.S. Mechanism of photosensitized oxidation. Science, 162, 963–970, 1968.

Friedrich E.G Jr. Relief for herpes vulvitis. Obstet. Gynecol., 41, 74, 1973.

Garzia, A. and Dansi, A. Gli idrocarburi policiclici come catalizzatori di ossidazione. Ann. Chim, 45, 31–39, 1955.

Garzia, A. and Dansi, A. Sul meccanismo di ossidazione di miscugli di benzopirene e cisteina. Il farmaco, 8, 449–454, 1953.

Gatti, M., Brazzelli, V., Bellosta, M., Vignini, R., Pizzala, R., and Borroni, G., Fotoprotezione e carotenoidi nella PUVA terapia della psoriasi, limiti e proposte. Dermatologia e Benessere, n. 1, anno II, 1–5, 1989.

Goeckerman W.H. Treatment of psoriasis. Northwest Med. 24, 229, 1925.

Griffin AC, Hakim RE, Knox J. The wave lenght effect upon erythemal and carcinogenic response in psoralen treated mice. J. Invest. Dermatol., 31, 289–290, 1958.

Heiwood, R., Palmer, A.K., Gregson, R.L. and Hummler, M. The toxicity of Beta-carotene. Toxicology 36: 91–100, 1985.

Honigsmann H, Konrad K, Gschnait F, Wolff K. Photochemotherapy of mycosis fungoides. In: proc Seventh Int Congr Photobiol, Rome, Italy, (Abstracts). 222, 1976.

Kaufman, R.H., Gardner, H.L., Braun, D., Wallis, C., Rawls, E., and Melnick, J. Herpes genitalis treated by photodynamic inactivation of virus. Am. J. Obstet. Gynecol, 117, 1144, 1973.

Kensler, T.W., and Trush, M.A. Role of oxygen radicals in tumor promotion. Env. Mutag. 6, 593–616, 1984.

Kohn Speyer, A.C. Effect of ultraviolet radiation on the incidence of tar cancer in mice. Lancet, 1, 1305–1306, 1929.

Kornhauser, A., Wamer, W., Giles, A. Effect of dietary beta-carotene on psoralen-induced phototoxicity. Ann. N.Y. Acad. Sci., 453, 91–104, 1985.

Krinsky, N.Carotenoid protection against oxidation. Pure and Applied Chemistry, 51, 649–660, 1979.

Kunze, E., Schauer, A., Eder, M.and Seefeldt, C. Early sequential lesions during development of experimental gastric cancer with special reference to dysplasias. J. Cancer Res. Clin. Oncol. 95, 247–264, 1979.

Lamola.A., Blumberg.E. The effectiveness of β-carotene and phytoene as systemic sunscreens. Proc. Am. Soc. Photobiol., 109, 1976.

Lawrence, D.J. and Bern, H.A. Vitamin A and mucous metaplasia. Ann. N.Y. Acad. Sci. 106: 646–653, 1963.

Macdonald, K., Holti, G., and Marks, J. Is there a place for β-carotene/canthaxanthin in photochemotherapy for psoriasis? Dermatologica, 169, 41–46, 1984.

Maisin, J. and De Jonghe, A. Au sujet de l'action de la lumière et de l'ozone sur certains corps cancérigènes. Compt. Rend. Soc. Biol., 117, 111–114, 1934.

Maltotsy, G. and Fabian, G. Measurement of the photodynamic effect of cancerogenic substances with biological indicators. Nature, 158, 877–878, 1943.

Mathews-Roth, M.M. Carotenoid pigments as antitumor agents. In "Current Chemotherapy and Infection disease". Nelson, J.D., and Grassi, C., Am. Soc. Microbiol., Washington, D.C, pp. 1503–1505, 1980.

Miller, E.C. Studies on the formation of protein-bound derivatives of 3,4-benzpyrene in the epidermal fraction of mouse skin. Cancer Res., 11, 100–108, 1951.

Miller, J.A. and Miller E.C. The carcinogenicities of fluoroderivatives of 10–methyl-1,2-benzanthracene II. Substitution of K region and the 3'-,6-,7-positions. Cancer Res., 23, 229–239, 1963.

Mortazawi SAM. Meladinine und UVA bei Vitiligo, Psoriasis, Parapsoriasis und Akne Vulgaris. Dermatol Monatsschr, 158, 908, 1972.

Morton, J.J., Luce-Clausen, E.M., and Mahoney, E.B. The effect of visible light on the development of tumours induced by benzpyrene in the skin of mice. Am. J. Roentgenol., 43, 896–898, 1940.

Morton, J.J.Luce-Clausen, E.M., and Mahoney, E.B. Visible light and skin tumours induced with benzpyrene in mice. Cancer Res., 2, 256–260, 1942.

Mottram, J.C., and Doniach, I. The photodynamic action of carcinogenic agents. Lancet, 1, 1156–1158, 1938.

Parrish J.A, Fitzpatrick T.B, Tanenbaum L, Pathak M.A. Photochemotherapy of psoriasis with oral methoxalen and longwave ultraviolet light. N. Engl. J. Med., 291, 1207, 1974.

Parrish, J.A, Wolff, K, Fitzpatrick T.B, et al. Oral psoralen photochemotherapy treatment of psoriasis. In: Proc Seventh Int Congr Photobiol, Rome, Italy (Abstracts). 225, 1976.

Pathak, M. A., and Joshi, P.C. Production of active oxygen species ($^1O_2$ and $^3O_2$ by psoralens and ultraviolet radiation (320–400 nm). Biochem. Biophys. Acta, 798, 115–126, 1984.

Peto, R., Doll, R., Buckley, J.D., and Sporn, M.B. Can dietary β-carotene materially reduce human cancer rates? Nature, 290, 201–208, 1981.

Prodi, G. and Maltoni, G. Azione dei raggi ultravioletti sulle carcinogenesi cutanee da 9,10-dimetil-1,2-benzantracene e 3,4-benzopirene nel ratto. Boll. Soc. Ital. Sper., 35, 245–248, 1959.

Rodighiero G. Biochemical and medical aspects of psoralens. Photochem. Photobiol. 24, 647, 1976.

Roenigk H.H. Photochemotherapy for psoriasis (PUVA). In Castellani A, ed: Research in Photobiology. New York: Plenum Press, 409–417, 1977.

Rusch, H.P., Kline, B.E., and Baumann, C.A. The nonadditive effect of ultraviolet light and other carcinogenic procedures. Cancer. Res., 2, 183–188, 1942.

Santamaria L. Photodynamic action and carcinogenicity. In Bucalossi P, Veronesi U, eds: Recent contribution to cancer research in Italy. Milano: CEA, 1960; 176.

Santamaria L., Bianchi A., Arnaboldi A., Daffara P. and Andreoni L. Photocarcinogenesis by methoxypsoralen, neutral red, proflavine, and long UV radiation. Cancer Det. and Prev., 8, 447–454, 1985.

Santamaria, L, Giordano, G.G, and Santamaria, R. Effetto fotodinamico dell'ematoporfirina sul siero di sangue. Studio elettroforetico. Atti Soc. It. Patol., 5 (2), 417–426, 1957.

Santamaria, L. and Fanelli, O. Photodynamic action of polycyclic hydrocarbons on isolated mitochondria in relation to their carcinogenicity. Progress in Photobiology. Proc. of the III Intl. Congr. on Photobiology, Elsevier Publ. Co., pp. 448–453, 1961.

Santamaria, L. Benazzo, L., Benazzo, M., and Bianchi, A. First clinical case-report (1980–88) of cancer chemoprevention with beta-carotene plus canthaxanthin supplemented to patients after radical treatment. Boll. Chim. Farm., 127, 57s–61s, 1988.

Santamaria, L. Problems of energy transfer in photodynamic reactions. Bull. Soc. Chim. Belg., 71, 889–905, 1962.

Santamaria, L., and Giordano, G.G. Effect of Long-wave ultraviolet radiation on polycyclic hydrocarbon carcinogenesis. In "The Biologic Effects of Ultraviolet Radiation", Urbach, F. (ed.), Pergamon Press Ltd., Oxford, 569–580, 1969.

Santamaria, L., and Prino, G. The photodynamic substances and their mechanism of action. U. Gallo and L. Santamaria Eds. Res. Prog. Org. Biol. Med,. Chem., 1, 259–336, North Holland Publ. Co., Amsterdam, 1964.

Santamaria, L., Bianchi, A, Andreoni, L., Santagati, G., Arnaboldi, A., and Bermond, P. 8-Methoxypsoralen photocarcinogenesis and its preventiom by dietary carotenoids. Preliminary results. Med. Biol. Env., 12, 533–537, 1984.

Santamaria, L., Bianchi, A., Arnaboldi, A., and Andreoni, L. Prevention of the benzo(a)pyrene photocarcinogenic effect by β-carotene and canthaxanthine. Preliminary study. Boll. Chim. Farm., 119, 745–748, 1980.

Santamaria, L., Bianchi, A., Arnaboldi, A., Andreoni, L., and Bermond, P. Dietary carotenoids block photocarcinogenic enhancement by benzo(a)pyrene and inhibit its carcinogenesis in the dark. Experientia, 39, 1043–1045., 1983.

Santamaria, L., Bianchi, A., Arnaboldi, A., Ravetto, C., Bianchi, L., Pizzala, R., Andreoni, L., Santagati, G., and Bermond, P. Chemoprevention of indirect and direct chemical carcinogenesis by carotenoids as oxygen radical quenchers. Ann. N.Y. Acad. Sci., 534, 584–596, 1988.

Santamaria, L., Bianchi, A., Ravetto, C., Arnaboldi, A., Santagati, G., and Andreoni, L. Prevention of gastric cancer induced by N-methyl-N'nitro-N-nitrosoguanidine.in rats fed supplemental carotenoids. J. Nutr. Growth Cancer, 4, 175–181, 1987.

Santamaria, L., Bianchi, L., Bianchi, A., Pizzala, R., Santagati, G., and Bermond, P. Photomutagenicity by 8-Methoxypsoralen with and without singlet oxygen involvement and its prevention by β-carotene. Relevance to the mechanism of 8-MOP photocarcinogenesis and to PUVA application. Med. Biol Env., 12, (1), 541–549, 1984.

Santamaria, L., Bianco, R., Torriani, C. Variazioni polarimetriche di siero di sangue umano, siero-albumina cristallizzata e peptidi in seguito ad effetto fotodinamico da ematoporfirina Atti Soc. It. Patol., 5 (2), 449–455, 1957.

Santamaria, L., Bucciarelli, G., and Bianco, R. Variazioni spettrofotometriche di siero di sangue, proteine purificate e peptidi in seguito a foto-ossidazione da ematoporfirina. Atti Soc. It. Patol., 5 (2), 469–475, 1957.

Santamaria, L., Calendi, E., and Monaco, L. Azione di idrocarburi policiclici cancerogeni e non cancerogeni su mitocondri in presenza di luce. Atti Soc. It. Patol., 6 (2), 829–831, 1959.

Santamaria, L., Fedi, N., Bartolo, A. Effetto fotodinamico dell' ematoporfirina sul siero di sangue. Studio Polarografico. Giorn. Bioch. 4, 267–277., 1955.

Santamaria, L., Fiornovelli, V., and Nava, C. Comportamento di mitocondri isolati esposti alla luce in presenza di derivati monometilici del 3,4-benzofenantrene e di ormoni steroidei. Atti Soc. It. Patol., 6 (2), 839–841, 1959.

Santamaria, L., Giordano, G.G., Alfisi, M., and Cascione, F. Effects of light on BP carcinogenesis. Nature, 210, 824–825, 1966.

Santamaria, L., Santamaria, R, Ciarfuglia, M. Tensione superficiale di soluzioni di siero di sangue fotossidato mediante ematoporfirina. Atti Soc. It. Patol., 5 (2), 475–461.

Sayre R.M., Black, H.S., Poh Agin: Beta carotene is not an oral sunscreen. Proc. Am. Soc. Photobiol., 113, 1981.

Seelig, M.G. and Cooper, Z.K. Light and tar cancer. Surg. Gynec. Obst., 56, 752–761, 1933.

Seifter, E., Rettura, G. Stratford, F. and Levinson, S.M. CH3BA tumor prevention and treatment with beta-carotene. Fed. Proc. 40: 652–656, 1981.

Smith, D.E., Santamaria, L., Smaller, B. Free radicals in photodynamic systems. In: M.S. Bloys et al. (eds.): Free Radicals in Biological Systems, pp. 305–310. New York, Academic Press, 1961.

Sporn, M.B., Dunlop, N.M., Newton, D.L. and Smith, J.M. Prevention of chemical carcinogenesis by vitamin A and its synthetic analogs (retinoids). Fed. Proc. 35: 1332–1338, 1976.

Temple, N.J. and Basu, T.K. Protective effect of beta-carotene against colon tumors in mice. J.N.C.I. 78 (6): 1211–1214, 1987.

Tronnier H., Schule D. First results of therapy with long wave UV after photosensibilisation of skin. In Schenk GO, ed: Proc Sixth Int Congr Photobiol, Bochum, 1972. Frankfurt: Deutsche Gesellschaft Lichtforschung, 340, 1974.

Uematzu S, Mizuno N. Methoxalen photochemotherapy of pustolosis palmaris and plantaris. In: Proc Seventh Int Congr Photobiol, Rome, Italy (Abstracts). 231, 1976.

Vles, F., De Coulon, A., and Ugo, A. Recherches sur le propriét's physico-chimiques des tissues en relation avec l'état normal ou pathologique de l'organisme. XXI partie. Influence de l'obscurité et de la lumére sur la cancérisation par le gourdon. Arch. Physique Biol., 12, 255–265, 1935.

Wolf, C., Steiner, A., and Honigsmann, H. Do oral carotenoids protect human skin against ultraviolet erythema, psoralen phototoxicity, and ultraviolet-induced DNA damage? J. Invest. Dermatol., 88 (1), 55–57, 1988.

Wollf, K., Fitzpatrick, T.B, Parrish, J.A et al. Photochemotherapy for psoriasis with orally administered methoxalen, Arch. Dermatol., 112, 943, 1976.

Wunderly, C.H., and Petzold, F.A. Die Losung cancerogener Kohlenwasserstoffe im Blutserum. Naturwiessenschaften, 39, 493–464, 1952.

UV and the Environment

UV and the Environment: An Introduction

EMANUEL RIKLIS

Nuclear Research Center-Negev
Beer-Sheva 84190, Israel

While space research is drawing increasing attention, the scope of interest in the biology of our environment on earth and in outer space is widening to the borderline between our atmosphere and outer space, with increasing interest in what impact might change in the near stratosphere have on life on earth.

In the introductory lecture to the Symposium on UV and the environment, delivered by Prof. Frederich Urbach, presenting data prepared by J. Kurylo, based on data compiled by Mack McFarland, recent knowledge of atmospheric ozone science has been described. Not only is radiation from outer space affecting life on earth, but also release of chemicals into the atmosphere is affecting the spectrum of ultraviolet radiation which reaches earth.

Ozone, which acts as a filter in the stratosphere appears to be depleted as a result of the release of halogenated CFCs which are accumulating in the troposphere and are slowly carried up to the stratosphere, where their photodecomposition releases chlorine which in turn attacks the ozone, converting it to molecular oxygen, thus leading to increased amounts of UV-B reaching the earth's surface. This could lead to an increase in skin cancer as well as to possible damage to crops and marine life. Generally speaking, a 1% decrease in ozone could result in a 2% increase in UV-B. The uncertainties in attempts to quantify ozone depletion theories are large as in the number of chemical compounds which may be responsible for the depletion of ozone and the creation of the 'ozone hole'. The acute worry from these events is based on the fact that CFC use is growing and it is accumulating, spreading and can be decomposed only at the expense of ozone. For that reason, on March 15, 1988, an international panel of scientists released the executive summary of the ozone panel report which supports a decrease in the production and release of chlorinated compounds. The exact effect of UV-B radiation on plant and animal species is full of uncertainties and more research is needed.

In space, not only UV-B radiation but rather UV-C and vacuum UV, in addition, of course, to heavy HZe particles are the source of concern. Research on the effects of UV-C under space-like irradiation conditions of dryness, vacuum and cold temperatures, did not start with the space flights era but earlier. As described in my own lecture in this session, already in 1963, while working on the discovery of DNA repair mechanisms, I have

Photobiology, Edited by E. Riklis
Plenum Press, New York, 1991

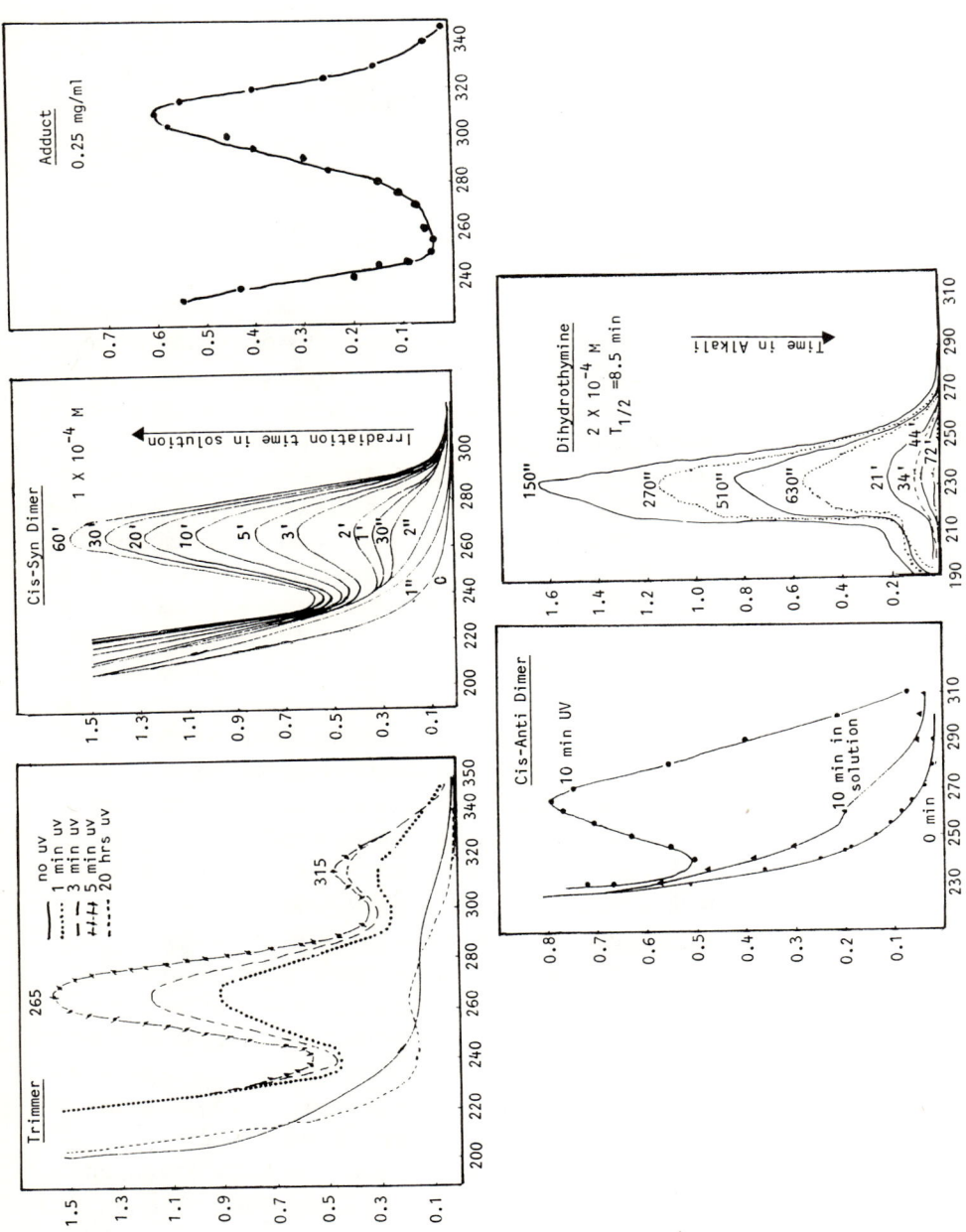

Figure 1. Major and minor thymine photoproducts: absorption spectra, photoproduction, photoreversibility, and dependence on dose and on thymine concentration.

shown that under conditions of dryness and vacuum no thymine dimers were formed, neither in DNA nor in a thin film of thymidine (Riklis, 1964; 1965). Instead, other photoproducts such as the 'spore product', the now well known 6–4 thymine photoproduct and others were formed. A detailed account of the different products formed and the conditions for their production in DNA or in frozen thymine solutions or dry thymine films was published long ago (Riklis, 1965, 1966, 1975, 1976; Cabantchick et al., 1970; Kraemer and Riklis, 1973) and therefore only the absorption spectra of some of these photoproducts which have been described in the Congress lecture will be shown here. The photoproducts whose formation has been described were: a thymine trimer, *cis*-syn thymine dimer, cytosine-thymine dimer, *cis*-anti thymine dimer, the adduct 5-hydroxy-6-4'(5'methyl-pyrimid 2'-1)dihydrothymine, the spore product 5-thyminyl-5,6-dihydrothymine, urea, dihydrothymine, n-propylurea. These were obtained under different conditions of thymine concentrations, cycles of freezing-thawing, exposure through filters to narrow spectral bands and conditions of dryness and vacuum vs. humidity and air or nitrogen as the ambience during exposure.

The interesting significance of these photoproducts is their similarity to photoproducts obtained in space flights or in space-like exposure conditions as described in the lecture of Dr. Gerda Horneck in this session and also published elsewhere (Horneck, 1981; Horneck et al., 1984; Lindberg et al. 1987). In these experiments, the effects of space solar UV light irradiation and space vacuum were studied in *B. subtilis* spores, with simulations performed in the laboratory. Of the few photoproducts obtained, in addition to the spore product, *trans*-syn thymine dimer and another minor photoproduct were formed, and the fact that mutation rate in the spores increased several fold under space ultrahigh vacuum may indicate the possible importance of minor photoproducts.

This short introduction describes the presentations of Urbach, Horneck and Riklis delivered at the symposium on UV and the environment. The other presentations in this symposium are given in full in the following articles.

References

Horneck, G. (1981). Survival of microorganisms in space: a review, *Adv. Space Res.* 1:39–48.

Horneck, G., Bucker, H., Reitz, G., Requardt, H., Dose, K., Martens, K.D., Martens, H.D. and Weber, P. (1984). Microorganisms in the space environment. Science 225:226–228.

Kraemer, J. and Riklis, E. (1973). Photoproduct formation in UV irradiated DNA at high temperature and high irradiation doses. *Intl. J. Radiat. Biol.* 23:75–81.

Lindberg, C. Horneck, G., Bucker, H., Reitz, G. and Requardt, H. (1987). Ultraviolet (180–300 nm) action spectra for inactivation and DNA photoproducts formation of *Bacillus subtilis* spores under ultrahigh vacuum. *Proc. 3rd Europ. Symp. on Life Sciences Research in Space, Graz, Austria*, pp. 197–200.

Riklis, E. (1964). Abstr. *Ann. Microbiol. Conf. Jerusalem*. Published in *Isr. J. Med. Sci.* 1, 321, 1965.

Riklis, E. (1965). Studies on mechanism of repair of ultraviolet-irradiated viral and bacterial DNA *in vivo* and *in vitro*. *Canad. J. Biochem.* 43:1207–1219.

Riklis, E. (1966). The photochemistryt of thymine dimerization by ultraviolet irradiation, NRCN-131.

Cabantchick, Z.I., Riklis, E., Schmidt, G.M.J. (1970). Photoproducts obtained in thymine solutions by UV light irradiation, NRCN-266.

Riklis, E. (1975). Photoproducts and their significance in radiation damage repair. *Suppl. Academia Brasileira de Ciencias* 45:221–226.

Riklis, E. (1976). Excision repair revisited, in DNA repair and late effects. *Proc. Symp. IGEGM* Vienna, Vol. 15:133–154.

Riklis, E. (1989). Photoproducts in DNA irradiated *in vitro* and *in vivo* under extreme environmental conditions. *Adv. Space Research* 9(10):223–234.

UV and Exobiology: Can Micro-Organisms Survive the Space Environment?

H.-D. MENNIGMANN

Institute of Microbiology University
D-6000 Frankfurt/M., FRG

The latest development in the evolutionary chain of events, at least on Earth, is Life. If one accepts that Life is an emanation of the characteristics of Matter one has probably also to accept that the same development inevitably can happen, or could have happened, again here and/or elsewhere provided the prevailing conditions are or were appropriate. But, what this chain of events did look like is in the first place *hypothesis*: we have no experimental means to say clearly how all this *did* happen; but, one can study how it *could* have happened and probably also how it could not. In other words: we can search for the *requirements* for development, evolution, distribution, and transfer of Life throughout the Universe; this is what is meant by the term *Exobiology*.

The recent advancements in leaving the boundaries of Earth have made possible new types of experiments to tackle the above problems, one of the oldest of which is the 'how' and 'where' of the Origin of Life. As far as *Biology* (in a very broad sense) is concerned, it will try to find answers to the question whether Life is, or has been, possible on other celestial bodies, and if so, could it have been transferred from one to the other.

The question of whether living matter could possibly survive a journey through Space has another implication. Some space transportation systems have already landed, or will land, on other celestial bodies; we are confronted by the moral and scientific imperative not to contaminate these by any terrestrial organisms. Thus, if such principles are valid, care has to be taken to avoid any *involuntary* contamination. Plans for *intentional* colonization of, for instance Mars, is a different matter.

Undoubtedly, Space is a harsh environment, and it is unnecessary to discuss here why *higher forms of Life* will not survive if brought there. Thus, the above question can be reduced to whether micro-organisms, their self-replicative parts, or very small genetic entities such as viruses and viroids, can do so. Therefore, in the following, an attempt will be made to determine the Space parameter most hostile to them.

From laboratory experiments it has been known for a long time that micro-organisms can survive very low temperatures. Thus, the low temperature of Space by itself should not do them any harm.

Photobiology, Edited by E. Riklis
Plenum Press, New York, 1991

We also know that even a high vacuum is not necessarily deleterious to cells. When it is, it is due to desiccation and its consequences on the integrity of biopolymers and cellular structures (Dose, 1986). Also, removal of other volatiles may be an important aspect. However, some micro-organisms can produce unique forms of dormant cells ("spores") endowed with special internal conditions and with additional surface layers so that they can endure prolonged periods of dryness. And even cells or spores that do survive show signs of vacuum-induced alterations like an increase in frequency of mutants, repair of DNA lesions, growth delay, and DNA-protein cross-links (Horneck et al. 1984).

Thus, due to the low pressure of 10^{-14} Pa and the low temperature of 4 K one would not expect to find any *active* form of life. However, as far as these parameters are concerned there should be good chances for micro-organisms to survive Space conditions in a *dormant* state if they happened to "get" there (Melosh, 1988).

In contrast to this, the radiation field so different from that of the Earth's surface can be expected to be deleterious enough to be the final determinant of survival of micro-organisms in Space. UV on the one hand can be assumed to be the most hazardous factor not only because of its quantity and quality, but also because of the synergistic effect together with vacuum as indicated by the considerable increase in UV sensitivity. On the other hand it is the easiest form of radiation to be protected against, so that survivors from UV would have a good chance to remain viable in Space for long periods of time. This, and other observations to be reported below, leaves us with the galactic ions ("HZE-particles") as the type of radiation that probably sets the ultimate limits to survival of micro-organisms in Space.

In the following, some past and future experiments will be described and discussed, in which attempts have been, or will be, made to contribute to our knowledge about the probabilities for living matter to exist in Space. As will be seen, these experiments are of two types: exposure of micro-organisms to, and their collection from, Space.

Early balloon and rocket born experiments

The first "collection type" experiments were performed by Imshenetsky and co-workers in the early seventies (Imshenetsky et al. 1976). They used meteorological rockets with a maximum altitude of about 100 km. On four successful flights they triggered sampling devices at various altitudes during the rocket ascent; a total of 31 micro-organisms was collected. These belonged to four sporeforming fungal and two non-sporeforming bacterial species. There is some indication of a gradient in the sense that the number of individuals collected decreases with altitude with an upper limit at about 80 km. However, the total number is too small as to draw any decisive conclusion with regards to an "upper limit of the biosphere". Nevertheless, it is certainly of some importance for further speculation to note that not only the usually more resistant spores, but also vegetative cells can survive at all because even at this relatively low altitude the environment is rather harsh.

Attempts have also been made to collect micro-organisms during spacecraft missions in low earth orbit, i.e. at higher altitudes. But, no single survivor was detected (Lorenz et al., 1976). This could be due to the even more deleterious environment. But, it should also be kept in mind that due to the enormous speed of the vehicle and despite considerable efforts to "cushion" this, the "landing" on the collection surface could have been lethal. Currently in the US more sophisticated devices are in a planning phase.

Most other experiments were of the type where organisms were sent up to be exposed to Space. What seems to have been the first experiment of this type was conducted by A. Stevens already in 1935 (cf. Parfenov and Lukin, 1973) in which seven fungal species were exposed to Space conditions during a 4-hour balloon flight at an altitude of 25 km. The death rate was observed to differ greatly between species. From this and most of the later similar reports it can only be deduced that Space conditions are more or less lethal, and no data are presented which would allow to draw any quantitative conclusion or, more importantly, which would have made possible a comparison with data from laboratory simulation experiments.

The first experiments of this type where sufficient quantitative data were presented were performed by Hotchin, Lorenz and co-workers in the mid-sixties (Lorenz et al. 1988). They exposed the bacterial virus T1 on six successful missions on balloons at an altitude of 34 km, six on sounding rockets at 70–150 km, and two on Gemini satellites.

The value of these experiments lies in the fact that the same group of investigators performed the same type of experiment with the same technique which allows one to at least *construct* survival kinetics despite the fact that the composite data are from several individual experiments. A few more details will be discussed in this case because on the one hand here we have a good example to demonstrate the difficulties of experimentation with biological objects in Space, and on the other hand it gives us a starting point to learn in by what follows why and how experiments became more and more sophisticated.

When plotting the survival data from all single experiments vs. incident photons of the wavelength region of 100–300 nm a biphasic curve resulted as is obtained for populations of two different apparent sensitivities.

But, there are good reasons to believe that this type of a curve typical for T1 survival is merely accidental: (1) The back-extrapolation value for the shallower part varied with the suspension medium from which the virus was dried. (2) Only from known and relevant laboratory experiments one would not expect the curve to have such a shape. Here, survival of dried T1 particles usually follows straight line kinetics. But, a dependence was noted on the humidity of the day of experimentation of whether one obtains a straight line or a biphasic curve (Hill and Rossi, 1954).

These early experiments already tell us a number of facts that are, or will be, taken up in later ones as will be described in this review. (1) UV is rather deleterious to the bacterial virus T1. Unfortunately, however, because of the difficulties in calculating the fluence it is not possible to express the sensitivity of dry and unprotected T1 towards solar electromagnetic radiation in quantitative terms. But, the results allow us to draw some important conclusions for later experiments which have been, or will be, flown.

(2) The experiments have clearly shown that biological material can survive Space conditions at least for a number of hours provided UV radiation is excluded. This is, among others, due to desiccation. This is the same line of arguments as used to explain the survival of a micro-organism inside the lunar TV camera after having been left on the lunar surface for 2.5 years.

(3) Already thin layers of material like broth exert a highly protective effect against inactivation by UV. From the experiments it is not obvious whether this is due to a simple shielding effect or to the highly hygroscopic nature of broth or to some other mechanism.

Experiments on the Apollo-16 mission

The next mission noteworthy in the present context was that of Apollo 16 in 1972. Here within the multi-user-facility "Microbial Ecology Evaluation Device" (MEED) the response towards UV irradiation of a great number of micro-organisms, in fluid suspension or mounted in a dry state, was studied; some of the dry samples were also exposed to Space vacuum (Taylor et al., 1972). The biological endpoints of primary interest were of quite different nature. Though in all experiments UV irradiation (mainly at about 260 nm) was the Space parameter under study only the ones by Bücker, Horneck, and Wollenhaupt (1973) and Spizizen and Isherwood (1975) are of relevance here because in all others the organisms were not exposed to conditions of free Space but were rather kept in closed cuvettes.

Unfortunately, however, published data allow only general conclusions to be drawn. With regards to the samples irradiated at a pressure of 1 bar, like more or less all authors, Spizizen and Isherwood found that on the whole there was fairly good agreement between the effects of on-ground and in-flight irradiation. For the vacuum exposed samples there are no data presented, but only the statement is made that they appeared to be more sensitive against UV; simulation experiments on ground seem to be missing.

In this latter respect, the experiment by Bücker et al. differs. These authors have also performed ground control experiments for both samples irradiated at 1 bar and *in vacuo* ($\approx 10^{-6}$ Torr). The results of the in-flight irradiation at 1 bar agree very well with these ground controls; in other words, as in the experiment by Spizizen and Isherwood, in-flight irradiations have not shown any major unexpected effects so that the ground irradiation experiments at this wavelength can be accepted as appropriate simulations. For the flight samples a synergistic action between UV irradiation and the Space vacuum was observed, the apparent sensitization being slightly smaller than on the ground.

With this experiment it has been shown convincingly and for the first time, that micro-organisms irradiated by solar UV while being exposed to Space vacuum are sensitized by it to more or less the same extent as in appropriate laboratory experiments. Since the total exposure time amounted to only 10 min and since the total energy was even reduced by spectral filters, these results show in addition, that unprotected microorganisms are very rapidly killed by solar UV and thus, in the absence of any protection, have no chance of survival under these conditions.

Experiments on the Spacelab-1 mission

The last group of experiments of interest here is the Space Environment Experiment ES 029 which was flown in 1984 on Spacelab-1. The hardware consisted of an exposure tray able to house 316 individual samples of dried spores of *Bacillus subtilis*. A time controlled shutter system and various combinations of different bandpass and neutral filters allowed the samples to be exposed to different parts of the UV region and to study fluence dependence of a number of biologically important phenomena (Horneck et al., 1984). Half of the tray was hermetically sealed so as to keep a pressure of 1 bar while the other half (covered as above, but without a hermetic seal) was to be exposed to the low ambient pressure.

In this review, emphasis is laid on *survival* of biological systems under the conditions of Space. Therefore, here and with regards to the other phenomena studied, it may suffice to state that both extent of mutant induction, significance of repair, growth delay, and formation of DNA lesions (i.e. thymine photoproducts and crosslinks to protein) are rather similar to the results obtained from terrestrial simulation experiments run in parallel.

The in-flight and simulation survival data for a fully repair-proficient strain of *Bacillus subtilis* show very good agreement for the vacuum exposed samples; this is not true for the 1-bar samples. The latter could be due to an artifact like a fall of pressure inside the exposure tray, an explanation which cannot be excluded with certainty. The argument against this interpretation is the observation (unpublished) that coincidence (for 254 nm irradiation) is better the more cellular repair proficiency is reduced.

The increase in sensitivity to radiation under Space vacuum reviewed above and its conformity, though of varying quality, with simulation experiments, is of importance for two reasons.

Firstly, the results from previous, more qualitative experiments are now substantiated by quantitative data. They prove unequivocally that even such a relatively resistant form of organism like a bacterial spore if unprotected, is inactivated by solar UV in a very short time and thus if exposed has not even a chance to remain viable during travel from one celestial body to another, however close.

Secondly, the aforementioned conformity is also of importance because it proves that the conditions of Space, at least to a certain extent (see discussion below on temperature), can reliably be simulated on the ground; and this in turn is of importance since opportunities for experimentation in Space are distinctly limited. However, many additional experiments need to be performed in order to fully evaluate the trend shown by previous missions and to ask the right questions for, maybe, more sophisticated experiments on future missions. The pursuit of these aims with laboratory experiments seems to be both justified and indispensable.

Implications for panspermia concepts and planetary quarantine

As outlined above, with or without the sensitizing effect of vacuum the survival of a micro-organism in Space is at best a matter of minutes; and it is mainly radiation in the region of 200 to 300 nm wavelength which is responsible for this effect. However, in all previous experiments the importance of one parameter, i.e. temperature in Space of about 4 K, has not been specifically investigated and one other, i.e. a protective coating, has barely been touched upon (see discussion above on early balloon and rocket-born experiments with bacteriophage T1).

Even if samples on an orbiter are exposed to Space in vented containers, actual pressure is not that of Space because of degassing of the orbiter etc.; this might affect induction of lesions due to low pressure (see above), but probably not UV sensitivity. Also, temperature is not the same as that of Space because experiment containers and supporting structures absorb solar infrared radiation. In the Spacelab-1 experiment temperature was measured inside the exposure tray; it varied between 17°C and 35°C with a mean of about 25°C during the whole mission, i.e. it just happened to be "room temperature". In future missions attempts should be made to study UV sensitivity at extremely low temperatures with the aid of an artificially-cooled exposure device.

The conjecture that temperature is probably the crucial parameter in the present context is fostered by recent laboratory experiments (Weber and Greenberg, 1985); these show that reduction of temperature down to about 10 K greatly reduces UV sensitivity of *Bacillus subtilis* spores in ultrahigh vacuum. From these data it has been calculated (Weber and Greenberg, 1985) to take about 150 years of solar UV irradiation in the diffuse interstellar medium (where UV radiation is not attenuated) to reduce survival of unprotected spores to 10 per cent. Once in Space this time span is more than sufficient for a micro-organism to travel at least to other planets of our solar system (Hotchin, 1968). Further reduction in sensitivity would occur if spores reside within dense interstellar clouds (where UV radiation is attenuated by several orders of magnitude) or if within such a cloud they accrete mantles of simple molecules which will then condensate into strongly UV absorbing polymers. Taking both of these possibilities together, a time span of 4.5 to 45 million years is calculated (Weber and Greenberg, 1985) to reduce survival down to 10 percent; the time for a particle of the size of a spore to travel from one solar system to another would be considerably shorter (Hotchin, 1968).

Notwithstanding these numbers it remains a matter of debate whether or not panspermia could "work". The available data do give an idea about the chances of survival of spores once in an environment as described above and only as far as UV radiation is concerned. However, the chances to get there *physically* by leaving a celestial body are really not too large (Imshenetsky et al., 1976; Melosh, 1988), and the same holds true, in consideration of the above data on unprotected spores, for getting there *alive*. Even if this did happen a spore would be confronted with other deleterious factors of which probably the most hazardous ones would be the HZE-particles (see above) against which shielding is impossible; the chances to survive these have been roughly calculated to be of the order of 10^5 to 10^6 years (Horneck, 1981).

In the present context panspermia is one area of interest, the other being *planetary quarantine* of which one particular aspect will be considered here. Outside contamination of a spacecraft may not pose too serious a problem because many micro-organisms will probably be thermally killed during take-off; in addition, their initial number is comparatively small. But, "it is known that orbiting spacecraft have already liberated relatively huge numbers of living microbial spores into terrestrial orbit in the form of fecal material discarded from spacecraft" (Hotchin, 1968). Of these most exist as "mantled" and "freeze-dried" organisms, thus being rendered highly UV-resistant. Since "human faeces contain 10^8–10^{14} viable microorganisms per person per day's excretion, of which approximately 10 per cent are spores" (Hotchin, 1968), and in view of their different starting position for leaving Earth's biosphere and the short travel time to other solar planets (Hotchin, 1968), they do not represent a negligible quantity. The question of whether and where any such organism could survive or even grow has to be left open here.

Outlook: Future experiments

The experiments reviewed have given a number of valuable answers, but several questions are left open and others are newly raised. Some of them, but not all, can be tackled in laboratory experiments, but finally, all have to be verified on Space missions. Key parameters pertinent to missions under way or in an advanced planning state would be: sensitivity of vegetative cells; sensitivity of other types of organisms in a resting state; stability, sensitivity, and polymerization of simple biomolecules and complex biopolymers; substances protective against desiccation effects; substances effective in shielding against radiation; significance of extremely low temperature; significance of UV radiation other then direct solar radiation; significance of long-duration exposure to Space conditions. These, singly or in combination, are subjects for one or the other of the following mission.

One set of experiments already in orbit is housed on the *Long Duration Exposure Facility* (LDEF). It was launched in April 1984; and retrieval was planned to take place one year later. But, it was in orbit until January 1990 and the experiments are awaiting evaluation.

Another set of experiments will be combined in the multi-user facility *Exobiology Radiation Assembly* (ERA) which is an ESA core-facility of the *European Retrievable Carrier* (EURECA) to be launched in 1991. Current plans foresee this to be retrieved after 6 to 9 months in orbit. ERA is a successor to the above-mentioned Spacelab-1 experiment, ES 029, because most of the parameters under study are the same; however, others have been added and the spectrum of test objects is now considerably increased. A similar device will be sent into orbit, probably in early 1992, on the German Spacelab Mission D2.

References

Bücker, H., G. Horneck, and H. Wollenhaupt (1973) Effects of space vacuum and solar UV–irradiation (254 nm) on colony forming ability of *Bacillus subtilis* spores. Proc. Symp. "Microbial Response to Space Environment" (G. Taylor, ed.); NASA TMX–58103, p. 87–103.

Dose, K. (1986) Survival under space vacuum: Biochemical aspects. *Adv. Space Res.* 12:307–312.

Hill, R.F. and H.H. Rossi (1954) The ultraviolet sensitivity and photoreactivability of T1 bacteriophage, I. Effect of irradiation conditions upon survival curves. *Radiation Res.* 1:282–293.

Horneck, G. (1981) Survival of microorganisms in space: A review. *Adv. Space Res.* 1:39–48.

Horneck, G., H. Bücker, G. Reitz, H. Requardt, K. Dose, K.D. Martens, H.D. Mennigmann, and P. Weber (1984) Microorganisms in the space environment. *Science* 225:226–228.

Hotchin, J. (1968) The microbiology of space. *J. British Interplanetary Soc.* 21:122–130.

Imshenetsky, A., S. Lysenko, G. Kazakov and N. Ramkova (1976) On micro–organisms of the stratosphere. *Life Sci. Space Res.* 14:359–362.

Lorenz, P.R., J. Hotchin, A.S. Markusen, G.B. Orlob, C.L. Hemenway, and D.S. Hallgren (1968) Survival of micro–organisms in space. *Space Life Sci.* 1:118–130.

Lorenz, P.R., G.B. Orlob, and C.L. Hemenway (1969) Survival of micro–organisms in space. *Space Life Sci.* 1:491–500.

Melosh, H.J. (1988) The rocky road to panspermia. *Nature* 232:687–688.

Parfenov, G.P. and A.A. Lukin (1973) Results and prospects of microbiological studies in outer space. *Space Life Sci.* 4:160–179.

Spizizen, J., J.E. Isherwood, and G.R. Taylor (1975) Effects of solar ultraviolet radiations on *Bacillus subtilis* spores and T_7 bacteriophage. *Life Sci. Space Res.* 13:143–149.

Taylor, G.R., C.E. Chassay, W.L. Ellis, B.G. Foster, P.A. Volz, J. Spizizen, H. Bücker, R.T. Wrenn, R.C. Simmonds, R.A. Long, M.B. Parson, E.V. Benton, J.V. Bailey, B.C. Wooley, and A.M. Heimpel (1972) Microbial response to space environment. *NASA SP–315*, p. 27.11–27.17.

Weber, P. and J.M. Greenberg (1985) Can spores survive in interstellar space? *Nature* 316:403–407.

Further Solar UV Spectral Measurements at the Dead Sea

A.P. KUSHELEVSKY

Department of Nuclear Engineering
Ben Gurion University
Beer Sheva, Israel

Abstract

Further UV solar spectral measurements were made during the month of February 1987 at the Dead Sea and at Beer Sheva. The spectra were similar to those measured previously during the summer with preferential attenuation at wavelengths close to 300 nm in the UV-B region. A moving edge spectrometric measurements with improved resolution confirmed this conclusion.

Introduction

The Dead Sea area is one of the most important health resorts in Israel for the treatment of psoriasis. Treatment is based on a combination of bathing in the Dead Sea and sunbathing in an open air solarium (Azizi et al., 1982).

While the role of the Dead Sea cannot be neglected, the success of the treatment at the Dead Sea (exceeding 80%) is credited primarily to the solar UV radiation which is filtered by the column of air over the Dead Sea, nearly 400 meters below sea level.

Previous measurements of the UV solar spectra at the Dead Sea (Kushelevsky and Slifkin, 1975) show that the shorter wavelength UV radiation is attenuated considerably more than UV radiation with longer wavelengths. This leads to a much 'softer' solar spectrum relatively low in erythemal UV-B radiation which allows the patients to sunbath for longer periods than at sea level with a lower risk of immediate and long-term damage to their skins.

The previous spectral measurements were carried out during the summer. The marked decrease in the UV-B content was assumed to be due to the haze over the Dead Sea which scatters and absorbs the incident sunlight as it was assumed that the scattering and absorption by the extra 400 meters of air above the area cannot explain, on its own, the relatively large attenuation of UV radiation.

Further measurements (Goldberg and Kushelevsky, 1977; Leibovici et al., 1987) using broad band UV sensitive radiometers, showing elevated UVA/UVB ratios at the Dead Sea, although confirming the spectral measurements throw a doubt on the role of the haze in selective attenuation, as those measurements also reveal elevated UVA/UVB

Photobiology, Edited by E. Riklis
Plenum Press, New York, 1991

ratios in the winter, when the haze over the area is much less than during the summer. Spectral measurements were therefore repeated during the winter and the results are discussed in this paper.

Experimental

A double monochromator spectroradiometer (IL700/760D/790) optimized for UV measurements using a solar blind UV sensitive photomultiplier as a detector, with extremely high stray light rejection, was used to measure the solar irradiance as a function of wavelength on a horizontal surface at Ein Bokek on the Dead Sea shore and at Beer Sheva (250 m above sea level). As in previous measurements the Beer-Sheva spectra was used as a standard against which the Dead Sea spectra were compared.

Measurements were made at 1 nm intervals using a 5 nm bandwidth. The Dead Sea/Beer Sheva irradiance ratios plotted against wavelength are shown in Figs. 1–3.

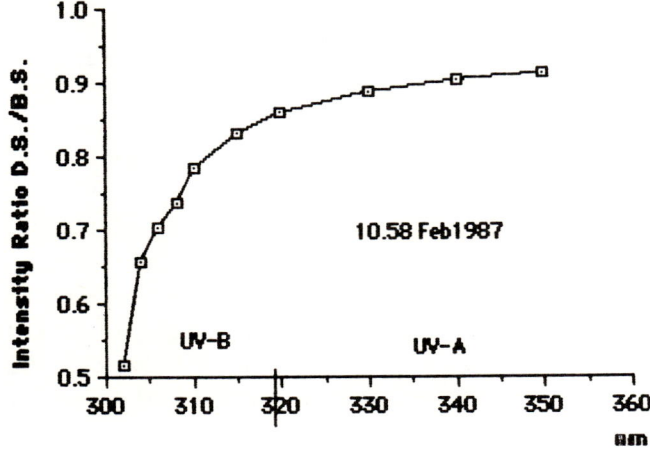

Figure 1. Intensity ratio D.S./B.S.

Improved resolution was obtained using a scanning edge measurement method (Table 1). This method consists of centering the spectroradiometer at a wavelength about a half of bandwidth below cut off so that no signal is observed and then increasing the wavelength setting so that the edge of the acceptance band overlaps the solar spectrum (Fig. 1) giving a signal which corresponds to the irradiance over the wavelength interval, equal in size to the step by which the central wavelength setting was increased. This, in essence, allows measurements to be made with a resolution independent of the bandwidth used. To be used for absolute radiometric measurements, however, this method requires the signal to be deconvoluted according to the spectroradiometer wavelength response function and the shape of the solar spectrum, but, for comparison purposes deconvolution is unnecessary.

Table 1. UV spectral intensities at Beer Sheva and the Dead Sea using moving edge method

TIME	λ(nm)	10nm b.w.		20 nm b.w.	
		BS	DS	BS	DS
10.58	300	0.70	0.5	2.25	1.95
	295	0.25	0.11	1.00	0.74
	290	0.03	–	0.06	0.02
11.33	300	0.95	0.70	2.53	2.35
	295	0.35	0.24	1.23	0.99
	290	0.06	0.03	0.43	0.28
11.57	300	0.95	0.81	2.25	2.25
	295	0.33	0.24	1.18	0.96
	290	0.04	0.01	0.33	0.24
13.00	300	0.84	0.64	2.37	1.83
	295	0.27	0.21	1.04	0.81
	290	0.05	0.04	0.36	0.25

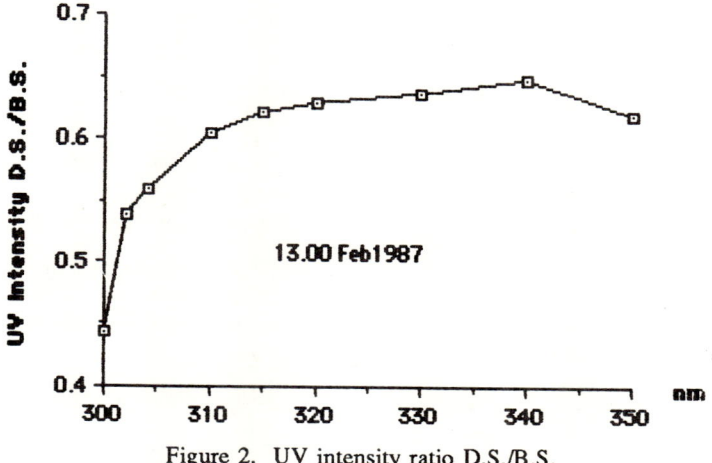

Figure 2. UV intensity ratio D.S./B.S.

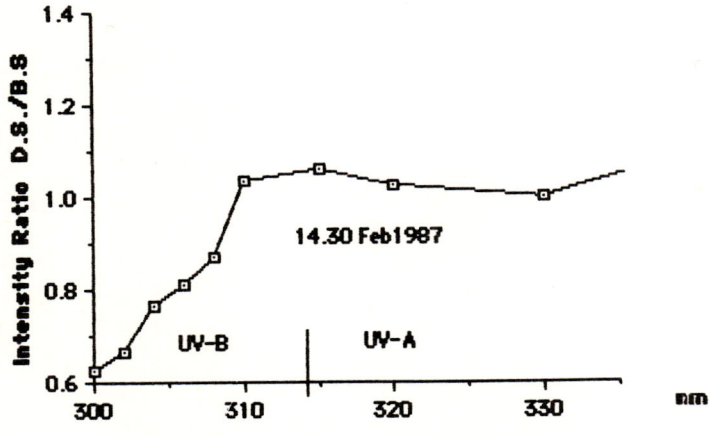

Figure 3. UV intensity ratio D.S./B.S.

Results and Discussion

The solar spectra measured at the Dead Sea using a 5 nm bandwidth normalized against Beer Sheva taken during the winter show the same pattern of attenuation at short wavelengths as the previous measurements taken in the summer.

The moving edge measurements shown in Table 1 corroborate these findings and show that the attenuation close to cut-off are even more significant than is suggested by the conventional spectral measurements.

These findings therefore suggest that the haze over the area, is not the most important factor causing the attenuation of short wavelength UV radiation as was previous thought, for otherwise we cannot explain the attenuation pattern at the Dead Sea in the winter.

This has important implications in scheduling patients to come to the Dead Sea for treatment during the winter to take advantage of the special solar UV spectra in the area.

References

Azizi, E., Kushelevsky, A.P., Avrach, W. and Schewach-Millet, M. (1982). Climate therapy for Psoriasis at the Dead Sea, Israel. *Isr. J. Med. Sci.* 18:267–270.

Kushelevsky, A.P. and Slifkin, M.A. (1975). UV measurements at the Dead Sea and at Beer Sheva, *Isr. J. Med. Sci.*, 11:488–490.

Goldberg, L.H. Kushelevsky, A.P. UV Light measurements at the Dead Sea. Proc. 2nd Int. Symp. on Psoriasis, Stanford (1976). Yorke: New York, pp. 461–463, 1977.

Leibovici, V., Sagi, E., Siladji, S., Greiter, F. and Holabar, K. (1987). Seasonal variation of UV radiation at the Dead Sea. *Dermatologia* 174:290–292.

Ozone and Ultraviolet Light are Additive Co-Carcinogens *In Vitro*

C. BOREK

Radiological Research Laboratory
Dept. of Radiation Oncology and
Department of Pathology/Cancer Center
Columbia University
New York, NY 10032 USA

Introduction

The identification of environmental agents which interact in an additive or synergistic fashion to induce neoplastic transformation is of relevance to human health.

Ozone (O_3) a reactive species of oxygen is a ubiquitous pollutant in urban atmosphere and a potential health hazard to man (NCR Report, Natl. Acad. Sci., 1977). Ozone is a key oxidant in photochemical smog and is formed by the action of ultraviolet light (UV light) on nitrogen oxides and hydrocarbons, emitted from automobiles (Borek, and Mehlman, 1983). Ozone is used as a disinfectant and is produced in UV lamps (NCR Report, Natl. Acad. Sci., 1977).

In earlier work we showed that ozone acts as a direct carcinogen *in vitro*, in mouse C3H/1OT-1/2 cells and in hamster embryo cells (Borek et al., 1986). Exposure of cells to 5 ppm ozone for 5 minutes induced neoplastic transformation and resulted in the activation of dominant transforming genes (Borek et al., 1989a). Our earlier studies also demonstrated that ozone acts as a co-carcinogen with ionizing radiation (Borek et al., 1986, 1989a). Exposure of cells to 3 or 4 Gy of gamma-rays, prior to ozone treatment resulted in a marked enhanced rate of transformation, which was statistically consistent with a synergistic interaction between the two agents (Borek et al., 1986, 1989a).

Ultraviolet light, a non-ionizing form of radiation acts as a carcinogen *in vitro* in a variety of cell systems including the hamster embryo and C3H/1OT-1/2 cells (Donger et al., 1981; Modal and Heidelberger, 1976; Chan and Little 1976; for review see Borek, 1987).

Our more recent studies (Borek et al., 1989b) were undertaken to investigate whether ozone acts with UV light to induce transformation at frequencies which are higher than those produced by each of the agents alone.

Photobiology, Edited by E. Riklis
Plenum Press, New York, 1991

Materials and Methods

Cells

We used primary cultures of hamster embryo cells and mouse C3H/1OT–1/2 cells at passage 10. The two cell systems have been used extensively in carcinogenesis studies and the culture conditions and transformation assays as well established (for reviews see Borek, 1987; 1984).

Ozone Exposure

Cells were exposed to 6 ppm of ozone for 10 min, as previously (Borek et al., 1986; 1989a).

For O_3 transformation and co-carcinogenesis experiments primary hamster cells and C3H/1OT–1/2 cells were seeded at 5×10^6 cells in 100 mm Falcon petri dishes in Dulbecco's medium containing 10% fetal calf serum (GIBCO). Twenty four hours after seeding, medium was replaced with 2 ml buffered saline and cells were exposed to 6 ppm O_3 for 10 min after which the saline was removed and replaced with complete medium. Ozone was formed from pure oxygen by an electric arc discharge apparatus (Borek et al., 1986). Before entering the chamber the O_3 was diluted with room air with the air intake regulated by a flow meter. The combined O_3-room air mixture was drawn into the exposure chamber by the way of a vacuum chamber. A steady state of O_3 was produced in the chamber 1 h prior to exposure. The concentration of O_3 inside the chamber measured in ppm was determined by a Pollution Control Industry UV monitor. Following exposure hamster embryo cells were cloned in 60 mm petri dishes at 700 cells per dish on 3×10^4 syngeneic feeder cells (Borek, 1986) and incubated in a humidified incubator at 37°C for 10 days. The C3H/1OT–1/2 cells were re-seeded out into 10 ml petri dishes to allow for 100–400 survivors and incubated for 6 weeks. Control cultures of both cell types not treated with O_3 but exposed to air and subjected to the same conditions were re-seeded as described above.

UV light Exposures to UV light was carried out by exposing hamster and mouse C3H/1OT–1/2 primary cultures to UV light at 254 nm at a dose rate of 4J/m^2.

Exposure to ozone and UV light

Combined exposures to UV and ozone were carried out by irradiating the cells with UV light (4J/m^2) 10 minutes prior to treating them with 6 ppm of ozone for 10 minutes. After single or combined treatments the cells were cloned as described above. Cells exposed to UV light and to air served as controls.C3H/1OT–1/2

Results and Discussions

Cell survival following the treatments was assessed by colony forming ability (Borek et al., 1987, 1989b). Transformation was scored in the hamster embryo and the mouse C3H/1OT–1/2 cell using morphological criteria (Borek 1987; 1984). In the C3H/1OT–1/2 cell system transformed foci included both type II and type III foci (Borek, 1987; 1989a). Hamster embryo transformed colonies were identified by their piled up morphology and irregular cell-cell orientation (Borek, 1987: 1984). The correlation between the morphological transformation of the cells and their malignant potential has been well established (reviewed Borek, 1987).

As previously we found that treatment of hamster embryo and mouse C3H/1OT–1/2 cells with O_3 resulted in enhanced cell transformation to control untreated cells, as illustrated in Figure 1. Also shown in Fig. 1 are transformation rates corresponding to a protocol where cells were first irradiated with UV light and then, exposed to O_3. Transformation was scored as transformants per surviving cell as previously (Borek, 1986, 1989a,b).

The results presented here indicate that O_3 acts in additive fashion with ultraviolet light to produce neoplastic transformation in hamster embryo cells and C3H/1OT–1/2 cells

Our earlier studies showed that ozone interaction with ionizing radiation is consistent with a synergistic interaction between the two agents and that the mechanisms underlying these transforming events was mediated in part via free radical mechanisms (Borek, 1986; 1989a,b).

The transforming mode of action of UV is wavelength dependent (Doniger, 1981). Most of the products produced at the wavelength are cyclobutane dimers and photoproducts (Cleaver et al., 1988), the latter being extremely unstable and able to convert to other products. While active oxygen production occurs mostly in the near UV range (reviewed in Cerutti, 1985), some free radical reaction occur even at 254. When ozone acts as a co-carcinogen with UV it may oxidize some of the UV induced

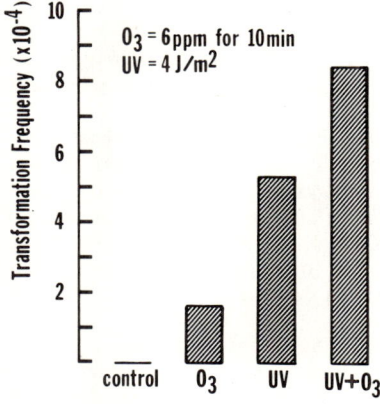

Figure 1. Transformation of mouse C3H1OT–1/2 by Ozone (O3) and ultraviolet light.

products as well as directly produce free radical intermediates by its interaction with a wide range of biological molecules (Borek, 1986, 1989a,b). Ongoing experiments indicate that Vitamin E, an antioxidant, inhibits transformation induced by UV and O_3 as well as markedly suppresses the co-carcinogenic action of UV and O_3. These results support the notion that similar to O_3 the carcinogenic effects of UV are mediated in part via free radical mechanisms and that the additive action of UV and O_3 as co-carcinogens is mediated in part via similar mechanisms.

Our findings underscore the putative long term toxic effects of additive interactions of ozone with ultraviolet light, two ubiquitous agents in our environment.

Acknowledgement

This article was supported by Grant No. CA-12536 from the National Cancer Institute, and by a contract from the National Foundation for Cancer Research.

References

Borek, C. (1984). *In vitro* cell cultures as tools in the study of free radicals and free radical modifiers in carcinogenesis. *Meth. Enzymol.* 105, 465–479.

Borek, C. (1987). The induction and regulation of radiogenic transformation *in vitro*: Cellular and molecular mechanisms. *International Encylopedia of Pharmacology and Therapeutics, Mechanisms of Cellular Transformation by Carcinogenic Agents*, edited by D. Grunberger, G.F. and G.F. Goff (New York: Pergamon Press), Section 126, pp. 151–155.

Borek, C., Mehlman, M.A. (1983). Evaluation of health effects, toxicity and biochemical mechanisms of ozone. *Biochemical Effects of Ozone and Related Photochemical Oxidants.* (Princeton, New Jersey: Senate Press), pp. 325–361.

Borek, C., Ong, A., and Mason, H. (1989a) Ozone and ultraviolet light act as additive co-carcinogens to induce *in vitro* neoplastic transformation. Teratogenesis, Carcinogenesis and Mutagenesis, 9:71–74.

Borek, C., Zaider, M., Ong, A., and Witz, G. (1986). Ozone acts alone and synergistically with ionizing radiation to induce *in vitro* neoplastic transformation. Carcinogenesis,. 7, 1611–1613.

Borek, C., Ong, A., and Zaider, M. (1989b). Ozone activities transforming genes *in vitro* and acts as a synergistic co-carcinogen with gamma-rays only if delivered after radiation. Carcinogenesis 10:1549–1551

Cerutti, P.A. (1985). Pro-oxidant states and tumor promotion. *Science*, 227, 375–381.Chan, G.L., Little, J.B. (1976). Introduction of oncogenic transformation *in vitro* by ultraviolet light. *Nature*, 264, 442–449.

Cleaver, J.E., Cortes, F., Karentz, D., Lutze, L.H., Morgan, W.F., Player, A.N., Vuksanovic, L., Mitchell, and D.L. (1988). The relative biological importance of cyclobutane and (6–4) pyrimidine-pyrimidone dimer photoproducts in human cells: Evidence from a xeroderma pigmentosum revertant. *Photochemistry and Photobiology*, 48, 41–49.

Doniger, J., Jacobson, E.D., Krell, K., DiPaolo, J.A., (1981). Ultraviolet light action spectra for neoplastic transformation of Syrian hamster cells correlate with spectrum for pyrimidine dimer formation in cellular DNA. *Proc. Natl. Acad. Sci.*, (USA) 78, 2378–2382.

Modal, S., Heidelberger, C., (1976). Transformation of C3H/1OT–1/2 CL mouse embryo fibroblasts by ultraviolet irradiation and a phorbol ester. *Nature*, 260, 710–711.

National Research Council Committee Report on Medical and Environmental Effects of Environmental Pollutants: Ozone and other Photochemical Oxidants. *Nat. Acad. Sci.*, (USA) Washington, 1977.

UV Risks and Regulations

Skin Photobiology and Regulations on UV Radiation

JAN C. VAN DER LEUN

University of Utrecht
Institute of Dermatology
Heidelberglaan 100
NL-3584 CX Utrecht
The Netherlands

The popularity of tanning with artificial sources of UV radiation has again been riding high during the past decade. The UV sources used have become bigger and bigger, too large for home use in many cases, but they appeared in beauty parlors, swimming pools and special tanning salons.

There are also worries, especially about the long-term consequences. In several countries, there is pressure from the public on politics and regulators to do something against this development. I will relate to you what happened in two countries.

First in country A; let us assume that everyone in this country carried the name of A. Regulator A was pondering on what to do. He hoped to devise something sensible and effective, and good luck had it that there was a well-known skin photobiologist in the country, by the name of A. The regulator consulted the skin photobiologist. The result, after some time, was a strict regulation; most UV sources for tanning were banned and the remaining types subjected to strict rules. The political system in A took care that the regulation came into effect soon, with the force of law.

In country B, the same problem arose. Regulator B also consulted with a skin photobiologist, B. Conclusion: one cannot expect much good from regulations,especially as one cannot regulate the sun. More effect is to be expected from education of the public and that should be done actively.

Regulators A and B meet in some international conference, and discuss why the outcome in their countries was so different. Each of them is convinced that his country took the right course; each of them has great confidence in his own expert.

The experts A and B meet, at a photobiology conference. They discuss the regulatory actions in their countries and wonder, why the outcome was so different. The two photobiologists have no appreciable difference of opinion on the risks involved in the tanning fashion, not even on the issue of what regulation could be effective, if any. They conclude that it might have been a good idea, to discuss the problem with each other in the first place. This was a piece of real history; the countries A and B were Sweden and the Netherlands, respectively.

Photobiology, Edited by E. Riklis
Plenum Press, New York, 1991

Against this background it appears better that such problems are dealt with by international committees. Ideally, the various countries might then receive the same consensus advice, and one might hope that they would all take similar action, perhaps even coordinated action.

Everything would be fine if there would be one international committee for one problem, in our case the risks of UV radiation. In the real world, there are many international committees, all dealing with basically the same problem, but approaching it from different sides.

The Comité Internationale de l'Éclairage (CIE) deals with illumination. But many lamps emit, besides light, also UV-radiation. The question arises, what damage is done to the eyes and the skins of people. Potential damages are identified, action spectra are defined and it is indicated how it can be estimated what damage is to be expected, and how such damage can be prevented or limited.

The International Electrotechnical Commission (IEC) deals with the safety of electrical equipment, and proposes regulations to prevent accidents. Not only electrical shock, but also burns and any type of damage that may result from the use of electrical equipment. Some types of electrical equipment emit unintentional UV-radiation, such as welding arcs, or intended, such as sunbeds. Therefore the IEC has a working group preparing safety rules for UV-emitting equipment.

There is a sensitive public awareness of the dangers of radiation, especially after the nuclear bombs and nuclear accidents. Almost every country has its radiation hygienists. National groups of radiation hygienists have together formed the International Radiation Protection Agency (IRPA). That has also established an International Non-Ionizing Radiation Committee (INIRC), which produces recommendations on all kinds of non-ionizing radiation, including one on UV radiation.

The World Health Organization (WHO) realizes that one of the threats to public health comes from radiations. The WHO produces Environmental Health Criteria Documents, and one of these is on UV radiation. The WHO Regional Office for Europe has produced a book on the Hazards of Non-Ionizing Radiation; it includes a chapter on UV radiation.

The United Nations Environment Programme (UNEP) has timely realized that one of the most acute threats to the global environment comes from a possible depletion of the ozone layer in the stratosphere; this would lead to an increased transmission of solar UV-B radiation to the earth's surface. UNEP formed a committee to study the consequences of increased UV-B irradiance, including the consequences for human health.

I do not claim completeness for this listing, these are just the committees and organizations within my own field of view. All these committees deal with basically the same problems, and derive their data from the same scientific investigations, yours and mine. There is some overlap of the people involved in all these efforts, but that is mainly limited to the experts. The participants closer to the organizational or regulatory side are different on each international committee. The committees never meet each other. They all have their own responsibilities and timetables.

Again and again some photobiologists are requested to make their knowledge available, and deal with the risks of UV radiation to people. There is a fair degree of willingness among the experts to spend some time and effort to this public service.

Perhaps we can do even better if we pay some more attention to these problems of risks and regulations among ourselves, and come to better founded opinions before these go into the machineries of committees. That is the reason why Dr. Liz Jacobson and I took the opportunity to organize this symposium as a part of this International Photobiology Congress, where so many experts are together. There is no point in forming another committee. We cannot even hope to coordinate the activities of the existing committees and organizations. We do hope, however, that the attention we pay together to the problems of risks and regulations will help us to come to a growing consensus and to offer a better service to the society in which we live.

Patterns of Human Exposure to Ultraviolet Radiation

FREDERICK URBACH

Temple University Medical Center, Philadelphia, USA

> *Dieu commande au Soleil d'animé
> la Nature, et la Lumière est un de
> ses mains.*
> Racine, Athalie I, 4

Exposure to ultraviolet radiation (UVR) occurs from both natural and artificial sources. The sun is the principal natural source. The known effects of UVR on man may be beneficial or detrimental, depending on a number of circumstances.

Artificial UVR sources are widely used in industry and, because of the germicidal properties of certain portions of the UVR spectrum they are also used in hospitals, biological laboratories, and schools. UVR is extensively used for therapeutic purposes, in the prevention of vitamin D deficiency, and for the treatment of skin diseases. Artificial UVR sources are available as consumer products for cosmetic purposes.

The migration of people between areas of different UVR exposure, whether for occupational or recreational reasons, gives rise to unforeseen exposures.

UVR can be classified into UV-A, UV-B and UV-C regions. Wavelengths in the UV-C region (200–280 nm) cause unpleasant, but usually not serious effects on the skin and eye. Although UV-C is very efficiently absorbed by nucleic acids, the overlying dead layers of skin absorb the radiation to such a degree that there is only mild erythema and, usually, no late sequelae, even after repeated exposures. Since solar UVR below 290 nm is effectively absorbed by stratospheric ozone, no such radiation reaches living organisms from natural sources.

Most observed biological effects of UV-B radiation (280–320 nm) are extremely detrimental to living organisms. However, living organisms are usually protected from excessive solar UV-B radiation by feathers, fur, or pigments that absorb the radiation before it reaches sensitive physiological targets. Other means of protection include behavioral patterns and the ability to tolerate certain UV-B radiation injury because of molecular and other repair mechanisms.

Much less is known about the biological effects of UV-A radiation (320–400 nm). It can augment the biological effects of UV-B, and doses of UV-A, which alone do not show any biological effect, can, in the presence of certain chemical agents, result in injury to tissues (phototoxicity, photoallergy, enhancement of photocarcinogenesis).

Photobiology, Edited by E. Riklis
Plenum Press, New York, 1991

Beneficial effects

It is now generally acknowledged that a long period of UVR deficiency may have a harmful effect on the human body. The best known manifestation of "UVR deficiency" is the development of vitamin D deficiency and rickets in children because of a disturbance in the phosphorus and calcium metabolism. The resultant effect on the bone-forming processes is accompanied by a sharp reduction in the defensive powers of the body, making it particularly vulnerable to many diseases. Appropriate measures to increase UV-B exposure by improving the architectural features of buildings (orientation of windows, use of UV-B transmitting window glass), the use of sun and sun-and-air bathing (solaria) and the development of artificial UVR sources and installations (photaria) have been shown to correct and protect disease states due to UVR deficiency. In fair-skinned people, all the beneficial effects can be obtained with daily suberythemal doses.

Harmful effects

These may be acute or chronic, and involve primarily the eyes and skin. The acute effects of UVR on the eyes consist of the development of photokeratitis and photoconjunctivitis, which are unpleasant but usually reversible and easily prevented by appropriate eyewear. Acute effects on the skin consist of solar erythema, "sunburn", which, if severe enough, may result in blistering and destruction of the surface of the skin with secondary infection and systemic effects, similar to a first or second degree heat burn. The skin has natural, adaptive protective mechanisms consisting of increased production of the skin pigment melanin, and thickening of the outer horny layer.

Chronic effects on the eye consist of the development of pterygium and squamous cell cancer of the conjunctiva and perhaps cataracts. Chronic skin changes due to UVR consist of "aging" (solar elastosis) and the induction of premalignant changes (actinic keratoses) and malignant skin tumors (non-melanoma and melanoma skin cancers). The evidence for a causal association of UV-B radiation with these chronic changes, particularly with skin cancer induction, has been reviewed in detail elsewhere.

Additional harmful effects (phototoxicity, photoallergy and enhanced photocarcinogenesis) are produced by the interaction of UVR and a variety of environmental and medicinal chemicals. This results in acute and chronic skin changes caused by UVR of wavelengths which are not normally of an injurious nature.

Solar Ultraviolet Radiation — Effect on Skin

In the UVR region, the sun emits like a black body of about 5,200°C. This radiation is first intercepted by the ozone made by the shorter (vacuum) ultraviolet radiation of the sun out of oxygen molecules. The ozone layer has a maximum density at about 25 km altitude, above this the ozone density is less because less and less oxygen atoms and nitrogen molecules are available for the triple collision which is needed for the formation of ozone. At layers below 30 km the change in the amount of ozone from normal photochemical processes becomes minute, and ozone becomes a permanent part of the stratosphere. Until 1965, the total amount of ozone in the tropical regions was essentially constant and smaller than at high altitudes where ozone amounts were higher and more variable with seasons. In the past 17 years, stratospheric ozone has decreased by an average of 2–3%, mainly because of man-made chemicals that attack ozone.

Ozone absorbs UVR primarily below 315nm. In addition, air molecules scatter UVR more than any other radiation, and therefore at least 50% of solar UVR reaches the ground scattered from the sky. The amount of UVR is also affected by other scattering and absorbing agents of the atmosphere — dust, aerosols, haze and clouds scatter, NO_x and SO_2 absorb UVR. The amount of UVR over continents as a rule is lower than over clear oceans or over tops of mountains. (For every 1,000 m elevation, UVB increases by about 15%). UVB reaching our body will also be altered by reflection from environmental surfaces (albedo). Blumthaler and Ambach (1988) recently reported mean values for albedo of total solar radiation and erythema effective weighted UVB for different surfaces (Table 1).

Table 1

Surface	Total Solar Radiation Albedo Mean %	Erythemic Solar Radiation Albedo Mean %	Remarks
Water	9.1	4.8	Clear Water
Field	11.5	2.2	Varying moisture
Rock	14.4	3.7	Various sizes
Stream Sand	23.8	9.8	Near stream
Grassland,corn	20.7	1.3	Varying height
New dry snow	87.0	94.4	High mountains
New wet snow	74.5	79.2	High mountains
Old dry snow	79.2	82.2	Varyingly dirty

UVB, as any other radiation, may be reflected by any surface by the combined effects of reflection, refraction and diffraction. Over UVB permeable materials such as water, there is mirror-like reflectivity which follows Fresnel's Law. This law says that the lower the angle of the sun is above the water, the higher the reflectivity. Consequently, the reflectivity for smooth water for UVR of high solar altitudes is negligibly small. It may become somewhat larger if the sea is rough. For very low angles the reflectivity may easily exceed 50%, but at these low levels the amount of UV B is small (less than 8% at sun angles of 30° above the horizon or less — Diffey et al, 1988).

The amount of UVB striking any surface varies of course with the direction, including the side turned away from the sun. All these exposures are markedly increased on the highly reflective surface such as snow (Büttner 1969, Figure 1).

Depending on latitude, it has been calculated that an outdoor worker may receive 25–40% of the ambient, flat surface UVR exposure, while indoor workers, including a few hours on weekends, may receive 10% of that. In terms of minimal erythema doses, that

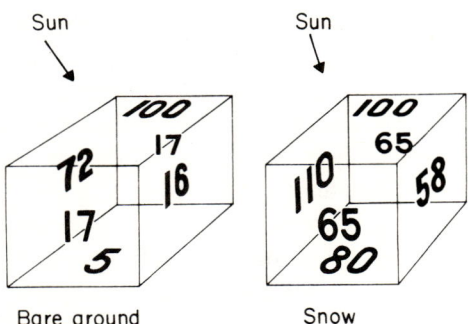

Bare ground   Snow

Figure 1

may represent 300–1,000 MED/year for outdoor workers, and 100 MED or less for indoor workers. Furthermore, a two-week vacation in a sunny climate may add 100+ MED's to the dose to indoor workers. (Cole et al, 1985; Diffey, 1987). For Sweden (60°N) the incidence of premalignant and malignant skin lesions is about 3.3 times higher in outdoor than in indoor workers. Diffey's calculations fit this difference for this northern country. The differences will of course be much higher in more southern latitudes such as the United States of America.

Finally, the distribution of non-melanoma skin cancers, which appear to be mainly due to chronic repeated solar UVB exposure, can be explained by purely geometric considerations. Their primary locations are on the head and neck (mostly nose, cheekbones, forehead, ears), upper back and dorsae of hands. Figure 2 shows the exposure of an upright person to solar UVR depending on sun angle. By the time where the sun angle is 30° above the horizontal, most of the body will be exposed, but the UVR content of the solar radiation is less than 8% of the overhead sun total (Diffey et al, 1988).

Table 2

Geographical Latitude	Annual Dose (MED)* Total	Contribution whenever solar altitude <30°
0°	7907	307
20°	6150	228
40°	3116	240
60°	926	132

*One MED (minimal erythema dose) is taken to be equivalent to an erythemally effective dose of 250 J/m^2 relative to 300 nm radiation.

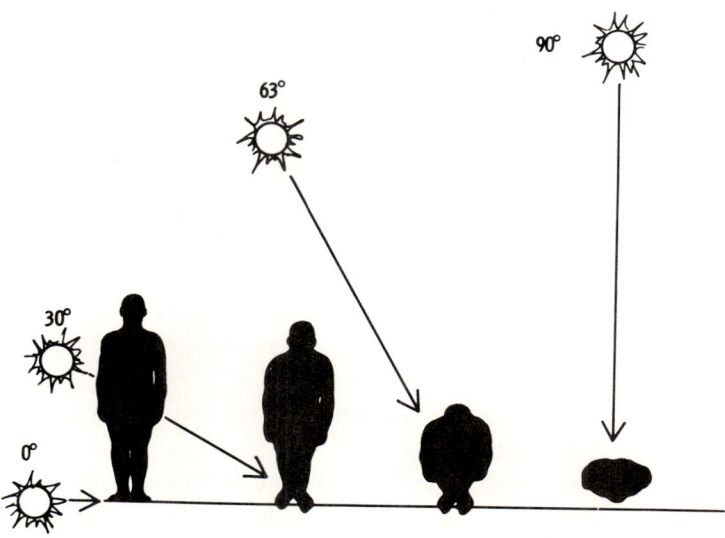

Figure 2. Surface area of direct sunlight exposure on a standing man facing the sun at different solar elevations

Solar Ultraviolet Radiation — Effect on Eyes

Exposure of the eye to UVB causes damage and necrosis of the corneal epithelium (photokeratitis-Lerman, 1980) and biochemical changes in lens protein *in vitro* which are believed to be precursors to cataract formation. The action spectrum for corneal effects peaks at the shortest wavelengths and found in sunlight, whereas for lens effects the most effective wavelengths are somewhat longer. The occurrence of photokeratitis in humans exposed to sunlight over reflective surfaces such as sand and snow is well-known. Cataracts have been associated with sunlight exposure in epidemiologic studies (Hiller et al, 1986).

Attempts have been made to quantitate ocular UVR exposure for individuals performing outdoor activities. The ocular dose of UVR depends on many factors, including ambient conditions, hours (and time of day) spent outdoors, use of protective clothing and behavior. The best and most recent study is that of Rosenthal et al (1988). Watermen (full-time Chesapeake Bay fisherman) received an average of 11% of flat surface ambient solar UVB to the eye area when not wearing hats (6.3–17.2%) and 7.2% (4.5–10.3%) when wearing hats. In contrast, groundsmen received 4.6% without and 2.0% with hats and carpenters 8.4% without and 5.2% with hats. It is interesting that in this study with human subjects the attenuation of exposure due to hats was substantially less (less than one-half) as compared with a mannikin study (Rosenthal et al, 1985).

These studies clearly show that surface reflectivity affects ocular ambient exposure. For the water areas, reflectivity was found to vary from 3–4%, in keeping with other reports. Also, wearing eyeglasses that absorb to 380 nm greatly decreased ocular ambient exposure of watermen to 1–2% (Rosenthal, 1985).

Exposure to Artificial Light Sources

Indoor Lighting

Indoor lighting has become a necessary part of our existence. For several decades, there has been emphasis on distributing enough luminous intensity to enhance visual acuity and provide efficient and safe illumination.

Fluorescent light sources are widely used because of superior economy. Luminaires with acrylic sheets as diffusers have been widely used to reduce glare. These sheets absorb virtually all UVR emitted from fluorescent lamps. More recently, fixtures with open louvres ("egg crate") have become prevalent because they increase brightness, but at the expense of allowing UVR to pass.

The exposed skin of an individual presents an irregular target for an overhead fluorescent source, primarily irradiating the head, since people in offices and stores are normally clothed. Figure 3 shows the spatial distribution of irradiance over the head of a mannequin in percent incidence of the top of the head. Measurements were made with an analog device, weighted to represent reasonably the human skin erythema action spectrum (Cole et al, 1985). As can be seen the orbit and submental areas receive very little UVR, the top of the ear, nose and temple the most.

Measurements of a number of commonly-used fluorescent lamps in various office and store configurations showed that they emit measurable amounts of UVB, and some

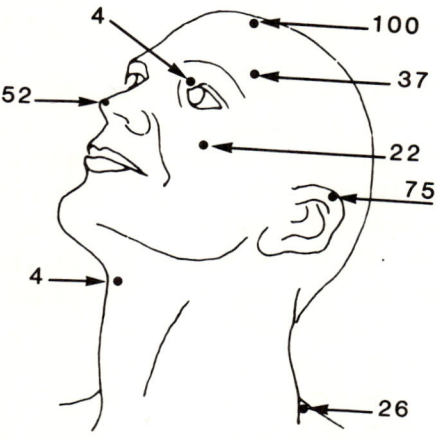

Figure 3

even small amounts of UVC. The visual brightness of such sources is not an adequate predictor of their UVR emission, and thus of their potential for biological hazard. It is possible that under some lighting conditions over a working lifetime, the risk of cumulative UVR damage might be significantly increased. Utilizing acrylic louvres lowers the risk to negligible (Cole et al, 1985).

Solaria

Most commercial tanning equipment in current use exploits the divergence in the erythema and pigmentation action spectra (Gange et al, 1986). These sources include Type I UVA fluorescent lamps (UVB content 1–2%, lower limit 315–320 nm). Type II UVA lamps (UVB content 0.05%, lower limit 330–340 nm) and optically filtered, metal halide doped mercury arc lamps (Mutzhas, 1986). The usual exposure to such lamps is 30 minutes, resulting in an exposure of about 4.5×10^5 J/M^2, or a pigmenting dose for skin types III and IV. Very fair persons can develop mild erythema from such a dose of UVR.

Diffey (1987) estimated that the use of solaria 10 times per year adds a negligible risk for developing skin cancer, but that use 3 times weekly from age 20 to age 50 (30 years) could double the likelihood of skin cancer development. We have estimated that once weekly use of type II lamps for 6 months (25 exposures/year) could, after 30 years, increase the incidence of skin cancer by 10%. Clearly, the increasing use of "sun" beds and "sun" canopies for home use gives considerable cause for concern, because no controls can be executed there.

It is clear from all studies that sunbathing during a two-week holiday in natural sunlight in very sunny climates presents a much more significant carcinogenic burden for the skin than moderate use of an UVA solarium. Although a UVA-induced tan does not provide as much protection against acute sunburn as a UVB-induced tan, a similar degree of tan induced by either waveband of radiation provides similar protection against DNA damage (Gange et al, 1985). However, there is no question that additional UVA exposure from solaria is additive to the damaging effect of solar UVB exposure.

Reference

Blumthaler, M. and Ambach, W. (1988), Solar UVB albedo of various surfaces. *Photochem. Photobiol.* 48 (1), 85–88.

Bittner, K.J.K. (1969), The effect of natural sunlight on human skin. In: The Biologic Effects of Ultraviolet Radiation, F. Urbach (ed), Pergamon Press, Oxford, pp. 237–249.

Cole, C., Forbes, P.D., Davies, R.E. and Urbach, F. (1985), Effect of indoor lighting on normal skin. In: The Medical and Biological Effects of Light. R.J. Wurtman, M.J. Baum and J.P. Potts, Jr., Eds., *Ann. NY Acad. Sciences* 453:305–316.

Diffey, B.L. (1984), Environmental exposure to UVB radiation. *Review of Environmental Health*, 4: (4)317–337.

Diffey, B.L. (1987), Analysis of the risk of skin cancer from sunlight and solaria in subjects living in Northern Europe. *Photodermatology* 4:118–126.

Diffey, B.L., Meanwell, E.F. and Loftus, M.J. (1988), Ambient ultraviolet radiation and skin cancer incidence. *Photodermatology* 5:175–178.

Gange, R.W., Blackett, A.D. Matzinger, E.A., Sutherland, B.M., and Kochevar, I.E. (1985), Comparative protection efficiency of UVA and UVB induced tans against erythema and formation of endonuclease sensitive sites in DNA in human skin. *J. Invest. Drm.* 85:362–364.

Gange, R.W., Park, Y.K., Auletta, M., Kagetsu, N., Blackett, A.D. and Parrish, J.A. (1986), Action spectra for cutaneous responses to ultraviolet radiation. In: Urbach, F. and Gange, R.W., eds., The Biologic Effects of UVA Radiation, Praeger, N.Y., pp. 57–65.

Hiller, S., Sperduto, R.D. and Ederer, F. (1986), Epidemiologic associations with nuclear, cortical and posterior sub capsular cataracts, *Am. J. Epidemiology* 124:916.

Lerman, S. (1980), Radiant energy and the eye. MacMillan, N.Y.

Mutzhas, M.F. (1986), UVA emitting light sources. In: Urbach, F. and Gange, R.W., eds., The Biologic Effects of UVA Radiation. Praeger, N.Y., pp. 10–23.

Rosenthal, F.S., Safran, M. and Taylor, H.R. (1985), The ocular dose of ultraviolet radiation from sunlight exposure. Photochem. *Photobiol.* 42(2), 163–171.

Rosenthal, F.S., Phoon, C., Bakalian, E. and Taylor, H.R. (1988), The ocular dose of ultraviolet radiation to outdoor workers.

Cutaneous Photosensitization: Hazard and Regulation

B. E. JOHNSON

Department of Dermatology
University of Dundee
Ninewells Hospital and Medical School
Dundee, DD1 9SY, Scotland, U.K.

The hazards of skin exposure to natural sunlight are well established. Of course, Government legislation against these hazards is impossible. However, legislation is possible to regulate exposure to artificial sources of ultraviolet (UV) radiation (Passchier and Bosnjakovic, 1987; Sykes and Jacobson 1989). I shall discuss another source of hazard for skin exposed to either natural sunlight or artificial sources of UV and visible radiation, for which regulation may be possible through Government legislation. This hazard is photosensitization (Blum, 1964; Johnson, 1984; Kochevar, 1987) a process in which reactions to normally harmless radiation are produced in a system by a specific radiation absorbing substance, the photosensitizer (Spikes 1977; Lamola 1974).

Photosensitizers (Table 1)

For cutaneous reactions, the photosensitizer may be (1) *exogenous*, a chemical introduced into the skin by topical, or parenteral administration in a therapeutic, domestic or industrial setting; (2) *endogenous*, a normal metabolite present in greater than normal concentrations or an abnormal metabolite.

Drugs (Table 2) are the major source of photosensitized reactions in the skin but plant materials, dyestuffs, polycyclic hydrocarbons in wood preservatives, coal tars and environmental pollutants, perfume and cosmetic constituents, sunscreen and printing ink materials, and even metal salts all contribute to the overall picture.

Cutaneous photosensitization

Photosensitized reactions in skin were classified as *phototoxic*, reactions derived from direct damage to cell and tissue constituents, and *photoallergic*, reactions for which immune system mechanisms, particularly those which are cell mediated, are involved (Epstein, 1939) More recently, activation of complement, usually associated with immune system reactions, has been shown to be a feature of porphyrin and drug induced cutaneous phototoxicity (Lim and Gigli, 1983).

Photobiology, Edited by E. Riklis
Plenum Press, New York, 1991

Table 1. A general classification of photosensitizers

Exogenous

Drugs: See Table 2.
Plant materials: Furocoumarins, Alpha-terthienyl, Polyacetylenes.
Dyestuffs: Thiazines; Methylene Blue, Toluidine Blue
             Xanthenes; Fluorescein, Eosin, Erythrosin, Rose Bengal.
             Anthraquinone based; Disperse Blue, Benzanthrone.
Polycyclic Hydrocarbons: Pitch, coal tars, Anthracene, Acridine, Fluoranthene.
Perfumes and cosmetics: Bergamot oil containing 5-MOP, Musk Ambrette, 6-methylcoumarin.
Sunscreens, Inks: Amyl o-dimethyl aminobenzoic acid.
Tatoos: Cadmium sulphide.
Miscellaneous: Cyclamate sweetener; blankophore fabric whitener; quinoxaline-n-dioxide.

Endogenous

Abnormal metabolites: Uroporphyrin, Coproporphyrin
Normal constituents: Protoporphyrin, Kynurenic acid, Tryptophan

Table 2. Drugs commonly reported as photosensitizing

Antibacterial:	Tetracyclines; sulphonamides; nalidixic acid; 4-quinolones.
Tranquillizer:	Phenothiazines (chlorpromazine).
Antidepressant:	protriptyline.
Diuretic:	Chlorthiazides; frusemide.
Antiarrhythmic/ Antihypertensive:	amiodarone, methyldopa, quinidine, propranolol.
Anti-inflammatory: (Non-Steroidal)	benoxaprofen, ibuprofen, azapropazone, naproxen, piroxicam, tiaprofenic acid.
Antifungal:	grizeofulvin.
Bacteriostat or Topical antifungal:	Halogenated salycilanilides, bithionol buclosamide, fentichlor, hexachlorophene

Commonly recognised molecular mechanisms of photosensitization vary involving reactive substrate, superoxide or hydroxyl radicals and hydrogen peroxide (type I) or excited singlet state oxygen (type II) both of which may produce damage in membrane or nucleic acid components of the cell, and also, photo-induced binding of photosensitizer to the target biomolecule, well illustrated by the interactions of linear furocoumarins such as 8-MOP with DNA. Toxic photoproduct formation is also included here even though the mechanism does not fulfill the exact requirements of photosensitization. This type of reaction was originally of interest in terms of the mechanisms of chlorpromazine photosensitized killing of Ehrlich tumour cells (Carraz and Beriel, 1962) or photohaemolysis with chlorpromazine (Johnson, 1984; Kochevar and Hom,1983) and protryptilene (Kochevar, 1980). It as gained in importance with the work of Beijersbergen van Henegowen and his colleagues who have shown systemic toxicity with photoproducts of chlordiazepoxide (Bakri et al, 1983; 1985) and the demonstration that the lipid soluble photoproduct of Benoxaprofen is more likely to be phototoxic than the parent compound (Kochevar et al, 1984).

It is evident that for any photosensitizer, the biologic effect observed may be the result of one or more preferred pathways rather than a single, well defined photo-sensitization reaction.

Phototoxicity

Phototoxicity then is the term used for skin reactions derived directly from photosensitized damage to cell and tissue components.

The molecular and cellular mechanisms which produce this damage vary depending mainly on the chemical structure of the photosensitizer, its lipid solubility, location in relation to the biomolecules with which interactions may take place, involvement of oxygen and and finally, the distribution of the photosensitizer in the skin.

In general, a phototoxic reaction should be produced in any subject in which sufficient photosensitizing agent is present within the skin which is then exposed to high enough doses of the appropriate radiation.

There is no such thing as a typical phototoxic reaction in the skin. However, it is possible to classify the major reaction patterns into four types (Table 3).

Table 3. Major patterns of cutaneous phototoxicity

Skin Reactions	Photosensitizers or Diseases
Prickling and burning during exposure; immediate erythema, oedema/urticaria with higher doses. Sometimes delayed erythema/hyper-pigmentation	Coal tar, Pitch, Anthraquinone based dyestuffs, benoxaprofen, amiodarone, chlorpromazine, haematoporphyrin, Erythropoietic Protoporphyria
Exaggerated sunburn	Drugs such as chlorpromazine, quinine, chlorthiazides, demethylchlortetracycline
Late onset erythema, blisters with slightly higher doses. Low exposure doses may lead to hyperpigmentation only	Psoralens, Phytophotodermatitis Berloque dermatitis
Increased skin fragility, blistering with friction	Nalidixic acid, frusemide, tetracycline, Porphyria Cutanea Tarda

For the immediate, prickling and burning reaction type, as seen in "Pitch Smarts", an oxygen dependent, membrane directed phototoxicity might be postulated while the severe reaction leading to a delayed erythema and blistering as obtained with 8-methoxypsoralen, is more likely mediated through changes in DNA. However, at the moment, there remains doubt concerning direct cause and effect, the overall pattern of skin response may derive from more than one molecular and cellular photosensitization event and, as seen with chlorpromazine, there may be some overlapping of reaction pattern for some photosensitizers.

In addition to these four major reaction types, some drug-induced photosensitivities may be manifest as photo-onycholysis, photosensitized damage to the nails, the development of small white bumps, known as milia, in exposed skin obtained with benoxaprofen for instance, and a so-called "lichenoid eruption" obtained with quinine. One feature of phototoxicity at the cellular level is photosensitized mutagenesis and over a longer term of chronic exposure, photosensitizers such as 8-MOP,and 5-MOP produce a photosensitized cutaneous carcinogenesis (Young et al, 1983).

Models for cutaneous photosensitization

Numerous models have been developed to examine the photosensitization process (Johnson et al, 1986). These models have served two purposes:
1. To act as a screening method for the phototoxic potential of drugs and other chemicals in the environment;
2. To provide information as to the mechanisms involved.

A screening model should be as simple as possible. In this regard, photohaemolysis, the photo-sensitized destruction of histidine and killing of Candida albicans are very useful allowing the differentiation between the phototoxicity obtained with extracts of two plant groups. The major psoralens, causal agents in "Phyto-photodermatitis" due to some plants of the Umbelliferae, Rutaceae, Moracae and Leguminosae are strongly positive in the Candida test but negative against histidine and in photohaemolysis. Extracts from plants of the Compositae such as Chrysanthemum, with which a more immediate and less striking skin reaction is obtained, are negative against Candida but positive in the photohaemolysis and histidine tests. Therefore, one of these simple tests on its own is not sufficient as a screen for phototoxicity. Where a specific intracellular target such as DNA is involved, the photohaemolysis model is obviously inadequate. In addition, the new generation of broad spectrum 4-quinolone antibiotics derived from nalidixic acid, have failed to reveal their phototoxic potential in either of these simple tests but are positive against mouse peritoneal macrophages and PHA stimulated human lymphocytes. It would certainly appear that a degree of sophistication is required for screening methods in terms of at least short term cell culture facilities and it may well be that the most reliable and useful test will be that which uses a standard fibroblast cell line.

Various animal models, the flank skin of hairless mice or plucked haired mice, the ears of haired mice, guinea pigs, rabbits and miniature swine, have been used to establish the phototoxic potential of a number of substances. The most reliable model so far appears to be the "mouse tail" test (Lunggren, 1984).

Photoallergy (Epstein, 1972)

Epstein (1939) introduced the term "photoallergy" to differentiate the reactions obtained with intradermal injections of sulfanilamide. A first reaction, obtained with all the subjects tested, resembled sunburn and occurred soon after exposure to sunlight. The second, obtained in a limited number of subjects appeared 10 days later after a second exposure and resembled the more severe reactions of cell mediated delayed hypersensitivity contact dermatitis. This second reaction, with histopathology like that of contact allergic dermatitis, was also obtained with 3,4'5'5-tetrachlorosalicylanilide (TCSA) incorporated in soaps as a bacteriostat in the 1960's giving rise to a minor epidemic of photo-sensitivity (Wilkinson, 1961,62; Herman and Sams, 1972). Its reproducibility in experimental animals with sensitization and challenge experiments and demonstrable lymphocyte mediation, appears to justify the term "photoallergy" (Harber and Baer, 1972; Ichikawa et al, 1981; Takigawa and Miyachi, 1982; Maurer, 1983; Barratt et al, 1987). The incidence of apparent photoallergic dermatitis fell dramatically after the withdrawal of halogenated phenolic compounds from the toiletries market (Smith and Epstein, 1977). More recently, episodes of similar photosensitivity have been associated with exposure to quinoxaline-n-dioxide, a foodstock animal growth promoter, and the fragrance materials, musk ambrette and 6-methyl coumarin. The commonly reported and confirmed photoallergic reactions follow the application of the photosensitizer by topical or intradermal routes rather than by mouth.

The nature of the allergen is obviously of interest. It is possible that photochemistry changes a normally harmless substance into a contact allergen. However, with TCSA, irradiation leads to the formation of a highly reactive intermediate which binds readily to protein, forming a complete antigen (Kochevar, 1979; Rickwood and Barratt, 1982).

Animal models for establishing the photoallergic potential of drugs and other substances are now well established although some variation in technique may be required for different compounds. (Maurer, 1983).

Relationship of photosensitization to persistent light reaction

The acute eczematous type of reaction obtained with photoallergy may be prolonged and, as with contact dermatitis, may be debilitating. Photoallergy is much rarer than phototoxicity and this is fortunate because in general, the reactions are more serious in terms of management than the majority of phototoxicity reactions. The development of a state of "persistent light reaction" where a characteristic photosensitivity dermatitis of chronic nature arises without the need for an exogenous photosensitizer, was reported to be associated with photosensitization by substances such as TCSA. The mechanisms involved are not known but, rather than being a prolongation of the TCSA photoallergic reaction state per se, a state of auto-sensitization might be induced through a phototoxic oxidation of histidine in cutaneous protein (Kochevar, 1979). Once this is established, it may be one factor among many, such as cell mediated contact dermatitis due to Compositae plant oleoresins (Addo et al, 1985) or fragrance materials (Addo et al, 1982) which are also phototoxic, in maintaining the chronic inflammation (Botcherby et al, 1984). An endogenous photosensitizer such as kynurenic acid which is very effective against histidine may well act as the source of this autosensitzation resulting in the exquisitely photosensitive condition known as Photosensitivity Dermatitis/Actinic Reticluoid (Frain-Bell, 1982).

Metabolism in drug induced photosensitivity

The phototoxic potential of many drugs which are known to be photosensitizing is revealed by in vitro tests. However, in some instances, a compound which is highly potent in vitro proves to be inactive *in vivo*. This anomoly was first observed in a study of the psoralens used in photo-chemotherapy for vitiligo when trimethylpsoralen (TMP) highly phototoxic in vitro and when applied topically was found to be minimally phototoxic when taken by mouth whereas 8-MOP retained a high degree of phototoxic potential when taken orally. The difference was explained in terms of both solubility of the different drugs and their metabolism, TMP rapidly producing inactive metabolites (Mandula and Pathak, 1976). A similar explanation may be valid for the relatively low incidence of cutaneous photosensitivity obtained with some non-steroidal anti-inflammatory drugs (NSAIDs) such as ketoprofen which is as potent as benoxaprofen in vitro but produces only an idiosynchratic photosensitivity in those patients for whom it is prescribed. It is clear that if metabolism is held responsible for converting a phototoxic agent into an inactive one, the reverse process may also occur. This may be seen with the synthetic retinoid, etretinate, a vitamin A analogue (Ferguson and Johnson, 1986) and the NSAIDs piroxicam (Kochevar et al, 1986) and sulindac. In each case, a major metabolite of the inactive parent drug is phototoxic. An intermediate position is held for amiodarone, an important anti-arrythmia drug, where the major desethyl metabolite is an order of magnitude more phototoxic than the parent compound (Hasan et al, 1984; Ferguson et al, 1985). With chlorpromazine, the situation is very complex and the desmethyl metabolites are more phototoxic than CPZ itself while others such as CPZ sulphoxide are inactive (Ljunggren and Moller, 1977).

Any study of drug induced photosensitivity should therefore include metabolic and pharmacokinetic studies.

Regulation and legislation

Legislation providing regulation of photosensitizing substances appears to be very limited. For some years now, the quantity of bergapten (5-MOP) in perfumes and cosmetic preparations is subject of a Hazardous Substances act in the United States due to the incidence of Berloque dermatitis (Marzulli and Maibach, 1970). Questions were raised in the British Parliament concerning the incorporation of 5-MOP in sunscreening agents as suntanning promoters but no legislation was required to stop the sale of such preparations and the latest research suggests that initial fears regarding them may not be merited (Young et al, 1988). Nonetheless, where abnormal cutaneous photosensitivity is seen to be associated with an exogenous agent which is susceptible to control, the the requirements for legislation should be considered.

Table 4. Government regulations for photosensitizing substances

AUSTRIA	No specific regulations. If known, drug induced photosensitization must be reported. Honigsmann, Vienna.
BELGIUM	No specific regulations. Drug photosensitization a reportable side-effect; part of information for patients. Roelandts, Leuven.
CANADA	No specific regulations. Phototoxicity and photoallergy included in recommended toxicity tests for new drugs. Ramsay, Toronto: Begin, Health & Welfare, Ottawa.
FINLAND	No specific regulations. If reported, special prohibition measures possible on recommendation of, for example, the Finnish Medical Board. Hannuksela, Oulu.
FRANCE	No specific regulations. Cesarini, Paris.
GERMANY (FDR)	No specific regulations. Problem recognised. Holzle and Plewig, Dusseldorf. Zesch, Berlin.
GERMANY (DDR)	Specific regulations for phototoxicity of new drugs and cosmetics. Screening tests required. Barth, Dresden.
GREECE	Pre-market; manufacturers required to report. National Drug Organisation check on data. Post-market; surveillance. Stratigos, Athens.
INDIA	Drug Controller of India provides regulation as with other side-effects. May require trials. Bhutani, New Delhi.
ITALY	As per European Economic Council directives. No specific regulations; photosensitization included in "irritant" and "sensitization" side effects. Santamaria, Pavia.
NETHERLANDS	No specific regulations. van Vloten, Utrecht.
NORWAY	No specific regulations. Notification as side-effect is required. Thune, Oslo.
SPAIN	No specific regulations. Certain cosmetic and toiletry constituents banned. Lecha; Mascaro, Barcelona.
SWEDEN	No specific regulations. New drugs of known photosensitizing family may require screen. Moller; Ljunggren, Malmo.
U.K.	No specific regulations. Committee on Safety of Medicines may require notification, may withdraw licence.
U.S.A.	No Federal Regulation but trials may be stopped. Regulation being considered. Felden, F.D.A., Rockville, MD.

The high incidence of phototoxic skin reactions obtained with the NSAID benoxaprofen and its use in litigation processes, after the withdrawal of the drug despite the manufacturer's warning of this effect, has concentrated minds on the importance of photosensitivity as a side effect of drugs (Anonymous, 1982; Gerber, 1988). This is one area in which regulation can certainly be established through legislation. However, a limited survey of Government legislation around the world (Table 4) shows that little attention is paid to this aspect of safety of medicines. In the majority of countries polled, the incidence of photosensitivity reactions in clinical trials or post market surveys are all that is required. Some advice may then be sought for confirmation and if the incidence of photosensitivity is high, certification may be withdrawn.

1051

Table 5. Product licence for drugs

Medicines Act 1968 - revised 1985 E.E.C. directive.

Expert reports required on:
1. Chemical and Pharmaceutical Documentation
    a. Composition.
    b. Method of preparation.
    c. Control of starting materials.
    d. Control tests on intermediates.
    e. Control tests on finished product.
    f. Stability.
    g. Other information.
2. Toxicological and Pharmaceutical Documentation
    a. Acute toxicity and toxicity with
    b. repeat adminstration.
    c. Foetal toxicity and fertility studies.
    d. Mutagenic potential.
    e. Carcinogenic potential.
    f. Pharmacodynamics.
    g. Pharmacokinetics.
3. Clinical Documentation
    a. Human pharmacology.
    b. Clinical documentation
    c. Other information.

Only in the German Democratic Republic is there a legislative requirement for new drug products to be tested for photosensitizing properties. The potential problem is recognised in Canada and a draft for new legislation includes indications of procedures for testing for photosensitization by topically applied agents using recognised aniaml models. In Spain, the halogenated salicylanilides and related substances are banned because of their photosensitizing properties. A summary of the procedures for product licencing in the United Kingdom required by the Medicines Act of 1968, revised according to European Council Directive 83/570/EEC (Table 5) shows that detailed toxicological tests at in vitro, animal model and clinical trial levels are established and recognised as of value in establishing the safety of a new drug. No mention of photosensitization is present nor, in the full text, are recommendations made for such tests.

It is not certain that photosensitization of the skin is a good reason for banning or withdrawing from the market a drug of proven and superior value unless an equally efficacious drug, without this particular side effect is available. So long as the nature of the photosensitizing action is known, it should be possible by a combination of judicious prescribing, advice about avoiding exposure to sunlight when high concentrations of the drug are circulating through or deposited in the skin, or application of an appropriate form of protection in the form of clothing or sunscreen, photosensitizing drugs might still be used. Legislation should therefore be directed at the product licencing stage of drug development in terms of acquisition of knowledge of hazard and provision of advised warning. The tests already required to determine levels of toxicity in general are established and at each stage, may be adapted for tests of photosensitizing potential. Once these requirements are fulfilled, photosensitization as a side effect of drug therapy may be treated on a risk benefit basis in the same way as any other side effect. Drug manufacturers such as I.C.I. and Roche already operate in-house testing to a certain extent. In the field of cosmetic fragrance materials and antibacterials, Unilever operate an advisory service in this respect. In addition, clinical expertise is available for the final trial stages.

There are two other sources of photosensitizing hazard for which regulations could be applied. The first may be illustrated by the anthraquinone dyestuff intermediate, benzanthrone (Walker, 1982; MacDonald et al, 1985). The incidence of photoxocity in process workers with this substance is high and might be regulated by some form of Factory Act or Safety at Work Act. However, there is no coverage of photosensitization as such in any form of such Acts. The problem is recognised but obviously not serious enough to attract the attention of either the Factory Inspectorate or the Trade Unions.

The second is seen as Phytophodermatitis, in the U.K. mainly associated with Giant Hogweed (Heracleum mantegazzianum). The problem of this severe cutaneous phototoxicity has excited Government at the local Town and Regional level. However, no legislation exists to cover hazardous plants in the U.K. unless they are harmful to livestock. Neither children nor council workers fall into this category and therefore, new legislation would be required to enforce the removal of the Giant Hogweed from the environment.

It is certainly not clear that plants such as the Giant Hogweed should be destroyed simply because they produce phototoxicity. Other plants constitute a greater hazard to man than that of photosensitization and it would seem that, as for drugs, knowledge of the hazard and education about its prevention are more appropriate requirements. This would no doubt be the view of the citizens of Tromso in northern Norway who regard their version of the Giant Hogweed, Heracleum laciniatum, with such affection as to call it the Tromso Palm (Kavli and Volden, 1984).

References

Addo, H.A., Ferguson, J., Johnson, B.E., and Frain-Bell, W., 1982, The relationship between, exposure to fragrance materials and persistent light reaction in the photosensitivity dermatitis with actinic reticuloid syndrome. *British Journal of Dermatology*, 107 ,261-274.

Addo, H.A., Sharma, S.C., Ferguson, J., Johnson, B.E. and Frain-Bell, W., 1985, a study of Compositae plant extract reactions in photosensitivity dermatitis. *Photodermatology*, 2, 68-79.

Anonymous, 1982, Benoxaprofen (editorial) *British Medical Journal*, 285, 459-460.

Bakri, A., Beijersbergen van Henegouwen, G.M.J. and Chanal, J.L., 1983, Photopharmacology of the tranquillizer chlordiazepoxide in relation to its phototoxicity. *Photochemistry and Photobiology*, 38, 177-183.

Bakri, A., Beijersbergen van Henegouwen, G and Chanal, J.L., 1985, Involvement of the N_4-oxide group in the phototoxicity of chlordiazepoxide in the rat. *Photodermatology*, 2, 205-212.

Barratt, M.D., Goodwin, B.F.J. and Lovell, W.W., 1987, Induction of photoallergy in guinea pigs by injection of photoallergen-protein conjugates. *Internation Archives of Allergy and applied Immunology*, 84, 385-389.

Blum, H.F., 1964, Photodynamic action and diseases caused by light. Hafner Publishing Company, New York.

Botcherby, P.K., Magnus, I.A., Marimo, B. and Gianelli, F., 1984, Actinic reticuloid - an idiopathic photodermatosis with cellular sensitivity to near ultraviolet radiation. *Photochemistry and Photobiology*, 39, 641-649.

Carraz, G. and Beriel, H., 1962, Photosensibilisants et radiosensibilitants. 2e memoire: les photomimetiques. *Therapie*, 17, 195-202.

Epstein, J.H., 1972, Photoallergy- a review. *Archives of Dermatology*, 106, 741-748.

Epstein, S., 1939, Photoallergy and primary phototoxicity to sulfanilamide. *Journal of Investigative Dermatology*, 2, 43-51.

Ferguson, J.F., Addo, H.A., Jones, S., Johnson, B.E. and Frain-Bell, W., 1985, A study of cutaneous photosensitivity induced by amiodarone. *British Journal of Dermatology*, 113, 537-549.

Ferguson, J. and Johnson, B.E., 1986, photosensitivity due to retinoids; clinical and laboratory studies. *British Journal of Dermatology*, 115, 275-293.

Frain-Bell, W., 1982, Photosensitivity dermatitis and actinic reticuloid. *Seminars in Dermatology*, 1, 161-168.

Gerber, P., 1988, Mass product-liability litigation. *The Medical Journal of Australia*, 148, 485-488.

Harber L.C. and Baer R.L., 1972, Pathogenic mechanisms of drug-induced photosensitivity. *Journal of investigative Dermatology*, 58, 327-342.

Hasan, T., Kochevar, I.E. and Abdulah, D., 1984, Amiodarone phototoxicity to human erythrocytes and lymphocytes. *Photochemistry and Photobiology*, 40, 715-719.

Herman, P.S. and Sams, W.M. Jr., 1972, Soap Photodermatitis, Photosensitivity to halogenated Salicylanilides. Charles C. Thomas, Springfield, Illinois.

Ichikawa, H., Armstrong, R.B. and Harber, L.C., 1981, Photoallergic contact dermatitis in guinea pigs: improved induction technique using Freund's complete adjuvant. *Journal of investigative Dermatology*, 76, 498-501

Johnson, B.E., 1974, Cellular mechanisms of chlorpromazine photosensitivity. *Proceedings of the Royal Society of Medicine*, 67, 871-872.

Johnson, B.E., 1987, Light sensitivity associated with drugs and chemicals. In: The physiology and pathophysiology of the skin. Jarrett, A. (ed) Academic Press, New York, pp2541-2606.

Johnson, B.E., Walker, E.M. and Hetherington, A.M., 1986, In vitro models for cutaneous phototoxicity. In: Skin Models, Models to Study Function and Disease of Skin. Marks, R and Plewig, G. (eds) Springer Verlag, pp 264-281.

Kavli, G. and Volden, G., 1984, Phytophotodermatitis. Photodermatology, 1, 65-75.

Kochevar, I.E., 1979, Photoallergic responses to chemicals. *Photochemistry and Photobiology*, 30, 437-440.

Kochevar, I.E., 1980, Possible mechanisms of toxicity due to photochemical products of protriptyline. *Toxicology and applied Pharmacology*, 54, 258-2 .

Kochevar, I.E., 1987, Mechanisms of drug photosensitization. *Photochemistry and Photobiology*, 45, 891-895.

Kochevar, I.E. and Hom, J., 1983, Photoproducts of chlorpromazine which cause red blood cell lysis. *Photochemistry and Photobiology*, 37, 163-168.

Kochevar, I.E., Wujek Hoover, K. and Gawienowski, M., 1984, Benoxaprofen photosensitization of cell membrane disruption. *Journal of Investigative Dermatology*, 82, 214-218.

Kochevar, I.E., Morison, W.L., Lamm, J.L., McAuliffe, D.J., Western, A. and Hood, A.F., 1986, Possible mechanism of Piroxicam induced photosensitivity. *Archives of Dermatology*, 122, 1283-1287.

Lamola, A.A., 1974, Fundamental aspects of spectroscopy and photochemistry of organic compounds; electronic energy transfer in biologic systems; and photosensitization. In: Sunlight and Man, Fitzpatrick, T.B. (ed) University of Tokyo Press, pp 17-55.

Lim, H.W. and Gigli, I., 1983, Complement-derived peptides in phototoxic reactions. In: Experimental and clinical photoimmunology. Daynes, R.A. and Spikes, J.D. (eds) C.R.C. Press Inc., Boca Raton, pp 81-93.

Ljunggren, B., 1984, The mouse tail phototoxicity test. *Photodermatology*, 1, 96-100.

Ljunggren, B. and Moller, H., 1977, Phenothiazine phototoxicity: an experimental study on chlorpromazine and its metabolites. *Journal of investigative Dermatology*, 68, 313-317.

MacDonald, K.J.S., Walker, S.A., Walker, E.M. and Johnson, B.E., 1985, The action spectrum for benzanthrone photosensitization of mouse macrophages. *Photodermatology*, 2, 237-240.

Mandula, B.B. and Pathak, M.A., 1976, Photochemotherapy: identification of a metabolite of 4,5',8-trimethylpsoralen. *Science*, 193, 1131-1134.

Marzulli, F.N. and Maibach, H.I., 1970, Perfume Phototoxicity. Journal of the Society of Cosmetic Chemists, 21, 695-715.

Maurer, T., 1983, Contact and Photocontact Allergens. Marcel Decker inc., New York and Basel.

Passchier, W.F. and Bosnjakovich, B.F.M., 1987, (eds) Human exposure to ultraviolet radiation. Risks and regulations.

Excerpta Medica, International Congress Series 744, Elsevier Science Publishers, Amsterdam, New York.

Rickwood, D.M. and Barratt, M.D., 1982, Evidence for a major strong binding site for tetrachlorosalicylanilide on human serum albumin. *Photochemistry and Photobiology*, 35, 643-647.

Smith, S.Z. and Epstein, J.H., 1977, Photocontact dermatitis to halogenated salicylanilides and related compounds. *Archives of Dermatology*, 113, 1372-1374

Spikes, J.D., 1977, Photosensitization. In: The Science of Photobiology. Smith K.C. (ed) Plenum, New York, pp 87-110.

Sykes, S.M. and Jacobson, E.D., 1989, UV regulatory strategies. This volume.

Takigawa, M. and Miyachi, Y., 1982, Mechanisms of contact photosensitivity to tetrachlorosalicylanilide under genetic restrictions of the major histocompatability complex. *Journal of investigative Dermatology*, 78, 108-115.

Walker, S.A.,1982, Photocontact Dermatitis. An investigation into photocontact dermatitis induced by benzanthrone. Thesis for M.D. degree, University of Edinburgh.

Wilkinson, D.S., 1961, Photodermatitis due to tetrachloro-salicylanilide. *British Journal of Dermatology*, 73, 213-219

Wilkinson, D.S., 1962, Further experiences with halogenated salicylanilides. *British Journal of Dermatology*. 74. 295-301.

Young, A.R., Magnus, I.A., Davies, A.C. and Smith, N.P., 1983, A comparison of the phototumorigenic potential of 8-MOP and 5-MOP in hairless albino mice exposed to solar simulated radiation. *British Journal of Dermatology*, 108, 507-518.

Young, A.R., Potten, C.S., Chadwick, C.A., Murphy, G.M. and Cohen, A.J., 1988, Inhibition of UV radiation-induced DNA damage by a 5-methoxypsoralen tan in human skin. *Pigment Cell Research*, 1, 350-354.

UV Regulatory Strategies

S.M. SYKES AND E.D. JACOBSON
Food and Drug Administration
Center for Devices and Radiological Health
5600 Fishers Ln Rockville, MD 20857, USA

Introduction

Ultraviolet Radiation (UVR) is hazardous to human health. These hazards were recently detailed at a symposium held in Amsterdam entitled, "Human Exposure to Ultraviolet Radiation: Risks and Regulations" (Passchier and Bosnjakovic, 1987). This symposium brought together international experts on UVR risk and protection and was the first of its kind to focus specifically on health risks from UVR and outline current methods for risk reduction. The purpose of the present symposium, "UV Risks and Regulations", has been to outline for a broader audience the results of the Amsterdam symposium along with new data and perspectives that have emerged in the last 20 months.

The regulation of human toxins is a complex, uncertain, and controversial process. The process is divided into two separate functions: risk assessment and risk management (Latin, 1988). Risk assessment is the scientific activity that develops probabilistic estimates of human hazard at various exposure levels. Risk management is the political activity that develops control programs to minimize the risk. The previous papers of this symposium have evaluated UVR bioeffects from a scientific perspective. The goal of the present report is to provide some perspectives on the bioeffects of UVR from a risk assessment standpoint and to analyze risk management processes used to reduce the hazard level.

Risk

Defined from a regulatory perspective, risk is the probability that some action will bring bodily harm (usually death or serious, debilitating injury). The key word in the definition for regulatory purposes is *probability*. Ideally, risks are compared and action priorities developed as a function of amount of risk.

In the U.S. the probability considered sufficient to trigger a regulatory response is referred to as the *de minimus* level. The legal doctrine of *de minimus non curat lex* holds that the law does not concern itself with trifling matters and that courts should be reluctant to apply the literal terms of a statute to mandate pointless results (Hallenbeck

Photobiology, Edited by E. Riklis
Plenum Press, New York, 1991

and Cunningham, 1986). U.S. regulatory agencies have set different *de minimus* levels depending on a number of factors that define acceptable risk from the perspective of that particular agency. The Environmental Protection Agency (EPA) and Food and Drug Administration (FDA), for example, define acceptable risk as one in a million, while the Occupational Safety and Health administration (OSHA) generally considers acceptable risk to be one in a thousand, though considerable variation exists between different occupational groups. In each case, however, sufficient probability of risk triggers action.

Risk Assessment

In 1982, the National Research Council of the National Academy of Sciences published a report that defined the process of risk assessment in the U.S. (NRC, 1983). Though practiced to different degrees by different regulatory agencies, risk assessment has nonetheless emerged as the common tool for the assessment of environmental toxins and a variety of risk situations (DHHS, 1986). It consists of four distinct parts: hazard identification, dose-response assessment, exposure assessment, and risk characterization (Russell and Gruber, 1987). Hazard identification refers to the qualitative evaluation of human and animal studies to determine the probable human consequences from exposure to the toxin of and studies to develop hazard predictive models. Exposure assessment characterizes the probable extent and pattern of human exposure to the toxin. The final step, risk characterization, uses the results of the three previous steps to express the quantitative risk to human health in probabilistic terms.

Quantitative estimation of risk is fundamental to rational risk management decision making. The development of regulatory responses to public health problems requires energy, lots of energy. Occasionally some obvious or perceived public health catastrophe, line the incidents with nuclear power plants at Chernobyl and Three Mile Island, stimulates the energy. More often the proponent of a given cause is a solitary person (at least initially), attempting to advise superiors and others of a risk situation without any widely perceived public health threat. In the process of generating interest, he transforms from a dispassionate analyst into a passionate advocate. Only a strong advocate will, however, be competing with other similar advocates for other actions for other toxins. Each is competing for the limited resources of the agency. The final decision maker is probably not an expert in any particular scientific area and needs a common basis for risk comparison. Risk assessment provides a method to address rationally the demands of competing advocates.

A full risk assessment of UVR is beyond the scope of the present report. Nonetheless there are a number of features of the risk assessment process as it applies to UVR that should be discussed in order to address adequately the myriad of risk management options. The following sections discuss some features of UVR that should be understood in order to bridge the gap between UVR bioeffects analysis and UVR risk assessment.

Hazard identification

Photobiologists traditionally divide the bioeffects of UVR into two groups, acute effects and long term effects. For risk assessment, however, this is not a particularly useful distinction. A more useful distinction is to analyze whether effects are stochastic or non-stochastic. At issue is whether or not a threshold exists. Stochastic effects are those for which the probability of an effect occurring is a function of dose, without threshold. Non-stochastic effects, conversely, are those for which the severity of effects varies with dose and for which a threshold exists. (Bosnjokavic, 1987). Thus erythema, for example, is a non-stochastic effect because it has a threshold value below which the effect will not occur and, when exceeded, will be more severe with increasing dose. Photocarcinogenesis, on the other hand, is thought to be a stochastic effect because the probability of its acquisition increases with dose, and no threshold dose has been found below which the effect will not occur.

The existence or absence of a threshold is fundamental to the development of risk management approaches. Sliney (1987) notes that there is no single concept more critical to the development of exposure limits than the existence of a threshold below which an adverse effect does not occur. The existence of a threshold indicates that if the dose is sufficiently reduced, the effect will be completely abolished. Conversely the absence of a threshold indicates that the incidence of an effect can only be reduced, not abolished. Thus the goals of the risk manager come a) eliminate the occurrence of undesired non-stochastic effects by insuring that exposures are below the threshold dose and b) to lower the probability of acquiring stochastic effects below the acceptable, *de minimus* level.

Unlike other environmental toxins, UV radiation is required for normal human function. This factor makes the risk assessment of UVR unique. Although the range of postulated beneficial effects of optical radiation is broad and a number of beneficial effects from UV have been postulated, one is known with certainty. Vitamin D_3 is essential for human health (Wolff, 1987). This vitamin regulates calcium absorption and has an influence on the immune system, the musculature, and the skin as well as the nervous and endocrine systems. In the United States, vitamin D_3 is often added to milk, but this practice may be of only moderate utility for some people. The major stimulus for vitamin D_3 production, and some consider the safest as well, in UVR exposure. Thus, the goal of the risk manager cannot be to reduce the amount of human exposure to UVR to zero. Rather risk managers must use the concept of a window of acceptable exposure to consider both hazardous and beneficial UVR bioeffects.

Dose-response evaluation

Two major factors separate the quantitative evaluation of UVR bioeffects from evaluations of other environmental toxins. First, humans are differentially sensitive to UV bioeffects. While individual sensitivity has to some extent been quantitiated. Six different skin types can be distinguished on the basis of the degree of pigmentation and sensitivity to erythema (Health Council of the Netherlands, 1986). Second, any

1059

quantitative analysis of UV bioeffects would be incomplete without consideration of the relative effects of component wavelengths. The lack of such data is a major shortcoming of many studies of UVR phenomenon. In this respect, the dose-response evaluation of UVR hazards is more complicated than dose-response analyses of other human toxins. These two factors elevate the quantitative analysis of UVR to a multidimensional process.

The term "action spectrum" is often applied to studies of the relative effects of different wavelengths of UVR, but in the analysis of medically relevant endpoints, such as erythema and carcinogenesis, the term is somewhat misapplied. Active spectrum is a photophysical concept which defines for each wavelength the fraction of the incident quanta which are not only absorbed but also photochemically active (Haber and Tevini, 1987). A spectrum of relative quantum yields is highly useful because it can be compared with the absorption spectra of suspected photopigments to suggest which may be involved in the biological reaction. Biomedical phenemona such as erythema and carcinogenesis are, however, end expressions of a host of biochemical interactions incrementally separated from the initial molecular absorption. Additionally, the optical properties of skin attenuate absorption considerably. Thus, erythema and carcinogenesis may share the same or similar action spectrum but not share the same initial photopigment. For risk assessment purposes the term "action spectrum" should probably be replaced by the term "effectiveness spectrum".

Human exposure evaluation

The quantitation of human exposure to a potential toxin is the weakest aspect of evaluatiing risk for regulatory purposes. This situation is particularly acute for UVR exposure because of the large element of personal choice. Bosnjakovic (1987) estimates that 90% of human UVR exposure is due to the natural source, the sun. The extent to which an individual is exposed to the sun is almost entirely at his own discretion.

The geometry of exposure is an additional factor that must be considered in human exposure evaluation. Marshall (1987) notes, for example, that because of the geometry of the head with deep eye sockets and bony eyebrows, the eyes will not receive an overhead exposure from midday sun. Even in those exposure situations when UVR falls directly on he cornea, reflection results in 50% merely reflected off the corneal surface. Thus, while the cornea is more sensitive to UVR injury than the skin, corneal burns from sunlight are rarely experienced in the absence of erythema because of geometric factors.

Personal choice and the geometry of exposure point not our the importance in relating experimental data to received dose rather than available dose, particularly in human studies. Retrospective epidemiology will always suffer from this shortcoming. Needed are more prospective studies of human exposure in a variety of circumstances that use personal dosimeters such as the polysulfone dosimeters recently discussed by Ferenczi, Hill and Scrimger (1987). Such data are only now beginning to become available.

Risk characterization

The goal of risk characterization is to merge the data from the previous three sections, hazard identification, dose-response analysis, and human exposure evaluation, to yield a quantitative probability of human hazard. In the case of UVR, this final stage of the analysis becomes particularly cumbersome. At each stage in the analysis a number of mediating factors have incrementally complicated the process. The combined effects of effectiveness spectra, individual sensitivity, and uncertainty over the extent of human exposure limit the accuracy of any quantitative estimation of risk.

For non-stochastic effects, thresholds have been established for Type II skin (Health Council of Netherlands, 1986). Figure 1 shows threshold curves derived for intentional exposure (boxes) and unintentional exposure (pluses). The curve for intentional exposure is essentially an erythema effectiveness spectrum, and the data have been normalized to a minimally effective erythemal dose of 200 J/m^2. This dose is considered to be the radiant exposure necessary to induce a minimal erythema in depending on the type. The curve for unintentional exposure is a combined keratitis/erythema curve with a peak at 270 rm. The data have been normalized to an effective radiant exposure at 30 J/m^2. It may be necessary to modify this curve as data for UVA cataracts become available.

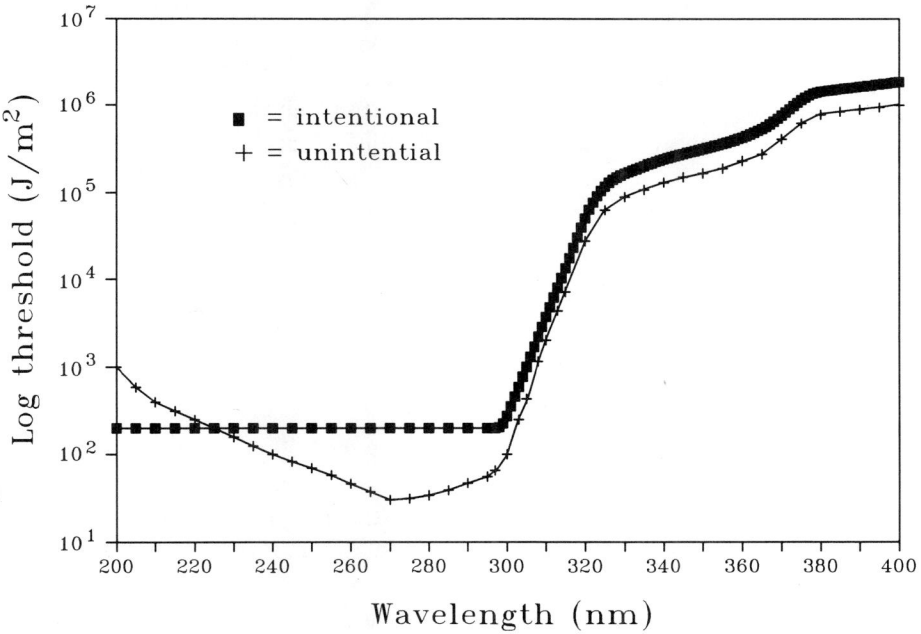

Figure 1. Human exposure limits recommended by a committee of the Health Council of the Netherlands (1986) for intentional (boxes) and unintentional (pluses) exposure to UVR as a function of wavelength (energy) (UVR Hazard Thresholds).

For the stochastic effect of skin cancer, the probabalistic computation has never been formally done. Bosnjakovic (1987) compared morbidity from exposure to ionizing radiation from natural sources and UV radiation from the sun in the Netherlands. He calculated an average individual mortality risk due to ionizing radiation of 2.5×10^{-5} per year. In contrast, the average individual mortality risk due to solar UV was in the range of 0.5 to 2.5×10^{-5} per year. He concluded that the stochastic risk due to UV radiation is probably quite comparable to that due to ionizing radiation and differs by less than a factor of four.

The increased risk of terminal skin cancer from indoor illumination sources can be derived from the analysis of Lytle et al., (1987). Using an exponential model of UV induced skin cancer incidence, the authors calculated than an annual indoor exposure of 75 minimal erythema doses (MEDs) might increase the incidence of non-melanoma skin cancer by 4.3 percent. A source emitting the maximum threshold limit value recommended by the American Conference of Governmental Industrial Hygienists (ACGIH) would yield 75 MEDs per year over 250 workdays (Cole et al., 1985). The overall incidence of non-melanoma skin cancer in the U.S. is 232 new cases per 100,000 people per year (Scotto et al., 1983). With a case fatality rate of 1%, the probability of death would be 2.32×10^{-5} per year, a number comparable with that derived by Bosjankovic. Using these estimates, as increase of 4.3 percent would calculate to an increased morbidity from non-melanoma skin cancer of approximately 1.0×10^{-6} per year. By comparison, Slaper and van der Leun (1987) estimate that for people receiving a normal sun exposure of 100 MED per year, an additional exposure of 100 MED per year between the ages of 15 and 45 increases risk by a factor of 3.4. They further note that excessive users of suntanning equipment could increase their risks by a factor of from 5 to 10. Risk factors of this order increase the probability of increased morbidity from skin cancer to as high as 2.3×10^{-4} per year, roughly the same risk or death from driving a motor vehicle in the United States (Wilson and Crouch, 1987). Calculation of lifetime risk, as opposed to annual risk, increases these estimates of nearly two orders of magnitude into the range of 10^{-2}.

Risk Management

Products

UVR, as such, can't be managed. It cannot be turned off at the source or legislated out of existance because the major offender, the sun, is not under terrestrial control. Rather we must place emphasis on the various manmade products that either emit or transmit UVR or adversely affect its environmental levels. Products such as sunlamps and germicidal lamps emit UVR and are in daily use in situations where individuals are exposed to high levels. Sunscreens variably transmit UVR and therefore attenuate UVR exposure. Sunglasses and intraocular lenses transmit UVR to varying degrees and must be included in the list of products possessing a UVR hazard component. Finally, and perhaps most significant, chlorofluorocarbons are now recognized as adversely affecting the earth's protective ozone layer, increasing the environmental exposure to UVR. Each of these products possess a different degree of UVR burden and each must be analyzed separately.

Goals

Risk management uses risk assessment data combined with analyses of economic factors and social considerations to formulate a risk reduction strategy within the statutory mandate and legal jurisdiction of the regulatory organization. This translation of scientific data to practical policies is a process that is not well understood by either advocates of certain positions or by the people the policy is designed to protect (Miller, 1987). Risk assessment uses formulas, data, and models and gives the impression of a logical process. Risk management uses none of these, and the formulation of strategies is highly controversial as a result.

While risk assessment is science and can conceivably be agreed on, risk management is policy and can, and probably should, vary with local circumstances. Wilson and Crouch (1987) note that the results, goals, and values of performing a risk assessment must be sharply contrasted with the cultural values assigned to the results. Such cultural values will influence decisions and may differ even for risk estimates that are identical in probability. The strong effect of societal value systems can be easily seen in the changing perceptions of tanning. Meijer, Lauxtermann, and Bosnjakovic (1987) quote the following line from a 1927 womens magazine: "If you want to be ugly and to displease men, do everything you can to to be tanned by the sun, so that the soft skin of your face becomes wrinkled." In recent years, in contrast, a tan has been considered a sign of vigor, and people flock to beaches each summer to develop a "healthy" tan. The cycle has most recently began to swing back in the other direction. In the last year or so have some sectors of the public begun to recognize the deleterious effects of too much UVR exposure. This year the Bain de Soleil model is shown in magazine ads wearing a shades-lighter "St. Tropez Tan".

Often there is a large component of individual discretion in the selection of risk situations. In radiation toxicology (and other toxicologic disciplines as well), exposure situations are classified as being either intentional (voluntary) or unintentional (involuntary) (Hallenbeck and Cunningham, 1986; Passchier and van der Leun, 1987). Intentional risks are those that we know about, or think we know about, and accept. Unintentional risks are, conversely, those that we don't know about and perhaps wouldn't except if we did. The distinction is fundamental to the development of risk management programs. For example, data or erythema, keratitis, and carcinogenesis have been combined into two hazard curves in the report of the Health Council of the Netherlands. One is intended for involuntary exposure and is a combined curve for both keratitis and erythema. The other is intended for voluntary exposures to tanning devices and is essentially an erythema effectiveness spectrum. The authors of the report felt that for voluntary exposure eye damage can be easily prevented if users are careful to shield their eyes during exposure.

In addition to societal values systems, economic factors must also be considered. There is no regulatory body dedicated solely to management of the risks from UVR. Numerous organizations exist, but each one manages UVR as well as any number of other potential health problems. Like all other government agencies, regulatory organizations have limited budgets and manpower and must concentrate on certain areas. Miller (1987) notes that any government organization charged with the responsibility of public health protection has more hazards, risks, and issues than can be addressed with the resources available. Strict regulatory approaches are unlikely to be effective if they can't be enforced. Miller observes that it is difficult for risk managers to tell risk assessors, and even the public, that a particular risk management program cannot be implemented because of competing priorities.

Some people tend to think of regulatory agencies as acting solely to develop standards and punishing those not in conformance. In practice this is far from reality. Regulatory agencies have come to realize that strict adversary relationships with regulated industries are often counterproductive to the goal of protecting public health. In responding to a real or suspected public health problem, a variety of approaches may be taken. The U.S. Food and Drug Administration, for example, uses education, regulation, enforcement, voluntary compliance, or some combination of all of these (Miller, 1987). The decision about which one to use is a complicated function of the risk assessment, level of public concern, and the nature of the hazard.

Methods

Miller (1987) suggests a useful model for analyzing the nature of the hazard. By breaking down the probability of injury, P, into its individual components, risk managers can get clues to effective risk reduction strategies. This is expressed pseudomathematically in the following formula:

$$P_{(injury)} = P_{(malfunction)} + P_{(user\ error)} = P_{(patient\ susceptibility)}$$

If product malfunction is the largest component of $P_{(injury)}$ then regulatory, enforcement, or voluntary actions by the manufacturer may be called for. If $P_{(injury)}$ is more a function of user error or patient susceptibility then education may be the stronger tool.

If risk assessment indicated that a problem was related to product malfunction, then risk managers must determine whether the product should remain on the market. If the product should be removed from the market, then a determination must be made if ceasing all future distribution is sufficient or if all units previously sold should be removed. If the product should remain on the market but in a modified version to address the risk, then the follow-up decision about modifying all existing units must be made. In this context, product modifications are not restricted to hardware changes but may include changes in instruction for use and restrictions on sale and distribution.

While many discussions of product-related risk solutions are specific to one

1064

manufacturer's product, the process of identifying options always includes the step of asking whether the particular problem is unique. If there is a suspicion that the particular problem is not unique, but is generic across all products of a type, then the options for dealing with it are expanded to include actions that would be directed at the entire industry. The development of generic performance standards can be particularly effective in this situation, though the process is often a slow one.

Another aspect of product related problems is understanding whether the design of the device or the implementation of that design by the manufacturing process if the root cause. If design related, than improving the manufacturing process if not likely to be a good approach. If the problem is in the manufacturing process, then quality control and quality assurance approaches become viable options.

In dealing with user error, education can be a particularly powerful tool. Van der Leun (1987) suggests that a 50% reduction in long term risk (skin cancer) can be achieved by about a 30% reduction in yearly UVB dose. As noted above, human UVR exposure comes from two sources, electronic product emissions and the sun. Since probably 90% of UV human exposure is from the sun (Passchier, 1987), even the total elimination of products that emit UVR would not bring about a 30% reduction in total dose. Electronic product emissions to provide an added UV exposure burden and must be minimized. However, to achieve a 50% reduction in long term risk, other approaches, in addition to the minimizing of electronic product emissions, must be taken. Chief among these is education.

Education is also a powerful tool when dealing with cases of individual sensitivity. Whether the differential sensitivity arises fro endogenous factors such as skin type or certain genetic diseases, or exogenous factors such as medications, the individual must be made aware of the adverse consequences of excessive UVR exposure.

Current standards

Three types of organizations relevant to risk management of UVR exist: those for scientific exchange, government regulatory organizations, and non-government standards and guidelines associations. For the scientific analysis of UVR bioeffects, two major organizations predominate. Association Internationale de Photobiologie (AIP), founded in 1928, and the American Society for Photobiology (ASP), found in 1972, both play a strong role in the dissemination of scientific information. In the regulatory area, most countries have a Ministry of Health or equivalent whose job entails formulating public health policy for that nation. In the U.S. this role is divided among several organizations. Three predominate in the handling of UVR hazards: the food and Drug Administration (FDA) for sunscreens and the regulation of medical devices and consumer products that emit UVR, the Environmental Protection Agency (EPA) for the regulation of environmental contaminants that affect solar UVR, and the Occupational Safety and Health Administration (OSHA) and its research arm, the National Institute of Occupational Safety and Health (NIOSH), for the regulation of safe levels of UVR in the workplace.

Because the resources of government organizations are limited, regulatory agencies are becoming more reliant on independent manufacturer and standards and guidelines organizations. Examples include the International Electrotechnical Commission (IEC), the Commission Internationale de l'Eclairage (CIE), the American National Standards Institute (ANSI), the International Non-Ionizing Radiation Committee (INIRC) of the International Radiation Protection Association (IRPA), and the American Council of Governmental Industrial Hygienists (ACGIH).

Each organization, whether government regulatory or non-government advisory, has its own goals and mandates. Government regulatory organizations in particular operate within the context of an often confusing set of laws and regulations. Sometimes the

Table 1. Current UVR standards (modified from Health Council of the Netherlands)

Country	Product	Standard	Features
Australia	Sunlamp	Standard 2635–1983	labeling required, limits on spectral composition, timer and emergency switch required, total UV irradiance limited to 200 W/m^2, 30 min maximum exposure, limits on irradiation regime in commercial establishments
Canada	Sunlamp	P.C. 1980–1652, part XI	labeling required, limits on spectral composition, timer required, 10 min maximum exposure
Germany	Sunlamp	Entwurf DIN 5050	classification required, prohibits emission of erythema-effective wavelengths above 325 rm and pigment-effective wavelengths below 300 rm
Sweden	Sunlamp	SSI FS 1982: 1	labeling required, limits on spectral composition, timer required, licence required, total UV irradiance limited to 200 W/m^2, 30 min maximum exposure
United Kingdom	Sunlamp	Guidance note GS 18	labeling required, timer and emergency switch required, exposure limit determined by manufacturer depending on properties of equipment
United States	Sunlamp	21 CFR part 1040.20	labeling required, limits on spectral composition, timer and emergency switch required, eye protection required, exposure limit determined by manufacturer depending on properties of equipment
United States	Mercury vapor	21 CFR part 1040.30	labeling required, protective housing required on some models

obvious is not so easy to execute. While FDA has the regulatory authority to assure that manufacturers of regulated products provide, for example, adequate instructions about proper maintenance and use, the agency has no regulatory control over the users of that information. In other countries, the opposite is true (Miller, 1987). It is not particularly important that they each arrive at the same conclusions, for each has its own purposes. It is important, however, that they use the same data set. Thus the role for scientific organizations like AIP and ASP can be viewed as working toward unifying the data set.

The single most regulated UVR emitting product in the world is the sunlamp. Sunlamp regulations exist in a least 6 countries including, Australia, Canada, Germany, Sweden the United Kingdom, and the United States. In addition, the U.S. has a performance standard for Mercury Vapor Lamps. Each of these standards is different (Table 1). The most strict, perhaps, is the Swedish regulation developed by the Swedish National Institute of Radiation Protection (NIRP) which effectively prohibits certain types of sunlamps. The U.S. standard is a more flexible performance standard that serves to insure that the user is protected from overexposure.

Probably the most used standard for UVR emitting products is not a standard at all. The threshold limit value developed by the ACGIH and recently adopted by IRPA is the major tool for estimating the risks of UV emitting products. This standard was designed to reduce the incidence of UV hazards in the workplace and is limited to the analysis of sources with exposures of eight hours per day or less. This standard is the backbone of the regulations recently recommended by the Health Council of the Netherlands (1986). The Council, however, extended the analytical capability of the ACGIH effort to better include the UCA and extend application beyond eight hours.

Conclusions

UVR is a known toxin and potent human carcinogen. While the possibility of adverse human effects has been known and appreciated for some time, the probability of severe effects is becoming more acute as lifespan increases past the latent periods for photocarcinogenesis and photocataractogenesis and as UVR availability and medical utilization escalates. The estimation and management of UVR risks is a multidimensional process that requires input from scientists, manufacturers and the general public. UVR risk management, should be a cooperative effort that promotes safety by insuring that the risks of products that emit UVR or attenuate its effects are well characterized by the scientist, minimized by the manufacturer, and understood by the public.

References

Bosnjakovic, B.F.M. (1987). Protection policies for ionizing and UV radiation: A confrontation of crucial determinants. In Human Exposure to Ultraviolet Radiation: Risks and Regulations (Edited by E.F. Passchier and B.F.M. Bosnjakovic). Elsevier Science Publishers, New York.

Cole, C., Forbes, P.D., Davies, R.E., and Urbach, F. (1985). Effect of indoor lighting on normal skin. *Ann. N.Y. Acad. Sci.* 453, 305–316.

DHHS (U.S. Department of Health and Human Services) (1986). Determining Risks to Health: Federal Policy and Practice. Auburn House Publishing Co., Dover, Mass.

Ferenczi, L.Z., Hill, G.B., and Scrimger, J.W. (1987). Quantitation of human exposure to UVR. In Human Exposure to Ultraviolet Radiation: Risks and Regulations (Edited by W.F. Passchier and B.F.M. Bosnjakovic). Elsevier Science Publishers, New York.

Hader, D.,and Tevini, M. (1987). General Photobiology. Pergamon Press, New York, 1987.

Hallenbeck, W.H., and Cunningham, K.M. (1986). Quantitative Risk Assessment for Environmental and Occupational Health. Lewis Publishers, Inc., Chelsea.

Health Council of the Netherlands (1986). UV Radiation. Pub.No. 1986–9E, The Hague, the Netherlands.

Latin, H. (1988). Good Science, Bad Regulation, and Toxic Risk Assessment. *Yale J. Reg.* 5, 89–148.

Lytle, C.D., Hitchins, V.M., and Beer, J.Z. (1987). Estimation of carcinogenic risk from lamps which emit ultraviolet radiation. In Human Exposure to Ultraviolet Radiation: Risks and Regulations (Edited by W.F. Passchier and B.F.M. Bosnjakovic). Elsevier Science Publishers, New York.

Marshall, J. (1987). Ultraviolet radiation and the eye. In Human Exposure to Ultraviolet Radiation: Risks and Regulations (Edited by W.F. Passchier and B.F.M. Bosnjakovic). Elsevier Science Publishers, New York.

Miller, E.A. (1987). Policies of risk reduction and protection. In Human Exposure to Ultraviolet Radiation: Risks and Regulations (Edited by W.F. Passchier and B.F.M. Bosnjakovic). Elsevier Science Publishers, New York.

NRC (National Research Council) (1983). Risk Assessment in the Federal Government: Managing the Process. National Academy Press, Washington, DC.

Passchier, W.F., and Bosnjakovic, B.F.M. (1987). Human Exposure to Ultraviolet Radiation: Risks and Regulations. Elsevier Science Publishers, New York.

Russell, M.M. (1987). Risk assessment in environmental policy-making. *Science* 235, 286–290.

Scotto, J., Fears, T.R., and Fraumeini, Jr. J.F., (1983). Incidence of Normelanoma Skin Cancer in the United States. U.S. Department of Health and Human Services, NIH Pub. No. 83–2433.

Slaper, H., and van der Leun, J.C. (1987). Human exposure to ultraviolet radiation: Quantitative modeling of skin cancer incidence. In Human Exposure to Ultraviolet Radiation: Risks and Regulations (Edited by W.F. Passchier and B.F.M. Bosnjakovic). Elsevier Science Publishers, New York.

Sliney, D.H. (1987). Unintentional exposure to ultraviolet radiation: Risk reduction and exposure limits. In Human Exposure to Ultraviolet Radiation: Risks and Regulations (Edited by W.F. Passchier and B.F.M. Bosnjakovic). Elsevier Science Publishers, New York.

Wilson, R., and Crouch, E.A.C. (1987). Risk assessment and comparisons: An introduction. *Science* 236, 267–270.

Wolff, F. (1987). Risk-benefit evaluation of UV-exposure. In Human Exposure to Ultraviolet Radiation: Risks and Regulations (Edited by W.F. Passchier and B.F.M. Bosnjakovic). Elsevier Science Publishers, New York.

Round Table: Photobiology in Developing Countries

K.K. ROHATGI-MUKHERJEE and E. RIKLIS

A Committee on Photobiology in Developing Countries was set up at the interim meeting of the AIP held in Grenoble in 1986 at the occasion of the founding meeting of the European Society for Photobiology. Members of the committee are K.K. Rohatgi-Mukherjee, Chairperson, S. Braslavsky and E. Riklis. It was assigned the task of identifying photobiology-related problems of importance to developing countries, and to suggest how the AIP may be of assistance in promoting activities in these countries. Members of the committee had knowledge of problems in the Latin America continent, S. Braslavsky being from Argentina and E. Riklis from having been Scientific Counsellor to Latin America while serving in Brazil, and K.K. Rohatgi-Mukherjee having knowledge of India and the Far East countries. Accordingly, a draft of proposed activities was sent to several organisations and scientists and contacts were maintained with the Academia Sinica in Shanghai, the Third World Academy of Sciences and other noted scientists, but no representative from these organisations arrived at the International Photobiology congress, which perhaps is an indication of one of the problems facing scientists in the developing countries.

Among the proposals for actions which could advance science in general and photobiology in particular in the developing countries were: organization of courses or short symposia with international participation, within the program of local scientific societies meetings; help by subscription to specialized journals; free access to scientific data banks.

The scientific topics which were considered suitable for development towards solving problems of importance in the developing countries are: Phototherapy of skin diseases, eye problems, industrial toxicology, agricultural problems, new methods of water purification using sunlight. For courses and seminars, more sophisticated topics were proposed, such as bio- and chemiluminescence, fluorescence markers, photodynamic therapy, and other topics related to medical and agricultural science.

The roundtable discussion on photobiology in developing countries, organized within the program of the 10th International Congress on Photobiology was led by K.K. Rohatgi-Mukherjee with the participation of, (in alphbetical order), Lars Bjorn, Silvia Braslavsky, Louis Caldas, Danuta Frackowiack, Emanuel Riklis, David Shugar, Kendrick Smith, Hiraku Takebe and Rex Tyrrell.

In the very lively discussion, the problems facing the developing countries and the

Photobiology, Edited by E. Riklis
Plenum Press, New York, 1991

standing of photobiology in all countries were described and ideas for improvement were put forth. It was noted that there are areas which are discussed in this congress that are not represented in Latin America, and few representatives from developing countries are normally present in international congresses. It was suggested that the senior members and laboratory directors should see to it that the young students come to the national and the international meetings, and one of the immediate tasks of AIP is to raise sums of money that will enable promising travel support to young scientists (Riklis).

Another pressing problem is the isolation in literature. Very few libraries receive major journals, and there are no databanks. Studying for a Ph.D. degree in the advanced countries only creates frustration upon going back to the conditions in the developing country, (Braslavsky). This has been overcome in, for example, the Biophysics Institute in Warsaw by encouraging the researchers who study in developed countries to maintain contacts with their places of specialisation and continue collaboration, that brings also grants and equipment. Journals are indeed very expensive and unattainable, but with Current Contents one can obtain all papers of interest directly from their authors. The Pan American Federation could be helpful also in this regard (Shugar).

It was noted that in some countries what was once flourishing science has declined to few still outstanding institutes which are the exception, not the rule. Some of this may be blamed on political situations, but a major problem is reflected in the fact that in an international meeting such as this one we see hundreds of participants from several countries but only one, two or none from other countries, and this is an indication of the problem - lack of resources. The AIP needs resources so it can plan participation of at least one from each developing country. As for the wider scope of developing photobiological scientific activities, the AIP as an international association should approach the bodies which have some resources in these countries, the National Research Councils, Academies of Science which are strong in the latin american countries, and certainly regional organizations like the Organization of American States, which is the regional organization of Latin American countries, and secure their support in the plans for development of scientific activities. National societies should be established and represent the AIP in front of local government and academia (Riklis).

The AIP should show that it has ideas and projects in order to be recognized as spokesman for the photobiology community, and indeed there are topics in Latin America of special importance, such as the rain forest problem. Need to develop a plan of what to do and with it get recognition from the organizations mentioned (Smith).

Indeed goals of AIP have been changed to do just that and therefore the formation of this committee and a committee for fund raising We need to update the list of photobiologists in the world and the AIP should and can be an information bank for people and their expertise. In the AIP secretariat there are already lists of thousands of photobiologists, members of the American society, the European, Japanese and other national societies (Tyrrell).

The ozone problem, as an example, is being studied in Argentina, yet when a NASA team came to do measurements there was no contact with local scientists. A local strong society could make this contact (Braslavsky).

This really depends on the persons directing this sort of activities in developing countries. An example of how contacts could be widened and visits of top scientists could be utilized to the maximum was shown when advanced seminars organized with the participation of Israeli scientists in one country in Latin America was followed by a tour of lectures and one day symposia given by the same scientists in all other countries of the region, thus enabling hundreds of local scientists and students to benefit from the visit of experts to one specific course. This required organization, and the collaboration of the national bodies, NRCs, Academies, universities as well as the regional organizations like the OAS. It could serve as a model for maximum utilization of cooperative efforts, and certainly the existence of a national society for that field of science could be very helpful (Riklis).

There are general economic problems, and perhaps help can be offered by other laboratories in forms which do not require special budgets, like offering scholarships from existing grants, and introduce students from developing countries to advanced instrumentation and techniques. (Frackowiack).

Problems of society establishment pale compared to fundamental problems of equipment, electricity, supplies, how to improve crops, etc, especially in countries in Africa.. we should put the efforts in these. Some of them do have universities and research teams but other need more basic support (Smith).

One could think of redirecting journals like current contents to developing countries after they are read by developed laboratories. (Bjorn).

The notion of creating and strengthening national photobiology societies was discussed: It was agreed that it is important to develop local photobiology societies. If such are active in the more developed countries like Argentina and Brazil they could organize courses for the less developed, and the same may be done in the East and in Africa (Braslavsky).

Regional societies can take care of this. Local societies should be based on strengthening existing capabilities (Riklis). Like the European Photobiology society. We should strengthen the local societies and activate new ones. In India it started because of AIP (Rohatgi-Mukherjee). The creation of the photobiology society in the USA was a gradual process, movements started groups of photobiology within existing biophysics or other societies, then grew to a separate society (Smith).

In Japan the radiation research was organized because of Hiroshima, and photobiologists joined that society. Photomedicine began 10 years ago, and only recently there is real organization of photobiology. In Korea and China there is some activity but in China they face problems of supplies, lack of foetal calf serum, sterile plastic dishes etc. Korea is richer but activity is still limited. Need to find good leaders to form photobiology group, people who have influence.

Above all that, photobiology should be recognized as a science. Even in the strong ASP we do not see the new young people. Members should identify as photobiologists and not in their original scientific discipline. We should protect photobiology (Smith).

To do this we need textbooks, list of requirements - the education committee is not meeting this year here so this committee should recommend it (Rohatgi-Mukherjee), and we need a laboratory manual for courses (Smith).

The round table discussion was concluded with the following DECISIONS:

The AIP (Association Internationale de Photobiologie) takes upon itself to create local groups and national photobiology societies and help activate them to develop existing capabilities. It is also recommended that the AIP should help regional activities to develop

It is recommended that the AIP is to prepare data bank of expertise in the AIP and of experts known to the AIP. These could be utilized to lead programs in developing countries.

It is recommended that efforts will be taken to make photobiology part of the curriculum as a mature field of science.

Author Index

Subject Index

Only principal keywords are listed in this Subject Index. The page numbers refer to the first page of each article.